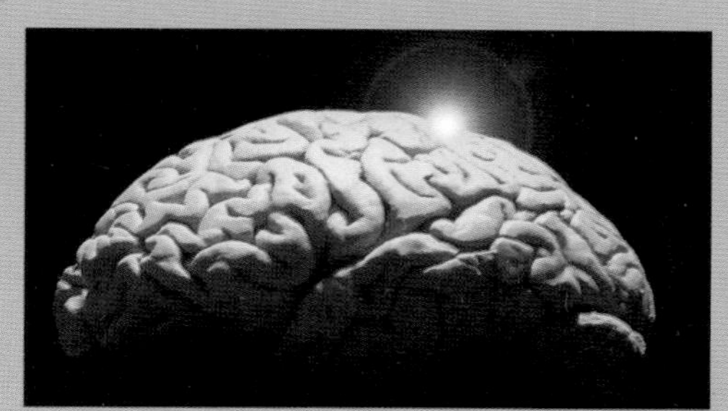

NEUROSCIENCE

Exploring the Brain

Second Edition

NEUROSCIENCE
Exploring the Brain

Second Edition

MARK F. BEAR, Ph.D.
Professor of Neuroscience
Howard Hughes Medical Institute
Brown University
Providence, Rhode Island

BARRY W. CONNORS, Ph.D.
Professor of Neuroscience
Brown University
Providence, Rhode Island

MICHAEL A. PARADISO, Ph.D.
Associate Professor of Neuroscience
Brown University
Providence, Rhode Island

Editor: Susan Katz
Managing Editor: Angela Heubeck
Development Editor: Betsy Dilernia
Marketing Manager: Christen DeMarco
Production Editor: Bill Cady
Illustrations: Duckwall Productions

351 West Camden Street
Baltimore, Maryland 21201-2436 USA

530 Walnut Street
Philadelphia, Pennsylvania 19106-3621 USA

Printed in the United States of America

Library of Congress Cataloging-in-Publication Data

Bear, Mark F.
Neuroscience: exploring the brain / Mark F. Bear, Barry W. Connors, Michael A. Paradiso—2nd ed.
p. cm.
Includes bibliographical references and index.
ISBN 0-7817-3255-7
1. Neurosciences. 2. Brain. I. Connors, Barry W. II. Paradiso, Michael A. III. Title.
[DNLM: 1. Brain. 2. Neurosciences. WL 300 B368n 2001]
QP355.2.B425 2001
612.8—dc21

00-046513

The publishers have made every effort to trace the copyright holders for borrowed material. If they have inadvertently overlooked any, they will be pleased to make the necessary arrangements at the first opportunity.

To purchase additional copies of this book, call our customer service department at **(800) 638-3030** or fax orders to **(301) 824-7390**. International customers should call **(301) 714-2324.**

Lippincott Williams & Wilkins customer service representatives are available from 8:30 am to 6:00 pm, EST, Monday through Friday, for telephone access. ***Or visit Lippincott Williams & Wilkins on the Internet: http://www.lww.com.***

02 03
3 4 5 6 7 8 9 10

to our parents

Naomi and Firman Bear
Rose and John Connors
Marie and Nicholas Paradiso

PREFACE

THE ORIGINS OF *NEUROSCIENCE: EXPLORING THE BRAIN*

For over twenty years, Brown University has offered a course called Neuroscience 1: An Introduction to the Nervous System. The course has been remarkably successful: Approximately one of every four Brown undergraduates takes it. For a few students, this is the beginning of a career in neuroscience; for others, it is the only science course he or she will take in college.

The success of introductory neuroscience reflects the fascination and curiosity everyone has for how we sense, move, feel, and think. We believe, however, that the success of our course also derives from the way it is taught and what is emphasized. A cornerstone of our philosophy is that we assume only minimal prior knowledge of biology, physics, and chemistry. The fundamentals required for understanding neuroscience are covered as the course progresses. This approach ensures that we can work up to advanced concepts with confidence that students are on track. We also strive to show that science is interesting, exciting, and fun. To this end, we include liberal amounts of commonsense metaphors, real-world examples, humor, and anecdotes. Finally, our course does not survey all of neurobiology. Instead, we focus on mammalian brains and, whenever possible, the human brain. In this sense, the course closely resembles what is taught to most second-year medical students, but without prerequisites. Similar courses are now offered at many colleges and universities by psychology, biology, and neuroscience departments.

The first edition of *Neuroscience: Exploring the Brain* was written to provide a suitable textbook for Neuroscience 1, incorporating the subject matter and philosophy that made introductory neuroscience successful here at Brown. It has been very gratifying to see that the book has gained popularity around the world, sometimes acting as the catalyst for new courses in introductory neuroscience. This enthusiastic response encouraged us to write a second edition. Not only have we updated the book with the latest discoveries in this fast-paced field, we have incorporated numerous suggestions for improvement from our students and colleagues.

NEW IN THE SECOND EDITION

Writing the second edition gave us the opportunity to review the research accomplishments of the past five years, and they are truly astonishing. Examples are the recent determination of the three-dimensional structure of a selectively permeable ion channel, important for understanding neuronal signaling, and the discovery of the hormone, leptin, which has revolutionized our understanding of how feeding behavior is regulated. The book has been revised to incorporate these and many other new findings. In addition to bringing the book up to date, we have expanded and added a number of new features.

More Connections With Real Life

A popular component of the first edition, *Of Special Interest* boxes illustrate how a knowledge of neuroscience can be applied. We have expanded this feature with an eye toward making more connections with real life, including substantially increased coverage of common nervous system disorders, such as Alzheimer's disease and mental retardation. In addition, we incorporated more discussion of neurological disorders in the text of the chapters, where such coverage helps illustrate important principles—for example, in the control of voluntary movement.

More Anatomy

Over the years, our students have consistently indicated that they would like to have expanded coverage of nervous system anatomy to give them a better appreciation of how all the different components fit together. We have responded by including *An Illustrated Guide to Human Neuroanatomy*, new to the second edition as an appendix to Chapter 7. This illustrated appendix provides a preview of the structures students will encounter in specific functional contexts in later chapters. To help students learn the new terminology, a Self-Quiz with labeling exercises is also provided.

More Behavioral Neuroscience

Interesting topics in neuroscience far outnumber the chapters that are appropriate for an introductory text. Feedback from our colleagues at other institutions indicated, however, a need to expand coverage of behavioral neuroscience. Based on this valuable input, we added three exciting new chapters that connect the brain and behavior: Motivation (Chapter 16), Sex and the Brain (Chapter 17), and Mental Illness (Chapter 21).

More Brain Food

Our goal was to provide a book that anyone—regardless of his or her science background—could begin on the first page and understand all the way through. Of course, neuroscience is a rigorous, quantitative scientific discipline. In the first edition, we covered advanced concepts in cellular neurophysiology using *Brain Food* boxes. We have expanded this feature in the second edition, providing greater coverage of advanced concepts and new technologies. Isolating this material from the main text gives instructors flexibility in making reading assignments that are appropriate for students' backgrounds.

New Discoverers

We authors are all active neuroscientists, and we want our readers to understand the allure of research. A unique feature of our book is the *Path of Discovery* boxes, in which famous neuroscientists tell stories about their own research. These essays serve several purposes: to give a flavor of the thrill of discovery; to show the importance of hard work and patience, as well as serendipity and intuition; to reveal the human side of science; and to entertain and amuse. We have continued this tradition in the new edition, with contributions from twenty-four esteemed scientists. Included in this illustrious group are three Nobel laureates: Erwin Neher, Torsten Wiesel, and Susumu Tonegawa. We are very grateful to the *Path of Discovery* authors for their time, effort, and enthusiasm.

AN OVERVIEW OF THE BOOK

Neuroscience: Exploring the Brain surveys the organization and function of the human nervous system. We present material at the cutting edge of neuroscience, in a way that is accessible to both science and nonscience students alike. The level of the material is comparable to an introductory college text in general biology.

The book is divided into four parts: Part I, Foundations; Part II, Sensory and Motor Systems; Part III, The Brain and Behavior; and Part IV, The Changing Brain. We begin Part I by introducing the modern field of neuroscience and tracing some of its historical antecedents. Then we take a close look at the structure and function of individual neurons, how they communicate chemically, and how these building blocks are arranged to form a nervous system. In Part II, we go inside the brain to examine the structure and function of the systems that serve the senses and command voluntary movements. In Part III, we explore the neurobiology of human behavior, including motivation, sex, mood, emotion, sleep, language, and attention. Finally, in Part IV, we look at how the environment modifies the brain, both during development and in adult learning and memory.

The human nervous system is examined at several different levels, ranging from the molecules that determine the functional properties of neurons, to the large systems in the brain that underlie cognition and behavior. Many disorders of the human nervous system are introduced as the book progresses, usually within the context of the specific neural system under discussion. Indeed, many insights into the normal functions of neural systems have come from the study of diseases that cause specific malfunctions of these systems. In addition, we discuss the actions of drugs and toxins on the brain, using this information to illustrate how different brain systems contribute to behavior and how drugs may alter brain function.

Organization of Part I: Foundations (Chapters 1–7)

The goal of Part I is to build a strong base of general knowledge in neurobiology. The chapters should be covered sequentially, although Chapters 1 and 6 can be skipped without a loss of continuity.

In Chapter 1, we use a historical approach to review some basic principles of nervous system function, and then we turn to the topic of how neuroscience research is conducted today. We directly confront the ethics of neuroscience research, particularly that which involves animals.

In Chapter 2, we focus mainly on the cell biology of the neuron. This is essential information for students inexperienced in biology, and we find that even those with a strong biology background find this review helpful. After touring the cell and its organelles, we go on to discuss the structural features that make neurons and their supporting cells unique, emphasizing the correlation of structure and function.

Chapters 3 and 4 are devoted to the physiology of the neuronal membrane. We cover the essential chemical, physical, and molecular properties that enable neurons to conduct electrical signals. Throughout, we appeal to students' intuition by using a commonsense approach, with liberal use of metaphors and real-life analogies.

Chapters 5 and 6 cover interneuronal communication, particularly chemical synaptic transmission. Chapter 5 presents the general principles of chemical synaptic transmission, and Chapter 6 discusses the neurotransmitters and their modes of action in greater detail. We also describe many of the modern methods used to study the chemistry of synaptic transmission. Later chapters do not assume an understanding of synaptic transmission at the depth of Chapter 6, however, so this can be skipped at the instructor's discretion. Most coverage of psychopharmacology appears in Chapter 15, after the general organization of the brain and its sensory and motor systems has been presented. In our experience, students wish to know where, in addition to how, drugs act on the nervous system and behavior.

Chapter 7 covers the gross anatomy of the nervous system. Here we focus on the common organizational plan of the mammalian nervous system by tracing the brain's embryological development. (Cellular aspects of development are covered in Chapter 22.) We show that the specializations of the human brain are simple variations on the basic plan that applies to all mammals.

The Chapter 7 appendix, *An Illustrated Guide to Human Neuroanatomy*, covers the surface and cross-sectional anatomy of the brain, the spinal cord, the autonomic nervous system, the cranial nerves, and the blood supply. A Self-Quiz will help students learn the terminology. We recommend that students become familiar with the anatomy in the *Illustrated Guide* before moving on to Part II.

Organization of Part II: Sensory and Motor Systems (Chapters 8–14)

Part II surveys the systems within the brain that control conscious sensation and voluntary movement. In general, these chapters do not need to be covered sequentially, except for Chapters 9 and 10 on vision and Chapters 13 and 14 on the control of movement.

We chose to begin Part II with a discussion of the chemical senses—smell and taste—in Chapter 8. These are good systems for illustrating the general principles and problems in the encoding of sensory information, and the transduction mechanisms have strong parallels with other systems.

Chapters 9 and 10 cover the visual system, an essential topic for all introductory neuroscience courses. Many details of visual system organization are presented, illustrating not only the depth of current knowledge but also the principles that apply across sensory systems.

Chapter 11 explores the auditory system, and Chapter 12 introduces the somatic sensory system. Audition and somatic sensation are such an important part of everyday life that it is hard to imagine teaching introductory neuroscience without discussing them. The vestibular sense of balance is now also covered in a separate section of Chapter 11. This placement offers instructors the option to skip the vestibular system at their discretion.

In Chapters 13 and 14, we discuss the motor systems of the brain. Considering how much of the brain is devoted to the control of movement, this more extensive treatment is clearly justified. We are well aware, however, that the complexities of the motor systems are daunting to students and instructors alike. We have tried to keep our discussion sharply focused, using numerous examples to connect with personal experience.

Organization of Part III: The Brain and Behavior (Chapters 15–21)

Part III explores how different neural systems contribute to different behaviors, focusing on the systems where the connection between the brain and behavior can be made most strongly. We cover the systems that control visceral function and homeostasis, simple motivated behaviors (such as eating and drinking), sex, mood, emotion, sleep, consciousness, language, and attention. Finally, we discuss what happens when these systems fail during mental illness.

Chapters 15–19 explore a number of neural systems that orchestrate widespread responses throughout the brain and the body. In Chapter 15, we focus on three systems that are characterized by their broad influence and their interesting neurotransmitter chemistry: the secretory hypothalamus, the autonomic nervous system, and the diffuse modulatory systems of the brain. We discuss how the behavioral manifestations of various drugs may result from disruptions of these systems.

In Chapter 16, we look at the physiological factors that motivate specific behaviors, focusing mainly on very recent research on the control of eating habits. Chapter 17 investigates the influence of sex on the brain and the influence of the brain on sexual behavior. Chapter 18 examines the neural systems believed to underlie emotional experience and expression, specifically emphasizing fear and anxiety, anger and aggression, reinforcement and reward.

In Chapter 19, we explore the systems that give rise to the rhythms of the brain, ranging from the rapid electrical rhythms of the brain during sleep and wakefulness, to the slow circadian rhythms controlling hormones, temperature, alertness, and metabolism. Part III ends with a discussion of the neuroscience of higher brain functions in Chapter 20 and of mental illness in Chapter 21.

Organization of Part IV: The Changing Brain (Chapters 22–24)

Part IV explores the cellular and molecular basis of brain development, and learning and memory, which represent two of the most exciting frontiers of modern neuroscience.

Chapter 22 examines the mechanisms used during brain development to ensure that the correct connections are made between neurons. The cellular aspects of development are discussed here rather than in Part I for several reasons. First, by this point in the book, students fully appreciate that normal brain function depends on its precise wiring. Because we use the visual system as a concrete example, the chapter also must follow a discussion of the visual pathways in Part II. Second, we explore aspects of experience-dependent development of the visual system that are regulated by the diffuse modulatory systems of the brain, so this chapter is placed after the early chapters of Part III. Finally, an exploration of the role of the sensory environment in brain development in Chapter 22 is followed in the next two chapters by discussions of how experience-dependent modifications of the brain form the basis for learning and memory. We see that many of the mechanisms are similar, illustrating the unity of biology.

Chapters 23 and 24 cover learning and memory. Chapter 23 focuses on the anatomy of memory, exploring how different parts of the brain contribute to storage of different types of information. Chapter 24 takes a deeper look into the molecular and cellular mechanisms of learning and memory, focusing on changes in synaptic connections.

HELPING STUDENTS LEARN

Neuroscience: Exploring the Brain is not an exhaustive study. It is intended to be a readable textbook that communicates to students the important principles of neuroscience clearly and effectively. To help students learn neuroscience, we include a number of features designed to enhance comprehension:

- **Chapter Outlines and Introductory and Concluding Remarks.** These preview the organization of each chapter, set the stage, and place the material into broader perspective.
- **Key Terms and Glossary.** Neuroscience has a language of its own, and to comprehend it, one must learn the vocabulary. In the text of each chapter, important terms are highlighted in boldface type. To facilitate review, these terms appear in a list at the end of each chapter, in the order in which they appeared in the text, along with page references. The same terms are assembled at the end of the book, with definitions, in a glossary.

- **Review Questions.** At the end of each chapter, we include a brief set of questions for review. These are specifically designed to provoke thought and help students integrate the material.
- **Internal Reviews of Neuroanatomical Terms.** In Chapter 7, where nervous system anatomy is discussed, the narrative is interrupted periodically with brief self-quiz vocabulary reviews to enhance understanding. In the Chapter 7 appendix, an extensive Self-Quiz is provided in the form of a workbook with labeling exercises.
- **References and Suggested Readings.** To guide study beyond the scope of the textbook, we provide selected readings that will lead students into the research literature associated with each chapter. Rather than including citations in the body of the chapters, where they would compromise the readability of the text, we have organized the references and suggested readings by chapter and listed them at the end of the book.
- **Full-Color Illustrations.** We believe in the power of illustrations—not those that "speak a thousand words," but those that each make a single point. The first edition of this book set a new standard for illustrations in a neuroscience text. The bar has been raised again, with many superb new illustrations for the second edition.

ACKNOWLEDGMENTS

First and foremost, we wish to thank four people who made extraordinary contributions to this edition of the book: Betsy Dilernia, Caitlin Duckwall, Jim McIlwain, and Suzanne Meagher. Betsy served as our development editor, once again keeping us in line with her purple pencil. We are especially grateful for the standard of excellence that she established and held us to. The clarity and consistency of the writing are due to her remarkable efforts. Caitlin produced the new art, and the results speak for themselves. Caitlin took our sometimes fuzzy concepts and made them a beautiful reality. Jim is a mentor, faculty colleague, and friend; he is also an award-winning professor of neuroscience. Jim read every word of our nascent manuscript and showed us how to improve it. Finally, we are forever indebted to Suzanne, who assisted us at every step. It is no exaggeration to say that without her incredible assistance, loyalty, and dedication to this project, the book would never have been completed. Suzanne, you are the best!

We again would like to acknowledge the architects and current trustees of the undergraduate neuroscience curriculum at Brown University. We thank Mitchell Glickstein, Ford Ebner, James McIlwain, Leon Cooper, James Anderson, Leslie Smith, John Donoghue, and John Stein for all they did to make undergraduate neuroscience great at Brown. We thank the staff of Lippincott Williams & Wilkins for believing in this project and guiding it towards successful completion. We gratefully acknowledge the research support provided to us over the years by the National Institutes of Health, the Whitehall Foundation, the Alfred P. Sloan Foundation, the Klingenstein Foundation, the Charles A. Dana Foundation, the National Science Foundation, the Keck Foundation, the Human Frontiers Science Program, the Office of Naval Research, and the Howard Hughes Medical Institute. We thank our colleagues in the Brown University Department of Neuroscience for their support of this project and helpful advice. We thank the anonymous but very helpful colleagues at other institutions who gave us comments on the first edition and reviewed the first draft of our manuscript for the second. We gratefully acknowledge the scientists who provided us with figures illustrating their research results. In addition, many students and colleagues helped us to improve the new edition by informing us about recent research, pointing out errors in the first edition and suggesting better ways to describe or illustrate concepts. We thank them all, including (but not limited to) Yael Amitai, Teresa Audesirk, Michael Beierlein, Steve Chamberlin, Richard Cantin, Z. H. Cho, Geoffrey Gold, Jennifer Hahn, Richard Huganir, David Glanzman, Robert Malenka, John Morrison, Saundra Patrick, Robert Patrick, Erik Sklar, John Stein, Nelson Spruston, J. Michael Walker, and Wes Wallace.

We thank our loved ones for standing by us despite the countless weekends and evenings lost to preparing this book.

Last, but not least, we wish to thank the thousands of students to whom we have had the privilege to teach neuroscience over the past two decades.

BRIEF CONTENTS

(continues)

PART IV THE CHANGING BRAIN

DETAILED CONTENTS

Chapter 4
The Action Potential

Chapter 5
Synaptic Transmission

Chapter 6
Neurotransmitter Systems

Chapter 7
The Structure of the Nervous System

PART II SENSORY AND MOTOR SYSTEMS

Chapter 8
The Chemical Senses

Chapter 9
The Eye

Chapter 10
The Central Visual System

Chapter 11
The Auditory and Vestibular Systems

Chapter 12
The Somatic Sensory System

Chapter 16
Motivation

Chapter 17
Sex and the Brain

Chapter 18
Brain Mechanisms of Emotion

Chapter 19
Rhythms of the Brain

Chapter 20
Language and Attention

Chapter 21 Mental Illness

PART IV THE CHANGING BRAIN

Chapter 22 Wiring the Brain

Chapter 23
Memory Systems

Chapter 24
Molecular Mechanisms of Learning and Memory

LIST OF BOXES

PATH OF DISCOVERY

OF SPECIAL INTEREST

BRAIN FOOD

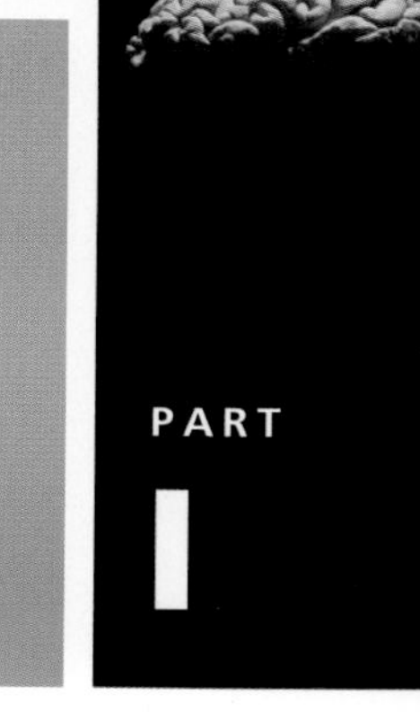

PART

I

FOUNDATIONS

CHAPTER

1

Introduction to Neuroscience

INTRODUCTION

Men ought to know that from nothing else but the brain come joys, delights, laughter and sports, and sorrows, griefs, despondency, and lamentations. And by this, in an especial manner, we acquire wisdom and knowledge, and see and hear and know what are foul and what are fair, what are bad and what are good, what are sweet and what are unsavory . . . . And by the same organ we become mad and delirious, and fears and terrors assail us . . . . All these things we endure from the brain when it is not healthy . . . . In these ways I am of the opinion that the brain exercises the greatest power in the man.

—Hippocrates, *On the Sacred Disease* (Fourth century B.C.)

It is human nature to be curious about how we see and hear; why some things feel good and others hurt; how we move; how we reason, learn, remember, and forget; the nature of anger and madness. These mysteries are starting to be unraveled by basic neuroscience research, and the conclusions of this research are the subject of this textbook.

The word "neuroscience" is young. The Society for Neuroscience, an association of professional neuroscientists, was founded as recently as 1970. The study of the brain, however, is as old as science itself. Historically, the scientists who devoted themselves to an understanding of the nervous system came from different scientific disciplines: medicine, biology, psychology, physics, chemistry, mathematics. The neuroscience revolution occurred when these scientists realized that the best hope for understanding the workings of the brain comes from an interdisciplinary approach, a combination of traditional approaches to yield a new synthesis, a new perspective. Most people involved in the scientific investigation of the nervous system today regard themselves as neuroscientists. Indeed, while the course you are now taking may be sponsored by the psychology or biology department at your university or college and may be called biopsychology or neurobiology, you can bet that your instructor is a neuroscientist.

The Society for Neuroscience is the largest and fastest-growing association of professional scientists in all of experimental biology. Far from being overly specialized, the field is as broad as nearly all of natural science, with the nervous system serving as the common point of focus. Understanding how the brain works requires knowledge about many things, from the structure of the water molecule to the electrical and chemical properties of the brain to why Pavlov's dog salivated when a bell rang. In this book, we will explore the brain with this broad perspective.

We begin the adventure with a brief tour of neuroscience. What have scientists thought about the brain over the ages? Who are the neuroscientists of today, and how do they approach studying the brain?

THE ORIGINS OF NEUROSCIENCE

You probably already know that the nervous system—the brain, spinal cord, and nerves of the body—is crucial for life and enables you to sense, move, and think. How did this view arise?

Evidence suggests that even our prehistoric ancestors appreciated that the brain was vital to life. The archeological record is rife with examples of hominid skulls, dating back a million years and more, bearing signs of fatal cranial damage, presumably inflicted by other hominids. As early as 7000 years ago, people were boring holes in each other's skulls (a process called trepanation), evidently with the aim not to kill but to cure (Figure 1.1). The skulls show signs of healing after the operation, indicating that this procedure was carried out on live subjects and was not merely a ritual conducted after death.

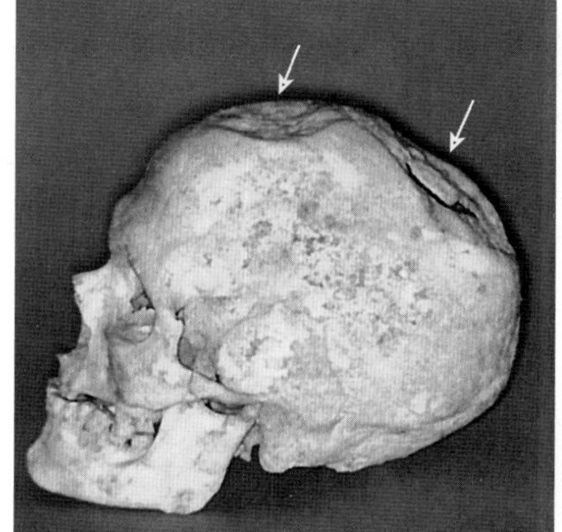

Figure 1.1
Evidence of prehistoric brain surgery. This skull of a man, over 7000 years old, was surgically opened when he was still alive. The arrows indicate two sites of trepanation. (Source: Alt et al., 1997, Fig. 1a.)

Some individuals apparently survived multiple skull surgeries. What early surgeons hoped to accomplish is not clear, although some have speculated that this procedure may have been used to treat headaches or mental disorders, perhaps by giving the evil spirits an escape route.

Recovered writings from the physicians of ancient Egypt, dating back almost 5000 years, indicate that they were well aware of many symptoms of brain damage. However, it is also very clear that the heart, not the brain, was considered to be the seat of the soul and the repository of memories. Indeed, while the rest of the body was carefully preserved for the afterlife, the brain of the deceased was simply scooped out through the nostrils and discarded! The view that the heart was the seat of consciousness and thought was not seriously challenged until the time of Hippocrates.

Views of the Brain in Ancient Greece

Consider the notion that the different parts of your body look different because they serve different purposes. The structures of the feet and hands are very different, and they perform very different functions: We walk on our feet and manipulate objects with our hands. Thus, we can say that there appears to be a very clear *correlation between structure and function*. Differences in appearance predict differences in function.

What can we glean about function from the structure of the head? Quick inspection and a few simple experiments (like closing your eyes) reveal that the head is specialized for sensing the environment. In the head are your eyes and ears, your nose and tongue. Even crude dissection shows that the nerves from these organs can be traced through the skull into the brain. What would you conclude about the brain from these observations?

If your answer is that the brain is the organ of sensation, then you have reached the same conclusion as several Greek scholars of the fourth century B.C. The most influential scholar was Hippocrates (460–379 B.C.), the father of Western medicine, who stated his belief that the brain not only was involved in sensation, but also was the seat of intelligence.

However, this view was not universally accepted. The famous Greek philosopher Aristotle (384–322 B.C.) clung to the belief that the heart was the center of intellect. What function did Aristotle reserve for the brain? He proposed it to be a radiator for the cooling of blood that was overheated by the seething heart. The rational temperament of humans was thus explained by the large cooling capacity of our brain.

Views of the Brain During the Roman Empire

The most important figure in Roman medicine was the Greek physician and writer Galen (A.D. 130–200), who embraced the Hippocratic view of brain function. As physician to the gladiators, he must have witnessed the unfortunate consequences of spinal and brain injury. However, Galen's opinions about the brain probably were influenced more by his many careful animal dissections. Figure 1.2 is a drawing of the brain of a sheep, one of Galen's favorite subjects. Two major parts are evident: the *cerebrum* in the front and the *cerebellum* in the back. (The structure of the brain is the subject of Chapter 7.) Just as we were able to deduce function from the structure of the hands and feet, Galen tried to deduce function from the structure of the cerebrum and the cerebellum. Poking the freshly dissected brain with a finger reveals the cerebellum to be rather hard and the cerebrum to be rather soft. From this observation, Galen suggested that the cerebrum must be the recipient of sensations and the cerebellum must command the muscles. Why did he propose this distinction? He recognized that to form memories, sensations must be imprinted onto the brain. Naturally, this must occur in the doughy cerebrum.

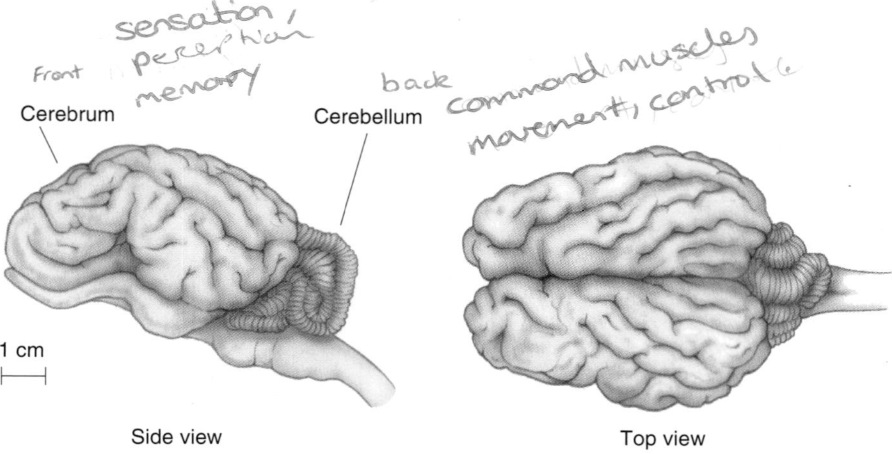

Figure 1.2
The brain of a sheep. Notice the location and appearance of the cerebrum and the cerebellum.

As improbable as his reasoning may seem, Galen's deductions were not that far from the truth. The cerebrum, in fact, is largely concerned with sensation and perception, and the cerebellum is primarily a movement control center. Moreover, the cerebrum is a repository of memory. We will see that this is not the only example in the history of neuroscience in which the right general conclusions were reached for the wrong reasons.

How does the brain receive sensations and move the limbs? Galen cut open the brain and found that it is hollow (Figure 1.3). In these hollow spaces, called *ventricles* (like the similar chambers in the heart), there is fluid. To Galen, this discovery fit perfectly with the prevailing theory that the body functioned according to a balance of four vital fluids, or humors. Sensations were registered and movements initiated by the movement of humors to or from the brain ventricles via the nerves, which were believed to be hollow tubes, like the blood vessels.

Views of the Brain From the Renaissance to the Nineteenth Century

Galen's view of the brain prevailed for almost 1500 years. More detail was added to the structure of the brain by the great anatomist Andreas Vesalius (1514–1564) during the Renaissance (Figure 1.4). However, ventricular localization of brain function remained essentially unchallenged. Indeed, the whole concept was strengthened in the early seventeenth century, when

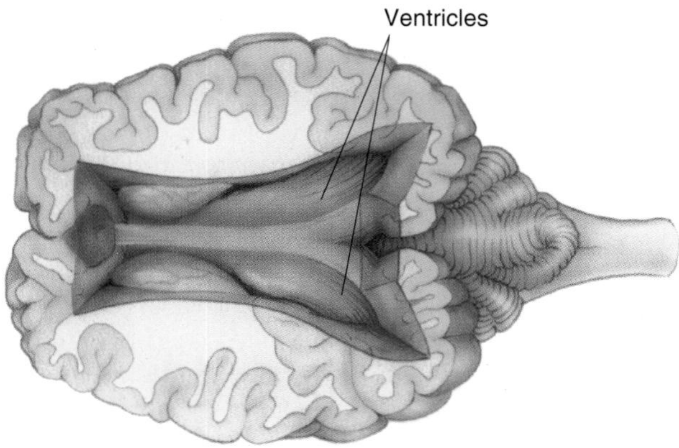

Figure 1.3
A dissected sheep brain showing the ventricles.

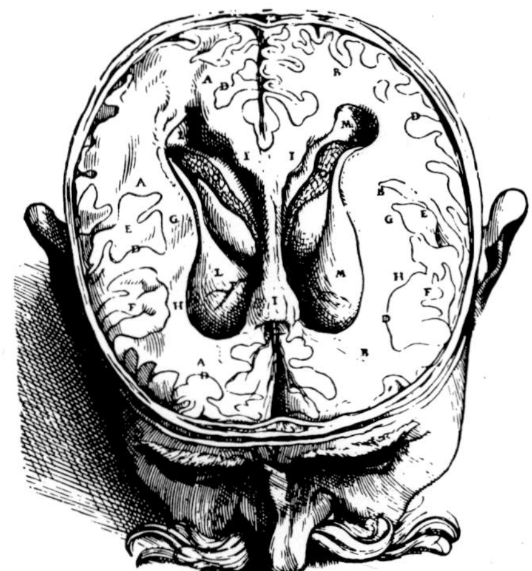

Figure 1.4
Human brain ventricles depicted during the Renaissance. This drawing is from *De humani corporis fabrica* by Vesalius (1543). The subject was probably a decapitated criminal. Great care was taken to be anatomically correct in depicting the ventricles. (Source: Finger, 1994, Fig. 2.8.)

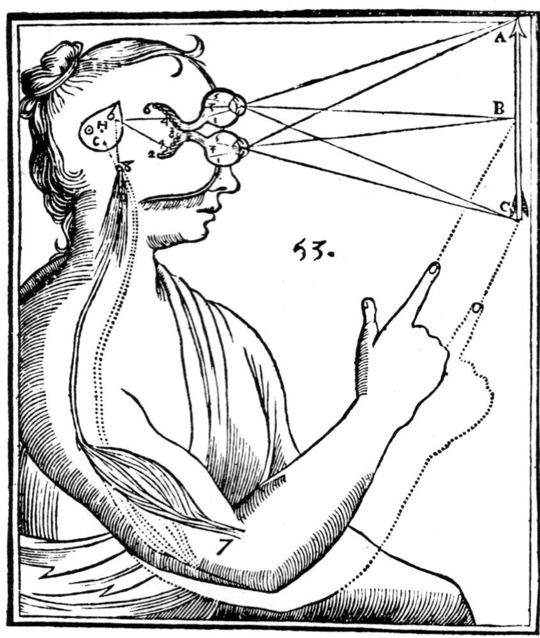

Figure 1.5
The brain according to Descartes. This drawing appeared in a 1662 publication by Descartes. Hollow nerves from the eyes project to the brain ventricles. The mind influences the motor response by controlling the pineal gland (H), which works like a valve to control the movement of animal spirits through the nerves that inflate the muscles. (Source: Finger, 1994, Fig. 2.16.)

French inventors began developing hydraulically controlled mechanical devices. These devices supported the notion that the brain could be machine-like in its function: Fluid forced out of the ventricles through the nerves might literally "pump you up" and cause the movement of the limbs. After all, don't the muscles bulge when they contract?

A chief advocate of this fluid-mechanical theory of brain function was the French mathematician and philosopher René Descartes (1596–1650). Although he thought this theory could explain the brain and behavior of other animals, it was inconceivable to Descartes that it could account for the full range of *human* behavior. He reasoned that unlike other animals, people possess intellect and a God-given soul. Thus, Descartes proposed that brain mechanisms control human behavior only to the extent that this behavior resembles that of the beasts. Uniquely human mental capabilities exist outside the brain in the "mind." Descartes believed that the mind is a spiritual entity that receives sensations and commands movements by communicating with the machinery of the brain via the pineal gland (Figure 1.5). Today, some people still believe that there is a "mind-brain problem," that somehow the human mind is distinct from the brain. However, as we shall see in Chapter 20, modern neuroscience research supports another conclusion: The mind has a physical basis, which is the brain.

Fortunately, other scientists during the seventeenth and eighteenth centuries broke away from Galen's tradition of focusing on the ventricles and began to give the substance of the brain a closer look. One of their observations was that brain tissue is divided into two parts: the *gray matter* and the *white matter* (Figure 1.6). What structure-function relationship did they propose? White matter, because it was continuous with the nerves of the body, was correctly believed to contain the fibers that bring information to and from the gray matter.

By the end of the eighteenth century, the nervous system had been completely dissected, and its gross anatomy had been described in detail. It was recognized that the nervous system has a central division, consisting of the brain and spinal cord, and a peripheral division, consisting of the network of nerves that course through the body (Figure 1.7). An important breakthrough in neuroanatomy was the observation that the same general pattern of

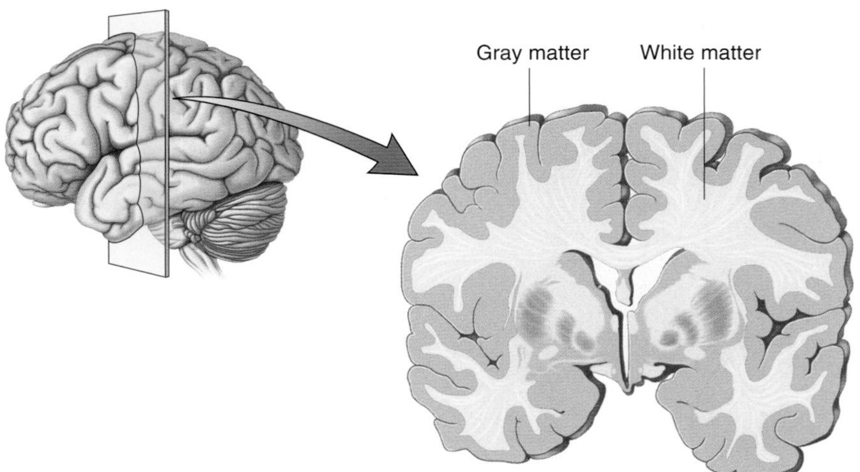

Figure 1.6
White matter and gray matter. The brain has been cut open to reveal these two types of tissue.

bumps (called *gyri*) and grooves (called *sulci* and *fissures*) could be identified on the surface of the brain in every individual (Figure 1.8). This pattern, which enables the parceling of the cerebrum into *lobes*, was the basis for speculation that different functions might be localized to the different bumps on the brain. The stage was now set for the era of cerebral localization.

Nineteenth-Century Views of the Brain

Let's review the state of understanding of the nervous system at the end of the eighteenth century:

- Injury to the brain can disrupt sensations, movement, and thought and can cause death.
- The brain communicates with the body via the nerves.
- The brain has different identifiable parts, which probably perform different functions.
- The brain operates like a machine and follows the laws of nature.

During the next 100 years, more would be learned about the function of the brain than had been learned in all of previously recorded history. This work provided the solid foundation on which twentieth-century neuroscience rests. Below we'll review four key insights gained during the nineteenth century.

Nerves as Wires. In 1751, Benjamin Franklin published a pamphlet entitled *Experiments and Observations on Electricity*, which heralded a new understanding of electrical phenomena. By the turn of the century, Italian scientist Luigi Galvani and German biologist Emil du Bois-Reymond had shown that muscles can be caused to twitch when nerves are stimulated electrically and that the brain itself can generate electricity. These discoveries finally displaced the notion that nerves communicate with the brain by the movement of fluid. The new concept was that the nerves are "wires" that conduct electrical signals to and from the brain.

Unresolved was whether the signals to the muscles causing movement use the same wires as those that register sensations from the skin. Bidirectional communication along the same wires was suggested by the observation that when a nerve in the body is cut, there is usually a loss of both sensation and movement in the affected region. However, it was also known that within each nerve of the body there are many thin filaments, or *nerve fibers*, each one of which could serve as an individual wire carrying information in a different direction.

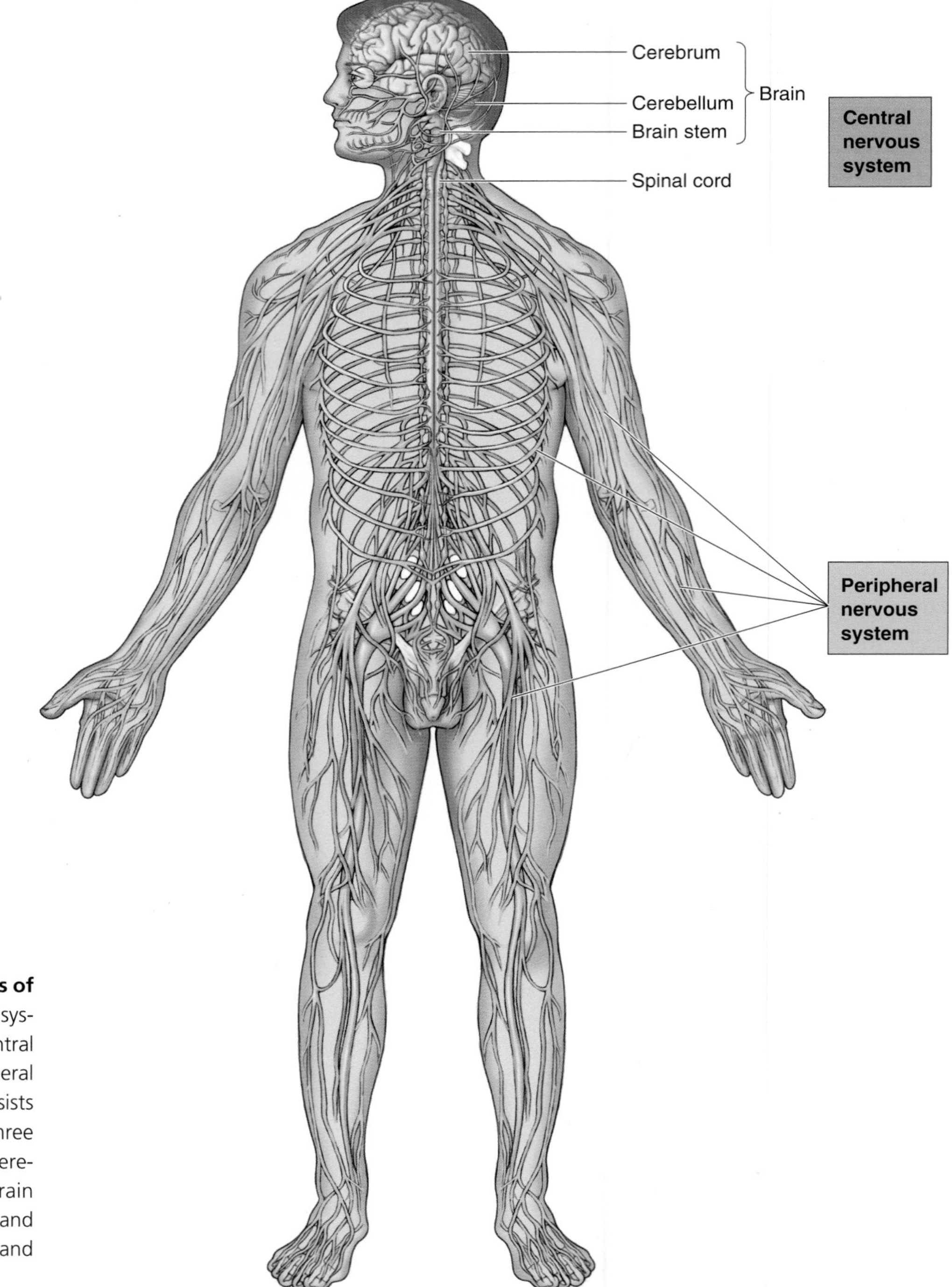

Figure 1.7
The basic anatomical subdivisions of the nervous system. The nervous system consists of two divisions, the central nervous system (CNS) and the peripheral nervous system (PNS). The CNS consists of the brain and spinal cord. The three major parts of the brain are the cerebrum, the cerebellum, and the brain stem. The PNS consists of the nerves and nerve cells that lie outside the brain and spinal cord.

This question was answered around 1810 by Scottish physician Charles Bell and French physiologist François Magendie. A curious anatomical fact is that just before the nerves attach to the spinal cord, the fibers divide into two branches, or roots. The dorsal root enters toward the back of the spinal cord, and the ventral root enters toward the front (Figure 1.9). Bell tested the possibility that these two spinal roots carry information in different directions by cutting each root separately and observing the consequences in experimental animals. He found that only cutting the ventral roots caused muscle paraly-

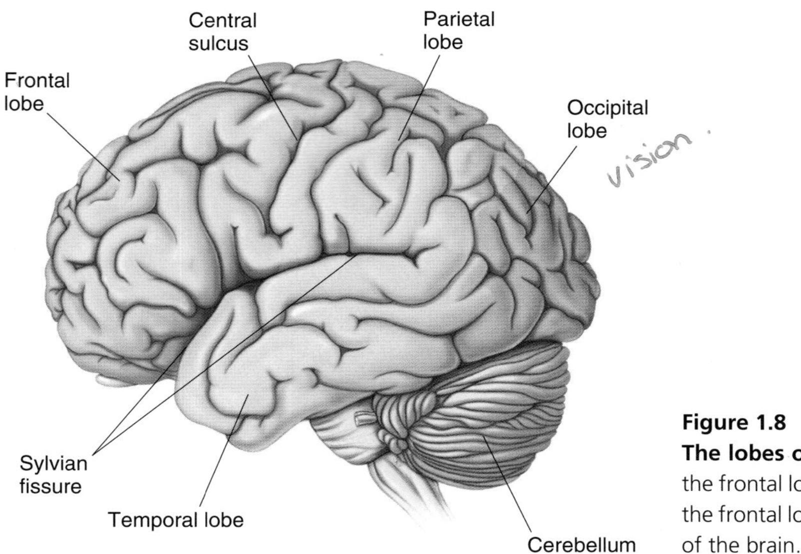

Figure 1.8
The lobes of the cerebrum. Notice the deep Sylvian fissure, dividing the frontal lobe from the temporal lobe, and the central sulcus, dividing the frontal lobe from the parietal lobe. The occipital lobe lies at the back of the brain. These landmarks can be found on all human brains.

sis. Later, Magendie was able to show that the dorsal roots carry sensory information into the spinal cord. Bell and Magendie concluded that within each nerve there is a mixture of many wires, some of which bring information into the brain and spinal cord and others that send information out to the muscles. In each sensory and motor nerve fiber, transmission is strictly one way. The two kinds of fibers are bundled together for most of their length, but they are anatomically segregated when they enter or exit the spinal cord.

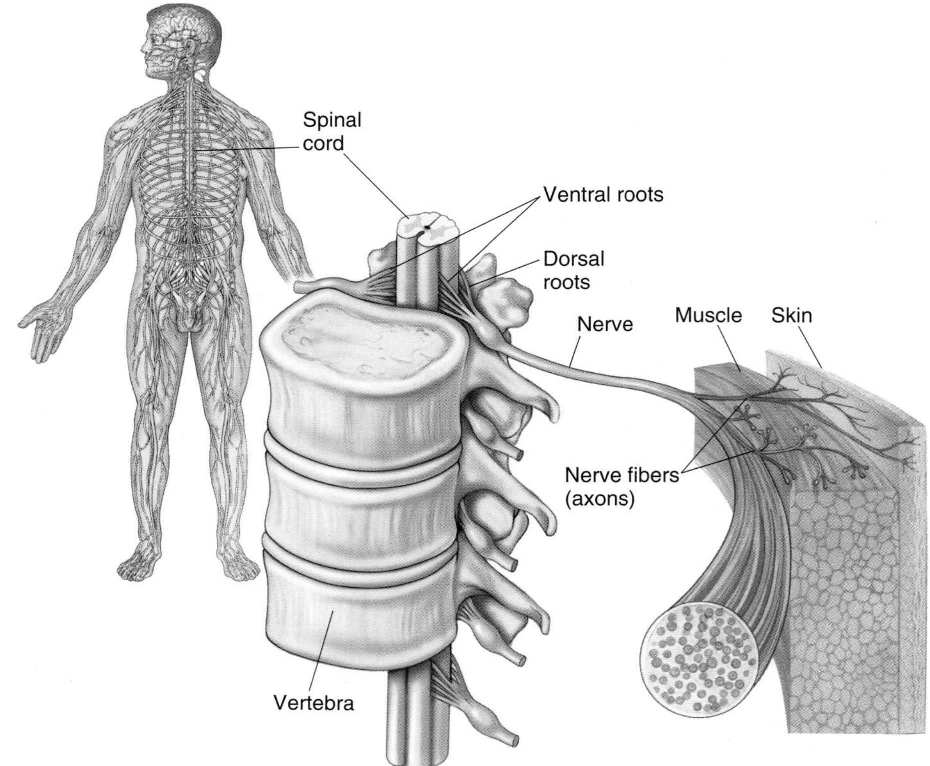

Figure 1.9
Spinal nerves and spinal nerve roots. Thirty-one pairs of nerves leave the spinal cord to supply the skin and the muscles. Cutting a spinal nerve leads to a loss of sensation and a loss of movement in the affected region of the body. Incoming sensory fibers and outgoing motor fibers divide into spinal roots where the nerves attach to the spinal cord. Bell and Magendie found that the ventral roots contain only motor fibers and the dorsal roots contain only sensory fibers.

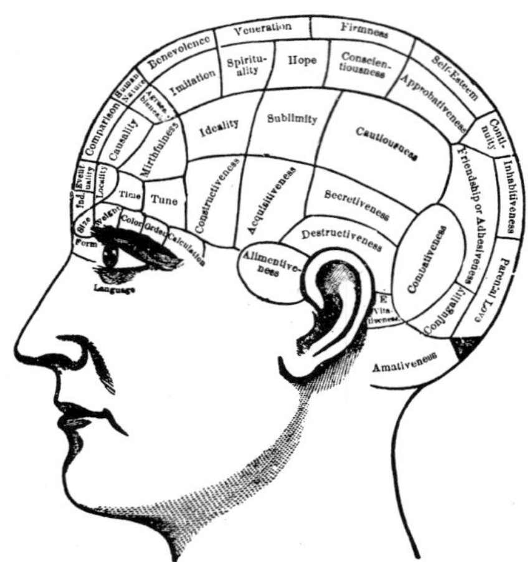

Figure 1.10
A phrenological map. According to Gall and his followers, different behavioral traits could be related to the size of different parts of the skull. (Source: Clarke and O'Malley, 1968, Fig. 118.)

Localization of Specific Functions to Different Parts of the Brain. If different functions are localized in different spinal roots, then perhaps different functions also are localized in different parts of the brain. In 1811, Bell proposed that the origin of the motor fibers is the cerebellum and the destination of the sensory fibers is the cerebrum.

How would you test this proposal? One way is to use the same approach that Bell and Magendie employed to identify the functions of the spinal roots: to destroy these parts of the brain and test for sensory and motor deficits. This approach, in which parts of the brain are systematically destroyed to determine their function, is called the *experimental ablation method*. In 1823, the esteemed French physiologist Marie-Jean-Pierre Flourens used this method in a variety of animals (particularly birds) to show that the cerebellum does indeed play a role in the coordination of movement. He also concluded that the cerebrum is involved in sensation and perception, as Bell and Galen before him had suggested. Unlike his predecessors, however, Flourens provided solid experimental support for his conclusions.

What about all those bumps on the brain's surface? Do they perform different functions as well? The idea that they do was irresistible to a young Austrian medical student named Franz Joseph Gall. Believing that bumps on the surface of the skull reflect the bumps on the surface of the brain, Gall proposed in 1809 that the propensity for certain personality traits, such as generosity, secretiveness, and destructiveness, could be related to the dimensions of the head (Figure 1.10). To support his claim, Gall and his followers collected and carefully measured the skulls of hundreds of people representing the extensive range of personality types, from the very gifted to the criminally insane. This new "science" of correlating the structure of the head with personality traits was called *phrenology*. Although the claims of the phrenologists were never taken seriously by the mainstream scientific community, they did capture the popular imagination of the time. In fact, a textbook on phrenology published in 1827 sold over 100,000 copies.

One of the most vociferous critics of phrenology was Flourens, the same man who had shown experimentally that the cerebellum and cerebrum perform different functions. His grounds for criticism were sound. For one thing, the shape of the skull is not correlated with the shape of the brain. In addition, Flourens performed experimental ablations showing that particular traits are not isolated to the portions of the cerebrum specified by phrenology. Flourens also concluded, however, that all regions of the cerebrum participate equally in all cerebral functions, a conclusion that later was shown to be erroneous.

Figure 1.11
Paul Broca (1824–1880). By carefully studying the brain of a man who had lost the faculty of speech after a brain lesion (see Figure 1.12), Broca became convinced that different functions could be localized to different parts of the cerebrum. (Source: Clarke and O'Malley, 1968, Fig. 121.)

The person usually credited with tilting the scales of scientific opinion firmly toward localization of function in the cerebrum was French neurologist Paul Broca (Figure 1.11). Broca was presented with a patient who could understand language but could not speak. Following the man's death in 1861, Broca carefully examined his brain and found a lesion in the left frontal lobe (Figure 1.12). Based on this case and several others like it, Broca concluded that this region of the human cerebrum was specifically responsible for the production of speech.

Solid experimental support for cerebral localization in animals quickly followed. German physiologists Gustav Fritsch and Eduard Hitzig showed in 1870 that applying small electrical currents to a circumscribed region of the exposed surface of the brain of a dog could elicit discrete movements. Scottish neurologist David Ferrier repeated these experiments with monkeys. In 1881, he showed that removal of this same region of the cerebrum causes paralysis of the muscles. Similarly, German physiologist Hermann Munk using experimental ablation presented evidence that the occipital lobe of the cerebrum was specifically required for vision.

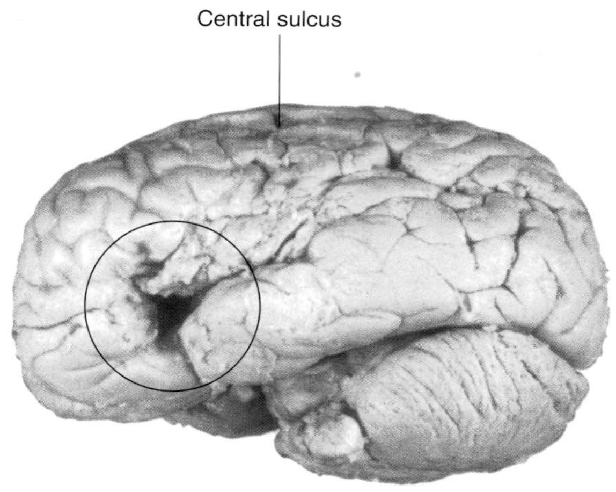

Figure 1.12
The brain that convinced Broca of localization of function in the cerebrum. This is the preserved brain of a patient who had lost the ability to speak before he died in 1861. The lesion that produced this deficit is circled. (Source: Corsi, 1991, Fig. III,4.)

As you will see in Part II of this book, we now know that there is a very clear division of labor in the cerebrum, with different parts performing very different functions. Today's maps of the functional divisions of the cerebrum rival even the most elaborate of those produced by the phrenologists. The big difference is that unlike the phrenologists, scientists today require solid experimental evidence before attributing a specific function to a portion of the brain. All the same, Gall seems to have had the right idea. It is natural to wonder why Flourens, the pioneer of brain localization of function, was misled into believing that the cerebrum acted as a whole and could not be subdivided. There are many reasons that this gifted experimentalist may have missed cerebral localization, but it seems clear that one reason was his visceral reaction against Gall and phrenology. He could not bring himself to agree even remotely with Gall, whom he viewed as a lunatic. This reminds us that science, for better or worse, was and still is a distinctly human endeavor.

The Evolution of Nervous Systems. In 1859, English biologist Charles Darwin (Figure 1.13) published *On the Origin of Species*. In this landmark work, he articulated a theory of evolution: that species of organisms evolved from a common ancestor. According to his theory, differences among species arise by a process Darwin called *natural selection*. As a result of the mechanisms of reproduction, the physical traits of the offspring are sometimes different from those of the parents. If these traits represent an advantage for survival, the offspring themselves will be more likely to reproduce, thus increasing the likelihood that the advantageous traits are passed on to the next generation. Over the course of many generations, this process has led to the development of traits that distinguish species today: flippers on harbor seals, paws on dogs, hands on raccoons, and so on. This single insight revolutionized biology. Today, scientific evidence ranging from anthropology to molecular genetics overwhelmingly supports the theory of evolution by natural selection.

Figure 1.13
Charles Darwin (1809–1882). Darwin proposed his theory of evolution, explaining how species evolve through the process of natural selection. (Source: The Bettman Archive.)

Darwin included behavior among the heritable traits that could evolve. For example, he noticed that many mammalian species show the same reaction when frightened: The pupils of the eyes get bigger, the heart races, hairs stand on end. This is as true for a human as it is for a dog. To Darwin, the similarities of this response pattern indicated that these different species evolved from a common ancestor, which possessed the same behavioral trait (advantageous presumably because it facilitated escape from predators). Because behavior reflects the activity of the nervous system, we can infer that the brain mechanisms that underlie this fear reaction may be similar, if not identical, across these species.

The idea that the nervous systems of different species evolved from common ancestors and may have common mechanisms is the rationale for relating the results of animal experiments to humans. Thus, for example, many of the details of electrical impulse conduction along nerve fibers were worked out first in the squid but are now known to apply equally well to humans. Most neuroscientists today use *animal models* of the process they wish to understand in humans. For example, rats show clear signs of addiction if they are given the chance to self-administer cocaine repeatedly. Consequently, rats are a valuable animal model for research focused on understanding how psychoactive drugs exert their effects on the nervous system.

On the other hand, many behavioral traits are highly specialized for the environment (or niche) a species normally occupies. For example, monkeys swinging from branch to branch have a keen sense of sight, while rats slinking through underground tunnels have poor vision but a highly evolved sense of touch using the whiskers on the snout. Adaptations are reflected in the structure and function of the brain of every species. By comparing the specializations of the brains of different species, neuroscientists have been able to identify which parts of the brain are specialized for different behavioral functions. Examples for monkeys and rats are shown in Figure 1.14.

The Neuron: The Basic Functional Unit of the Brain. Refinement of the microscope in the early 1800s gave scientists their first opportunity to examine animal tissues at high magnifications. In 1839, German zoologist Theodor Schwann proposed what came to be known as the *cell theory*: All tissues are composed of microscopic units called cells.

Although cells in the brain had been identified and described, there was still controversy about whether the individual "nerve cell" was actually the

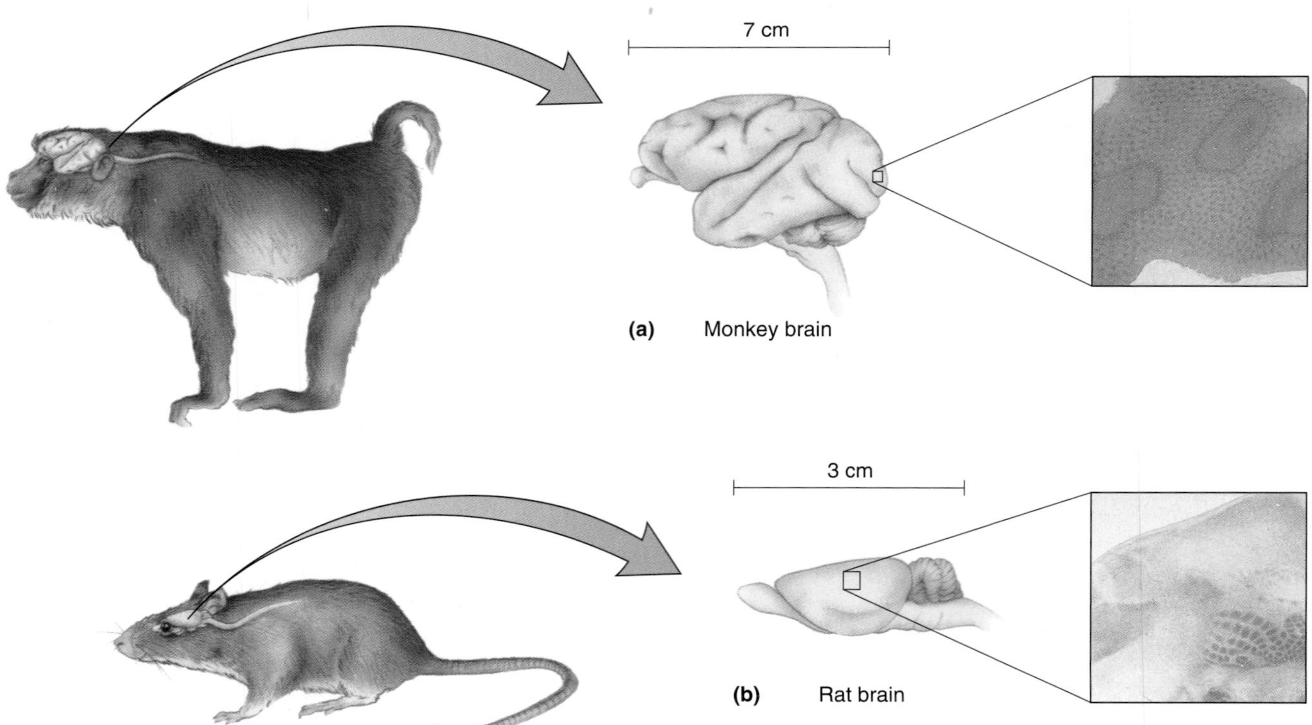

Figure 1.14
Different brain specializations in monkeys and rats. (a) The brain of the macaque monkey has a highly evolved sense of sight. The boxed region receives information from the eyes. When this region is sliced open and stained to show metabolically active tissue, a mosaic of "blobs" appears. The neurons within the blobs are specialized to analyze colors in the visual world. **(b)** The brain of a rat has a highly evolved sense of touch to the face. The boxed region receives information from the whiskers. When this region is sliced open and stained to show the location of the neurons, a mosaic of "barrels" appears. Each barrel is specialized to receive input from a single whisker on the rat's face. (Photomicrographs courtesy of Dr. S. H. C. Hendry.)

basic unit of brain function. Nerve cells usually have a number of thin projections, or processes, that extend from a central cell body (Figure 1.15). Initially, scientists could not decide whether the processes from different cells fuse together like the blood vessels of the circulatory system. If this were true, then the "nerve net" of connected nerve cells would represent the elementary unit of brain function.

Chapter 2 presents a brief history of how this issue was resolved. Suffice it to say that by 1900, the individual nerve cell, now called the neuron, was recognized to be the basic functional unit of the nervous system.

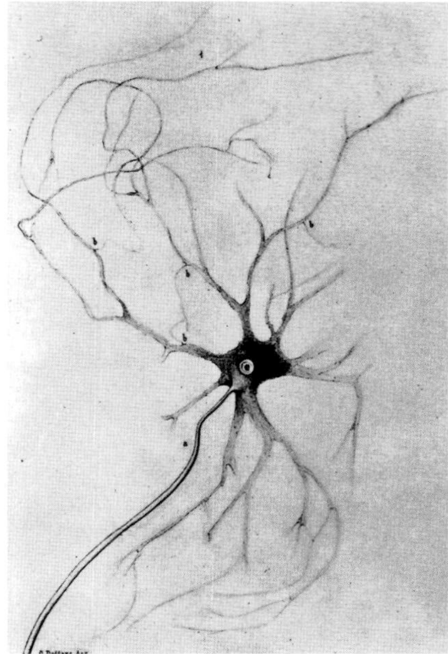

Figure 1.15
An early depiction of a nerve cell. Published in 1865, this drawing by German anatomist Otto Deiters shows a nerve cell, or neuron, and its many projections, called neurites. For a time it was thought that the neurites from different neurons might fuse together like the blood vessels of the circulatory system. We now know that neurons are distinct entities that communicate using chemical signals. (Source: Clarke and O'Malley, 1968, Fig. 16.)

NEUROSCIENCE TODAY

The history of modern neuroscience is still being written, and the accomplishments, to date, form the basis for this textbook. We will discuss the most recent developments throughout the book. Now let's take a look at how brain research is conducted today and why its continuation is important to society.

Levels of Analysis

History has clearly shown that understanding how the brain works is a big challenge. To reduce the complexity of the problem, neuroscientists break it into smaller pieces for systematic experimental analysis. This is called the *reductionist approach*. The size of the unit of study defines what is often called the *level of analysis*. In ascending order of complexity, these levels are molecular, cellular, systems, behavioral, and cognitive.

Molecular Neuroscience. The brain has been called the most complex piece of matter in the universe. Brain matter consists of a fantastic variety of molecules, many of which are unique to the nervous system. These different molecules play many different roles that are crucial for brain function: messengers that allow neurons to communicate with one another, sentries that control what materials can enter or leave neurons, conductors that orchestrate neuron growth, archivists of past experiences. The study of the brain at this most elementary level is called molecular neuroscience.

Cellular Neuroscience. The next level of analysis is cellular neuroscience, which focuses on studying how all those molecules work together to give the neuron its special properties. Among the questions asked at this level are: How many different types of neurons are there, and how do they differ in function? How do neurons influence other neurons? How do neurons become "wired together" during fetal development? How do neurons perform computations?

Systems Neuroscience. Constellations of neurons form complex circuits that perform a common function: vision, for example, or voluntary movement. Thus, we can speak of the "visual system" and the "motor system," each of which possesses its own distinct circuitry within the brain. At this level of analysis, called systems neuroscience, neuroscientists study how different neural circuits analyze sensory information, form perceptions of the external world, make decisions, and execute movements.

Behavioral Neuroscience. How do neural systems work together to produce integrated behaviors? For example, are different forms of memory accounted for by different systems? Where in the brain do "mind-altering" drugs act, and what is the normal contribution of these systems to the regulation of mood and behavior? What neural systems account for gender-specific behaviors? Where in the brain do dreams come from? These questions are studied in behavioral neuroscience.

Cognitive Neuroscience. Perhaps the greatest challenge of neuroscience is understanding the neural mechanisms responsible for the higher levels of human mental activity, such as self-awareness, mental imagery, and language. Research at this level, called cognitive neuroscience, studies how the activity of the brain creates the mind.

Neuroscientists

"Neuroscientist" sounds impressive, kind of like "rocket scientist." But we were all students once, just like you. For whatever reason—maybe our eyesight was poor, or perhaps a family member suffered a loss of speech after a stroke, and we wanted to know why—we came to share a thirst for knowledge of how the brain works. Perhaps you will, too.

Being a neuroscientist is rewarding, but it does not come easily. Many years of training are required. One perhaps begins by helping out in a research lab during or after college and then going to graduate school to earn an advanced degree, either a Ph.D. or an M.D. (or both). This usually is followed by several years of postdoctoral research to learn new techniques or ways of thinking under the direction of an established neuroscientist. Finally, the "young" neuroscientist is ready to set up shop at a university, institute, or hospital.

Broadly speaking, neuroscience research (and neuroscientists) may be divided into two types: *clinical* and *experimental*. Clinical research is mainly conducted by physicians (M.D.s). The main medical specialties associated with the human nervous system are neurology, psychiatry, neurosurgery, and neuropathology (Table 1.1). Many who conduct clinical research continue in the tradition of Broca, attempting to deduce from the behavioral effects of brain damage the functions of various parts of the brain. Others conduct studies to assess the benefits and risks of new types of treatment.

Despite the obvious value of clinical research, the foundation for all medical treatments of the nervous system was and continues to be laid by experimental neuroscientists, who may hold either an M.D. or a Ph.D. The experimental approaches to studying the brain are so broad that they include almost every conceivable methodology. Thus, despite the interdisciplinary nature of neuroscience, expertise in a particular methodology is still often used to distinguish one neuroscientist from another. Thus, there are *neuroanatomists*, who use sophisticated microscopes to trace connections in the brain; *neurophysiologists*, who use electrodes, amplifiers, and oscilloscopes to measure the brain's electrical activity; *neuropharmacologists*, who use "de-

Table 1.1 Medical Specialists Associated With the Nervous System

SPECIALIST	DESCRIPTION
Neurologist	An M.D. trained to diagnose and treat diseases of the nervous system
Psychiatrist	An M.D. trained to diagnose and treat disorders of mood and personality
Neurosurgeon	An M.D. trained to perform surgery on the brain and spinal cord
Neuropathologist	An M.D. or Ph.D. trained to recognize the changes in nervous tissue that result from disease

Table 1.2 Types of Experimental Neuroscientists

TYPE	DESCRIPTION
Computational neuroscientist	Uses mathematics and computers to construct models of brain functions
Developmental neurobiologist	Analyzes the development and maturation of the brain
Molecular neurobiologist	Uses the genetic material of neurons to understand the structure and function of brain molecules
Neuroanatomist	Studies the structure of the nervous system
Neurochemist	Studies the chemistry of the nervous system
Neuroethologist	Studies the neural basis of species-specific animal behaviors in natural settings
Neuropharmacologist	Examines the effects of drugs on the nervous system
Neurophysiologist	Measures the electrical activity of the nervous system
Neuropsychologist	Studies the neural basis of human behavior
Physiological psychologist (biological psychologist, psychobiologist)	Studies the biological basis of animal behavior
Psychophysicist	Quantitatively measures perceptual abilities

signer drugs" to study the chemistry of brain function; *molecular neurobiologists*, who probe the genetic material of neurons to find clues about the structure of brain molecules; and so on. Table 1.2 lists some of the types of experimental neuroscientists. Ask your instructor what type or types he or she is.

The Scientific Process

Neuroscientists of all stripes endeavor to establish truths about the nervous system. Regardless of the level of analysis they choose, they work according to the scientific process, which consists of four essential steps: observation, replication, interpretation, and verification.

Observation. Observations typically are made during experiments designed to test a particular hypothesis. For example, Bell hypothesized that the ventral roots contain the nerve fibers that control the muscles. To test this idea, he performed the experiment in which he cut these fibers and then observed whether or not muscular paralysis resulted. Other types of observation derive from carefully watching the world around us, or from introspection, or from human clinical cases. For example, Broca's careful observations led him to correlate left frontal lobe damage with the loss of the ability to speak.

Replication. Whether the observation is experimental or clinical, it is essential that it be replicated before it can be accepted by the scientist as fact. Replication simply means repeating the experiment on different subjects or making similar observations in different patients, as many times as necessary to rule out the possibility that the observation occurred by chance.

Interpretation. Once the scientist believes the observation is correct, he or she makes an interpretation. Interpretations depend on the state of knowledge (or ignorance) at the time the observation was made and on the preconceived notions (the "mind set") of the scientist who made it. As such, interpretations do not always withstand the test of time. For example, at the time he made his observations, Flourens was unaware that the cerebrum of a bird is fundamentally different from that of a mammal. Thus, he wrongly concluded from experimental ablations in birds that there was no localization of certain functions in the cerebrum of mammals. Moreover, as mentioned before, his profound distaste for Gall surely also colored his interpretation. The point is that the correct interpretation often remains unrecognized until long after the original observations were made. Indeed, major breakthroughs are sometimes made when old observations are reinterpreted in a new light.

Verification. The final step of the scientific process is verification. This step is distinct from the replication performed by the original observer. Verification means that the observation is sufficiently robust that it will be seen by any competent scientist who precisely follows the protocols of the original observer. Successful verification generally means that the observation is accepted as fact. However, not all observations can be verified. Sometimes this is due to inaccuracies in the original report or insufficient replication. But failure to verify usually stems from the fact that additional variables, such as temperature or time of day, contributed to the original result. Thus, the process of verification, if affirmative, establishes new scientific fact, and, if negative, suggests new interpretations for the original observation.

Occasionally, one reads in the popular press about a case of "scientific fraud." Researchers face keen competition for limited research funds and feel considerable pressure to "publish or perish." In the interest of expediency, a few have actually published "observations" that were never made. Fortunately, such instances of fraud are rare, thanks to the scientific process. Before long, other scientists find they are unable to verify the fraudulent observations and begin to question how they could have been obtained in the first place. The material you will learn in this book stands as a strong testament to the success of the scientific process.

The Use of Animals in Neuroscience Research

Most of what we know about the nervous system has come from experiments on animals. In most cases, the animals are killed so their brains can be examined neuroanatomically, neurophysiologically, and/or neurochemically. The fact that animals are sacrificed for the pursuit of human knowledge raises questions about the ethics of animal research.

The Animals. Let's begin by putting the issue in perspective. Throughout history, humans have considered animals and animal products as renewable natural resources to be used for food, clothing, transportation, recreation, sport, and companionship. The animals used for research, education, and testing have always been a small fraction of the total used for other purposes. For example, in the United States today, the number of animals used for all types of biomedical research is less than 1% of the number that are killed for food alone.[1] The number used specifically for neuroscience research is much smaller still.

Neuroscience experiments are conducted using many different species ranging from snails to monkeys. The choice of animal species is generally dictated by the question under investigation, the level of analysis, and the extent to which the knowledge gained at this level can be related to humans. As a rule, the more basic the process under investigation, the more distant can be the evolutionary relationship with humans. Thus, experiments aimed at un-

[1]According to the National Academy of Sciences Institute of Medicine, 1991.

derstanding the molecular basis of nerve impulse conduction can be carried out on a distantly related species, such as the squid. On the other hand, understanding the neural basis of movement and perceptual disorders in humans has required experiments on more closely related species, such as the macaque monkey. Today, more than half of the animals used for neuroscience research are rodents—mice and rats—that are bred specifically for this purpose.

Animal Welfare. In the developed world today, most educated adults have a concern for animal welfare. Neuroscientists share this concern and work to ensure that animals are well treated. It must also be appreciated, however, that society has not always placed such value on animal welfare, as reflected in some of the scientific practices of the past. For example, in his experiments early in the nineteenth century, Magendie used unanesthetized puppies (for which he was later criticized by his scientific rival Bell). Before passing judgment, consider that the philosophy of Descartes was very influential in French society at that time. Animals of all types were believed to be simple automata, biological machines that lacked any semblance of emotion. As disturbing as this now seems, it is also worth bearing in mind that humans scarcely had any more respect for one another than they did for animals during this period (slavery was still practiced in the United States, for example). Fortunately, some things have changed quite dramatically since then. Heightened awareness of animal welfare in recent years has led to significant improvements in how animals are treated in biomedical research. Unfortunately, other things have changed little. Humans around the world continue to abuse one another in myriad ways (child abuse, violent crime, ethnic cleansing, and so on).

Today, neuroscientists accept certain moral responsibilities toward their animal subjects:

1. Animals are used only for worthwhile experiments that promise to advance our knowledge of the nervous system.
2. All necessary steps are taken to minimize pain and distress experienced by the experimental animals (use of anesthetics, analgesics, etc.).
3. All possible alternatives to the use of animals are considered.

Adherence to this ethical code is monitored in a number of ways. First, research proposals must pass a review by the Institutional Animal Care and Use Committee (IACUC). Members of this committee include a veterinarian, scientists in other disciplines, and nonscientist community representatives. After passing the IACUC review, proposals are evaluated for scientific merit by a panel of expert neuroscientists. This step ensures that only the most worthwhile projects are carried out. Then, when neuroscientists attempt to publish their observations in the professional journals, the papers are carefully reviewed by other neuroscientists for scientific merit and for animal welfare concerns. Reservations about either issue can lead to rejection of the papers, which in turn can lead to a loss of funding for the research. In addition to these monitoring procedures, federal law sets strict standards for the housing and care of laboratory animals.

Animal Rights. Most people accept the necessity for animal experimentation to advance knowledge, as long as it is performed humanely and with the proper respect for the animal's welfare. However, a vocal and increasingly violent minority seeks the total abolition of animal use for human purposes, including experimentation. These people subscribe to a philosophical position called "animal rights." According to this way of thinking, animals have the same legal and moral rights as humans do.

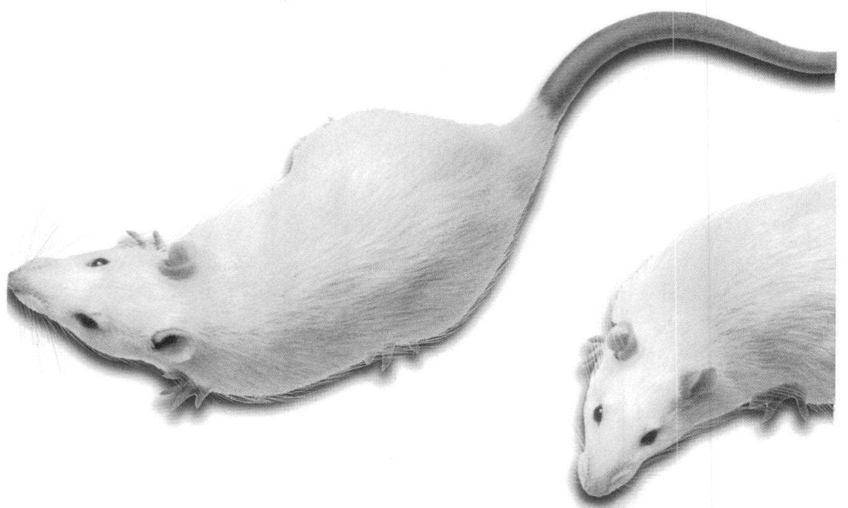
They've Saved More People Than 911.

Figure 1.16
Our debt to animal research. This poster counters the claims of animal rights activists by raising public awareness of the benefits of animal research. (Source: National Foundation for Biomedical Research.)

If you are an animal lover, you may be sympathetic to this position. But consider the following questions. Are you willing to deprive yourself and your family of medical procedures that were developed using animals? Is the death of a mouse equivalent to the death of a human being? Is keeping a pet the moral equivalent of slavery? Is eating meat the moral equivalent of murder? Is it unethical to take the life of a pig to save the life of a child? Is controlling the rodent population in the sewers or the roach population in your home morally equivalent to the Holocaust? If your answer is no to any of these questions, then you do not subscribe to the philosophy of animal rights. *Animal welfare*—a concern that all responsible people share—must not be confused with animal rights.

Animal rights activists have vigorously pursued their agenda against animal research, sometimes with alarming success. They have manipulated public opinion with repeated allegations of cruelty in animal experiments that are grossly distorted or blatantly false. They have vandalized laboratories, destroying years of hard-won scientific data and hundreds of thousands of dollars of equipment (that you, the taxpayer, had paid for). Using threats of violence, they have driven some researchers out of science altogether.

Fortunately, the tide is turning. Thanks to the efforts of a number of people, scientists and nonscientists alike, the false claims of the extremists have been exposed, and the benefits to humankind of animal research have been extolled (Figure 1.16). Considering the staggering toll in terms of human suffering that results from disorders of the nervous system, neuroscientists take the position that it is immoral *not* to wisely use all the resources nature has provided, including animals, to gain an understanding of how the brain functions in health and in disease.

The Cost of Ignorance: Nervous System Disorders

Modern neuroscience research is expensive, but the cost of ignorance about the brain is far greater. Table 1.3 lists some of the disorders that affect the ner-

Table 1.3 Some Major Disorders of the Nervous System

DISORDER	DESCRIPTION
Alzheimer's disease	A progressive degenerative disease of the brain, characterized by dementia and always fatal
Cerebral palsy	A motor disorder caused by damage to the cerebrum at the time of birth
Depression	A serious disorder of mood characterized by insomnia, loss of appetite, and feelings of dejection
Epilepsy	A condition characterized by periodic disturbances of brain electrical activity that can lead to seizures, loss of consciousness, and sensory disturbances
Multiple sclerosis	A progressive disease that affects nerve conduction, characterized by episodes of weakness, lack of coordination, and speech disturbance
Parkinson's disease	A progressive disease of the brain that leads to difficulty in initiating voluntary movement
Schizophrenia	A severe psychotic illness characterized by delusions, hallucinations, and bizarre behavior
Spinal injury	A loss of feeling and movement caused by traumatic damage to the spinal cord
Stroke	A loss of brain function caused by disruption of the blood supply, usually leading to permanent sensory, motor, or cognitive deficit

vous system. It is likely that your family has felt the impact of one or more of these. Let's look at a few brain disorders and examine their effects on society.[2]

Alzheimer's disease and Parkinson's disease are both characterized by progressive degeneration of specific neurons in the brain. Parkinson's disease, which results in a crippling impairment of voluntary movement, currently affects approximately 500,000 Americans. Alzheimer's disease leads to dementia, a state of confusion characterized by the loss of ability to learn new information and to recall previously acquired knowledge. The U.S. National Institutes of Health (NIH) estimates that dementia affects 10% of people over age 65 and 50% of people over age 85. The number of Americans with dementia totals well over 3 million. Indeed, it is now recognized that dementia is not an inevitable outcome of aging, as was once believed, but is a sign of brain disease. Alzheimer's disease progresses mercilessly, robbing its victims first of their minds, then of control over basic bodily functions, and finally of their lives; the disease is always fatal. In the United States, the annual cost of care for people with dementia is approximately $90 billion.

Depression and schizophrenia are disorders of mood and thought. Depression is characterized by overwhelming feelings of dejection, worthlessness, and guilt. Fifteen million Americans will experience a major depressive illness at some time in their lives. Depression is the leading cause of suicide, which claims 30,000 lives each year in the United States. Schizophrenia is a severe personality disorder characterized by delusions, hallucinations, and bizarre behavior. This disease often strikes at the prime of life—adolescence or early adulthood—and can persist for life. Over 2 million Americans suffer from schizophrenia. The National Institute of Mental Health (NIMH) estimates that mental disorders, such as depression and schizophrenia, cost the United States in excess of $130 billion annually.

Stroke is the third leading cause of death in the United States. Stroke victims who do not die, some 100,000 per year, are likely to be permanently disabled. The annual cost of stroke nationwide is $25 billion. Alcohol and drug addiction affects virtually every family in the United States. The cost in terms of treatment, lost wages, and other consequences approaches $150 billion per year. These few examples only scratch the surface. *More Americans are hospitalized with neurological and mental disorders than with any other major disease group, including heart disease and cancer.*

[2]Source of statistics for disorders in this section: U.S. Office of Science and Technology Policy, 1991.

The economic costs of brain dysfunction are enormous, but they pale in comparison with the staggering emotional toll on victims and their families. The prevention and treatment of brain disorders require an understanding of normal brain function, and this basic understanding is the goal of neuroscience. Neuroscience research has already contributed to the development of increasingly effective treatments for Parkinson's disease, depression, and schizophrenia. New strategies are being tested to rescue dying neurons in people with Alzheimer's disease and those who have had a stroke. Major progress has been made in understanding how drugs and alcohol affect the brain and how they lead to addictive behavior. The material in this book demonstrates that a lot is known about the function of the brain. But what we know is insignificant compared with what is still left to be learned.

CONCLUDING REMARKS

In this chapter, we have emphasized that neuroscience is a distinctly human endeavor. The historical foundations of neuroscience were laid by many people over many generations. Men and women today are working at all levels of analysis, using all types of technology, to shed light on the workings of the brain. The fruits of this labor form the basis for this textbook.

The goal of neuroscience is to understand how nervous systems function. Many important insights can be gained from a vantage point outside the head. Because the brain's activity is reflected in behavior, careful behavioral measurements inform us of the capabilities and limitations of brain function. Computer models that reproduce the brain's computational properties can help us understand how these properties might arise. From the scalp we can measure brain waves, which tell us something about the electrical activity of different parts of the brain during various behavioral states. New computer-assisted imaging techniques enable researchers to examine the structure of the living brain as it sits in the head. And using even more sophisticated imaging methods, we are beginning to see what different parts of the human brain become active under different conditions. But none of these noninvasive methods, old or new, can substitute for experimentation with living brain tissue. We cannot make sense of remotely detected signals without being able to see how they are generated and what their significance is. To understand *how* the brain works, we must open the head and examine what's inside—neuroanatomically, neurophysiologically, and neurochemically.

The pace of neuroscience research today is truly breathtaking and raises hopes that soon we will have new treatments for the wide range of nervous system disorders that debilitate and cripple millions of people annually. In recognition of the progress and promise of brain research, the U.S. Congress designated the 1990s the "Decade of the Brain." (An esteemed colleague of ours has suggested that while this was a good idea, Congress was perhaps overly optimistic; he suggests that we designate the new century as the "Century of the Brain.") Despite the progress during the past decade and the centuries preceding it, we still have a long way to go before we fully understand how the brain performs all of its amazing feats. But this is the fun of being a neuroscientist: Because our ignorance of brain function is so vast, a startling new discovery lurks around virtually every corner.

REVIEW QUESTIONS

1. What are brain ventricles, and what functions have been ascribed to them over the ages?
2. What experiment did Bell perform to show that the nerves of the body contain a mixture of sensory and motor fibers?
3. What did Flourens' experiments suggest were the functions of the cerebrum and the cerebellum?
4. What is the meaning of the term animal model?
5. A region of the cerebrum is now called Broca's area. What function do you think this region performs, and why?
6. What are the different levels of analysis in neuroscience research? What types of question do researchers ask at each level?
7. What are the steps in the scientific process? Describe each one.

CHAPTER

2 Neurons and Glia

INTRODUCTION

All tissues and organs in the body consist of cells. The specialized functions of cells and how they interact determine the functions of organs. The brain is an organ—to be sure, the most sophisticated and complex organ that nature has devised. But the basic strategy for unraveling its function is no different from that used to investigate the pancreas or the lung. We must begin by learning how brain cells work individually and then see how they are assembled to work together. In neuroscience, there is no need to separate *mind* from *brain*; once we fully understand the individual and concerted actions of brain cells, we will understand the origins of our mental abilities. The organization of this book reflects this "neurophilosophy." We begin with the cells of the nervous system—their structure, function, and means of communication. In later chapters, we will explore how these cells are assembled into circuits that mediate sensation, perception, movement, speech, and emotion.

In this chapter, we focus on the structure of the different types of cells in the nervous system: *neurons* and *glia*. These are broad categories, within which are many types of cells that differ based on their structure, chemistry, and function. Nonetheless, the distinction between neurons and glia is important. Although there are many neurons in the human brain (about 100 billion), glia outnumber neurons by tenfold. Based on these numbers, it might appear that we should focus our attention on glia for insights into the cellular functions of the nervous system. However, neurons are the most important cells for the unique functions of the brain. It is the neurons that sense changes in the environment, communicate these changes to other neurons, and command the body's responses to these sensations. Glia are thought to contribute to brain function mainly by insulating, supporting, and nourishing neighboring neurons. If the brain were a chocolate-chip cookie and the neurons were chocolate chips, the glia would be the cookie dough that fills all the other space and ensures that the chips are suspended in their appropriate locations. Indeed, the term *glia* is derived from the Greek word for "glue," giving the impression that the main function of these cells is to keep the brain from running out of our ears! As we shall see later in the chapter, the simplicity of this view is probably a good indication of the depth of our ignorance about glial function. However, we still are confident that neurons perform the bulk of information processing in the brain. Therefore, we will focus 90% of our attention on 10% of brain cells: the neurons.

Neuroscience, like other fields, has a language all its own. To use this language, you must learn the vocabulary. After you have read this chapter, take a few minutes to review the key terms list and make sure you understand the meaning of each term. Your neuroscience vocabulary will grow as you work your way through the book.

THE NEURON DOCTRINE

To study the structure of brain cells, scientists have had to overcome several obstacles. The first was the small size. Most cells are in the range of 0.01–0.05 mm in diameter. The tip of an unsharpened pencil lead is about 2 mm across; neurons are 40–200 times smaller. (For a review of the metric system, see Table 2.1.) This size is at or beyond the limit of what can be seen by the naked eye. Therefore, progress in cellular neuroscience was not possible before the development of the compound microscope in the late seventeenth century. Even then, obstacles remained. To observe brain tissue using a microscope, it was necessary to make very thin slices, ideally not much thicker than the diameter of the cells. However, brain tissue has a consistency like a bowl of Jello: not firm enough to make thin slices. Thus, the study of the anatomy of

Table 2.1 Units of Size in the Metric System

UNIT	ABBREVIATION	METER EQUIVALENT	REAL-WORLD EQUIVALENT
Kilometer	km	10^3 m	About two-thirds of a mile.
Meter	m	1 m	About 3 feet.
Centimeter	cm	10^{-2} m	Thickness of your little finger.
Millimeter	mm	10^{-3} m	Thickness of your toenail.
Micrometer	μm	10^{-6} m	Near the limit of resolution for the light microscope.
Nanometer	nm	10^{-9} m	Near the limit of resolution for the electron microscope.

Figure 2.1
Nissl-stained neurons. A thin slice of brain tissue has been stained with Cresyl violet, a Nissl stain. The clumps of deeply stained material around the cell nuclei are Nissl bodies. (Source: Hammersen, 1980, Fig. 493.)

brain cells had to await the development of a method to harden the tissue without disturbing its structure and an instrument that could produce very thin slices. Early in the nineteenth century, scientists discovered how to harden, or "fix," tissues by immersing them in formaldehyde, and they developed a special device called a microtome to make very thin slices.

These technical advances spawned the field of **histology**, the microscopic study of the structure of tissues. But scientists studying brain structure faced yet another obstacle. Freshly prepared brain has a uniform, cream-colored appearance under the microscope; the tissue has no differences in pigmentation to enable histologists to resolve individual cells. Thus, the final breakthrough in neurohistology was the introduction of stains that could selectively color some, but not all, parts of the cells in brain tissue.

One stain, still used today, was introduced by the German neurologist Franz Nissl in the late nineteenth century. Nissl showed that a class of basic dyes would stain the nuclei of all cells and also stain clumps of material surrounding the nuclei of neurons (Figure 2.1). These clumps are called *Nissl bodies*, and the stain is known as the **Nissl stain**. The Nissl stain is extremely useful for two reasons. First, it distinguishes neurons and glia from one another. Second, it enables histologists to study the arrangement, or **cytoarchitecture**, of neurons in different parts of the brain. (The prefix *cyto-* is from the Greek word for "cell.") The study of cytoarchitecture led to the realization that the brain consists of many specialized regions. We now know that each region performs a different function.

Figure 2.2
Camillo Golgi (1843–1926). (Source: Finger, 1994, Fig. 3.22.)

The Golgi Stain

The Nissl stain, however, does not tell the whole story. A Nissl-stained neuron looks like little more than a lump of protoplasm containing a nucleus. Neurons are much more than that, but how much more was not recognized until the publication of the work of Italian histologist Camillo Golgi (Figure 2.2). In 1873, Golgi discovered that by soaking brain tissue in a silver chromate solution, now called the **Golgi stain**, a small percentage of neurons became darkly colored in their entirety (Figure 2.3). This revealed that the neuronal cell body, the region of the neuron around the nucleus that is shown with the Nissl stain, is actually only a small fraction of the total structure of the neuron. Notice in Figures 2.1 and 2.3 how different histological stains can provide strikingly different views of the same tissue. Today, neurohistology remains an active field in neuroscience, along with its credo: "The gain in brain is mainly in the stain."

The Golgi stain shows that neurons have at least two distinguishable parts: a central region that contains the cell nucleus, and numerous thin tubes that radiate away from the central region. The swollen region containing the cell nucleus has several names that are used interchangeably: **cell body**,

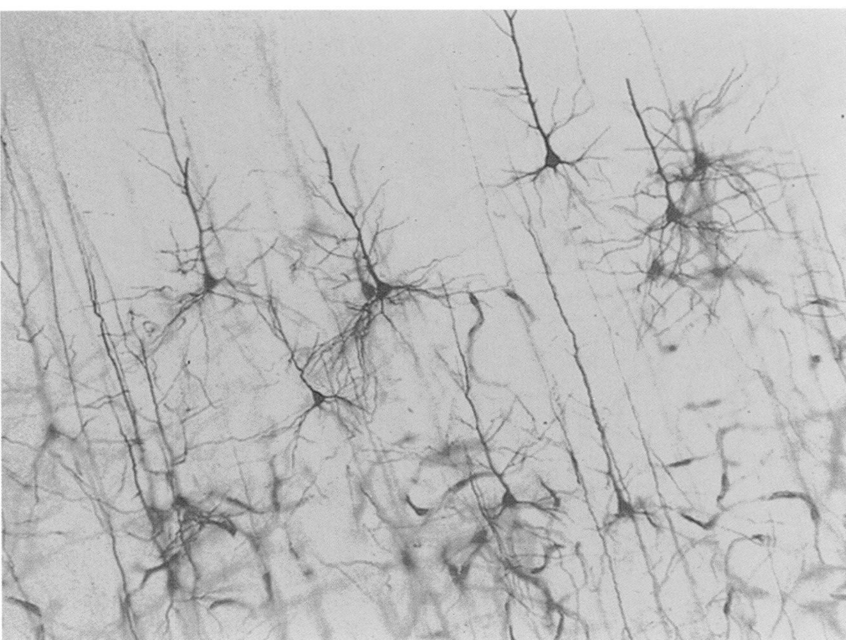

Figure 2.3
Golgi-stained neurons. (Source: Hubel, 1988, p. 126.)

soma (plural: somata), and **perikaryon** (plural: perikarya). The thin tubes that radiate away from the soma are called **neurites** and are of two types: **axons** and **dendrites** (Figure 2.4).

The cell body usually gives rise to a single axon. The axon is of uniform diameter throughout its length, and if it branches, the branches generally extend at right angles. Because axons can travel over great distances in the body (a meter or more), it was immediately recognized by the histologists of the day that axons must act like "wires" that carry the output of the neurons. Dendrites, on the other hand, rarely extend more than 2 mm in length. Many dendrites extend from the cell body and generally taper to a fine point. Early histologists recognized that because dendrites come in contact with many axons, they must act as the antennae of the neuron to receive incoming signals, or input.

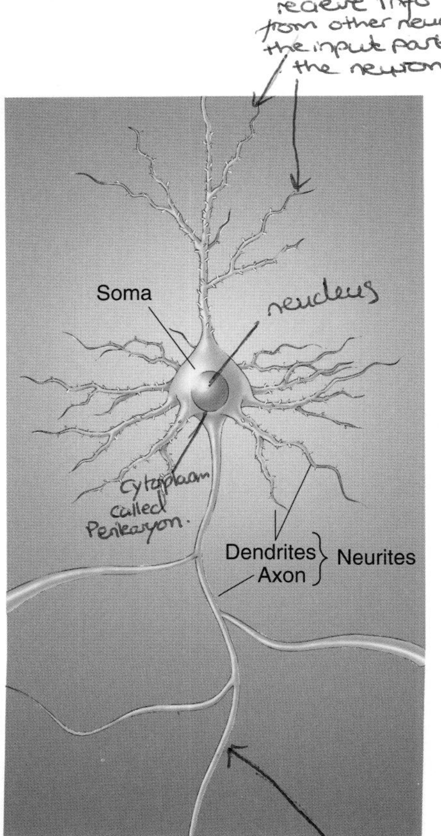

Figure 2.4
The basic parts of a neuron.

Cajal's Contribution

Golgi invented the stain, but it was a Spanish contemporary of Golgi who used it to greatest effect. Santiago Ramón y Cajal was a skilled histologist and artist who learned about Golgi's method in 1888 (Figure 2.5). In a remarkable series of publications over the next 25 years, Cajal used the Golgi stain to work out the circuitry of many regions of the brain (Figure 2.6). Ironically, Golgi and Cajal drew completely opposite conclusions about neurons. Golgi championed the view that the neurites of different cells are fused together to form a continuous reticulum, or network, similar to the arteries and veins of the circulatory system. According to this reticular theory, the brain is an exception to the cell theory, which states that the individual cell is the elementary functional unit of all animal tissues. Cajal, on the other hand, argued forcefully that the neurites of different neurons are not continuous with one another and must *communicate by contact, not continuity*. This idea that the neuron adhered to the cell theory came to be known as the **neuron doctrine**. Although Golgi and Cajal shared the Nobel Prize in 1906, they remained rivals to the end.

Figure 2.5
Santiago Ramón y Cajal (1852–1934). (Source: Finger, 1994, Fig. 3.26.)

The scientific evidence over the next 50 years weighed heavily in favor of the neuron doctrine, but final proof had to wait until the development of the electron microscope in the 1950s (Box 2.1). With the increased resolving power of the electron microscope, it was finally possible to show that the neurites of different neurons are not continuous with one another. Thus, our starting point in the exploration of the brain must be the individual neuron.

THE PROTOTYPICAL NEURON

As we have seen, the neuron consists of several parts: the soma, the dendrites, and the axon. The inside of the neuron is separated from the outside by the limiting skin, the *neuronal membrane*, which lies like a circus tent on an intricate internal scaffolding, giving each part of the cell its special three-dimensional appearance. Let's explore the inside of the neuron and learn about the functions of the different parts (Figure 2.7).

The Soma

We begin our tour at the soma, the roughly spherical central part of the neuron. The cell body of the typical neuron is about 20 μm in diameter. The watery fluid inside the cell, called the **cytosol**, is a salty, potassium-rich solution that is separated from the outside by the neuronal membrane. Within the soma are a number of membrane-enclosed structures collectively called **organelles**.

The cell body of the neuron contains the same organelles that are found in all animal cells. The most important ones are the nucleus, the rough endoplasmic reticulum, the smooth endoplasmic reticulum, the Golgi apparatus, and the mitochondria. Everything contained within the confines of the cell membrane, including the organelles but excluding the nucleus, is referred to collectively as the **cytoplasm**.

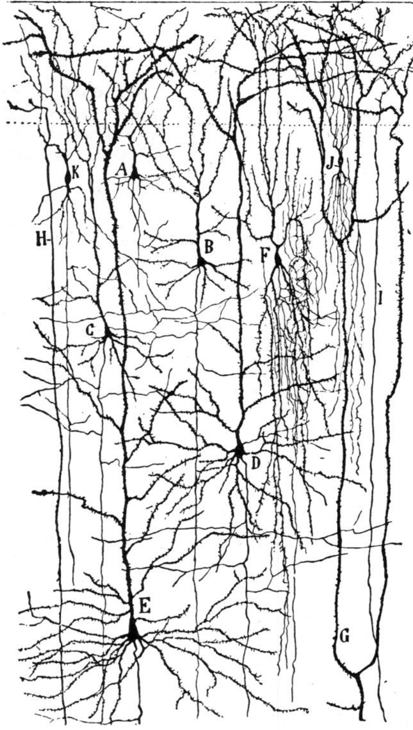

Figure 2.6
One of Cajal's many drawings of brain circuitry. The letters label the different elements Cajal identified in an area of the human cerebral cortex that controls voluntary movement. We will learn more about this part of the brain in Chapter 14. (Source: DeFelipe and Jones, 1988, Fig. 90.)

The Nucleus. Its name derived from the Latin word for "nut," the **nucleus** of the cell is spherical, centrally located, and about 5–10 μm across. It is contained within a double membrane called the *nuclear envelope*. The nuclear envelope is perforated by pores that measure about 0.1 μm across.

Within the nucleus are **chromosomes**, which contain the genetic material, **DNA (deoxyribonucleic acid)**. Your DNA was passed on to you from your parents, and it contains the blueprint for your entire body. The DNA in each of your neurons is the same, and it is the same as the DNA in the cells of your liver and kidney. What distinguishes a neuron from a liver cell is the specific parts of the DNA that are used to assemble the cell.

Each chromosome contains an uninterrupted double-stranded braid of DNA 2 nm wide. If the DNA from the 46 human chromosomes were laid out straight, end to end, it would measure more than 2 m in length. If we were to consider this total length of DNA as analogous to the string of letters that makes up this book, the genes would be analogous to the individual words. Each gene is a segment of DNA that can measure anywhere from 0.1 μm to several micrometers in length.

The "reading" of the DNA is known as **gene expression**, and the final product of gene expression is the synthesis of remarkable molecules called proteins. Proteins exist in a wide variety of shapes and sizes, perform many different functions, and bestow upon neurons virtually all of their unique characteristics. **Protein synthesis**, the assembly of protein molecules, occurs in the cytoplasm. Because the DNA never leaves the nucleus, there must be an intermediary that carries the genetic message to the sites of protein synthesis in the cytoplasm. This function is performed by another long molecule

Box 2.1

OF SPECIAL INTEREST

Advances in Microscopy

The human eye can distinguish two points only if they are separated by more than about one-tenth of a millimeter (100 μm). Thus, we can say that 100 μm is near the *limit of resolution* for the unaided eye. Neurons have a diameter of about 20 μm, and neurites can measure as small as a fraction of a micrometer. The light microscope, therefore, was a necessary development before neuronal structure could be studied. But this type of microscopy has a theoretical limit imposed by the properties of microscope lenses and visible light. With the standard light microscope, the limit of resolution is about 0.1 μm. However, the space between neurons measures only 0.02 μm (20 nm). No wonder two esteemed scientists, Golgi and Cajal, disagreed about whether neurites were continuous from one cell to the next. This question could not be answered until the electron microscope was developed and applied to biological specimens, which only occurred within the past 50 years or so.

The electron microscope uses an electron beam instead of light to form images, dramatically increasing the resolving power. The limit of resolution for an electron microscope is about 0.1 nm—a million times better than the unaided eye. Our insights into the fine structure of the inside of neurons—the *ultrastructure*—have all come from electron microscopic examination of the brain.

Today, microscopes on the leading edge of technology use laser beams to illuminate the tissue and computers to create digital images (Figure A). Unlike the traditional methods of light and electron microscopy, which require tissue fixation, these new techniques give neuroscientists their first chance to peer into brain tissue that is still alive.

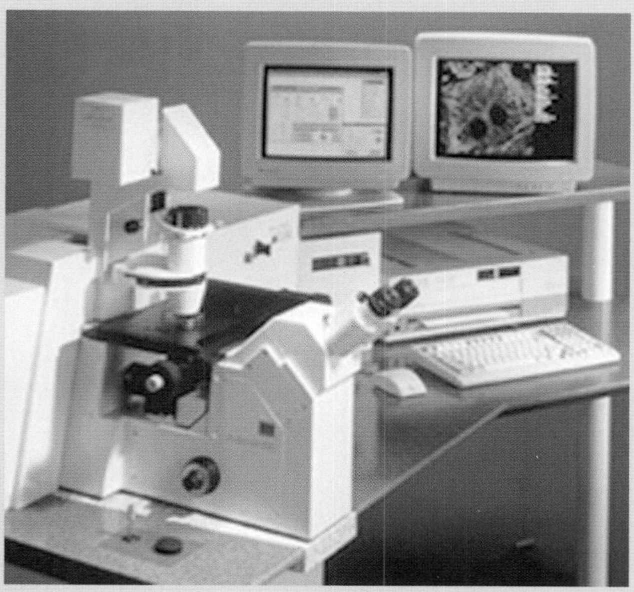

Figure A
Laser microscope and computer. (Source: Carl Zeiss, Inc.)

called **messenger ribonucleic acid**, or **mRNA**. Messenger RNA consists of four different nucleic acids strung together in various sequences to form a chain. The detailed sequence of the nucleic acids in the chain represents the information in the gene, just as the sequence of letters gives meaning to a written word. The process of assembling a piece of mRNA that contains the information of a gene is called **transcription**, and the resulting mRNA is called the *transcript* (Figure 2.8).

Messenger RNA transcripts emerge from the nucleus via pores in the nuclear envelope and travel to the sites of protein synthesis elsewhere in the neuron. At these sites, a protein molecule is assembled much as the mRNA molecule was: by linking together many small molecules into a chain. In the case of protein, the building blocks are the **amino acids**, of which there are 20 different kinds. This assembling of proteins from amino acids under the direction of the mRNA is called **translation**.

The scientific study of this process, which begins with the DNA of the nucleus and ends with the synthesis of protein molecules in the cell, is called

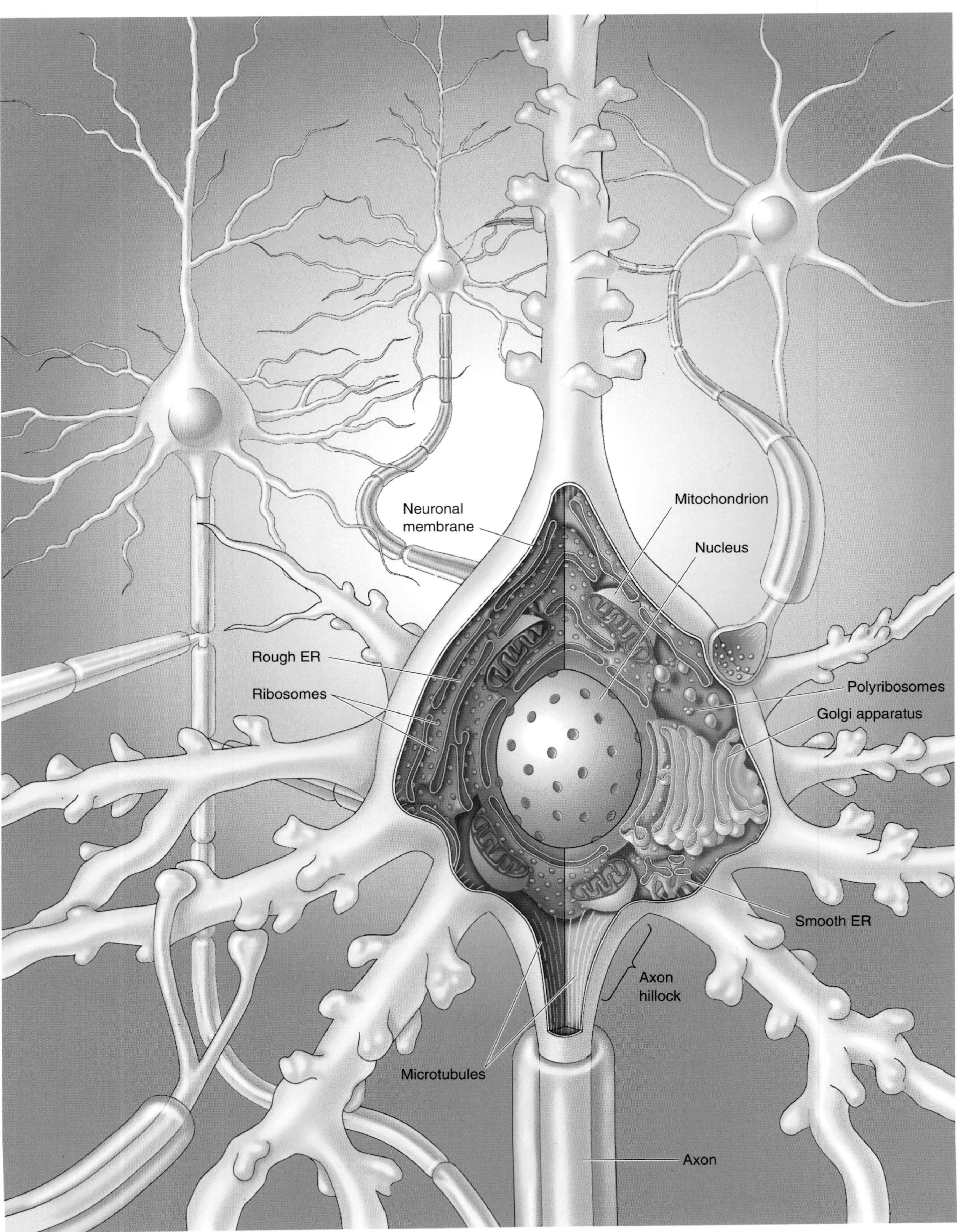
Neuronal membrane
Mitochondrion
Nucleus
Rough ER
Ribosomes
Polyribosomes
Golgi apparatus
Smooth ER
Axon hillock
Microtubules
Axon

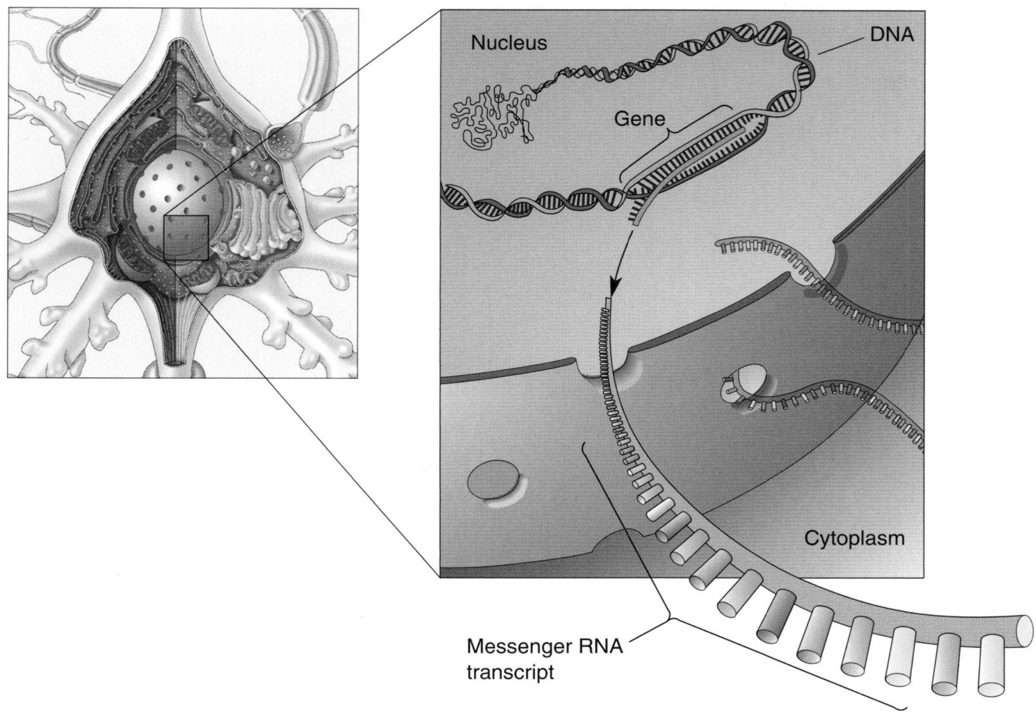

Figure 2.8
Gene transcription. Messenger RNA molecules carry the genetic instructions for protein assembly from the nucleus to the cytoplasm.

molecular biology. The "central dogma" of molecular biology is summarized as follows:

$$\text{DNA} \xrightarrow{\text{Transcription}} \text{mRNA} \xrightarrow{\text{Translation}} \text{Protein}$$

An emerging new field within neuroscience is called *molecular neurobiology*. Molecular neurobiologists use the information contained in the genes to determine the structure and functions of neuronal proteins.

Rough Endoplasmic Reticulum. Not far from the nucleus are enclosed stacks of membrane dotted with dense globular structures called **ribosomes**, which measure about 25 nm in diameter. The stacks are called **rough endoplasmic reticulum**, or **rough ER** (Figure 2.9). Rough ER abounds in neurons, far more than in glia or most other non-neuronal cells. In fact, we have already been introduced to rough ER by another name: Nissl bodies. This organelle is stained with the dyes that Nissl introduced 100 years ago.

Rough ER is a major site of protein synthesis in neurons. RNA transcripts bind to the ribosomes, and the ribosomes translate the instructions contained in the mRNA to assemble a protein molecule. Thus, ribosomes take raw material in the form of amino acids and manufacture proteins using the blueprint provided by the mRNA (Figure 2.10a).

Not all ribosomes are attached to the rough ER. Many are freely floating and are called free ribosomes. Several free ribosomes may appear to be attached by a thread; these are called **polyribosomes**. The thread is a single

Figure 2.7
The internal structure of a typical neuron.

Figure 2.9
Rough endoplasmic reticulum, or rough ER.

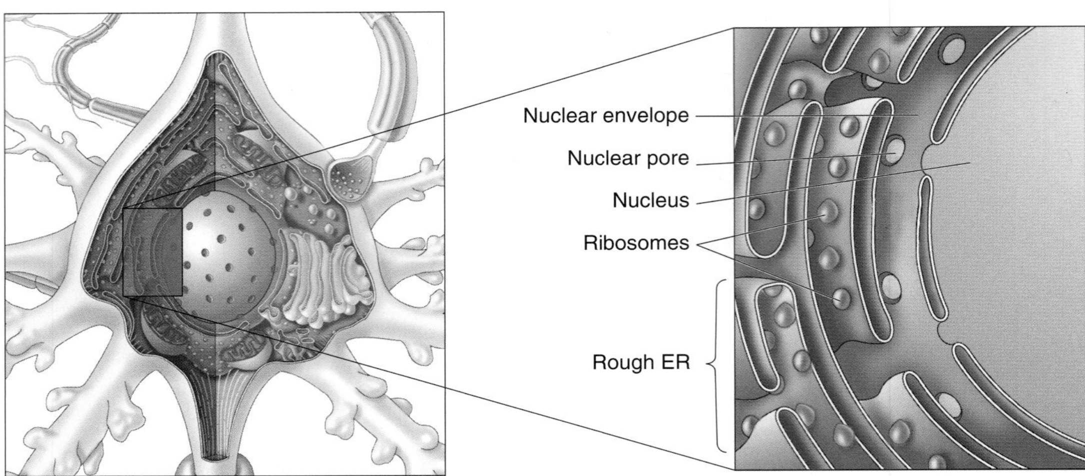

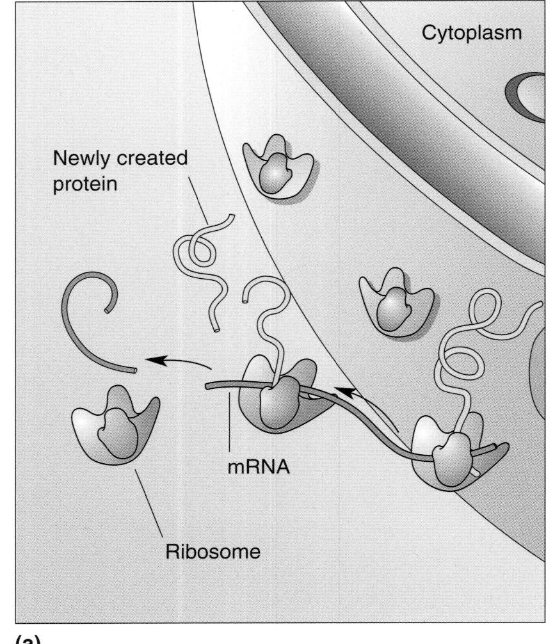

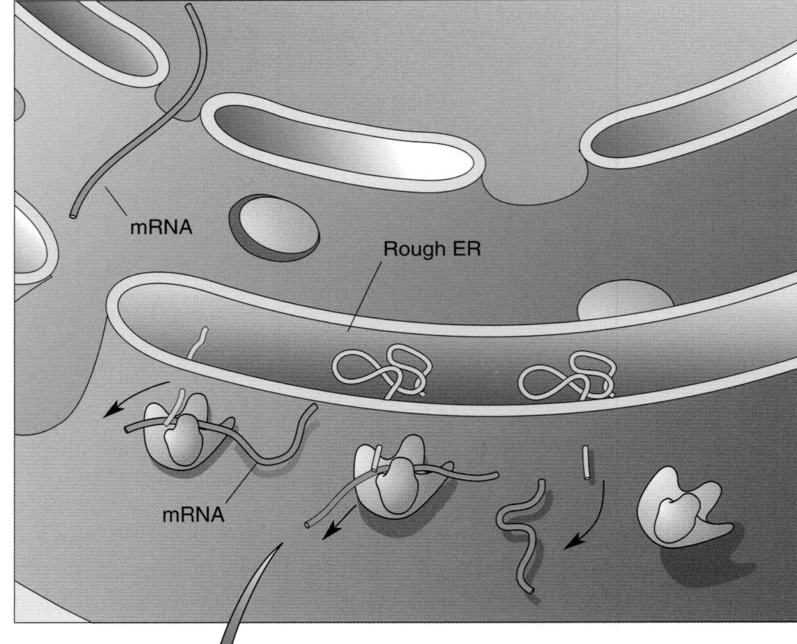

Figure 2.10
Protein synthesis on a free ribosome and on rough ER. Messenger RNA (mRNA) binds to a ribosome, initiating protein synthesis. **(a)** Proteins synthesized on free ribosomes are destined for the cytosol. **(b)** Proteins synthesized on the rough ER are destined to be enclosed by or inserted in membrane. Membrane-associated proteins are inserted into the membrane as they are assembled.

strand of mRNA, and the associated ribosomes are working on it to make multiple copies of the same protein.

What is the difference between proteins synthesized on the rough ER and those synthesized on the free ribosomes? The answer appears to lie in the intended fate of the protein molecule. If it is destined to reside within the cytosol of the neuron, then the protein's mRNA transcript shuns the ribosomes of the rough ER and gravitates toward the free ribosomes. However, if the protein is destined to be inserted into the membrane of the cell or an organelle, then it is synthesized on the rough ER. As the protein is being assembled, it is threaded back and forth through the membrane of the rough ER, where it is trapped (Figure 2.10b). It is not surprising that neurons are so well endowed with rough ER, because, as we shall see in later chapters, special membrane proteins are what give these cells their remarkable information-processing abilities.

Smooth Endoplasmic Reticulum and the Golgi Apparatus. The remainder of the cytosol of the soma is crowded with stacks of membranous organelles that look a lot like rough ER without the ribosomes, so much so that one type is called **smooth endoplasmic reticulum**, or **smooth ER**. Smooth ER is actually quite heterogeneous and performs different functions in different locations. Some smooth ER is continuous with rough ER and is believed to be a site where the proteins that jut out from the membrane are carefully folded, giving them their three-dimensional structure. Other types of smooth ER play no direct role in the processing of protein molecules but instead regulate the internal concentrations of substances such as calcium. (This organelle is particularly prominent in muscle cells, where it is called sarcoplasmic reticulum, as we will see in Chapter 13.)

The stack of membrane-enclosed disks in the soma that lies farthest from the nucleus is the **Golgi apparatus**, first described in 1898 by Camillo Golgi (Figure 2.11). This is a site of extensive "post-translational" chemical processing of proteins. One important function of the Golgi apparatus is believed to be the sorting of certain proteins that are destined for delivery to different parts of the neuron, such as the axon and the dendrites.

The Mitochondrion. Another very abundant organelle in the soma is the **mitochondrion** (plural: mitochondria). In neurons, these sausage-shaped structures measure about 1 μm in length. Within the enclosure of their outer membrane are multiple folds of inner membrane called *cristae* (singular: crista). Between the cristae is an inner space called *matrix*.

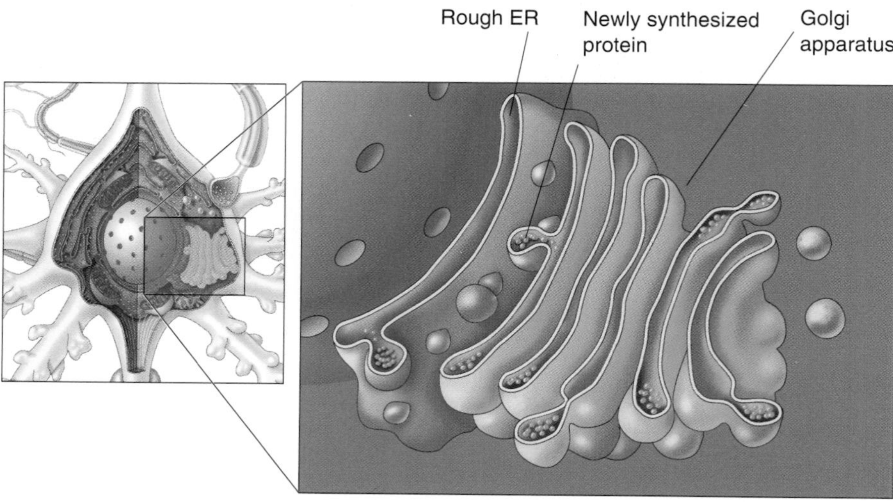

Figure 2.11
The Golgi apparatus.

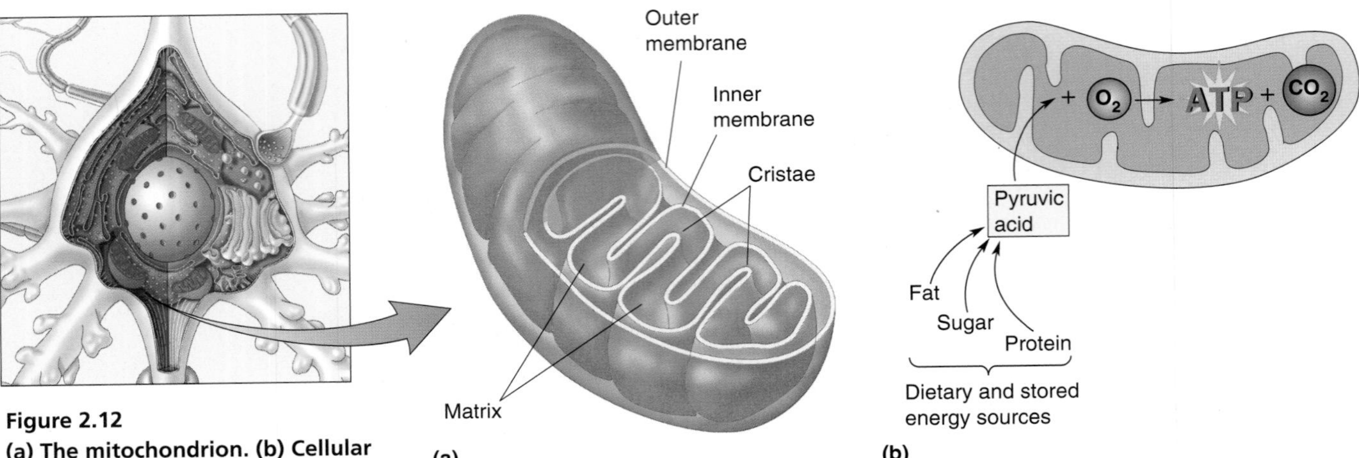

Figure 2.12
(a) The mitochondrion. (b) Cellular respiration. ATP is the energy currency that fuels biochemical reactions in neurons.

The mitochondria are the site of *cellular respiration* (Figure 2.12). When a mitochondrion "takes a breath," it pulls inside pyruvic acid (derived from sugars and digested proteins and fats) and oxygen, both of which are floating in the cytosol. Within the inner compartment of the mitochondrion, pyruvic acid enters into a complex series of biochemical reactions called the *Krebs cycle*, named after the German-British scientist Hans Krebs, who first proposed it in 1937. The biochemical products of the Krebs cycle provide energy that, in an additional series of reactions within the cristae (called the electron-transport chain), results in the addition of phosphate to adenosine diphosphate (ADP), yielding **adenosine triphosphate (ATP)**, the cell's energy source. When the mitochondrion "exhales," 17 ATP molecules are released for every molecule of pyruvic acid that had been taken in.

ATP is the energy currency of the cell. The chemical energy stored in ATP is used to fuel most of the biochemical reactions of the neuron. For example, as we shall see in Chapter 3, special proteins in the neuronal membrane use the energy released by the breakdown of ATP into ADP to pump certain substances across the membrane to establish concentration differences between the inside of the neuron and the outside.

The Neuronal Membrane

The **neuronal membrane** serves as a barrier to enclose the cytoplasm inside the neuron and to exclude certain substances that float in the fluid that bathes the neuron. The membrane is about 5 nm thick and is studded with proteins. As mentioned earlier, some of the membrane-associated proteins pump substances from the inside to the outside. Others form pores that regulate which substances can gain access to the inside of the neuron. An important characteristic of neurons is that the protein composition of the membrane varies depending on whether it is in the soma, the dendrites, or the axon.

The function of neurons cannot be understood without understanding the structure and function of the membrane and its associated proteins. In fact, this topic is so important that we'll spend a good deal of the next four chapters looking at how the membrane endows neurons with the remarkable ability to transfer electrical signals throughout the brain and body.

The Cytoskeleton

Earlier we compared the neuronal membrane to a circus tent that was draped on an internal scaffolding. This scaffolding is called the **cytoskeleton**, and it gives the neuron its characteristic shape. The "bones" of the cytoskeleton are

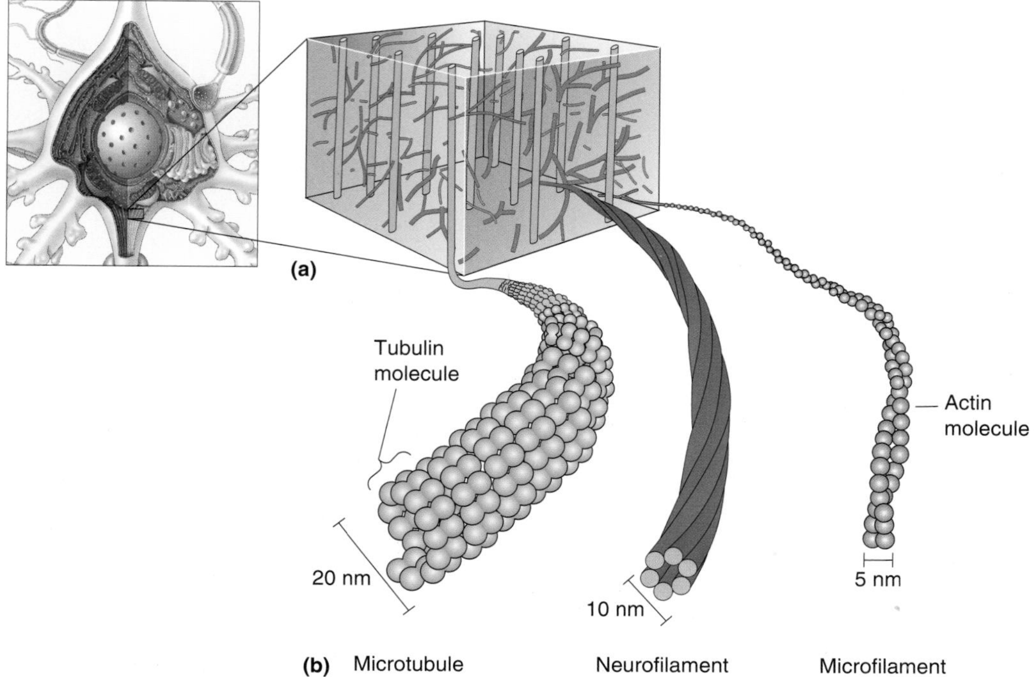

Figure 2.13
(a) The cytoskeleton. (b) Components of the cytoskeleton. The arrangement of microtubules, neurofilaments, and microfilaments gives the neuron its characteristic shape.

the microtubules, microfilaments, and neurofilaments (Figure 2.13). By drawing an analogy with a scaffolding, we do not mean that the cytoskeleton is static. On the contrary, elements of the cytoskeleton are dynamically regulated and are very likely in continual motion. Your neurons are probably squirming around in your head even as you read this sentence.

Microtubules. Measuring 20 nm in diameter, **microtubules** are big and run longitudinally down neurites. A microtubule appears as a straight, thick-walled hollow pipe. The wall of the pipe is composed of smaller strands that are braided like rope around the hollow core. Each of the smaller strands consists of the protein *tubulin*. A single tubulin molecule is small and globular; the strand consists of tubulins stuck together like meatballs on a barbecue skewer. The process of joining small proteins to form a long strand is called *polymerization*; the resulting strand is called a *polymer*. Polymerization and depolymerization of microtubules and, therefore, of neuronal shape can be regulated by various signals within the neuron.

One class of proteins that participate in the regulation of microtubule assembly and function are *microtubule-associated proteins*, or *MAPs*. Among other functions (many of which are unknown), MAPs anchor the microtubules to one another and to other parts of the neuron. Pathological changes in an axonal MAP, called *tau*, have been implicated in the dementia that accompanies Alzheimer's disease (Box 2.2).

Microfilaments. Measuring only 5 nm in diameter, **microfilaments** are about the same thickness as the cell membrane. Found throughout the neuron, they are particularly numerous in the neurites. Microfilaments are braids of two thin strands, and the strands are polymers of the protein *actin*.

Box 2.2

OF SPECIAL INTEREST

Alzheimer's Disease and the Neuronal Cytoskeleton

The neurites are the most remarkable structural feature of a neuron. Their elaborate branching patterns, critical for information processing, reflect the organization of the underlying cytoskeleton. It is therefore no surprise that a devastating loss of brain function can result when the cytoskeleton of neurons is disrupted. An example is *Alzheimer's disease*, which is characterized by disruption of the cytoskeleton of neurons in the cerebral cortex, a region of the brain crucial for cognitive function. This disorder and its underlying brain pathology were first described in 1907 by the German physician A. Alzheimer in a paper entitled "A Characteristic Disease of the Cerebral Cortex." Below are excerpts from the English translation.

> One of the first disease symptoms of a 51-year-old woman was a strong feeling of jealousy toward her husband. Very soon she showed rapidly increasing memory impairments; she could not find her way about her home, she dragged objects to and fro, hid herself, or sometimes thought that people were out to kill her, then she would start to scream loudly.
>
> During institutionalization her gestures showed a complete helplessness. She was disoriented as to time and place. From time to time she would state that she did not understand anything, that she felt confused and totally lost. Sometimes she considered the coming of the doctor as an official visit and apologized for not having finished her work, but other times she would start to yell in the fear that the doctor wanted to operate on her; or there were times that she would send him away in complete indignation, uttering phrases that indicated her fear that the doctor wanted to damage her woman's honor. From time to time she was completely delirious, dragging her blankets and sheets to and fro, calling for her husband and daughter, and seeming to have auditory hallucinations. Often she would scream for hours and hours in a horrible voice.
>
> Mental regression advanced quite steadily. After four and a half years of illness the patient died. She was completely apathetic in the end, and was confined to bed in a fetal position. (Bick et al., 1987, pp. 1–2.)

Following her death, Alzheimer examined the woman's brain under the microscope. He made particular note of changes in the "neurofibrils," elements of the cytoskeleton that are stained by a silver solution.

> The Bielschowsky silver preparation showed very characteristic changes in the neurofibrils. However, inside an apparently normal-looking cell, one or more single fibers could be observed that became prominent through their striking thickness and specific impregnability. At a more advanced stage, many fibrils arranged parallel showed the same changes. Then they accumulated forming dense bundles and gradually advanced to the surface of the cell. Eventually, the nucleus and cytoplasm disappeared, and only a tangled bundle of fibrils indicated the site where once the nèuron had been located.
>
> As these fibrils can be stained with dyes different from the normal neurofibrils, a chemical transformation of the fibril substance must have taken place. This might be the reason why the fibrils survived the destruction of the cell. It seems that the transformation of the fibrils goes hand in hand with the storage of an as yet not closely examined pathological product of the metabolism in the neuron. About one-quarter to one-third of all the neurons of the cerebral cortex showed such alterations. Numerous neurons, especially in the upper cell layers, had totally disappeared. (Bick et al., 1987, pp. 2–3.)

The severity of the dementia in Alzheimer's disease is well correlated with the number and distribution of what are now commonly known as *neurofibrillary tangles*, the "tombstones" of dead and dying neurons (Figure A). Indeed, as Alzheimer speculated, tangle formation in the cerebral cortex is very likely to cause the symptoms of the disease. Electron microscopy reveals that the major components of the tangles are *paired helical filaments*, long fibrous proteins braided together like strands of a rope (Figure B). It is now understood that these filaments consist of the microtubule-associated protein tau.

(Box 2.2, continued)

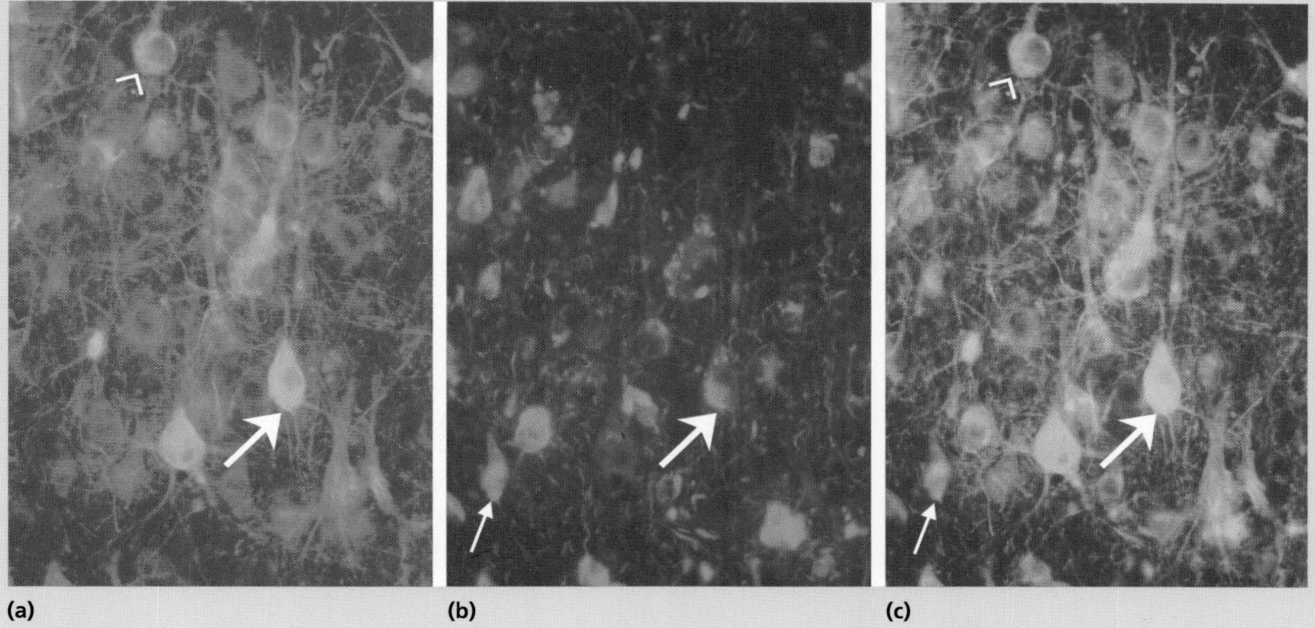

Figure A

Neurons in a human brain with Alzheimer's disease. Normal neurons contain neurofilaments but no neurofibrillary tangles. **(a)** Brain tissue stained by a method that makes neuronal neurofilaments fluoresce green, showing viable neurons. **(b)** The same region of the brain stained to show the presence of tau within neurofibrillary tangles, revealed by red fluorescence. **(c)** Superimposition of images in parts a and b. The neuron indicated by the arrowhead contains neurofilaments but no tangles and, therefore, is healthy. The neuron indicated by the large arrow has neurofilaments but also has started to show accumulation of tau and, therefore, is diseased. The neuron indicated by the small arrow in parts b and c is dead because it contains no neurofilaments. The remaining tangle is the tombstone of a neuron killed by Alzheimer's disease. (Source: Courtesy of Dr. John Morrison and modified from Vickers et al., 1994.)

Tau normally functions as a bridge between the microtubules in axons, ensuring that they run straight and parallel to one another. In Alzheimer's disease, the tau detaches from the microtubules and accumulates in the soma. This disruption of the cytoskeleton causes the axons to wither, thus impeding the normal flow of information in the affected neurons.

What causes the changes in tau? Attention is focused on another protein that accumulates in the brains of Alzheimer's patients, called *amyloid*. The field of Alzheimer's disease research moves very fast, but the consensus today is that the abnormal secretion of amyloid by neurons is the first step in the process that leads to neurofibrillary tangle formation and dementia. Current hope for therapeutic intervention is focused on strategies to reduce the depositions of amyloid in the brain. The need for effective therapy is urgent: In the United States alone, over 3 million people are afflicted with this tragic disease.

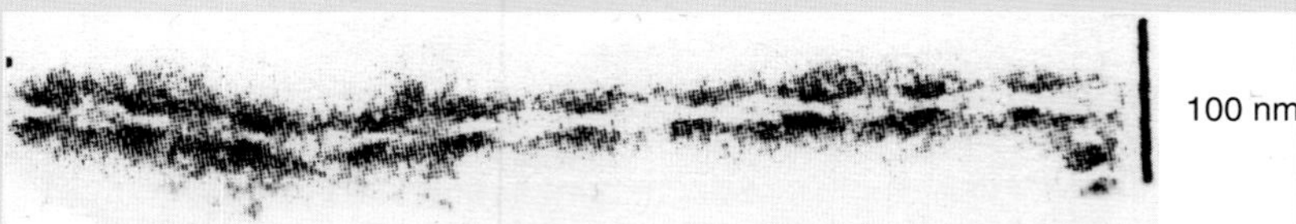

Figure B

Paired helical filaments of a tangle. (Source: Goedert, 1996, Fig. 2b.)

Actin is one of the most abundant proteins in cells of all types, including neurons, and is believed to play a role in changing cell shape. Indeed, as we shall see in Chapter 13, actin filaments are critically involved in the mechanism of muscle contraction.

Like microtubules, actin microfilaments are constantly undergoing assembly and disassembly, and this process is regulated by signals in the neuron. Besides running longitudinally down the core of the neurites like microtubules, microfilaments are also closely associated with the membrane. They are anchored to the membrane by attachments with a meshwork of fibrous proteins that line the inside of the membrane like a spider web.

Neurofilaments. With a diameter of 10 nm, **neurofilaments** are intermediate in size between microtubules and microfilaments. In fact, they exist in all cells of the body with the name *intermediate filament*; only in neurons are they called neurofilaments. The difference in name actually does reflect subtle differences in structure from one tissue to the next. An example of an intermediate filament from another tissue is keratin, which, when bundled together, makes up hair.

Of the types of fibrous structure we have discussed, neurofilaments most closely resemble the bones and ligaments of the skeleton. A neurofilament consists of multiple subunits (building blocks) that are organized like a chain of sausages. The internal structure of each subunit consists of three protein strands woven together. Unlike microfilaments and microtubules, these strands consist of individual long protein molecules, each of which is coiled in a tight, springlike configuration. This structure makes neurofilaments mechanically very strong.

The Axon

So far, we've explored the soma, organelles, membrane, and cytoskeleton. However, none of these structures is unique to neurons; they are found in all the cells in our body. But now we encounter the axon, a structure found only in neurons that is highly specialized for the transfer of information over distances in the nervous system.

The axon begins with a region called the **axon hillock**, which tapers to form the initial segment of the axon proper (Figure 2.14). Two noteworthy features distinguish the axon from the soma:

1. No rough ER extends into the axon, and there are few, if any, free ribosomes.
2. The protein composition of the axon membrane is fundamentally different from that of the soma membrane.

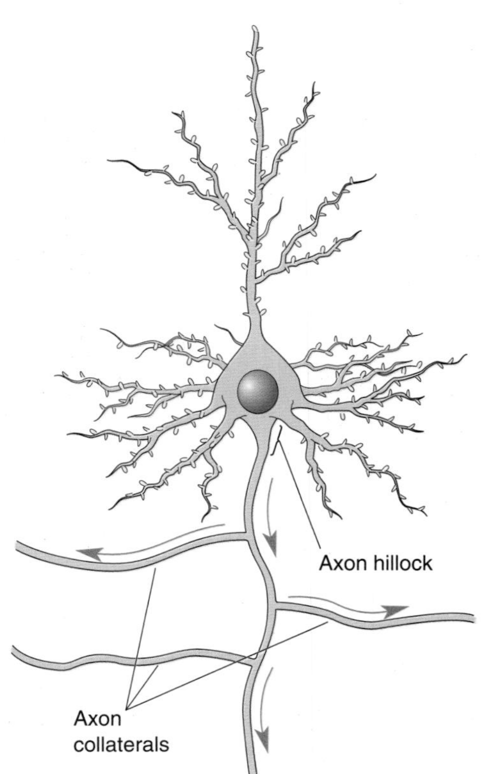

Figure 2.14
The axon and axon collaterals. The axon functions like a telegraph wire to send electrical impulses to distant sites in the nervous system. The arrows indicate the direction of information flow.

These structural differences translate into functional distinctions. Because there are no ribosomes, there is no protein synthesis in the axon. This means that all proteins in the axon must originate in the soma. And it is the different proteins in the axonal membrane that enable it to serve as the "telegraph wire" that sends information over great distances.

Axons may extend from less than a millimeter to over a meter long. Axons often branch, and these branches are called **axon collaterals**. Occasionally, an axon collateral will return to communicate with the same cell that gave rise to the axon or with the dendrites of neighboring cells. These axon branches are called *recurrent collaterals*.

The diameter of an axon is variable, ranging from less than 1 μm to about 25 μm in humans and to as large as 1 mm in the squid. This variation in axon size is important. As will be explained in Chapter 4, the speed of the electrical signal that sweeps down the axon—the *nerve impulse*—varies depending on axonal diameter. The thicker the axon, the faster the impulse travels.

The Axon Terminal. All axons have a beginning (the axon hillock), a middle (the axon proper), and an end. The end is called the **axon terminal** or **terminal bouton** (French for "button"), reflecting the fact that it usually appears as a swollen disk (Figure 2.15). The terminal is a site where the axon comes in contact with other neurons (or other cells) and passes information on to them. This point of contact is called the **synapse**, a word derived from the Greek, meaning "to fasten together." Sometimes axons have many branches at their ends, and each branch forms a synapse on dendrites or cell bodies in the same region. These branches are collectively called the **terminal arbor**. Sometimes axons form synapses at swollen regions along their length and then continue on to terminate elsewhere. Such swellings are called *boutons en passant* ("buttons in passing"). In either case, when a neuron makes synaptic contact with another cell, it is said to innervate that cell, or to provide **innervation**.

The cytoplasm of the axon terminal differs from that of the axon in several ways:

1. Microtubules do not extend into the terminal.
2. The terminal contains numerous small bubbles of membrane, called **synaptic vesicles**, that measure about 50 nm in diameter.

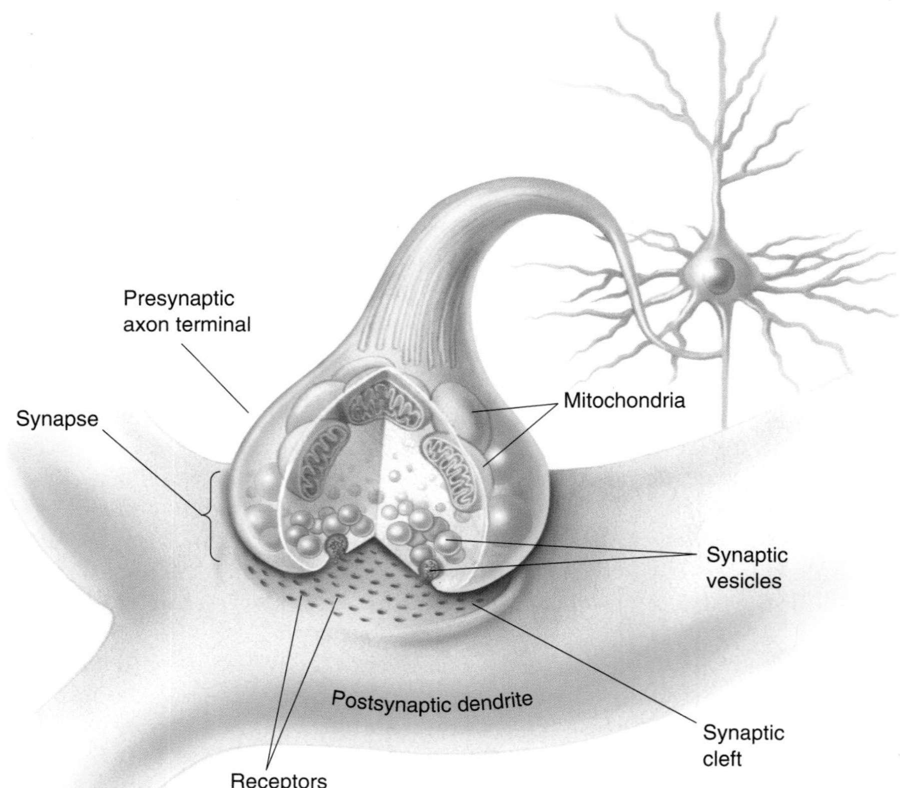

Figure 2.15
The axon terminal and the synapse. Axon terminals form synapses with the dendrites or somata of other neurons. When a nerve impulse arrives in the presynaptic axon terminal, neurotransmitter molecules are released from synaptic vesicles into the synaptic cleft. Neurotransmitter then binds to specific receptor proteins, causing the generation of electrical or chemical signals in the postsynaptic cell.

3. The inside surface of the membrane that faces the synapse has a particularly dense covering of proteins.
4. It has numerous mitochondria, indicating a high energy demand.

The Synapse. Although Chapters 5 and 6 are devoted entirely to how information is transferred from one neuron to another at the synapse, we provide a preview here. The synapse has two sides: *presynaptic* and *postsynaptic* (Figure 2.15). These names indicate the usual direction of information flow, which is from "pre" to "post." The presynaptic side generally consists of an axon terminal, while the postsynaptic side may be the dendrite or soma of another neuron. The space between the presynaptic and postsynaptic membranes is called the **synaptic cleft**. The transfer of information at the synapse from one neuron to another is called **synaptic transmission**.

At most synapses, information in the form of electrical impulses traveling down the axon is converted in the terminal into a chemical signal that crosses the synaptic cleft. On the postsynaptic membrane, this chemical signal is converted again into an electrical one. The chemical signal is called a **neurotransmitter**, and it is stored in and released from the synaptic vesicles within the terminal. As we will see, different neurotransmitters are used by different types of neurons.

This electrical-to-chemical-to-electrical transformation of information makes possible many of the brain's computational abilities. Modification of this process is involved in memory and learning, and synaptic transmission dysfunction accounts for certain mental disorders. The synapse is also the site of action for nerve gas and for most psychoactive drugs.

Axoplasmic Transport. As mentioned, one feature of the cytoplasm of axons, including the terminal, is the absence of ribosomes. Since ribosomes are the protein factories of the cell, their absence means that the proteins of the axon must be synthesized in the soma and then shipped down the axon. Indeed, in the mid-nineteenth century English physiologist Augustus Waller showed that axons cannot be sustained when separated from their parent cell body. The degeneration of axons that occurs when they are cut is now called *Wallerian degeneration.* Because it can be detected with certain staining methods, Wallerian degeneration is one way to trace axonal connections in the brain.

Wallerian degeneration occurs because the normal flow of materials from the soma to the axon terminal is interrupted. This movement of material down the axon is called **axoplasmic transport**, first demonstrated directly by the experiments of American neurobiologist Paul Weiss and his colleagues in the 1940s. They found that if they tied a thread around an axon, material accumulated on the side of the axon closest to the soma. When the knot was untied, the accumulated material continued down the axon at a rate of 1–10 mm per day.

This was a remarkable discovery, but it is not the whole story. If all material moved down the axon by this transport mechanism alone, it would not reach the ends of the longest axons for at least half a year—too long a wait to feed hungry synapses. In the late 1960s, methods were developed to track the movements of protein molecules down the axon into the terminal. These methods entailed injecting the somata of neurons with radioactive amino acids. Recall that amino acids are the building blocks of proteins. The "hot" amino acids were assembled into proteins, and the arrival of radioactive proteins in the axon terminal was measured to calculate the rate of transport. Bernice Grafstein of Rockefeller University discovered that this *fast axoplasmic transport* (so named to distinguish it from *slow axoplasmic transport* described by Weiss) occurred at a rate as high as 1000 mm per day.

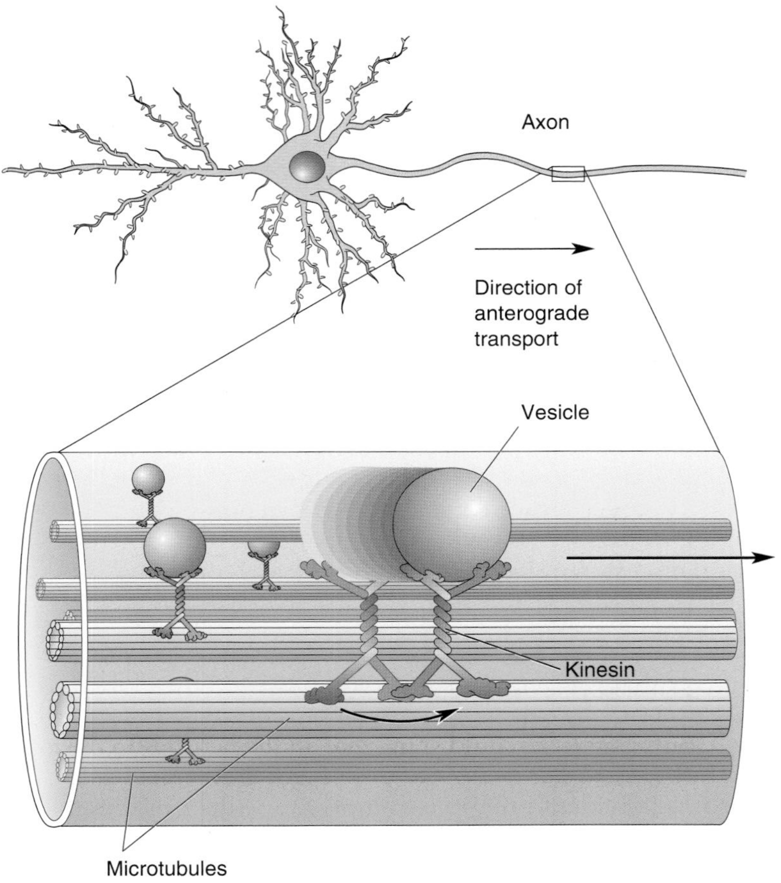

Figure 2.16
A mechanism for the movement of material on the microtubules of the axon. Trapped in membrane-enclosed vesicles, material is transported from the soma to the axon terminal by the action of the protein kinesin, which "walks" along microtubules at the expense of ATP.

Much is now known about how axoplasmic transport works. Material is enclosed within vesicles, which then "walk down" the microtubules of the axon. The "legs" are provided by a protein called *kinesin*, and the process is fueled by ATP (Figure 2.16). Kinesin moves material only from the soma to the terminal. All movement of material in this direction is called **anterograde transport**.

In addition to anterograde transport, there is a mechanism for the movement of material up the axon from the terminal to the soma. This process is believed to provide signals to the soma about changes in the metabolic needs of the axon terminal. Movement in this direction, from terminal to soma, is called **retrograde transport**. The molecular mechanism is similar to anterograde transport, except the "legs" for retrograde transport are provided by a different protein called *dynein*. Both anterograde and retrograde transport mechanisms have been exploited by neuroscientists to trace connections in the brain (Box 2.3).

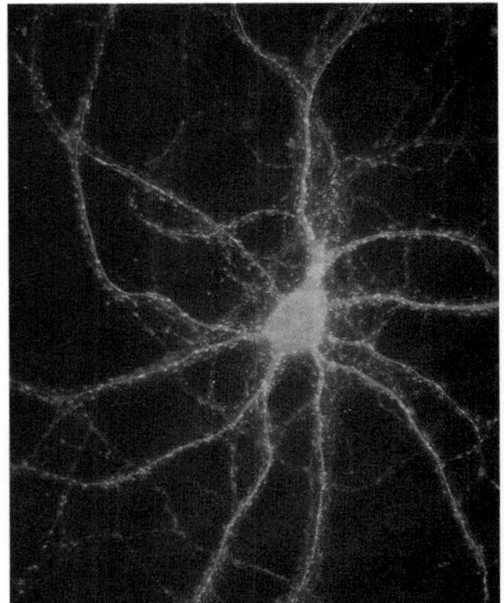

Figure 2.17
Dendrites receiving synaptic inputs from axon terminals. A neuron has been made to fluoresce green, using a method that reveals the distribution of a microtubule-associated protein. Axon terminals have been made to fluoresce orange-red, using a method to reveal the distribution of synaptic vesicles. The axons and cell bodies that contribute these axon terminals are not visible in this photomicrograph. (Source: *Neuron* 10 [Suppl.], 1993, cover figure.)

Dendrites

The term *dendrite* is derived from the Greek for "tree," reflecting the fact that these neurites resemble the branches of a tree as they extend from the soma. The dendrites of a single neuron are collectively called a **dendritic tree**; each branch of the tree is called a *dendritic branch*. The wide variety of shapes and sizes of dendritic trees are used to classify different groups of neurons.

Because dendrites function as the antennae of the neuron, they are covered with thousands of synapses (Figure 2.17). The dendritic membrane under the

Figure 2.18
Dendritic spines. This is a computer reconstruction of a segment of dendrite, showing the variable shapes and sizes of spines. Each spine is postsynaptic to one or two axon terminals. (Source: Harris and Stevens, 1989, cover figure.)

synapse (the *postsynaptic* membrane) has many specialized protein molecules called **receptors** that detect the neurotransmitters in the synaptic cleft.

The dendrites of some neurons are covered with specialized structures called **dendritic spines** that receive some types of synaptic input. Spines look like little punching bags that hang off the dendrite (Figure 2.18). The unusual morphology of spines has fascinated neuroscientists ever since their discovery by Cajal. They are believed to isolate various chemical reactions that are triggered by some types of synaptic activation. Spine structure is sensitive to the type and amount of synaptic activity. Unusual changes in spines have been shown to occur in the brains of individuals with cognitive impairments (Box 2.4).

For the most part, the cytoplasm of dendrites resembles that of axons. It is filled with cytoskeletal elements and mitochondria. One interesting difference was discovered by University of Virginia neuroscientist Oswald Steward. He found that polyribosomes can be observed in dendrites, often right under spines. Steward's research suggests that synaptic transmission can actually direct local protein synthesis in some neurons (Box 2.5). In Chapter 24, we will see that synaptic regulation of protein synthesis is crucial for information storage by the brain.

CLASSIFYING NEURONS

It is unlikely that we can ever hope to understand how each of the hundred billion neurons in the nervous system uniquely contributes to the function of the brain. But what if we could show that all the neurons in the brain can be divided into a small number of categories and that within each category, all neurons function identically? The complexity of the problem might then be reduced to understanding the unique contribution of each category rather than each cell. It is with this hope that neuroscientists have devised schemes for classifying neurons.

Classification Based on the Number of Neurites

Neurons can be classified according to the total number of neurites (axons and dendrites) that extend from the soma (Figure 2.19). A neuron that has a single neurite is said to be **unipolar.** If there are two neurites, the cell is **bipolar**, and if there are three or more, the cell is multipolar. Most neurons in the brain are **multipolar**.

Classification Based on Dendrites

Dendritic trees can vary widely from one type of neuron to another. Some have inspired elegant names like "double bouquet cells." Others have less interesting names, such as "alpha cells." Classification is often unique to a particular part of the brain. For example, in the cerebral cortex (the structure that lies just under the surface of the cerebrum), there are two broad classes: **pyramidal cells** (pyramid-shaped) and **stellate cells** (star-shaped) (Figure 2.20).

Another simple way to classify neurons is according to whether their dendrites have spines. Those that do are called **spiny** and those that do not are called **aspinous**. These dendritic classification schemes can overlap. For example, in the cerebral cortex, all pyramidal cells are spiny. Stellate cells, on the other hand, can be either spiny or aspinous.

Classification Based on Connections

Information is delivered to the nervous system by neurons that have neurites in the sensory surfaces of the body, such as the skin and the retina of the eye.

Box 2.3

OF SPECIAL INTEREST

Hitching a Ride on "Retrorail"

Fast anterograde transport of proteins in axons was shown by injecting the soma with radioactive amino acids. The success of this method immediately suggested a way to trace connections in the brain. For example, to determine where neurons in the eye send their axons, the eye was injected with radioactive proline, an amino acid. The proline was incorporated into proteins in the somata that were then transported to the axon terminals. By use of a technique called autoradiography, the location of radioactive axon terminals could be detected, thereby revealing the extent of the connection between the eye and the brain.

It was subsequently discovered that retrograde transport could also be exploited to work out connections in the brain. Strangely enough, the enzyme horseradish peroxidase (HRP) is selectively taken up by axon terminals and then transported retrogradely to the soma. A chemical reaction can then be initiated to visualize the location of the HRP in slices of brain tissue. This method is commonly used to trace connections in the brain (Figure A).

Some viruses also exploit retrograde transport to infect neurons. For example, the oral type of herpesvirus enters axon terminals in the lips and mouth and is then transported back to the parent cell bodies. Here the virus usually remains dormant until physical or emotional stress occurs (as on a first date), at which time it replicates and returns to the nerve ending, causing a painful cold sore. Similarly, the rabies virus enters the nervous system by retrograde transport through axons in the skin. However, once inside the soma, the virus wastes no time in replicating madly, killing its neuronal host. The virus is then taken up by other neurons within the nervous system, and the process repeats itself again and again, usually until the victim dies.

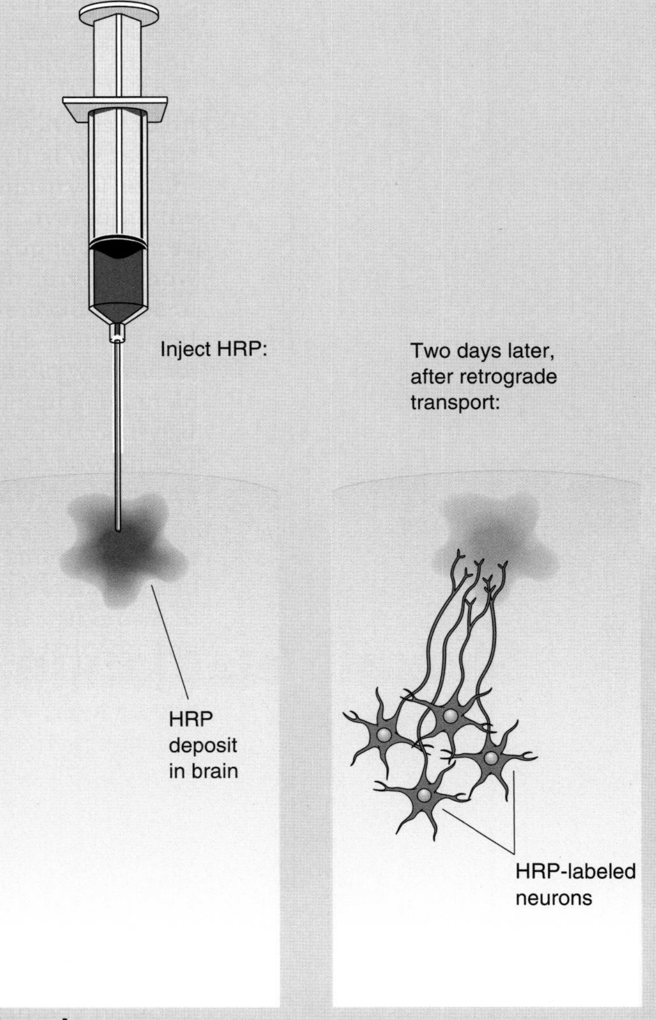

Figure A

Cells with these connections are called **primary sensory neurons**. Other neurons have axons that form synapses with the muscles and command movements; these are called **motor neurons**. But most neurons in the nervous system form connections only with other neurons. According to this classification scheme, these cells are all called **interneurons**.

Classification Based on Axon Length

Some neurons have long axons that extend from one part of the brain to the other; these are called **Golgi type I neurons**, or projection neurons. Other neurons have short axons that do not extend beyond the vicinity of the cell body; these are called **Golgi type II neurons**, or local circuit neurons. In the cerebral cortex, for example, pyramidal cells usually have long axons that extend to other parts of the brain and are therefore Golgi type I neurons. In contrast, stellate cells have axons that never extend beyond the cerebral cortex and are therefore Golgi type II neurons.

Classification Based on Neurotransmitter

The classification schemes presented so far are based on the morphology of neurons as revealed by a Golgi stain. Newer methods that enable neuroscientists to identify which neurons contain specific neurotransmitters have resulted in a scheme for classifying neurons based on their chemistry. For example, the motor neurons that command voluntary movements all release the neurotransmitter *acetylcholine* at their synapses. These cells are therefore also classified as cholinergic, meaning that they use this particular neurotransmitter. Collections of cells that use a common neurotransmitter make up the brain's neurotransmitter systems (see Chapters 6 and 15).

GLIA

We have devoted most of our attention in this chapter to the neurons. While this decision is justified by the state of current knowledge, some neuroscien-

Box 2.4

OF SPECIAL INTEREST

Mental Retardation and Dendritic Spines

The elaborate architecture of a neuron's dendritic tree is a good reflection of the complexity of its synaptic connections with other neurons. Brain function depends on these highly precise synaptic connections, which are formed during the fetal period and are refined during infancy and early childhood. Not surprisingly, this very complex developmental process is vulnerable to disruption. *Mental retardation* is said to have occurred if a disruption of brain development results in subaverage cognitive functioning that impairs adaptive behavior.

The use of standardized tests indicates that intelligence in the general population is distributed along a bell-shaped (Gaussian) curve. By convention, the mean intelligence quotient (IQ) is set to be 100. About two-thirds of the total population falls within 15 points (one standard deviation) of the mean, and 95% of the population falls within 30 points (two standard deviations). People with intelligence scores below 70 are considered to be mentally retarded if the cognitive impairment affects the person's ability to adapt his or her behavior to the setting in which he or she lives. Some 2–3% of humans fit this description.

Mental retardation has many causes. The most severe forms are associated with genetic disorders. An example is a condition called *phenylketonuria* (*PKU*). The basic abnormality is a deficit in the liver enzyme that metabolizes the dietary amino acid phenylalanine. Infants born with PKU have an abnormally high level of the amino acid in the blood and brain. If the condition goes untreated, brain growth is stunted and severe mental retardation results. Another example is *Down syndrome*, which occurs when the fetus has an extra copy of chromosome 21, thus disrupting normal gene expression during brain development.

A second known cause of mental retardation is accidents during pregnancy and childbirth. Examples are maternal infections with German measles (rubella) and asphyxia during childbirth. A third cause of mental retardation is poor nutrition during pregnancy. An example is *fetal alcohol syndrome*, a constellation of developmental abnormalities that occur in children born to alcoholic mothers. A fourth cause, thought to account for the majority of cases, is environmental impoverishment—the lack of good nutrition, socialization, and sensory stimulation—during infancy.

While some forms of mental retardation have very clear physical correlates (e.g., stunted growth; abnormalities in the structure of the head, hands, and body), most cases have only

Figure 2.19
Classification of neurons based on the number of neurites.

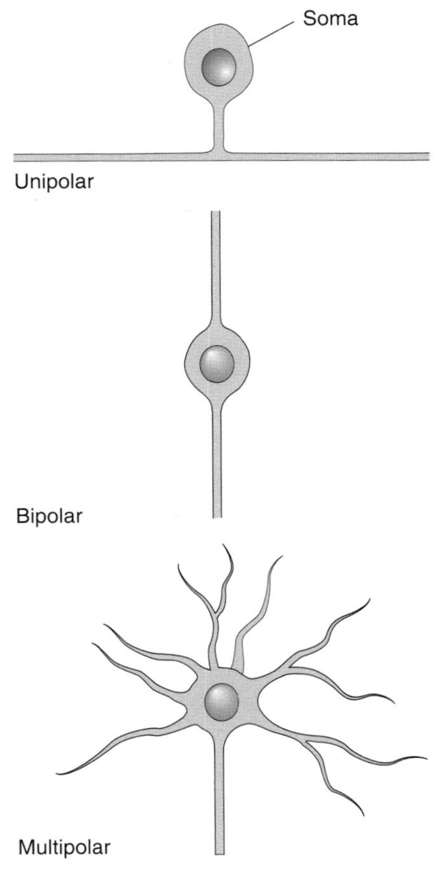

tists consider glia to be the "sleeping giants" of neuroscience. One day, they suppose, it will be shown that glia contribute much more importantly to information processing in the brain than is currently appreciated. At present, however, the evidence indicates that glia contribute to brain function mainly by supporting neuronal functions. Although their role may be subordinate, without glia, the brain could not function properly.

Astrocytes

The most numerous glia in the brain are called **astrocytes** (Figure 2.21). These cells fill the spaces between neurons. The space that remains between the neurons and the astrocytes in the brain measures only about 20 nm wide. Consequently, astrocytes probably influence whether a neurite can grow or retract. And when we speak of fluid "bathing" neurons in the brain, it is more like a sponge bath than a session in a hot tub.

An essential role of astrocytes is regulating the chemical content of this *extracellular space*. For example, astrocytes envelop synaptic junctions in the brain, thereby restricting the spread of neurotransmitter molecules that have been released. Astrocytes also have special proteins in their membranes that actively remove many neurotransmitters from the synaptic cleft. A recent and unexpected discovery is that astrocytic membranes also possess neurotransmitter receptors that, like the receptors on neurons, can trigger electrical and biochemical events inside the glial cell. Besides regulating neurotransmitters, astrocytes also tightly control the extracellular concentration of sev-

(Box 2.4, continued)

behavioral manifestations. The brains of these individuals appear grossly normal. How, then, do we account for the profound cognitive impairment? An important clue came in the 1970s from the research of Miguel Marin-Padilla, working at Dartmouth College, and Dominick Purpura, working at the Albert Einstein College of Medicine in New York City. Using the Golgi stain, they studied the brains of retarded children and discovered remarkable changes in dendritic structure. The dendrites of retarded children had many fewer dendritic spines, and the spines that they did have were unusually long and thin (Figure A). The extent of the spine changes was well correlated with the degree of mental retardation.

Dendritic spines are an important target of synaptic input. Purpura pointed out that the dendritic spines of mentally retarded children resemble those of the normal human fetus. He suggested that mental retardation reflects the failure of normal circuits to form in the brain. In the three decades since this seminal work was published, it has been established that normal synaptic development, including maturation of the dendritic spines, depends critically on the environment during infancy and early childhood. An impoverished environment during an early "critical period" of development can lead to profound changes in the circuits of the brain. However, there is some good news. Many of the deprivation-induced changes in the brain can be reversed if intervention occurs early enough. In Chapter 22, we will take a closer look at the role of experience in brain development.

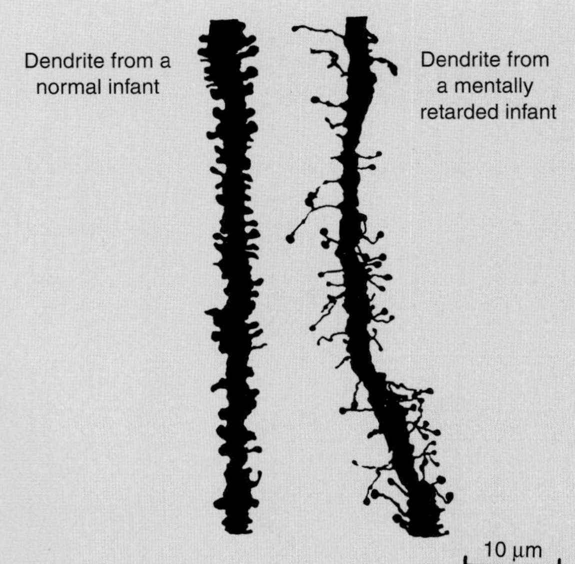

Figure A
Normal and abnormal dendrites. (Source: Purpura, 1974, Fig. 2A.)

Box 2.5

PATH OF DISCOVERY

The Story of Dendritic Protein Synthesis

BY OSWALD STEWARD

It's relatively rare that a single observation causes one to completely redirect one's research. But such was the case when I noticed the selective localization of polyribosomes beneath postsynaptic sites on dendrites, what I now call synapse-associated polyribosome complexes, or SPRCs (Figure A). A polyribosome is a cluster of ribosomes bound to mRNA, the machinery that cells use to synthesize protein. The observation of polyribosomes closely associated with synapses on dendrites triggered an idea that was, at the time, counter to prevailing dogma—that neurons can locally synthesize certain key proteins at synaptic sites on dendrites.

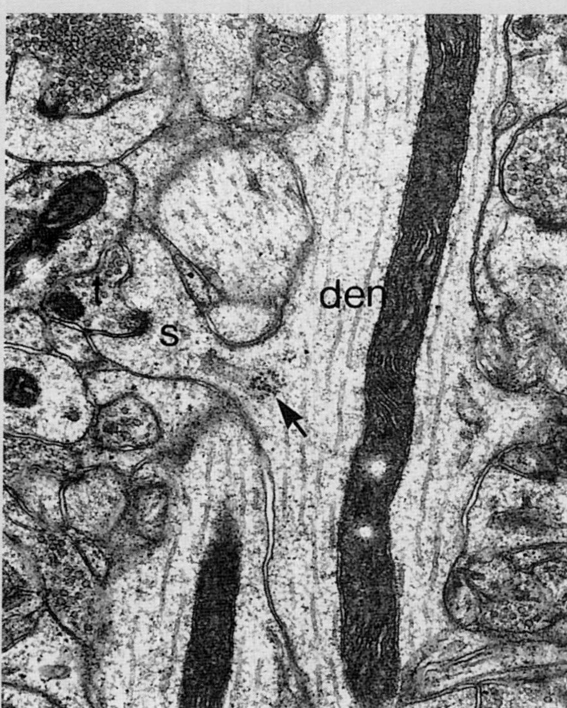

Figure A
A synapse-associated polyribosome complex. This electron micrograph shows a dendrite (den) with a SPRC (arrow). The SPRC is closely associated with a dendritic spine (s), which is in contact with an axon terminal (t). (Source: Courtesy of Dr. Oswald Steward.)

Oswald Steward

I discovered SPRCs and formulated the basic hypothesis for dendritic protein synthesis in 1979. The observation made such an impact that I started a scientific diary that I have kept to this day. I had been studying how neurons modify connections after injury, focusing on the examples of axonal "sprouting" that occur in a region of the brain called the hippocampus. I was trying to define the cellular and molecular mechanisms of this growth. At that time, one could not study gene expression in individual neurons (as is now possible using molecular biological techniques). Instead, those of us thinking in such terms measured the overall level of protein synthetic activity by neurons. In an experiment designed to measure neuronal protein synthesis during sprouting, we were surprised to find evidence of protein synthesis within the part of the hippocampus that contained dendrites but few neuronal cell bodies. To find out what elements were responsible for this protein synthesis, we examined these parts with the electron microscope to determine where polyribosomes were. I still recall the first striking images of clusters of polyribosomes in and near dendritic spine synapses. It wasn't so much the observation as the idea that immediately came to mind—that the ability to synthesize certain proteins at individual synaptic sites would provide a mechanism that would allow neurons to modify the molecular composition of *individual synapses* on a moment-by-moment basis.

eral substances that have the potential to interfere with proper neuronal function. For example, astrocytes regulate the concentration of potassium ions in the extracellular fluid.

Myelinating Glia

Unlike astrocytes, the primary function of **oligodendroglial** and **Schwann cells** is clear. These glia provide layers of membrane that insulate axons. Boston University anatomist Alan Peters, a pioneer in the electron micro-

(Box 2.5, continued)

This idea had powerful face validity to me, yet it ran counter to the prevailing view that neuronal protein synthesis occurred only in the cell body. With the nagging concern that I might have misidentified the polyribosomes, I carried a sheaf of my electron micrographs to the American Association of Anatomy meeting. For the entire meeting, I watched for Alan Peters, first author of the definitive book *Fine Structure of the Nervous System*. When I finally ran into him, we spread out my micrographs on a hallway floor. I was delighted when he confirmed my identification of the polyribosomes. He also explained why the selective positioning of the polyribosomes hadn't been noticed before: "In electron microscopy, you may not notice something unless you're looking specifically for it." I had noticed the polyribosomes because I was looking for what could account for the protein synthesis in this part of the hippocampus.

Looking back, I believe the entire concept that a unique subset of proteins was synthesized locally at individual synapses emerged almost fully formed when I saw the original micrograph. This was an "Aha! experience," the basis for which I can't fully explain. However, about a year before the discovery of SPRCs, I was hosting a visit by Sir John Eccles to the University of Virginia. Eccles had won the Nobel Prize for his studies of synaptic transmission. He and I discussed several things about the practice of science, and I still remember one comment in particular: "You can generate enormous amounts of data and never have an impact on the field. The trick is to develop a *story*." His comment was fresh in my mind when I first saw the SPRCs, and it may have helped generate that extra creative energy for the SPRC story to come together in such an explosive way.

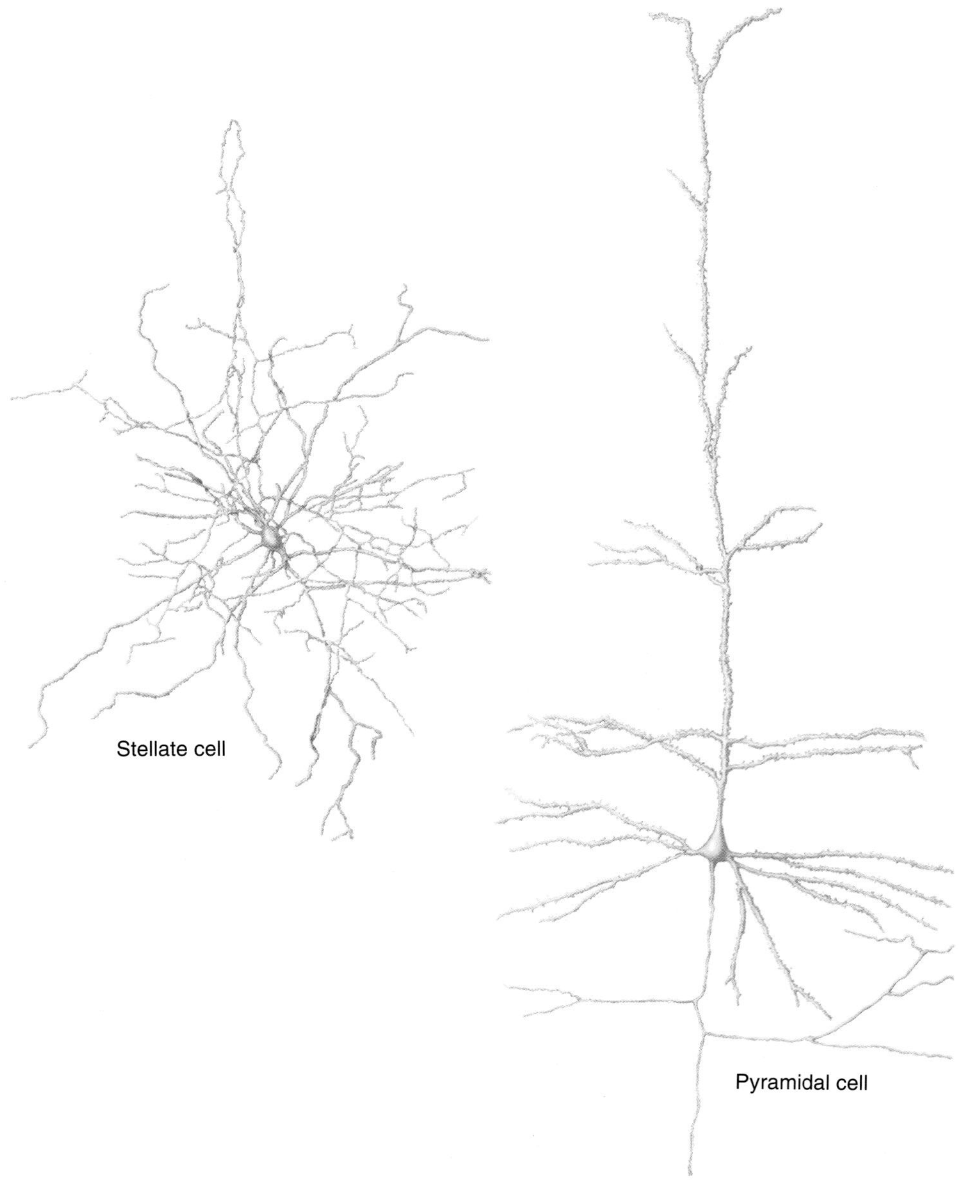

Figure 2.20
Classification of neurons based on dendritic tree structure. Pyramidal cells and stellate cells, distinguished by the arrangement of their dendrites, are two types of neurons found in the cerebral cortex.

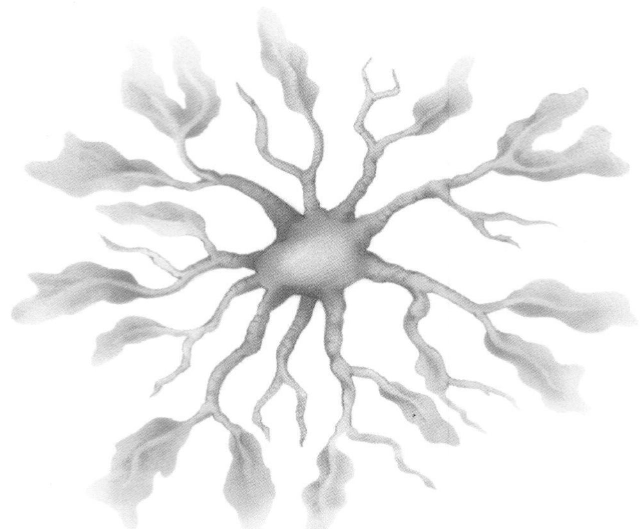

Figure 2.21
An astrocyte. Astrocytes fill most of the space in the brain that is not occupied by neurons and blood vessels.

scopic study of the nervous system, showed that this wrapping, called **myelin**, spirals around axons in the brain (Figure 2.22). Because the axon fits inside the spiral wrapping like a sword in its scabbard, the name *myelin sheath* describes the entire covering. The sheath is interrupted periodically, leaving a short length where the axonal membrane is exposed. This region is called a **node of Ranvier** (Figure 2.23).

We will see in Chapter 4 that myelin serves to speed the propagation of nerve impulses down the axon. Oligodendroglia and Schwann cells differ in their location and some other characteristics. For example, oligodendroglia are found only in the central nervous system (brain and spinal cord), while Schwann cells are found only in the peripheral nervous system (parts outside the skull and vertebral column). Another difference is that one oligoden-

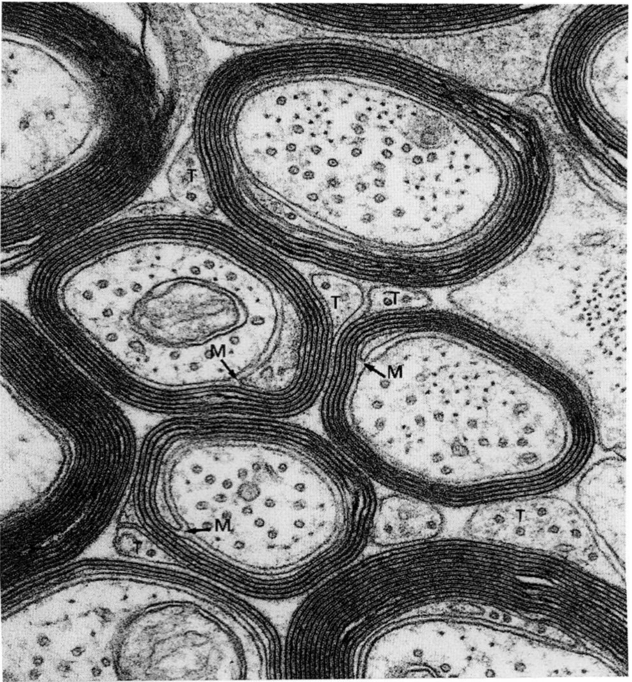

Figure 2.22
Myelinated optic nerve fibers cut in cross section. (Source: Courtesy of Dr. Alan Peters.)

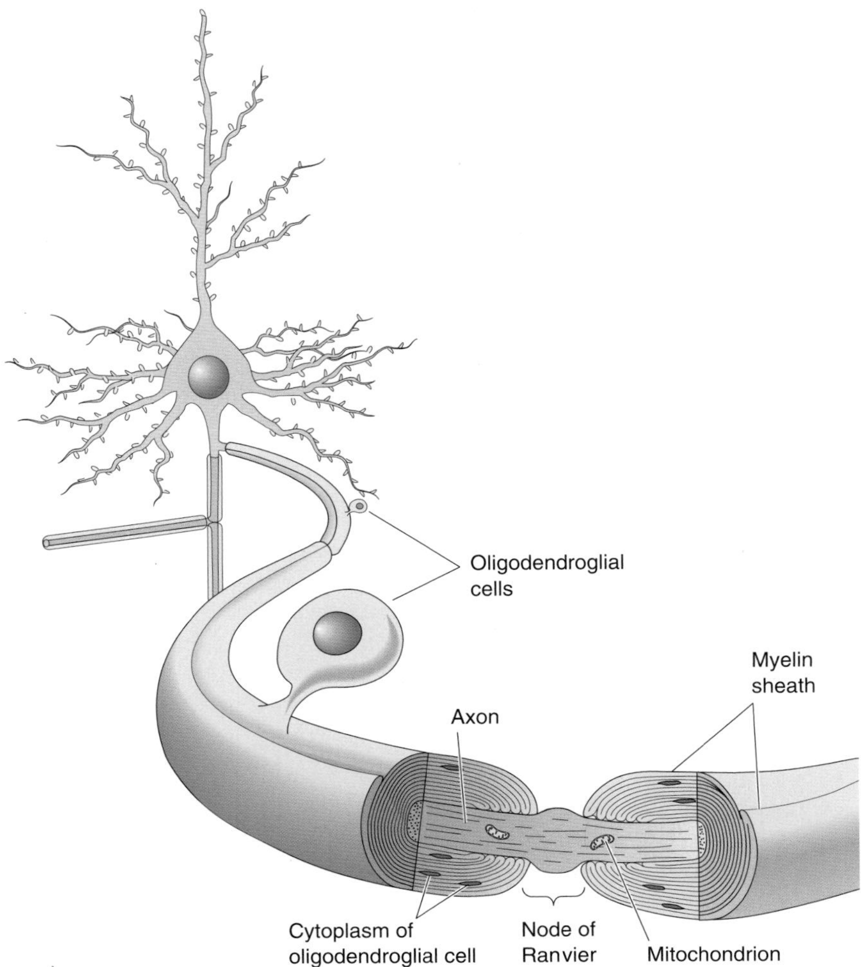

Figure 2.23
An oligodendroglial cell. Like the Schwann cells found in the nerves of the body, oligodendroglia provide myelin sheaths around axons in the brain and spinal cord. The myelin sheath of an axon is interrupted periodically at the nodes of Ranvier.

droglial cell will contribute myelin to several axons, while each Schwann cell myelinates only a single axon.

Other Non-Neuronal Cells

Even if we eliminated every neuron, every astrocyte, and every oligodendroglial cell, other cells would still remain in the brain. For the sake of completeness, we must mention these other cells. First, special cells called **ependymal cells** provide the lining of fluid-filled ventricles within the brain, and they also play a role in directing cell migration during brain development. Second, a class of cells called **microglia** function as phagocytes to remove debris left by dead or degenerating neurons and glia. Finally, the vasculature of the brain—arteries, veins, and capillaries—would still be there.

CONCLUDING REMARKS

Learning about the structural characteristics of the neuron provides insight into how neurons and their different parts work, because structure correlates with function. For example, the absence of ribosomes in the axon correctly predicts that proteins in the axon terminal must be provided by the soma via axoplasmic transport. A large number of mitochondria in the axon terminal correctly predicts a high energy demand. The elaborate structure of the dendritic tree appears ideally suited for the receipt of incoming information, and indeed, this is where most of the synapses are formed with the axons of other neurons.

From the time of Nissl it has been recognized that an important feature of neurons is the rough ER. What does this tell us about neurons? We have said that rough ER is a site of the synthesis of proteins destined to be inserted into the membrane. We will now see how the various proteins in the neuronal membrane give rise to the unique capabilities of neurons to transmit, receive, and store information.

KEY TERMS

The Neuron Doctrine
histology (p. 24)
Nissl stain (p. 24)
cytoarchitecture (p. 24)
Golgi stain (p. 24)
cell body (p. 24)
soma (p. 25)
perikaryon (p. 25)
neurite (p. 25)
axon (p. 25)
dendrite (p. 25)
neuron doctrine (p. 25)

The Prototypical Neuron
cytosol (p. 26)
organelle (p. 26)
cytoplasm (p. 26)
nucleus (p. 26)
chromosome (p. 26)
DNA (deoxyribonucleic acid) (p. 26)
gene expression (p. 26)
protein synthesis (p. 26)
mRNA (messenger ribonucleic acid) (p. 27)
transcription (p. 27)
amino acid (p. 27)
translation (p. 27)
ribosome (p. 29)
rough endoplasmic reticulum (rough ER) (p. 29)
polyribosome (p. 29)
smooth endoplasmic reticulum (smooth ER) (p. 31)
Golgi apparatus (p. 31)
mitochondrion (p. 31)
ATP (adenosine triphosphate) (p. 32)
neuronal membrane (p. 32)
cytoskeleton (p. 32)
microtubule (p. 33)
microfilament (p. 33)
neurofilament (p. 36)
axon hillock (p. 36)
axon collateral (p. 36)
axon terminal (p. 37)
terminal bouton (p. 37)
synapse (p. 37)
terminal arbor (p. 37)
innervatation (p. 37)
synaptic vesicle (p. 37)
synaptic cleft (p. 38)
synaptic transmission (p. 38)
neurotransmitter (p. 38)
axoplasmic transport (p. 38)
anterograde transport (p. 39)
retrograde transport (p. 39)
dendritic tree (p. 39)
receptor (p. 40)
dendritic spine (p. 40)

Classifying Neurons
unipolar neuron (p. 40)
bipolar neuron (p. 40)
multipolar neuron (p. 40)
pyramidal cell (p. 40)
stellate cell (p. 40)
spiny neuron (p. 40)
aspinous neuron (p. 40)
primary sensory neuron (p. 41)
motor neuron (p. 41)
interneuron (p. 41)
Golgi type I neuron (p. 41)
Golgi type II neuron (p. 42)

Glia
astrocyte (p. 43)
oligodendroglial cell (p. 44)
Schwann cell (p. 44)
myelin (p. 46)
node of Ranvier (p. 46)
ependymal cell (p. 47)
microglial cell (p. 47)

REVIEW QUESTIONS

1. State the neuron doctrine in a single sentence. To whom is this insight credited?
2. Which parts of a neuron are shown by a Golgi stain that are not shown by a Nissl stain?
3. What are three physical characteristics that distinguish axons from dendrites?
4. Which of the following structures are unique to neurons and which are not: nucleus, mitochondria, rough ER, synaptic vesicle, Golgi apparatus?
5. What are the steps by which the information in the DNA of the nucleus directs the synthesis of a membrane-associated protein molecule?
6. Colchicine is a drug that causes microtubules to break apart (depolymerize). What effect would this drug have on anterograde transport? What would happen in the axon terminal?
7. Classify the cortical pyramidal cell based on (a) the number of neurites, (b) the presence or absence of dendritic spines, (c) connections, and (d) axon length.
8. What is myelin? What does it do? Which cells provide it in the central nervous system?

CHAPTER

3 The Neuronal Membrane at Rest

INTRODUCTION

Consider the problem your nervous system confronts when you step on a thumbtack. Your reactions are automatic: You shriek with pain as you jerk up your foot. For this simple response to occur, breaking of the skin must be translated into neural signals that travel rapidly and reliably up the long sensory nerves of your leg. In the spinal cord, these signals are transferred to interneurons. Some of these neurons connect with the parts of your brain that interpret the signals as being painful. Others connect to the motor neurons that control the leg muscles that withdraw your foot. Thus, even this simple reflex, depicted in Figure 3.1, requires the nervous system to *collect, distribute,* and *integrate* information. A goal of cellular neurophysiology is to understand the biological mechanisms that underlie these functions.

The neuron solves the problem of conducting information over a distance by using electrical signals that sweep along the axon. In this sense, axons act like telephone wires. The analogy stops here, however, because the type of signal used by the neuron is constrained by the special environment of the nervous system. In a copper telephone wire, information can be transferred over long distances at a high rate (about half the speed of light) because telephone wire is a superb conductor of electrons, is well insulated, and is suspended in air (air being a poor conductor of electricity). Electrons will, therefore, move within the wire instead of radiating away. In contrast, electrical charge in the cytosol of the axon is carried by electrically charged atoms (ions) instead of free electrons. This makes cytosol far less conductive than copper wire. Also, the axon is not especially well insulated and is bathed in salty extracellular fluid, which conducts electricity. Thus, like water flowing down a leaky garden hose, electrical current passively conducting down the axon would not go very far before it would leak out.

Fortunately, the axonal membrane has properties that enable it to conduct a special type of signal—the nerve impulse, or **action potential**—that overcomes these biological constraints. In contrast to passively conducted electrical signals, action potentials do not diminish over distance; they are signals of fixed size and duration. Information is encoded in the frequency of action

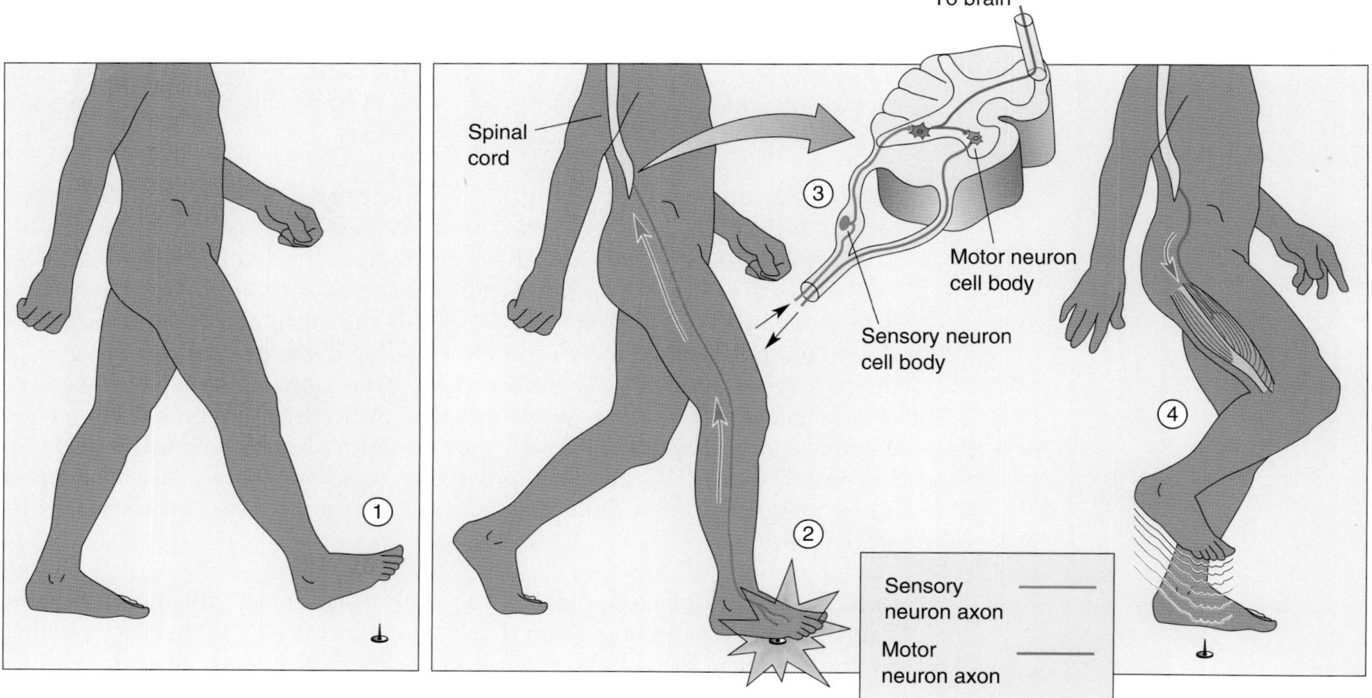

Figure 3.1

A simple reflex. ① A person steps on a thumbtack. ② The breaking of the skin is translated into signals that travel up sensory nerve fibers (the direction of information flow, indicated by the arrows). ③ In the spinal cord, the information is distributed to interneurons. Some of these neurons send axons to the brain, where the painful sensation is registered. Others synapse on motor neurons, which send descending signals to the muscles. ④ The motor commands lead to muscle contraction and withdrawal of the foot.

potentials of individual neurons, as well as in the distribution and number of neurons firing action potentials in a given nerve. This type of code is partly analogous to Morse code sent down a telegraph wire; information is encoded in the pattern of electrical impulses. Cells capable of generating and conducting action potentials, which include both nerve and muscle cells, are said to have **excitable membrane**. The "action" in action potentials occurs at the cell membrane.

When a cell with excitable membrane is not generating impulses, it is said to be at rest. In the resting neuron, the cytosol along the inside surface of the membrane has a negative electrical charge compared with that of the outside. This difference in electrical charge across the membrane is called the **resting membrane potential** (or resting potential). The action potential is simply a brief reversal of this condition, and for an instant—about a thousandth of a second—the inside of the membrane becomes positively charged with respect to the outside. Therefore, to understand how neurons signal one another, we must learn how the neuronal membrane at rest separates electrical charge, how electrical charge can be rapidly redistributed across the membrane during the action potential, and how the impulse can propagate reliably along the axon.

In this chapter, we begin our exploration of neuronal signaling by tackling the first question: How does the resting membrane potential arise? Understanding the resting potential is very important because it forms the foundation for understanding the rest of neuronal physiology. And knowledge of neuronal physiology is central to understanding the capabilities and limitations of brain function.

THE CAST OF CHEMICALS

We begin our discussion of the resting membrane potential by introducing the three main players: the salty fluids on either side of the membrane, the membrane itself, and the proteins that span the membrane. Each of these has certain properties that contribute to establishing the resting potential.

Cytosol and Extracellular Fluid

Water is the main ingredient of the fluid inside the neuron, the intracellular fluid or cytosol, and the fluid that bathes the neuron, the extracellular fluid. Electrically charged atoms—ions—are dissolved in this water, and they are responsible for the resting and action potentials.

Water. For our purpose here, the most important property of the water molecule (H_2O) is its uneven distribution of electrical charge (Figure 3.2a). The two hydrogen atoms and the oxygen atom are bonded together covalently, which means they share electrons. The oxygen atom, however, has a greater affinity for electrons than does the hydrogen atom. As a result, the shared electrons spend more time associated with the oxygen atom than with the two hydrogen atoms. Therefore, the oxygen atom acquires a net negative charge (because it has extra electrons), and the hydrogen atoms acquire a net positive charge. Thus, H_2O is said to be a polar molecule, held together by *polar covalent bonds*. This electrical polarity makes water an effective solvent of other charged or polar molecules; that is, other polar molecules tend to dissolve in water.

Ions. Atoms or molecules that have a net electrical charge are known as **ions**. Table salt is a crystal of sodium (Na^+) and chloride (Cl^-) ions held together by the electrical attraction of oppositely charged atoms. This attraction is

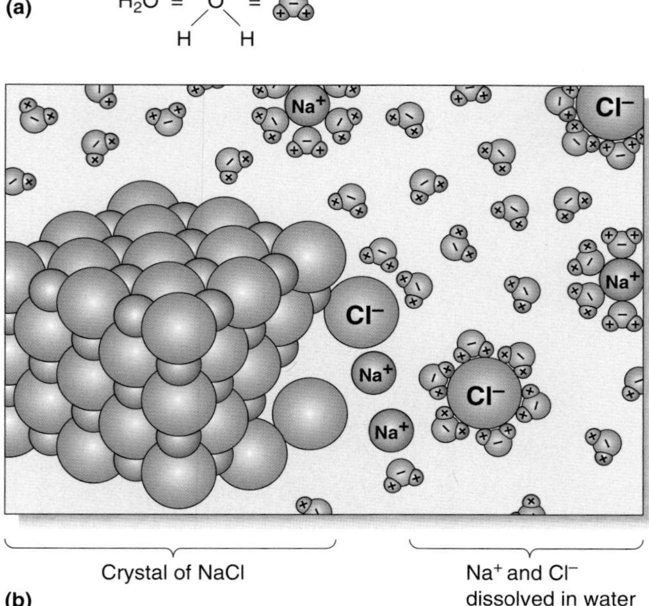

Figure 3.2
Water is a polar solvent. (a) Representations of the atomic structure of the water molecule. The oxygen atom has a net negative electrical charge, and the hydrogen atoms have a net positive electrical charge, making water a polar molecule. **(b)** A crystal of sodium chloride (NaCl) dissolves in water because the polar water molecules have a stronger attraction for the electrically charged sodium and chloride ions than the ions do for one another.

called an *ionic bond*. Salt dissolves readily in water because the charged portions of the water molecule have a stronger attraction for the ions than they have for each other (Figure 3.2b). As each ion breaks away from the crystal, it is surrounded by a sphere of water molecules. Each positively charged ion (Na^+, in this case) will be covered by water molecules oriented so that the oxygen atom (the negative pole) will be facing the ion. Likewise, each negatively charged ion (Cl^-) will be surrounded by the hydrogen atoms of the water molecules. These clouds of water that surround each ion are called spheres of hydration, and they effectively insulate the ions from one another.

The electrical charge of an atom depends on the difference between the number of protons and electrons. When this difference is 1, the ion is said to be *monovalent*; when the difference is 2, the ion is *divalent*; and so on. Ions with a net positive charge are called **cations**; ions with a negative charge are called **anions**. Remember that ions are the major charge carriers involved in the conduction of electricity in biological systems, including the neuron. The ions of particular importance for cellular neurophysiology are the monovalent cations Na^+ (sodium) and K^+ (potassium), the divalent cation Ca^{2+} (calcium), and the monovalent anion Cl^- (chloride).

The Phospholipid Membrane

As we have seen, substances with uneven electrical charges dissolve in water because of the polarity of the water molecule. These substances, including ions and polar molecules, are said to be "water-loving," or *hydrophilic*. However, compounds whose atoms are bonded by *nonpolar covalent bonds* have no basis for chemical interactions with water. A nonpolar covalent bond occurs when the shared electrons are distributed evenly in the molecule so that no portion acquires a net electrical charge. Such compounds do not dissolve in water and are said to be "water-fearing," or *hydrophobic*. One familiar example of a hydrophobic substance is olive oil, and as you know, oil and water don't mix. Another example is *lipid*, a class of water-insoluble biological molecules important to the structure of cell membranes. The lipids of the neuronal membrane contribute to the resting and action potentials by forming a barrier to water-soluble ions and, indeed, to water itself.

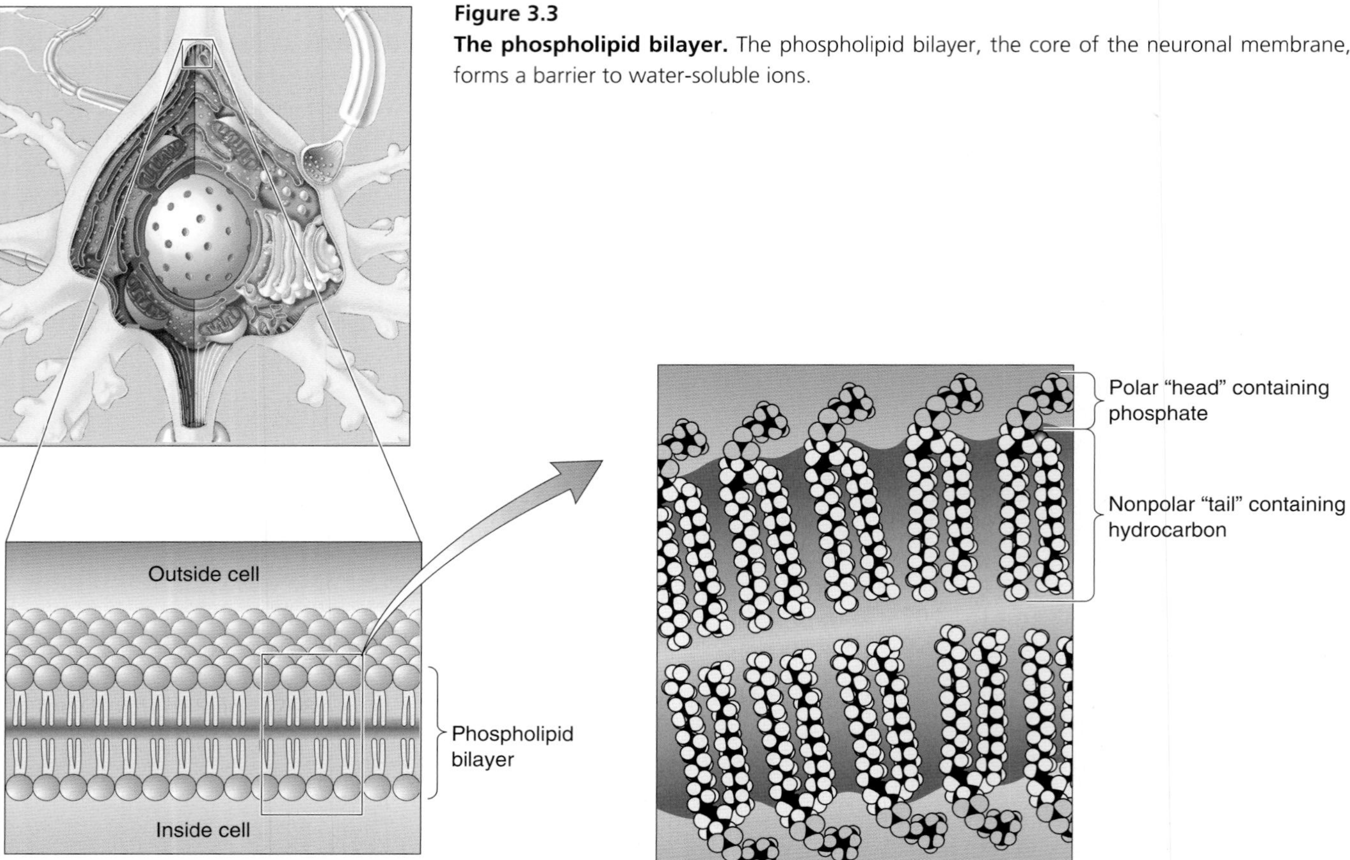

Figure 3.3
The phospholipid bilayer. The phospholipid bilayer, the core of the neuronal membrane, forms a barrier to water-soluble ions.

The Phospholipid Bilayer. The main chemical building blocks of cell membranes are *phospholipids*. Like other lipids, phospholipids contain long nonpolar chains of carbon atoms bonded to hydrogen atoms. In addition, however, a phospholipid has a polar phosphate group (a phosphorus atom bonded to three oxygen atoms) attached to one end of the molecule. Thus, phospholipids are said to have a polar "head" (containing phosphate) that is hydrophilic and a nonpolar "tail" (containing hydrocarbon) that is hydrophobic.

The neuronal membrane consists of a sheet of phospholipids two molecules thick. A cross section through the membrane, shown in Figure 3.3, reveals that the hydrophilic heads face the outer and inner watery environments and the hydrophobic tails face each other. This stable arrangement, called a **phospholipid bilayer**, effectively isolates the cytosol of the neuron from the extracellular fluid.

Protein

The type and distribution of protein molecules distinguish neurons from other types of cells. The *enzymes* that catalyze chemical reactions in the neuron, the *cytoskeleton* that gives a neuron its special shape, the *receptors* that are sensitive to neurotransmitters—all are made up of protein molecules. The resting and action potentials depend on special proteins that span the phospholipid bilayer. These proteins provide routes for ions to cross the neuronal membrane.

Protein Structure. To perform their many functions in the neuron, different proteins have widely different shapes, sizes, and chemical characteristics. To understand this diversity, let's briefly review protein structure.

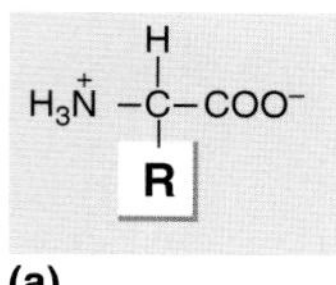

(a)

Figure 3.4
Amino acids, the building blocks of protein. (a) Every amino acid has in common a central alpha carbon, an amino group (NH_3^+), and a carboxyl group (COO^-). Amino acids differ from one another based on a variable R group. **(b)** The 20 amino acids that are used by neurons to make proteins. Noted in parentheses are the common abbreviations used for the various amino acids.

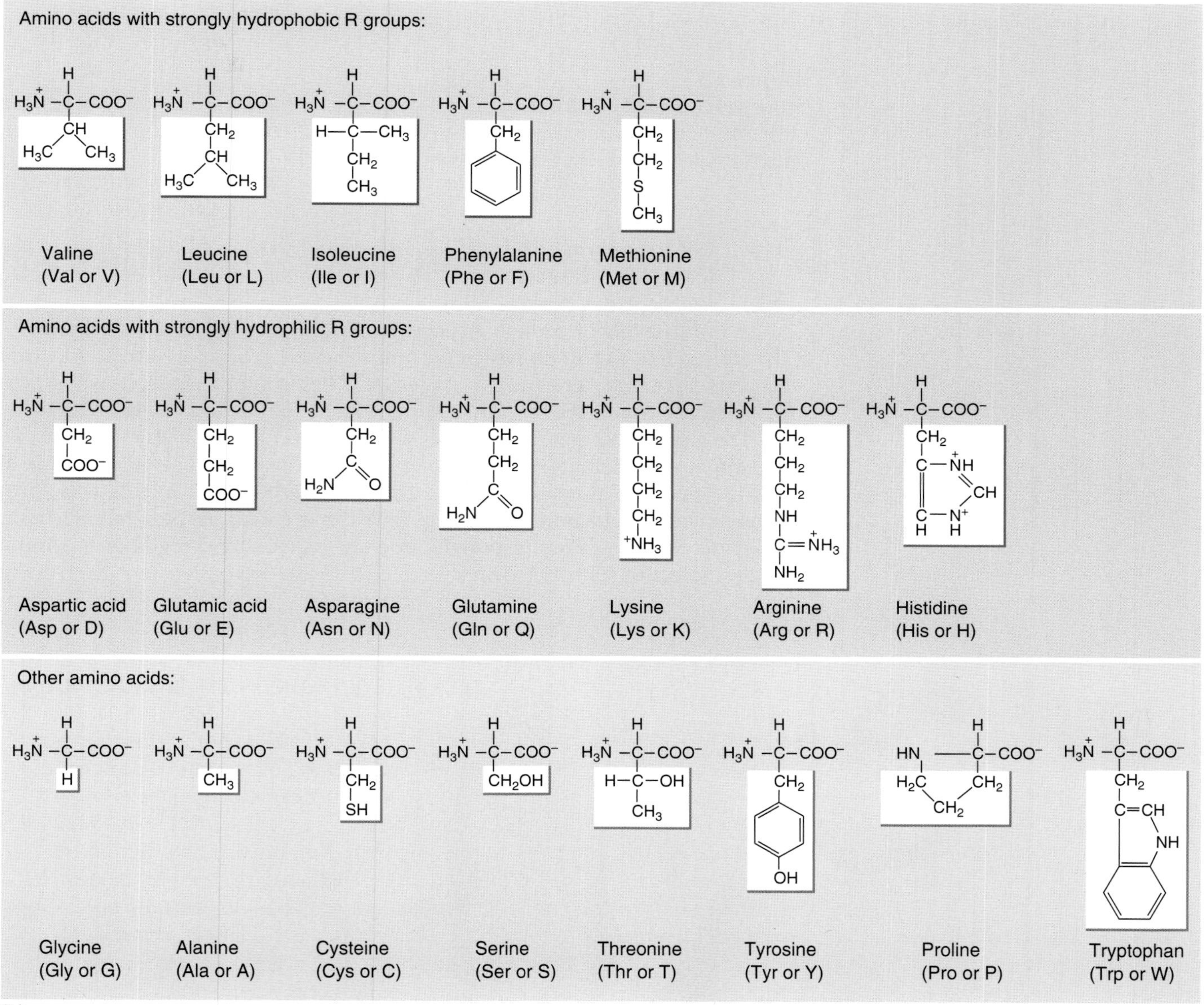

Amino acids with strongly hydrophobic R groups:

Valine (Val or V)
Leucine (Leu or L)
Isoleucine (Ile or I)
Phenylalanine (Phe or F)
Methionine (Met or M)

Amino acids with strongly hydrophilic R groups:

Aspartic acid (Asp or D)
Glutamic acid (Glu or E)
Asparagine (Asn or N)
Glutamine (Gln or Q)
Lysine (Lys or K)
Arginine (Arg or R)
Histidine (His or H)

Other amino acids:

Glycine (Gly or G)
Alanine (Ala or A)
Cysteine (Cys or C)
Serine (Ser or S)
Threonine (Thr or T)
Tyrosine (Tyr or Y)
Proline (Pro or P)
Tryptophan (Trp or W)

(b)

As mentioned in Chapter 2, proteins are molecules assembled from various combinations of 20 different amino acids. The basic structure of an amino acid is shown in Figure 3.4a. All amino acids have a central carbon atom (the alpha carbon), which is covalently bonded to four molecular groups: a hydrogen atom, an amino group (NH_3^+), a carboxyl group (COO^-), and a variable group called the *R group* (R for residue). The differences between amino acids result from differences in the size and nature of these R groups (Figure 3.4b). The properties of the R group determine the chemical relationships in which each amino acid can participate.

Proteins are synthesized in the ribosomes of the neuronal cell body. In this process, amino acids assemble into a chain connected by **peptide bonds,**

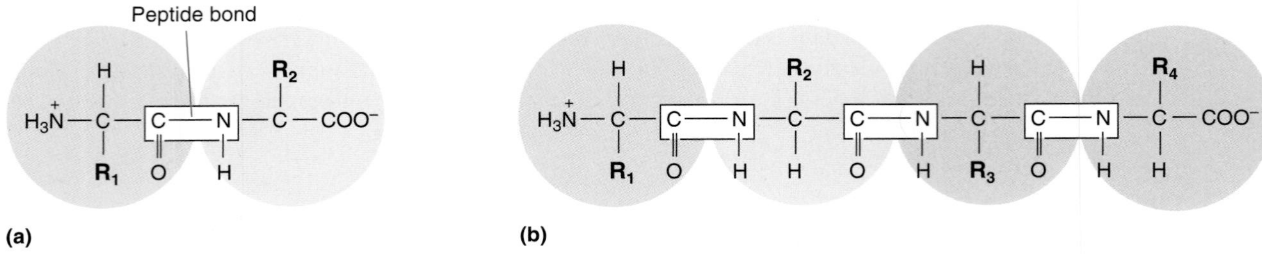

Figure 3.5
The peptide bond and a polypeptide. (a) Peptide bonds attach amino acids together. The bond forms between the carboxyl group of one amino acid and the amino group of another. **(b)** A polypeptide is a single chain of amino acids.

which join the amino group of one amino acid to the carboxyl group of the next (Figure 3.5a). Proteins made of a single chain of amino acids are also called **polypeptides** (Figure 3.5b).

The four levels of protein structure are shown in Figure 3.6. The *primary structure* is like a chain in which the amino acid R groups are linked together by peptide bonds. As a protein molecule is being synthesized, however, the polypeptide chain can coil into a spiral-like configuration called an *alpha helix*. The alpha helix is an example of what is called the *secondary structure* of a protein molecule. Interactions among the R groups can cause the molecule to change its three-dimensional conformation even further. In this way, proteins can bend, fold, and assume a globular shape. This shape is called *tertiary structure*. Finally, different polypeptide chains can bond together to form a larger molecule; such a protein is said to have *quaternary structure*. Each of the

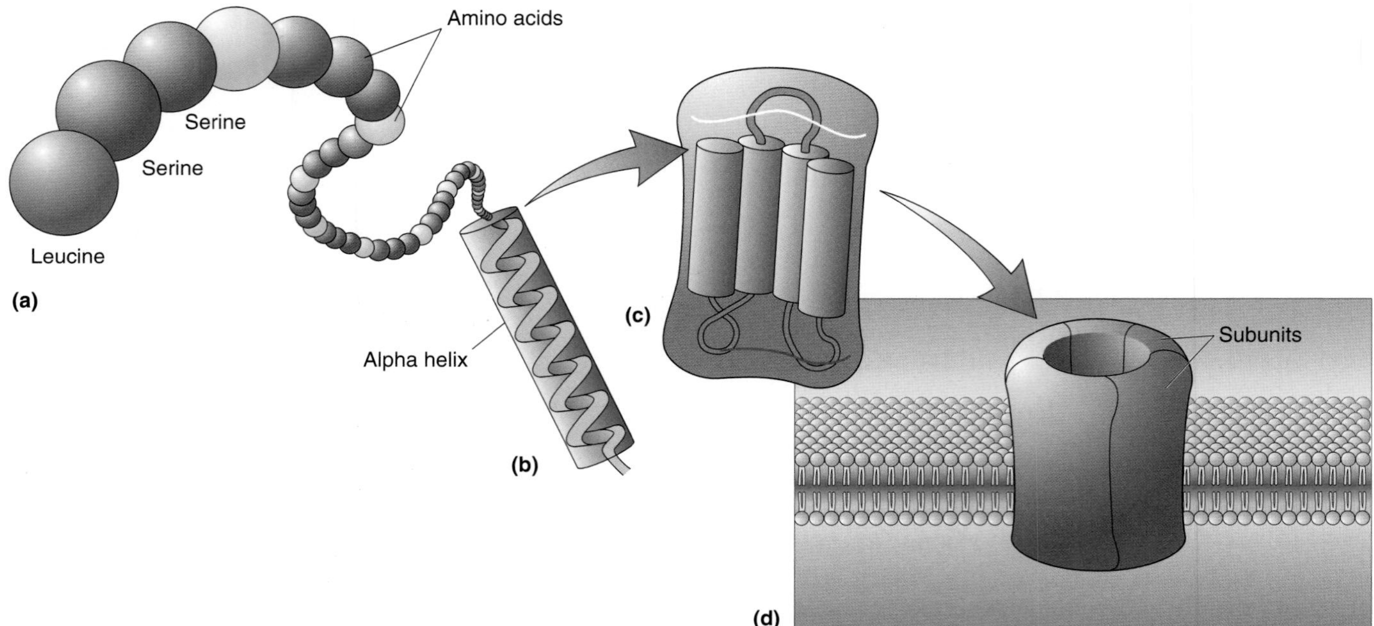

Figure 3.6
Protein structure. (a) Primary structure: the sequence of amino acids in the polypeptide. **(b)** Secondary structure: coiling of a polypeptide into an alpha helix. **(c)** Tertiary structure: three-dimensional folding of a polypeptide. **(d)** Quaternary structure: various polypeptides bonded together to form a larger protein.

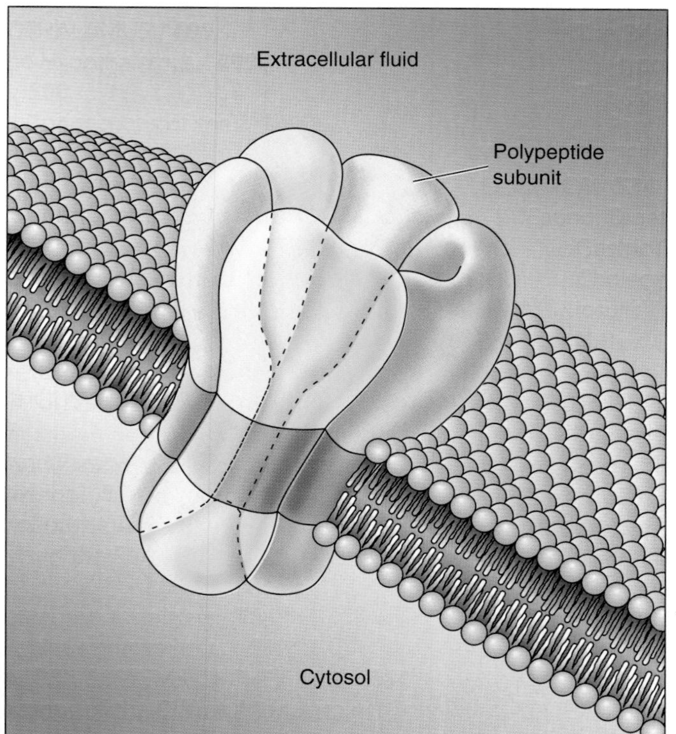

Figure 3.7
A membrane ion channel. Ion channels consist of membrane-spanning proteins that assemble to form a pore. In this example, the channel protein has five polypeptide subunits. Each subunit has a hydrophobic surface region (shaded) that readily associates with the phospholipid bilayer.

different polypeptides contributing to a protein with quaternary structure is called a *subunit*.

Channel Proteins. The exposed surface of a protein may be chemically heterogeneous. Regions where nonpolar R groups are exposed are hydrophobic and tend to associate readily with lipid. Regions with exposed polar R groups are hydrophilic and tend to avoid a lipid environment. Therefore, it is not difficult to imagine classes of rod-shaped proteins with polar groups exposed at either end but with only hydrophobic groups showing on their middle surfaces. This type of protein could be suspended in a phospholipid bilayer, with its hydrophobic portion inside the membrane and its hydrophilic ends exposed to the watery environments on either side.

Ion channels are made from just these sorts of membrane-spanning protein molecules. Typically, a functional channel across the membrane requires that four to six similar protein molecules assemble to form a pore between them (Figure 3.7). The subunit composition varies from one type of channel to the next, and this is what specifies their different properties. One important property of most ion channels, specified by the diameter of the pore and the nature of the R groups lining it, is **ion selectivity**. Potassium channels are selectively permeable to K^+. Likewise, sodium channels are permeable almost exclusively to Na^+, calcium channels to Ca^{2+}, and so on. Another important property of many channels is **gating**. Channels with this property can be opened and closed—gated—by changes in the local microenvironment of the membrane.

You will learn much more about channels as you work your way through this book. *Understanding ion channels in the neuronal membrane is key to understanding cellular neurophysiology.*

Ion Pumps. In addition to those that form channels, other membrane-spanning proteins come together to form **ion pumps**. Recall from Chapter 2 that

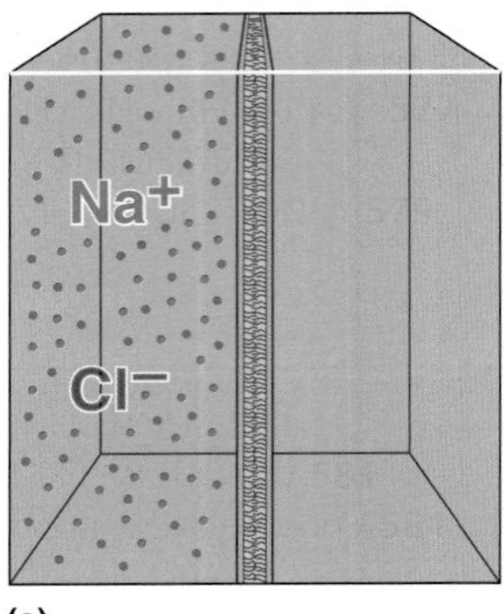

(a)

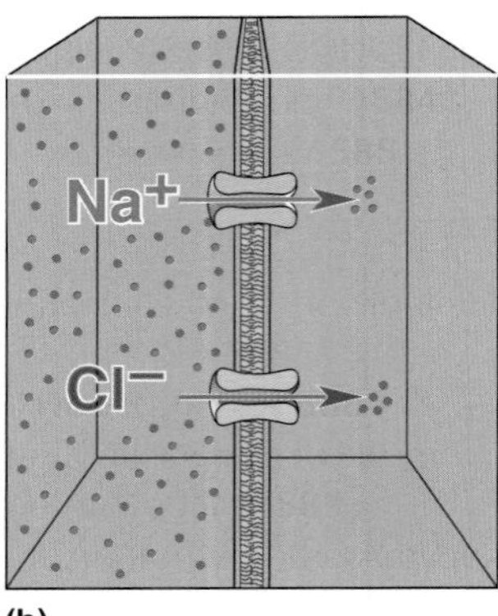

(b)

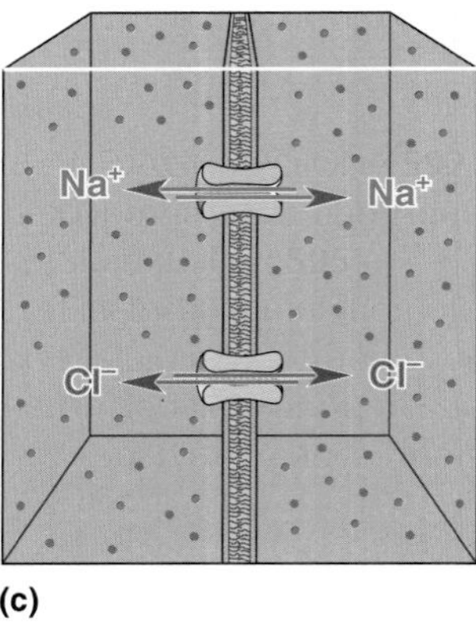

(c)

Figure 3.8
Diffusion. (a) NaCl has been dissolved on the left side of an impermeable membrane. The sizes of the letters Na^+ and Cl^- indicate the relative concentrations of these ions. **(b)** Channels inserted in the membrane allow passage of Na^+ and Cl^-. Because there is a large concentration gradient across the membrane, there will be a net movement of Na^+ and Cl^- from the region of high concentration to the region of low concentration, from left to right. **(c)** In the absence of any other factors, the net movement of Na^+ and Cl^- across the membrane ceases when they are equally distributed on the two sides of the permeable membrane.

adenosine triphosphate (ATP) is the energy currency of cells. Ion pumps are enzymes that use the energy released by the breakdown of ATP to transport certain ions across the membrane. We will see that these pumps play a critical role in neuronal signaling by transporting Na^+ and Ca^{2+} from the inside of the neuron to the outside.

THE MOVEMENT OF IONS

A channel across a membrane is like a bridge across a river (or in the case of a gated channel, like a drawbridge): It provides a path to cross from one side to the other. The existence of a bridge does not necessarily compel us to cross it, however. The bridge we cross during the weekday commute may lie unused on the weekend. The same can be said of membrane ion channels. The existence of an open channel in the membrane does not necessarily mean that there will be a net movement of ions across the membrane. Such movement also requires that external forces be applied to drive them across. Because the functioning nervous system requires the movement of ions across the neuronal membrane, it is important that we understand these forces. Ionic movements through channels are influenced by two factors: diffusion and electricity.

Diffusion

Ions and molecules dissolved in water are in constant motion. This temperature-dependent random movement tends to distribute the ions evenly throughout the solution so that there is a net movement of ions from regions of high concentration to regions of low concentration; this movement is called **diffusion**. As an example, consider adding a teaspoon of milk to a cup of hot tea. The milk tends to spread evenly through the tea. If the thermal energy of the solution is reduced, as with iced tea, the diffusion of milk molecules takes noticeably longer.

Although ions typically do not pass through a phospholipid bilayer directly, diffusion causes ions to be pushed through channels in the membrane. For example, if NaCl is dissolved in the fluid on one side of a permeable membrane (i.e., with channels that permit Na^+ and Cl^- passage), the Na^+ and Cl^- ions cross until they are evenly distributed in the solutions on both sides (Figure 3.8). As with the previous example, the net movement is from the region of high concentration to the region of low concentration. (For a review of how concentrations are expressed, see Box 3.1.) Such a difference in concentration is called a **concentration gradient**. Thus, we say that ions flow down a concentration gradient. Driving ions across the membrane by diffusion, therefore, happens when (1) the membrane possesses channels permeable to the ions, and (2) there is a concentration gradient across the membrane.

Electricity

Besides diffusion down a concentration gradient, another way to induce a net movement of ions in a solution is to use an electrical field, because ions are electrically charged particles. Consider the situation in Figure 3.9, in which wires from the two terminals of a battery are placed in a solution containing dissolved NaCl. Remember, *opposite charges attract and like charges repel.* Consequently, there will be a net movement of Na^+ toward the negative terminal (the cathode) and of Cl^- toward the positive terminal (the anode). The movement of electrical charge is called **electrical current**, represented by the symbol I and measured in units called amperes (amps). According to the convention established by Benjamin Franklin, current is defined as being posi-

Box 3.1

B R A I N F O O D

A Review of Moles and Molarity

Concentrations of substances are expressed as the number of molecules per liter of solution. The number of molecules is usually expressed in *moles*. One mole is 6.02×10^{23} molecules. A solution is said to be 1 Molar (M) if it has a concentration of 1 mole per liter. A 1 millimolar (mM) solution has 0.001 moles per liter. The abbreviation for concentration is a pair of brackets. Thus, we read [NaCl] = 1 mM as "the concentration of the sodium chloride solution is 1 millimolar."

tive in the direction of positive-charge movement. In this example, therefore, positive current flows in the direction of Na^+ movement, from the anode to the cathode.

Two important factors determine how much current will flow: electrical potential and electrical conductance. **Electrical potential**, also called **voltage**, is the force exerted on a charged particle, and it reflects the difference in charge between the anode and the cathode. More current flows as this difference increases. Voltage is represented by the symbol V and is measured in units called volts. As an example, the difference in electrical potential between the terminals of a car battery is 12 volts; that is, the electrical potential at one terminal is 12 volts more positive than that at the other.

Electrical conductance is the relative ability of an electrical charge to migrate from one point to another. It is represented by the symbol g and measured in units called siemens (S). Conductance depends on the number of particles available to carry electrical charge and on the ease with which these particles can travel through space. A term that expresses the same property in a different way is **electrical resistance**, the relative inability of an electrical charge to migrate. It is represented by the symbol R and measured in units called ohms (Ω). Resistance is simply the inverse of conductance (i.e., $R = 1/g$).

There is a simple relationship between potential (V), conductance (g), and the amount of current (I) that will flow. This relationship, known as **Ohm's law**, may be written $I = gV$: Current is the product of the conductance and the potential difference. Notice that if the conductance is zero, no current will flow even when the potential difference is very large. Likewise, when the potential difference is zero, no current will flow even when the conductance is very large.

Consider the situation illustrated in Figure 3.10a, in which NaCl has been dissolved in equal concentrations on either side of a phospholipid bilayer. If we drop wires from the two terminals of a battery into the solution on either side, we will generate a large potential difference across this membrane. No current will flow, however, because there are no channels to allow migration of Na^+ and Cl^- across the membrane; the conductance of the membrane is zero. Driving an ion across the membrane electrically, therefore, requires that (1) the membrane possesses channels permeable to that ion and (2) there is an electrical potential difference across the membrane (Figure 3.10b).

The stage is now set. We have electrically charged ions in solution on either side of the neuronal membrane. Ions can cross the membrane only by way of protein channels. The protein channels can be highly selective for spe-

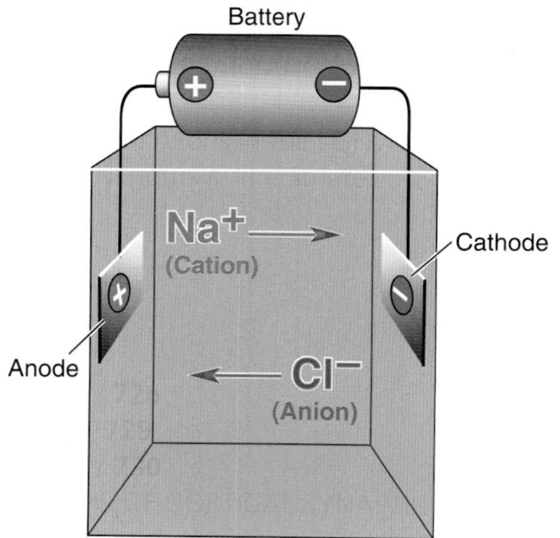

Figure 3.9
The movement of ions influenced by an electrical field.

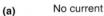

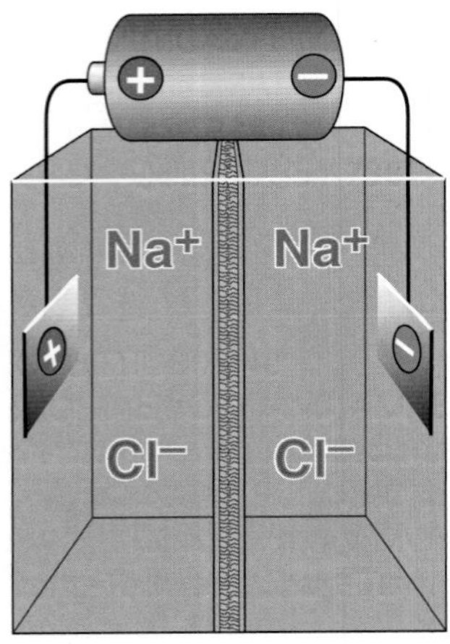

(a) No current

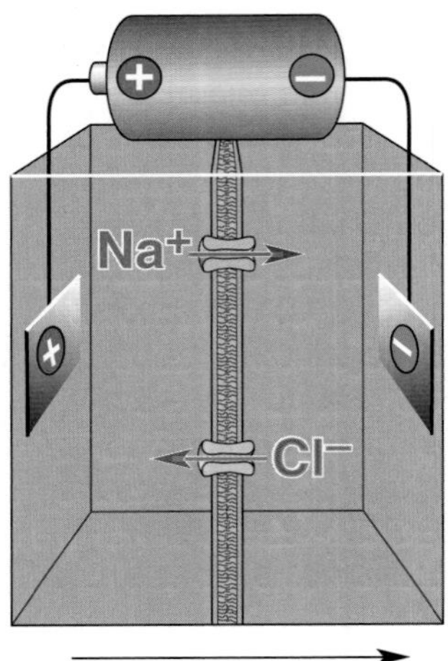

(b)

Figure 3.10
Electrical current flow across a membrane. (a) Voltage applied across a phospholipid bilayer causes no electrical current because there are no channels to allow the passage of electrically charged ions from one side to the other; the conductance of the membrane is zero. **(b)** Inserting channels in the membrane allows ions to cross. Electrical current flows in the direction of cation movement (from left to right in this example).

cific ions. The movement of any ion through its channel depends on the concentration gradient and the difference in electrical potential across the membrane. Let's use this knowledge to explore the resting membrane potential.

THE IONIC BASIS OF THE RESTING MEMBRANE POTENTIAL

The **membrane potential** is the voltage across the neuronal membrane at any moment, represented by the symbol V_m. Sometimes V_m is "at rest"; at other times it is not (such as during an action potential). V_m can be measured by inserting a microelectrode into the cytosol. A typical **microelectrode** is a thin glass tube with an extremely fine tip (diameter 0.5 μm) that will penetrate the membrane of a neuron with minimal damage. It is filled with an electrically conductive salt solution and connected to a voltmeter. The voltmeter measures the electrical potential difference between the tip of this microelectrode and a wire placed outside the cell (Figure 3.11). This method reveals that electrical charge is unevenly distributed across the neuronal membrane. The inside of the neuron is electrically negative with respect to the outside. This steady difference, the resting potential, is maintained whenever a neuron is not generating impulses.

The resting potential of a typical neuron is about −65 millivolts (1 mV = 0.001 volts). Stated another way, for a neuron at rest, $V_m = -65$ mV. This negative resting membrane potential inside the neuron is an absolute requirement for a functioning nervous system. To understand the negative membrane potential, we look to the ions that are available and how they are distributed inside and outside the neuron.

Equilibrium Potentials

Consider a hypothetical cell in which the inside is separated from the outside by a pure phospholipid membrane with no proteins. Inside this cell, a concentrated potassium salt solution is dissolved, yielding K^+ and A^- (A^- for anion, any molecule with a negative charge). Outside the cell is a solution with the same salt but diluted twentyfold with water. Although a large con-

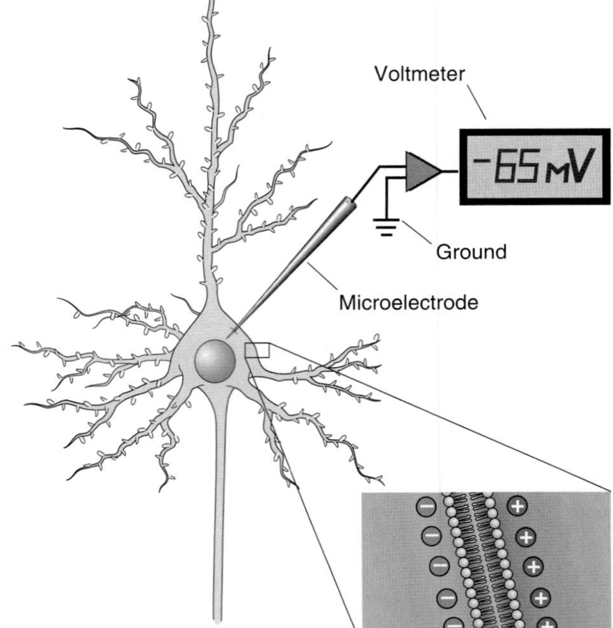

Figure 3.11
Measuring the resting membrane potential. A voltmeter measures the difference in electrical potential between the tip of a microelectrode inside the cell and a wire in the extracellular fluid. Typically, the inside of the neuron is about −65 mV with respect to the outside. This potential is caused by the uneven distribution of electrical charge across the membrane (enlargement).

centration gradient exists between the inside of the cell and the outside, there will be no net movement of ions because the phospholipid bilayer, having no channel proteins, is impermeable to charged, hydrophilic atoms. Under these conditions, a microelectrode would record no potential difference between the inside and the outside of the cell. In other words, V_m would be equal to 0 mV because the ratio of K^+ to A^- on each side of the membrane equals 1; both solutions are electrically neutral (Figure 3.12a).

Consider how this situation would change if potassium channels were inserted into the phospholipid bilayer. Because of the selective permeability of these channels, K^+ would be free to pass across the membrane, but A^- would not. Initially, diffusion rules: K^+ ions pass through the channels out of the cell, down the steep concentration gradient. Because A^- is left behind, however, the inside of the cell immediately begins to acquire a net negative charge, and an electrical potential difference is established across the membrane (Figure 3.12b). As the inside fluid acquires more and more net negative charge, the electrical force starts to pull positively charged K^+ ions back through the channels into the cell. When a certain potential difference is reached, the electrical force pulling K^+ ions inside exactly counterbalances the force of diffusion pushing them out. Thus, an *equilibrium* state is reached in which the diffusional and electrical forces are equal and opposite, and the net movement of K^+ across the membrane ceases (Figure 3.12c). The electrical potential difference that exactly balances an ionic concentration gradient is called an **ionic equilibrium potential**, or simply **equilibrium potential**; it is represented by the symbol E_{ion}. In this example, the equilibrium potential will be about −80 mV.

The example in Figure 3.12 demonstrates that generating a steady electrical potential difference across a membrane is a relatively simple matter. All that is required is an ionic concentration gradient and selective ionic permeability. Before moving on to the situation in real neurons, however, we can use this example to make four important points.

1. *Large changes in membrane potential are caused by minuscule changes in ionic concentrations.* In Figure 3.12, channels were inserted and K^+ ions flowed out of the cell until the membrane potential went from 0 mV to the equilibrium potential of −80 mV. How much does this ionic redistribution affect the K^+ concentration on either side of the membrane? Not much. For a cell with a 50-μm diameter containing 100 mM K^+, it can be calculated that the concentration change required to take the membrane from 0 to −80 mV is about 0.00001 mM. That is, when the channels were inserted and the K^+ flowed out until equilibrium was reached, the internal K^+ concentration went from 100 mM to 99.99999 mM—a negligible drop in concentration.
2. *The net difference in electrical charge occurs at the inside and outside surfaces of the membrane.* Because the phospholipid bilayer is so thin (less than 5 nm thick), it is possible for ions on one side to interact electrostatically with

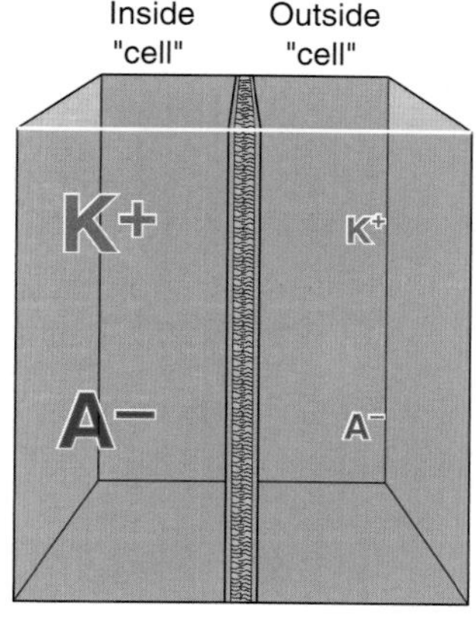

(a)

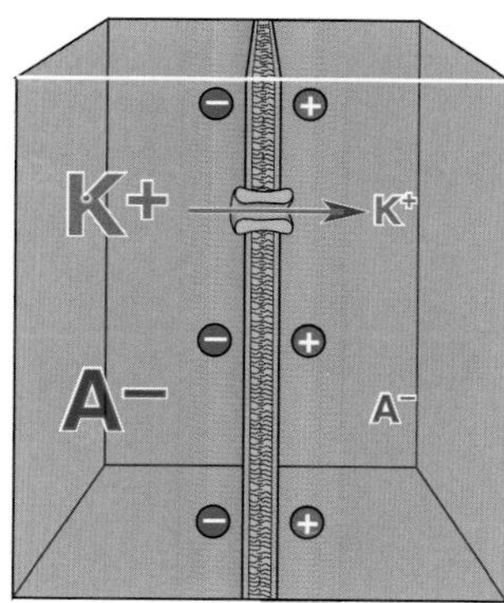

(b)

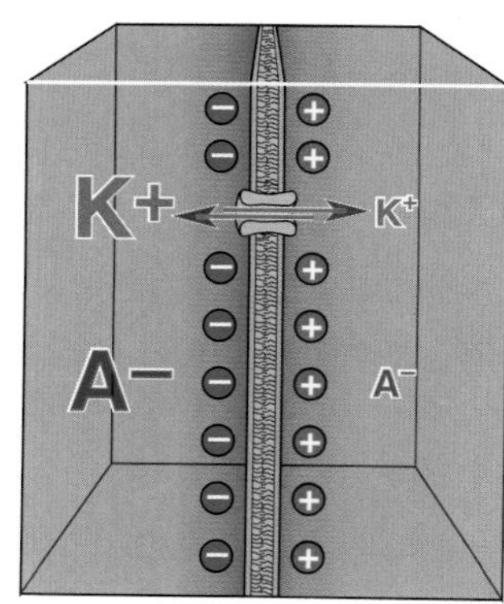

(c)

Figure 3.12
Establishing equilibrium in a selectively permeable membrane. (a) An impermeable membrane separates two regions: one of high salt concentration (inside) and the other of low salt concentration (outside). **(b)** Inserting a channel that is selectively permeable to K^+ into the membrane initially results in a net movement of K^+ ions down their concentration gradient, from left to right. **(c)** A net accumulation of positive charge on the outside and negative charge on the inside retards the movement of positively charged K^+ ions from the inside to the outside. Equilibrium is established such that there is no net movement of ions across the membrane, leaving a charge difference between the two sides.

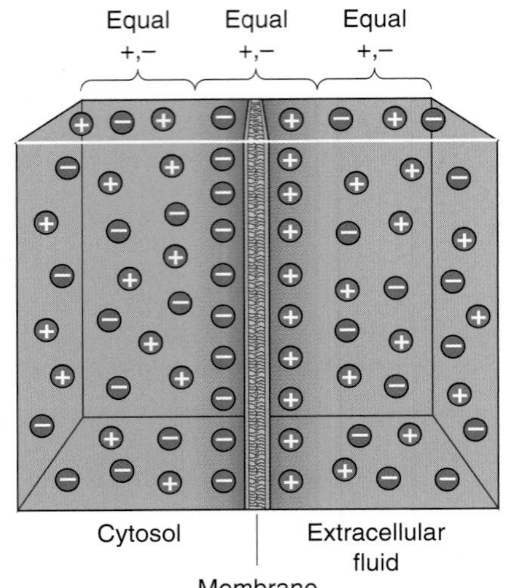

Figure 3.13
The distribution of electrical charge across the membrane. The uneven charges inside and outside the neuron line up along the membrane because of electrostatic attraction across this very thin barrier. Notice that the bulk of the cytosol and extracellular fluid is electrically neutral.

ions on the other side. Thus, the negative charges inside the neuron and the positive charges outside the neuron tend to be mutually attracted to the cell membrane. Consider how, on a warm summer evening, mosquitoes are attracted to the outside face of a windowpane when the inside lights are on. Similarly, the net negative charge inside the cell is not distributed evenly in the cytosol but, rather, is localized at the inner face of the membrane (Figure 3.13). In this way, the membrane is said to store electrical charge, a property called *capacitance*.

3. *Ions are driven across the membrane at a rate proportional to the difference between the membrane potential and the equilibrium potential.* Notice from our example in Figure 3.12 that when the channels were inserted, there was a net movement of K^+ only as long as the electrical membrane potential differed from the equilibrium potential. The difference between the real membrane potential and the equilibrium potential ($V_m - E_{ion}$) for a particular ion is called the **ionic driving force**. We'll talk more about this in Chapters 4 and 5 when we discuss the movement of ions across the membrane during the action potential and synaptic transmission.
4. *If the concentration difference across the membrane is known for an ion, an equilibrium potential can be calculated for that ion.* In our example in Figure 3.12, we assumed that K^+ was more concentrated inside the cell than outside it. From this knowledge, we were able to deduce that the equilibrium potential would be negative if the membrane were selectively permeable to K^+. Let's consider another example, in which Na^+ is more concentrated *outside* the cell (Figure 3.14). If the membrane contained sodium channels, Na^+ would flow down the concentration gradient *into* the cell. The entry of positively charged ions would cause the cytosol on the inner surface of the membrane to acquire a net positive charge. The positively charged interior of the cell would now repel Na^+ ions, tending to push them back out through their channels. At a certain potential difference, the electrical force pushing Na^+ ions out would exactly counterbalance the force of diffusion pushing them in. In this example, the membrane potential at equilibrium would be positive on the inside.

The examples in Figures 3.12 and 3.14 illustrate that if we know the ionic concentration difference across the membrane, we can figure out the equilibrium potential for any ion. Prove it to yourself. Assume that Ca^{2+} is more concentrated on the outside of the cell than on the inside and that the membrane is selectively permeable to Ca^{2+}. See if you can figure out whether the inside of the cell would be positive or negative at equilibrium. Try it again, assuming that the membrane is selectively permeable to Cl^- and that Cl^- is more concentrated outside the cell than inside it. (Pay attention here; note the charge of the ion.)

The Nernst Equation. The preceding examples show that each ion has its own equilibrium potential—the steady electrical potential that would be achieved if the membrane were permeable only to that ion. Thus, we can speak of the potassium equilibrium potential, E_K; the sodium equilibrium potential, E_{Na}; the calcium equilibrium potential, E_{Ca}; and so on. And knowing the electrical charge of the ion and the concentration difference across the membrane, we can easily deduce whether the inside of the cell would be positive or negative at equilibrium. In fact, the *exact* value of an equilibrium potential in mV can be calculated using an equation derived from the principles of physical chemistry, the **Nernst equation**, which takes into consideration the charge of the ion, the temperature, and the ratio of the external and internal ion concentrations. Using the Nernst equation, we can calculate the value of the equilibrium potential for any ion. For example, if K^+ is concentrated twentyfold on the inside of a cell, the Nernst equation tells us that $E_K = -80$ mV (Box 3.2).

The Distribution of Ions Across the Membrane

It should now be clear that the neuronal membrane potential depends on the ionic concentrations on either side of the membrane. Estimates of these concentrations appear in Figure 3.15. The important point is that *K^+ is more concentrated on the inside than the outside, and Na^+ and Ca^{2+} are more concentrated on the outside than the inside.*

How do these concentration gradients arise? Ionic concentration gradients are established by the actions of ion pumps in the neuronal membrane. Two ion pumps are especially important in cellular neurophysiology: the sodium-potassium pump and the calcium pump. The **sodium-potassium pump** is an enzyme that breaks down ATP in the presence of internal Na^+. The chemical energy released by this reaction drives the pump, which exchanges internal Na^+ for external K^+. The actions of this pump ensure that K^+ is concentrated inside the neuron and that Na^+ is concentrated outside. Notice that the pump pushes these ions across the membrane against their concentration gradients (Figure 3.16). This work requires the expenditure of metabolic energy. Indeed, it has been estimated that the sodium-potassium pump expends as much as 70% of the total amount of ATP utilized by the brain.

The **calcium pump** is also an enzyme that actively transports Ca^{2+} out of the cytosol across the cell membrane. Additional mechanisms decrease intracellular [Ca^{2+}] to a very low level (0.0002 mM); these include intracellular calcium-binding proteins and organelles, such as mitochondria and types of endoplasmic reticulum, that sequester cytosolic calcium ions.

Ion pumps are the unsung heroes of cellular neurophysiology. They work in the background to ensure that the ionic concentration gradients are established and maintained. These proteins may lack the glamour of a gated ion channel, but without ion pumps, the resting membrane potential would not exist, and the brain would not function.

Relative Ion Permeabilities of the Membrane at Rest

The pumps establish ionic concentration gradients across the neuronal membrane. With knowledge of these ionic concentrations, we can use the Nernst equation to calculate equilibrium potentials for the different ions (Figure 3.15). Remember, though, that an equilibrium potential for an ion is the membrane potential that results if a membrane is *selectively permeable* to that ion alone. In reality, however, neurons are not permeable to only a single type of ion. How do we incorporate this detail into our thinking?

Let's consider a few scenarios involving K^+ and Na^+. If the membrane of a neuron were permeable only to K^+, the membrane potential would equal

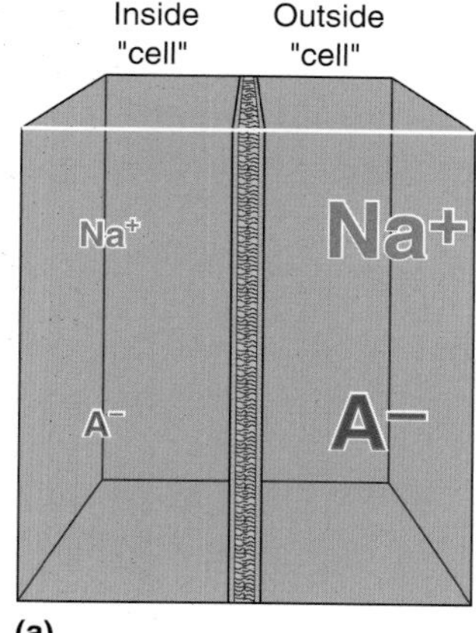

(a)

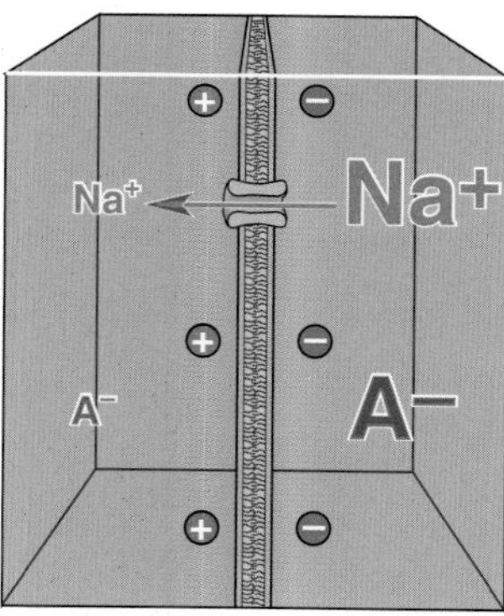

(b)

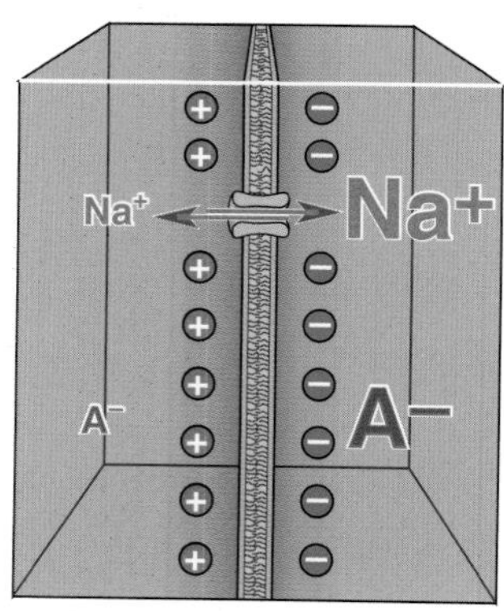

(c)

Figure 3.14
Another example of establishing equilibrium in a selectively permeable membrane. **(a)** An impermeable membrane separates two regions: one of high salt concentration (outside) and the other of low salt concentration (inside). **(b)** Inserting a channel that is selectively permeable to Na^+ into the membrane initially results in a net movement of Na^+ ions down their concentration gradient, from right to left. **(c)** A net accumulation of positive charge on the inside and negative charge on the outside retards the movement of positively charged Na^+ ions from the outside to the inside. Equilibrium is established such that there is no net movement of ions across the membrane, leaving a charge difference between the two sides; in this case, the inside of the cell is positively charged with respect to the outside.

Box 3.2

BRAIN FOOD

The Nernst Equation

The equilibrium potential for an ion can be calculated using the Nernst equation:

$$E_{ion} = 2.303 \frac{RT}{zF} \log \frac{[ion]_o}{[ion]_i}$$

where E_{ion} equals ionic equilibrium potential, R equals gas constant, T equals absolute temperature, z equals charge of the ion, F equals Faraday's constant, log equals base 10 logarithm, $[ion]_o$ equals ionic concentration outside the cell, and $[ion]_i$ equals ionic concentration inside the cell. The Nernst equation can be derived from the principles of physical chemistry. Let's see if we can make some sense of it.

Remember that equilibrium is the balance of two influences: diffusion, which pushes an ion down its concentration gradient, and electricity, which causes an ion to be attracted to opposite charges and repelled by like charges. Increasing the thermal energy of each particle increases diffusion and, therefore, increases the potential difference achieved at equilibrium. Thus, E_{ion} is proportional to T. On the other hand, increasing the electrical charge of each particle decreases the potential difference needed to balance diffusion. Therefore, E_{ion} is inversely proportional to the charge of the ion (z). We need not worry about R and F in the Nernst equation because they are constants.

At body temperature (37°C), the Nernst equation for the important ions—K^+, Na^+, Cl^-, and Ca^{2+}—simplifies to this:

$$E_K = 61.54 \text{ mV} \log \frac{[K^+]_o}{[K^+]_i}$$

$$E_{Na} = 61.54 \text{ mV} \log \frac{[Na^+]_o}{[Na^+]_i}$$

$$E_{Cl} = -61.54 \text{ mV} \log \frac{[Cl^-]_o}{[Cl^-]_i}$$

$$E_{Ca} = 30.77 \text{ mV} \log \frac{[Ca^{2+}]_o}{[Ca^{2+}]_i}$$

Therefore, to calculate the equilibrium potential for a certain type of ion at body temperature, all we need to know is the ionic concentrations on either side of the membrane. For instance, in the example we used in Figure 3.12, we stipulated that K^+ was twentyfold more concentrated inside the cell than outside:

$$\text{If } \frac{[K^+]_o}{[K^+]_i} = \frac{1}{20}$$

$$\text{and } \log \frac{1}{20} = -1.3$$

$$\text{then } E_K = 61.54 \text{ mV} \times -1.3 = -80 \text{ mV}.$$

Notice that there is no term in the Nernst equation for permeability or ionic conductance. Thus, calculating the value of E_{ion} does not require knowledge of the selectivity or the permeability of the membrane for the ion. There is an equilibrium potential for each ion in the intracellular and extracellular fluid. E_{ion} is the membrane potential that would just balance the ion's concentration gradient, so that no net ionic current would flow if the membrane were permeable to that ion.

E_K, which according to Figure 3.15 is −80 mV. On the other hand, if the membrane of a neuron were permeable only to Na^+, the membrane potential would equal E_{Na}, 62 mV. If the membrane were equally permeable to K^+ and Na^+, however, the resulting membrane potential would be some average of E_{Na} and E_K. What if the membrane were 40 times more permeable to K^+ than to Na^+? The membrane potential again would be between E_{Na} and E_K but much closer to E_K than to E_{Na}. This approximates the situation in real neurons. The resting membrane potential of −65 mV approaches, but does not achieve, the potassium equilibrium potential of −80 mV. This difference

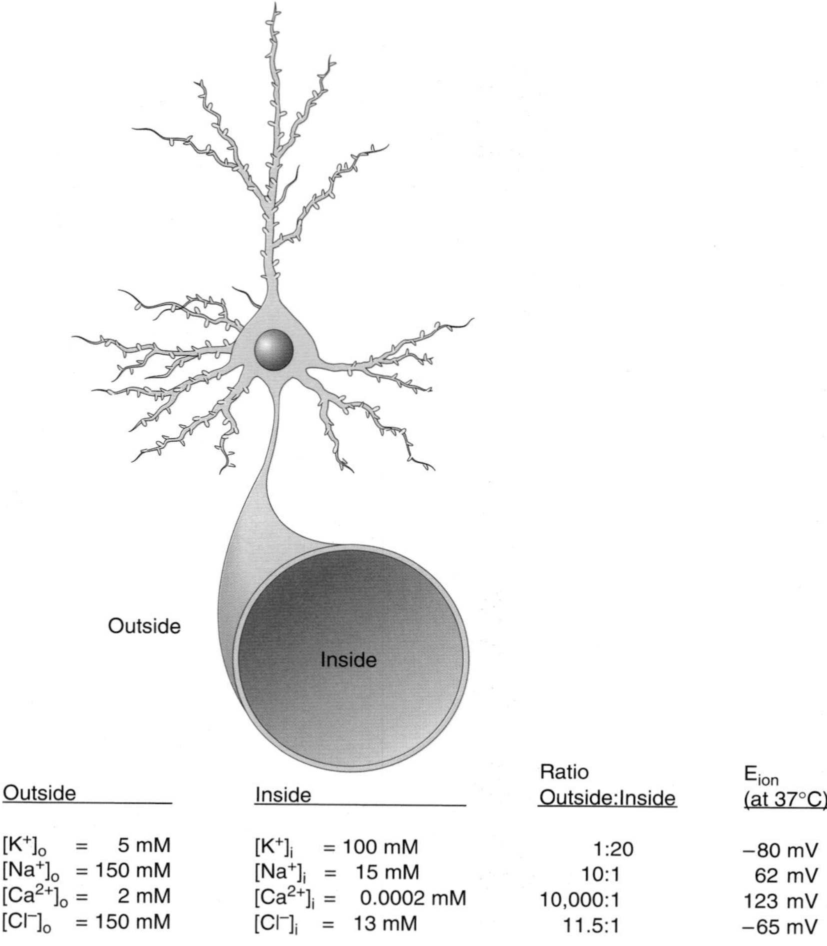

Outside	Inside	Ratio Outside:Inside	E_{ion} (at 37°C)
$[K^+]_o$ = 5 mM	$[K^+]_i$ = 100 mM	1:20	−80 mV
$[Na^+]_o$ = 150 mM	$[Na^+]_i$ = 15 mM	10:1	62 mV
$[Ca^{2+}]_o$ = 2 mM	$[Ca^{2+}]_i$ = 0.0002 mM	10,000:1	123 mV
$[Cl^-]_o$ = 150 mM	$[Cl^-]_i$ = 13 mM	11.5:1	−65 mV

Figure 3.15
Approximate ion concentrations on either side of a neuronal membrane. E_{ion} is the membrane potential that would be achieved at body temperature if the membrane were selectively permeable to that ion.

arises because although the membrane at rest is highly permeable to K^+, there is also a steady leak of Na^+ into the cell.

The resting membrane potential can be calculated using the **Goldman equation**, a mathematical formula that takes into consideration the relative permeability of the membrane to different ions. If we concern ourselves only with K^+ and Na^+, use the ionic concentrations in Figure 3.15, and assume that the resting membrane permeability is fortyfold greater to K^+ than to Na^+, then the Goldman equation predicts a resting membrane potential of −65 mV, the observed value (Box 3.3).

The Wide World of Potassium Channels. As we have seen, the selective permeability of potassium channels is a key determinant of the resting membrane potential and therefore of neuronal function. What is the molecular basis for this ionic selectivity? Selectivity for K^+ ions derives from the arrangement of amino acid residues that line the pore regions of the channels. Thus, it was a major breakthrough in 1987 when Lily and Yuh Nung Jan and their students at the University of California at San Francisco succeeded in determining the amino acid sequences of a family of potassium channels (Box 3.4). The search was conducted using the fruit fly *Drosophila melanogaster*. While these insects may be annoying in the kitchen, they are extremely valuable in the laboratory because their genes can be studied and manipulated in ways that are not possible in mammals.

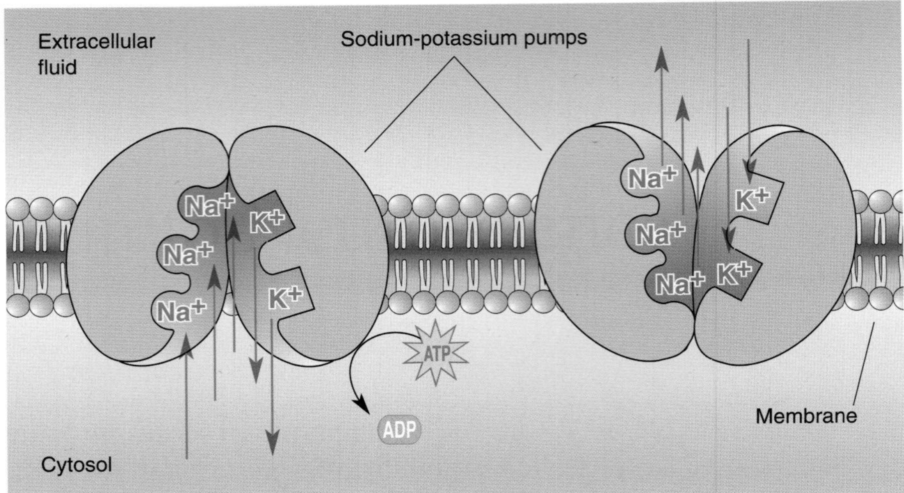

Figure 3.16
The sodium-potassium pump. This ion pump is a membrane-associated protein that transports ions across the membrane against their concentration gradients at the expense of metabolic energy. (ADP = adenosine diphosphate.)

Normal flies, like humans, can be put to sleep with ether vapors. While conducting research on anesthetized insects, investigators discovered that flies of one mutant strain responded to the ether by shaking their legs, wings, and abdomen. This strain of fly was designated *Shaker*. Detailed studies soon showed that the odd behavior was explained by a defect in a particular type of potassium channel. Using molecular biological techniques, the Jans mapped the gene that was mutated in *Shaker*. Knowledge of the DNA

Box 3.3

BRAIN FOOD

The Goldman Equation

If the membrane of a real neuron were permeable only to K^+, the resting membrane potential would equal E_K, about −80 mV. But it does not; the measured resting membrane potential of a typical neuron is about −65 mV. This discrepancy is explained because real neurons at rest are not exclusively permeable to K^+; there is also some Na^+ permeability. Stated another way, the *relative permeability* of the resting neuronal membrane is quite high to K^+ and low to Na^+. If the relative permeabilities are known, it is possible to calculate the membrane potential at equilibrium by using the Goldman equation. Thus, for a membrane permeable only to Na^+ and K^+ at 37°C:

$$V_m = 61.54 \text{ mV} \log \frac{P_K [K^+]_o + P_{Na} [Na^+]_o}{P_K [K^+]_i + P_{Na} [Na^+]_i}$$

where V_m is the membrane potential, P_K and P_{Na} are the relative permeabilities to K^+ and Na^+, respectively, and the other terms are the same as for the Nernst equation.

If the resting membrane ion permeability to K^+ is 40 times greater than it is to Na^+, then solving the Goldman equation using the concentrations in Figure 3.15 yields:

$$V_m = 61.54 \text{ mV} \log \frac{40\,(5) + 1\,(150)}{40\,(100) + 1\,(15)}$$

$$= 61.54 \text{ mV} \log \frac{350}{4015}$$

$$= -65 \text{ mV}$$

Box 3.4

PATH OF DISCOVERY

Shaker Flies and Their Defective Potassium Channels

BY LILY AND YUH NUNG JAN

Having gone to graduate school at Caltech with an intention to study theoretical high-energy physics, we had the good fortune of being lured to biology by one of the great pioneers of molecular genetics, Max Delbrück. We stayed at Caltech for our postdoctoral training, attracted by Seymour Benzer's idea of using genetics, which had been so well developed in the fruit fly *Drosophila melanogaster*, in the study of nervous system function. It was more or less chance that led to our discovery of an association between potassium channels and mutations in a gene called *Shaker*.

In 1974, to prepare for our postdoctoral studies in Seymour's lab, we went to Cold Spring Harbor Laboratory to take summer courses in neurobiology. During a break, we went to Yale University to visit Doug Kankel, a former postdoctoral fellow with Seymour, and helped ourselves to a bottle of fruit flies. That fly bottle proved handy when we returned to the laboratory course. Surrounded by wonderful teachers like JacSue Kehoe, Philippe Ascher, and Enrico Stefani, we made electrophysiological recordings from the *Drosophila* larval muscles and repeated a number of the experiments we had just learned in the frog and other preparations. After returning to Caltech, we thought that perhaps larval muscle recordings could be used as an assay system for genetic mutations that affect synaptic transmission. Before getting to the mutants, however, we spent several months characterizing neuromuscular transmission in normal (wild-type) *Drosophila* larvae, to make sure it worked in much the same way as in the frog.

Electrophysiological recording was impractical as a primary screen of the thousands of different flies in which mutations had been produced. Figuring that mutations affecting synaptic transmission are likely to result in behavioral defects if they do not kill the flies altogether, we began by recording from mutants in Seymour's stock collection that showed peculiar behaviors. Shortly after we started this mutant screen, on April 28, 1975, we came upon KS133, a *Shaker* mutant isolated by Seymour's group (so named because of its propensity to shake under ether anesthesia). We found unusual electrophysiological properties but struggled to understand exactly what was wrong. We soon realized that we needed a more sophisticated method to study these flies. We were fortunate to collaborate with Mike Dennis, who had taught us in the Cold Spring Harbor summer course. Mike invited us to his laboratory at the University of California at San Francisco (UCSF) and taught us how to use a compound microscope to guide our recordings from the muscle. Those week-long visits with Mike remain vivid in our memories.

Lily and Yuh Nung Jan

Our studies soon implicated a defect in the potassium channels of the *Shaker* mutants. This defect could arise either because the *Shaker* gene codes for a potassium channel or because the *Shaker* gene in some other way controls potassium channel expression or function. This question was resolved 10 years later, after we set up our own lab at UCSF. In 1987, together with our students Diane Papzian, Tom Schwarz, Bruce Tempel, and Leslie Timpe, we succeeded in determining the full DNA sequence of the *Shaker* gene. This work finally revealed that the protein encoded by the *Shaker* gene forms a potassium channel (Figure A).

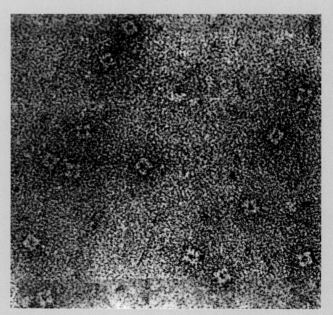

Figure A

Shaker potassium channels in the cell membrane of the fruit fly *Drosophila*, viewed from above with an electron microscope. (Source: Li et al., 1994, Fig. 2.)

sequence of what is now called the *Shaker* potassium channel enabled researchers to find the genes for other potassium channels based on sequence similarity. This analysis has revealed the existence of a very large number of different potassium channels, including those responsible for the maintenance of the resting membrane potential in neurons.

Most potassium channels have four subunits that are arranged like the staves of a barrel to form a pore (Figure 3.17). Despite their diversity, the subunits of different potassium channels have common structural features that bestow selectivity for K^+ ions. Of particular interest is a region called the *pore loop*, which contributes to the *selectivity filter* that makes the channel permeable mostly to K^+ ions (Figure 3.18). Mutations involving only a single amino acid in this region can severely disrupt neuronal function.

An example of this is seen in a strain of mice called *Weaver*. These animals have difficulty maintaining posture and moving normally. The defect has been traced to the mutation of a single amino acid in the pore loop of a potassium channel found in specific neurons of the cerebellum, a region of the brain important for motor coordination. As a consequence of the mutation, Na^+ as well as K^+ ions can pass through the channel. Increased sodium permeability causes the membrane potential of the neurons to become less negative, thus disrupting neuronal function. (Indeed, the absence of the normal negative membrane potential in these cells is believed to be the cause of their untimely death.) In recent years it has become increasingly clear that many

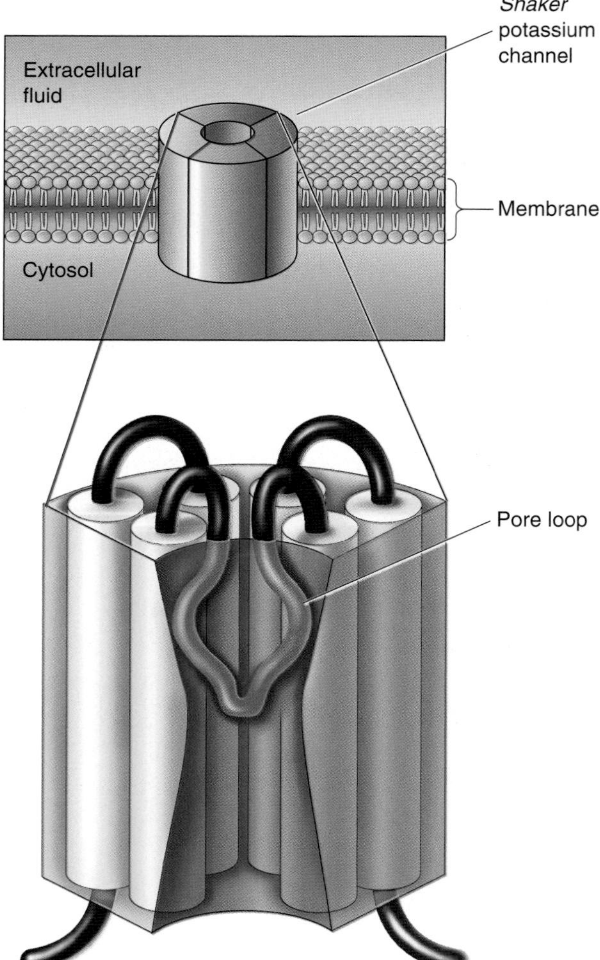

Figure 3.17
Structure of a potassium channel. The *Shaker* potassium channel has four subunits arranged like staves of a barrel to form a pore. Enlargement: The tertiary structure of the protein subunit contains a pore loop, a part of the polypeptide chain that makes a hairpin turn within the plane of the membrane. The pore loop is a critical part of the filter that makes the channel selectively permeable to K^+ ions.

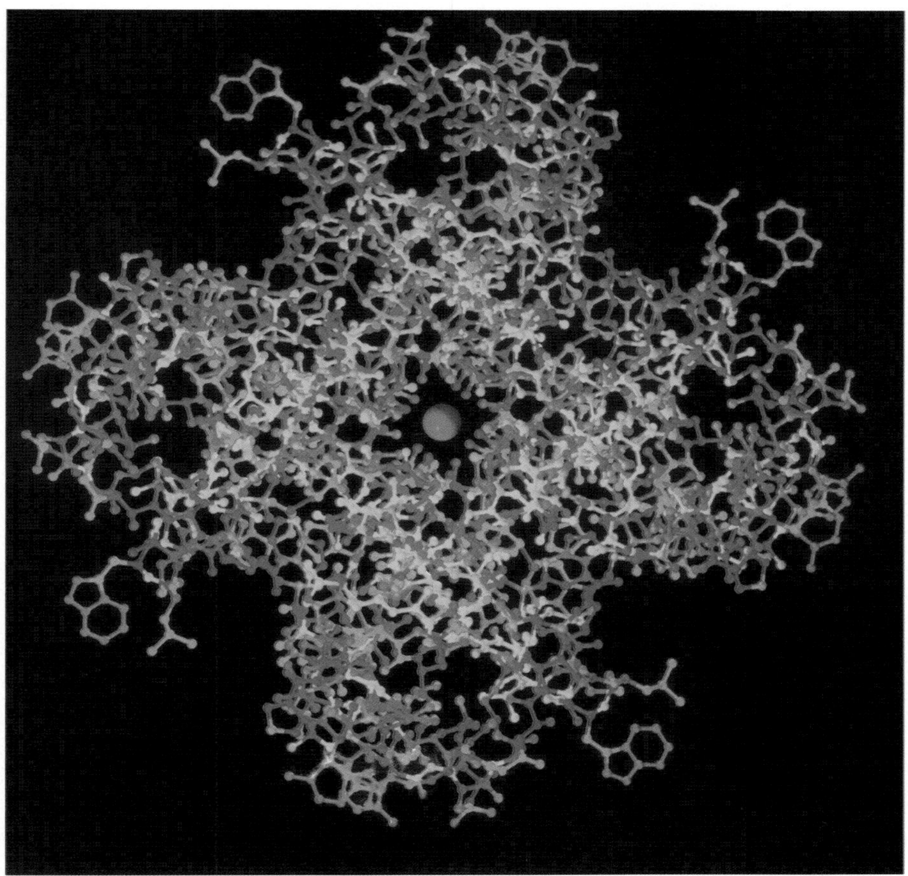

Figure 3.18
A view of the potassium channel pore. The atomic structure of potassium-selective ion channels has recently been solved. Here we are looking into the pore from the outside. The red ball in the middle is a K^+ ion. (Source: Doyle et al., 1998.)

inherited neurological disorders in humans, such as certain forms of epilepsy, may be explained by mutations of specific potassium channels.

The Importance of Regulating the External Potassium Concentration

Because the neuronal membrane at rest is mostly permeable to K^+, the membrane potential is close to E_K. Another consequence of high K^+ permeability is that the membrane potential is particularly sensitive to changes in the concentration of extracellular potassium. This relationship is shown in Figure 3.19. A tenfold change in the K^+ concentration outside the cell, $[K^+]_o$, from 5 to 50 mM, would take the membrane potential from −65 to −17 mV. A change in membrane potential from the normal resting value (−65 mV) to a less negative value is called a **depolarization** of the membrane. Therefore, *increasing extracellular potassium depolarizes neurons.*

The sensitivity of the membrane potential to $[K^+]_o$ has led to the evolution of mechanisms that tightly regulate extracellular potassium concentrations in the brain. One of these is the **blood-brain barrier**, a specialization of the walls of brain capillaries that limits the movement of potassium (and other blood-borne substances) into the extracellular fluid of the brain.

Glia, particularly astrocytes, also possess efficient mechanisms to take up extracellular K^+ whenever concentrations rise, as they normally do during

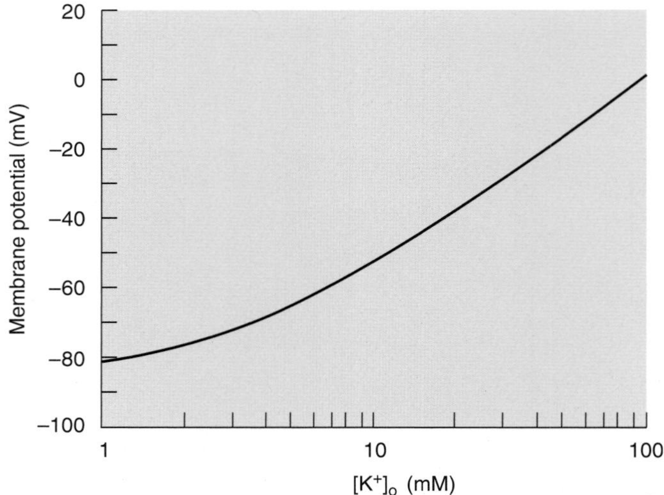

Figure 3.19

The dependence of membrane potential on external potassium concentration. Because the neuronal membrane at rest is mostly permeable to potassium, a tenfold change in $[K^+]_o$ from 5 to 50 mM causes a 48-mV depolarization of the membrane. This function was calculated using the Goldman equation (Box 3.3).

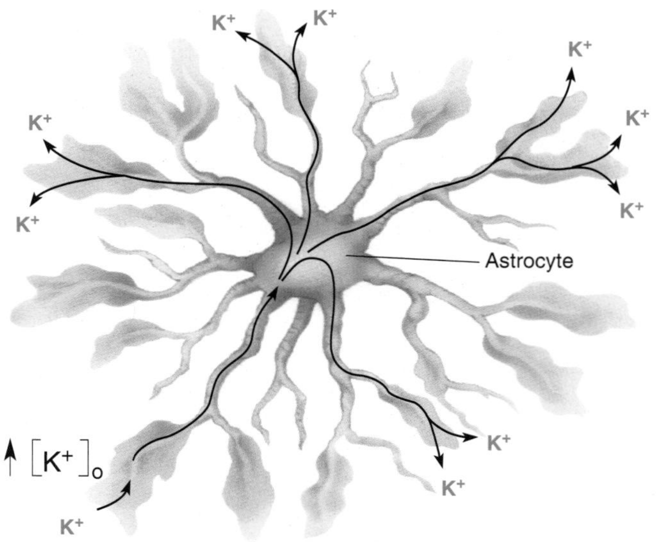

Figure 3.20

Potassium spatial buffering by astrocytes. When brain $[K^+]_o$ increases as a result of local neural activity, K^+ enters astrocytes via membrane channels. The extensive network of astrocytic processes helps dissipate the K^+ over a large area.

Box 3.5

OF SPECIAL INTEREST

Death by Lethal Injection

On June 4, 1990, Dr. Jack Kevorkian shocked the medical profession by assisting in the suicide of Janet Adkins. Adkins, a 54-year-old happily married mother of three, had been diagnosed with Alzheimer's disease, a progressive brain disorder that always results in senile dementia and death. Mrs. Adkins was a member of the Hemlock Society, which advocates euthanasia as an alternative to death by terminal illness. Dr. Kevorkian agreed to help Mrs. Adkins take her own life. In the back of a 1968 Volkswagen van at a campsite in Oakland County, Michigan, she was hooked to an intravenous line that infused a harmless saline solution. To choose death, Mrs. Adkins switched the solution to one that contained an anesthetic solution, followed automatically by potassium chloride. The anesthetic caused Mrs. Adkins to become unconscious by suppressing the activity of neurons in part of the brain called the reticular formation. However, cardiac arrest and death were caused by the KCl injection. The ionic basis of the resting membrane potential explains why the heart stopped beating.

Recall that the proper functioning of excitable cells (including those of cardiac muscle) requires that their membranes be maintained at the resting potential whenever they are not generating impulses. The negative resting potential is a result of selective ionic permeability to K^+ and to the metabolic pumps that concentrate potassium inside the cell. However, as Figure 3.19 shows, membrane potential is very sensitive to changes in the extracellular concentration of potassium. A tenfold rise in extracellular K^+ would wipe out the resting potential. Although neurons in the brain are somewhat protected from large changes in $[K^+]_o$, other excitable cells in the body, such as muscle cells, are not. Without negative resting potentials, cardiac muscle cells can no longer generate the impulses that lead to contraction, and the heart immediately stops beating. Intravenous potassium chloride is, therefore, a lethal injection.

periods of neural activity. Remember that astrocytes fill most of the space between neurons in the brain. Astrocytes have membrane potassium pumps that concentrate K^+ in their cytosol, and they also have potassium channels. When $[K^+]_o$ increases, K^+ ions enter the astrocyte through the potassium channels, causing the astrocyte membrane to depolarize. The entry of K^+ ions increases the internal potassium concentration, $[K^+]_i$, which is believed to be dissipated over a large area by the extensive network of astrocytic processes. This mechanism for the regulation of $[K^+]_o$ by astrocytes is called *potassium spatial buffering* (Figure 3.20).

Not all excitable cells are protected from increases in potassium, however. Muscle cells, for example, do not have a blood-brain barrier or glial buffering mechanisms. Consequently, although the brain is relatively protected, elevations of $[K^+]$ in the blood can have serious consequences for body physiology (Box 3.5).

CONCLUDING REMARKS

We have now explored the resting membrane potential. The activity of the sodium-potassium pump produces and maintains a large K^+ concentration gradient across the membrane. The neuronal membrane at rest is highly permeable to K^+ because of the membrane potassium channels. The movement of K^+ ions across the membrane, down their concentration gradient, leaves the inside of the neuron negatively charged.

The electrical potential difference across the membrane can be thought of as a battery whose charge is maintained by the work of the ion pumps. In the next chapter, we'll see how this battery runs our brain.

KEY TERMS

Introduction
action potential (p. 51)
excitable membrane (p. 52)
resting membrane potential (p. 52)

The Cast of Chemicals
ion (p. 52)
cation (p. 53)
anion (p. 53)
phospholipid bilayer (p. 54)
peptide bond (p. 55)
polypeptide (p. 56)
ion channel (p. 57)
ion selectivity (p. 57)
gating (p. 57)
ion pump (p. 57)

The Movement of Ions
diffusion (p. 58)
concentration gradient (p. 58)
electrical current (p. 58)
electrical potential (p. 59)
voltage (p. 59)
electrical conductance (p. 59)
electrical resistance (p. 59)
Ohm's law (p. 59)

The Ionic Basis of the Resting Membrane Potential
membrane potential (p. 60)
microelectrode (p. 60)
ionic equilibrium potential (p. 61)
equilibrium potential (p. 61)
ionic driving force (p. 62)
Nernst equation (p. 62)
sodium-potassium pump (p. 63)
calcium pump (p. 63)
Goldman equation (p. 65)
depolarization (p. 69)
blood-brain barrier (p. 69)

REVIEW QUESTIONS

1. What two functions do proteins in the neuronal membrane perform to establish and maintain the resting membrane potential?
2. On which side of the neuronal membrane are Na^+ ions more abundant?
3. When the membrane is at the potassium equilibrium potential, in which direction (in or out) is there a net movement of potassium ions?
4. There is a much greater K^+ concentration inside the cell than outside. Why, then, is the resting membrane potential negative?
5. When the brain is deprived of oxygen, the mitochondria within neurons cease producing ATP. What effect would this have on the membrane potential? Why?

CHAPTER

4

The Action Potential

INTRODUCTION

Now we come to the signal that conveys information over distances in the nervous system—the action potential. As we saw in Chapter 3, the cytosol in the neuron at rest is negatively charged with respect to the extracellular fluid. The action potential is a rapid reversal of this situation such that, for an instant, the inside of the membrane becomes positively charged with respect to the outside. The action potential is also often called a spike, a nerve impulse, or a discharge.

The action potentials generated by a cell are all similar in size and duration, and they do not diminish as they are conducted down the axon. Keep in mind the big picture: The *frequency* and *pattern* of action potentials constitute the code used by neurons to transfer information from one location to another. In this chapter, we discuss the mechanisms that are responsible for the action potential and how it propagates down the axonal membrane.

PROPERTIES OF THE ACTION POTENTIAL

Action potentials have certain universal properties, features that are shared by axons in the nervous systems of every beast, from a squid to a college student. Let's begin by exploring some of these properties. What does the action potential look like? How is it initiated? How rapidly can a neuron generate action potentials?

The Ups and Downs of an Action Potential

In Chapter 3, we saw that the membrane potential, V_m, can be determined by inserting a microelectrode in the cell. A voltmeter is used to measure the electrical potential difference between the tip of this intracellular microelectrode and another placed outside the cell. When the neuronal membrane is at rest, the voltmeter reads a steady potential difference of about −65 mV. During the action potential, however, the membrane potential briefly becomes positive. Because this occurs so rapidly—100 times faster than the blink of an eye—a special type of voltmeter, called an *oscilloscope*, is used to study action potentials. The oscilloscope records the voltage as it changes over time (Box 4.1).

An action potential as it would appear on the display of an oscilloscope is shown in Figure 4.1. This graph represents a plot of membrane potential versus time. Notice that the action potential has certain identifiable parts. The first part, called the **rising phase**, is characterized by a rapid depolarization of the membrane. This change in membrane potential continues until V_m reaches a peak value of about 40 mV. The part of the action potential where the inside of the neuron is positively charged with respect to the outside is called the **overshoot**. The **falling phase** of the action potential is a rapid repolarization until the membrane is actually more negative than the resting potential. The last part of the falling phase is called the **undershoot**, or **afterhyperpolarization**. Finally, there is a gradual restoration of the resting potential. From beginning to end, the action potential lasts about 2 milliseconds (msec).

The Generation of an Action Potential

In Chapter 3, we said that breaking of the skin by a thumbtack was sufficient to generate action potentials in a sensory nerve. Let's use this example to see how an action potential begins.

The perception of sharp pain when a thumbtack enters your foot is caused by the generation of action potentials in certain nerve fibers in the skin. (We'll

Box 4.1

BRAIN FOOD

Methods of Recording Action Potentials

Methods for studying nerve impulses may be broadly divided into two types: intracellular and extracellular (Figure A). *Intracellular recording* requires impaling the neuron or axon with a microelectrode. The small size of most neurons makes this method challenging, explaining why so many of the early studies on action potentials were performed on the neurons of invertebrates, which can be 50–100 times larger than mammalian neurons. Fortunately, recent technical advances have made even the smallest vertebrate neurons accessible to intracellular recording methods, and these studies have confirmed that much of what was learned in invertebrates is directly applicable to humans.

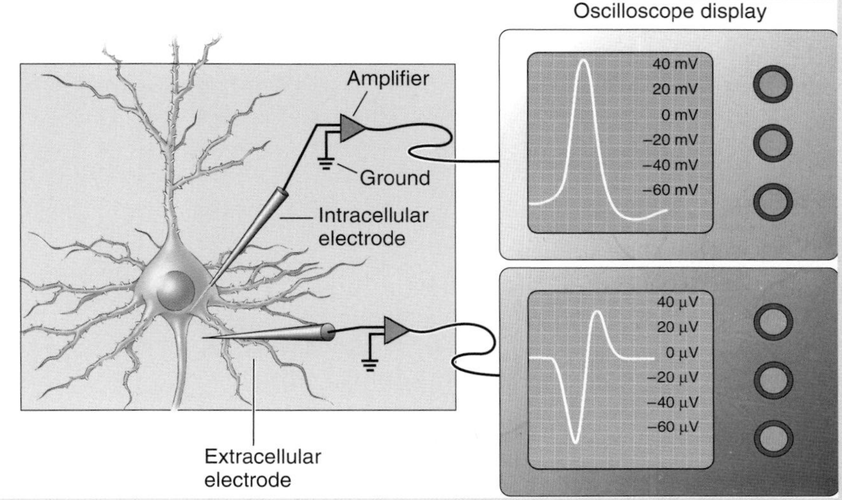

Figure A

The goal of intracellular recording is simple: to measure the potential difference between the tip of the intracellular electrode and another electrode placed in the solution bathing the neuron (continuous with the earth, and thus called ground). The intracellular electrode is filled with a concentrated salt solution (often KCl) having a high electrical conductivity. The electrode is connected to an amplifier that compares the potential difference between this electrode and ground. This potential difference can be displayed using an oscilloscope. The oscilloscope sweeps a beam of electrons from left to right across a phosphor screen. Vertical deflections of this beam can be read as changes in voltage. The oscilloscope is really just a sophisticated voltmeter that can record rapid changes in voltage, such as an action potential.

As we see in this chapter, the action potential is characterized by a sequence of ionic movements across the neuronal membrane. These electrical currents can be detected without impaling the neuron by placing an electrode near the membrane. This is the principle behind *extracellular recording*. Again, we measure the potential difference between the tip of the recording electrode and ground. The electrode can be a fine glass capillary filled with a salt solution, but it is often simply a thin insulated metal wire. Normally, in the absence of neural activity, the potential difference between the extracellular recording electrode and ground is zero. When the action potential arrives at the recording position, however, positive charges flow away from the recording electrode into the neuron. Then, as the action potential passes by, positive charges flow out across the membrane toward the recording electrode. Thus, the extracellular action potential is characterized by a brief, alternating voltage difference between the recording electrode and ground. (Notice the different voltage changes produced by the action potential recorded with intracellular and extracellular recordings.) These changes in voltage can be seen using an oscilloscope, but they can also be *heard* by connecting the output of the amplifier to a loudspeaker. Each impulse makes a distinctive popping sound. Indeed, recording the activity of an active sensory nerve sounds just like making popcorn.

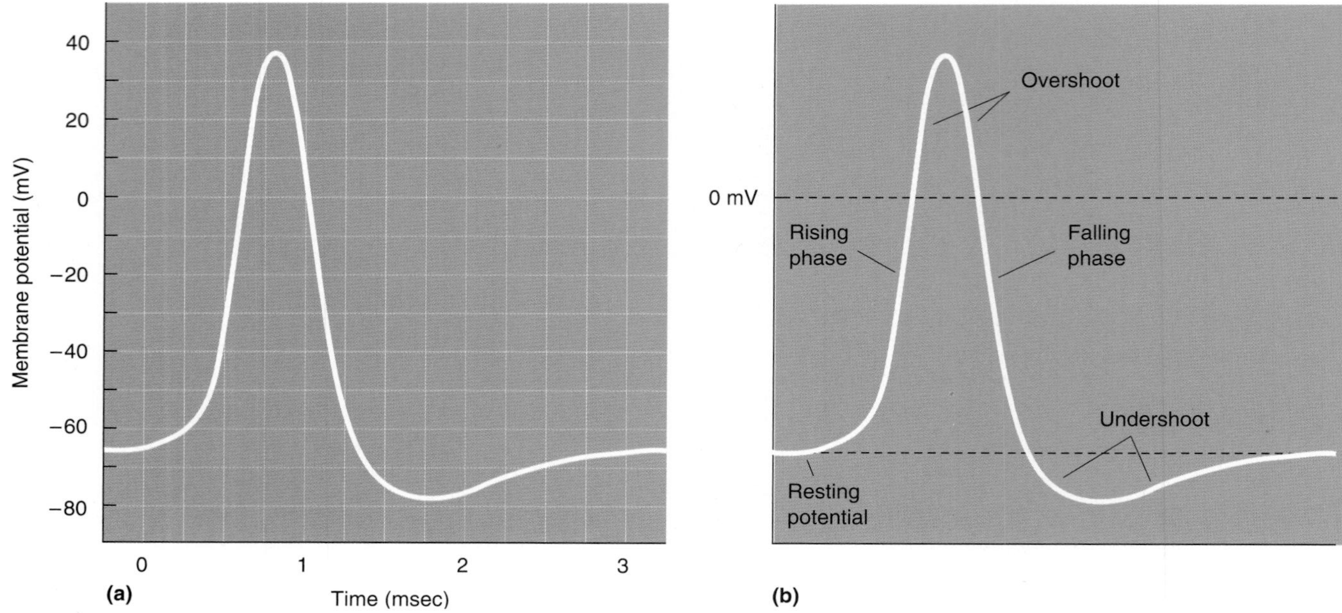

Figure 4.1
An action potential. (a) An action potential displayed by an oscilloscope. **(b)** The parts of an action potential.

learn more about pain in Chapter 12.) The membrane of these fibers is believed to possess a type of gated sodium channel that opens when the nerve ending is stretched. The initial chain of events is, therefore, (1) the thumbtack enters the skin, (2) the membrane of the nerve fibers in the skin is stretched, and (3) Na^+-permeable channels open. Because of the large concentration gradient and the negative charge of the cytosol, Na^+ ions enter the fiber through these channels. The entry of Na^+ depolarizes the membrane; that is, the cytoplasmic (inside) surface of the membrane becomes less negative. If this depolarization, called a *generator potential*, achieves a critical level, the membrane will generate an action potential. The critical level of depolarization that must be crossed in order to trigger an action potential is called **threshold**. *Action potentials are caused by depolarization of the membrane beyond threshold.*

The depolarization that causes action potentials arises in different ways in different neurons. In our example above, depolarization was caused by the entry of Na^+ through specialized ion channels that were sensitive to membrane stretching. In interneurons, depolarization is usually caused by Na^+ entry through channels that are sensitive to neurotransmitters released by other neurons. In addition to these natural routes, neurons can be depolarized by injecting electrical current through a microelectrode, a method commonly used by neuroscientists to study action potentials in different cells.

Generating an action potential by depolarizing a neuron is something like taking a photograph by pressing the shutter button on a camera. Applying increasing pressure on the button has no effect until it crosses a threshold value, and then "click"—the shutter opens and one frame of film is exposed. Applying increasing depolarization to a neuron has no effect until it crosses threshold, and then "pop"—one action potential. For this reason, action potentials are said to be "all-or-none."

The Generation of Multiple Action Potentials

In the above example, we likened the generation of an action potential by depolarization to taking a photograph by pressing the shutter button on a cam-

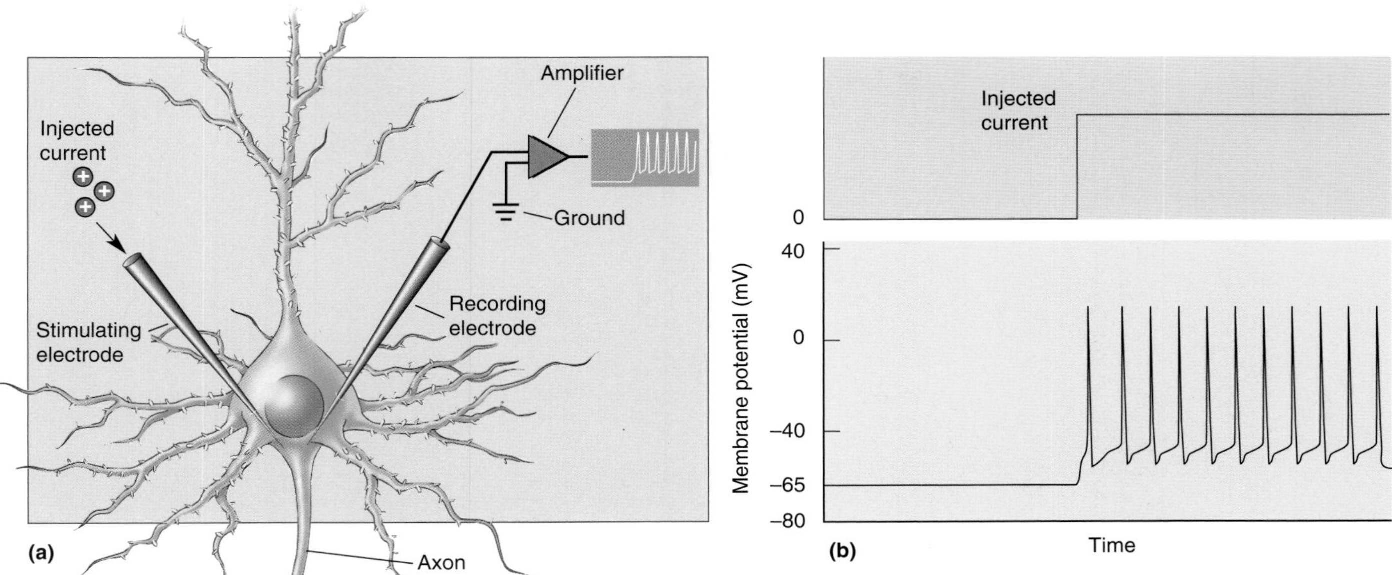

Figure 4.2
The effect of injecting a positive charge into a neuron. (a) The axon hillock is impaled by two electrodes, one for recording the membrane potential relative to ground and the other for stimulating the neuron with electrical current. **(b)** When electrical current is injected into the neuron (top trace), the membrane is depolarized sufficiently to fire action potentials (bottom trace).

era. But what if the camera were one of those fancy motor-driven models that fashion and sports photographers use? In that case, continued pressure on the shutter button beyond threshold would cause the camera to shoot frame after frame. The same thing is true for a neuron. If, for example, we pass continuous depolarizing current into a neuron through a microelectrode, we will generate not one but many action potentials in succession (Figure 4.2).

The rate of action potential generation depends on the magnitude of the continuous depolarizing current. If we pass enough current through a microelectrode to depolarize just to threshold but not far beyond, we might find that the cell generates action potentials at a rate of something like 1 per second, or 1 hertz (1 Hz). If we crank up the current a little more, however, we will find that the rate of action potential generation increases, say, to 50 impulses per second (50 Hz). Thus, the *firing frequency* of action potentials reflects the magnitude of the depolarizing current. This is one way that stimulation intensity is encoded in the nervous system (Figure 4.3).

Although firing frequency increases with the amount of depolarizing current, there is a limit to the rate at which a neuron can generate action potentials. The maximum firing frequency is about 1000 Hz; once an action potential is initiated, it is impossible to initiate another for about 1 msec. This period of time is called the **absolute refractory period**. In addition, it can be relatively difficult to initiate another action potential for several milliseconds after the end of the absolute refractory period. During this **relative refractory period**, the amount of current required to depolarize the neuron to action potential threshold is elevated above normal.

We will now take a look at how the movement of ions across the membrane through specialized protein channels causes a neural signal with these properties.

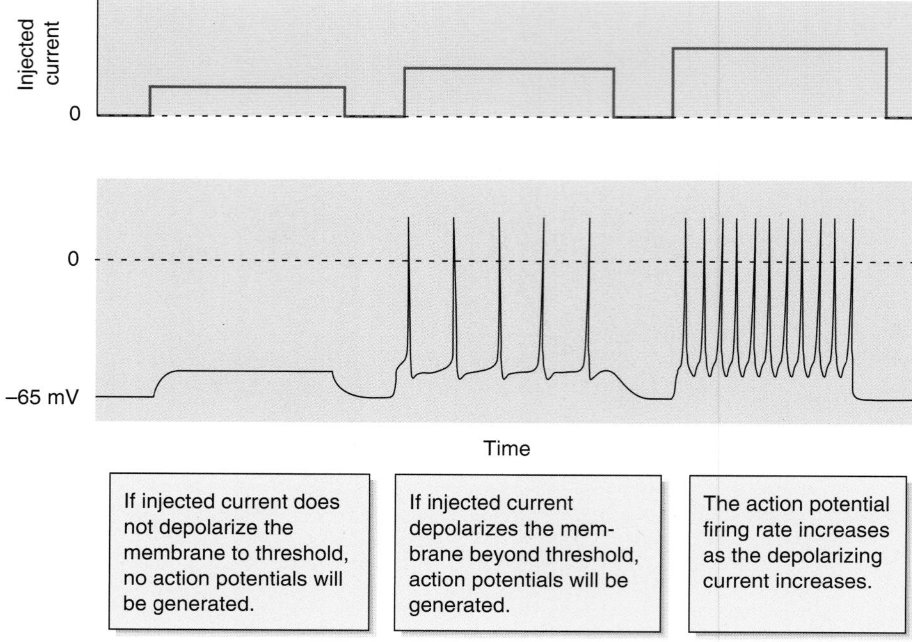

Figure 4.3
The dependence of action potential firing frequency on the level of depolarization.

THE ACTION POTENTIAL IN THEORY

The action potential is a dramatic redistribution of electrical charge across the membrane. *Depolarization of the cell during the action potential is caused by the influx of sodium ions across the membrane, and repolarization is caused by the efflux of potassium ions.* Let's apply some of the concepts introduced in Chapter 3 to help us understand how ions are driven across the membrane and how these ionic movements affect the membrane potential.

Membrane Currents and Conductances

Consider the ideal neuron illustrated in Figure 4.4. The membrane of this cell has three types of protein molecules: sodium-potassium pumps, potassium channels, and sodium channels. The pumps work continuously to establish and maintain concentration gradients. As in all of our previous examples, we'll assume that K^+ is concentrated twentyfold inside the cell and that Na^+ is concentrated tenfold outside the cell. According to the Nernst equation, at 37°C, $E_K = -80$ mV and $E_{Na} = 62$ mV. Let's use this cell to explore the factors that govern the movement of ions across the membrane.

We begin by assuming that both the potassium channels and the sodium channels are closed and that the membrane potential, V_m, is equal to 0 mV (Figure 4.4a). Now let's open the potassium channels only (Figure 4.4b). As we learned in Chapter 3, K^+ ions will flow out of the cell, down their concentration gradient, until the inside becomes negatively charged and $V_m = E_K$ (Figure 4.4c). Here we want to focus our attention on the movement of K^+ that took the membrane potential from 0 mV to -80 mV. Consider these three points:

1. The net movement of K^+ ions across the membrane is an electrical current. We can represent this current with the symbol I_K.

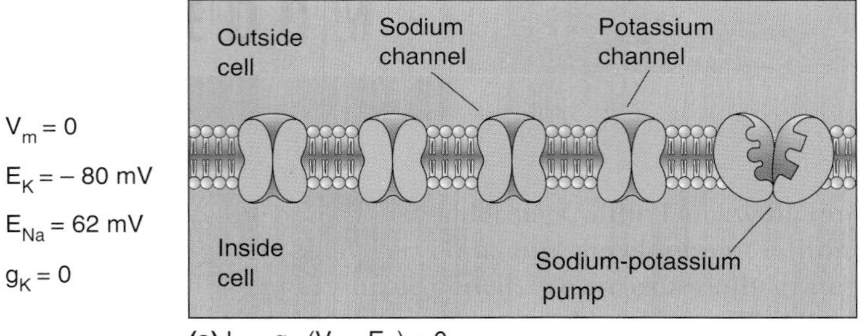

(a) $I_K = g_K (V_m - E_K) = 0$

$0 > V_m > E_K$

$E_K = -80$ mV

$E_{Na} = 62$ mV

$g_K > 0$

K^+ K^+

(b) $I_K = g_K (V_m - E_K) > 0$

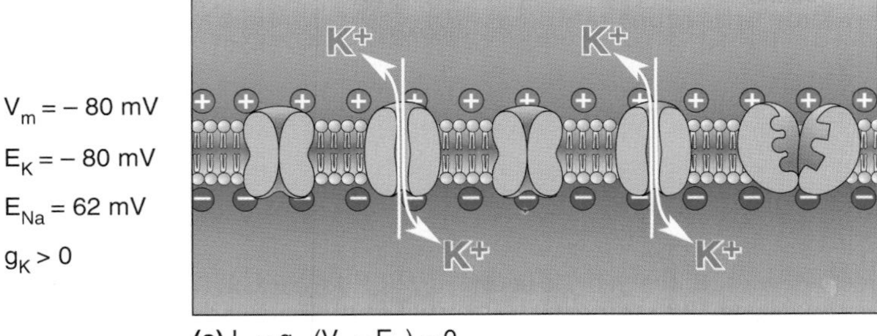

(c) $I_K = g_K (V_m - E_K) = 0$

Figure 4.4
Membrane currents and conductances. Here is an ideal neuron with sodium-potassium pumps, potassium channels, and sodium channels. The pumps establish ionic concentration gradients so that K^+ is concentrated inside the cell and Na^+ is concentrated outside the cell. **(a)** Initially we assume that all channels are closed and the membrane potential equals 0 mV. **(b)** Now we open the potassium channels, and K^+ flows out of the cell. This movement of K^+ is an electrical current, I_K, and it flows as long as the membrane conductance to K^+ ions, g_K, is greater than zero, and the membrane potential is not equal to the potassium equilibrium potential. **(c)** At equilibrium, there is no net potassium current because although g_K is greater than zero, the membrane potential at equilibrium equals E_K. At equilibrium, an equal number of K^+ ions enters and leaves.

2. The number of open potassium channels is proportional to an electrical conductance. We can represent this conductance with the symbol g_K.
3. Membrane potassium current, I_K, will flow only as long as $V_m \neq E_K$. The driving force on K^+ is defined as the difference between the real membrane potential and the equilibrium potential, and it can be written as $V_m - E_K$.

There is a simple relationship between the ionic driving force, ionic conductance, and the amount of ionic current that will flow. For K^+ ions, this may be written:

$$I_K = g_K (V_m - E_K).$$

More generally, we write:

$$I_{ion} = g_{ion} (V_m - E_{ion}).$$

If this sounds familiar, that is because it is simply an expression of Ohm's law, $I = gV$, which we learned about in Chapter 3.

Now let's take another look at our example. Initially, we began with $V_m =$ 0 mV and no ionic membrane permeability (Figure 4.4a). There is a large driving force on K^+ ions because $V_m \neq E_K$; in fact, $(V_m - E_K) = 80$ mV. However, because the membrane is impermeable to K^+, the potassium conductance, g_K, equals zero. Consequently, $I_K = 0$. Potassium current only flows when we stipulate that the membrane has open potassium channels, and therefore $g_K > 0$. Now K^+ ions flow out of the cell—as long as the membrane potential differs from the potassium equilibrium potential (Figure 4.4b). Notice that the current flow is in the direction that takes V_m toward E_K. When $V_m = E_K$, the membrane is at equilibrium and no net current will flow. In this condition, although there is a large potassium conductance, g_K, there is no longer any net driving force on the K^+ ions (Figure 4.4c).

The Ins and Outs of an Action Potential

Let's pick up the action where we left off in the last section. The membrane of our ideal neuron is permeable only to K^+, and $V_m = E_K = -80$ mV. What's happening with the Na^+ ions concentrated outside the cell? Because the membrane potential is so negative with respect to the sodium equilibrium potential, there is a very large driving force on Na^+ ($[V_m - E_{Na}] = [-80 \text{ mV} - 62 \text{ mV}] = -142$ mV). Nonetheless, there can be no net Na^+ current as long as the membrane is impermeable to Na^+. But now let's open the sodium channels and see what happens to the membrane potential.

At the instant we change the ionic permeability of the membrane, g_{Na} is high, and as we have said, there is a large driving force pushing on Na^+. Thus, we have what it takes to generate a large sodium current, I_{Na}, across the membrane. Na^+ ions pass through the membrane sodium channels in the direction that takes V_m toward E_{Na}; in this case, the sodium current, I_{Na}, is inward across the membrane. Assuming the membrane permeability is now far greater to sodium than it is to potassium, this influx of Na^+ depolarizes the neuron until V_m approaches E_{Na}, 62 mV.

Notice that something remarkable happened here. Simply by switching the dominant membrane permeability from K^+ to Na^+, we were able to rapidly reverse the membrane potential. In theory, then, the rising phase of the action potential could be explained if, in response to depolarization of the membrane beyond threshold, membrane sodium channels opened. This would allow Na^+ to enter the neuron, causing a massive depolarization until the membrane potential approached E_{Na}.

How could we account for the falling phase of the action potential? Simply assume that sodium channels quickly close and the potassium channels remain open, so the dominant membrane ion permeability switches back from Na^+ to K^+. Then K^+ would flow out of the cell until the membrane potential again equals E_K. Notice that if g_K increased during the falling phase, the action potential would be even briefer.

Our model for the ins and outs and ups and downs of the action potential in an ideal neuron is shown in Figure 4.5. The rising phase of the action potential is explained by an inward sodium current, and the falling phase is explained by an outward potassium current. The action potential, therefore, can be accounted for simply by the movement of ions through channels that are gated by changes in the membrane potential. If you understand this concept, you understand a lot about the ionic basis of the action potential. What's left now is to see how this actually happens—in a real neuron.

THE ACTION POTENTIAL IN REALITY

Let's quickly review our theory of the action potential. When the membrane is depolarized to threshold, there is a transient increase in g_{Na}. The increase in g_{Na} allows the entry of Na^+ ions, which depolarizes the neuron. And the increase in g_{Na} must be brief to account for the short duration of the action potential. Restoring the negative membrane potential would be further aided by a transient increase in g_K during the falling phase, allowing K^+ ions to leave the depolarized neuron faster.

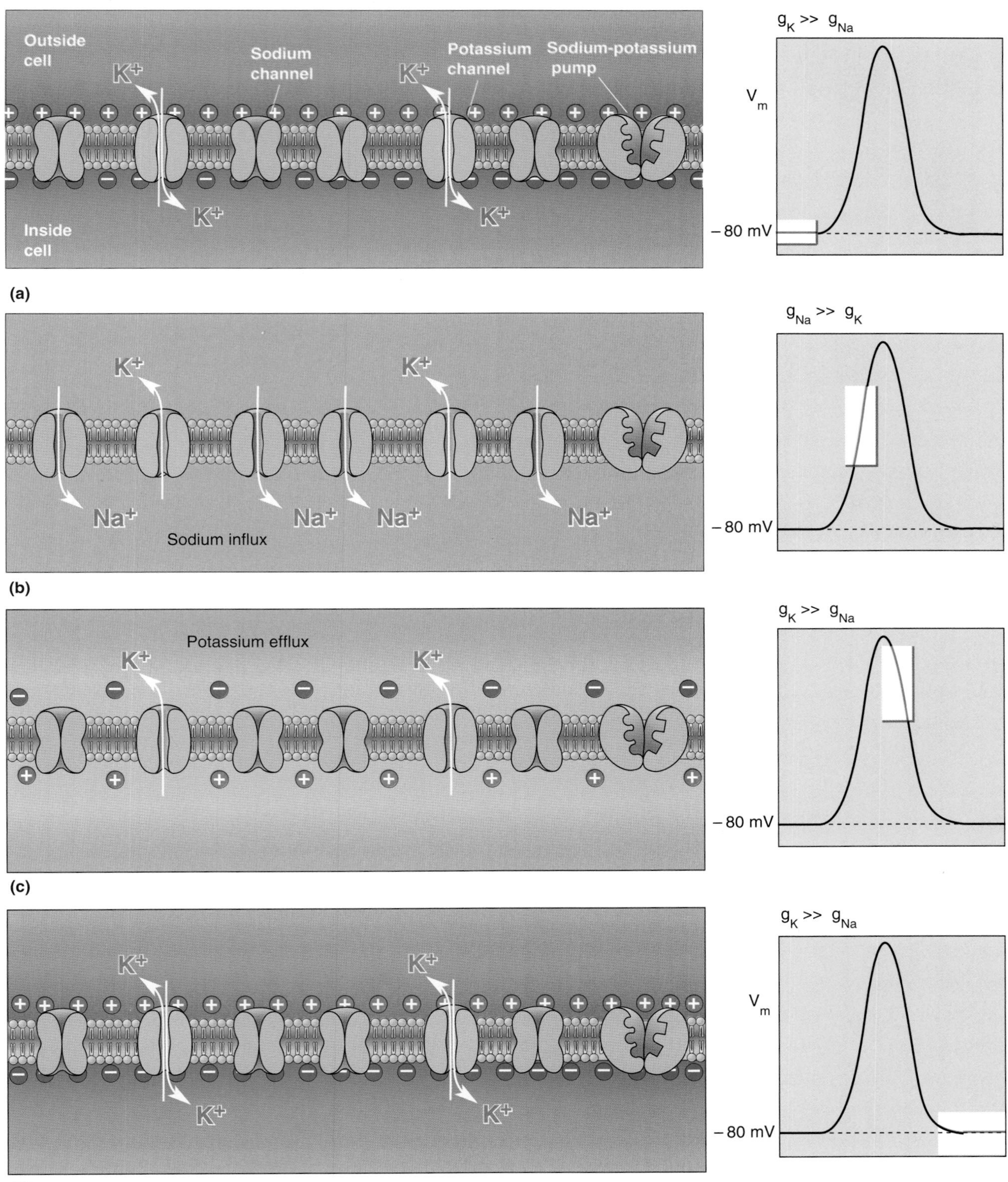

Figure 4.5

Flipping the membrane potential by changing the relative ionic permeability of the membrane. (a) The ideal neuron, introduced in Figure 4.4. We begin by assuming that the membrane is permeable only to K^+ and that $V_m = E_K$. **(b)** We stipulate that the membrane sodium channels open so that $g_{Na} >> g_K$. There is a large driving force on Na^+, so Na^+ ions rush into the cell, taking V_m toward E_{Na}. **(c)** Now we close the sodium channels so that $g_K >> g_{Na}$. Because the membrane potential is positive, there is a large driving force on K^+ ions. The efflux of K^+ takes V_m back toward E_K. **(d)** The resting state is restored: $V_m = E_K$.

Testing this theory is simple enough in principle. All one has to do is measure the sodium and potassium conductances of the membrane during the action potential. In practice, however, such a measurement proved to be quite difficult in real neurons. The key technical breakthrough was the introduction of a device called the **voltage clamp**, invented by the American physiologist Kenneth C. Cole. The decisive experiments using it were performed by Cambridge University physiologists Alan Hodgkin and Andrew Huxley around 1950. The voltage clamp enabled Hodgkin and Huxley to "clamp" the membrane potential of an axon at any value they chose. They could then deduce the changes in membrane conductance that occur at different membrane potentials by measuring the currents that flowed across the membrane. In an elegant series of experiments, Hodgkin and Huxley showed that the rising phase of the action potential was indeed caused by a transient increase in g_{Na} and an influx of Na^+ ions and that the falling phase was associated with an increase in g_K and an efflux of K^+ ions. Their accomplishments earned them the Nobel Prize in 1963.

To account for the transient changes in g_{Na}, Hodgkin and Huxley proposed the existence of sodium "gates" in the axonal membrane. They hypothesized that these gates are "activated"—opened—by depolarization above threshold, and "inactivated"—closed and locked—when the membrane acquires a positive membrane potential. These gates are "deinactivated"—unlocked and enabled to be opened again—only after the membrane potential returns to a negative value.

It is a tribute to Hodgkin and Huxley that their hypotheses about membrane gates predated by more than 20 years the direct demonstration of voltage-gated channel proteins in the neuronal membrane. We have a new understanding of gated membrane channels, thanks to two recent scientific breakthroughs. First, new molecular biological techniques have enabled neuroscientists to determine the detailed structure of these proteins. Second, new neurophysiological techniques have enabled neuroscientists to measure the ionic currents that pass through single channels. We will now explore the action potential from the perspective of these membrane ion channels.

The Voltage-Gated Sodium Channel

The **voltage-gated sodium channel** is aptly named. The protein forms a pore in the membrane that is highly selective to Na^+ ions, and the pore is opened and closed by changes in the electrical potential of the membrane.

Sodium Channel Structure. The voltage-gated sodium channel is created from a single long polypeptide. The molecule has four distinct domains, numbered I–IV; each domain consists of six transmembrane alpha helices, numbered S1–S6 (Figure 4.6). The four domains are believed to clump together to form a pore between them. The pore is closed at the negative resting membrane potential. When the membrane is depolarized to threshold, however, the molecule twists into a configuration that allows the passage of Na^+ through the pore (Figure 4.7).

Like the potassium channel, the sodium channel has pore loops that are assembled into a selectivity filter. This filter makes the sodium channel 12 times more permeable to Na^+ than it is to K^+. Apparently, the Na^+ ions are stripped of most but not all of their associated water molecules as they pass into the channel. The retained water, which serves as a sort of molecular chaperone for the ion, is necessary for the ion to pass the selectivity filter. The ion-water complex can then be used to select Na^+ and exclude K^+ (Figure 4.8).

The sodium channel is gated by a change in voltage across the membrane. It has now been established that the voltage sensor resides in segment S4 of

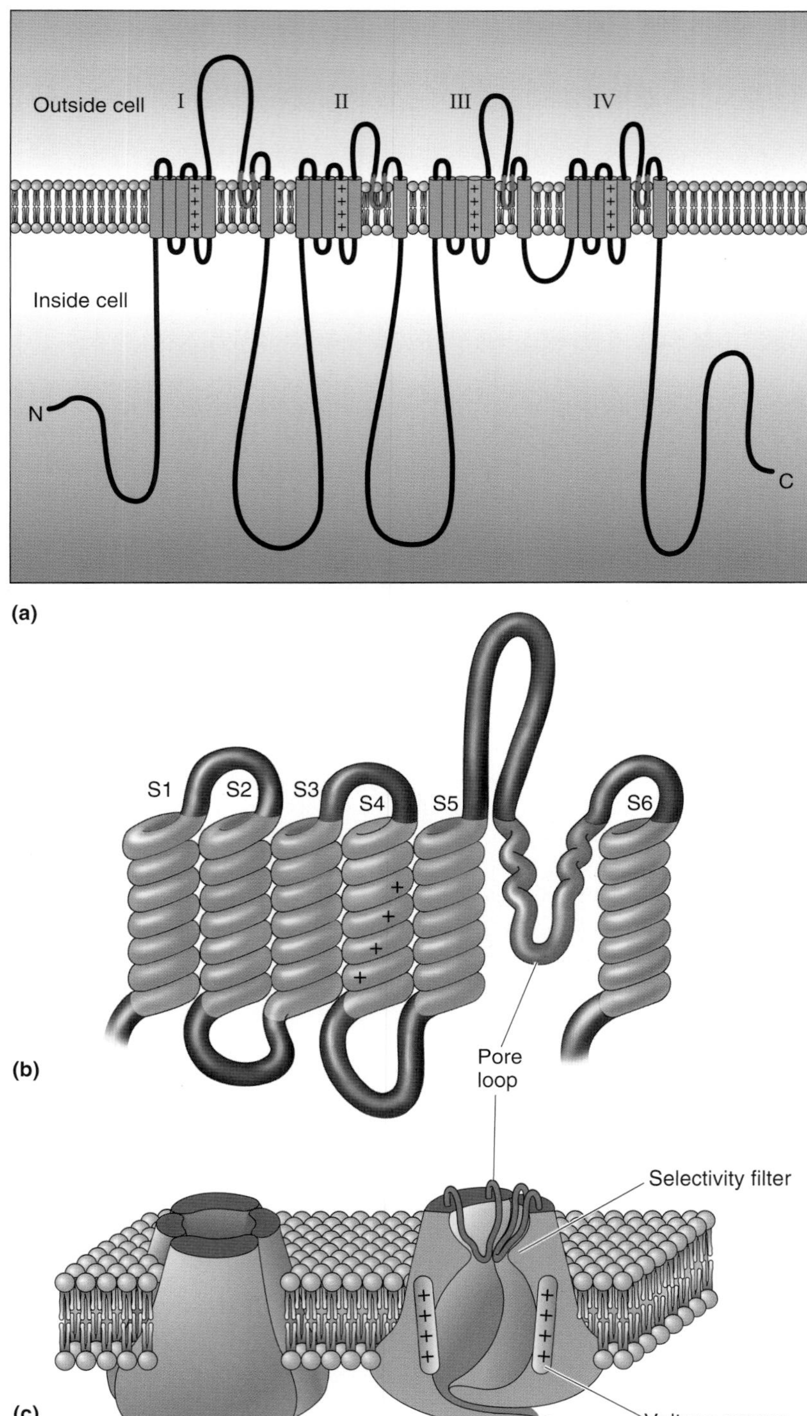

Figure 4.6

Structure of the voltage-gated sodium channel. (a) How the sodium channel polypeptide chain is believed to be woven into the membrane. The molecule consists of four domains, I–IV. Each domain consists of six alpha helices (represented by the blue cylinders), which pass back and forth across the membrane. **(b)** An expanded view of one domain showing the voltage sensor of alpha helix S4 and the pore loop (red), which contributes to the selectivity filter. **(c)** A view of the molecule showing how the domains may arrange themselves to form a pore between them. (Source: Adapted from Armstrong and Hille, 1998, Fig. 1.)

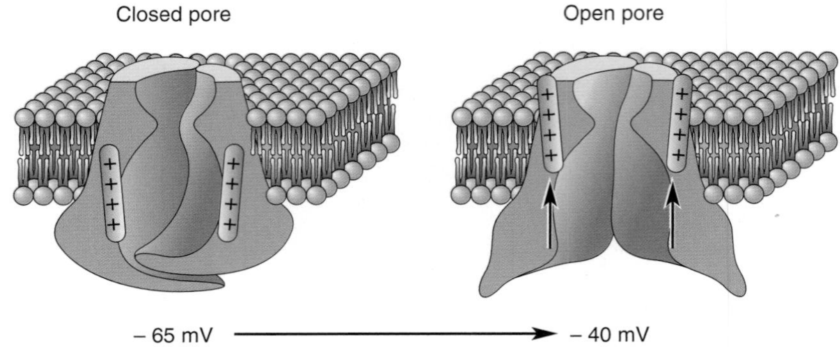

Figure 4.7
A hypothetical model for changing the configuration of the sodium channel by depolarizing the membrane.

the molecule. In this segment, positively charged amino acid residues are regularly spaced along the coils of the helix. Thus, the entire segment can be forced to move by changing the membrane potential. Depolarization pushes S4 away from the inside of the membrane, and this conformational change in the molecule causes the gate to open.

Functional Properties of the Sodium Channel. Research performed around 1980 in the laboratory of Erwin Neher at the Max Planck Institute in Göttingen, Germany, revealed the functional properties of the voltage-gated sodium channel. A new method called the **patch clamp** was used to study the ionic currents passing through individual ion channels (Box 4.2). The patch-clamp method entails sealing the tip of an electrode to a very small *patch* of neuronal membrane. This patch then can be torn away from the neuron, and the ionic currents across it can be measured as the membrane potential is *clamped* at any value the experimenter selects. With luck, the patch will contain only a single channel, and the behavior of this channel can be studied. Patch clamping enabled Neher and his colleagues to study the functional properties of the voltage-gated sodium channel (Box 4.3).

Changing the membrane potential of a patch of axonal membrane from −80 to −65 mV has little effect on the voltage-gated sodium channels. They remain closed because depolarization of the membrane has not yet reached threshold. Changing the membrane potential from −65 to −40 mV, however, causes these channels to pop open. As shown in Figure 4.9, voltage-gated sodium channels have a characteristic pattern of behavior:

1. They open with little delay.
2. They stay open for about 1 msec and then close (inactivate).
3. They cannot be opened again by depolarization until the membrane potential returns to a negative value near threshold.

A hypothetical model for how conformational changes in the voltage-gated sodium channel could account for these properties is illustrated in Figure 4.9c.

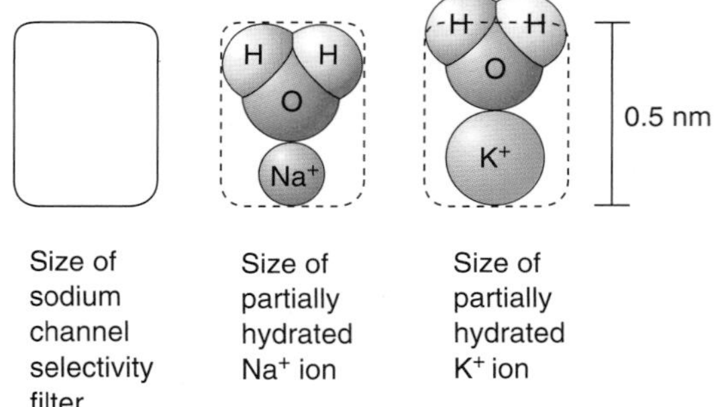

Figure 4.8
Dimensions of the sodium channel selectivity filter. Water accompanies the ions as they pass through the channel. Hydrated Na^+ fits; hydrated K^+ does not. (Source: Adapted from Hille, 1992, Figs. 5, 6.)

Box 4.2

BRAIN FOOD

The Patch-Clamp Method

The very existence of voltage-gated channels in the neuronal membrane was merely conjecture until the development of methods to study individual channel proteins. A revolutionary new method, the patch clamp, was developed by German neuroscientists Bert Sakmann and Erwin Neher in the mid-1970s. In recognition of their contribution, Sakmann and Neher were awarded the 1991 Nobel Prize.

Patch clamping enables one to record ionic currents through single channels (Figure A). The first step is gently lowering the fire-polished tip of a glass recording electrode, 1–5 μm in diameter, onto the membrane of the neuron (part a), and then applying suction through the electrode tip (part b). A tight seal forms between the walls of the electrode and the underlying patch of membrane. This "gigaohm" seal (so named because of its high electrical resistance, $>10^9\ \Omega$) leaves the ions in the electrode only one path to take, through the channels in the underlying patch of membrane. If the electrode is withdrawn from the cell, the membrane patch can be torn away (part c), and ionic currents can be measured as steady voltages are applied across the membrane (part d).

With a little luck, one can resolve currents flowing through single channels. If the patch contains a voltage-gated sodium channel, for example, then changing the membrane potential from −65 to −40 mV will cause the channel to open, and current (I) will flow through it (part e). The amplitude of the measured current at a constant membrane voltage reflects the channel conductance, and the duration of the current reflects the time the channel is open.

Patch-clamp recordings reveal that most channels flip between two conductance states that can be interpreted as open or closed. The time they remain open can vary, but the single-channel conductance value stays the same and is therefore said to be unitary. Ions can pass through single channels at an astonishing rate—well over a million per second.

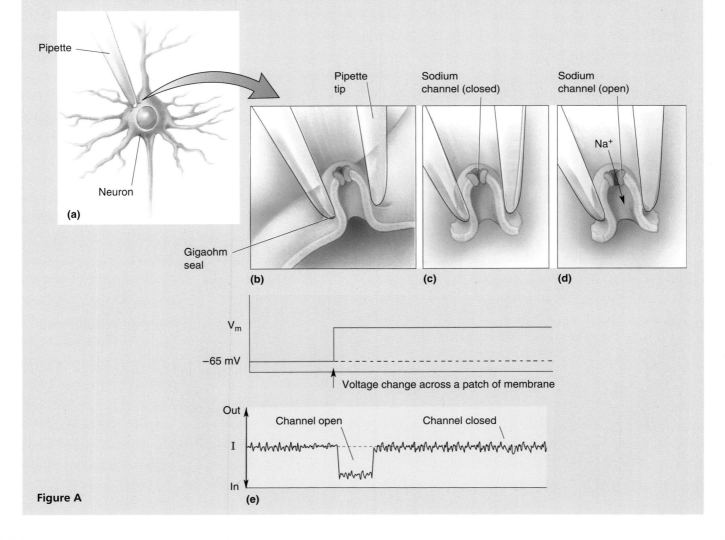

Figure A

Box 4.3

PATH OF DISCOVERY

The Challenge of Resolving Voltage-Gated Channels

BY ERWIN NEHER

Soon after Bert Sakmann and I succeeded in resolving the single-channel currents that can be activated in muscle cells by the neurotransmitter acetylcholine, the question arose whether voltage-gated channels were amenable to the same approach. For some time we were in a deadlocked position. The acetylcholine-gated channels we studied really had favorable properties: The currents were large in amplitude, and the channels stayed open for a long time. Even so, our techniques were so crude in the early 1970s (Figure A) that we could hardly resolve them. Voltage-gated channels seemed to be beyond the reach of our method because of their rapid switching and small amplitude.

Erwin Neher

Nevertheless, one morning at breakfast during a symposium at Yale University, Franco Conti persuaded me to try recording from the squid giant axon. So I went to the Marine Biological Station at Camogli, Italy, for a number of very pleasant visits. Franco and I inserted L-shaped measuring pipettes longitudinally into squid giant axons, attempting to press the pipette against the membrane from the inside for a patch-clamp measurement. I learned to appreciate Franco's art in dissecting and handling the fragile axon. Nevertheless, we experienced lots of frustration from the difficulty of the task, except for a few exciting recordings that did show regular voltage-dependent, open-close fluctuations compatible with the properties of voltage-gated potassium channels. The records of these experiments made it into the pages of *Nature,* but by and large the approach was too difficult to be pursued further.

Things changed when Fred Sigworth joined the laboratory. Fred had just completed his Ph.D. thesis, in which he described a method of analysis to determine channel current fluctuations even if they could not be fully resolved. He set out to record such fluctuations in muscle cells. While he set up his experiments, I managed to achieve what is now known as the "gigaohm" seal, or "gigaseal." With the gigaseal, all those problems of resolution improved by an order of magnitude. Fred, trained in electronic engineering, quickly improved the electronics to match the improved recording conditions. Within a few weeks, he called to tell me he had recorded a whole tape of well-resolved voltage-gated sodium channel currents.

Figure A

Scanning electron micrograph of the tip of a patch pipette from the "early days." The opening of this pipette is 0.5 μm in diameter.

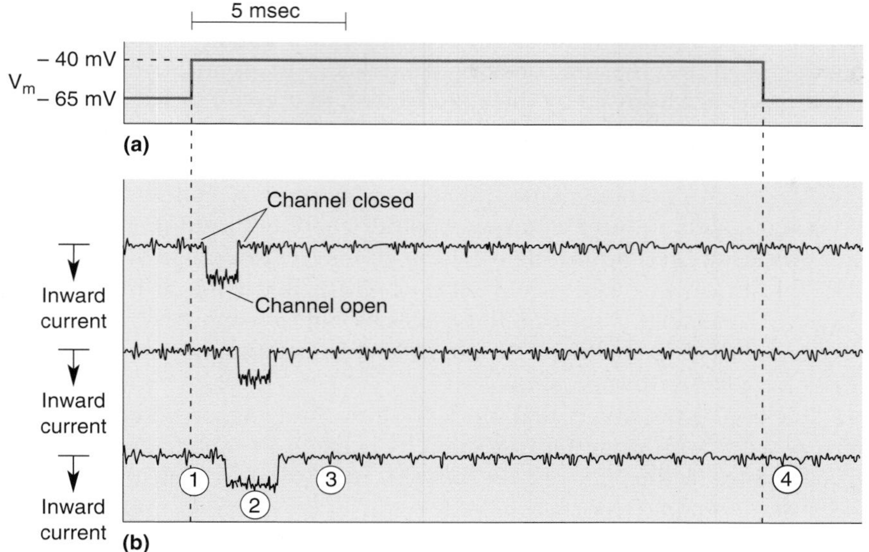

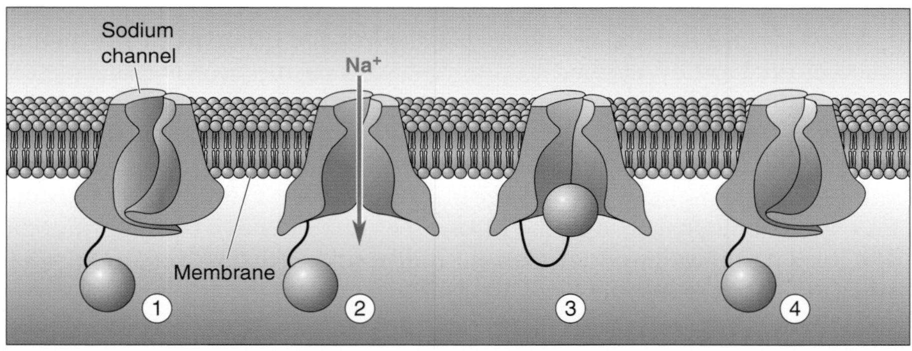

(c)

Figure 4.9
The opening and closing of sodium channels upon membrane depolarization. (a) This trace shows the electrical potential across a patch of membrane. When the membrane potential is changed from −65 to −40 mV, the sodium channels pop open. **(b)** These traces show how three different channels respond to the voltage step. Each line is a record of the electrical current that flows through a single channel. ① At −65 mV, the channels are closed, so there is no current. ② When the membrane is depolarized to −40 mV, the channels briefly open and current flows inward, represented by the downward deflection in the current traces. Although there is some variation from channel to channel, all of them open with little delay and stay open for less than 1 msec. After they open once, they close and stay closed as long as the membrane is maintained at a depolarized V_m. ③ The closure of the sodium channel by steady depolarization is called inactivation. ④ To deinactivate the channels, the membrane must be returned to −65 mV. **(c)** A model for how changes in the conformation of the sodium channel protein might yield its functional properties. ① The closed channel ② opens upon membrane depolarization. ③ Inactivation occurs when a globular portion of the protein swings up and occludes the pore. ④ Deinactivation occurs when the globular portion swings away and the pore closes by movement of the transmembrane domains.

A single channel does not an action potential make. The membrane of an axon may contain thousands of sodium channels per square micrometer (μm^2), and the concerted action of all these channels is required to generate what we measure as an action potential. Nonetheless, it is interesting to see how many of the properties of the action potential can be explained by the properties of the voltage-gated sodium channel. For example, the fact that single channels do not open until a critical level of membrane depolarization is reached explains the action potential threshold. The rapid opening of the channels in response to depolarization explains why the rising phase of the action potential occurs so quickly. And the short time the channels stay open before inactivating (about 1 msec) partly explains why the action potential is so brief. Furthermore, inactivation of the channels can account for the absolute refractory period: Another action potential cannot be generated until the channels are deinactivated.

The Effects of Toxins on the Sodium Channel. Currents through the sodium channel, as well as action potentials, can be completely blocked with **tetrodotoxin** (**TTX**). This potent toxin clogs the Na^+-permeable pore by binding tightly to a specific site on the outside of the channel. Originally isolated from the ovaries of the Japanese puffer fish, TTX is usually fatal if ingested. Nonetheless, puffer fish are considered a delicacy in Japan. Sushi chefs train for years, and are licensed by the government, to prepare puffer fish in such a way that eating them causes numbness around the mouth. Talk about adventuresome eating!

TTX is one of a number of natural toxins that interfere with the function of the voltage-gated sodium channel. Another channel-blocking toxin is *saxitoxin*, produced by dinoflagellates of the genus *Gonyaulax*. Saxitoxin is concentrated in clams, mussels, and other shellfish that feed on these marine protozoa. Occasionally, the dinoflagellates bloom, causing what is known as a "red tide." Eating shellfish at these times can be fatal, because of the unusually high concentration of the toxin.

In addition to the toxins that block sodium channels, certain compounds interfere with nervous system function by causing the channels to open inappropriately. In this category is *batrachotoxin*, isolated from the skin of a species of Colombian frog. Batrachotoxin causes the channels to open at more negative potentials and to stay open much longer than usual, thus scrambling the information encoded by the action potentials. Toxins produced by lilies (*veratridine*) and buttercups (*aconitine*) have a similar mechanism of action. Sodium channel inactivation is also disrupted by toxins from scorpions and sea anemones.

What can we learn from these toxins? First, the different toxins disrupt channel function by binding to different sites on the protein. Information about toxin binding and its consequences has helped researchers deduce the three-dimensional structure of the sodium channel. Second, the toxins can be used as experimental tools to study the consequences of blocking action potentials. For example, as we shall see in later chapters, TTX is commonly used for experiments that require the blocking of impulses in a nerve or muscle. The third and most important lesson from studying toxins? Be careful what you put in your mouth!

Voltage-Gated Potassium Channels

Hodgkin and Huxley's experiments indicated that the falling phase of the action potential was explained only partly by the inactivation of g_{Na}. They also found a transient increase in g_K that functioned to speed the restoration of a negative membrane potential after the spike. They proposed the existence of membrane potassium gates that, like sodium gates, open in response to depolarization of the membrane. Unlike sodium gates, however, potassium gates do not open immediately upon depolarization; it takes about 1 msec for them to open. Because of this delay and because this potassium conductance serves to rectify, or reset, the membrane potential, they called this conductance the *delayed rectifier*.

We now know that there are many types of **voltage-gated potassium channels**. Most of them open when the membrane is depolarized, and they function to diminish any further depolarization by giving K^+ ions a path to leave the cell across the membrane. The known voltage-gated potassium channels have a similar structure. The channel proteins consist of four separate polypeptide subunits that come together to form a pore between them. Like the sodium channel, these proteins are sensitive to changes in the electrical field across the membrane. When the membrane is depolarized, the subunits are believed to twist into a shape that allows K^+ ions to pass through the pore.

Putting the Pieces Together

We can now use what we've learned about ions and channels to explain the key properties of the action potential (Figure 4.10):

- *Threshold*. Threshold is the membrane potential at which enough voltage-gated sodium channels open so that the relative ionic permeability of the membrane favors sodium over potassium.
- *Rising phase*. When the inside of the membrane has a negative electrical potential, there is a large driving force on Na^+ ions. Therefore, Na^+ ions rush into the cell through the open sodium channels, causing rapid depolarization of the membrane.
- *Overshoot*. Because the relative permeability of the membrane greatly favors sodium, the membrane potential goes to a value close to E_{Na}, which is greater than 0 mV.
- *Falling phase*. The behavior of two types of channel contributes to the falling phase. First, the voltage-gated sodium channels inactivate. Second, the voltage-gated potassium channels finally open (triggered to do so 1 msec earlier by the depolarization of the membrane). There is a great driving force on K^+ ions when the membrane is strongly depolarized. Therefore, K^+ ions rush out of the cell through the open channels, causing the membrane potential to become negative again.
- *Undershoot*. The open voltage-gated potassium channels add to the resting potassium membrane permeability. Because there is very little sodium permeability, the membrane potential goes toward E_K, causing hyperpolarization relative to the resting membrane potential until the voltage-gated potassium channels close again.
- *Absolute refractory period*. Sodium channels inactivate when the membrane becomes strongly depolarized. They cannot be activated again, and another action potential cannot be generated until the membrane potential goes sufficiently negative to deinactivate the channels.
- *Relative refractory period*. The membrane potential stays hyperpolarized until the voltage-gated potassium channels close. Therefore, more depolarizing current is required to bring the membrane potential to threshold.

We've seen that channels and the movement of ions through them can explain the properties of the action potential. But it is important to remember that the sodium-potassium pump also is working quietly in the background. Imagine that the entry of Na^+ during each action potential is like a wave coming over the bow of a boat making way in heavy seas. Like the continuous action of the boat's bilge pump, the sodium-potassium pump works all the time to transport Na^+ back across the membrane. The pump maintains the ionic concentration gradients that drive Na^+ and K^+ through their channels during the action potential.

ACTION POTENTIAL CONDUCTION

To transfer information from one point to another in the nervous system, it is necessary that the action potential, once generated, be conducted down the axon. This process is like the burning of a fuse. Imagine your favorite cartoon character holding a stick of dynamite with a burning match held under the fuse. The fuse ignites when it gets hot enough (beyond some threshold). The flame heats up the segment of fuse immediately ahead of it until it ignites. In this way, the flame steadily works its way down the fuse. Note that the fuse lit at one end only burns in one direction; the flame cannot turn back on itself because the combustible material just behind it is spent.

Propagation of the action potential along the axon is similar to the propagation of the flame along the fuse. When the axon is depolarized sufficiently

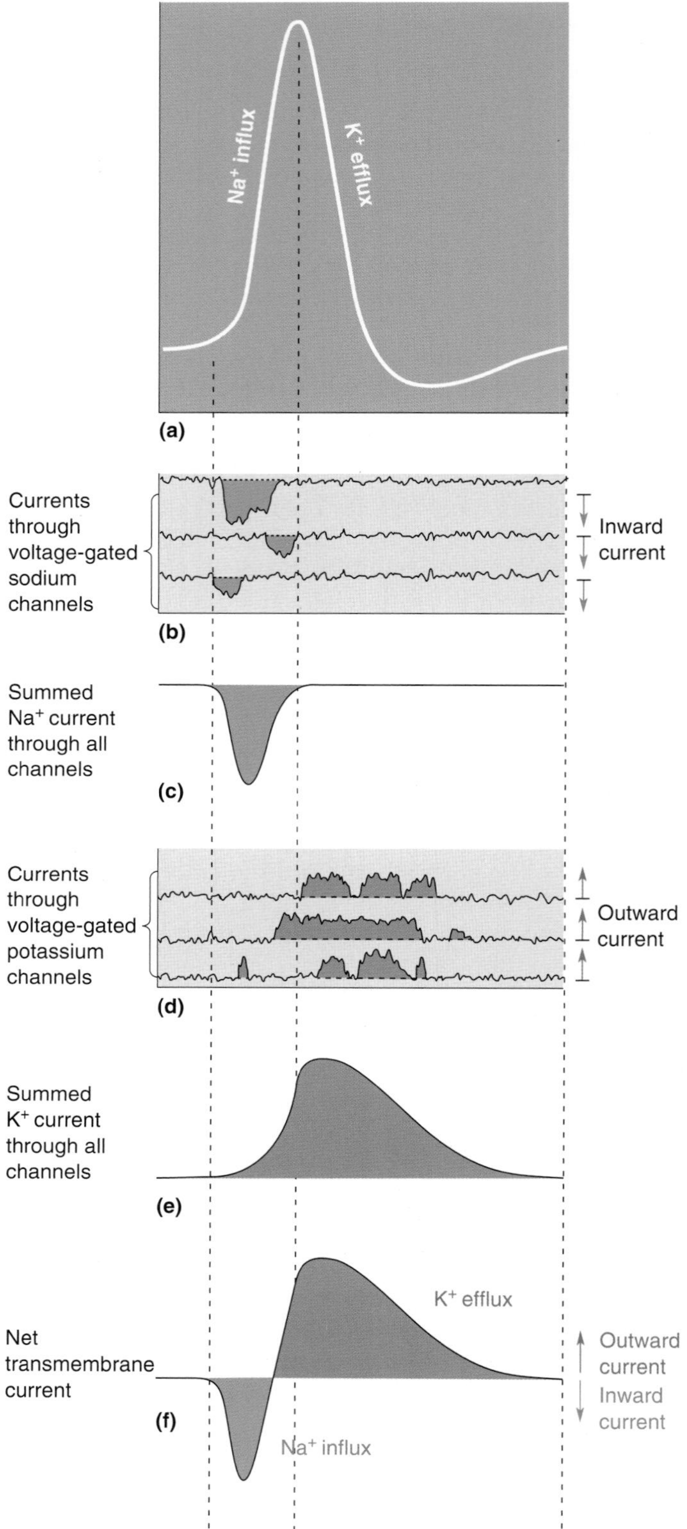

Figure 4.10

The molecular basis of the action potential. (a) The membrane potential as it changes in time during an action potential. The rising phase of the action potential is caused by the influx of Na^+ ions through hundreds of voltage-gated sodium channels. The falling phase is caused by sodium channel inactivation and the efflux of K^+ ions through voltage-gated potassium channels. **(b)** The inward currents through three representative voltage-gated sodium channels. Each channel opens with little delay when the membrane is depolarized to threshold. The channels stay open for no more than 1 msec and then inactivate. **(c)** The summed Na^+ current flowing through all the sodium channels. **(d)** The outward currents through three representative voltage-gated potassium channels. Voltage-gated potassium channels open about 1 msec after the membrane is depolarized to threshold and stay open as long as the membrane is depolarized. The high potassium permeability causes the membrane to hyperpolarize briefly. When the voltage-gated potassium channels close, the membrane potential relaxes back to the resting value, around −65 mV. **(e)** The summed K^+ current flowing through all the potassium channels. **(f)** The net transmembrane current during the action potential (the sum of parts c and e).

to reach threshold, voltage-gated sodium channels open, and the action potential is initiated. The influx of positive charge depolarizes the segment of membrane immediately before it until it reaches threshold and generates its own action potential (Figure 4.11). In this way, the action potential works its way down the axon until it reaches the axon terminal, thereby initiating synaptic transmission (the subject of Chapter 5).

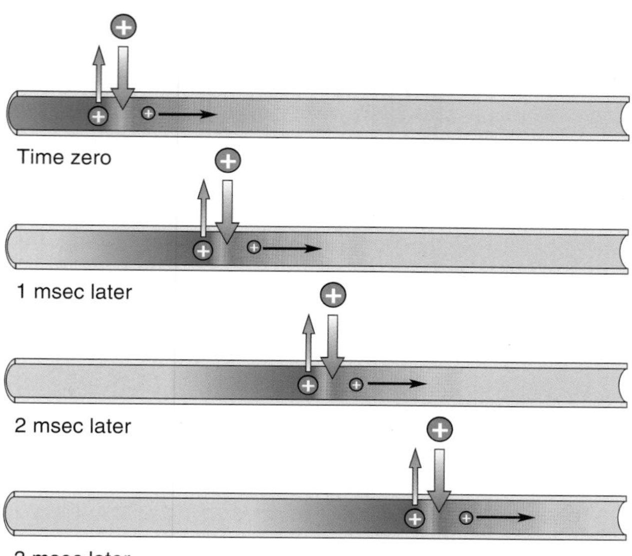

Figure 4.11
Action potential conduction. The entry of positive charge during the action potential causes the membrane just ahead to depolarize to threshold.

An action potential initiated at one end of an axon only propagates in one direction; it does not turn back on itself. This is because the membrane just behind it is refractory as a result of inactivation of the sodium channels. But just like the fuse, an action potential can be generated by depolarization at either end of the axon and therefore propagate in either direction. (Normally, action potentials conduct only in one direction; this is called *orthodromic* conduction. Backward propagation, sometimes elicited experimentally, is called *antidromic* conduction.) Because the axonal membrane is excitable (capable of generating action potentials) along its entire length, the impulse will propagate without decrement. The fuse works the same way because it is combustible along its entire length. Unlike the fuse, however, the axon can regenerate its firing ability.

Action potential conduction velocities vary, but 10 m/sec is a typical rate. Remember that from start to finish the action potential lasts about 2 msec. From this, we can calculate the length of membrane that is engaged in the action potential at any instant in time:

$$10 \text{ m/sec} \times 2 \times 10^{-3} \text{ sec} = 2 \times 10^{-2} \text{ m}.$$

Therefore, an action potential traveling at 10 m/sec occurs over a 2-cm length of axon.

Factors Influencing Conduction Velocity

Remember that the inward Na^+ current during the action potential depolarizes the patch of membrane just ahead. If this patch reaches threshold, it will fire an action potential, and the action potential will "burn" on down the membrane. The speed with which the action potential propagates down the axon depends on how far the depolarization ahead of the action potential spreads, which in turn depends on certain physical characteristics of the axon.

Imagine that the influx of positive charge into an axon during the action potential is like turning on the water to a leaky garden hose. There are two paths the water can take: one, down the inside of the hose; the other, across the hose through the leaks. How much water goes along each path depends on their relative resistance; most of the water will take the path of least resistance. If the hose is narrow and the leaks are numerous and large, most of the water will flow out through the leaks. If the hose is wide and the leaks are few and tiny, most of the water will flow down the inside of the hose. The same principles apply to positive current spreading down the axon ahead of the action potential. There are two paths that positive charge can take: one, down the inside of the axon; the other, across the axonal membrane. If the axon is narrow and there are many open membrane pores, most of the current will flow out across the membrane. If the axon is wide and there are few open membrane pores, most of the current will flow down inside the axon. The farther the current goes down the axon, the farther ahead of the action potential the membrane will be depolarized, and the faster the action potential will propagate. As a rule, therefore, action potential conduction velocity increases with increasing axonal diameter.

As a consequence of this relationship between axonal diameter and conduction velocity, neural pathways that are especially important for survival have evolved unusually large axons. An example is the giant axon of the squid, which is part of a pathway that mediates an escape reflex in response to strong sensory stimulation. The squid giant axon can be 1 mm in diameter, so large that originally it was thought to be part of the animal's circulatory system. Neuroscience owes a debt to British zoologist J. Z. Young, who in 1939 called attention to the squid giant axon as an experimental preparation for studying the biophysics of the neuronal membrane. Hodgkin and Huxley used this preparation to elucidate the ionic basis of the action potential, and the giant axon continues to be used today for a wide range of neurobiological studies.

Axonal size and the number of voltage-gated channels in the membrane also affect axonal excitability. Smaller axons require greater depolarization to reach action potential threshold and are more sensitive to blockage by local anesthetics (Box 4.4).

Myelin and Saltatory Conduction. The good thing about fat axons is that they conduct action potentials faster; the bad thing about them is that they take up a lot of space. If all of the axons in your brain were the diameter of a squid giant axon, your head would be too big to fit through a barn door. Fortunately, vertebrates evolved another solution for increasing action potential conduction velocity: ensheathing the axon with insulation called myelin (see Chapter 2). Myelin consists of many wraps of membrane provided by glial support cells—Schwann cells in the peripheral nervous system (outside the brain and spinal cord) and oligodendroglia in the central nervous system. Just as wrapping the leaky garden hose with duct tape facilitates water flow down the inside of the hose, myelin facilitates current flow down the inside of the axon, thereby increasing action potential conduction velocity (Box 4.5).

The myelin sheath does not extend continuously along the entire length of the axon. There are breaks in the insulation where ions cross the membrane to generate action potentials. As you'll recall from Chapter 2, these breaks in the myelin sheath are the nodes of Ranvier (Figure 4.12). Voltage-gated sodium channels are concentrated in the membrane of the nodes. The distance between nodes is usually 0.2–2.0 mm, depending on the size of the axon (thicker axons have larger internodal distances).

Imagine that the action potential traveling along the axon membrane is like you traveling down a sidewalk. Action potential conduction without myelin is like walking down the sidewalk in small steps, heel-to-toe, using every inch of the sidewalk to creep along. Conduction with myelin, in contrast, is like skipping down the sidewalk. In myelinated axons, action potentials skip from node to node (Figure 4.13). This type of action potential propagation is called **saltatory conduction** (from the Latin meaning "to leap").

Box 4.4

OF SPECIAL INTEREST

Local Anesthesia

Although you've tried to tough it out, you just can't take it anymore. You finally give in to the pain of the toothache and head for the dentist. Fortunately, the worst part of having a cavity filled is the pinprick in the gum caused by the injection needle. After the injection, your mouth is numb and you daydream while the dentist drills and repairs your tooth. What was injected, and how did it work?

Local anesthetics are drugs that temporarily block action potentials in axons. They are called "local" because they are injected directly into the tissue where anesthesia—the absence of sensation—is desired. Small axons, firing a lot of action potentials, are most sensitive to conduction block by local anesthetics.

The first local anesthetic introduced into medical practice was cocaine. The chemical was originally isolated from the leaves of the coca plant in 1860 by the German physician Albert Niemann. According to the custom of the pharmacologists of his day, Niemann tasted the new compound and discovered that it caused his tongue to go numb. It was soon learned that cocaine had toxic and addictive properties. (The mind-altering effect of cocaine was studied by another well-known physician of that era, Sigmund Freud. Cocaine alters mood by a mechanism distinct from its local anesthetic action, as we shall see in Chapter 15.)

The search for a suitable synthetic substitute for cocaine led to the development of lidocaine, which is now the most widely used local anesthetic. Lidocaine can be dissolved into a jelly and smeared onto the mucous membranes of the mouth (and elsewhere) to numb the nerve endings (called *topical anesthesia)*; it can be injected directly into a tissue (*infiltration anesthesia*) or a nerve (*nerve block*); it can even be infused into the cerebrospinal fluid bathing the spinal cord (*spinal anesthesia)*, where it can numb large parts of the body.

Lidocaine and other local anesthetics prevent action potentials by binding to the voltage-gated sodium channels. The binding site for lidocaine has been identified as the S6 alpha helix of domain IV of the protein (Figure A). Lidocaine cannot gain access to this site from the outside. The anesthetic first must cross the axonal membrane and then pass through the open gate of the channel to find its binding site inside the pore. This explains why active nerves are blocked faster (the sodium channel gates are open more often). The bound lidocaine interferes with the flow of Na^+ that normally results from depolarizing the channel.

Smaller axons are affected by local anesthetics before larger axons because their action potentials have less of a safety margin; more of the voltage-gated sodium channels must function to ensure that the action potential doesn't fizzle out as it conducts down the axon. This increased sensitivity of small axons to local anesthetics is fortuitous in clinical practice. As we will discover in Chapter 12, it is the smaller fibers that convey information about painful stimuli like a toothache.

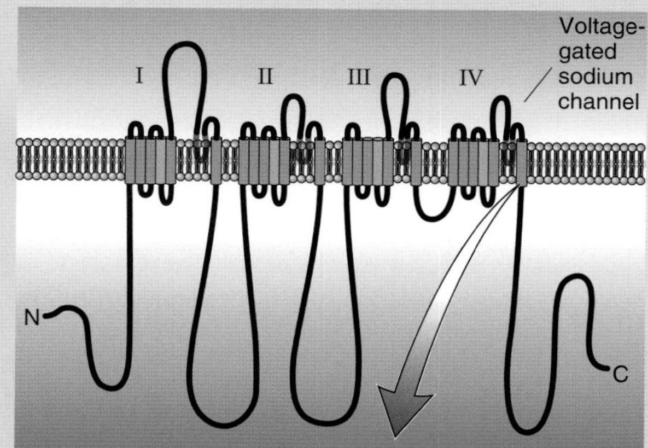

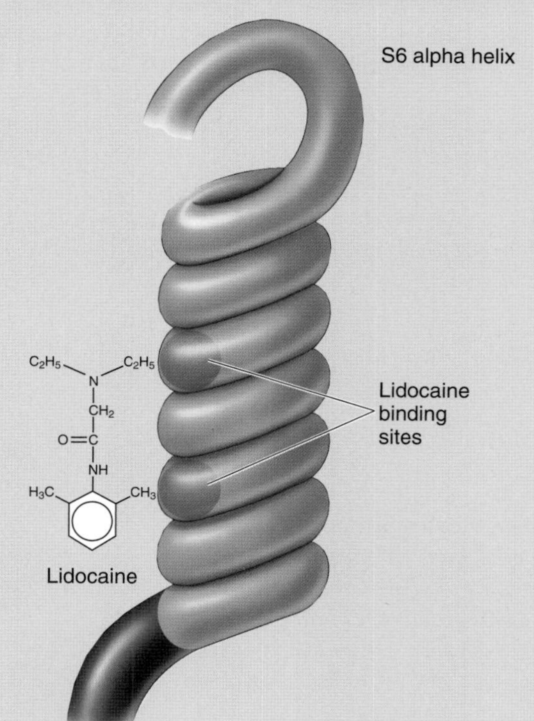

Figure A
Lidocaine's mechanism of action. (Source: Adapted from Hardman, et al., 1996, Fig. 15-3.)

Box 4.5

OF SPECIAL INTEREST

Multiple Sclerosis, a Demyelinating Disease

The critical importance of myelin for the normal transfer of information in the human nervous system is revealed by the neurological disorder known as *multiple sclerosis* (*MS*). Victims of MS often complain of weakness, lack of coordination, and impaired vision and speech. The disease is capricious, usually marked by remissions and relapses that occur over many years. Although the precise cause of MS is still poorly understood, the cause of the sensory and motor disturbances is now quite clear. MS attacks the myelin sheaths of bundles of axons in the brain, spinal cord, and optic nerves. The name is derived from the Greek word for "hardening," which describes the lesions that develop around bundles of axons; and the sclerosis is multiple because the disease attacks many sites in the nervous system at the same time.

Lesions in the brain can now be viewed noninvasively using new methods such as magnetic resonance imaging (MRI). However, neurologists have been able to diagnose MS for many years by taking advantage of the fact that myelin serves the nervous system by increasing the velocity of axonal conduction. One simple test involves stimulating the eye with a checkerboard pattern, then measuring the elapsed time until an electrical response occurs from the scalp over the part of the brain that is a target of the optic nerve. People who have MS are characterized by a marked slowing of the conduction velocity of their optic nerve.

Another demyelinating disease is *Guillain-Barré syndrome*, which attacks the myelin of the peripheral nerves that innervate muscle and skin. This disease may follow minor infectious illnesses and inoculations, and it appears to result from an anomalous immunological response against one's own myelin. The symptoms stem directly from the slowing and/or failure of action potential conduction in the axons that innervate the muscles. This conduction deficit can be demonstrated clinically by stimulating the peripheral nerves electrically through the skin, then measuring the time it takes to evoke a response (a twitch of a muscle, for instance). Both MS and Guillain-Barré syndrome are characterized by a profound slowing of the response time because saltatory conduction is disrupted.

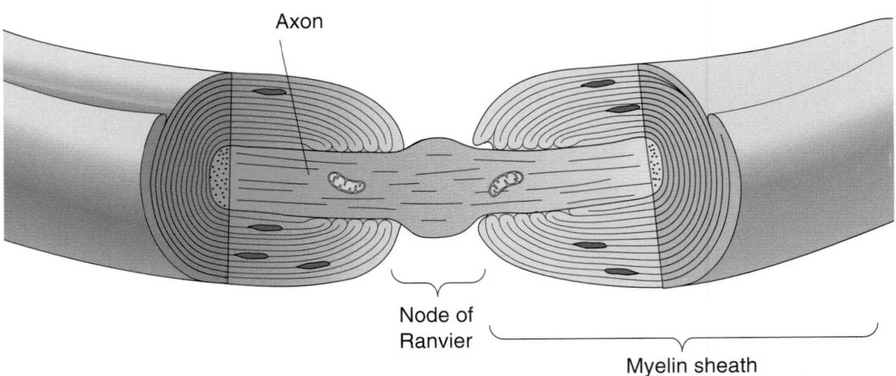

Figure 4.12
The myelin sheath and node of Ranvier. The electrical insulation provided by myelin helps speed action potential conduction from node to node. Voltage-gated sodium channels are concentrated in the axonal membrane at the nodes of Ranvier.

ACTION POTENTIALS, AXONS, AND DENDRITES

Action potentials of the type discussed in this chapter are a feature mainly of axons. As a rule, the membranes of dendrites and neuronal cell bodies do not generate sodium-dependent action potentials because they have very few voltage-gated sodium channels. Only membrane that contains these specialized protein molecules is capable of generating action potentials, and this type of excitable membrane is usually found only in axons. Therefore, the part of the neuron where an axon originates from the soma, the axon hillock, is often also called the **spike-initiation zone**. In a typical neuron in the brain

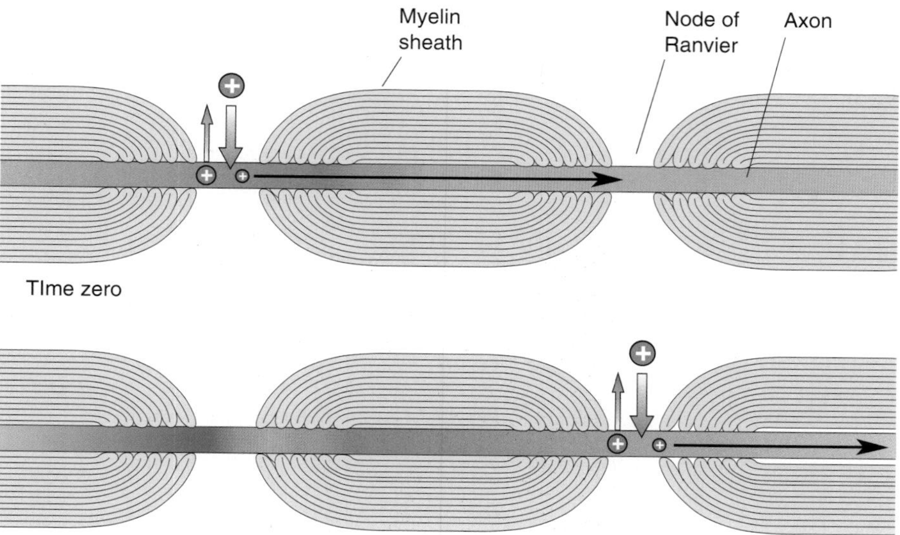

Figure 4.13
Saltatory conduction. Myelin allows current to spread farther and faster between nodes, thus speeding action potential conduction. Compare this figure with Figure 4.11.

or spinal cord, the depolarization of the dendrites and soma caused by synaptic input from other neurons leads to the generation of action potentials if the membrane of the *axon hillock* is depolarized beyond threshold (Figure 4.14a). In most sensory neurons, however, the spike-initiation zone occurs near the *sensory nerve endings*, where the depolarization caused by sensory stimulation leads to the generation of action potentials that propagate up the sensory nerves (Figure 4.14b).

In Chapter 2, we learned that axons and dendrites differ in their morphology. We now see that they are functionally different and that this difference in function is specified at the molecular level by the type of protein in the neuronal membrane. Differences in the types and density of membrane ion channels also can account for the characteristic electrical properties of different types of neuron (Box 4.6).

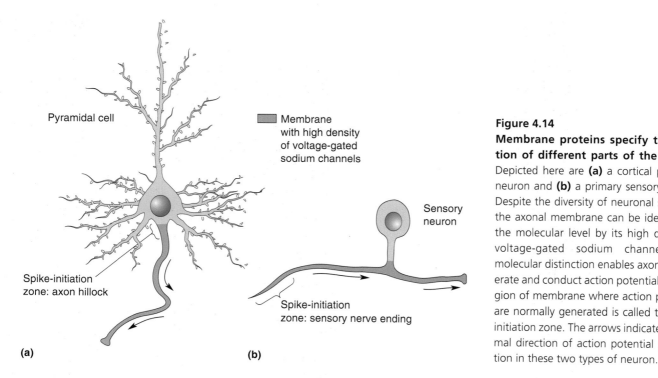

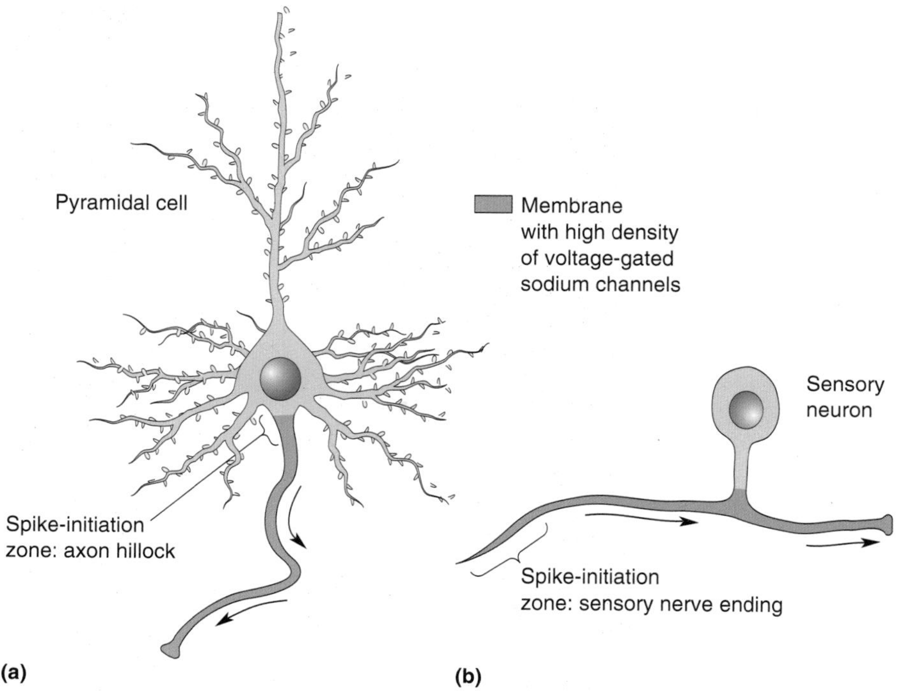

Figure 4.14
Membrane proteins specify the function of different parts of the neuron. Depicted here are **(a)** a cortical pyramidal neuron and **(b)** a primary sensory neuron. Despite the diversity of neuronal structure, the axonal membrane can be identified at the molecular level by its high density of voltage-gated sodium channels. This molecular distinction enables axons to generate and conduct action potentials. The region of membrane where action potentials are normally generated is called the spike-initiation zone. The arrows indicate the normal direction of action potential propagation in these two types of neuron.

Box 4.6

OF SPECIAL INTEREST

The Eclectic Electric Behavior of Neurons

Neurons are not all alike; they vary in shape, size, and connections. Neurons also differ from one another in their electrical properties. A few examples of the diverse behavior of neurons are shown in Figure A.

The cerebral cortex has two major types of neurons, as defined by morphology: aspinous stellate cells and spiny pyramidal cells. A stellate cell typically responds to a steady depolarizing current injected into its soma by firing action potentials at a relatively steady frequency throughout the stimulus (part a). However, most pyramidal cells cannot sustain a steady firing rate. Instead, they fire rapidly at the beginning of the stimulus and then slow down, even if the stimulus remains strong (part b). This slowing over time is called *adaptation*, and it is a very common property among excitable cells. Another firing pattern is the burst, a rapid cluster of action potentials followed by a brief pause. Some cells, including a particular subtype of large pyramidal neuron in the cortex, can even respond to a steady input with rhythmic, repetitive bursts (part c). Variability of firing patterns is not confined to the cerebral cortex. Surveys of many areas of the brain imply that neurons have just as large an assortment of electrical behaviors as morphologies.

What accounts for the diverse behavior of different types of neurons? Ultimately, each neuron's physiology is determined by the properties and numbers of the ion channels in its membrane. There are many more types of ion channels than the few described in this chapter, and each has distinctive properties. For example, some potassium channels activate only very slowly. A neuron with a high density of these potassium channels will show adaptation because during a prolonged stimulus, more and more of the slow potassium channels will open and the outward currents they progressively generate will tend to hyperpolarize the membrane. When you realize that a single neuron may express more than a dozen types of ion channels, the source of diverse firing behavior becomes clear. It is the complex interactions between multiple ion channels that create the eclectic electric signature of each class of neuron.

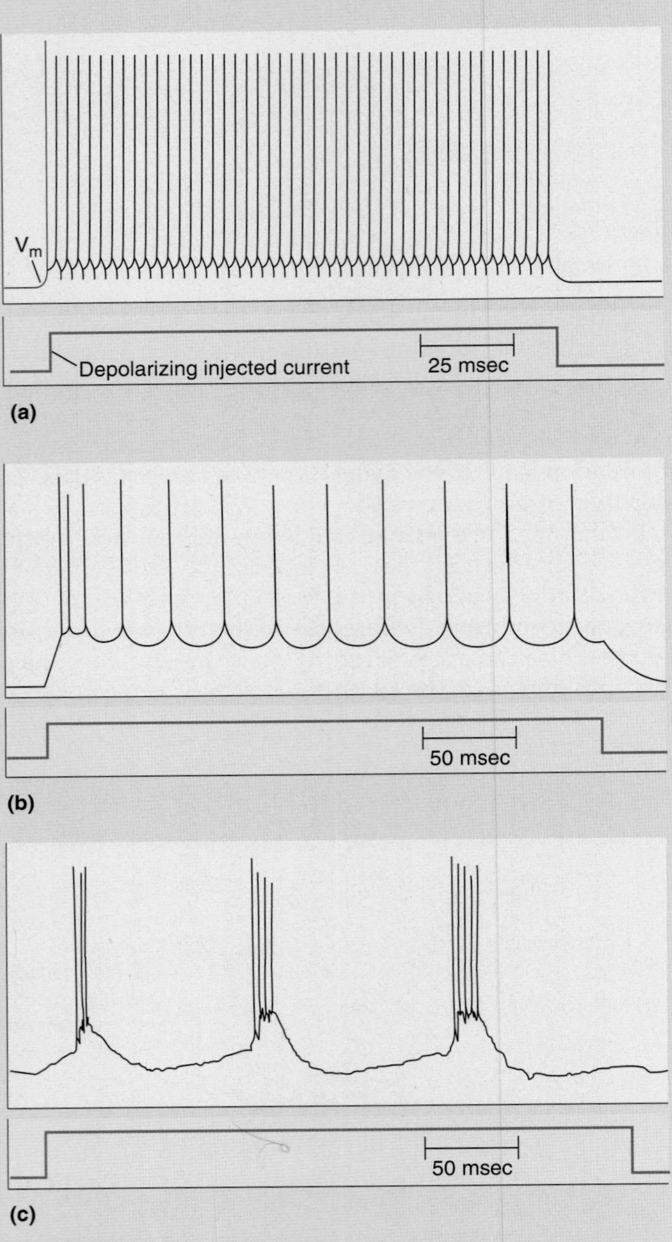

Figure A
The diverse behavior of neurons. (Source: Adapted from Agmon and Connors, 1992.)

CONCLUDING REMARKS

Let's return briefly to the example in Chapter 3 of stepping on a thumbtack. The breaking of the skin caused by the tack stretches the sensory nerve endings of the foot. Special ion channels that are sensitive to the stretching of the membrane open and allow positively charged sodium ions to enter the nerve endings. This influx of positive charge depolarizes the membrane of the spike-initiation zone to threshold, and the action potential is generated. The positive charge that enters during the rising phase of the action potential spreads down the axon and depolarizes the membrane ahead to threshold. In this way the action potential is continuously regenerated as it sweeps like a wave up the sensory axon. We now come to the step where this information is distributed and integrated by other neurons in the central nervous system. This transfer of information from one neuron to another is called *synaptic transmission*, the subject of the next two chapters.

It should come as no surprise that synaptic transmission, like the action potential, depends on specialized proteins in the neuronal membrane. Thus, a picture of the brain as a complicated mesh of interacting neuronal membranes begins to emerge. Consider that a typical neuron with all its neurites has a membrane surface area of about 250,000 μm^2. The surface area of the 100 billion neurons that make up the human brain comes to 25,000 m^2—roughly the size of four soccer fields. This expanse of membrane, with its myriad specialized protein molecules, constitutes the fabric of our minds.

KEY TERMS

Properties of the Action Potential
rising phase (p. 74)
overshoot (p. 74)
falling phase (p. 74)
undershoot (p. 74)
after-hyperpolarization (p. 74)
threshold (p. 76)
absolute refractory period (p. 77)
relative refractory period (p. 77)

The Action Potential in Reality
voltage clamp (p. 82)
voltage-gated sodium channel (p. 82)
patch clamp (p. 84)
tetrodotoxin (TTX) (p. 87)
voltage-gated potassium channel (p. 88)

Action Potential Conduction
saltatory conduction (p. 92)

Action Potentials, Axons, and Dendrites
spike-initiation zone (p. 94)

REVIEW QUESTIONS

1. Define membrane potential (V_m) and sodium equilibrium potential (E_{Na}). Which of these, if any, changes during the course of an action potential?
2. What ions carry the early inward and late outward currents during the action potential?
3. Why is the action potential referred to as "all-or-none"?
4. Some voltage-gated K^+ channels are known as delayed rectifiers because of the timing of their opening during an action potential. What would happen if these channels took much longer than normal to open?
5. Imagine we have labeled tetrodotoxin (TTX) so that it can be seen with a microscope. If we wash this TTX onto a neuron, what parts of the cell would you expect to be labeled? What would be the consequence of applying TTX to this neuron?
6. How does action potential conduction velocity vary with axonal diameter? Why?

CHAPTER 5

Synaptic Transmission

INTRODUCTION

In Chapters 3 and 4, we discussed how mechanical energy, such as a thumbtack entering your foot, can be converted into a neural signal. First, specialized ion channels of the sensory nerve endings allow a positive charge to enter the axon. If this depolarization reaches threshold, then action potentials are generated. Because the axonal membrane is excitable and has voltage-gated sodium channels, action potentials can propagate without decrement up the long sensory nerves. For this information to be processed by the rest of the nervous system, it is necessary that these neural signals be passed on to other neurons, e.g., the motor neurons that control muscle contraction, as well as neurons in the brain and spinal cord that lead to a coordinated reflex response. By the end of the nineteenth century, it was recognized that this transfer of information from one neuron to another occurs at specialized sites of contact. In 1897, English physiologist Charles Sherrington gave these sites their name: synapses. The process of information transfer at a synapse is called *synaptic transmission.*

The physical nature of synaptic transmission was argued for almost a century. One attractive hypothesis, especially considering the speed of synaptic transmission, was that it was simply electrical current flowing from one neuron to the next. In 1959, Harvard University physiologists Edwin Furshpan and David Potter finally proved the existence of such **electrical synapses**. We now know, however, that electrical synapses account for only a very small fraction of the total number of synapses in the mammalian brain.

An alternative hypothesis, also dating to the 1800s, was that chemical neurotransmitters transfer information from one neuron to another at the synapse. In 1921, Otto Loewi, then the head of the Pharmacology Department at the University of Graz in Austria, provided solid support for the concept of **chemical synapses**. Loewi showed that electrical stimulation of axons innervating the frog's heart caused the release of a chemical and that this chemical could mimic the effects of neuron stimulation on the heartbeat (Box 5.1). Later, Bernard Katz and his colleagues at University College, London, conclusively demonstrated that fast transmission at the synapse between a motor neuron axon and skeletal muscle was chemically mediated. By 1951, John Eccles of the Australian National University was able to study the physiology of synaptic transmission within the mammalian central nervous system (CNS) with a new tool, the glass microelectrode. These experiments indicated that many CNS synapses also use a chemical transmitter. We now know that most synapses in the brain are chemical synapses. During the past decade, new methods of studying the molecules involved in synaptic transmission have revealed that synapses are far more complex than most neuroscientists anticipated.

Synaptic transmission is a large and fascinating topic. It is impossible to understand the actions of psychoactive drugs, the causes of mental disorders, the neural bases of learning and memory—indeed, any of the operations of the nervous system—without knowledge of synaptic transmission. Therefore, we've devoted several chapters to this topic, mainly focusing on chemical synapses. In this chapter, we begin by exploring the basic mechanisms of synaptic transmission. What do different types of synapse look like? How are neurotransmitters synthesized and stored, and how are they released in response to an action potential in the axon terminal? How do neurotransmitters act on the postsynaptic membrane? How do single neurons integrate the inputs provided by the thousands of synapses that impinge upon them?

Box 5.1

OF SPECIAL INTEREST

Otto Loewi and *Vagusstoff*

One of the more colorful stories in the history of neuroscience was contributed by Otto Loewi, who, working in Austria in the 1920s, showed definitively that synaptic transmission between nerve and heart is chemically mediated. The heart is supplied with two types of innervation; one type speeds the beating of the heart, and the other slows it. The latter type of innervation is supplied by the vagus nerve. Loewi isolated a frog heart with the vagal innervation left intact, stimulated the nerve electrically, and observed the expected effect, the slowing of the heartbeat. The critical demonstration that this effect was chemically mediated came when he took the solution that bathed this heart, applied it to a second isolated frog heart, and found that the beating of this one also slowed. The idea for this experiment had actually come to Loewi in a dream. Below is his own account:

> In the night of Easter Sunday, 1921, I awoke, turned on the light, and jotted down a few notes on a tiny slip of paper. Then, I fell asleep again. It occurred to me at six o'clock in the morning that during the night I had written down something most important, but I was unable to decipher the scrawl. That Sunday was the most desperate day in my whole scientific life. During the next night, however, I awoke again, at three o'clock, and I remembered what it was. This time I did not take any risk; I got up immediately, went to the laboratory, made the experiment on the frog's heart, described above, and at five o'clock the chemical transmission of the nervous impulse was conclusively proved . . . . Careful consideration in daytime would undoubtedly have rejected the kind of experiment I performed, because it would have seemed most unlikely that if a nervous impulse released a transmitting agent, it would do so not just in sufficient quantity to influence the effector organ, in my case the heart, but indeed in such an excess that it could partly escape into the fluid which filled the heart, and could therefore be detected. Yet the whole nocturnal concept of the experiment was based on this eventuality, and the result proved to be positive, contrary to expectation.
>
> (Loewi, 1953, pp. 33, 34)

The active compound, which Loewi called *vagusstoff*, turned out to be acetylcholine. As we shall see in this chapter, acetylcholine is also a transmitter at the synapse between nerve and skeletal muscle. Here, unlike at the heart, acetylcholine causes excitation and contraction of the muscle.

TYPES OF SYNAPSES

We introduced the synapse in Chapter 2. A synapse is the specialized junction where an axon terminal contacts another neuron or cell type. The normal direction of information flow is from the axon terminal to the target neuron; thus, the axon terminal is said to be *presynaptic* and the target neuron is said to be *postsynaptic*. Let's take a closer look at the different types of synapse.

Electrical Synapses

As we have said, most mammalian synapses are chemical, but there is a simpler, evolutionarily ancient form of electrical synapse that allows the direct transfer of ionic current from one cell to the next. Electrical synapses occur at specialized sites called **gap junctions**. At a gap junction, the presynaptic and postsynaptic membranes are separated by only about 3 nm, and this narrow gap is spanned by special proteins called *connexins*. Six connexins combine to form a channel called a *connexon*, which allows ions to pass directly from the cytoplasm of one cell to the cytoplasm of the other (Figure 5.1). Most gap junctions allow ionic current to pass equally well in both directions; therefore, unlike most chemical synapses, electrical synapses are bidirectional. Because electrical current can pass through these channels, cells connected by

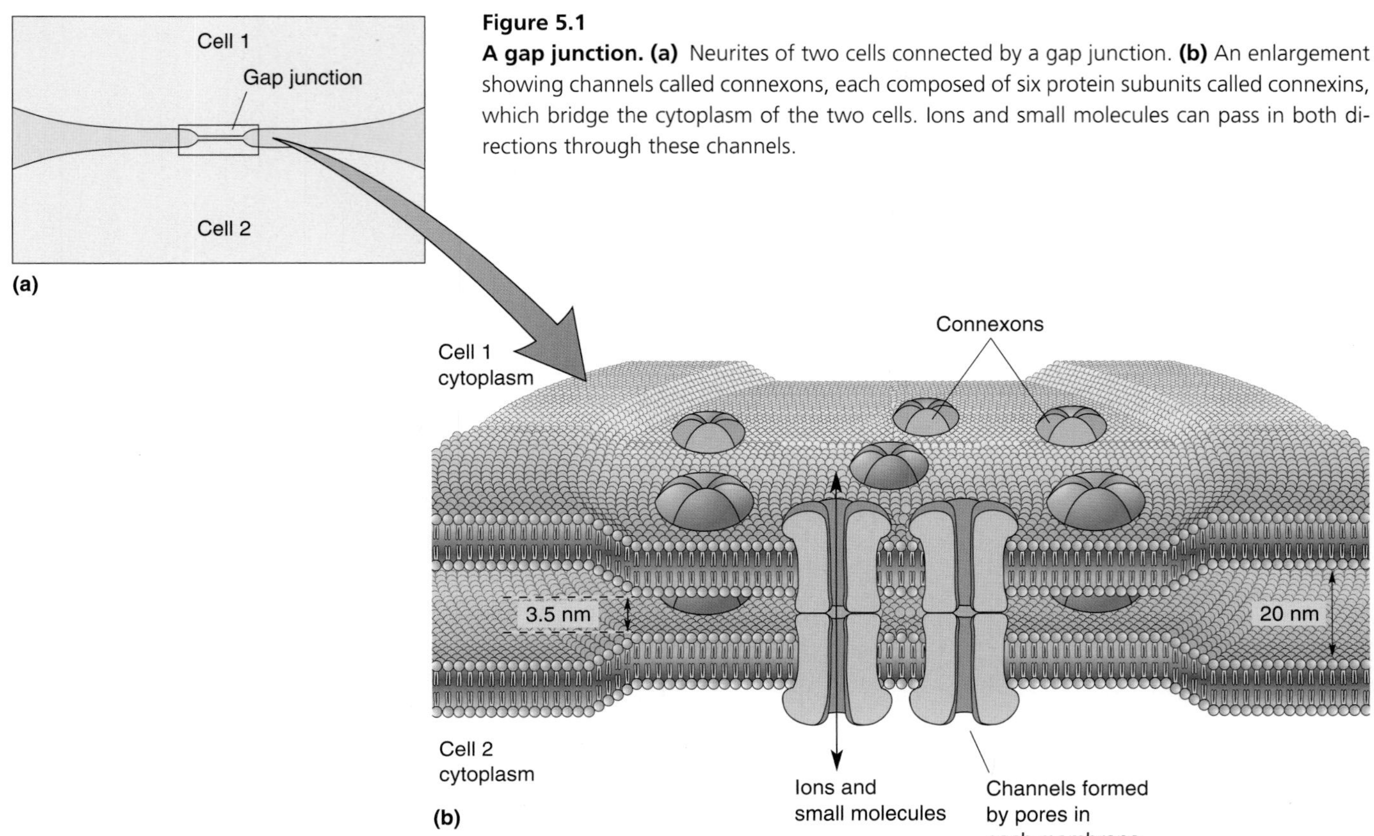

Figure 5.1
A gap junction. (a) Neurites of two cells connected by a gap junction. **(b)** An enlargement showing channels called connexons, each composed of six protein subunits called connexins, which bridge the cytoplasm of the two cells. Ions and small molecules can pass in both directions through these channels.

gap junctions are said to be *electrotonically coupled*. The pore of a connexon is among the largest known. Its diameter is about 2 nm, big enough for all major cellular ions and many small organic molecules to pass through.

Transmission at electrical synapses is very fast and, if the synapse is large, fail-safe. Thus, an action potential in the presynaptic neuron can produce, almost instantaneously, an action potential in the postsynaptic neuron. In invertebrate species, electrical synapses are commonly found between sensory and motor neurons in neural pathways mediating escape reflexes. This enables an animal to beat the hastiest retreat possible when faced with a dangerous situation. In the adult mammalian CNS, electrical synapses are mainly found in specialized locations where normal function requires that the activity of neighboring neurons be highly synchronized.

Although gap junctions between adult mammalian neurons are relatively rare, they are very common in some non-neural cells, including glia, epithelial cells, smooth and cardiac muscle cells, liver cells, and some glandular cells. They also occur frequently between neurons during early embryonic stages. There is evidence that during brain development, gap junctions allow neighboring cells to share both electrical and chemical signals that may help coordinate their growth and maturation.

Chemical Synapses

As a rule, synaptic transmission in the mature human nervous system is chemical, so we will focus exclusively on chemical synapses. Before we discuss the different types of chemical synapse, let's take a look at some of their universal characteristics (Figure 5.2).

The presynaptic and postsynaptic membranes at chemical synapses are separated by a *synaptic cleft* that is 20–50 nm wide, 10 times the width of the

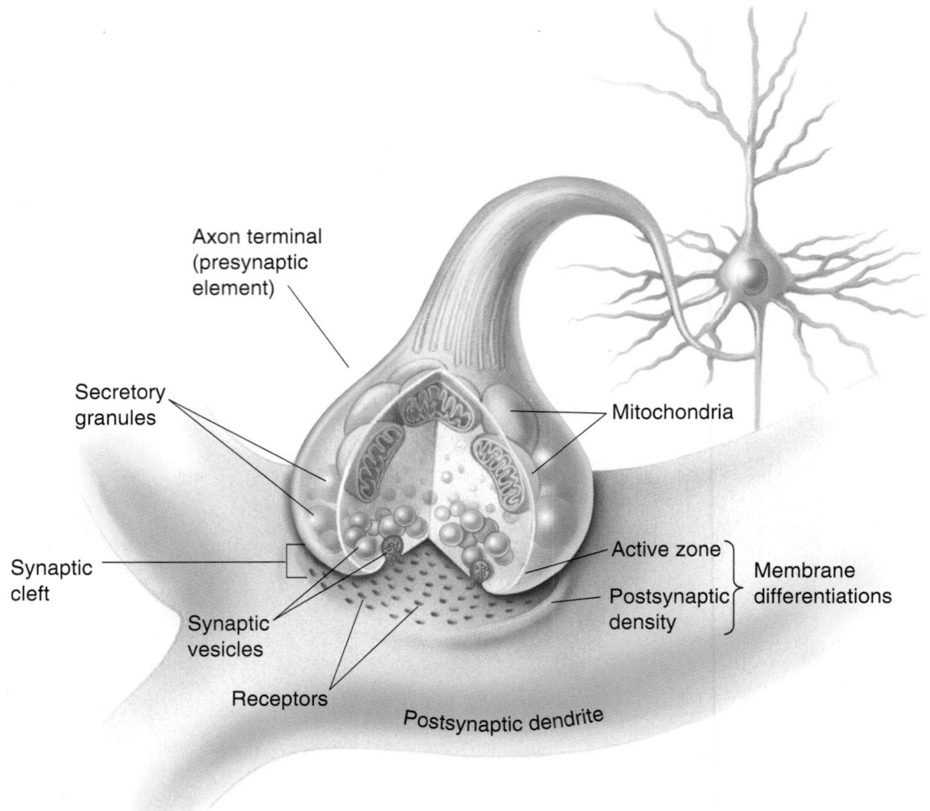

Figure 5.2
The parts of a chemical synapse.

separation at gap junctions. The cleft is filled with a matrix of fibrous extracellular protein. One function of this matrix is to make the presynaptic and postsynaptic membranes adhere to each other. The presynaptic side of the synapse, also called the *presynaptic element,* is usually an axon terminal. The terminal typically contains dozens of small membrane-enclosed spheres, about 50 nm in diameter, called *synaptic vesicles* (Figure 5.3a). These vesicles store neurotransmitter, the chemical used to communicate with the postsynaptic neuron. Many axon terminals also contain larger vesicles, about 100 nm in diameter, called **secretory granules**. Secretory granules contain soluble protein that appears dark in the electron microscope, so they are sometimes called **large, dense-core vesicles** (Figure 5.3b).

Dense accumulations of protein adjacent to and within the membranes on either side of the synaptic cleft are collectively called **membrane differentiations**. On the *presynaptic* side, proteins jutting into the cytoplasm of the terminal along the intracellular face of the membrane look like a field of pyramids. The pyramids and the membrane associated with them are the actual sites of neurotransmitter release, called **active zones**. Synaptic vesicles are clustered in the cytoplasm adjacent to the active zones.

The protein thickly accumulated in and just under the *postsynaptic* membrane is called the **postsynaptic density**. The postsynaptic density contains the neurotransmitter receptors, which convert the *intercellular* chemical signal (i.e., neurotransmitter) into an *intracellular* signal (i.e., a change in membrane potential, or a chemical change) in the postsynaptic cell. As we shall see, the nature of this postsynaptic response can be quite varied, depending on the type of protein receptor that is activated by the neurotransmitter.

CNS Synapses. In the CNS, different types of synapse may be distinguished by which part of the neuron is postsynaptic to the axon terminal. If the post-

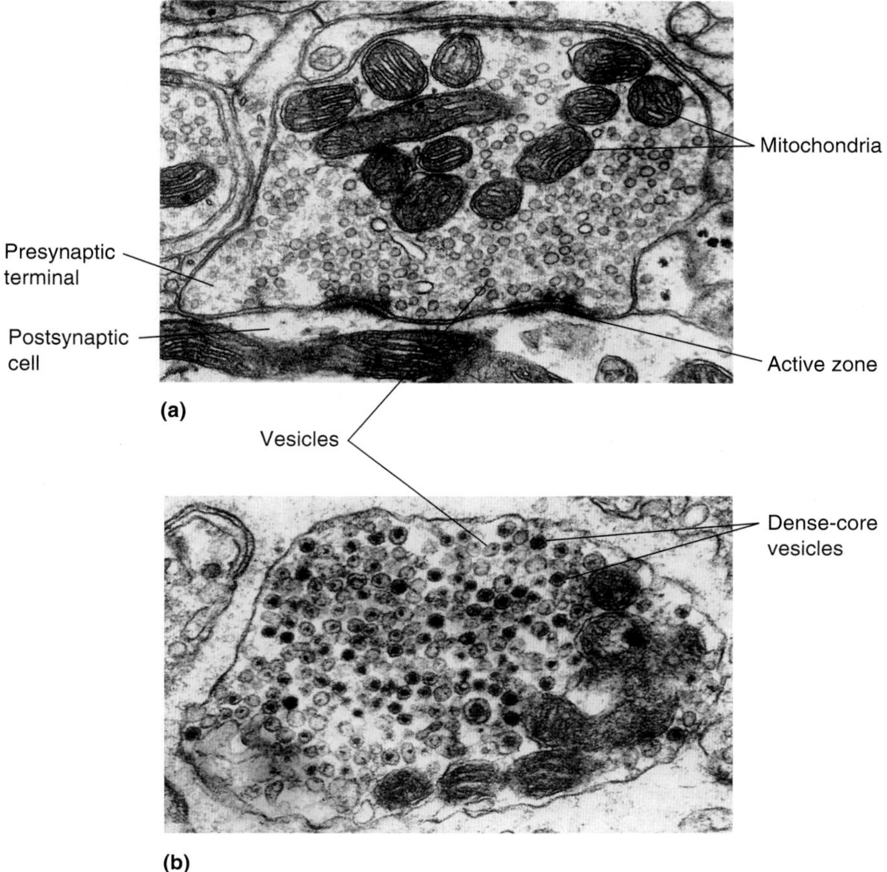

Figure 5.3
Chemical synapses, as seen with the electron microscope. (a) Fast excitatory synapse in the CNS. (Source: Adapted from Heuser and Reese, 1977, p. 262.) **(b)** A synapse in the PNS, with numerous dense-core vesicles. (Source: Adapted from Heuser and Reese, 1977, p. 278.)

synaptic membrane is on a dendrite, the synapse is said to be *axodendritic*. If the postsynaptic membrane is on the cell body, the synapse is said to be *axosomatic*. In some cases, the postsynaptic membrane is on another axon, and these synapses are called *axoaxonic* (Figure 5.4). In certain specialized neurons, *dendrites* actually form synapses with one another; these are called *dendrodendritic* synapses. The size and shape of CNS synapses also vary widely (Figure 5.5).

CNS synapses may be further divided into two general categories based on the appearance of their presynaptic and postsynaptic membrane differentiations under the powerful magnification of the electron microscope. Synapses in which the membrane differentiation on the postsynaptic side is thicker than that on the presynaptic side are called asymmetrical synapses, or **Gray's type I synapses**; those in which the membrane differentiations are of similar thickness are called symmetrical synapses, or **Gray's type II synapses** (Figure 5.6). As we shall see later in the chapter, these structural differences can be correlated with functional differences. Gray's type I synapses are usually excitatory, while Gray's type II synapses are usually inhibitory.

The Neuromuscular Junction. Synaptic junctions also exist outside the brain and spinal cord. For example, axons of the autonomic nervous system innervate glands, smooth muscle, and the heart. Chemical synapses also occur between the axons of motor neurons of the spinal cord and skeletal muscle. Such a synapse is called a **neuromuscular junction**, and it has many of the structural features of chemical synapses in the CNS (Figure 5.7).

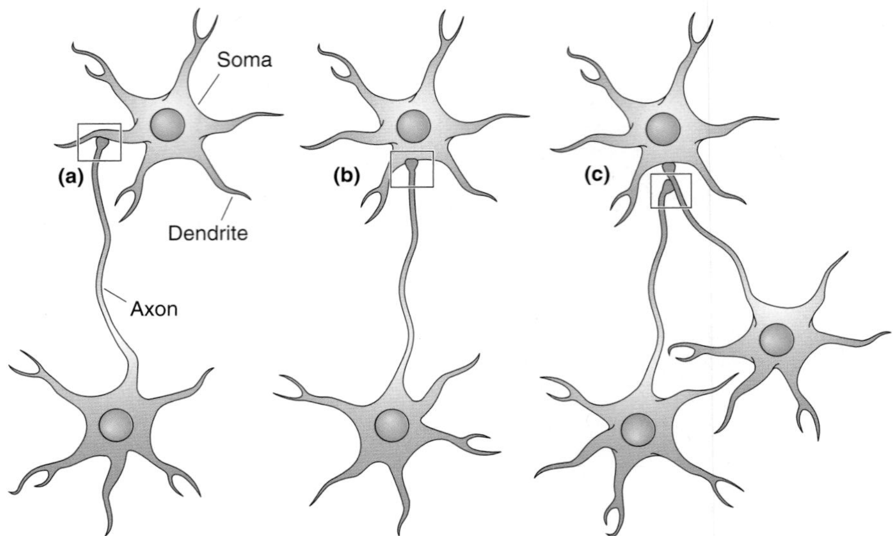

Figure 5.4
Synaptic arrangements in the CNS. (a) Axodendritic synapse. **(b)** Axosomatic synapse. **(c)** Axoaxonic synapse.

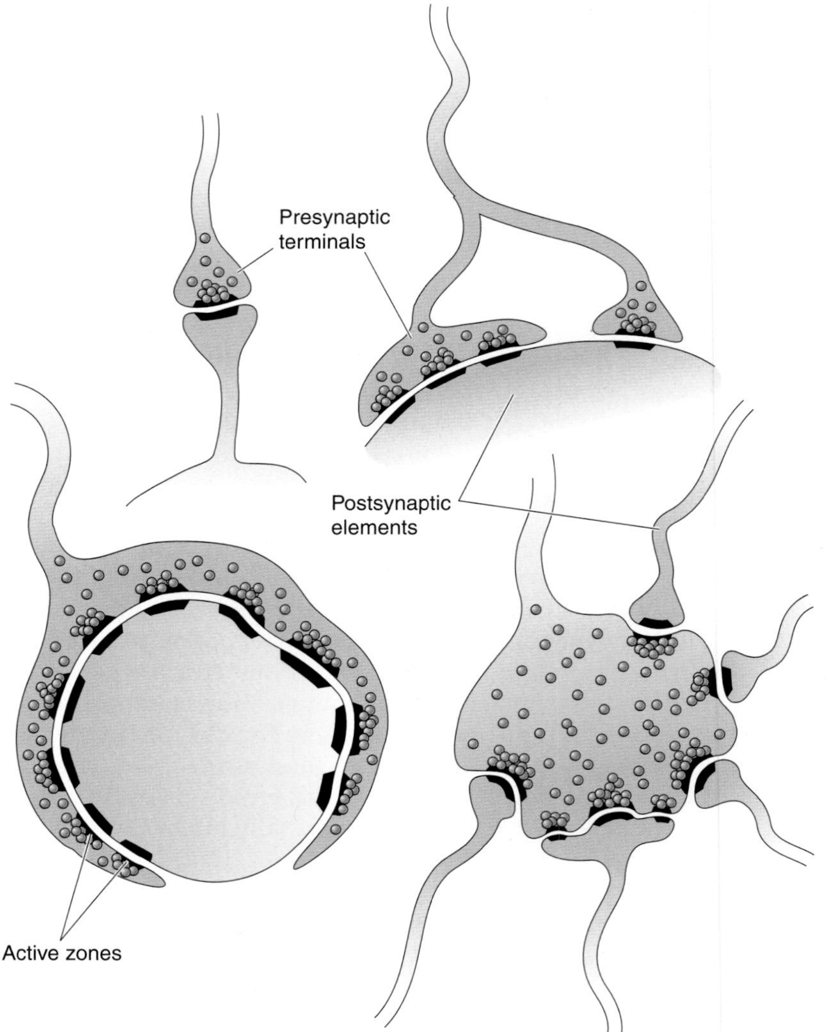

Figure 5.5
Various sizes of CNS synapses. Notice that larger synapses have more active zones.

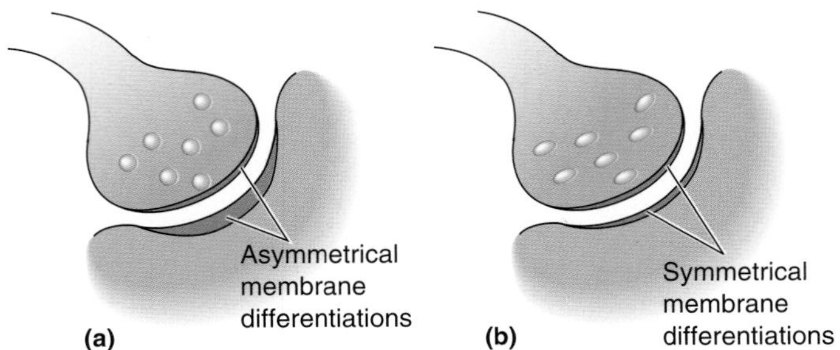

Figure 5.6
Two categories of CNS synaptic membrane differentiations. (a) A Gray's type I synapse is asymmetrical and usually excitatory. **(b)** A Gray's type II synapse is symmetrical and usually inhibitory.

Neuromuscular synaptic transmission is fast and reliable. An action potential in the motor axon always causes an action potential in the muscle cell it innervates. This reliability is accounted for, in part, by structural specializations of the neuromuscular junction. Its most important specialization is its size; it is one of the largest synapses in the body. The presynaptic terminal also contains a large number of active zones. In addition, the postsynaptic membrane, also called the **motor end-plate**, contains a series of shallow folds. The presynaptic active zones are precisely aligned with these junctional folds, and the postsynaptic membrane of the folds is packed with neurotransmitter receptors. This structure ensures that many neurotransmitter molecules are focally released onto a large chemoreceptive membrane surface.

Because neuromuscular junctions are more accessible to the scientist than CNS synapses, much of what we know about the mechanisms of synaptic transmission was first established here. Neuromuscular junctions are also of considerable clinical significance; diseases, drugs, and poisons that interfere with this chemical synapse have direct effects on vital bodily functions.

PRINCIPLES OF CHEMICAL SYNAPTIC TRANSMISSION

Consider the basic requirements of chemical synaptic transmission. There must be a mechanism for synthesizing neurotransmitter and packing it into the synaptic vesicles, a mechanism causing vesicles to spill their contents into the synaptic cleft in response to a presynaptic action potential, a mechanism for producing an electrical or biochemical response to neurotransmitter in the postsynaptic neuron, and a mechanism for removing neurotransmitter from the synaptic cleft. And to be useful for sensation, perception, and the control of movement, all of these things must occur very rapidly. No wonder physiologists initially were skeptical about the existence of chemical synapses in the brain!

Fortunately, thanks to several decades of research on the topic, we now understand how many of these aspects of synaptic transmission are so efficiently carried out. Here we'll present a general survey of the basic principles. In Chapter 6, we will take a more detailed look at the individual neurotransmitters and their modes of postsynaptic action.

Neurotransmitters

Since the discovery of chemical synaptic transmission, the search has been on to identify neurotransmitters in the brain. Our current understanding is that most neurotransmitters fall into one of three chemical categories: (1) *amino acids*, (2) *amines*, and (3) *peptides* (Table 5.1). Some representatives of these categories are shown in Figure 5.8. The amino acid and amine neurotransmitters are all small organic molecules containing a nitrogen atom, and they are stored in and released from synaptic vesicles. Peptide neurotransmitters are

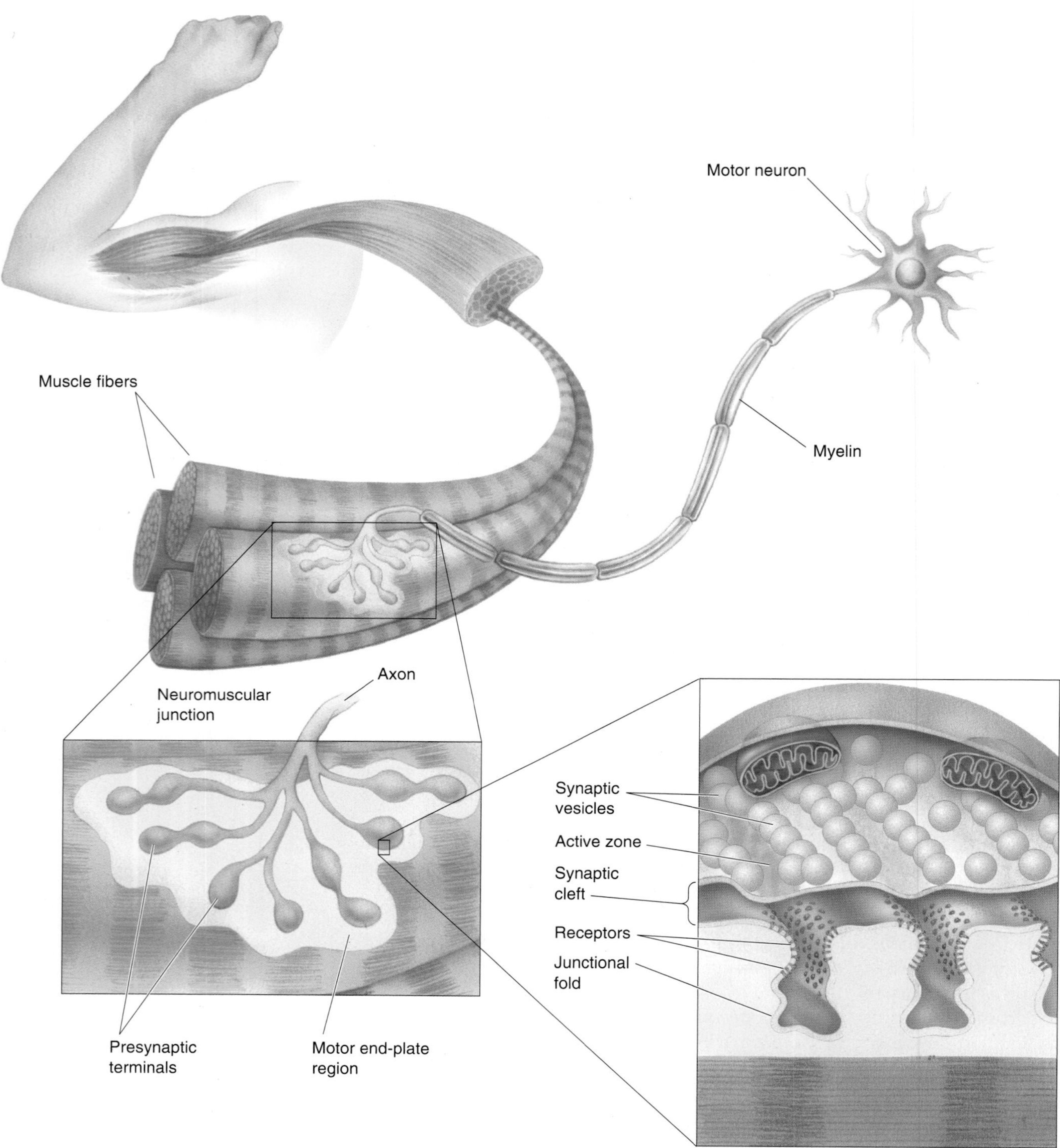

Figure 5.7

The neuromuscular junction. The postsynaptic membrane, called the motor end-plate, contains junctional folds with numerous neurotransmitter receptors.

Table 5.1 The Major Neurotransmitters

AMINO ACIDS	AMINES	PEPTIDES
Gamma-amino butyric acid (GABA)	Acetylcholine (ACh)	Cholecystokinin (CCK)
Glutamate (Glu)	Dopamine (DA)	Dynorphin
Glycine (Gly)	Epinephrine	Enkephalins (Enk)
	Histamine	*N*-Acetylaspartylglutamate (NAAG)
	Norepinephrine (NE)	Neuropeptide Y
	Serotonin (5-HT)	Somatostatin
		Substance P
		Thyrotropin-releasing hormone
		Vasoactive intestinal polypeptide (VIP)

large molecules stored in and released from secretory granules. As mentioned earlier, secretory granules and synaptic vesicles are frequently observed in the same axon terminals. Consistent with this observation, peptides often exist in the same axon terminals that contain amine or amino acid neurotransmitters. And, as we'll discuss in a moment, these different neurotransmitters are released under different conditions.

Different neurons in the brain release different neurotransmitters. Fast synaptic transmission at most CNS synapses is mediated by the amino acids **glutamate (Glu)**, **gamma-aminobutyric acid (GABA)**, and **glycine (Gly)**. The amine **acetylcholine (ACh)** mediates fast synaptic transmission at all neuromuscular junctions. Slower forms of synaptic transmission in the CNS and in the periphery are mediated by transmitters from all three chemical categories.

Neurotransmitter Synthesis and Storage

Chemical synaptic transmission requires that neurotransmitters be synthesized and ready for release. Different neurotransmitters are synthesized in different ways. For example, glutamate and glycine are among the 20 amino

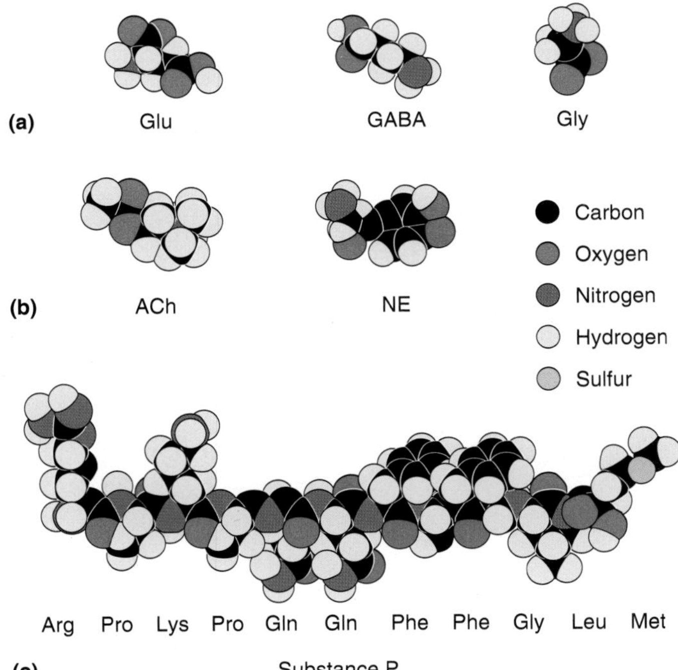

Figure 5.8
Representative neurotransmitters. (a) The amino acid neurotransmitters glutamate, GABA, and glycine. **(b)** The amine neurotransmitters acetylcholine and norepinephrine. **(c)** The peptide neurotransmitter substance P. (For the abbreviations and chemical structures of amino acids in substance P, see Figure 3.4b.)

Nucleus
Precursor peptide
Rough ER
Ribosome
Active peptide neurotransmitter
Golgi apparatus
Secretory granules
Synaptic vesicles
(a)

Precursor molecule
① Synthetic enzyme
Neurotransmitter molecule
② Transporter protein
Synaptic vesicle
(b)

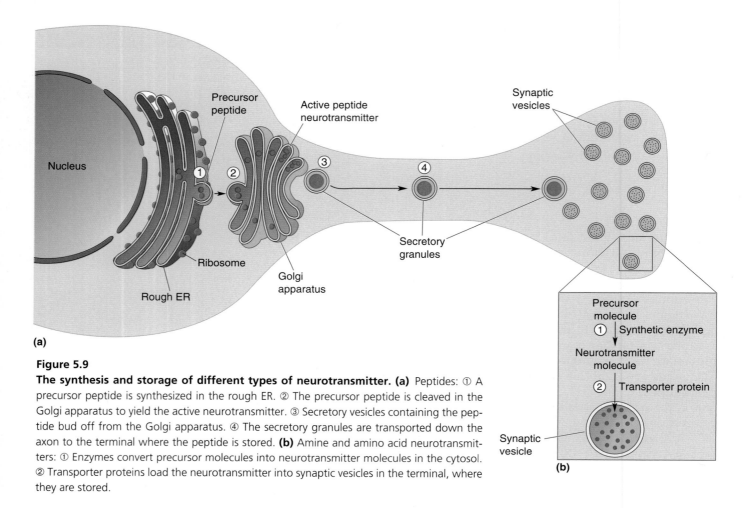

Figure 5.9
The synthesis and storage of different types of neurotransmitter. (a) Peptides: ① A precursor peptide is synthesized in the rough ER. ② The precursor peptide is cleaved in the Golgi apparatus to yield the active neurotransmitter. ③ Secretory vesicles containing the peptide bud off from the Golgi apparatus. ④ The secretory granules are transported down the axon to the terminal where the peptide is stored. **(b)** Amine and amino acid neurotransmitters: ① Enzymes convert precursor molecules into neurotransmitter molecules in the cytosol. ② Transporter proteins load the neurotransmitter into synaptic vesicles in the terminal, where they are stored.

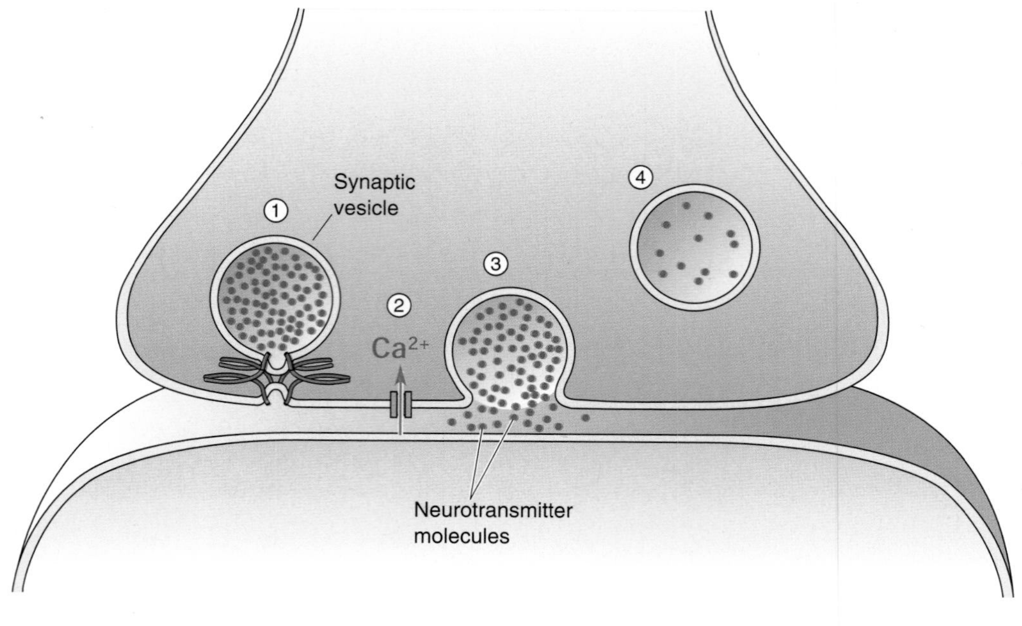

Figure 5.10
The release of neurotransmitter by exocytosis. ① A synaptic vesicle loaded with neurotransmitter, in response to ② an influx of Ca^{2+} through voltage-gated calcium channels, ③ releases its contents into the synaptic cleft by the fusion of the vesicle membrane with the presynaptic membrane, and ④ is eventually recycled by the process of endocytosis.

acids that are the building blocks of protein (see Figure 3.4b); consequently, they are abundant in all cells of the body, including neurons. In contrast, GABA and the amines are made only by the neurons that release them. These neurons contain specific enzymes that synthesize the neurotransmitters from various metabolic precursors. The synthesizing enzymes for both amino acid and amine neurotransmitters are transported to the axon terminal, where they locally and rapidly direct transmitter synthesis.

Once synthesized in the cytosol of the axon terminal, the amino acid and amine neurotransmitters must be taken up by the synaptic vesicles. Concentrating these neurotransmitters inside the vesicle is the job of **transporters**, special proteins embedded in the vesicle membrane.

Quite different mechanisms are used to synthesize and store peptides in secretory granules. As we learned in Chapters 2 and 3, peptides are formed when amino acids are strung together by the ribosomes of the cell body. In the case of peptide neurotransmitters, this occurs in the rough endoplasmic reticulum (ER). Generally, the peptide synthesized in the rough ER is cleaved in the Golgi apparatus to yield the active neurotransmitter. Secretory granules containing the peptide bud off from the Golgi apparatus and are carried to the axon terminal by axoplasmic transport. Figure 5.9 compares the synthesis and storage of amine and amino acid neurotransmitters with that of peptide neurotransmitters.

Neurotransmitter Release

Neurotransmitter release is triggered by the arrival of an action potential in the axon terminal. The depolarization of the terminal membrane causes **voltage-gated calcium channels** in the active zones to open. These membrane channels are very similar to the sodium channels we discussed in Chapter 4, except that they are permeable to Ca^{2+} ions instead of Na^+ ions. There is a large inward driving force on Ca^{2+}. Remember that the internal calcium ion concentration, $[Ca^{2+}]_i$, at rest is very low, only 0.0002 mM, so Ca^{2+} will flood the cytoplasm of the axon terminal as long as the calcium channels are open. The resulting elevation in $[Ca^{2+}]_i$ is the signal that causes neurotransmitter to be released from synaptic vesicles.

The vesicles release their contents by a process called **exocytosis**. The membrane of the synaptic vesicle fuses to the presynaptic membrane at the active zone, allowing the contents of the vesicle to spill out into the synaptic cleft (Figure 5.10). Studies of a giant synapse in the squid nervous system showed that exocytosis can occur very rapidly—within 0.2 msec of the Ca^{2+} influx into the terminal. Exocytosis is fast because Ca^{2+} enters at the active zone, precisely where synaptic vesicles are ready and waiting to release their contents. In this local "microdomain" around the active zone, calcium can achieve very high concentrations (greater than 0.1 mM).

The precise mechanism for how increased $[Ca^{2+}]_i$ stimulates exocytosis is poorly understood, but it is under intensive investigation (Box 5.2). The speed of neurotransmitter release suggests that the vesicles involved are those already "docked" at the active zones. Docking is believed to involve interactions between proteins in the synaptic vesicle membrane and the active zone (Box 5.3). In the presence of high $[Ca^{2+}]_i$, these proteins alter their conformation so that the lipid bilayers of the vesicle and presynaptic membranes fuse, forming a pore that allows the neurotransmitter to escape into the cleft. The mouth of this exocytotic fusion pore continues to expand until the membrane of the vesicle is fully incorporated into the presynaptic membrane (Figure 5.11). The vesicle membrane is later recovered by the process of **endocytosis**, and the recycled vesicle is refilled with neurotransmitter (Figure 5.10). During periods of prolonged stimulation, vesicles are mobilized from

Box 5.2

PATH OF DISCOVERY

The Functional Anatomy of Neurotransmitter Release

BY THOMAS SÜDHOF

When I first learned how synapses work, I was immediately fascinated by what looked like an unnecessarily complicated process. At first it seemed illogical that one neuron would send a signal to another neuron by synthesizing neurotransmitters, packing them into synaptic vesicles, secreting them upon stimulation, and then having them activate postsynaptic receptors. The process seemed overly complicated. This complexity was also attractive, however, because it suggested that something important was happening. Only after considerable reading did I realize that synaptic transmission *had* to be as complicated as it is in order to achieve the unique speed and plasticity that are so important for information processing in the brain.

While looking at electron micrographs of a synapse, I was impressed by the elaborate presynaptic machinery, which generates buckets of synaptic vesicles, fills them with neurotransmitter, and moves them around. In spite of copious information about the morphology and physiology of synapses, little was known about their molecules, let alone their mechanisms. At that time, I was a student in Germany in the laboratory of V. P. Whittaker, who was the first to purify and characterize synaptic vesicles. Working at the dawn of the era of molecular biology, I realized how little we knew about the kind of stuff synapses are made of. Analysis of postsynaptic neurotransmitter receptors promised to be relatively straightforward. But on the presynaptic side, the key molecules were unidentified, the molecules of synaptic vesicles and of active zones were unknown, and the nature of the adhesion and recognition between pre- and postsynaptic plasma membranes was mysterious. This was when I developed an interest in the molecular architecture of presynaptic terminals.

After completing postdoctoral training with M. S. Brown and J. L. Goldstein on a different topic (the biology of cholesterol), I felt better equipped to study the molecular design of synapses. I set out in 1986, using the techniques I had learned from my teachers. My initial approach was guided by the idea that synaptic vesicles are simple: small organelles with a limited number of protein components whose only function is to mediate the storage and release of neurotransmitters.

Collaborating with Reinhard Jahn, I started with the hypothesis that all functions involved in neurotransmitter release must act on synaptic vesicles directly or indirectly. We characterized the major proteins of synaptic vesicles, cloning the first, synaptophysin, in 1987; cloning other proteins followed quickly over the next 3 years. Important results also came from several other laboratories, most notably those of H. Betz, R. Scheller, K. Buckley, P. Greengard, and R. Kelly. Our original plan was rather unimaginative and descriptive: We first wanted to purify and clone the various proteins, then study their structures, and finally analyze their functions by knocking them out in mice. Today, 15 years after the project started, the molecular and genetic description of synaptic vesicles is probably more nearly complete than that of any other organelle.

Thomas Südhof

Not surprisingly, the synapse has yielded all kinds of proteins from among about 20 protein families. Some proteins are permanently associated with the vesicles, while others dynamically cycle on and off the vesicles during exocytosis and endocytosis. One of the most interesting findings is that synaptic vesicles are economically organized. Only a single vesicle protein, synaptobrevin, seems to be essential for vesicle fusion, and another single protein, synaptotagmin, is required for fast calcium triggering of fusion. Most of the remaining vesicle proteins are involved in neurotransmitter transport into the vesicles or in regulating the fusion reaction.

The study of synapses has expanded enormously, extending to other synaptic structures and components, especially the active zone. With luck, the definition of synaptic vesicles will be complete in the next decade, and we will understand how all of its components function. I am still fascinated with vesicles, their movements in the nerve terminal, and their precisely regulated and timed fusion. One of the wonders of nature is the cell's ability to make such vesicles in abundance and of uniform size, to cluster them at synapses, and to use them to store and release neurotransmitters. This amazing process will continue to raise interesting questions.

Box 5.3

BRAIN FOOD

How to SNARE a Vesicle

Yeasts are single-celled organisms beloved for their ability to make dough rise and grape juice ferment. Remarkably, the humble yeasts have some close similarities to the chemical synapses in our brain. Recent research has shown that the proteins controlling secretion in both yeast cells and synapses are only minor variations of each other. Apparently, these molecules are so generally useful that they have been conserved across more than a billion years of evolution, and they are found in all eukaryotic cells.

The trick to fast synaptic function is to deliver neurotransmitter-filled vesicles to just the right place, the presynaptic membrane, and then cause them to fuse at just the right time, when an action potential delivers a pulse of high Ca^{2+} concentration to the cytosol. This process of exocytosis is a special case of a more general cellular problem, *membrane trafficking*. Cells have many types of membranes, including those enclosing the whole cell, the nucleus, endoplasmic reticulum, Golgi apparatus, and various types of vesicles. To avoid intracellular chaos, each of these membranes has to be moved and delivered to specific locations within the cell. After delivery, one type of membrane often has to fuse with another type. A common molecular machinery has evolved for the delivery and fusion of all of these membranes, and small variations in these molecules determine how and when membrane trafficking takes place.

The specific binding and fusion of membranes seem to depend on the SNARE[1] family of proteins, which were first found in yeast. Each SNARE peptide has a lipid-loving end that embeds itself within the membrane and a longer tail that projects into the cytosol. Vesicles have "v-SNAREs," and the outer membrane has "t-SNAREs" (for "target" membrane). The cytosolic ends of these two complementary types of SNAREs can bind tightly to one another, allowing a vesicle to "dock" very close to a presynaptic membrane and nowhere else (Figure A).

Although SNARE-to-SNARE complexes form the main connection between vesicle and target membranes, a large and bewildering array of other proteins stick to the "SNAREpin." The functions of most of them are still unknown, but synaptotagmin, a vesicle protein, may be the critical Ca^{2+} sensor that rapidly triggers vesicle fusion and thus transmitter release. On the presynaptic membrane side, calcium channels may form part of the docking complex. By placing the calcium channels very close to the docked vesicles, inflowing Ca^{2+} can trigger transmitter release with astonishing speed—within about 60 μsec in a mammalian synapse at body temperature.

There is a long way to go before we understand all of the molecules involved in synaptic transmission. In the meantime, we can count on yeasts to provide delightful brain food (and drink) for thought.

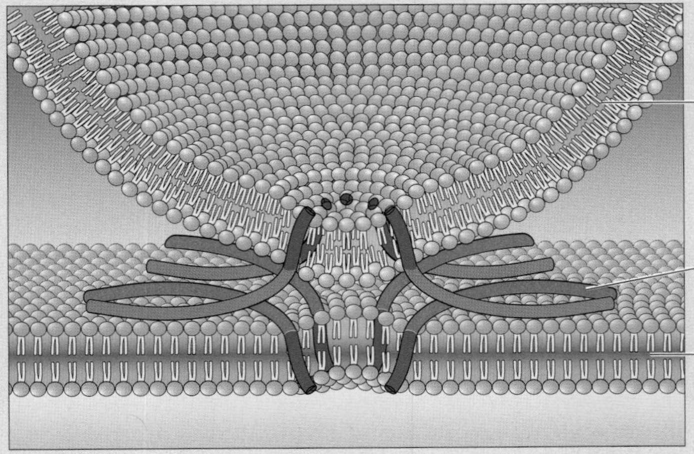

Figure A
Cellular SNAREpins
(Source: Adapted from Weber et al., 1998, Fig. 5.)

[1] SNARE stands for SNAP receptor; SNAP stands for soluble NSF attachment protein; NSF stands for *N*-ethylmaleimide-sensitive factor. Protein names can be like Russian babushka dolls, with one name hiding another, which hides another, and so on.

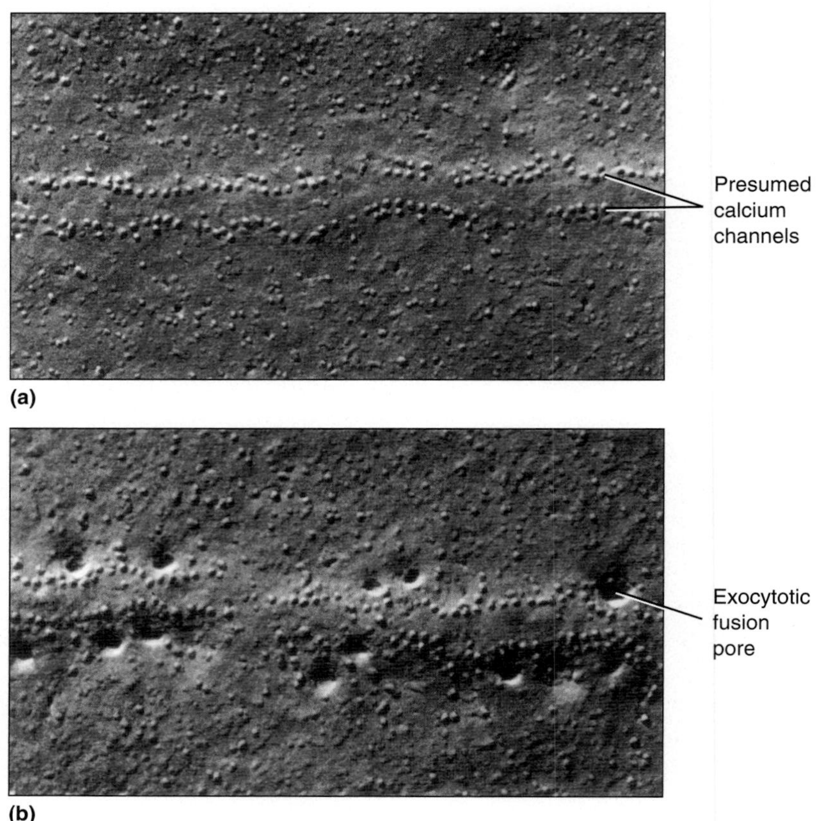

Figure 5.11
A "receptor's eye" view of neurotransmitter release. (a) Extracellular surface of the active zone at the frog neuromuscular junction. The particles are believed to be calcium channels. **(b)** The presynaptic terminal has been stimulated to release neurotransmitter. The exocytotic fusion pores are where synaptic vesicles have fused with the presynaptic membrane and released their contents. (Source: Heuser and Reese, 1973.)

a "reserve pool" that is bound to the cytoskeleton of the axon terminal. The release of these vesicles from the cytoskeleton and their docking to the active zone are also triggered by elevations of $[Ca^{2+}]_i$.

Secretory granules also release peptide neurotransmitters by exocytosis in a calcium-dependent fashion but typically not at the active zones. Because the sites of granule exocytosis occur at a distance from the sites of Ca^{2+} entry, peptide neurotransmitters are usually not released in response to every action potential invading the terminal. Instead, the release of peptides generally requires high-frequency trains of action potentials, so that the $[Ca^{2+}]_i$ throughout the terminal can build to the level required to trigger release away from the active zones. Unlike the fast release of amino acid and amine neurotransmitters, the release of peptides is a leisurely process, taking 50 msec or more.

Neurotransmitter Receptors and Effectors

Neurotransmitters released into the synaptic cleft affect the postsynaptic neuron by binding to specific receptor proteins that are embedded in the postsynaptic density. The binding of neurotransmitter to the receptor is like inserting a key in a lock; this causes conformational changes in the protein, and the protein can then function differently. Although there are well over 100 different neurotransmitter receptors, they can be divided into two types: transmitter-gated ion channels and G-protein-coupled receptors.

Transmitter-Gated Ion Channels. Receptors known as **transmitter-gated ion channels** are membrane-spanning proteins consisting of four or five subunits that come together to form a pore between them (Figure 5.12). In the absence of neurotransmitter, the pore is usually closed. When neurotransmitter

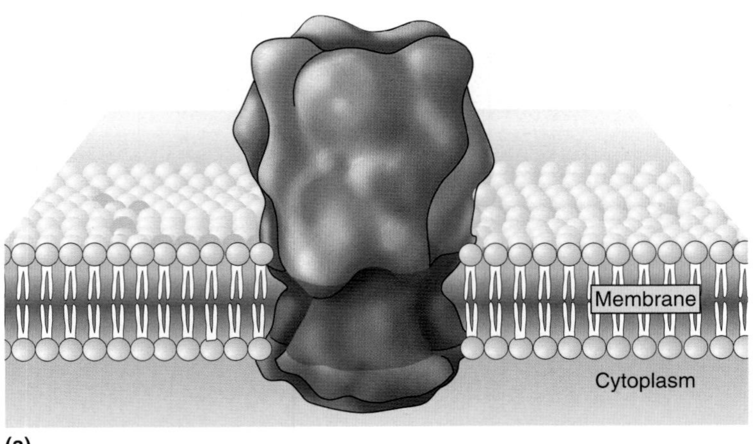

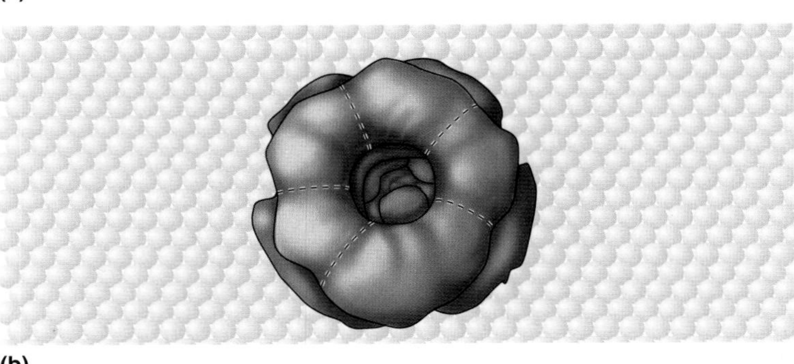

Figure 5.12
The structure of a transmitter-gated ion channel. (a) Side view of an ACh-gated ion channel as it is believed to appear. **(b)** Top view of the channel showing the pore. Broken lines indicate the approximate borders between the five subunits. (Source: Adapted from Unwin, 1993.)

binds to specific sites on the extracellular region of the channel, it induces a conformational change—just a slight twist of the subunits—which within microseconds causes the pore to open. The functional consequence of this depends on which ions can pass through the pore.

Transmitter-gated channels generally do not show the same degree of ion selectivity as do voltage-gated channels. For example, the ACh-gated ion channels at the neuromuscular junction are permeable to both Na^+ and K^+ ions. Nonetheless, as a rule, if the open channels are permeable to Na^+, the net effect will be to depolarize the postsynaptic cell from the resting membrane potential (Box 5.4). Because it tends to bring the membrane potential toward threshold for generating action potentials, this effect is said to be *excitatory*. A transient postsynaptic membrane depolarization caused by the presynaptic release of neurotransmitter is called an **excitatory postsynaptic potential** (**EPSP**) (Figure 5.13). Synaptic activation of ACh-gated and glutamate-gated ion channels causes EPSPs.

If the transmitter-gated channels are permeable to Cl^-, the net effect will be to hyperpolarize the postsynaptic cell from the resting membrane potential (because the chloride equilibrium potential is negative; see Chapter 3). Because it tends to bring the membrane potential away from threshold for generating action potentials, this effect is said to be *inhibitory*. A transient hyperpolarization of the postsynaptic membrane potential caused by the presynaptic release of neurotransmitter is called an **inhibitory postsynaptic potential** (**IPSP**) (Figure 5.14). Synaptic activation of glycine-gated or GABA-gated ion channels causes an IPSP. We'll discuss EPSPs and IPSPs in more detail when we explore the principles of synaptic integration.

G-Protein-Coupled Receptors. Fast chemical synaptic transmission is mediated by amino acid and amine neurotransmitters acting on transmitter-gated ion channels. However, all three types of neurotransmitter, acting on

Box 5.4

BRAIN FOOD

Reversal Potentials

In Chapter 4, we saw that when the membrane voltage-gated sodium channels open during the action potential, Na^+ ions enter the cell, causing the membrane potential to depolarize rapidly until it approaches the sodium equilibrium potential, E_{Na}, about 40 mV. Unlike the voltage-gated channels, however, many transmitter-gated ion channels are not permeable to a single type of ion. For example, the ACh-gated ion channel at the neuromuscular junction is permeable to both Na^+ and to K^+. Let's explore the functional consequence of activating these channels.

In Chapter 3, we learned that the membrane potential, V_m, can be calculated using the Goldman equation, which takes into account the relative permeability of the membrane to different ions (Box 3.3). If the membrane were equally permeable to Na^+ and K^+, as it would be if the ACh-gated channels were open, then V_m would have a value between E_{Na} and E_K, around 0 mV. Therefore, ionic current would flow through the channels in a direction that brings the membrane potential toward 0 mV. If the membrane potential were <0 mV before ACh was applied, as is usually the case, the direction of net current flow through the ACh-gated ion channels would be *inward*, causing a depolarization. If the membrane potential were >0 mV before ACh was applied, however, the direction of net current flow through the ACh-gated ion channels would be *outward*, causing the membrane potential to become less positive.

Ionic current flow at different membrane voltages can be graphed, as shown in Figure A. Such a graph is called an I-V plot (I: current; V: voltage). The critical value of membrane potential at which the direction of current flow reverses is called the *reversal potential*. In this case, the reversal potential would be 0 mV. The experimental determination of a reversal potential, therefore, helps tell us which types of ions the membrane is permeable to.

If by changing the relative permeability of the membrane to different ions a neurotransmitter causes V_m to move toward a value that is more positive than action potential threshold, the neurotransmitter action is termed excitatory. As a rule, neurotransmitters that open a channel permeable to Na^+ are excitatory. If a neurotransmitter causes V_m to take on a value that is more negative than action potential threshold, the neurotransmitter action is termed inhibitory. Neurotransmitters that open a channel permeable to Cl^- tend to be inhibitory, as are neurotransmitters that open a channel permeable only to K^+.

At positive membrane potentials, ACh causes outward current

Out

Membrane current

Membrane voltage

I-V plot during ACh application

–60 mV

60 mV

Reversal potential

In

At negative membrane potentials, ACh causes inward current

Figure A

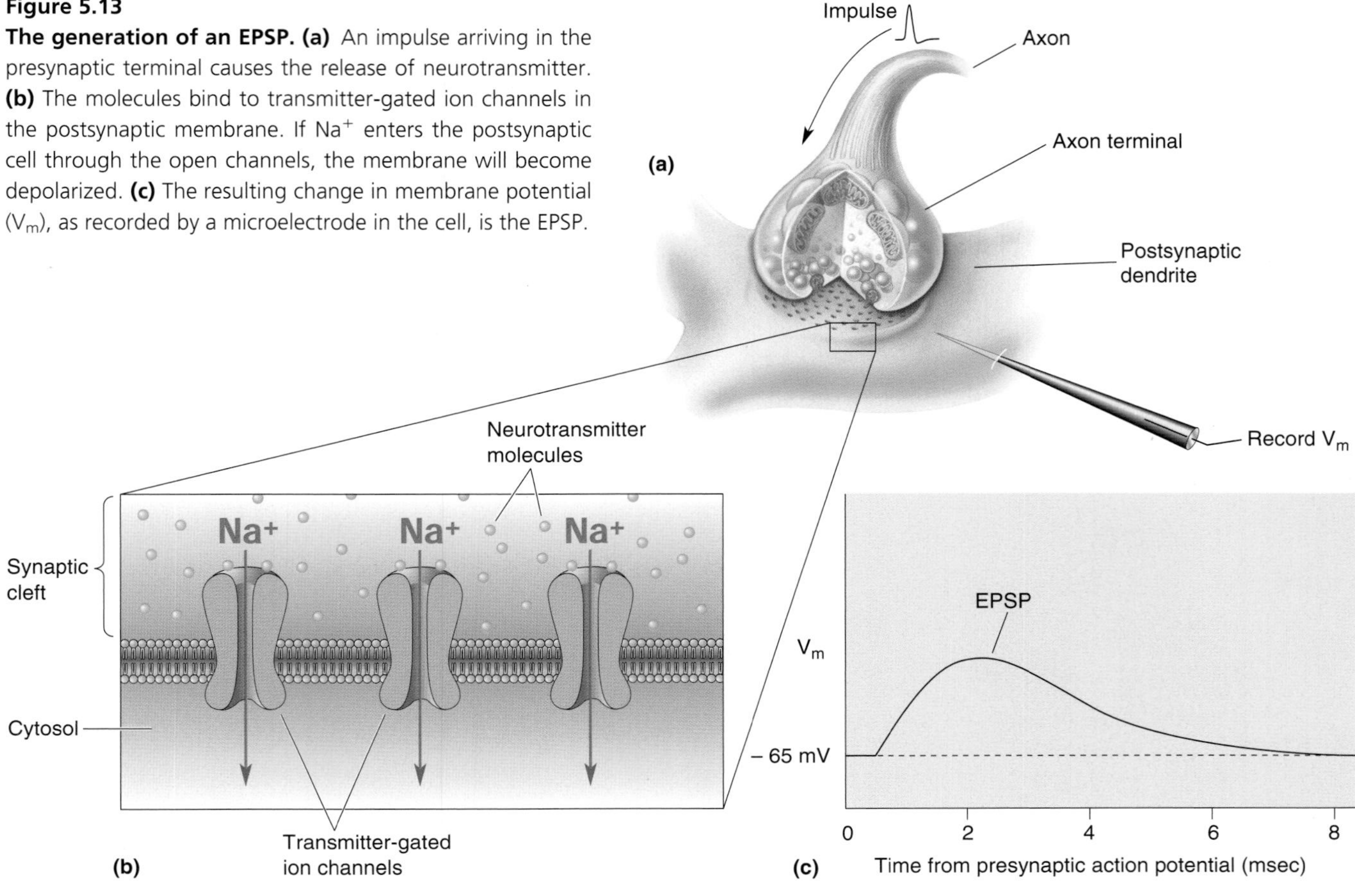

Figure 5.13
The generation of an EPSP. (a) An impulse arriving in the presynaptic terminal causes the release of neurotransmitter. **(b)** The molecules bind to transmitter-gated ion channels in the postsynaptic membrane. If Na^+ enters the postsynaptic cell through the open channels, the membrane will become depolarized. **(c)** The resulting change in membrane potential (V_m), as recorded by a microelectrode in the cell, is the EPSP.

G-protein-coupled receptors, can also have slower, longer-lasting, and much more diverse postsynaptic actions. This type of transmitter action involves three steps:

1. Neurotransmitter molecules bind to receptor proteins embedded in the postsynaptic membrane.
2. The receptor proteins activate small proteins, called **G-proteins,** which are free to move along the intracellular face of the postsynaptic membrane.
3. The activated G-proteins activate "effector" proteins.

Effector proteins can be G-protein-gated ion channels in the membrane (Figure 5.15a), or they can be enzymes that synthesize molecules called **second messengers** that diffuse away in the cytosol (Figure 5.15b). Second messengers can activate additional enzymes in the cytosol that can regulate ion channel function and alter cellular metabolism. Because G-protein-coupled receptors can trigger widespread metabolic effects, they are often called **metabotropic receptors**.

We'll discuss the different neurotransmitters, their receptors, and their effectors in more detail in Chapter 6. You should be aware, however, that the same neurotransmitter can have different postsynaptic actions, depending on what receptors it binds to. An example is the effect of ACh on the heart and on skeletal muscles. ACh slows the rhythmic contractions of the heart by causing a slow hyperpolarization of the cardiac muscle cells. In contrast, in skeletal muscle, ACh induces contraction by causing a rapid depolarization of the muscle fibers. These different actions are explained by different receptors. In the heart, the ACh receptor is coupled by a G-protein to a

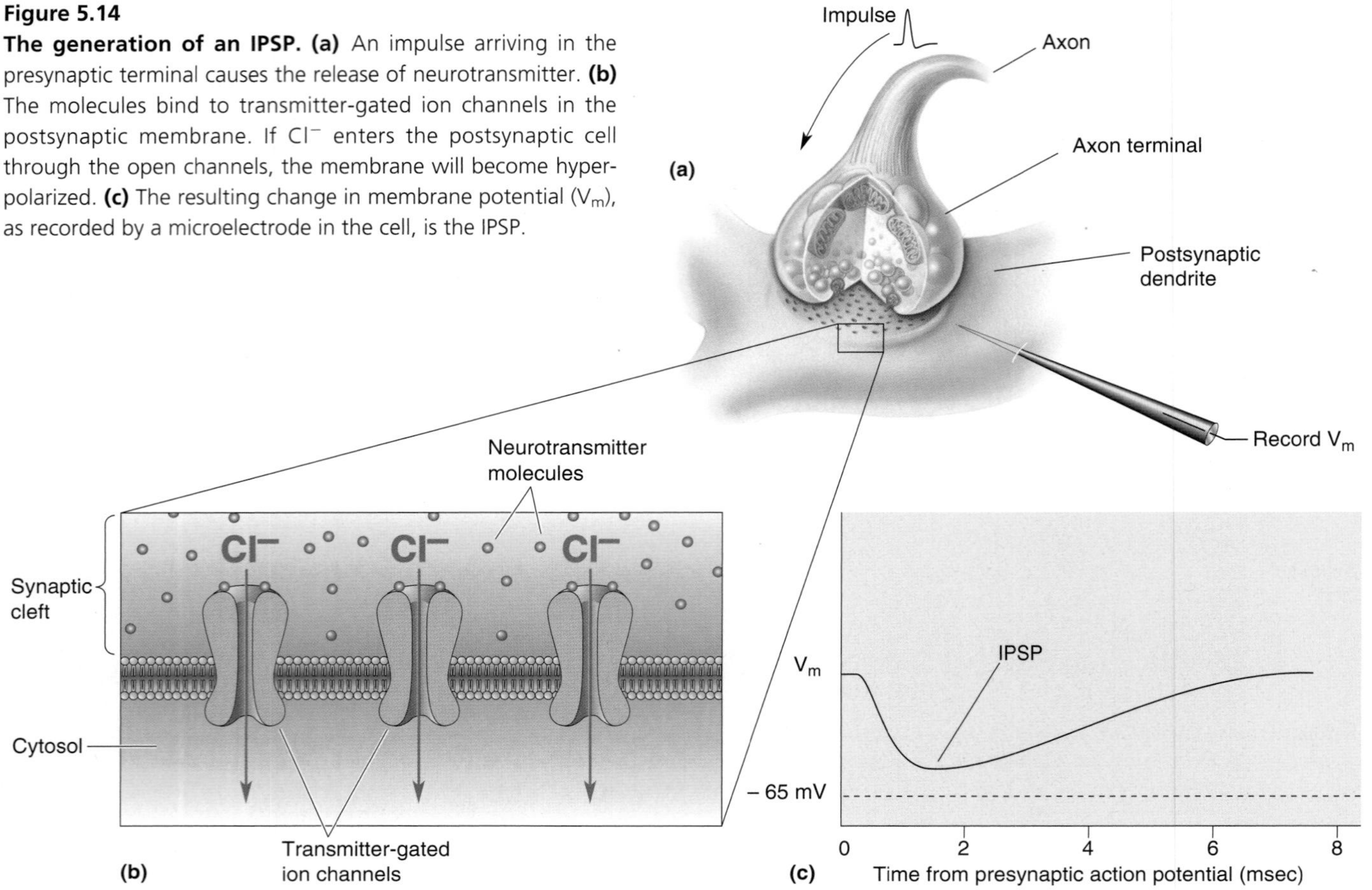

Figure 5.14
The generation of an IPSP. (a) An impulse arriving in the presynaptic terminal causes the release of neurotransmitter. **(b)** The molecules bind to transmitter-gated ion channels in the postsynaptic membrane. If Cl^- enters the postsynaptic cell through the open channels, the membrane will become hyperpolarized. **(c)** The resulting change in membrane potential (V_m), as recorded by a microelectrode in the cell, is the IPSP.

potassium channel. The opening of the potassium channel hyperpolarizes the cardiac muscle fibers. In skeletal muscle, the receptor is an ACh-gated ion channel permeable to Na^+. The opening of this channel depolarizes the muscle fibers.

Autoreceptors. Besides being a part of the postsynaptic density, neurotransmitter receptors are commonly found in the membrane of the presynaptic axon terminal. Presynaptic receptors that are sensitive to the neurotransmitter released by the presynaptic terminal are called **autoreceptors**. Typically, autoreceptors are G-protein-coupled receptors that stimulate second messenger formation. The consequences of activating these receptors vary, but a common effect is inhibition of neurotransmitter release and, in some cases, neurotransmitter synthesis. Such autoreceptors appear to function as a sort of safety valve to reduce release when the concentration of neurotransmitter in the synaptic cleft gets too high.

Neurotransmitter Recovery and Degradation

Once the released neurotransmitter has interacted with postsynaptic receptors, it must be cleared from the synaptic cleft to permit another round of synaptic transmission. One way this happens is by simple diffusion of the transmitter molecules away from the synapse. For most of the amino acid and amine neurotransmitters, however, diffusion is aided by their reuptake into the presynaptic axon terminal. Reuptake occurs by the action of specific neurotransmitter transporter proteins in the presynaptic membrane. Once inside the cytosol of the terminal, the transmitters may be enzymatically destroyed or reloaded into synaptic vesicles. Neurotransmitter transporters

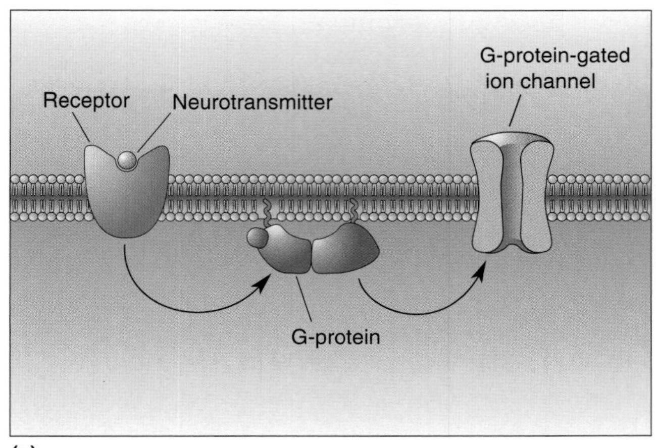

(a)

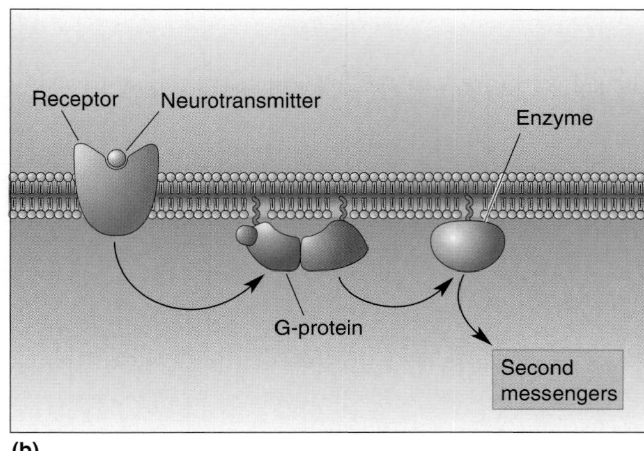

(b)

Figure 5.15
Transmitter actions at G-protein-coupled receptors. The binding of neurotransmitter to the receptor leads to the activation of G-proteins. Activated G-proteins activate effector proteins, which may be **(a)** ion channels or **(b)** enzymes that generate intracellular second messengers.

also exist in the membranes of glia surrounding the synapse, which assist in the removal of neurotransmitter from the cleft.

Another way neurotransmitter action can be terminated is by enzymatic destruction in the synaptic cleft itself. This is how ACh is removed at the neuromuscular junction, for example. The enzyme acetylcholinesterase (AChE) is deposited in the cleft by the muscle cells. AChE cleaves the ACh molecule, rendering it inactive at the ACh receptors.

The importance of transmitter removal from the cleft should not be underestimated. At the neuromuscular junction, for example, uninterrupted exposure to high concentrations of ACh after several seconds leads to a process called *desensitization*, in which the transmitter-gated channels close despite the continued presence of ACh. This desensitized state can persist for many seconds, even after the neurotransmitter is removed. The rapid destruction of ACh by AChE normally prevents desensitization from occurring. If the AChE is inhibited, as it is by various nerve gases used as chemical weapons, however, the ACh receptors will become desensitized and neuromuscular transmission will fail.

Neuropharmacology

Each of the steps of synaptic transmission we have discussed so far—neurotransmitter synthesis, loading into synaptic vesicles, exocytosis, binding and activation of receptors, reuptake, and degradation—is chemical and, therefore, can be affected by specific drugs and toxins (Box 5.5). The study of the effects of drugs on nervous system tissue is called **neuropharmacology**.

Earlier we mentioned that nerve gases can interfere with synaptic transmission at the neuromuscular junction by inhibiting the enzyme AChE. This interference represents one class of drug action, which is to inhibit the normal function of specific proteins involved in synaptic transmission; such drugs are called **inhibitors**. Inhibitors of neurotransmitter receptors, called **receptor antagonists**, bind to the receptors and block (antagonize) the normal action of the transmitter. An example of a receptor antagonist is curare, an arrow-tip poison traditionally used by South American Indians to paralyze their prey. Curare binds tightly to the ACh receptors on skeletal muscle cells and blocks the actions of ACh, thereby preventing muscle contraction.

Box 5.5

OF SPECIAL INTEREST

Bacteria, Spiders, Snakes, and You

What do the bacteria *Clostridium botulinum*, black widow spiders, cobras, and humans have in common? They all produce toxins that attack the chemical synaptic transmission that occurs at the neuromuscular junction. Botulism is caused by several kinds of botulinum neurotoxins that are produced by the growth of *C. botulinum* in improperly canned foods. (The name comes from the Latin word for "sausage" because of the early association of the disease with poorly preserved meat.) Botulinum toxins are potent blockers of neuromuscular transmission; it has been estimated that as few as 10 molecules of the toxins are enough to inhibit a cholinergic synapse. Botulinum toxins are extraordinarily specific enzymes that destroy certain of the SNARE proteins in the presynaptic terminals, which are critical for transmitter release (Box 5.3). Ironically, this specific action of the toxins made them important tools in the early research on SNAREs.

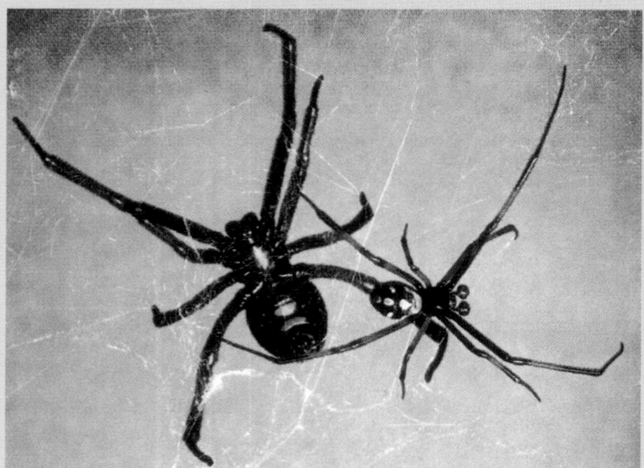

Figure A
Black widow spiders. (Source: Matthews, 1995, p. 174.)

Although its mechanism of action is different, black widow spider venom also exerts deadly effects by affecting transmitter release (Figure A). The venom first increases and then eliminates ACh release at the neuromuscular junction. Electron microscopic examination of synapses poisoned with black widow spider venom reveals that the axon terminals are swollen and the synaptic vesicles are missing. The action of the venom, a protein molecule, is not entirely understood. Venom binds with proteins on the outside of the presynaptic membrane, perhaps forming a membrane pore that depolarizes the terminal and allows Ca^{2+} to enter and trigger rapid and total depletion of transmitter. In some cases, the venom can induce transmitter release even without the need for Ca^{2+}.

The bite of the Taiwanese cobra also results in the blockade of neuromuscular transmission in its victim by yet another mechanism. One of the active compounds in the snake's venom, called α-bungarotoxin, is a peptide molecule that binds so tightly to the postsynaptic nicotinic ACh receptors that it takes days to be removed. Often there is not time for its removal, however, because cobra toxin prevents the activation of nicotinic receptors by ACh, thereby paralyzing the respiratory muscles of its victims.

We humans have synthesized a large number of chemicals that poison synaptic transmission at the neuromuscular junction. Originally motivated by the search for chemical warfare agents, this effort led to the development of a new class of compounds called *organophosphates*. These irreversible inhibitors of AChE, by preventing the degradation of ACh, probably kill their victims by desensitizing ACh receptors. The organophosphates used today as insecticides, like parathion, are toxic to humans only in high doses.

Other drugs bind to receptors, but instead of inhibiting them, they mimic the actions of the naturally occurring neurotransmitter. These drugs are called **receptor agonists**. An example of a receptor agonist is nicotine, derived from the tobacco plant. Nicotine binds to and activates the ACh receptors in skeletal muscle. In fact, the ACh-gated ion channels in muscle are also called **nicotinic ACh receptors**, to distinguish them from other types of ACh receptors, such as those in the heart, that are not activated by nicotine. There are also nicotinic ACh receptors in the CNS, and these are involved in the addictive effects of tobacco use.

The immense chemical complexity of synaptic transmission makes it especially susceptible to the medical corollary of Murphy's Law, which states that if a physiological process can go wrong, it will go wrong. When chemical synaptic transmission goes wrong, the nervous system malfunctions. Defective neurotransmission is believed to be the root cause of a large number of neurological and psychiatric disorders. The good news is that thanks to our growing knowledge of the neuropharmacology of synaptic transmission, clinicians have new and increasingly effective therapeutic drugs for treating these disorders. We'll discuss the synaptic basis of some psychiatric disorders and their neuropharmacological treatment in Chapter 21.

PRINCIPLES OF SYNAPTIC INTEGRATION

Most CNS neurons receive thousands of synaptic inputs that activate different combinations of transmitter-gated ion channels and G-protein-coupled receptors. The postsynaptic neuron integrates all of these complex ionic and chemical signals and gives rise to a simple form of output: action potentials. The transformation of many synaptic inputs to a single neuronal output constitutes a neural computation. The brain performs billions of neural computations every second we are alive. As a first step toward understanding how neural computations are performed, let's explore some basic principles of synaptic integration.

The Integration of EPSPs

The most elementary postsynaptic response is the opening of a single transmitter-gated channel (Figure 5.16). Inward current through these channels depolarizes the postsynaptic membrane, causing the EPSP. The postsynaptic membrane of one synapse may have a few tens to several thousands of transmitter-gated channels; how many are activated during synaptic transmission depends mainly on how much neurotransmitter is released.

Quantal Analysis of EPSPs. The elementary unit of neurotransmitter release is the contents of a single synaptic vesicle. Vesicles each contain about the same number of transmitter molecules (several thousand); the total amount of transmitter released is some multiple of this number. Consequently, the amplitude of the postsynaptic EPSP is some multiple of the response to the contents of a single vesicle. Stated another way, postsynaptic EPSPs at a given synapse are *quantized*; they are multiples of an indivisible unit, the *quantum*, that reflects the number of transmitter molecules in a single synaptic vesicle and the number of postsynaptic receptors available at the synapse.

At many synapses, exocytosis of vesicles occurs at some very low rate in the absence of presynaptic stimulation. The size of the postsynaptic response to this spontaneously released neurotransmitter can be measured electro-

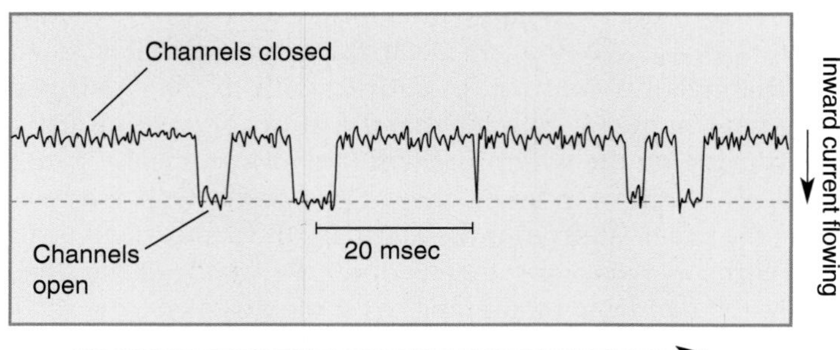

Figure 5.16
A patch-clamp recording from a transmitter-gated ion channel. Ionic current passes through the channels when the channels are open. In the presence of neurotransmitter, they rapidly alternate between open and closed states. (Source: Adapted from Neher and Sakmann, 1992.)

physiologically. This tiny response is a **miniature postsynaptic potential**, often called simply a "mini." Each mini is generated by the transmitter contents of one vesicle. The amplitude of the postsynaptic EPSP evoked by a presynaptic action potential, then, is simply an integer multiple (1×, 2×, 3×, etc.) of the mini-amplitude.

Quantal analysis, a method of comparing the amplitudes of miniature and evoked postsynaptic potentials, can be used to determine how many vesicles release neurotransmitter during normal synaptic transmission. Quantal analysis of transmission at the neuromuscular junction reveals that a single action potential in the presynaptic terminal triggers the exocytosis of about 200 synaptic vesicles, causing an EPSP of 40 mV or more. At many CNS synapses, in striking contrast, the contents of only a *single vesicle* are released in response to a presynaptic action potential, causing an EPSP of only a few tenths of a millivolt.

EPSP Summation. The difference between excitatory transmission at neuromuscular junctions and CNS synapses is not surprising. The neuromuscular junction has evolved to be fail-safe; it must work every time, and the best way to ensure this is to generate an EPSP of a huge size. On the other hand, if every CNS synapse were, by itself, capable of triggering an action potential in its postsynaptic cell (as the neuromuscular junction can), then a neuron would be little more than a simple relay station. Instead, most neurons perform more sophisticated computations, requiring that many EPSPs add together to produce a significant postsynaptic depolarization. This is what is meant by integration of EPSPs.

EPSP summation represents the simplest form of synaptic integration in the CNS. There are two types of summation: spatial and temporal. **Spatial summation** is the adding together of EPSPs generated simultaneously at many different synapses on a dendrite. **Temporal summation** is the adding together of EPSPs generated at the same synapse if they occur in rapid succession, within 1–15 msec of one another (Figure 5.17).

The Contribution of Dendritic Properties to Synaptic Integration

Even with the summation of several EPSPs out on a dendrite, the depolarization still may not be enough to cause a neuron to fire an action potential. The current entering at the sites of synaptic contact must spread down the dendrite and through the soma and cause the membrane of the spike-initiation zone to be depolarized beyond threshold before an action potential can be generated. The effectiveness of an excitatory synapse in triggering an action potential, therefore, depends on how far the synapse is from the spike-initiation zone and on the properties of the dendritic membrane.

Dendritic Cable Properties. To simplify the analysis of how dendritic properties contribute to synaptic integration, let's assume that dendrites function as cylindrical cables that are electrically passive, that is, lacking voltage-gated ion channels (in contrast, of course, with axons). Using an analogy introduced in Chapter 4, imagine that the influx of positive charge at a synapse is like turning on the water that will flow down a leaky garden hose (the dendrite). The water can take either of two paths: One is down the inside of the hose; the other is through the leaks. By the same token, synaptic current can take either of two paths: One is down the inside of the dendrite; the other is across the dendritic membrane. At some distance from the site of current influx, the EPSP amplitude may approach zero because of the dissipation of the current across the membrane.

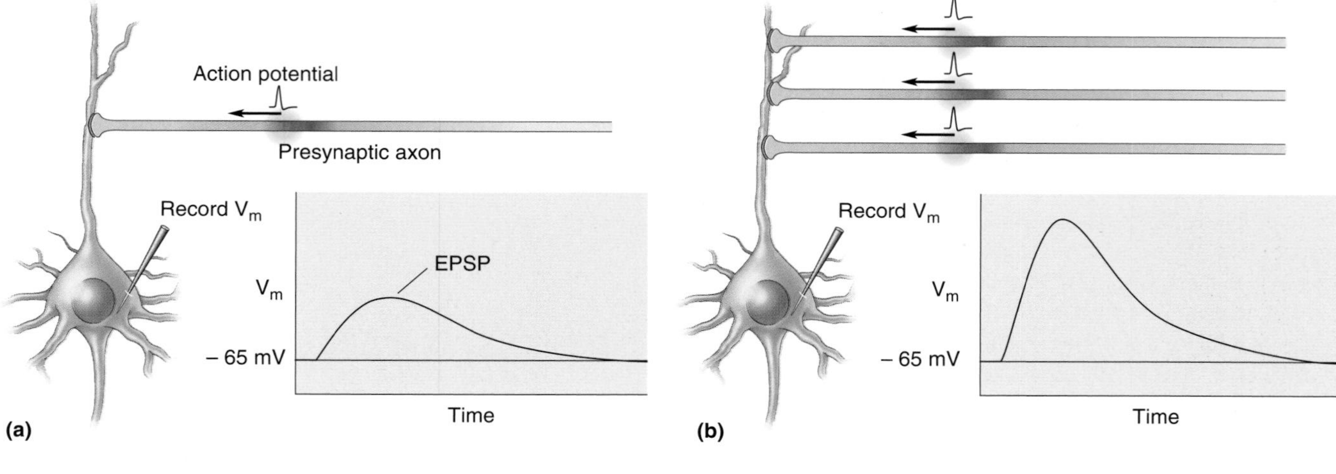

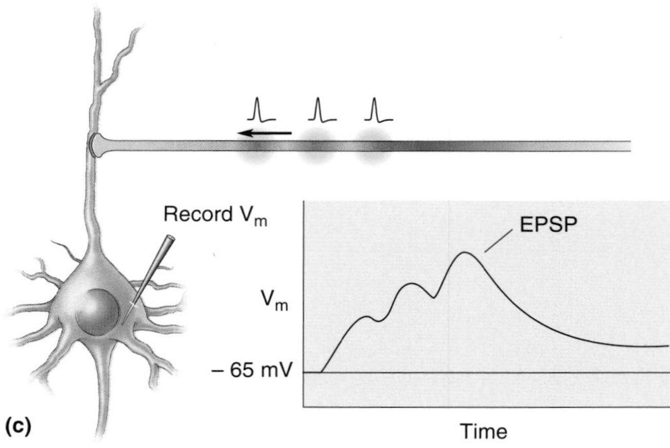

Figure 5.17
EPSP summation. (a) A presynaptic action potential triggers a small EPSP in a postsynaptic neuron. **(b)** Spatial summation of EPSPs: When two or more presynaptic inputs are active at the same time, their individual EPSPs add together. **(c)** Temporal summation of EPSPs: When the same presynaptic fiber fires action potentials in quick succession, the individual EPSPs add together.

The decrease in depolarization as a function of distance along a dendritic cable is plotted in Figure 5.18. To simplify the mathematics, in this example we will assume that the dendrite is infinitely long, unbranched, and uniform in diameter. Notice that the amount of depolarization falls off exponentially with increasing distance. Depolarization of the membrane at a given distance (V_x) can be described by the equation $V_x = V_o/e^{x/\lambda}$, where V_o is depolarization at the origin (just under the synapse), e (= 2.718 . . .) is the base of natural logarithms, x is the distance from the synapse, and λ is a constant that depends on the properties of the dendrite. Notice that when $x = \lambda$, $V_x = V_o/e$. Put another way, $V_\lambda \approx 0.37\ (V_o)$. This distance λ, where the depolarization is 37% of that at the origin, is called the dendritic **length constant**. (Remember that this analysis is an oversimplification. Real dendrites have finite lengths and branches and tend to taper, all of which also affect the spread of current and thus the effectiveness of synaptic potentials.)

The length constant is an index of how far depolarization can spread down a dendrite or axon. The longer the length constant, the more likely it is that EPSPs generated at distant synapses will depolarize the membrane at the axon hillock. The value of λ in our idealized, electrically passive dendrite depends on two factors: (1) the resistance to current flowing longitudinally down the dendrite, called the **internal resistance** (r_i), and (2) the resistance to current flowing across the membrane, called the **membrane resistance** (r_m). Most current will take the path of least resistance; therefore, the value of λ will increase as membrane resistance increases, because more depolarizing current will flow down the inside of the dendrite. The value of λ will decrease

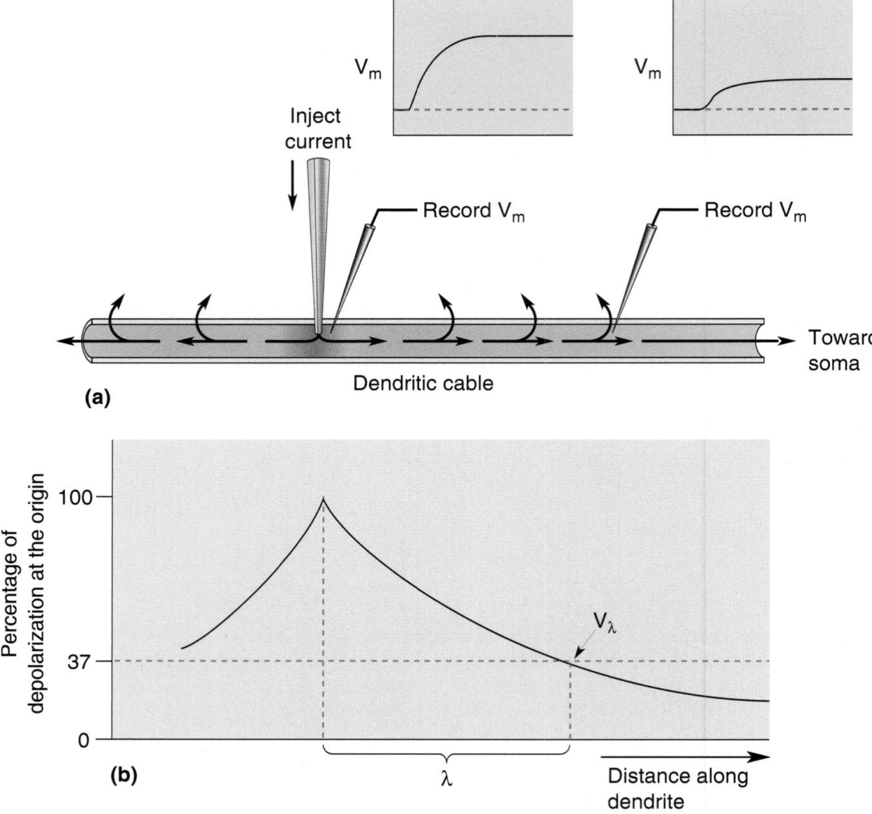

Figure 5.18
Decreasing depolarization as a function of distance along a long dendritic cable. (a) Current is injected into a dendrite, and the depolarization is recorded. As this current spreads down the dendrite, much of it dissipates across the membrane. Therefore, the depolarization measured at a distance from the site of current injection is smaller than that measured right under it. **(b)** A plot of membrane depolarization as a function of distance along the dendrite. At the distance λ, one length constant, the membrane depolarization (V_λ), is 37% of that at the origin.

as internal resistance increases, because more current will flow across the membrane. Just as water will flow farther down a wide hose with few leaks, synaptic current will flow farther down a wide dendrite (low r_i) with few open membrane channels (high r_m).

The internal resistance depends only on the diameter of the dendrite and the electrical properties of the cytoplasm; consequently, it is relatively constant in a mature neuron. The membrane resistance, in contrast, depends on the number of open ion channels, which changes from moment to moment depending on what other synapses are active. The dendritic length constant, therefore, is not constant at all! As we will see in a moment, fluctuations in the value of λ are an important factor in synaptic integration.

Excitable Dendrites. Our analysis of dendritic cable properties made another important assumption: The dendrite's membrane is electrically passive and lacks voltage-gated channels. Some dendrites in the brain have nearly passive and inexcitable membranes and thus follow the simple cable equations. The dendrites of spinal motor neurons, for example, are very close to passive. However, many other neuronal dendrites are decidedly not passive (Figure 5.19). A variety of neurons have dendrites with significant numbers of voltage-gated sodium, calcium, and potassium channels. There are usually not enough of these dendritic channels to generate fully propagating action potentials, as in axons. But the voltage-gated channels in dendrites can act as important amplifiers of small postsynaptic potentials generated far out on dendrites. EPSPs that would diminish to near nothingness in a long, passive dendrite may nevertheless be large enough to trigger the opening of voltage-gated sodium channels, which in turn would add current to boost the synaptic signal along toward the soma.

Paradoxically, dendritic sodium channels in some cells may also serve to carry electrical signals in the other direction—from the soma outward along

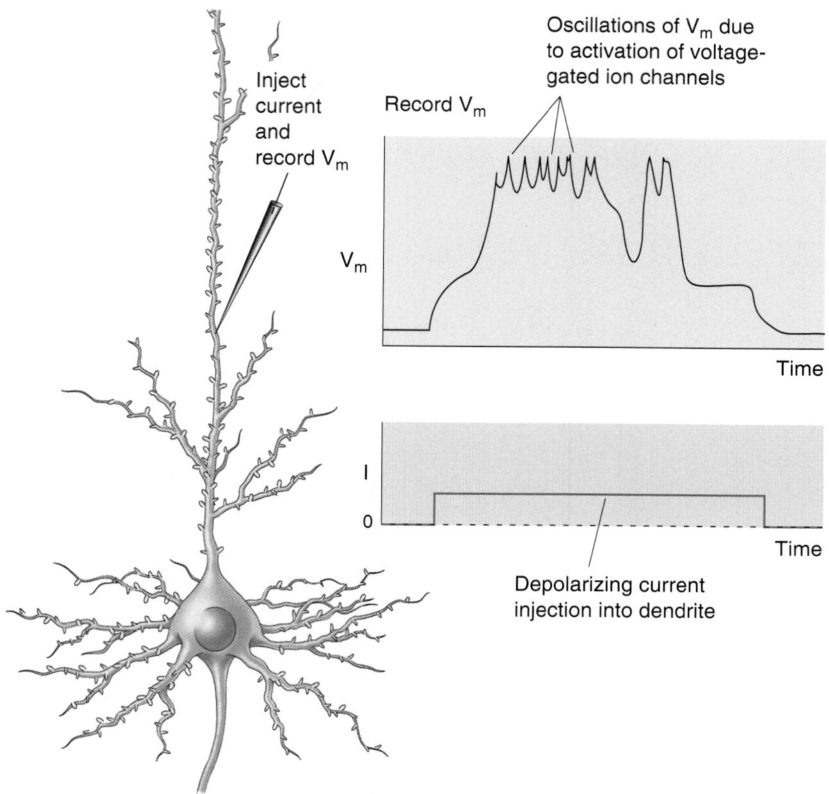

Figure 5.19
Excitable dendrites. Pyramidal cells in the cerebral cortex often have long vertical apical dendrites. When these dendrites are injected with depolarizing current, many generate complex oscillations of membrane potential because of the voltage-gated sodium and calcium channels in their membrane.

dendrites. This may be a mechanism by which synapses on dendrites are informed that a spike occurred in the soma, and it has relevance for hypotheses about the cellular mechanisms of learning that will be discussed in Chapter 24.

Inhibition

So far, we've seen that whether an EPSP contributes to the action potential output of a neuron depends on several factors, including the number of coactive excitatory synapses, the distance of the synapse from the spike-initiation zone, and the properties of the dendritic membrane. Of course, not all synapses in the brain are excitatory. The action of some synapses is to take the membrane potential away from action potential threshold; these are called inhibitory synapses. Inhibitory synapses exert powerful control over a neuron's output (Box 5.6).

IPSPs and Shunting Inhibition. The postsynaptic receptors under most inhibitory synapses are very similar to those under excitatory synapses; they're transmitter-gated ion channels. The only important differences are that they bind different neurotransmitters (either GABA or glycine) and that they allow different ions to pass through their channels. The transmitter-gated channels of most inhibitory synapses are permeable to only one natural ion, Cl^-. Opening of the chloride channel allows Cl^- ions to cross the membrane in a direction that brings the membrane potential toward the chloride equilibrium potential, E_{Cl}, about −65 mV. If the membrane potential were less negative than −65 mV when the transmitter was released, activation of these channels would cause a hyperpolarizing IPSP.

Notice that if the resting membrane potential were already −65 mV, no IPSP would be visible after chloride channel activation because the value of the membrane potential would already equal E_{Cl} (i.e., the reversal potential

Box 5.6

OF SPECIAL INTEREST

Startling Mutations

A flash of lightning . . . a thunderclap . . . a tap on the shoulder when you *think* you're alone! If you are not expecting them, any of these stimuli can make you jump, grimace, hunch your shoulders, and breathe faster. We all know the brief but dramatic nature of the startle response.

Luckily, when lightning strikes twice or a friend taps our shoulder again, we tend to be much less startled the second time. We quickly habituate and relax. For an unfortunate minority of mice, cows, dogs, horses, and people, however, life is a succession of exaggerated startle responses. Even normally benign stimuli, such as hands clapping or a touch to the nose, may trigger an uncontrollable stiffening of the body, an involuntary shout, flexion of the arms and legs, and a fall to the ground. Worse yet, these overdone responses don't adapt when the stimuli are repeated. The clinical term for startle disease is *hyperekplexia*,, and the first recorded cases were in members of a community of French-Canadian lumberjacks in 1878. Hyperekplexia is an inherited condition, seen worldwide, and its sufferers are known by colorful local names: the "Jumping Frenchmen of Maine" (Quebec), "myriachit" (Siberia), "latah" (Malaysia), and "Ragin' Cajuns" (Louisiana).

We now know the molecular basis for two general types of startle diseases. Remarkably, both involve defects of inhibitory glycine receptors. The first type, identified in humans and in a mutant mouse called *spasmodic*, is caused by a mutation of a gene for the glycine receptor. The change is the smallest one possible—the abnormal receptors have only one amino acid (of more than 400) coded wrong—but the result is a chloride channel that opens less frequently when exposed to the neurotransmitter glycine. The second type of startle disease is seen in the mutant mouse *spastic* and in a strain of cattle. In these animals, normal glycine receptors are expressed, but in fewer than normal numbers. The two forms of startle disease thus take different routes to the same unfortunate end: The transmitter glycine is less effective at inhibiting neurons in the spinal cord and brain stem.

Most neural circuits depend on a delicate balance of synaptic excitation and inhibition for normal functioning. If excitation is increased or inhibition reduced, then a turbulent and hyperexcitable state may result. An impairment of glycine function yields exaggerated startles; reduced GABA function can lead to the seizures of epilepsy (as discussed in Chapter 19). How can such diseases be treated? There is often a clear and simple logic. Drugs that enhance inhibition can be very helpful.

for that synapse; see Box 5.4). If there is no visible IPSP, is the neuron really inhibited? The answer is yes. Consider the situation illustrated in Figure 5.20, with an excitatory synapse on a distal segment of dendrite and an inhibitory synapse on a proximal segment of dendrite, near the soma. Activation of the excitatory synapse leads to the influx of positive charge into the dendrite. This current depolarizes the membrane as it flows toward the soma. At the site of the active inhibitory synapse, however, the membrane potential is approximately equal to E_{Cl}, -65 mV. Positive current therefore flows outward across the membrane at this site to bring V_m to -65 mV. This synapse acts as an electrical shunt, preventing the current from flowing through the soma to the axon hillock. This type of inhibition is called **shunting inhibition**. The actual physical basis of shunting inhibition is the *inward movement of negatively charged chloride ions,* which is formally equivalent to *outward positive current* flow. Shunting inhibition is like cutting a big hole in the leaky garden hose—all of the water flows down this path of least resistance out of the hose before it gets to the nozzle, where it can "activate" the flowers in your garden.

Thus, you can see that the action of inhibitory synapses also contributes to synaptic integration. The IPSPs can be subtracted from EPSPs, making the postsynaptic neuron less likely to fire action potentials. In addition, shunting

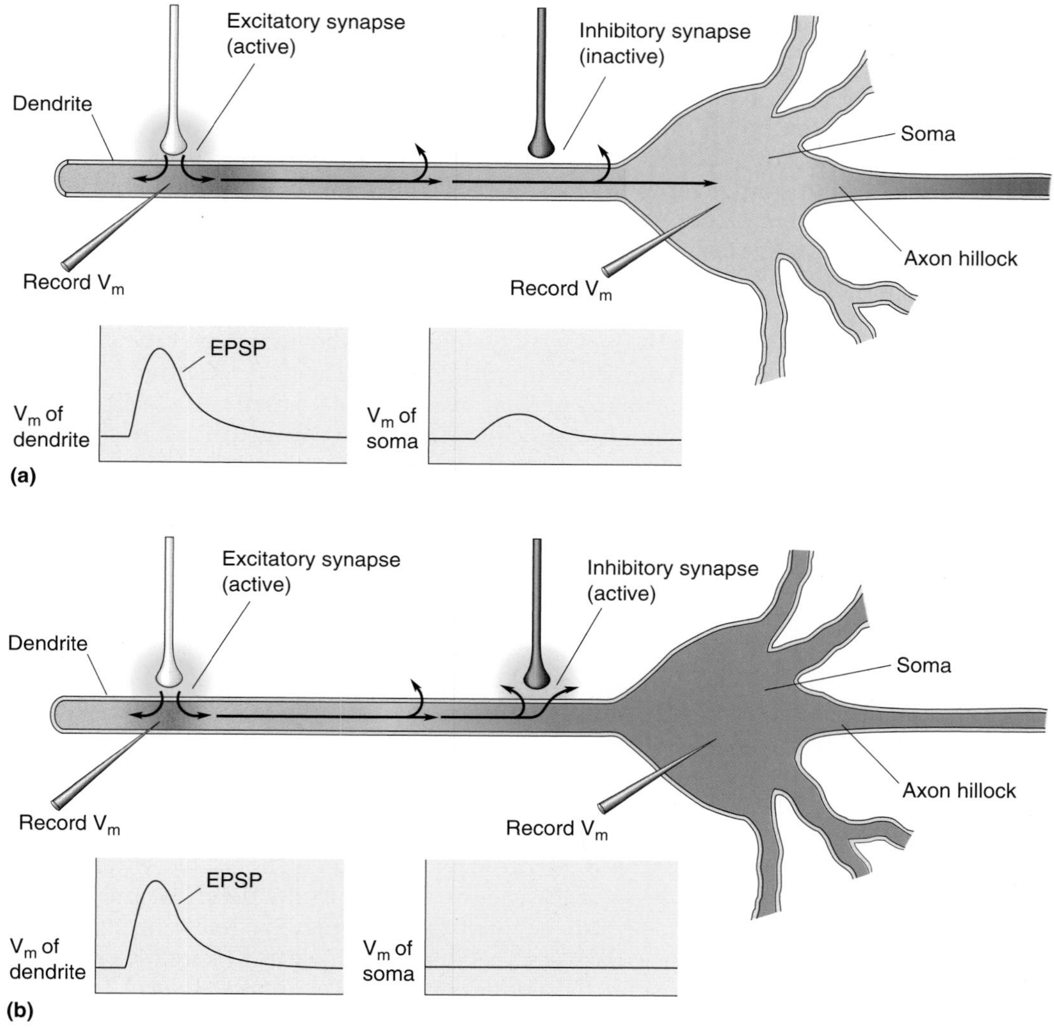

Figure 5.20
Shunting inhibition. A neuron receiving one excitatory and one inhibitory input. **(a)** Stimulation of the excitatory input causes inward postsynaptic current that spreads to the soma, where it can be recorded as an EPSP. **(b)** When the inhibitory and excitatory inputs are stimulated together, the depolarizing current leaks out before it reaches the soma.

inhibition acts to drastically reduce r_m and, consequently, λ, thus allowing positive current to flow out across the membrane instead of internally down the dendrite toward the spike-initiation zone.

The Geometry of Excitatory and Inhibitory Synapses. Inhibitory synapses in the brain that use GABA or glycine as a neurotransmitter always have a morphology characteristic of Gray's type II (Figure 5.6). This structure contrasts with excitatory synapses that use glutamate, which always have a Gray's type I morphology. This correlation between structure and function has been useful for working out the geometric relationships among excitatory and inhibitory synapses on individual neurons. In addition to being spread over the dendrites, inhibitory synapses on many neurons are found clustered on the soma and near the axon hillock, where they are in an especially powerful position to influence the activity of the postsynaptic neuron.

Modulation

Most of the postsynaptic mechanisms we've discussed so far involve transmitter receptors that are themselves ion channels. To be sure, synapses with transmitter-gated channels carry the bulk of the specific information that is processed by the nervous system. Many synapses with G-protein-coupled neurotransmitter receptors, however, are not directly associated with an ion channel. Synaptic activation of these receptors does not directly evoke EPSPs and IPSPs but instead *modifies* the effectiveness of EPSPs generated by other synapses with transmitter-gated channels. This type of synaptic transmission is called **modulation**. We'll give you a taste for how modulation influences synaptic integration by exploring the effects of activating one type of G-protein-coupled receptor in the brain, the norepinephrine β receptor.

The binding of the amine neurotransmitter **norepinephrine (NE)** to the β receptor triggers a cascade of biochemical events within the cell. In short, the β receptor activates a G-protein that, in turn, activates an effector protein, the intracellular enzyme adenylyl cyclase. **Adenylyl cyclase** catalyzes the chemical reaction that converts adenosine triphosphate (ATP), the product of oxidative metabolism in the mitochondria, into a compound called **cyclic adenosine monophosphate**, or **cAMP**, that is free to diffuse within the cytosol. Thus, the *first* chemical message of synaptic transmission (the release of NE into the synaptic cleft) is converted by the β receptor into a *second* message (cAMP); cAMP is an example of a second messenger.

The effect of cAMP is to stimulate another enzyme called a protein kinase. **Protein kinases** catalyze a chemical reaction called **phosphorylation**, the transfer of phosphate groups (PO_3) from ATP to specific sites on cell proteins (Figure 5.21). The significance of phosphorylation is that it can change the conformation of a protein, thereby changing that protein's activity.

In some neurons, one of the proteins that is phosphorylated when cAMP rises is a particular type of potassium channel in the dendritic membrane. Phosphorylation causes this channel to close, thereby reducing the membrane K^+ conductance. By itself, this does not cause any dramatic effects on the neuron. But consider the wider consequence: *decreasing the K^+ conductance increases the dendritic membrane resistance and therefore increases the length constant.* It is like wrapping the leaky garden hose in duct tape; more water can flow down the inside of the hose and less leaks out the sides. As a consequence of increasing λ, distant or weak excitatory synapses will become more effective in depolarizing the spike-initiation zone beyond threshold; the cell will become *more excitable*. Thus, the binding of NE to β receptors produces

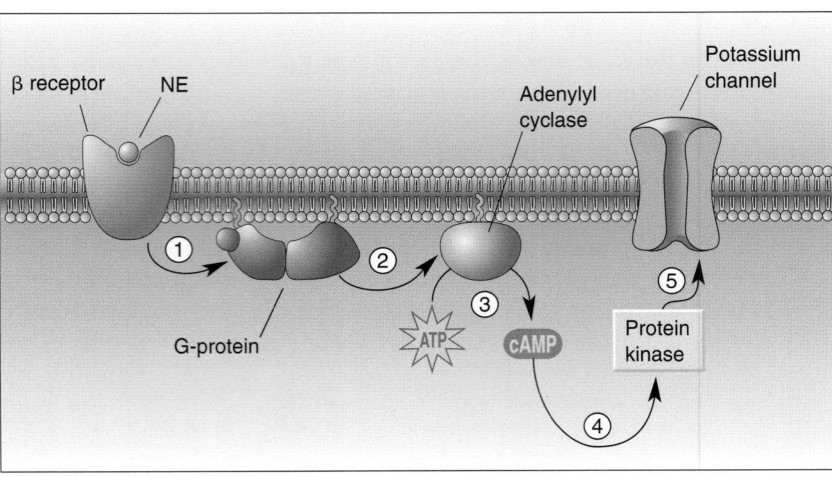

Figure 5.21
Modulation by the NE β receptor. ① The binding of NE to the receptor activates a G-protein in the membrane. ② The G-protein activates the enzyme adenylyl cyclase. ③ Adenylyl cyclase converts ATP into the second messenger cAMP. ④ cAMP activates a protein kinase. ⑤ The protein kinase causes a potassium channel to close by attaching a phosphate group to it.

little change in membrane potential but greatly increases the response produced by another neurotransmitter at an excitatory synapse. Because this effect involves several biochemical intermediaries, it can last far longer than the presence of the modulatory transmitter itself.

We have described one particular G-protein-coupled receptor and the consequences of activating it in one type of neuron. But it is important to recognize that other types of receptor can lead to the formation of other types of second messenger molecules. Activation of any of these receptor types will initiate a distinct cascade of biochemical reactions in the postsynaptic neuron that do not always include phosphorylation and decreases in membrane conductance. In fact, cAMP in a different cell type with different enzymes may produce functionally opposite changes in the excitability of cells.

In Chapter 6, we will describe more examples of synaptic modulation and their mechanisms. You can already see, however, that modulatory forms of synaptic transmission offer an almost limitless number of ways that information encoded by presynaptic impulse activity can be transformed and used by the postsynaptic neuron.

CONCLUDING REMARKS

In this chapter, we have covered the basic principles of chemical synaptic transmission. The action potential that arose in the sensory nerve when you stepped on that thumbtack in Chapter 3 and swept up the axon in Chapter 4 has now reached the axon terminal in the spinal cord. The depolarization of the terminal triggered the presynaptic entry of Ca^{2+} through voltage-gated calcium channels, which then stimulated exocytosis of the contents of synaptic vesicles. Liberated neurotransmitter diffused across the synaptic cleft and attached to specific receptors in the postsynaptic membrane. The transmitter (probably glutamate) caused transmitter-gated channels to open, which allowed positive charge to enter the postsynaptic dendrite. Because the sensory nerve was firing action potentials at a high rate and because many synapses were activated at the same time, the EPSPs summed to bring the spike-initiation zone of the postsynaptic neuron to threshold, and this cell then generated action potentials. If the postsynaptic cell were a motor neuron, this activity would cause the release of ACh at the neuromuscular junction and muscle contraction. If the postsynaptic cell were an interneuron that used GABA as a neurotransmitter, the activity of the cell would result in inhibition of its synaptic targets. If this cell used a modulatory transmitter such as NE, the activity could cause lasting changes in the excitability or metabolism of its synaptic targets. It is this rich diversity of chemical synaptic interactions that allows complex behaviors (such as shrieking with pain as you jerk up your foot) to emerge from simple stimuli (such as stepping on a thumbtack).

Although we surveyed chemical synaptic transmission in this chapter, we did not cover the *chemistry* of synaptic transmission in any detail. In Chapter 6, we'll take a closer look at the chemical "nuts and bolts" of different neurotransmitter systems. In Part III, after we've examined the sensory and motor systems, we'll explore the contributions of several different neurotransmitters to nervous system function and behavior. You'll see that the chemistry of synaptic transmission warrants all of this attention because defective neurotransmission is the basis for many neurological and psychiatric disorders. And virtually all psychoactive drugs, both therapeutic and illicit, exert their effects at chemical synapses.

In addition to providing explanations for aspects of neural information processing and the effects of drugs, knowledge of chemical synaptic transmission is the key to understanding the neural basis of learning and memory.

Memories are established by modification of the effectiveness of chemical synapses in the brain. Material discussed in this chapter suggests possible sites of modification, ranging from changes in presynaptic Ca^{2+} entry and neurotransmitter release to alterations in postsynaptic receptors or excitability. As we shall see in Chapter 24, all of these changes are likely to contribute to the storage of information by the nervous system.

KEY TERMS

Introduction
electrical synapse (p. 99)
chemical synapse (p. 99)

Types of Synapses
gap junction (p. 100)
secretory granule (p. 102)
large, dense-core vesicle (p. 102)
membrane differentiation (p. 102)
active zone (p. 102)
postsynaptic density (p. 102)
Gray's type I synapse (p. 103)
Gray's type II synapse (p. 103)
neuromuscular junction (p. 103)
motor end-plate (p. 105)

Principles of Chemical Synaptic Transmission
glutamate (Glu) (p. 107)
gamma-aminobutyric acid (GABA) (p. 107)
glycine (Gly) (p. 107)
acetylcholine (ACh) (p. 107)
transporter (p. 109)
voltage-gated calcium channel (p. 109)
exocytosis (p. 109)
endocytosis (p. 109)
transmitter-gated ion channel (p. 112)
excitatory postsynaptic potential (EPSP) (p. 113)
inhibitory postsynaptic potential (IPSP) (p. 113)
G-protein-coupled receptor (p. 115)
G-protein (p. 115)
second messenger (p. 115)
metabotropic receptor (p. 115)
autoreceptor (p. 116)
neuropharmacology (p. 117)
inhibitor (p. 117)
receptor antagonist (p. 117)
receptor agonist (p. 118)
nicotinic ACh receptor (p. 118)

Principles of Synaptic Integration
miniature postsynaptic potential (p. 120)
quantal analysis (p. 120)
EPSP summation (p. 120)
spatial summation (p. 120)
temporal summation (p. 120)
length constant (p. 121)
internal resistance (p. 121)
membrane resistance (p. 121)
shunting inhibition (p. 124)
modulation (p. 126)
norepinephrine (NE) (p. 126)
adenylyl cyclase (p. 126)
cyclic adenosine monophosphate (cAMP) (p. 126)
protein kinase (p. 126)
phosphorylation (p. 126)

REVIEW QUESTIONS

1. What is meant by quantal release of neurotransmitter?
2. You apply ACh and activate nicotinic receptors on a muscle cell. Which way will current flow through the receptor channels when $V_m = -60$ mV? When $V_m = 0$ mV? When $V_m = 60$ mV? Why?
3. In this chapter we discussed a GABA-gated ion channel that is permeable to Cl^-. GABA also activates a G-protein-coupled receptor called the $GABA_B$ receptor, which causes potassium-selective channels to open. What effect would $GABA_B$ receptor activation have on the membrane potential?
4. You think you have discovered a new neurotransmitter, and you are studying its effect on a neuron. The reversal potential for the response caused by the new chemical is −60 mV. Is this substance excitatory or inhibitory? Why?
5. A drug called strychnine, isolated from the seeds of a tree native to India and commonly used as rat poison, blocks the effects of glycine. Is strychnine an agonist or an antagonist of the glycine receptor?
6. How does nerve gas cause respiratory paralysis?
7. Why is an excitatory synapse on the soma more effective in evoking action potentials in the postsynaptic neuron than an excitatory synapse on the tip of a dendrite?
8. What are the steps that lead to increased excitability in a neuron when NE is released presynaptically?

CHAPTER 6

Neurotransmitter Systems

INTRODUCTION

The normal human brain is an orderly set of chemical reactions. As we have seen, some of the brain's most important chemical reactions are those associated with synaptic transmission. Chapter 5 introduced the general principles of chemical synaptic transmission, using a few specific neurotransmitters as examples. In this chapter, we will explore in more depth the variety and elegance of the major neurotransmitter systems.

Neurotransmitter systems begin with neurotransmitters. In Chapter 5 we discussed the three major classes: *amino acids*, *amines*, and *peptides*. Even a partial list of the known transmitters, such as that appearing in Table 5.1, has over 20 molecules. Each of these molecules can define a particular transmitter system. In addition to the molecule itself, a neurotransmitter system includes all the molecular machinery responsible for transmitter synthesis, vesicular packaging, reuptake and degradation, and transmitter action (Figure 6.1).

The first molecule positively identified as a neurotransmitter by Otto Loewi in the 1920s was acetylcholine, or ACh (see Box 5.1). To describe the cells that produce and release ACh, British pharmacologist Henry Dale introduced the term **cholinergic**. (Dale shared the 1936 Nobel Prize with Loewi, in recognition of his neuropharmacological studies of synaptic transmission.) Dale termed the neurons that use the amine neurotransmitter norepinephrine (NE) **noradrenergic** (NE is known as noradrenaline in Great Britain). The convention of using the suffix *-ergic* continued when additional transmitters were identified. Therefore, today we speak of **glutamatergic** synapses that use glutamate, **GABAergic** synapses that use gamma-

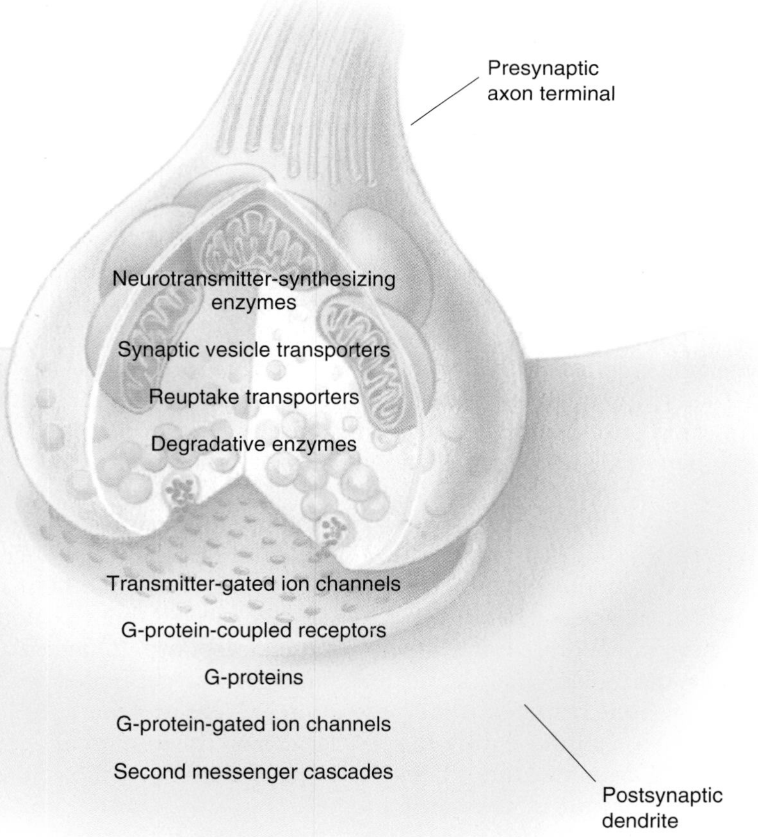

Figure 6.1
Elements of neurotransmitter systems.

aminobutyric acid (GABA), **peptidergic** synapses that use peptides, and so on. These adjectives are also used to identify the various neurotransmitter systems. Thus, for example, ACh and all the molecular machinery associated with it are collectively called the *cholinergic system*.

With this terminology in hand, we can begin our exploration of the neurotransmitter systems. We start with a discussion of the experimental strategies that have been used to study neurotransmitter systems. Then we will look at the synthesis and metabolism of specific neurotransmitters and explore how these molecules exert their postsynaptic effects. In Part III, after we have learned more about the structural and functional organization of the nervous system, we'll take another look at specific neurotransmitter systems in the context of their individual contributions to the regulation of brain function and behavior.

STUDYING NEUROTRANSMITTER SYSTEMS

The first step in studying a neurotransmitter system is usually identifying the neurotransmitter. This is no simple task; the brain contains uncountable different chemicals. How can we decide which few chemicals are used as transmitters?

Over the years, neuroscientists have established certain criteria that must be met for a molecule to be considered a neurotransmitter:

1. The molecule must be synthesized and stored in the presynaptic neuron.
2. The molecule must be released by the presynaptic axon terminal upon stimulation.
3. The molecule, when experimentally applied, must produce a response in the postsynaptic cell that mimics the response produced by the release of neurotransmitter from the presynaptic neuron.

Let's begin by exploring some of the strategies and methods that are used to satisfy these criteria.

Localization of Transmitters and Transmitter-Synthesizing Enzymes

The investigator often begins with little more than a hunch that a particular molecule may be a neurotransmitter. This idea may be based on observing that the molecule is concentrated in brain tissue or that the application of the molecule to certain neurons alters their action potential firing rate. Whatever the inspiration, the first step in confirming the hypothesis is to show that the molecule is, in fact, localized in and synthesized by particular neurons. Many methods have been used to satisfy this criterion for different neurotransmitters. Two of the most important techniques used today are immunocytochemistry and *in situ* hybridization.

Immunocytochemistry. The method of **immunocytochemistry** is used to anatomically localize particular molecules to particular cells. The principle behind the method is quite simple (Figure 6.2). Once the neurotransmitter candidate has been chemically purified, it is injected into the bloodstream of an animal, where it stimulates an immune response. (Often, to evoke a response, the molecule is chemically coupled to a larger molecule.) One aspect of the immune response is the generation of large proteins called antibodies. Antibodies can bind tightly to specific sites on the foreign molecule—in this case, the transmitter candidate. These specific antibody molecules can be recovered from a blood sample of the immunized animal and chemically tagged with a colorful marker that can be seen with a microscope. When

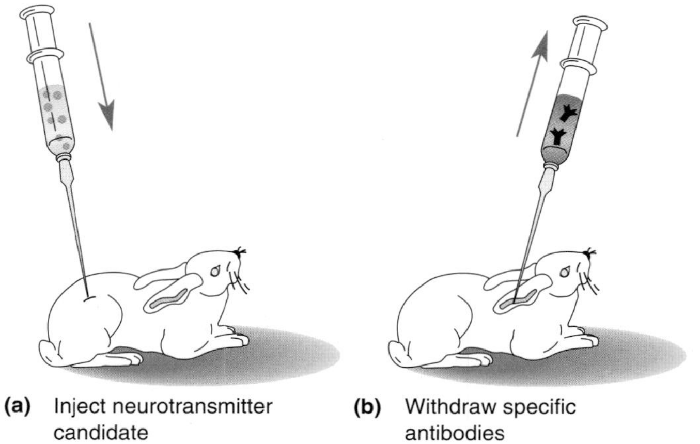

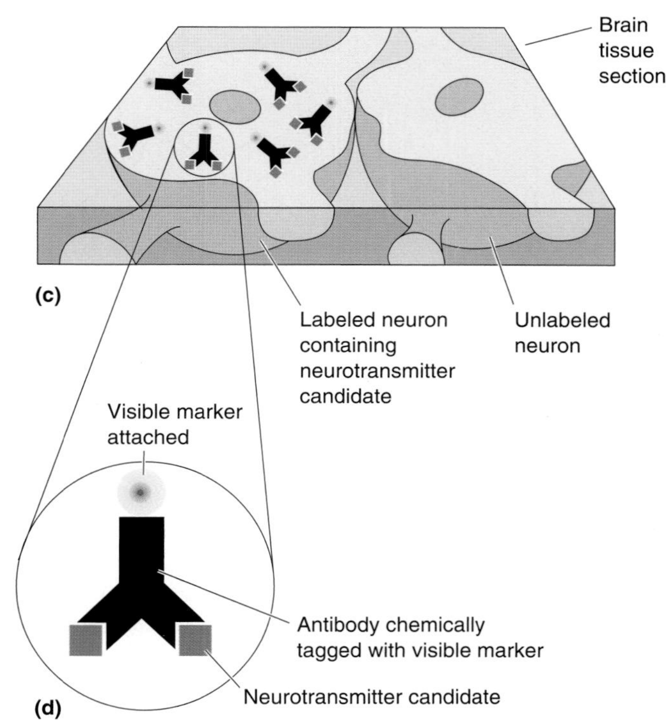

Figure 6.2
Immunocytochemistry. This method uses labeled antibodies to identify the location of molecules within cells. **(a)** The molecule of interest, a neurotransmitter candidate, is injected into an animal, causing an immune response and the generation of antibodies. **(b)** Blood is withdrawn from the animal, and the antibodies are isolated from the serum. **(c)** The antibodies are tagged with a visible marker and applied to sections of brain tissue. The antibodies label only cells that contain the neurotransmitter candidate. **(d)** A close-up of the complex that includes the neurotransmitter candidates, an antibody, and its visible marker.

these labeled antibodies are applied to a section of brain tissue, they will color just those cells that contain the transmitter candidate (Figure 6.3).

Immunocytochemistry can be used to localize any molecule for which a specific antibody can be generated, including the synthesizing enzymes for transmitter candidates. Demonstration that the transmitter candidate and its synthesizing enzyme are contained in the same neuron—or better yet, in the same axon terminal—can help satisfy the criterion that the molecule be localized in and synthesized by a particular neuron.

***In Situ* Hybridization.** The method known as ***in situ* hybridization** is also useful for confirming that a cell synthesizes a particular protein or peptide. Recall from Chapter 2 that proteins are assembled by the ribosomes according to instructions from specific messenger ribonucleic acid (mRNA) molecules. There is a unique mRNA molecule for every polypeptide synthesized by a neuron. The mRNA transcript consists of four different nucleic acids linked together in various sequences to form a strand. Each nucleic acid has the unusual property that it will bind tightly to one other complementary nucleic acid. Thus, if the sequence of nucleic acids in a strand of mRNA is known, it is possible to construct in the lab a complementary strand that will stick, like a strip of Velcro, to the mRNA molecule. The complementary strand is called a *probe*, and the process by which the probe bonds to the mRNA molecule is called *hybridization* (Figure 6.4). To see if the mRNA for a particular peptide is localized in a neuron, we chemically label the appropriate probe so it can be detected, apply it to a section of brain tissue, allow time for the probes to stick to any complementary mRNA strands, then wash

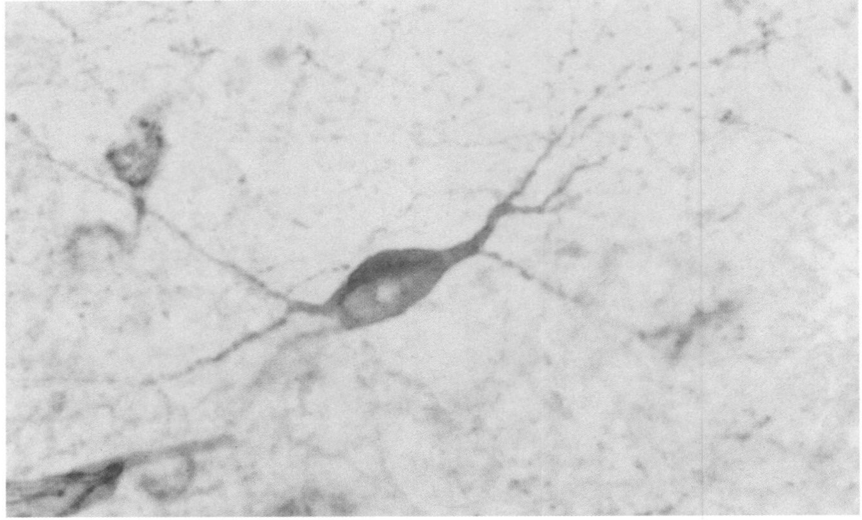

Figure 6.3
Immunocytochemical localization of a peptide neurotransmitter in neurons. (Source: Courtesy of Dr.Y. Amitai and S. L. Patrick.)

away all the extra probes that have not stuck. Finally, we search for neurons that contain the label.

For *in situ* hybridization, the probes are usually labeled by making them radioactive. Since we cannot see radioactivity, hybridized probes are detected by laying the brain tissue on a sheet of special film that is sensitive to radioactive emissions. After exposure to the tissue, the film is developed like a photograph, and negative images of the radioactive cells are visible as clusters of small dots (Figure 6.5). This technique for viewing the distribution of radioactivity is called **autoradiography**.

In summary, immunocytochemistry is a method for viewing the location of specific molecules, including proteins, in sections of brain tissue. *In situ* hybridization is a method for localizing specific mRNA transcripts for proteins. Together, these methods enable us to see whether a neuron contains and synthesizes a transmitter candidate.

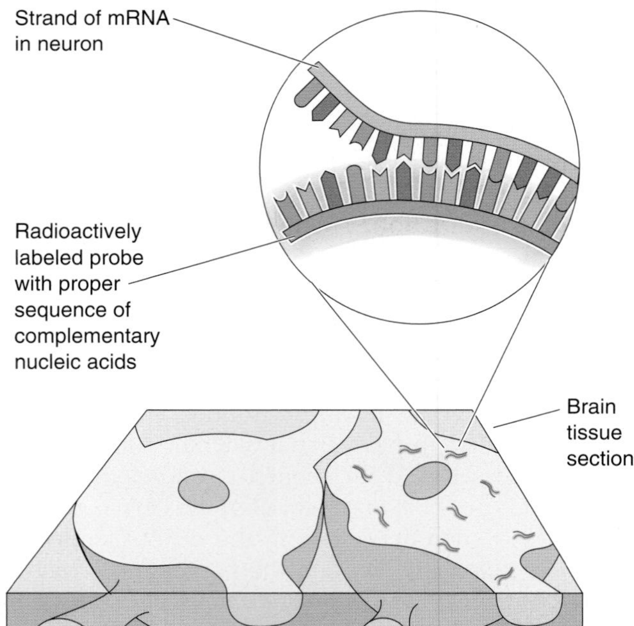

Figure 6.4
***In situ* hybridization.** Strands of mRNA consist of nucleotides arranged in a specific sequence. Each of these nucleotides will stick to one other complementary nucleotide. In the method of *in situ* hybridization, a synthetic probe containing a sequence of complementary nucleotides that will allow it to stick to the mRNA is constructed. If the probe is labeled, the location of cells containing the mRNA will be revealed.

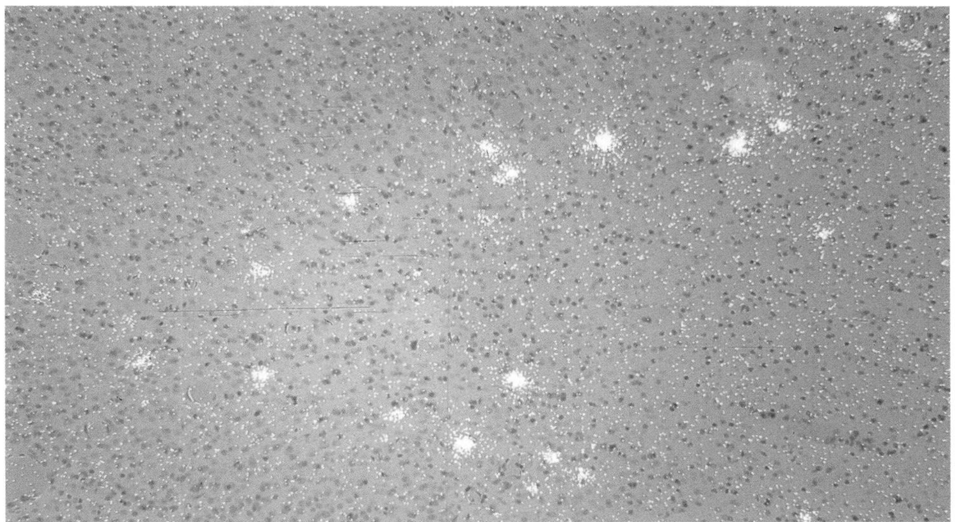

Figure 6.5
***In situ* hybridization of the mRNA for a peptide neurotransmitter in neurons.** Only neurons with the proper mRNA are labeled, with clusters of white dots. (Source: Courtesy of Dr. S. H. C. Hendry.)

Studying Transmitter Release

Once we are satisfied that a transmitter candidate is synthesized by a neuron and localized to the presynaptic terminal, we must show that it is actually released upon stimulation. In some cases, a specific set of cells or axons can be stimulated while taking samples of the fluids bathing their synaptic targets. The biological activity of the sample can then be tested to see if it mimics the effect of the intact synapses, and then the sample can be chemically analyzed to reveal the structure of the active molecule. This general approach helped Loewi and Dale identify ACh as a transmitter at many peripheral synapses.

Unlike the peripheral nervous system (PNS), the nervous system outside the brain and spinal cord, most regions of the central nervous system (CNS) contain a diverse mixture of intermingled synapses using different neurotransmitters. This usually makes it impossible to stimulate a single population of synapses, containing only a single neurotransmitter. Researchers must be content with stimulating many synapses in a region of the brain and collecting and measuring all the chemicals that are released. One way this is done is by using brain slices that are kept alive *in vitro*. To stimulate release, the slices are bathed in a solution containing a high K^+ concentration. This treatment causes a large membrane depolarization (see Figure 3.19), thereby stimulating transmitter release from the axon terminals in the tissue. Because transmitter release requires the entry of Ca^{2+} into the axon terminal, it must also be shown that the release of the neurotransmitter candidate from the tissue slice after depolarization occurs only when Ca^{2+} ions are present in the bathing solution.

Even when it has been shown that a transmitter candidate is released upon depolarization in a calcium-dependent fashion, we still cannot be sure that the molecules collected in the fluids were released from the axon terminals; they may have been released as a secondary consequence of synaptic activation. These technical difficulties make the second criterion—that a transmitter candidate must be released by the presynaptic axon terminal upon stimulation—the most difficult to satisfy unequivocally in the CNS.

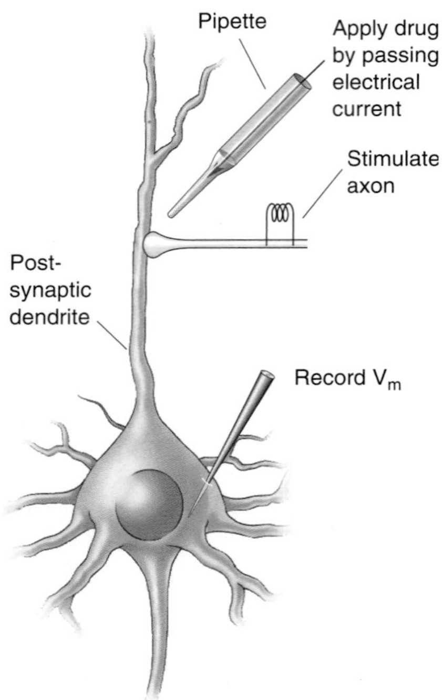

Figure 6.6
Microionophoresis. This method enables a researcher to apply very small amounts of drugs or neurotransmitter candidates to the surface of neurons. The responses generated by the drug can be compared to those generated by synaptic stimulation.

Studying Synaptic Mimicry

Establishing that a molecule is localized in, synthesized by, and released from a neuron is still not sufficient to qualify it as a neurotransmitter. A third criterion must be met: The molecule must evoke the same response as that produced by the release of naturally occurring neurotransmitter from the presynaptic neuron.

To assess the postsynaptic actions of a transmitter candidate, a method called **microionophoresis** is often used. Most neurotransmitter candidates can be dissolved in solutions that will cause them to acquire a net electrical charge. A glass pipette with a very fine tip just a few micrometers across is filled with the ionized solution. The tip of the pipette is carefully positioned next to the postsynaptic membrane of the neuron, and the transmitter candidate is ejected in very small amounts by passing electrical current through the pipette. A microelectrode in the postsynaptic neuron can be used to measure the effects of the transmitter candidate on the membrane potential (Figure 6.6).

If ionophoretic application of the molecule causes electrophysiological changes that mimic the effects of transmitter released at the synapse *and* if the other criteria of localization, synthesis, and release have been met, then the molecule and the transmitter usually are considered to be the same chemical.

Studying Receptors

Each neurotransmitter exerts its postsynaptic effects by binding to specific receptors. As a rule, no two neurotransmitters bind to the same receptor; however, one neurotransmitter can bind to many different receptors. Each of the different receptors that a neurotransmitter binds to is called a **receptor subtype**. For example, in Chapter 5 we learned that ACh acts on two different cholinergic receptor *subtypes*: One type is present in skeletal muscle, and the other is present in heart muscle. Both subtypes are also present in many other organs and within the CNS.

Researchers have tried almost every method of biological and chemical analysis to study the different receptor subtypes of the various neurotransmitter systems. Three approaches have proved to be particularly useful: neuropharmacological analysis of synaptic transmission, ligand-binding methods, and, most recently, molecular analysis of receptor proteins.

Neuropharmacological Analysis. Much of what we know about receptor subtypes was first learned using neuropharmacological analysis. For instance, skeletal muscle and heart muscle respond differently to various cholinergic drugs. *Nicotine*, derived from the tobacco plant, is a receptor agonist in skeletal muscle but has no effect in the heart. On the other hand, *muscarine*, derived from a poisonous species of mushroom, has no effect on skeletal muscle but is an agonist at the cholinergic receptor subtype in the heart. (Recall that ACh slows the heart rate; muscarine is poisonous because it causes a precipitous drop in heart rate and blood pressure.) Thus, two ACh receptor subtypes can be distinguished by the actions of different drugs. In fact, the receptors were given the names of their agonists: **nicotinic receptors** in skeletal muscle and **muscarinic receptors** in the heart. Nicotinic and muscarinic receptors also exist in the brain.

Another way to distinguish receptor subtypes is to use selective antagonists. The arrow-tip poison *curare* inhibits the action of ACh at nicotinic receptors (therefore causing paralysis), and *atropine*, derived from belladonna plants, antagonizes ACh at muscarinic receptors (Figure 6.7). (The eyedrops an ophthalmologist uses to dilate your pupils are related to atropine.)

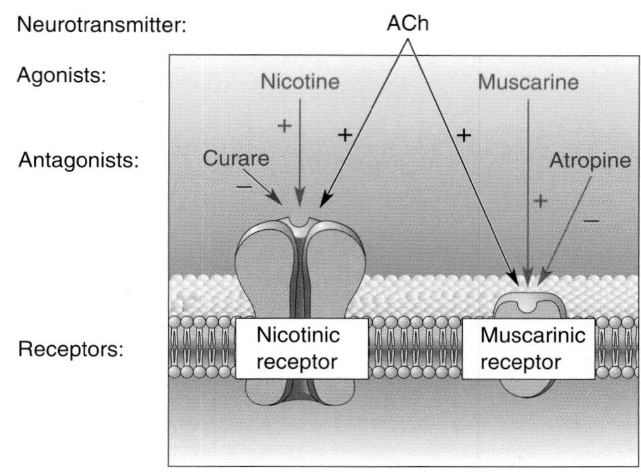

Figure 6.7
The neuropharmacology of cholinergic synaptic transmission. Sites on transmitter receptors can bind the transmitter itself (ACh), an agonist that mimics the transmitter, or an antagonist that blocks the effects of the transmitter and agonists.

Different drugs were also used to distinguish several subtypes of glutamate receptors. Three subtypes are **AMPA receptors**, **NMDA receptors**, and **kainate receptors**, each named for a different chemical agonist. (AMPA stands for α-amino-3-hydroxy-5-methyl-4-isoxazole propionate, and NMDA stands for *N*-methyl-D-aspartate.) Glutamate activates all three subtypes, but AMPA acts only at the AMPA receptor, NMDA acts only at the NMDA receptor, and so on (Figure 6.8).

Similar pharmacological analyses were used to split the NE receptors into two subtypes, α and β, and to divide GABA receptors into $GABA_A$ and $GABA_B$ subtypes. The same can be said for virtually all the neurotransmitter systems. Thus, selective drugs have been extremely useful for categorizing receptor subclasses (Table 6.1). In addition, neuropharmacological analysis has been invaluable for assessing the contributions of neurotransmitter systems to brain function.

Ligand-Binding Methods. As we said, the first step in studying a neurotransmitter system is usually identifying the neurotransmitter. With the discovery in the 1970s that many drugs interact selectively with neurotransmitter receptors, however, researchers realized that they could use these compounds to begin analyzing receptors, even before the neurotransmitter itself had been identified. A pioneer of this approach was Solomon Snyder at Johns Hopkins University, who was interested in studying compounds called opiates. Opiates are a broad class of drugs that are both medically im-

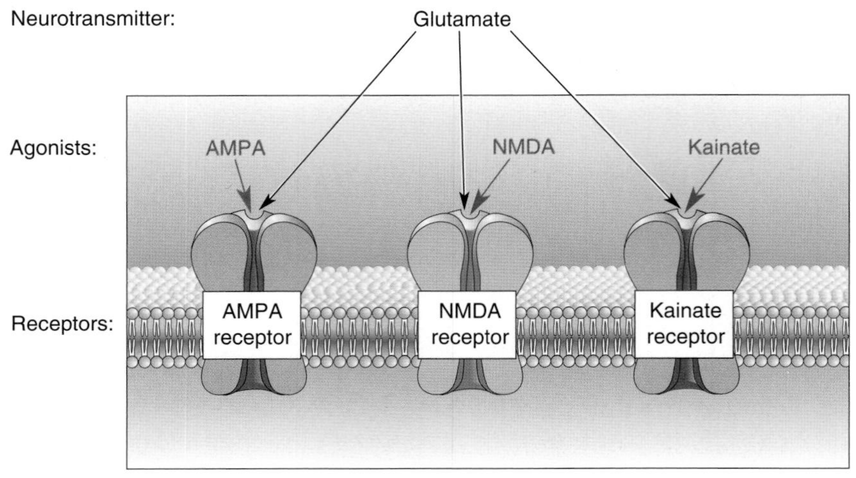

Figure 6.8
The neuropharmacology of glutamatergic synaptic transmission. There are three main subtypes of glutamate receptors, each of which binds glutamate and each of which is activated selectively by a different agonist.

Table 6.1 The Neuropharmacology of Some Receptor Subtypes

NEUROTRANSMITTER	RECEPTOR SUBTYPE	AGONIST	ANTAGONIST
Acetylcholine (ACh)	Nicotinic receptor	Nicotine	Curare
	Muscarinic receptor	Muscarine	Atropine
Norepinephrine (NE)	α receptor	Phenylephrine	Phenoxybenzamine
	β receptor	Isoproterenol	Propranolol
Glutamate (Glu)	AMPA	AMPA	CNQX
	NMDA	NMDA	AP5
GABA	$GABA_A$	Muscimol	Bicuculline
	$GABA_B$	Baclofen	Phaclofen

portant and commonly abused. Their effects include pain relief, euphoria, depressed breathing, and constipation.

The question Snyder and his student Candace Pert originally set out to answer was how heroin, morphine, and other opiates exert their effects on the brain. They hypothesized that opiates might be agonists at specific receptors in neuronal membranes. To test this idea, they radioactively labeled opiate compounds and applied them in small quantities to neuronal membranes that had been isolated from different parts of the brain. If receptors existed in the membrane, the labeled opiates should bind tightly to them. This is just what they found. The radioactive drugs labeled specific sites on the membranes of some, but not all, neurons in the brain (Figure 6.9). Following the discovery of opiate receptors, the search was on to identify endogenous opiates, or *endorphins*, the naturally occurring neurotransmitters that act on these receptors. Two peptides called *enkephalins* were soon isolated from the brain, and they eventually proved to be opiate neurotransmitters.

Any chemical compound that binds to a specific site on a receptor is called a *ligand* for that receptor (from the Latin word meaning "to bind"). The technique of studying receptors using radioactively labeled ligands is called the **ligand-binding method**. Notice that a ligand for a receptor can be an agonist, an antagonist, or the chemical neurotransmitter itself. Specific ligands were invaluable for isolating neurotransmitter receptors and determining their chemical structure (Box 6.1). Ligand-binding methods have been enormously important for mapping the anatomical distribution of different neurotransmitter receptors in the brain.

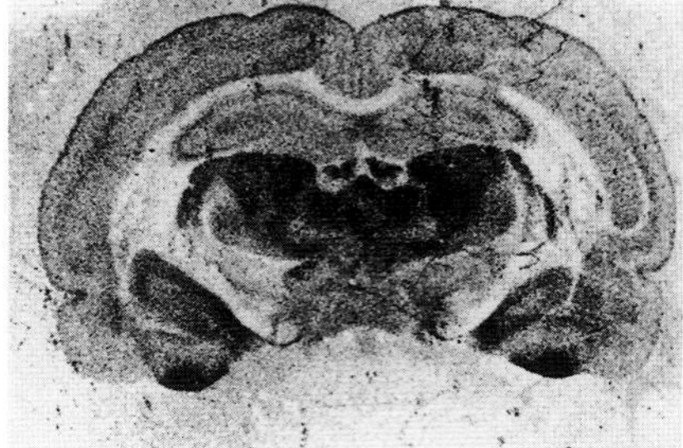

Figure 6.9
Opiate receptor binding to a slice of rat brain. Special film was exposed to a brain section that had radioactive opiate receptor ligands bound to it. The dark regions contain more receptors than do the light ones. (Source: Snyder, 1986, p. 44.)

Box 6.1

PATH OF DISCOVERY

A Brief History of the Nicotinic Receptor

BY JEAN-PIERRE CHANGEUX

The concept of receptor is almost as old as the concept of enzyme, but enzymes were identified at the molecular level many years before receptors were. One key to receptor identification was the use of drugs and poisons, whose nervous system effects had been known almost since the origins of humanity. In ingenious experiments reported in 1857, the French physiologist Claude Bernard localized the paralyzing action of curare to the site where peripheral motor fibers communicate with skeletal muscle. In 1905, inspired by Bernard's results, the British pharmacologist John Newport Langeley proposed that the surface of the muscle underlying nerve endings has a receptive substance (today called the receptor) that combines with nicotine and curare.

Although the concept of receptor had been clearly stated, doubts remained that receptors would ever be identified. Indeed, in 1943 the Nobel laureate Sir Henry Dale even challenged the receptor theory itself. Nevertheless, biochemists such as David Nachmansohn remained confident. After fleeing Nazi Germany for Paris in the late 1930s, Nachmansohn discovered that the electric organ of an eel (the *Torpedo*) is a rich source of cholinergic synapses. Nachmansohn already believed that the receptor was a protein, and the catalytic site of enzymes became the prevailing model for the receptor binding site of drugs. The important difference is that an enzyme leads to a catalytic product, whereas a neurotransmitter receptor results in a "puncture" that permits ions to flow through the membrane. But this scheme did not explain how the binding of transmitter to receptor leads to the opening of an ion pathway.

Studies of bacterial enzymes and vertebrate oxygen carriers such as hemoglobin provided a plausible molecular answer. Those proteins were called "allosteric," in that they carry at least two alien categories of binding sites—one for the regulatory ligand and the other for the substrate of the biological activity. By 1959 I was a graduate student in Jacques Monod's laboratory, where I worked on a bacterial regulatory enzyme, L-threonine deaminase. In my dissertation, I carefully suggested that "membrane phenomena, . . . such as synaptic transmission," might involve "mechanisms analogous to those described with allosteric proteins." For neurotransmitter receptors, the regulatory site would be the drug neurotransmitter binding site, and the active site would be the ion channel. Reversible changes of the receptor's shape would throw a "molecular switch" between the channel-shut and the channel-open states. The physiological signal (acetylcholine) or the drug (nicotine) would then stabilize the open channel state; curare would keep it closed.

Jean-Pierre Changeux

To evaluate the merit of these ideas, I needed an identified neurotransmitter receptor in large quantities; the eel's electric organ seemed to have considerable advantages. A postdoctoral fellow, Michiki Kasai, and I separated micron-sized fragments of membranes from the electric tissue. These fragments had the remarkable ability to close up into microsacs that could be filled with radioactive sodium or potassium ions. Adding acetylcholine dramatically enhanced the flow of ions into and out of these sacs, similar to the response of muscle membrane to acetylcholine, indicating that the receptor was present and functional. Even after several more years of difficult work learning how to isolate the receptor, however, I knew that studies with the usual drugs would not convince the scientific community.

By chance, in 1970, Chen Yuan Lee of the National Taiwan University offered us a new and better tool. Lee had been working for decades on toxins from the venom of poisonous snakes (cobra or *Bungarus*). One toxin, α-bungarotoxin, happened to block neuromuscular transmission in a highly selective and almost irreversible manner. Would α-bungarotoxin work on the rather primitive electric fish systems?

The result was unexpected but wonderful: α-Bungarotoxin completely blocked the response of single cells from the electrical organ, of the microsacs prepared from the electrical tissue, and of the reversible binding of ligands to the molecules prepared from these membranes. This step was decisive. The acetylcholine receptor protein was purified and its biochemical identification achieved. Studies since then have strongly supported the view that a reversible "allosteric switch" between a resting closed state and an active open state accounts for signal transduction at the synapse.

Molecular Analysis. There has been an explosion of information about neurotransmitter receptors in the past 20 years, thanks to the new methods for studying protein molecules. Information obtained with these methods has enabled us to divide the neurotransmitter receptor proteins into two groups: transmitter-gated ion channels and G-protein-coupled (metabotropic) receptors (see Chapter 5).

Molecular neurobiologists have determined the structure of the polypeptides that make up many proteins, and these studies have led to some startling conclusions. Receptor subtype diversity was expected from the actions of different drugs, but the extent of the diversity was not appreciated until researchers determined how many different polypeptides could serve as subunits of functional receptors.

Consider as an example the $GABA_A$ receptor, a transmitter-gated chloride channel. Each channel requires five subunits, and there are five major classes of subunit proteins, designated α, β, γ, δ, and ρ. At least six different polypeptides (designated $\alpha 1$–6) can substitute for one another as an α subunit. Three different polypeptides (designated $\beta 1$–3) can substitute as a β subunit, and two different polypeptides ($\gamma 1$ and $\gamma 2$) can be used as a γ subunit. Although this is probably not the final tally, let's use these numbers to make an interesting calculation. If it takes five subunits to form a $GABA_A$ receptor-gated channel and there are 13 possible subunits to choose from, then there are 74,269 possible combinations of subunits. This means there are 74,269 potential subtypes of $GABA_A$ receptors!

Recognize, however, that most of the possible subunit combinations are never manufactured by neurons and, even if they were, would not work properly. Nonetheless, it is clear that receptor classifications like those appearing in Table 6.1, while still useful, underestimate the diversity of receptor subtypes in the brain.

NEUROTRANSMITTER CHEMISTRY

Research using methods such as those discussed above has led to the conclusion that the major neurotransmitters are amino acids, amines, and peptides. Evolution is conservative and opportunistic, and it often puts common and familiar things to new uses. This seems to be true about the evolution of neurotransmitters. For the most part, they are similar or identical to the basic chemicals of life, the same substances that cells in all species, from bacteria to giraffes, use for metabolism. Amino acids, the building blocks of protein, are essential to life. Most of the known neurotransmitter molecules are either (1) amino acids, (2) amines derived from amino acids, or (3) peptides constructed from amino acids. ACh is an exception; it is derived from acetyl-coenzyme A (acetyl CoA), a ubiquitous product of cellular respiration in mitochondria, and choline, which is important for fat metabolism throughout the body.

Amino acid and amine transmitters are generally each stored in and released by separate sets of neurons. The convention established by Dale classifies neurons into mutually exclusive groups by neurotransmitter (cholinergic, glutamatergic, GABAergic, and so on). The idea that a neuron has only one neurotransmitter is often called **Dale's principle**. Strictly speaking, many peptide-containing neurons violate Dale's principle, because these cells usually release more than one neurotransmitter: an amino acid or amine *and* a peptide. Nonetheless, the amino acid and amine neurotransmitters can be used to assign most neurons to distinct, nonoverlapping classes. Let's take a look at the biochemical mechanisms that differentiate these neurons.

Cholinergic Neurons

Acetylcholine (ACh) is the neurotransmitter at the neuromuscular junction and therefore is synthesized by all the motor neurons in the spinal cord and brain stem. Other cholinergic cells contribute to the functions of specific circuits in the PNS and CNS, as we will see in Chapter 15.

ACh synthesis requires a specific enzyme, *choline acetyltransferase (ChAT)* (Figure 6.10). Like all presynaptic proteins, ChAT is manufactured in the soma and transported to the axon terminal. Only cholinergic neurons contain ChAT, so this enzyme is a good marker for cells that use ACh as a neurotransmitter. ChAT synthesizes ACh in the cytosol of the axon terminal, and the neurotransmitter is concentrated in synaptic vesicles by the actions of an ACh **transporter** (Box 6.2).

ChAT transfers an acetyl group from acetyl CoA to choline (Figure 6.11a). The source of choline is the extracellular fluid, where it exists in low micromolar concentrations. Choline is taken up by the cholinergic axon terminals via a specific transporter. Because the availability of choline limits how much ACh can be synthesized in the axon terminal, the transport of choline into the neuron is said to be the **rate-limiting step** in ACh synthesis. For certain diseases in which a deficit in cholinergic synaptic transmission has been noted, dietary supplements of choline are sometimes prescribed to boost ACh levels in the brain.

Cholinergic neurons also manufacture the ACh degradative enzyme *acetylcholinesterase (AChE).* AChE is secreted into the synaptic cleft and is associated with cholinergic axon terminal membranes. AChE is also manufactured by some noncholinergic neurons, however, so this enzyme is not as useful a marker for cholinergic synapses as ChAT.

AChE degrades ACh into choline and acetic acid (Figure 6.11b). This happens very quickly, because AChE has one of the fastest catalytic rates among all known enzymes. Much of the resulting choline is taken up by the cholin-

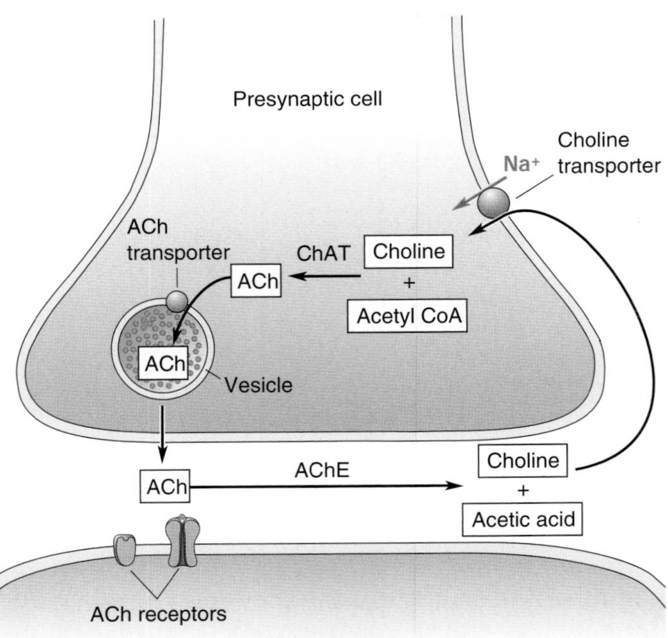

Figure 6.10
The life cycle of ACh.

BRAIN FOOD

Pumping Ions and Transmitters

Neurotransmitters may lead an exciting life, but the most mundane part of it would seem to be the steps that recycle them back from the synaptic cleft and eventually into a vesicle. Where synapses are concerned, the exotic proteins of exocytosis and the innumerable transmitter receptors get most of the publicity. Yet the neurotransmitter *transporters* are very interesting for at least two reasons: They succeed at an extraordinarily difficult job, and they are the molecular site at which many important psychoactive drugs act.

The hard work of transporters is pumping transmitter molecules across membranes so effectively that they become highly concentrated in very small places. There are two general types of neurotransmitter transporter. One type, the *neuronal membrane transporter,* shuttles transmitter from the extracellular fluid, including the synaptic cleft, and concentrates it up to 10,000 times higher within the cytosol of the presynaptic terminal. A second type, the *vesicular transporter*, then crams transmitter into vesicles at concentrations that may be 100,000 times higher than in the cytosol. Inside cholinergic vesicles, for example, ACh may reach the incredible concentration of 1000 mM, or 1 molar—in other words, about twice the concentration of salt in seawater!

How do transporters achieve such dramatic feats of concentration? Concentrating a chemical is like carrying a weight uphill; both are extremely unlikely to occur unless energy is applied to the task. Recall from Chapter 3 that ion pumps in the plasma membrane use ATP as their source of energy to transport Na^+, K^+, and Ca^{2+} against their concentration gradients. These ion gradients are essential for setting the resting potential and for powering the ionic currents that underlie action and synaptic potentials. Notice, however, that once ionic gradients are established across a membrane, they can themselves be tapped as sources of energy. Just as the energy spent in pulling up the weights on a cuckoo clock can be reclaimed to turn the gears and hands of the clock (as the weights slowly fall down again), transporters use transmembrane gradients of Na^+ or H^+ as an energy source for moving transmitter molecules up steep concentration gradients. The transporter lets one transmembrane gradient, that of Na^+ or H^+, run down a bit in order to build up another gradient, that of the transmitter.

The transporters themselves are large proteins that span membranes. There can be several transporters for one transmitter (e.g., at least four subtypes are known for GABA). Figure A shows how they work. Plasma membrane transporters use a *cotransport* mechanism, carrying two Na^+ ions along with one transmitter molecule. By contrast, vesicular membrane transporters use a *countertransport* mechanism that trades a transmitter molecule from the cytosol for a H^+ from inside the vesicle. Vesicle membranes have ATP-driven H^+ pumps that keep their contents very acidic, or high in protons (i.e., H^+ ions)

What is the relevance of all this to drugs and disease? Many psychoactive drugs, such as amphetamines and cocaine, potently block certain transporters. By altering the normal recycling process of various transmitters, the drugs lead to chemical imbalances in the brain that can have profound effects on mood and behavior. It is also possible that defects in transporters can lead to psychiatric or neurological disease; certainly some of the drugs that are therapeutically useful in psychiatry work by blocking transporters. The numerous links between transmitters, drugs, disease, and treatment are tantalizing but complex and will be discussed further in Chapters 15 and 21.

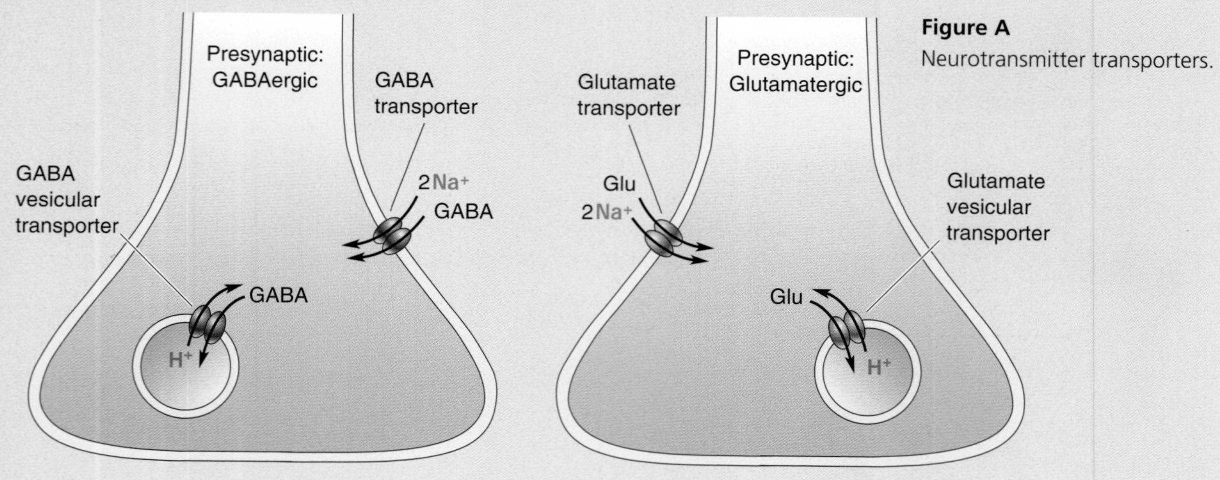

Figure A
Neurotransmitter transporters.

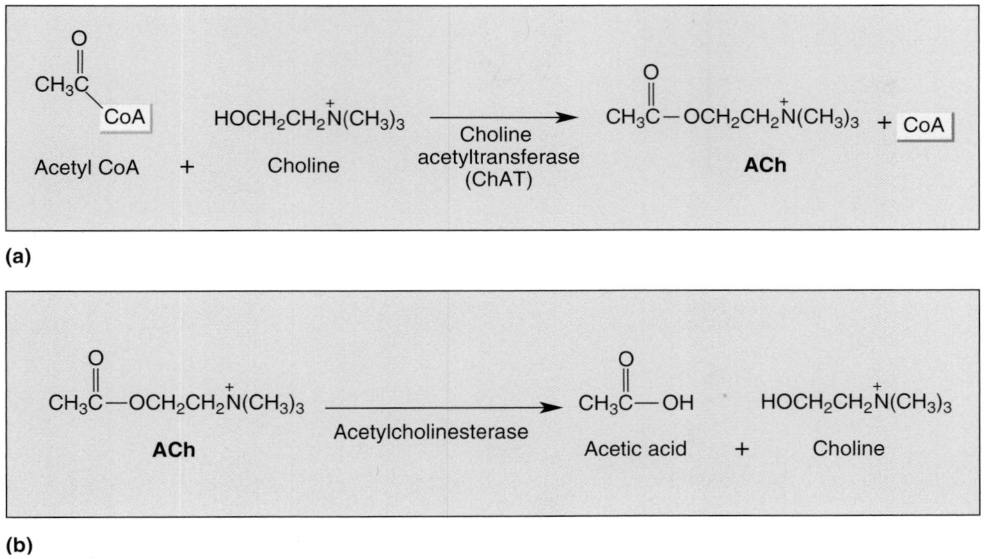

$$CH_3C(=O)\text{–CoA} + HOCH_2CH_2\overset{+}{N}(CH_3)_3 \xrightarrow[\text{(ChAT)}]{\text{Choline acetyltransferase}} CH_3C(=O)\text{–}OCH_2CH_2\overset{+}{N}(CH_3)_3 + \text{CoA}$$

Acetyl CoA + Choline → ACh

(a)

$$CH_3C(=O)\text{—}OCH_2CH_2\overset{+}{N}(CH_3)_3 \xrightarrow{\text{Acetylcholinesterase}} CH_3C(=O)\text{—}OH + HOCH_2CH_2\overset{+}{N}(CH_3)_3$$

ACh → Acetic acid + Choline

(b)

Figure 6.11
Acetylcholine. (a) ACh synthesis. **(b)** ACh degradation.

ergic axon terminal and reused for ACh synthesis. In Chapter 5, we mentioned that AChE is the target of many nerve gases and insecticides. Inhibition of AChE prevents the breakdown of ACh, disrupting transmission at cholinergic synapses on skeletal muscle and heart muscle. Acute effects include marked decreases in heart rate and blood pressure; death from the irreversible inhibition of AChE is, however, typically a result of respiratory paralysis.

Catecholaminergic Neurons

The amino acid tyrosine is the precursor for three different amine neurotransmitters that contain a chemical structure called a *catechol* (Figure 6.12a). These neurotransmitters are collectively called **catecholamines**. The catecholamine neurotransmitters are **dopamine** (**DA**), **norepinephrine** (**NE**), and **epinephrine**, also called **adrenaline.** (Figure 6.12b). Catecholaminergic neurons are found in regions of the nervous system involved in the regulation of movement, mood, attention, and visceral function (discussed further in Chapter 15).

All catecholaminergic neurons contain the enzyme *tyrosine hydroxylase (TH)*, a catalyst in the first step in catecholamine synthesis, the conversion of tyrosine to a compound called **dopa** (L-dihydroxyphenylalanine) (Figure 6.13a). The activity of TH is rate-limiting for catecholamine synthesis. The enzyme's activity is regulated by various signals in the cytosol of the axon terminal. For example, decreased catecholamine release by the axon terminal causes the catecholamine concentration in the cytosol to rise, thereby inhibiting TH. This type of regulation is called *end-product inhibition.* On the other hand, during periods when catecholamines are released at a high rate, the elevation in $[Ca^{2+}]_i$ that accompanies neurotransmitter release triggers an increase in the activity of TH, so transmitter supply keeps up with demand. In addition, prolonged periods of stimulation actually cause the synthesis of more mRNA that codes for the enzyme.

Dopa is converted into the neurotransmitter dopamine by the enzyme *dopa decarboxylase* (Figure 6.13b). Dopa decarboxylase is abundant in catecholaminergic neurons, so the amount of dopamine synthesized primarily depends on the amount of dopa available. In a movement disorder called *Parkinson's disease*, dopaminergic neurons in the brain slowly degenerate and eventually die. One strategy for treating Parkinson's disease is the adminis-

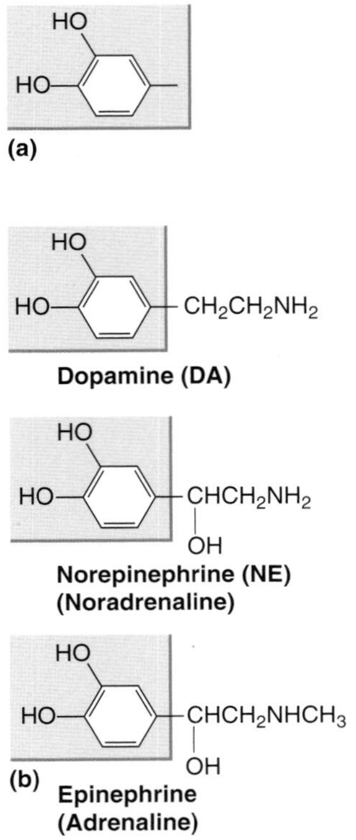

HO, HO (catechol ring)

(a)

HO, HO — $CH_2CH_2NH_2$
Dopamine (DA)

HO, HO — $CHCH_2NH_2$ (OH)
Norepinephrine (NE) (Noradrenaline)

HO, HO — $CHCH_2NHCH_3$ (OH)
Epinephrine (Adrenaline)

(b)

Figure 6.12
The catecholamines. (a) The catechol group. **(b)** Catecholamine neurotransmitters.

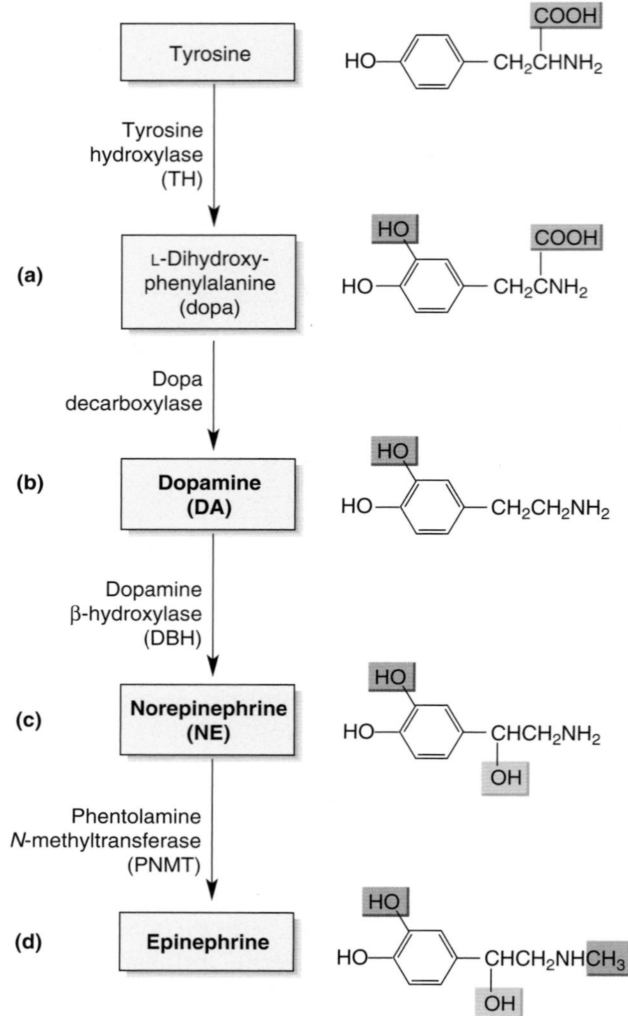

Figure 6.13
The synthesis of catecholamines from tyrosine. The catecholamine neurotransmitters are in boldface type.

tration of dopa, which causes an increase in DA synthesis in the surviving neurons, increasing the amount of DA available for release. (We will learn more about dopamine and movement in Chapter 14.)

Neurons that use NE as a neurotransmitter contain, in addition to TH and dopa decarboxylase, the enzyme *dopamine β-hydroxylase (DBH)*, which converts dopamine to norepinephrine (Figure 6.13c). Note that DBH is not found in the cytosol but is found within the synaptic vesicles. Thus, in noradrenergic axon terminals, DA is transported from the cytosol to the synaptic vesicles, and there it is made into NE.

The last in the line of catecholamine neurotransmitters is epinephrine (adrenaline). Adrenergic neurons contain the enzyme *phentolamine N-methyltransferase (PNMT)*, which converts NE to epinephrine (Figure 6.13d). Curiously, PNMT is in the cytosol of adrenergic axon terminals. Thus, NE must first be synthesized in the vesicles and released into the cytosol for conversion into epinephrine, and then the epinephrine must again be transported into vesicles for release. In addition to serving as a neurotransmitter in the brain, epinephrine is released by the adrenal gland into the bloodstream. As we shall see in Chapter 15, circulating epinephrine acts at receptors throughout the body to produce a coordinated visceral response.

There is no fast extracellular degradative enzyme analagous to AChE for the catecholamine systems. Instead, the actions of catecholamines in the synaptic cleft are terminated by selective uptake of the neurotransmitters back into the axon terminal via Na^+-dependent transporters. This step is sensitive to a number of different drugs. For example, amphetamine and cocaine block catecholamine uptake, therefore prolonging the actions of the neurotransmitter in the cleft. Once inside the axon terminal, the catecholamines may be reloaded into synaptic vesicles for reuse or may be enzymatically destroyed by the action of *monoamine oxidase (MAO)*, an enzyme found on the outer membrane of mitochondria.

Serotonergic Neurons

The amine neurotransmitter **serotonin**, also called *5-hydroxytryptamine* and abbreviated **5-HT**, is derived from the amino acid tryptophan. **Serotonergic** neurons are relatively few in number, but as we shall see in Part III, they appear to play an important role in the brain systems that regulate mood, emotional behavior, and sleep.

The synthesis of serotonin occurs in two steps, just like the synthesis of dopamine (Figure 6.14). Tryptophan is converted first into an intermediary called 5-HTP (5-hydroxytryptophan) by the enzyme *tryptophan hydroxylase*. The 5-HTP is then converted to 5-HT by the enzyme *5-HTP decarboxylase*. Serotonin synthesis appears to be limited by the availability of tryptophan in the extracellular fluid bathing neurons. The source of brain tryptophan is the blood, and the source of blood tryptophan is the diet (grains, meat, and dairy products are particularly rich in tryptophan). Thus, a dietary deficiency of tryptophan can quickly lead to a depletion of serotonin in the brain.

Following release from the axon terminal, 5-HT is removed from the synaptic cleft by the action of a specific transporter. The process of serotonin reuptake, like catecholamine reuptake, is sensitive to a number of different drugs. For example, several clinically useful antidepressant drugs, including fluoxetine (trade name Prozac), are selective inhibitors of serotonin reuptake. Once it is back in the cytosol of the serotonergic axon terminal, the transmitter is either reloaded into synaptic vesicles or degraded by MAO.

Amino Acidergic Neurons

The amino acids **glutamate (Glu)**, **glycine (Gly)**, and **gamma-aminobutyric acid (GABA)** serve as neurotransmitters at most CNS synapses (Figure 6.15).

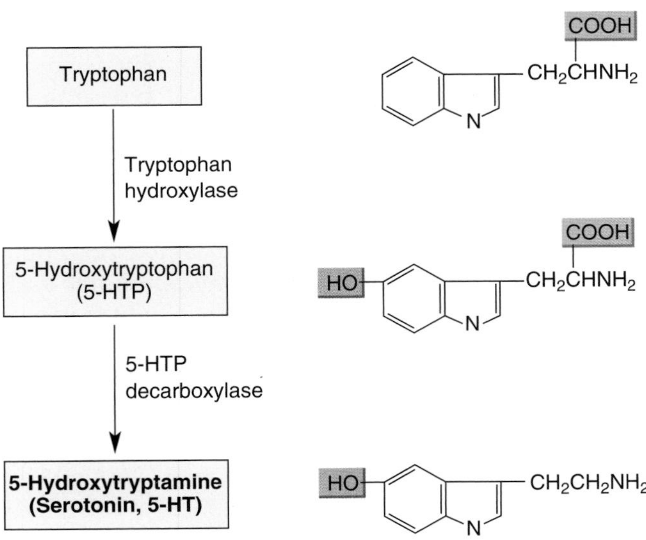

Figure 6.14
The synthesis of serotonin from tryptophan.

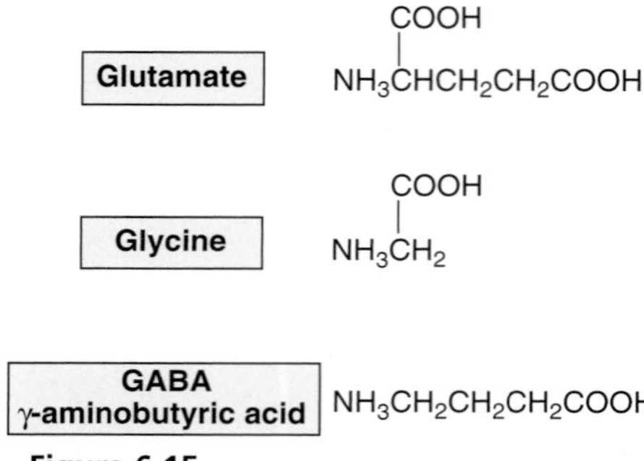

Figure 6.15
Amino acid neurotransmitters.

Of these, only GABA is unique to those neurons that use it as a neurotransmitter; the others are among the 20 amino acids that make up proteins.

Glutamate and glycine are synthesized from glucose and other precursors using enzymes that exist in all cells. Differences among neurons in the synthesis of these amino acids are therefore quantitative rather than qualitative. For example, the average glutamate concentration in the cytosol of glutamatergic axon terminals has been estimated to be about 20 mM, two or three times higher than that in nonglutamatergic cells. The more important distinction between glutamatergic and nonglutamatergic neurons, however, is the transporter that loads the synaptic vesicles. In glutamatergic axon terminals, but not in other types, the glutamate transporter concentrates glutamate until it reaches a value of about 50 mM in the synaptic vesicles.

Because GABA is not one of the 20 amino acids used to construct proteins, it is synthesized only by those neurons that use it as a neurotransmitter. The precursor for GABA is glutamate, and the key synthesizing enzyme is *glutamic acid decarboxylase (GAD)* (Figure 6.16). GAD, therefore, is a good marker for GABAergic neurons. Immunocytochemical studies have shown that GABAergic neurons are distributed widely in the brain. GABAergic neurons are the major source of synaptic inhibition in the nervous system.

The synaptic actions of the amino acid neurotransmitters are terminated by selective uptake into the presynaptic terminals and glia, once again via specific Na^+-dependent transporters. Inside the terminal or glial cell, GABA is metabolized by the enzyme *GABA transaminase.*

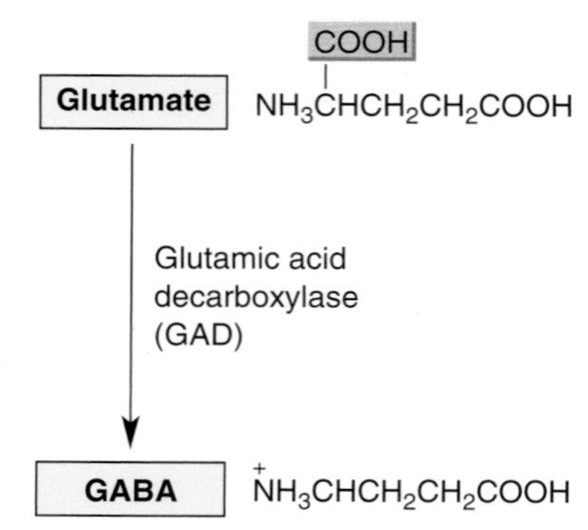

Figure 6.16
The synthesis of GABA from glutamate.

Other Neurotransmitter Candidates and Intercellular Messengers

In addition to the amines and amino acids, a few other small molecules may serve as chemical messengers between neurons. For instance, researchers have recently focused on adenosine triphosphate (ATP), a key molecule in cellular metabolism, as a possible neurotransmitter. ATP is concentrated in vesicles at many synapses in the CNS and PNS, and it is released into the cleft by presynaptic spikes in a Ca^{2+}-dependent manner, just as the classic transmitters are. ATP is often packaged in vesicles along with another classic transmitter. For example, catecholamine-containing vesicles may have 100 mM of ATP, an enormous quantity, in addition to 400 mM of the catecholamine itself.

ATP directly excites some neurons by gating a cation channel. In this sense, some of the neurotransmitter functions of ATP may be similar to those of glutamate. The *nucelotide receptors*, such as the ATP receptor, are the most recently discovered type of transmitter-gated ion channel. There is also a large class of G-protein-coupled nucleotide receptors.

The most exotic chemical messenger to be proposed for intercellular communication is actually a gaseous molecule, **nitric oxide** (**NO**). Carbon monoxide (CO) has also been suggested as a messenger, although evidence thus far is meager. These are the same nitric oxide and carbon monoxide that are major air pollutants from internal combustion engines. NO is synthesized from the amino acid arginine by many cells of the body, and it has powerful biological effects. In the nervous system, NO may have unique functions. It seems to be released from postsynaptic neurons by nonvesicular means and acts on presynaptic terminals. Communication in this direction, from "post" to "pre," is called *retrograde communication*; thus, NO is said to be a **retrograde messenger**. Because NO is small and membrane-permeable, it can diffuse much more freely than other transmitter molecules, even penetrating through one cell to affect another beyond it. Its influence, rather than being confined to the site of the cells that released it, may spread throughout a small region of local tissue. On the other hand, NO is evanescent and breaks down very rapidly. The functions of gaseous transmitters are being extensively studied and hotly debated.

Before leaving the topic of neurotransmitter chemistry, we will point out once again that many of the chemicals we call neurotransmitters may also be present in high concentrations in non-neural parts of the body. A chemical may serve dual purposes, mediating communication in the nervous system but doing something entirely different elsewhere. Amino acids, of course, are used to make proteins throughout the body. ATP is the energy source for all cells. NO is released from endothelial cells and causes the smooth muscle of blood vessels to relax. (One consequence in males is penile erection.) The cells with the highest levels of ACh are not in the brain but in the cornea of the eye, where there are no ACh receptors. Likewise, the highest serotonin levels are not in neurons but in blood platelets. These observations underscore the importance of rigorous analysis before a chemical is assigned a neurotransmitter role (Box 6.3).

The operation of a neurotransmitter system is like a play with two acts. Act I is presynaptic and culminates in the transient elevation of neurotransmitter concentration in the synaptic cleft. We are now ready to move on to Act II, the generation of electrical and biochemical signals in the postsynaptic neuron. The main players are transmitter-gated channels and G-protein-coupled receptors.

TRANSMITTER-GATED CHANNELS

In Chapter 5, we learned that ACh and the amino acid neurotransmitters mediate fast synaptic transmission by acting on transmitter-gated ion channels. These channels are magnificent minuscule machines. A single channel can be a sensitive detector of chemicals and voltage, it can regulate the flow of surprisingly large currents with great precision, it can sift and select between very similar ions, and it can be regulated by other receptor systems. Yet each channel is only about 11 nm long, just barely visible with the best computer-enhanced electron microscopic methods. Although the secrets of these channels are now being revealed, we still have much to learn.

The Basic Structure of Transmitter-Gated Channels

The most thoroughly studied transmitter-gated ion channel is the nicotinic ACh receptor in skeletal muscle. It is a pentamer, an amalgam of five protein subunits arranged like the staves of a barrel to form a single pore through the membrane (Figure 6.17a). Four different types of polypeptides are used as subunits for the nicotinic receptor, and they are designated α, β, γ, and δ. A complete mature channel is made from two α units, and one each of β, γ, and δ (abbreviated $\alpha_2\beta\gamma\delta$). There is one ACh binding site on each of the α subunits; the simultaneous binding of ACh to both sites is required for the channel to open (Figure 6.17b).

Although each subunit type has a different primary structure, there are stretches where the different polypeptide chains have a similar sequence of amino acids. For example, each subunit polypeptide has four separate segments that will coil into alpha helices (Figure 6.17a). Because the amino acid residues of these segments are hydrophobic, the four alpha helices are believed to be where the polypeptide is threaded back and forth across the membrane similar to the pore loops of potassium and sodium channels (see Chapters 3 and 4).

The primary structures of the subunits of many other transmitter-gated

Box 6.3

OF SPECIAL INTEREST

Are Cannabinoids Neurotransmitters?

Most neurotransmitters were discovered long before their receptors, but new techniques have tended to reverse this tradition. We now have receptors in search of transmitters. This is the beginning of one such story. It is too soon to write the ending.

Cannabis sativa is the botanical name for hemp, a fibrous plant used through the ages for making rope and cloth. These days, cannabis is much more popular as dope than rope; it is widely, and usually illegally, sold as marijuana or hashish. The Chinese first recognized the potent psychoactive properties of cannabis 4000 years ago, but Western society learned of its intoxicating properties only in the nineteenth century, when Napolean's troops returned to France with Egyptian hashish. The French physician Jacques-Joseph Moreau introduced hashish to Parisian literary society and wrote the first medical description of cannabis' effects. At low doses, these effects can be euphoria, feelings of calm and relaxation, altered sensations, reduced pain, increased laughter, talkativeness, hunger, and lightheadedness, as well as decreased problem-solving ability, short-term memory, and psychomotor performance (i.e., the skills necessary for driving). High doses of cannabis can cause profound personality changes and even hallucinations. In recent years, forms of cannabis have been approved for limited medicinal use in the United States, primarily to treat nausea and vomiting in cancer patients undergoing chemotherapy and to stimulate appetite in some AIDS patients.

The active ingredient in cannabis is an oily chemical called Δ^9-tetrahydrocannabinol, or THC. During the late 1980s, it became apparent that THC can bind to specific G-protein-coupled "cannabinoid" receptors in the brain, particularly in motor control areas, the cerebral cortex, and pain pathways. At about the same time, a group at the National Institute of Mental Health cloned the gene for an unknown G-protein-coupled receptor. Further work showed that the mystery receptor was the cannabinoid receptor, and that activating it leads to G-protein-mediated responses. Two types of cannabinoid receptors are now known: CB1 receptors are in the brain, and CB2 receptors are in other organs of the body.

Remarkably, the brain has more CB1 receptors than *any* other G-protein-coupled receptor. What are they doing there? We don't know yet, but we are quite certain they did not evolve to bind the THC from hemp. The natural ligand for a receptor is never the synthetic drug, plant toxin, or snake venom that might have helped us identify that receptor in the first place. It is much more likely that the cannabinoid receptors exist to bind some signaling molecule made by the brain, just as the neurotransmitter receptors described in this chapter do. Is there a THC-like neurotransmitter? We don't know this either, but there are a few candidate molecules that have so far not met all the criteria needed to achieve neurotransmitter status. The most promising is anandamide (from *ananda*, the Sanskrit word for "internal bliss"). J. Michael Walker and his group at Brown University recently showed that painful sensory stimuli trigger the release of anandamide from neurons in a pain center of the brain; furthermore, electrically stimulating this same pain center suppressed pain by activating cannabinoid receptors.

As the search for new transmitters continues, the hunt is also on for further subtypes of CB receptors and for more selective compounds that bind to them. Cannabinoids are potentially useful for relief of nausea, pain suppression, muscle relaxation, treatment of seizures, and decreasing the intraocular pressure of glaucoma. Cannabinoid therapies might be more practical if new drugs can be developed that retain the therapeutic benefits without causing psychoactive side effects.

channels in the brain have now been deduced, and there are obvious similarities (Figure 6.18). All contain the four hydrophobic segments that are thought to span the membrane. Most of the channels are thought to be pentameric complexes, with close similarities to the nicotinic ACh receptor. Possible exceptions are the glutamate-gated channels. Recent evidence indicates that some glutamate receptors are tetramers, having four subunits that comprise a functional channel. It is also likely that the M2 region of the glutamate subunits does not span the membrane but instead forms a hairpin that both enters and exits from the inside of the membrane.

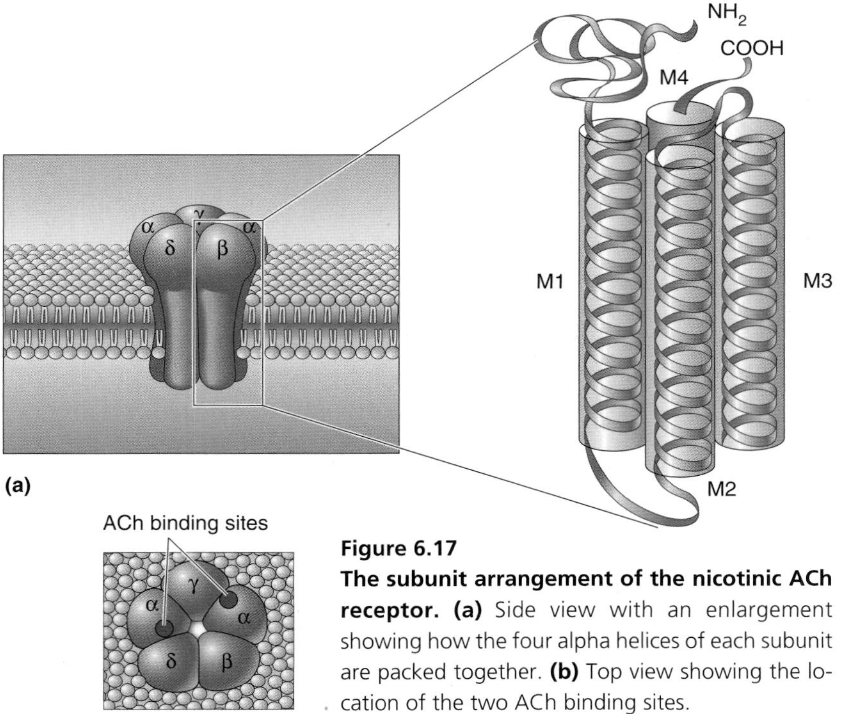

ACh binding sites

(b)

Figure 6.17
The subunit arrangement of the nicotinic ACh receptor. (a) Side view with an enlargement showing how the four alpha helices of each subunit are packed together. **(b)** Top view showing the location of the two ACh binding sites.

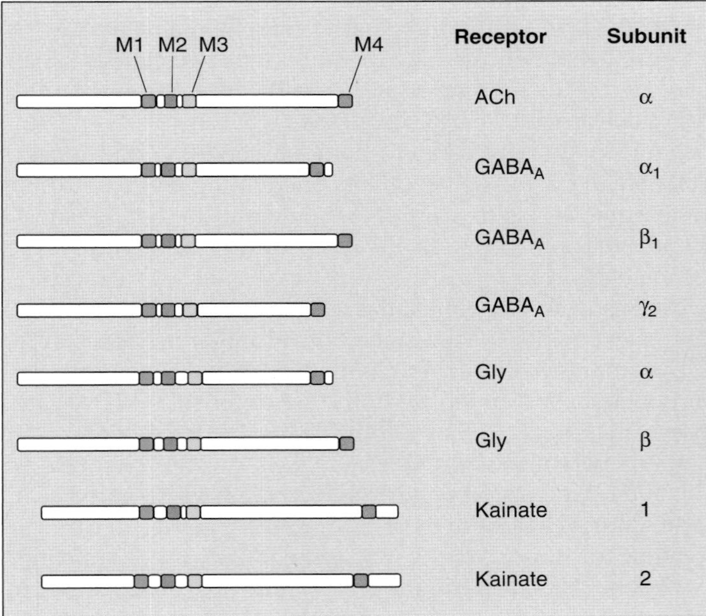

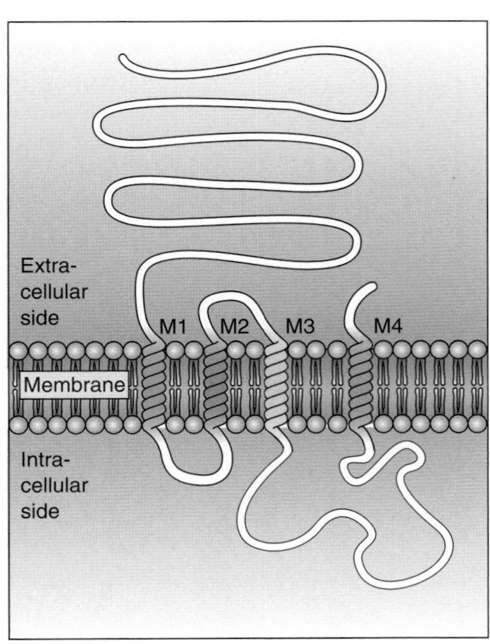

(a) (b)

Figure 6.18
Similarities in the structure of subunits for different transmitter-gated ion channels. (a) If the polypeptides for various channel subunits were stretched out in a line, this is how they would compare. They have in common the four regions called M1–M4, which are segments where the polypeptides will coil into alpha helices to span the membrane. Kainate receptors are subtypes of glutamate receptors. **(b)** M1–M4 regions of the ACh α subunit, as they are believed to be threaded through the membrane.

The most interesting variations among channel structures are the ones that account for their differences. Different transmitter binding sites let one channel respond to Glu while another responds to GABA; certain amino acids around the narrow ion pore allow only Na^+ and K^+ to flow through some channels, Ca^{2+} through others, and only Cl^- through yet others.

Amino Acid-Gated Channels

Amino acid-gated channels mediate most of the fast synaptic transmission in the CNS. Let's take a closer look at their functions, because they are central to topics as diverse as sensory systems, memory, and disease. Several properties of these channels distinguish them from one another and define their functions within the brain.

- The *pharmacology* of their binding sites describes which transmitters affect them and how drugs interact with them.
- The *kinetics* of the transmitter binding process and channel gating determine the duration of their effect.
- The *selectivity* of the ion channels determines whether they produce excitation or inhibition and whether Ca^{2+} enters the cell in significant amounts.
- The *conductance* of open channels helps determine the magnitude of their effects.

All these properties are a direct result of the molecular structure of the channels.

Glutamate-Gated Channels. As we discussed previously, three glutamate receptor subtypes bear the names of their selective agonists: AMPA, NMDA, and kainate. Each of these is a glutamate-gated ion channel. The AMPA-gated and NMDA-gated channels mediate the bulk of fast excitatory synaptic transmission in the brain. Kainate receptors also exist throughout the brain, but their functions are not clearly understood.

AMPA-gated channels are permeable to both Na^+ and K^+, and most of them are not permeable to Ca^{2+}. The net effect of activating them at normal, negative membrane potentials is to admit Na^+ ions into the cell, causing a rapid and large depolarization. Thus, AMPA receptors at CNS synapses mediate excitatory transmission in much the same way as nicotinic receptors mediate synaptic excitation at neuromuscular junctions.

AMPA receptors coexist with NMDA receptors at many synapses in the brain, so most glutamate-mediated excitatory postsynaptic potentials (EPSPs) have components contributed by both (Figure 6.19). NMDA-gated channels also cause excitation of a cell by admitting Na^+, but they differ from AMPA receptors in two very important ways: (1) NMDA-gated channels are permeable to Ca^{2+}, and (2) inward ionic current through NMDA-gated channels is voltage-dependent. We'll discuss each of these properties in turn.

It is hard to overstate the importance of intracellular Ca^{2+} to cell functions. We have already seen that Ca^{2+} can trigger presynaptic neurotransmitter release. Postsynaptically, Ca^{2+} can activate many enzymes, regulate the opening of a variety of channels, and affect gene expression; in excessive amounts, Ca^{2+} can even trigger the death of a cell (Box 6.4). Thus, activation of NMDA receptors can, in principle, cause widespread and lasting changes in the postsynaptic neuron. Indeed, as we will see in Chapter 24, Ca^{2+} entry through NMDA-gated channels may cause the changes that lead to long-term memory.

When the NMDA-gated channel opens, Ca^{2+} and Na^+ enter the cell (and K^+ leaves), but the magnitude of this inward ionic current depends on the

Figure 6.19

The coexistence of NMDA and AMPA receptors in the postsynaptic membrane of a CNS synapse. (a) An impulse arriving in the presynaptic terminal causes the release of glutamate. **(b)** Glutamate binds to AMPA-gated and NMDA-gated channels in the postsynaptic membrane. **(c)** Entry of Na^+ through the AMPA channels and Na^+ and Ca^{2+} through the NMDA channels causes an EPSP.

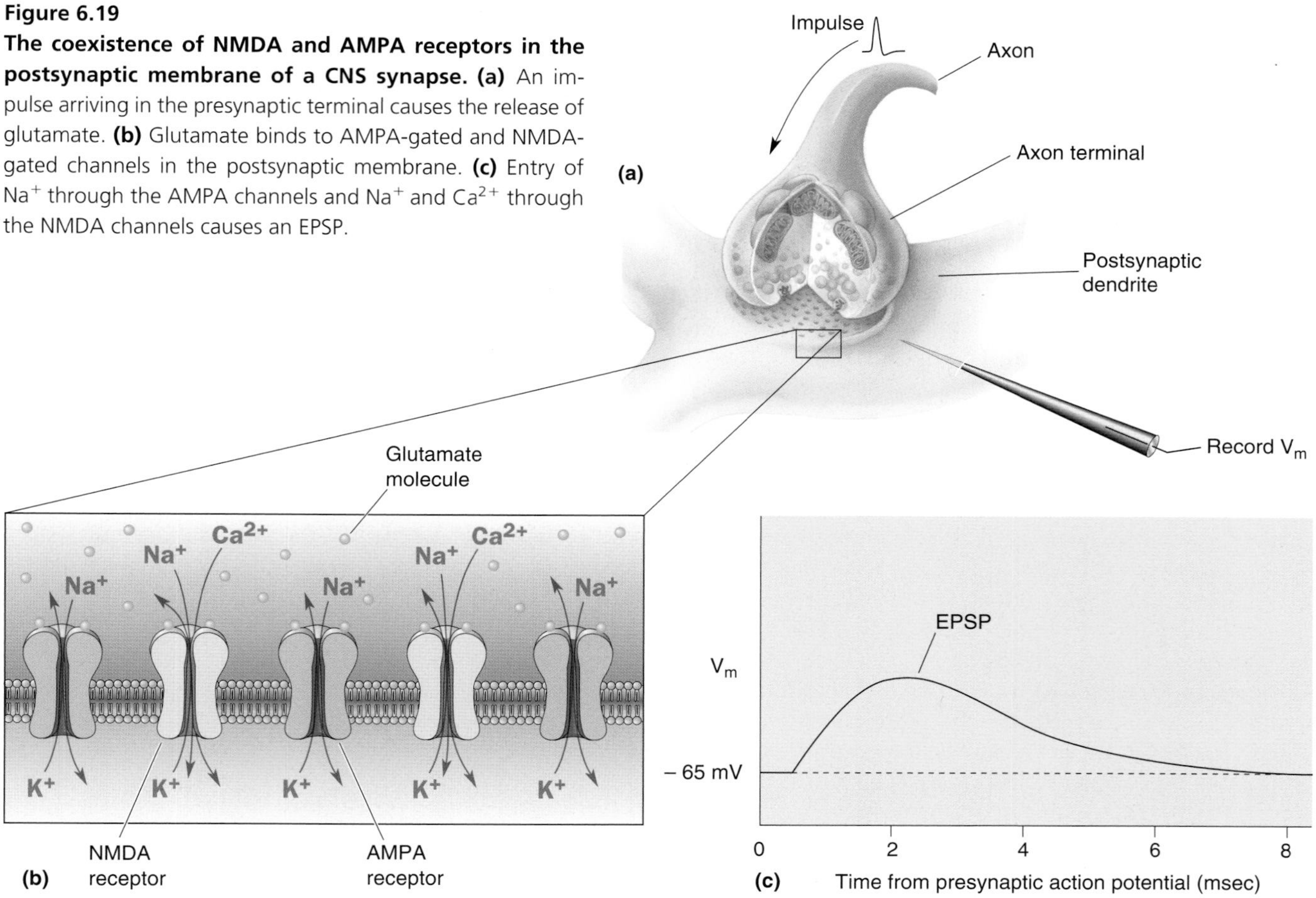

postsynaptic membrane potential in an unusual way for an unusual reason. When glutamate binds to the NMDA receptor, the pore opens as usual. At normal negative resting membrane potentials, however, the channel becomes clogged by Mg^{2+} ions, and the "magnesium block" prevents other ions from passing freely through the NMDA channel. Mg^{2+} pops out of the pore only when the membrane is depolarized, which usually follows activation of AMPA channels at the same and neighboring synapses. Thus, inward ionic current through the NMDA channel is *voltage-dependent*, in addition to being transmitter-gated. Both glutamate and depolarization must coincide before the channel will pass current (Figure 6.20). This property has a significant impact on synaptic integration at many locations in the CNS.

GABA-Gated and Glycine-Gated Channels. GABA mediates most synaptic inhibition in the CNS, and glycine mediates most of the rest. Both the $GABA_A$ receptor and the glycine receptor gate a chloride channel. Surprisingly, inhibitory $GABA_A$ and glycine receptors have a structure very similar to that of excitatory nicotinic ACh receptors, despite the fact that the first two are selective for anions, while the last is selective for cations. Each receptor has α subunits that bind the transmitter and β subunits that do not.

Synaptic inhibition must be tightly regulated in the brain. Too much causes loss of consciousness and coma; too little leads to a seizure. The need to control inhibition may explain why the $GABA_A$ receptor has, in addition to its GABA binding site, several other sites where chemicals can dramatically modulate its function. For example, two classes of drugs, **benzodiazepines** (such as the tranquilizer diazepam, or Valium) and **barbiturates**

Box 6.4

OF SPECIAL INTEREST

The Brain's Exciting Poisons

Neurons of the brain rarely regenerate, so each dead neuron is one less we have for thinking. One of the fascinating ironies of neuronal life and death is that glutamate, the most essential neurotransmitter in the brain, is also one of the biggest killers of neurons. A large percentage of the brain's synapses release glutamate, and it is stored in large quantities. Even the cytosol of nonglutamatergic neurons has a very high glutamate concentration, greater than 3 mM. An ominous observation is that when you apply this same amount of glutamate to the outside of isolated neurons, they die within minutes.

The voracious metabolic rate of the brain demands a continuous supply of oxygen and glucose. If blood flow ceases, as in cardiac arrest, neural activity will stop within seconds, and permanent damage will result within a few minutes. Disease states such as cardiac arrest, stroke, brain trauma, seizures, and oxygen deficiency can initiate a vicious cycle of excess glutamate release. Whenever neurons cannot generate enough ATP to keep their ion pumps working hard, membranes depolarize and Ca^{2+} leaks into cells. The entry of Ca^{2+} triggers the synaptic release of glutamate. Glutamate further depolarizes neurons, which further raises intracellular Ca^{2+} and causes still more glutamate to be released. At this point, there may even be a *reversal* of the glutamate transporter, further contributing to the cellular leakage of glutamate.

When glutamate reaches high concentrations, it kills neurons by overexciting them, a process called *excitotoxicity*. Glutamate simply activates its several types of receptors, which allow excessive amounts of Na^+, K^+, and Ca^{2+} to flow across the membrane. The NMDA subtype of the glutamate-gated channel is a critical player in excitotoxicity because it is the main route for Ca^{2+} entry. Neuron damage or death occurs because of swelling resulting from water uptake and stimulation by Ca^{2+} of intracellular enzymes that degrade proteins, lipids, and nucleic acids. Neurons literally digest themselves.

Excitotoxicity has recently been implicated in several progressive neurodegenerative human diseases, such as *amyotrophic lateral sclerosis* (ALS, also known as Lou Gehrig's disease), in which spinal motor neurons slowly die, and *Alzheimer's disease*, in which brain neurons slowly die. The effects of various environmental toxins mimic aspects of these diseases. Eating large quantities of a certain type of chickpea can cause lathyrism, a degeneration of motor neurons. The pea contains an excitotoxin called β-oxalylaminoalanine, which activates glutamate receptors. A toxin called domoic acid, found in contaminated mussels, is also a glutamate receptor agonist. Ingesting small amounts of domoic acid causes seizures and brain damage. And another plant excitotoxin, β-methylaminoalanine, may cause a hideous condition that combines signs of ALS, Alzheimer's disease, and Parkinson's disease in individual patients on the island of Guam.

As researchers sort out the tangled web of excitotoxins, receptors, enzymes, and neurological disease, new strategies for treatment will emerge. Already, glutamate receptor antagonists that can obstruct these excitotoxic cascades and minimize neuronal suicide show clinical promise. Genetic manipulations may eventually thwart neurodegenerative conditions in susceptible people.

(including phenobarbital, and other sedatives and anticonvulsants), each bind to their own distinct site on the outside face of the $GABA_A$ channel (Figure 6.21). By themselves, these drugs do very little to the channel. But when GABA is present, benzodiazepines increase the frequency of channel openings, while barbiturates increase the duration of channel openings. The result in each case is more inhibitory Cl^- current, stronger inhibitory postsynaptic potentials, and the behavioral consequences of enhanced inhibition. The actions of benzodiazepines and barbiturates are selective for the $GABA_A$ receptor, and the drugs have no effect on glycine receptor function. Some of this selectivity can be understood in molecular terms; only receptors with the γ type of $GABA_A$ subunit, in addition to α and β subunits, respond to benzodiazepines.

Another popular drug that strongly enhances the function of the $GABA_A$ receptor is ethanol, the form of alcohol imbibed in beverages. Ethanol has complex actions that include effects on NMDA, glycine, nicotinic ACh, and serotonin receptors. Its effects on $GABA_A$ channels depend on their specific structure. Evidence indicates that particular α, β, and γ subunits are necessary for constructing an ethanol-sensitive $GABA_A$ receptor, similar to the structure that is benzodiazepine-sensitive. This explains why ethanol enhances inhibition in some brain areas but not others. By understanding this molecular and anatomical specificity, we can begin to appreciate how drugs like ethanol lead to such powerful and addictive effects on behavior.

These myriad drug effects present an interesting paradox. Surely the $GABA_A$ receptor did not evolve modulatory binding sites just for the benefit of our modern drugs. The paradox has motivated researchers to search for endogenous ligands, natural chemicals that might bind to benzodiazepine and barbiturate sites and serve as regulators of inhibition. There is now substantial evidence that natural benzodiazepine-like ligands exist, although identifying them and understanding their functions are proving difficult. Other good candidates as natural modulators of $GABA_A$ receptors are the *neurosteroids*, natural metabolites of steroid hormones that are synthesized from cholesterol primarily in the gonads and adrenal glands but also in glial cells of the brain. Although some neurosteroids suppress inhibitory function, others enhance it, and they seem to do so by binding to their own site on the $GABA_A$ receptor (Figure 6.21), distinct from those of the other drugs we've mentioned. The functions of natural neurosteroids are also obscure, but they suggest a means by which brain and body physiology could be regulated in parallel by the same chemicals.

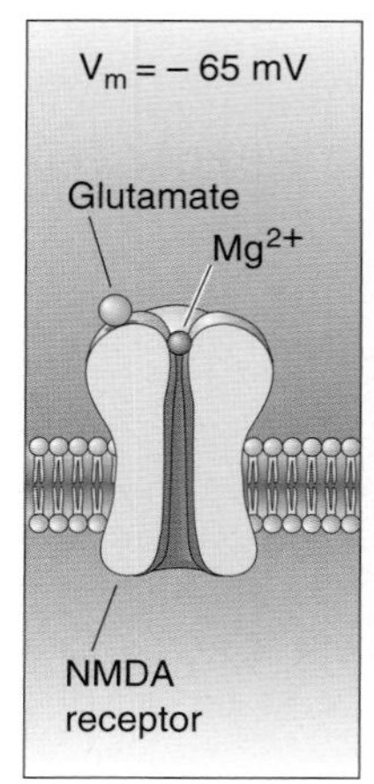

(a) Glutamate

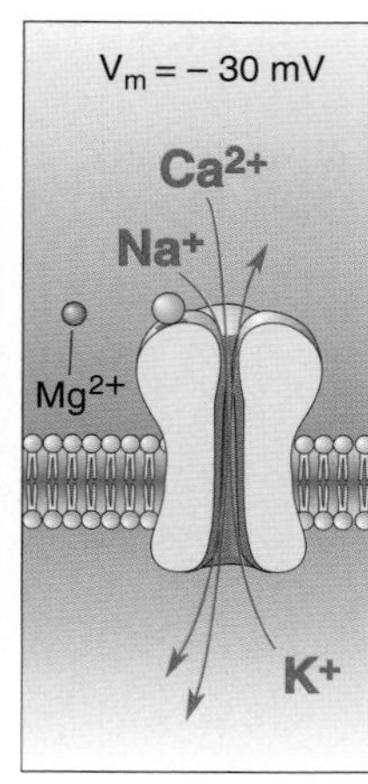

(b) Glutamate and depolarization

Figure 6.20
The flow of inward ionic current through the NMDA-gated channel. (a) Glutamate alone causes the channel to open, but at the resting membrane potential, the pore becomes blocked by Mg^{2+} ions. **(b)** Depolarization of the membrane relieves the Mg^{2+} block and allows Na^+ and Ca^{2+} to enter.

G-PROTEIN-COUPLED RECEPTORS AND EFFECTORS

There are multiple subtypes of G-protein-coupled receptors in every known neurotransmitter system. In Chapter 5, we learned that transmission at these receptors involves three steps: (1) binding of the neurotransmitter to the receptor protein, (2) activation of G-proteins, and (3) activation of effector systems. Let's focus on each of these steps.

The Basic Structure of G-Protein-Coupled Receptors

Most G-protein-coupled receptors are simple variations on a common plan, consisting of a single polypeptide containing seven membrane-spanning alpha helices (Figure 6.22). Two of the extracellular loops of the polypeptide form the transmitter binding sites. Structural variations in this region deter-

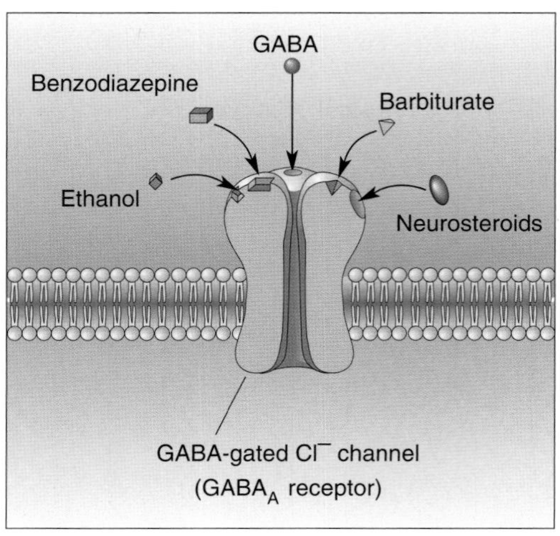

Figure 6.21
The binding of drugs to the $GABA_A$ receptor. The drugs by themselves do not open the channel, but they change the effect that GABA has when it binds to the channel at the same time as the drug.

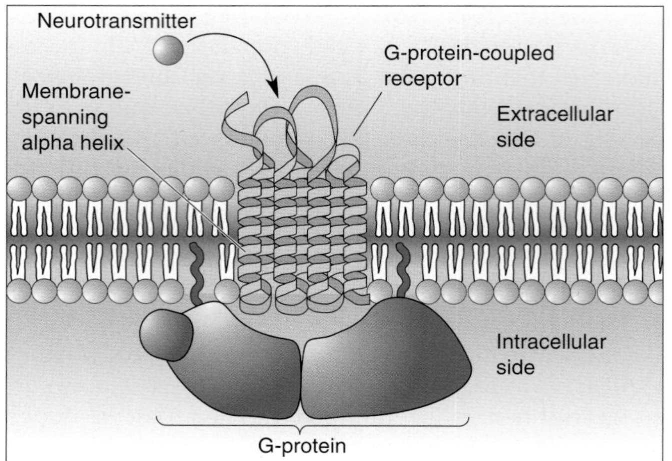

Figure 6.22
The basic structure of a G-protein-coupled receptor. Most metabotropic receptors have seven membrane-spanning alpha helices, a transmitter binding site on the extracellular side, and a G-protein binding site on the intracellular side.

mine which neurotransmitters, agonists, and antagonists bind to the receptor. Two of the intracellular loops can bind to and activate G-proteins. Structural variations here determine which G-proteins and, consequently, which effector systems are activated in response to transmitter binding.

A partial list of G-protein-coupled receptors appears in Table 6.2. About 100 such receptors have been described. Most of these were unknown only a few years ago, before the powerful methods of molecular neurobiology were applied to the problem.

The Ubiquitous G-Proteins

G-proteins are the common link in most signaling pathways that start with a neurotransmitter receptor and end with effector proteins. G-protein is short for guanosine triphosphate (GTP)-binding protein, which is actually a diverse family of about 20 types. There are many more transmitter receptors than G-proteins, so some types of G-proteins can be activated by many receptors.

All G-proteins have the same basic mode of operation (Figure 6.23):

1. Each G-protein has three subunits, termed α, β, and γ. In the resting state, a guanosine diphosphate (GDP) molecule is bound to the G_α subunit, and the whole complex floats around on the inner surface of the membrane.
2. If this GDP-bound G-protein bumps into the proper type of receptor *and* if that receptor has a transmitter molecule bound to it, the G-protein releases its GDP and exchanges it for a GTP that it picks up from the cytosol.

Table 6.2 Some G-Protein-Coupled Neurotransmitter Receptors

NEUROTRANSMITTER	RECEPTOR(S)
Acetylcholine (ACh)	Muscarinic receptors (M_1, M_2, M_3, M_4, M_5)
Glutamate (Glu)	Metabotropic glutamate receptors (mGluR1–7)
GABA	$GABA_B$ receptor
Serotonin (5-HT)	$5\text{-}HT_{1(A, B, C, D\alpha, D\beta, E, F)}$ $5\text{-}HT_{2, 2F}$, $5\text{-}HT_4$, $5\text{-}HT_{5\alpha, 5\beta}$
Dopamine (DA)	D_1, $D_{2S, Sl}$, D_3, D_4, D_5
Norepinephrine (NE)	α_1, α_2, β_1, β_2, β_3
Enkephalin	μ, δ, κ
Cannabinoid	CB1, CB2

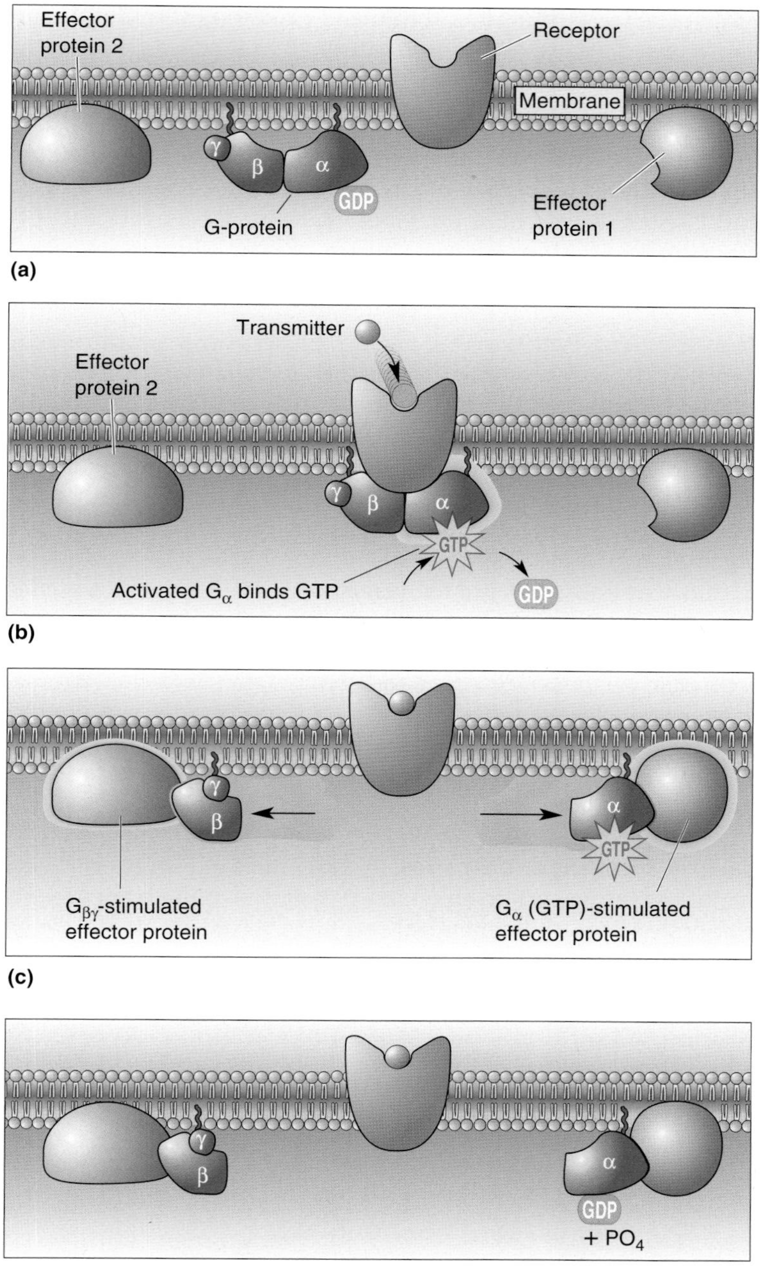

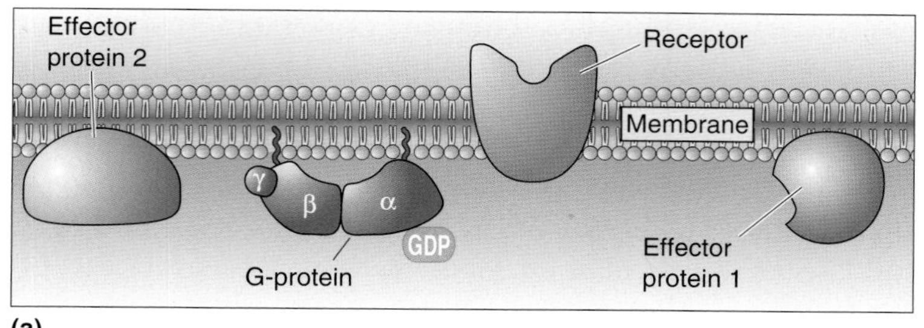

(a)

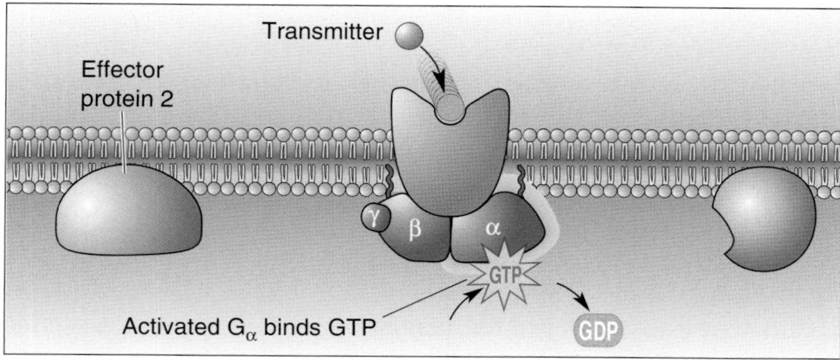

(b)

γ
β
α
GTP
$G_{\beta\gamma}$-stimulated effector protein
G_α (GTP)-stimulated effector protein

(c)

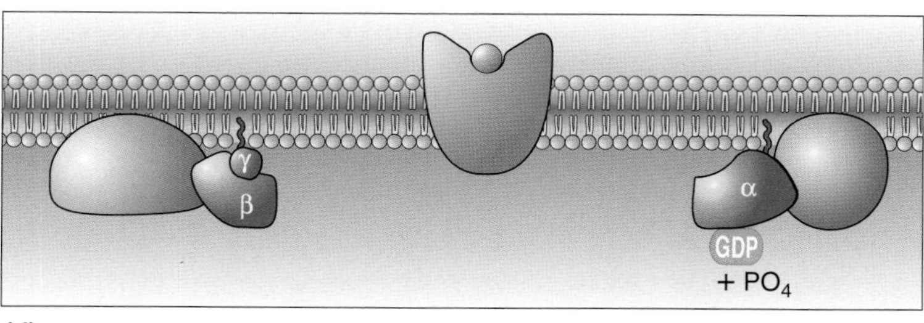

(d)

Figure 6.23
The basic mode of operation of G-proteins. (a) In its inactive state, the α subunit of the G-protein binds GDP. **(b)** When activated by a G-protein-coupled receptor, the GDP is exchanged for GTP. **(c)** The activated G-protein splits, and both the G_α (GTP) subunit and the $G_{\beta\gamma}$ subunit become available to activate effector proteins. **(d)** The G_α subunit slowly removes phosphate (PO_4) from GTP, converting GTP to GDP and terminating its own activity.

3. The activated GTP-bound G-protein splits into two parts: the G_α subunit plus GTP, and the $G_{\beta\gamma}$ complex. Both can then move on to stimulate various effector proteins.
4. The G_α subunit is itself an enzyme that breaks down GTP into GDP. Therefore, G_α eventually terminates its own activity by converting the bound GTP to GDP.
5. The G_α and $G_{\beta\gamma}$ subunits come back together, allowing the cycle to begin again.

The first G-proteins that were discovered had the effect of stimulating effector proteins. Subsequently, it was found that other G-proteins could inhibit these same effectors. Thus, the simplest scheme for subdividing the G-proteins is G_s, designating that the G-protein is stimulatory, and G_i , designating that the G-protein is inhibitory.

G-Protein-Coupled Effector Systems

In Chapter 5, we learned that activated G-proteins exert their effects by binding to either of two types of effector proteins: G-protein-gated ion channels and G-protein-activated enzymes. Because the effects do not involve any other chemical intermediaries, the first route is sometimes called the shortcut pathway.

The Shortcut Pathway. A variety of neurotransmitters use the shortcut pathway, from receptor to G-protein to ion channel. One example is the muscarinic receptors in the heart. These ACh receptors are coupled via G-proteins to potassium channels, explaining why ACh slows the heart rate (Figure 6.24). In this case, the $\beta\gamma$ subunits migrate laterally along the membrane until they bind to the right type of potassium channel. Another example is neuronal $GABA_B$ receptors, also coupled by the shortcut pathway to potassium channels.

Shortcut pathways are the fastest of the G-protein-coupled systems, having responses beginning within 30–100 msec of neurotransmitter binding. Although not quite as fast as a transmitter-gated channel, which uses no intermediary between receptor and channel, this is faster than the second messenger cascades we will describe next. The shortcut pathway is also very localized, compared with other effector systems. As the G-protein diffuses within the membrane, it apparently cannot move very far, so only channels nearby can be affected.

Second Messenger Cascades. G-proteins can also exert their effects by directly activating certain enzymes. Activation of these enzymes can trigger an elaborate series of biochemical reactions, a cascade that often ends in the activation of other "downstream" enzymes that alter neuronal function.

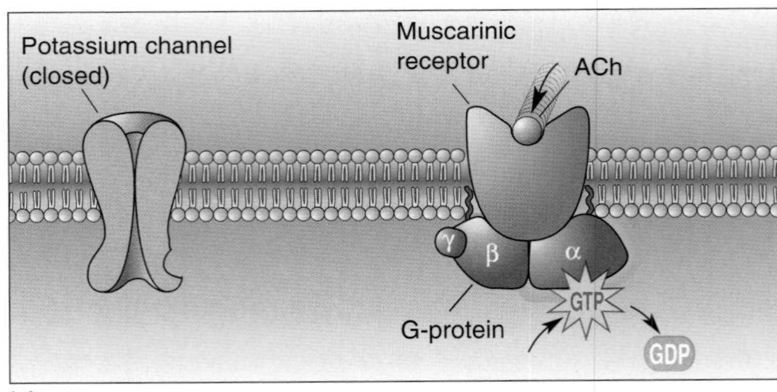

(a)

(b)

Figure 6.24
The shortcut pathway. (a) G-proteins in heart muscle are activated by ACh binding to muscarinic receptors. **(b)** The activated $G_{\beta\gamma}$ subunit directly gates a potassium channel.

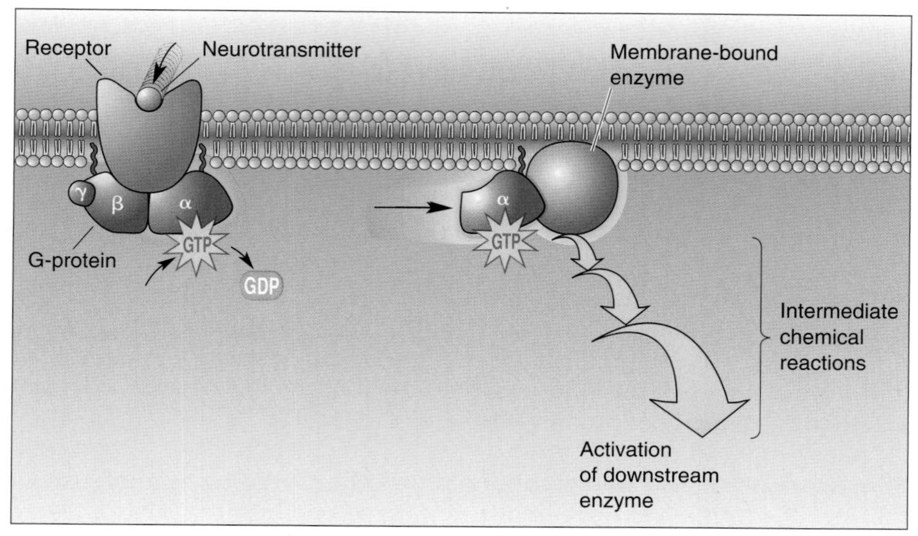

Figure 6.25
The components of a second messenger cascade.

Between the first enzyme and the last are several *second messengers*. The whole process that couples the neurotransmitter, via multiple steps, to activation of a downstream enzyme is called a **second messenger cascade** (Figure 6.25).

In Chapter 5, we introduced the cAMP second messenger cascade initiated by the activation of the NE β receptor (Figure 6.26a). It begins with the β receptor activating the stimulatory G-protein, G_s, which proceeds to stimulate the membrane-bound enzyme adenylyl cyclase. Adenylyl cyclase converts ATP to cAMP. The subsequent rise of cAMP in the cytosol activates a specific downstream enzyme called **protein kinase A (PKA)**.

Many biochemical processes are regulated with a push-pull method, one to stimulate them and one to inhibit them, and cAMP production is no exception. The activation of a second type of NE receptor, called the α_2 receptor, leads to the activation of G_i (inhibitory G-protein). G_i suppresses the activity of adenylyl cyclase, and this effect can take precedence over the stimulatory system (Figure 6.26b).

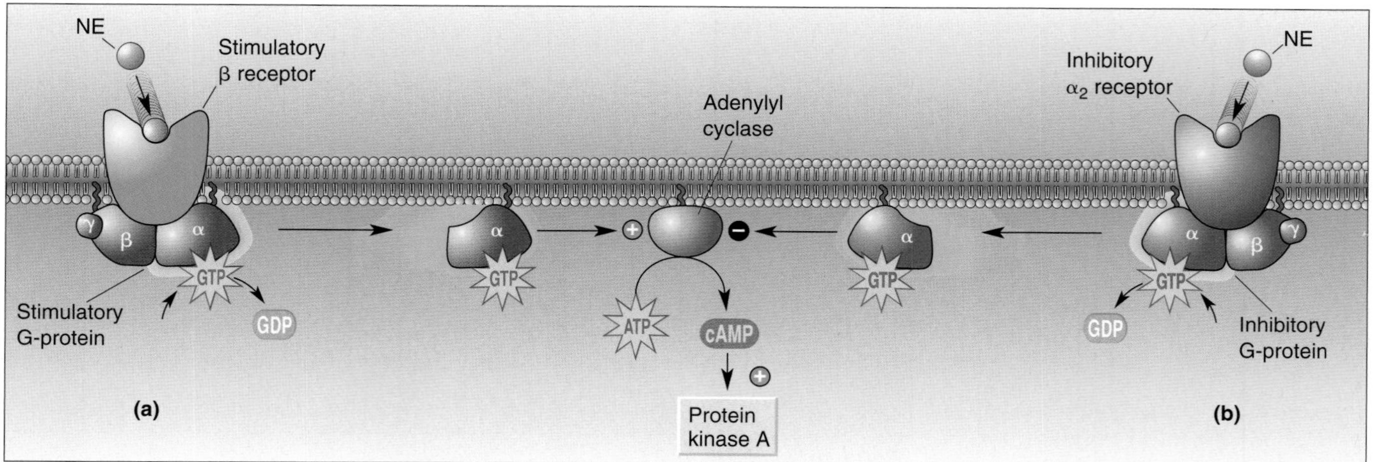

Figure 6.26
The stimulation and inhibition of adenylyl cyclase by different G-proteins. (a) The binding of NE to the β receptor activates G_s, which in turn activates adenylyl cyclase. Adenylyl cyclase generates cAMP, which activates the downstream enzyme protein kinase A. **(b)** The binding of NE to the α_2 receptor activates G_i, which inhibits adenylyl cyclase.

Some messenger cascades can branch. Figure 6.27 shows how the activation of various G-proteins can stimulate **phospholipase C (PLC)**, an enzyme that floats in the membrane like adenylyl cyclase. PLC acts on a membrane phospholipid (PIP_2, or phosphatidylinositol-4,5-bisphosphate), splitting it to form the two molecules that serve as second messengers: **diacylglycerol (DAG)** and **inositol-1,4,5-triphosphate (IP_3)**. DAG, which is lipid-soluble, stays within the plane of the membrane, where it activates a downstream enzyme, **protein kinase C (PKC)**. At the same time, the water-soluble IP_3 diffuses away in the cytosol and binds to specific receptors on the smooth ER and other membrane-enclosed organelles in the cell. These receptors are IP_3-gated calcium channels, so IP_3 causes the organelles to discharge their stores of Ca^{2+}. As we have said, elevations in cytosolic Ca^{2+} can trigger widespread and long-lasting effects. One effect is activation of the enzyme **calcium-calmodulin-dependent protein kinase**, or **CaMK**.

Phosphorylation and Dephosphorylation. The preceding examples show that key downstream enzymes in many of the second messenger cascades are *protein kinases* (PKA, PKC, CaMK). As mentioned in Chapter 5, protein kinases transfer phosphate from ATP floating in the cytosol to proteins, a reaction called *phosphorylation*. The addition of phosphate groups to a protein changes its conformation slightly, thereby changing its biological activity. Phosphorylation of ion channels, for example, can strongly influence the probability that they will open or close.

Consider the consequence of activating β receptors on cardiac muscle cells. The subsequent rise in cAMP activates PKA, which phosphorylates the cell's voltage-gated calcium channels, and this *enhances* their activity. More Ca^{2+} flows, and the heart beats more strongly. By contrast, the stimulation of β adrenergic receptors in many neurons seems to have no effect on calcium channels but causes *inhibition* of certain potassium channels. Reduced K^+ conductance causes a slight depolarization, increases the length constant, and makes the neuron more excitable (see Chapter 5).

If transmitter-stimulated kinases were allowed to phosphorylate without some method of reversing the process, all proteins would quickly become saturated with phosphates, and further regulation would become impossi-

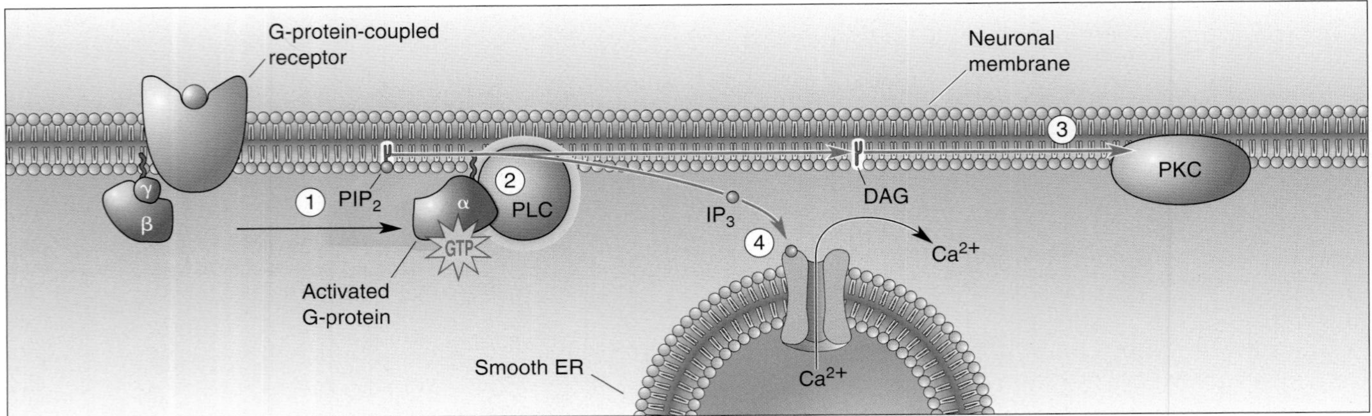

Figure 6.27
Second messengers generated by the breakdown of PIP_2, a membrane phospholipid. ① Activated G-proteins stimulate the enzyme phospholipase C (PLC). ② PLC splits PIP_2 into DAG and IP_3. ③ DAG stimulates the downstream enzyme protein kinase C (PKC). ④ IP_3 stimulates the release of Ca^{2+} from intracellular stores. The Ca^{2+} can go on to stimulate various downstream enzymes.

ble. Enzymes called **protein phosphatases** save the day; they act rapidly to remove phosphate groups. The degree of channel phosphorylation at any moment, therefore, depends on the dynamic balance of phosphorylation by kinases and dephosphorylation by phosphatases (Figure 6.28).

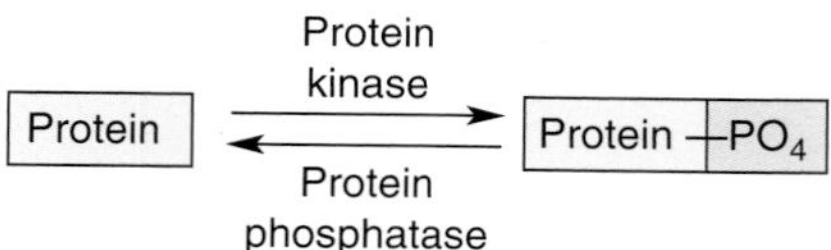

Figure 6.28
Protein phosphorylation and dephosphorylation.

The Function of Signal Cascades. Synaptic transmission using transmitter-gated channels is simple and fast. Transmission involving G-protein-coupled receptors is complex and slow. What are the advantages of having such long chains of command? One important advantage is *signal amplification*: The activation of one G protein-coupled receptor can lead to the activation of not one, but many, ion channels (Figure 6.29).

Signal amplification can occur at several places in the cascade. A single neurotransmitter molecule, bound to one receptor, can activate perhaps 10–20 G-proteins; each G-protein can activate an adenylyl cyclase, which can make many cAMP molecules that can spread to activate many kinases; each kinase can then phosphorylate many channels. If all cascade components were tied together in a clump, signaling would be severely limited. The use of small messengers that can diffuse quickly (such as cAMP) also allows sig-

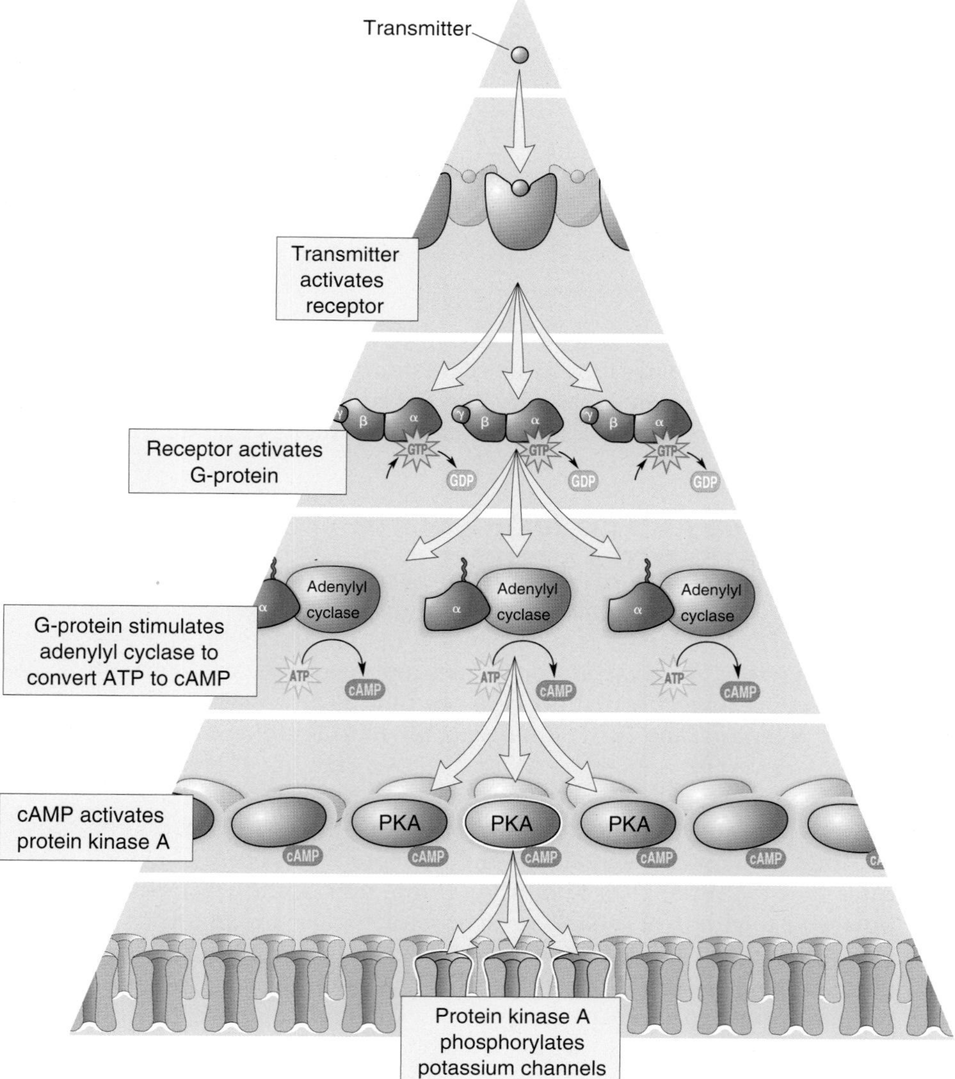

Figure 6.29
Signal amplification by G-protein-coupled second messenger cascades. When a transmitter activates a G-protein-coupled receptor, there can be amplification of the messengers at several stages of the cascade, so that ultimately many channels are affected.

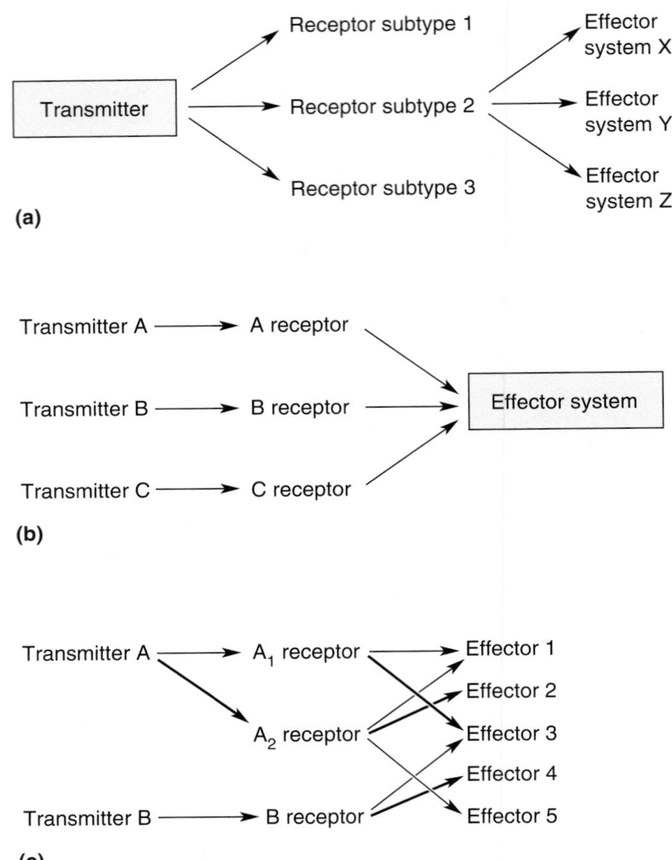

Figure 6.30
Divergence and convergence in neurotransmitter signaling systems. (a) Divergence. **(b)** Convergence. **(c)** Integrated divergence and convergence.

naling at a distance, over a wide stretch of cell membrane. Signal cascades also provide many sites for further regulation, as well as interaction between cascades. Finally, signal cascades can generate very long-lasting chemical changes in cells, which may form the basis for, among other things, a lifetime of memories.

DIVERGENCE AND CONVERGENCE IN NEUROTRANSMITTER SYSTEMS

Glutamate is the most common excitatory neurotransmitter in the brain, while GABA is the pervasive inhibitory neurotransmitter. But this is only part of the story, because any single neurotransmitter can have many different effects. A molecule of glutamate can bind to any of several kinds of glutamate receptors, and each of these can mediate a different effect. The ability of one transmitter to activate more than one subtype of receptor and cause more than one type of postsynaptic response is called *divergence*.

Divergence is the rule among neurotransmitter systems. Every known neurotransmitter can activate multiple receptor subtypes, and evidence indicates that the number of known receptors will continue to escalate as the powerful methods of molecular neurobiology are applied to each system. Because of the multiple receptor subtypes, one transmitter can affect different neurons (or even different parts of the same neuron) in very different ways. Divergence also occurs at points beyond the receptor level, depending on which G-proteins and which effector systems are activated. Divergence may occur at any stage in the cascade of transmitter effects (Figure 6.30a).

Neurotransmitters can also exhibit *convergence* of effects. Multiple transmitters, each activating their own receptor type, can converge to affect the same effector systems (Figure 6.30b). Convergence in a single cell can occur at the level of the G-protein, the second messenger cascade, or the type of ion channel. Neurons integrate divergent and convergent signaling systems, resulting in a complex map of chemical effects (Figure 6.30c). The wonder is that it ever works; the challenge is to understand how.

CONCLUDING REMARKS

Neurotransmitters are the essential links between neurons and between neurons and other effector cells, such as muscle cells and glandular cells. But it is important to think of transmitters as one link in a chain of events, inciting chemical changes both fast and slow, divergent and convergent. You can envision the many signaling pathways onto and within a single neuron as a kind of information network. This network is in delicate balance, shifting its effects dynamically as the demands on a neuron vary with changes in the organism's behavior.

The signaling network within a single neuron resembles in some ways the neural networks of the brain itself. It receives a variety of inputs, in the form of transmitters bombarding it at different times and places. These inputs cause an increased drive through some signal pathways and a decreased drive through others, and the information is recombined to yield a particular output that is more than a simple summation of the inputs. Signals regulate signals, chemical changes can leave lasting traces of their history, drugs can shift the balance of signaling power, and in a literal sense, the brain and its chemicals are one.

KEY TERMS

Introduction
cholinergic (p. 131)
noradrenergic (p. 131)
glutamatergic (p. 131)
GABAergic (p. 131)
peptidergic (p. 132)

Studying Neurotransmitter Systems
immunocytochemistry (p. 132)
in situ hybridization (p. 133)
autoradiography (p. 134)
microionophoresis (p. 136)
receptor subtype (p. 136)
nicotinic receptor (p. 136)
muscarinic receptor (p. 136)
AMPA receptor (p. 137)
NMDA receptor (p. 137)
kainate receptor (p. 137)
ligand-binding method (p. 138)

Neurotransmitter Chemistry
Dale's principle (p. 140)
acetylcholine (ACh) (p. 141)
transporter (p. 141)
rate-limiting step (p. 141)
catecholamines (p. 143)
dopamine (DA) (p. 143)
norepinephrine (NE) (p. 143)
epinephrine (adrenaline) (p. 143)
dopa (p. 143)
serotonin (5-HT) (p. 145)
serotonergic (p. 145)
glutamate (Glu) (p. 145)
glycine (Gly) (p. 145)
gamma-aminobutyric acid (GABA) (p. 145)
nitric oxide (NO) (p. 146)
retrograde messenger (p. 146)

Transmitter-Gated Channels
benzodiazepine (p. 151)
barbiturate (p. 151)

G-Protein-Coupled Receptors and Effectors
second messenger cascade (p. 157)
protein kinase A (PKA) (p. 157)
phospholipase C (PLC) (p. 158)
diacylglycerol (DAG) (p. 158)
inositol-1,4,5-triphosphate (IP_3) (p. 158)
protein kinase C (PKC) (p. 158)
calcium-calmodulin-dependent protein kinase (CaMK) (p. 158)
protein phosphatase (p. 159)

REVIEW QUESTIONS

1. If you could place microelectrodes into both a presynaptic and a postsynaptic neuron, how would you determine whether the synapse between them was chemically or electrically mediated?
2. List the criteria that are used to determine whether a chemical serves as a neurotransmitter. What are the various experimental strategies you could use to show that ACh fulfills the criteria of a neurotransmitter at the neuromuscular junction?
3. What are three methods that could be used to show that a neurotransmitter receptor is synthesized or localized in a particular neuron?
4. Compare and contrast the properties of (a) AMPA and NMDA receptors, and (b) $GABA_A$ and $GABA_B$ receptors.
5. Synaptic inhibition is an important feature of the circuitry in the cerebral cortex. How would you determine whether GABA or Gly, or both, or neither is the inhibitory neurotransmitter of the cortex?
6. Glutamate activates a number of different metabotropic receptors. The consequence of activating one subtype is the *inhibition* of cAMP formation. A consequence of activating a second subtype is *activation* of protein kinase C. Propose mechanisms for these different effects.
7. Do convergence and divergence of neurotransmitter effects occur in single neurons?
8. Ca^{2+} ions are considered to be second messengers. Why?

CHAPTER 7

The Structure of the Nervous System

INTRODUCTION

In previous chapters, we saw how individual neurons function and communicate. Now, we are ready to assemble them into a nervous system that sees, hears, feels, moves, remembers, and dreams. Just as an understanding of neuronal structure was necessary for understanding neuronal function, we must understand nervous system structure to understand brain function.

Neuroanatomy has challenged generations of students, and for good reason: The human brain is extremely complicated. Our brain is, however, merely a variation on a plan that is common to the brains of all mammals (Figure 7.1). The human brain appears complicated because it is distorted as a result of the selective growth of some parts within the confines of the skull. But once the basic mammalian plan is understood, these specializations of the human brain become transparent.

We begin by introducing the general organization of the mammalian brain and the terms used to describe it. Then we take a look at how the three-dimensional structure of the brain arises during embryological and fetal development. Following the course of development makes it easier to understand how the parts of the adult brain fit together. Finally, we explore the cerebral neocortex, a structure that is unique to mammals and proportionately the largest in humans. An Illustrated Guide to Human Neuroanatomy follows the chapter as an appendix.

The neuroanatomy presented in this chapter provides the canvas on which we will paint the sensory and motor systems in Chapters 8–14. Because you will encounter a lot of new terms, self-quizzes within the chapter provide an opportunity for review.

GROSS ORGANIZATION OF THE MAMMALIAN NERVOUS SYSTEM

The nervous system of all mammals has two divisions: the central nervous system (CNS) and the peripheral nervous system (PNS). In this section, we identify some of the important components of the CNS and the PNS. We also discuss the membranes that surround the brain and the fluid-filled ventricles within the brain. Then, we explore some new methods of examining the structure of the living brain. But first, we need to review some anatomical terminology.

Anatomical References

Getting to know your way around the brain is like getting to know your way around a city. To describe your location in the city, you use points of reference such as north, south, east, and west, up and down. The same is true for the brain; only the terms, called *anatomical references*, are different.

Consider the nervous system of a rat (Figure 7.2a). We begin with the rat, because it is a simplified version that has all of the general features of mammalian nervous system organization. In the head lies the brain, and the spinal cord runs down inside the backbone toward the tail. The direction pointing toward the rat's nose is called **anterior** or **rostral** (from the Latin for "beak"). The direction pointing toward the rat's tail is called **posterior** or **caudal** (from the Latin for "tail"). The direction pointing up is called **dorsal** (from the Latin for "back"), and the direction pointing down is called **ventral** (from the Latin for "belly"). Thus, the rat spinal cord runs anterior to posterior. The top side of the spinal cord is the dorsal side, and the bottom side is the ventral side.

If we look down on the nervous system, we see that it may be divided into two equal halves (Figure 7.2b). The right side of the brain and spinal cord is the mirror image of the left side. This characteristic is known as *bilateral symmetry*. With just a few exceptions, most structures within the nervous system come in pairs, one on the right side and the other on the left. The line running down the middle of the nervous system is called the **midline**, and this gives

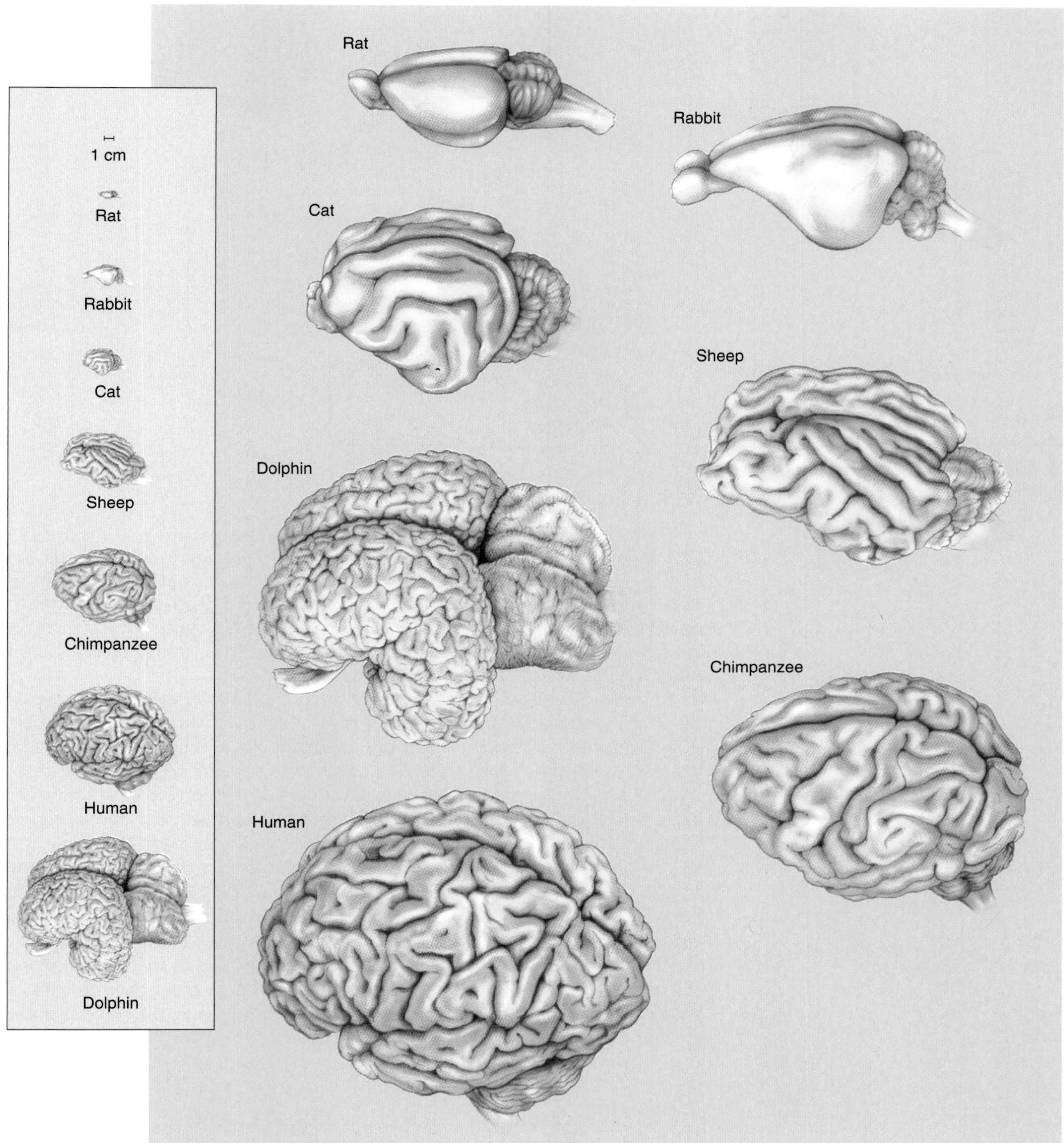

Figure 7.1
Mammalian brains. Despite differences in complexity, the brains of all of these species have many features in common. The brains have been drawn to appear approximately the same size; their relative sizes are shown in the inset on the left.

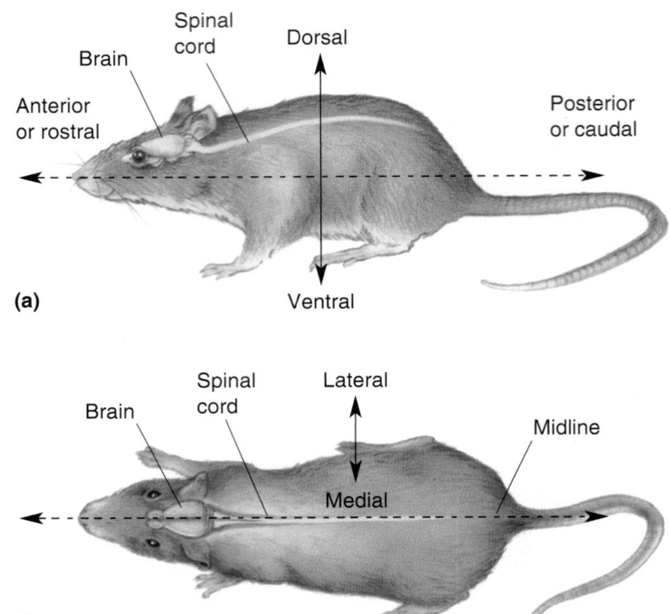

Figure 7.2
Basic anatomical references in the nervous system of a rat. (a) Side view. **(b)** Top view.

us another way to describe direction. Structures closer to the midline are **medial**; structures farther away from the midline are **lateral**. In other words, the nose is medial to the eyes, the eyes are medial to the ears, and so on. In addition, two structures that are on the same side are said to be **ipsilateral** to each other; for example, the right ear is ipsilateral to the right eye. If the structures are on opposite sides of the midline, they are said to be **contralateral** to each other; the right ear is contralateral to the left ear.

To view the internal structure of the brain, it is usually necessary to slice it up. In the language of anatomists, a slice is called a *section*; to slice is *to section*. Although one could imagine an infinite number of ways we might cut into the brain, the standard approach is to make cuts parallel to one of the three *anatomical planes of section*. The plane of the section resulting from splitting the brain into equal right and left halves is called the **midsagittal plane** (Figure 7.3a). Sections parallel to the midsagittal plane are in the **sagittal plane**.

The two other anatomical planes are perpendicular to the sagittal plane and to one another. The **horizontal plane** is parallel to the ground (Figure 7.3b). A single section in this plane could pass through both the eyes and the ears. Thus, horizontal sections split the brain into dorsal and ventral parts. The **coronal plane** is perpendicular to the ground and to the sagittal plane (Figure 7.3c). A single section in this plane could pass through both eyes or both ears but not through all four at the same time. Thus, the coronal plane splits the brain into anterior and posterior parts.

Self-Quiz

Take a few moments right now and be sure you understand the meaning of these terms:

anterior	ventral	contralateral
rostral	midline	midsagittal plane
posterior	medial	sagittal plane
caudal	lateral	horizontal plane
dorsal	ipsilateral	coronal plane

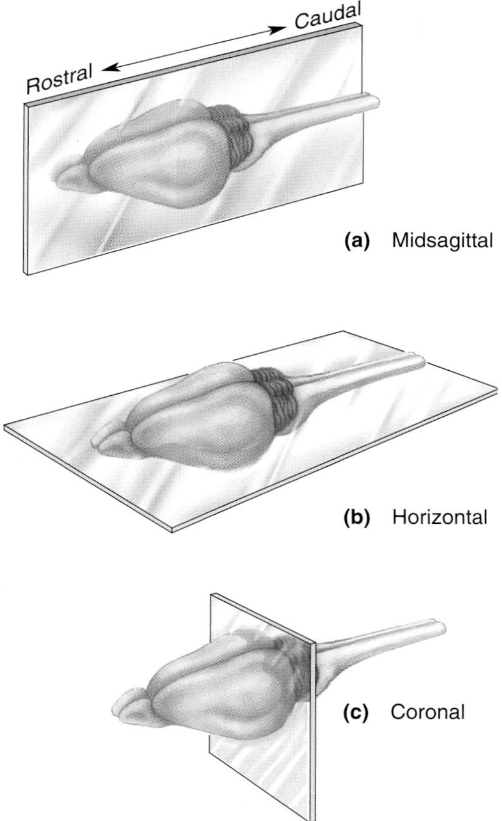

Figure 7.3
Anatomical planes of section.

The Central Nervous System

The **central nervous system**, or **CNS**, consists of the parts of the nervous system that are encased in bone: the **brain** and the **spinal cord**. The brain lies entirely within the skull. A side view of the rat brain reveals three parts that are common to all mammals: the cerebrum, the cerebellum, and the brain stem (Figure 7.4a).

The Cerebrum. The rostral-most and largest part of the brain is the **cerebrum**. Figure 7.4b shows the rat cerebrum as it appears when viewed from above. Notice that it is clearly split down the middle into two **cerebral hemispheres** separated by the deep *sagittal fissure*. In general, the *right* cerebral hemisphere receives sensations from, and controls movements of, the *left* side of the body. Similarly, the *left* cerebral hemisphere is concerned with sensations and movements on the *right* side of the body.

The Cerebellum. Lying behind the cerebrum is the **cerebellum** (Latin for "little brain"). While the cerebellum is, in fact, dwarfed by the large cerebrum, it actually contains as many neurons as both cerebral hemispheres combined. The cerebellum is primarily a movement control center that has extensive connections with the cerebrum and the spinal cord. In contrast to the cerebral hemispheres, the left side of the cerebellum is concerned with movements of the left side of the body, and the right side of the cerebellum is concerned with movements of the right side of the body.

The Brain Stem. The remaining part of the brain is the brain stem, best observed in a midsagittal view of the brain (Figure 7.4c). The **brain stem** forms

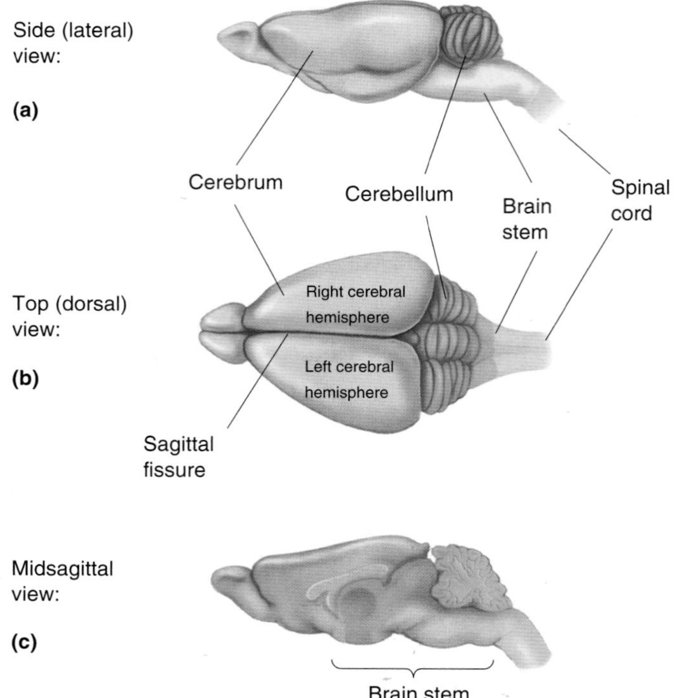

Figure 7.4
The brain of a rat. (a) Side (lateral) view. **(b)** Top (dorsal) view. **(c)** Midsagittal view.

the stalk from which the cerebral hemispheres and the cerebellum sprout. The brain stem is a complex nexus of fibers and cells that, in part, serves to relay information from the cerebrum to the spinal cord and cerebellum and from the spinal cord and cerebellum to the cerebrum. However, the brain stem is also the region that regulates vital functions, such as breathing, consciousness, and the control of body temperature. Indeed, while the brain stem is considered the most primitive part of the mammalian brain, it is also the most important to life. One can survive damage to the cerebrum and cerebellum, but damage to the brain stem usually means rapid death.

The Spinal Cord. The spinal cord is encased in the bony vertebral column and is attached to the brain stem. The spinal cord is the major conduit of information from the skin, joints, and muscles to the brain and from the brain to the skin, joints, and muscles. A transection of the spinal cord results in anesthesia (lack of feeling) in the skin and paralysis of the muscles in parts of the body caudal to the cut. Paralysis in this case does not mean that the muscles cannot function but, rather, that they cannot be controlled by the brain.

The spinal cord communicates with the body via the **spinal nerves**, which are part of the peripheral nervous system (discussed below). Spinal nerves exit the spinal cord through notches between each vertebra of the vertebral column. Each spinal nerve attaches to the spinal cord by means of two branches, the **dorsal root** and the **ventral root** (Figure 7.5). Recall from Chapter 1 that François Magendie showed that the dorsal root contains axons bringing information *into* the spinal cord, such as those that signal the accidental entry of a thumbtack into your foot (see Figure 3.1). Charles Bell showed that the ventral root contains axons carrying information *away from* the spinal cord, for example, to the muscles that jerk your foot away in response to the pain of the thumbtack.

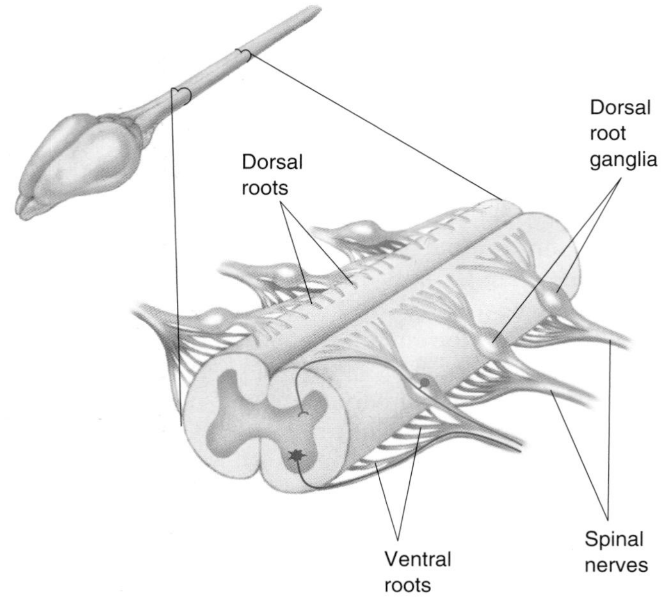

Figure 7.5
The spinal cord. The spinal cord runs inside the vertebral column. Axons enter and exit the spinal cord via the dorsal and ventral roots, respectively. These roots come together to form the spinal nerves that course through the body.

The Peripheral Nervous System

All of the parts of the nervous system other than the brain and spinal cord constitute the **peripheral nervous system**, or **PNS**. The PNS can be divided into two parts: the somatic PNS and the visceral PNS.

The Somatic PNS. All of the spinal nerves that innervate the skin, the joints, and the muscles that are under voluntary control are part of the **somatic PNS**. The somatic motor axons, which command muscle contraction, derive from motor neurons in the ventral spinal cord. The cell bodies of the motor neurons lie within the CNS, but their axons are mostly in the PNS.

The somatic sensory axons, which innervate and collect information from the skin, muscles, and joints, enter the spinal cord via the dorsal roots. The somata of these neurons lie outside the spinal cord in clusters called **dorsal root ganglia**. There is a dorsal root ganglion for each spinal nerve (Figure 7.5).

The Visceral PNS. The **visceral PNS**, also called the involuntary, vegetative, or **autonomic nervous system** (**ANS**), consists of the neurons that innervate the internal organs, blood vessels, and glands. Visceral sensory axons bring information about visceral function to the CNS, such as the pressure and oxygen content of the blood in the arteries. Visceral motor fibers command the contraction and relaxation of muscles that form the walls of the intestines and the blood vessels (called smooth muscles), the rate of cardiac muscle contraction, and the secretory function of various glands. For example, the visceral PNS controls blood pressure by regulating the heart rate and the diameter of the blood vessels.

We will return to the structure and function of the ANS in Chapter 15. For now, remember that when one speaks of an emotional reaction that is beyond voluntary control, such as "butterflies in the stomach" or blushing, it usually is mediated by the visceral PNS (the ANS).

Afferent and Efferent Axons. Our discussion of the PNS is a good place to introduce two terms that are used to describe axons in the nervous system. Derived from the Latin, **afferent** ("carry to") and **efferent** ("carry from") indicate whether the axons are transporting information *toward* or *away from* a particular point. Consider the axons in the PNS relative to a point of reference in the CNS. The somatic or visceral sensory axons bringing information *into* the CNS are afferents. The axons that emerge *from* the CNS to innervate the muscles and glands are efferents.

The Cranial Nerves

In addition to the nerves that arise from the spinal cord and innervate the body, there are 12 pairs of **cranial nerves** that arise from the brain stem and innervate (mostly) the head. Each cranial nerve has a name and a number associated with it (originally numbered by Galen, about 1800 years ago, from anterior to posterior).

Some of the cranial nerves are part of the CNS, others are part of the somatic PNS, and still others are part of the visceral PNS. Many cranial nerves contain a complex mixture of axons that perform various functions. The cranial nerves and their functions are summarized in the chapter appendix.

The Meninges

The CNS, the part of the nervous system encased in the skull and vertebral column, does not come in direct contact with the overlying bone. It is protected by three membranes collectively called the **meninges** (singular: meninx), from the Greek for "covering." The three membranes are the dura mater, the arachnoid membrane, and the pia mater (Figure 7.6).

The outermost covering is the **dura mater**, which is Latin for "hard mother," an accurate description of the dura's leather-like consistency. The dura forms a tough, inelastic bag that surrounds the brain and spinal cord. Just under the dura lies the **arachnoid membrane**, from the Greek for "spider." This meningeal layer has an appearance and consistency resembling those of a spider web. While there normally is no space between the dura and the arachnoid, if the blood vessels passing through the dura are ruptured, blood can collect here and form a *subdural hematoma*. The buildup of fluid in this subdural space can disrupt brain function by compressing parts of the CNS. The disorder is treated by drilling a hole in the skull and draining the blood.

The **pia mater**, the "gentle mother," is a thin membrane that adheres closely to the surface of the brain. Along the pia run many blood vessels that ultimately dive into the substance of the underlying brain. The pia is separated from the arachnoid by a fluid-filled space. This *subarachnoid space* is filled with salty clear liquid called **cerebrospinal fluid**, or **CSF**. Thus, in a sense, the brain floats inside the head in this thin layer of CSF.

The Ventricular System

In Chapter 1, we noted that the brain is hollow. The fluid-filled caverns and canals inside the brain constitute the **ventricular system**. The fluid that runs in this system is CSF, the same as the fluid in the subarachnoid space. CSF is produced by a special tissue, the choroid plexus, in the ventricles of the cerebral hemispheres. CSF flows from the paired ventricles of the cerebrum to a series of connected, unpaired cavities at the core of the brain stem (Figure 7.7). CSF exits the ventricular system and enters the subarachnoid space by way of small openings, or apertures, near where the cerebellum attaches to the brain stem. In the subarachnoid space, CSF is absorbed by the blood vessels at special structures called arachnoid villi. If the normal flow of CSF is disrupted, brain damage can result (Box 7.1).

In a moment we will return to this subject to fill in some details about the

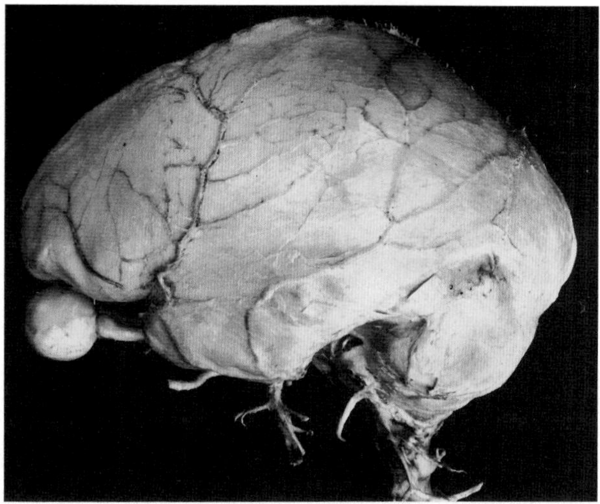

(a)

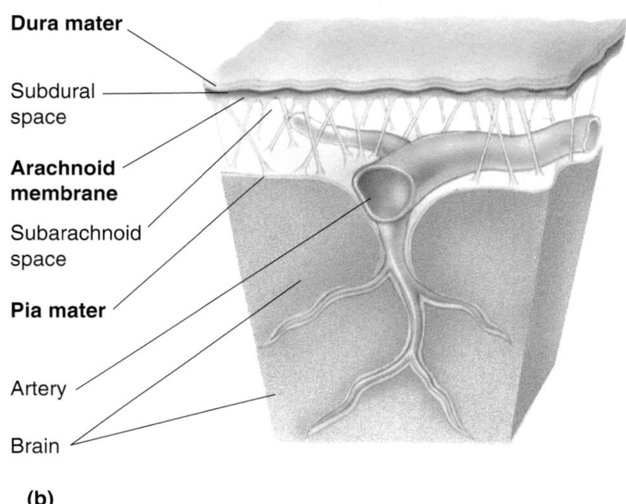

(b)

Figure 7.6
The meninges. (a) The skull has been removed to show the tough outer meningeal membrane, the dura mater. (Source: Gluhbegoric and Williams, 1980.) **(b)** Illustrated in cross section, the three meningeal layers protecting the brain and spinal cord are the dura mater, the arachnoid membrane, and the pia mater.

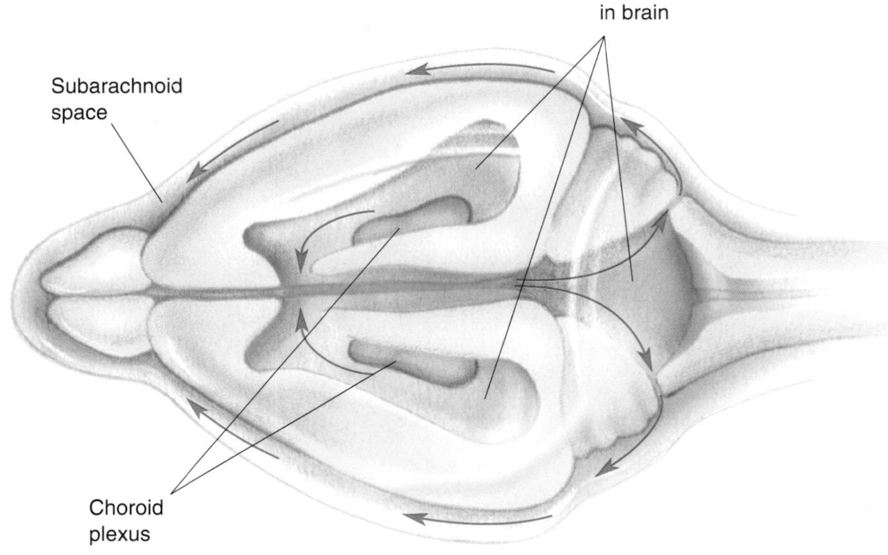

Figure 7.7
The ventricular system in a rat brain. CSF is produced in the ventricles of the paired cerebral hemispheres and flows through a series of unpaired ventricles at the core of the brain stem. CSF escapes into the subarachnoid space via small apertures near the base of the cerebellum. CSF is absorbed into the blood in the subarachnoid space.

Box 7.1

OF SPECIAL INTEREST

Water on the Brain

If the flow of CSF from the choroid plexus through the ventricular system to the subarachnoid space is impaired, the fluid will back up and cause a swelling of the ventricles. This condition is called *hydrocephalus*, meaning "water head."

Occasionally, babies are born with hydrocephalus. Because the skull is soft and not completely formed, however, the head expands to accommodate the increase in intracranial fluid, sparing the brain from damage. Often this condition goes unnoticed until the size of the head reaches enormous proportions.

In adults, hydrocephalus is a much more serious situation because the skull cannot expand, and intracranial pressure increases as a result. The soft brain tissue is compressed, impairing function and leading to death if left untreated. Typically, this "obstructive" hydrocephalus is also accompanied by severe headache caused by the distension of nerve endings in the meninges. Treatment consists of inserting a tube into the swollen ventricle and draining off the excess fluid (Figure A).

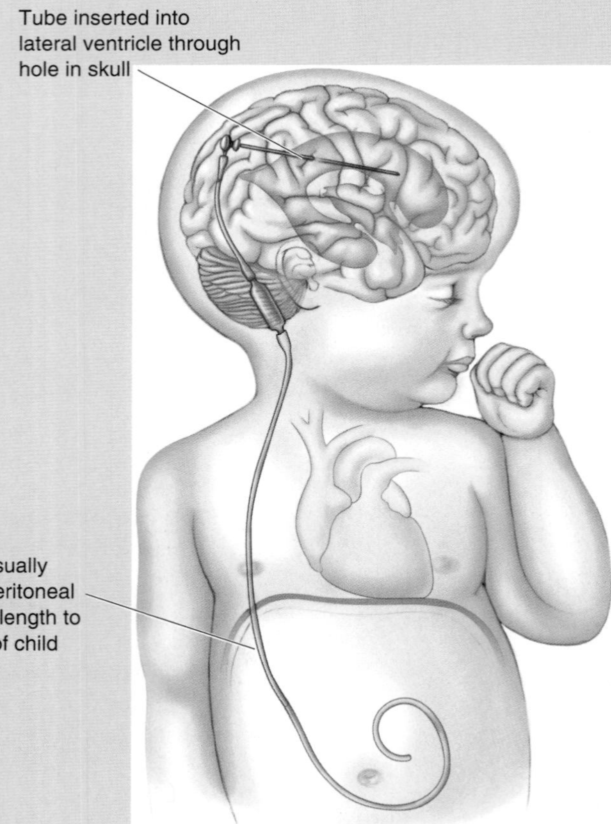

Figure A

ventricular system. As we will see, understanding the organization of the ventricular system holds the key to understanding how the mammalian brain is organized.

Imaging the Living Brain

For centuries, anatomists have investigated the structure of the brain by removing it from the head, sectioning it in various planes, staining the sections, and examining the stained sections. Much has been learned by this approach, but there are some limitations. Most obviously, the brain removed from the head is dead. This, to say the least, limits the usefulness of this method for examining the brain and for diagnosing neurological disorders in living individuals. Neuroanatomy has been revolutionized by the introduction of exciting new methods that enable one to produce images of the living brain. Here we briefly introduce them.

Computed Tomography. Some types of electromagnetic radiation, like X-rays, penetrate the body and are absorbed by various "radiopaque" tissues. Thus, using X-ray-sensitive film, one can make two-dimensional images of the shadows formed by the radiopaque structures within the body. This technique works well for the bones of the skull but not for the brain. The brain is a complex three-dimensional volume of slight and varying radiopacity, so little information can be gleaned from a single two-dimensional X-ray image.

An ingenious solution, called *computed tomography*, or *CT*, was developed by Godfrey Hounsfield and Allan Cormack, who shared the Nobel Prize in 1979. The goal of CT is to generate an image of a slice of brain. (The word *tomography* is derived from the Greek word for "cut.") To accomplish this, an X-ray source is rotated around the head within the plane of the desired cross section. On the other side of the head, in the trajectory of the X-ray beam, are sensitive electronic sensors of X-irradiation. The information about relative radiopacity obtained with different "viewing" angles is fed to a computer that executes a mathematical algorithm on the data. The end result is a digital reconstruction of the position and amount of radiopaque material within the plane of the slice. CT scans noninvasively revealed, for the first time, the gross organization of gray and white matter and the position of the ventricles in the living brain.

Magnetic Resonance Imaging. While still used widely, CT is gradually being replaced by a newer imaging method, *magnetic resonance imaging*, or *MRI*. The advantages of MRI are that it yields a much more detailed map of the brain than does CT, it does not require X-irradiation, and images of brain slices can be made in any plane desired. MRI uses information about how hydrogen atoms in the brain respond to perturbations of a strong magnetic field (Box 7.2). The electromagnetic signals emitted by the atoms are detected by an array of sensors around the head and fed to a powerful computer that constructs a map of the brain. The information from an MRI scan can be used to build a strikingly detailed image of the whole brain.

Functional Brain Imaging. CT and MRI are extremely valuable for detecting structural changes in the living brain, such as brain tumors and brain swelling after a head injury. Nonetheless, much of what goes on in the brain—healthy or diseased—is chemical and electrical in nature and not observable by simple inspection of the brain's anatomy. Amazingly, however, even these secrets are beginning to yield to the newest imaging techniques.

The two "functional imaging" techniques now in widespread use are *positron emission tomography*, or *PET*, and *functional magnetic resonance imag-*

Box 7.2

BRAIN FOOD

Magnetic Resonance Imaging

Magnetic resonance imaging (MRI) is a general technique that can be used for determining the amount of certain atoms at various locations in the body. It has become an important tool in neuroscience because it can be used noninvasively to obtain a detailed picture of the nervous system, particularly the brain.

In the most common form of MRI, the hydrogen atoms are quantified—for instance, those in water or fat in the brain. An important fact of physics is that when a hydrogen atom is put into a magnetic field, its nucleus (which consists of a single proton) can exist in either of two states: a high-energy state or a low-energy state. Because hydrogen atoms are abundant in the brain, there are many protons in each state.

The key to MRI is making the protons jump from one state to the other. Energy is added to the protons by passing an electromagnetic wave (i.e., a radio signal) through the head while it is positioned between the poles of a large magnet. When the radio signal is set at just the right frequency, the protons in the low-energy state absorb energy from the signal and hop to the high-energy state. The frequency at which the protons absorb energy is called the resonant frequency (hence the name magnetic resonance). When the radio signal is turned off, some of the protons fall back down to the low-energy state, emitting a radio signal of their own at a particular frequency. This signal can be picked up by a radio receiver. The stronger the signal, the more hydrogen atoms between the poles of the magnet.

If we used this procedure, we would simply get a measurement of the total amount of hydrogen in the head. However, it is possible to measure hydrogen amounts at a fine spatial scale by taking advantage of the fact that the frequency at which protons emit energy is proportional to the size of the magnetic field. In the MRI machines used in hospitals, the magnetic fields vary from one side of the magnet to the other. This gives a spatial code to the radio waves emitted by the protons: High-frequency signals come from hydrogen atoms near the strong side of the magnet, and low-frequency signals come from the weak side of the magnet. The last step in the MRI process is to orient the gradient of the magnet at many different angles relative to the head and measure the amount of hydrogen. It takes about 15 minutes to make all of the measurements for a typical brain scan. A sophisticated computer program is used to make a single image from the measurements, resulting in a picture of the distribution of hydrogen atoms in the head.

Figure A is an MRI image of a lateral view of the brain in a living human. In Figure B, another MRI image, a slice has been made in the brain. Notice how clearly you can see the white and gray matter. This differentiation makes it possible to see the effects of demyelinating diseases on white matter in the brain. MRI images also reveal lesions in the brain, because tumors and inflammation generally increase the amount of extracellular water.

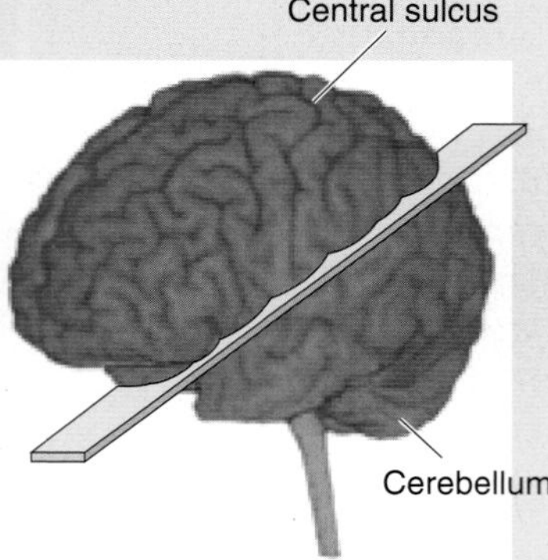

Figure A

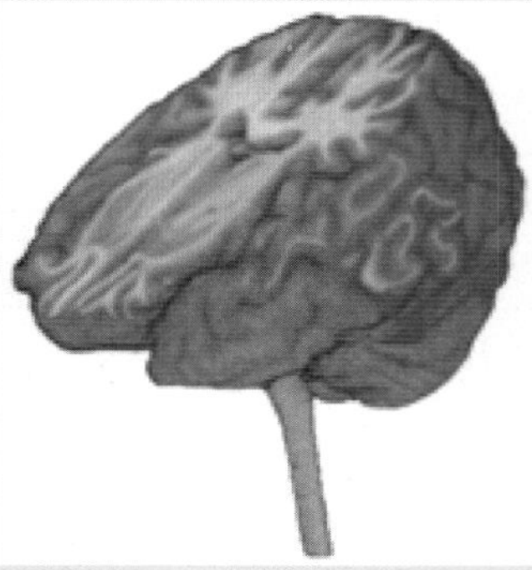

Figure B

ing, or *fMRI*. While the technical details differ, both methods detect changes in regional blood flow and metabolism within the brain (Box 7.3). The principle is simple. Neurons that are active demand more glucose and oxygen. The brain vasculature responds to neural activity by directing more blood to the active regions. Thus, by detecting changes in blood flow, PET and fMRI reveal the regions of brain that are most active under various circumstances.

The advent of imaging techniques has offered neuroscientists the extraordinary opportunity of peering into the living, thinking brain. As you can imagine, however, even the most sophisticated brain images are useless unless you know what you are looking at. Next, let's take a closer look at how the brain is organized.

Self-Quiz

Take a few moments right now and be sure you understand the meaning of these terms:

- central nervous system (CNS)
- brain
- spinal cord
- cerebrum
- cerebral hemispheres
- cerebellum
- brain stem
- spinal nerve
- dorsal root
- ventral root
- peripheral nervous system (PNS)
- somatic PNS
- dorsal root ganglia
- visceral PNS
- autonomic nervous system (ANS)
- afferent
- efferent
- cranial nerve
- meninges
- dura mater
- arachnoid membrane
- pia mater
- cerebrospinal fluid (CSF)
- ventricular system

UNDERSTANDING CNS STRUCTURE THROUGH DEVELOPMENT

The entire CNS is derived from the walls of a fluid-filled tube that is formed at an early stage of embryonic development. The tube itself becomes the adult ventricular system. Thus, by examining how this tube changes during the course of fetal development, we can understand how the brain is organized and how the different parts fit together. In this section, we focus on development as a way to understand the structural organization of the brain. In Chapter 22, we will revisit the topic of development to see how neurons are born, how they find their way to their final locations in the CNS, and how they make the appropriate synaptic connections with one another.

As you work your way through this section and through the rest of the book, you will encounter many names used by anatomists to refer to groups of related neurons and axons. Some common names used for describing collections of neurons and axons are given in Tables 7.1 and 7.2. Take a few moments to familiarize yourself with these new terms before continuing.

Anatomy by itself can be pretty dry. It really comes alive only after the functions of structures are understood. The remainder of this book is devoted to explaining the functional organization of the nervous system. We have punctuated this section with a preview of some structure–function relationships, however, to provide you with a general sense of how the different parts individually and collectively contribute to the function of the CNS.

Formation of the Neural Tube

The embryo begins as a flat disk with three distinct layers of cells called endoderm, mesoderm, and ectoderm. The *endoderm* ultimately gives rise to the

Box 7.3

BRAIN FOOD

Functional Imaging of Brain Activity: PET and fMRI

Until recently, "mind reading" has been beyond the reach of science. With the introduction of *positron emission tomography* (*PET*) and *functional magnetic resonance imaging* (*fMRI*), however, it is possible to observe and measure changes in brain activity associated with the planning and execution of specific tasks.

PET imaging was developed in the 1970s by two groups of physicists, one at Washington University, led by M. Ter-pogossian and M. E. Phelps, and a second at UCLA, led by Z. H. Cho. The basic procedure is simple. A radioactive solution containing atoms that emit positrons (positively charged electrons) is introduced into the bloodstream. Positrons, emitted wherever the blood goes, interact with electrons to produce photons of electromagnetic radiation. The locations of the positron-emitting atoms are found by detectors that pick up the photons.

One powerful application of PET is the measurement of metabolic activity in the brain. In a technique developed by Louis Sokoloff and his colleagues at the National Institute of Mental Health, a positron-emitting isotope of fluorine or oxygen is attached to 2-deoxyglucose (2-DG). This radioactive 2-DG is injected into the bloodstream, and it travels to the brain. Metabolically active neurons, which normally use glucose, also take up the 2-DG. The 2-DG is phosphorylated by enzymes inside the neuron, and this modification prevents the 2-DG from leaving. Thus, the amount of radioactive 2-DG accumulated in a neuron and the number of positron emissions indicate the level of neuronal metabolic activity.

In a typical PET application, a person's head is placed in an apparatus surrounded by detectors (Figure A). By use of computer algorithms, the photons resulting from positron emissions reaching each of the detectors are recorded. With this information, levels of activity for populations of neurons at various sites in the brain can be calculated. Compiling these measurements produces an image of the brain activity pattern. The researcher monitors brain activity while the subject performs a task, such as moving a finger or reading aloud. Different tasks "light up" different brain areas. To obtain a picture of the activity induced by a particular behavioral or thought task, a subtraction technique is used. Even in the absence of any sensory stimulation, the PET image will contain a great deal of brain activity. To create an image of the brain activity resulting from a specific task, such as a person looking at a picture, this background activity is subtracted out (Figure B).

Although PET imaging has proven to be a valuable technique, it has significant limitations. Because the spatial resolution is only 5–10 mm^3, the images show the activity of many thousands of cells. Also, obtaining a single PET brain scan may take one to many minutes. This, along with concerns about radiation exposure, limit the number of obtainable scans from one person in a reasonable time. Thus, the work of S. Ogawa at Bell Labs, showing that MRI techniques could be used to measure local changes in blood oxygen levels that result from brain activity, was an important advance.

The fMRI method takes advantage of the fact that oxyhemoglobin (the oxygenated from of hemoglobin in the blood) has a different magnetic resonance from that of deoxyhemoglobin (hemoglobin that has donated its oxygen). More ac-

lining of many of the internal organs (viscera). From the *mesoderm* arise the bones of the skeleton and the muscles. The nervous system and the skin derive entirely from the *ectoderm*.

Our focus is on changes in the part of the ectoderm that gives rise to the nervous system: the *neural plate*. At this early stage (about 3 weeks of gestation in humans), the brain consists only of a flat sheet of cells (Figure 7.8a). The next event of interest is the formation of a groove in the neural plate, called the *neural groove*, that runs rostral to caudal (Figure 7.8b). The walls of the groove, the *neural folds*, subsequently move together and fuse dorsally, forming the **neural tube** (Figure 7.8c). *The entire central nervous system develops from the walls of the neural tube.* As the neural folds come together, some neural ectoderm is pinched off and comes to lie just lateral to the neural tube.

(Box 7.3, *continued*)

tive regions of the brain receive more blood, and this blood donates more of its oxygen. Functional MRI detects the locations of increased neural activity by measuring the ratio of oxyhemoglobin to deoxyhemoglobin. It has emerged as the method of choice for functional brain imaging because the scans can be made rapidly (50 msec), have good spatial resolution (3 mm^3), and are completely noninvasive.

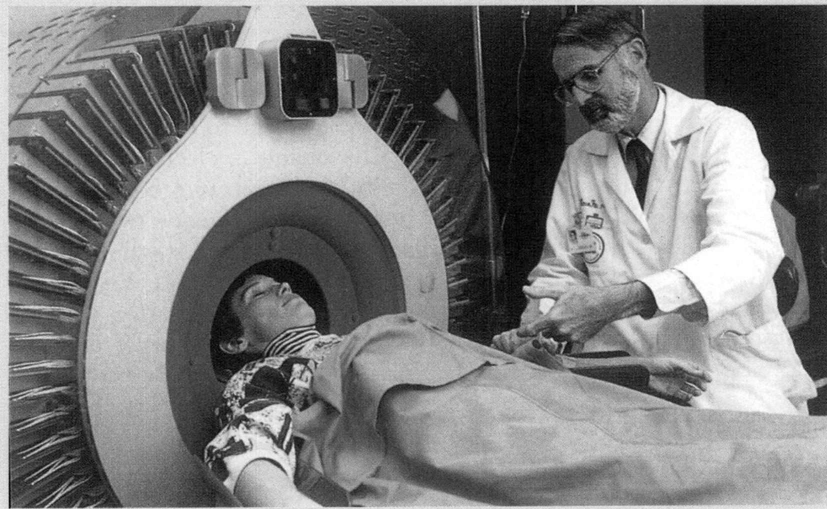

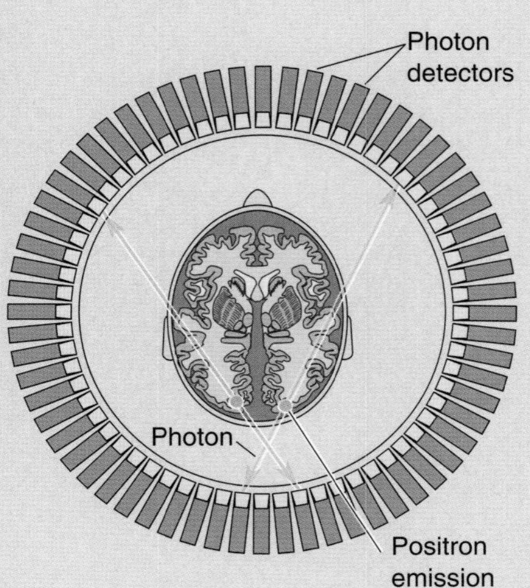

Figure A
The PET procedure. (Source: Posner and Raichle, 1994, p. 61.)

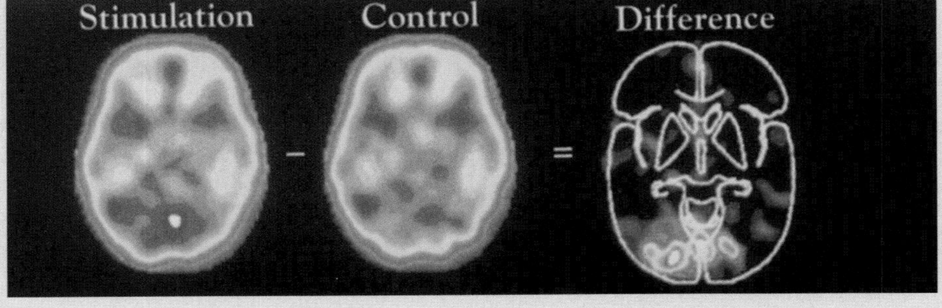

Figure B
A PET image. (Source: Posner and Raichle, 1994, p. 65.)

This tissue is called the **neural crest** (Figure 7.8d). *All neurons with cell bodies in the peripheral nervous system derive from the neural crest.*

The neural crest develops in close association with the underlying mesoderm. The mesoderm at this stage of development forms prominent bulges on either side of the neural tube called *somites*. From these somites, the 33 individual vertebrae of the spinal column and the related skeletal muscles will develop. The nerves that innervate these skeletal muscles are therefore called *somatic* motor nerves.

The process by which the neural plate becomes the neural tube is called **neurulation**. Neurulation occurs very early in embryonic development, about 22 days after conception in humans. A common birth defect is the failure of appropriate closure of the neural tube. Fortunately, recent research suggests that most cases of neural tube defects can be avoided by ensuring proper maternal nutrition during this period (Box 7.4).

Table 7.1 Collections of Neurons

NAME	DESCRIPTION AND EXAMPLE
Gray matter	A generic term for a collection of neuronal cell bodies in the CNS. When a freshly dissected brain is cut open, neurons appear gray.
Cortex	Any collection of neurons that form a thin sheet, usually at the brain's surface. *Cortex* is Latin for "bark." Example: *cerebral cortex*, the sheet of neurons found just under the surface of the cerebrum.
Nucleus	A clearly distinguishable mass of neurons, usually deep in the brain (not to be confused with the nucleus of a cell). *Nucleus* is from the Latin word for "nut." Example: *lateral geniculate nucleus*, a cell group in the brain stem that relays information from the eye to the cerebral cortex.
Substantia	A group of related neurons deep within the brain, usually with less distinct borders than those of nuclei. Example: *substantia nigra* (from the Latin for "black substance"), a brain stem cell group involved in the control of voluntary movement.
Locus (plural: loci)	A small, well-defined group of cells. Example: *locus ceruleus* (Latin for "blue spot"), a brain stem cell group involved in the control of wakefulness and behavioral arousal.
Ganglion (plural: ganglia)	A collection of neurons in the PNS. *Ganglion* is from the Greek for "knot." Example: the *dorsal root ganglia*, which contain the cell bodies of sensory axons entering the spinal cord via the dorsal roots. Only one cell group in the CNS goes by this name: the *basal ganglia*, which are structures lying deep within the cerebrum that control movement.

Three Primary Brain Vesicles

The process by which structures become more elaborate and specialized during development is called **differentiation**. The first step in the differentiation of the brain is the development at the rostral end of the neural tube of three swellings called the primary vesicles (Figure 7.9). *The entire brain derives from the three primary vesicles of the neural tube.*

The rostral-most vesicle is called the *prosencephalon*. Pro is Greek for "before"; *encephalon* is from the Greek for "brain." Thus, the prosencephalon is also called the **forebrain**. Behind the prosencephalon lies another vesicle called the *mesencephalon*, or **midbrain**. Caudal to this is the third primary vesicle, the *rhombencephalon*, or **hindbrain**. The rhombencephalon connects with the caudal neural tube, which gives rise to the spinal cord.

Table 7.2 Collections of Axons

NAME	DESCRIPTION AND EXAMPLE
Nerve	A bundle of axons in the PNS. Only one collection of CNS axons, the *optic nerve*, is called a nerve.
White matter	A generic term for a collection of CNS axons. When a freshly dissected brain is cut open, axons appear white.
Tract	A collection of CNS axons having a common site of origin and a common destination. Example: *corticospinal tract*, which originates in the cerebral cortex and ends in the spinal cord.
Bundle	A collection of axons that run together but that do not necessarily have the same origin and destination. Example: *medial forebrain bundle*, which connects cells scattered within the cerebrum and brain stem.
Capsule	A collection of axons that connect the cerebrum with the brain stem. Example: *internal capsule*, which connects the brain stem with the cerebral cortex.
Commissure	Any collection of axons that connect one side of the brain with the other side.
Lemniscus	A tract that meanders through the brain like a ribbon. Example: *medial lemniscus*, which brings touch information from the spinal cord through the brain stem.

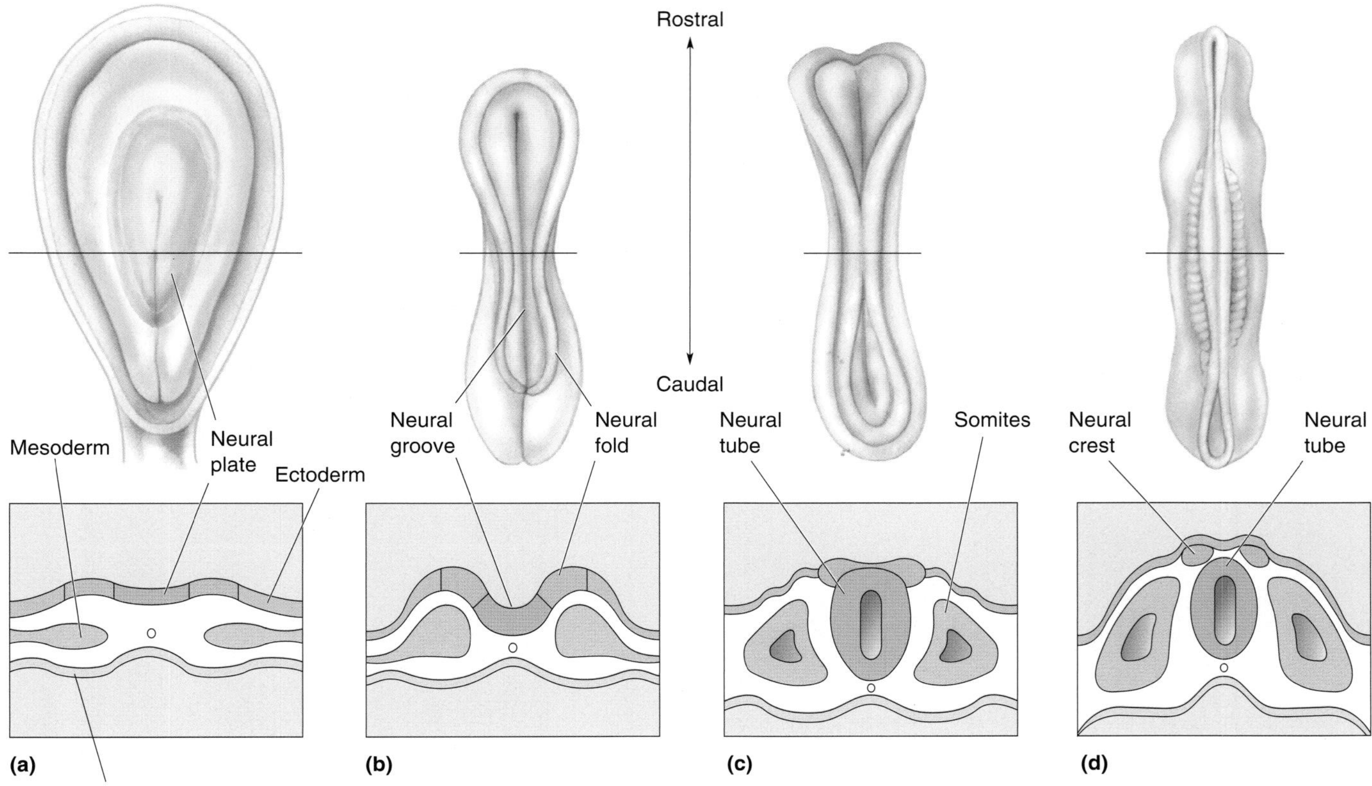

Figure 7.8
Formation of the neural tube and neural crest. These schematic illustrations follow the early development of the nervous system in the embryo. The drawings above are dorsal views of the embryo; those below are cross sections. **(a)** The primitive embryonic CNS begins as a thin sheet of ectoderm. **(b)** The first important step in the development of the nervous system is the formation of the neural groove. **(c)** The walls of the groove, called neural folds, come together and fuse, forming the neural tube. **(d)** The bits of neural ectoderm that are pinched off when the tube rolls up are called the neural crest, from which the PNS will develop. The somites are mesoderm that will give rise to much of the skeletal system and the muscles.

Differentiation of the Forebrain

The next important developments occur in the forebrain, where secondary vesicles sprout off on both sides of the prosencephalon. The secondary vesicles are the *optic vesicles* and the *telencephalic vesicles*. The unpaired structure that remains after the secondary vesicles have sprouted off is called the **diencephalon**, or "between brain" (Figure 7.10). Thus, the forebrain at this stage consists of the two optic vesicles, the two telencephalic vesicles, and the diencephalon.

The optic vesicles grow and invaginate (fold in) to form the optic stalks and the optic cups, which will ultimately become the *optic nerves* and the two *retinas* in the adult (Figure 7.11). The important point is that the retina, at the back of the eye, and the optic nerve connecting the eye to the diencephalon are part of the brain, not the PNS.

Differentiation of the Telencephalon and Diencephalon. The telencephalic vesicles together form the **telencephalon**, or "endbrain," consisting of the two cerebral hemispheres. The telencephalon continues to develop in

Box 7.4

OF SPECIAL INTEREST

Nutrition and the Neural Tube

Neural tube formation is a crucial event in the development of the nervous system. It occurs early—only 3 weeks after conception—when the mother is often unaware that she is pregnant. Failure of the neural tube to close correctly is a common birth defect, occurring in approximately 1 in every 500 live births. A recent discovery of enormous public health importance is that many neural tube defects can be traced to a deficiency of the vitamin *folic acid* (or *folate*) in the maternal diet during the weeks immediately after conception. It has been estimated that dietary supplementation of folic acid during this period could reduce the incidence of neural tube defects by 90%.

Formation of the neural tube is a complex process (Figure A). It depends on a precise sequence of changes in the three-dimensional shape of individual cells and on changes in the adhesion of each cell to its neighbors. The timing of neurulation also must be coordinated with changes in non-neural ectoderm and the mesoderm. At the molecular level, successful neurulation depends on specific sequences of gene expression that are controlled, in part, by the position and local chemical environment of the cell. It is no wonder that this process is highly sensitive to chemicals or chemical deficiencies in the maternal circulation.

The fusion of the neural folds to form the neural tube occurs first in the middle, then anteriorly and posteriorly (Figure B). Failure of the anterior neural tube to close results in a condition called *anencephaly*. Anencephaly, characterized by degeneration of the forebrain and skull, is always fatal. Failure of the posterior neural tube to close results in a condition called *spina bifida*. In its most severe form, spina bifida is characterized by the failure of the posterior spinal cord to form from the neural plate (*bifida* is from the Latin word meaning "cleft in two parts"). Less severe forms are characterized by defects in the meninges and vertebrae overlying the posterior spinal cord. Spina bifida, while usually not fatal, does require extensive and costly medical care. The medical expenses associated with spina bifida in the United States alone are estimated to be $200 million annually.

Folic acid plays an essential role in a number of metabolic pathways, including the biosynthesis of DNA that naturally

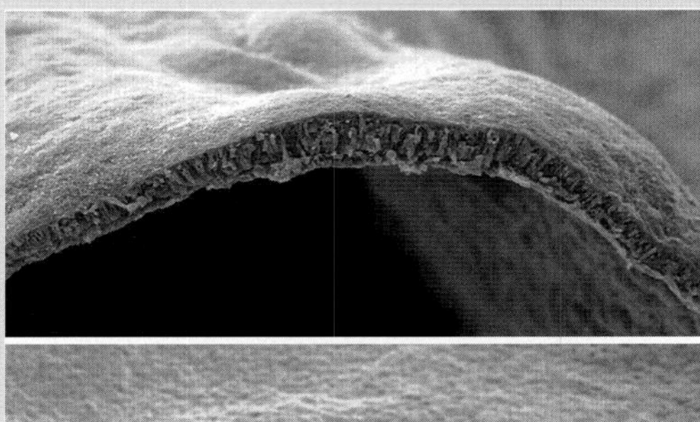

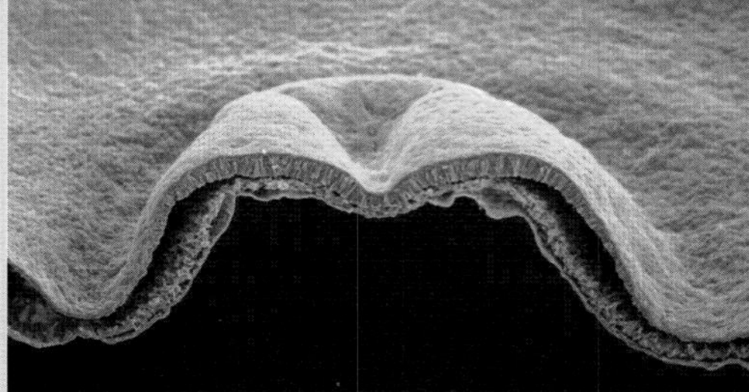

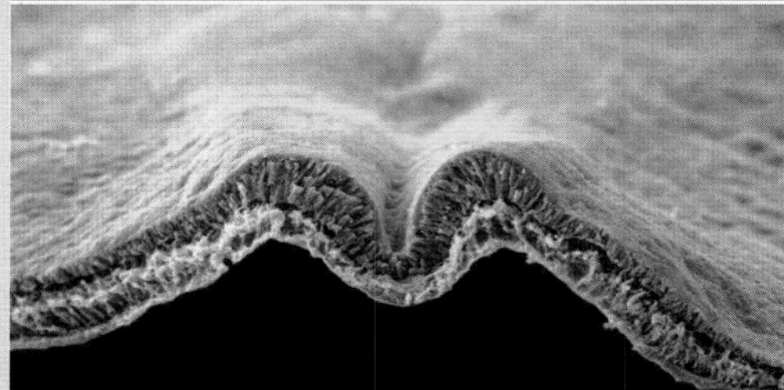

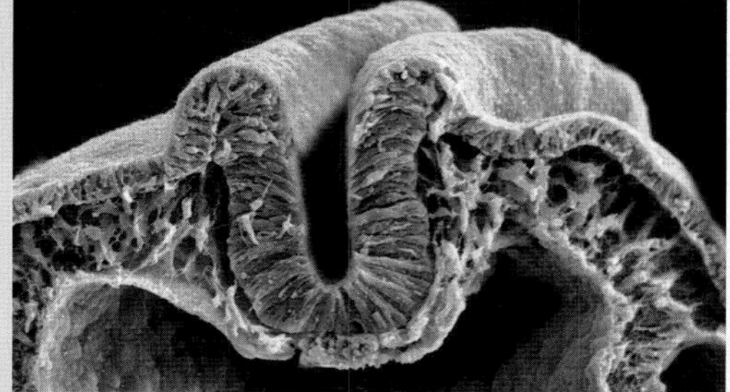

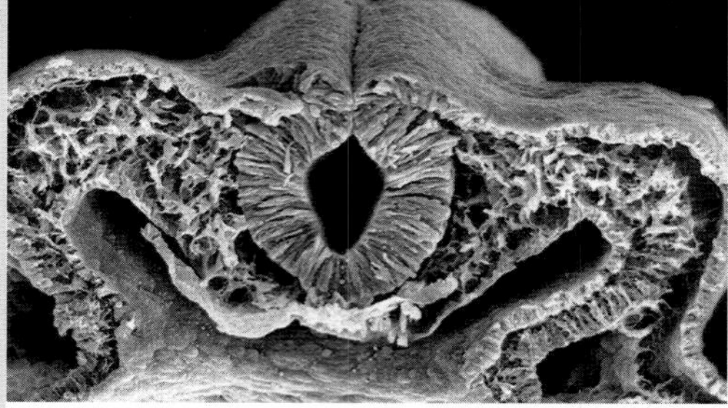

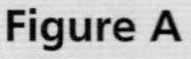

Figure A
Scanning electron micrographs of neurulation
(Source: Smith and Schoenwolf, 1997.)

(Box 7.4, continued)

must occur during development as cells divide. Although we do not precisely understand why folic acid deficiency increases the incidence of neural tube defects, one can easily imagine how it could alter the complex choreography of neurulation. The name is derived from the Latin word for "leaf," reflecting the fact that folic acid was first isolated from spinach leaves. Besides green leafy vegetables, good dietary sources of folic acid are liver, yeast, eggs, beans, and oranges. Many breakfast cereals are now fortified with folic acid. Nonetheless, the folic acid intake of the average American is only half of what is recommended to prevent birth defects (0.4 mg/day). The U.S. Centers for Disease Control and Prevention recommends that women take multivitamins containing 0.4 mg of folic acid before planning pregnancy.

Figure B
(a) Neural tube closure. **(b)** Neural tube defects.

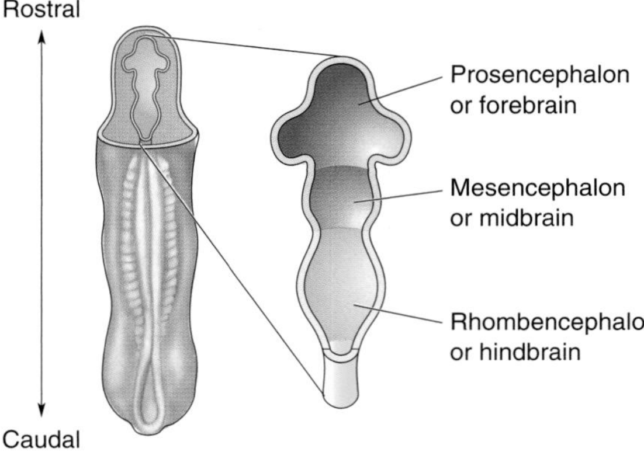

Figure 7.9
The three primary brain vesicles. The rostral end of the neural tube differentiates to form the three vesicles that will give rise to the entire brain. This view is from above, and the vesicles have been cut horizontally so that we can see the inside of the neural tube.

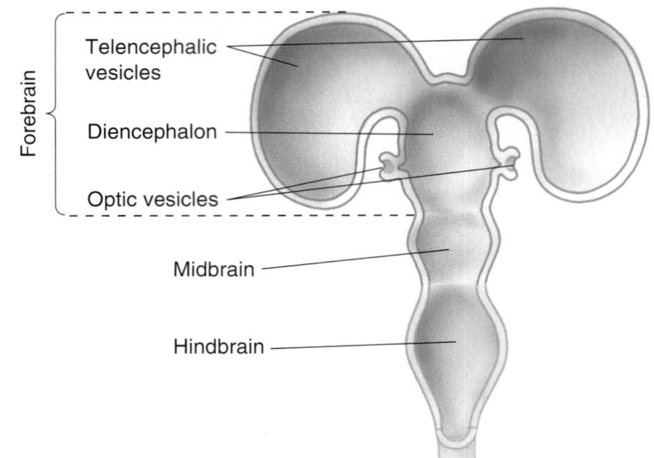

Figure 7.10
The secondary brain vesicles of the forebrain. The forebrain differentiates into the diencephalon and the paired telencephalic and optic vesicles. The optic vesicles develop into the eyes.

four ways. (1) The telencephalic vesicles grow posteriorly so that they lie over and lateral to the diencephalon (Figure 7.12a). (2) Another pair of vesicles sprout off the ventral surfaces of the cerebral hemispheres, giving rise to the **olfactory bulbs** and related structures that participate in the sense of smell (Figure 7.12b). (3) The cells of the walls of the telencephalon divide and differentiate into various structures. (4) White matter systems develop, carrying axons to and from the neurons of the telencephalon.

Figure 7.13 shows a coronal section through the primitive mammalian forebrain, illustrating how the different parts of the telencephalon and diencephalon differentiate and fit together. Notice that the two cerebral hemispheres lie above and on either side of the diencephalon and that the ventral-medial surfaces of the hemispheres have fused with the lateral surfaces of the diencephalon (Figure 7.13a).

The fluid-filled spaces within the cerebral hemispheres are called the **lateral ventricles**, and the space at the center of the diencephalon is called the **third ventricle** (Figure 7.13b). The paired lateral ventricles are a key landmark in the adult brain: Whenever you see paired fluid-filled ventricles in a brain section, you know that the tissue surrounding them is in the telencephalon. The elongated, slitlike appearance of the third ventricle in cross section is also a useful feature for identifying the diencephalon.

Notice in Figure 7.13 that the walls of the telencephalic vesicles appear swollen because of the proliferation of neurons. These neurons form two different types of gray matter in the telencephalon: the **cerebral cortex** and the **basal telencephalon**. Likewise, the diencephalon differentiates into two structures: the **thalamus** and the **hypothalamus** (Figure 7.13c). The thalamus, nestled deep inside the forebrain, gets its name from the Greek word for "inner chamber."

The neurons of the developing forebrain extend axons to communicate with other parts of the nervous system. These axons bundle together to form three major white matter systems: the cortical white matter, the corpus callosum, and the internal capsule (Figure 7.13d). The **cortical white matter** contains all of the axons that run to and from the neurons in the cerebral cortex. The **corpus callosum** is continuous with the cortical white matter and forms an axonal bridge that links cortical neurons of the two cerebral hemispheres. The cortical white matter is also continuous with the **internal capsule**, which links the cortex with the brain stem, particularly the thalamus.

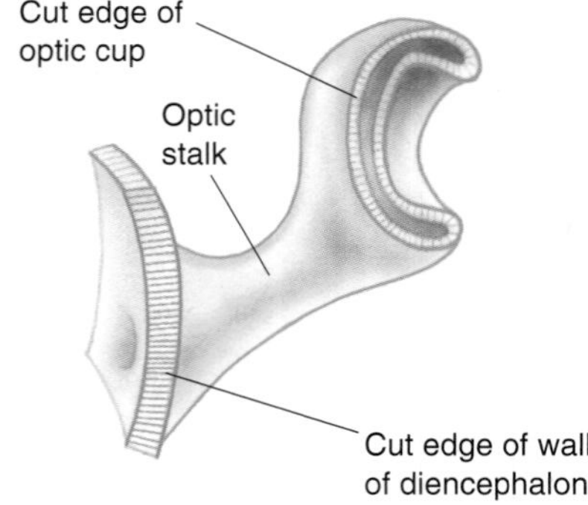

Figure 7.11
Early development of the eye. The optic vesicle differentiates into the optic stalk and the optic cup. The optic stalk will become the optic nerve, and the optic cup will become the retina.

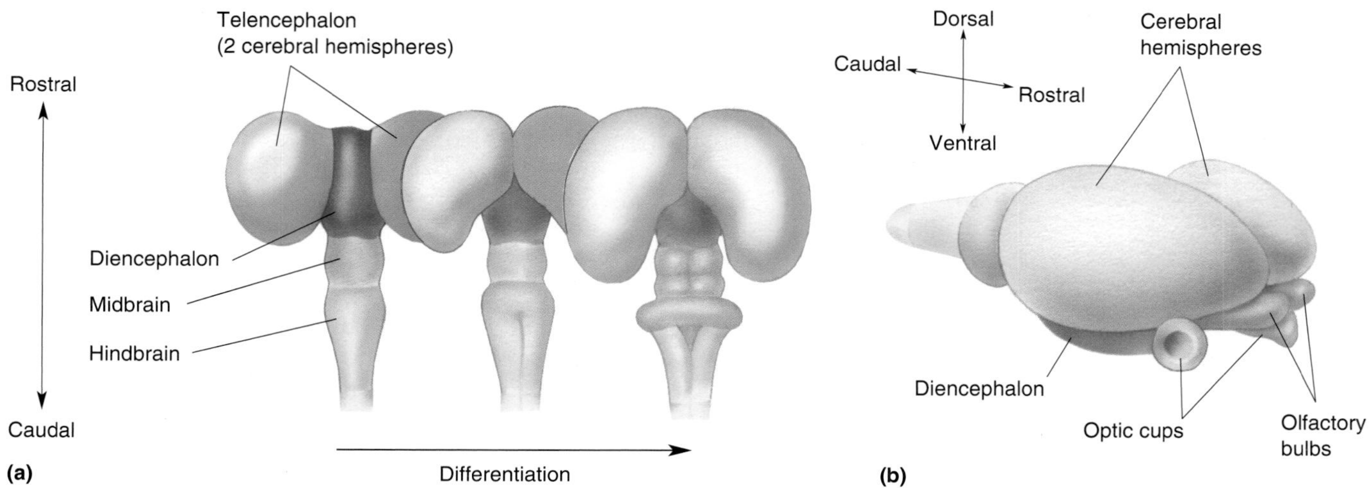

Figure 7.12
Differentiation of the telencephalon. (a) As development proceeds, the cerebral hemispheres swell and grow posteriorly and laterally to envelop the diencephalon. **(b)** The olfactory bulbs sprout off the ventral surfaces of each telencephalic vesicle.

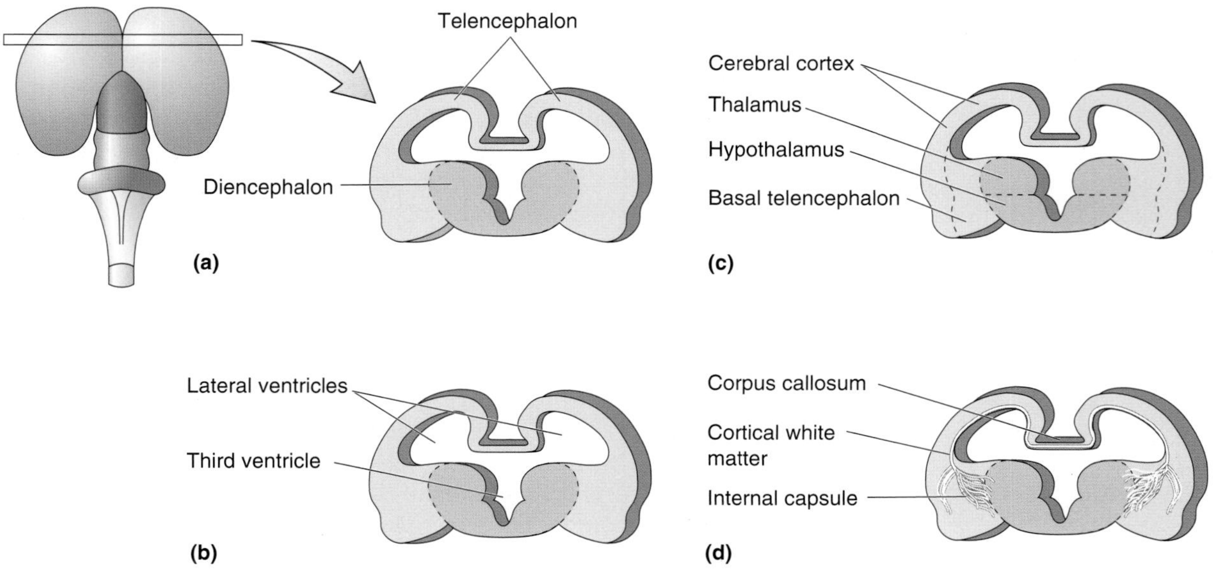

Figure 7.13
Structural features of the forebrain. (a) A coronal section through the primitive forebrain, showing the two main divisions: the telencephalon and the diencephalon. **(b)** Ventricles of the forebrain. **(c)** Gray matter of the forebrain. **(d)** White matter systems of the forebrain.

Forebrain Structure–Function Relationships. The forebrain is the seat of perceptions, conscious awareness, cognition, and voluntary action. All of this depends on extensive interconnections with the sensory and motor neurons of the brain stem and spinal cord.

Arguably the most important structure in the forebrain is the cerebral cortex. As we will see later in this chapter, the cortex is the brain structure that has expanded the most over the course of human evolution. Cortical neurons receive sensory information, form perceptions of the outside world, and command voluntary movements.

Neurons in the olfactory bulbs receive information from cells that sense chemicals in the nose (odors) and relay this information caudally to a part of the cerebral cortex for further analysis. Information from the eyes, ears, and skin is also brought to the cerebral cortex for analysis. Each of the sensory pathways serving vision, audition (hearing), and somatic sensation, however, relays (i.e., synapses upon neurons) in the thalamus en route to the cortex. Thus, the thalamus is often called the gateway to the cerebral cortex (Figure 7.14).

Thalamic neurons send axons to the cortex via the internal capsule. As a general rule, the axons of each internal capsule carry information to the cortex about the contralateral side of the body. Therefore, if a thumbtack entered the *right* foot, it would be relayed to the *left* cortex by the *left* thalamus via axons in the *left* internal capsule. But how is the right foot to know what the left foot is doing? One important way is by communication between the hemispheres via the axons in the corpus callosum.

Cortical neurons also send axons through the internal capsule back to the brain stem. Some cortical axons course all the way to the spinal cord, forming the corticospinal tract. This is one important way cortex can command voluntary movement. Another way is by communicating with neurons in the basal ganglia, a collection of cells in the basal telencephalon. The term *basal* is used to describe structures deep in the brain; the basal ganglia lie deep within the cerebrum. The functions of the basal ganglia are poorly understood, but it is known that damage to these structures disrupts the ability to initiate voluntary movement. Other structures contributing to other brain functions are also present in the basal telencephalon. For example, in Chapter 18 we'll discuss a structure called the amygdala that is involved in fear and emotion.

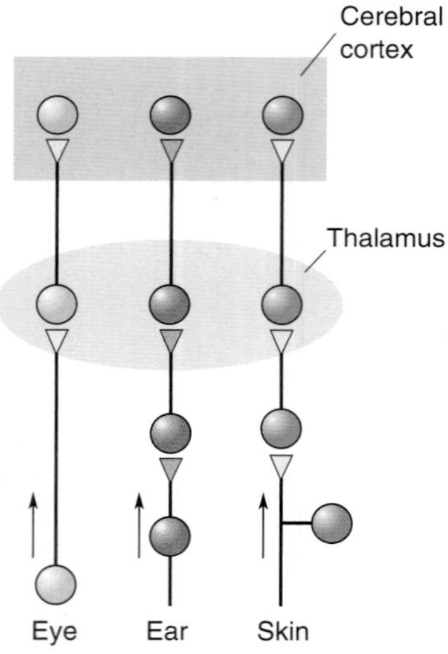

Figure 7.14
The thalamus: gateway to the cerebral cortex. The sensory pathways from the eye, ear, and skin all relay in the thalamus before terminating in the cerebral cortex. The arrows indicate the direction of information flow.

Although the hypothalamus lies just under the thalamus, functionally it is more closely related to certain telencephalic structures, such as the amygdala. The hypothalamus performs many primitive functions and therefore has not changed much over the course of mammalian evolution. "Primitive" does not mean unimportant or uninteresting, however. The hypothalamus controls the visceral (autonomic) nervous system, which regulates bodily functions in response to the needs of the organism. For example, when you are threatened, the hypothalamus orchestrates the body's visceral fight-or-flight response. Hypothalamic commands to the ANS will lead to (among other things) an increase in the heart rate, increased blood flow to the muscles for escape, and even the standing of your hair on end. Conversely, when you're relaxing after Sunday brunch, the hypothalamus ensures that the brain is well nourished via commands to the ANS, which will increase peristalsis (movement of material through the gastrointestinal tract) and redirect blood to your digestive system. The hypothalamus also plays a key role in motivating animals to find food, drink, and sex in response to their needs. Aside from its connections to the ANS, the hypothalamus also directs bodily responses via connections with the pituitary gland located below the diencephalon. This gland communicates to many parts of the body by releasing hormones into the bloodstream.

Self-Quiz

Listed below are derivatives of the forebrain that we have discussed. Be sure you know what each of these terms means.

Primary Vesicle	Secondary Vesicle	Some Adult Derivatives
Forebrain (prosencephalon)	Optic vesicle	Retina
		Optic nerve
	Thalamus (diencephalon)	Dorsal thalamus
		Hypothalamus
		Third ventricle
	Telencephalon	Olfactory bulb
		Cerebral cortex
		Basal telencephalon
		Corpus callosum
		Cortical white matter
		Internal capsule

Differentiation of the Midbrain

Unlike the forebrain, the midbrain differentiates relatively little during subsequent brain development (Figure 7.15). The dorsal surface of the mesencephalic vesicle becomes a structure called the **tectum** (Latin for "roof"). The floor of the midbrain becomes the **tegmentum**. The CSF-filled space in between constricts into a narrow channel called the **cerebral aqueduct**. The aqueduct connects rostrally with the third ventricle of the diencephalon. Because it is small and circular in cross section, the cerebral aqueduct is a good landmark for identifying the midbrain.

Midbrain Structure–Function Relationships. For such a seemingly simple structure, the functions of the midbrain are remarkably diverse. Besides serving as a conduit for information passing from the spinal cord to the forebrain and from the forebrain to the spinal cord, the midbrain contains neurons that contribute to sensory systems, the control of movement, and several other functions.

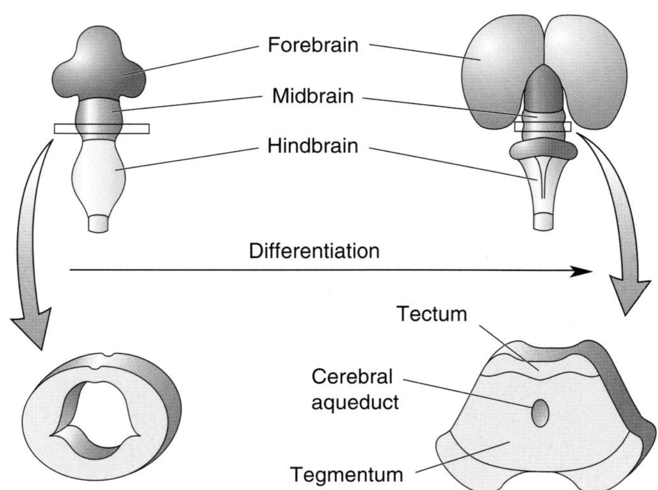

Figure 7.15
Differentiation of the midbrain. The midbrain differentiates into the tectum and the tegmentum. The CSF-filled space at the core of the midbrain is the cerebral aqueduct. (Drawings are not to scale.)

The midbrain contains axons descending from the cerebral cortex to the brain stem and the spinal cord. For example, the corticospinal tract courses through the midbrain en route to the spinal cord. Damage to this tract in the midbrain on one side impairs voluntary control of movement on the opposite side of the body.

The tectum differentiates into two structures: the superior colliculus and the inferior colliculus. The *superior colliculus* receives direct input from the eye, so it is also called the optic tectum. One function of the optic tectum is to control eye movements, which it does via synaptic connections with the motor neurons that innervate the eye muscles. Some of the axons that supply the eye muscles originate in the midbrain, bundling together to form cranial nerves III and IV (Table 7.1).

The *inferior colliculus* also receives sensory information, but from the ear instead of the eye. The inferior colliculus serves as an important relay station for auditory information en route to the thalamus.

The tegmentum is one of the most colorful regions of the brain because it contains both the substantia nigra (the black substance) and the red nucleus. These two cell groups are involved in the control of voluntary movement. Other cell groups scattered in the midbrain have axons that project widely throughout much of the CNS and function to regulate consciousness, mood, pleasure, and pain.

Differentiation of the Hindbrain

The hindbrain differentiates into three important structures: the **cerebellum**, the **pons**, and the **medulla oblongata**—also called simply the **medulla**. The cerebellum and pons derive from the rostral half of the hindbrain (called the metencephalon); the medulla derives from the caudal half (called the myelencephalon). The CSF-filled tube becomes the **fourth ventricle**, which is continuous with the cerebral aqueduct of the midbrain.

At the three-vesicle stage, the rostral hindbrain in cross section is a simple tube. In subsequent weeks, the tissue along the dorsal-lateral wall of the tube, called the rhombic lip, grows dorsally and medially until it fuses with its twin on the other side. The resulting flap of brain tissue grows into the cerebellum. The ventral wall of the tube differentiates and swells to form the pons (Figure 7.16).

Less dramatic changes occur during the differentiation of the caudal half of the hindbrain into the medulla. The ventral and lateral walls of this region swell, leaving the roof covered only with a thin layer of non-neuronal ependymal cells (Figure 7.17). Along the ventral surface of each side of the medulla runs a major white matter system. Cut in cross section, these bundles of axons appear somewhat triangular, explaining why they are called the *medullary pyramids*.

Hindbrain Structure–Function Relationships. Like the midbrain, the hindbrain is an important conduit for information passing from the forebrain to the spinal cord and from the spinal cord to the forebrain. In addition, neurons of the hindbrain contribute to the processing of sensory information, the control of voluntary movement, and regulation of the autonomic nervous system.

The cerebellum, the "little brain," is an important movement control center. It receives massive axonal inputs from the spinal cord and the pons. The spinal cord inputs provide information about where the body is in space. The inputs from the pons relay information from the cerebral cortex, specifying the goals of intended movements. The cerebellum compares these types of information and calculates the sequences of muscle contractions that are re-

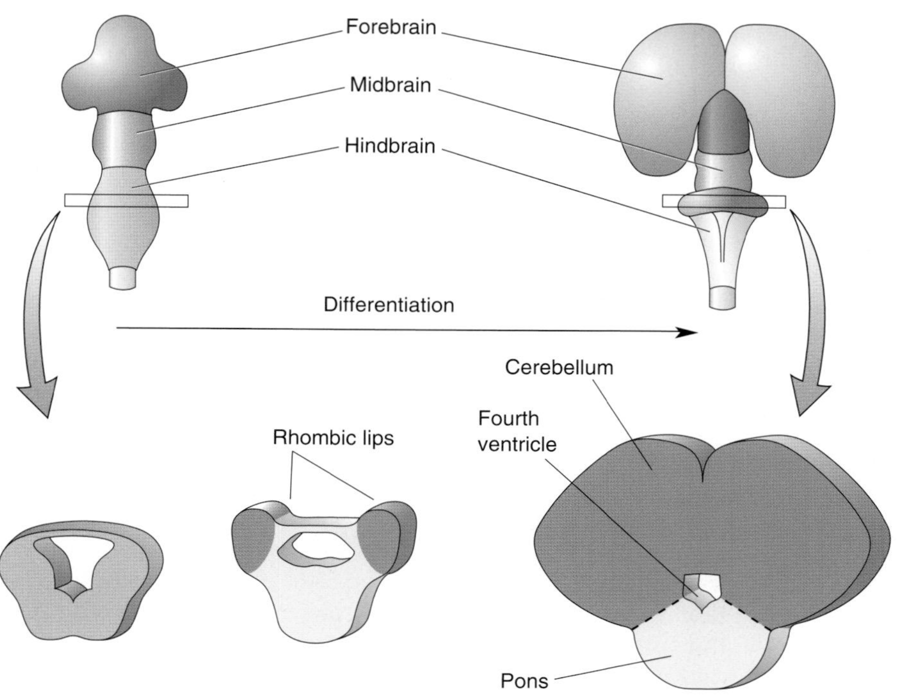

Figure 7.16
Differentiation of the rostral hindbrain. The rostral hindbrain differentiates into the cerebellum and pons. The cerebellum is formed by the growth and fusion of the rhombic lips. The CSF-filled space at the core of the hindbrain is the fourth ventricle. (Drawings are not to scale.)

quired to achieve the movement goals. Damage to the cerebellum results in uncoordinated and inaccurate movements.

Of the descending axons passing through the midbrain, over 90%—about 20 million axons in the human—synapse on neurons in the pons. The pontine cells relay all of this information to the cerebellum on the opposite site. Thus, the pons serves as a massive switchboard connecting the cerebral cortex to the cerebellum. (The word *pons* is Latin for "bridge.") The pons bulges out from the ventral surface of the brain stem to accommodate all this circuitry.

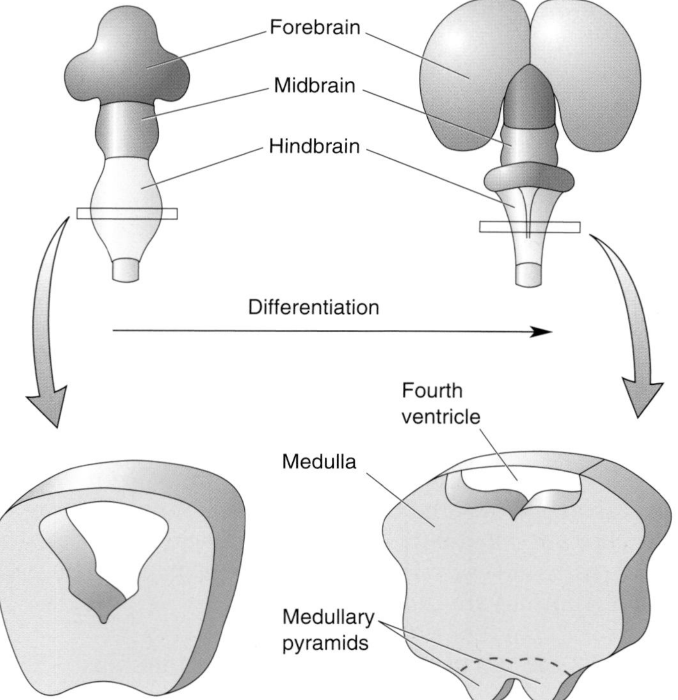

Figure 7.17
Differentiation of the caudal hindbrain. The caudal hindbrain differentiates into the medulla. The medullary pyramids are bundles of axons coursing caudally toward the spinal cord. The CSF-filled space at the core of the medulla is the fourth ventricle. (Drawings are not to scale.)

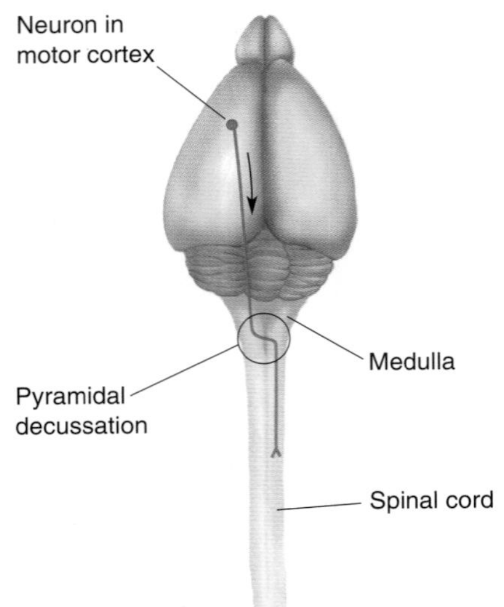

Figure 7.18
The pyramidal decussation. The corticospinal tract crosses from one side to the other in the medulla.

The axons that do not terminate in the pons continue caudally and enter the medullary pyramids. Most of these axons originate in the cerebral cortex and are part of the corticospinal tract. Thus, "pyramidal tract" is often used as a synonym for corticospinal tract. Near where the medulla joins with the spinal cord, each pyramidal tract crosses from one side of the midline to the other. A crossing of axons from one side to the other is called a *decussation*, and this one is called the *pyramidal decussation*. The crossing of axons in the medulla explains why the cortex of one side of the brain controls movements on the opposite side of the body (Figure 7.18).

In addition to the white matter systems passing through, the medulla contains neurons that perform many sensory and motor functions. For example, the axons of the auditory nerves, bringing auditory information from the ears, synapse on cells in the cochlear nuclei of the medulla. The cochlear nuclei project axons to a number of structures, including the tectum of the midbrain (inferior colliculus, discussed earlier). Damage to the cochlear nuclei leads to deafness.

Other sensory functions of the medulla include touch and taste. The medulla contains neurons that relay somatic sensory information from the spinal cord to the thalamus. Destruction of the cells leads to anesthesia (loss of feeling). Other neurons relay gustatory (taste) information from the tongue to the thalamus. And among the motor neurons in the medulla are cells that control the tongue muscles via cranial nerve XII. (So think of the medulla next time you stick out your tongue!)

Self-Quiz

Listed below are derivatives of the midbrain and hindbrain that we have discussed. Be sure you know what each of these terms means.

Primary Vesicle	Some Adult Derivatives
Midbrain (mesencephalon)	Tectum
	Tegmentum
	Cerebral aqueduct
Hindbrain (rhombencephalon)	Cerebellum
	Pons
	Fourth ventricle
	Medulla

Differentiation of the Spinal Cord

As shown in Figure 7.19, the transformation of the caudal neural tube into the spinal cord is straightforward, compared with the differentiation of the brain. With the expansion of the tissue in the walls, the cavity of the tube constricts to form the tiny CSF-filled **spinal canal**.

Cut in cross section, the gray matter of the spinal cord (where the neurons are) has the appearance of a butterfly. The upper part of the butterfly's wing is the **dorsal horn**, and the lower part is the **ventral horn**. The gray matter between the dorsal and ventral horns is called the *intermediate zone*. Everything else is white matter, consisting of columns of axons that run up and down the spinal cord. Thus, the bundles of axons running along the dorsal surface of the cord are called the *dorsal columns*, the bundles of axons lateral to the spinal gray matter on each side are called the *lateral columns*, and the bundles on the ventral surface are called the *ventral columns*.

Spinal Cord Structure–Function Relationships. In general, dorsal horn cells receive sensory inputs from the dorsal root fibers, ventral horn cells pro-

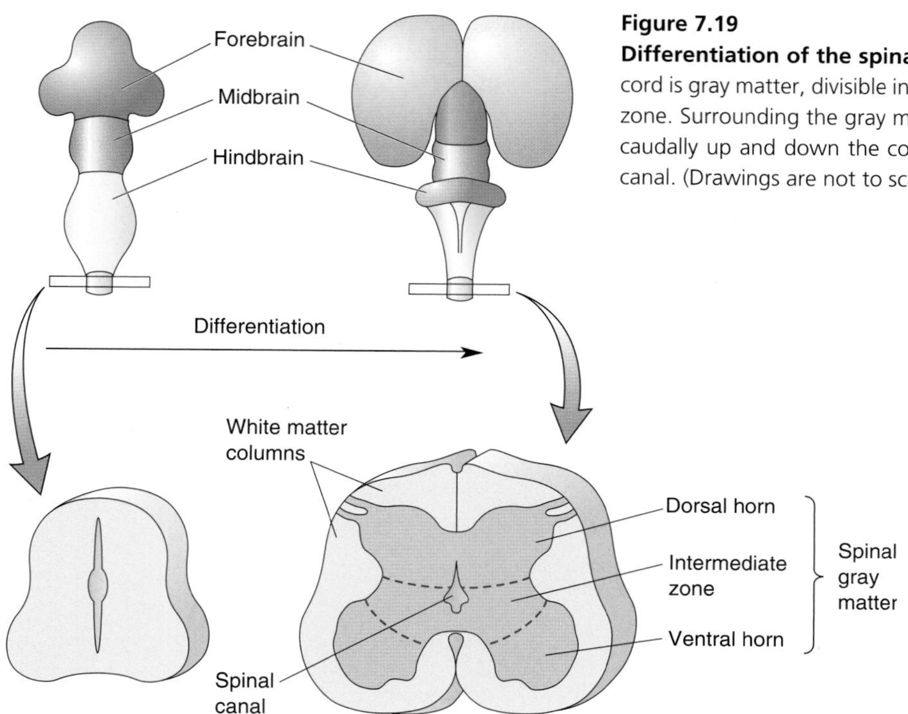

Figure 7.19
Differentiation of the spinal cord. The butterfly-shaped core of the spinal cord is gray matter, divisible into dorsal and ventral horns and an intermediate zone. Surrounding the gray matter are white matter columns running rostrocaudally up and down the cord. The narrow CSF-filled channel is the spinal canal. (Drawings are not to scale.)

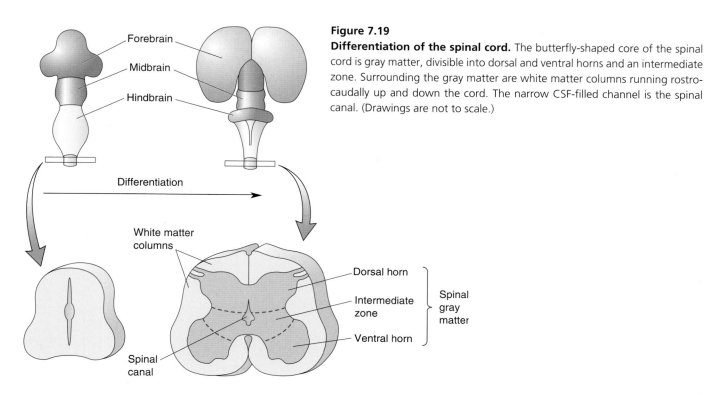

ject axons into the ventral roots that innervate muscles, and intermediate zone cells are interneurons that shape motor outputs in response to sensory inputs and descending commands from the brain.

The large dorsal column contains axons that carry somatic sensory (touch) information up the spinal cord toward the brain. It is a superhighway that speeds information from the ipsilateral side of the body up to nuclei in the medulla. The postsynaptic neurons in the medulla give rise to axons that decussate and ascend to the thalamus on the contralateral side. This crossing of axons in the medulla explains why touching the left side of the body is sensed by the right side of the brain.

The lateral column contains the axons of the descending corticospinal tract, which also cross from one side to the other in the medulla. These axons innervate the neurons of the intermediate zone and ventral horn and communicate the signals that control voluntary movement.

There are at least a half-dozen tracts that run in the columns of each side of the spinal cord. Most are one-way, taking information either to or from the brain. Thus, the spinal cord is the major conduit of information from the skin, joints, and muscles to the brain and from the brain to the skin, joints, and muscles. The spinal cord is also much more than that, however. The neurons of the spinal gray matter begin the analysis of sensory information, play a critical role in coordinating movements, and orchestrate simple reflexes, such as jerking your foot away from a thumbtack.

Putting the Pieces Together

We have discussed the development of different parts of the CNS: the telencephalon, diencephalon, midbrain, hindbrain, and spinal cord. Now let's put all of the individual pieces together to make a whole central nervous system.

Figure 7.20 is a highly schematic illustration that captures the basic organizational plan of the CNS of all mammals, including humans. The paired hemispheres of the telencephalon surround the lateral ventricles. Dorsal to the lateral ventricles, at the surface of the brain, lies the cortex. Ventral and lateral to the lateral ventricles lies the basal telencephalon. The lateral ventricles are continuous with the third ventricle of the diencephalon. Surrounding this ventricle are the thalamus and the hypothalamus. The third ventricle is continuous with the cerebral aqueduct. Dorsal to the aqueduct is the tectum. Ventral to the aqueduct is the midbrain tegmentum. The aqueduct connects with the fourth ventricle, which lies at the core of the hindbrain. Dorsal to the fourth ventricle sprouts the cerebellum. Ventral to the fourth ventricle lie the pons and the medulla.

You should see by now that finding your way around the brain is easy if you can identify which parts of the ventricular system are in the neighborhood (Table 7.3). Even in the complicated human brain, the ventricular system holds the key to understanding brain structure.

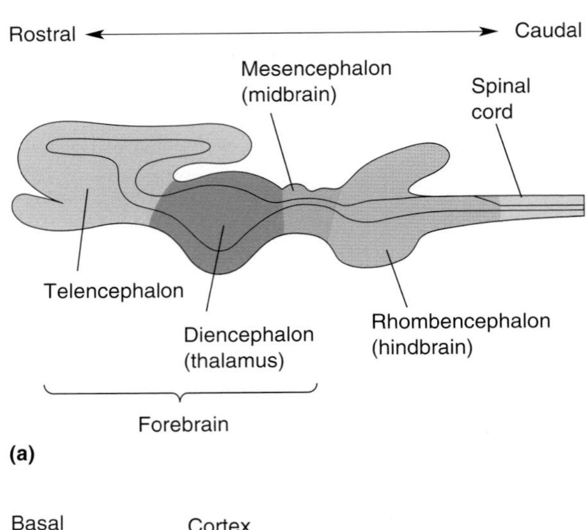

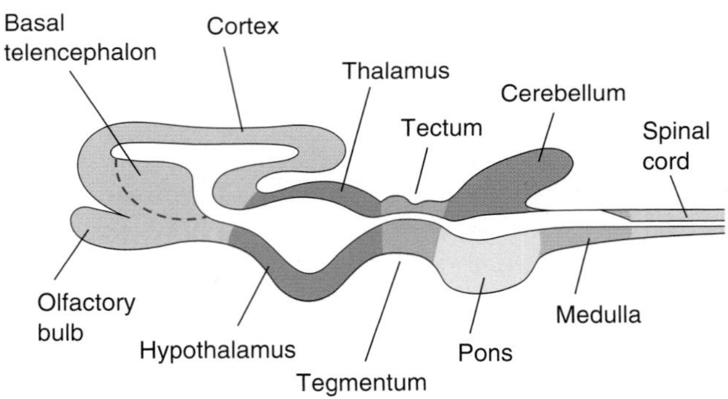

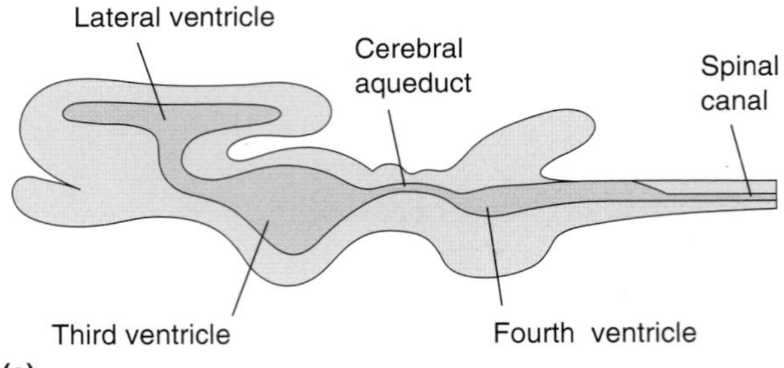

Figure 7.20
The brainship Enterprise. (a) The basic plan of the mammalian brain with the major subdivisions indicated. **(b)** Major structures within each division of the brain. Note that the telencephalon consists of two hemispheres, although only one is illustrated. **(c)** The ventricular system.

Table 7.3 The Ventricular System of the Brain

COMPONENT	RELATED BRAIN STRUCTURES
Lateral ventricles	Cerebral cortex
	Basal telencephalon
Third ventricle	Thalamus
	Hypothalamus
Cerebral aqueduct	Tectum
	Midbrain tegmentum
Fourth ventricle	Cerebellum
	Pons
	Medulla

Special Features of the Human CNS

So far, we've explored the basic plan of the CNS as it applies to all mammals. Figure 7.21 compares the brains of the rat and the human. You can see immediately that there are indeed many similarities but also some obvious differences.

Let's start by reviewing the similarities. The dorsal view of both brains reveals the paired hemispheres of the telencephalon (Figure 7.21a). A midsagittal view of the two brains shows the telencephalon extending rostrally from the diencephalon (Figure 7.21b). The diencephalon surrounds the third ventricle, the midbrain surrounds the cerebral aqueduct, and the cerebellum, pons, and medulla surround the fourth ventricle. Notice how the pons swells below the cerebellum and how structurally elaborate the cerebellum is.

Now let's consider some of the structural differences between rat and human brains. Figure 7.21a reveals a striking one: the many convolutions on the surface of the human cerebrum. The grooves in the surface of the cerebrum are called **sulci** (singular: sulcus), and the bumps are called **gyri** (singular: gyrus). Remember that the thin sheet of neurons that lies just under the surface of the cerebrum is the cerebral cortex. Sulci and gyri result from the tremendous expansion of the surface area of the cerebral cortex during human fetal development. The adult human cortex, measuring about 1100 cm^2, must fold and wrinkle to fit within the confines of the skull. This increase in cortical surface area is one of the "distortions" of the human brain. Clinical and experimental evidence indicates that the cortex is the seat of uniquely human reasoning and cognition. Without the cerebral cortex, a person would be blind, deaf, dumb, and unable to initiate voluntary movement. We will take a closer look at the structure of the cerebral cortex in a moment.

The side views of the rat and human brains in Figure 7.21c reveal further differences in the forebrain. One is the small size of the olfactory bulb in the human relative to the rat. On the other hand, notice again the growth of the cerebral hemisphere in the human. See how the cerebral hemisphere of the human brain arcs posteriorly, ventrolaterally, and then anteriorly to resemble a ram's horn. The tip of the "horn" lies right under the temporal bone (temple) of the skull, so this portion of the brain is called the **temporal lobe**. Three other lobes named after skull bones also describe the parts of the human cerebrum. The portion of the cerebrum lying just under the frontal bone of the forehead is called the **frontal lobe**. The deep **central sulcus** marks the posterior border of the frontal lobe, caudal to which lies the **parietal lobe**, under the parietal bone. Caudal to that, at the back of the cerebrum under the occipital bone, lies the **occipital lobe** (Figure 7.22).

Despite the disproportionate growth of the cerebrum, the human brain still follows the basic mammalian brain plan laid out during embryonic development. Again, the ventricles are key. Although the ventricular system is

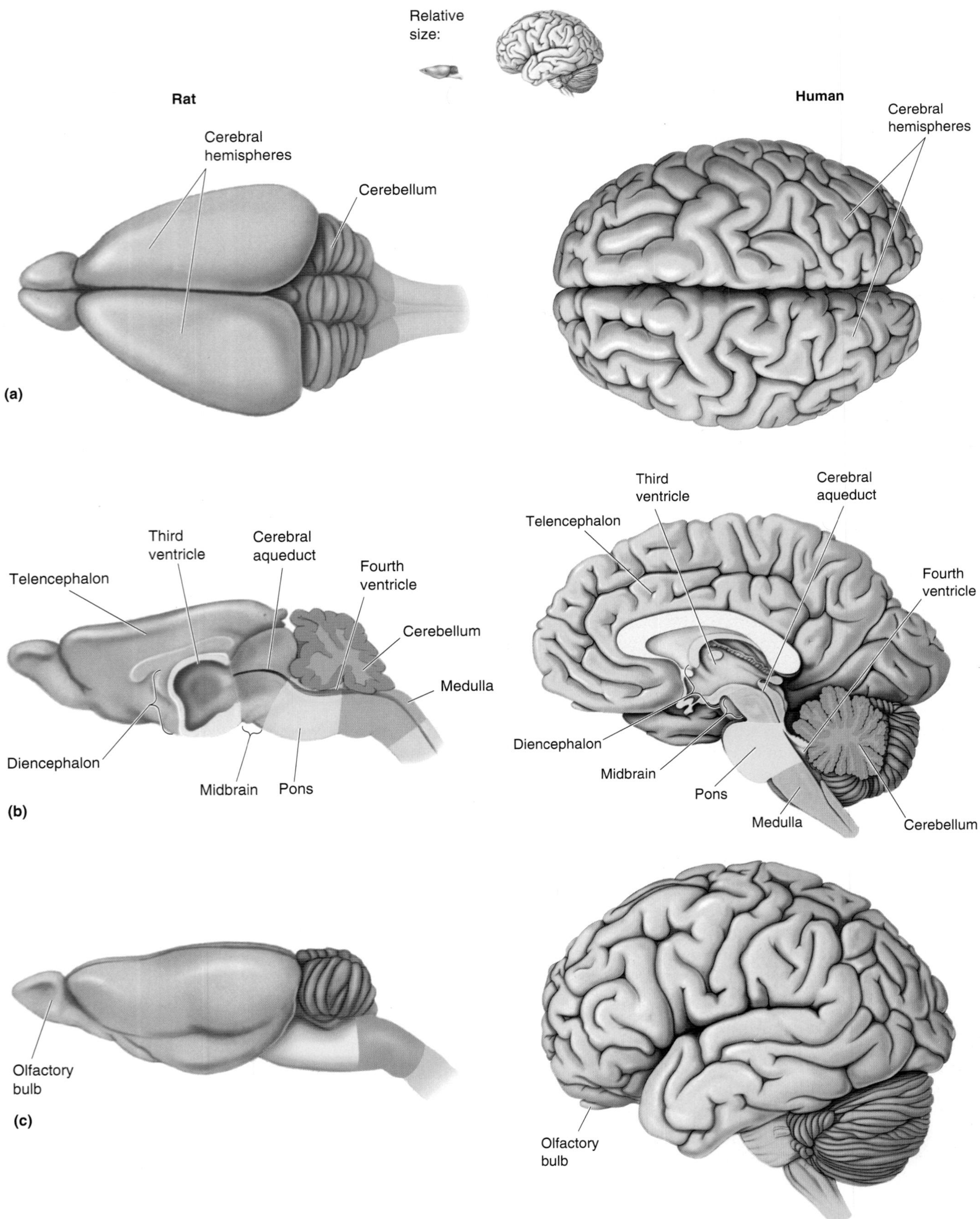

Relative size:
Rat
Human
Cerebral hemispheres
Cerebellum
Cerebral hemispheres
(a)
Third ventricle
Cerebral aqueduct
Telencephalon
Fourth ventricle
Cerebellum
Medulla
Diencephalon
Midbrain
Pons
(b)
Third ventricle
Cerebral aqueduct
Telencephalon
Fourth ventricle
Diencephalon
Midbrain
Pons
Medulla
Cerebellum
Olfactory bulb
(c)
Olfactory bulb

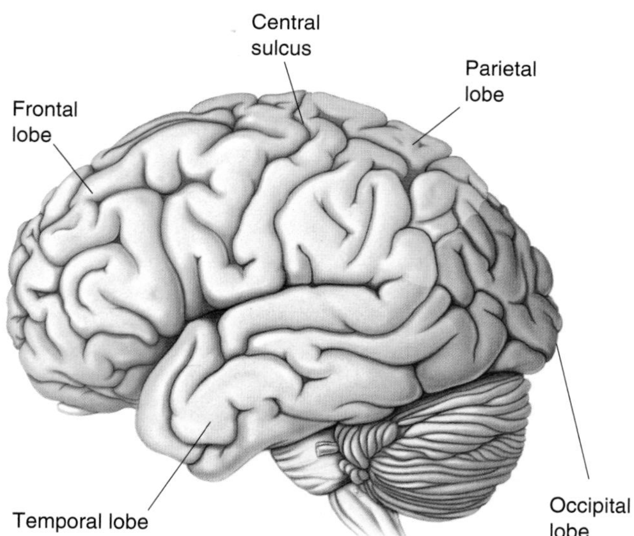

Figure 7.22
The lobes of the human cerebrum.

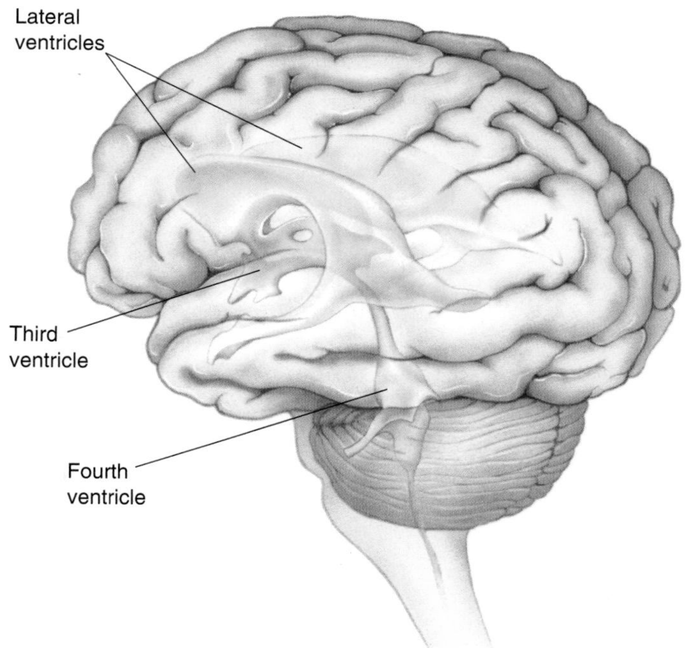

Figure 7.23
The human ventricular system. Although the ventricles are distorted by the growth of the brain, the basic relationships of the ventricles to the surrounding brain are the same as those illustrated in Figure 7.20.

distorted, particularly by the growth of the temporal lobes, the relationships that relate the brain to the ventricles still hold (Figure 7.23).

A GUIDE TO THE CEREBRAL CORTEX

Because of its prominence in the human brain, the cerebral cortex deserves further description. As we will see repeatedly in subsequent chapters, the systems in the brain that are responsible for sensations, perceptions, voluntary movement, learning, speech, and cognition all converge in this remarkable organ.

Figure 7.21
The rat brain and human brain compared. (a) Dorsal view. **(b)** Midsagittal view. **(c)** Lateral view. (Brains are not drawn to the same scale.)

Types of Cerebral Cortex

Cerebral cortex in the brain of all vertebrate animals has several common features, as shown in Figure 7.24. First, the cell bodies of cortical neurons are always arranged in layers, or sheets, most of which lie parallel to the surface of the brain. Second, the layer of neurons closest to the surface (the most superficial cell layer) is separated from the pia by a zone that lacks neurons; it is called the molecular layer, or simply *layer I*. Third, at least one cell layer contains pyramidal cells that emit large dendrites, called *apical dendrites*, that extend up to layer I, where they form multiple branches. Thus, we can say that the cerebral cortex has a characteristic cytoarchitecture that distinguishes it, for example, from the nuclei of the basal telencephalon or the thalamus.

Figure 7.25 shows a Nissl-stained coronal section through the caudal telencephalon of a rat brain. You don't need to be Cajal to see that different types of cortex are characterized by their cytoarchitecture. Medial to the lateral ventricle is a piece of cortex that is folded onto itself in a peculiar shape. This structure is called the **hippocampus** (from the Greek for "seahorse"); despite its bends, it has only a single cell layer. Connected to the hippocampus ventrally and laterally is another type of cortex that has only two cell layers. It is called the **olfactory cortex,** because it is continuous with the olfac-

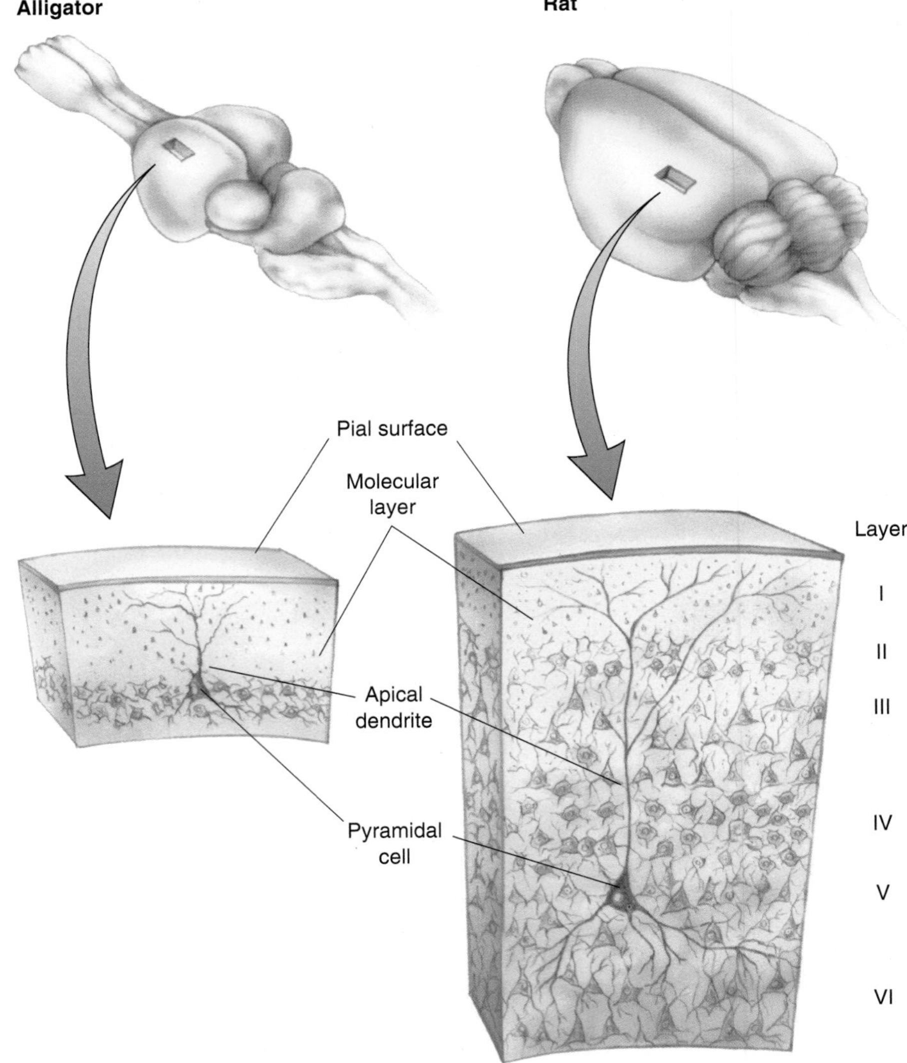

Figure 7.24
General features of cerebral cortex. On the left is the structure of cortex in an alligator; on the right is the structure of cortex in a rat. In both species, the cortex lies just under the pia of the cerebral hemisphere, contains a molecular layer, and has pyramidal cells arranged in layers.

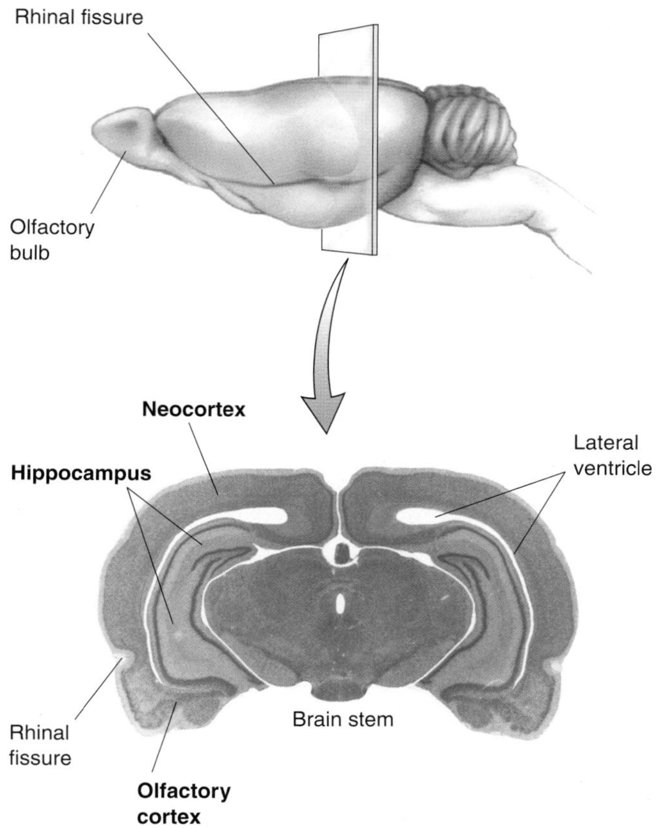

Figure 7.25
Three types of cortex in a mammal. In this section of a rat brain, the lateral ventricles lie between the neocortex and the hippocampus on each side. The ventricles are not obvious because they are very long and thin in this region. Below the telencephalon lies the brain stem. What region of brain stem is this, based on the appearance of the fluid-filled space at its core?

tory bulb, which sits further anterior. The olfactory cortex is separated by a sulcus, called the *rhinal fissure*, from another more elaborate type of cortex, the **neocortex**, which has many cell layers. Unlike the hippocampus and olfactory cortex, *neocortex is found only in mammals*. Thus, when we said previously that the cerebral cortex has expanded over the course of human evolution, we really meant that the neocortex has expanded. Similarly, when we said that the thalamus is the gateway to the cortex, we meant that it is the gateway to the neocortex. Most neuroscientists (ourselves included) are such neocortical chauvinists that the term *cortex*, if left unqualified, usually refers to the cerebral neocortex.

In Chapter 8, we will discuss the olfactory cortex in the context of the sense of smell. Further discussion of the hippocampus is reserved until later in this book, when we explore its role in the limbic system (Chapter 18) and in memory and learning (Chapters 23 and 24). The neocortex will figure prominently in our discussions of vision, audition, somatic sensation, and the control of voluntary movement in Part II, so let's examine its structure in more detail.

Areas of Neocortex

Just as cytoarchitecture can be used to distinguish cerebral cortex from the basal telencephalon and the neocortex from the olfactory cortex, it can be used to divide the neocortex into zones. This is precisely what the famous German neuroanatomist Korbinian Brodmann did at the beginning of the twentieth century. He constructed a **cytoarchitectural map** of the neocortex (Figure 7.26). In this map, each area of cortex having a common cytoarchitecture is given a number. Thus, we have "area 17" at the tip of the occipital lobe, "area 4" just anterior to the central sulcus in the frontal lobe, and so on.

Figure 7.26
Brodmann's cytoarchitectural map of the human cerebral cortex.

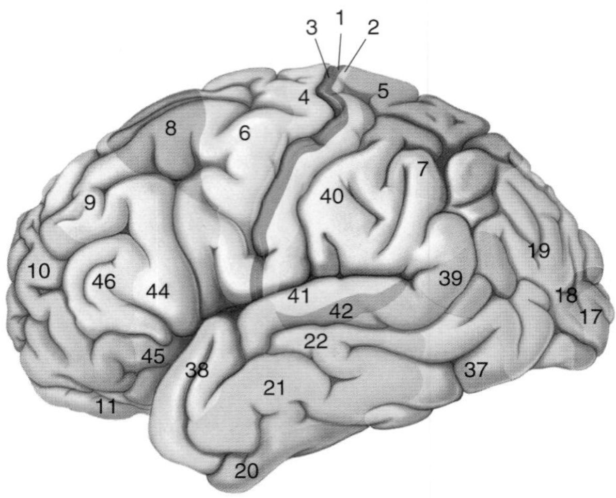

What Brodmann guessed but could not show was that cortical areas that look different perform different functions. We now have evidence that this is true. For instance, we can say that area 17 is visual cortex, because it receives signals from a nucleus of the thalamus that is connected to the retina. Indeed, without area 17, a human is blind. Similarly, we can say that area 4 is motor cortex, because neurons in this area project axons directly to the motor neurons of the ventral horn that command muscles to contract. Notice that the different functions of these two areas are specified by their different connections.

Neocortical Evolution and Structure–Function Relationships. A problem that has fascinated neuroscientists since the time of Brodmann is how neocortex has changed over the course of mammalian evolution. The brain is a soft tissue, so there is no fossil record of the cortex of our early mammalian ancestors. Nonetheless, considerable insight can be gained by comparing the cortex of different living species (Figure 7.1). The surface area of the cortex varies tremendously among species; for example, a comparison of mouse, monkey, and human cortex reveals differences in size on the order of 1:100: 1000. On the other hand, there is little difference in the thickness of the neocortex among mammals, varying by no more than a factor of two. Thus, we can conclude that the amount of cortex has changed over the course of evolution, but its basic structure has not.

Brodmann proposed that neocortex expanded by the insertion of new areas. Jon Kaas and his students at Vanderbilt University have sought to address this issue by studying the structure and function of different cortical areas in many species (Box 7.5). Their research suggests that the primordial neocortex consisted mainly of three types of cortex—cortex that also exists to some degree in all living species. The first type consists of *primary sensory areas*, which are first to receive signals from the ascending sensory pathways. For example, area 17 is designated as primary visual cortex, or V1, because it receives input from the eyes via a direct path: retina to thalamus to cortex. The second type of neocortex consists of *secondary sensory areas*, so designated because of their heavy interconnections with the primary sensory areas. The third type of cortex consists of *motor areas*, which are intimately concerned with the control of voluntary movement. These cortical areas receive inputs from thalamic nuclei that relay information from the basal telencephalon and the cerebellum, and they send outputs to motor control neurons in the brain stem and spinal cord. For example, because cortical area 4 sends outputs directly to motor neurons in the ventral horn of the spinal cord, it is designated

Box 7.5

PATH OF DISCOVERY

Evolution of the Evolution of the Neocortex

BY JON KAAS

The neocortex varies so much in size and shape across mammals that it is natural to wonder how such variation comes about in evolution. My enduring interest in brain evolution developed because I had the opportunity to train with two outstanding scientists who studied cortical variation. As a graduate student, I worked with Irving Diamond at Duke University. I studied the function of auditory cortex in cats as his other students explored cortical organization in such unusual species as tree shrews, squirrels, hedgehogs, and opossums. I continued with my postdoctoral training at the University of Wisconsin with Clinton Woolsey, who had illustrated many neuroscience textbooks with his depictions of sensory and motor areas in monkeys, cats, and rabbits.

At the time, the available methods greatly limited understanding cortical organization across species. Cortex had long been divided into areas of supposed functional significance by regional differences in the appearance of brain sections stained for cell bodies or myelinated fibers, but only a few cortical areas were histologically distinct enough to be recognized with widespread agreement. Woolsey advanced our understanding of cortical organization by recording from the brain surface during sensory stimulation and by electrically stimulating the brain to evoke movements. These procedures revealed much about the organization of primary sensory and motor representations, but more precise mapping methods were needed to distinguish the various cortical representations from each other. In addition, many parts of sensory cortex failed to respond to stimuli because of the anesthetics in common use.

Because I was using surface recordings and traditional anesthetics, my early efforts to study visual cortex in monkeys were not very productive. Fortunately, just across the laboratory hallway, Wally Welker was pioneering the use of microelectrodes in mapping somatosensory cortex in raccoons, beavers, and other mammals. Systematic progressions of microelectrode recordings across cortex revealed the organization of somatosensory representations with great precision. Another member of Woolsey's lab, Vincente Montero, had already used microelectrodes to map visual cortex in rats. Thus, I started to apply these new methods in studies of visual cortex organization in hedgehogs, tree shrews, and squirrels in collaboration with Bill Hall, Herb Killackey, and Irving Diamond at Duke. The results clearly demonstrated that we could learn a lot by using these microelectrode mapping methods.

Jon Kaas

I returned with great optimism to studies of visual cortex organization in primates. I was lucky enough to team up with a new arrival at Woolsey's lab, John Allman, then a graduate student in anthropology. We enjoyed immediate success in a series of exhausting experiments on visual cortex in owl monkeys (Figure A). A vast extent of previously unexplored cortex proved to consist of a patchwork of systematic maps (or representations) of the retina. The first such map that we described we called the middle temporal area, MT, for its location in the temporal lobe. MT corresponded to a densely myelinated section of cortex that surprisingly had been overlooked in earlier histological studies of the cortex. MT is now one of the most investigated visual areas, and it has been identified in all primate species studied, including humans. We went on to show that visual cortex in monkeys includes a number of distinct visual areas.

(*continues on next page*)

as primary motor cortex, or M1. Kaas' analysis suggests that the common mammalian ancestor had about 20 areas that could be assigned to these three categories.

Figure 7.27 shows views of the brain of a rat, a cat, and a human, with the primary sensory and motor areas identified. It is plain to see that when we speak of the expansion of the cortex in mammalian evolution, what has expanded is the region that lies between these areas. Research by Kaas and others has shown that much of the "in-between" cortex reflects expansion of the number of secondary sensory areas devoted to the analysis of sensory infor-

(Box 7.5, *continued*)

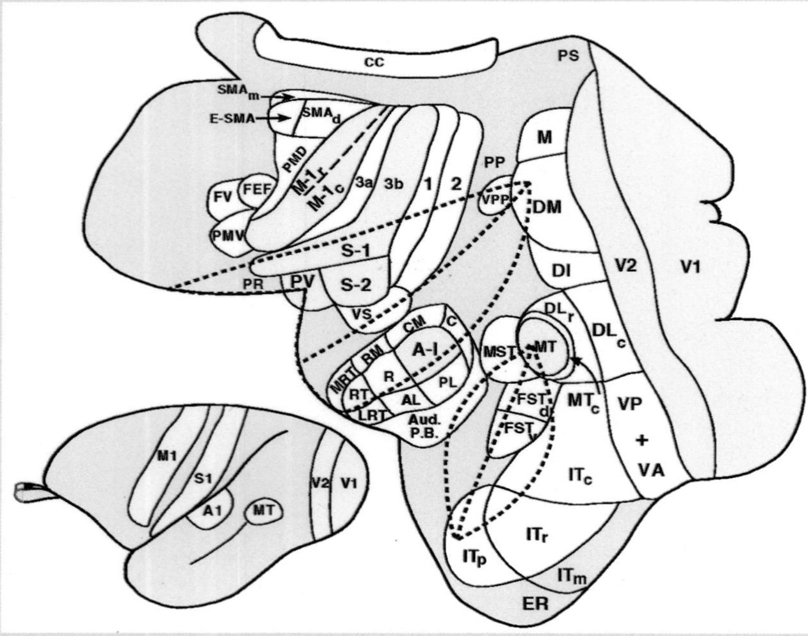

Figure A
Some proposed subdivisions of cortex in owl monkeys. Lower left: Several areas (yellow) on a lateral view of the brain. Upper right: A view of the cortex as a flattened sheet, where fissures have been opened (broken lines). Visual areas are grouped caudally (right), auditory areas are grouped in the temporal lobe (lower middle), and somatosensory and motor areas are rostral (left). (Source: Courtesy of Jon Kaas.)

After starting my laboratory at Vanderbilt University, I expanded the range of these studies by using the new techniques for revealing the connections of the brain. I also expanded into studies of auditory and somatosensory cortex with students and Mike Merzenich, a friend from my postdoctoral days at Wisconsin. We studied a variety of primate species, as well as tree shrews, squirrels, and other mammals.

These research efforts led to a number of major conclusions. Vast amounts of cortex in monkeys and other advanced mammals with large brains are devoted to sensory processing. Investigators today are still trying to define cortical areas and determine their numbers, but there is widespread agreement on the identities of some areas, and the numbers for some mammals clearly are high. Monkeys, for example, have some 20–30 visual areas, 10–15 auditory areas, and 8–15 somatosensory areas. In addition, most mammals share a limited set of homologous sensory and motor areas, including the primary visual, auditory, and somatosensory fields. These areas have been retained from the first mammals, where they occupied most of the neocortex, and they continue to occupy most of the neocortex in species of living mammals with small brains and little neocortex. Additional sensory processing areas have emerged independently in various lines of mammalian evolution. Although we are not certain how new areas emerge in evolution, we know that mammals commonly advance in brain function by increasing the number of brain areas rather than just increasing brain size.

An explosion of new and powerful research methods is now making dreams possible. I expect such major advances over the next few years that our current understanding of cortical organization, function, and variation will seem rather primitive. The real exploration has just begun.

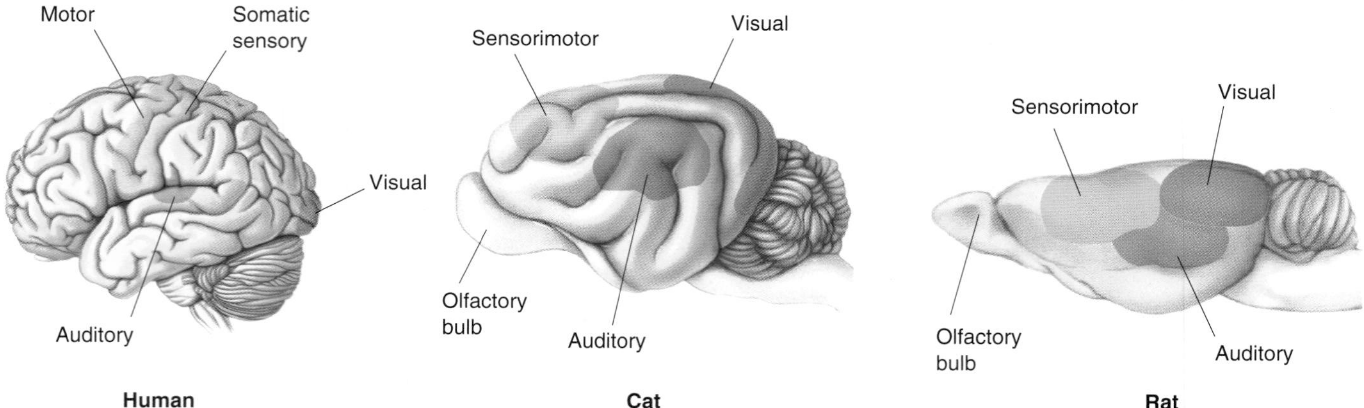

Figure 7.27
A lateral view of the cerebral cortex in three species. Notice the expansion of the human cortex that is neither strictly primary sensory nor strictly motor.

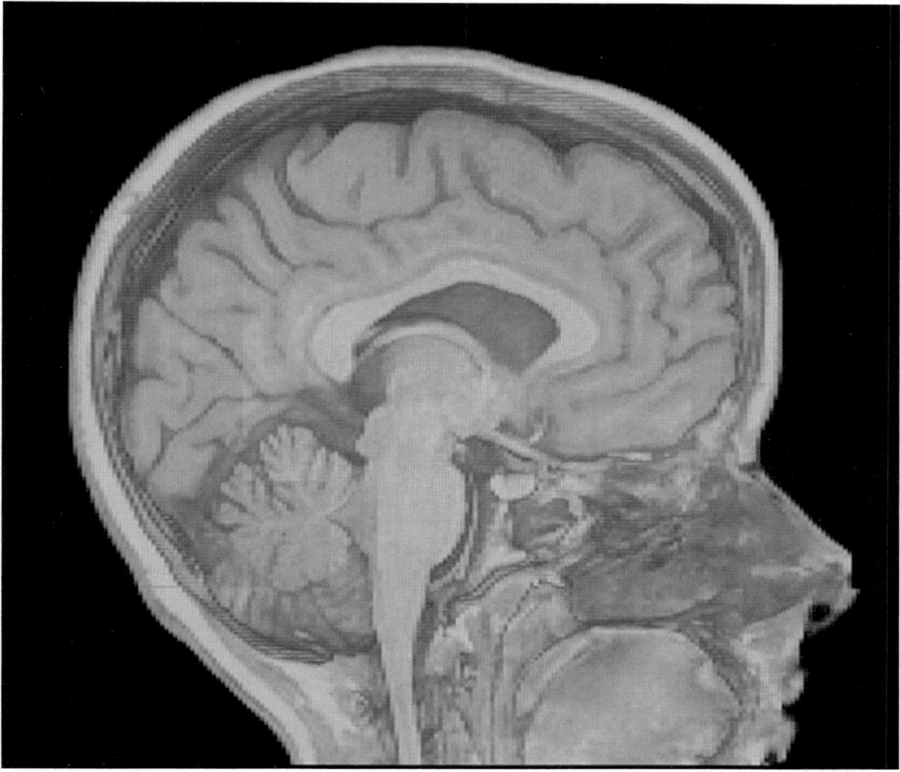

Figure 7.28
An MRI scan of the living human brain. How many structures can you label? (Source: Courtesy of Dr. J. N. Sanes.)

mation. For example, in primates that depend heavily on vision, like humans, the number of secondary visual areas has been estimated to be between 20 and 40. Even after we have assigned primary sensory, motor, and secondary sensory functions to large regions of cortex, however, a considerable amount of area remains in the human brain, particularly in the frontal and temporal lobes. These are the *association areas* of cortex. Association cortex is a more recent development, a noteworthy characteristic of the primate brain. The emergence of the "mind"—our unique ability to interpret behavior (our own and that of others) in terms of unobservable mental states, such as desires, intentions, and beliefs—correlates best with the expansion of the frontal cortex. Indeed, as we will see in Chapter 18, lesions of the frontal cortex can profoundly alter an individual's personality.

CONCLUDING REMARKS

Although we have covered a lot of new ground in this chapter, we have only scratched the surface of neuroanatomy. Clearly, the brain deserves its status as the most complex piece of matter in the universe. What we have presented here is a shell, or scaffold, of the nervous system and some of its contents.

Understanding neuroanatomy is necessary for understanding how the brain works. This statement is just as true for an undergraduate first-time neuroscience student as it is for a neurologist or a neurosurgeon. In fact, neuroanatomy has taken on new relevance with the advent of methods to image the living brain (Figure 7.28).

An Illustrated Guide to Human Neuroanatomy appears as an appendix to this chapter. Use the Guide as an atlas to locate various structures of interest.

Labeling exercises are also provided to help you learn the names of the parts of the nervous system you will encounter in this book.

In Part II, Sensory and Motor Systems, the anatomy presented in Chapter 7 and its appendix will come alive as we explore how the brain goes about the tasks of smelling, tasting, seeing, hearing, sensing touch, and moving.

KEY TERMS

Gross Organization of the Mammalian Nervous System

anterior (p. 164)
rostral (p. 164)
posterior (p. 164)
caudal (p. 164)
dorsal (p. 164)
ventral (p. 164)
midline (p. 164)
medial (p. 166)
lateral (p. 166)
ipsilateral (p. 166)
contralateral (p. 166)
midsagittal plane (p. 166)
sagittal plane (p. 166)
horizontal plane (p. 166)
coronal plane (p. 166)
central nervous system (CNS) (p. 167)
brain (p. 167)
spinal cord (p. 167)
cerebrum (p. 167)
cerebral hemispheres (p. 167)
cerebellum (p. 167)
brain stem (p. 167)
spinal nerve (p. 168)
dorsal root (p. 168)
ventral root (p. 168)
peripheral nervous system (PNS) (p. 169)
somatic PNS (p. 169)
dorsal root ganglion (p. 169)
visceral PNS (p. 169)
autonomic nervous system (ANS) (p. 169)
afferent (p. 169)
efferent (p. 169)
cranial nerve (p. 170)
meninges (p. 170)
dura mater (p. 170)
arachnoid membrane (p. 170)
pia mater (p. 170)
cerebrospinal fluid (CSF) (p. 170)
ventricular system (p. 170)

Understanding CNS Structure Through Development

neural tube (p. 176)
neural crest (p. 177)
neurulation (p. 177)
differentiation (p. 178)
forebrain (p. 178)
midbrain (p. 178)
hindbrain (p. 178)
gray matter (p. 178)
cortex (p. 178)
nucleus (p. 178)
substantia (p. 178)
locus (p. 178)
ganglion (p. 178)
nerve (p. 178)
white matter (p. 178)
tract (p. 178)
bundle (p. 178)
capsule (p. 178)
commissure (p. 178)
lemniscus (p. 178)
diencephalon (p. 179)
telencephalon (p. 179)
olfactory bulb (p. 182)
lateral ventricle (p. 182)
third ventricle (p. 182)
cerebral cortex (p. 182)
basal telencephalon (p. 182)
thalamus (p. 182)
hypothalamus (p. 182)
cortical white matter (p. 182)
corpus callosum (p. 182)
internal capsule (p. 182)
tectum (p. 185)
tegmentum (p. 185)
cerebral aqueduct (p. 185)
pons (p. 186)
medulla oblongata (p. 186)
fourth ventricle (p. 186)
spinal canal (p. 188)
dorsal horn (p. 188)
ventral horn (p. 188)
sulcus (p. 191)
gyrus (p. 191)
temporal lobe (p. 191)
frontal lobe (p. 191)
central sulcus (p. 191)
parietal lobe (p. 191)
occipital lobe (p. 191)

A Guide to the Cerebral Cortex

hippocampus (p. 194)
olfactory cortex (p. 194)
neocortex (p. 195)
cytoarchitectural map (p. 195)

REVIEW QUESTIONS

1. Are the dorsal root ganglia in the central or the peripheral nervous system?
2. Is the myelin sheath of optic nerve axons provided by Schwann cells or oligodendroglia? Why?
3. Imagine that you are a neurosurgeon about to remove a tumor lodged deep inside the brain. The top of the skull has been removed. What now lies between you and the brain? Which layer or layers must be cut before you reach the CSF?
4. What is the fate of tissue derived from the embryonic neural tube? Neural crest?
5. Name the three main parts of the hindbrain. Which of these are also part of the brain stem?
6. Where is CSF produced? What path does it take before it is absorbed into the bloodstream? Name the parts of the CNS it will pass through in its voyage from brain to blood.
7. What are three features that characterize the structure of cerebral cortex?

AN ILLUSTRATED GUIDE TO HUMAN NEUROANATOMY

INTRODUCTION

As we will see in the remainder of the book, a fruitful way to explore the nervous system is to divide it into functional systems. Thus, the *olfactory system* consists of those parts of the brain that are devoted to the sense of smell, the *visual system* includes those parts that are devoted to vision, and so on. While this functional approach to investigating nervous system structure has many merits, it can make the "big picture"—how all these systems fit together inside the box we call the brain—difficult to see. The goal of this Illustrated Guide is to help you learn in advance about some of the anatomy that will be discussed in the subsequent chapters. Here we concentrate on naming the structures and seeing how they are related physically; their functional significance is discussed in the remainder of the book.

The Guide is organized into six main parts. The first part covers the surface anatomy of the brain, the structures that can been seen by inspection of the whole brain, and those parts that are visible when the two cerebral hemispheres are separated by a cut in the midsagittal plane. Next, we explore the cross-sectional anatomy of the brain, using a series of slabs that contain structures of interest. The brief third and fourth parts cover the spinal cord and the autonomic nervous system. The fifth part of the Guide illustrates the cranial nerves and summarizes their diverse functions. The last part illustrates the blood supply of the brain.

The nervous system has an astonishing number of bits and pieces. In this Guide, we focus on those structures that will appear later in the book when we discuss the various functional systems. Nonetheless, even this abbreviated atlas of neuroanatomy yields a formidable new vocabulary. Therefore, to help you learn the terminology, an extensive self-quiz review is provided at the end in the form of a workbook with labeling exercises.

SURFACE ANATOMY OF THE BRAIN (next page)

Imagine you hold in your hands a human brain that has been dissected from the skull. It is wet and spongy and weighs about 1.4 kg (3 pounds). Looking down on the brain's dorsal surface reveals the convoluted surface of the cerebrum. Flipping the brain over shows the complex ventral surface that normally rests on the floor of the skull. Holding the brain up and looking at its side—the lateral view—shows the "ram's horn" shape of the cerebrum coming off the stalk of the brain stem. The brain stem is shown more clearly if we slice the brain right down the middle and view its medial surface. In the part of the Guide that follows, we will name the important structures that are revealed by such an inspection of the brain. Notice the magnification of the drawings: 1× is life-size, 2× is twice life-size, 0.6× is 60% of life-size, and so on.

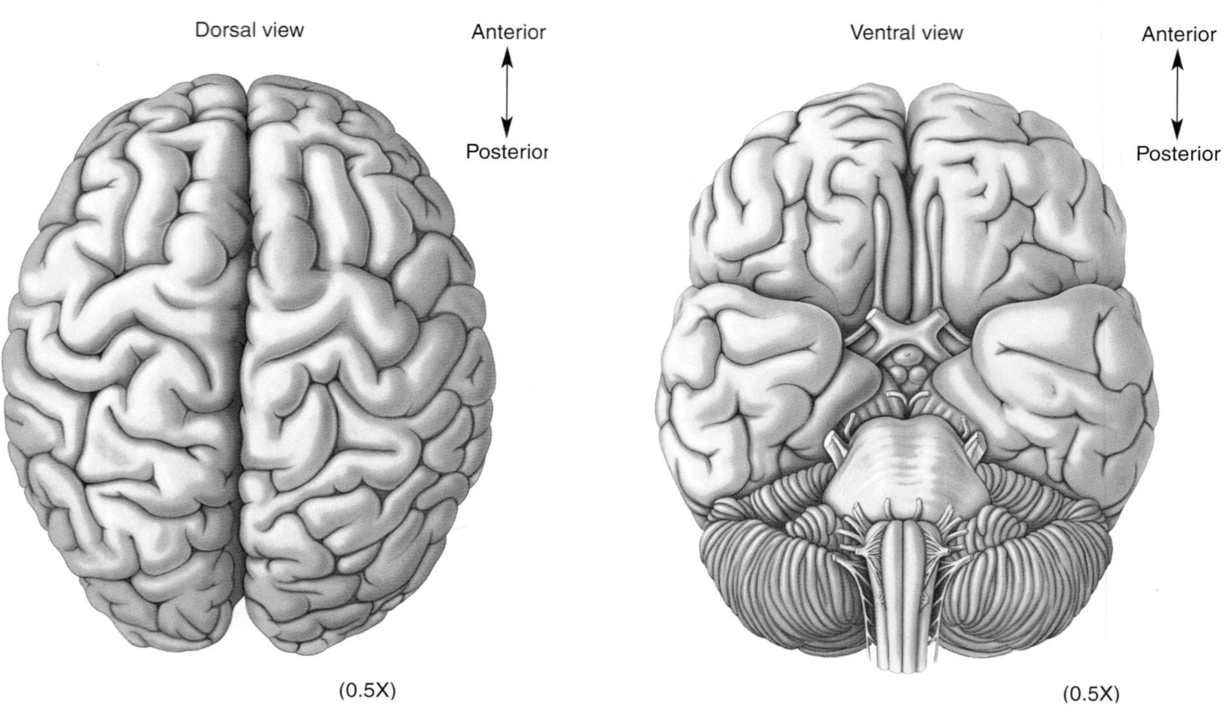
Dorsal view
Anterior
Posterior
(0.5X)
Ventral view
Anterior
Posterior
(0.5X)

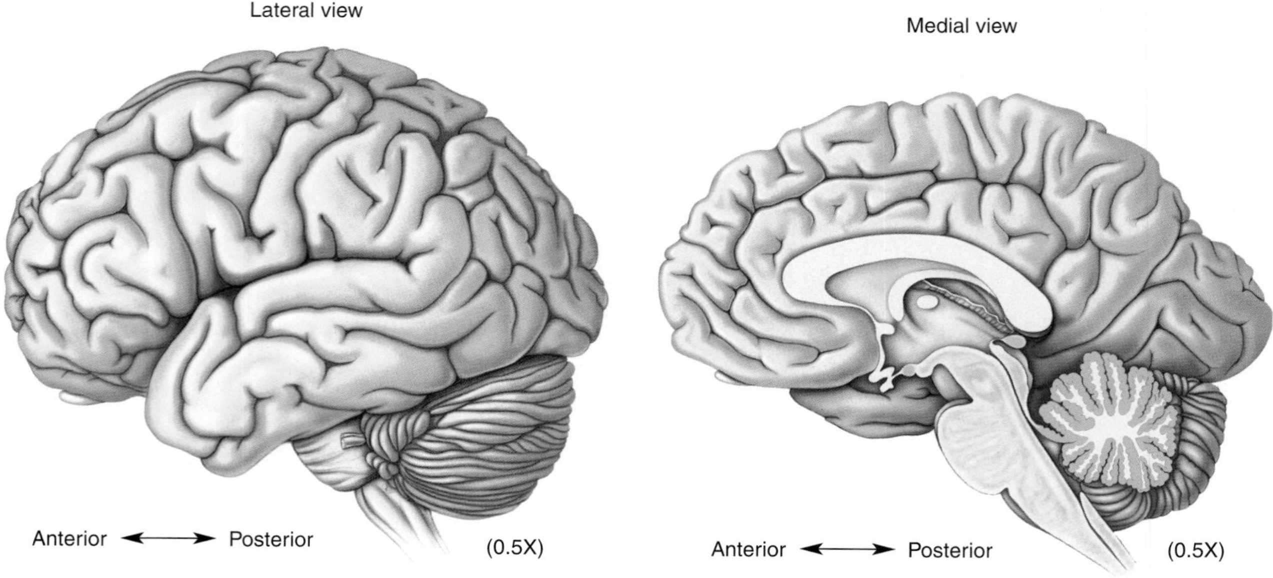
Lateral view
Anterior
Posterior
(0.5X)
Medial view
Anterior
Posterior
(0.5X)

The Lateral Surface of the Brain

(a) Gross Features. This is a life-size drawing of the brain. Gross inspection reveals the three major parts: the large cerebrum, the brain stem that forms its stalk, and the rippled cerebellum. The diminutive olfactory bulb of the cerebrum can also be seen in this lateral view.

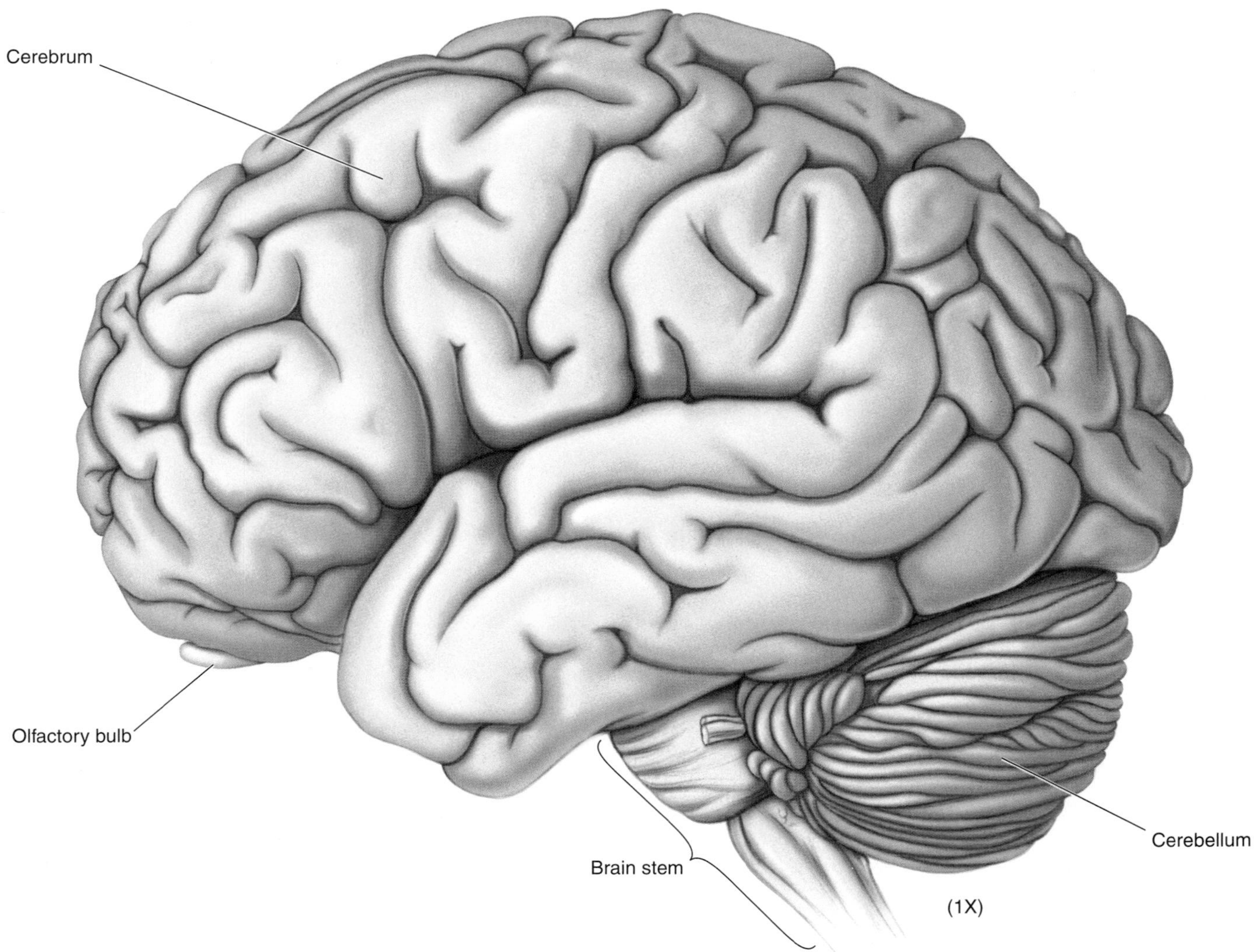

(b) Selected Gyri, Sulci, and Fissures. The cerebrum is noteworthy for its convoluted surface. The bumps are called gyri, and the grooves are called sulci or, if they are especially deep, fissures. The precise pattern of gyri and sulci can vary considerably from individual to individual, but many features are common to all human brains. Some of the important landmarks are labeled here. The postcentral gyrus lies immediately posterior to the central sulcus, and the precentral gyrus lies immediately anterior to the central sulcus. The neurons of the postcentral gyrus are involved in somatic sensation (touch, Chapter 12), and those of the precentral gyrus control voluntary movement (Chapter 14). Neurons in the superior temporal gyrus are involved in audition (hearing, Chapter 11).

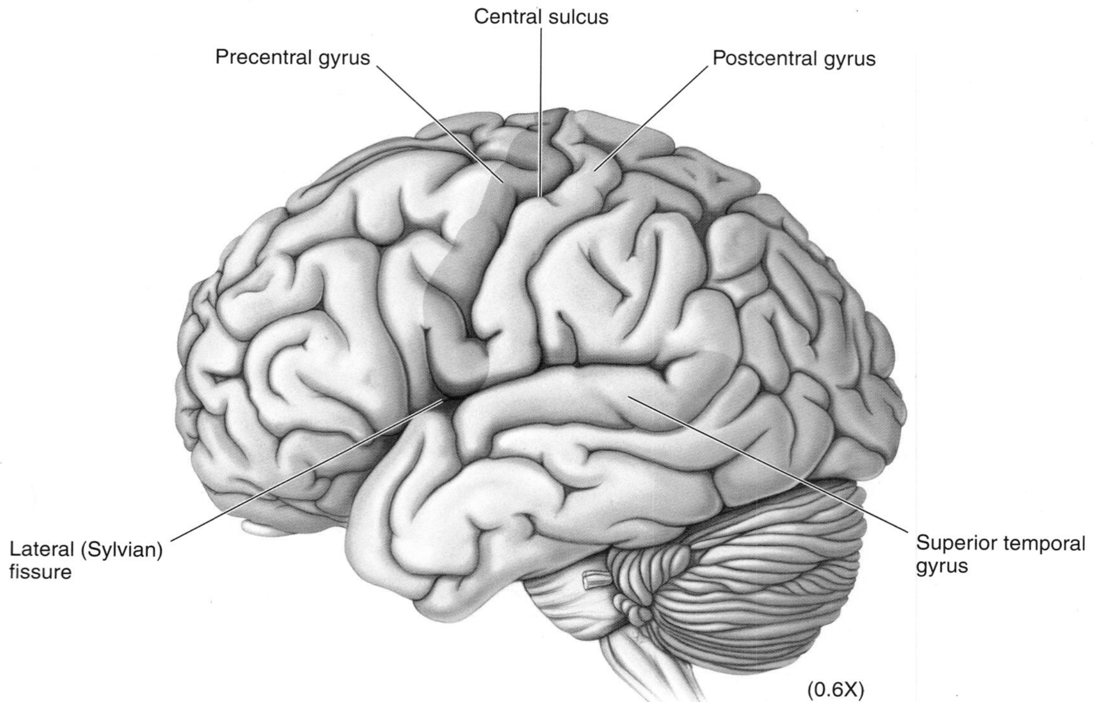

(c) Cerebral Lobes and the Insula. By convention, the cerebrum is subdivided into lobes named after the bones of the skull that lie over them. The central sulcus divides the frontal lobe from the parietal lobe. The temporal lobe lies immediately ventral to the deep lateral (Sylvian) fissure. The occipital lobe lies at the very back of the cerebrum, bordering both parietal and temporal lobes. A buried piece of the cerebral cortex, the insula (Latin for "island"), is revealed if the margins of the lateral fissure are gently pulled apart (inset). The insula borders and separates the temporal and frontal lobes.

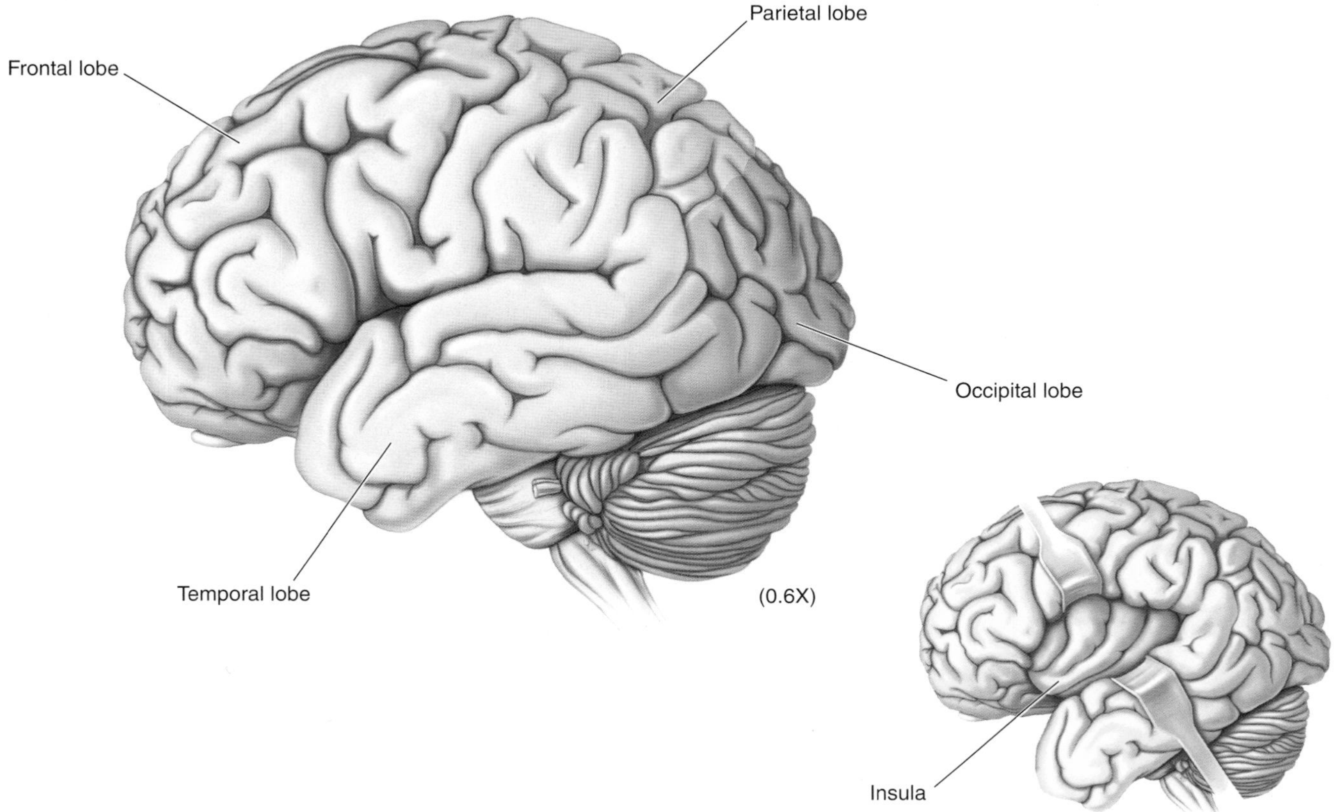

(d) Major Sensory, Motor, and Association Areas of Cortex (next page). The cerebral cortex is organized like a patchwork quilt. The various areas, first identified by Brodmann, differ from one another in microscopic structure and function. Visual areas 17, 18, and 19 (Chapter 10) are in the occipital lobe, somatic sensory areas 3, 1, and 2 (Chapter 12) are in the parietal lobe, and auditory areas 41 and 42 (Chapter 11) are in the temporal lobe. On the inferior surface of the parietal lobe (the operculum) and buried in the insula is gustatory area 43, devoted to the sense of taste (Chapter 8).

In addition to the analysis of sensory information, the cerebral cortex plays an important role in the control of voluntary movement. The major motor control areas— primary motor cortex (area 4), the supplementary motor area, and the premotor area —lie in the frontal lobe, anterior to the central sulcus (Chapter 14). In the human brain, large expanses of cortex cannot be simply assigned to sensory or motor functions. These constitute the association areas of cortex. Some of the more important areas are the prefrontal cortex (Chapters 20 and 23), the posterior parietal cortex (Chapters 12, 20, and 23), and the inferotemporal cortex (Chapter 23).

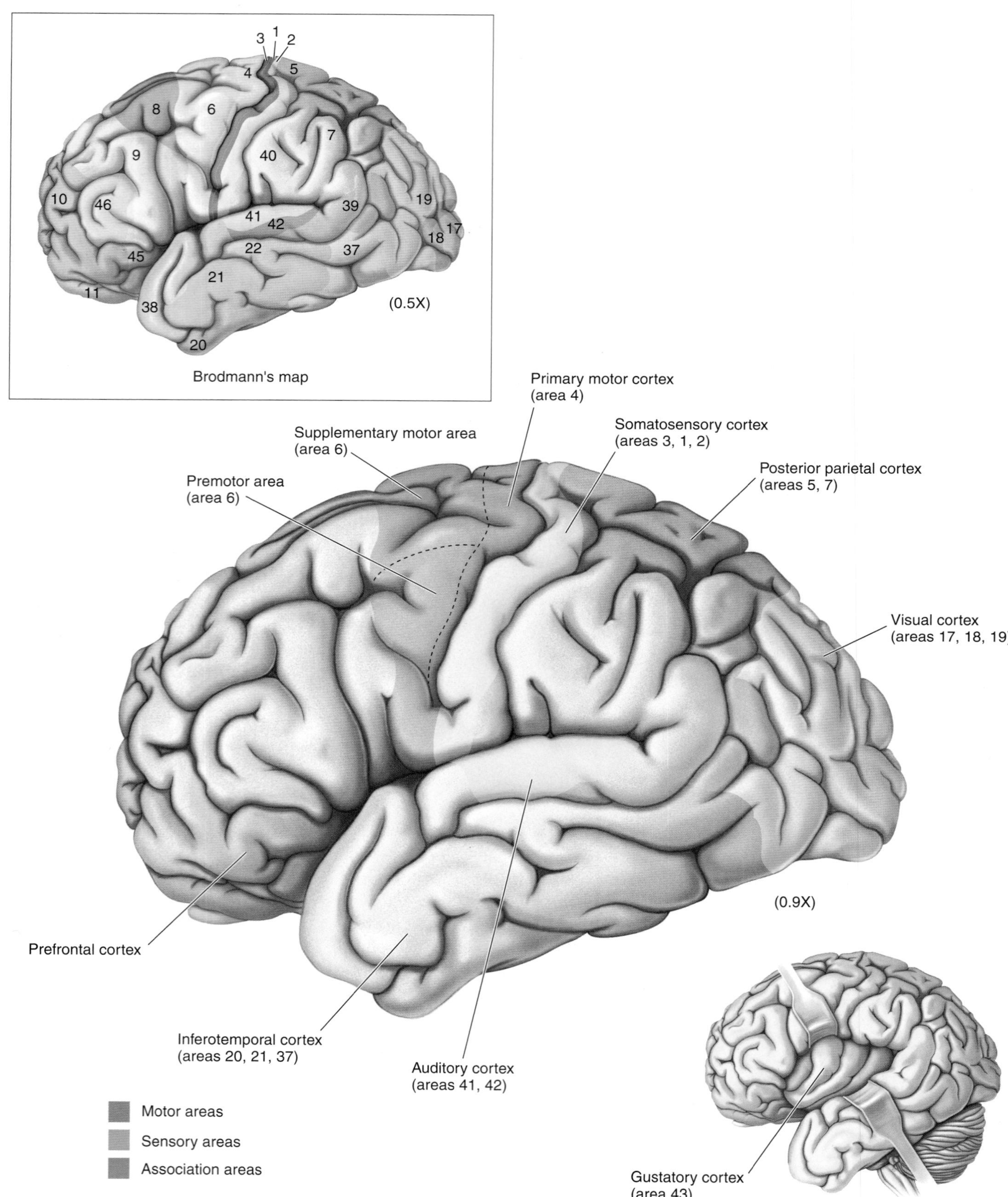
3
1
2
4
5
8
6
7
9
40
10
46
41
42
39
19
17
18
45
22
37
21
11
38
(0.5X)
20
Brodmann's map
Primary motor cortex
(area 4)
Supplementary motor area
(area 6)
Somatosensory cortex
(areas 3, 1, 2)
Premotor area
(area 6)
Posterior parietal cortex
(areas 5, 7)
Visual cortex
(areas 17, 18, 19)
(0.9X)
Prefrontal cortex
Inferotemporal cortex
(areas 20, 21, 37)
Auditory cortex
(areas 41, 42)
Motor areas
Sensory areas
Association areas
Gustatory cortex
(area 43)

The Medial Surface of the Brain

(a) Brain Stem Structures. Splitting the brain down the middle exposes the medial surface of the cerebrum, shown in this life-size illustration. This view also shows the midsagittal cut surface of the brain stem, consisting of the diencephalon (thalamus and hypothalamus), the midbrain (tectum and tegmentum), the pons, and the medulla. (Some anatomists define the brain stem as consisting only of the midbrain, pons, and medulla.)

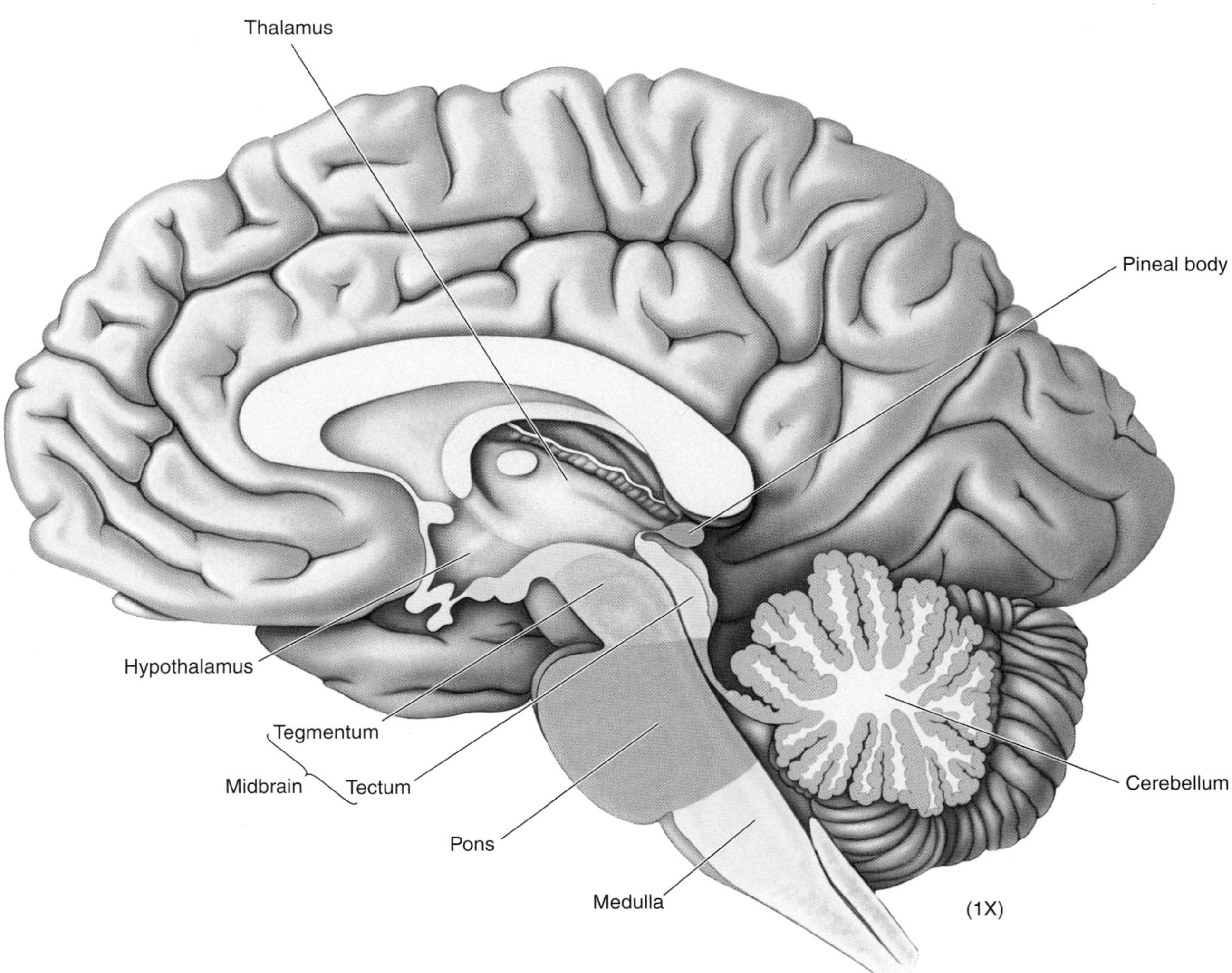

(b) Forebrain Structures (facing page). Shown here are the important forebrain structures that can be observed by viewing the medial surface of the brain. Notice the cut surface of the corpus callosum, a huge bundle of axons that connects the two sides of the cerebrum. The unique contributions of the two cerebral hemispheres to human brain function can be studied in patients whose callosum has been sectioned (Chapter 20). The fornix is another prominent fiber bundle that connects the hippocampus on each side with the hypothalamus. *Fornix* is Latin for "arch." Some of the axons in the fornix regulate memory storage (Chapter 23).

In the lower illustration, the brain has been tilted slightly to show the positions of the amygdala and hippocampus. These are "phantom views" of these structures, since they cannot be observed directly from the surface. Both lie deep to the overlying cortex. We will see them again in cross section later in the Guide. The amygdala (Latin for "almond") is an important structure for regulating emotional states (Chapter 18), and the hippocampus is important for memory (Chapters 23 and 24).

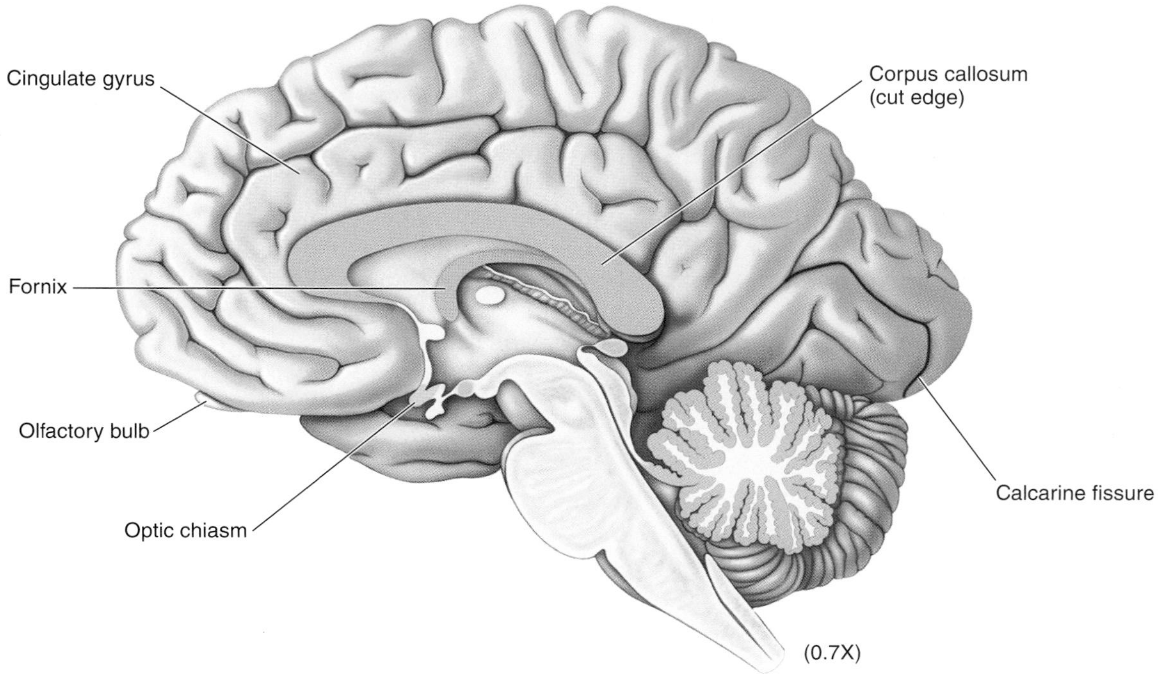

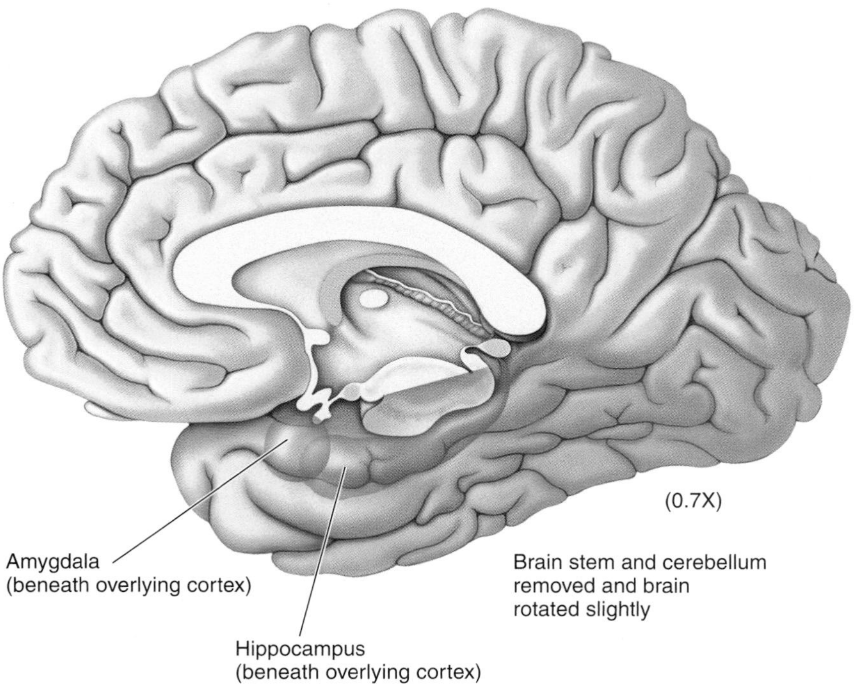

Brain stem and cerebellum removed and brain rotated slightly

(c) Ventricles. The lateral walls of the unpaired parts of the ventricular system—the third ventricle, the cerebral aqueduct, the fourth ventricle, and the spinal canal—can be observed in the medial view of the brain. These are handy landmarks, because the thalamus and hypothalamus lie next to the third ventricle; the midbrain lies next to the aqueduct; the pons, cerebellum, and medulla lie next to the fourth ventricle; and the spinal cord forms the walls of the spinal canal. The lateral ventricles are paired structures that sprout like antlers from the third ventricle. A phantom view of the right lateral ventricle, which lies under the cortex, is shown in the lower illustration. The two cerebral hemispheres surround the two lateral ventricles. Notice how a coronal section of the brain at the thalamus–midbrain junction intersects the "horns" of the lateral ventricle of each hemisphere twice.

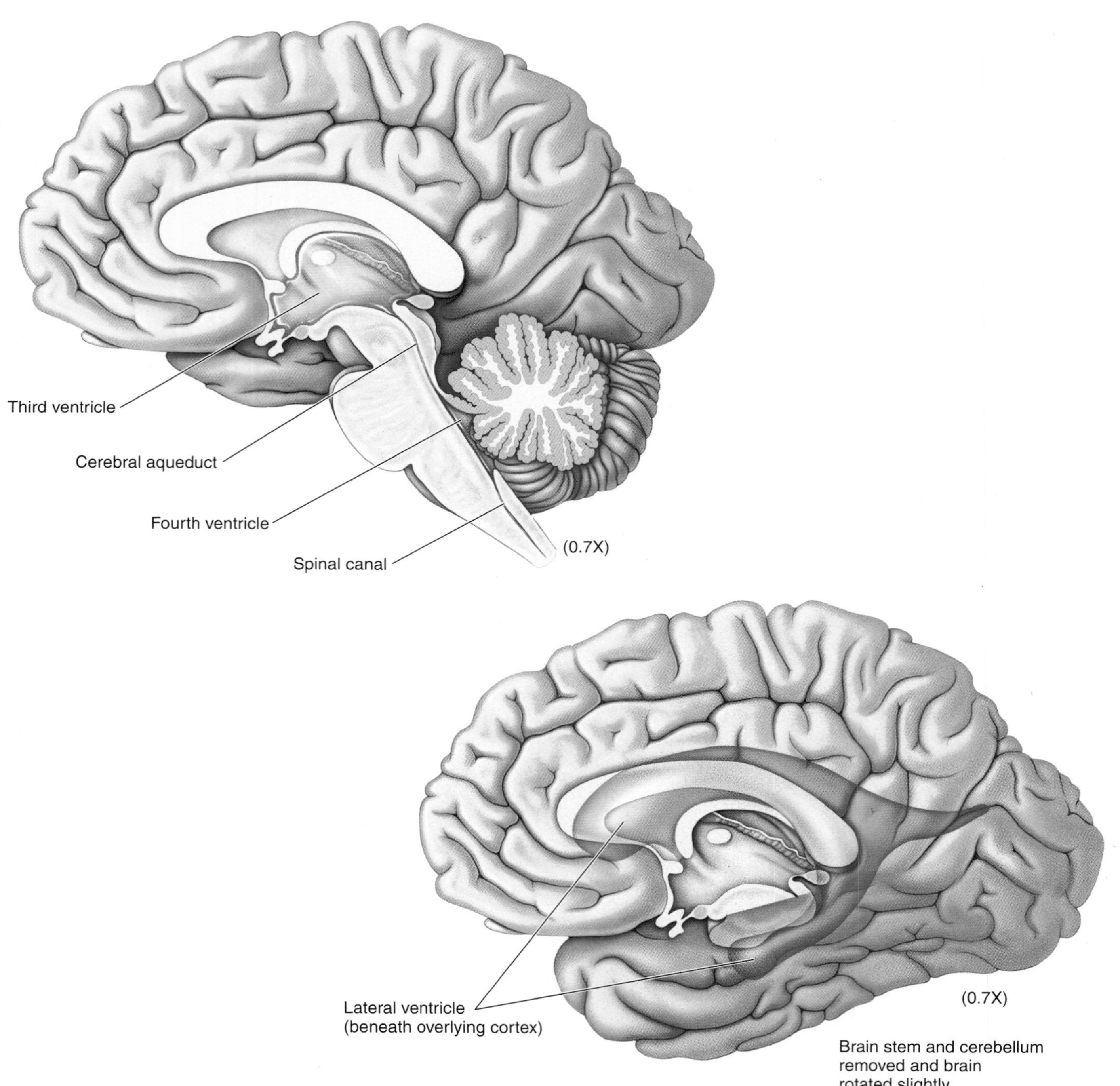

The Ventral Surface of the Brain

The underside of the brain has many distinct anatomical features. Notice the nerves emerging from the brain stem; these are the cranial nerves, which will be illustrated in more detail later in the Guide. Also notice the X-shaped optic chiasm just anterior to the hypothalamus. The chiasm is where many axons from the eyes decussate or cross from one side to the other. The bundles of axons anterior to the chiasm, which emerge from the backs of the eyes, are the optic nerves. The bundles lying posterior to the chiasm, which disappear into the thalamus, are called the optic tracts (Chapter 10). The paired mammillary bodies (from the Latin for "nipple") are a prominent feature of the ventral surface of the brain. These nuclei of the hypothalamus are part of the circuitry that stores memory (Chapter 23) and are a major target of the axons of the fornix (seen in the medial view). Notice also the olfactory bulbs (Chapter 8) and the midbrain, pons, and medulla.

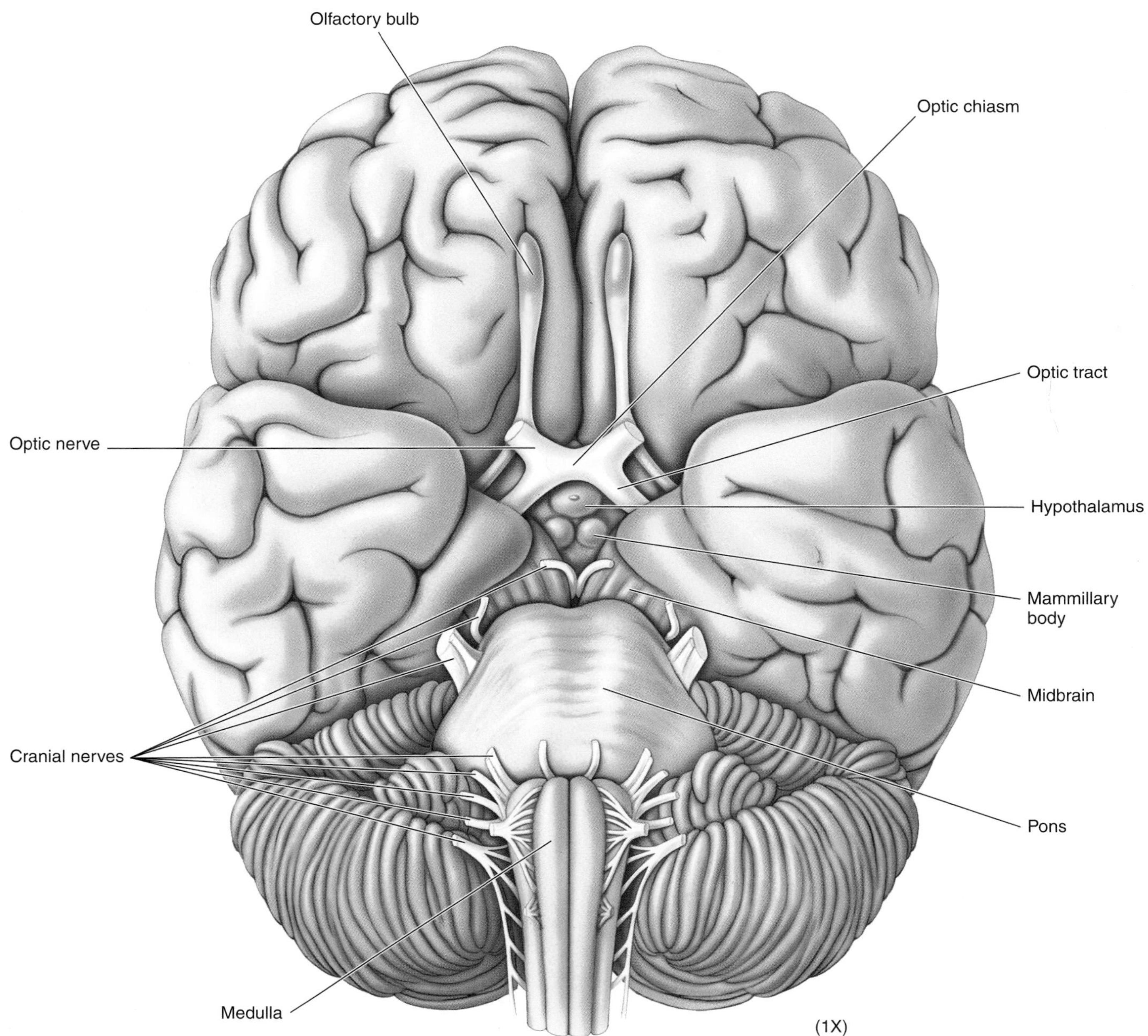

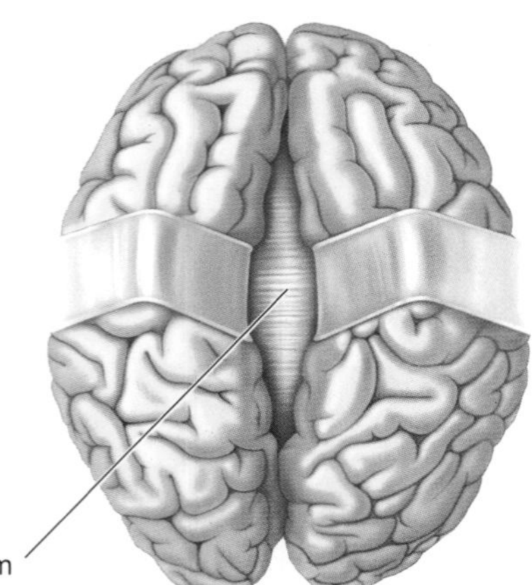

The Dorsal Surface of the Brain

(a) Cerebrum. The dorsal view of the brain is dominated by the large cerebrum. Notice the paired cerebral hemispheres. These are connected by the axons of the corpus callosum (Chapter 20), which can be seen if the hemispheres are retracted slightly. The medial view of the brain, illustrated previously, showed the callosum in cross section.

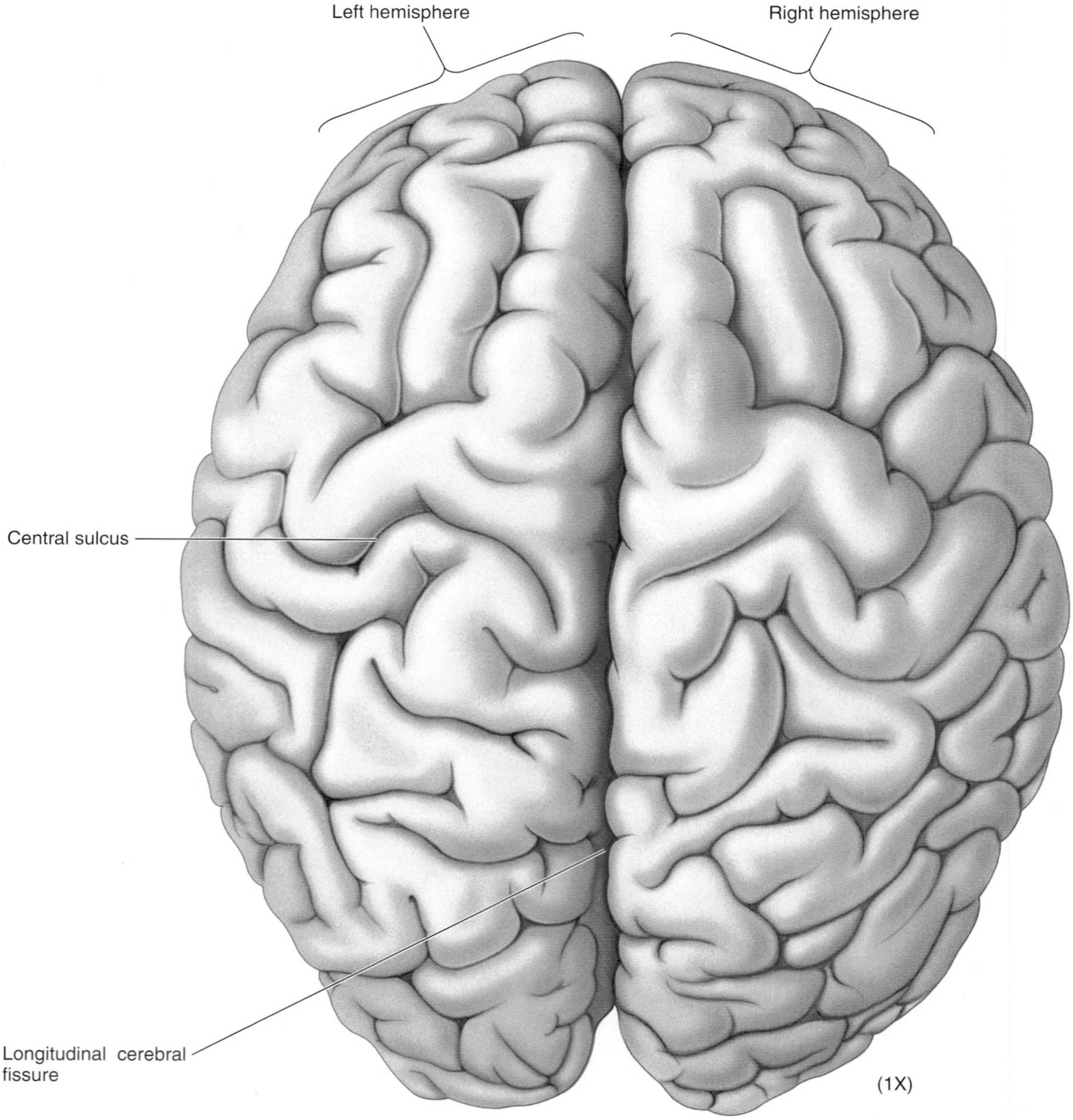

(b) Cerebrum Removed. The cerebellum dominates the dorsal view of the brain if the cerebrum is removed and the brain is tilted slightly forward. The cerebellum, an important motor control structure (Chapter 14), is divided into two hemispheres and a midline region called the vermis (Latin for "worm").

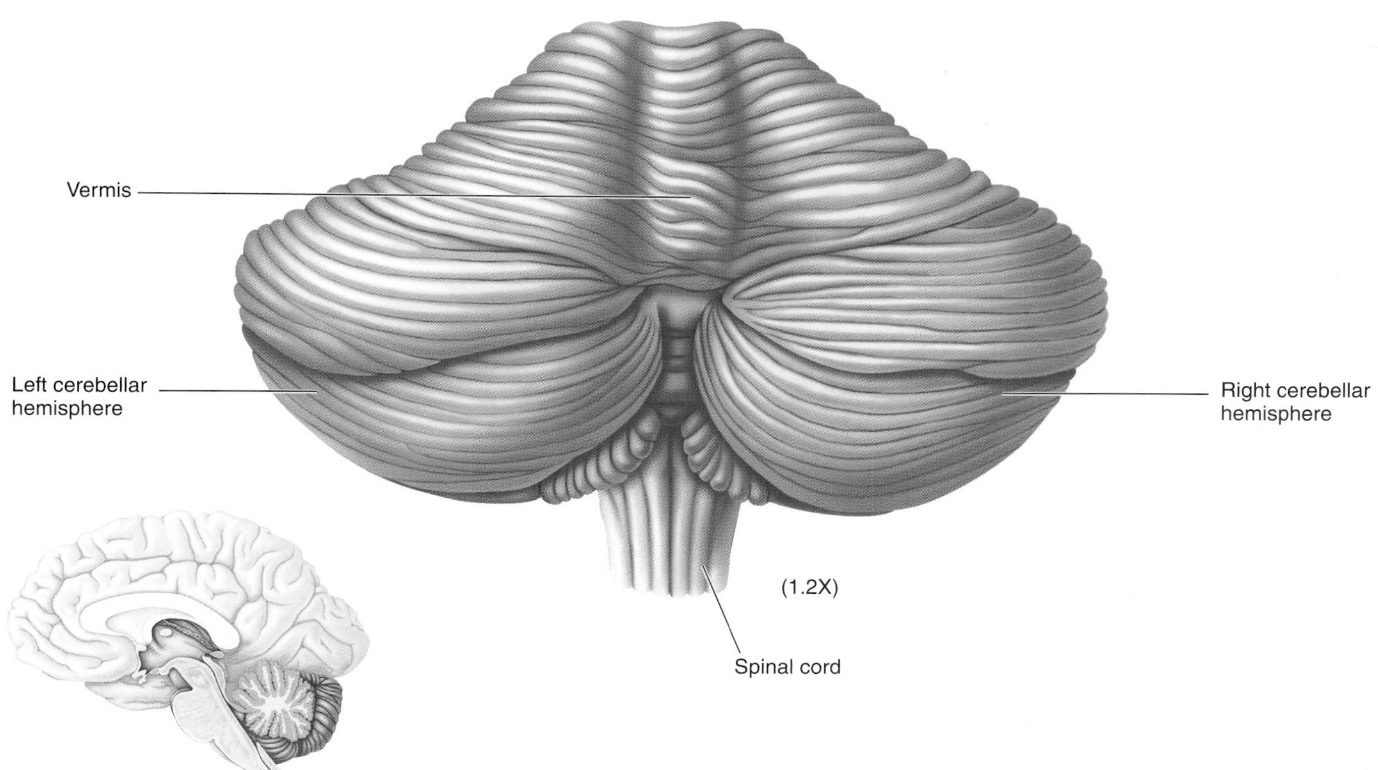

(c) Cerebrum and Cerebellum Removed. The top surface of the brain stem is exposed when both the cerebrum and the cerebellum are removed. The major divisions of the brain stem are labeled on the left side, and some specific structures are labeled on the right side. The pineal body, lying atop the thalamus, secretes melatonin and is involved in the regulation of sleep and sexual behavior (Chapters 17 and 19). The superior colliculus receives direct input from the eyes (Chapter 10) and is involved in the control of eye movements (Chapter 14), while the inferior colliculus is an important component of the auditory system (Chapter 11). *Colliculus* is Latin for "mound." The cerebellar peduncles are the large bundles of axons that connect the cerebellum and the brain stem (Chapter 14).

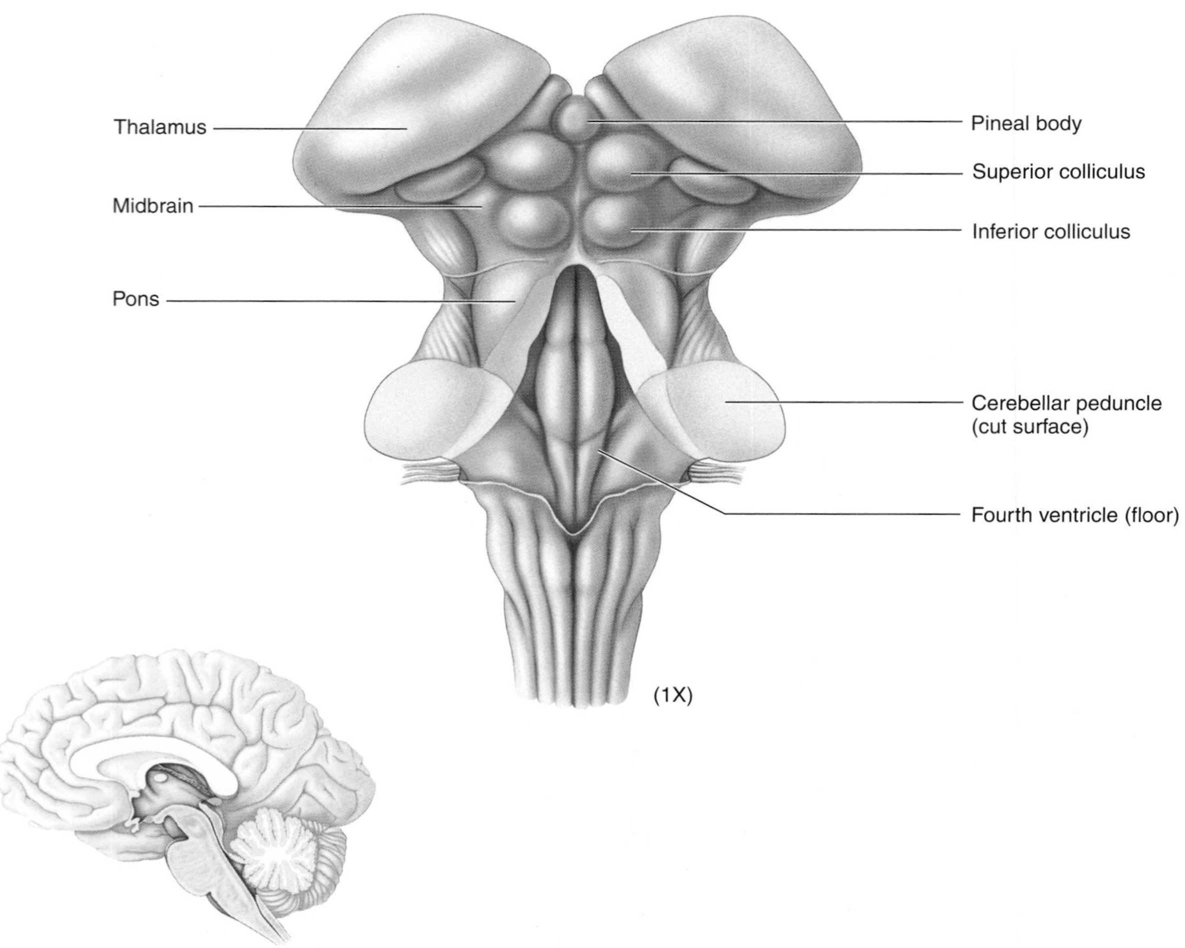

CROSS-SECTIONAL ANATOMY OF THE BRAIN

Understanding the brain requires that we peer inside it, and this is accomplished by making cross sections. Cross sections can be made physically with a knife or, in the case of noninvasive imaging of the living brain, digitally with an MRI or a CT scan. For learning the internal organization of the brain, the best approach is to make cross sections that are perpendicular to the axis defined by the embryonic neural tube, called the *neuraxis.* The neuraxis bends as the human fetus grows, particularly at the junction of the midbrain and thalamus. Consequently, the best plane of section depends on exactly where along the neuraxis we are looking.

In this part of the Guide, we look at drawings of a series of cross-sectional slabs of the brain, showing the internal structure of the forebrain (cross sections 1–3), the midbrain (cross sections 4 and 5), the pons and cerebellum (cross section 6), and the medulla (cross sections 7–9). The drawings are schematic, in that structures within the slab are sometimes projected onto the slab's visible surface.

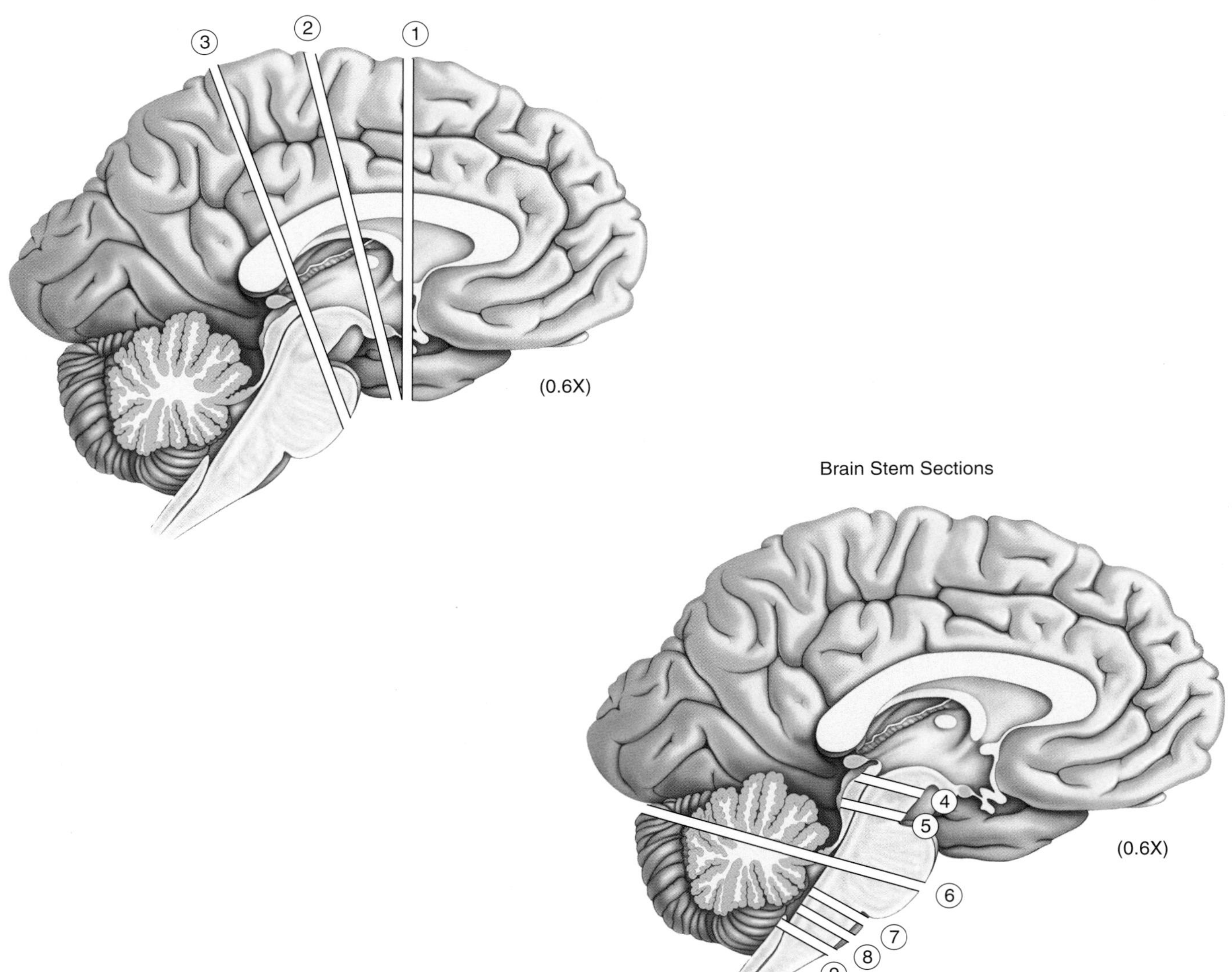

Cross Section 1: Forebrain at Thalamus–Telencephalon Junction

(a) Gross Features. The telencephalon surrounds the lateral ventricles, and the thalamus surrounds the third ventricle. In this section, the lateral ventricles can be seen sprouting from the slitlike third ventricle. The hypothalamus, forming the floor of the third ventricle, is a vital control center for many basic bodily functions (Chapters 15–17). The insula (Chapter 8) lies at the base of the lateral (Sylvian) fissure, here separating the frontal lobe from the temporal lobe. The heterogeneous region lying deep within the telencephalon, medial to the insula and lateral to the thalamus, is called the basal forebrain.

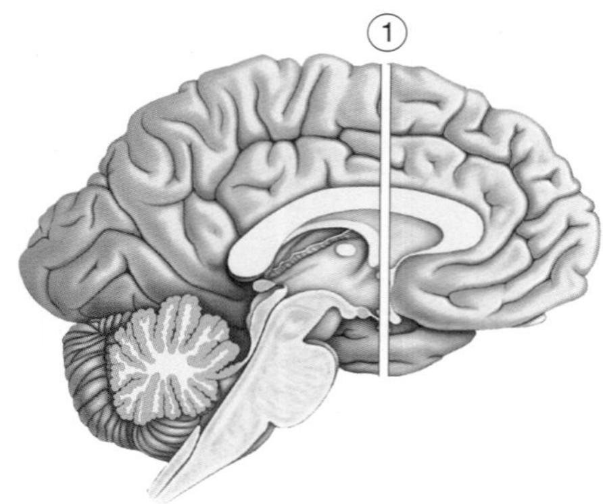

Frontal lobe

Lateral ventricle

Thalamus

Insula

Lateral (Sylvian) fissure

Third ventricle

Temporal lobe

(1X)

Basal forebrain

Hypothalamus

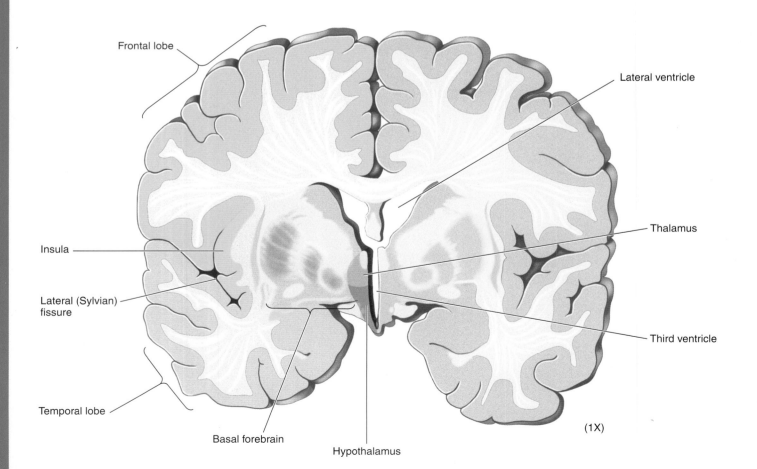

(b) Selected Cell and Fiber Groups. Here we take a more detailed look at the structures of the forebrain. The internal capsule is the large collection of axons connecting the cortical white matter with the thalamus, and the corpus callosum is the enormous sling of axons connecting the cerebral cortex of the two hemispheres. The fornix, shown first in the medial view of the brain, is shown here in cross section, where it loops around the stalk of the lateral ventricle. The neurons of the closely associated septal area (from *saeptum*, which is Latin for "partition") contribute axons to the fornix and are involved in memory storage (Chapter 23). Three important collections of neurons in the basal telencephalon are also shown: the caudate nucleus, the putamen, and the globus pallidus. Collectively, these structures, called the "basal ganglia," are an important part of the brain systems that control movement (Chapter 14).

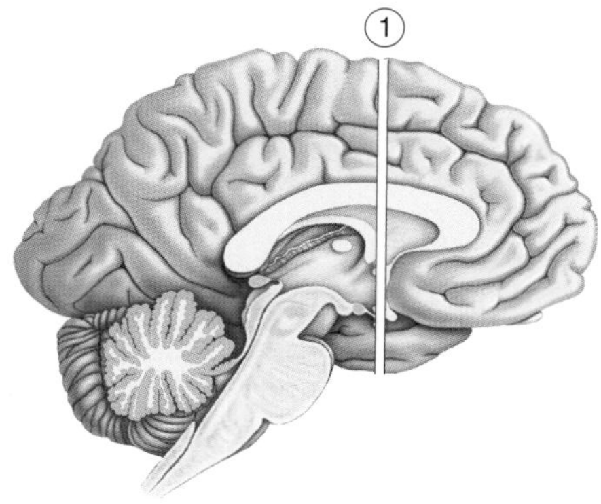

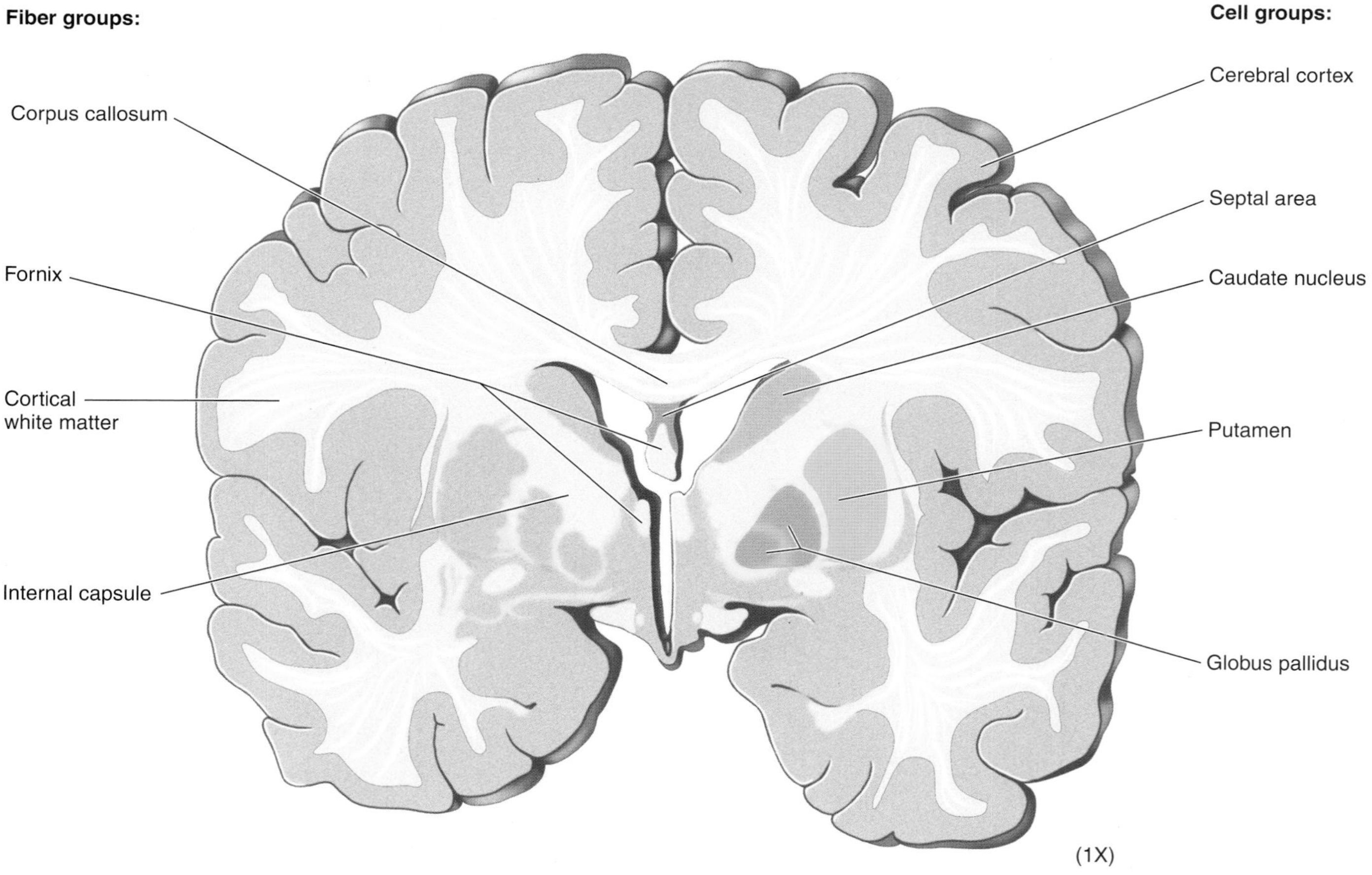

Fiber groups:

Corpus callosum

Fornix

Cortical white matter

Internal capsule

Cell groups:

Cerebral cortex

Septal area

Caudate nucleus

Putamen

Globus pallidus

(1X)

Cross Section 2: Forebrain at Mid-Thalamus

(a) Gross Features. As we look slightly caudal in the neuraxis, we see the heart-shaped thalamus (from the Greek for "inner chamber") surrounding the small third ventricle at the brain's core. Just ventral to the thalamus lies the hypothalamus. The telencephalon is organized much like what we saw in cross section 1. Because we are looking slightly posterior, the lateral fissure here separates the parietal lobe from the temporal lobe.

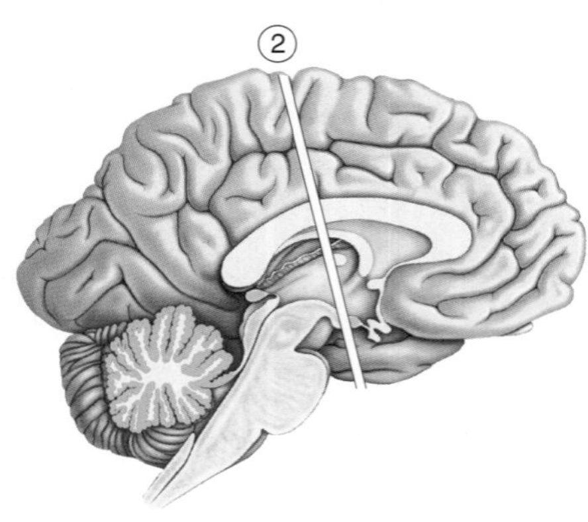

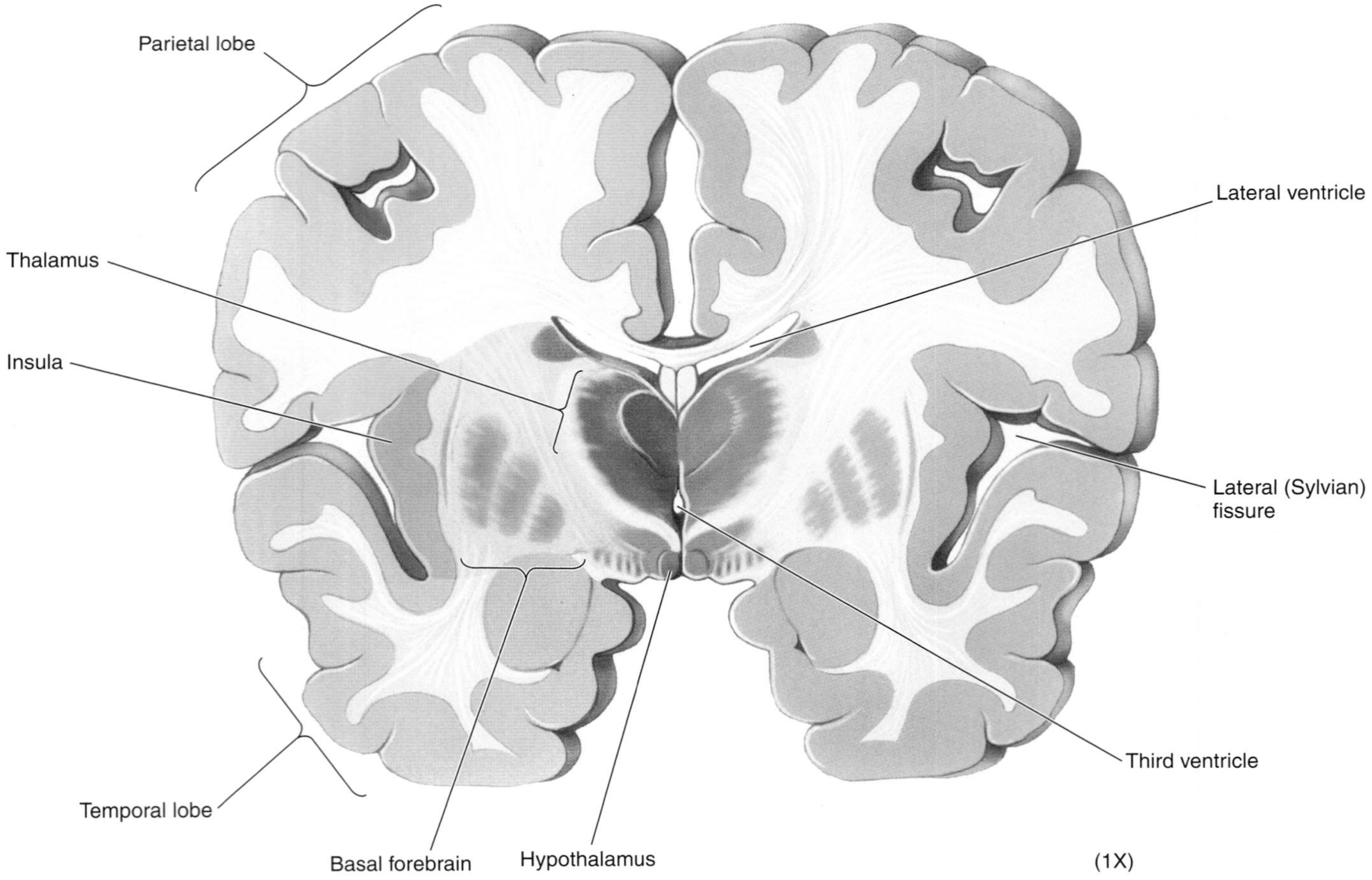

(b) Selected Cell and Fiber Groups. Many important cell and fiber groups appear at this level of the neuraxis. One structure apparent in the telencephalon is the amygdala, involved in the regulation of emotion (Chapter 18) and memory (Chapter 23). The thalamus is divided into separate nuclei, two of which—the ventral posterior nucleus and the ventral lateral nucleus—are labeled. The thalamus provides much of the input to the cerebral cortex, with different thalamic nuclei projecting axons to different areas of cortex. The ventral posterior nucleus, a part of the somatic sensory system (Chapter 12), projects to the cortex of the postcentral gyrus. The ventral lateral nucleus and closely related ventral anterior nucleus (not shown) are parts of the motor system (Chapter 14); they project to the motor cortex of the precentral gyrus. Visible below the thalamus are the subthalamus and the mammillary bodies of the hypothalamus. The subthalamus is a part of the motor system (Chapter 14), while the mammillary bodies receive information from the fornix and contribute to the regulation of memory (Chapter 23). Because this section also encroaches on the midbrain, a little of the substantia nigra ("black substance") can be seen near the base of the brain stem. The substantia nigra is also a part of the motor system (Chapter 14). Parkinson's disease results from the degeneration of this structure.

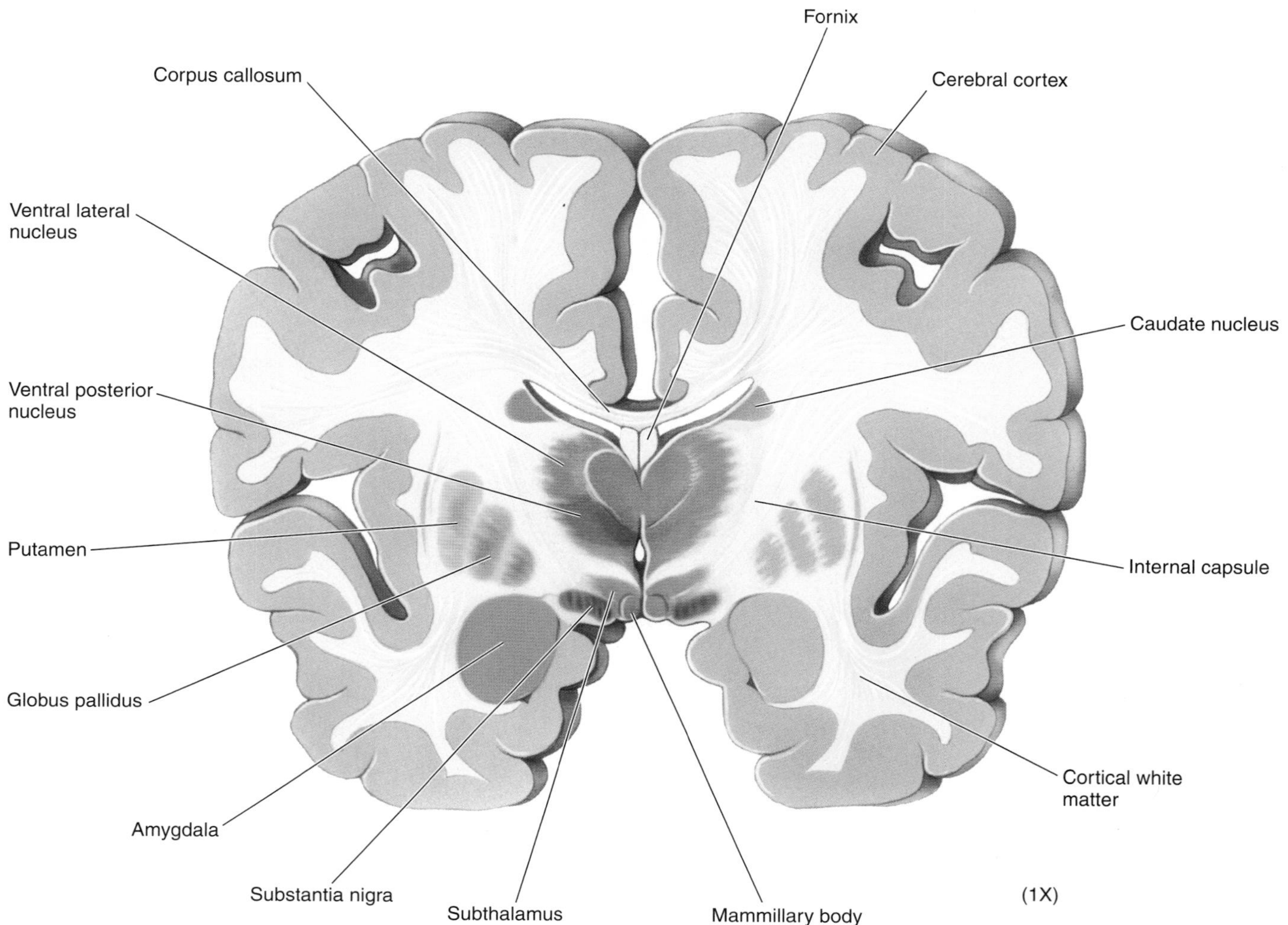

Cross Section 3: Forebrain at Thalamus–Midbrain Junction

(a) Gross Features. The neuraxis bends sharply at the junction of the thalamus and the midbrain. This cross section is taken at a level where the teardrop-shaped third ventricle communicates with the cerebral aqueduct. The brain surrounding the third ventricle is thalamus, and the brain around the cerebral aqueduct is midbrain. The lateral ventricles of each hemisphere appear twice in this section. You can see why by reviewing the phantom view of the ventricle, shown previously.

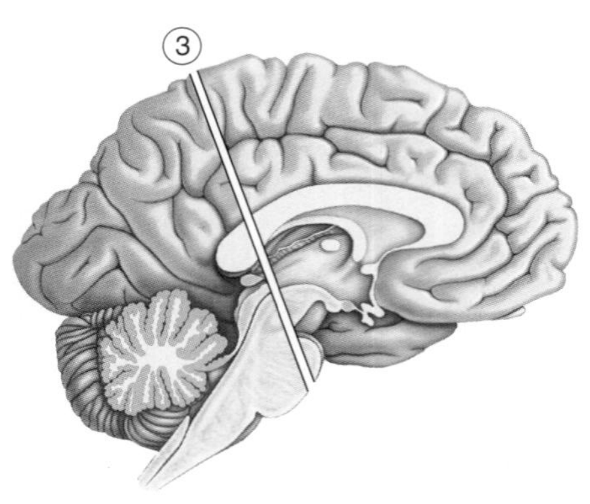

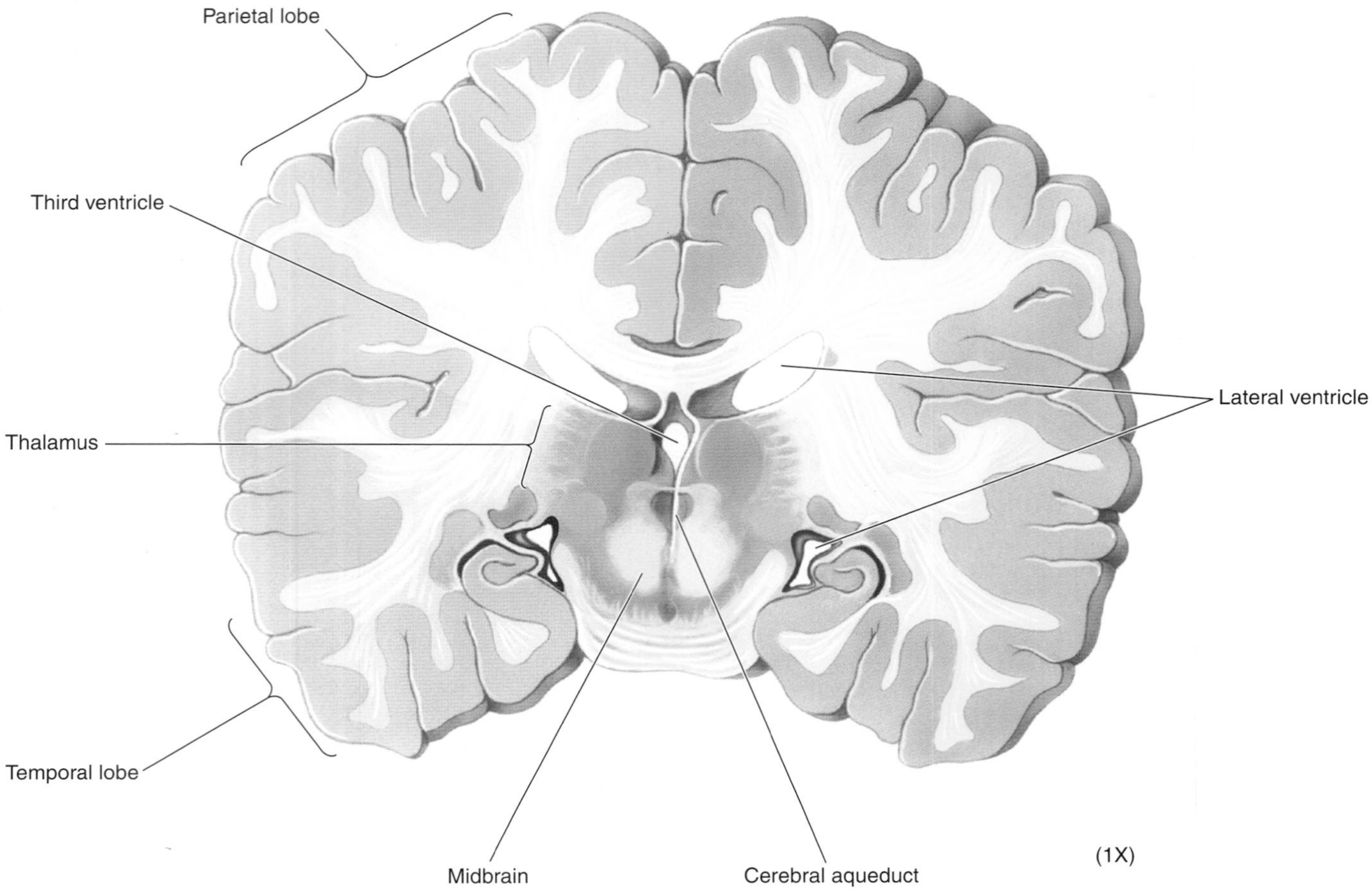

Parietal lobe
Third ventricle
Thalamus
Lateral ventricle
Temporal lobe
Midbrain
Cerebral aqueduct
(1X)

(b) Selected Cell and Fiber Groups. This section contains two more important nuclei of the thalamus: the medial and lateral geniculate nuclei. Geniculate is from the Latin for "knee." The lateral geniculate nucleus relays information to the visual cortex (Chapter 10), and the medial geniculate nucleus relays information to the auditory cortex (Chapter 11). Also notice the location of the hippocampus, a relatively simple form of cerebral cortex bordering the lateral ventricle of the temporal lobe. The hippocampus (from the Greek for "seahorse") plays an important role in learning and memory (Chapters 23 and 24).

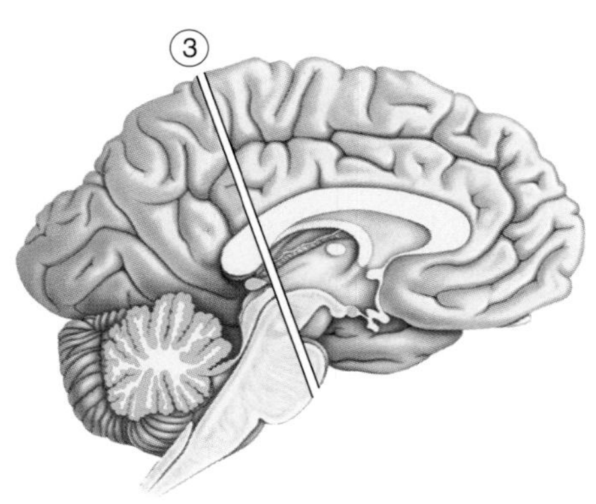

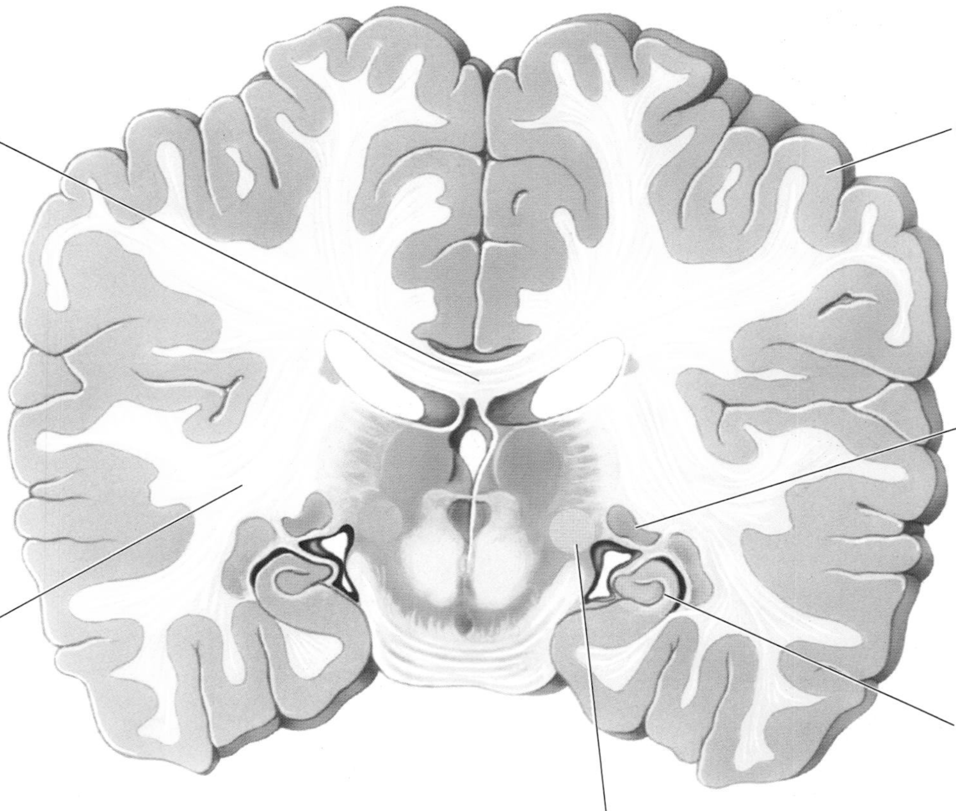

Corpus callosum
Cerebral cortex
Lateral geniculate nucleus
Cortical white matter
Hippocampus
Medial geniculate nucleus
(1X)

Cross Section 4: Rostral Midbrain

We are now at the midbrain. The plane of section has been angled relative to the forebrain sections, so that it remains perpendicular to the neuraxis. The core of the midbrain is the small cerebral aqueduct. Here, the roof of the midbrain, also called the tectum (Latin for "roof"), consists of the paired superior colliculi. As discussed earlier, the superior colliculus is a part of the visual system (Chapter 10), and the substantia nigra is a part of the motor system (Chapter 14). The red nucleus is also a motor control structure (Chapter 14), while the periaqueductal gray is important in the control of somatic pain sensations (Chapter 12).

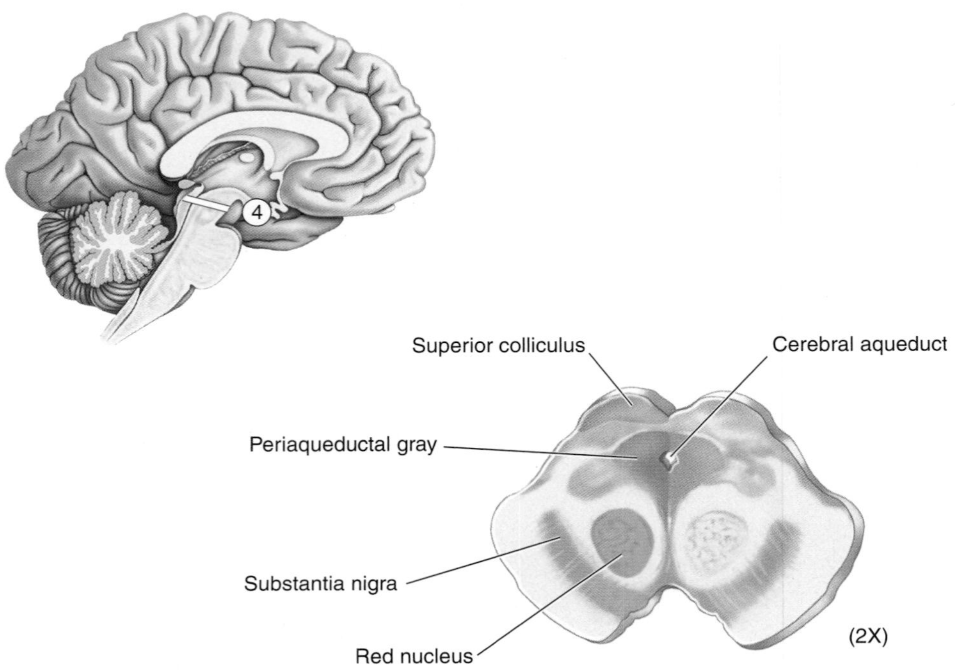

Cross Section 5: Caudal Midbrain

The caudal midbrain appears similar to the rostral midbrain. At this level, however, the roof is formed by the inferior colliculi (part of the auditory system; Chapter 11) instead of by the superior colliculi. Review the dorsal view of the brain stem to see how the superior and inferior colliculi are situated relative to each other.

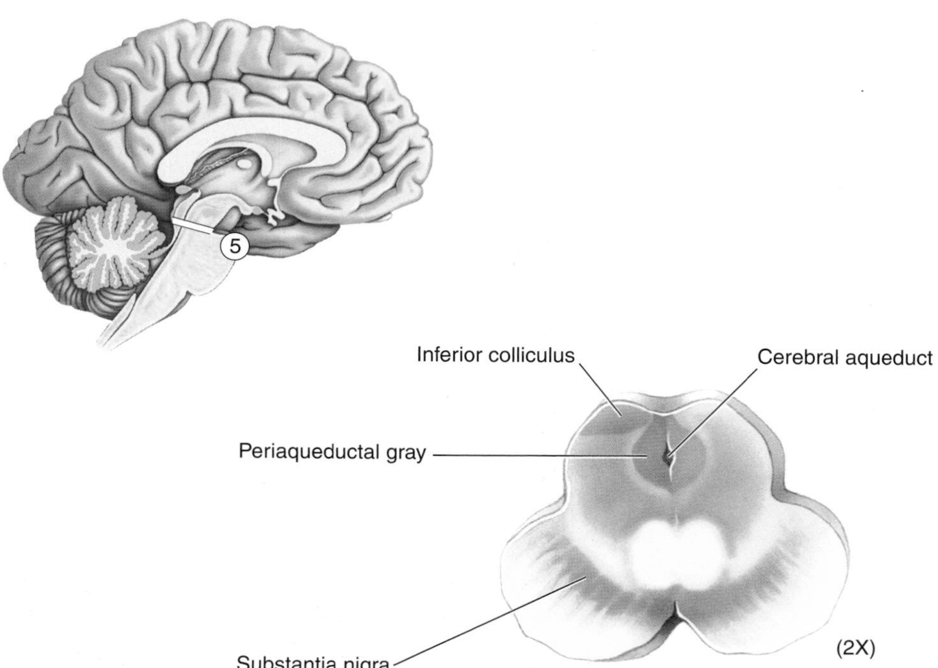

Cross Section 6: Pons and Cerebellum

This section shows the pons and cerebellum, parts of the rostral hindbrain that border the fourth ventricle. As mentioned earlier, the cerebellum is important in the control of movement. Much of the input to the cerebellar cortex derives from the pontine nuclei, while the output of the cerebellum is from neurons of the deep cerebellar nuclei (Chapter 14). The reticular formation (from *reticulum*, which is Latin for "net") runs from the midbrain to the medulla at the core of the brain stem, just under the cerebral aqueduct and fourth ventricle. One function of the reticular formation is to regulate sleep and wakefulness (Chapter 19). In addition, a function of the pontine reticular formation is to control body posture (Chapter 14).

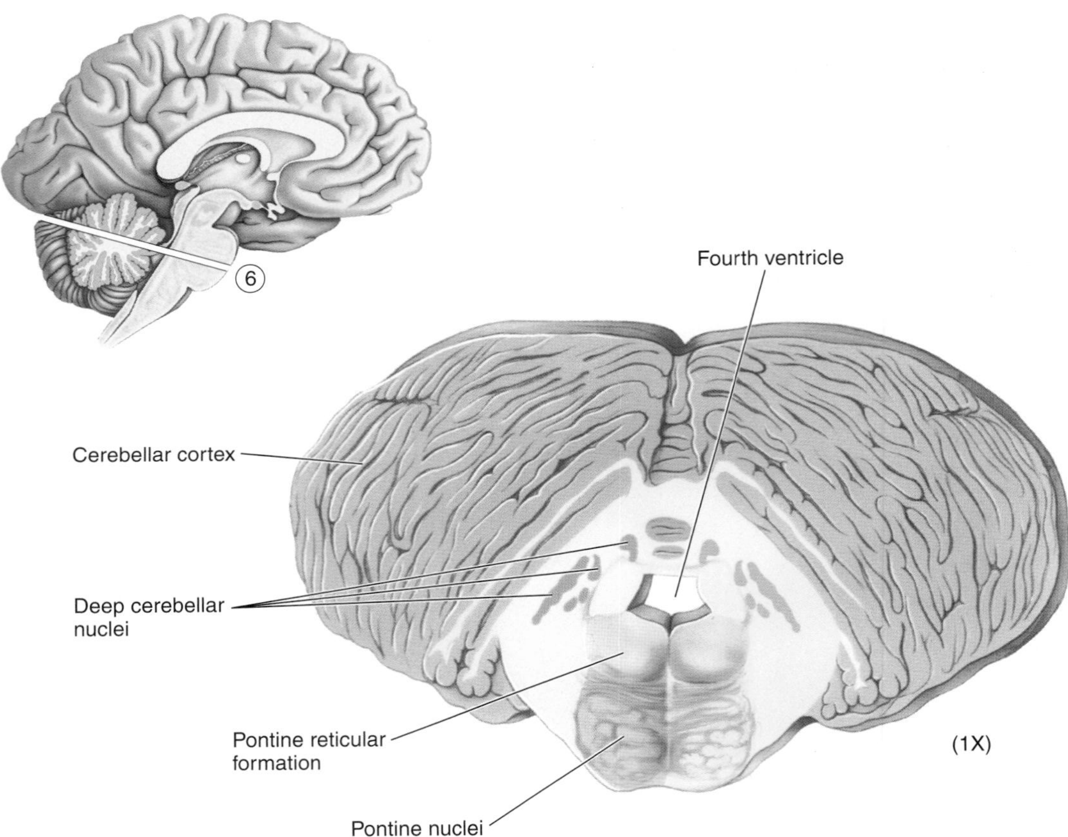

Cross Section 7: Rostral Medulla

As we move further caudally along the neuraxis, the brain surrounding the fourth ventricle becomes the medulla. The medulla is a complex region of the brain. Here we focus only on those structures whose functions are discussed later in the book. At the very floor of the medulla lie the medullary pyramids, huge bundles of axons descending from the forebrain toward the spinal cord. The pyramids contain the corticospinal tracts, which are involved in the control of voluntary movement (Chapter 14). Several nuclei that are important for hearing are also found in the rostral medulla: the dorsal and ventral cochlear nuclei and the superior olive (Chapter 11). Also shown are the inferior olive, important for motor control (Chapter 14), and the raphe nucleus, important for the modulation of pain, mood, and wakefulness (Chapters 12, 19, and 21).

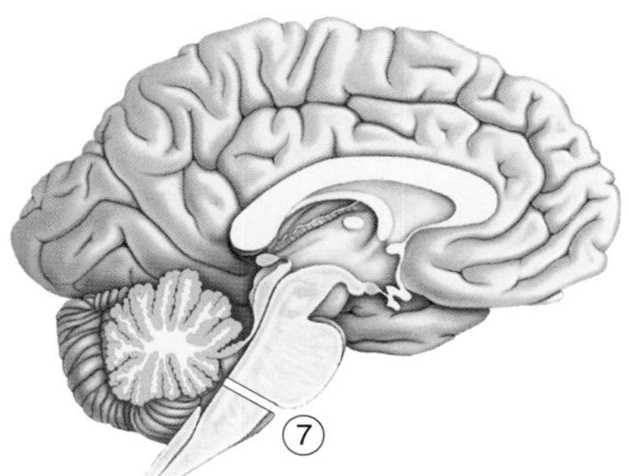

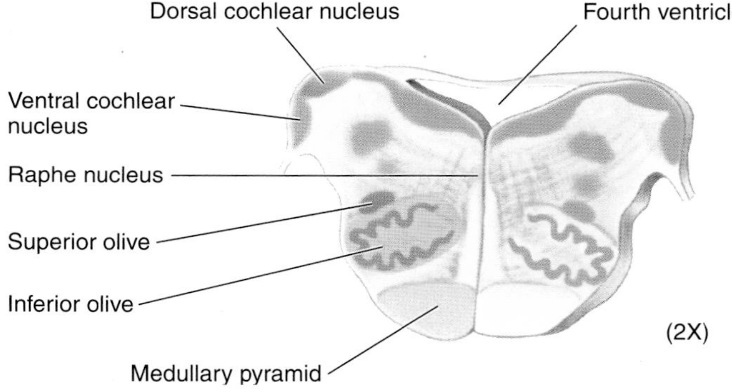

Cross Section 8: Mid-Medulla

The mid-medulla contains some of the same structures labeled in cross section 7. Notice also the medial lemniscus. *Lemniscus* is Latin for "ribbon." The medial lemniscus contains axons bringing information about somatic sensation to the thalamus (Chapter 12). The gustatory nucleus, serving the sense of taste (Chapter 8), is a part of the larger nucleus of the solitary tract, which regulates aspects of visceral function (Chapters 15 and 16). The vestibular nuclei serve the sense of balance (Chapter 11).

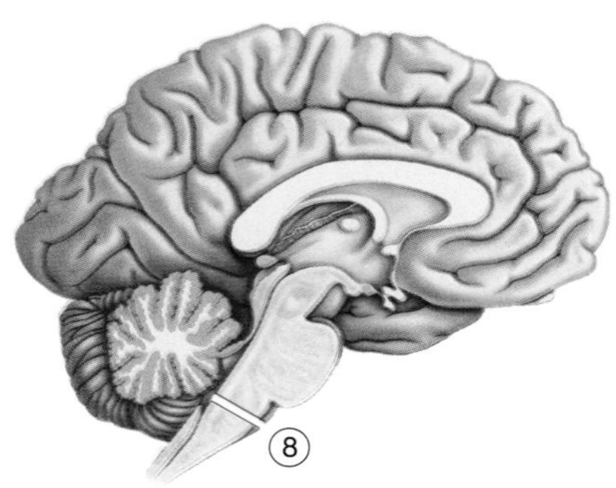

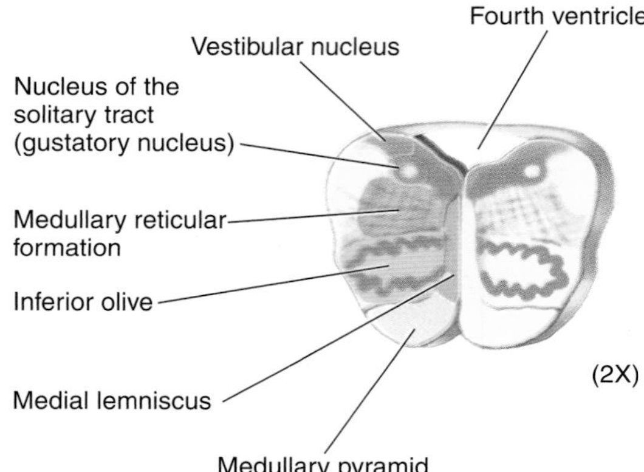

Cross Section 9: Medulla–Spinal Cord Junction

As the medulla disappears, so does the fourth ventricle, now replaced by the beginning of the spinal canal. Notice the dorsal column nuclei, which receive somatic sensory information from the spinal cord (Chapter 12). Axons arising from the neurons in each dorsal column nucleus cross to the other side of the brain (decussate) and ascend to the thalamus via the medial lemniscus.

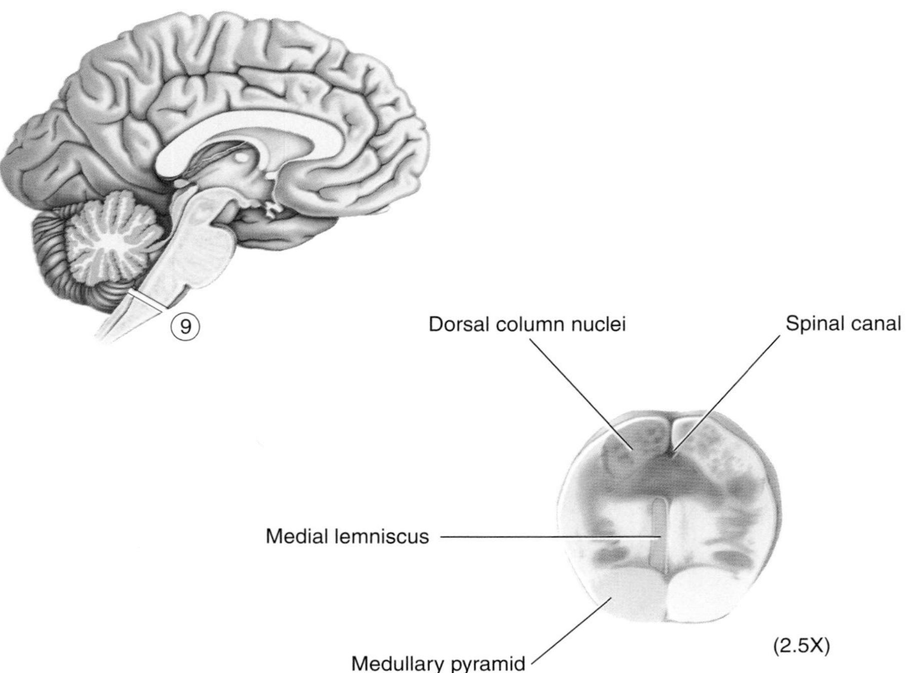

THE SPINAL CORD

Dorsal Surface of Spinal Cord and Spinal Nerves

The spinal cord lies within the vertebral column. The spinal nerves, a part of the somatic PNS, communicate with the cord via notches between the vertebrae. The vertebrae are described according to their location. In the neck, they are called cervical vertebrae and are numbered from C1 to C7. The vertebrae attached to ribs are called thoracic vertebrae and are numbered from T1 to T12. The five vertebrae of the lower back are called lumbar, and those within the pelvic area are called sacral.

The spinal nerves and the associated segments of the spinal cord adopt the names of the vertebrae; eight cervical nerves are associated with seven cervical vertebrae. Also, the spinal cord in the adult human ends at about the level of the third lumbar vertebra. This disparity arises because the spinal cord does not grow after birth, whereas the spinal column does. The bundles of spinal nerves streaming down within the lumbar and sacral vertebral column are called the cauda equina, Latin for "horse's tail."

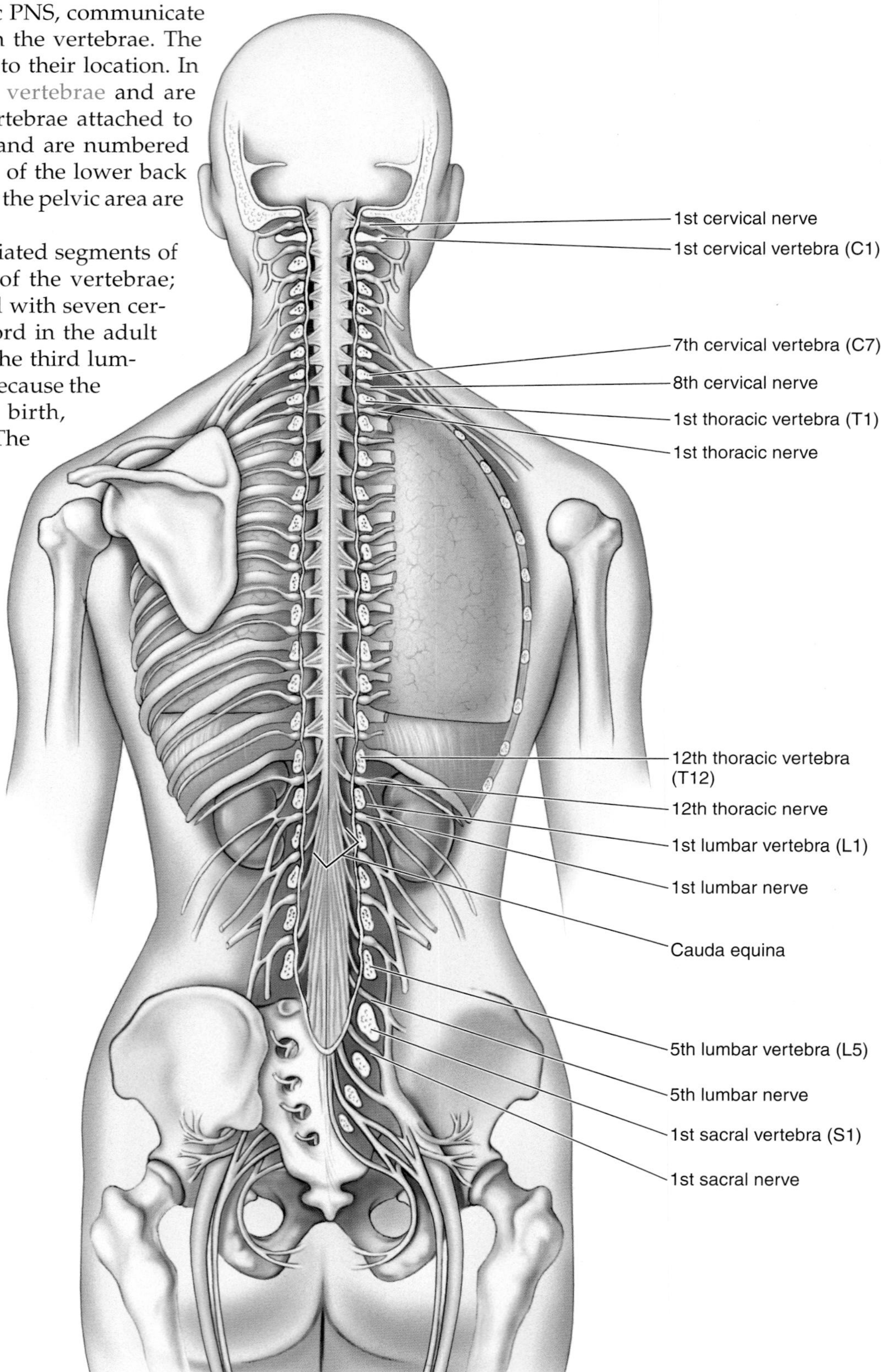

Ventral-Lateral Surface

This view shows how the spinal nerves attach to the spinal cord and how the spinal meninges are organized. As the nerve passes into the vertebral notch (not shown), it splits into two roots. The dorsal root carries sensory axons whose cell bodies lie in the dorsal root ganglia. The ventral root carries motor axons arising from the gray matter of the ventral spinal cord. The butterfly-shaped core of the spinal cord is gray matter consisting of neuronal cell bodies. The gray matter is divided into the dorsal, lateral, and ventral horns. Notice how the organization of gray and white matter in the spinal cord differs from that in the forebrain. In the forebrain, the gray matter surrounds the white matter; in the spinal cord, it is the other way around. The thick shell of white matter, containing the long axons that run up and down the cord, is divided into three columns: the dorsal columns, the lateral columns, and the ventral columns.

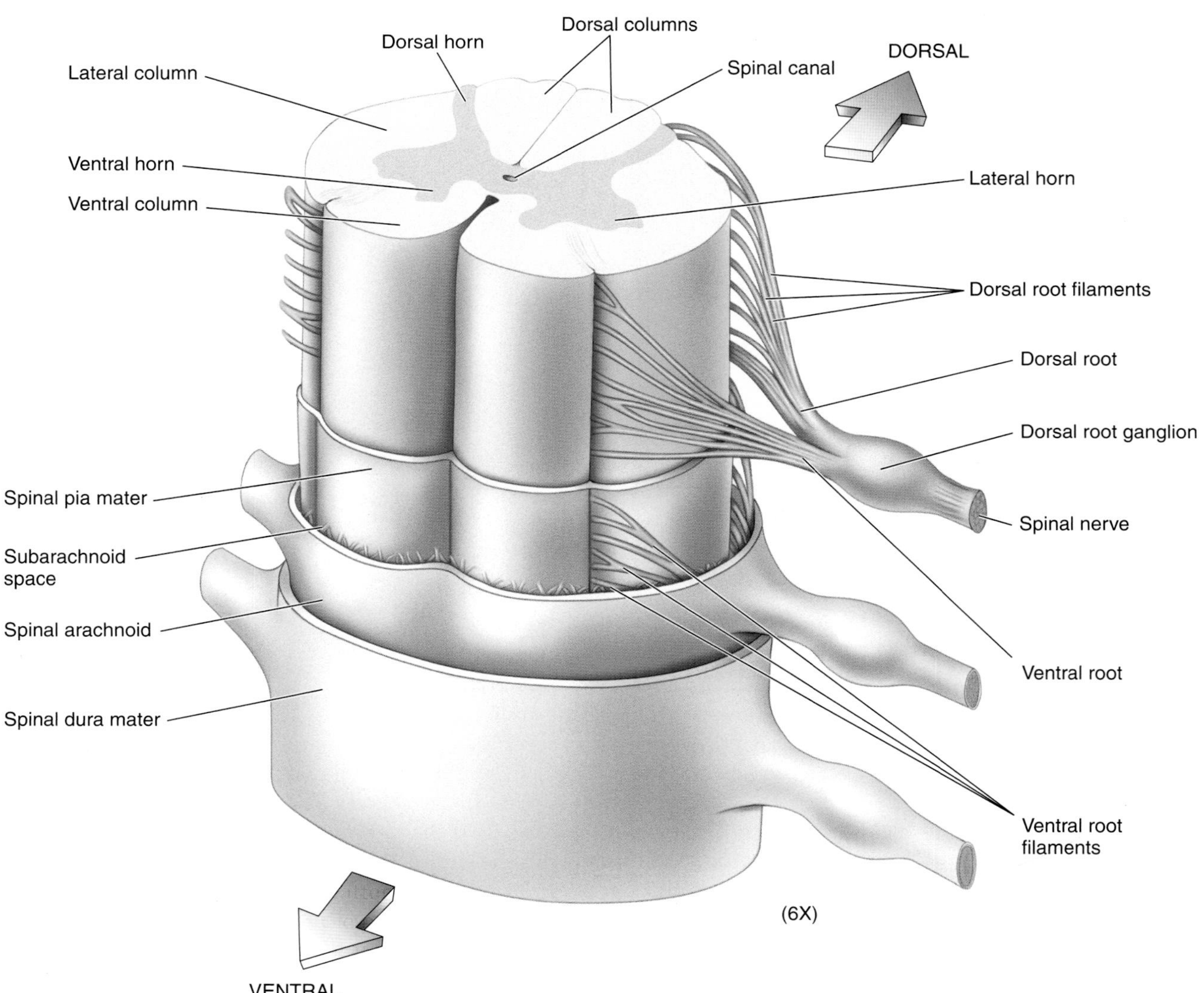

Cross-Sectional Anatomy

Illustrated in this view are some of the important tracts of axons running up and down the spinal cord. On the left side, the major ascending sensory pathways are indicated. The entire dorsal column consists of sensory axons ascending to the brain. This pathway is important for the conscious appreciation of touch. The spinothalamic tract carries information about painful stimuli and temperature. The somatic sensory system is covered in Chapter 12. On the right side are some of the descending tracts important for the control of movement (Chapter 14). The names of the tracts describe their origins and terminations (e.g., the vestibulospinal tract originates in the vestibular nuclei of the medulla and terminates in the spinal cord). The descending tracts contribute to two pathways: the lateral and ventromedial pathways. The lateral pathway carries the commands for voluntary movements, especially of the extremities. The ventromedial pathway participates mainly in the maintenance of posture and certain reflex movements.

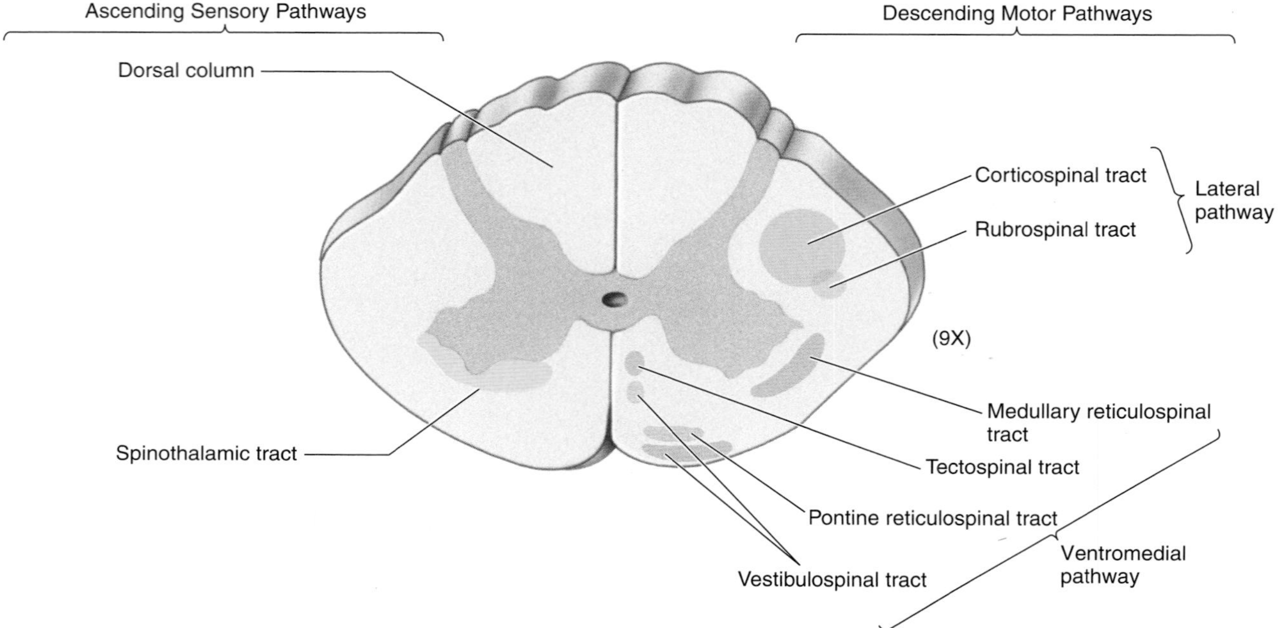

THE AUTONOMIC NERVOUS SYSTEM (facing page)

In addition to the somatic PNS, which is devoted largely to the voluntary control of movement and conscious skin sensations, there is the visceral PNS, devoted to the regulation of the internal organs, glands, and vasculature. Because this regulation occurs automatically and is not under direct conscious control, this system is called the autonomic nervous system, or ANS. The two most important divisions of the ANS are the sympathetic and parasympathetic divisions.

The figure illustrates the cavity of the body as it appears when it has been sectioned sagittally at the level of the eye. Notice the vertebral column, which is encased in a thick wall of connective tissue. The spinal nerves can be seen emerging from the column. The sympathetic division of the ANS includes a chain of ganglia that runs along the side of the vertebral column. These ganglia communicate with the spinal nerves, with one another, and with a large number of internal organs. The parasympathetic division of the ANS is organized quite differently. Much of the parasympathetic innervation of the viscera arises from the vagus nerve, one of the cranial nerves emerging from the medulla. The other major source of parasympathetic fibers is the sacral spinal nerves. (The functional organization of the ANS is discussed in Chapter 15.)

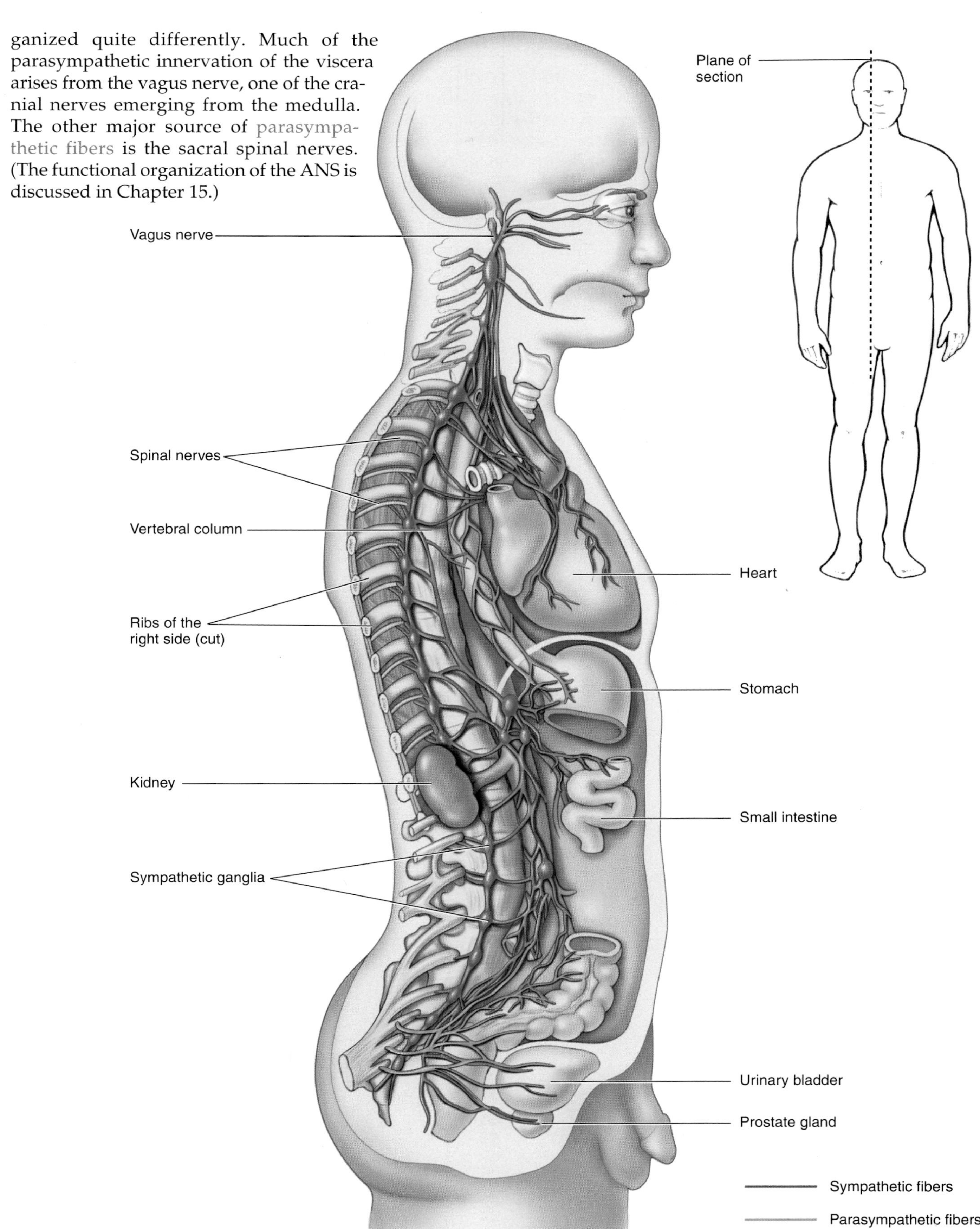

THE CRANIAL NERVES

Twelve pairs of cranial nerves emerge from the base of the brain. The first two "nerves" are actually parts of the CNS, serving olfaction and vision. The rest are like the spinal nerves, in that they contain axons of the PNS. As the illustration shows, however, a single nerve often has fibers performing many functions. Knowledge of the nerves and their diverse functions is a valuable aid in the diagnosis of a number of neurological disorders. It is important to recognize that the cranial nerves have associated cranial nerve nuclei in the midbrain, pons, and medulla. Examples are the cochlear and vestibular nuclei, which receive information from cranial nerve VIII. Most of the cranial nerve nuclei were not illustrated or labeled in the brain stem cross sections, however, because their functions are not discussed explicitly in this book.

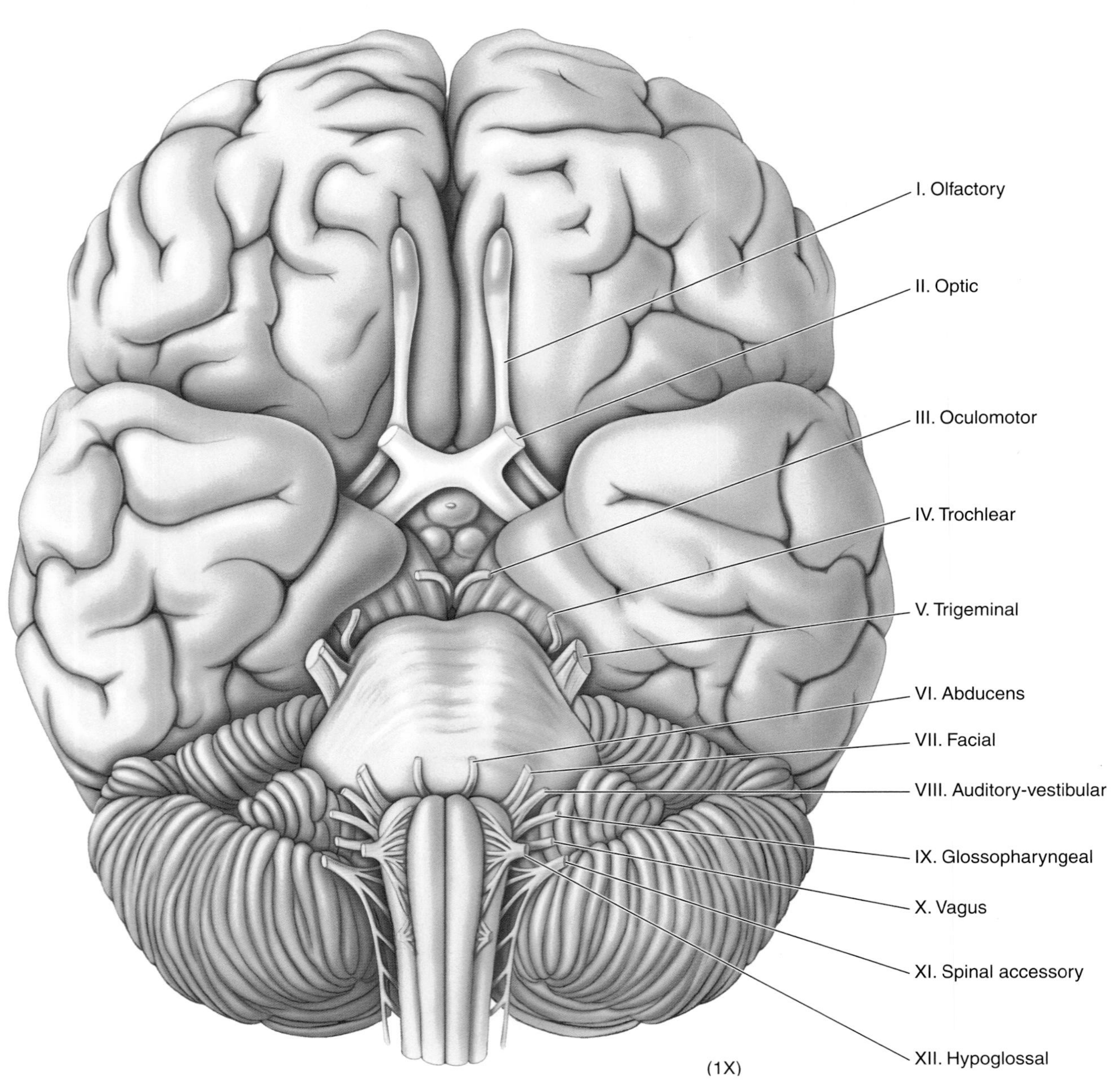

Nerve Number and Name	Types of Axons	Important Functions
I. Olfactory	Special sensory	Sensation of smell
II. Optic	Special sensory	Sensation of vision
III. Oculomotor	Somatic motor Visceral motor	Movements of the eye and eyelid Parasympathetic control of pupil size
IV. Trochlear	Somatic motor	Movements of the eye
V. Trigeminal	Somatic sensory Somatic motor	Sensation of touch to the face Movement of muscles of mastication (chewing)
VI. Abducens	Somatic motor	Movements of the eye
VII. Facial	Somatic sensory Special sensory	Movement of muscles of facial expression Sensation of taste in anterior two-thirds of the tongue
VIII. Auditory-vestibular	Special sensory	Sensation of hearing and balance
IX. Glossopharyngeal	Somatic motor Visceral motor Special sensory Visceral sensory	Movement of muscles in the throat (oropharynx) Parasympathetic control of the salivary glands Sensation of taste in posterior one-third of the tongue Detection of blood pressure changes in the aorta
X. Vagus	Visceral motor Visceral sensory Somatic motor	Parasympathetic control of the heart, lungs, and abdominal organs Sensation of pain associated with viscera Movement of muscles in the throat (oropharynx)
XI. Spinal accessory	Somatic motor	Movement of muscles in the throat and neck
XII. Hypoglossal	Somatic motor	Movement of the tongue

THE BLOOD SUPPLY OF THE BRAIN

Ventral View

Two pairs of arteries supply blood to the brain: the vertebral arteries and the internal carotid arteries. The vertebral arteries converge near the base of the pons to form the unpaired basilar artery. The vertebral and basal arteries supply blood to the brain stem and cerebellum. At the level of the midbrain, the basilar artery splits into the right and left superior cerebellar arteries and the posterior cerebral arteries. The posterior cerebral arteries send branches, called posterior communicating arteries, that connect them to the internal carotids. The internal carotids branch to form the middle cerebral arteries and the anterior cerebral arteries. The anterior cerebral arteries of each side are connected by the anterior communicating artery. Thus, the posterior cerebral and communicating arteries, the internal carotids, and the anterior cerebral and communicating arteries form a ring of connected arteries at the brain's base. This ring is called the *circle of Willis*.

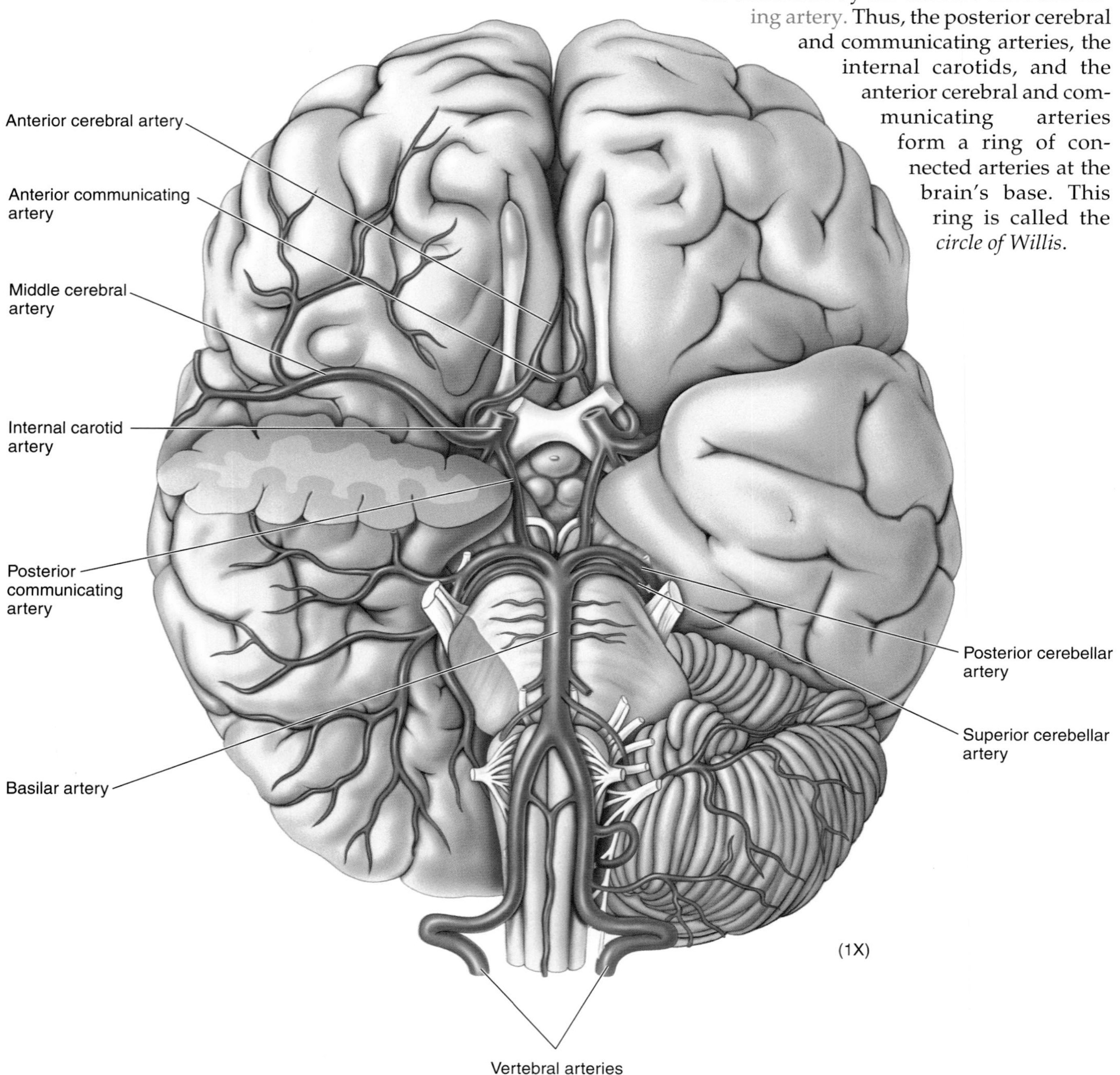

Lateral View

Most of the lateral surface of the cerebrum is supplied by the middle cerebral artery. This artery also feeds the deep structures of the basal forebrain.

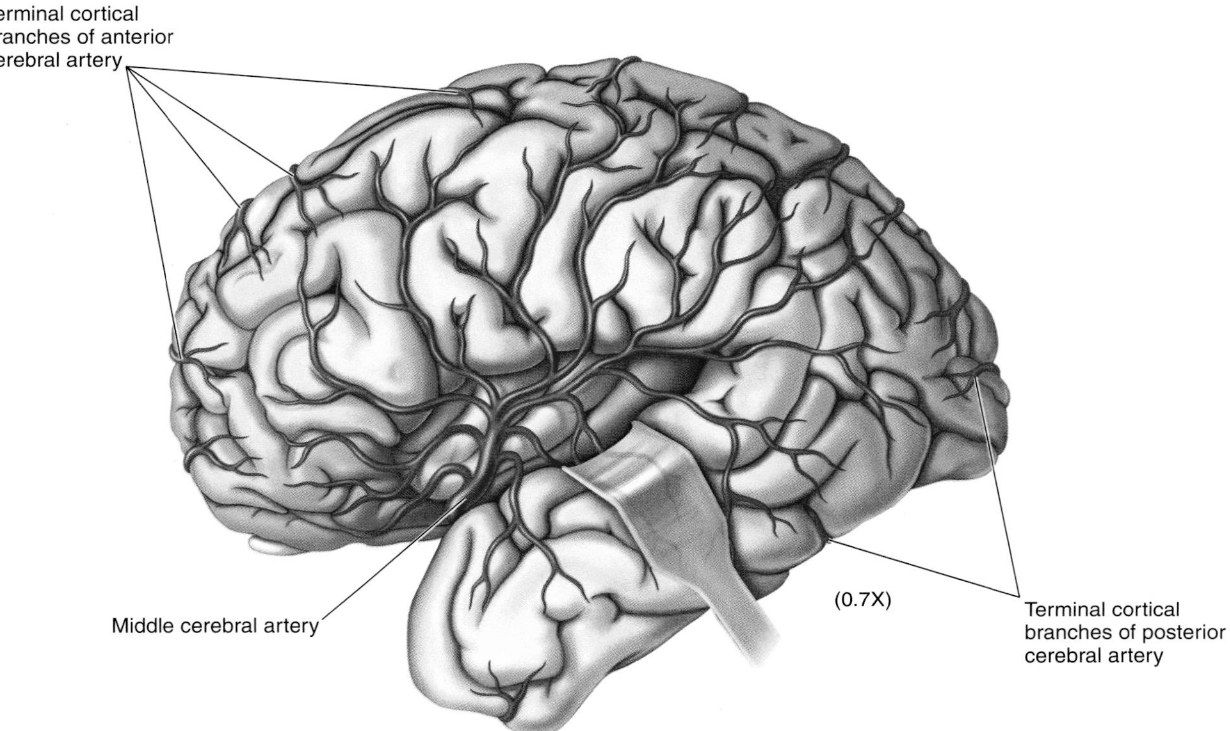

Medial View (Brain Stem Removed)

Most of the medial wall of the cerebral hemisphere is supplied by the anterior cerebral artery. The posterior cerebral artery feeds the medial wall of the occipital lobe and the inferior part of the temporal lobe.

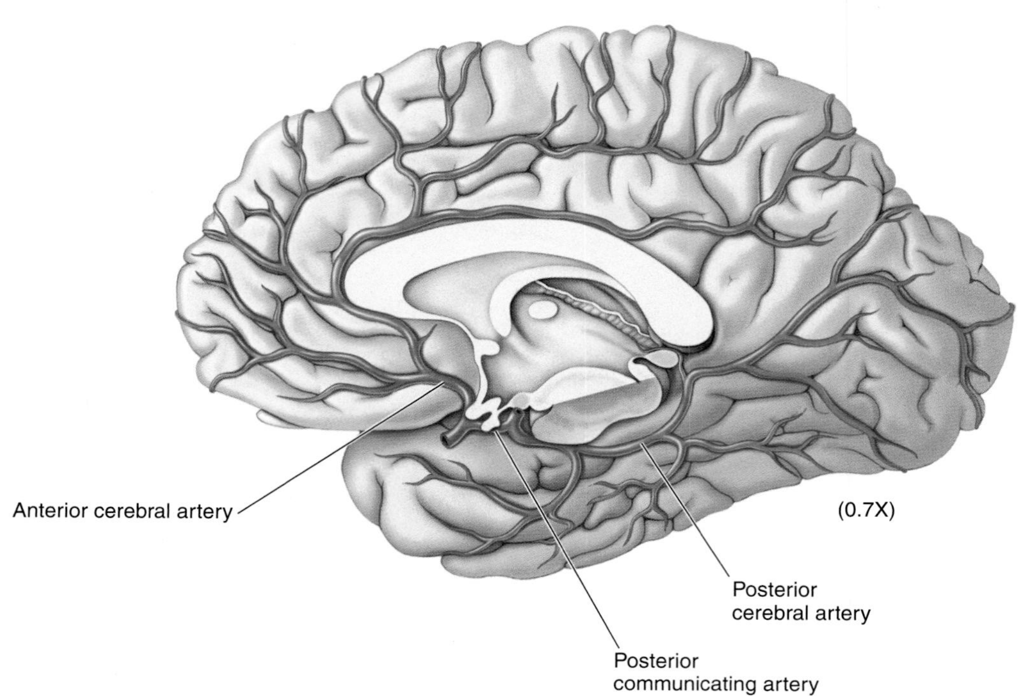

Self-Quiz

This review workbook is designed to help you learn the neuroanatomy that has been presented. Here we have reproduced the images from the Guide; instead of labels, however, numbered leader lines (arranged clockwise) point to the structures of interest. Test your knowledge by filling in the appropriate names in the spaces provided. To review what you have learned, quiz yourself by putting your hand over the names. This technique greatly facilitates the learning and retention of anatomical terms. Mastery of the vocabulary of neuroanatomy will serve you well as you learn about the functional organization of the brain in the remainder of the book.

SELF-QUIZ

The Lateral Surface of the Brain

(a) Gross Features

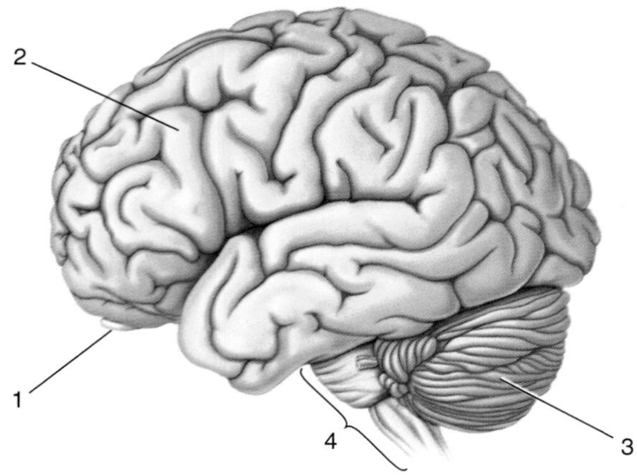

1. ______________________________

2. ______________________________

3. ______________________________

4. ______________________________

(b) Selected Gyri, Sulci, and Fissures

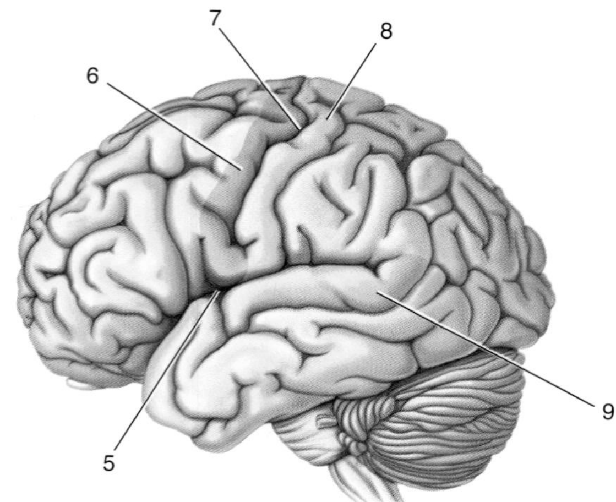

5. ______________________________

6. ______________________________

7. ______________________________

8. ______________________________

9. ______________________________

(c) Cerebral Lobes and the Insula

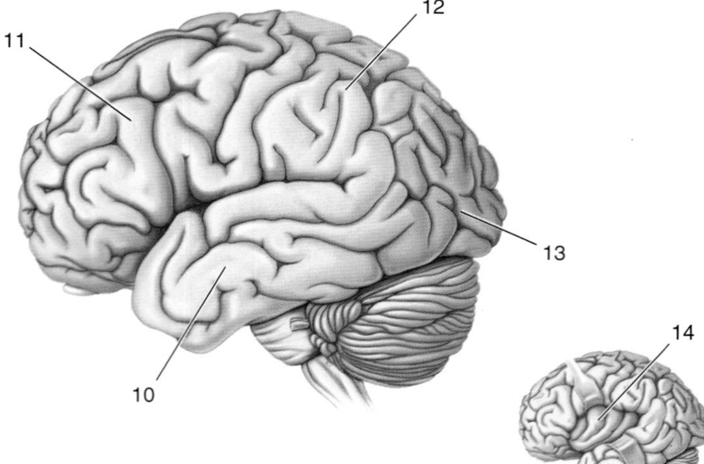

10. ______________________________

11. ______________________________

12. ______________________________

13. ______________________________

14. ______________________________

SELF-QUIZ

The Lateral Surface of the Brain *(continued)*

(d) Major Sensory, Motor, and Association Areas of Cortex

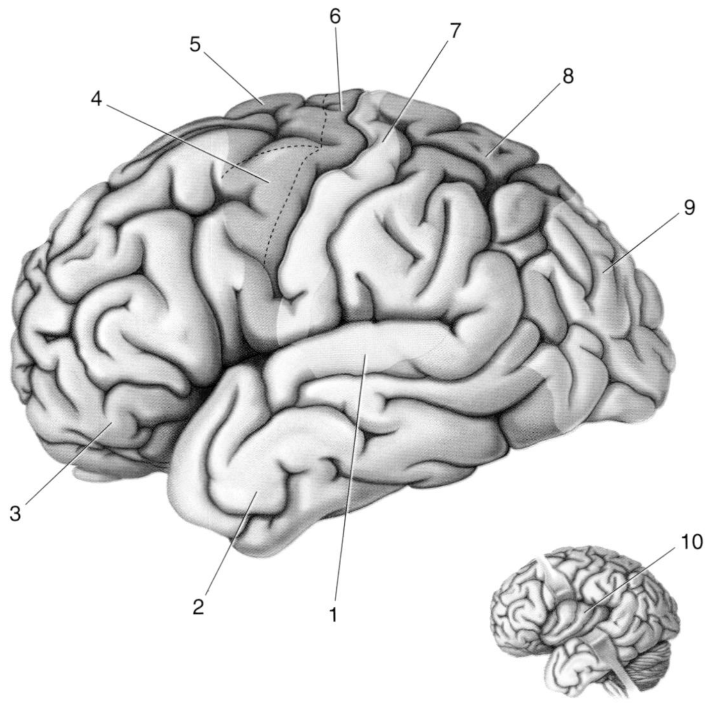

1. ______________________________
2. ______________________________
3. ______________________________
4. ______________________________
5. ______________________________
6. ______________________________
7. ______________________________
8. ______________________________
9. ______________________________
10. ______________________________

SELF-QUIZ

The Medial Surface of the Brain

(a) Brain Stem Structures

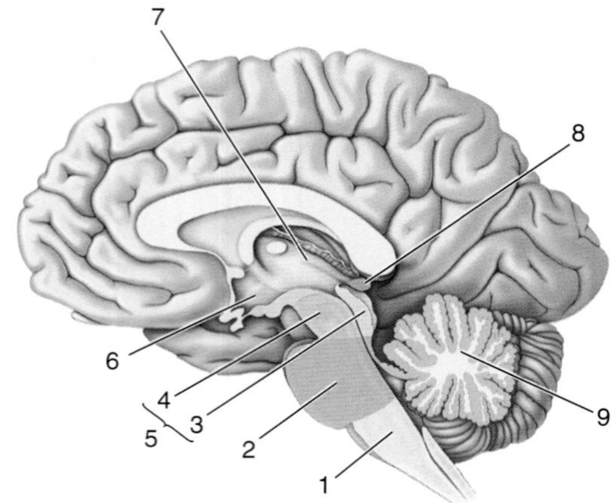

1. ____________________
2. ____________________
3. ____________________
4. ____________________
5. ____________________
6. ____________________
7. ____________________
8. ____________________
9. ____________________

(b) Forebrain Structures

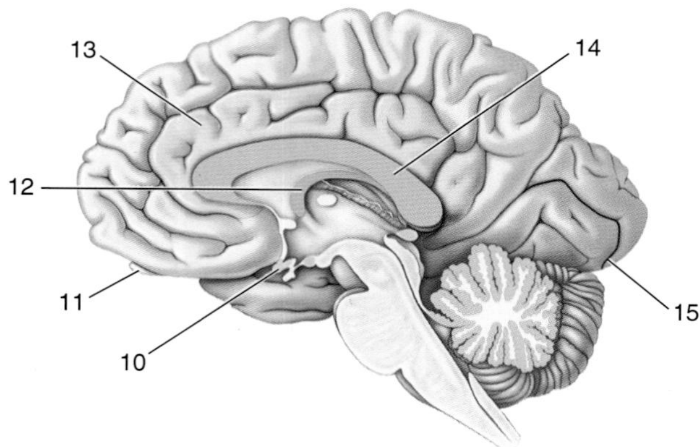

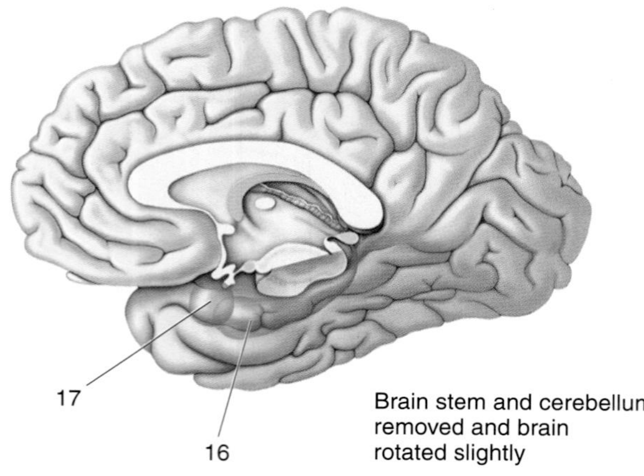

Brain stem and cerebellum removed and brain rotated slightly

10. ____________________
11. ____________________
12. ____________________
13. ____________________
14. ____________________
15. ____________________
16. ____________________
17. ____________________

SELF-QUIZ

The Medial Surface of the Brain

(c) Ventricles

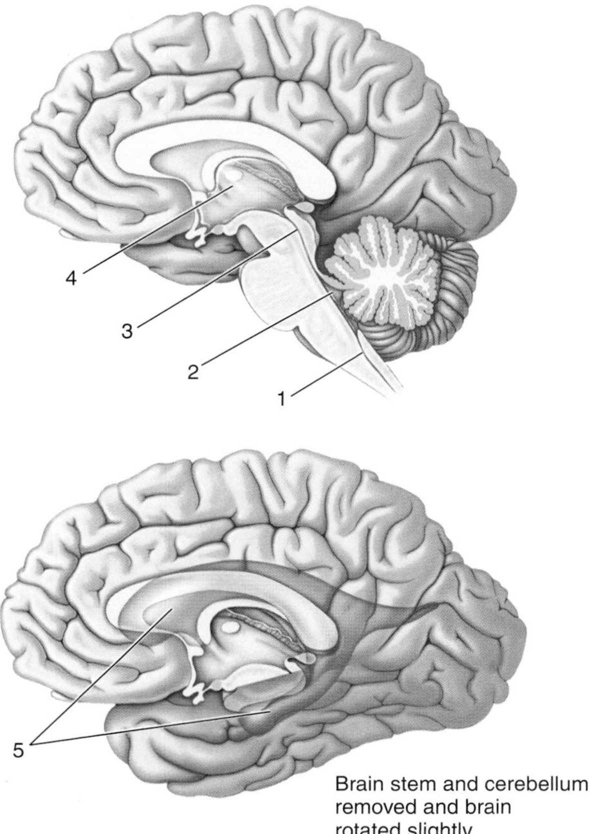

Brain stem and cerebellum removed and brain rotated slightly

1. ______________________
2. ______________________
3. ______________________
4. ______________________
5. ______________________

The Ventral Surface of the Brain

(a) Gross Features

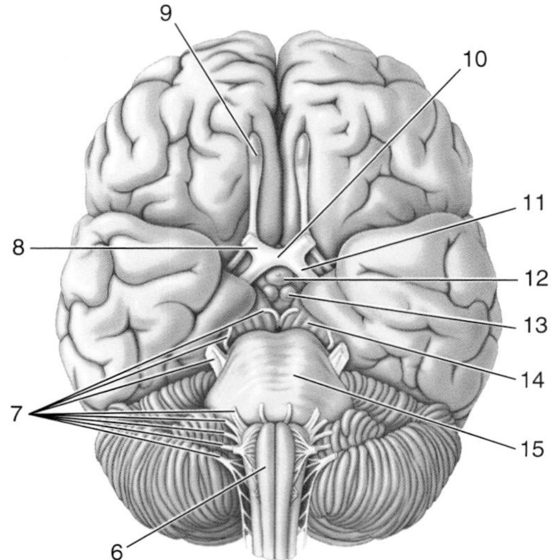

6. ______________________
7. ______________________
8. ______________________
9. ______________________
10. ______________________
11. ______________________
12. ______________________
13. ______________________
14. ______________________
15. ______________________

SELF-QUIZ

The Dorsal Surface of the Brain

(a) Cerebrum

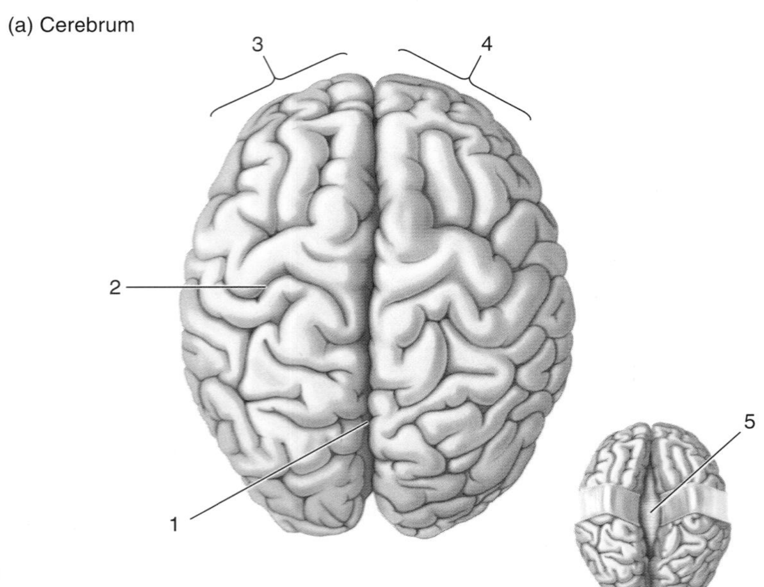

1. ______________________

2. ______________________

3. ______________________

4. ______________________

5. ______________________

(b) Cerebrum Removed

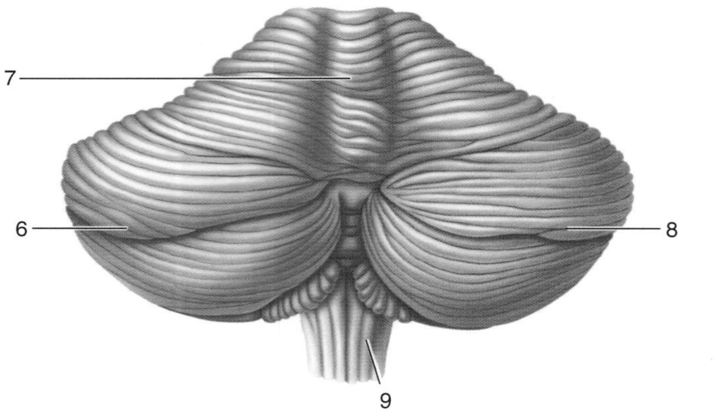

6. ______________________

7. ______________________

8. ______________________

9. ______________________

(c) Cerebrum and Cerebellum Removed

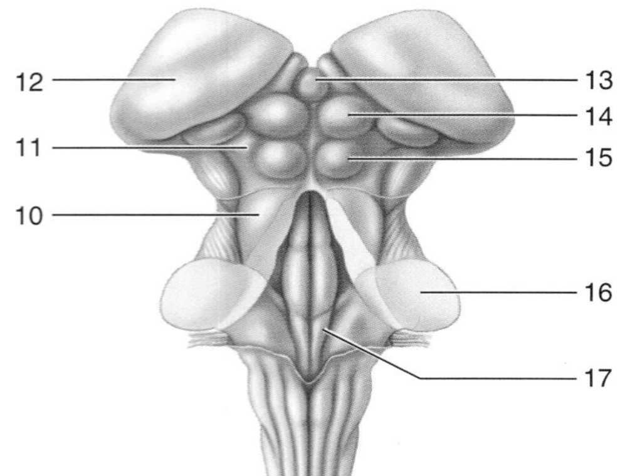

10. ______________________

11. ______________________

12. ______________________

13. ______________________

14. ______________________

15. ______________________

16. ______________________

17. ______________________

SELF-QUIZ

Forebrain at Thalamus–Telencephalon Junction

(a) Gross Features

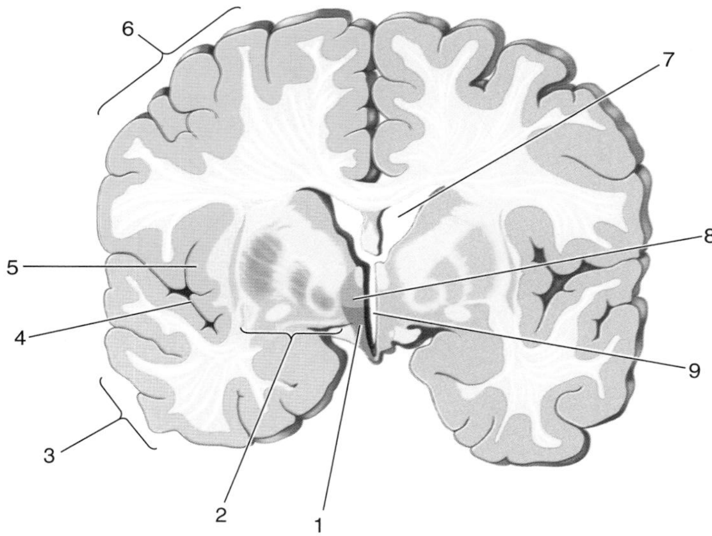

(b) Selected Cell and Fiber Groups

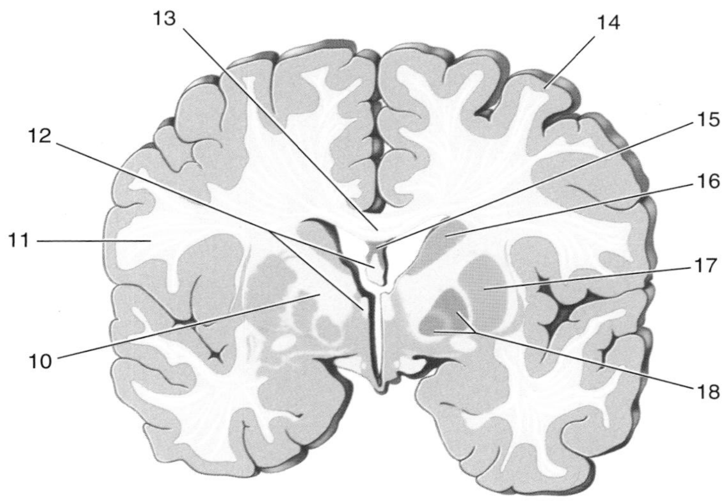

1. ______________________

2. ______________________

3. ______________________

4. ______________________

5. ______________________

6. ______________________

7. ______________________

8. ______________________

9. ______________________

10. ______________________

11. ______________________

12. ______________________

13. ______________________

14. ______________________

15. ______________________

16. ______________________

17. ______________________

18. ______________________

SELF-QUIZ

Forebrain at Mid-Thalamus

(a) Gross Features

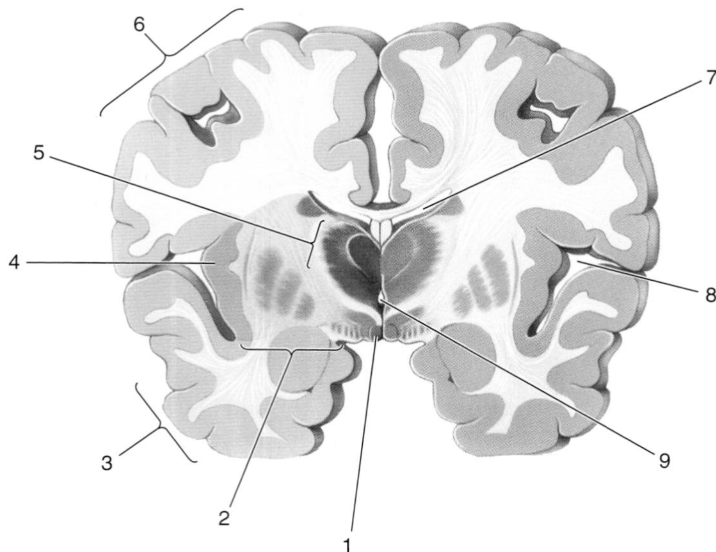

(b) Selected Cell and Fiber Groups

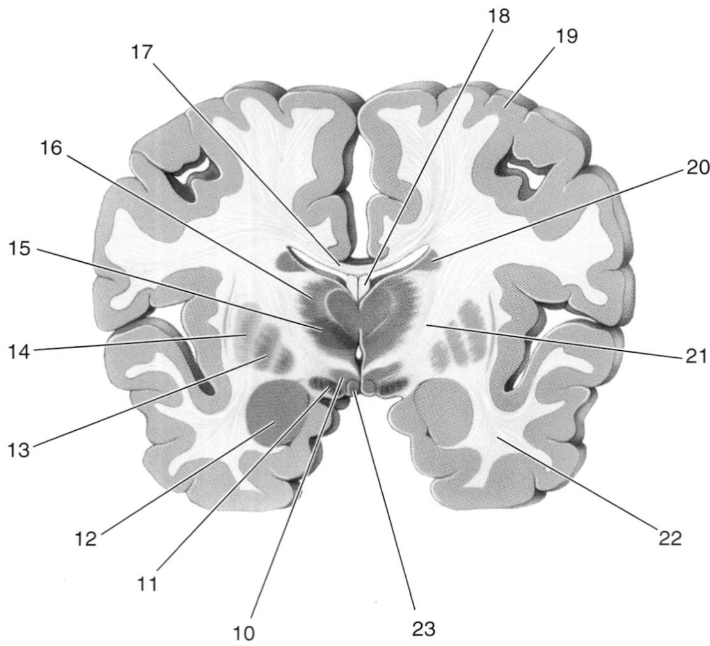

1. ____________________
2. ____________________
3. ____________________
4. ____________________
5. ____________________
6. ____________________
7. ____________________
8. ____________________
9. ____________________
10. ____________________
11. ____________________
12. ____________________
13. ____________________
14. ____________________
15. ____________________
16. ____________________
17. ____________________
18. ____________________
19. ____________________
20. ____________________
21. ____________________
22. ____________________
23. ____________________

SELF-QUIZ

Forebrain at Thalamus–Midbrain Junction

(a) Gross Features

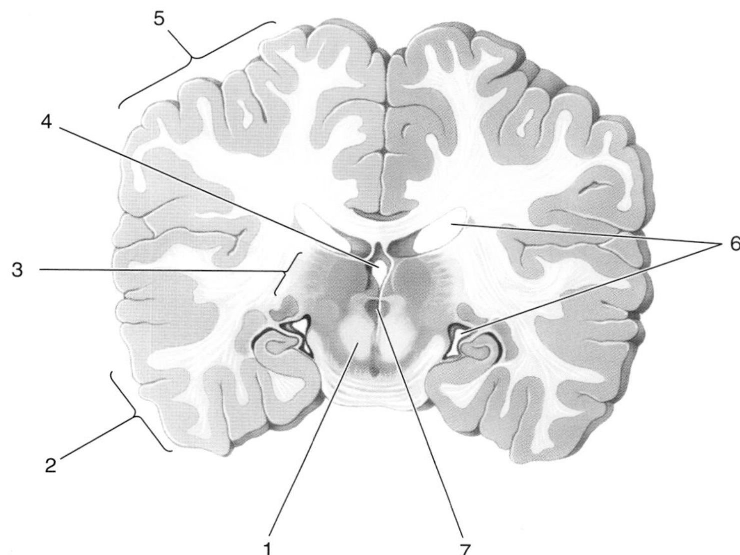

1. ______________________
2. ______________________
3. ______________________
4. ______________________
5. ______________________
6. ______________________
7. ______________________

(b) Selected Cell and Fiber Groups

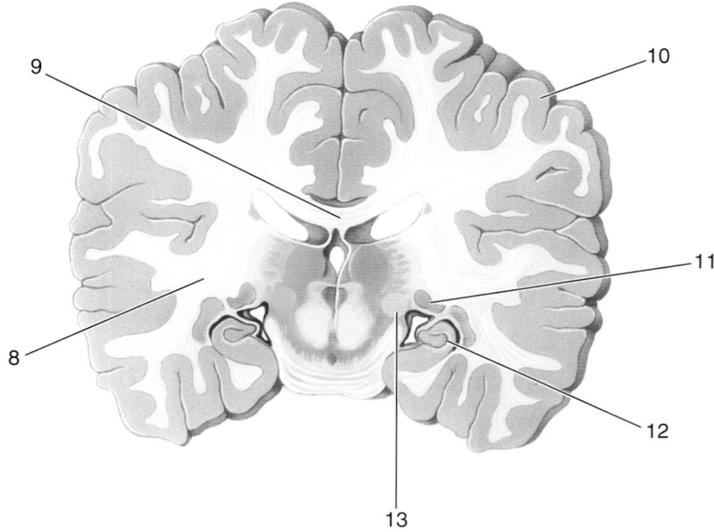

8. ______________________
9. ______________________
10. ______________________
11. ______________________
12. ______________________
13. ______________________

SELF-QUIZ

Rostral Midbrain

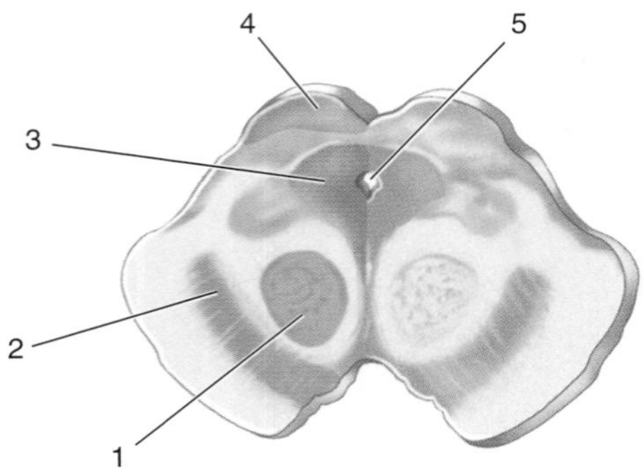

1. ______________________________
2. ______________________________
3. ______________________________
4. ______________________________
5. ______________________________

Caudal Midbrain

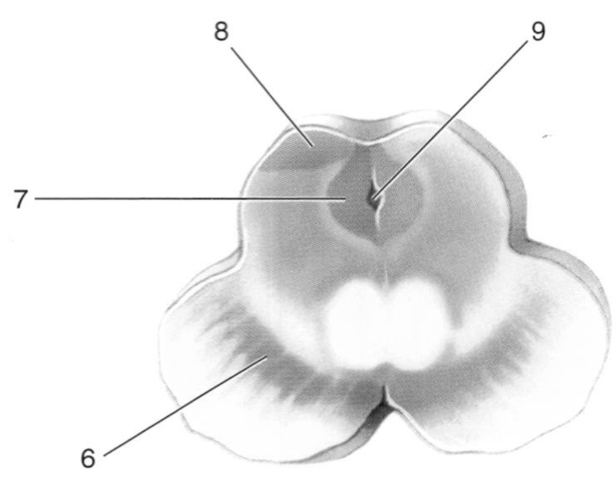

6. ______________________________
7. ______________________________
8. ______________________________
9. ______________________________

Pons and Cerebellum

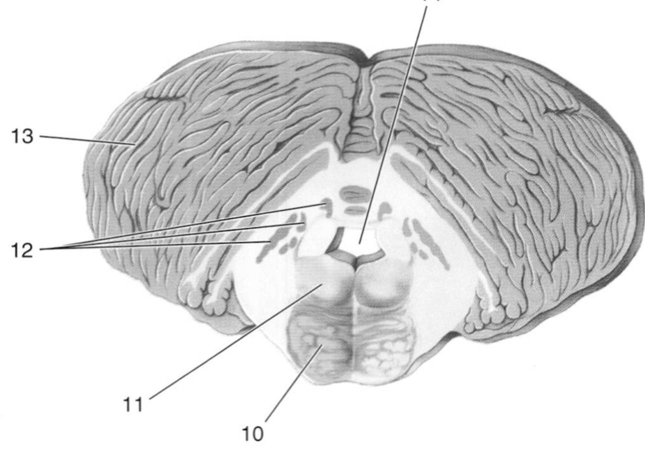

10. ______________________________
11. ______________________________
12. ______________________________
13. ______________________________
14. ______________________________

SELF-QUIZ

Rostral Medulla

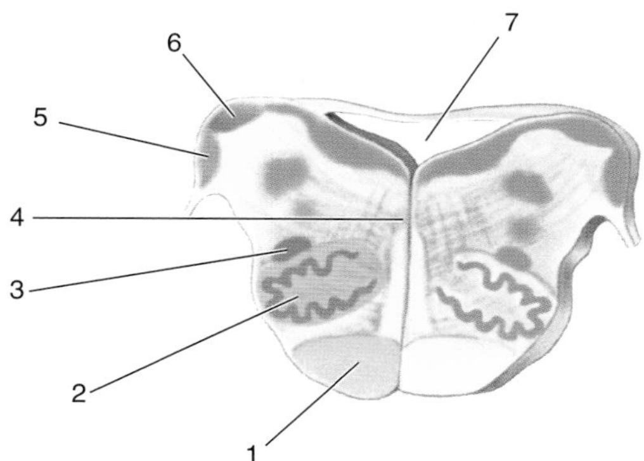

Mid-Medulla

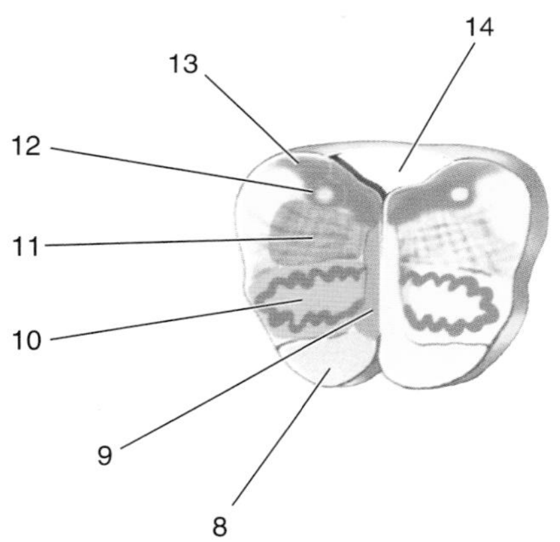

Medulla–Spinal Cord Junction

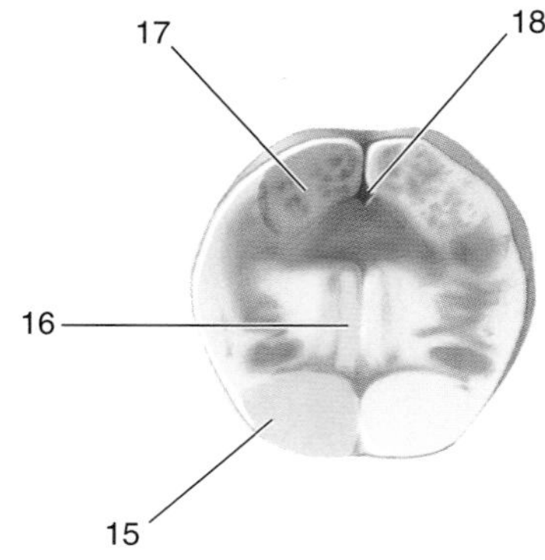

1. ____________________

2. ____________________

3. ____________________

4. ____________________

5. ____________________

6. ____________________

7. ____________________

8. ____________________

9. ____________________

10. ____________________

11. ____________________

12. ____________________

13. ____________________

14. ____________________

15. ____________________

16. ____________________

17. ____________________

18. ____________________

SELF-QUIZ

Spinal Cord, Ventral-Lateral Surface

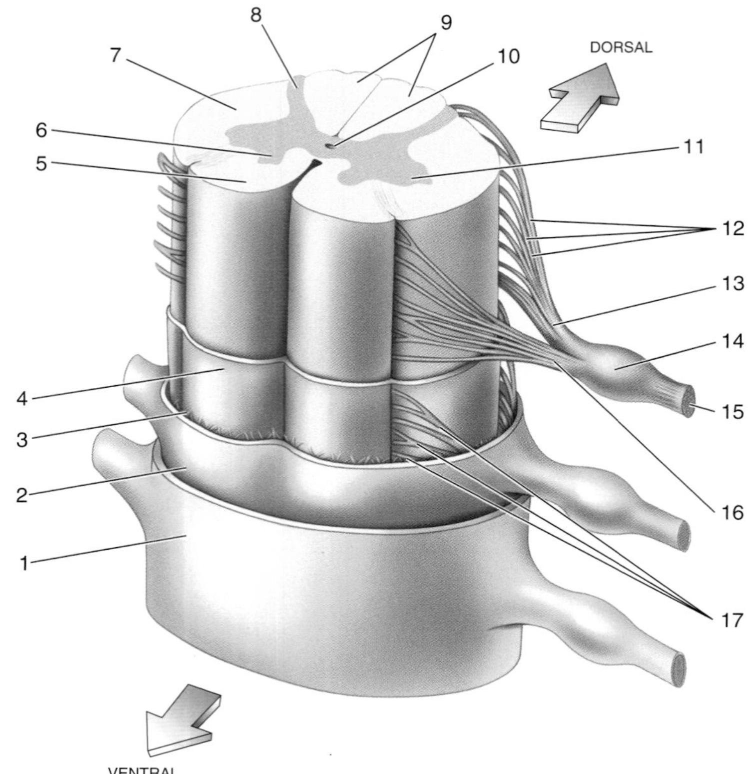

Spinal Cord, Cross-Sectional Anatomy

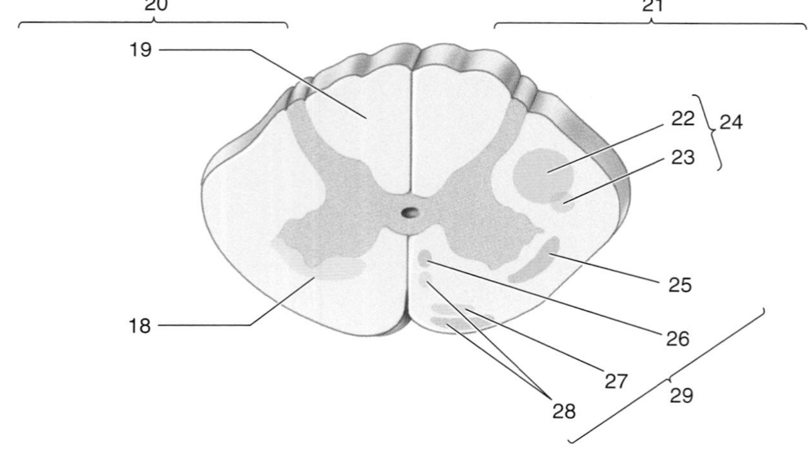

1. ______
2. ______
3. ______
4. ______
5. ______
6. ______
7. ______
8. ______
9. ______
10. ______
11. ______
12. ______
13. ______
14. ______
15. ______
16. ______
17. ______
18. ______
19. ______
20. ______
21. ______
22. ______
23. ______
24. ______
25. ______
26. ______
27. ______
28. ______
29. ______

SELF-QUIZ

The Cranial Nerves

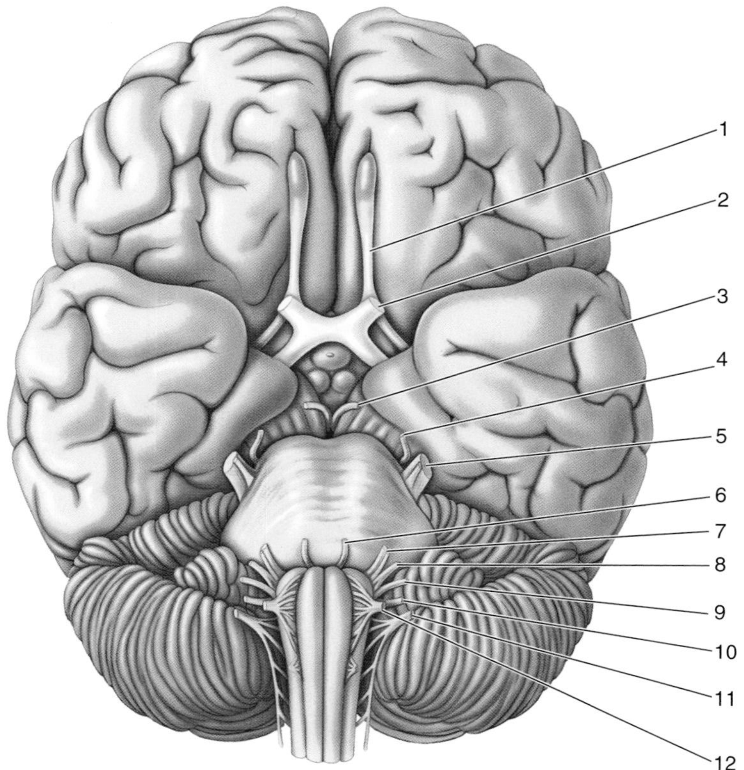

1. ____________________

2. ____________________

3. ____________________

4. ____________________

5. ____________________

6. ____________________

7. ____________________

8. ____________________

9. ____________________

10. ____________________

11. ____________________

12. ____________________

SELF-QUIZ

The Blood Supply of the Brain

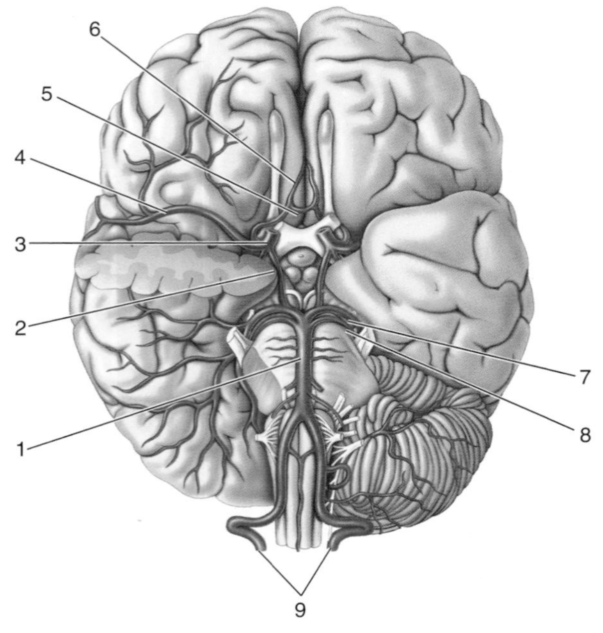

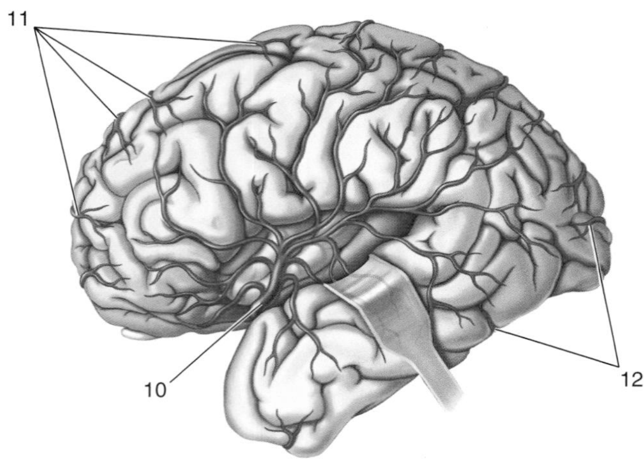

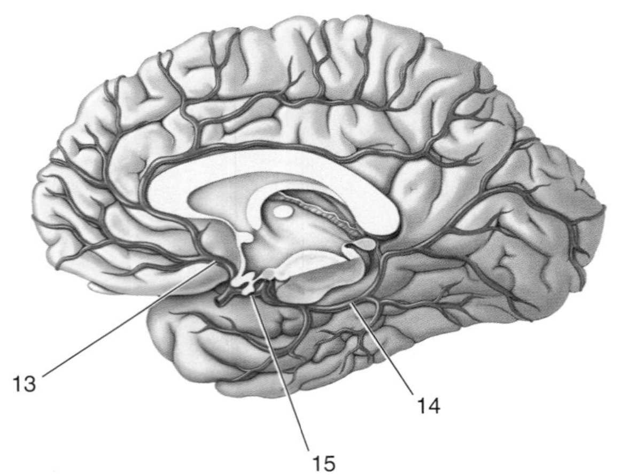

1. ______________________________

2. ______________________________

3. ______________________________

4. ______________________________

5. ______________________________

6. ______________________________

7. ______________________________

8. ______________________________

9. ______________________________

10. ______________________________

11. ______________________________

12. ______________________________

13. ______________________________

14. ______________________________

15. ______________________________

PART

II

SENSORY AND MOTOR SYSTEMS

CHAPTER

8

The Chemical Senses

INTRODUCTION

Life evolved in a sea of chemicals. From the beginning, organisms have floated or swum in water containing chemical substances that signal food, poison, or sex. In this sense, things have not changed much in 3 billion years. Animals, including humans, depend on the chemical senses to help identify nourishment (the sweetness of honey, the aroma of pizza), noxious substances (the bitterness of plant poisons), or the suitability of a potential mate. Chemical sensation is the oldest and most common of the sensory systems. Even brainless bacteria can detect and tumble toward a favorable food source.

Multicellular organisms must detect chemicals in both their internal and their external environments. The variety of chemical detection systems has expanded considerably over the course of evolution. Humans live in a sea of air full of volatile chemicals; we put chemicals into our mouths for a variety of reasons, and we carry a complex sea within us in the form of blood and other fluids that bathe our cells. We have specialized detection systems for the chemicals in each milieu. The mechanisms of chemical sensation that originally evolved to detect environmental substances now serve as the basis for chemical communication between cells and organs, using hormones and neurotransmitters. Virtually every cell in every organism is responsive to many chemicals.

In this chapter, we consider the most familiar of our chemical senses: taste, or **gustation**, and smell, or **olfaction**. Although taste and smell reach our awareness most often, they are not the only important chemical senses we have. Many types of chemically sensitive cells, or **chemoreceptors,** are distributed throughout the body. For example, some nerve endings in skin and mucous membranes warn us of irritating chemicals. A wide range of chemoreceptors report subconsciously and consciously about our internal state: Nerve endings in the digestive organs detect many types of ingested substances, receptors in arteries of the neck measure carbon dioxide and oxygen levels in our blood, and sensory endings in muscles respond to acidity, giving us the burning feeling that comes with exertion and oxygen debt.

Gustation and olfaction have a similar task: the detection of environmental chemicals. In fact, only by using both senses can the nervous system perceive flavor. Gustation and olfaction have unusually strong and direct connections with our most basic internal needs, including thirst, hunger, emotion, sex, and certain forms of memory. The systems of gustation and olfaction are, however, separate and different, from the structures and mechanisms of their chemoreceptors, to the gross organization of their central connections, to their effects on behavior. The neural information from each system is processed in parallel and is merged only at rather high levels in the cerebral cortex.

TASTE

Humans evolved as omnivores (from the Latin *omnis*, "all," and *vorare*, "to eat"), opportunistically eating the plants and animals they could gather, scavenge, or kill. A sensitive and versatile system of taste was necessary to distinguish between new sources of food and potential toxins. Some of our taste preferences are inborn. We have an innate preference for sweetness, satisfied by mother's milk. Bitter substances are instinctively rejected; indeed, many kinds of poisons are bitter. Experience can strongly modify our instincts, however, and we can learn to tolerate and even enjoy the bitterness of such substances as coffee and quinine. The body also has the capacity to recognize a deficiency of certain key nutrients and develop an appetite for them. For example, when deprived of essential salt, we may crave salty foods.

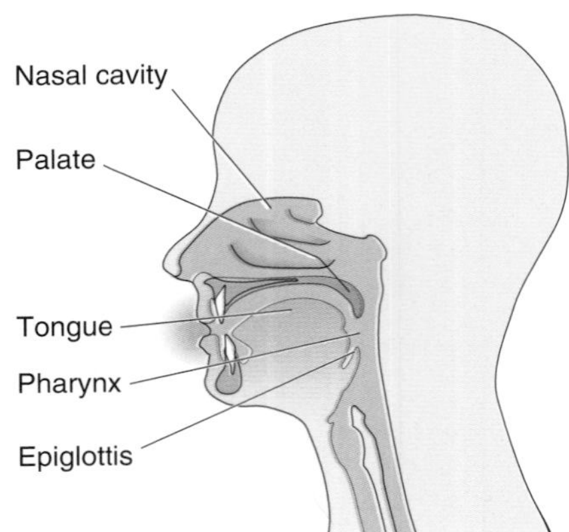

Figure 8.1
Anatomy of the mouth, throat, and nasal passages. Taste is primarily a function of the tongue, but regions of the pharynx, palate, and epiglottis have some sensitivity. Notice how the nasal passages are placed so that odors from ingested food can enter through the nose or the pharynx, thereby easily contributing to perceptions of flavor through olfaction.

The Basic Tastes

Although the number of chemicals is endless and the variety of flavors seems immeasurable, it is likely that we can recognize only a few basic tastes. Most neuroscientists put the number at four or five. The four basic taste qualities are saltiness, sourness, sweetness, and bitterness. The more unfamiliar fifth taste quality is *umami*, meaning "delicious" in Japanese; it is defined by the taste of the amino acid glutamate (monosodium glutamate, or MSG, being the familiar culinary form). There are several kinds of evidence for the basic tastes, the most compelling of which is behavioral. When people are asked to taste many pairs of chemicals and compare them, they describe each taste as one of four categories: salt, sour, sweet, or bitter.

The correspondence between chemistry and taste is obvious in some cases. Most acids taste sour, and most salts taste salty. But the chemistry of substances can vary considerably, while their basic taste remains the same. Many substances are sweet, from familiar sugars (like fructose, present in fruits and honey, and sucrose, which is white table sugar), to certain proteins (monellin), to artificial sweeteners (saccharin and aspartame, the second of which is made from two amino acids). Surprisingly, sugars are the least sweet of all of these; gram for gram, the artificial sweeteners and proteins are 10,000–100,000 times sweeter than sucrose. Bitter substances range from simple ions, such as K^+ (KCl actually evokes both bitter and salty tastes) and Mg^{2+}, to complex organic molecules, such as quinine and caffeine.

How, then, do we perceive the countless flavors of foods, such as chocolate, strawberries, and barbecue sauce? First, each food activates a different combination of the basic tastes, helping make it unique. Second, most foods have a distinctive flavor as a result of their taste *and* smell, occurring simultaneously. For example, without the sense of smell, a bite of onion can be easily mistaken for a bite of apple. Third, other sensory modalities may contribute to a unique food-tasting experience. Texture and temperature are important, and pain sensations are essential to the hot, spicy flavor of foods laced with capsaicin, the key ingredient in hot peppers.

The Organs of Taste

Experience tells us that we taste with our tongue, but other areas of the mouth, such as the palate, pharynx, and epiglottis, are also involved (Figure 8.1). Odors from the food we are eating can also pass via the pharynx into the nasal cavity, where they can be detected by olfactory receptors. The tip of the tongue is most sensitive to sweetness, the back of it to bitterness, and the sides to saltiness and sourness, as shown in Figure 8.2. This does not mean, however, that we taste sweetness only with the tip of our tongue. Most of the tongue is sensitive to all basic tastes. The tongue map implies only that certain regions of the tongue are more sensitive to the basic tastes than are other regions.

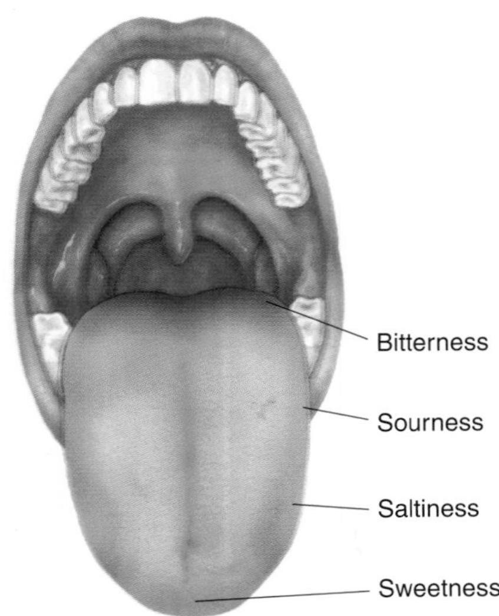

Figure 8.2
The tongue map: regions of the lowest thresholds for the basic tastes. The center of the tongue lacks taste buds and is relatively insensitive to taste.

Scattered about the surface of the tongue are small projections called **papillae** (Latin for "bumps"; singular: papilla), which are shaped like ridges (*foliate papillae*), pimples (*vallate papillae*), or mushrooms (*fungiform papillae*) (Figure 8.3a). Facing a mirror, stick your tongue out and shine a flashlight on it, and you will see your papillae easily—small, rounded ones at the front and sides, and large ones in the back. Each papilla has one to several hundred **taste buds**, visible only with a microscope (Figure 8.3b). Each taste bud has 50–150 **taste receptor cells,** or taste cells, arranged within the bud like the sections of an orange. Taste cells are only about 1% of the tongue epithelium. Taste buds also have basal cells that surround the taste cells plus a set of gustatory afferent axons (Figure 8.3c). A person typically has

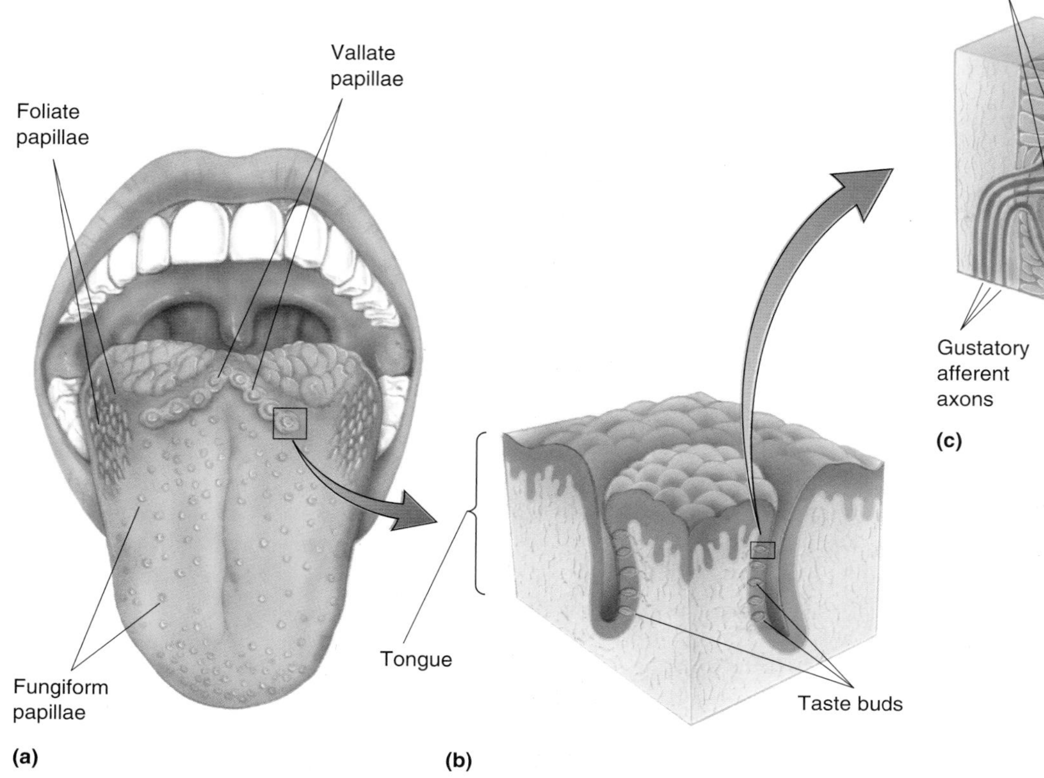

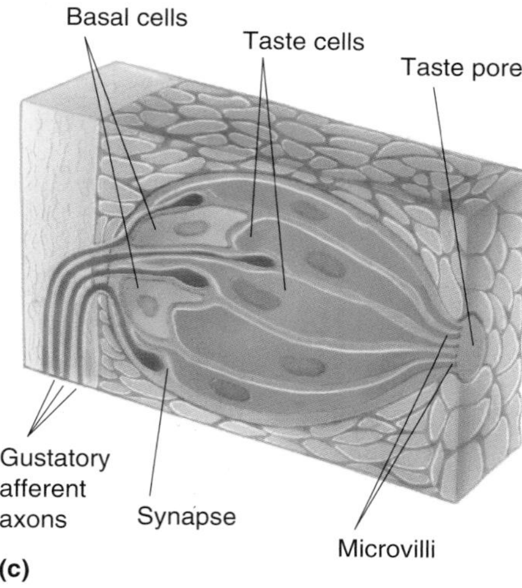

Figure 8.3

The tongue, its papillae, and its taste buds. (a) Papillae are the taste-sensitive structures. The largest and most posterior are the vallate papillae. Foliate papillae are elongated. Fungiform papillae are relatively large toward the back of the tongue and much smaller along the sides and tip. **(b)** Cross-sectional view of a vallate papilla showing the locations of taste buds. **(c)** A taste bud is a cluster of taste cells (receptor cells), gustatory afferent axons and their synapses with taste cells, and basal cells. Microvilli at the apical end of the taste cells extend into the taste pore, the site where chemicals dissolved in saliva can interact directly with taste cells.

2000–5000 taste buds, although exceptional persons have as few as 500 or as many as 20,000.

By using tiny droplets, it is possible to expose a single papilla on a person's tongue to low concentrations of various basic taste stimuli (something almost purely sour, like vinegar, or almost purely sweet, like a sucrose solution). Concentrations too low are not tasted, but at some critical concentration, the stimulus evokes a perception of taste—this is the *threshold* concentration. At concentrations just above threshold, most papillae tend to be sensitive to only one basic taste; there are sour-sensitive papillae and sweet-sensitive papillae, for example. When the concentrations of the taste stimuli are increased, however, most papillae become less selective. Whereas a papilla might have responded only to sweet when all stimuli were weak, it may also respond to sour and salt if they are made stronger.

This lack of specificity is a common phenomenon in sensory systems. Most sensory receptors are surprisingly indiscriminate about the things that excite them. This presents a paradox: If single taste receptors show only small differences in response to ice cream and bananas, how can we distinguish reliably between differences as subtle as two brands of chocolate? The answer lies in the brain.

Taste Receptor Cells

The chemically sensitive part of a taste receptor cell is its small membrane region, called the *apical end*, near the surface of the tongue. The apical ends have thin extensions called *microvilli* that project into the *taste pore*, a small opening on the surface of the tongue where the taste cell is exposed to the contents of the mouth (Figure 8.3c). By standard histological criteria, taste receptor cells are not neurons. They do, however, form synapses with the endings of the gustatory afferent axons near the bottom of the taste bud. Taste receptor cells also make both electrical and chemical synapses onto some of the basal cells; some basal cells synapse onto the sensory axons, and these may form a simple information-processing circuit within each taste bud. Cells of the taste bud undergo a constant cycle of growth, death, and regeneration; the lifespan of one taste cell is about 2 weeks. This process depends on an influence of the sensory nerve, because if the nerve is cut, the taste buds degenerate.

When a taste receptor cell is activated by an appropriate chemical, its membrane potential changes, either depolarizing or hyperpolarizing. This voltage shift is called the **receptor potential** (Figure 8.4a). If the receptor potential is depolarizing and large enough, most taste receptor cells, like neurons, may fire action potentials. In any case, depolarization of the receptor membrane causes voltage-gated calcium channels to open; Ca^{2+} enters the cytoplasm, triggering the release of transmitter molecules. This is basic synaptic transmission from taste cell to sensory axon. The identity of the taste receptor's transmitter is unknown, but we do know that it excites the postsynaptic sensory axon and causes it to fire action potentials (Figure 8.4b), which communicate the taste signal into the brain stem.

More than 90% of receptor cells respond to two or more of the basic tastes, emphasizing that even the first cell in the taste process is relatively unselective to chemical types. An example is cell 2 in Figure 8.4a, which gives strong depolarizing responses to both salt (NaCl) and sour (HCl, hydrochloric acid) stimuli. Taste cells and their gustatory axons differ widely in their response preferences, however. Each of the gustatory axons in Figure 8.4b is influenced by four basic tastes, but each has a clearly different bias.

Figure 8.5 shows the results of similar recordings from four gustatory axons in a rat. One responds strongly only to salt, one only to sweet, and two to all but sweet. Why does one cell respond only to a single chemical type, while another responds to three or four types? The answer is that the responses depend on the particular transduction mechanisms in each cell.

Mechanisms of Taste Transduction

The process by which an environmental stimulus causes an electrical response in a sensory receptor cell is called **transduction** (from the Latin *transducere*, "to lead across"). The nervous system has myriad transduction mechanisms, which make it sensitive to chemicals, pressures, sounds, and light. The nature of the transduction mechanism determines the specific sensitivity of a sensory system. We see because our eyes have photoreceptors. If our tongue had photoreceptors, we might see with our mouth.

Some sensory systems have a single basic type of receptor cell that uses one transduction mechanism (e.g., the auditory system). Taste transduction involves many processes, however, and each basic taste may use one or more of these mechanisms. Taste stimuli, or *tastants*, may (1) directly pass through ion channels (salt and sour), (2) bind to and block ion channels (sour and bitter), (3) bind to and open ion channels (some amino acids), or (4) bind to membrane receptors that activate second messenger systems that, in turn,

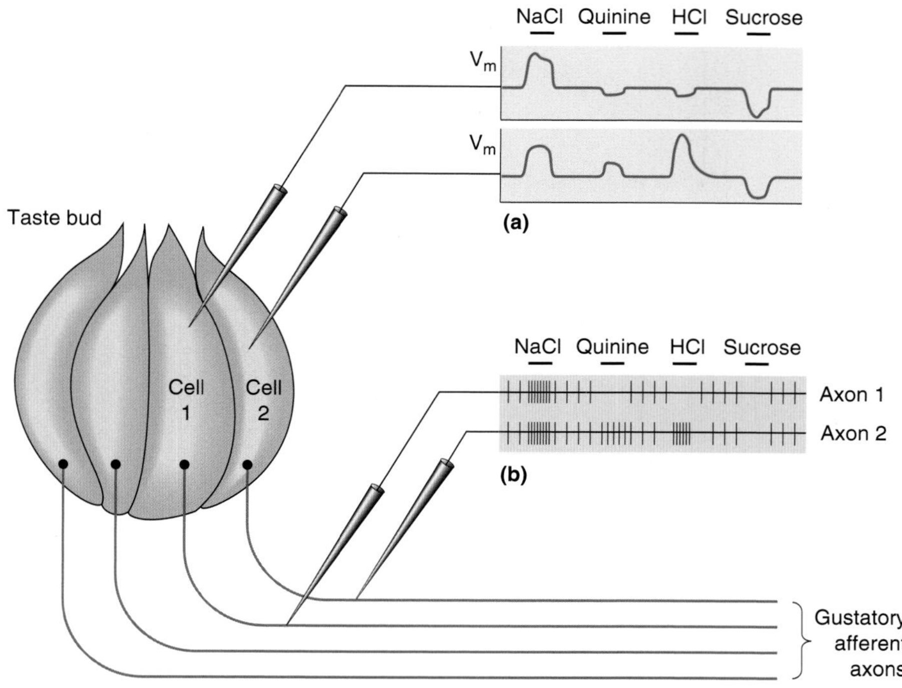

Figure 8.4
Responsiveness of taste cells and gustatory axons. (a) Two different cells were exposed to salt (NaCl), bitter (quinine), sour (HCl), and sweet (sucrose) stimuli, and their membrane potential was recorded with electrodes. Notice the different sensitivities of the two cells. **(b)** In this case, the action potential discharge of the sensory axons was recorded. This is an example of extracellular recording of action potentials. Each vertical deflection in the record is a single action potential.

open or close ion channels (sweet, bitter, and umami). These are familiar processes, the functional building blocks of signaling in all neurons and synapses.

As explained in Chapter 1, much of what we know about how our brain works comes from studying animals other than humans. Our understanding of taste transduction is no exception. Important clues have come from the cells of catfish, mudpuppies (a large, voracious salamander), mice, and rats. Each animal has certain experimental advantages (very large taste cells, for example), and the information they provide tells us where to start looking when we study human cells directly. In the case of taste (and much of the rest of neuroscience), the human story is still sparse and must be supplemented by information obtained from animals.

Saltiness. The prototypical salty chemical is table salt (NaCl), which apart from water is the major component of the ocean, blood, and chicken soup.

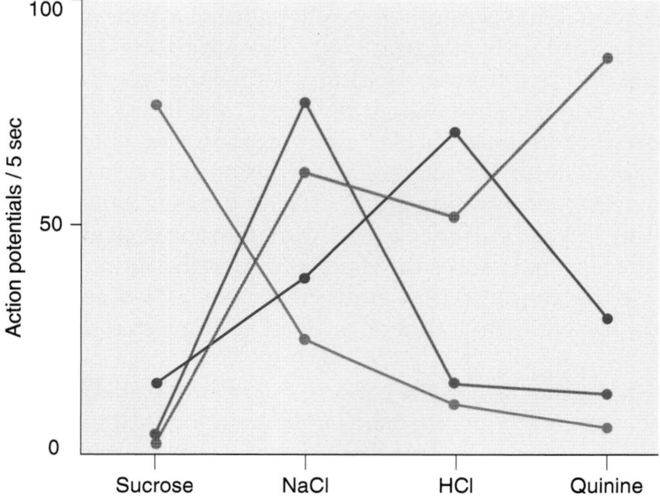

Figure 8.5
Action potential firing rates of four primary gustatory nerve axons in a rat. The taste stimuli were sweet (sucrose), salt (NaCl), sour (HCl), and bitter (quinine). Each colored line represents measurements from a single axon. Notice the differences in selectivity between axons. (Source: Adapted from Sato, 1980, p. 23.)

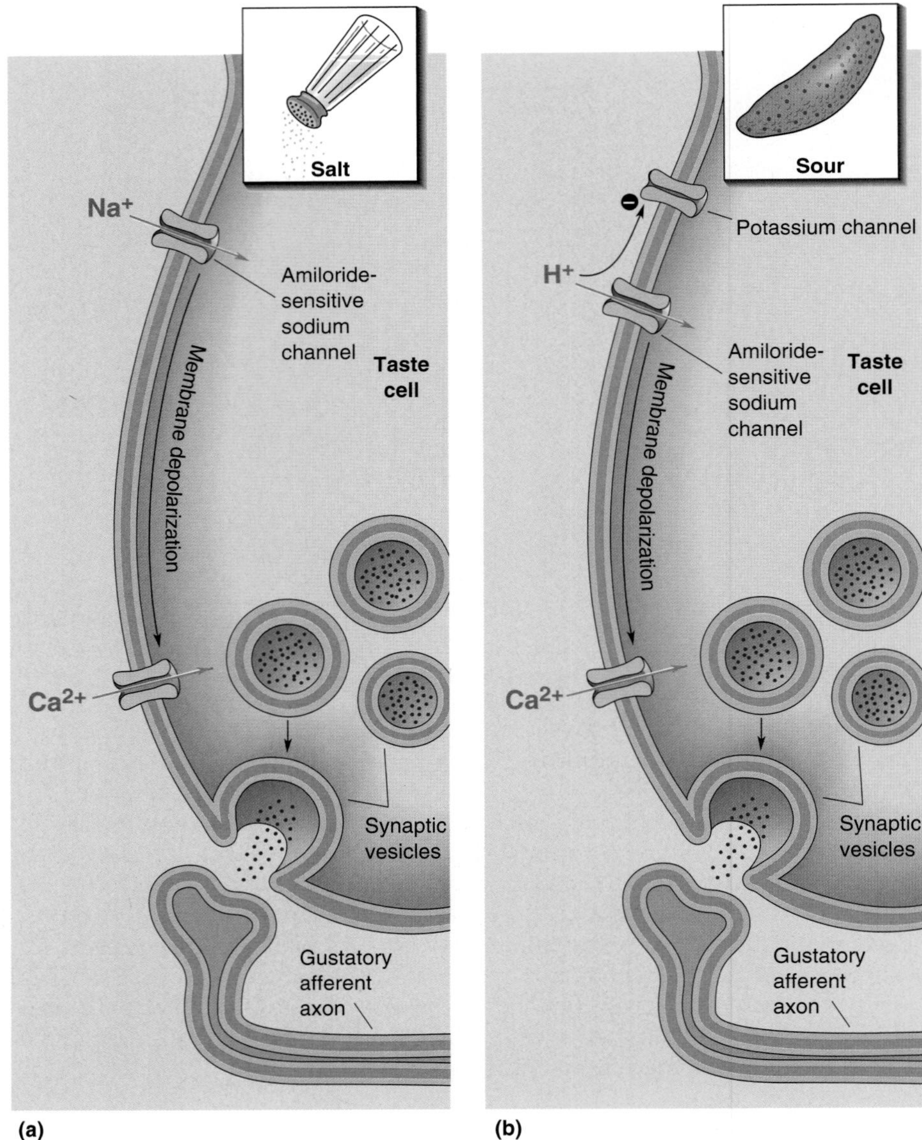

Figure 8.6
Transduction mechanisms of (a) salt and (b) sour tastants. Tastants can interact directly with ion channels, by either passing through them (Na^+ and H^+) or blocking them (H^+ blocking the potassium channel). Then, the membrane voltage influences calcium channels on the basal membrane, which in turn influence the intracellular [Ca^{2+}] and transmitter release.

The taste of salt is mostly the taste of the cation Na^+. Salt-sensitive taste cells have a Na^+-selective channel that is common in other epithelial cells and that is blocked by the drug amiloride (Figure 8.6a). The amiloride-sensitive sodium channel is quite different from the sodium channel that generates action potentials; the taste channel is insensitive to voltage, and it stays open all the time. When you sip chicken soup, the Na^+ concentration rises outside the receptor cell, and the gradient for Na^+ across the membrane is made steeper. Na^+ then diffuses down its concentration gradient, which means it flows into the cell, and the resulting inward current causes the membrane to depolarize. This is similar to the process that causes the rising phase of the action potential, except in that case the concentration gradient for Na^+ stays constant, while the conductance for Na^+ (the number of open sodium channels) increases temporarily.

The anions of salts affect the taste of the cations. For example, NaCl tastes saltier than Na acetate, apparently because the larger an anion is, the more it *inhibits* the salt taste of the cation. The mechanisms of anion inhibition are poorly understood. Another complication is that as the anions become larger, they tend to take on tastes of their own. Sodium saccharin tastes sweet because the Na^+ concentrations are far too low for us to taste the saltiness, and the saccharin potently activates sweetness receptors.

Sourness. Foods taste sour because of their high acidity (otherwise known as low pH). Acids, such as HCl, dissolve in water and generate hydrogen ions (protons, or H^+). Thus, protons are the causative agents of acidity and sourness, and they are known to affect sensitive taste receptors in two ways (Figure 8.6b). First, H^+ can permeate the amiloride-sensitive sodium channel, the same channel that mediates the taste of salt. This causes an inward H^+ current and depolarizes the cell. (The cell would not be able to distinguish H^+ from Na^+ if this were the only transduction mechanism available to it.) Second, H^+ can bind to and *block* K^+-selective channels. When the K^+ permeability of a membrane is decreased, it depolarizes. These may not be the only mechanisms of sour taste transduction, because changes of pH can affect virtually all cellular processes.

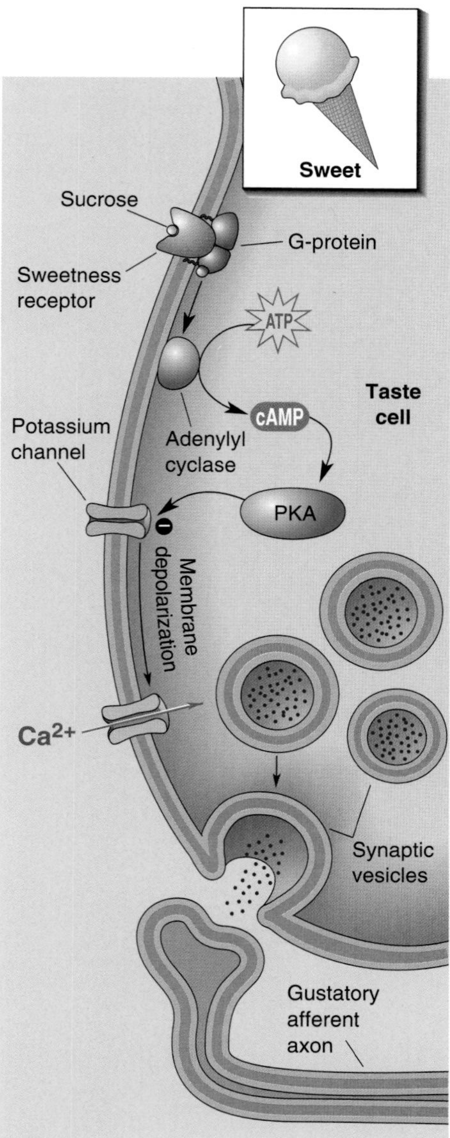

Figure 8.7
Transduction mechanisms for sweet tastants. Tastants bind directly to G-protein-coupled membrane receptors and trigger the synthesis of cAMP, which leads to the blockade of a potassium channel, depolarization, Ca^{2+} entry, and transmitter release.

Sweetness. There are many sweet tastants, and several mechanisms are sensitive to them. Some molecules become sweet when they bind to specific receptor sites and activate a cascade of second messengers in certain taste cells (Figure 8.7). In one case, the cascade is similar to that caused by the activation of the norepinephrine receptor in some neurons (see Chapter 5). It involves a G-protein-coupled membrane receptor triggering the formation of cAMP within the cytoplasm, which activates protein kinase A (PKA), which phosphorylates a K^+-selective channel (apparently a different one from that involved in sourness), causing a blockade. Once again, the blocking of K^+ channels causes depolarization of the receptor cell.

Some sweet stimuli can activate a second messenger pathway that involves IP_3, similar to the $Bitter_2$ mechanism described next. Finally, there may also be a second transduction mechanism for sweetness that does not involve second messengers. In this case, a set of cation channels may be gated directly by sugars.

Bitterness. Bitter taste receptors are poison detectors. Perhaps because poisons are so chemically diverse, there are several mechanisms for bitter taste transduction (Figure 8.8). Some bitter compounds, such as calcium and quinine ($Bitter_1$ in Figure 8.8), can bind directly to K^+-selective channels and block them (similar to the sourness mechanism).

There are also specific membrane receptor proteins for bitter substances (as in $Bitter_1$ in Figure 8.8), which activate G-protein-coupled second messenger cascades that are different from those of the sweetness mechanism. One type of bitter receptor triggers an increased production of the intracellular messenger inositol triphosphate (IP_3). IP_3 pathways are ubiquitous signaling systems in cells throughout the body, as described in Chapter 6. An unusual feature of the IP_3 bitterness system is that it modulates transmitter release without changing the membrane potential of the receptor cell, because it directly triggers the release of Ca^{2+} from intracellular storage sites. Yet another bitterness receptor seems to *reduce* cAMP levels by stimulating the enzyme that breaks cAMP down.

Amino Acids. "Amino acids" may not be the answer at the tip of your tongue when asked to list your favorite tastes, but recall that proteins are made from amino acids and that they are also excellent energy sources. In short, amino acids are the foods your mother would want you to eat. Many amino acids also taste good, although some taste bitter. Not surprisingly,

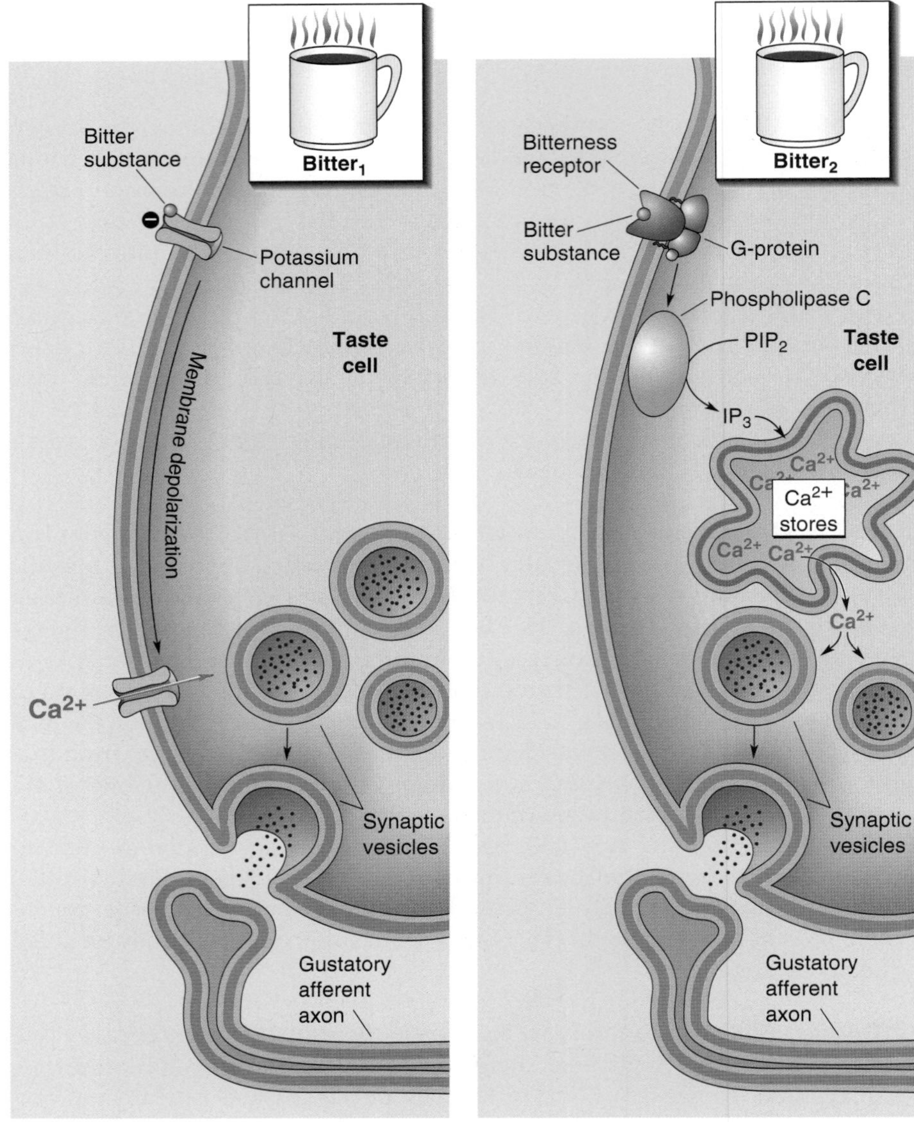

Figure 8.8
Transduction mechanisms for bitter tastants. Tastants can either block a potassium channel ($Bitter_1$) or bind directly to a G-protein-coupled membrane receptor ($Bitter_2$) that triggers IP_3 synthesis and the release of Ca^{2+} from internal storage sites. PIP_2 is the membrane lipid phosphatidylinositol-4,5-bisphosphate.

there are numerous transduction mechanisms for amino acids. The best studied are those for the umami taste, which comes from glutamate or aspartate. There seem to be at least two pathways. Glutamate can directly activate an ion channel that is permeable to the cations Na^+ and Ca^{2+} (Figure 8.9). The resulting inward current causes depolarization, which opens voltage-gated calcium channels and triggers transmitter release; the Ca^{2+} that flows directly through the glutamate channel may also contribute to transmitter release. This is another familiar process, because glutamate-gated channels are part of the most prevalent excitatory neurotransmitter system in the brain.

In the second umami pathway, glutamate binds to a G-protein-coupled receptor (a particular subtype of metabotropic glutamate receptor; see Chapter 5). This receptor probably decreases cAMP levels, which in turn modify some unknown channel. Other amino acids have still other transduction pathways. For example, arginine and proline may gate their own channels. Amino acids that taste bitter (leucine, for example) may trigger other second messenger-mediated systems.

Central Taste Pathways

The main flow of taste information is from taste buds to the primary gustatory axons, into the brain stem, up to the thalamus, and finally to the cerebral cortex (Figure 8.10). Three cranial nerves carry primary gustatory axons and bring taste information to the brain. The anterior two-thirds of the tongue and the palate send axons into the chorda tympani, which is a branch of cranial nerve VII, the *facial nerve*. The posterior third of the tongue is innervated by a branch of cranial nerve IX, the *glossopharyngeal nerve*. The regions around the throat, including the glottis, epiglottis, and pharynx, send taste axons to a branch of cranial nerve X, the *vagus nerve*. These nerves are involved in a variety of other sensory and motor functions, but their taste axons all enter the brain stem, bundle together, and synapse within the slender **gustatory nucleus**, a part of the *solitary nucleus* in the medulla.

Taste pathways diverge from the gustatory nucleus. The conscious experience of taste is presumably mediated by the cerebral cortex. The path to the neocortex via the thalamus is a common one for sensory information. Neurons of the gustatory nucleus synapse on a subset of small neurons in the **ventral posterior medial (VPM) nucleus**, which is a portion of the thalamus that deals with sensory information from the head. The VPM taste neurons then send axons to the **primary gustatory cortex** in Brodmann's area 36 and the insula-operculum regions of cortex. The taste pathways to the thalamus and cortex are primarily ipsilateral to the cranial nerves that supply them. Lesions within the VPM thalamus or the gustatory cortex—as a result of a stroke, for example—can cause *ageusia*, the loss of taste perception.

Gustation is important to basic behaviors such as the control of feeding and digestion, both of which involve additional taste pathways. Gustatory nucleus cells project to a variety of brain stem regions, largely in the medulla, that are involved in swallowing, salivation, gagging, vomiting, and basic physiological functions such as digestion and respiration. In addition, gustatory information is distributed to the hypothalamus and related parts of the basal telencephalon. These structures seem to be involved in the palatability of foods and the forces that motivate us to eat (Box 8.1). Localized lesions of the hypothalamus or amygdala, a nucleus of the basal telencephalon, can cause an animal to overeat chronically, ignore food, or alter its preferences for food types, as we will see in Chapter 16.

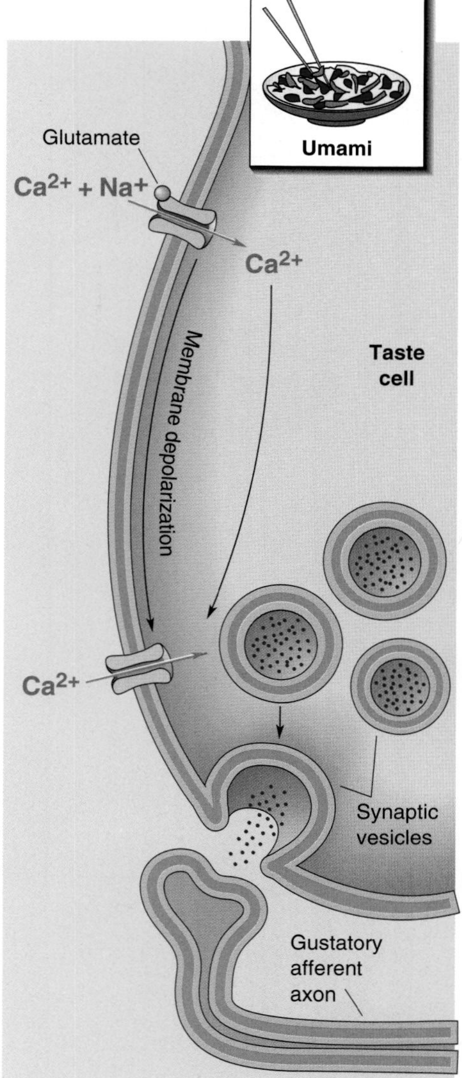

Figure 8.9

Transduction mechanisms for umami tastants (glutamate). Some amino acid tastants can bind to a cation-permeable channel, leading to a change in membrane current and potential, hence the direct entry of Ca^{2+}.

The Neural Coding of Taste

If you were going to design a system for coding tastes, you might begin with many specific taste receptors for many basic tastes (e.g., sweet, sour, salty, bitter, chocolate, banana, mango, beef, Swiss cheese). Then you would connect each receptor type, by separate sets of axons, to neurons in the brain that also responded to only one specific taste. All the way up to the cortex you would expect to find specific neurons responding to "sweet" and "chocolate," and the flavor of chocolate ice cream would involve the rapid firing of these cells and very few of the "salty," "sour," and "banana" cells. This concept is the *labeled line hypothesis*, and it seems simple and rational. Unfortunately, it is incompatible with several facts. As we have seen, individual taste receptor cells tend to be *broadly tuned* to stimuli; that is, they are not very specific in their responses. Primary taste axons are even less specific than receptor cells, and most central taste neurons continue to be broadly responsive all the way into the cortex. In other words, the response of a single taste cell is often ambiguous about the food being tasted; the labels on the taste lines are uncertain rather than distinct.

Cells in the taste system are broadly tuned for several reasons. If one taste receptor cell has several transduction mechanisms, it responds to several types of tastants, although it may still respond most strongly to one or two.

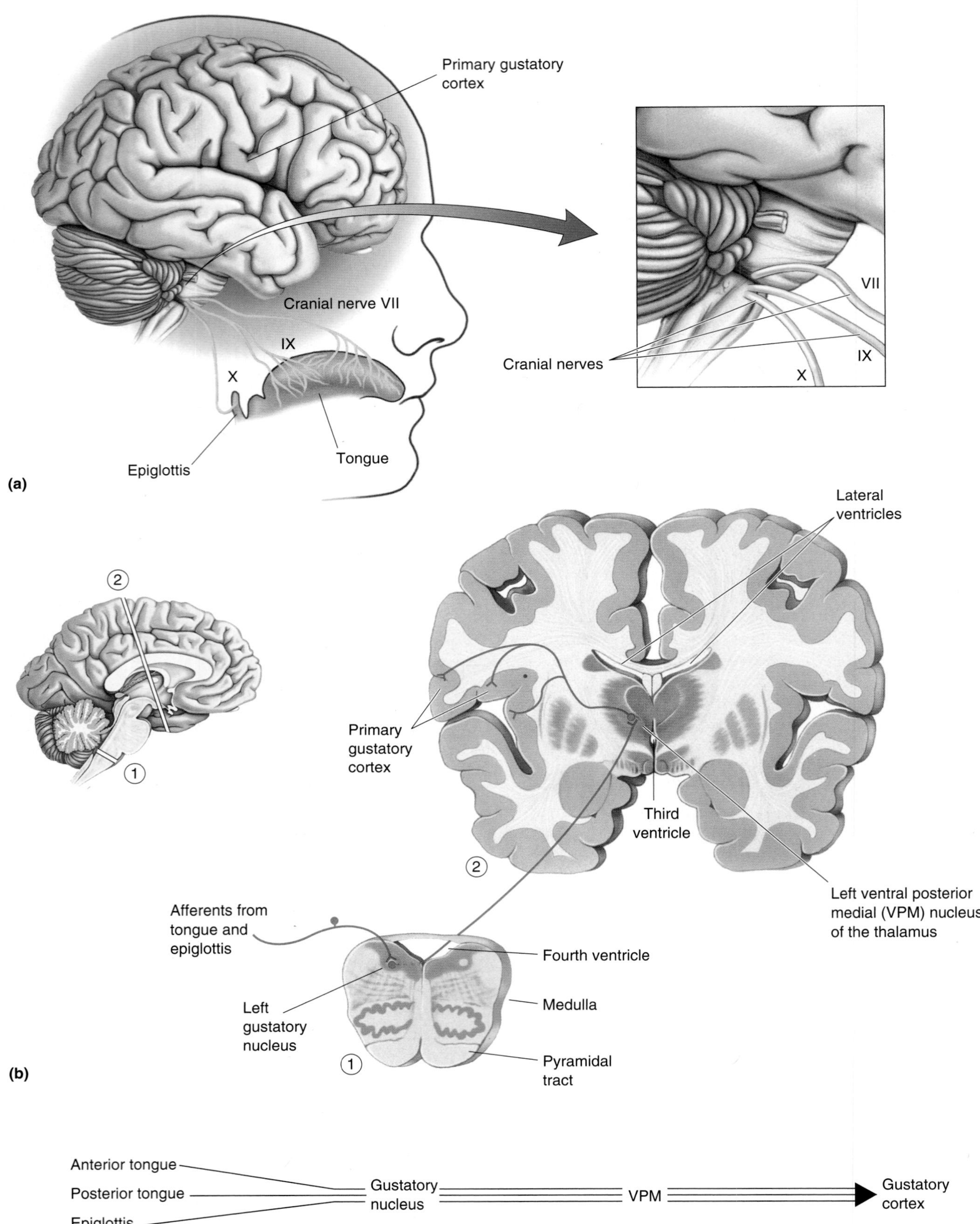
Primary gustatory cortex
Cranial nerve VII
IX
X
Tongue
Epiglottis
(a)
VII
IX
X
Cranial nerves
Lateral ventricles
2
1
Primary gustatory cortex
Third ventricle
2
Left ventral posterior medial (VPM) nucleus of the thalamus
Afferents from tongue and epiglottis
Fourth ventricle
Left gustatory nucleus
Medulla
1
Pyramidal tract
(b)
Anterior tongue
Posterior tongue
Epiglottis
Gustatory nucleus
VPM
Gustatory cortex
(c)

Box 8.1

OF SPECIAL INTEREST

Memories of a Very Bad Meal

When one of us was 14 years old, he ended an entertaining day at an amusement park by snacking on one of his favorite New England foods—fried clams. Within an hour he became nauseated, vomited, and had a most unpleasant bus ride home. Presumably, the clams had spoiled. Sadly, for years afterward he could not even imagine eating another fried clam, and the smell of them alone was repulsive. The fried clam aversion was quite specific. It did not affect his enjoyment of other foods, and he felt no prejudice for amusement park rides, buses, or the friends he had been with the day he got sick.

By the time the author reached his thirties, he could happily dine on fried clams again. He also read about research that John Garcia, working at Harvard Medical School, had done just about the same time as the original bad-clam experience. Garcia fed rats a sweet liquid, and in some cases he then gave them a drug that made them briefly feel ill. After even one such trial, rats that had received the drug avoided the sweet stimulus forever. The rats' aversion was specific for the taste stimulus; they did not avoid sound or light stimuli under the same conditions.

Extensive research has shown that *flavor aversion learning* results in a particularly robust form of associative memory. It is most effective for food stimuli (taste and smell both contribute); it requires remarkably little experience (as little as one trial); and it can last a very long time—more than 50 years in some people. And learning occurs even when there is a very long delay between the food and the nausea. This is obviously a useful form of learning in the wild. An animal can't afford to be a slow learner when new foods might be poisonous. For modern humans, this can backfire; many perfectly good fried clams have remained uneaten. Food aversion can be a more serious problem for patients undergoing radiation therapy or chemotherapy for cancer, when the nausea induced by their treatments makes many foods unpalatable. On the other hand, taste aversion learning has also been used to prevent coyotes from stealing domestic sheep and to help people reduce their dependence on alcohol and cigarettes.

In addition, receptor cell inputs converge onto afferent axons. Each receptor synapses onto a primary taste axon that also receives input from several other receptors, in that papilla as well as its neighbors. This means that one axon may combine the taste information from several papillae. If one of those receptors is mostly sensitive to sour stimuli and another is mostly sensitive to salt stimuli, then the axon will respond to salt *and* sour. This pattern continues into the brain: Neurons of the gustatory nucleus receive synapses from many axons of different taste specificities, and they may become less selective for tastes than the primary taste axons.

All of this mixing of taste information may seem to be an inefficient way to design a coding system. Why not use many taste cells that are highly specific? In part, the answer may be that we would need an enormous variety of receptor types, and even then we could not respond to new tastes. So when you taste chocolate ice cream, how does the brain sort through its apparently ambiguous information about the flavor to make clear distinctions between

Figure 8.10

Central taste pathways. (a) Taste information from different regions of the tongue and mouth cavity is carried by three cranial nerves—VII, IX, and X—to the medulla. **(b)** Gustatory axons enter the small gustatory nucleus within the medulla. Axons from the gustatory nucleus synapse on neurons of the thalamus, and these in turn project to regions of the cerebral cortex that include parts of the postcentral gyrus, the operculum, and the insular cortex. The enlargements show the locations of the planes of section through ① the medulla and ② the forebrain. **(c)** Summary of the central taste pathways.

chocolate and thousands of other possibilities? The likely answer is a scheme that includes features of roughly labeled lines and **population coding**, in which the responses of a large number of broadly tuned neurons, rather than a small number of precisely tuned neurons, are used to specify the properties of a particular stimulus, such as a taste.

Population coding schemes seem to be used throughout the sensory and motor systems of the brain, as we shall see in later chapters. In the case of taste, receptors are not sensitive to all tastes; most respond broadly—to salt and sour but not to bitter and sweet, for example. Only with a large population of taste cells with different response patterns can the brain distinguish between alternative tastes. One food activates a certain subset of neurons, with some of them firing very strongly, some moderately, and some not at all or perhaps even inhibited below their spontaneous firing rates (i.e., their nonstimulated rates); a second food excites some of the cells activated by the first food but also others; and the overall patterns of discharge rates are distinctly different. The relevant population may even include neurons activated by the olfactory, temperature, and textural features of a food; certainly the creamy coldness of chocolate ice cream contributes to our ability to distinguish it from chocolate cake.

Box 8.2

OF SPECIAL INTEREST

Human Pheromones?

Smells are surer than sounds and sights to make your heartstrings crack.

—Rudyard Kipling

Odors can certainly sway emotions and arouse memories, but just how important are they to human behavior? Each of us has a distinctive set of odors that mark our identity as surely as our fingerprints or genes do. In fact, variations in body odor are probably genetically determined. Bloodhounds have great difficulty distinguishing between the smells of identical twins but not between those of fraternal siblings. For some animals, odor identity is essential: When her lamb is born, the ewe establishes a long-term memory of its specific smell and develops an enduring bond based largely on olfactory cues. In a newly inseminated female mouse, the smell of a strange male (but not the smell of her recent mate, which she remembers) will trigger an abortion of the pregnancy.

Humans can recognize the scents of other humans. Infants as young as 6 days old show a clear preference for the smell of their own mother's breast over that of other nursing mothers. The mothers, in turn, can usually identify the odor of their own infant from among several choices.

About 30 years ago, researcher Martha McClintock reported that women who spend a lot of time together (college roommates, for example) often find that their menstrual cycles synchronize. This effect is probably mediated by pheromones. In 1998, McClintock and Kathleen Stern, working at the University of Chicago, found that odorless compounds from one group of women (the "donors") could influence the timing of the menstrual cycles of other women (the "recipients"). Body chemicals were collected by placing cotton pads under the arms of the donors for at least 8 hours. The pads were then wiped under the noses of the recipients, who agreed not to wash their faces for 6 hours. The recipients were not told the source of the chemicals on the pads and did not consciously perceive any odor from them except the alcohol used as a carrier. Nevertheless, depending on the donor's time in her menstrual cycle, the recipient's cycle either shortened or lengthened. These dramatic results are the best evidence yet that humans can communicate with pheromones.

Many animals use the *accessory olfactory system* to detect pheromones and mediate a variety of social behaviors involving mother, mate, territory, and food. The accessory system runs parallel to the primary olfactory system. It consists of a separate chemically sensitive region in the nasal cavity, the *vomeronasal organ*, which projects to the *accessory olfactory bulb*, and from there provides input to the hypothalamus. It had been thought for a long time that the vomeronasal organ in mature humans was absent or vestigial, but recent research indicates that it is present in adults (Figure A). Its precise function in humans is not yet clear.

Napoleon Bonaparte once wrote to his love Josephine, ask-

SMELL

Olfaction brings both good news and bad news. It combines with taste to help us identify foods, and it increases our enjoyment of many of them. But it can also warn of potentially harmful substances (spoiled meat) or places (smoke-filled rooms). In olfaction, the bad news may outweigh the good; by some estimates, we can smell several hundred thousand substances, but only about 20% of them smell pleasant. Practice helps in olfaction, and professional perfumers and whiskey blenders can actually distinguish between as many as 100,000 odors.

Smell is also a mode of communication. Chemicals released by the body, called **pheromones** (from the Greek *pherein*, "to carry," and *horman*, "to excite"), are important signals for reproductive behaviors, and they may also be used to mark territories, identify individuals, and indicate aggression or submission. Although systems of pheromones are well developed in many animals, their importance to humans is not clear (Box 8.2). Research by Martha McClintock, now at the University of Chicago, provides some of the best evidence yet that humans may indeed communicate with pheromones (Box 8.3).

(Box 8.2, *continued*)

ing her not to bathe for the two weeks until they would next meet, so he could enjoy her natural aromas. The scent of a woman may, indeed, be a source of arousal for sexually experienced males, presumably because of learned associations. But there is not yet any hard evidence for human pheromones that might mediate sexual attraction (for members of either sex) via innate mechanisms. Considering the commercial implications of such a substance, you can be sure the search will continue.

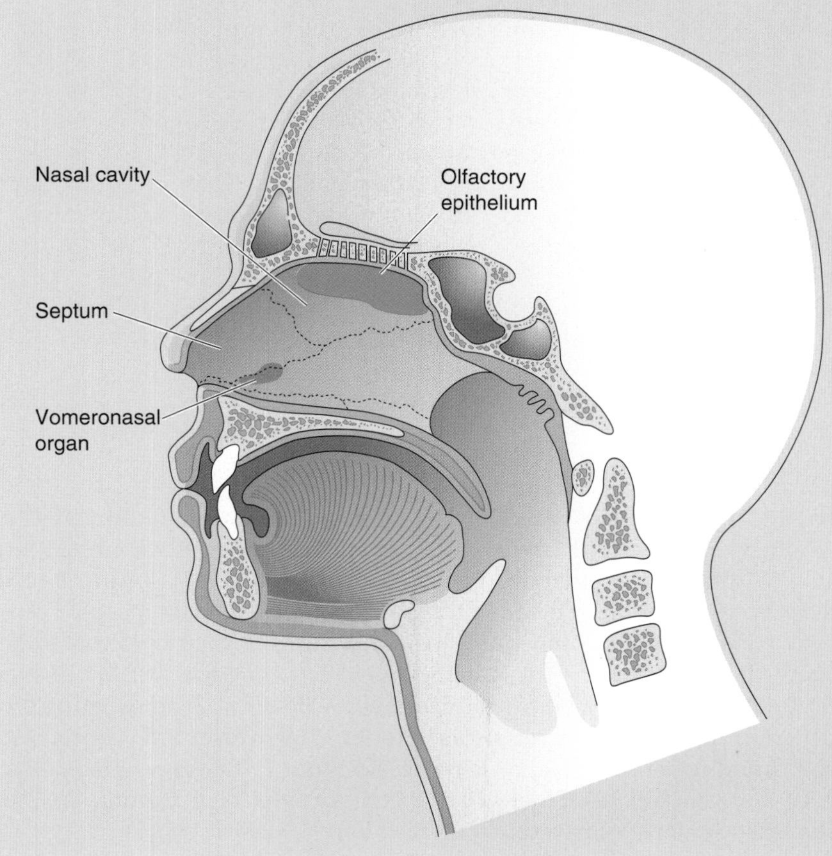

Figure A
Location of the vomeronasal organs, two small pits on either side of the nasal septum.

Box 8.3

PATH OF DISCOVERY

Tales of Two Species: Hormones, Pheromones, and Behavior

BY MARTHA McCLINTOCK

I did my first study in humans on menstrual synchrony and suppression. I had observed these phenomena among my dorm mates in college, and I did the study to convince a skeptical professor that the phenomena could, in fact, occur and were statistically significant. Fortunately, no one told me that it is difficult to get good compliance in human studies. In fact, all 135 dorm members gave me permission to study them, and each gave me complete data for a year!

In graduate school at the University of Pennsylvania, I switched to research with rats, because that was the species used by the people who could give me rigorous training in behavioral neuroscience and endocrinology. (I also worked on vision in goldfish and rhesus monkeys.) Furthermore, there is a huge amount of literature on rats, and I could draw on this published research when I needed to know the basic parameters of systems related to whatever problem I was studying; I didn't have to discover everything myself. Given that I am interested in studying complex systems—namely, the interaction between social, behavioral, and physiological levels of analysis—the existing body of literature made the enterprise feasible. Finally, I was interested in studying feral animals and comparing their behavior to the documented results from studies of laboratory animals. Fortunately, a former student had left his wild rat equipment available for my use, so I was able to study pheromonal and mating systems of both wild and laboratory rats.

Since moving to the University of Chicago, I have continued to do research on both rats and humans. I have found that my initial focus on menstrual synchrony was too narrow and that other manifestations of pheromonal regulation of the timing of ovarian function include asynchrony and cycle stability. Moreover, menstrual synchrony is but one manifestation of a broader phenomenon: Pheromones from a variety of reproductive states, such as pregnancy and lactation, can regulate ovarian function throughout life. It is now clear that there is a great deal more to the story than I had originally reported.

Martha McClintock

In addition, I have branched out into other research areas, studying phenomena that are not the same or even parallel in the two species. Rats are powerful subjects for studies of endocrine mechanisms in the context of sexual and natural selection. Humans are essential for studying the more sophisticated aspects of chemosensory and pheromonal communication, as well as the effects of beliefs and emotions on endocrine function. Nonetheless, all of these studies share the theme of behavioral regulation of fertility, with chemosensory communication being one of the major routes by which social interactions can "get under the skin" and affect neural and endocrine function.

The Organs of Smell

We do not smell with our nose. Rather, we smell with a small, thin sheet of cells high up in the nasal cavity called the **olfactory epithelium** (Figure 8.11a). The olfactory epithelium has three main cell types (Figure 8.11b). *Olfactory receptor cells* are the site of transduction. Unlike taste receptor cells, olfactory receptors are genuine neurons, with axons of their own that penetrate into the central nervous system. *Supporting cells* are similar to glia; among other things, they help produce mucus. *Basal cells* are the source of new receptor cells. Olfactory receptors (similar to taste receptors) continually grow, die, and regenerate in a cycle that lasts about 4–8 weeks. In fact, olfactory receptor cells are one of the very few types of neurons in the nervous system that are regularly replaced throughout life.

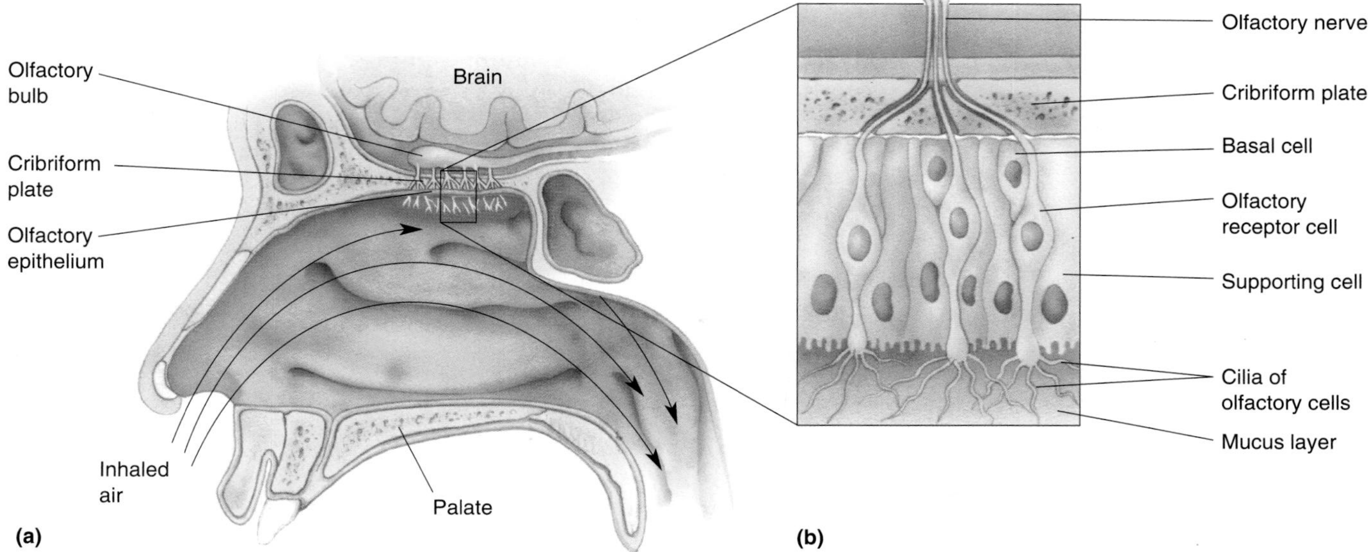

Figure 8.11

(a) Location and (b) structure of the olfactory epithelium. The epithelium consists of a layer of olfactory receptor cells, supporting cells, and basal cells. Odorants dissolve in the mucus layer and contact the cilia of the olfactory cells. Axons of the olfactory cells penetrate the bony cribriform plate on their way to the central nervous system (CNS).

Sniffing brings air through the convoluted nasal passages, but only a small percentage of that air passes over the olfactory epithelium. The epithelium exudes a thin coating of mucus, which flows constantly and is replaced about every 10 minutes. Chemical stimuli in the air, called *odorants*, dissolve in the mucus layer before they reach receptor cells. Mucus consists of a water base with dissolved mucopolysaccharides (long chains of sugars); a variety of proteins including antibodies, enzymes, and odorant binding proteins; and salts. The antibodies are critical because olfactory cells can be a direct route by which some viruses (such as the rabies virus) and bacteria enter the brain. Also important are odorant binding proteins, which are small and soluble and may help concentrate odorants in the mucus.

The size of the olfactory epithelium is one indicator of an animal's olfactory acuity. Humans are relatively weak smellers (although even we can detect some odorants at concentrations as low as a few parts per trillion). The surface area of the human olfactory epithelium is only about 10 cm^2. The olfactory epithelium of certain dogs may be over 170 cm^2, and dogs have more than 100 times more receptors in each square centimeter than humans. By sniffing the aromatic air above the ground, dogs can detect the few molecules left by someone walking there hours before. Humans may only be able to smell the dog when he licks their face.

Olfactory Receptor Neurons

Olfactory receptor neurons have a single thin dendrite that ends with a small knob at the surface of the epithelium (Figure 8.11b). Waving from the knob, within the mucus layer, are several long, thin cilia. Odorants dissolved in the mucus bind to the surface of the cilia and activate the transduction process. On the opposite side of the olfactory receptor cell is a very thin unmyelinated axon. Collectively, the olfactory axons constitute the *olfactory nerve* (cranial nerve I). The olfactory axons do not all come together as a single nerve bundle, as in other cranial nerves. Instead, after leaving the epithelium, small clusters of the axons penetrate a thin sheet of bone called the *cribriform plate,* then course into the **olfactory bulb** (Figure 8.11a). The olfactory axons are fragile, and during traumatic injury, such as a blow to the head, the forces between the cribriform plate and surrounding tissue can permanently sever the olfactory axons. The result is *anosmia,* the inability to smell.

Olfactory Transduction. Although taste receptors use many molecular signaling systems for transduction, olfactory receptors probably use only one (Figure 8.12). All of the transduction molecules are in the thin cilia. The olfactory pathway can be summarized as follows:

Odorants →
- Binding to membrane odorant receptor proteins →
  - G-protein (G_{olf}) stimulation →
    - Activation of adenylyl cyclase →
      - Formation of cyclic AMP →
        - Binding of cAMP to specific cation channel →
          - Opening of cation channels and influx of Na^+ and Ca^{2+} →
            - Opening of Ca^{2+}-activated chloride channels
              - Current flow and membrane depolarization (receptor potential).

Once the cation-selective cAMP-gated channels open, current flows inward, and the membrane of the olfactory neuron depolarizes (Figures 8.12 and 8.13). Besides Na^+, the cAMP-gated channel allows substantial amounts of Ca^{2+} to enter the cilia. In turn, the Ca^{2+} triggers a Ca^{2+}-activated chloride current that may amplify the olfactory receptor potential. (This is a switch from the usual effect of Cl^- currents, which inhibit neurons; in olfactory cells,

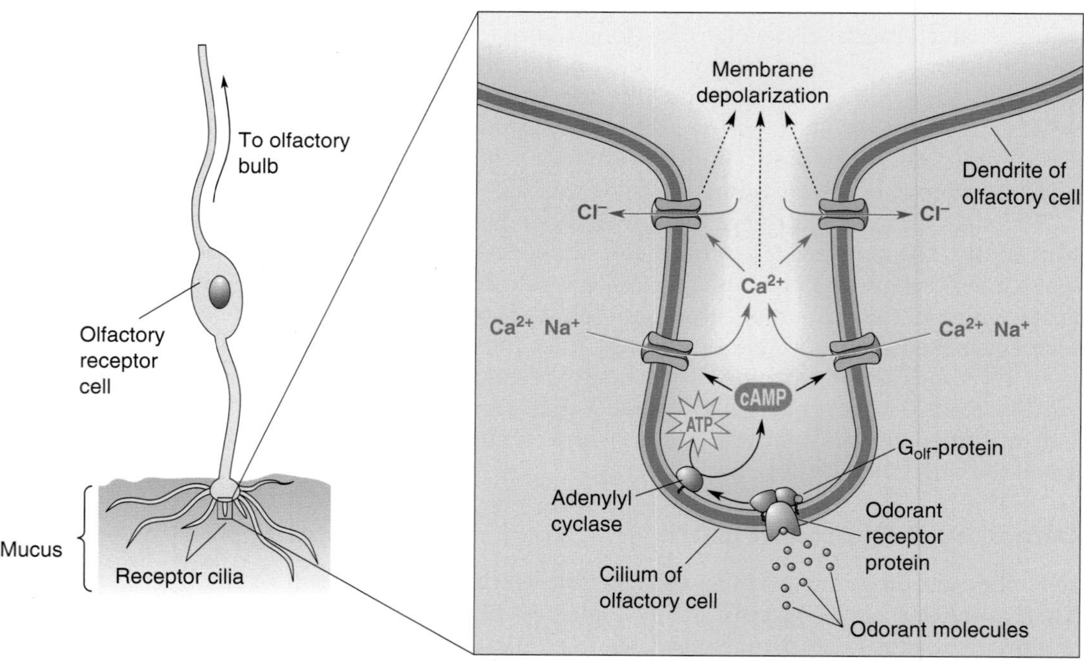

Figure 8.12
Transduction mechanisms of vertebrate olfactory receptor cells. This drawing shows a single cilium of an olfactory receptor cell and the signaling molecules of olfactory transduction that it contains. G_{olf} is a special form of G-protein found only in olfactory receptor cells.

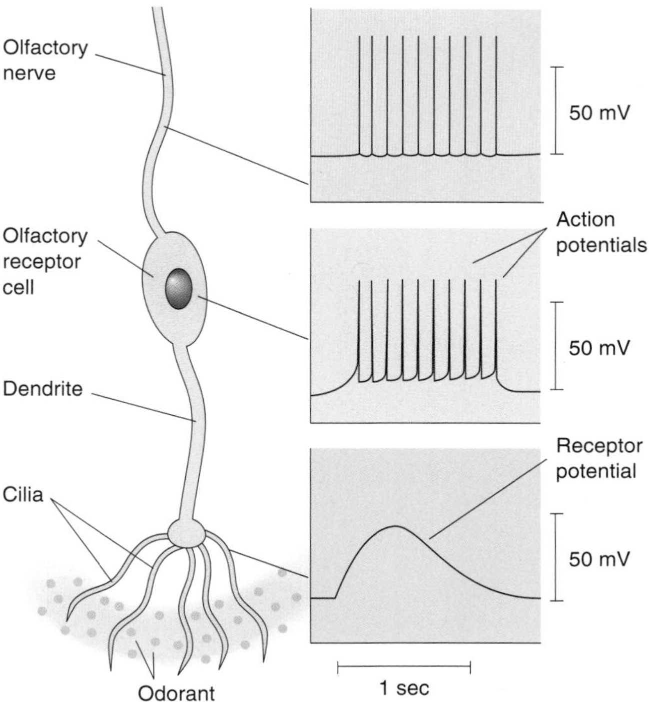

Figure 8.13
Voltage recordings from an olfactory receptor cell during stimulation. Odorants generate a slow receptor potential in the cilia; the receptor potential propagates down the dendrite and triggers a series of action potentials within the soma of the olfactory receptor cell. Finally, the action potentials (but not the receptor potential) propagate continuously down the olfactory nerve axon.

the internal Cl^- concentration must be unusually high so that a Cl^- current tends to depolarize rather than hyperpolarize the membrane.) If the resulting receptor potential is large enough, it will exceed the threshold for action potentials in the cell body, and spikes will propagate out along the axon into the central nervous system (CNS) (Figure 8.13).

The olfactory response may terminate for several reasons. Odorants diffuse away, scavenger enzymes in the mucus layer often break them down, and cAMP in the receptor cell may activate other signaling pathways that end the transduction process. Even in the continuing presence of an odorant, the strength of a smell usually fades. This is because the response of the receptor cell itself adapts to an odorant within about a minute.

This signaling pathway has two unusual features: the receptor binding proteins at the beginning and the cAMP-gated channels near the end. The receptor proteins have odorant binding sites on their extracellular surface. Because you can discriminate thousands of different odorants, you might guess that there are many types of odorant receptor proteins. Your guess would be right, but the surprise is that the number of receptor proteins is very large indeed. Researchers Linda Buck and Richard Axel, working at Columbia University in 1991, found that there are about 1000 odorant receptor genes, making it the largest family of genes yet discovered. Each receptor gene has a slightly different structure from the next, so each receptor protein encoded by these genes differs in its ability to bind odorants. Another surprising fact is that each olfactory receptor cell seems to express only one of the 1000 types of receptor genes. It follows that there are about 1000 types of receptor cells, each identified by the particular receptor gene it has chosen to express. The olfactory receptor sheet is organized into a few large zones, and each zone contains receptor cells that express a different subset of the receptor genes (Figure 8.14). Within each zone, individual receptor types are scattered randomly (Figure 8.15a).

Olfactory receptor proteins belong to the superfamily of proteins whose members have seven transmembrane alpha helices. This superfamily also in-

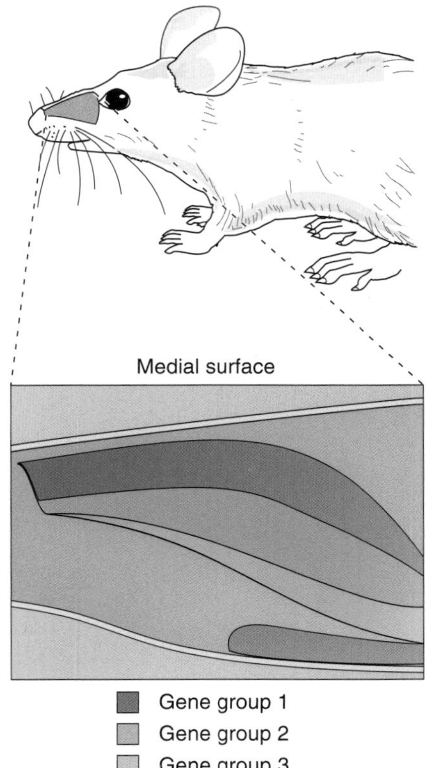

Figure 8.14
Maps of the expression of different olfactory receptor proteins on the olfactory epithelium of a mouse. Three groups of genes were mapped in this case, and each had a separate nonoverlapping zone of distribution. (Source: Adapted from Ressler et al., 1993, p. 602.)

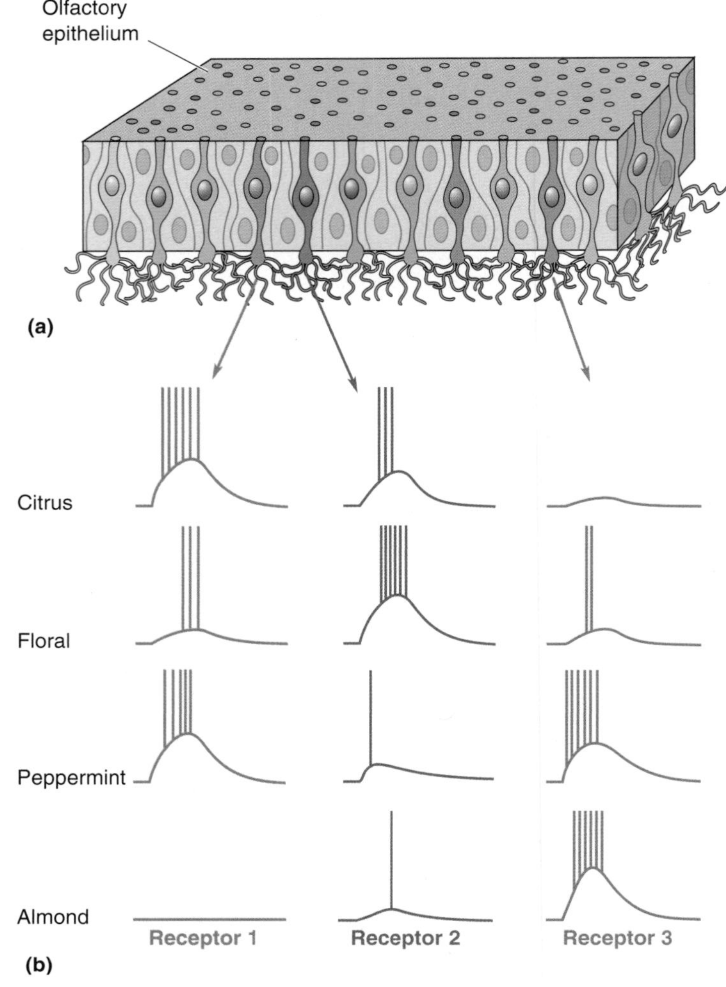

Figure 8.15
Broad tuning of single olfactory receptor cells. (a) Each receptor cell expresses a single olfactory receptor protein (here coded by cell color), and different cells are randomly scattered within a region of the epithelium. **(b)** Microelectrode recordings from three cells show that each one responds to many odors, but with differing preferences. By measuring responses from all three cells, each of the four odors can be clearly distinguished.

cludes a variety of neurotransmitter receptors that were described in Chapter 6. All proteins in the superfamily are coupled to G-proteins, which in turn relay a signal to other second messenger systems within the cell (olfactory receptor cells use a particular type of G-protein called G_{olf}). There is growing evidence that the only second messenger mediating olfactory transduction in vertebrates is cAMP. Some of the most compelling studies have used genetic engineering to produce mice in which critical proteins of the olfactory cAMP pathway have been knocked out (G_{olf}, for example); these mice are inevitably anosmic for a wide variety of odors.

In neurons, cAMP is a common second messenger, but the way it acts in olfactory transduction is quite unusual. Tadashi Nakamura and Geoffrey Gold, working at Yale University in 1987, showed that a population of channels in the cilia of olfactory cells responds directly to cAMP; that is, the channels are cAMP-gated. In Chapter 9, we will see that cyclic nucleotide-gated channels are also used for visual transduction. This is another demonstration that biology is conservative, that evolution recycles its good ideas: smelling and seeing use some very similar molecular mechanisms.

How are the 1000 types of receptor cells used to discriminate between tens of thousands of odors? As with taste, olfaction involves a population coding scheme. Each receptor protein binds different odorants more or less readily, so its receptor cell is more or less sensitive to those odorants (Figure 8.15b).

Some cells are more particular about the chemical structure of the odorants they respond to, but in general each receptor is quite broadly tuned. A corollary of this is that each odorant activates many of the 1000 types of receptors. The concentration of odorant is also important, and more odorant tends to generate stronger responses. Thus, each olfactory cell yields very ambiguous information about odorant type and strength. It is the job of the central olfactory pathways to look at the full package of information arriving from the receptor sheet—the population code—and use it to classify the odors further.

Central Olfactory Pathways

Olfactory receptor neurons send axons into the two olfactory bulbs (Figure 8.16). The bulbs are a neuroscientist's wonderland, full of neural circuits with fascinating dendritic arrangements, unusual reciprocal synapses, and high levels of many neurotransmitters. The input layer of each bulb contains about 2000 spherical structures called **glomeruli** (singular: glomerulus), each about 50–200 μm in diameter. Within each glomerulus, the endings of about 25,000 primary olfactory axons (axons from the receptor cells) converge and terminate on the dendrites of about 100 second-order olfactory neurons.

Recent studies revealed that the mapping of receptor cells onto glomeruli is astonishingly precise. Each glomerulus receives receptor axons from a large region of the olfactory epithelium. When molecular labeling methods are used to tag each receptor neuron expressing one particular receptor gene of the mouse—in this case, a gene called P2—we can see that the P2-labeled axons all converge onto only two glomeruli in each bulb, one of which is shown in Figure 8.17a. No axons seem to be out of place; such accuracy challenges our knowledge about axonal pathfinding during development (see Chapter 22).

This precision mapping is also consistent across the two olfactory bulbs; each bulb has only two P2-targeted glomeruli in symmetrical positions (Figure 8.17b). The positions of the P2 glomeruli within each bulb are even consistent from one mouse to another. Finally, it seems that each glomerulus receives input from *only* receptor cells of one particular type. This means that

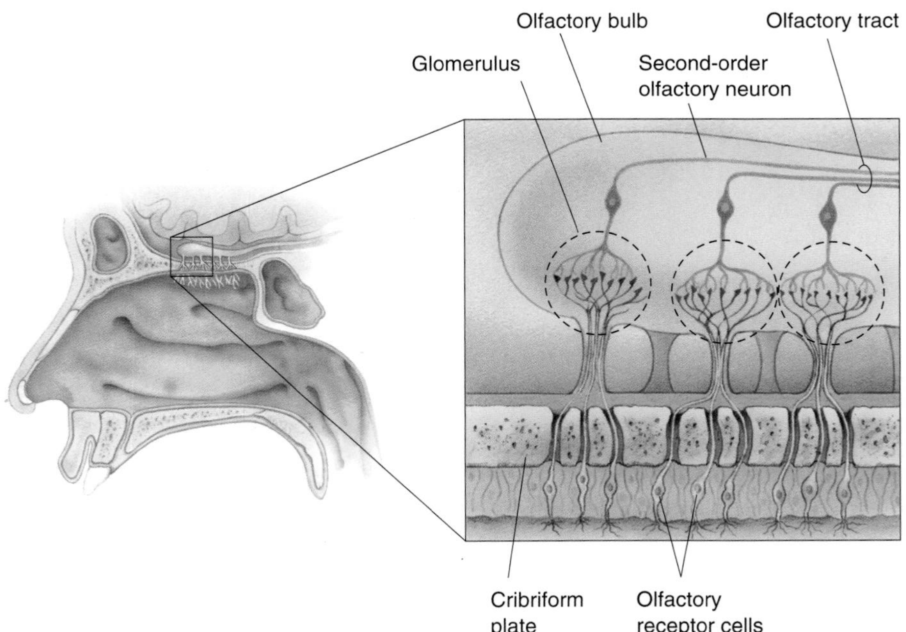

Figure 8.16
Structure of an olfactory bulb. Axons of olfactory receptor cells penetrate the cribriform plate and enter the olfactory bulb. After multiple branching, each olfactory axon synapses on second-order neurons within a spherical glomerulus. The second-order neurons send axons through the olfactory tract further into the brain.

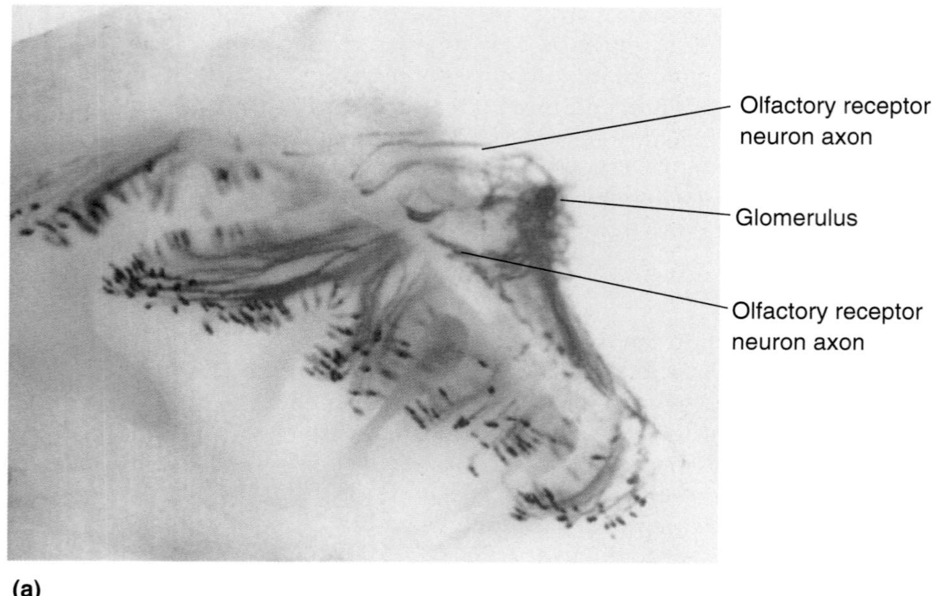

(a)

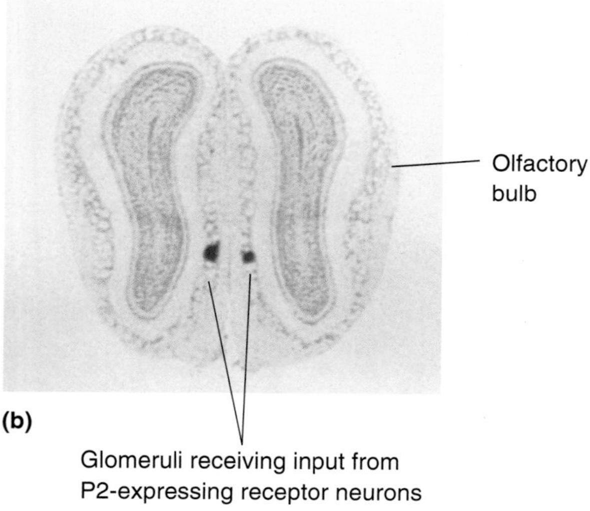

(b)

Figure 8.17

Convergence of olfactory nerve axons onto the olfactory bulb. Olfactory receptor neurons expressing a particular receptor gene all send their axons to the same glomeruli. **(a)** In a mouse, receptor neurons expressing the P2 receptor gene were labeled blue, and every neuron sent its axon to the same glomerulus in the olfactory bulb. **(b)** When the two bulbs were cut in cross section, it was possible to see that the P2-containing receptor axons project to symmetrically placed glomeruli in each bulb. (Source: Adapted from Mombaerts et al., 1996, p. 680.)

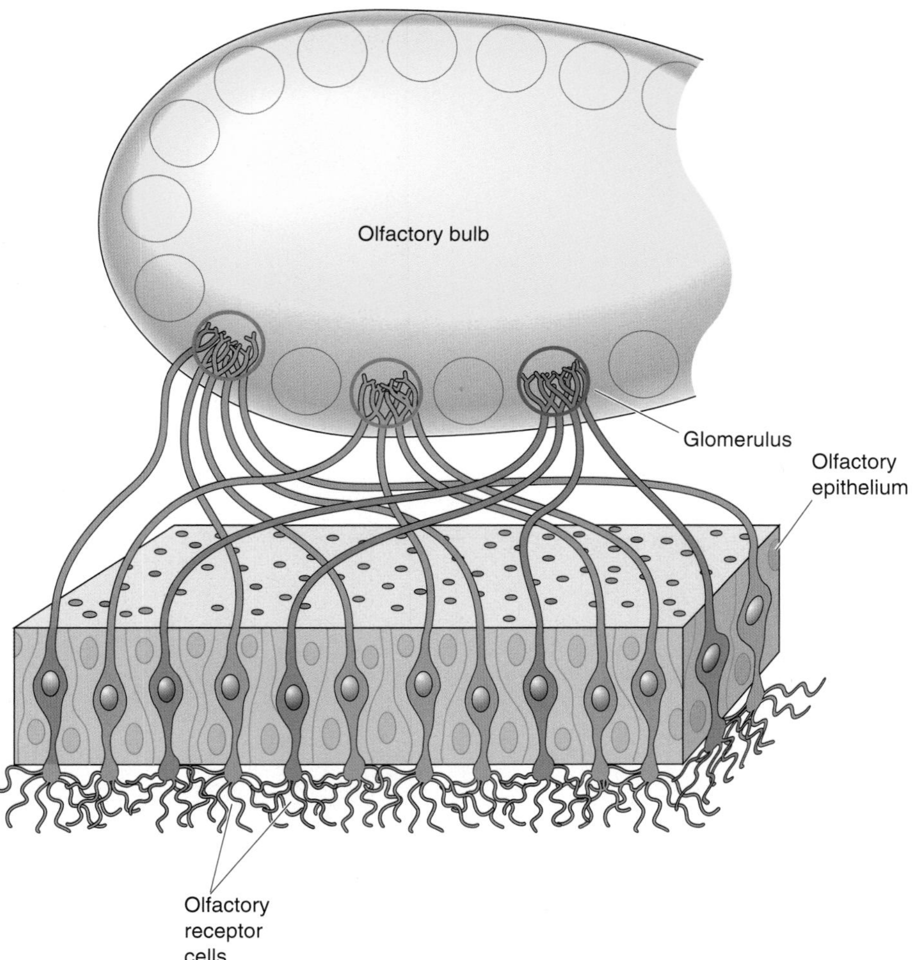

Figure 8.18
Specific mapping of olfactory receptor neurons onto glomeruli. Each glomerulus receives input only from receptor cells expressing a particular receptor protein gene. Receptor cells expressing a particular gene are color-coded.

the array of glomeruli within a bulb is a very orderly map of the receptor genes expressed in the olfactory epithelium (Figure 8.18) and, by implication, a map of odor information.

Olfactory information is modified by inhibitory and excitatory interactions within and among the glomeruli and between the two bulbs. Neurons in the bulbs are also subject to modulation from systems of axons that descend from higher areas of the brain. While it is obvious that the elegant circuitry of the olfactory bulbs has important functions, it is not entirely clear what those functions are. It is likely that they begin to segregate odorant signals into broad categories, independent of their strength and possible interference from other odorants. The precise identification of an odor probably requires further processing in the next stages of the olfactory system.

The output axons of the olfactory bulbs course through the olfactory tracts. Each olfactory tract projects directly into the primitive regions of the cerebral cortex; from here information passes to the thalamus and finally on to the neocortex. This anatomy makes olfaction unique. All other sensory systems *first* pass through the thalamus before projecting to the cerebral cortex. Among other things, the olfactory arrangement gives an unusually direct and widespread influence on the parts of the forebrain that have roles in emotion, motivation, and certain kinds of memory (see Chapters 16, 18, 23, and 24). A partial summary of olfactory pathways begins with axons from the olfactory bulbs, which project to several areas of the olfactory cortex,

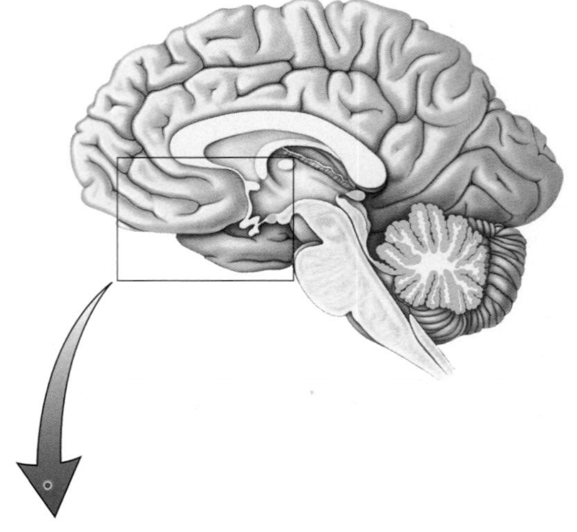

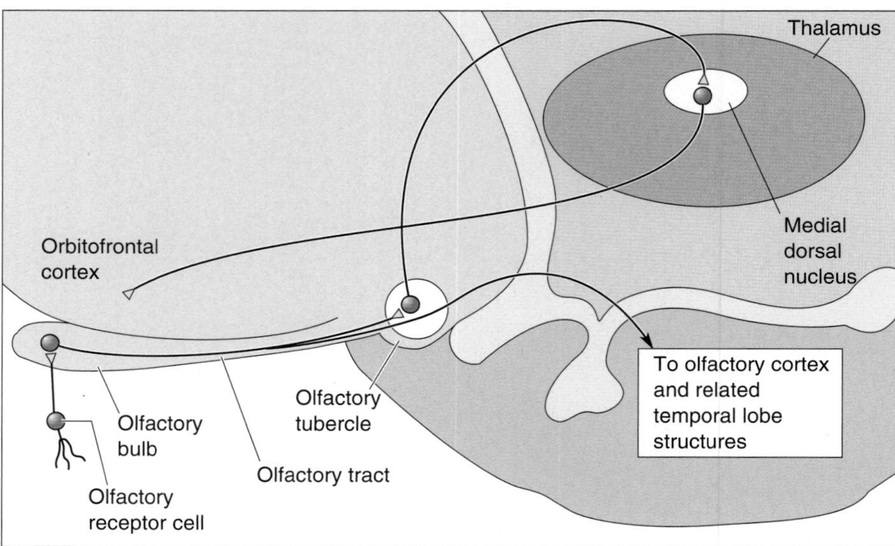

Figure 8.19
Central olfactory pathways. Axons of the olfactory tract branch and enter many regions of the forebrain, including the olfactory cortex. The neocortex is reached only by a pathway that synapses in the medial dorsal nucleus of the thalamus.

which in turn relay the information on to other structures. So many structures receive olfactory connections that listing those parts of the brain to which olfactory projections do *not* have access might be easier. As a general principle, several parallel pathways mediate different olfactory functions, including odor discrimination and conscious perception, motivational and emotional features, behaviors as disparate as reproduction and feeding, and imprinting and memory. Conscious perceptions of smell may be mediated by a path from the *olfactory tubercle* to the *medial dorsal nucleus* of the thalamus to the *orbitofrontal cortex* (situated right behind the eyes) (Figure 8.19).

Spatial and Temporal Representations of Olfactory Information

In olfaction, we are faced with a paradox similar to the one in gustation. Individual receptors are broadly tuned to their stimuli; that is, each cell is sensitive to a wide variety of chemicals. When we smell those same chemicals, however, we can easily tell them apart. How is the whole brain doing what single olfactory cells cannot? We will discuss three important ideas: (1) Each odor is represented by the activity of a large population of neurons; (2)

the neurons responsive to particular odors may be organized into spatial maps; and (3) the timing of action potentials may be an essential code for particular odors.

Olfactory Population Coding. As in gustation, the olfactory system uses the responses of a large population of receptors to encode a specific stimulus. A simplistic example was shown in Figure 8.15b. When presented with a citrus smell, none of the three receptor cells can clearly distinguish it from the other odors. But by looking at the *combination* of responses from all three cells, the brain could distinguish the citrus smell unambiguously from floral, peppermint, and almond. By using such population coding, you can imagine how an olfactory system with 1000 different receptors might be able to recognize tens of thousands of odors.

Olfactory Maps. A **sensory map** is an orderly arrangement of neurons that correlates with certain features of the environment. Microelectrode recordings show that many receptor neurons respond to the presentation of a single odorant and that these cells are distributed across a wide area of the olfactory epithelium (Figure 8.15). This is consistent with the widespread distribution of each receptor gene. We have seen, however, that the axons of each receptor cell type synapse upon particular glomeruli in the olfactory bulbs. This arrangement yields a sensory map according to which neurons in a specific place in the bulb respond to particular odors. The maps of regions activated by one chemical stimulus can be visualized with special recording methods. Such experiments reveal that while many bulb neurons are activated by one odor, the neurons' positions form complex but reproducible *spatial* patterns, as shown in Figure 8.20. Thus, the smell of a particular chemical is converted into a specific map within the "neural space" of the bulbs, and the form of the map depends on the nature and concentration of the odorant.

You will see in subsequent chapters that every sensory system uses spatial maps, perhaps for many purposes. In most cases, the maps correspond obviously to features of the sensory world. For example, in the visual system, there are maps of visual space, and in the somatic sensory system, there are maps of the body surface. The maps of the chemical senses are unusual in that the stimuli themselves have no meaningful spatial properties. Although seeing a skunk walking in front of you may tell you *what* and *where* it is, smell by itself can reveal only the *what*. (By moving the head about, it is possible to localize smells only crudely.) Because the olfactory system does not have to map the spatial pattern of an odor in the same way that the visual system has to map the spatial patterns of light, neural odor maps may be available for other purposes, such as discrimination between a huge number of different chemicals.

But does the brain actually use neural odor maps to distinguish between chemicals? We don't know. For a map to be useful, there must be something that reads and understands it. With practice and very specialized goggles, we

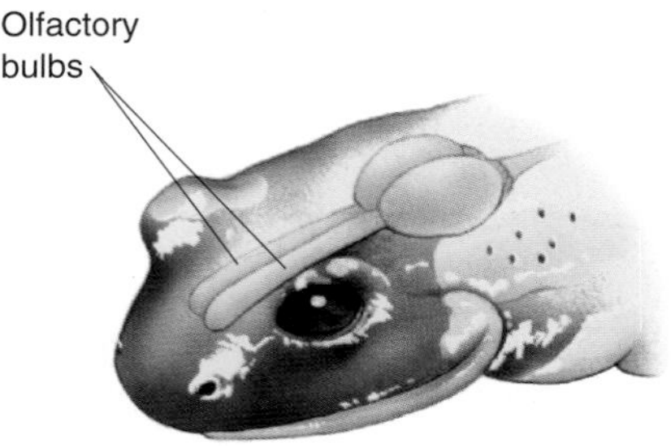

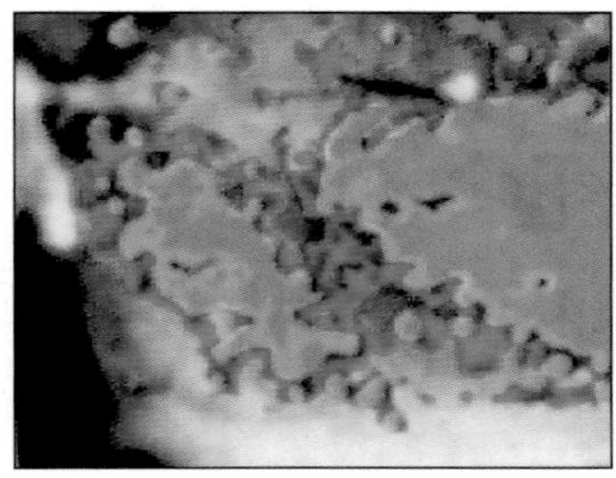
(a)

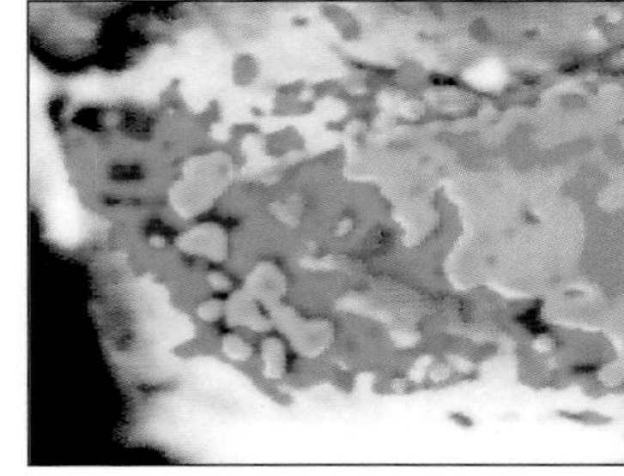
(b)

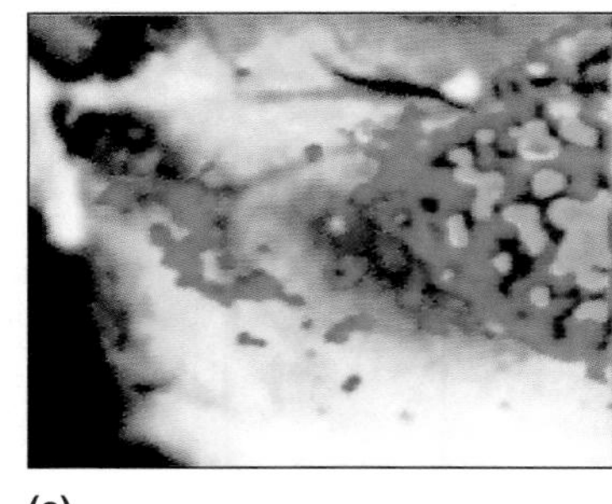
(c)

Figure 8.20
Maps of neural activation of the olfactory bulb. The activity of many olfactory neurons in a salamander olfactory bulb was recorded with a specialized optical method. The cells were stained with dyes that are sensitive to membrane voltage, and neural activity was signaled by changes in the amount of light emitted by the dye. The colors on the maps represent various levels of neural activity; hotter colors imply more activity. Different olfactants evoked different spatial patterns of neural activation in the bulb: **(a)** amyl acetate (banana), **(b)** limonene (citrus), **(c)** ethyl *n*-butyrate (pineapple). (Source: Adapted from Kauer, 1991, p. 82.)

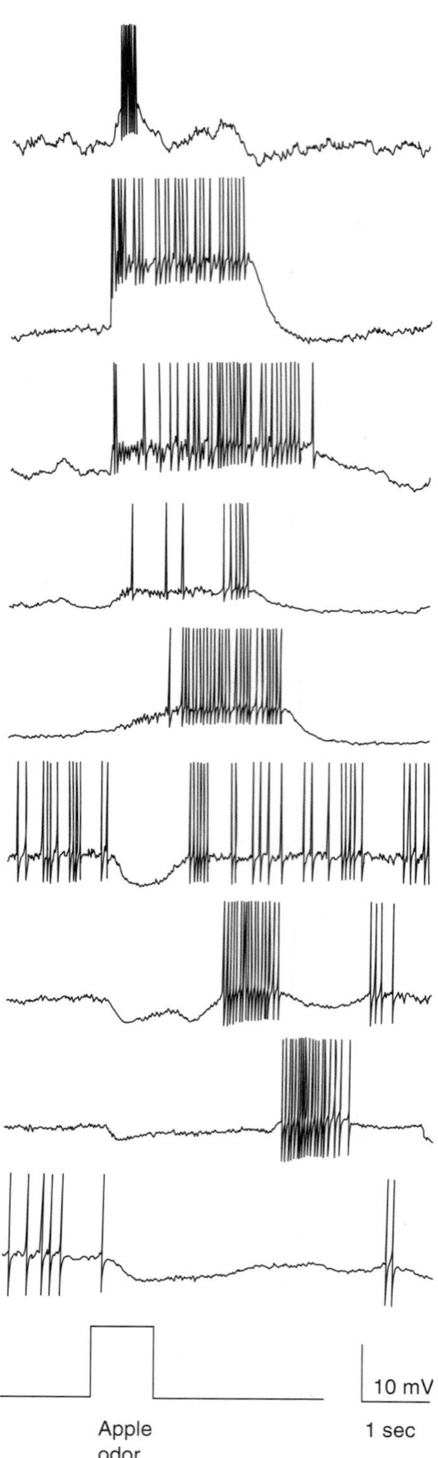

Figure 8.21
Temporal spiking patterns. The odor of apple produces a range of temporal spiking patterns in nine olfactory neurons. These recordings are from neurons in the antenna lobe of a locust. (Source: Laurent et al., 1996, p. 3839.)

might be able to read the "alphabet" of odors mapped on the surface of the olfactory bulb with our eyes. This may roughly approximate what higher regions of the olfactory system do, but so far there is no evidence that the olfactory cortex has this capability. An alternative idea is that spatial maps do not encode odors at all but are simply the most efficient way for the nervous system to form appropriate connections between related sets of neurons (e.g., receptor cells and glomerular cells). With orderly mapping, the lengths of axons and dendrites can be minimized. Neurons with similar functions that must interconnect can also find one another more easily if they are neighbors. The spatial map that results may be simply a side effect of these developmental requirements rather than a fundamental mechanism of sensory coding itself.

Temporal Coding in the Olfactory System. There is growing evidence that the *temporal* patterns of spiking are essential features of olfactory coding. Compared to many sounds and sights, odors are inherently slow stimuli, so the rapid timing of action potentials is not necessary to encode the timing of odors. **Temporal coding,** which depends on the timing of spikes, may instead encode the quality of odors. Hints about the possible importance of timing are easy to find. Researchers have known for many decades that the olfactory bulb and cortex generate oscillations of activity when odors are presented to the receptors, but the relevance of these rhythms is still unknown. Temporal patterns are also evident in the spatial odor maps, as they sometimes change shape during the presentation of a single odor.

Recent work by Gilles Laurent and his colleagues, working at the California Institute of Technology, has provided some of the most convincing evidence for temporal odor codes. Recording from the olfactory systems of insects, which have a neural organization somewhat analogous to the vertebrate olfactory system, the researchers found that one odor generates a wide range of temporal spiking patterns in various central olfactory neurons (Figure 8.21). By analyzing the detailed timing of spikes within cells and between groups of cells, they showed that odor information is encoded by the number, temporal pattern, rhythmicity, and cell-to-cell synchrony of spikes.

As with the spatial maps, however, demonstrating that information is carried by spike timing is only a first step; proving that the brain actually *uses* that information is much more difficult. In a fascinating experiment with honeybees, Laurent and his colleagues were able to disrupt the rhythmic synchrony of odor responses without otherwise affecting their spiking responses. This loss of synchronous spiking was associated with a loss of the bees' ability to discriminate between similar odors although not between broad categories of odors. The implication is that the bee analyzes an odor not only by keeping track of *which* olfactory neurons fire but also by *when* they fire. It will be interesting to see whether similar processes occur in a mammalian olfactory system.

CONCLUDING REMARKS

The chemical senses are a good place to begin learning about sensory systems. Smell and taste are the most basic of sensations. Gustation and olfaction use a variety of transduction mechanisms to recognize the astronomical number of chemicals we encounter in the environment. Yet the molecular mechanisms of transduction are very similar to the signaling systems used in every cell of the body for functions as diverse as neurotransmission and fertilization. We will see that the transduction mechanisms in other sensory systems, although highly specialized, also derive from common cellular processes. Remarkable parallels, such as the molecular similarity between the sensory cells of smelling and seeing, have been discovered.

Common sensory principles also extend to the level of neural systems. Most sensory cells are broadly tuned for their stimuli. This means that the nervous system must use population codes to represent and analyze sensory information, resulting in our remarkably precise and detailed perceptions. Populations of neurons are often arranged in sensory maps within the brain. And the timing of action potentials may be used to represent sensory information in ways that are not yet understood. In the chapters that follow, we will see trends in the anatomy and physiology of systems dealing with light, sound, and pressure.

KEY TERMS

Introduction
gustation (p. 255)
olfaction (p. 255)
chemoreceptor (p. 255)

Taste
papilla (p. 256)
taste bud (p. 256)
taste receptor cell (p. 256)
receptor potential (p. 258)
transduction (p. 258)
gustatory nucleus (p. 263)
ventral posterior medial (VPM) nucleus (p. 263)
primary gustatory cortex (p. 263)
population coding (p. 266)

Smell
pheromone (p. 267)
olfactory epithelium (p. 268)
olfactory bulb (p. 270)
glomerulus (p. 273)
sensory map (p. 277)
temporal coding (p. 278)

REVIEW QUESTIONS

1. Most tastes are some combination of the five basic tastes. What other sensory factors can help define the specific perceptions associated with a particular food?
2. The transduction of saltiness is accomplished, in part, by a Na^+-permeable channel. Why would a sugar-permeable membrane channel be a poor mechanism for the transduction of sweetness?
3. Some tastants apparently cause a hyperpolarization of the receptor cell membrane. Do the transduction mechanisms outlined in this chapter suggest ways that the cells might hyperpolarize? If not, suggest two mechanisms that may account for a negative shift in the membrane potential.
4. Why would the size of an animal's olfactory epithelium and, consequently, the number of receptor cells be related to its olfactory acuity?
5. Receptor cells of the gustatory and olfactory systems undergo a constant cycle of growth, death, and maturation. Therefore, the connections they make with the brain must be continually renewed as well. Can you propose a set of mechanisms that would allow the connections to be remade in a specific way again and again over the course of an entire lifetime?
6. If the olfactory system does use some kind of spatial mapping to encode specific odors, how may the rest of the brain read the map?

CHAPTER

9 The Eye

INTRODUCTION

Vision is remarkable because it lets us detect things as tiny and close as a mosquito on the tip of our nose or as immense and far away as a galaxy near the fringes of the universe. Sensitivity to light enables animals, including humans, to detect prey, predators, and mates. The success of the visual process requires that light reflected off distant objects be localized in reference to the individual and the environment; that some form of object identification based on size, shape, color, and experience take place; that movement of objects be detected; and that recognition of objects be possible in the full range of lighting conditions normally experienced by the individual in its habitat.

Light is electromagnetic energy that is emitted in the form of waves. We live in a turbulent sea of electromagnetic radiation. Like any ocean, this sea has large waves and small waves, short wavelets and long rollers. The waves crash into objects and are absorbed, scattered, reflected, and bent. Because of the nature of electromagnetic waves and their interactions with the environment, the visual system can extract information about the world. This is a big job, and it requires a lot of neural machinery. However, the mastery of vision over the course of vertebrate evolution has had surprising rewards. It has provided new ways to communicate, given rise to brain mechanisms for predicting the trajectory of objects and events in time and space, allowed for new forms of mental imagery and abstraction, and led to the creation of a world of art. The significance of vision is perhaps best attested to by the fact that about half of the human cerebral cortex is involved with analyzing the visual world.

The mammalian visual system begins with the eye. At the back of the eye is the **retina**, which contains photoreceptors specialized to convert light energy into neural activity. The rest of the eye acts like a camera, forming crisp, clear images of the world on the retina. Like a quality 35-mm camera, the eye automatically adjusts to differences in illumination and automatically focuses itself on objects of interest. The eye has some additional features not yet available on cameras, such as the ability to track moving objects (by eye movement) and the ability to keep its transparent surfaces clean (by tears and blinking).

While much of the eye functions like a camera, the retina is much more than film. In fact, as mentioned in Chapter 7, the retina is actually part of the brain. (Think about that the next time you look deep into someone's eyes.) In a sense, each eye has two overlapping retinas: one specialized for low light levels that we encounter from dusk to dawn; and another specialized for higher light levels and for the detection of color, from sunrise to sunset. Regardless of the time of day, however, the output of the retina is not a faithful reproduction of the intensity of the light falling on it. Rather, the retina is specialized to detect *differences* in the intensity of light falling on different parts of it, almost regardless of the absolute intensity. This capability probably accounts for the fact that print looks black on a white page both outside on a sunny day and inside a dimly lit room. Image processing is well under way in the retina before any visual information reaches the rest of the brain.

Axons of retinal neurons are bundled into optic nerves, which distribute visual information in the form of action potentials to several brain structures that perform various functions. Some targets of the optic nerves are involved in regulating biological rhythms, which are synchronized with the light-dark daily cycle; others are involved in the control of eye position and optics. The first synaptic relay in the pathway that serves visual perception occurs, however, in a cell group of the dorsal thalamus called the *lateral geniculate nucleus*, or *LGN*. From the LGN, visual information ascends to the cerebral cortex, where it is interpreted and remembered.

In this chapter, we explore the eye and the retina. We shall see how light carries information to our visual system, how the eye forms images on the retina, and how the retina converts light energy into neural signals that can be used to extract information about luminance and color differences. In Chapter 10, we will pick up the visual pathway at the back of the eye and take it through the thalamus to the cerebral cortex.

PROPERTIES OF LIGHT

The visual system uses light to form images of the world around us. Let's briefly review the physical properties of light and its interactions with the environment.

Light

Light is the electromagnetic radiation that is visible to our eyes. Electromagnetic radiation can be described as a wave of energy. Like any wave, electromagnetic radiation has a *wavelength*, the distance between successive peaks or troughs; *frequency*, the number of waves per second; and *amplitude*, the difference between wave trough and peak (Figure 9.1).

The energy content of electromagnetic radiation is proportional to its frequency. Radiation emitted at a high frequency (short wavelengths) has the highest energy content; examples are gamma radiation and X-rays, with wavelengths less than 10^{-9} m (<1 nm). Conversely, radiation emitted at lower frequencies (longer wavelengths) has less energy; examples are radar and radio waves, which have wavelengths greater than 1 mm. Only a small part of the electromagnetic spectrum is detectable by our visual system; visible light consists of wavelengths of 400–700 nm (Figure 9.2). As first shown by Isaac Newton early in the eighteenth century, the mix of wavelengths in this range emitted by the sun appears to humans as white, whereas light of a single wavelength appears as one of the colors of the rainbow. A "hot" color, such as red or orange, consists of light with a longer wavelength and, hence, has *less* energy than a "cool" color, such as blue or violet. Clearly, colors are themselves "colored" by the brain, based on our subjective experiences.

Optics

In a vacuum, a wave of electromagnetic radiation travels in a straight line and thus can be described as a *ray*. Light rays in our environment also travel in straight lines until they interact with the atoms and molecules of the atmosphere and objects on the ground. These interactions include reflection, absorption, and refraction (Figure 9.3). The study of light rays and their interactions is called *optics*.

Reflection is the bouncing of light rays off a surface. The manner in which a ray of light is reflected depends on the angle at which it strikes the surface. A ray striking a mirror perpendicularly is reflected 180 degrees back upon itself, a ray striking the mirror at a 45 degree angle is reflected 90 degrees, and so on. Most of what we see is light reflected off objects in our environment.

Absorption is the transfer of light energy to a particle or surface. You can feel this energy transfer on your skin on a sunny day, as visible light is ab-

Figure 9.1
Characteristics of electromagnetic radiation.

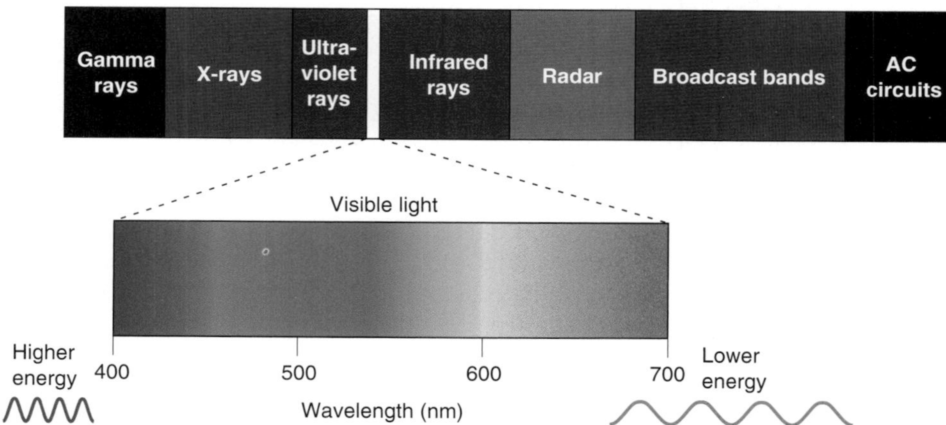

Figure 9.2
The electromagnetic spectrum. Only electromagnetic radiation with wavelengths of 400–700 nm is visible to the naked human eye. Within this visible spectrum, different wavelengths appear as different colors.

sorbed and warms you up. Surfaces that appear black absorb the energy of all visible wavelengths. Some compounds absorb light energy only in a limited range of wavelengths, then reflect the remaining wavelengths. This property is the basis for the colored pigments of paints. For example, a blue pigment absorbs long wavelengths but reflects a range of short wavelengths centered on 430 nm that are perceived as blue. As we will see in a moment, light-sensitive photoreceptor cells in the retina contain pigments and use the energy absorbed from light to generate changes in membrane potential.

Images are formed in the eye by **refraction**, the bending of light rays that can occur when they travel from one transparent medium to another. Consider a ray of light passing from the air into a pool of water. If the ray strikes the water surface perpendicularly, it will pass through in a straight line. If light strikes the surface at an angle, however, it will bend toward a line that is perpendicular to the surface. This bending of light occurs because the speed of light differs in the two media: Light passes through air more rapidly than through water, and the greater the difference between the speed of light in the two media, the greater the angle of refraction. The transparent media in the eye bend light rays to form images on the retina.

THE STRUCTURE OF THE EYE

The eye is an organ specialized for the detection, localization, and analysis of light. Here we introduce the structure of this remarkable organ in terms of its gross anatomy, ophthalmoscopic appearance, and cross-sectional anatomy.

Gross Anatomy of the Eye

When you look into someone's eyes, what are you really looking at? The main structures are shown in Figure 9.4. The **pupil** is the opening that allows light to enter the eye and reach the retina; it appears dark because of the light-absorbing pigments in the retina. The pupil is surrounded by the **iris**, whose pigmentation provides what we call the eye's color. The iris contains two muscles that can vary the size of the pupil; one makes it smaller when it contracts, while the other makes it larger. The pupil and iris are covered by the glassy transparent external surface of the eye, the **cornea**. The cornea lacks blood vessels and is nourished by the fluid behind it, the **aqueous humor**, and, on its outside surface, by the tear film that is continuously replenished by the blinking of the eyelids. The cornea is continuous with the **sclera**, the "white of the eye," which forms the tough wall of the eyeball. Inserted into the sclera are three pairs of muscles, the **extraocular muscles**, which move

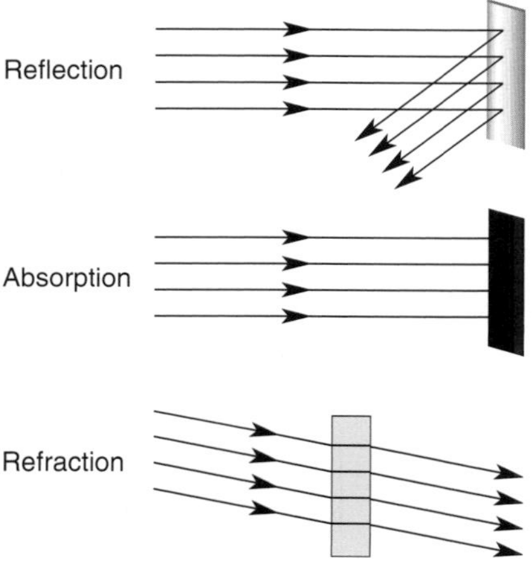

Figure 9.3
Some interactions between light and the environment.

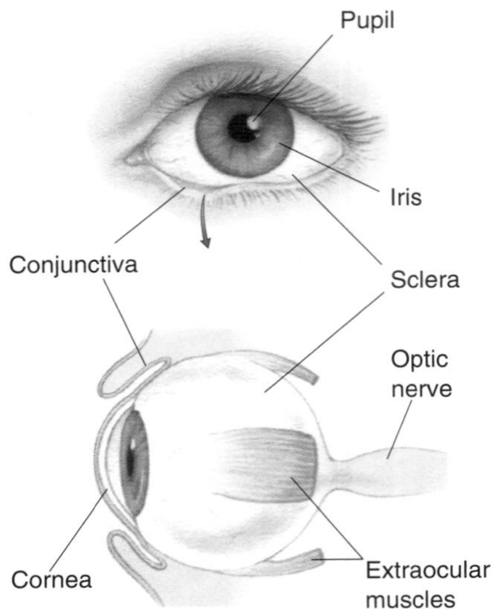

Figure 9.4
Gross anatomy of the human eye.

the eyeball in the bony orbits of the skull. These muscles normally are not visible because they lie behind the **conjunctiva**, a membrane that folds back from the inside of the eyelids and attaches to the sclera. The **optic nerve**, carrying axons from the retina, exits the eye at the back, passes through the orbit, and reaches the brain at its base, near the pituitary gland.

Ophthalmoscopic Appearance of the Eye. Another view of the eye is afforded by the ophthalmoscope, a device that enables one to peer into the eye through the pupil to the retina (Figure 9.5). The most obvious feature of the retina viewed through an ophthalmoscope is the blood vessels on its surface. These retinal vessels originate from a pale circular region called the **optic disk**, which is also where the optic nerve fibers exit the retina. The optic disk is sometimes called the optic nerve head.

It is interesting to note that the sensation of light cannot occur at the optic disk because there are no photoreceptors here, nor can it occur where the large blood vessels exist because the vessels cast shadows on the retina. Nevertheless, our perception of the visual world appears seamless. We are not aware of any holes in our field of vision because the brain fills in our perception of these areas. (They are not *actually* filled in—we fool ourselves into thinking they are.) There are tricks by which we can demonstrate the "blind" retinal regions (Box 9.1).

At the middle of each retina is a darker region with a yellowish hue. This is the **macula** (Latin for "spot"), the part of the retina for central (as opposed to peripheral) vision. Besides its color, the macula is distinguished by the relative absence of large blood vessels. Notice in Figure 9.5 that the vessels arc from the optic disk to the macula; this is also the trajectory of the optic nerve fibers from the macula en route to the optic nerve head. The relative absence of large blood vessels in this region of the retina is one of the specializations that improves the quality of central vision. Another specialization of the central retina that can sometimes be discerned with the ophthalmoscope is the **fovea**, a dark spot about 2 mm in diameter. *Fovea* is Latin for "pit," and the retina is thinner in the fovea than elsewhere. Because it marks the center of the retina, the fovea is a convenient anatomical reference point. Thus, the part of the retina that lies closer to the nose than the fovea is called nasal, the part

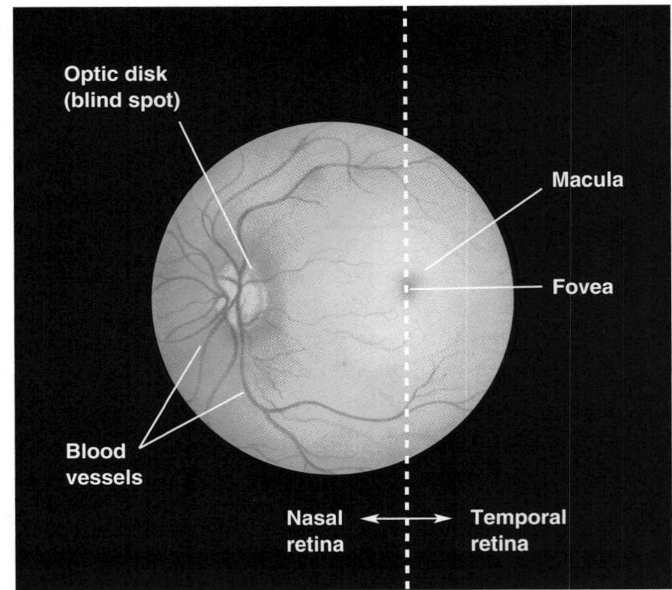

Figure 9.5
Ophthalmoscopic appearance of the retina.

Box 9.1

OF SPECIAL INTEREST

Demonstrating the Blind Regions of Your Eye

A look through an ophthalmoscope leads to the startling revelation that there is a sizable hole in the retina. The region where the optic nerve axons exit the eye and the retinal blood vessels enter the eye, the optic disk, is devoid of photoreceptors. Moreover, the blood vessels coursing across the retina are opaque and block the light from falling on photoreceptors beneath them. Although we normally do not notice them, these blind regions can be demonstrated. Look at Figure A. Hold the book about 1.5 feet away, close your right eye, and fixate on the cross with your left eye. Move the book (or your head) around slightly, and eventually you will find a position where the black circle disappears. At this position, the spot is imaged on the optic disk of the left eye. This region of visual space is called the blind spot for the left eye.

The blood vessels are a little tricky to demonstrate, but give this a try. Get a standard household flashlight. In a dark or dimly lit room, close your left eye (it helps to hold the eye closed with your finger so you can open your right eye wider). Look straight ahead with the open right eye, and shine the flashlight at an angle into the corner of the eye from the side. Jiggle the light back and forth, up and down. If you're lucky, you'll see an image of your own retinal blood vessels. This is possible because the illumination of the eye at this oblique angle causes the retinal blood vessels to cast long shadows on the adjacent regions of retina. For the shadows to be visible, they must be swept back and forth on the retina, hence the jiggling of the light.

If we have all of these light-insensitive regions in the retina, why does the visual world appear uninterrupted and seamless? The detailed answer to this question is still unknown, but mechanisms in the visual cortex apparently "fill in" the missing regions. Filling in can be demonstrated with the stimulus shown in Figure B. Again, fixate on the cross with your left eye. You will find that where the break in the line falls is imaged on the blind spot, your perception will be of a continuous, uninterrupted line.

Figure A

Figure B

that lies near the temple is called temporal, the part of the retina above the fovea is called superior, and that below it is called inferior.

Cross-Sectional Anatomy of the Eye

A cross-sectional view of the eye shows the path taken by light as it passes through the cornea toward the retina (Figure 9.6). This view reveals the **lens**, another transparent surface located behind the iris. The lens is suspended by ligaments (called zonule fibers) attached to the **ciliary muscles**, which are attached to the sclera and form a ring inside the eye. When they contract, the middle of the ring gets smaller, and the tension on the suspensory ligaments decreases. When the tension of these ligaments decreases, the lens tends to become thicker because of its natural elasticity. Conversely, relaxation of the ciliary muscles causes the ring to get bigger, and consequently, the tension on the suspensory ligaments increases. This has the effect of stretching the lens

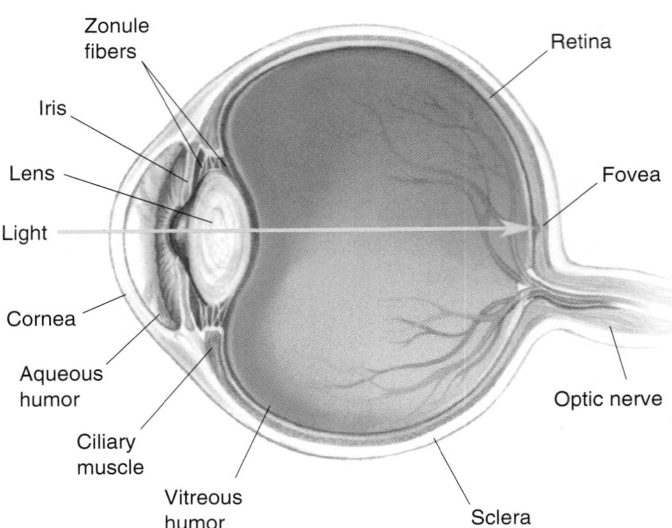

Figure 9.6
The eye in cross section.

into a flat shape. As we shall see, these changes in the shape of the lens enable our eyes to adjust their focus to different viewing distances.

The lens also divides the interior of the eye into two compartments containing slightly different fluids. The aqueous humor, introduced previously, is the watery fluid that lies between the cornea and the lens. The more viscous, jellylike **vitreous humor** lies between the lens and the retina; its pressure keeps the globe of the eye spherical.

Although the eyes do a remarkable job of delivering precise visual information to the rest of the brain, a variety of disorders can compromise this ability (Box 9.2).

IMAGE FORMATION BY THE EYE

The eye collects the light rays emitted by or reflected off objects in the environment and focuses them onto the retina to form images. Let's see how the different structures in the eye contribute to this function.

Refraction by the Cornea

Consider the light emitted from a distant source, perhaps a bright star at night. We see the star as a point of light because the eye focuses the star's light to a point on the retina. The light rays striking the surface of the eye from a distant star are virtually parallel, so they must be bent by the process of refraction.

Recall that as light passes into a medium where its speed is slowed, it will bend toward a line that is perpendicular to the border between the media. This is precisely the situation as light strikes the cornea and passes from the air into the aqueous humor. As shown in Figure 9.7, the light rays that strike the curved surface of the cornea bend so that they converge on the back of the eye; those that enter the center of the eye pass straight to the retina. The distance from the refractive surface to the point where parallel light rays converge is called the *focal distance*. Focal distance depends on the curvature of the cornea: The tighter the curve, the shorter the focal distance. The reciprocal of the focal distance in meters is a unit of measurement called the **diopter**. The cornea has a refractive power of about 42 diopters, which means that parallel light rays striking the corneal surface will be focused 0.024 m (2.4 cm) behind it, about the distance from cornea to retina.

Box 9.2

OF SPECIAL INTEREST

Eye Disorders

Once you know the basic structure of the eye, you can understand how a partial or complete loss of vision results from abnormalities in various components. For example, if there is an imbalance in the extraocular muscles of the two eyes, the eyes will point in different directions. Such a misalignment or lack of coordination between the two eyes is called *strabismus*, and there are two varieties. In *esotropia*, the directions of gaze of the two eyes cross, and the person is said to be cross-eyed. In *exotropia*, the directions of gaze diverge, and the person is said to be wall-eyed (Figure A). In most cases, strabismus of either type is congenital and should be corrected during early childhood. Treatment usually involves the use of prismatic glasses or surgery to the extraocular muscles to realign the eyes. Without treatment, conflicting images are sent to the brain from the two eyes, degrading depth perception and, more importantly, causing the person to suppress input from one eye. The dominant eye will be normal, but the suppressed eye will become *amblyopic*, meaning that it has poor visual acuity. If medical intervention is delayed until adulthood, the condition cannot be corrected.

A common eye disorder among older adults is *cataract*, a clouding of the lens (Figure B). Many people over age 65 have some degree of cataract; if it significantly impairs vision, surgery is usually required. In a cataract operation, the lens is removed and replaced with an artificial plastic lens. Although the artificial lens cannot adjust its focus like the normal lens, it provides a clear image, and glasses can be used for near and far vision (see Box 9.3).

Glaucoma, a progressive loss of vision associated with elevated intraocular pressure, is a leading cause of blindness. Glaucoma is associated with increased pressure of the aqueous humor, and pressure in the aqueous plays a crucial role in maintaining the shape of the eye. As this pressure increases, the entire eye is stressed, ultimately damaging the relatively weak point where the optic nerve leaves the eye. The axons in the optic nerve are compressed, and vision is gradually lost from the periphery inward. Unfortunately, by the time a person notices a loss of more central vision, the damage is advanced and a significant portion of the eye is permanently blind. For this reason, early detection and treatment with medication or surgery to reduce intraocular pressure are essential.

The light-sensitive retina at the back of the eye is the site of numerous disorders that pose a significant risk of blindness. You may have heard of a professional boxer having a *detached retina*. As the name implies, the retina pulls away from the underlying wall of the eye because of a blow to the head or shrinkage of the vitreous humor. Once the retina has started to detach, fluid from the vitreous space flows through small tears in the retina resulting from the trauma, and this causes more of the retina to separate. Symptoms of retinal detachment include abnormal perception of shadows and flashes of light. Treatment often involves laser surgery to scar the edge of the retinal tear, thereby reattaching the retina to the back of the eye.

In *retinitis pigmentosa*, there is a progressive degeneration of the photoreceptors. At first, there is a loss of peripheral vision and night vision. Subsequently, total blindness may result. The cause of this disease is unknown, but in some forms, it clearly has a strong genetic component. There is no effective treatment. In contrast to the tunnel vision experienced by patients with retinitis pigmentosa, people with *macular degeneration* lose only central vision. Tiny yellowish deposits called drusen build up under the macula, and they are often associated with photoreceptor degeneration. In this common, incurable disorder of older people, peripheral vision may remain normal, but the ability to read, watch television, and recognize faces is lost.

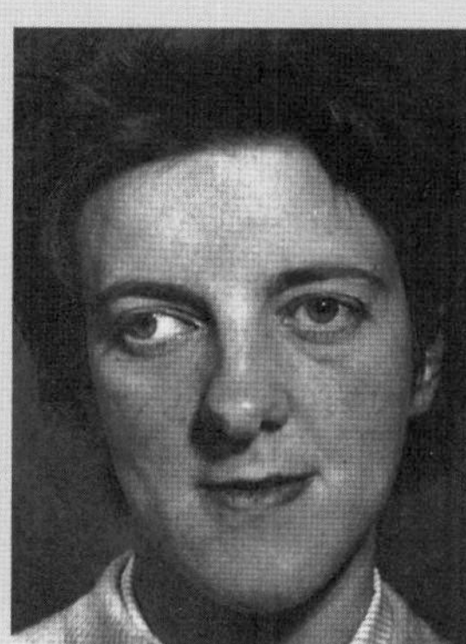

Figure A
Extropia. (Source: Newell, 1965, p. 330.)

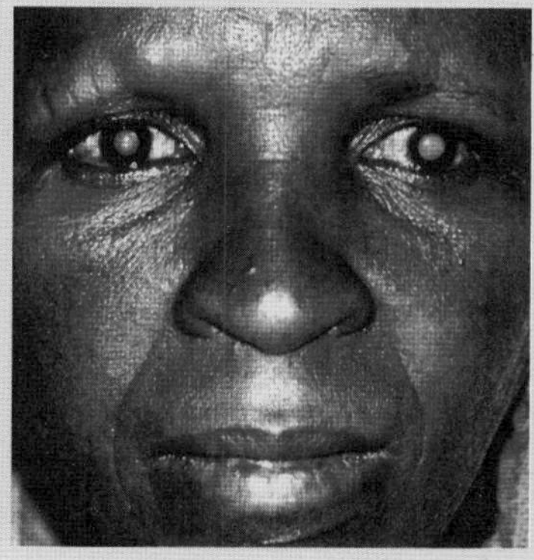

Figure B
Cataract. (Source: Schwab, 1987, p. 22.)

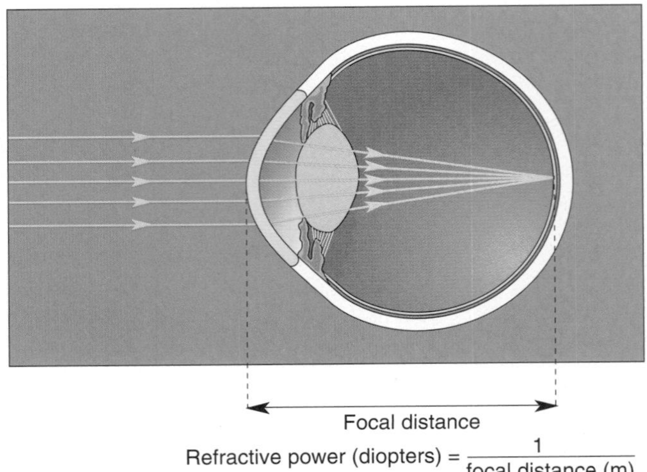

Figure 9.7
Refraction by the cornea.

Remember that refractive power depends on the slowing of light at the air-cornea interface. If we replace air with a medium that passes light at about the same speed as the eye, the refractive power of the cornea will be eliminated. This is why things look blurry when you open your eyes underwater; the water-cornea interface has very little focusing power. A scuba mask restores the air-cornea interface and, consequently, the refractive power of the eye.

Accommodation by the Lens

Although the cornea performs most of the eye's refraction, the lens contributes another dozen or so diopters to the formation of a sharp image at a distant point. More importantly, however, the lens is involved in forming crisp images of objects located closer than about 9 m from the eye. As objects approach, the light rays originating at a point can no longer be considered to be parallel. Rather, these rays diverge, and greater refractive power is required to bring them into focus on the retina. This additional focusing power is provided by changing the shape of the lens, a process called **accommodation** (Figure 9.8).

During accommodation, the ciliary muscles contract (relieving the tension on the suspensory ligaments), and the lens, because of its natural elasticity, becomes rounder. This rounding increases the curvature of the lens surfaces, thereby increasing their refractive power. The ability to accommodate

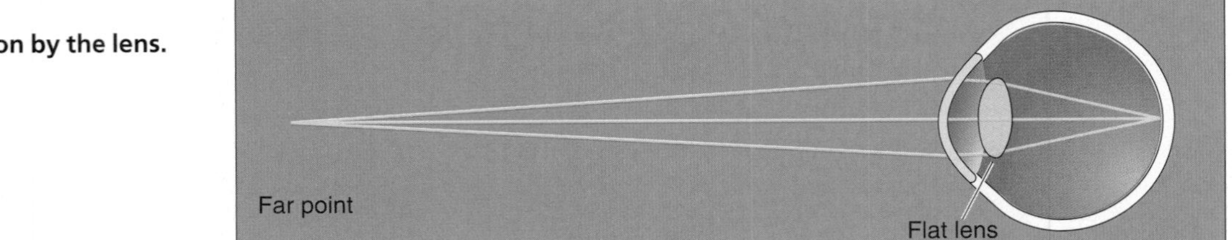

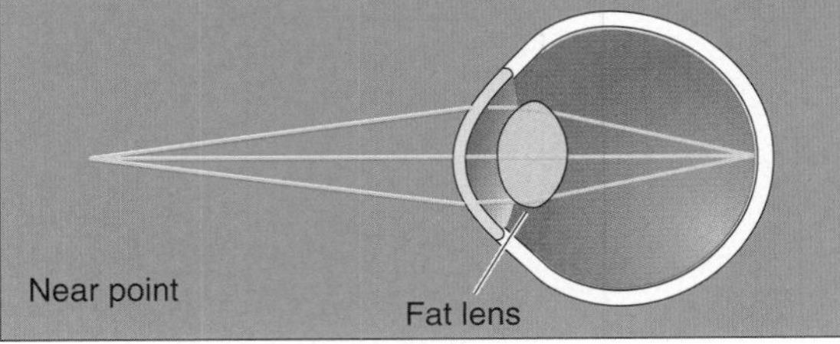

Figure 9.8
Accommodation: refraction by the lens.

changes with age. An infant can focus objects just beyond his or her nose, whereas many middle-aged adults cannot focus objects closer than about arm's length. Fortunately, artificial lenses can compensate for this and other defects of the eye's optics (Box 9.3).

The Pupillary Light Reflex

In addition to the cornea and the lens, the pupil contributes to the optical qualities of the eye by continuously adjusting for different ambient light levels. To check this for yourself, stand in front of a bathroom mirror with the lights out for a few seconds and then watch your pupils change size when the light goes on. This **pupillary light reflex** involves connections between the retina and neurons in the brain stem that synapse on the motor neurons that control the muscle that constricts the pupil. An interesting property of this reflex is that it is *consensual*; that is, shining a light into only one eye causes constriction of the pupils of both eyes. It is unusual, indeed, for the pupils not to be the same size; the lack of a consensual pupillary light reflex is often taken as a sign of a serious neurological disorder involving the brain stem.

Constriction of the pupil increases the depth of focus, just like decreasing the aperture size (increasing the f-stop) on a camera lens. To understand why this is true, consider two points in space, one close and the other far away. When the eye accommodates to the closer point, the image of the farther point on the retina no longer forms a point but rather a blurred circle. Decreasing the aperture—constricting the pupil—reduces the size of this blurred circle so that its image more closely approximates a point. In this way, distant objects appear to be less out of focus.

The Visual Field

The structure of the eyes and where they sit in our head limit how much of the world we can see at any one time. Let's investigate the extent of the space seen by one eye. Holding a pencil horizontally in your right hand, close your left eye and look at a point straight ahead. Keeping your eye fixated on this point, slowly move the pencil to the right (toward your right ear) across your field of view until the whole pencil disappears. Repeat this exercise, moving the pencil to the left, and then up and down. The points where you can no longer see the pencil mark the limits of the **visual field** for your right eye. Now look at the middle of the pencil as you hold it horizontally in front of you. Figure 9.9 shows how the light reflected off this pencil falls on your retina. Notice that the image is inverted; the left visual field is imaged on the right side of the retina, and the right visual field is imaged on the left side of the retina.

Visual Acuity

The ability of the eye to distinguish between two nearby points is called **visual acuity**. Acuity depends on several factors, but especially on the spacing of photoreceptors in the retina and the precision of the eye's refraction.

Distance across the retina can be described in terms of degrees of **visual angle**. A right angle subtends (spans) 90 degrees; the moon, for example, subtends an angle of about 0.5 degrees (Figure 9.10). We can speak of the eye's ability to resolve points that are separated by a certain number of degrees of visual angle. The Snellen eye chart, which we have all read at the doctor's office, tests our ability to discriminate characters that each subtend various angles at a viewing distance of 20 feet. Your vision is 20/20 when you can recognize a letter that subtends an angle of 0.083 degrees (equivalent to 5 minutes of arc, where 1 minute is a sixtieth of a degree).

Box 9.3

OF SPECIAL INTEREST

Vision Correction With Lenses and Surgery

When the ciliary muscles are relaxed and the lens is flat, the eye is said to be *emmetropic* if parallel light rays from a distant point source are focused sharply on the back of the retina. (The word is from the Greek *emmetros*, "in proper measure," and *ope*, "sight.") Stated another way, the emmetropic eye focuses parallel light rays on the retina without the need for accommodation (Figure A).

Now consider what happens when the eyeball is too short from front to back (Figure B). The light rays are focused at some point *behind* the retina. Without accommodation, all of the point image that the retina can see is a blurry circle. This condition is known as *hyperopia*, or farsightedness, because much of the accommodative power of the lens is needed to image far points and the lens cannot accommodate enough to image near points. Farsightedness can be corrected by placing a convex glass or plastic lens in front of the eye (Figure C). The curved front edge of the lens, like the cornea, bends light toward the center of the retina. Because the light again passes from glass into air, the back of the lens also bends the light (light going from glass to air speeds up and is bent *away* from the perpendicular).

If the eyeball is too long rather than too short, parallel rays converge before the retina, cross, and again are imaged on the retina as a blurry circle (Figure D). This condition is known as *myopia*, or nearsightedness. Thus, for the nearsighted eye to see distant points clearly, artificial concave lenses must be used to move the point image back onto the retina (Figure E).

Some eyes have irregularities in the curvature of the cornea or lens that lead to various amounts of refraction depending on whether light rays enter the eye in a horizontal or a vertical axis. This condition, known as *astigmatism*, can be corrected by using an artificial lens that is curved more along one axis than others.

Even if you are fortunate enough to have perfectly shaped eyeballs and a symmetrical refractive system, you probably will not escape *presbyopia* (from the Greek meaning "old eye"). This condition, a hardening of the lens that accompanies the aging process, is thought to be explained by the fact that new lens cells are generated throughout life, while no cells are lost. The hardening denies the lens of its elasticity, leaving it unable to thicken sufficiently during accommodation and unable to flatten sufficiently during relaxation. The correction for presbyopia, first introduced by none other than Benjamin Franklin, is a bifocal lens. These lenses are concave on top to assist far vision and convex on the bottom to assist near vision.

In hyperopia and myopia, the amount of refraction provided by the cornea is either too little or too great for the length of the eyeball. But what if we could change the amount of refraction the cornea provides? In the early 1970s, a group of Russian ophthalmologists developed *radial keratotomy* for the correction of myopia. In this procedure, a number of very fine radial incisions are made through the peripheral portion of the cornea. The cuts relax and flatten the central cornea, thus reducing the amount of refraction and minimizing the myopia. More recently, techniques have been developed using lasers to reshape the cornea. In *photorefractive keratectomy* (PRK), a laser is used to reshape the outer surface of the cornea by vaporizing thin layers. In *laser in situ keratomileusis* (LASIK), a thin flap of the cornea is lifted so that the laser can reshape the cornea from the inside. On the horizon are surgical procedures to alter corneal shape in a manner that is reversible.

MICROSCOPIC ANATOMY OF THE RETINA

Now that we have an image formed on the retina, we can get to the neuroscience of vision: the conversion of light energy into neural activity. To begin our discussion of image processing in the retina, we must introduce the cellular architecture of this bit of brain.

The basic system of retinal information processing is shown in Figure 9.11. The most direct pathway for visual information flow is from **photoreceptors** to **bipolar cells** to **ganglion cells**. The ganglion cells fire action potentials in response to light, and these impulses propagate down the optic nerve to the rest of the brain. Besides the cells in this direct path from photoreceptor to brain, retinal processing is influenced by two additional cell types. **Horizontal cells** receive input from the photoreceptors and project neurites

(Box 9.3, *continued*)

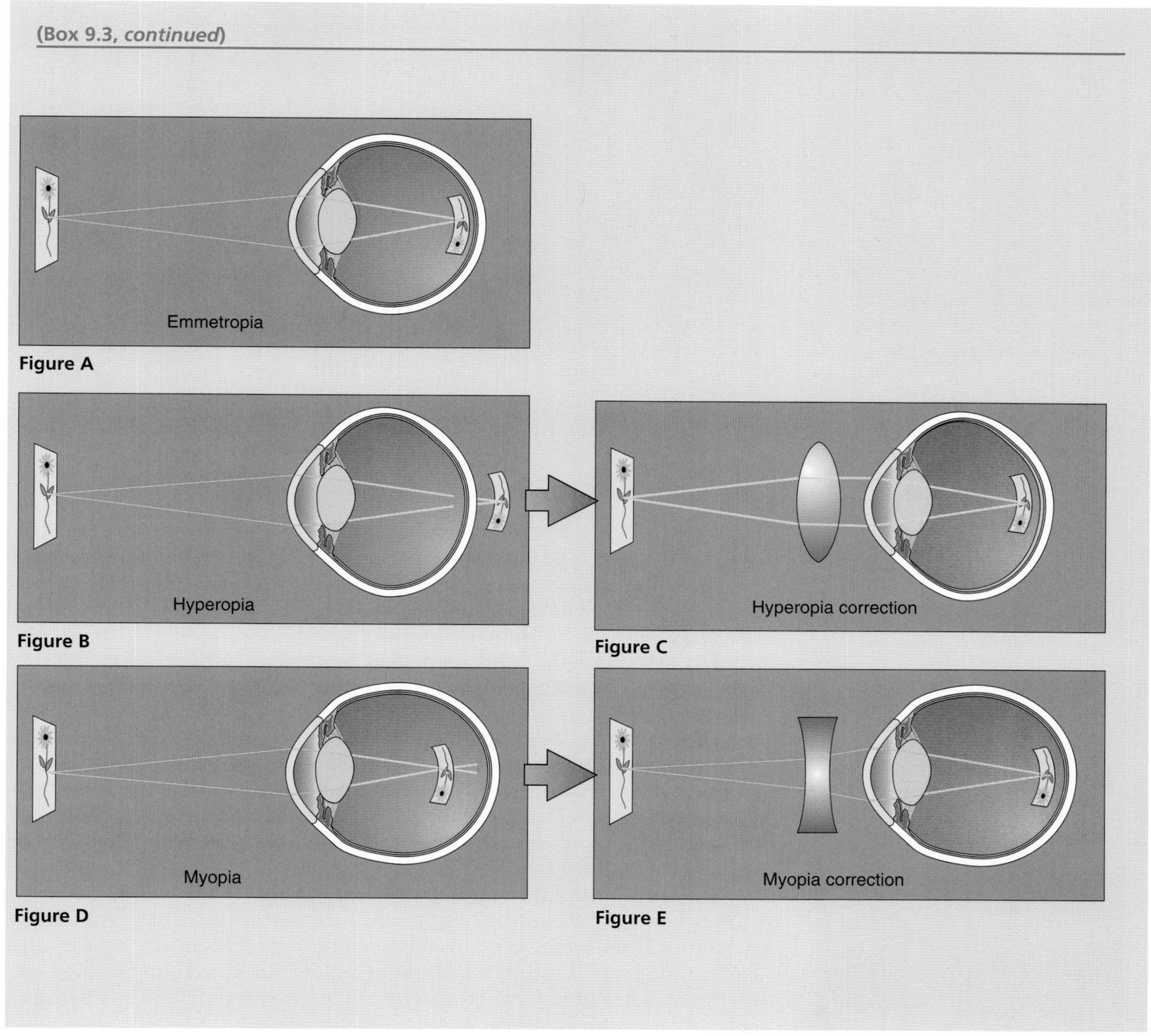

Figure A

Figure B

Figure C

Figure D

Figure E

laterally to influence surrounding bipolar cells and photoreceptors. **Amacrine cells** receive input from bipolar cells and project laterally to influence surrounding ganglion cells, bipolar cells, and other amacrine cells.

There are two important points to remember here:

1. *The only light-sensitive cells in the retina are the photoreceptors.* All other cells are influenced by light only via direct and indirect synaptic interactions with the photoreceptors.
2. *The ganglion cells are the only source of output from the retina.* No other retinal cell type projects an axon through the optic nerve.

Now let's take a look at how the different cell types are arranged in the retina.

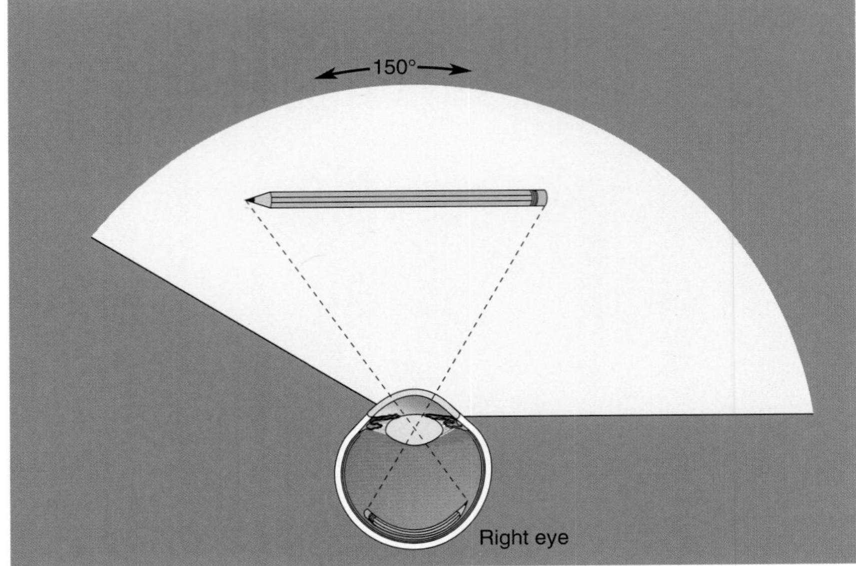

Figure 9.9
The visual field for one eye. The visual field is the total amount of space that can be viewed by the retina when the eye is fixated straight ahead. Notice how the image of an object in the visual field (pencil) is inverted on the retina.

The Laminar Organization of the Retina

Figure 9.12 shows that the retina has a *laminar organization*: Cells are organized in layers. Notice that the layers are seemingly inside-out; light must pass from the vitreous humor through the ganglion cells and bipolar cells *before* it reaches the photoreceptors. Because these cells are relatively transparent, image distortion is minimal.

The cell layers of the retina are named in reference to the middle of the eyeball. Thus, the innermost layer is the **ganglion cell layer**, which contains the cell bodies of the ganglion cells. Next is the **inner nuclear layer**, which contains the cell bodies of the bipolar cells, the horizontal and amacrine cells. The next layer is the **outer nuclear layer**, which contains the cell bodies of the photoreceptors. Finally, the **layer of photoreceptor outer segments** contains the light-sensitive elements of the retina. The outer segments are embedded in a pigmented epithelium that is specialized to absorb any light that passes entirely through the retina. This minimizes the reflection of light within the eye that would blur an image and decrease acuity.

Between the ganglion cell layer and the inner nuclear layer is the **inner plexiform layer**, which contains the synaptic contacts between bipolar cells, amacrine cells, and ganglion cells. Between the outer and inner nuclear layers is the **outer plexiform layer**, where the photoreceptors make synaptic contact with the bipolar and horizontal cells.

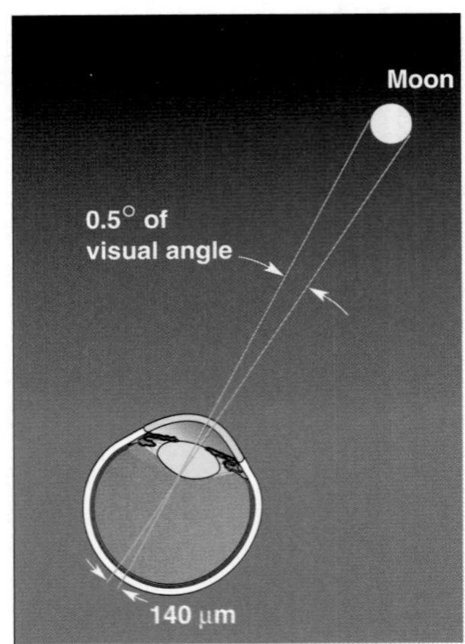

Figure 9.10
Visual angle. Distances across the retina can be expressed as degrees of visual angle.

Photoreceptor Structure

The conversion of electromagnetic radiation into neural signals occurs in the 125 million photoreceptors at the back of the retina. Every photoreceptor has four regions: an outer segment, an inner segment, a cell body, and a synaptic terminal. The outer segment contains a stack of membranous disks. Light-sensitive *photopigments* in the disk membranes absorb light, thereby triggering changes in the photoreceptor membrane potential (discussed later). Figure 9.13 shows the two types of photoreceptor in the retina, easily distinguished by the appearance of their outer segments. **Rod photoreceptors** have a long, cylindrical outer segment containing many disks. **Cone photoreceptors** have a shorter, tapering outer segment containing relatively few membranous disks.

The structural differences between rods and cones correlate with important functional differences. For example, the greater number of disks and the higher photopigment concentration in rods make the rods 1000 times more sensitive to light than the cones. Indeed, under nighttime lighting, or *scotopic* conditions, only rods contribute to vision. Conversely, under daytime lighting, or *photopic* conditions, cones do the bulk of the work. For this reason, the retina is said to be *duplex*: a scotopic retina using only rods and a photopic retina using mainly cones.

Rods and cones differ in other respects as well. All rods contain the same photopigment, but there are three types of cone, each containing a different pigment. The variations among pigments make the different cones sensitive to different wavelengths of light. As we shall see in a moment, only the cones, not the rods, are responsible for our ability to see color.

Figure 9.11
The basic system of retinal information processing. Information about light flows from the photoreceptors to bipolar cells to ganglion cells, which project axons out of the eye in the optic nerve. Horizontal cells and amacrine cells modify the responses of bipolar cells and ganglion cells via lateral connections.

Regional Differences in Retinal Structure

Retinal structure varies from the fovea to the retinal periphery. In general, the peripheral retina has a higher ratio of rods to cones (Figure 9.14). It also has a higher ratio of photoreceptors to ganglion cells. The combined effect of this arrangement is that the peripheral retina is more sensitive to light because (1) rods are specialized for low light and (2) there are more photoreceptors feed-

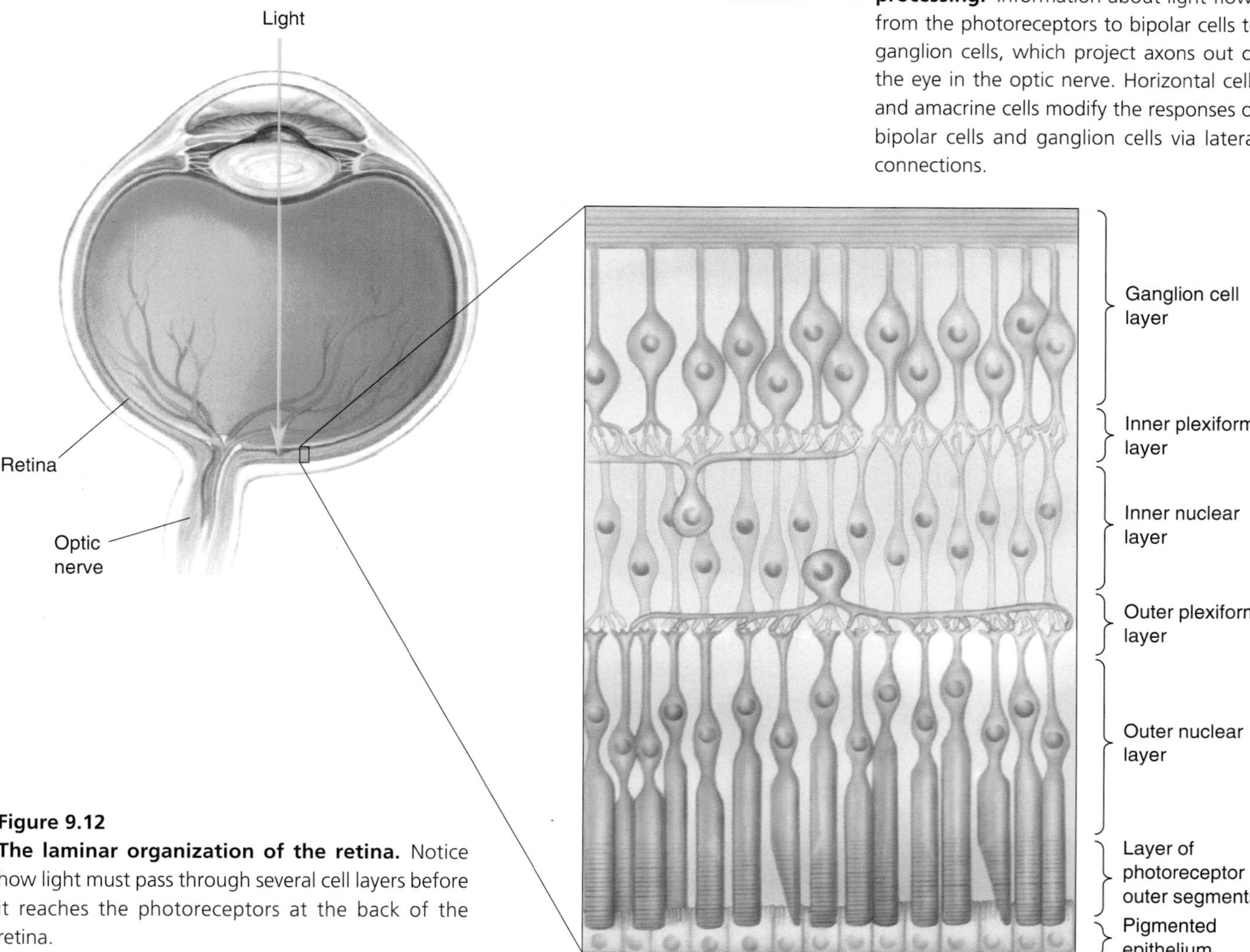

Figure 9.12
The laminar organization of the retina. Notice how light must pass through several cell layers before it reaches the photoreceptors at the back of the retina.

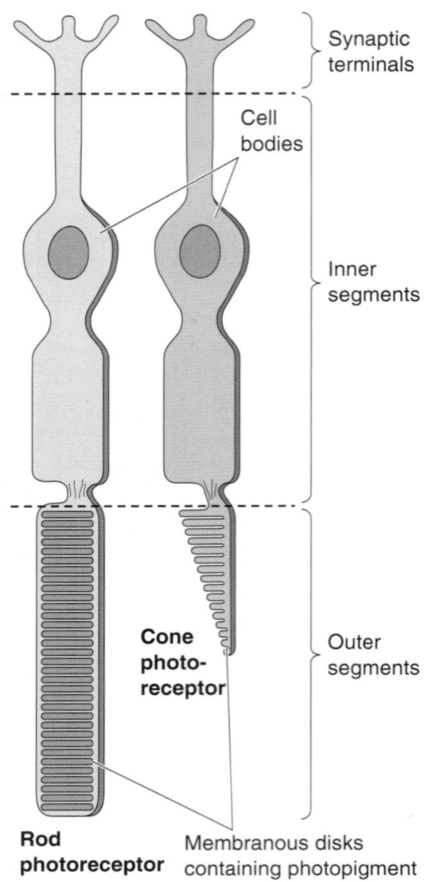

Figure 9.13
A rod and a cone.

ing information to each ganglion cell. You can prove this to yourself on a starry night. (It's fun! Try it with a friend.) First, spend about 20 minutes in the dark getting oriented, and then gaze at a bright star. Fixating on this star, search your peripheral vision for a dim star. Then move your eyes to look at this dim star. You will find that the faint star disappears when it is imaged on the central retina (when you look straight at it) but reappears when it is imaged on the peripheral retina (when you look slightly to the side of it).

The same characteristics that enable the peripheral retina to detect faint stars at night make it relatively poor at resolving fine details in daylight. This is because daytime vision requires cones and because good visual acuity requires a low ratio of photoreceptors to ganglion cells. The region of retina most highly specialized for high-resolution vision is the fovea. Recall that the fovea is a thinning of the retina at the center of the macula. In cross section, the fovea appears as a pit in the retina. Its pitlike appearance is due to the lateral displacement of ganglion cells from the fovea, allowing light to strike the photoreceptors without passing through the other retinal cell layers (Figure 9.15). This structural specialization maximizes visual acuity at the fovea by pushing aside other cells that might scatter light and blur the image. The fovea also is unique because it contains no rods; all of its photoreceptors are cones.

PHOTOTRANSDUCTION

The photoreceptors convert, or *transduce*, light energy into changes in membrane potential. We begin our discussion of phototransduction with rods, which outnumber cones in the human retina by 20 to 1. Most of what has been learned about phototransduction by rods has proven to be applicable to cones as well.

Phototransduction in Rods

As we discussed in Part I, one way information is represented in the nervous system is as changes in the membrane potential of neurons. Thus, we look for a mechanism by which the absorption of light energy can be transduced into a change in the photoreceptor membrane potential. In many respects, this process is analogous to the transduction of chemical signals into electrical signals that occurs during synaptic transmission. At a G-protein-coupled neurotransmitter receptor, for example, the binding of transmitter to the receptor activates G-proteins in the membrane, which in turn stimulate various effector enzymes (Figure 9.16a). These enzymes alter the intracellular concentration of cytoplasmic second messenger molecules, which directly or indirectly change the conductance of membrane ion channels, thereby altering membrane potential. Similarly, in the photoreceptor, light stimulation of the photopigment activates G-proteins, which in turn activate an effector enzyme that changes the cytoplasmic concentration of a second messenger molecule. This change causes a membrane ion channel to close, and the membrane potential is thereby altered (Figure 9.16b).

Recall from Chapter 3 that a typical neuron at rest has a membrane potential of about -65 mV, close to the equilibrium potential for K^+. In contrast, in complete darkness, the membrane potential of the rod outer segment is about -30 mV. This depolarization is caused by the steady influx of Na^+ through special channels in the outer segment membrane (Figure 9.17a). This movement of positive charge across the membrane is called the **dark current**. In 1985, a team of Russian scientists led by Evgeniy Fesenko found that these sodium channels are stimulated to open—are gated—by an intracellular second messenger called **cyclic guanosine monophosphate**, or **cGMP**. Evidently, cGMP is continually produced in the photoreceptor by the en-

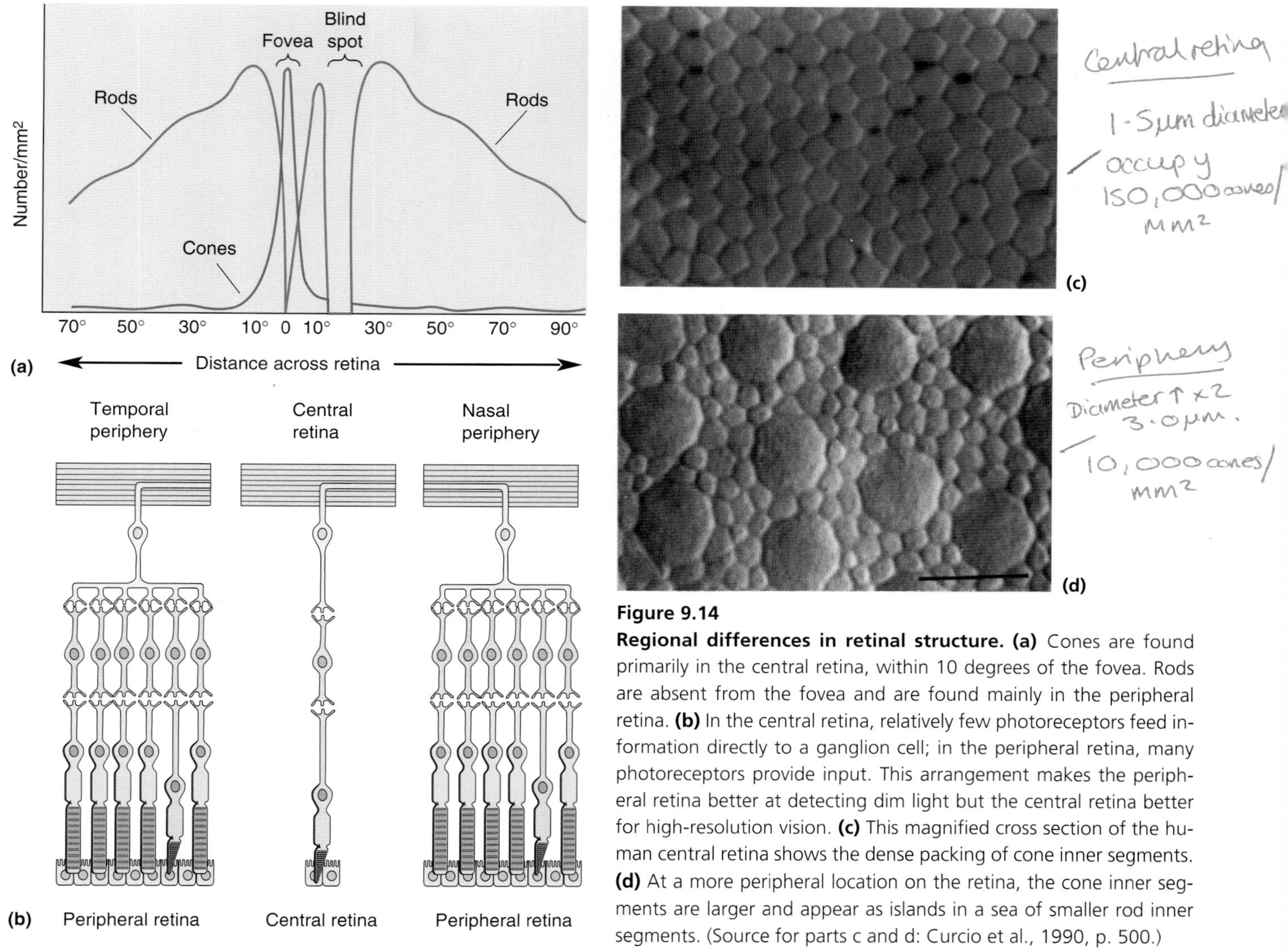

Blind spot
Fovea
Rods
Rods
Number/mm²
Cones
70° 50° 30° 10° 0 10° 30° 50° 70° 90°
(a) Distance across retina

Temporal periphery
Central retina
Nasal periphery

(b) Peripheral retina Central retina Peripheral retina

(c)

(d)

Figure 9.14
Regional differences in retinal structure. (a) Cones are found primarily in the central retina, within 10 degrees of the fovea. Rods are absent from the fovea and are found mainly in the peripheral retina. **(b)** In the central retina, relatively few photoreceptors feed information directly to a ganglion cell; in the peripheral retina, many photoreceptors provide input. This arrangement makes the peripheral retina better at detecting dim light but the central retina better for high-resolution vision. **(c)** This magnified cross section of the human central retina shows the dense packing of cone inner segments. **(d)** At a more peripheral location on the retina, the cone inner segments are larger and appear as islands in a sea of smaller rod inner segments. (Source for parts c and d: Curcio et al., 1990, p. 500.)

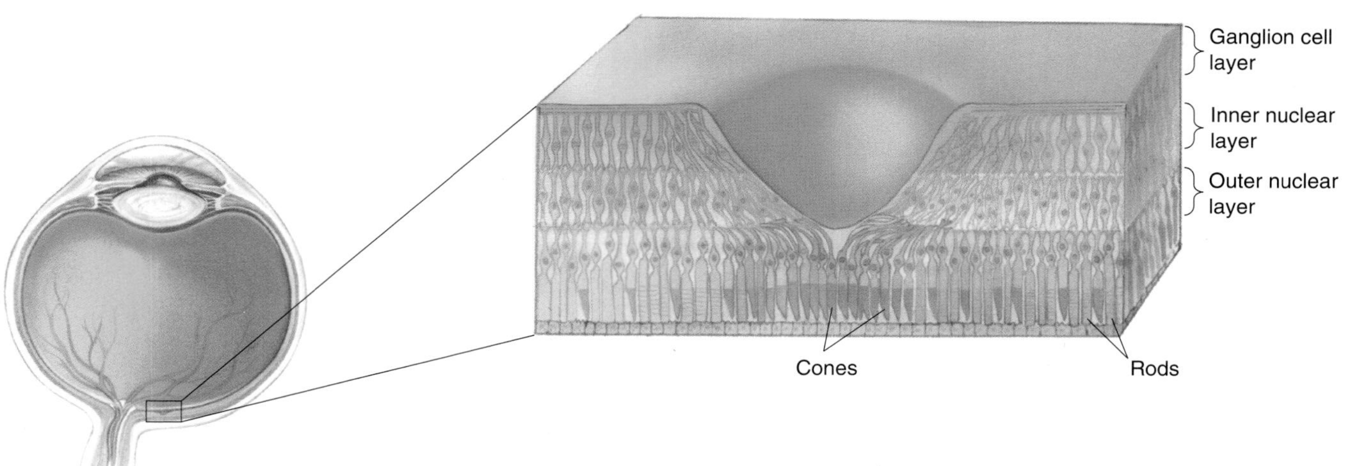

Figure 9.15
The fovea in cross section. The ganglion cell layer and the inner nuclear layer are displaced laterally to allow light to strike the foveal photoreceptors directly.

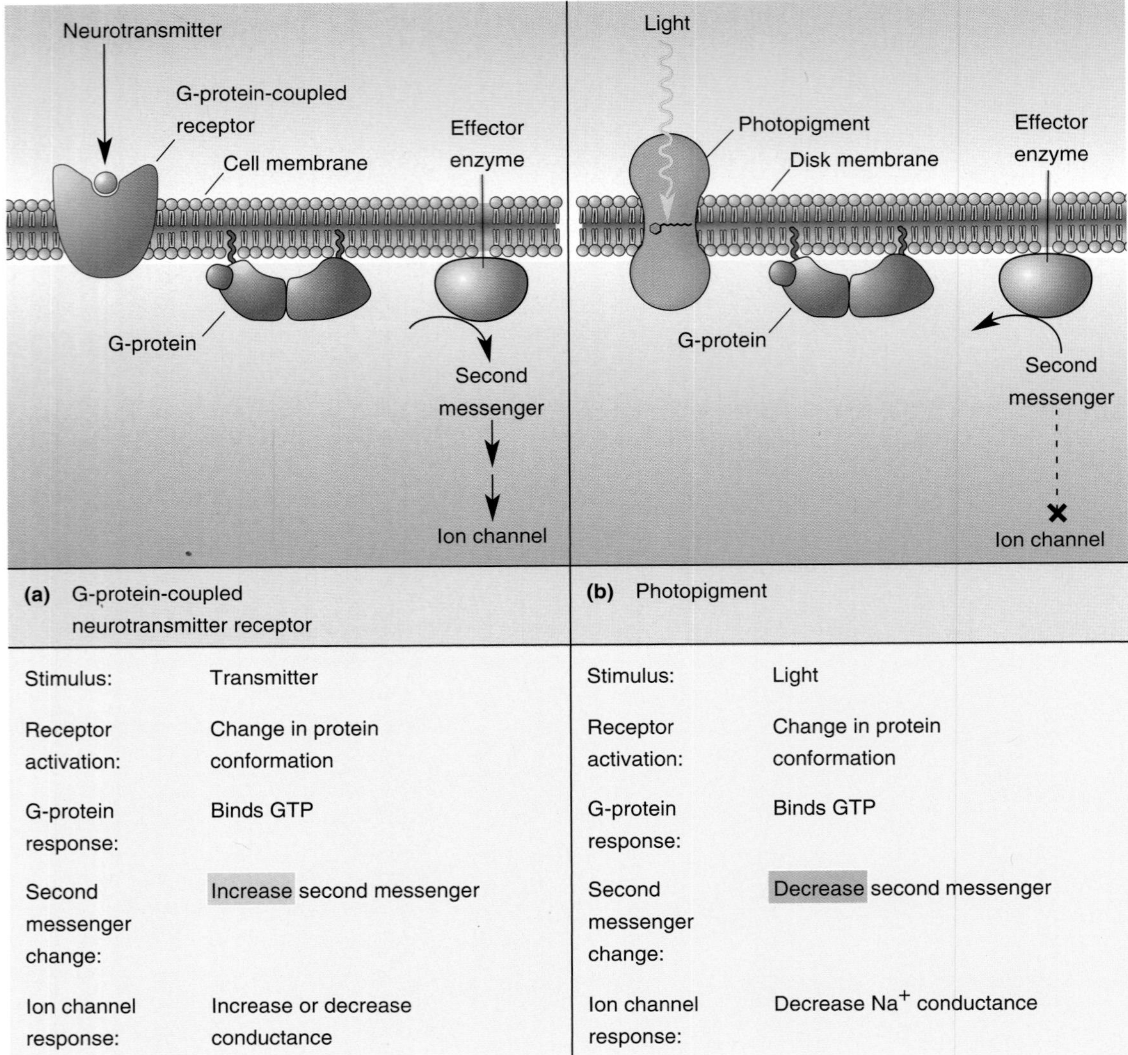

Figure 9.16
A comparison of events triggered by the activation of (a) a G-protein-coupled neurotransmitter receptor and (b) a photopigment.

zyme guanylate cyclase, keeping the Na^+ channels open. Light reduces cGMP; this causes the Na^+ channels to close, and the membrane potential becomes *more negative* (Figure 9.17b). Thus, *photoreceptors hyperpolarize in response to light.*

The hyperpolarizing response to light is initiated by the absorption of electromagnetic radiation by the photopigment in the membrane of the stacked disks in the rod outer segments. In the rods, this pigment is called **rhodopsin**. Rhodopsin can be thought of as a receptor protein with a prebound chemical agonist. The receptor protein is called *opsin,* and it has the seven transmembrane alpha helices typical of G-protein-coupled receptors throughout the body. The prebound agonist is called *retinal,* a derivative of vitamin A. The absorption of light causes a change in the conformation of retinal so that it activates the opsin (Figure 9.18). This process is called bleaching because it changes the wavelengths absorbed by the rhodopsin (the photopigment literally changes color from purple to yellow). The bleaching of rhodopsin stimulates a G-protein called **transducin** in the disk membrane, which in turn activates the effector enzyme **phosphodiesterase** (**PDE**). PDE breaks down the

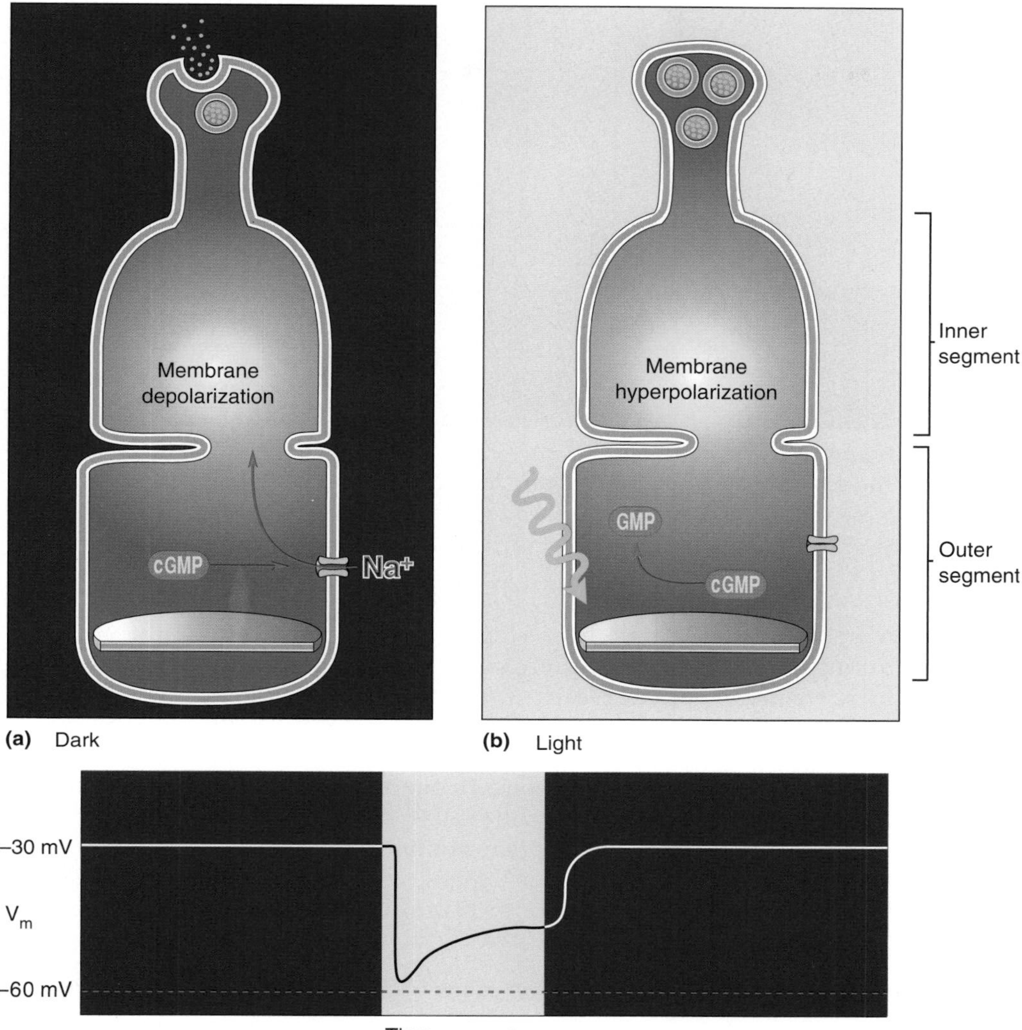

Figure 9.17
The hyperpolarization of photoreceptors in response to light. Photoreceptors are continuously depolarized in the dark because of an inward sodium current, the dark current. **(a)** Sodium enters the photoreceptor through a cGMP-gated channel. **(b)** Light leads to the activation of an enzyme that destroys cGMP, thereby shutting off the Na^+ current and hyperpolarizing the cell.

cGMP that is normally present in the cytoplasm of the rod (in the dark). The reduction in cGMP causes the Na^+ channels to close and the membrane to hyperpolarize.

One of the interesting functional consequences of using a biochemical cascade for transduction is signal *amplification*. Many G-proteins are activated by each photopigment molecule, and each PDE enzyme breaks down more than one cGMP molecule. This amplification gives our visual system the ability to detect as little as a single photon, the elementary unit of light energy. The complete sequence of events of phototransduction in rods is illustrated in Figure 9.19.

Phototransduction in Cones

Prolonged illumination of the rods causes cGMP levels to fall to the point where the response to light becomes *saturated*; additional light causes no

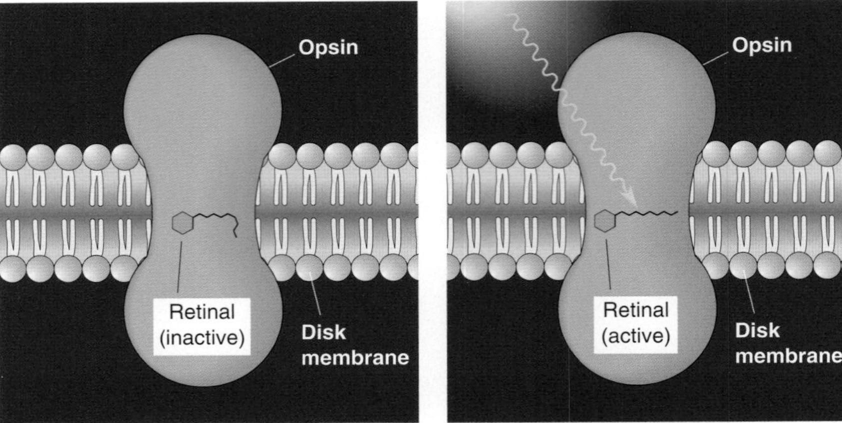

Figure 9.18
The activation of rhodopsin by light. Rhodopsin consists of a protein with seven transmembrane alpha helices, called an opsin, and a small molecule derived from vitamin A, called retinal. Retinal undergoes a change in conformation when it absorbs light, activating the opsin.

more hyperpolarization. This is the situation in bright sunlight. Thus, vision during the day depends entirely on the cones, whose photopigments require more energy to become bleached.

The process of phototransduction in cones is virtually the same as in rods; the only major difference is in the type of opsins in the membranous disks of the cone outer segments. The cones in our retinas contain one of three opsins that give the photopigments different spectral sensitivities. Thus, we can speak of "blue" cones, which are maximally activated by light with a wavelength of about 430 nm; "green" cones, which are maximally activated by light with a wavelength of about 530 nm; and "red" cones, which are maxi-

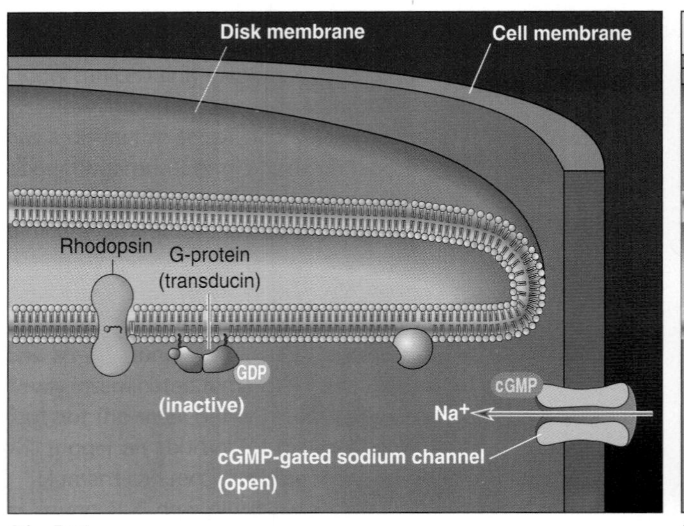

(a) Dark

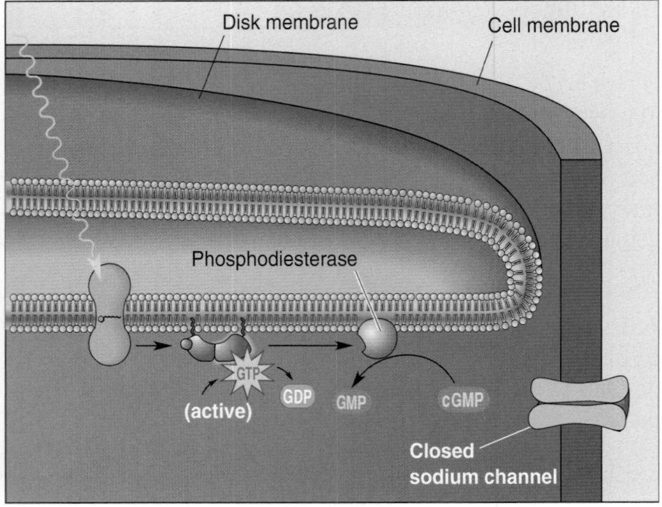

(b) Light

Figure 9.19
The light-activated biochemical cascade in a photoreceptor. (a) In the dark, cGMP gates a sodium channel, causing an inward Na^+ current and depolarization of the cell. **(b)** The activation of rhodopsin by light energy causes the G-protein (transducin) to exchange GDP for GTP (see Chapter 6), which in turn activates the enzyme phosphodiesterase (PDE). PDE breaks down cGMP and shuts off the dark current.

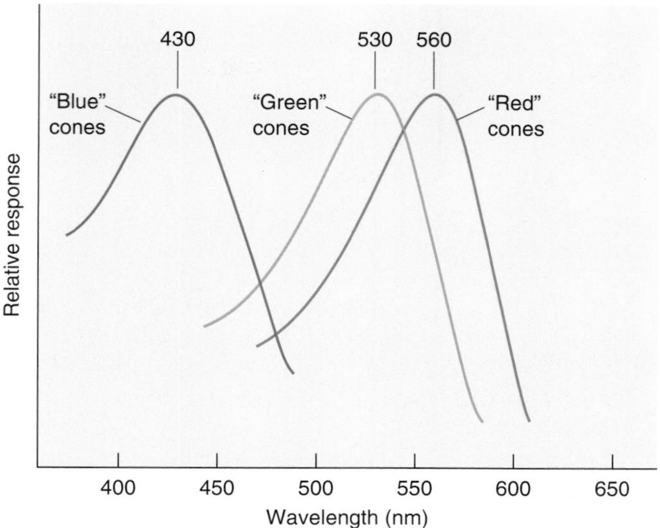

Figure 9.20
The spectral sensitivity of the three types of cone pigments.

mally activated by light with a wavelength of about 560 nm (Figure 9.20). Recent research by Jeremy Nathans and his colleagues at Stanford University and Johns Hopkins University has identified the genes that code for the red, green, and blue visual pigments, giving us considerable insight into the genetic basis of color vision (Box 9.4).

Color Detection. The color that we perceive is largely determined by the relative contributions of blue, green, and red cones to the retinal signal. The fact that our visual system detects colors in this way was actually predicted almost 200 years ago by British physicist Thomas Young. Young showed in 1802 that every color of the rainbow, including white, could be created by mixing the proper ratio of red, green, and blue light (Figure 9.21). He proposed, quite correctly, that at each point in the retina there exists a cluster of three receptor types, each type being maximally sensitive to either blue, green, or red. Young's ideas were later championed by Hermann von Helmholtz, an influential nineteenth-century German physiologist. (Among his accomplishments is the invention of the ophthalmoscope in 1851.) This theory of color vision came to be known as the **Young-Helmholtz trichromacy theory**. According to the theory, the brain assigns colors based on a comparison of the readout of the three cone types. When all types of cones are equally active, as in broad-spectrum light, we perceive "white." Various forms of color blindness result when one or more of the cone photopigment types is missing (Box 9.5).

If cones are entirely responsible for the perception of color, we should not be able to perceive color differences when cones are inactive. This inference is correct, and you can demonstrate it to yourself. Go outside on a dark night and try to distinguish the colors of various objects. It is difficult to detect colors at night because only the rods, with a single type of photopigment, are activated in dim light. (Bright neon signs are still seen as colored because they emit sufficient light to affect the cones.) The peak sensitivity of the rods is to a wavelength of about 500 nm, perceived as blue-green under photopic conditions. This fact is the basis for two points of view about the design of automobile dashboard indicator lights. One view is that the lights should be dim blue-green to take advantage of the spectral sensitivity of the rods. An alternate view is that the lights should be bright red, because this wavelength affects mainly cones, leaving the rods unsaturated, resulting in better night vision.

Box 9.4

PATH OF DISCOVERY

Genes and Vision

BY JEREMY NATHANS

I can trace my interest in visual pigments and human vision to the happy circumstance of being in the right place at the right time. In 1981, I was a second-year M.D.-Ph.D. student at Stanford and had the good fortune to hear Denis Baylor and Lubert Stryer deliver an extraordinary pair of research seminars on photoreceptors and their role in vision. Baylor described his elucidation of the electrical responses of single photoreceptor cells, and Stryer described his discovery of the light-triggered enzyme cascade that mediates phototransduction. These two talks opened my eyes to the remarkable performance of photoreceptor cells and to the beauty of vision research.

Among the papers I read at that time, George Wald's 1967 Nobel lecture stood out as especially interesting. In that paper, Wald reviewed what was then known about visual pigments, the light-absorbing proteins that initiate phototransduction. Wald also described his work on color vision and summarized the evidence that rod and cone pigments form an evolutionarily related family of proteins. Most intriguing was Wald's suggestion that the common inherited variations in human color vision might simply reflect variations within the genes that encode the three cone pigments.

Between 1967 and 1981, little progress was made in testing Wald's ideas. Cones are present as only a minor cell population in most mammalian retinas (including the human retina), and there is a corresponding paucity of cone visual pigments. Indeed, as of this writing (1998), no one has purified a cone pigment from a mammalian retina. Purifying and analyzing human cone pigments to directly test Wald's ideas represented a daunting technical challenge. While reading Wald's paper, however, I realized that with the passage of 15 years, his ideas could be tested in a way that could not have been envisioned in 1967: by isolating the cone pigment genes using recombinant DNA technology.

At this point I benefited from a second happy coincidence of time and place: One year earlier, I had joined the laboratory of David Hogness, a superb molecular biologist and one of the creators of recombinant DNA technology. Although Hogness worked on fruit fly development, he found the idea of looking at color vision and its variation intriguing, and he gave me his full support for this somewhat unconventional research project.

Jeremy Nathans

Our strategy was first to isolate the gene encoding rhodopsin, the abundant and better characterized visual pigment in rods, and then to use the cloned rhodopsin gene as a hybridization probe to identify related sequences in human genomic DNA. (In accordance with the Socratic dictum "Know thyself," I used my own DNA for these experiments.) If cone and rod visual pigments were derived from a common ancestral pigment, their genes might share enough sequence similarity that one strand from the cloned rod pigment gene could form a stable, albeit imperfect, double helix with the complementary strand from a cone pigment gene. After 4 years of hard work, we were able to assemble the complete amino acid sequences of human rod and cone pigments, as deduced from their gene sequences. In retrospect, we realized that the sequence similarity between rod and cone pigment genes was just barely high enough for our strategy to work. With the visual pigment genes in hand, we were able to show that variant red-green color vision results from rearrangements in the red and green cone pigment genes.

In more recent years, I have continued to pursue the biology of visual pigments and their role in human vision in collaboration with my students Isabel Chiu, Shannath Merbs, Hui Sun, Ching-Hwa Sung, Yanshu Wang, Chuck Weitz, and Don Zack. We have also extended the research to include studies of clinically significant visual impairment, such as retinitis pigmentosa.

While much has been learned about visual pigments and their role in vision, much remains unknown. Now, as a mentor to younger scientists, I hope to create that special combination of right time and place for those who would tackle these still unanswered questions.

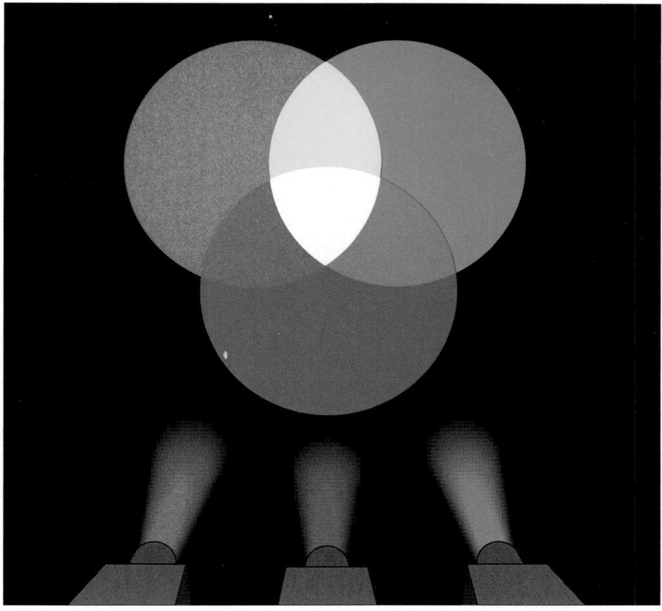

Figure 9.21
Mixing colored lights. The mixing of red, green, and blue light causes equal activation of the three types of cones, and the perception of "white" results.

Dark and Light Adaptation

This transition from all-cone daytime vision to all-rod nighttime vision is not instantaneous; it takes about 20–25 minutes (hence the time needed to get oriented in the star-gazing exercise above). This phenomenon is called **dark adaptation**, or getting used to the dark. Sensitivity to light actually increases a millionfold during this period. Dark adaptation is explained by a number of factors, including dilation of the pupils, regeneration of unbleached rhodopsin, and adjustment of the functional circuitry of the retina so that information from more rods is available to each ganglion cell. Because of this tremendous increase in sensitivity, when the dark-adapted eye goes back into bright light, it is temporarily saturated. This explains what happens when you first go outside on a bright day. Over the next 5–10 minutes, the eyes undergo **light adaptation**, reversing the changes in the retina that accompanied dark adaptation. This light-dark adaptation in the duplex retina gives our visual system the ability to operate in light intensities ranging from moonless midnight to bright high noon.

RETINAL PROCESSING

Well before the discovery of how photoreceptors work, researchers were able to explain some of the ways the retina processes visual images. Since about 1950, neuroscientists have studied the action potential discharges of retinal ganglion cells as the retina is stimulated with light. The pioneers of this approach were neurophysiologists Keffer Hartline, Stephen Kuffler, and Horace Barlow, with Hartline and Kuffler working in the United States and Barlow working in England. Their research uncovered which aspects of a visual image were encoded as ganglion cell output. Early studies of horseshoe crabs and frogs gave way to research on cats and monkeys. It was found that similar principles are involved in retinal processing across a wide range of species.

Progress in understanding how ganglion cell properties are generated by synaptic interactions in the retina has been slower. This is because *only ganglion cells fire action potentials;* all other cells in the retina (except some amacrine cells) respond to stimulation with graded changes in membrane

Box 9.5

OF SPECIAL INTEREST

Genetic Mistakes and the Colors We See

The color we perceive is largely determined by the relative amounts of light absorbed by the red, green, and blue visual pigments in our cones. For example, the perception of yellow light (Figure A) can be matched by an appropriate mixture of red and green light (Figure B). Because we use a three-color system, humans are *trichromats*. However, not all normal trichromats perceive colors the same. For example, if a population of people is asked to choose the wavelength of light that most appears green without being yellowish or bluish, there will be some variation in the choice. But there are significant abnormalities of color vision that extend well beyond this range of normal trichromatic vision.

Most abnormalities in color vision are the result of small genetic mistakes that lead to the loss of one visual pigment or a shift in the spectral sensitivity of one type of pigment. The most common abnormalities involve red-green color vision, and they are much more common in men than women. The reason for this pattern is that the genes responsible for the red and green pigments are on the X chromosome, whereas the gene that codes for the blue pigment is on chromosome 7. Men have abnormal red-green vision if there is a defect on the single X chromosome they inherit from their mother. Women have abnormal red-green vision only if both parents contribute abnormal X chromosomes.

About 6% of men have a red or green pigment that absorbs somewhat different wavelengths of light than the pigments of the rest of the population. These men are referred to as *anomalous trichromats* because they require somewhat different mixtures of red, green, and blue to see intermediate colors (and white) than other people do. Most anomalous trichromats have normal genes to encode the blue and either the red or green pigments, but they also have a hybrid gene that encodes a protein with an abnormal absorption spectrum between that of normal red and green pigments. For example, a person with an anomalous green pigment can match a yellow light with a red-green mixture containing less red than a normal trichromat (Figure C). Anomalous trichromats perceive the full spectrum of colors that normal trichromats perceive, but in rare instances they disagree about the precise color of an object (e.g., blue versus greenish blue).

About 2% of men actually lack either the red or the green pigment, making them red-green color-blind. Because this leaves them with a two-color system, they are referred to as *dichromats*. People lacking the green pigment are less sensitive to green and confuse certain red and green colors that appear different to trichromats. A "green dichromat" can match a yellow light with either red or green light; no mixture is needed (Figure D). In contrast to the roughly 8% of men who are either missing one pigment or have an anomalous pigment, only about 1% of women have such color abnormalities.

People without one color pigment are considered color-blind, but they actually perceive quite a colorful world. Estimates of the number of people lacking all color vision are

potential. The detection of such graded changes requires technically challenging intracellular recording methods, whereas action potentials can be detected using simple extracellular recording methods (see Box 4.1). It was not until the early 1970s that John Dowling and Frank Werblin of Harvard University were able to show how ganglion cell responses are built from the interactions of horizontal and bipolar cells.

The most direct path for information flow in the retina is from a cone photoreceptor to bipolar cell to ganglion cell. At each synaptic relay, the responses are modified by the lateral connections of horizontal cells and amacrine cells. We'll first focus on how information is transformed as it passes from photoreceptors to bipolar cells, and then we'll explore ganglion cell output in the last section.

Transformations in the Outer Plexiform Layer

Photoreceptors, like other neurons, release neurotransmitter when depolarized. As we have seen, photoreceptors are depolarized in the dark and are *hy-*

(Box 9.5, *continued*)

quite variable, but less than about 0.001% of the population is thought to have this condition. In one type, both red and green cone pigments are missing, in many cases because mutations of the red and green genes make them nonfunctional. These people are blue cone *monochromats* and live in a world that varies only in lightness, like a trichromat's perception of a black-and-white movie.

Recent research has shown that, precisely speaking, there may not be such a thing as normal color vision. In a group of males classified as normal trichromats, it was found that some require slightly more red than others to perceive yellow in a red-green mixture. This difference, which is tiny compared with the deficits discussed above, results from a single alteration of the red pigment gene. The 60% of males that have the amino acid serine at site 180 in the red pigment gene are more sensitive to long-wavelength light than the 40% that have the amino acid alanine at this site. Imagine what would happen if a woman had different red gene varieties on her two X chromosomes. Both red genes should be expressed, leading to different red pigments in two populations of cones. In principle, such women should have a form of tetrachromatic color vision, a rarity among all animals.

Figure A
Yellow light.

Figure B
Color mixture for a normal trichromat to perceive yellow.

Figure C
Color mixture for an anomalous trichromat to perceive yellow.

Figure D
Color mixture for a "green dichromat" to perceive yellow. (Source: Adapted from Nathans, 1989.)

perpolarized by light. We thus have the counterintuitive situation in which photoreceptors actually release fewer transmitter molecules in the light than in the dark. We can reconcile this apparent paradox, however, if we take the point of view that *dark* rather than *light* is the preferred stimulus for a photoreceptor. Thus, when a shadow passes across a photoreceptor, it responds by depolarizing and releasing neurotransmitter. The transmitter substance of the photoreceptor is thought to be the amino acid glutamate.

In the outer plexiform layer, each photoreceptor is in synaptic contact with two types of retinal neuron: bipolar cells and horizontal cells. Remember that bipolar cells create the direct pathway from photoreceptors to ganglion cells; horizontal cells feed information laterally in the outer plexiform layer to influence the activity of neighboring bipolar cells.

Bipolar Cell Receptive Fields. Bipolar cells can be categorized into two classes, based on their responses to the glutamate released by photoreceptors. In **OFF bipolar cells**, glutamate-gated cation channels mediate a classical depolarizing excitatory postsynaptic potential (EPSP) from the influx of Na^+ ions. **ON bipolar cells** respond to glutamate by hyperpolarizing (evidently using G-protein-coupled receptors). Notice that the names OFF and ON refer to whether these cells depolarize in response to light *off* (more glutamate) or to light *on* (less glutamate).

Each bipolar cell receives direct synaptic input from a cluster of photoreceptors. The number of photoreceptors in this cluster ranges from one at the center of the fovea to thousands in the peripheral retina. In addition to these direct connections with photoreceptors, bipolar cells are connected via horizontal cells to a circumscribed ring of photoreceptors that surrounds this central cluster. The **receptive field** of a bipolar cell (or any other cell in the visual system) is *the area of retina that, when stimulated with light, changes the cell's membrane potential.* The receptive field of a bipolar cell consists of two parts: a circular area of retina providing direct photoreceptor input, called the *receptive field center*, and a surrounding area of retina providing input via horizontal cells, called the *receptive field surround* (Figure 9.22a). Receptive field dimensions can be measured in millimeters across the retina or, more commonly, in degrees of visual angle. One millimeter on the retina corresponds to a visual angle of about 3.5 degrees. Bipolar cell receptive field diameters range from a fraction of a degree in the central retina to several degrees in the peripheral retina.

The response of a bipolar cell's membrane potential to light in the receptive field center is opposite to that of light in the surround. For example, if illumination of the center causes depolarization of the bipolar cell (an ON response), then illumination of the surround will cause an antagonistic hyperpolarization of the bipolar cell (Figure 9.22b, c). Likewise, if the cell is depolarized by a spot turning from light to dark in the center of its receptive field (an OFF response), it will be hyperpolarized by the same dark stimulus applied to the surround. Thus, these cells are said to have antagonistic **center-surround receptive fields**.

The center-surround receptive field organization is passed on from bipolar cells to ganglion cells via synapses in the inner plexiform layer. The lateral connections of the amacrine cells in the inner plexiform layer also contribute to the elaboration of ganglion cell receptive fields, but so far the precise contributions of these connections remain poorly understood.

RETINAL OUTPUT

The sole source of output from the retina to the rest of the brain is the action potentials arising from the million or so ganglion cells. The activity of these cells can be recorded electrophysiologically not only in the retina but also in the optic nerve, where their axons travel.

Ganglion Cell Receptive Fields

Most retinal ganglion cells have the concentric center-surround receptive field organization discussed above for bipolar cells. ON-center and OFF-center ganglion cells receive input from the corresponding type of bipolar cell. Thus, an ON-center ganglion cell will be depolarized and respond with a barrage of action potentials when a small spot of light is projected onto the middle of its receptive field. Likewise, an OFF-center cell will respond to a small dark spot presented to the middle of its receptive field. In both types of cell, however, the response to stimulation of the center is canceled by the response to stimulation of the surround (Figure 9.23). The surprising implication is that most retinal ganglion cells are not particularly responsive to changes in illumination that include both the receptive field center and the receptive

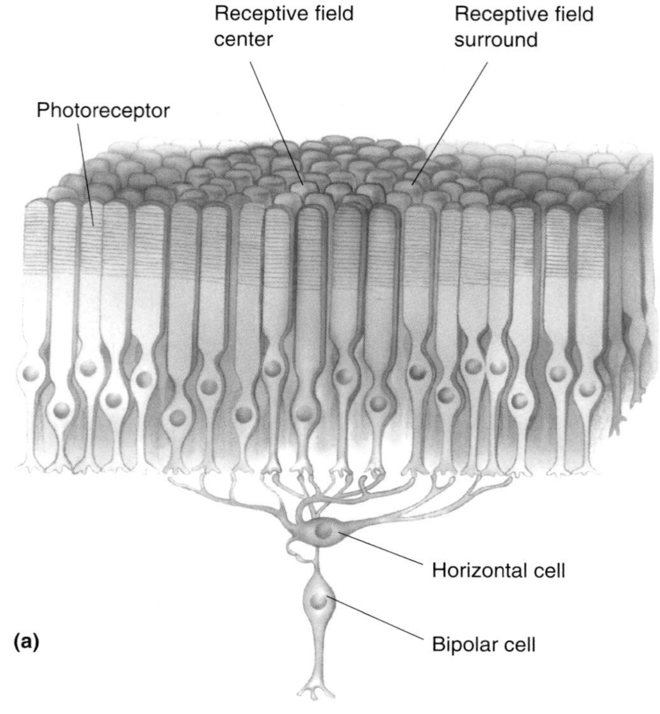

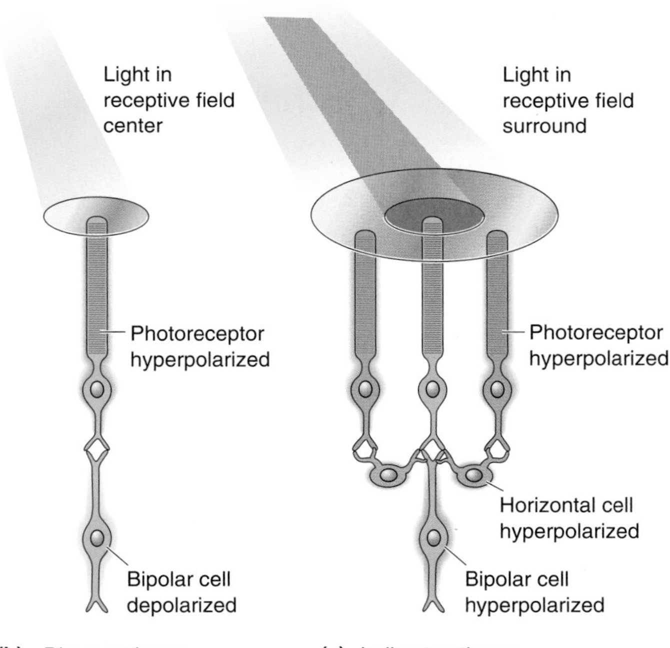

Figure 9.22

Direct and indirect pathways from photoreceptor to bipolar cell. (a) Bipolar cells receive direct synaptic input from a cluster of photoreceptors constituting the receptive field center. In addition, they receive indirect input from surrounding photoreceptors via horizontal cells. **(b)** An ON-center bipolar cell is depolarized by light in the receptive field center via the direct pathway. **(c)** Light in the receptive field surround hyperpolarizes the ON-center bipolar cell via the indirect pathway. Because of the intervening horizontal cell, the effect of light on the surround photoreceptors is always opposite the effect of light on the center photoreceptors.

field surround. Rather, it appears that the ganglion cells are mainly responsive to *differences* in illumination that occur within their receptive fields.

To illustrate this point, consider the response generated by an OFF-center cell as a dark shadow crosses its receptive field (Figure 9.24). Remember that in such a cell, dark in the center of the receptive field causes the cell to depolarize, whereas dark in the surround causes the cell to hyperpolarize. In uniform illumination, the center and surround cancel to yield some low level of response (Figure 9.24a). When the shadow enters the surround region of the

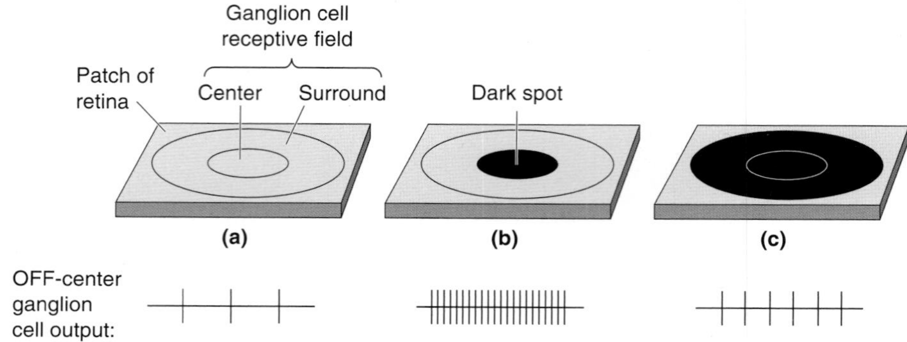

Figure 9.23

A center-surround ganglion cell receptive field. (a, b) An OFF-center ganglion cell responds with a barrage of action potentials when a dark spot is imaged on its receptive field center. **(c)** If the spot is enlarged to include the receptive field surround, the response is greatly reduced.

receptive field without encroaching on the center, it has the effect of hyperpolarizing the neuron, leading to a decrease in the cell's firing rate (Figure 9.24b). As the shadow begins to include the center, however, the partial inhibition by the surround is overcome, and the cell response increases (Figure 9.24c). But when the shadow finally fills the entire surround, the center response is again canceled (Figure 9.24d). Notice that the cell response in this example is only slightly different in uniform light and in uniform dark; the response is modulated mainly by the presence of a light-dark border in its receptive field.

Now let's consider the output of *all* of the OFF-center ganglion cells that are stimulated by a stationary light-dark border imaged on the retina. The responses fall into the four categories illustrated in Figure 9.24. Thus, the cells that register the presence of this shadow are those with receptive field centers and surrounds that are differentially affected by the border. The population of cells with receptive field centers "viewing" the light side of the border will be inhibited (Figure 9.24b). The population of cells with centers "viewing" the dark side of the border will be excited (Figure 9.24c). In this way, the difference in illumination at a light-dark border is not faithfully represented by the difference in the output of ganglion cells on either side of the border. Instead, *the center-surround organization of the receptive fields leads to a neural response that exaggerates the contrast at borders.*

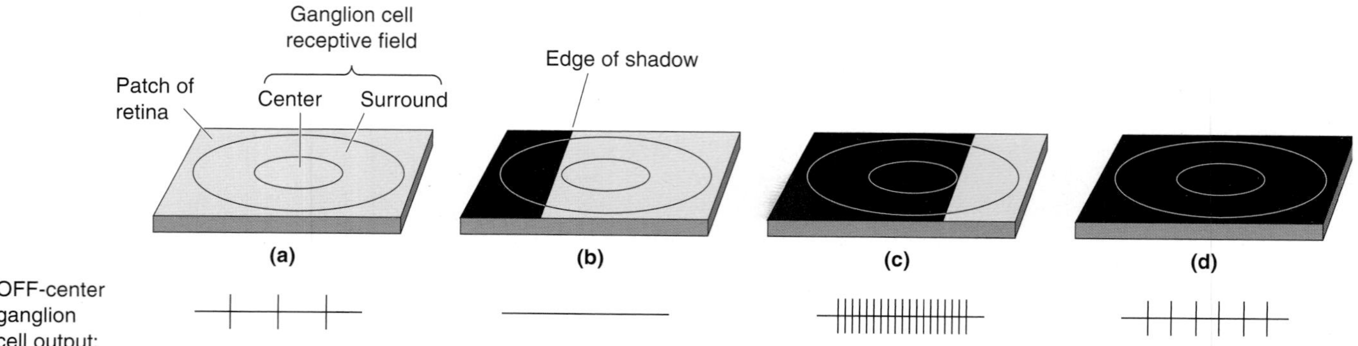

Figure 9.24

Ganglion cell output in response to a light-dark border falling within its receptive field.

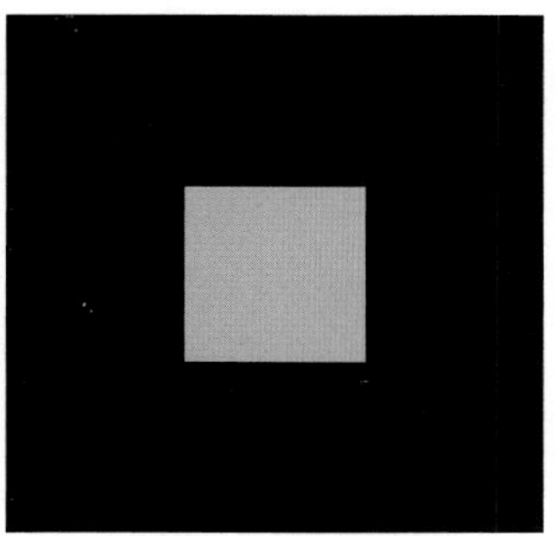

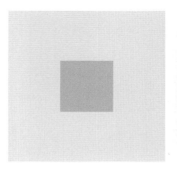

Figure 9.25
The influence of contrast on the perception of light and dark. The central boxes are identical shades of gray, but because the border is lighter on the left, the left box appears darker.

From the organization of ganglion cell receptive fields we can infer that our visual system is specialized to detect local spatial variations rather than the absolute magnitudes of light falling on the retina. Thus, the perception of lightness or darkness is not absolute but relative. This is demonstrated by the two framed boxes in Figure 9.25. Even though the central boxes are shaded identically, the left box *appears* darker than the right box because the left one is framed by a lighter background shading. Similarly, the print on this page appears black on a white background both in a dimly lit room and outside at high noon. This perceptual constancy cannot be explained by the absolute light intensities reflected by the print and the page. In fact, the "white" page reflects less light in the dorm than the "black" print does outside at high noon. Again, what matters is not the absolute light intensity, but the *difference* in light intensities that arise from the print and the page, and this difference remains constant despite the change in illumination.

Types of Ganglion Cells

Most ganglion cells in the mammalian retina have a center-surround receptive field with either an ON or an OFF center. They can be further categorized on the basis of their appearance, connectivity, and electrophysiological properties. For example, the ganglion cells in the cat retina can be divided into three categories (α, β, and γ) on the basis of the size of their somata and dendritic arbors. Another classification scheme (W, X, Y) developed for the cat was based on the visual response properties of ganglion cells. As it turned out, α cells and Y cells were one and the same, as were β cells and X cells.

Studies of the cat retina helped guide subsequent research on the primate retina. In the macaque monkey retina and human retina, two major types of ganglion cells are distinguished: large **M-type ganglion cells** and smaller **P-type ganglion cells**. (M stands for *magno*, from the Latin for "large"; P stands for *parvo*, from the Latin for "small."). Figure 9.26 shows the relative sizes of M and P ganglion cells at the same location on the retina. P cells constitute about 90% of the ganglion cell population, M cells about 5%, and the remaining 5% is made up of a variety of **nonM-nonP ganglion cell** types that are less well characterized.

The visual response properties of M cells differ from those of P cells in several ways. They have larger receptive fields, conduct action potentials more rapidly in the optic nerve, and are more sensitive to low-contrast stimuli. In addition, M cells respond to stimulation of their receptive field centers with a transient burst of action potentials, while P cells respond with a sustained discharge as long as the stimulus is on (Figure 9.27). The prevailing view is that M cells are particularly important for the detection of stimulus movement, while P cells are more sensitive to stimulus form and fine detail.

Color-Opponent Ganglion Cells. Another important distinction between ganglion cell types is that P cells and some nonM-nonP cells are sensitive to

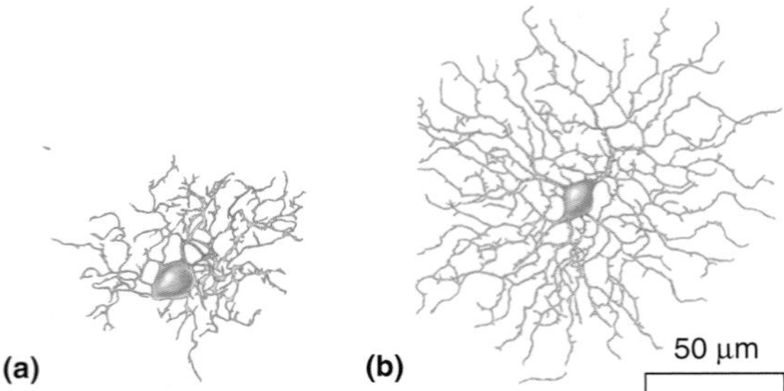

Figure 9.26
P and M ganglion cells in the macaque monkey retina. (a) A small P cell from the peripheral retina. **(b)** An M cell from a similar retinal location is significantly larger. (Source: Watanabe and Rodieck, 1989, pp. 437, 439.)

differences in the wavelength of light. The majority of these color-sensitive neurons are called **color-opponent cells**, reflecting the fact that the response to one wavelength in the receptive field center is canceled by showing another wavelength in the receptive field surround. In P cells, the colors that cancel each other are red and green, and in the nonM-nonP cells, the opponent colors are blue and yellow. Consider, for example, a cell with a red ON center and a green OFF surround (Figure 9.28). The center of the receptive field is fed mainly by red cones; therefore, the cell responds to red light by firing action potentials. Even a red light that bathes the entire receptive field is an effective stimulus, because the surround is less sensitive to this wavelength. The response to red is canceled only by green light on the surround. Shorthand notation for such a cell is R^+G^-, meaning simply that it is excited by red in the receptive field center and inhibited by green in the surround. What would be the response to white light on the entire receptive field? Because white light contains all visible wavelengths, both center and surround would be equally activated, thereby canceling the response of the cell.

Blue-yellow color opponency works the same way. Consider a cell with a blue ON center and a yellow OFF surround (B^+Y^-). Blue light drives blue cones that feed the receptive field center, while yellow light activates both red and green cones that feed the surround. Again, diffuse blue light would be an effective stimulus for this cell, but yellow on the surround would can-

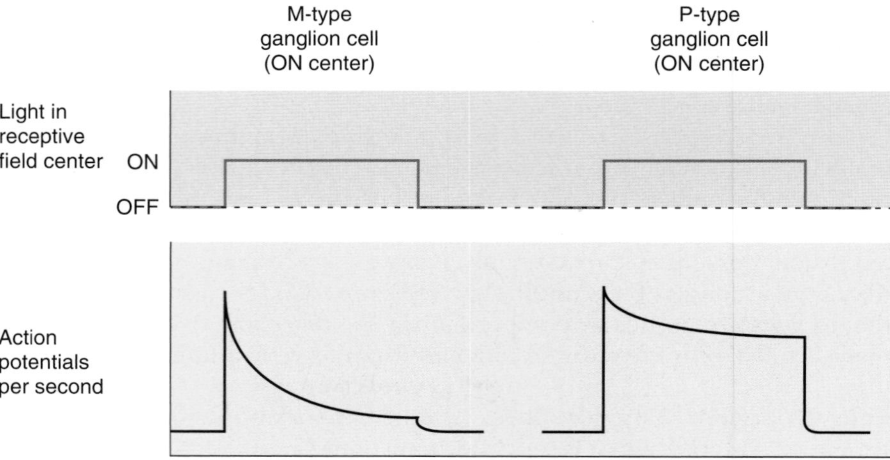

Figure 9.27
Different responses to light of M-type and P-type ganglion cells.

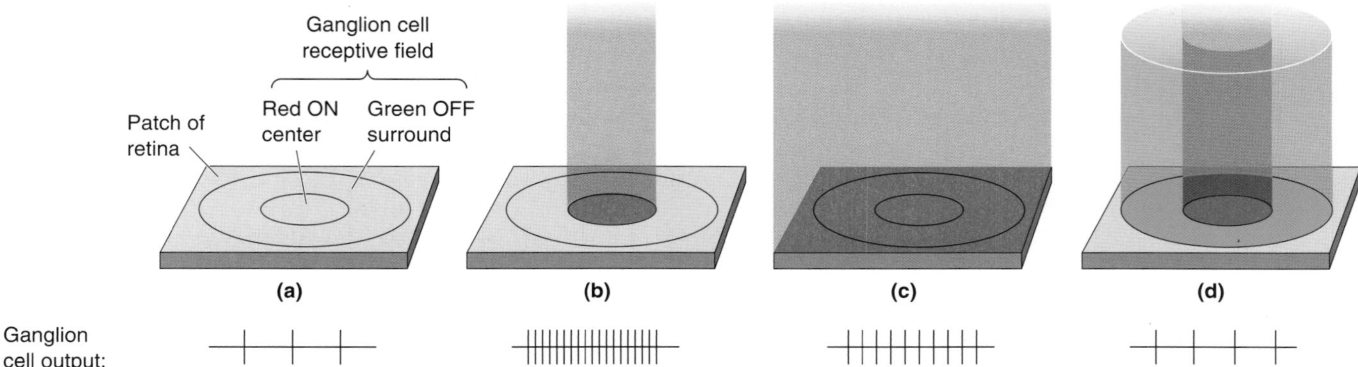

Figure 9.28
A color-opponent center-surround receptive field of a P-type ganglion cell.

cel the response, as would diffuse white light. The lack of color opponency in M cells is accounted for by the fact that both the center and surround of the receptive field receive input from more than one type of cone.

Perceived color is based on the relative activity of ganglion cells whose receptive field centers receive input from red, green, and blue cones. Demonstrate this to yourself by fixating on the cross in the middle of the red box in Figure 9.29 for a minute or so. This will have the effect of fatiguing some of the red cones. Then look at the white box. The activation of the green cones by the white light is unopposed, and you see a green square. Similarly, if you fixate on the blue box, you will see yellow when you shift your gaze to the white box. Thus, it appears that the ganglion cells provide a stream of information to the brain that is involved in the spatial comparison of three opposing processes: light versus dark, red versus green, and blue versus yellow.

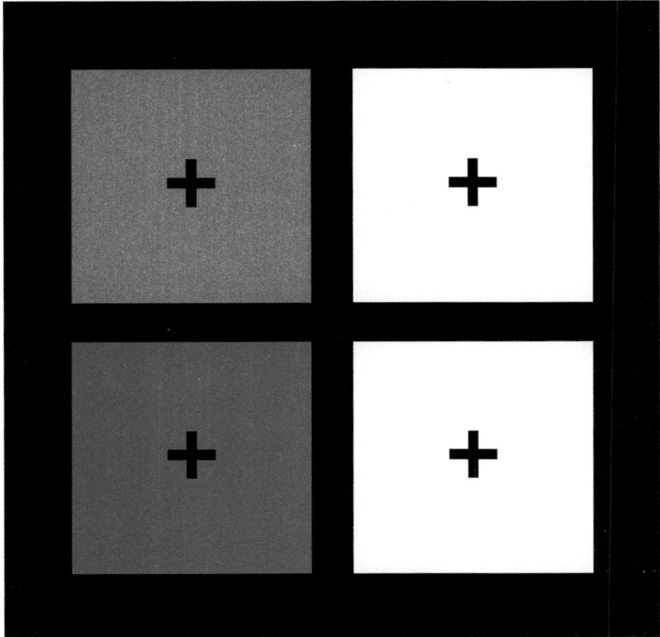

Figure 9.29
Color opponency revealed. Fixate on the cross in the red box on the left for 60 seconds, then shift your gaze to the cross in the white box. What color do you see? Try it again with the blue box.

Parallel Processing

One of the important concepts that emerges from our discussion of the retina is the idea of **parallel processing** in the visual system. Here's why. First, we view the world with not one but two eyes that provide two parallel streams of information. In the central visual system these streams are compared to give information about *depth*, the distance of an object from the observer. Second, there appear to be independent streams of information about light and dark that arise from the ON-center and OFF-center ganglion cells in each retina. Third, ganglion cells of both ON and OFF varieties have different types of receptive fields and response properties. M cells can detect subtle contrasts over their large receptive fields and are likely to contribute to low-resolution vision. P cells have small receptive fields that are well suited for the discrimination of fine detail. P cells and nonM-nonP cells are specialized for the separate processing of red-green and blue-yellow information.

CONCLUDING REMARKS

In this chapter, we have seen how light emitted by or reflected off objects in space can be imaged by the eye onto the retina. Light energy is first converted into membrane potential changes in the mosaic of photoreceptors. The transduction mechanism in photoreceptors is very similar to that in olfactory receptor cells, both of which involve cyclic nucleotide-gated ion channels. Photoreceptor membrane potential is converted into a chemical signal (the neurotransmitter glutamate), which is again converted into membrane potential changes in the postsynaptic bipolar and horizontal cells. This process of electrical-to-chemical-to-electrical signaling repeats again and again, until the presence of light or dark or color is finally converted to a change in the action potential firing frequency of the ganglion cells.

The information from the 125 million photoreceptors is funneled into 1 million ganglion cells. In the central retina, particularly the fovea, relatively few photoreceptors feed each ganglion cell, whereas in the peripheral retina, thousands of receptors do. Thus, the mapping of visual space onto the array of optic nerve fibers is not uniform. Rather, in "neural space," there is an overrepresentation of the central few degrees of visual space, and signals from individual cones are more important. This specialization ensures high acuity in central vision but also requires that the eye move to bring the images of objects of interest onto the fovea.

As we shall see in the next chapter, there is good reason to believe that the different types of information that arise from different types of ganglion cells are, at least in the early stages, processed independently. Parallel streams of information—for example, from the right and left eyes—remain segregated at the first synaptic relay in the lateral geniculate nucleus of the thalamus. The same can be said for the M-cell and P-cell synaptic relays in the LGN. In the visual cortex, it appears that parallel paths may process different visual attributes. For example, the distinction in the retina between neurons that do and do not convey information about color is preserved in the visual cortex. In general, each of the more than two dozen visual cortical areas may be specialized for the analysis of different types of retinal output.

KEY TERMS

Introduction
vision (p. 281)
retina (p. 281)

Properties of Light
refraction (p. 283)

The Structure of the Eye
pupil (p. 283)
iris (p. 283)
cornea (p. 283)
aqueous humor (p. 283)
sclera (p. 283)
extraocular muscle (p. 283)
conjunctiva (p. 284)
optic nerve (p. 284)
optic disk (p. 284)
macula (p. 284)
fovea (p. 284)
lens (p. 285)
ciliary muscle (p. 285)
vitreous humor (p. 286)

Image Formation by the Eye
diopter (p. 286)
accommodation (p. 288)
pupillary light reflex (p. 289)
visual field (p. 289)
visual acuity (p. 289)
visual angle (p. 289)

Microscopic Anatomy of the Retina
photoreceptor (p. 290)
bipolar cell (p. 290)
ganglion cell (p. 290)
horizontal cell (p. 290)
amacrine cell (p. 291)
ganglion cell layer (p. 292)
inner nuclear layer (p. 292)
outer nuclear layer (p. 292)
layer of photoreceptor outer segments (p. 292)
inner plexiform layer (p. 292)
outer plexiform layer (p. 292)
rod photoreceptor (p. 292)
cone photoreceptor (p. 292)

Phototransduction
dark current (p. 294)
cyclic guanosine monophosphate (cGMP) (p. 294)
rhodopsin (p. 296)
transducin (p. 296)
phosphodiesterase (PDE) (p. 296)
Young-Helmholtz trichromacy theory (p. 299)
dark adaptation (p. 301)
light adaptation (p. 301)

Retinal Processing
OFF bipolar cell (p. 303)
ON bipolar cell (p. 304)
receptive field (p. 304)
center-surround receptive field (p. 304)

Retinal Output
M-type ganglion cell (p. 307)
P-type ganglion cell (p. 307)
nonM-nonP ganglion cell (p. 307)
color-opponent cell (p. 308)
parallel processing (p. 310)

REVIEW QUESTIONS

1. What physical property of light is most closely related to the perception of color?
2. Name eight structures in the eye that light passes through before it strikes the photoreceptors.
3. Why is a scuba mask necessary for clear vision under water?
4. What is myopia, and how is it corrected?
5. Give three reasons explaining why visual acuity is best when images fall on the fovea.
6. How does the membrane potential change in response to a spot of light in the receptive field center of a photoreceptor? Of an ON bipolar cell? Of an OFF-center ganglion cell? Why?
7. What happens in the retina when you "get used to the dark?" Why can't you see color at night?
8. In what way is retinal output *not* a faithful reproduction of the visual image falling on the retina?
9. In retinitis pigmentosa, early symptoms include the loss of peripheral vision and night vision. Loss of what type of cells could lead to such symptoms prior to any blindness?

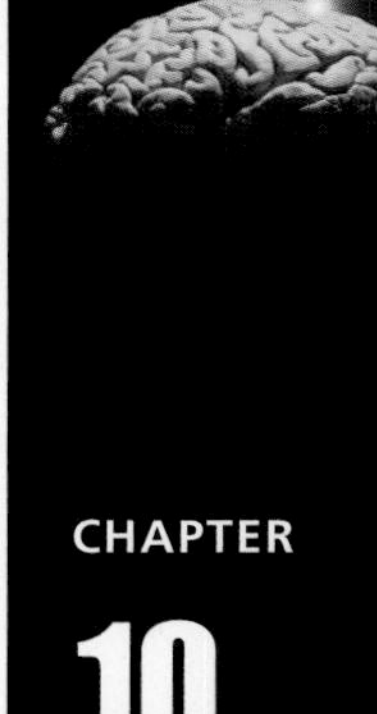

CHAPTER

10

The Central Visual System

INTRODUCTION

Although our visual system provides us with a unified picture of the world around us, this picture has multiple facets. Objects we see have shape and color. They have position in space, and sometimes they move. For us to see each of these properties, neurons somewhere in the visual system must be sensitive to them. Moreover, because we have two eyes, we actually have two visual images in our head, and somehow they must be merged.

In Chapter 9, we saw that in many ways the eye acts like a camera. But starting with the retina, the rest of the visual system is far more elaborate, far more interesting, and capable of doing far more than any camera. For example, we saw that the retina does not simply pass along information about the patterns of light and dark that fall on it. Rather, the retina extracts information about different facets of the visual image. What we perceive about the world around us, therefore, depends on what information is extracted by the retina and how this information is analyzed and interpreted by the rest of the CNS. An example is color. There is no such thing as color in the physical world; there is simply a spectrum of visible wavelengths of light that are reflected by objects around us. Based on the information extracted by the three types of cone photoreceptors, however, our brain somehow synthesizes a rainbow of colors and fills our world with it.

In this chapter, we explore how the information extracted by the retina is analyzed by the central visual system. The pathway serving conscious visual perception includes the *lateral geniculate nucleus (LGN)* of the thalamus and the primary visual cortex, also called *area 17*, *V1*, or *striate cortex*. We will see that the information funneled through this geniculocortical pathway is segregated into separate parallel processing channels that are specialized for the analysis of different stimulus attributes. The striate cortex then feeds this information to over two dozen different extrastriate cortical areas in the temporal and parietal lobes, and many of these appear to be specialized for different types of analysis.

Much of what we know about the central visual system was first worked out in the domestic cat and then extended to the rhesus monkey, *Macaca mulatta*. The macaque monkey, as it is also called, relies heavily on vision for survival in its habitat, as do we humans. In fact, tests of the performance of this primate's visual system show that in virtually all respects, it rivals that of humans. Thus, although most of this chapter concerns the organization of the macaque visual system, most neuroscientists agree that it approximates very closely the situation in our own brain.

Although visual neuroscience cannot yet explain many aspects of visual perception (some interesting examples are shown in Figure 10.1), significant progress has been made in answering a more basic question: How do neurons represent the different facets of the visual world? By examining those stimuli that make different neurons in the visual cortex respond and how these response properties arise, we begin to see how the brain portrays the visual world around us.

THE RETINOFUGAL PROJECTION

The neural pathway that leaves the eye, beginning with the optic nerve, is often called the **retinofugal projection**. The suffix *-fugal*, from the Latin for "flee," is commonly used in neuroanatomy to describe a pathway that is directed away from a structure. Thus, a centrifugal projection goes away from the center, a corticofugal projection goes away from the cortex, and the retinofugal projection goes away from the retina.

We begin our tour of the central visual system by looking at how the retinofugal projection courses from each eye to the brain stem on each side

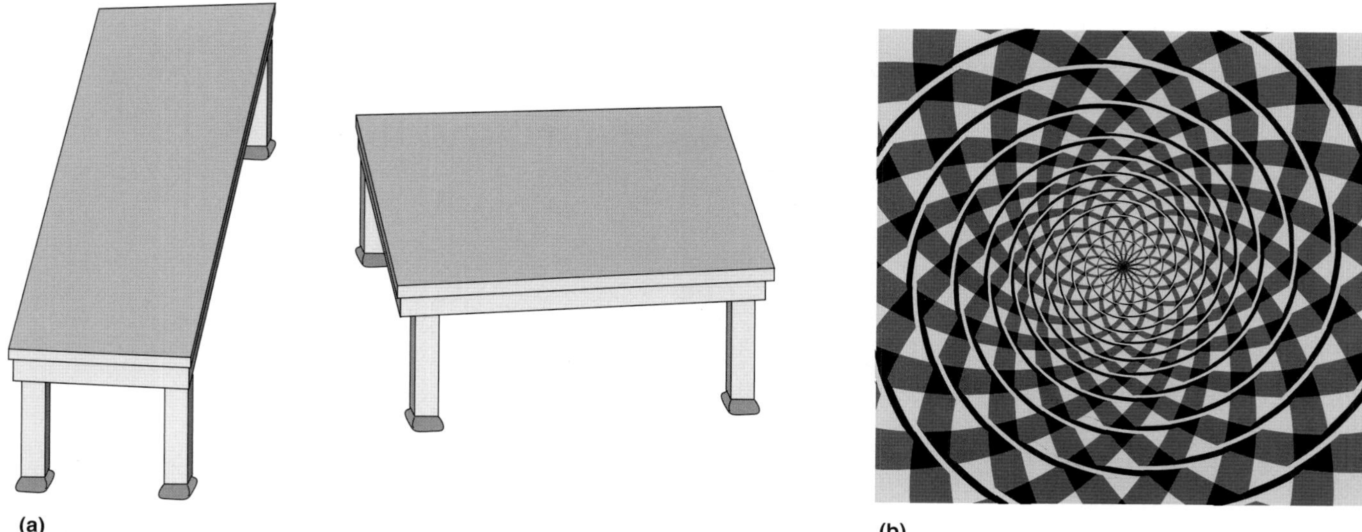

Figure 10.1
Perceptual illusions. (a) The two tabletops have identical dimensions and are imaged on similar-sized patches of retina, but the perceived sizes are quite different. **(b)** This is an illusory spiral. Try tracing it with your finger.

and how the task of analyzing the visual world initially is divided among and organized within certain structures of the brain stem. Then we'll focus on the major arm of the retinofugal projection that mediates conscious visual perception.

The Optic Nerve, Optic Chiasm, and Optic Tract

The ganglion cell axons "fleeing" the retina pass through three structures before they form synapses in the brain stem. The components of this retinofugal projection are, in order, the optic nerve, the optic chiasm, and the optic tract (Figure 10.2). The **optic nerves** exit the left and right eyes at the optic disks, travel through the fatty tissue behind the eyes in their bony orbits, then pass through holes in the floor of the skull. The optic nerves from both eyes combine to form the **optic chiasm**, which lies at the base of the brain, just anterior to where the pituitary gland dangles down. At the optic chiasm, the axons originating in the nasal retinas cross from one side to the other. The crossing of a fiber bundle from one side of the brain to the other is called a **decussation**. Because only the axons originating in the nasal retinas cross, we say that a partial decussation of the retinofugal projection occurs at the optic chiasm. Following the partial decussation at the optic chiasm, the axons of the retinofugal projections form the **optic tracts**, which run just under the pia along the lateral surfaces of the diencephalon.

Right and Left Visual Hemifields

To understand the significance of the partial decussation of the retinofugal projection at the optic chiasm, let's review the concept of the visual field introduced in Chapter 9. The full visual field is the entire region of space (measured in degrees of visual angle) that can be seen with both eyes looking straight ahead. Fix your gaze on a point straight ahead. Now imagine a vertical line passing through the fixation point, dividing the visual field into left and right halves. By definition, objects appearing to the left of the midline are

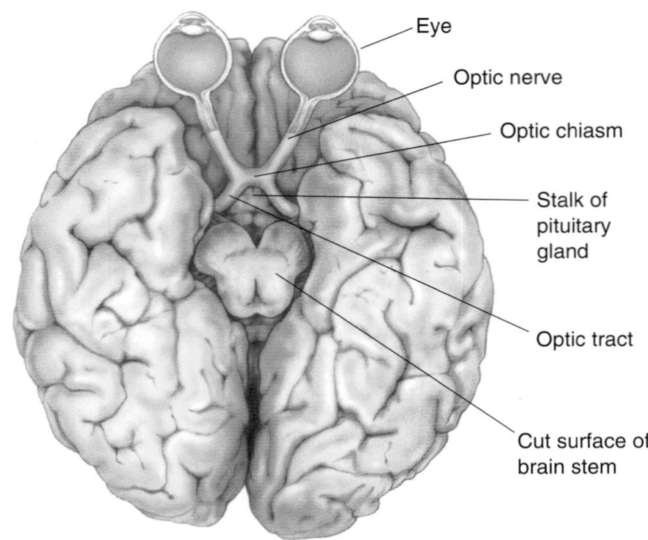

Figure 10.2
Retinofugal projection. This view of the base of the brain shows the optic nerves, optic chiasm, and optic tracts.

in the left **visual hemifield**, and objects appearing to the right of the midline are in the right visual hemifield (Figure 10.3).

By looking straight ahead with both eyes open and then alternately closing one eye and then the other, you will see that the central portion of both visual hemifields is viewed by both retinas. This region of space is therefore called the **binocular visual field**. Notice that objects in the binocular region of the left visual hemifield will be imaged on the nasal retina of the left eye and on the temporal retina of the right eye. Because the fibers from the nasal portion of the left retina cross to the right side at the optic chiasm, all of the information about the left visual hemifield is directed to the right side of the brain. Remember this rule of thumb: Optic nerve fibers cross in the optic chiasm so that *the left visual hemifield is "viewed" by the right hemisphere and the right visual hemifield is "viewed" by the left hemisphere.*

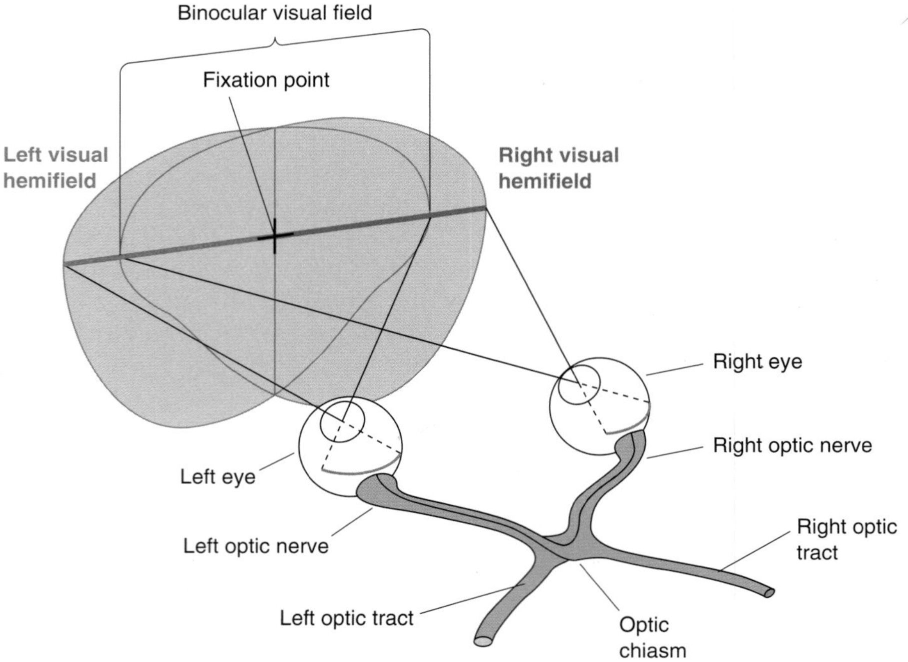

Figure 10.3
Right and left visual hemifields. Ganglion cells in both retinas that are responsive to visual stimuli in the right visual hemifield project axons into the left optic tract. Similarly, ganglion cells viewing the left visual hemifield project into the right optic tract.

Targets of the Optic Tract

A small number of optic tract axons peel off to form synaptic connections with cells in the hypothalamus, and another 10% or so continue past the thalamus to innervate the midbrain. But most of them innervate the **lateral geniculate nucleus (LGN)** of the dorsal thalamus. The neurons in the LGN give rise to axons that project to the primary visual cortex. This projection from LGN to cortex is called the **optic radiation**. Lesions anywhere in the retinofugal projection from eye to LGN to visual cortex cause blindness in humans. Therefore, we know that this pathway mediates conscious visual perception (Figure 10.4).

From our knowledge of how the visual world is represented in the retinofugal projection, we can predict the types of perceptual deficits that result from its destruction at different levels, as might occur from a traumatic injury to the head, a tumor, or an interruption of the blood supply. As shown

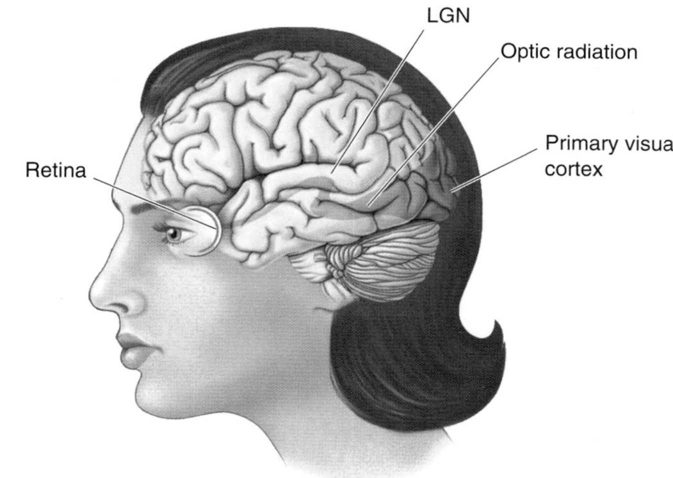

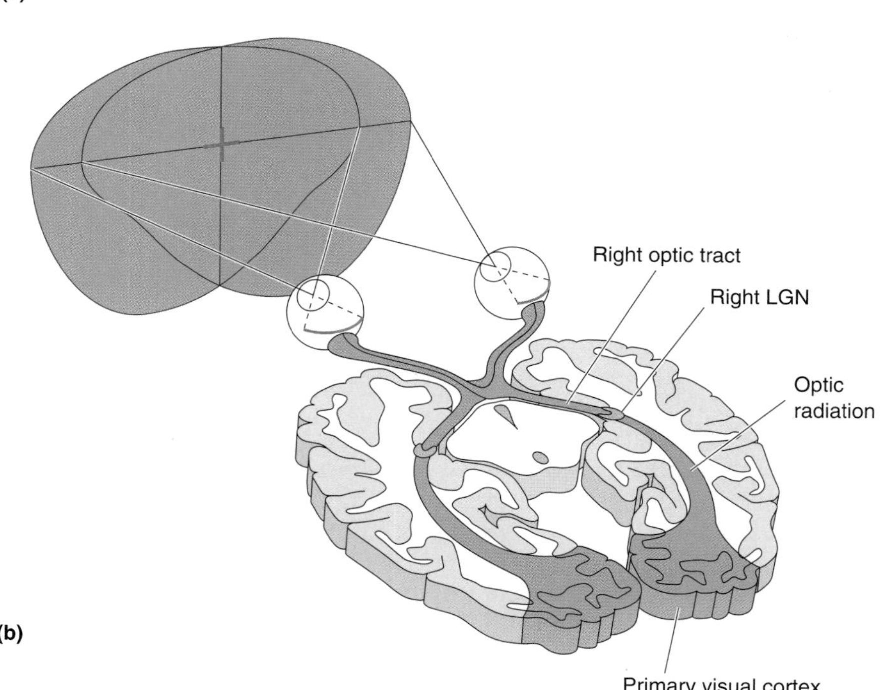

Figure 10.4
The visual pathway that mediates conscious visual perception. (a) A side view of the human brain with the retinogeniculocortical pathway shown inside (blue). **(b)** A horizontal section through the brain exposing the same pathway.

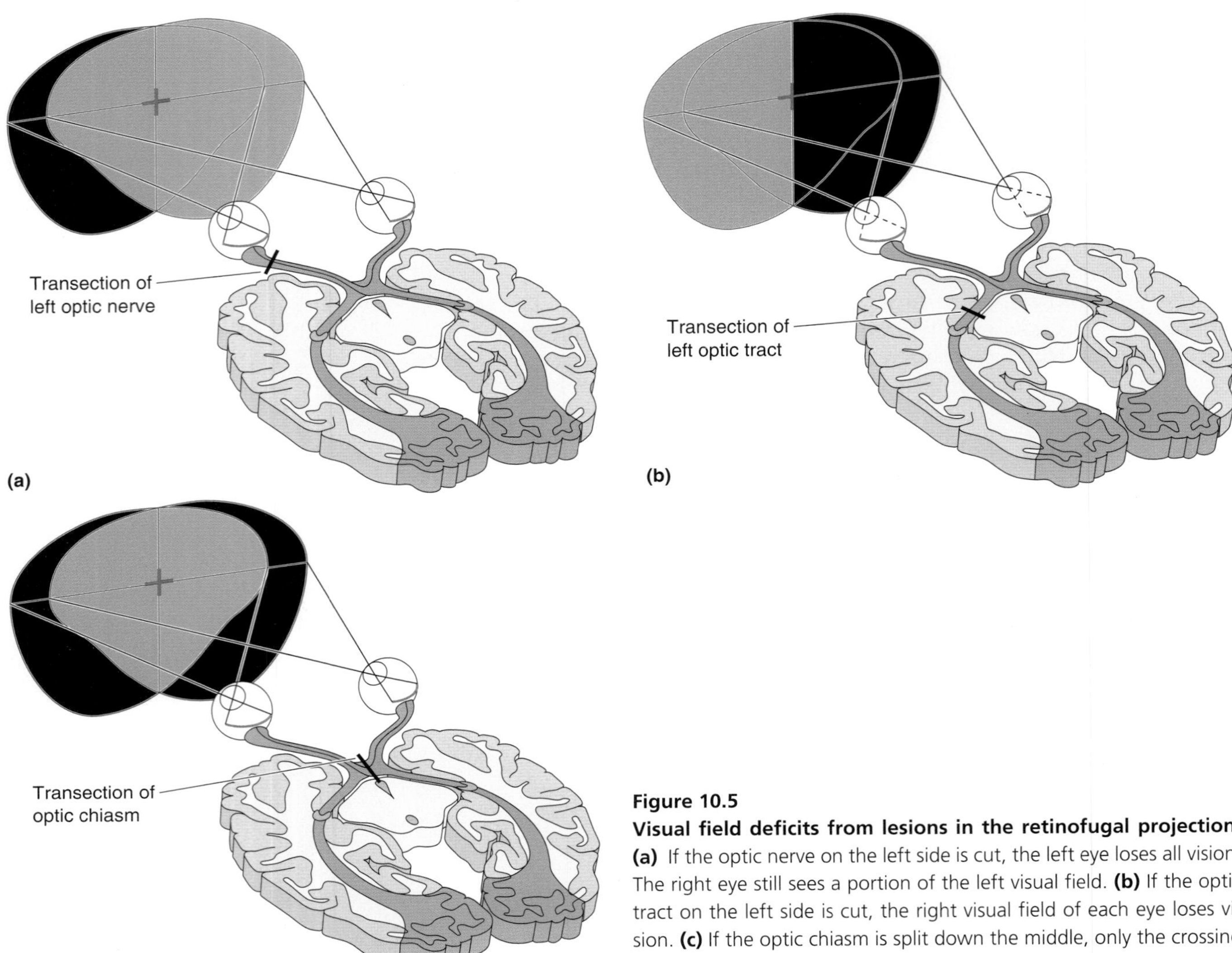

Figure 10.5
Visual field deficits from lesions in the retinofugal projection. **(a)** If the optic nerve on the left side is cut, the left eye loses all vision. The right eye still sees a portion of the left visual field. **(b)** If the optic tract on the left side is cut, the right visual field of each eye loses vision. **(c)** If the optic chiasm is split down the middle, only the crossing fibers are damaged, and both eyes lose peripheral vision.

in Figure 10.5, while a transection of the left optic nerve would render a person blind in the left eye only, a transection of the left optic tract would lead to blindness in the right visual field as viewed through either eye. A midline transection of the optic chiasm would affect only the fibers that cross the midline. Because these fibers originate in the nasal portions of both retinas, blindness would result in the regions of the visual field viewed by the nasal retinas, that is, the peripheral visual fields on both sides (Box 10.1). Because unique deficits result from lesions at different sites, neurologists and neuro-ophthalmologists can locate sites of damage by assessing visual field deficits.

Nonthalamic Targets of the Optic Tract. As we have said, some retinal ganglion cells send axons to innervate structures other than the LGN. Direct projections to part of the hypothalamus play an important role in synchronizing a variety of biological rhythms, including sleep and wakefulness, with the daily dark-light cycle (see Chapter 19). Direct projections to a part of the midbrain called the *pretectum* control the size of the pupil and certain types of eye movement. And about 10% of the ganglion cells in the retina project to a part of the midbrain tectum called the **superior colliculus** (Latin for "little hill") (Figure 10.6).

Box 10.1

OF SPECIAL INTEREST

David and Goliath

Many of you are familiar with the story of David and Goliath, which appears in the Hebrew scriptures (Old Testament). The armies of the Philistines and the Israelites were gathered for battle when Goliath, a Philistine, came forth and challenged the Israelites to settle the dispute by sending out their best man to face him in a fight to the death. Goliath, it seems, was a man of great proportions, measuring over "six cubits" in height. If you consider that a cubit is the distance from the elbow to the tip of the middle finger, about 20 inches, this guy was over 10 feet tall! Goliath was armed to the teeth with body armor, a javelin, and a sword. To face this giant, the Israelites sent David, a young and diminutive shepherd, armed only with a sling and five smooth stones. Here's how the action is described in the Revised Standard Version of the Bible (1 Samuel 17:48):

When the Philistine arose and came and drew near to meet David, David ran quickly toward the battle line to meet the Philistine. And David put his hand in his bag and took out a stone, and slung it, and struck the Philistine on his forehead; the stone sank into his forehead, and he fell on his face to the ground.

Now why, you might ask, are we giving a theology lesson in a neuroscience textbook? The answer is that our understanding of the visual pathway offers an explanation (besides divine intervention) for why Goliath was at a disadvantage in this battle. Body size is regulated by the secretion of growth hormone from the anterior lobe of the pituitary gland. In some cases, the anterior lobe becomes hypertrophied (swollen) and produces excessive amounts of the hormone, resulting in body growth to unusually large proportions. Such individuals, who are called pituitary giants, can measure well over 8 feet tall.

Pituitary hypertrophy also disrupts normal vision. Recall that the optic nerve fibers from the nasal retinas cross in the optic chiasm, which butts up against the stalk of the pituitary. Any enlargement of the pituitary compresses these crossing fibers and results in a loss of peripheral vision called bitemporal hemianopia, or tunnel vision. (See if you can figure out why this is so from what you know about the visual pathway.) We can speculate that David was able to draw close and smite Goliath because when David raced to the battle line, the pituitary giant had completely lost sight of him.

While 10% may not sound like much of a projection, bear in mind that in primates this is about 150,000 neurons, which is equivalent to the total number of retinal ganglion cells in a cat! In fact, the tectum of the midbrain is the major target of the retinofugal projection in all nonmammalian vertebrates (fish, amphibians, birds, and reptiles). In these vertebrate groups, the superior colliculus is called the **optic tectum**. This is why the projection from the retina to the superior colliculus is often called the **retinotectal projection**, even in mammals.

Retinotopy. The retinotectal projection illustrates a general organizational feature of the central visual system called retinotopy. **Retinotopy** is an organization whereby neighboring cells in the retina feed information to neighboring places in their target structures—in this case, the superior colliculus. In this way, the two-dimensional surface of the retina is mapped onto the two-dimensional surface of the colliculus.

There are two important points to remember about retinotopy. First, the mapping of the visual field onto a retinotopically organized structure is often distorted, because visual space is not sampled uniformly by the cells in the retina. Recall from Chapter 9 that there are many more ganglion cells with receptive fields in or near the fovea than in the periphery. Thus, the representation of the visual field is distorted in the superior colliculus: The central few degrees of the visual field are overrepresented, or *magnified*, in the retinotopic

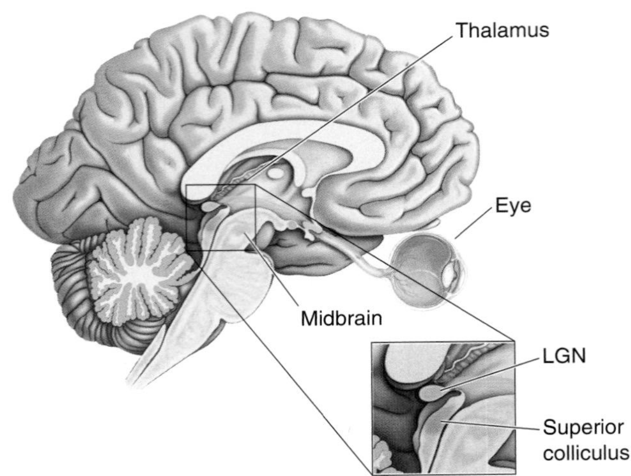

Figure 10.6
Superior colliculus.

map (Figure 10.7). Second, a discrete point of light can activate many cells in the retina and often many more cells in the target structure because of the overlap of receptive fields. The image of a point of light on the retina actually activates a large population of superior colliculus neurons; every neuron that contains that point in its receptive field would be activated. Thus, a position on the superior colliculus in retinotopic coordinates is the site of a peak in the *distribution* of postsynaptic activity that arises when the retina is stimulated with a point of light at the corresponding retinal location.

In the superior colliculus, a patch of neurons activated by a point of light via indirect connections with motor neurons in the brain stem commands eye and head movements to bring the image of this point in space onto the fovea. This branch of the retinofugal projection is thereby involved in orienting the eyes in response to new stimuli in the visual periphery.

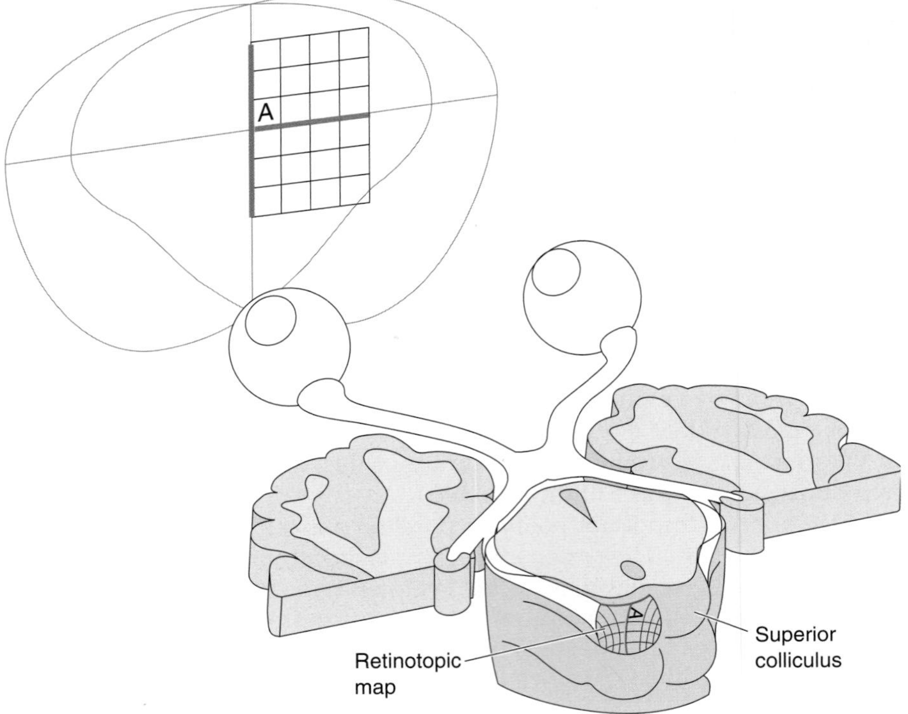

Figure 10.7
A retinotopic map of the superior colliculus. There is a map of visual space in the superior colliculus, but the map is distorted, with more tissue devoted to analysis of the central visual field. Similar maps are found in the LGN and primary visual cortex.

We will return to the superior colliculus when we discuss motor systems in Chapter 14. As we shall now see, however, the basic principles of retinotopy also apply to the LGN and primary visual cortex.

THE LATERAL GENICULATE NUCLEUS

The right and left lateral geniculate nuclei, located in the dorsal thalamus, are the major targets of the two optic tracts. Viewed in cross section, each LGN appears to be arranged in six distinct layers of cells (Figure 10.8). By convention, the layers are numbered 1 through 6, starting with the most ventral, layer 1. In three dimensions, the layers of the LGN are arranged like a stack of six pancakes, one on top of the other. The pancakes do not lie flat, however; they are bent around the optic tract like a knee joint. This shape explains the name "geniculate," from the Latin *geniculatus*, meaning "like a little knee."

The LGN is the gateway to the visual cortex and, therefore, to conscious visual perception. Here we will explore the structure and function of this thalamic nucleus.

The Segregation of Input by Eye and by Ganglion Cell Type

LGN neurons receive synaptic input from the retinal ganglion cells, and most geniculate neurons project an axon to primary visual cortex via the optic radiation. The segregation of LGN neurons into layers suggests that various types of retinal information are being kept separate at this synaptic relay, and indeed this is the case: Axons arising from the M-type, P-type, and nonM-nonP ganglion cells in the two retinas synapse on cells in different LGN layers.

Recall from our rule of thumb that the *right* LGN receives information about the left visual field. The left visual field is viewed by both the nasal left

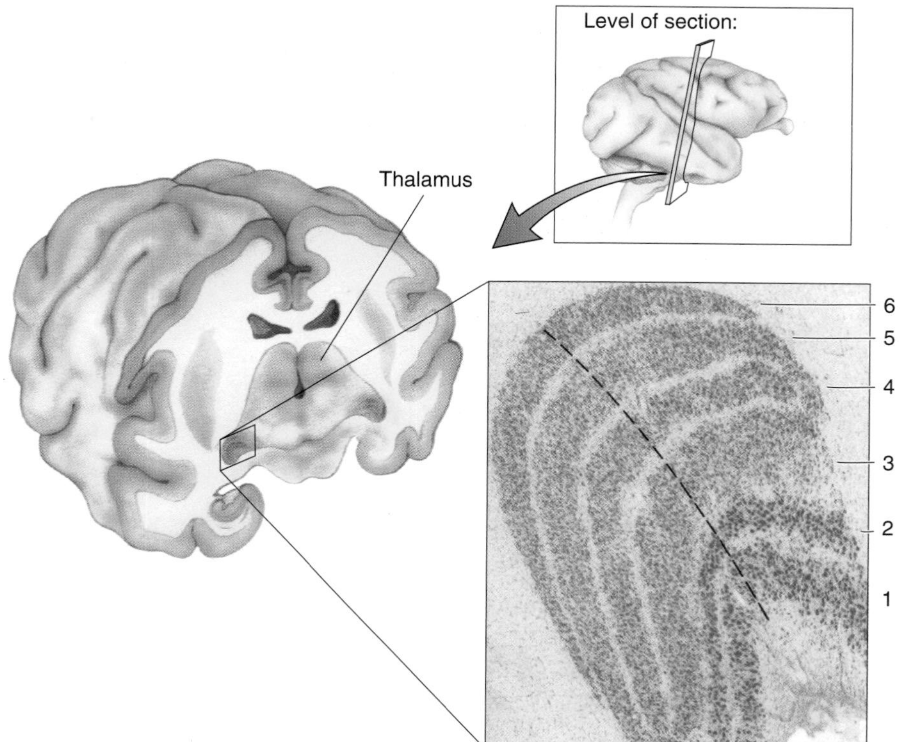

Figure 10.8
LGN of the macaque monkey. The tissue has been stained to show cell bodies, which appear as dots. Notice particularly the six layers and the larger size of the cells in the two ventral layers (layers 1 and 2). (Source: Adapted from Hubel, 1988, p. 65.)

retina and the temporal right retina. At the LGN, input from the two eyes is kept separate. In the right LGN, the right eye (ipsilateral) axons synapse on LGN cells in layers 2, 3, and 5. The left eye (contralateral) axons synapse on cells in layers 1, 4, and 6 (Figure 10.9).

A complete retinotopic map exists in each LGN layer, just as in the superior colliculus. And despite the segregation of left-eye and right-eye fibers, these different maps occur in perfect register. In other words, if we stuck a toothpick perpendicularly through the layers of the LGN, it would pass through cells responding to the same points in visual space viewed alternately by the right and left eyes.

A closer look at the LGN in Figure 10.8 reveals that the two ventral layers, 1 and 2, contain larger neurons, and the four more dorsal layers, 3 through 6, contain smaller cells. The ventral layers are therefore called **magnocellular LGN layers**, and the dorsal layers are called **parvocellular LGN layers**. Recall from Chapter 9 that ganglion cells in the retina may also be classified into magnocellular and parvocellular groups. As it turns out, P-type ganglion cells in the retina project exclusively to the parvocellular LGN, and M-type ganglion cells in the retina project entirely to the magnocellular LGN.

In addition to the neurons in the six principal layers of the LGN, numerous tiny neurons lie just ventral to each layer. Cells in these **koniocellular layers** (*konio-* is from the Greek for "dust") receive input from the nonM-nonP types of retinal ganglion cells and project to visual cortex. Note that the koniocellular layers are not uniquely numbered, because historically, the six thick layers were numbered before cells in the koniocellular layers were discovered. In Chapter 9 we saw that in the retina, M-type, P-type, and nonM-nonP–type ganglion cells respond differently to light and color. In the LGN, the different information derived from the three categories of retinal ganglion cells from the two eyes remains segregated.

The anatomical organization of the LGN supports the idea that the retina

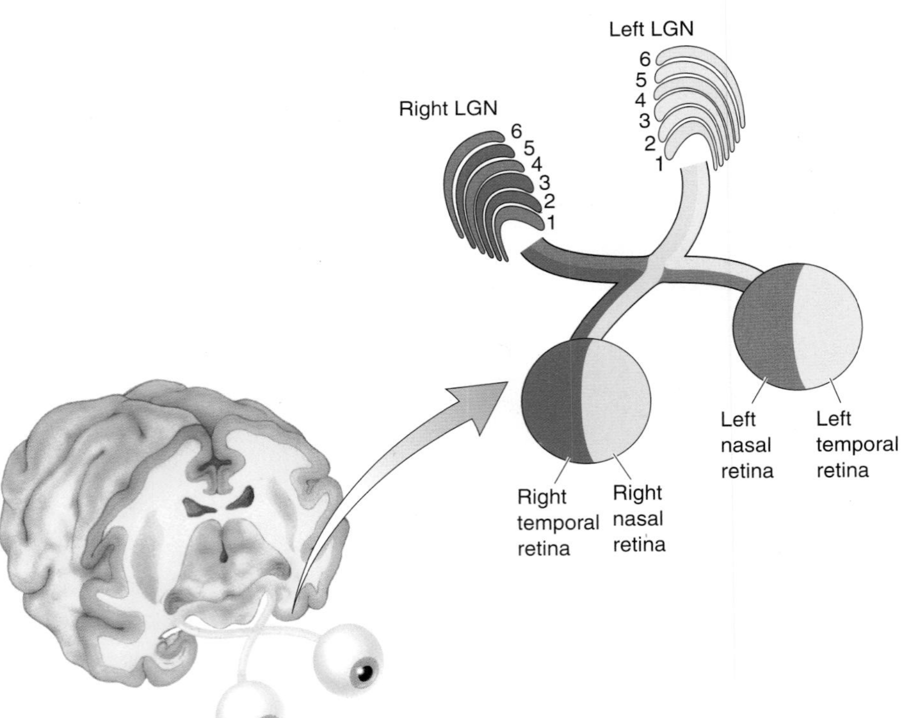

Figure 10.9
Retinal inputs to the LGN layers.

gives rise to streams of information that are processed in parallel. This organization is summarized in Figure 10.10.

Receptive Fields

Insertion of a microelectrode into the LGN makes possible the study of action potential discharges of geniculate neurons in response to visual stimuli, just as was done in the retina. The surprising conclusion of such studies is that the visual receptive fields of LGN neurons are almost identical with those of the ganglion cells that feed them. Thus, for example, magnocellular LGN neurons have relatively large center-surround receptive fields, respond to stimulation of their receptive field centers with a transient burst of action potentials, and are insensitive to differences in wavelength. All in all, they are just like M-type ganglion cells. Likewise, parvocellular LGN cells, like P-type retinal ganglion cells, have relatively small center-surround receptive fields, respond to stimulation of their receptive field centers with a sustained increase in the frequency of action potentials, and many show red-green color opponency. Although the small size of the cells in the koniocellular layers has impeded progress in identifying their response properties, recent work suggests that cells in the layers ventral to principal layers 3 and 4 exhibit blue-yellow color opponency, whereas cells in other koniocellular layers are not color-selective. Within all layers of the LGN, ON-center and OFF-center cells are intermixed.

Nonretinal Inputs to the LGN

What makes the similarity of LGN and ganglion cell receptive fields so surprising is that the retina is not the main source of synaptic input to the LGN. The major input, constituting about 80% of the excitatory synapses, comes from primary visual cortex. Thus, one might reasonably expect that this cor-

Retinal output: Eye	Retinal output: Ganglion cell type	LGN: LGN cell type	LGN: Layer
Contralateral	P-type	Parvocellular	6
	nonM-nonP	Koniocellular	
Ipsilateral	P-type	Parvocellular	5
	nonM-nonP	Koniocellular	
Contralateral	P-type	Parvocellular	4
	nonM-nonP	Koniocellular	
Ipsilateral	P-type	Parvocellular	3
	nonM-nonP	Koniocellular	
Ipsilateral	M-type	Magnocellular	2
	nonM-nonP	Koniocellular	
Contralateral	M-type	Magnocellular	1
	nonM-nonP	Koniocellular	

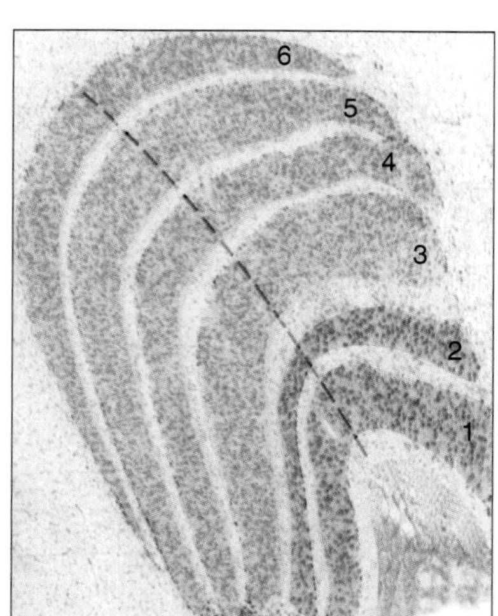

Figure 10.10
A summary of ganglion cell inputs to the different LGN layers. There is a thin koniocellular layer (shown in pink) ventral to each of the six principal layers.

ticofugal feedback pathway would significantly alter the qualities of the visual responses recorded in the LGN. So far, however, a role for this massive input has not been clearly identified.

The LGN also receives synaptic inputs from neurons in the brain stem whose activity is related to alertness and attentiveness (see Chapters 15 and 19). Have you ever "seen" a flash of light when you are startled in a dark room? This perceived flash may be a result of the direct activation of LGN neurons by this pathway. Usually, however, this input does not directly evoke action potentials in LGN neurons. But it can powerfully modulate the magnitude of LGN responses to visual stimuli. (Recall modulation from Chapters 5 and 6.) Thus, the LGN is more than a simple relay from retina to cortex; it is the first site in the ascending visual pathway where what we see is influenced by how we feel.

ANATOMY OF THE STRIATE CORTEX

The LGN has a single major synaptic target: primary visual cortex. Recall from Chapter 7 that the cortex may be divided into a number of distinct areas based on their connections and cytoarchitecture. Primary visual cortex is Brodmann's **area 17**, which lies in the occipital lobe of the primate brain. Much of area 17 lies on the medial surface of the hemisphere, surrounding the calcarine fissure (Figure 10.11). Other terms used interchangeably to describe the primary visual cortex are **V1** and **striate cortex**. (The term *striate* refers to the fact that area V1 has a uniquely dense stripe of myelinated afferent axons running into it, traveling parallel to the surface.)

We have seen that the axons of different types of retinal ganglion cells synapse on anatomically segregated neurons in the LGN. In this section, we look at the anatomy of the striate cortex and then trace the connections different LGN cells make with cortical neurons. In a later section, we will explore how this information is analyzed by cortical neurons. As we did in the LGN, in striate cortex we'll see a close correlation between structure and function.

Figure 10.11
Cortical area 17. Top views are lateral; bottom views are medial.

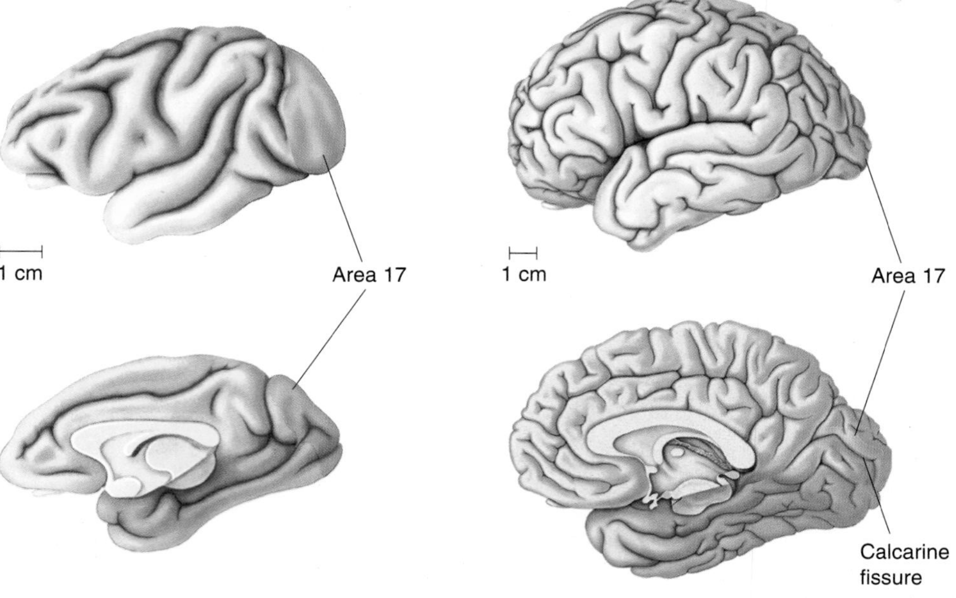

Lamination of the Striate Cortex

The neocortex, in general, and striate cortex, in particular, have neuronal cell bodies arranged into about a half-dozen layers. These layers can be seen clearly in a Nissl stain of the cortex, which, as described Chapter 2, leaves a deposit of dye (usually blue or violet) in the soma of each neuron. Starting at the white matter (containing the cortical input and output fibers), the cell layers are named by Roman numerals VI, V, IV, III, and II. Layer I, just under the pia mater, is largely devoid of neurons and consists almost entirely of axons and dendrites of cells in other layers. The full thickness of the striate cortex from white matter to pia is about 2 mm, the height of the lowercase letter m (Figure 10.12).

As Figure 10.12 shows, the lamination of striate cortex has been somewhat forced into a six-layer scheme. There are actually at least nine distinct layers of neurons. But to maintain Brodmann's convention that neocortex has six layers, we combine three sublayers into layer IV, labeled IVA, IVB, and IVC. Layer IVC is further divided into two tiers called IVCα and IVCβ. The anatomical segregation of neurons into layers suggests that there is a division of labor in the cortex that is similar to what we saw in the LGN. We can learn a lot about how the cortex handles visual information by examining the structure and connections of its different layers.

The Cells of Different Layers. Many neuronal shapes have been identified in striate cortex, but here we'll focus on two principal types, defined by the appearance of their dendritic trees (Figure 10.13). *Spiny stellate cells* are small neurons with spine-covered dendrites that radiate from the cell body (recall dendritic spines from Chapter 2). They are seen primarily in the two tiers of layer IVC. Outside of layer IVC are many *pyramidal cells*. These neurons are characterized by one or more thick apical dendrites that branch as they ascend to the pia and by multiple basal dendrites that extend horizontally. (Pyramidal cell dendrites are also covered with spines.)

A pyramidal cell in one layer may have dendrites extending into other layers. Remember that *only pyramidal cells send axons out of striate cortex* to form

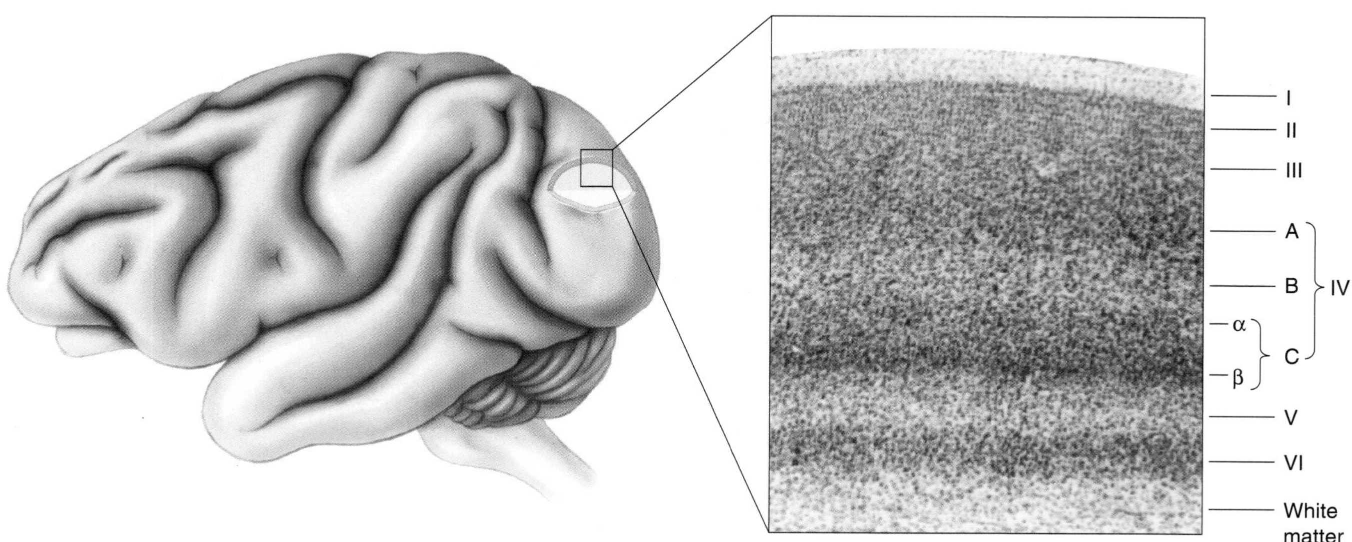

Figure 10.12
Cytoarchitecture of the striate cortex. The tissue has been Nissl-stained to show cell bodies, which appear as dots. (Source: Adapted from Hubel, 1988, p. 97.)

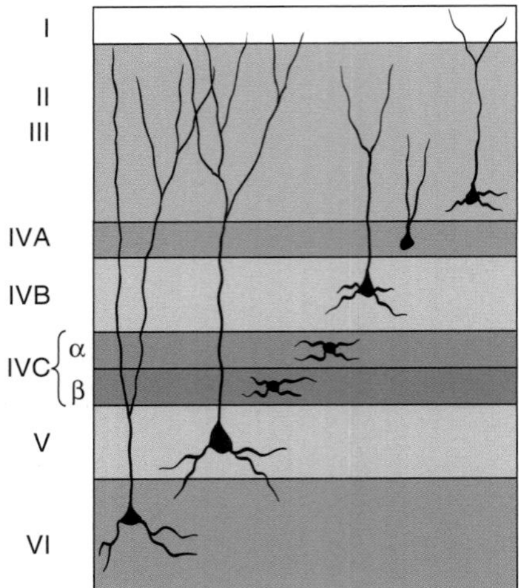

Figure 10.13
Dendritic morphology of some cells in striate cortex. Notice particularly that pyramidal cells are found in layers III, IVB, V, and VI and that spiny stellate cells are found in layer IVC.

connections with other parts of the brain. The axons of stellate cells only make local connections within the cortex.

Input-Output Organization of Different Layers. Axons from the LGN terminate in several cortical layers but mainly in layer IVC. This axonal projection maintains the retinotopy of the LGN, so neighboring cells in layer IV receive their inputs (via the LGN relay) from neighboring regions of the retina (Figure 10.14). As in the LGN and superior colliculus, many more cortical neurons receive input from the central part of the retina than from more peripheral parts.

Layer IVC stellate cells send axons up mainly to layers IVB and III. Here, an axon may form synapses with the dendrites of pyramidal cells of all layers. As previously mentioned, the pyramidal cells send axons out of striate cortex into the white matter. The pyramidal cells in different layers innervate different structures. Layer III and layer IVB pyramidal cells send their axons to other cortical areas. Layer V pyramidal cells send axons all the way down to the superior colliculus and pons. Layer VI pyramidal cells give rise to the massive axonal projection to the LGN (Figure 10.15). Pyramidal cell axons in all layers also branch and form local connections in the cortex.

Most intracortical connections occur along a radial line that runs across the layers, from white matter to layer I, perpendicular to the cortical surface. This pattern of *radial connections* maintains the retinotopic organization established in layer IV. Therefore, a cell in layer VI, for example, receives information from the same part of the retina as does a cell above it in layer IV (Figure 10.16a). The axons of some layer III pyramidal cells, however, extend collateral branches that make *horizontal connections* within layer III (Figure 10.16b). Radial and horizontal connections play different roles in the analysis of the visual world, as we shall see later in the chapter.

LGN Input to Layer IVC

Now we want to focus on the details of how the different types of LGN neurons innervate the cells in cortical layer IVC. We've seen that the output of the LGN is divided into streams of information, for example, from the magnocellular and parvocellular layers serving the right and left eyes. These streams remain anatomically segregated in layer IVC.

The segregation of magnocellular and parvocellular LGN inputs is straightforward. Magnocellular LGN neurons project to layer IVCα, and parvocellular LGN neurons project to layer IVCβ. Imagine that the two tiers of layer IVC are pancakes, stacked one (α) on top of the other (β). Because the input from the LGN to the cortex is arranged topographically, we see that layer IVC contains two overlapping retinotopic maps, one from the magnocellular LGN (IVCα) and the other from the parvocellular LGN (IVCβ).

How are the left eye and right eye LGN inputs segregated when they reach layer IVC of striate cortex? The answer was provided by a ground-breaking experiment performed in the early 1970s at Harvard Medical School by neuroscientists David Hubel and Torsten Wiesel. They injected a radioactive amino acid into one eye of a monkey (Figure 10.17). This amino acid was incorporated into proteins by the ganglion cells, and the proteins were transported down the ganglion cell axons into the LGN (recall anterograde transport from Chapter 2). Here the radioactive proteins spilled out of the ganglion cell axon terminals and were taken up by nearby LGN neurons. But not all LGN cells took up the radioactive material; only cells that were postsynaptic to the inputs from the injected eye incorporated the labeled protein. These cells then transported the radioactive proteins to their axon terminals in layer IVC of striate cortex. The location of the radioactive axon terminals

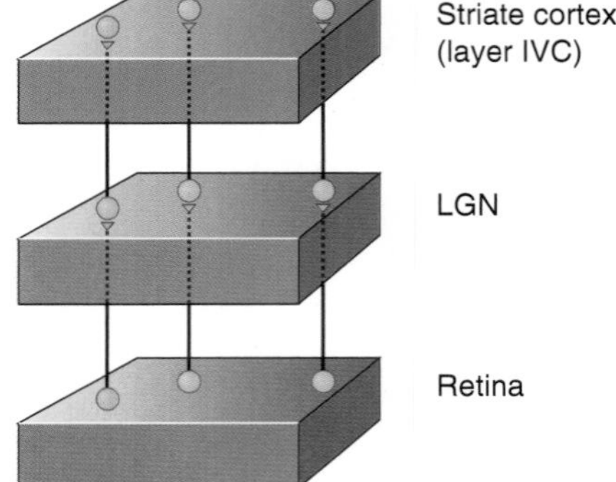

Figure 10.14
Retinotopy is preserved in the LGN projection to layer IVC.

was visualized by first placing a film of emulsion over thin sections of striate cortex and later developing the emulsion like a photograph, a process called *autoradiography* (introduced in Chapter 6). The resulting collection of silver grains on the film marked the location of the radioactive LGN inputs.

In sections cut perpendicular to the cortical surface, Hubel and Wiesel observed that the distribution of axon terminals relaying information from the injected eye was not continuous in layer IVC but was split up into a series of equally spaced patches, with each about 0.5 mm wide (Figure 10.18a). These patches were termed **ocular dominance columns**. In later experiments, the cortex was sectioned tangentially, parallel to layer IV. This revealed that the left eye and right eye inputs to layer IV are laid out as a series of alternating bands, like the stripes of a zebra (Figure 10.18b).

Putting the anatomy together, you can see that any chunk of layer IVC measuring 0.5 mm (the thickness of layer IVC) × 1 mm × 1 mm would contain a full complement of segregated inputs from both magnocellular and parvocellular LGN layers from both the left and the right eye.

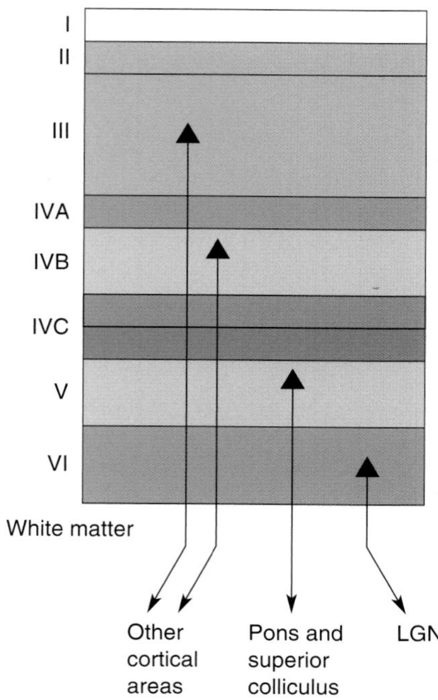

Figure 10.15
Organization of outputs from striate cortex.

Innervation of Other Cortical Layers

Layer IVC neurons project axons radially up to layers IVB and III, where, for the first time, information from the left eye and right eye begins to mix. Even

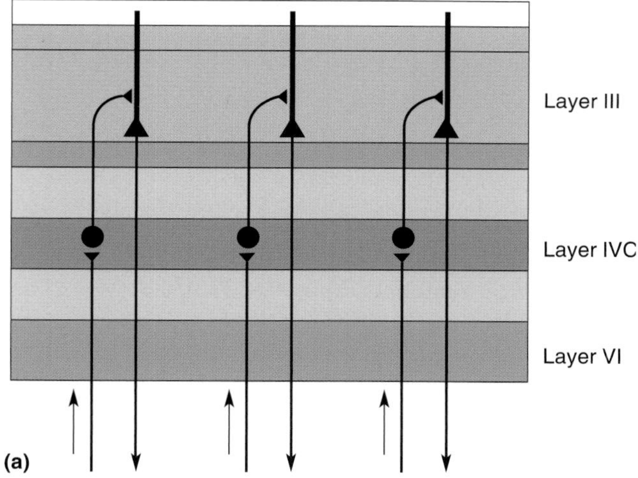

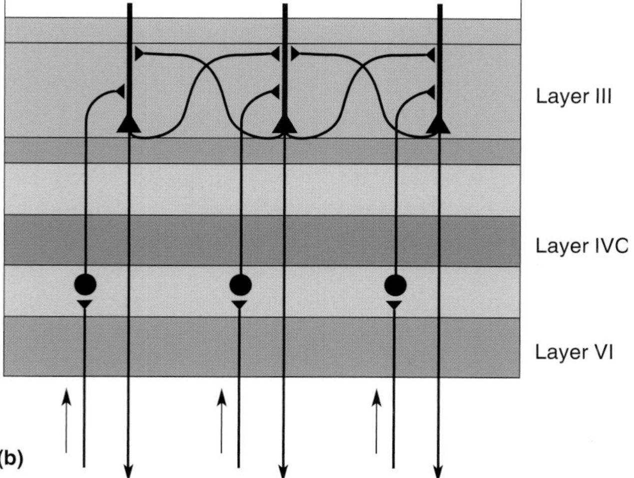

Figure 10.16
Patterns of intracortical connections. **(a)** Radial connections. **(b)** Horizontal connections.

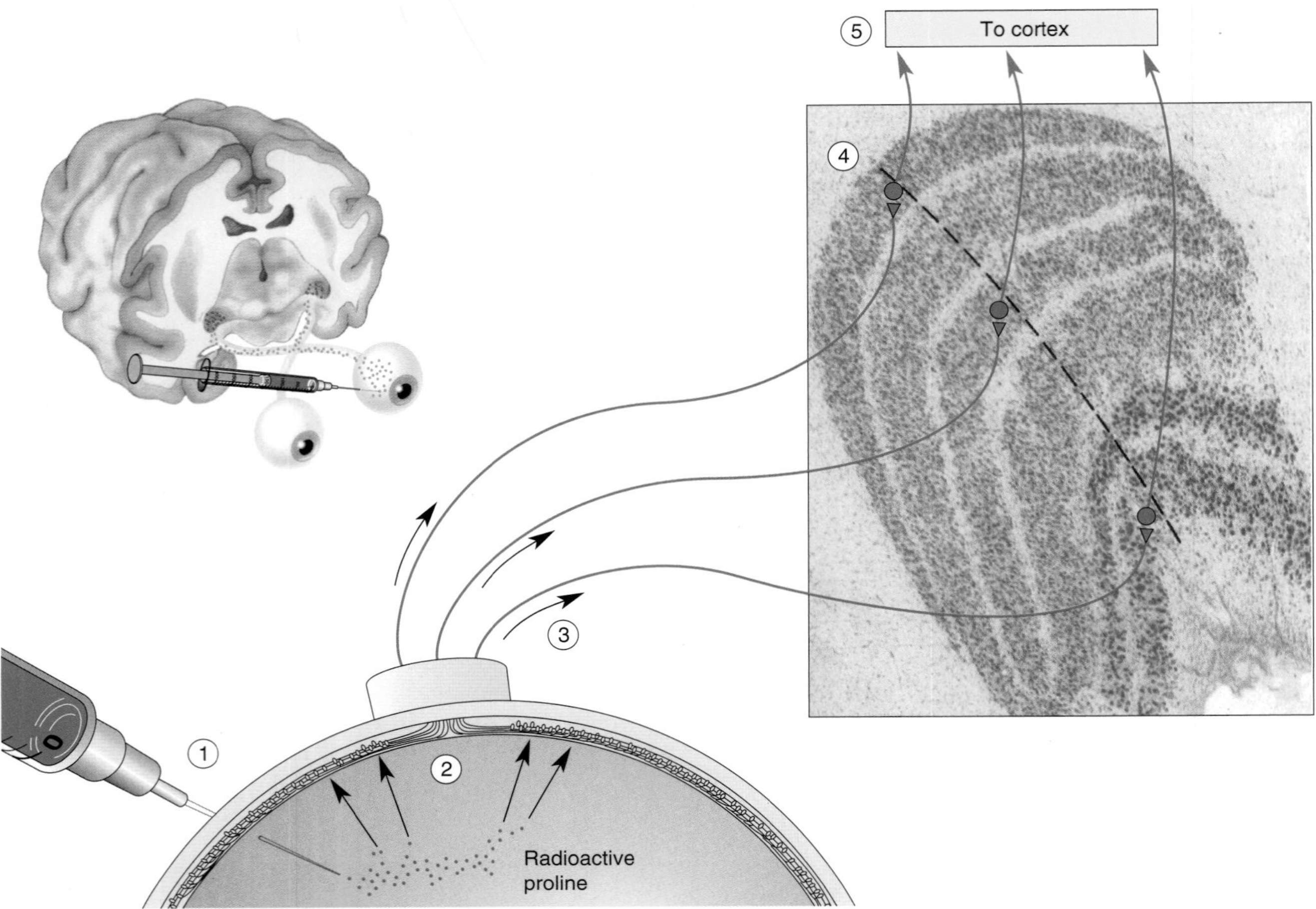

Figure 10.17

Transneuronal autoradiography. Radioactive proline is ① injected into one eye, where it is ② taken up by retinal ganglion cells and incorporated into proteins that are ③ transported down the axons to the LGN. Some radioactivity spills out of the retinal terminals and ④ is taken up by LGN neurons that then ⑤ transport it to striate cortex. The location of radioactivity can be determined by using autoradiography.

so, there continues to be considerable anatomical segregation of the magnocellular and parvocellular processing streams. Layer IVCα, which receives magnocellular LGN input, projects mainly to cells in layer IVB. Layer IVCβ, which receives parvocellular LGN input, projects mainly to layer III.

Blobs. In addition to the parvocellular input relayed from layer IVCβ, recent research indicates that a subset of cells in layer III receives direct input from the LGN. These layer III cells are found in what are known as "blobs." The blob story began in 1978, when Margaret Wong-Riley, at the University of California, San Francisco, discovered an inhomogeneity in the anatomy of the upper cortical layers. She used a staining procedure designed to reveal the presence of **cytochrome oxidase**, a mitochondrial enzyme used for cell metabolism, and found that it is not uniformly distributed in layers II and III. Rather, the cytochrome oxidase staining in cross sections of striate cortex appears as a colonnade, a series of pillars at regular intervals, running the full thickness of layers II and III and layers V and VI (Figure 10.19a). When the cortex is sliced tangentially through layer III, these pillars appear like the spots of a leopard (Figure 10.19b). These pillars of cytochrome oxidase-rich neurons, after some debate, eventually were given the name **blobs**. The blobs

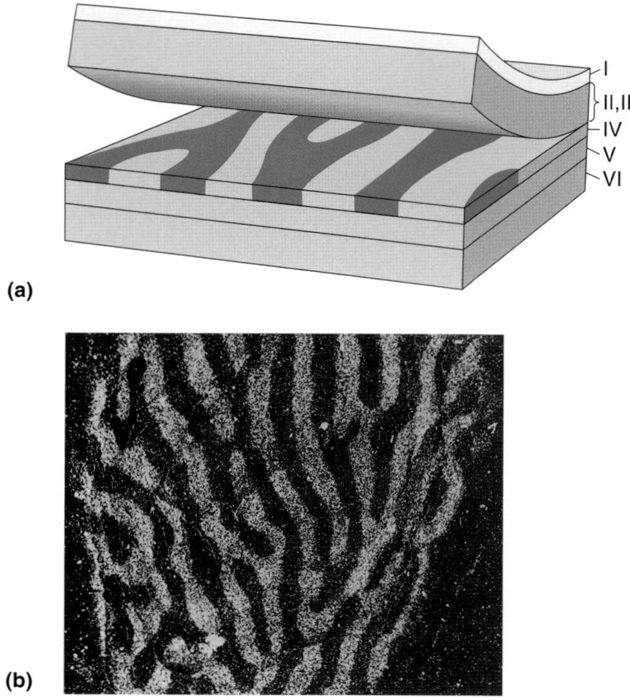

Figure 10.18
Ocular dominance columns in striate cortex. (a) Organization of ocular dominance columns in layer IV of the striate cortex of a macaque monkey. The distribution of LGN axons serving one eye is darkly shaded. In cross section, these eye-specific zones appear as patches, with each about 0.5 mm wide, in layer IV. When the superficial layers are peeled back, allowing a view of the ocular dominance columns in layer IV from above, these zones take on the appearance of zebra stripes. **(b)** An autoradiograph of a histological section of layer IV viewed from above. Two weeks prior to the experiment, one eye of this monkey was injected with radioactive proline. In the autoradiograph, the radioactive LGN terminals appear bright on a dark background. (Source: LeVay et al., 1980.)

are in rows, with each blob centered on an ocular dominance stripe in layer IV. Between the blobs are "interblob" regions.

It was soon shown that both blobs and interblobs receive parvocellular LGN input via a relay in layer IVCβ. In 1994, Johns Hopkins University neuroscientists Stewart Hendry and Takashi Yoshioka discovered that the blob cells in macaque striate cortex also receive direct LGN innervation from the koniocellular layers.

Parallel Pathways

Figure 10.20 summarizes the connections of striate cortex and its LGN input in terms of anatomically distinct parallel pathways that arise in the retina. The magnocellular pathway begins with the M-type ganglion cells of the retina. These cells project axons to the magnocellular layers of the LGN. The magnocellular layers of the LGN project to layer IVCα of striate cortex. Layer IVCα neurons project to layer IVB (Figure 10.20a).

The parvocellular pathway originates with the P-type ganglion cells of the retina, which project to the parvocellular layers of the LGN. The parvocellular LGN sends axons to layer IVCβ of striate cortex. Layer IVCβ neurons project to layer III blob and interblob regions (Figure 10.20b).

The koniocellular pathway arises from the subset of ganglion cells that are neither M-type nor P-type. These nonM-nonP cells project to the koniocellular layers of the LGN. The koniocellular LGN projects directly to the layer III blobs (Figure 10.20c). Notice that the blobs are a site of convergence of parvocellular and koniocellular inputs.

Notice also in Figure 10.20 that the information from the two eyes remains segregated all the way through layer IVC. Inputs from the two eyes begin to mix in layer IVB and in the interblob region of layer III. The blob cells, however, receive input mainly from one eye or the other. Keep these anatomically defined parallel pathways in mind as we look at how neurons in striate cortex respond to visual stimuli in the next section.

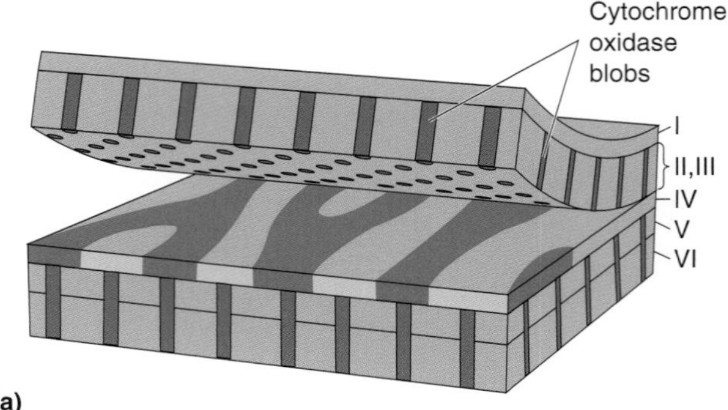

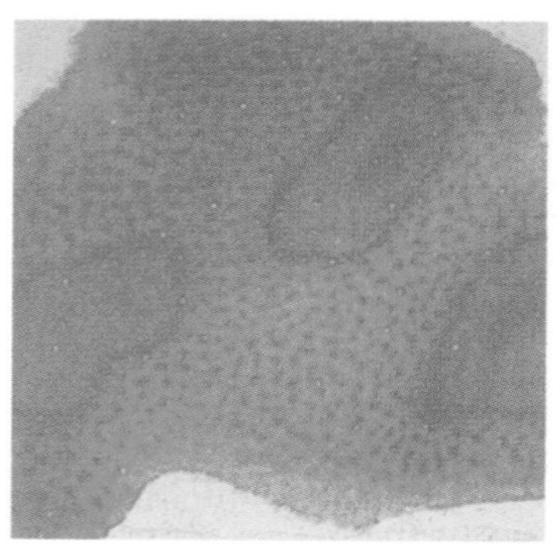

Figure 10.19
Cytochrome oxidase blobs. Distribution of the enzyme cytochrome oxidase in the cells of striate cortex. **(a)** The organization of cytochrome oxidase blobs in striate cortex of a macaque monkey. **(b)** A photograph of a histological section of layer III, stained for cytochrome oxidase and viewed from above. (Source: Courtesy of Dr. S. H. C. Hendry.)

PHYSIOLOGY OF THE STRIATE CORTEX

Beginning in the early 1960s, Hubel and Wiesel were the first to systematically explore the physiology of striate cortex with microelectrodes. They were students of Stephen Kuffler, who was then at Johns Hopkins University and later moved with them to Harvard. They extended Kuffler's innovative methods of receptive field mapping to the central visual pathways. After showing that LGN neurons behave much like retinal ganglion cells, they turned their attention to striate cortex, initially in cats and later in monkeys. (Here we will be concerned only with the monkey cortex.) The work on the physiology of striate cortex that continues today is built on the solid foundation provided by Hubel and Wiesel's pioneering studies. Their contributions to our understanding of the cerebral cortex were recognized with the Nobel Prize in 1981.

Instead of presenting the physiological results in the order in which they were discovered, we will discuss the physiology of cortical neurons in light of the anatomy described earlier. The latest research suggests that functionally there are three relatively independent channels of visual information processing. The channel that begins with the magnocellular retinal ganglion cells and leads to layer IVB of striate cortex we will call the **magnocellular channel**, or simply the **M channel**. The channel that begins with the parvocellular retinal ganglion cells and leads to the interblob regions of layer III we will call the **parvocellular-interblob channel**, or the **P-IB channel**. We will call the channel that converges on the blobs of layer III and passes through the parvocellular and koniocellular layers of the LGN the **blob channel**. We'll see that each of these channels appears to process different facets of vision.

The M Channel

Recall that cells in the magnocellular LGN are activated by only one eye (i.e., they are monocular), have center-surround receptive fields, respond transiently to visual stimulation, and are insensitive to the wavelength of light. At the level of the first synaptic relay in layer IVCα, we see an interesting elaboration of receptive field structure. While some cells in this layer have circular receptive fields, the receptive fields of many layer IVCα neurons are elongated along a particular axis, with an ON-center or OFF-center region flanked on one or both sides by an antagonistic surround (Figure 10.21). One gets the impression that these cells receive a converging input from three or more LGN cells with receptive fields that are aligned along one axis (Figure 10.21c). Hubel and Wiesel called neurons of this type **simple cells**. Simple cells respond best to a thin slit of light or dark that is aligned with the long axis of the receptive field, and they respond poorly, if at all, to a stimulus that is perpendicular to this axis (Figure 10.22). Thus, we say that layer IVCα cells possess **orientation selectivity**.

Layer IVCα cells project up to layer IVB, where the cells also have orientation-selective simple receptive fields. One important difference between the cells in the two layers is that while cells in layer IVCα respond only to stimulation of one eye or the other (just as the anatomy of ocular dominance columns predicts), many cells in layer IVB respond to stimulation of both eyes. Such cells are said to have binocular receptive fields. The construction of **binocular receptive fields** is essential in binocular animals, such as humans. Without binocular neurons, we would be unable to use the inputs from both eyes to form a single image of the world around us.

Besides binocularity, another important physiological property of layer IVB cells is **direction selectivity**. Figure 10.23 shows how a direction-selective cell might respond to a moving stimulus. The cell responds to an elongated stimulus swept across the receptive field, but only in a particular direction of movement. Sensitivity to the direction of stimulus motion is a

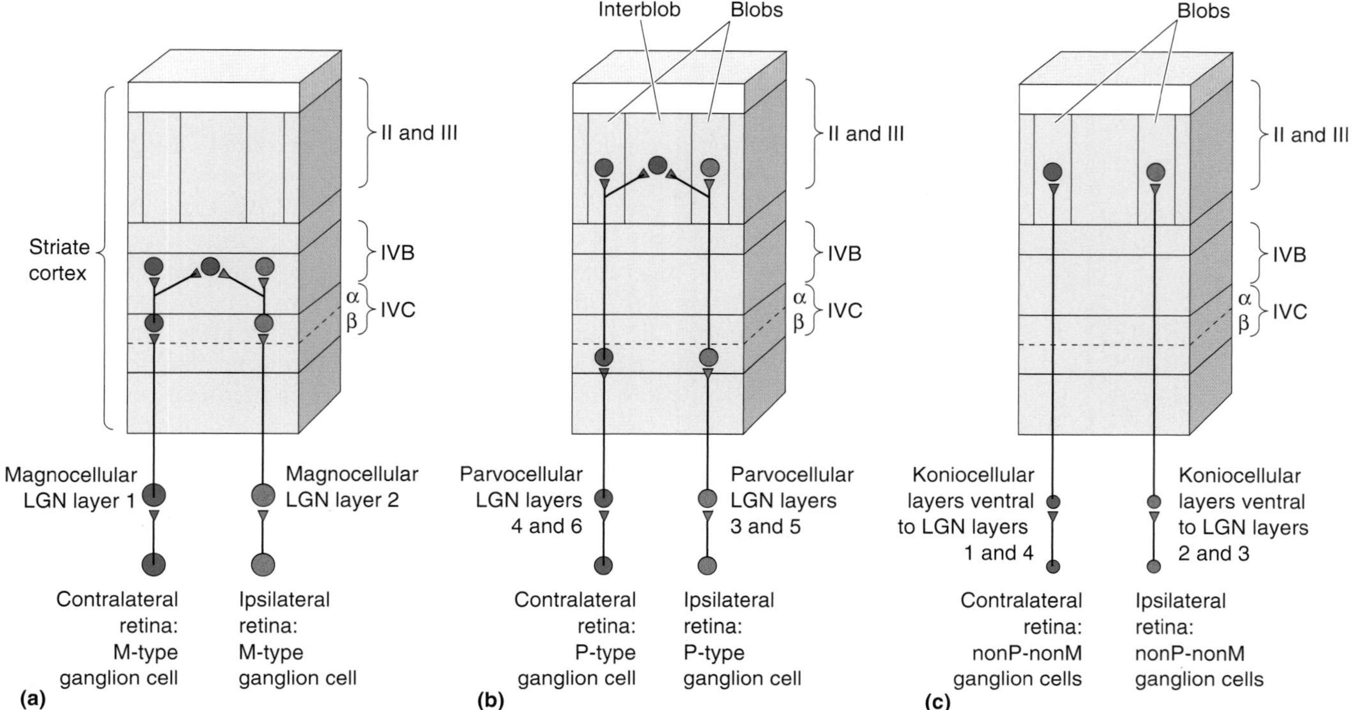

Figure 10.20
Three parallel pathways from the retina to striate cortex. (a) Magnocellular pathway. **(b)** Parvocellular pathway. **(c)** Koniocellular pathway.

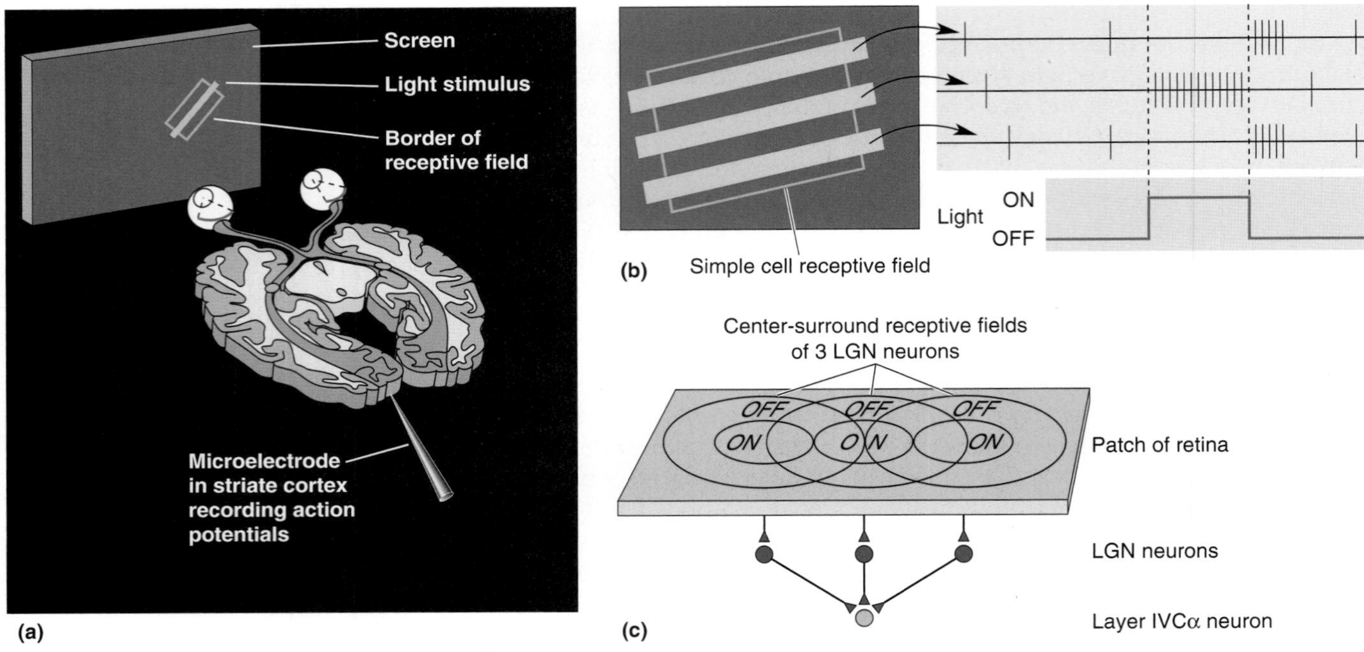

Figure 10.21
A simple cell receptive field. (a) The responses of a neuron in layer IVCα are monitored as visual stimuli are presented to its receptive field. The visual stimulus is a light bar. **(b)** The response of a simple cell to various stimuli. It responds best to an oriented bar of light, and the response can be ON or OFF, depending on where the bar lies in the receptive field. **(c)** A possible construction of a simple cell receptive field by the convergence of three LGN cell axons with center-surround receptive fields.

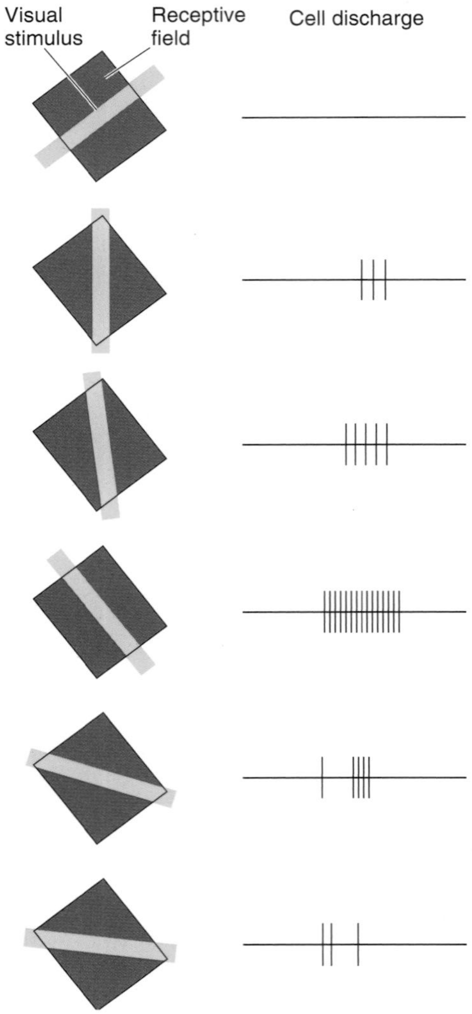

hallmark of neurons in the M channel. For this reason, the M channel is thought to be specialized for the *analysis of object motion*.

The P-IB Channel

Neurons in the parvocellular layers of the LGN have small, circular monocular receptive fields. Most of these cells show red-green center-surround opponency. In all of these respects, the cells of layer IVCβ act just like the LGN neurons that feed them. Recall that IVCβ cells relay information up to layers II and III and that this projection targets both blob and interblob regions.

We will discuss blobs in a moment; for now, let's concentrate on the interblob cells in layer III. Hubel and Wiesel called most of these **complex cells**, because their receptive fields appeared to be more complex than those of simple cells. Unlike simple cells, complex cell receptive fields do not have distinct ON and OFF regions (Figure 10.24). Instead, complex cells give ON and OFF responses to stimuli throughout the receptive field. Complex cells are mostly binocular, relatively insensitive to the wavelength of light, and highly selective to stimulus orientation (even more so than simple cells).

The analysis of stimulus orientation appears to be one of the most important functions of striate cortex. Presumably this analysis is required to discriminate and identify objects on the basis of their shape. Because virtually all interblob cells are orientation-selective and have small receptive fields, the P-IB channel is thought to be devoted to the *analysis of object shape*.

Visual stimulus | Receptive field | Cell discharge

Figure 10.22
Responses of an orientation-selective neuron.

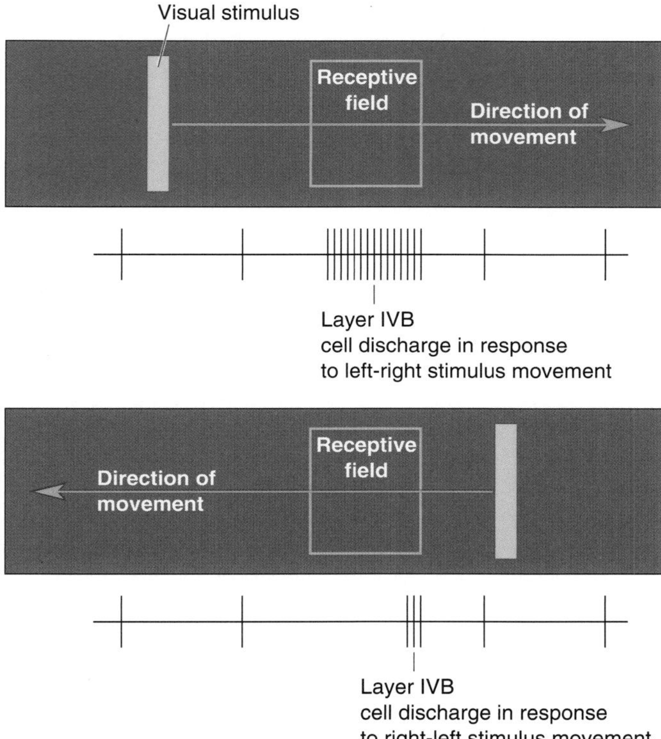

Figure 10.23
Responses of a direction-selective neuron.

Orientation Columns

We have mentioned that many cortical neurons exhibit orientation selectivity. Because this property may be generated independently in the M channel and P-IB channel, however, you may wonder whether the orientation selectivity of neurons in different layers is related. From the earliest work of Hubel and Wiesel, the answer to this question has been an emphatic yes. As a microelectrode is advanced *radially* (perpendicular to the surface) from one layer to the next, the preferred orientation remains the same for all selective neurons encountered. This includes all of the cells of layers V and VI having both simple and complex receptive fields. Hubel and Wiesel called such a radial column of cells an **orientation column**.

Perhaps predictably, as an electrode passes *tangentially* (parallel to the surface) through the cortex in a single layer, the preferred orientation shifts. We now know, from the use of a technique called optical imaging, that there is a mosaiclike pattern of optimal orientations in striate cortex (Box 10.2). If an

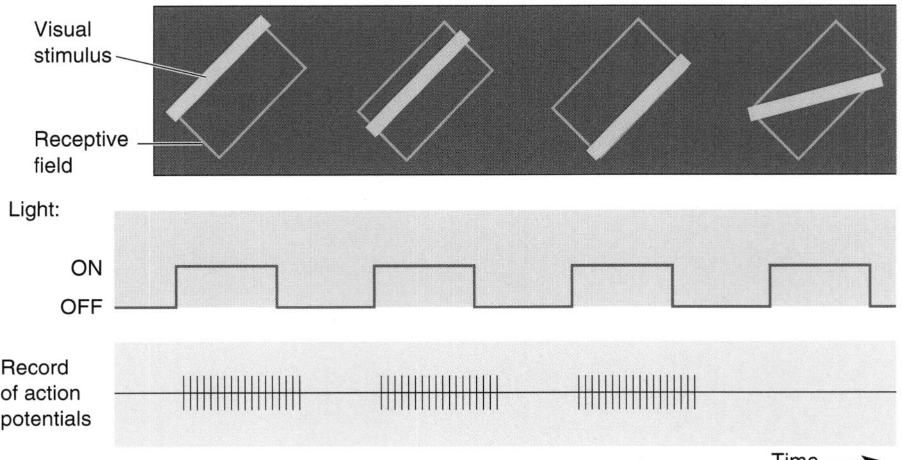

Figure 10.24
A complex cell receptive field. Like a simple cell, a complex cell responds best to an oriented bar of light. Responses occur to both light ON and light OFF, however, regardless of position in the receptive field.

Box 10.2

BRAIN FOOD

Optical Imaging of Neural Activity

Most of what we know about the response properties of neurons in the visual system and of every other system in the brain has been learned from intracellular and extracellular recordings with microelectrodes. These recordings give very precise information about the activity of one or a few cells. Unless one inserts thousands of electrodes, however, it is not possible to observe patterns of activity across large populations of neurons.

But what if there were a way to simultaneously record signals from thousands of neurons simply by aiming a camera at the surface of the brain? Incredibly, one can observe brain activity with this optical recording approach, and the resulting images have given us new insight into the organization of the cerebral cortex. In one version of optical recording, a voltage-sensitive dye is applied to the surface of the brain. The molecules in the dye bind to cell membranes, and they change their optical properties in proportion to variations in membrane potential. The change is detected with either an array of photodetectors or a video camera. If this technique is used to record from a single neuron, the output of the optical detector is very similar to an intracellular recording. In recordings from the cerebral cortex, the activity of individual neurons cannot be resolved, and the optical signal represents a summation of the changes in membrane potential of the neurons and glial cells in an area about 100 μm across.

A second way to study cortical activity optically is to image what are called intrinsic signals. When neurons are active, numerous changes occur in the neurons themselves and in the surrounding tissue. Examples of such changes are ion movement, neurotransmitter release, and alterations in blood volume and oxygenation. Because these factors are correlated with the level of neural activity and have (very small) effects on the reflection of light from the brain, they are called intrinsic signals for optical recording. Thus, when intrinsic signals are used to study brain activity, membrane potentials or action potentials are not directly measured. To record intrinsic signals, light is projected onto the brain, and a sensitive video camera records the reflected light. With the wavelengths of light usually used for illumination, the intrinsic signal is dominated by changes associated with activity-dependent increases in blood volume or blood oxygen saturation. One disadvantage of optical recording with intrinsic signals is that its reliance on slow vascular changes makes it incapable of the millisecond temporal resolution possible with voltage-sensitive dyes. Intrinsic signals, however, have the advantage of being free of the toxic side effects and damage that occur when the cortex is bathed in a dye.

Figure A shows the vasculature in a portion of primary visual cortex. Figure B shows ocular dominance columns in the same patch of striate cortex obtained by imaging areas in which blood flow changes occurred during visual stimulation. This figure is actually a subtraction of two images, one made when only the right eye was visually stimulated minus another made when only the left eye was stimulated. Consequently, the dark bands represent cells dominated by the left eye, and the light bands represent cells dominated by the right eye.

Figure C is a color-coded representation of preferred orientation in the same patch of striate cortex. Four optical images were recorded while bars of light at four orientations were swept across the visual field. Each location in the figure is colored according to the orientation that produced the greatest response at each location on the brain (blue, horizontal; red, 45 degrees; yellow, vertical; turquoise, 135 degrees). Consistent with earlier results obtained with electrodes (Figure 10.25), in some regions the orientation changes progressively along a straight line. The optical recording technique, however, reveals that cortical organization based on orientation is much more complex than an idealized pattern of parallel "columns."

(Box 10.2, *continued*)

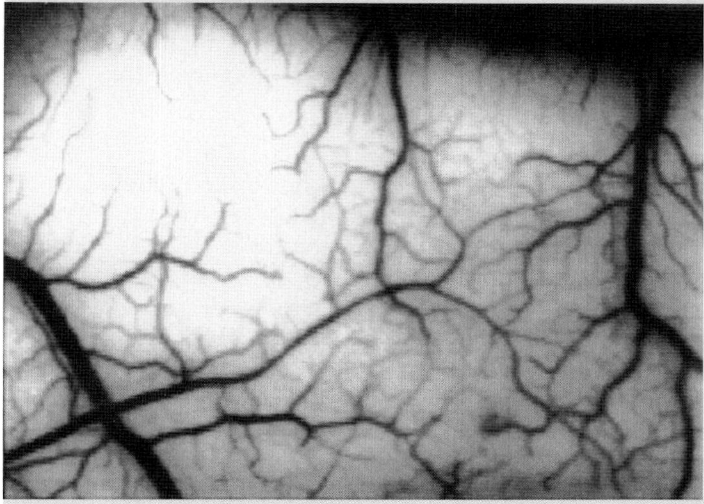

Figure A
Vasculature on the surface of primary visual cortex. (Source: Ts'o et al., 1990, Fig. 1A.)

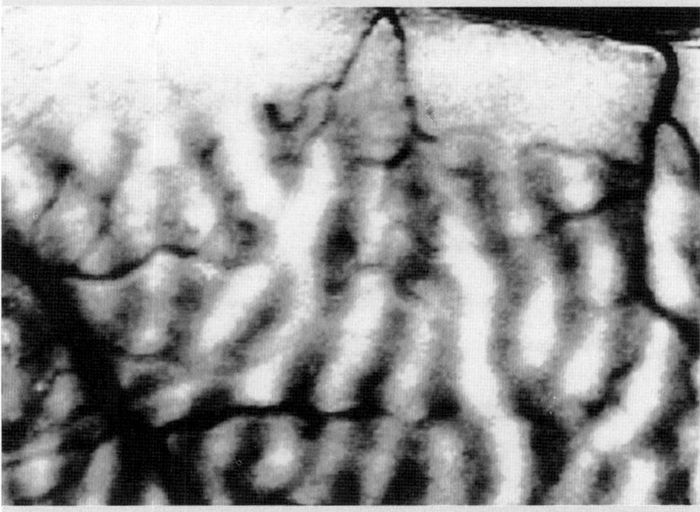

Figure B
Ocular dominance columns. (Source: Ts'o et al., 1990, Fig. 1B.)

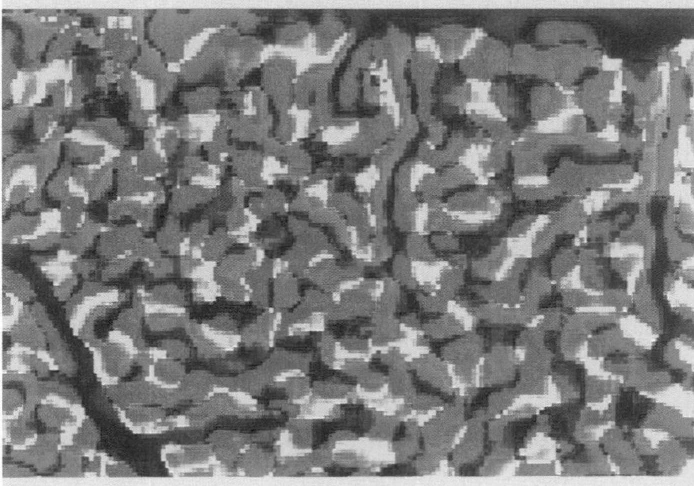

Figure C
A map of preferred orientations. (Source: Ts'o et al., 1990, Fig. 1C.)

electrode is passed at certain angles through this mosaic, the preferred orientation rotates like the sweep of the minute hand of a clock, from the top of the hour to 10 past to 20 past, and so on (Figure 10.25). If the electrode is moved at other angles, more sudden shifts in preferred orientation occur. Hubel and Wiesel found that a complete 180-degree shift in preferred orientation required an average traverse of about 1 mm within layer III.

Physiology of the Blobs

After Wong-Riley's discovery of the cytochrome oxidase blobs, Hubel, working with Harvard Medical School neurobiologist Margaret Livingstone, recorded from neurons in the blobs and found that they are very different from the interblob cells in layer III. Most blob cells are wavelength-sensitive and monocular, and they lack orientation selectivity. The visual responses of blob cells resemble those of their major sources of input: The neurons in layer IVCβ and the koniocellular layers of the LGN.

The receptive fields of most blob neurons are circular. Some have the color-opponent center-surround organization observed in the parvocellular and koniocellular layers of the LGN. Other blob cell receptive fields have red-green or blue-yellow color opponency in the center of their receptive fields, with no surround regions at all. Still other cells have both a color-opponent center and a color-opponent surround and are called double-opponent cells. For present purposes, remember that blobs contain most of the color-sensitive neurons outside of layer IVC. Thus, the blob channel appears to be specialized for the *analysis of object color*. Without it, we might be color-blind.

Putting the Pieces Together

We have seen that the anatomy and physiology of the central visual pathways, from retina to striate cortex, are consistent with the idea that there are several relatively independent parallel processing channels. Each one appears to be specialized for the analysis of a different facet of the visual scene: the M channel for the analysis of object motion, the P-IB channel for the analysis of object shape, and the blob channel for the analysis of object color.

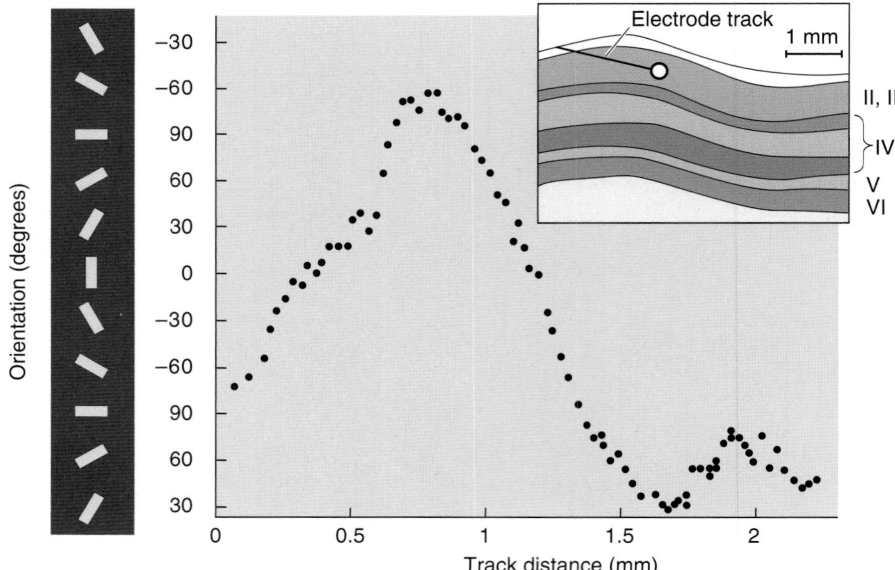

Figure 10.25
Systematic variation of orientation preferences across striate cortex. As an electrode is advanced tangentially across layer III of striate cortex, the orientation preference of the neurons encountered is recorded and plotted. There is a periodic, regular shift in preferred orientation. (Source: Adapted from Hubel and Wiesel, 1968.)

Each point in the visual world is analyzed by a circumscribed patch of cells in the cortex. Hubel and Wiesel showed that the image of a point in space falls within the receptive fields of neurons across a traverse of about 2 mm of layer III. For a complete analysis, this 2 × 2 mm patch of active neurons must include representatives from each of the processing channels from right and left eyes.

Fortunately, a 2 × 2 mm chunk of cortex would contain two complete sets of ocular dominance columns, 16 blobs, and, in the cells between blobs, a complete sampling (twice over) of all 180 degrees of possible orientations. Thus, Hubel and Wiesel have argued that a 2 × 2 mm chunk of striate cortex is both necessary and sufficient to analyze the image of a point in space, *necessary* because its removal would leave a blind spot for this point in the visual field, and *sufficient* because it contains all the neural machinery required to analyze the participation of this point in oriented and/or colored contours viewed through either eye. Such a unit of brain tissue has come to be called a **cortical module**. Striate cortex is constructed of perhaps a thousand of these modules (Figure 10.26). We can think of a visual scene being simultaneously processed by these modules, with each "looking" at a portion of the scene.

BEYOND STRIATE CORTEX

Striate cortex is called V1, for "visual area one," because it is the first cortical area to receive information from the LGN. Beyond V1 lie another two dozen distinct areas of cortex, each of which contains a representation of the visual world. The contributions to vision of these *extrastriate areas* are still being vigorously debated. The emerging picture, however, is that there are two large-scale cortical *streams* of visual processing, one stretching dorsally from striate cortex toward the parietal lobe and serving the analysis of visual motion and the other projecting ventrally toward the temporal lobe and serving the recognition of objects (Figure 10.27). These processing streams have primarily been studied in the macaque monkey brain. Recent work with fMRI, however, has begun to identify areas in the human brain that have properties analogous to those of brain areas in the macaque (Figure 10.28).

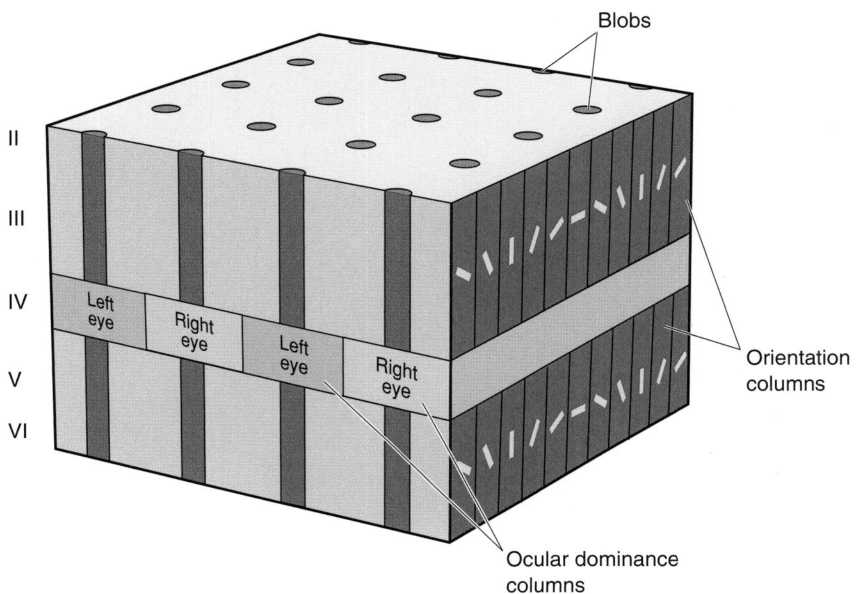

Figure 10.26
A cortical module.

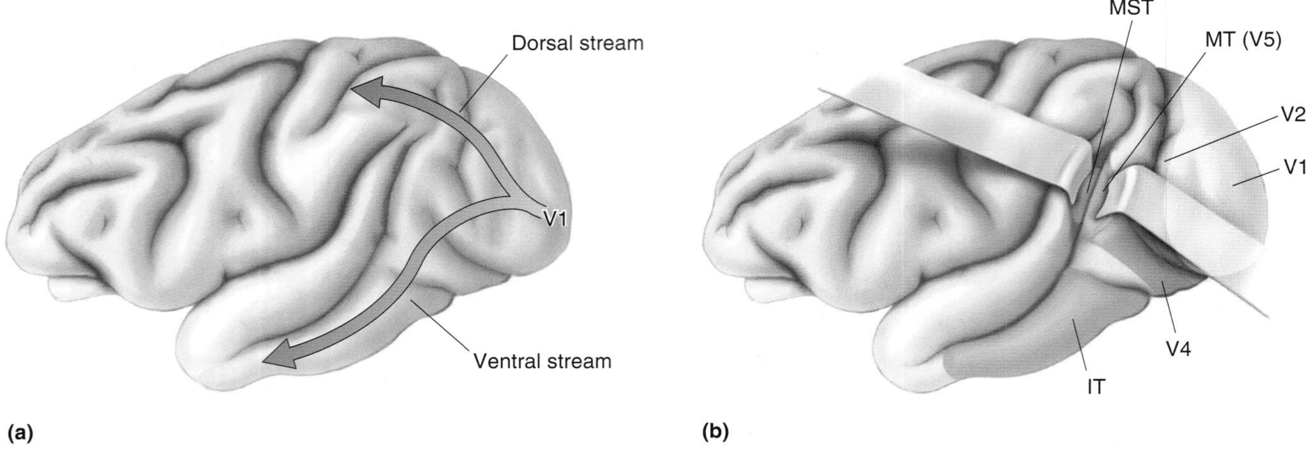

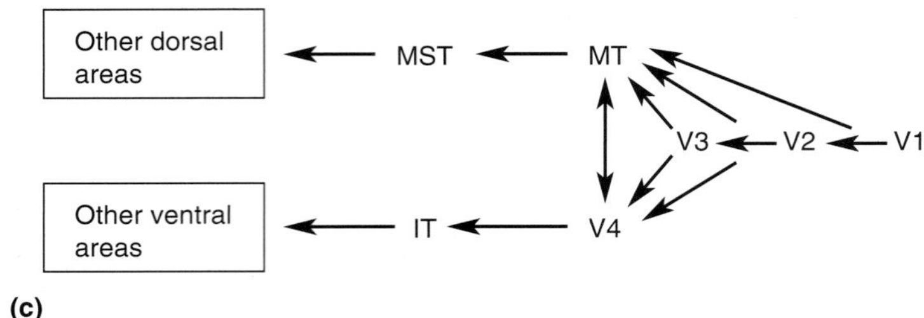

Figure 10.27
Beyond striate cortex in the macaque brain. (a) Dorsal and ventral visual processing streams. **(b)** Extrastriate visual areas. **(c)** Flow of information in the dorsal and ventral streams.

Figure 10.28
Visual areas in the human brain in both medial and lateral views. (Source: Courtesy of Dr. M. Sereno.)

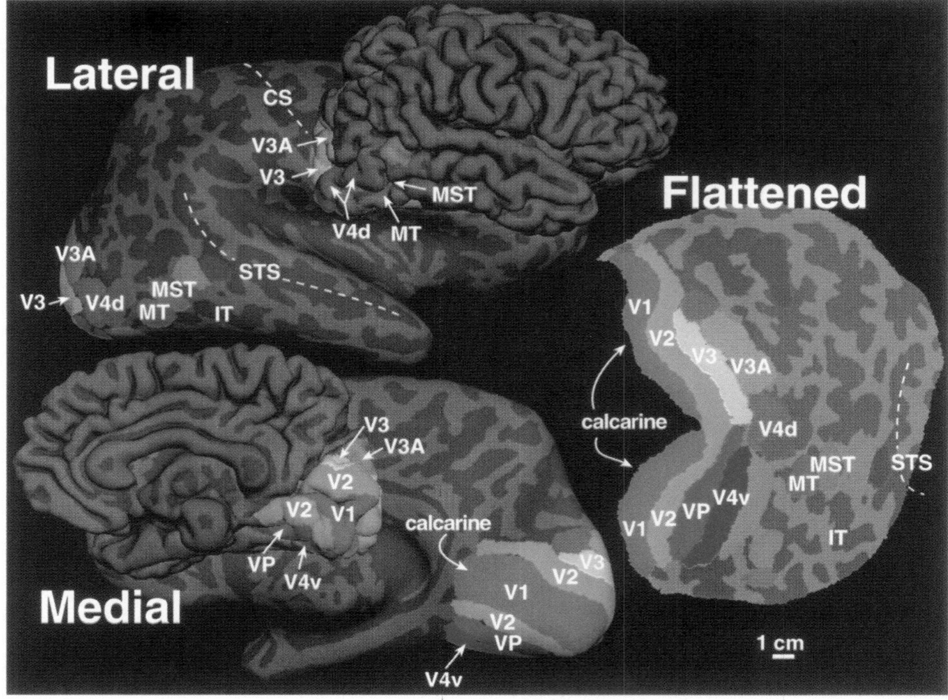

The Dorsal Stream

The cortical areas composing the dorsal stream are not arranged in a strict serial hierarchy, but there does appear to be a progression of areas in which more complex or specialized visual representations develop. Projections from V1 extend to areas designated V2 and V3, but we will skip further ahead in the dorsal stream.

Area MT. There is strong evidence that in an area known as V5, or MT (because of its location in the middle temporal lobe in some monkeys), a specialized processing of object motion takes place. **Area MT** receives retinotopically organized input from a number of other cortical areas, such as V2 and V3, and is directly innervated by cells in layer IVB of striate cortex. Recall that layer IVB is part of the magnocellular channel, which is characterized by cells with large receptive fields, transient responses to light, and direction selectivity. Neurons in area MT have large receptive fields that respond to stimulus movement in a narrow range of directions. Area MT is most notable for the fact that almost all of the cells are direction-selective, unlike areas earlier in the dorsal stream or anywhere in the ventral stream. The neurons in MT also respond to types of motion, such as drifting spots of light, that are not good stimuli for cells in other areas—it appears that the motion of the objects is more important than their structure. Further specialization for motion processing is evident in the organization of MT. This cortical area is arranged into direction-of-motion columns analogous to the orientation columns in V1. Presumably, the perception of movement at any point in space depends on a comparison of the activity across columns spanning a full 360-degree range of preferred directions.

Most intriguing is the finding by Anthony Movshon and his colleagues at New York University that some MT cells seem to respond to the perceived direction of motion, even when this direction is not the same as the actual directions in which objects move (Box 10.3). In one experiment used to explore this point, bars of light are moved simultaneously in two directions. Although the bars physically move in directions 45 degrees to the left and right of vertical, their simultaneous motion leads to the perception of a *plaid* that moves as a unit straight up (Figure 10.29). Presented with such a stimulus, the V1 cells that respond most strongly are those tuned for directions of motion 45 degrees from vertical; the neurons tuned for vertical motion do not respond. This is logical, given the physical movement directions of the stimulus bars. In MT, however, some cells that prefer vertical motion to simple stimuli also respond strongly to the plaid. Even though the bars are actually moving at 45-degree angles, the MT cells respond most when a plaid appears to move up. Testing with a variety of stimuli shows that the activity of these neurons seems to correlate with the perceived direction of motion rather than the physical direction. This exciting finding suggests that MT makes a transformation that is critical to building our perception of motion.

In complementary studies, William Newsome and his colleagues at Stanford University have shown that weak electrical stimulation in MT of the monkey appears to alter the perceived direction in which small dots of light move. For example, if electrical stimulation is applied to cells in a direction column preferring rightward movement, the monkey makes behavioral decisions suggesting that it has perceived motion in that direction. This is perhaps the closest visual neurophysiology has come to finding the basis for visual perception.

Dorsal Areas and Motion Processing. Beyond area MT, in the parietal lobe, are areas with additional types of specialized movement sensitivity. For ex-

Box 10.3

PATH OF DISCOVERY

Seeking the Brain's Representation of Coherent Motion

BY ANTHONY MOVSHON

From my first exposure to vision research, I have been fascinated by the relationship between neuronal activity and perceptual states. Perceptual phenomena have often given me inspiration for physiological experiments. I sometimes have a gut feeling that a particular perceptual phenomenon has a brain representation that I can pin down. Perhaps the best example of this in my career began in 1979, shortly after Ted Adelson had joined my laboratory.

We created patterns, which we called "plaids," by superimposing two sinusoidal gratings of different orientations. When we moved the two gratings across the screen, the plaids moved coherently in a direction different from either grating, and we saw them unambiguously as a single object. We quickly learned that by changing the width and spacing (the spatial frequency) of the stripes of one grating, we could abolish the coherent percept. Now the two gratings slid transparently across one another, with each moving its own way, and obstinately refused to cohere. What made the effect especially tantalizing was that the change in coherence was purely a perceptual response: All of the plaids we generated could have been painted on a single rigidly moving piece of paper, yet our brains reliably chose to see some plaids as having one coherent motion and others as having two transparent ones.

The simplest interpretation was that the brain represented the motion of the plaid in two ways. Somewhere, perhaps at an early processing level, the motions of each component grating would be separately encoded. Later in the system, these independent signals would be combined to compute the whole pattern's motion. Our perceptions could then be understood in terms of the relative activity of the neurons that signaled components and the neurons that signaled patterns; our subsequent psychophysical experiments supported this two-stage idea.

The perceptual effects were so compelling that I soon began to think hard about their physiological basis. With Martin Gizzi, then a graduate student, I soon established that direction-selective neurons in the cat's primary visual cortex signaled only the motion of the component gratings. If, for example, a cell responded best to a vertical grating moving to the right, its responses were unchanged if we made a plaid by adding a horizontal grating moving down; the cell behaved simply as though the second grating weren't there. Where might we find the cells that supported the perception of coherence? During the next couple of years, we recorded from a variety of visual areas in the cat. In every case cells responded reliably to the motion of the component gratings in our plaids and never to the motion of the pattern as a whole.

Anthony Movshon

In 1981, I found myself at dinner with Bill Newsome, then a postdoctoral fellow at NIH. Bill firmly maintained that cells in the cortical area he studied—area MT of the macaque—would do the honorable thing and respond to the coherent motion of the plaids. We agreed to collaborate and soon found that about half of the cells in MT responded to the components of plaid patterns as primary cortex cells did. But the other half responded to the coherent plaids in exactly the way that humans perceived them: If such a pattern direction-selective cell responded best to a rightward-moving grating, it also preferred rightward-moving plaids. Like human observers, pattern direction-selective cells did not respond to components within coherent patterns.

These MT cells also had another property that explained why coherence broke down when the components of a plaid were made different: Like other cortical cells, each one responds only to a particular range of spatial frequencies. Shown a plaid with gratings of different spatial frequency, pattern direction-selective cells cannot detect the grating of the "wrong" frequency, and therefore, like human observers, lose their response to the coherent motion.

Many years later Jack Scannell and Malcolm Young found pattern direction-selective cells in an area of the cat's cortex, of which I was unaware when our experiments were done. By switching to the macaque, I was able to confirm my original conjecture about the neural encoding of pattern motion. Had I not had that chance, I might have spent many dry years probing the cat's cortex. These are the perils of what the vision scientist Bela Julesz calls "psychoanatomy": You may be able to come up with a good idea about how the brain does something, but you also need a sense of adventure and a dose of good fortune to put your electrodes in the right place to prove it.

ample, in an area known as *MST*, there are cells selective for linear motion (as in MT), radial motion (either inward or outward from a central point), and circular motion (either clockwise or counterclockwise). We do not know how the visual system makes use of the neurons with complex motion-sensitive properties in MST or the "simpler" direction-selective cells in V1, MT, and other areas. Three roles have been proposed, however:

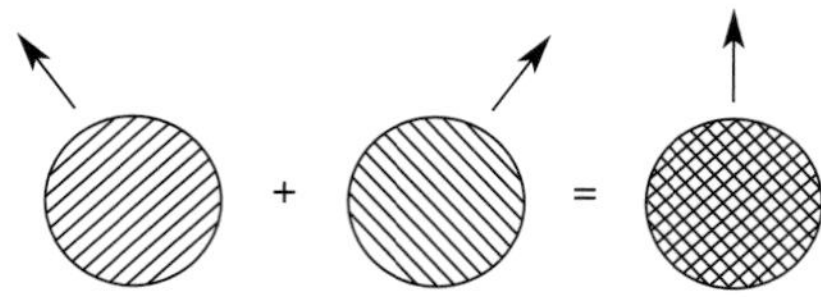

Figure 10.29
A plaid stimulus used to study motion perception and visual area MT. The stimulus is constructed by adding two sets of bars moving at 45-degree angles to each side of vertical. The plaid appears to move straight up.

1. *Navigation:* As we move through our environment, objects stream past our eyes, and the directions and speeds of objects in our peripheral vision provide valuable information that can be used for navigation.
2. *Directing eye movements:* Our ability to sense and analyze motion must also be used when we follow objects with our eyes and when we quickly move our eyes to objects in our peripheral vision that catch our attention.
3. *Motion perception:* We live in a world filled with motion, and survival sometimes depends on our interpretation of moving objects.

Striking evidence that cortical areas in the vicinity of MT and MST are critical for motion perception in humans comes from extremely rare cases in which brain lesions selectively disrupt the perception of motion. The clearest case was reported in 1983 by Josef Zihl and his colleagues at the Max Planck Institute for Psychiatry in Munich, Germany. Zihl studied a woman who had a stroke at the age of 43, bilaterally damaging portions of extrastriate visual cortex known to be particularly responsive to motion (Figure 10.30). Although some ill effects of the stroke, such as difficulty naming objects, were evident, neuropsychological testing showed the patient to be generally normal and with relatively normal vision, except for one serious deficit: She appeared to be incapable of visually perceiving motion. For example, she complained that when she poured coffee into a cup, it appeared to be frozen until suddenly it was overflowing. More ominously, she had trouble crossing the street. Clearly, this seemingly minor loss of motion perception had profound ramifications for the woman's lifestyle. The implication of this case is that motion perception may be based on specialized mechanisms located beyond striate cortex.

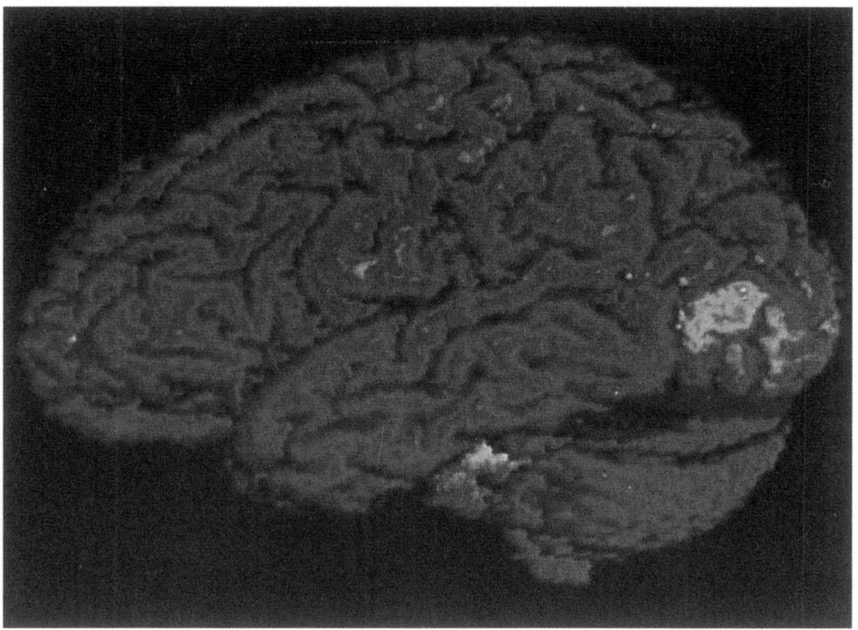

Figure 10.30
Human brain activity in response to visual motion. In this PET image, an area on the lateral surface of the occipital lobe (shown in red and yellow) is particularly active. (Source: Zeki, 1993, Plate 2.)

The Ventral Stream

In parallel with the dorsal stream, a progression of areas from V1, V2, and V3 running ventrally toward the temporal lobes appears specialized for the analysis of visual attributes other than motion.

Area V4. One of the most-studied areas in the ventral stream is **area V4** (Figures 10.27 and 10.28). V4 receives input from the blob and interblob regions of striate cortex via a relay in V2. Neurons in area V4 have larger receptive fields than cells in striate cortex, and many of the cells are both orientation-selective and color-selective. Although there is a good deal of ongoing speculation concerning the function of V4, this area appears to be important for perception of both shape and color. If this area is damaged in monkeys, perceptual deficits involving both shape and color result.

A rare clinical syndrome in humans known as *achromatopsia* is characterized by a partial or complete loss of color vision despite the presence of normal functional cones in the retina. People with this condition describe their world as drab, consisting of only shades of gray. Imagine how unappetizing a gray banana would be! Because achromatopsia is associated with cortical damage in the occipital and temporal lobes without damage to V1, the LGN, or the retina, the syndrome suggests that there is specialized color processing in the ventral stream. Consistent with the coexistence of color-sensitive and shape-sensitive cells in the ventral stream, achromatopsia is usually accompanied by deficits in form perception. Some researchers have proposed that V4 is a particularly critical area for color and form perception, but there is still controversy about whether the lesions associated with achromatopsia correspond to a human V4 area.

Area IT. Beyond V4 in the ventral stream are cortical areas that contain neurons with complicated spatial receptive fields. A major output of V4 is an area in the inferior temporal lobe known as **area IT**. A wide variety of colors and abstract shapes have been found to be good stimuli for cells in IT. As we will see in Chapter 23, this area appears to be important for both visual perception and visual memory. One of the most intriguing findings concerning IT is that a small percentage of the neurons respond strongly to pictures of faces. These cells may also respond to stimuli other than faces, but faces produce a particularly vigorous response, and some faces are more effective stimuli than others. This finding in monkeys appears consistent with images made using fMRI in humans, which indicate that there is a small area in the human brain that is more responsive to faces than to other stimuli (Figure 10.31). The finding of face-selective cells has sparked much interest, in part because of a syndrome called *prosopagnosia*—difficulty recognizing faces even though vision is otherwise normal. This rare syndrome usually results from a stroke and is associated with damage to extrastriate visual cortex.

Could it be that our brains contain a group of cells highly specialized for face recognition? The answer is not known. While most scientists agree that faces are particularly good stimuli for a small percentage of cells, this does not mean that these cells are not involved in processing other types of information.

FROM SINGLE NEURONS TO PERCEPTION

Visual *perception*—the task of identifying and assigning meaning to objects in space—obviously requires the concerted action of many cortical neurons. But which neurons in which cortical areas are responsible for what we perceive? How is the simultaneous activity of widely separated cortical neurons integrated, and where does this integration take place? Neuroscience research is only beginning to tackle these challenging questions. Sometimes, however, basic observations about receptive fields can give us insight into how we perceive (Box 10.4).

From Photoreceptors to Grandmother Cells

Comparing the receptive field properties of neurons at different points in the visual system may provide insight into the basis of perception. The receptive fields of photoreceptors are simply small patches on the retina, whereas those of retinal ganglion cells have a center-surround structure. The ganglion cells are sensitive to variables such as contrast and the wavelength of light. In striate cortex, we encounter simple and complex receptive fields that have several new properties, including orientation selectivity and binocularity. We have seen that in extrastriate cortical areas, cells are selectively responsive to more complex shapes, object motion, and even faces. It appears that the visual system consists of a hierarchy of areas in which receptive fields become increasingly more complex, moving away from V1. Perhaps our perception of specific objects is based on the excitation of a small number of specialized neurons in some ultimate perceptual area that has not yet been identified. Is our recognition of our grandmother based on the responses of five or ten cells with receptive field properties so highly refined that the cells respond only to one person? The closest approximation to this is the face-selective neurons in area IT. Even these fascinating cells do not respond to only one face, however.

While it is by no means settled, there are several arguments against the idea that perception is based on extremely selective receptive fields such as those of "grandmother cells." First, recordings have been made from most parts of the monkey brain, but there is no evidence that a portion of cortex has cells tuned to each of the millions of objects that we all recognize. Second, such great selectivity appears to be counter to the general principle of broad tuning that pertains throughout the nervous system. Photoreceptors respond to a range of wavelengths; simple cells respond to many orientations; cells in MT respond to motion in a range of directions; and face cells usually respond to many faces. Moreover, cells that are selective for one property—orientation, color, or whatever—are always sensitive to other properties. For example, we can focus on the orientation selectivity of V1 neurons and the way in which this may relate to the perception of form, neglecting the fact that the same cells may selectively respond to size, direction of motion, and so on. Finally, it may be too "risky" for the nervous system to rely on extreme selectivity. A blow to the head might kill all five grandmother cells, and in an instant we would lose our ability to recognize her.

Parallel Processing

If we do not rely on grandmother cells, how does perception work? One alternative hypothesis is formulated around the observation that parallel processing is used throughout the visual system (and other brain systems). We encountered parallel processing in Chapter 9, when we discussed ON and OFF and M and P ganglion cells. In this chapter, we saw M, P-IB, and blob parallel channels in V1. Extending away from V1 are dorsal and ventral streams of processing, and the different areas in these two streams are biased, or specialized, for various stimulus properties. Perhaps the brain uses a "division of labor" principle for perception. Within a given cortical area, many broadly tuned cells may serve to represent features of objects. At a bigger scale, a relatively large group of cortical areas may contribute to perception, some dealing mainly with color or form, others mainly with motion.

Box 10.4

OF SPECIAL INTEREST

Depth Perception, Random Dots, and the Shopping Mall

If you venture into a shopping mall and visit a bookstore or novelty shop, you will see books or posters showing patterns of dots or splotches of color that supposedly contain pictures in 3-D if you contort your eyes just the right way. These fascinating and sometimes frustrating pictures are a sophisticated version of stereograms, which have been around since the nineteenth century. Our two eyes always see slightly different images of the world because of the distance between them in the head. The closer objects are to the head, the greater the difference in the two images. You can easily demonstrate this to yourself by holding a finger up in front of your eyes and viewing it at various distances alternately with the left or right eye closed.

Long before anything was known about binocular neurons in visual cortex, stereograms were a popular form of recreation. They became readily available with the development of photography (Figure A). Two photographs were taken with lenses separated by a distance roughly the same as that of human eyes. If you look at the left photograph with your left eye and the right photograph with your right eye (by relaxing the eye muscles or with a stereoscope), your brain combines the images and interprets the different views as cues for distance.

In 1960, Bela Julesz, working at the Bell Telephone Laboratories, invented random-dot stereograms (Figure B). In principle, these paired images of random dots are the same as the nineteenth-century stereograms. The big difference is that no image can be seen with normal binocular viewing. To see the image in 3-D, you must direct your left and right eyes to left and right images. The principle in constructing the stereo images is to create a background of randomly spaced dots, and wherever an area should be closer or farther away in the fused image, the dots shown to one eye are horizontally shifted relative to those in the other eye. Imagine looking at a white index card covered with random black dots while you hold it in front of a large piece of white paper covered with similar dots. If you look at it with first one eye closed and then the other, the dots on the index card will shift horizontally more than those on the more distant piece of paper. The pair of stereo images captures this difference in viewpoint and erases any other indication that there is a square in front, such as the edge of the index card. Random-dot stereograms shocked many scientists, because in 1960 it was commonly thought that depth was perceived only after the images in each eye were separately recognized.

While stereograms elicit a striking image in 3-D, their use was limited by the fact that many observers needed a stereoscope to fuse the two images. This problem was remedied in the 1970s, when Christopher Tyler created autostereograms at the Smith-Kettlewell Eye Research Institute. An autostereogram is a single image that, when properly viewed, gives the perception of objects in 3-D (Figure C). These colorful images containing 3-D pictures you can see at the mall are based on an old illusion called the wallpaper effect. If you look at wallpaper that contains a repeating pattern, you can cross (or diverge) your eyes and view one piece of the pattern with one eye and the next cycle of the pattern with the other eye. The effect makes the wallpaper appear to be closer (or farther away).

In an autostereogram, the wallpaper effect is combined with random-dot stereograms by creating vertical strips containing a repeating pattern of random dots. The depth information is encoded in the positions of the dots and the repetition rate of the pattern. To see the image in an autostere-

Consider the pattern of activity that may arise in striate cortex when you view a simple geometric shape like a stop sign (Figure 10.32). Because the sign takes up a substantial portion of the visual field, neural activity in striate cortex is distributed among many cortical modules. The readout of each module provides information about orientation and color at a fraction of the contrast border along the edge of the sign. Since we do not know how perception works, perhaps this pattern of activity and the associated patterns in extrastriate cortex are all there is to perception. Many scientists find this idea unsatisfactory, however. They argue that somehow the activity of the mod-

(Box 10.4, *continued*)

ogram, you must either cross or diverge your eyes so that they look at strips offset by a repetition cycle. For example, to see the 3-D skull in Figure C, you have to relax your eye muscles so that the left eye looks at the left dot on top and the right eye at the right dot. You will know you are getting close when you see three dots at the top of the image. Relax and keep looking, and the picture will become visible.

One of the fascinating things about stereograms is that you often must look at them for tens of seconds or even minutes, while your eyes become "properly" misaligned and your visual cortex "figures out" the correspondence between the left and right eye views. We do not know what is going on in the brain during this period in which it struggles to make sense of the splotchy figure.

Figure A
A nineteenth-century stereogram. (Source: Horibuchi, 1994, p. 38.)

Figure B
A random-dot stereogram and the perception that results from binocularly fusing the images. (Source: Julesz, 1971, p. 21.)

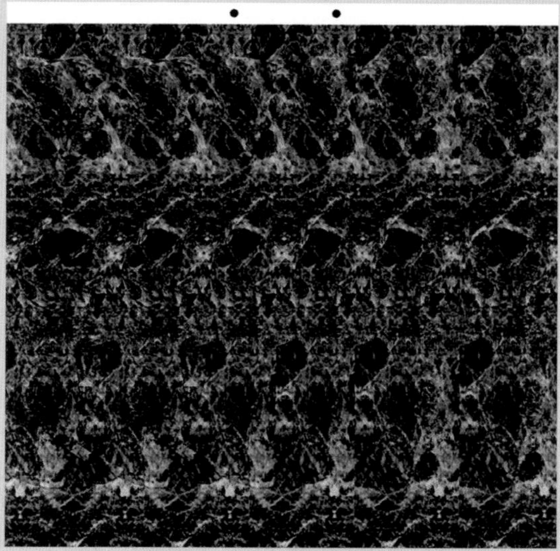

Figure C
An autostereogram. (Source: Horibuchi, 1994, p. 54.)

ules must be compared to make a decision that this activity represents a single object—a stop sign—and not a jumble of line segments.

One scheme for perceptually linking the neurons that respond to a stop sign (or one's grandmother) takes advantage of the fact that the cortical modules in striate cortex and subsets of cells in extrastriate cortical areas activated by a single object are active simultaneously. (In Chapter 8, we saw how the olfactory system may use the timing of neuronal activity to represent sensory information.) Thus, the contours of an object could be perceptually bound together if the activity of a single module were in some way sensitive to the ac-

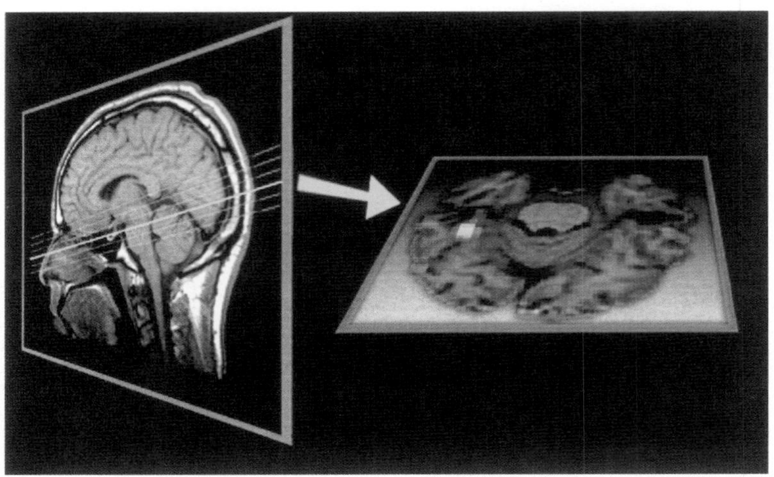

Figure 10.31
Human brain activity elicited by pictures of faces. Using fMRI, brain activity was recorded first in response to faces and second in response to nonface stimuli. The colored area in the brain section on the right showed significantly greater responses to faces. (Source: Courtesy of Drs. I. Gauthier, J. C. Gore, and M. Tarr.)

tivity of other modules that were active at precisely the same time. In fact, there is an anatomical basis for cross talk between modules: the horizontal connections that connect the cells of layer III. Similar connections even cross from one side of the brain to the other via the corpus callosum to bind together the cells that respond to images falling on the seam between the right and left visual fields. These connections selectively wire up the cells that have common receptive field characteristics; e.g., cells responding more to horizontal lines connect to other cells that share the same orientation preference.

In 1989, neuroscientists Charles Gray and Wolf Singer, working together at the Max Planck Institute in Frankfurt, Germany, discovered that visual cortical cells in widely separated modules often fire synchronously when activated by the contours of a single common object. Work by these scientists and others has gone on to show that objects often elicit synchronous activity even across widely separated cortical areas. This work suggests that the internal

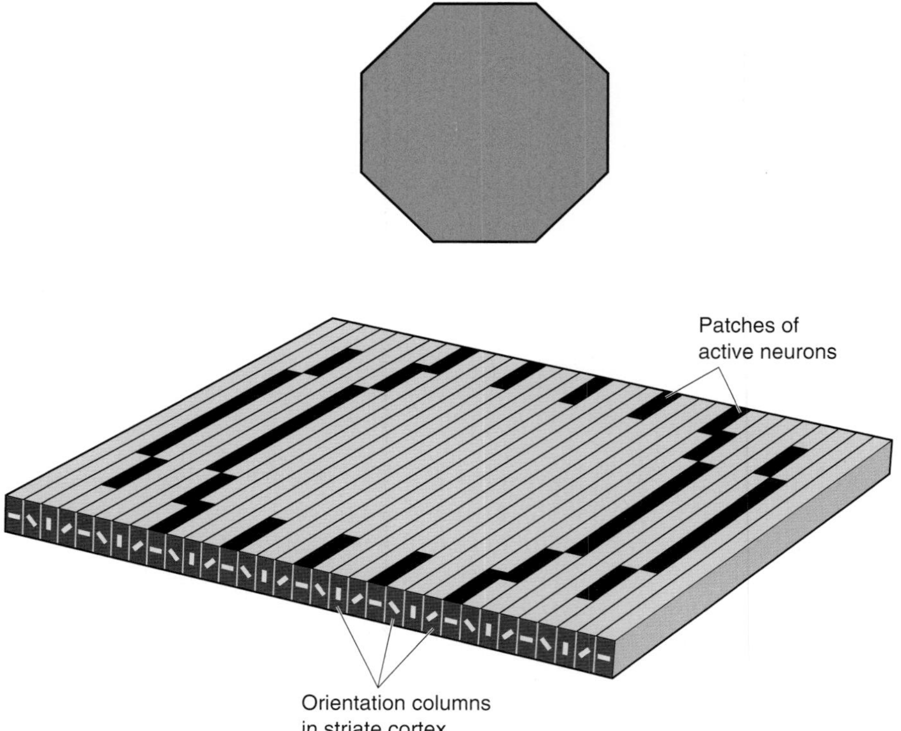

Figure 10.32
A highly schematic illustration of neurons in striate cortex responding to the shape of a stop sign. The actual pattern would show much more widespread activity because of the broad orientation tuning of the neurons

neural representation of a stop sign or one's grandmother may be the synchronous activity of a large population of cells distributed across the cortex.

Despite a lack of consensus about the significance of synchronous activity and the particular manner in which objects are represented, most visual neuroscientists believe that the representation of objects involves activity distributed across many cortical neurons. Because objects have many properties such as color, orientation, and direction of motion, it may also be necessary to compare or integrate activity across multiple cortical areas to determine perception.

CONCLUDING REMARKS

In this chapter, we have outlined the organization of the sensory pathway from eye to thalamus to cortex. We saw that vision actually involves the perception of numerous different properties of objects—color, form, movement—and that these properties are processed in parallel by different cells of the visual system. This processing of information evidently requires a strict segregation of inputs at the thalamus, some limited convergence of information in striate cortex, and finally a massive divergence of information as it is passed on to higher cortical areas. The distributed nature of the cortical processing of visual information is underscored when you consider that the output of a million ganglion cells can recruit the activity of well over a billion cortical neurons throughout the occipital, parietal, and temporal lobes! Somehow this widespread cortical activity is combined to form a single seamless perception of the visual world.

Heed the lessons learned from the visual system. As we shall see in later chapters, the principles of organization in this system—parallel processing, topographic mappings of sensory surfaces, synaptic relays in the dorsal thalamus, cortical modules, and multiple cortical representations—are also features of the sensory systems devoted to hearing and touch.

KEY TERMS

The Retinofugal Projection
retinofugal projection (p. 314)
optic nerve (p. 315)
optic chiasm (p. 315)
decussation (p. 315)
optic tract (p. 315)
visual hemifield (p. 316)
binocular visual field (p. 316)
lateral geniculate nucleus (LGN) (p. 317)
optic radiation (p. 317)
superior colliculus (p. 318)
optic tectum (p. 319)
retinotectal projection (p. 319)
retinotopy (p. 319)

The Lateral Geniculate Nucleus
magnocellular LGN layer (p. 322)
parvocellular LGN layer (p. 322)
koniocellular layer (p. 322)

Anatomy of the Striate Cortex
area 17 (p. 324)
V1 (p. 324)
striate cortex (p. 324)
ocular dominance column (p. 327)
cytochrome oxidase (p. 328)
blob (p. 328)

Physiology of the Striate Cortex
magnocellular channel (M channel) (p. 330)
parvocellular-interblob channel (P-IB channel) (p. 330)
blob channel (p. 330)
simple cell (p. 331)
orientation selectivity (p. 331)
binocular receptive field (p. 331)
direction selectivity (p. 331)
complex cell (p. 332)
orientation column (p. 333)
cortical module (p. 337)

Beyond Striate Cortex
area MT (p. 339)
area V4 (p. 342)
area IT (p. 342)

REVIEW QUESTIONS

1. Following a bicycle accident, you are disturbed to find that you cannot see anything in the left visual field. Where has the retinofugal pathway been damaged?
2. What is the source of most of the input to the *left* LGN?
3. A worm has eaten part of one lateral geniculate nucleus. You can no longer perceive color in the right visual field of the right eye. What layer or layers of which LGN have been damaged?
4. List the chain of connections that link a cone in the retina to a blob cell in striate cortex.
5. How are receptive fields transformed at each of the synaptic relays that connect an M-type retinal ganglion cell to a neuron in striate cortical layer IVB?
6. Which pathway, magnocellular or parvocellular, provides a greater percentage of the input to striate cortex? What are two analyses of the visual world that are thought to involve mainly this pathway? What about the other pathway?
7. What is meant by parallel processing in the visual system? Give two examples.
8. If a child is born cross-eyed and the condition is not corrected before age 10, binocular depth perception will be lost forever. This is explained by a modification in the circuitry of the visual system. From your knowledge of the central visual system, where do you think the circuitry has been modified?
9. In what ways is area MT more specialized for the detection of visual motion than area V1?
10. For many years it was thought that depth perception involved recognition of objects in each eye separately, followed by binocular integration. How do the stereograms discussed in Box 10.4 disprove this hypothesis? What areas of the brain are possible sites for binocular integration?

The Auditory and Vestibular Systems

CHAPTER **11**

Box 11.1

OF SPECIAL INTEREST

Infrasound

Most people are familiar with ultrasound (sound above the 20 kHz limit of our hearing) because it has everyday applications, from ultrasonic cleaners to medical imaging. Some animals can hear these high frequencies. For instance, dog whistles work because dogs can hear up to about 40 kHz. Less well known is infrasound, or sound at low frequencies, below about 20 Hz. Some animals can hear these frequencies; one is the elephant, which can detect 15 Hz tones at sound levels inaudible to humans. Whales produce low-frequency sounds, which are thought to be a means of communication over distances of kilometers. Low-frequency vibrations are also produced by the earth, and it is thought that some animals may sense an impending earthquake by hearing such sound.

Even though we usually cannot hear very low frequencies with our ears, they are present in our environment and can have unpleasant subconscious effects. Infrasound is produced by such devices as air conditioners, boilers, aircraft, and automobiles. Though even intense infrasound from these machines does not cause hearing loss, it can cause dizziness, nausea, and headache. Many cars produce low-frequency sound at highway speeds, making sensitive people carsick. At very high levels, low-frequency sound may also produce resonances in body cavities such as the chest and stomach that can damage internal organs. You may want to think twice before standing directly in front of a large speaker at a concert!

In addition to mechanical equipment, our body generates inaudible low-frequency sound. When muscle changes length, individual fibers vibrate, producing low-intensity sound at about 25 Hz. While we cannot normally hear these sounds, you can demonstrate them to yourself by carefully putting your thumbs in your ears and making a fist with each hand. As you tighten your fist, you can hear a low rumbling sound produced by the contraction of your forearm muscles. Other muscles, including your heart, produce inaudible sounds at frequencies near 20 Hz.

It's probably just as well that we aren't more aware of infrasound. It would be hard to get any work done if we had to listen to the sounds of our bodies along with the drone of machinery.

faintest sound that can be heard. If we were any more sensitive, we would hear a constant roar from the random movement of air molecules.

Real-world sounds rarely consist of simple periodic sound waves at one frequency and intensity. It is the simultaneous combination of different frequency waves at different intensities that gives different musical instruments and human voices their unique tonal qualities.

THE STRUCTURE OF THE AUDITORY SYSTEM

Before exploring how variations in air pressure are translated into neural activity, let's quickly survey the structure of the auditory system. The components of the ear are shown in Figure 11.3. The visible portion of the ear consists primarily of cartilage covered by skin, forming a sort of funnel called the **pinna** (Latin for "wing"), which helps collect sounds from a wide area. It has also been suggested that the convolutions in the pinna play a role in localizing sounds, something we discuss later in the chapter. In humans, the pinna is more or less fixed in position, but animals such as cats and horses have considerable muscular control over the position of their pinna and can orient it toward a source of sound.

The entrance to the internal ear is called the **auditory canal**, and it extends about 2.5 cm (1 inch) inside the skull before it ends at the **tympanic membrane**, also known as the *eardrum*. Connected to the medial surface of the tympanic membrane is a series of bones called **ossicles** (from the Latin for "little bones"; the ossicles are indeed the smallest bones in the body). Located

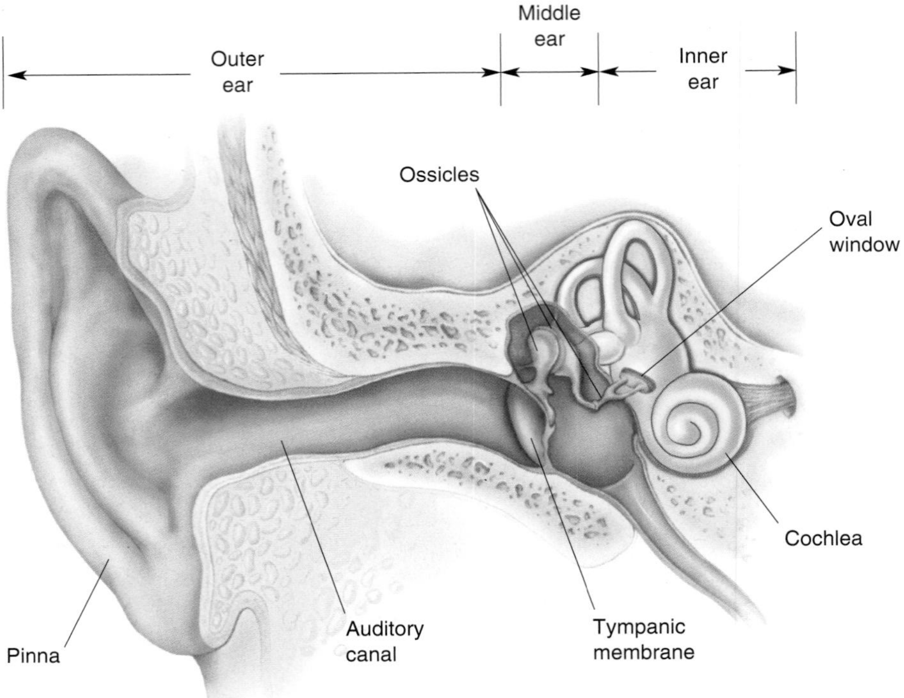

Figure 11.3
The outer, middle, and inner ear.

in a small air-filled chamber, the ossicles transfer movements of the tympanic membrane into movements of a second membrane covering a hole in the bone of the skull called the **oval window**. Behind the oval window is the fluid-filled **cochlea**, which contains the apparatus for transforming the physical motion of the oval window membrane into a neuronal response. Thus, the first stages of the basic auditory pathway look like this:

Sound wave moves the tympanic membrane →
Tympanic membrane moves the ossicles →
Ossicles move the membrane at the oval window →
Motion at the oval window moves fluid in the cochlea →
Movement of fluid in the cochlea causes a response in sensory neurons.

All of the structures from the pinna inward are considered components of the ear, and it is conventional to refer to the ear as having three main divisions. The structures from the pinna to the tympanic membrane make up the **outer ear**; the tympanic membrane and the ossicles constitute the **middle ear**; and the apparatus medial to the oval window is the **inner ear**.

Once a neural response to sound is generated in the inner ear, the signal is transferred to and processed by a series of nuclei in the brain stem. Output from these nuclei is sent to a relay in the thalamus, the **medial geniculate nucleus**, or **MGN**. Finally, the MGN projects to **primary auditory cortex**, or **A1**, in the temporal lobe. In one sense, the auditory pathway is more complex than the visual pathway because there are more intermediate stages between the sensory receptors and cortex. The systems have analogous components, however. Each starts with sensory receptors, which connect to early integration stages (in the retina for vision and in the brain stem for audition), then to a thalamic relay and sensory cortex (Figure 11.4).

THE MIDDLE EAR

The outer ear funnels sound to the middle ear, an air-filled cavity containing the first elements that move in response to sound. In the middle ear, variations in air pressure are converted into movements of the ossicles. In this section, we'll explore how the middle ear performs an essential transformation of sound energy.

Components of the Middle Ear

The structures within the middle ear are the tympanic membrane, the ossicles, and two tiny muscles that attach to the ossicles. The tympanic membrane is somewhat conical, with the point of the cone extending into the cavity of the middle ear. There are three ossicles, each named (from the Latin) after an object it slightly resembles (Figure 11.5). The ossicle attached to the tympanic membrane is the **malleus** ("hammer"), which forms a rigid connection with the **incus** ("anvil"). The incus forms a flexible connection with the **stapes** ("stirrup"). The flat bottom portion of the stapes, the *footplate,* moves in and out like a piston at the oval window, thus transmitting sound vibrations to the fluids of the cochlea in the inner ear.

The air in the middle ear is continuous with the air in the nasal cavities via the **Eustachian tube**, although this tube is usually closed by a valve. When you're in an ascending airplane or a car heading up a mountain, the pressure of the surrounding air decreases. As long as the valve on the Eustachian tube

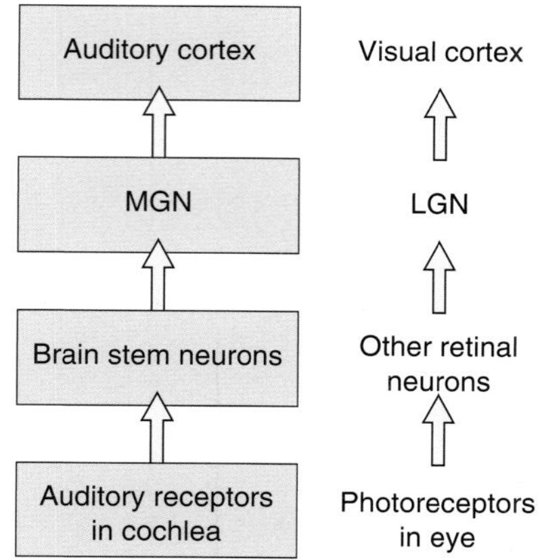

Figure 11.4
Auditory and visual pathways compared. Following the sensory receptors, both systems have early integration stages, a thalamic relay, and a projection to neocortex.

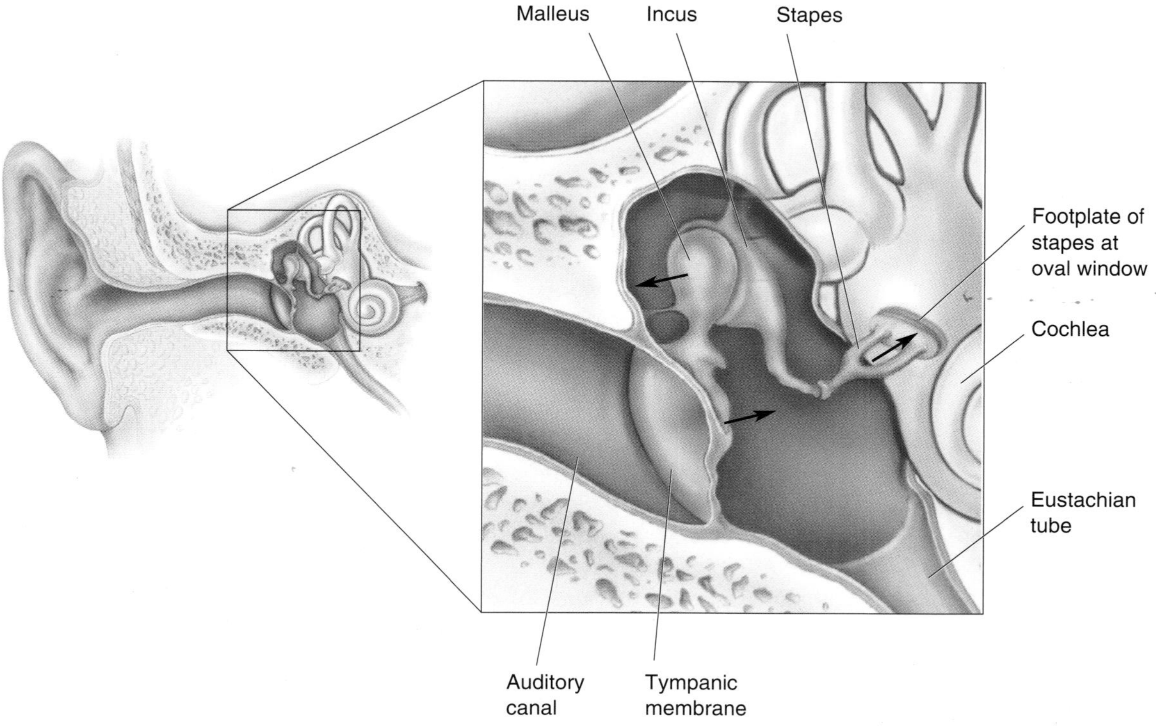

Figure 11.5
The middle ear. As the arrows indicate, when air pressure pushes the tympanic membrane, the bottom of the malleus is pushed inward, and the lever action of the ossicles makes the footplate of the stapes push inward at the oval window. The pressure pushing at the oval window is greater than that at the tympanic membrane, in part because the surface area of the footplate of the stapes is smaller than the surface area of the tympanic membrane.

is closed, however, the air in the middle ear stays at the pressure of the air before you started to climb. Because the pressure inside the middle ear is higher than the air pressure outside, the tympanic membrane bulges out, and you feel unpleasant pressure or pain in the ear. The pain can be relieved by yawning or swallowing, either of which opens the Eustachian tube, thereby equalizing the air pressure in the middle ear with the ambient air pressure.

Sound Force Amplification by the Ossicles

Sound waves move the tympanic membrane, and the ossicles move another membrane at the oval window. Why isn't the ear arranged so that sound waves directly move the membrane at the oval window? The problem is that the cochlea is filled with fluid, not air. If sound waves impinged directly on the oval window, the membrane would barely move, and all but 0.1% of the sound energy would be reflected because of the pressure the cochlear fluid exerts at the back of the oval window. If you've ever noticed how quiet it is under water, you know how well water reflects sound coming from above. The fluid in the inner ear resists being moved much more than air does (i.e., fluid has greater inertia), so more pressure is needed to vibrate the fluid than the air can provide. The ossicles provide this necessary amplification in pressure.

To understand the process, consider the definition of pressure. The pressure on a membrane is defined as the force pushing it divided by its surface area. The pressure at the oval window becomes greater than the pressure at the tympanic membrane if (1) the force on the oval window membrane is greater than that on the tympanic membrane or (2) the surface area of the oval window is smaller than the area of the tympanic membrane. The middle ear uses both mechanisms. It increases pressure at the oval window by altering both the force and the surface area. The force at the oval window is greater because the ossicles act like levers. Sound causes large movements of the tympanic membrane, which are transformed into smaller but stronger vibrations of the oval window. And the surface area of the oval window is much smaller than that of the tympanic membrane. These factors combine to make the pressure at the oval window about 20 times greater than the pressure at the tympanic membrane, and this increase is sufficient to move the fluid in the inner ear.

The Attenuation Reflex

Two muscles attached to the ossicles have a significant effect on sound transmission to the inner ear. The *tensor tympani muscle* is anchored to bone in the cavity of the middle ear at one end and attaches to the malleus at the other end (Figure 11.6). The *stapedius muscle* also extends from a fixed anchor of bone and attaches to the stapes. When these muscles contract, the chain of ossicles becomes much more rigid, and sound conduction to the inner ear is greatly diminished. The onset of a loud sound causes these muscles to contract, a response called the **attenuation reflex**. Sound attenuation is much greater at low frequencies than at high frequencies.

A number of functions have been proposed for this reflex. One function may be to adapt the ear to continuous sound at high intensities. Loud sounds that would otherwise saturate the response of the receptors in the inner ear could be reduced to a level below saturation by the attenuation reflex, thus increasing the dynamic range we can hear. The attenuation reflex also protects the inner ear from loud sounds that would otherwise damage it. Unfortunately, the reflex has a delay of 50–100 msec from the time that sound reaches the ear, so it doesn't offer much protection from very sudden loud

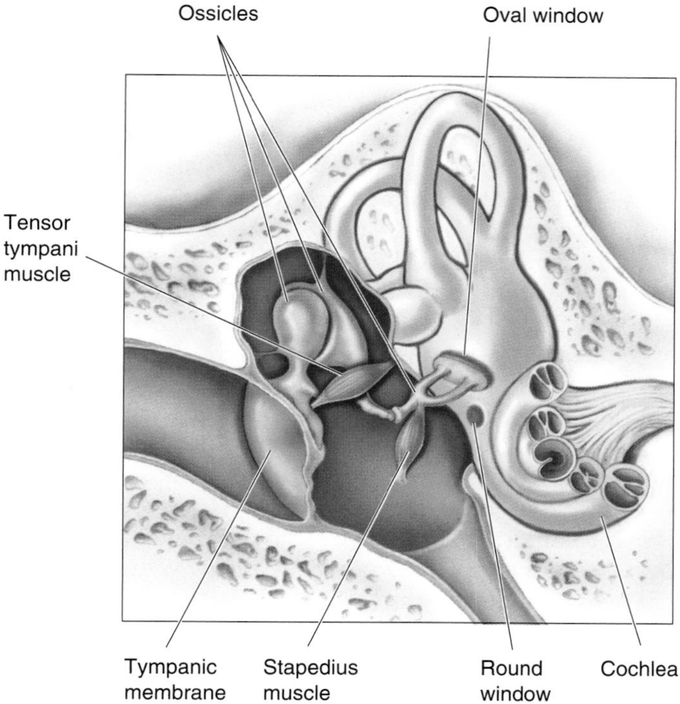

Figure 11.6
The middle and inner ear. Both the stapedius muscle and the tensor tympani muscle are attached to the wall of the middle ear at one end and to the ossicles at the other end.

sounds; damage may already be done by the time the muscles contract. This is why, despite the best efforts of your attenuation reflex, a loud explosion (or music on your Walkman) can still damage your cochlea. Because the attenuation reflex suppresses low frequencies more than high frequencies, it tends to make high-frequency sounds easier to discern in an environment with a lot of low-frequency noise. This capability enables us to understand speech more easily in a noisy environment than we could without the reflex. It is thought that the attenuation reflex is also activated when we speak, so we don't hear our own voices as loudly as we otherwise would.

THE INNER EAR

Although considered part of the ear, not all of the inner ear is concerned with hearing. The inner ear consists of the cochlea, which is part of the auditory system, and the labyrinth, which is not. The labyrinth is an important part of the *vestibular system*, which helps maintain the body's equilibrium. The vestibular system is discussed later in the chapter. Here we are concerned only with the cochlea and the role it plays in transforming sound into a neural signal.

Anatomy of the Cochlea

The cochlea (Latin for "snail") has a spiral shape resembling a snail's shell. Figure 11.6 shows the cochlea cut in half. One can approximate the structure of the cochlea by wrapping a drinking straw two and a half times around the sharpened tip of a pencil. In the cochlea, the hollow tube (represented by the straw) has walls made of bone. The central pillar of the cochlea (represented by the pencil) is a conical bony structure called the *modiolus*. The actual dimensions are much smaller than the straw-and-pencil model, the cochlea's hollow tube being about 32 mm long and 2 mm in diameter. Rolled up, the human cochlea is about the size of a pea. At the base of the cochlea are two membrane-covered holes: the oval window, which as we have seen is below the footplate of the stapes, and the **round window**.

If the cochlea is cut in cross section, we can see that the tube is divided into three fluid-filled chambers: the **scala vestibuli**, the **scala media**, and the **scala tympani** (Figure 11.7). The three scalae wrap around inside the cochlea like a spiral staircase (*scala* is from the Latin for "stairway"). **Reissner's membrane** separates the scala vestibuli from the scala media, and the **basilar membrane** separates the scala tympani from the scala media. Sitting upon the basilar membrane is the **organ of Corti**, which contains auditory receptor neurons; hanging over this organ is the **tectorial membrane**. At the apex of the cochlea the scala media is closed off, and the scala tympani becomes continuous with the scala vestibuli at a hole in the membranes called the **helicotrema** (Figure 11.8). At the base of the cochlea, the scala vestibuli meets the oval window and the scala tympani meets the round window.

The fluid in the scala vestibuli and scala tympani, called **perilymph**, has an ionic content similar to that of cerebrospinal fluid: low K^+ (7 mM) and high Na^+ (140 mM) concentrations. The scala media is filled with **endolymph**, which is unusual in that it has ionic concentrations similar to that of intracellular fluid, high K^+ (150 mM) and low Na^+ (1 mM), even though it is extracellular. This difference in ion content is generated by active transport processes taking place at the *stria vascularis*, the endothelium lining one wall of the scala media (Figure 11.7). The stria vascularis reabsorbs sodium and secretes potassium against their concentration gradients. Because of the ionic concentration differences and the permeability of Reissner's membrane, the endolymph has an electrical potential that is about 80 mV more positive than that of the perilymph; this is called the *endocochlear potential*. We shall see that

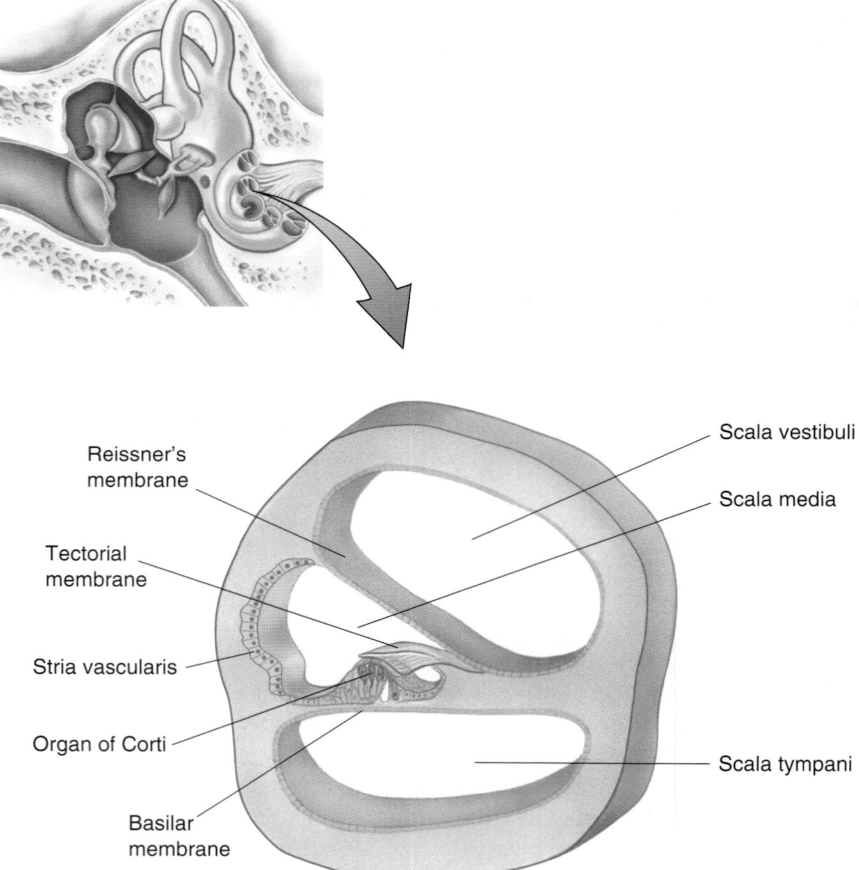

Figure 11.7
The three scalae of the cochlea. Viewed in cross section, the cochlea contains three small parallel chambers. These chambers, the scalae, are separated by Reissner's membrane and the basilar membrane. The organ of Corti contains the auditory receptors; it sits upon the basilar membrane and is covered by the tectorial membrane.

the endocochlear potential is important because it enhances auditory transduction.

Physiology of the Cochlea

Despite its structural complexity, the basic operation of the inner ear is fairly simple. Look at Figure 11.8 and imagine what happens when the ossicles move the membrane that covers the oval window. Inward motion at the oval window pushes perilymph into the scala vestibuli. If the membranes inside the cochlea were completely rigid, the increase in fluid pressure at the oval window would reach up the scala vestibuli, through the helicotrema, and back down the scala tympani to the round window. Because the fluid pressure has nowhere else to escape, the membrane at the round window would bulge out in response to the inward movement of the membrane at the oval window. Although this isn't exactly what happens, the description conveys the very important point that any motion at the oval window must be accompanied by a complementary motion at the round window. Such movement must occur because the cochlea is filled with incompressible fluid held in a solid bony container. The consequence of pushing in at the oval window is a bit like pushing in one end of a tubular water balloon—the other end has to bulge out. The only reason this description of the events in the cochlea is not quite accurate is that some structures inside the cochlea are not rigid. Most important, the basilar membrane is flexible and bends in response to sound.

The Response of the Basilar Membrane to Sound. The basilar membrane has two structural properties that determine the way it responds to sound. First, the membrane is wider at the apex than at the base by a factor of about

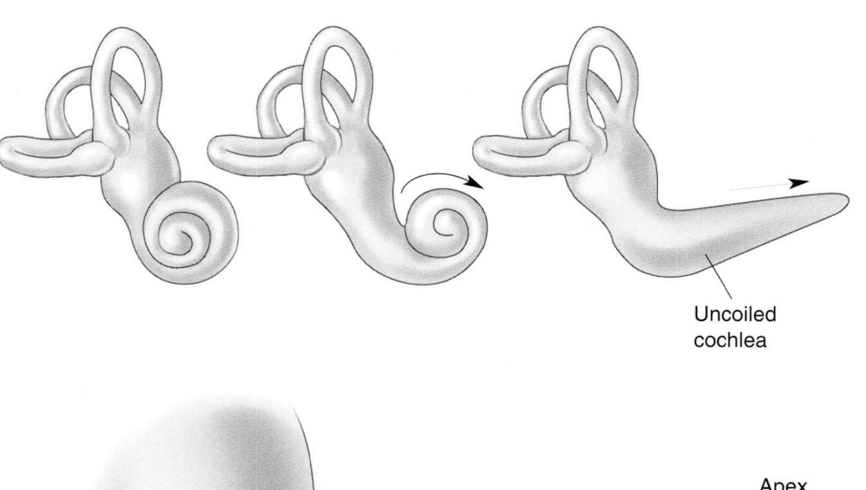

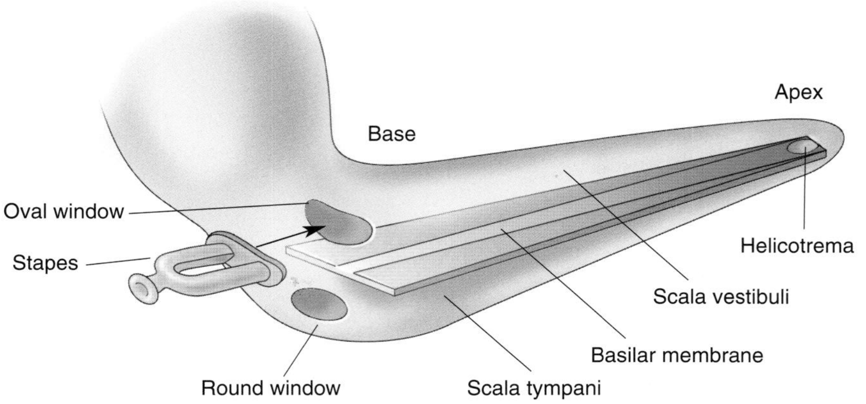

Figure 11.8
The basilar membrane in an uncoiled cochlea. Although the cochlea narrows from base to apex, the basilar membrane widens toward the apex. The helicotrema is a hole at the apex of the basilar membrane, which connects the scala vestibuli and scala tympani.

5. Second, the stiffness of the membrane decreases from base to apex, the base being about 100 times stiffer. Think of it as a flipper of the sort used for swimming, with a narrow, stiff base and a wide, floppy apex. When sound pushes the footplate of the stapes at the oval window, perilymph is displaced within the scala vestibuli, and endolymph is displaced within the scala media because Reissner's membrane is very flexible. Sound can also pull the footplate, reversing the pressure gradient. Sound causes a continual push-pull motion of the footplate.

We owe much of our understanding of the response of the basilar membrane to the research of Hungarian-American biophysicist Georg von Békésy. Von Békésy determined that the movement of the endolymph makes the basilar membrane bend near its base, starting a wave that propagates toward the apex. The wave that travels up the basilar membrane is similar to the wave that runs along a rope if you hold one end in your hand and give it a snap (Figure 11.9). The distance the wave travels up the basilar membrane depends on the frequency of the sound. If the frequency is high, the stiffer base of the membrane will vibrate a good deal, dissipating most of the energy, and the wave will not propagate very far (Figure 11.10a). Low-frequency sounds, however, generate waves that travel all the way up to the floppy apex of the membrane before most of the energy is dissipated (Figure 11.10b). The response of the basilar membrane establishes a place code in which different locations of membrane are maximally deformed at different sound frequencies (Figure 11.10c). As we shall see, the differences in the traveling waves produced by different sound frequencies are responsible for the neural coding of pitch.

The Organ of Corti and Associated Structures. Everything we have discussed to this point involves the mechanical transformations of sound energy that occur in the middle and inner ear. Now we come to the point in the system where neurons are first involved. The auditory receptor cells, which convert mechanical energy into a change in membrane polarization, are in the organ of Corti (named for the Italian anatomist who first identified it). The organ of Corti consists of hair cells, the rods of Corti, and various supporting cells.

The auditory receptors are called **hair cells** because each one has about 100 hairy-looking **stereocilia** (singular: stereocilium) extending from its top. The hair cells and stereocilia are shown in Figure 11.11 as they appear when viewed with a scanning electron microscope. The critical event in the transduction of sound into a neural signal is the bending of these cilia. For this reason, we want to examine the organ of Corti in more detail to see how flexing of the basilar membrane leads to bending of the stereocilia.

The hair cells are sandwiched between the basilar membrane and a thin sheet of tissue called the **reticular lamina** (Figure 11.12). The *rods of Corti* span

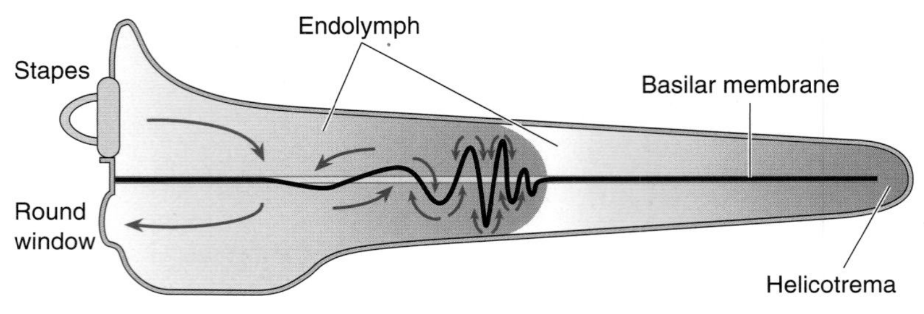

Figure 11.9
A traveling wave in the basilar membrane. As the stapes moves in and out, it causes endolymph to flow, as shown by the arrows. This generates a traveling wave in the basilar membrane. (The size of the wave is magnified about 1 million times in this illustration.) At this frequency, 3000 Hz, the fluid and membrane movement end abruptly about halfway between the base and apex. (Source: Adapted from Nobili, Mammano, and Ashmore, 1998, Fig. 1.)

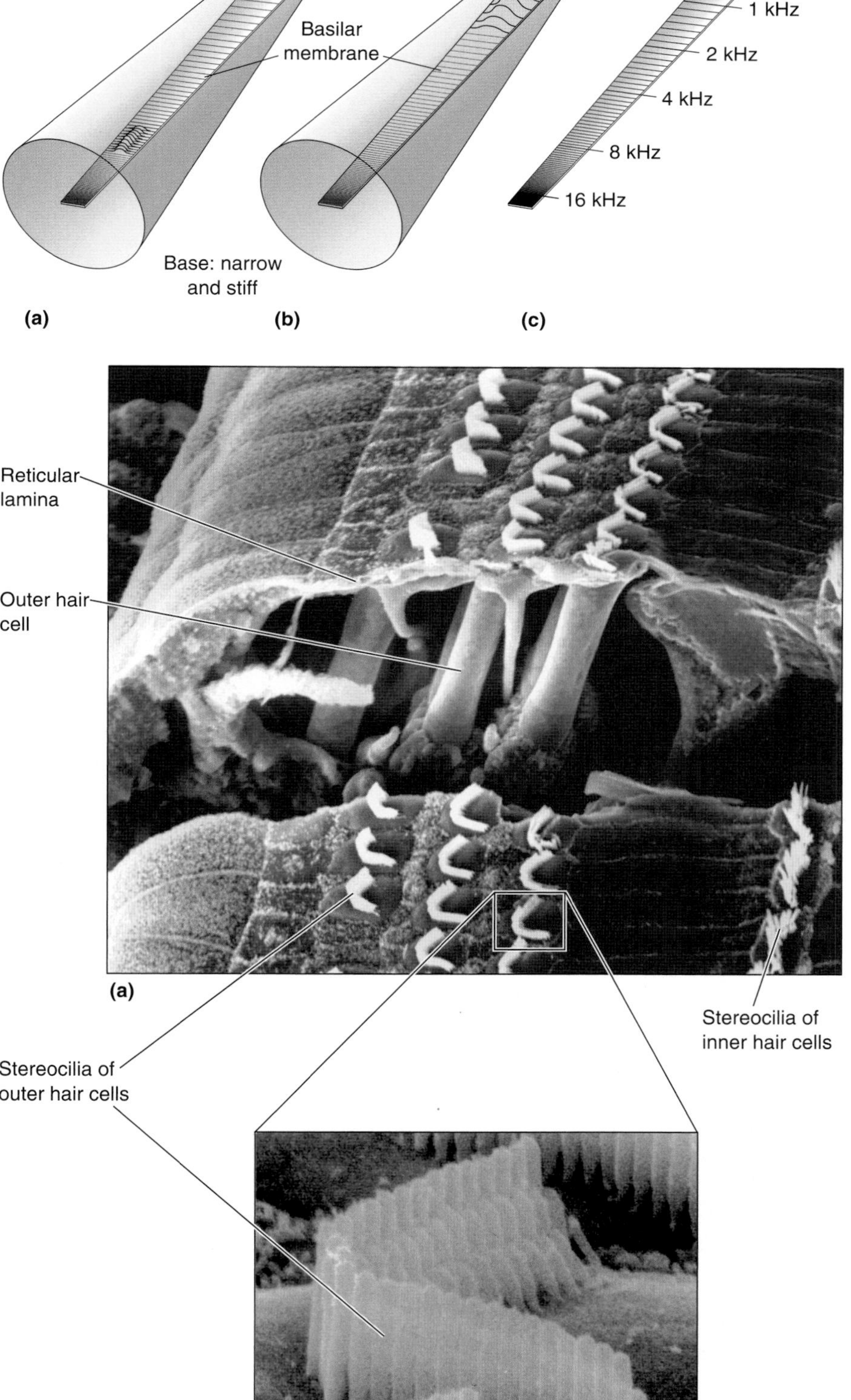

Figure 11.10
The response of the basilar membrane to sound. The cochlea is again shown uncoiled. **(a)** High-frequency sound produces a traveling wave, which dissipates near the narrow and stiff base of the basilar membrane. **(b)** Low-frequency sound produces a wave that propagates all the way to the apex of the basilar membrane before dissipating. (The bending of the basilar membrane is greatly exaggerated for the purpose of illustration.) **(c)** There is a place code on the basilar membrane for the frequency that produces the maximum amplitude deflection.

Figure 11.11
Hair cells viewed through the scanning electron microscope. (a) Hair cells and their stereocilia. **(b)** A higher-resolution view of the stereocilia on an outer hair cell. The stereocilia are approximately 5 mm long. (Source: Courtesy of I. Hunter-Duvar and R. Harrison, The Hospital for Sick Children, Toronto, Ontario, Canada.)

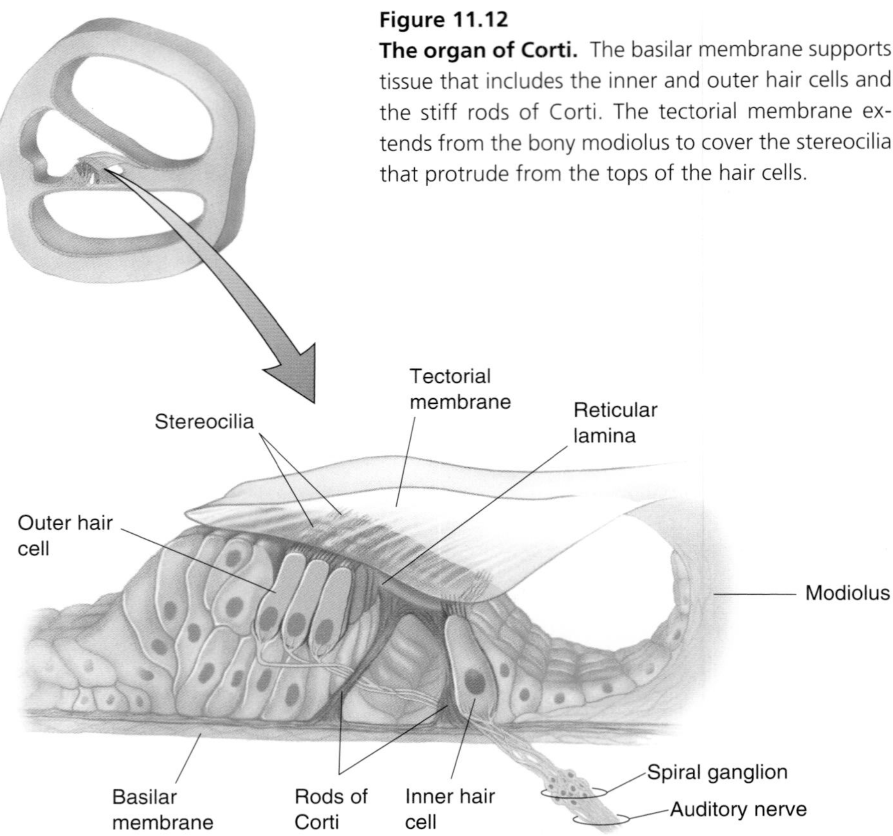

Figure 11.12
The organ of Corti. The basilar membrane supports tissue that includes the inner and outer hair cells and the stiff rods of Corti. The tectorial membrane extends from the bony modiolus to cover the stereocilia that protrude from the tops of the hair cells.

these two membranes and provide structural support. Hair cells between the modiolus and the rods of Corti are called **inner hair cells** (about 3500 form a single row), and cells farther out than the rods of Corti are called **outer hair cells** (humans have about 15,000–20,000 arranged in three rows). The stereocilia at the tops of the hair cells extend above the reticular lamina into the endolymph, and their tips end either in the gelatinous substance of the tectorial membrane (the outer hair cells) or just below the tectorial membrane (the inner hair cells). To keep the membranes within the organ of Corti straight in your mind, remember that the *basilar* is at the *base* of the organ of Corti, the *tectorial* forms a *roof* over the structure, and the *reticular* is in the *middle*, holding onto the hair cells.

Hair cells form synapses on neurons whose cell bodies are in the **spiral ganglion** within the modiolus. Spiral ganglion cells are bipolar, with neurites extending to the bases and sides of the hair cells, where they receive synaptic input. Axons from the spiral ganglion enter the **auditory-vestibular nerve** (cranial nerve VIII), which projects to the cochlear nuclei in the medulla.

Transduction by Hair Cells. When the basilar membrane moves in response to a motion at the stapes, the entire foundation supporting the hair cells moves, because the basilar membrane, rods of Corti, reticular lamina, and hair cells are all rigidly connected. These structures move as a unit, pivoting up toward the tectorial membrane or away from it. When the basilar membrane moves up, the reticular lamina moves up and in toward the modiolus. Conversely, downward motion of the basilar membrane causes the reticular lamina to move down and away from the modiolus. When the reticular lamina moves inward or outward relative to the modiolus, it also moves in or out with respect to the tectorial membrane. Because the tectorial membrane holds the tips of the outer hair cell stereocilia, the lateral motion of the reticular lamina relative to the tectorial membrane bends the stereocilia on the outer hair cells one way or the other (Figure 11.13). The tips of stereocilia from inner hair cells are also bent similarly, probably because they are pushed by moving endolymph. The stereocilia contain aligned actin filaments that make them stiff, so they bend as rigid rods. Cross-linked filaments connect the stereocilia on each hair cell, making all of the cilia on a hair cell move together as a unit. Putting it all together, you can imagine a sound wave making the basilar membrane jiggle between the two positions shown in Figure 11.13a and b and the hair cell cilia thus bending back and forth against the tectorial membrane.

Until recently, progress in our understanding of how hair cells convert the bending of stereocilia into neural signals was slow. Because the cochlea is encased in bone, it is quite difficult to record from the hair cells. In the 1980s, A. J. Hudspeth and his colleagues, then at the California Institute of Technology, pioneered a new approach in which hair cells are isolated from the inner ear and studied *in vitro*. The *in vitro* technique has revealed much about the transduction mechanism. Recordings from hair cells indicate that when the stereocilia bend in one direction, the hair cell depolarizes, and when they bend in the other direction, the cell hyperpolarizes (Figure 11.14a). When a sound wave causes the stereocilia to bend back and forth, the hair cell generates a receptor potential that alternately hyperpolarizes and depolarizes from the resting potential of −70 mV (Figure 11.14b).

To appreciate just how efficiently the ear works, take a moment to notice the scale on the X axis of Figure 11.14a. Its units are nanometers; recall that 1 nm equals 10^{-9} m. The graph shows that the receptor potential of the hair cell is saturated by the time the tips of its stereocilia have moved about 20 nm to the side; this is what an extremely loud sound may do. But the softest sound you can hear moves the stereocilia only 0.3 nm to each side. This is an astoundingly small distance—about the diameter of a large atom! Since each stereocilium is about 500 nm (or 0.5 mm) in diameter, this soft sound only has to jiggle the stereocilia about one-thousandth of their diameter in order to produce perceptible noise. How does the hair cell transduce such infinitesimally small amounts of sound energy?

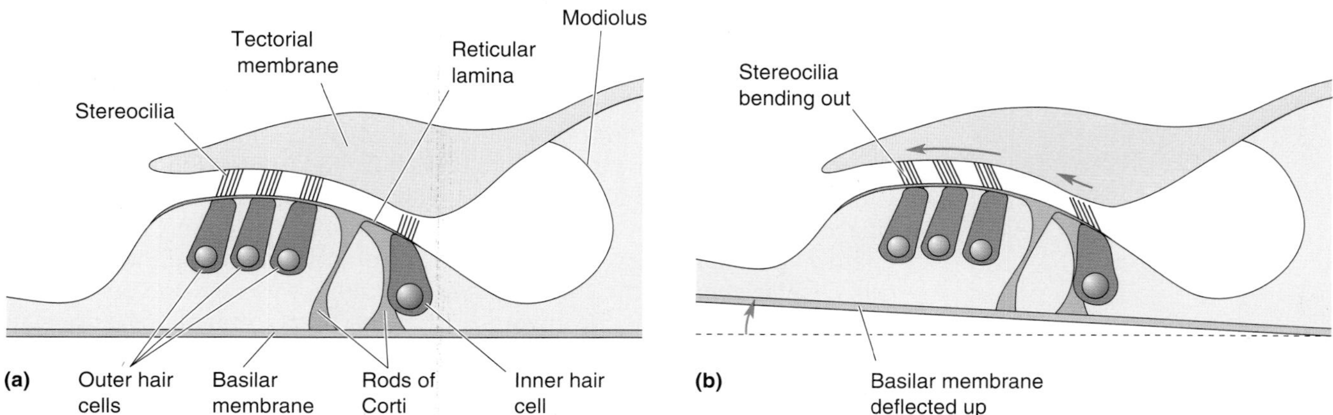

Figure 11.13

The bending of the stereocilia produced by the upward motion of the basilar membrane. (a) At rest, the hair cells are held between the reticular lamina and the basilar membrane, and the tips of the outer hair cell stereocilia are attached to the tectorial membrane. **(b)** When sound causes the basilar membrane to deflect upward, the reticular lamina moves up and inward toward the modiolus, causing the stereocilia to bend outward.

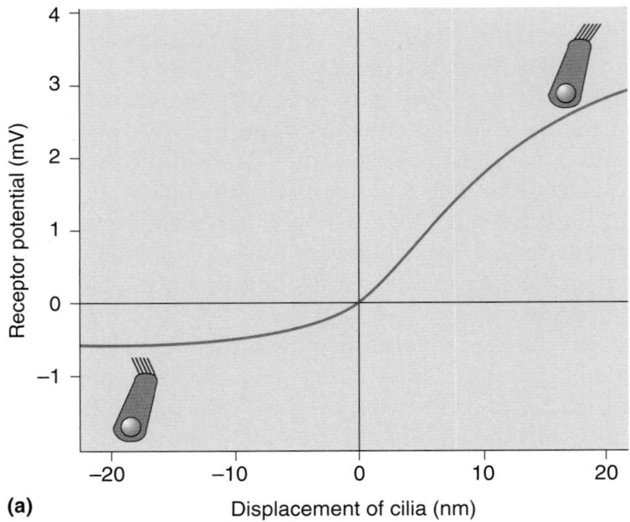

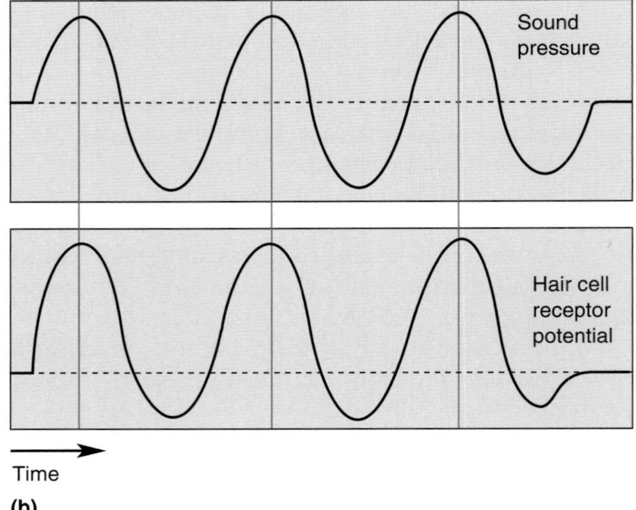

Figure 11.14
Hair cell receptor potentials. (a) The hair cell depolarizes or hyperpolarizes depending on the direction in which the stereocilia bend. **(b)** The hair cell receptor potential closely follows the air pressure changes during a low-frequency sound.

Changes in hair cell receptor potential result from the opening of potassium channels on the tips of the stereocilia when the cilia bend. Figure 11.15 shows how these interesting channels are believed to function. Each channel is linked by an elastic filament to the wall of the adjacent cilium. When the cilia are straight, the tension on this filament holds the channel in a partially opened state, allowing a small leak of K^+ from the endolymph into the hair cell. Displacement of the cilia in one direction increases tension on the linking filament, increasing the inward K^+ current. Displacement in the opposite direction relieves tension on the linking filament, thereby allowing the channel to close completely, preventing inward K^+ movement. The entry of K^+ into the hair cell causes a depolarization, which in turn activates voltage-gated calcium channels (Figure 11.15b). The entry of Ca^{2+} triggers the release of neurotransmitter, probably glutamate, which activates the spiral ganglion fibers lying postsynaptic to the hair cell.

It is interesting that the opening of K^+ channels produces a depolarization in the hair cell, whereas the opening of K^+ channels *hyperpolarizes* most neurons. The reason that hair cells respond differently from neurons is the un-

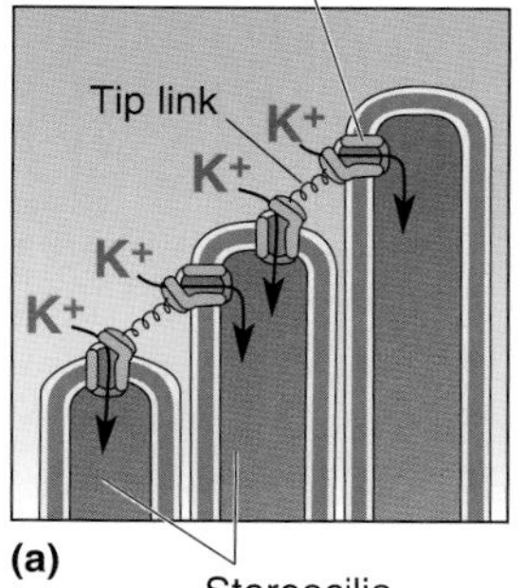

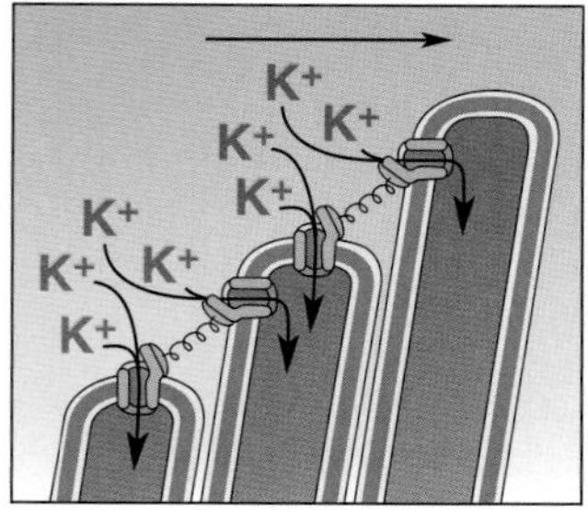

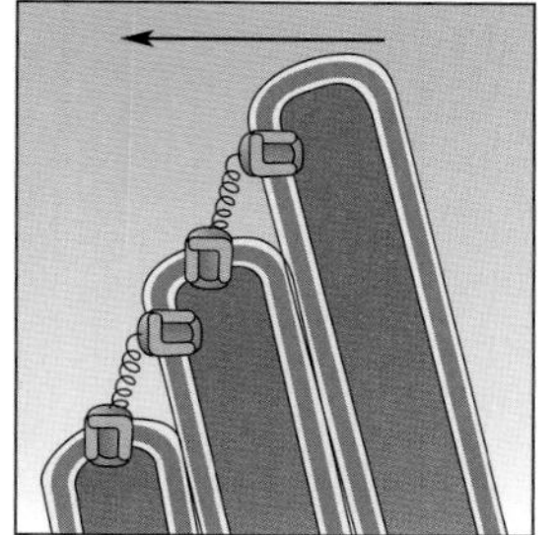

Figure 11.15
Depolarization of a hair cell. (a) Potassium channels at the tips of the stereocilia open when the tip links joining the stereocilia are stretched. **(b)** The entry of potassium depolarizes the hair cell, which opens voltage-gated Ca^{2+} channels. Incoming calcium leads to the release of neurotransmitter from synaptic vesicles, which then diffuses to the postsynaptic spiral ganglion neurite.

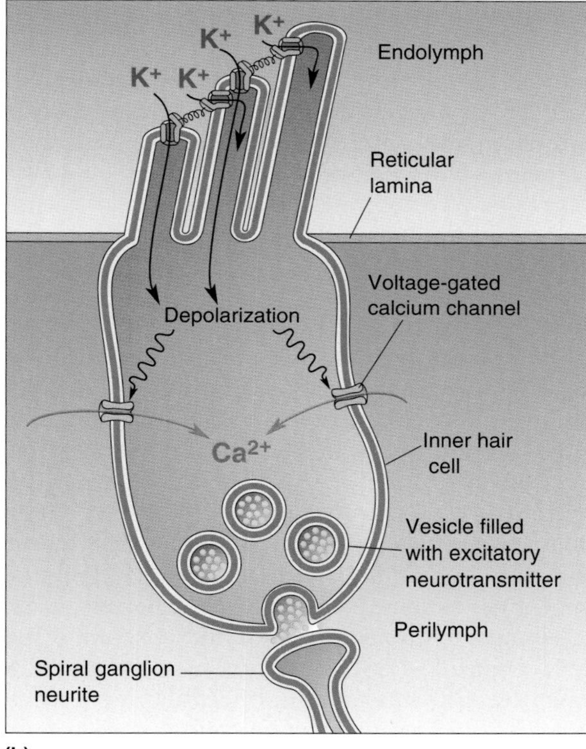

usually high K^+ concentration in endolymph, which yields a K^+ equilibrium potential of 0 mV compared with the equilibrium potential of −80 mV in typical neurons. Another reason that K^+ is driven into hair cells is the +80 mV endocochlear potential, which helps create a 125 mV gradient across the membranes of the stereocilia.

The Innervation of Hair Cells. The auditory nerve consists of the axons of neurons whose cell bodies are located in the spiral ganglion. Thus, the spiral ganglion neurons, which are the first in the auditory pathway to fire action potentials, provide all the auditory information sent to the brain. Curiously, there is a significant difference in the spiral ganglion innervation of the inner and outer hair cells. It is estimated that the number of neurons in the spiral ganglion is in the range of 35,000–50,000. Despite the fact that inner hair cells are outnumbered by outer hair cells by a factor of 3 to 1, more than 95% of the spiral ganglion neurons communicate with the relatively small number of in-

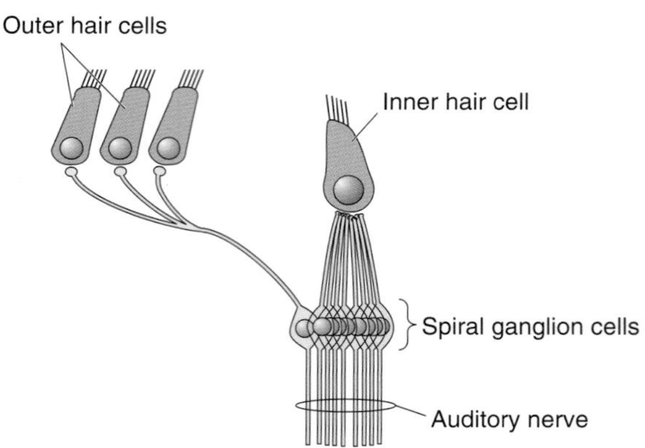

Figure 11.16
The innervation of hair cells by neurons from the spiral ganglion.

ner hair cells, and less than 5% receive synaptic input from the more numerous outer hair cells (Figure 11.16). Consequently, one spiral ganglion fiber receives input from only one inner hair cell; moreover, each inner hair cell feeds about 10 spiral ganglion neurites. The situation is the opposite with outer hair cells. Because they outnumber their spiral ganglion cells, one spiral ganglion fiber synapses with numerous outer hair cells.

Simply based on these numbers, we can infer that the majority of the information leaving the cochlea comes from inner hair cells. If that's true, what are the outer hair cells for?

Amplification by Outer Hair Cells. Given that outer hair cells far outnumber inner hair cells, it seems paradoxical that most of the cochlear output is derived from inner hair cells. Ongoing research suggests, however, that outer hair cells play a very important role in sound transduction. Ironically, one clue to the nature of this role was the discovery that the ear not only transduces sound but can create it, too (Box 11.2).

Outer hair cells seem to act like tiny motors that amplify the movement of the basilar membrane during low-intensity sound stimuli. The keys to this function are *motor proteins*, found in the membranes of outer hair cells (Figure 11.17a). Motor proteins can change the length of outer hair cells, and it has been found that outer hair cells respond to sound with both a receptor potential and a change in length (Figure 11.17b). The nature of motor proteins is unknown; they do not resemble any other system of cellular movement. The hair cells' motor is driven by the receptor potential, and it does not use adenosine triphosphate (ATP) as an energy source. It is also extremely fast, as it must be able to keep up with the movements induced by high-frequency sounds.

Because outer hair cells are attached to the basilar membrane and reticular lamina, when motor proteins change the length of the hair cell, the basilar membrane is pulled toward or pushed away from the reticular lamina and tectorial membrane. This is why the word *motor* is used—the outer hair cells actively change the physical relationship between the cochlear membranes.

The motor effect of outer hair cells makes a significant contribution to the traveling wave that propagates down the basilar membrane. This was demonstrated in 1991 by Mario Ruggero and Nola Rich at the University of Minnesota, who administered the chemical furosemide into experimental animals. Furosemide temporarily decreases the transduction that normally results from the bending of stereocilia on hair cells, and these researchers found that it significantly reduced the movement of the basilar membrane in response to sound (Figure 11.17c, d). This effect of furosemide is believed to result from inactivation of the outer hair cell motor proteins. It is sometimes

OF SPECIAL INTEREST

Noisy Ears: Otoacoustic Emissions

Sensory systems are supposed to detect stimulus energy in the environment, not generate it. Can you imagine eyes glowing in the dark or noses smelling like roses? How about ears buzzing loudly? The truth is that retinas don't radiate light, and olfactory receptors don't emit odors, but some ears can definitely generate sounds loud enough for a bystander to hear. Such sounds are called *otoacoustic emissions*. In one early description, a man sitting next to his dog realized the animal was humming; after some anxious investigation, he discovered that the sound came from one of the dog's ears.

The ears of all vertebrates, including humans, can emit sounds. Presenting a short sound stimulus, such as a click, to a normal human ear causes an "echo" that can be picked up with a microphone in the auditory canal. This echo is not usually noticed because it is too faint to be heard over other sounds in the environment. Ears that emit relatively loud sounds spontaneously, in the absence of any incoming sound, have often sustained cochlear damage as the result of exposure to extremely loud sounds (explosions, machines, rock bands), drugs, or disease. If spontaneous otoacoustic emissions are loud enough, they may cause tinnitus, a ringing in the ears (Box 11.6).

The mechanism that causes the ear to generate its *own* sounds—the cochlear amplifier—is the same one that functions to improve its detection of environmental sounds, but operating in reverse. Normal outer hair cells, stimulated with a click, react with a quick movement that drives the cochlear fluids and membranes, which move the ossicles and ultimately vibrate the tympanic membrane to produce sound in the outside air (the echo). Spontaneous emissions occur because the sensitivity of the cochlear amplifier is very high. Most people with normal hearing can perceive them in an exceptionally quiet environment. Damaged regions of the cochlea can somehow facilitate the spontaneous movement of some outer hair cells, so that they vibrate all the time. Strangely enough, most people are unaware that their ears are broadcasting sounds. Apparently, their central auditory neurons recognize the spontaneous cochlear activity as noise and suppress the perception of it. The benefit is that they are spared from an otherwise maddening tinnitus, but the cost is partial hearing loss in the affected frequency range.

said that the outer hair cells constitute a **cochlear amplifier**. When the outer hair cells amplify the response of the basilar membrane, the stereocilia on the inner hair cells bend more, and the increased transduction process in the inner hair cells produces a greater response in the auditory nerve. Through this feedback system, therefore, outer hair cells contribute significantly to the output of the cochlea. Without the cochlear amplifier, the peak movement of the basilar membrane would be about 100-fold smaller.

The effect of outer hair cells on the response of inner hair cells can be modified by neurons outside the cochlea. In addition to the spiral ganglion afferents that project from the cochlea to the brain stem, there are about 1000 efferent fibers projecting *from* the brain stem *toward* the cochlea. These efferents diverge widely, synapsing onto outer hair cells and releasing acetylcholine. Stimulation of these efferents changes the shape of the outer hair cells, thereby affecting the responses of inner hair cells. In this way, descending input from the brain to the cochlea can regulate auditory sensitivity.

The amplifying effect of outer hair cells explains how certain antibiotics (e.g., kanamycin) that damage hair cells can lead to deafness. After excessive exposure to antibiotics, many inner hair cells have reduced responses to sound. The antibiotic almost exclusively damages outer hair cells, however, not inner hair cells. For this reason, deafness produced by antibiotics is thought to be a consequence of damage to the cochlear amplifier (i.e., outer hair cells), demonstrating just how essential a role this amplifier plays.

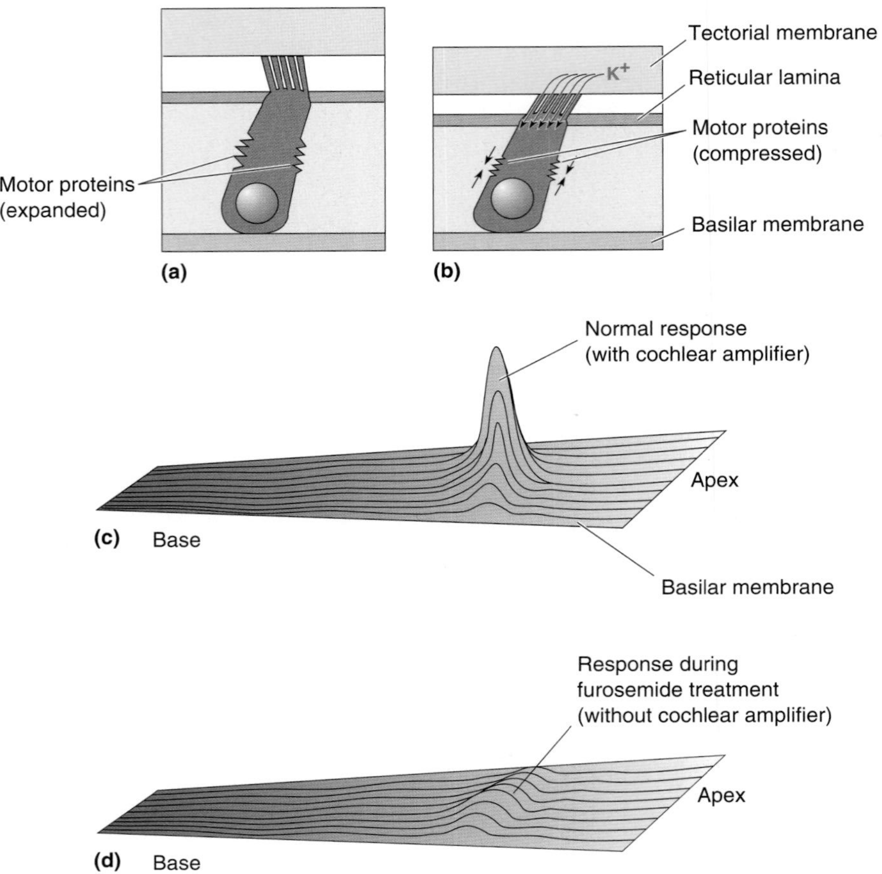

Figure 11.17
Amplification by outer hair cells. (a) Motor proteins in the membranes of outer hair cells. **(b)** When potassium enters the hair cell, motor proteins are activated and shorten the hair cell. **(c)** The conformational change of the hair cell increases the bending of the basilar membrane and is therefore called cochlear amplification. **(d)** Furosemide decreases hair cell transduction, consequently reducing the bending of the basilar membrane. (Source: Adapted from Ashmore and Kolston, 1994, Fig. 2, 3.)

CENTRAL AUDITORY PROCESSES

The auditory pathway appears more complex than the visual pathway because there are more synapses at nuclei intermediate between the sensory organ and the cortex. Also, in contrast to the visual system, there are many more alternative pathways by which signals can get from one nucleus to the next. Nonetheless, the amount of information processing in the two systems is not that different when you consider that the cells and synapses of the auditory system in the brain stem are analogous to interactions in the layers of the retina. We now look at auditory circuitry, focusing on the transformations of auditory information that occur along the way.

The Anatomy of Auditory Pathways

Afferents from the spiral ganglion enter the brain stem in the auditory-vestibular nerve. At the level of the medulla, the axons innervate the **dorsal cochlear nucleus** and **ventral cochlear nucleus** ipsilateral to the cochlea where the axons originate. Each axon branches so that it synapses on neurons in both cochlear nuclei. From this point on, the system gets more complicated, and the connections are less well understood, because there are multiple parallel pathways. Rather than trying to describe all of these connections, we follow one particularly important pathway from the cochlear nuclei to auditory cortex (Figure 11.18). Cells in the ventral cochlear nucleus send out axons that project to the **superior olive** (also called the *superior olivary nucleus*) on both sides of the brain stem. Axons of the olivary neurons ascend in the *lateral lemniscus* (a lemniscus is a collection of axons) and innervate the **infe-**

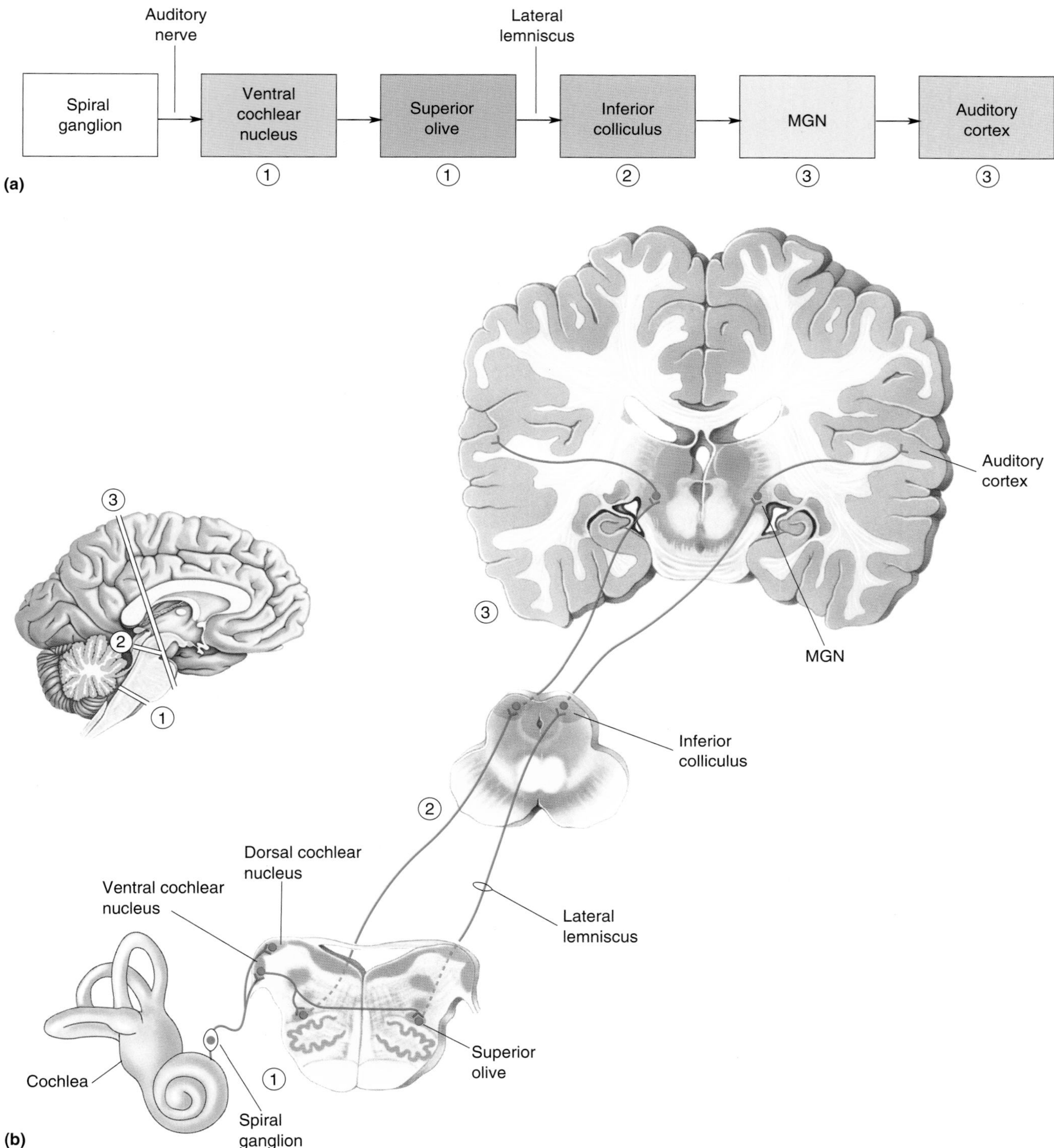

Figure 11.18

Auditory pathways. Neural signals can travel from the spiral ganglion to auditory cortex via numerous pathways. Here, a primary pathway is shown **(a)** schematically and **(b)** through brain stem cross sections. Notice that only the connections from one side are illustrated.

rior colliculus of the midbrain. Many efferents of the dorsal cochlear nucleus follow a route similar to the pathway from the ventral cochlear nucleus, but the dorsal path bypasses the superior olive. Although there are other routes from the cochlear nuclei to the inferior colliculus with additional intermediate relays, *all ascending auditory pathways converge onto the inferior colliculus.* The neurons in the inferior colliculus send out axons to the medial geniculate nucleus (MGN) of the thalamus, which in turn projects to auditory cortex.

Before moving on to the response properties of auditory neurons, we should make several points:

1. Projections and brain stem nuclei other than the ones described contribute to the auditory pathways. For instance, the inferior colliculus sends axons not only to the MGN but also to the superior colliculus, where the integration of auditory and visual information occurs, and to the cerebellum.
2. There is extensive feedback in the auditory pathways. For instance, brain stem neurons send axons that contact outer hair cells, and auditory cortex sends axons to the MGN and inferior colliculus.
3. Other than the cochlear nuclei, auditory nuclei in the brain stem receive input from both ears. This explains the clinically important fact that the only way in which brain stem damage can produce deafness in one ear is if the cochlear nuclei (or auditory-vestibular nerve) on one side are destroyed.

Response Properties of Neurons in the Auditory Pathway

To understand the transformations of auditory signals that occur in the brain stem, we must first consider the nature of the input from the neurons in the spiral ganglion of the cochlea. Because most spiral ganglion cells receive input from a single inner hair cell at a particular location on the basilar membrane, they fire action potentials only in response to sound within a limited frequency range. After all, hair cells are excited by deformations of the basilar membrane, and each portion of the membrane is maximally sensitive to a particular range of frequencies.

Figure 11.19 shows the results of an experiment in which action potentials were recorded from a single auditory nerve fiber (i.e., the axon of a spiral

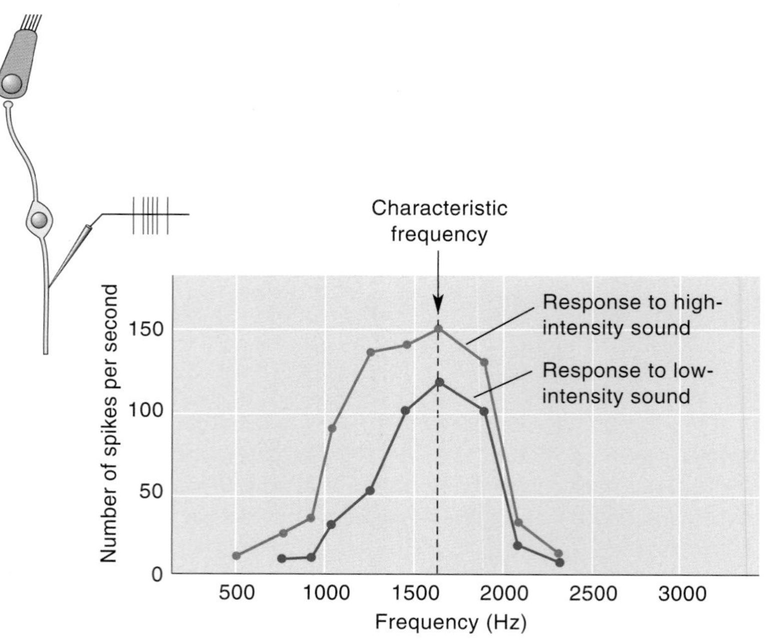

Figure 11.19
The response of an auditory nerve fiber to various sound frequencies. This neuron is frequency tuned, and its greatest response is at the characteristic frequency. (Source: Adapted from Rose, Hind, Anderson, and Brugge, 1971, Fig. 2.)

ganglion cell). The graph represents the firing rate in response to sounds at different frequencies. The neuron is most responsive to sound at one frequency, called the neuron's **characteristic frequency**, and is less responsive at neighboring frequencies. This type of frequency tuning is typical of neurons at each of the relays from cochlea to cortex.

As one ascends the auditory pathway in the brain stem, the response properties of the cells become more diverse and complex, just as in the visual pathway. For instance, some cells in the cochlear nuclei are especially sensitive to sounds varying in frequency with time (think of the sound of a trombone as it slides from a low note to a high note). In the MGN there are cells that respond to fairly complex sounds such as vocalizations, as well as other cells that show simple frequency selectivity, as in the auditory-vestibular nerve. An important development in the superior olive is that cells receive input from cochlear nuclei on both sides of the brain stem. As discussed below, such *binaural* neurons are probably important for sound localization.

ENCODING SOUND INTENSITY AND FREQUENCY

If you stop reading this book for a moment, you can focus on the many sounds around you. You can probably hear sounds you have been ignoring, and you can selectively pay attention to different sounds occurring at the same time. We are usually bathed in an amazing diversity of sounds—from chattering people to cars to the radio to electrical noises—and our brain must be able to analyze the important ones while ignoring the noise. We cannot yet account for the perception of each of these sounds by pointing to particular neurons in the brain. Most sounds have certain features in common, however, including intensity, frequency, and a point of origin. Each of these features is represented differently in the auditory pathway.

Stimulus Intensity

Information about sound intensity is coded in two interrelated ways: the firing rates of neurons and the number of active neurons. As a stimulus gets more intense, the basilar membrane vibrates with greater amplitude, causing the membrane potential of the activated hair cells to be more depolarized or hyperpolarized. This causes the nerve fibers with which the hair cells synapse to fire action potentials at greater rates. In Figure 11.19, the auditory-vestibular nerve fiber fires faster to the same sound frequencies when the intensity is increased. In addition, more intense stimuli produce movements of the basilar membrane over a greater distance (Figure 11.20), which leads to the activation of more hair cells. In a single auditory-vestibular nerve fiber, this increase in the number of activated hair cells causes a broadening of the frequency range to which the fiber responds. The loudness we perceive is thought to be correlated with the number of active neurons in the auditory-vestibular nerve (and throughout the auditory pathway) and their firing rates.

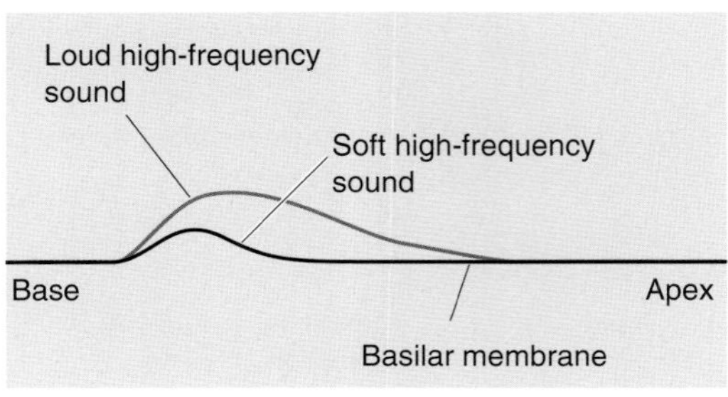

Figure 11.20
The influence of sound intensity on the response of the basilar membrane. At the same sound frequency, the maximum bending of the basilar membrane will occur closer to the apex as the intensity is increased.

Stimulus Frequency, Tonotopy, and Phase Locking

From the hair cells in the cochlea through the various nuclei leading to auditory cortex, most neurons are sensitive to stimulus frequency. They are most sensitive at their characteristic frequency. How is frequency represented in the central nervous system?

Tonotopy. Frequency sensitivity is largely a consequence of the mechanics of the basilar membrane, because different portions of the membrane are maximally deformed by sound of different frequencies. Moving from the base to the apex of the cochlea, the frequency that produces a maximal deformation of the basilar membrane progressively decreases. There is a corresponding representation of frequency in the auditory nerve; auditory-vestibular nerve fibers connected to hair cells near the apical basilar membrane have low characteristic frequencies, and those connected to hair cells near the basal basilar membrane have high characteristic frequencies (Figure 11.21). When neurons in the auditory-vestibular nerve synapse in the cochlear nuclei, they do so in an organized pattern based on characteristic frequency. Nearby neurons have similar characteristic frequencies, and a systematic relationship exists between position in the cochlear nucleus and characteristic frequency. In other words, there is a map of the basilar membrane within the cochlear nuclei. Systematic organization of characteristic frequency within an auditory structure is called **tonotopy**, analogous to retinotopy in the visual system. Tonotopic maps exist on the basilar membrane within each of the auditory-vestibular nerve relay nuclei, the MGN, and auditory cortex.

Because of the tonotopy throughout the auditory system, the location of active neurons in auditory nuclei is one indication of the frequency of the sound. There are two reasons, however, frequency must be coded in some way other than the site of maximal activation in tonotopic maps. One reason is that these maps do not contain neurons with very low characteristic frequencies, below about 200 Hz. As a result, the site of maximal activation may be the same for a 50 Hz tone as for a 200 Hz tone, so there must be some other way to distinguish them. The second reason something other than tonotopy is needed is that the region of the basilar membrane maximally displaced by a sound depends on its intensity in addition to its frequency (Figure 11.20). At a fixed frequency, a more intense sound will produce a maximal deformation at a point further up the basilar membrane than a less intense sound.

Phase Locking. The main source of information about sound frequency that complements information derived from tonotopic maps is the timing of neu-

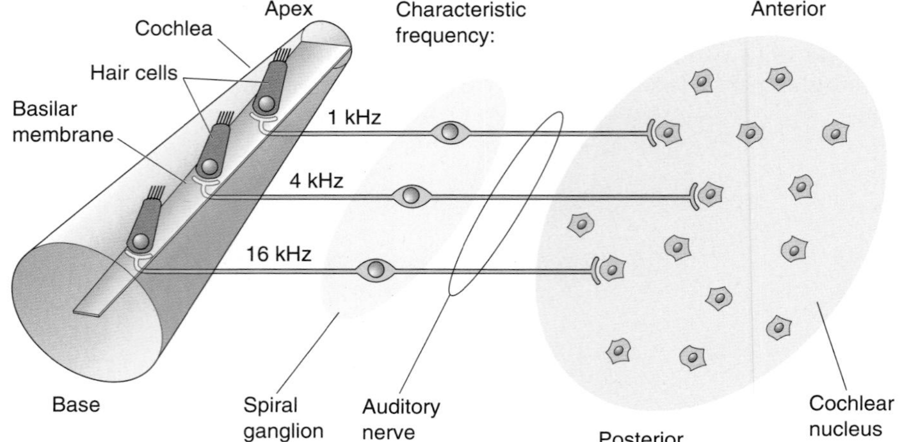

Figure 11.21
Tonotopic maps on the basilar membrane and cochlear nucleus. Going from the base to the apex of the cochlea, the basilar membrane resonates with increasingly lower frequencies. This tonotopy is preserved in the auditory nerve and cochlear nucleus. In the cochlear nucleus, there are bands of cells with similar characteristic frequencies; characteristic frequencies increase progressively from posterior to anterior.

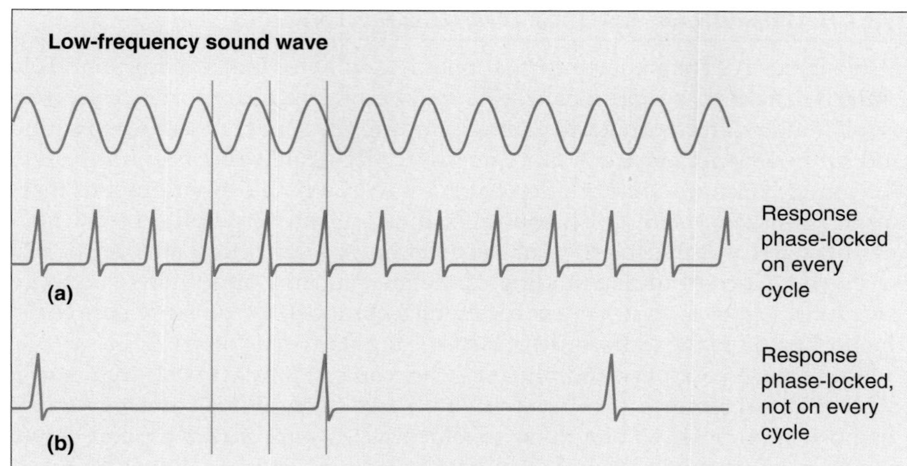

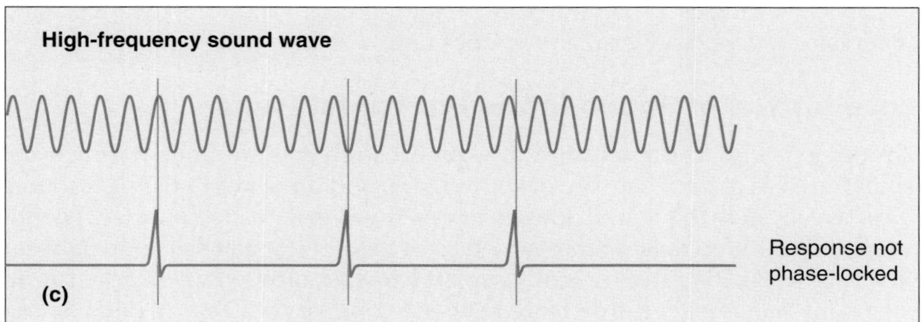

Figure 11.22
Phase locking in the response of auditory nerve fibers. Sound at a low frequency can elicit a phase-locked response, either **(a)** on every cycle of the stimulus or **(b)** on some fraction of the cycles. At high frequencies, **(c)** the response does not have a fixed phase relationship to the stimulus.

ral firing. Recordings made from neurons in the auditory nerve show **phase locking**, the consistent firing of a cell at the same phase of a sound wave (Figure 11.22). Think of a sound wave as a sinusoidal variation in air pressure. A phase-locked neuron fires action potentials at the peaks, the troughs, or some other constant location on the wave. At low frequencies, some neurons fire action potentials every time the sound has a particular phase (Figure 11.22a). This makes it easy to determine the frequency of the sound; it is the same as the frequency of the neuron's action potentials.

Phase locking can occur even if an action potential has not fired on every cycle (Figure 11.22b). For instance, a neuron may respond to a 1000 Hz sound with an action potential on only perhaps 25% of the cycles of the input, but those action potentials always occur at the same phase of the sound. If you have a group of such neurons, each responding to different cycles of the input signal, it is possible to have a response to every cycle (by some member of the group) and thus a measure of sound frequency. It is likely that intermediate sound frequencies are represented by the pooled activity of a number of neurons, each of which fires in a phase-locked manner; this is called the **volley principle**. Phase locking occurs with sound waves up to about 4 kHz. Above this point, the action potentials fired by a neuron are at random phases of the sound wave (Figure 11.22c) because the intrinsic variability in the timing of the action potential becomes comparable to the time interval between successive cycles of the sound. In other words, the sound waves cycle too fast for the action potentials of single neurons to represent their timing accurately. Above 4 kHz, frequencies are represented by tonotopy alone.

To summarize, here is how different frequencies are represented: At very low frequencies, phase locking is used; at intermediate frequencies, both phase locking and tonotopy are useful; and at high frequencies, tonotopy must be relied on to indicate sound frequency.

MECHANISMS OF SOUND LOCALIZATION

While the use of frequency information is essential for interpreting sounds in our environment, sound localization can be of critical importance for survival. If a predator is about to eat you, finding the source of a sudden sound and running away are much more important than analyzing the subtleties of the sound. Humans are not often eaten by wild animals anymore, but there are other situations in which sound localization can be helpful. If you carelessly try to cross the street, your localization of a car's horn may be all that saves you. Current understanding of the mechanisms underlying sound localization suggests that we use different techniques for locating sources in the horizontal plane (left–right) and vertical plane (up–down).

If you close your eyes and plug one ear, you can locate a bird singing as it flies overhead almost as well as with both ears open. But if you try to locate the horizontal position of a duck quacking as it swims across a pond, you'll find that you're much less able using only one ear. Thus, good horizontal localization requires a comparison of the sounds reaching the two ears, whereas good vertical localization does not.

Localization of Sound in the Horizontal Plane

An obvious cue to the location of a sound source is the time at which the sound arrives at each ear. We have two ears, and if we aren't facing a source directly, it takes the sound longer to reach one ear than the other. For instance, if a sudden noise comes at you from the right, it reaches your right ear first (Figure 11.23a); it arrives at your left ear later, after what is known as an *interaural time delay.* If the distance between your ears is 20 cm, sound coming from the right, perpendicular to your head, reaches your left ear 0.6 msec af-

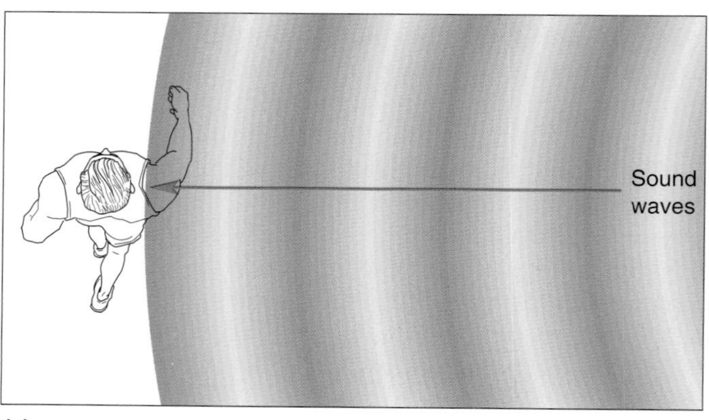

(a)

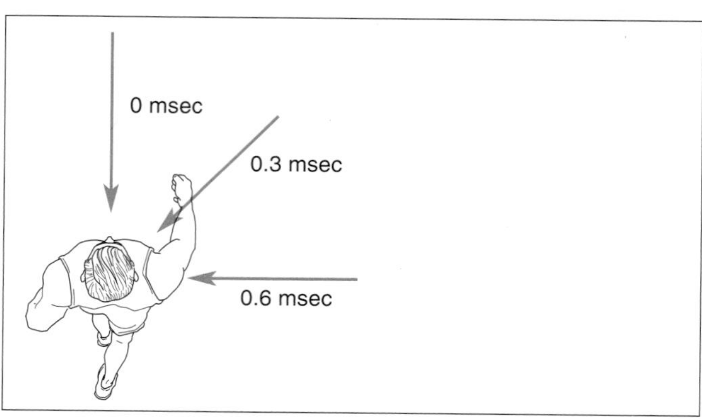

(b)

Figure 11.23
Interaural time delay as a cue to the location of sound. (a) Sound waves coming from the right side reach the right ear first, and there is a large interaural delay before the sound propagates to the left ear. **(b)** If the sound comes from straight ahead, there is no interaural delay. Delays for three sound directions are shown.

ter reaching your right ear. If the sound comes from straight ahead, there is no interaural delay; and at angles between straight ahead and perpendicular, the delay is between 0 and 0.6 msec (Figure 11.23b). Sounds from the left side yield delays opposite to those on the right. Thus, there is a simple relationship between location and interaural delay. Detected by specialized neurons in the brain stem, the delay enables us to locate the source of the sound in the horizontal plane.

If we don't hear the onset of a sound because it is a continuous tone rather than a sudden noise, we cannot know the initial arrival times of the sound at the two ears. Thus, continuous tones pose more of a problem for sound localization because they are always present at both ears. We can, however, use arrival time to localize continuous sound in a slightly different manner from localizing a sudden sound. The only thing that can be compared between continuous tones is the time at which the same *phase* of the sound wave reaches each ear. Imagine you are exposed to a 200 Hz sound coming from the right. At this frequency, one cycle of the sound covers 172 cm, which is much more than the 20 cm distance between your ears. After a peak in the sound pressure wave passes the right ear, you must wait 0.6 msec, the time it takes sound to travel 20 cm, before detecting a peak at the left ear. Of course, if the sound is straight ahead, peaks in the continuous tone reach the ears simultaneously. Because the sound wave is much longer than the distance between the ears, we can reliably use the interaural delay of the peak in the wave to determine sound location.

Things are more complicated with continuous tones at high frequencies. Suppose the sound coming from the right now has a frequency of 20,000 Hz, which means that one cycle of the sound covers 1.7 cm. After a peak reaches the right ear, does it still take 0.6 msec before a peak arrives at the left ear? No, it takes a much shorter time, because many peaks of such a high-frequency wave fit between your ears. No longer is there a simple relationship between the direction the sound comes from and the arrival times of the peaks at the two ears. Interaural arrival time simply is not useful for locating sounds with frequencies so high that one cycle of the sound wave is smaller than the distance between your ears (i.e., greater than about 2000 Hz).

Fortunately, the brain has another process for sound localization at high frequencies: *interaural intensity difference*. An interaural intensity difference exists between the two ears because your head effectively casts a sound shadow (Figure 11.24). There is a direct relationship between the direction of the sound and the extent to which your head shadows the sound to one ear. If sound comes directly from the right, the left ear hears a significantly lower intensity (Figure 11.24a). With sound coming from straight ahead, the same intensity reaches the two ears (Figure 11.24b), and with sound coming from intermediate directions, there are intermediate intensity differences (Figure 11.24c). Neurons sensitive to differences in intensity can use this information to locate the sound. Intensity information cannot be used to locate sounds at lower frequencies because sound waves at these frequencies diffract around the head, and the intensities at the two ears are roughly equivalent. There is no sound shadow at low frequencies.

Let's summarize the two processes for localizing sound in the horizontal plane. With sounds in the range of 20–2000 Hz, the process involves *interaural time delay*. From 2000–20,000 Hz, *interaural intensity difference* is used. Together these two processes constitute the **duplex theory of sound localization**.

The Sensitivity of Binaural Neurons to Sound Location. From our discussion of the auditory pathway, recall that neurons in the cochlear nuclei receive afferents only from the ipsilateral auditory nerve. Thus, all of these cells are *monaural*, meaning that they respond to sound presented to one ear only.

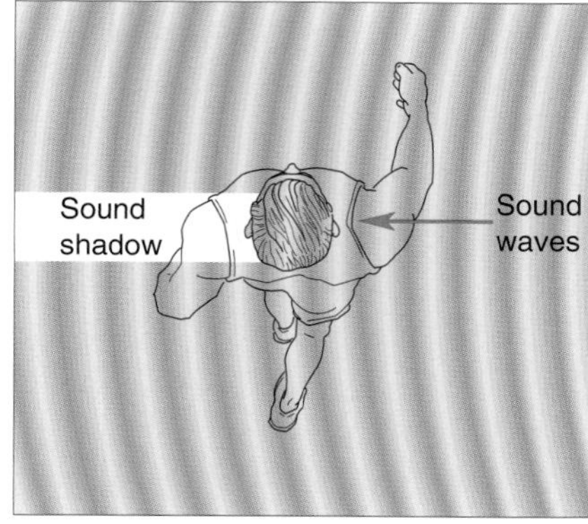

(a)

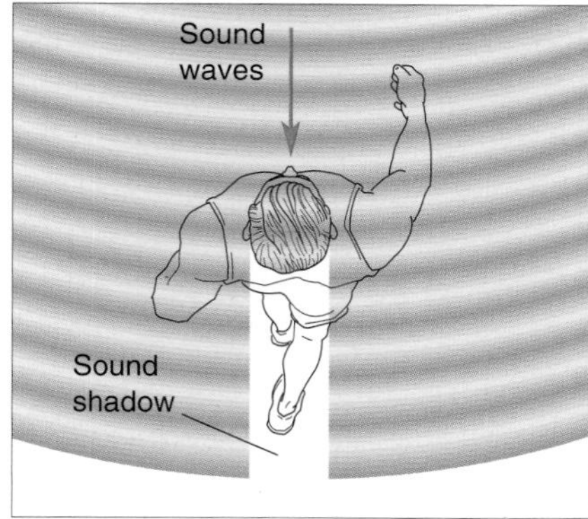

(b)

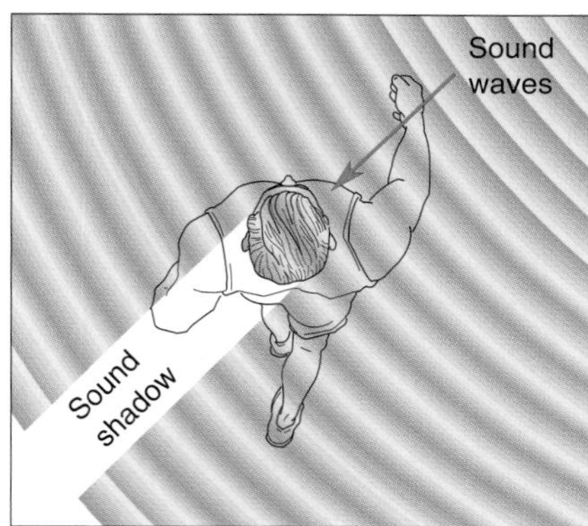

(c)

Figure 11.24
Interaural intensity difference as a cue to sound location. (a) With high-frequency sound, the head casts a sound shadow to the left when sound waves come from the right. Lower-intensity sound in the left ear is a cue that the sound came from the right. **(b)** If the sound comes from straight ahead, a sound shadow is cast behind the head, but the sound reaches the two ears with the same intensity. **(c)** Sound coming from an oblique angle partially shadows the left ear.

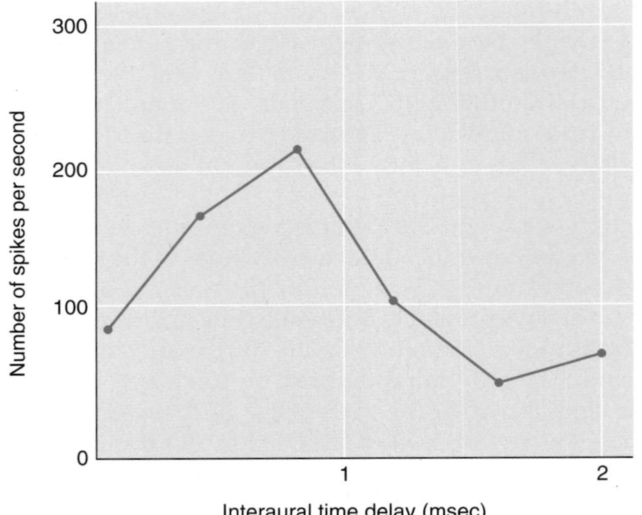

Figure 11.25
Responses of a neuron in the superior olive sensitive to interaural time delay. This neuron has an optimal delay of about 1 msec.

At all later stages of processing in the auditory system, however, there are *binaural* neurons whose responses are influenced by sound at both ears. The response properties of binaural neurons imply that they play an important role in sound localization in the horizontal plane.

The first structure with binaural neurons is the superior olive. While there is some controversy concerning the relationship between the activity of such neurons and the behavioral localization of sound, there are several compelling correlations. Neurons in the superior olive receive input from cochlear nuclei on both sides of the brain stem. Cells in the cochlear nuclei that project to the superior olive typically have responses phase-locked to lower-frequency sound input. Consequently, an olivary neuron receiving spikes from the left and right cochlear nuclei can compute interaural time delay. Recordings made in the superior olive show that each neuron typically gives its greatest response to a particular interaural delay (Figure 11.25). Because interaural delay varies with sound location, each of these neurons may be encoding a particular position in the horizontal plane.

How can a neural circuit produce neurons sensitive to interaural delay? The key is to use axons as *delay lines* and to measure small time differences precisely. A sound hitting the left ear triggers action potentials in the left cochlear nucleus, which propagate along afferent axons into the superior olive (Figure 11.26). Within 0.6 msec of hitting the left ear, that sound reaches the right ear (assuming the sound comes directly from the left) and triggers action potentials in axons from the right cochlear nucleus. Because of the way the axons and neurons are arranged in the olive, however, the action potentials from each side take different lengths of time to arrive at the various postsynaptic neurons in the olive. For example, the axon from the left cochlear nucleus has a longer path to travel to neuron 3 in Figure 11.26 than the axon from the right cochlear nucleus; therefore, the arrival of the spike from the left side is delayed just enough that it coincides with the arrival of the spike from the right side. By arriving at precisely the same time, action potentials from the two sides produce excitatory postsynaptic potentials (EPSPs) that summate, yielding a larger EPSP that excites olivary neuron 3 more strongly than an EPSP from each ear could alone. When an interaural delay is more or less than 0.6 msec, the spikes do not arrive together, and thus the EPSPs they trigger do not summate as much.

Other neurons in the superior olive are tuned to other interaural times because of systematic differences in the arrangement of axonal delay lines. To

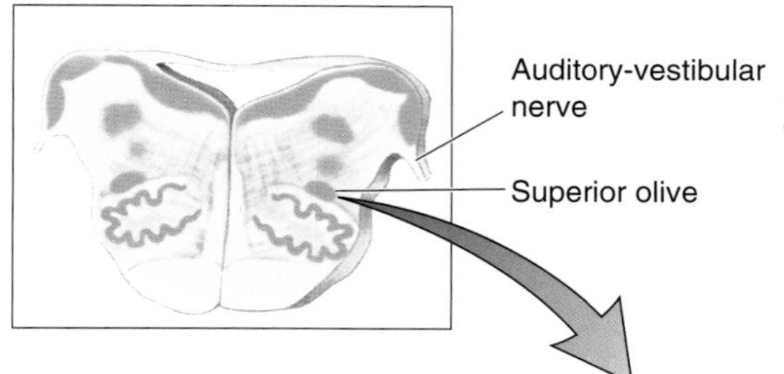

Sound from the left side initiates activity in the left cochlear nucleus; activity is then sent to the superior olive.

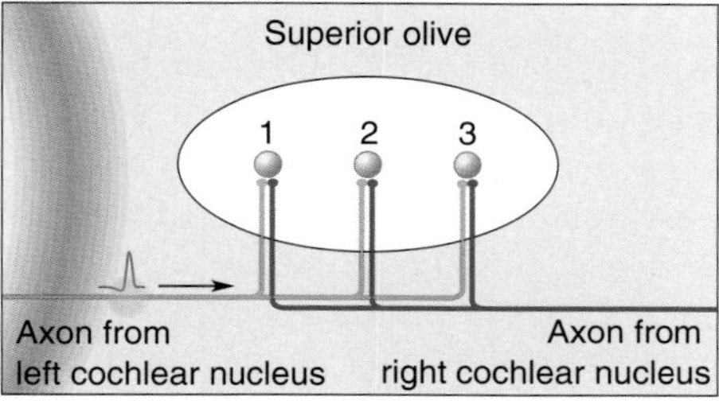

Very soon, the sound reaches the right ear, initiating activity in the right cochlear nucleus. Meanwhile, the first impulse has traveled farther along its axon.

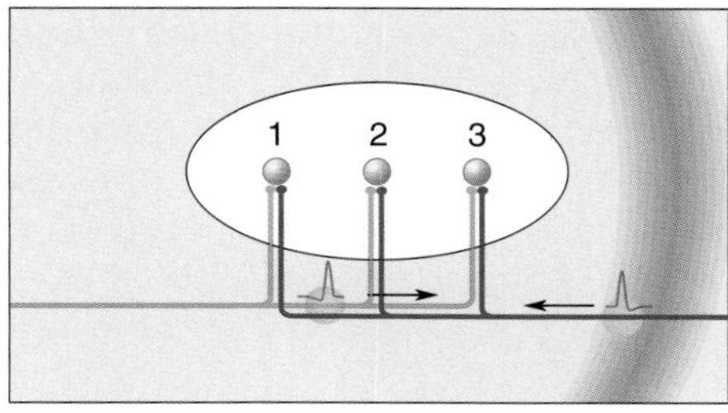

Both impulses reach olivary neuron 3 at the same time, and summation of synaptic potentials generates an action potential.

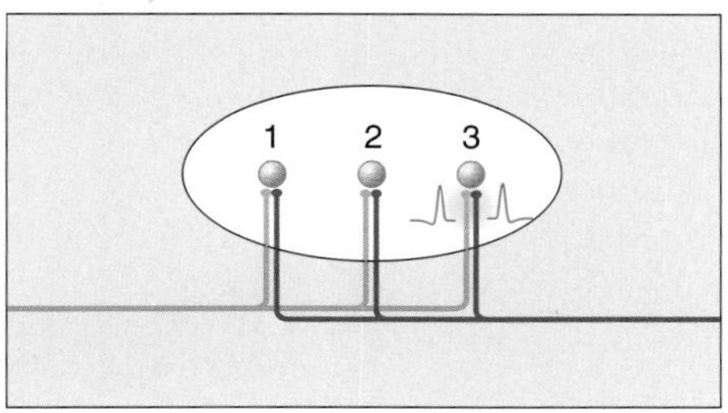

Figure 11.26
Delay lines and neuronal sensitivity to interaural delay.

measure timing differences as accurately as possible, many neurons and synapses of the auditory system are specially adapted for rapid operation; their action potentials and EPSPs are much faster than those of most other neurons in the brain (Box 11.3). There are limits to auditory time measurement of this type, however. Phase locking is essential for a precise comparison of the timing of inputs, and because phase locking occurs only at relatively low frequencies, it makes sense that interaural delays are useful only for localizing sounds of relatively low frequency.

In addition to their sensitivity to interaural delay, neurons in the superior olive are sensitive to the other sound location cue, interaural intensity. The responses of two types of binaural neurons are shown in Figure 11.27. One type of neuron (type EE) is moderately excited by sound presented to either ear but gives a maximal response only when both ears are stimulated. The other type of neuron (type EI) is excited by sound in one ear but inhibited by sound in the other ear. These EI neurons can be particularly sensitive to differences in intensity at the two ears because the effect of added excitation from one ear is compounded by the effect of decreased inhibition from the other ear. Presumably, such a mechanism contributes to horizontal localization of high-frequency sound by encoding differences in interaural intensity.

Box 11.3

BRAIN FOOD

Auditory Neurons, Fast and Sure

Neurons that process sound information are the extreme timing machines of the brain. They are designed for speed, to preserve and analyze the very rapid neural signals that encode small but meaningful variations in sound signals. For example, trained musicians can distinguish between two tones of 1000 Hz and 1001 Hz. This requires that they detect a difference of only 1 μsec in the sound wavelengths. A single action potential lasts about 1000 times longer! People can discriminate the direction of a sound source in the horizontal plane with a precision of 2 degrees; this demands that they discriminate the 11 μsec difference between the time it takes a sound to reach their two ears. Auditory information must pass through three sets of synapses before timing comparisons are made in the superior olive, so the transmission lines have to be swift and faithful.

It is instructive to examine the physiological features that help auditory cells work as rapidly and precisely as neural physiology will allow. Speed starts with the hair cells, which have fast transduction channels with no second messengers to delay the process. The synapses that hair cells make onto auditory nerves are extraordinarily fast when turning on and off, so they can follow the repetitive nature of sound precisely. Their speed comes in part from the way they handle Ca^{2+}, the trigger for transmitter release. Hair cell calcium channels activate and inactivate exceptionally quickly, they are clustered tightly adjacent to vesicle release sites, and special cytoplasmic proteins avidly bind Ca^{2+} after its influx, so that transmitter release ends promptly. Myelin greatly increases the conduction velocity of auditory nerve axons, as it does for many axons in the brain, but the auditory nerve is unique in having myelin on its cell bodies as well.

Many of the synapses at the brain stem end of the auditory-vestibular nerve, in the cochlear nucleus, and farther up the auditory pathway in the superior olive are simply huge (Figure A). Their size ensures the release of many vesicles of glutamate, hence a large and reliable EPSP with very little jitter in its timing. On the postsynaptic side, the glutamate encounters a unique "auditory" α-amino-3-hydroxy-5-methyl-4-isoxazole propionic acid (AMPA) type of glutamate receptor that is found nowhere else in the brain. The synaptic current generated by this receptor turns on about as quickly as it does at other synapses, but it also turns off in a blazing few tenths of a millisecond. Again, rapidly resetting the synapse to prepare for the next signal is just as important as the ability to turn it on quickly.

Finally, many auditory neurons in the brain stem have a shape and physiology optimized for speed of electrical conduction. Dendrites are few and thick, and synapses tend to be close to, on, or even enveloping the soma, so that EPSPs are not slowed and weakened by dendritic cable properties. Cell membranes have numerous potassium channels that reset the membrane potential very rapidly after the onset of an EPSP or an action potential. Speed and reliability are of the essence in audition, and auditory neurons are optimized for both.

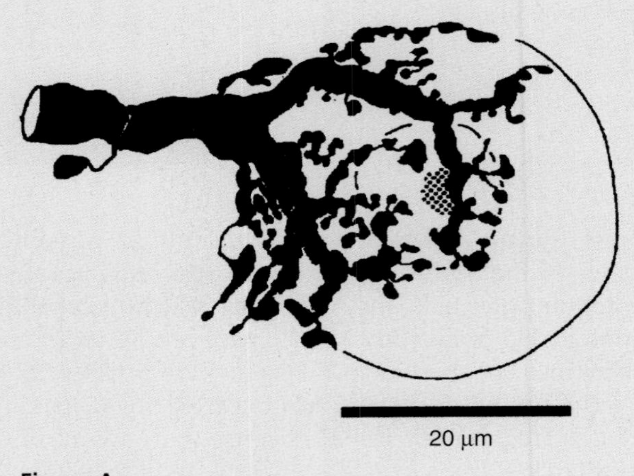

Figure A

Localization of Sound in the Vertical Plane

Comparing inputs to both ears is not very useful for localizing sounds in the vertical plane because as a sound source moves up and down, neither the interaural delay nor the interaural intensity changes. This is why, as mentioned earlier, plugging one ear affects localizing sounds in the vertical plane much less than localizing sounds in the horizontal plane. To impair vertical sound localization seriously, one must place a tube in the auditory canal to bypass the pinna. The sweeping curves of the outer ear are essential for assessing the

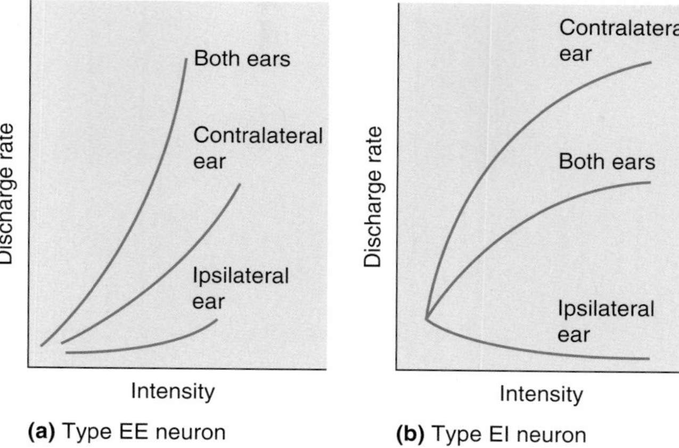

Figure 11.27
Binaural responses in the superior olive. (a) At all sound intensities, this EE neuron gives a stronger response to *bilateral* stimulation than to *unilateral* stimulation. **(b)** The response of this EI neuron increases with increasing sound intensity in the contralateral ear but decreases as intensity rises in the ipsilateral ear. The response to bilateral stimulation is reduced from the contralateral response.

elevation of a source of sound. The bumps and ridges apparently produce reflections of the entering sound. The delays between the direct path and the reflected path change as a sound source moves vertically (Figure 11.28). The combined sound, direct and reflected, is subtly different when it comes from above or below. In addition, the outer ear allows higher-frequency sounds to enter the auditory canal more effectively when they come from an elevated source. Consistent with the idea that reflections off the outer ear are important is the finding that vertical localization is seriously impaired if the convolutions of the pinna are covered.

Some animals are extremely good at vertical sound localization, even though they do not have a pinna. For example, a barn owl can swoop down on a squeaking mouse in the dark, locating accurately by sound, not sight (Box 11.4). Although owls do not have a pinna, they use the same techniques we use for horizontal localization (interaural differences) because their ears are at different heights on their head. Some animals have a more "active" system for sound localization than humans and owls. Certain bats emit sounds that reflect from objects, and the bats use these echoes to locate objects without sight. Many bats use reflected sound, analogous to the sonar used by ships, to hunt insects. In 1989, James Simmons at Brown University made the startling discovery that bats can discriminate time delays that differ by as little as 0.00001 msec. This finding challenges our understanding of how the

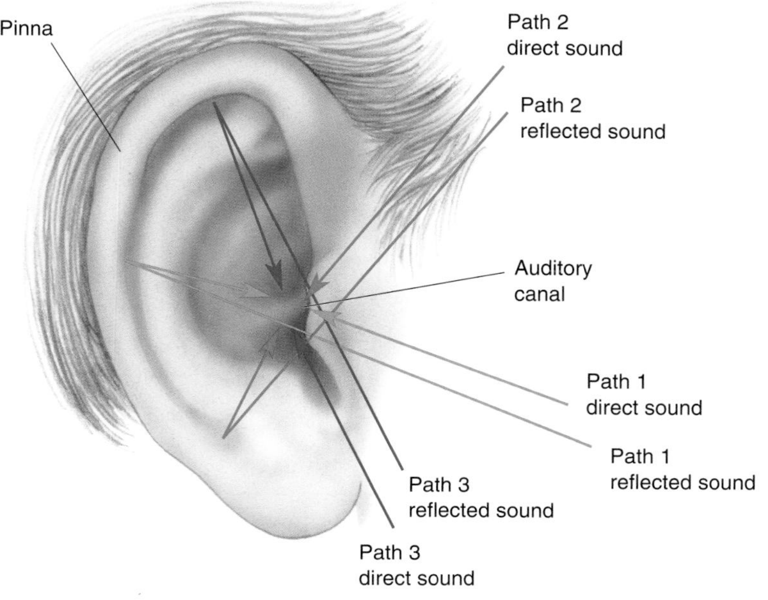

Figure 11.28
Vertical sound localization based on reflections from the pinna.

Box 11.4

PATH OF DISCOVERY

A Search for the Impossible

BY MASAKAZU KONISHI

When I moved to the California Institute of Technology in 1975, Jack Pettigrew suggested that we work together on the visual system of the barn owls I had brought from Princeton. We studied the stimulus selectivity of neurons in the Wulst, a forebrain visual area that is very large in owls. This experience was one of the turning points in my research career.

As soon as my new soundproof room became available, I decided to use it to look for auditory neurons with spatial receptive fields analogous to those of the Wulst neurons. Jack's laboratory had an elaborate optical device to cast visual stimuli on a screen. Jack and I conceived of an acoustic equivalent of this setup that could move a small loudspeaker to any place equally distant from the center of the subject's head—in this case, an owl.

The next important step was to decide where to look for space-specific neurons in the owl's brain. We had some success looking in a forebrain structure called Field L, which is the auditory equivalent of the Wulst, but progress was slow. At about this time, Eric Knudsen joined my lab as a postdoctoral fellow, and I suggested to him the possibility of finding a map of auditory space in the owl's brain. Although he liked the idea, neither of us thought deeply about it, because if we had, we would have rejected the notion right away. With Eric aboard we moved faster, collecting more neurons with desirable properties in Field L. Disappointingly, the neurons did not form a map of auditory space.

Eric suggested we try the midbrain auditory area next, because he had once worked on the homologous region in catfish. During our first recording session in this area, we saw the neurons' preferences for sound source position shift as we moved the electrode to adjoining locations. This finding was very exciting and encouraged us immensely. We worked very hard for the next 3 months, and finally Eric was able to reconstruct an auditory space map from the data we collected from 19 neurons in one owl.

Figure A
Masakazu Konishi

Receptive fields and maps occur in other sensory systems because the sensory surface (e.g., the skin, the retina) on which space is mapped projects topographically to higher centers. In the auditory system, the inner ear maps sound frequency instead of space. The auditory system is, therefore, expected to have neither spatial receptive fields nor maps. Arnold Starr had pointed this out in a review published well before we started our search. If we had read his paper, we might not have undertaken this project. Naiveté is sometimes useful.

The discovery of the space map marked the end of our wishful expeditions and the beginning of more logical approaches. How to explain spatial receptive fields became the next question. When Andy Moiseff joined my lab as a postdoctoral fellow, our plan was to map the receptive fields of auditory neurons and then determine what binaural cues they require. Andy's former mentors told him this would be impossible, but we managed to do it despite the skepticism. This and subsequent studies showed that auditory neurons have spatial receptive fields because they are tuned to combinations of interaural time and intensity differences, which define the horizontal and vertical directions, respectively, of the owl's auditory space.

As our work continued to evolve, we could see a beautiful picture emerging. We were very lucky from the beginning, because we would have given up the project, had we not found the map.

nervous system, using action potentials lasting almost a millisecond, can perform such fine temporal discriminations.

AUDITORY CORTEX

Axons leaving the MGN project to auditory cortex via the internal capsule in an array called the *acoustic radiation*. Primary auditory cortex (A1) corresponds to Brodmann's area 41 in the temporal lobe (Figure 11.29a). The structure of A1 and the secondary auditory areas is in many ways similar to corresponding areas of the visual cortex. Layer I contains few cell bodies, and layers II and III contain mostly small pyramidal cells. Layer IV, where the medial geniculate axons terminate, is composed of densely packed granule cells. Layers V and VI contain mostly pyramidal cells that tend to be larger than those in the superficial layers. Let's look at how these cortical neurons respond to sound.

Neuronal Response Properties

In general, neurons in monkey (and presumably human) A1 are relatively sharply tuned for sound frequency and possess characteristic frequencies covering the audible spectrum of frequencies. In electrode penetrations made perpendicular to the cortical surface in monkeys, the cells encountered tend to have similar characteristic frequencies, suggesting a columnar organization on the basis of frequency. In the tonotopic representation in A1, low frequencies are represented rostrally and laterally, whereas high frequencies are represented caudally and medially (Figure 11.29b). Roughly speaking, *isofrequency bands* run mediolaterally across A1. In other words, strips of neurons running across A1 contain neurons that have fairly similar characteristic frequencies.

In the visual system, it is possible to describe large numbers of cortical neurons as having some variation on a general receptive field that is either simple or complex. So far, it has not been possible to sort the diverse auditory receptive fields into a similarly small number of categories. As at earlier stages in the auditory pathway, neurons have different temporal response patterns; some have a transient response to a brief sound, and others have a sustained response.

In addition to the frequency tuning that occurs in most cells, some neurons are intensity tuned, giving a peak response to a particular sound intensity. Even within a vertical column perpendicular to the cortical surface, there may be considerable diversity in the degree of tuning to sound frequency. Some neurons are sharply tuned for frequency, and others are barely tuned at all; the degree of tuning does not seem to correlate well with cortical layers. Other sounds that produce responses in cortical neurons include clicks, bursts of noise, frequency-modulated sounds, and animal vocalizations. Clarifying the role of these neurons that respond to seemingly complex stimuli is one of the challenges researchers face (Box 11.5).

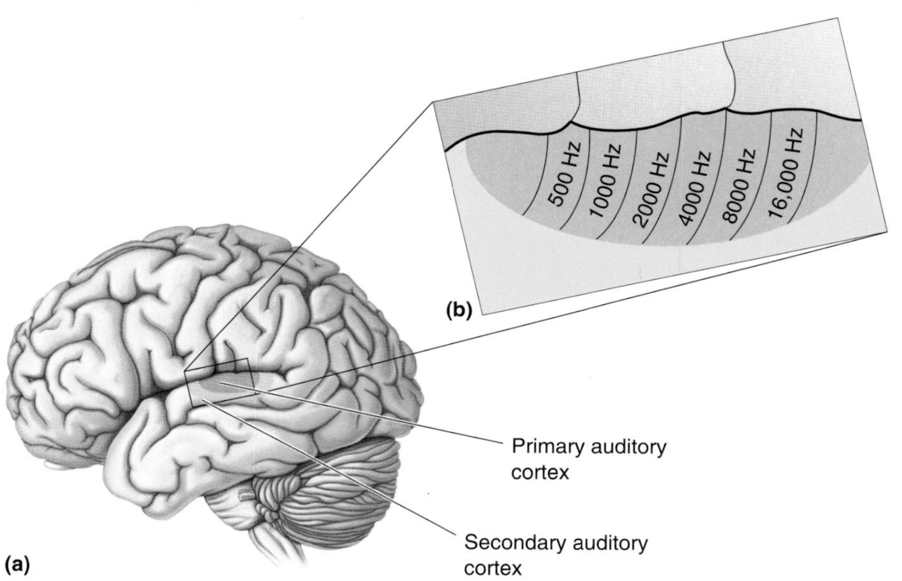

Figure 11.29
Primary auditory cortex in humans. (a) Primary auditory cortex (purple) and secondary auditory areas (yellow) on the superior temporal lobe. **(b)** Tonotopic organization within primary auditory cortex. The numbers are characteristic frequencies.

Box 11.5

OF SPECIAL INTEREST

How Does Auditory Cortex Work? Consult a Specialist

The function of an animal's brain is to help it stay alive and reproduce. Different species have vastly different habits and needs, and some animals have evolved a sensory system specialized for processing its favorite stimuli. The exaggerated systems of the sensory specialists, such as barn owls and bats, sometimes help us understand how we sensory generalists work.

Barn owls find their prey (a scurrying mouse, for example) in the dark by listening very carefully. They are particularly adept at identifying and localizing faint sounds, and some of the neural mechanisms of sound localization were first understood in owls. Bats have a unique and more active auditory technique. They find their food (a fluttering moth, for example) by echolocating it. The bats emit brief calls and listen for the faint echoes reflected from the target. Bats require their cortex for proper echolocation. Studying bat cortex can provide insight into how auditory cortex works in bats, and it may also enlighten us about the human cortex.

The most interesting stimuli for echolocating bats are their own calls and echoes. A bat's language is very limited. To echolocate, most bats scream loudly at ultrasonic frequencies (20-100 kHz) using an essentially one-word vocabulary. The call of the mustached bat (*Pteronotus parnellii*) is very brief, no more than 20 msec long; it consists of a steady constant-frequency (CF) part followed by a sweep of descending frequency—the frequency-modulated (FM) part. Figure A is a graph of the bat's call and echo, showing the frequencies of the sounds plotted against time. As it flies, the bat rapidly and continually repeats the call. By listening to its own calls and their echoes and carefully comparing them in many ways, the

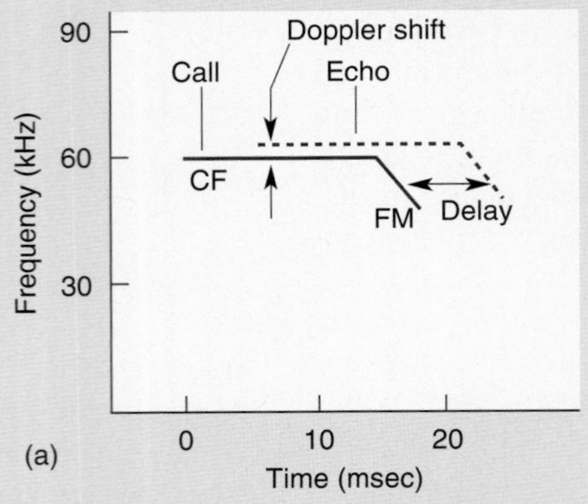

Figure A
A bat's call and echo (Source: Adapted from Suga, 1995, p. 302.)

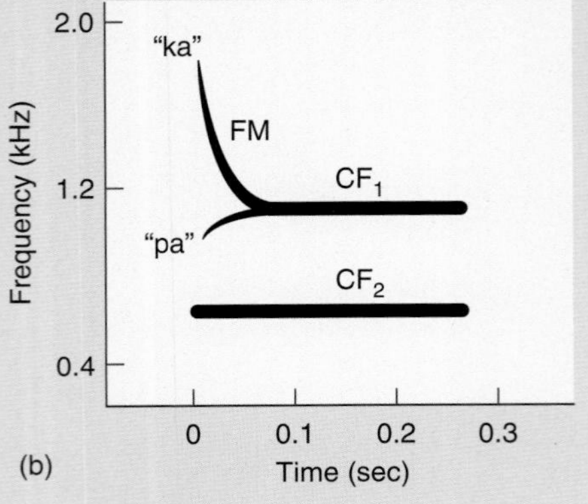

Figure B
A human's spoken words. (Source: Adapted from Suga, 1995, p. 296.)

(Box 11.5, *continued*)

bat builds a remarkably detailed auditory image of the nearby world. For example, the *delay* between the call and its echo depends on the distance to a reflecting target (1 msec of delay for each 17 cm of distance). If the target is moving toward or away from the bat, the frequency of the echo is *Doppler-shifted* higher or lower (think of the shifting pitch of an ambulance siren as it passes by you; a 1 kHz shift corresponds to a speed of about 3 m/sec). A moth's beating wings cause a *rhythm* in the echoes, and that helps the bat know there's a particular kind of insect in front of it and not something less edible. Many other subtle changes in the echo's frequency, timing, loudness, and pattern tell the bat about other features of the targets.

Nobuo Suga at Washington University has studied in great detail the processing of call-echo information by the auditory cortex of the mustached bat. Suga found that the bat's cortex is a melange of distinct auditory areas. Many are specialized for detecting particular features important for echolocating, and others seem more generalized. For example, a large region is devoted to processing Doppler shifts of echoes right around 60 kHz, the loudest part of the bat's call; this area processes information about target velocity and location. Three separate areas detect call-echo delays and yield information about target distance. We are beginning to understand the neural mechanisms that extract such information.

The basic features of a bat's calls and a human's spoken words are similar, although human speech is much slower and lower in pitch. Human syllables consist of particular combinations of CF periods, FM sweeps, brief pauses, and bursts of noise. For example, the syllable "ka" differs from "pa" because their initial FM sweeps bend in different directions (Figure B). It is likely that the neuronal circuits that process speech sounds in human auditory cortex use principles similar to those in the bat's cortex. Interpreting those speech sounds as words and understanding the concepts they imply are in the realm of language. The brain mechanisms of language will be discussed in Chapter 20.

Given the wide variety of response types that neurophysiologists encounter in studying auditory cortex, you can understand why it is reassuring to see some sort of organization or unifying principle. One organizational principle already discussed is the tonotopic representation in many auditory areas. A second organizational principle is the presence in auditory cortex of columns of cells with similar binaural interaction. As at lower levels in the auditory system, one can distinguish EE and EI neurons (Figure 11.27). Recall that EE cells respond more to stimulation of both ears than to either ear separately, while EI cells are inhibited if both ears are stimulated. If one makes an electrode penetration perpendicular to the cortical surface, most of the cells encountered are of one type, either EE or EI. In cat cortex, there are alternating patches of EE and EI cells within an isofrequency band, and evidence suggests that the binaural interaction columns (sometimes called *summation and suppression columns*) are roughly at right angles to the isofrequency bands (Figure 11.30). As we discussed for the superior olive, neurons sensitive to interaural time delays and interaural intensity differences probably play a role in sound localization.

In addition to A1, other cortical areas located on the superior surface of the temporal lobe respond to auditory stimuli. Some of these higher auditory areas are tonotopically organized, and others do not seem to be. As in visual cortex, there is a tendency for the stimuli that evoke a strong response to be more complex than at lower levels in the system. An example of specialization is Wernicke's area, which we will discuss in Chapter 20. Destruction of this area does not interfere with the sensation of sound, but it seriously impairs the ability to interpret spoken language.

The Effects of Auditory Cortical Lesions and Ablation

While deafness results from the bilateral ablation of auditory cortex, it is more often the consequence of damage to the ear (Box 11.6). A surprising de-

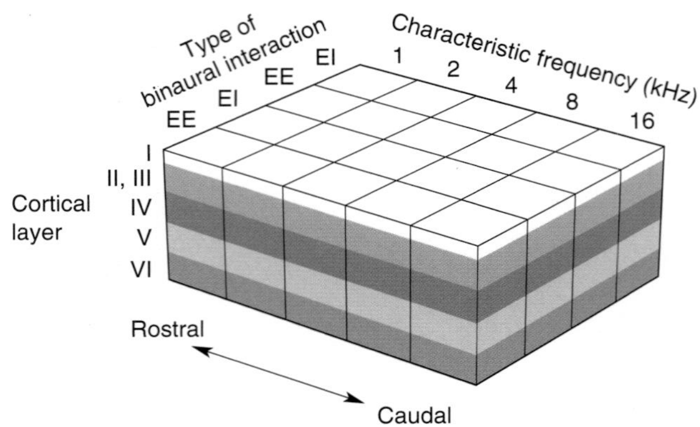

Figure 11.30
A hypothetical "ice cube" model of auditory cortex. In this model, characteristic frequencies increase moving rostral to caudal, and alternating bands of EE and EI cells run mediolateral.

Box 11.6

OF SPECIAL INTEREST

Auditory Disorders

Although the effects of cortical lesions provide important information about the role of auditory cortex in perception, the perceptual deficit we all associate with the auditory system, deafness, usually results from problems in or near the cochlea. Deafness is conventionally subdivided into two categories: conduction deafness and nerve deafness.

Hearing loss resulting from a disturbance in sound conductance from the outer ear to the cochlea is called *conduction deafness*. This sensory deficit can be caused by something as simple as excessive wax in the ear or by more serious diseases of the ossicles. A number of diseases cause binding of the ossicles to the bone of the middle ear, impairing the transfer of sound. Fortunately, most of the mechanical problems in the middle ear that interfere with sound conductance can be treated surgically.

Nerve deafness is deafness associated with the loss of neurons in either the auditory nerve or the hair cells of the cochlea. Nerve deafness sometimes results from tumors affecting the inner ear. It can also be caused by drugs that are toxic to hair cells, such as quinine and certain antibiotics. Exposure to loud sounds, such as explosions and loud music, is another cause of injury to the cochlea. Depending on the degree of cell loss, various treatments are possible. If the cochlea or auditory nerve on one side is completely destroyed, deafness in that ear is absolute. Partial loss of hair cells is more common, however. In these cases, a hearing aid can be used to amplify the sound for the remaining hair cells.

In more severe cases involving considerable damage to the cochlea but an intact auditory-vestibular nerve, some hearing can be restored by implanting an artificial electronic cochlea. This device has an electrode that is placed over or within the cochlea to stimulate the auditory-vestibular nerve electrically. The artificial cochlea is wired to a microphone, which receives the sound. The frequency coding usually provided by the mechanics of the basilar membrane and hair cells can be partially achieved by the use of multiple electrodes that stimulate different portions of the auditory nerve in response to sounds of various frequencies. The success of these devices is quite variable, ranging from restoration of the ability to understand speech to only a crude perception of changes in sounds.

With deafness, a person hears less sound than normal. With the hearing disorder *tinnitus*, a person hears noises in the ears in the absence of any sound stimulus (Box 11.2). The subjective sensation can take many forms, including buzzing, humming, and whistling. You may have had a mild and temporary form of tinnitus after being at a party with really loud music; your brain may have had fun, but your hair cells are in shock! Tinnitus is a relatively common disorder that can seriously interfere with concentration and work if it persists. You can imagine how distracting it would be if you constantly heard whispering or humming or the crinkling of paper.

Tinnitus can be a symptom of any of a number of neurological problems. Although it frequently accompanies diseases involving the cochlea or auditory-vestibular nerve, it may result from exposure to loud sounds or abnormal vasculature of the neck. Although clinical treatment of tinnitus is only partially successful, the annoyance of the noise can often be lessened by using a device that constantly produces a sound in the affected ear or ears. For unknown reasons, the constant real sound is less annoying than the sound of the tinnitus that gets blocked out.

gree of normal auditory function is retained after unilateral lesions in auditory cortex. This is in marked contrast to the visual system, in which a unilateral cortical lesion of striate cortex leads to complete blindness in one visual hemifield. The reason for greater preservation of function after lesions in auditory cortex is that both ears send output to cortex in both hemispheres. In humans, the primary deficit that results from a unilateral loss of A1 is the inability to determine the source of a sound. It may be possible to determine which side of the head a sound comes from, but there is little ability to locate the sound more precisely. Performance on such tasks as frequency or intensity discrimination is nearly normal.

Studies in experimental animals indicate that smaller lesions can produce rather specific localization deficits. Because of the tonotopic organization of A1, it is possible to make a restricted cortical lesion that destroys neurons with characteristic frequencies within a limited frequency range. Interestingly, there is a deficit in localization only for sounds roughly corresponding to the characteristic frequencies of the missing cells. This finding reinforces the idea that information in different frequency bands may be processed in parallel by tonotopically organized structures.

THE VESTIBULAR SYSTEM

Strangely enough, listening to music and riding a bicycle both involve sensations that are transduced by hair cells. The vestibular system reports on the position and movement of the head, gives us our sense of balance and equilibrium, and helps coordinate movements of the head and eyes and adjustments to body posture. When the vestibular system operates normally, we are usually unaware of it. When its function is disrupted, however, the results can include the unpleasant feelings we usually associate with motion sickness: vertigo and nausea plus a sense of disequilibrium and uncontrollable eye movements.

The Vestibular Labyrinth

The vestibular and auditory systems both use hair cells to transduce movements. Common biological structures often have common origins. In this case, the organs of mammalian balance and hearing both evolved from the *lateral line organs* present in all aquatic vertebrates. Lateral lines consist of small pits or tubes along an animal's sides. Each pit contains clusters of hair-like sensory cells whose cilia project into a gelatinous substance that is open to the water in which the animal swims. The purpose of lateral lines in many animals is to sense vibrations or pressure changes in the water. In some cases, they are also sensitive to temperature or electrical fields. Lateral lines were lost as reptiles evolved, but the exquisite mechanical sensitivity of hair cells was adopted and adapted for use in the structures of the inner ear that derived from the lateral line.

In mammals, all hair cells are contained within sets of interconnected chambers called the **vestibular labyrinth** (Figure 11.31a). We have already discussed the auditory portion of the labyrinth, the spiraling cochlea (Figure 11.6). The vestibular labyrinth includes two types of structures with different functions: the **otolith organs**, which detect the force of gravity and tilts of the head, and the **semicircular canals**, which are sensitive to head rotation. The ultimate purpose of each structure is to transmit mechanical energy, derived from head movement, to its hair cells. Each is sensitive to different kinds of movement not because their hair cells differ but because of the specialized structures within which the hair cells reside.

The otolith organs are a pair of relatively large chambers, the *saccule* and

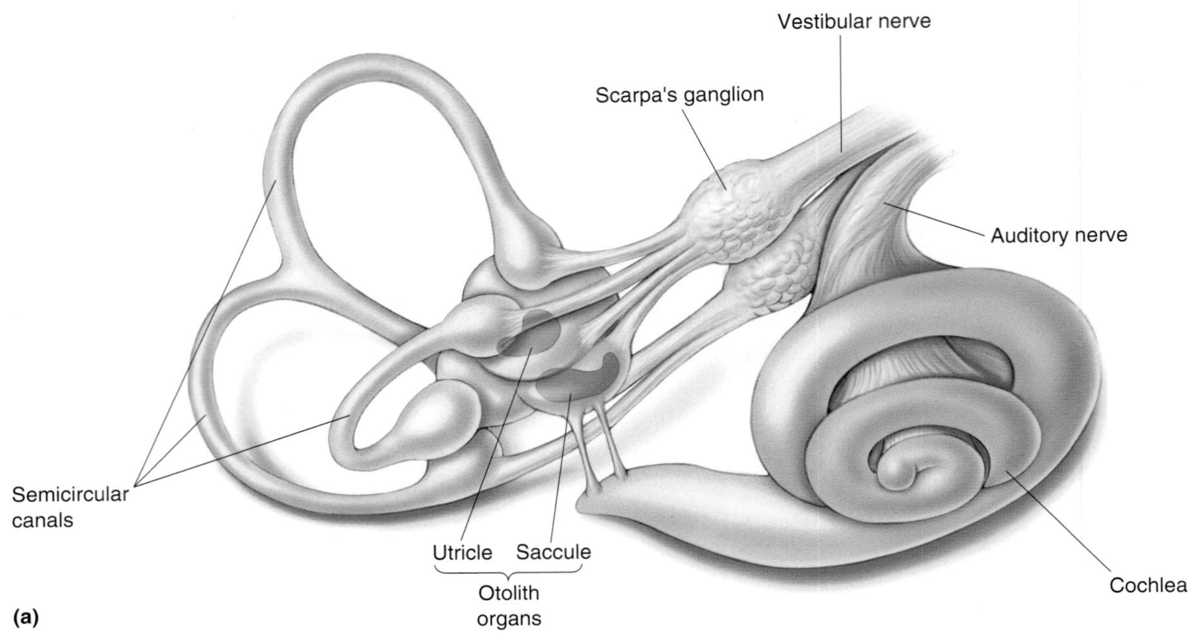

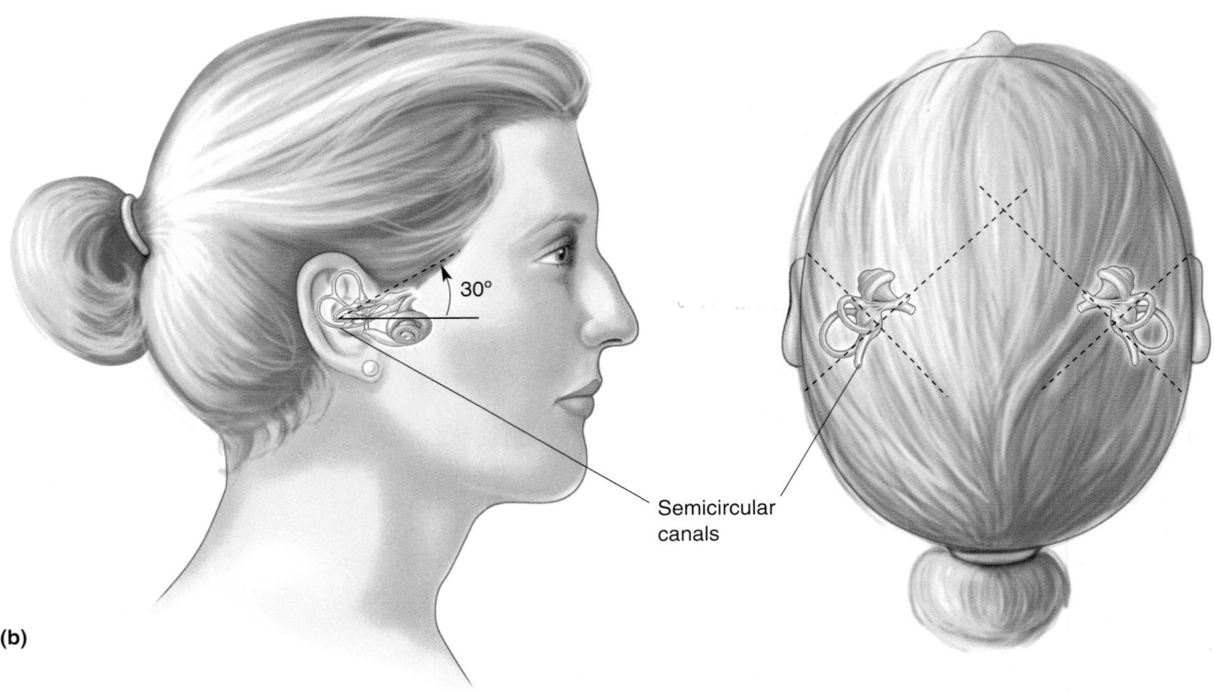

Figure 11.31
The vestibular labyrinth. (a) Locations of the otolith organs (utricle and saccule) and semicircular canals. **(b)** A vestibular labyrinth resides on each side of the head, with the semicircular canals arranged in parallel planes.

the *utricle*, near the center of the labyrinth. The semicircular canals are the three arcing structures of the labyrinth. They lie in approximately orthogonal planes, which means that there is an angle of about 90 degrees between any pair of them (Figure 11.31b). A set of vestibular organs resides on each side of the head, and they are mirror images of each other.

Each hair cell of the vestibular organs makes an excitatory synapse with the end of a sensory axon from the vestibular nerve, a branch of cranial nerve VIII (the auditory-vestibular nerve). There are about 20,000 vestibular nerve axons on each side of the head, and their cell bodies lie in *Scarpa's ganglion*.

The Otolith Organs

The saccule and utricle detect changes of head angle and *linear acceleration* of the head. When you tilt your head, the angle between your otolith organs and the direction of the force of gravity changes. Linear acceleration also generates force in proportion to the mass of an object. Forces due to linear acceleration are the sort you encounter when you ride in an elevator or a car as it starts or stops. In contrast, when a car or elevator moves at constant velocity, acceleration is zero, so there is no force apart from gravitational force. This is why you can fly steadily at 600 mph in a jet yet feel perfectly still; the sudden bouncing you feel during air turbulence, however, is another good example of the forces generated by linear acceleration and of movements detected by your otolith organs.

Each otolith organ contains a sensory epithelium called a **macula**, which is vertically oriented within the saccule and horizontally oriented within the utricle when the head is upright. (The vestibular macula and the retinal macula are entirely different structures.) The vestibular macula contains hair cells that lie among a bed of supporting cells with their cilia projecting into a gelatinous cap (Figure 11.32). Movements are transduced by hair cells in the maculae when the hair bundles are deflected. The unique feature of the otolith organs is the tiny crystals of calcium carbonate called *otoliths*, 1–5 μm in diameter. (*Otolith* is Greek for "ear stone.") Otoliths encrust the surface of the macula's gelatinous cap near the tips of the hair bundles, and they are the key to the tilt sensitivity of the macula. The otoliths have a higher density than the endolymph that surrounds them. When the angle of the head changes or when the head accelerates, a force is exerted on the otoliths; this exerts a force in the same direction on the cap, which moves slightly, and the cilia of the hair cells bend. Not just any deflection will do, however. Each hair cell has one especially tall cilium, the *kinocilium*. The bending of hairs toward the kinocilium results in a depolarizing, excitatory receptor potential. Bending the hairs away from the kinocilium hyperpolarizes and inhibits the cell. The cell is exquisitely sensitive to direction. If the hairs are perpendicular to their preferred direction, they barely respond. The transduction mechanism of vestibular hair cells is essentially the same as that of auditory hair cells (Figure 11.15). As with auditory hair cells, only tiny hair movements are necessary. The response saturates when the hairs are bent less than 0.5 μm, about the diameter of one cilium.

The head can tilt and move in any direction, but the hair cells of the utricle and saccule are oriented to transduce all of them effectively. The saccular maculae are oriented more or less vertically, while the utricular maculae are mostly horizontal (Figure 11.33). On each macula the direction preference of the hair cells varies systematically. There are enough hair cells in each macula to cover a full range of directions. The mirror-image orientation of the saccule and utricle on each side of the head means that when a given head movement excites hair cells on one side, it tends to inhibit hair cells in the corresponding location on the other. Thus, any tilt or acceleration of the head ex-

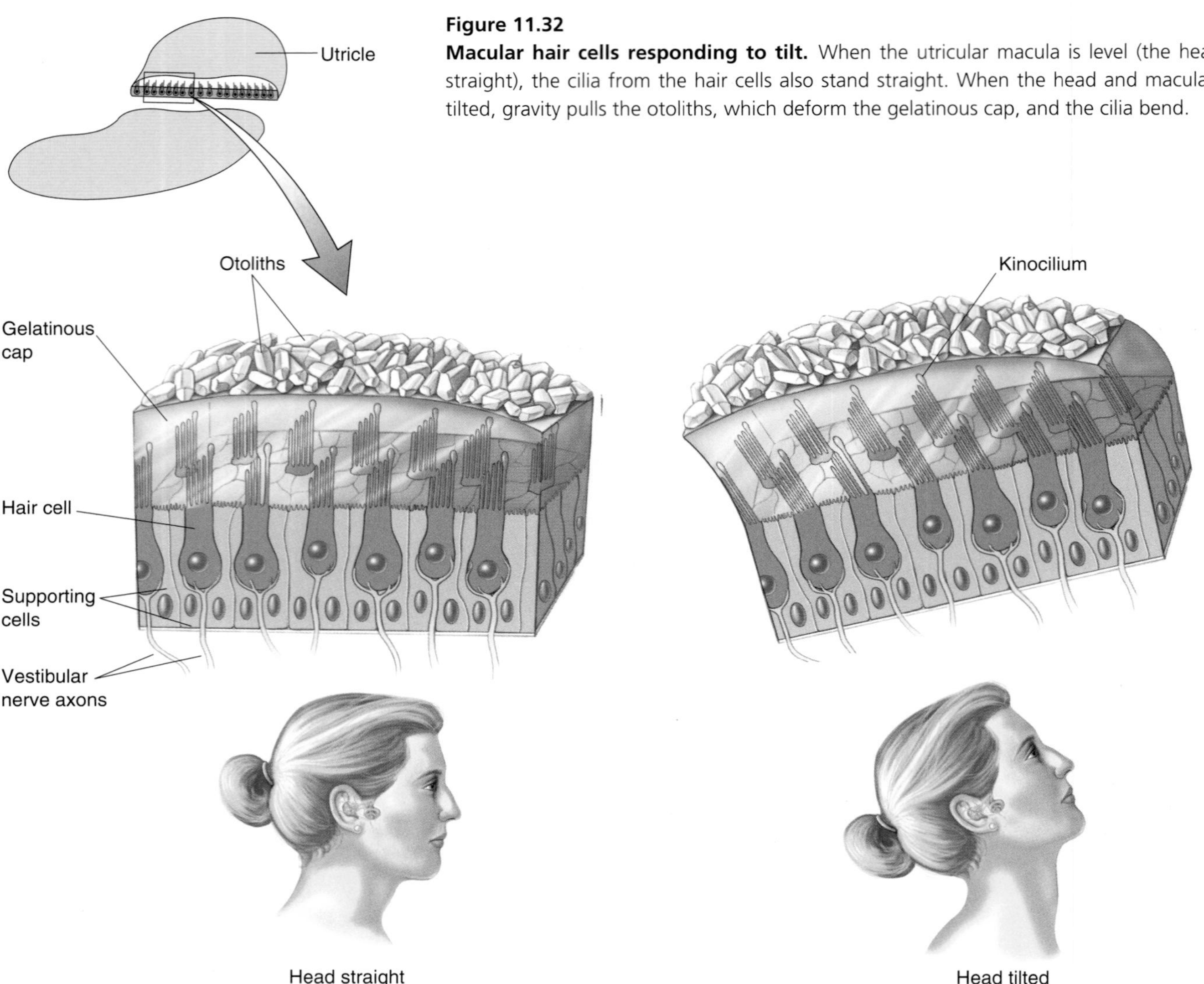

Figure 11.32
Macular hair cells responding to tilt. When the utricular macula is level (the head is straight), the cilia from the hair cells also stand straight. When the head and macula are tilted, gravity pulls the otoliths, which deform the gelatinous cap, and the cilia bend.

cites some hair cells, inhibits others, and has no effect on the rest. The central nervous system, by simultaneously using the information encoded by the full population of otolithic hair cells, can unambiguously interpret all possible linear movements.

The Semicircular Canals

The semicircular canals detect turning movements of the head, such as shaking your head from side to side or nodding. As with the otolith organs, the semicircular canals sense acceleration, but of a different kind. *Angular acceleration* is generated by sudden rotational movements, and it is the primary stimulus for the semicircular canals.

The hair cells of the semicircular canals are clustered within a sheet of cells, the *crista,* within a bulge along the canal called the **ampulla** (Figure 11.34a). The cilia project into the gelatinous *cupula,* which spans the lumen of the canal within the ampulla. All of the hair cells in an ampulla have their kinocilia oriented in the same direction, which means that they all get excited or inhibited together. The semicircular canals are filled with endolymph, the same fluid that fills the cochlea. Bending of the cilia occurs when the canal is suddenly rotated about its axis like a wheel; as the wall of the canal and the

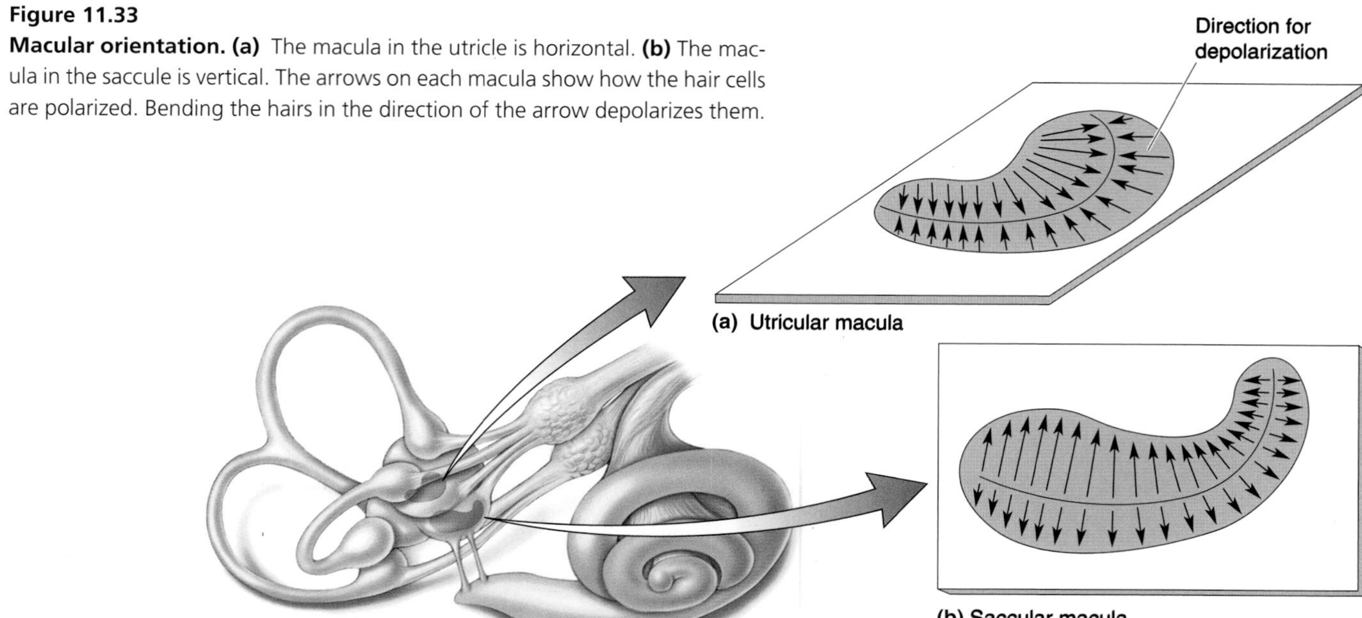

Figure 11.33
Macular orientation. (a) The macula in the utricle is horizontal. **(b)** The macula in the saccule is vertical. The arrows on each macula show how the hair cells are polarized. Bending the hairs in the direction of the arrow depolarizes them.

cupula begin to spin, the endolymph tends to stay behind because of inertia. The sluggish endolymph exerts a force upon the cupula, much like wind upon a sail (Figure 11.34b). This bows the cupula, which bends the cilia, which (depending on the direction of the rotation) either excites or inhibits the release of neurotransmitter from the hair cells onto the vestibular nerve axons.

If head rotation is maintained at a constant velocity, the friction of endolymph with the canal walls eventually makes the two move together, and this reduces and then eliminates the bending of the cupula after 15–30 seconds. Such adaptation to rotation can be seen clearly in the firing rates of vestibular axons from the canals (Figure 11.35). (This sort of prolonged head rotation is not something you encounter very often unless you have a taste for certain amusement park rides.) When rotation of the head and its canals finally stops, the inertia of the endolymph causes the cupula to bend in the other direction, generating an opposite response from the hair cells and a temporary sensation of counterrotation.

Together, the three semicircular canals on one side of the head help sense all possible head rotation angles. This is further ensured because each canal is paired with another on the opposite side of the head (Figure 11.31b). Each member of a pair sits within the same orientation plane as its partner and responds to rotation about the same axis. While rotation excites the hair cells of one canal, however, it inhibits the hair cells of its contralateral partner canal. Vestibular axons fire at high rates even at rest, so their activity can be driven either up or down depending on the direction of rotation. This "push-pull" arrangement—each rotation causing excitation on one side and inhibition on the other (Figure 11.35)—optimizes the ability of the brain to detect rotational movements.

Central Vestibular Pathways and Vestibular Reflexes

The central vestibular pathways coordinate and integrate information about head and body movement and use it to control the output of motor neurons that adjust head, eye, and body positions. Primary vestibular axons from cra-

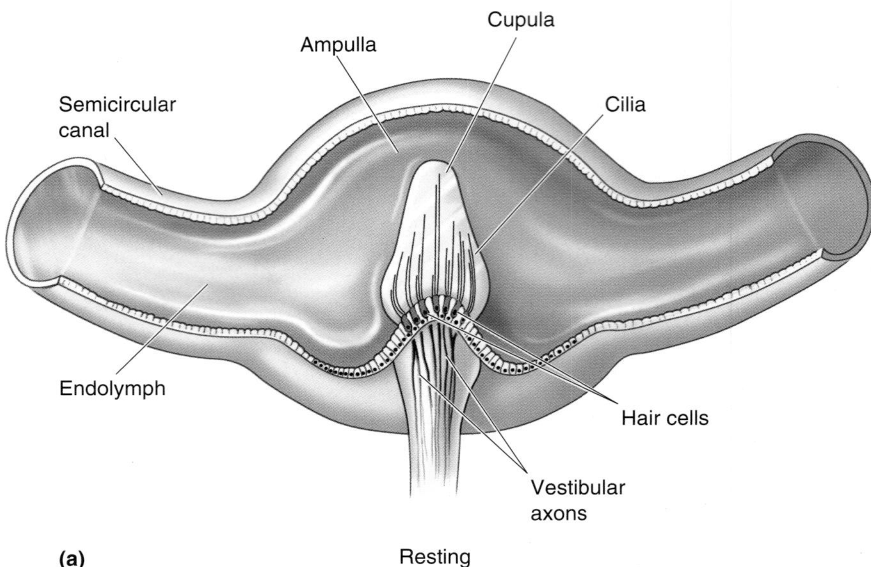

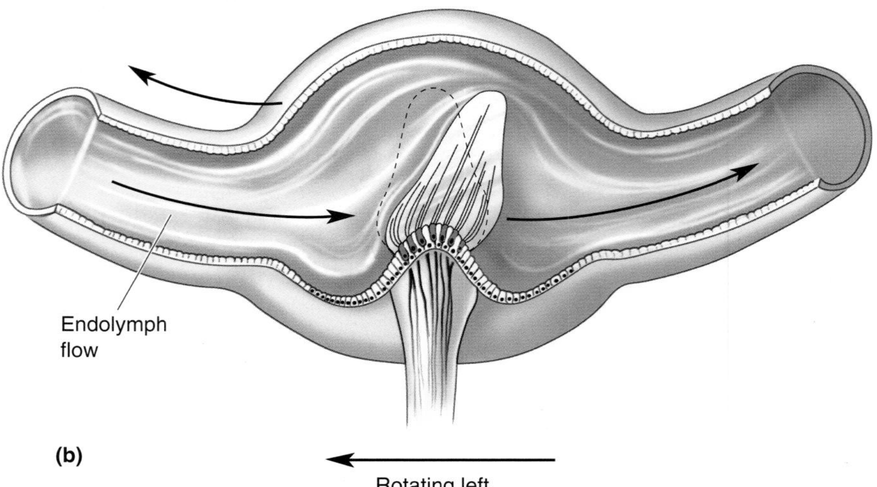

Figure 11.34
A cross section through the ampulla of a semicircular canal. (a) The cilia of hair cells penetrate into the gelatinous cupula, which is bathed in the endolymph that fills the canals. **(b)** When the canal rotates leftward, the endolymph lags behind and applies force to the cupula, bending the cilia within it.

nial nerve VIII make direct connections to the **vestibular nucleus** on the same side of the brain stem and to the cerebellum (Figure 11.36). The vestibular nuclei also receive inputs from other parts of the brain, including the cerebellum, and the visual and somatic sensory systems, thereby combining incoming vestibular information with data about the motor system and other sensory modalities.

The vestibular nucleus, which has several subdivisions, in turn projects to a variety of targets above it in the brain stem and below it into the spinal cord (Figure 11.36). For example, axons from the otolith organs project to the lateral vestibular nucleus, which then projects via the *vestibulospinal tract* to excite spinal motor neurons controlling muscles in the legs that help to maintain posture (see Chapter 14). This pathway helps the body maintain an upright posture even on the rolling deck of a boat. Axons from the semicircular canals project to the medial vestibular nucleus, which sends axons via the *medial longitudinal fasciculus* to excite motor neurons of trunk and neck muscles that orient the head. This pathway helps the head stay straight even as the body cavorts around below it.

Similar to the other sensory systems, the vestibular system makes connections to the thalamus and then to the neocortex. The vestibular nuclei send axons into the *ventral posterior (VP) nucleus* of the thalamus, which projects to regions close to the representation of the face in the primary somatosensory and primary motor areas of cortex (see Chapters 12 and 14). Although its functions are unclear, at the cortical level there is considerable integration of information about movements of the body, the eyes, and the visual scene. It is likely that the cortex continually maintains a representation of body position and orientation in space that is essential for our perception of equilibrium and for planning and executing complex, coordinated movements.

The Vestibulo-ocular Reflex. One particularly important function of the central vestibular system is to keep your eyes pointed in a particular direction even while you are dancing like a fool. This is accomplished by the **vestibulo-ocular reflex (VOR)**. Recall that accurate vision requires the image to remain stable on the retinas despite movement of the head (see Chapter 9). Each eye can be moved by a set of six extraocular muscles. The VOR works by sensing rotations of the head and immediately commands a compensatory movement of the eyes in the opposite direction. The movement helps keep your line of sight tightly fixed on a visual target. Since the VOR is a reflex triggered by *vestibular* input, it works amazingly well even in the dark or when your eyes are closed.

Imagine driving down a very bumpy road. Since the VOR makes constant adjustments, your view of the world ahead is quite stable because eye movement compensates for each bump and its consequent movement of your head. To appreciate how effective your VOR is, compare the stability of a passing object during the bumpy drive as you look at it first with your eyes alone and then through the viewfinder of a camera. Unless you have a very sophisticated camera with the electromechanical equivalent of a VOR, you'll find that your view jumps around hopelessly because your arms are not nearly quick or accurate enough to move the camera with each bump.

The effectiveness of the VOR depends on connections from the semicircular canals to the vestibular nucleus, to the cranial nerve nuclei that excite the

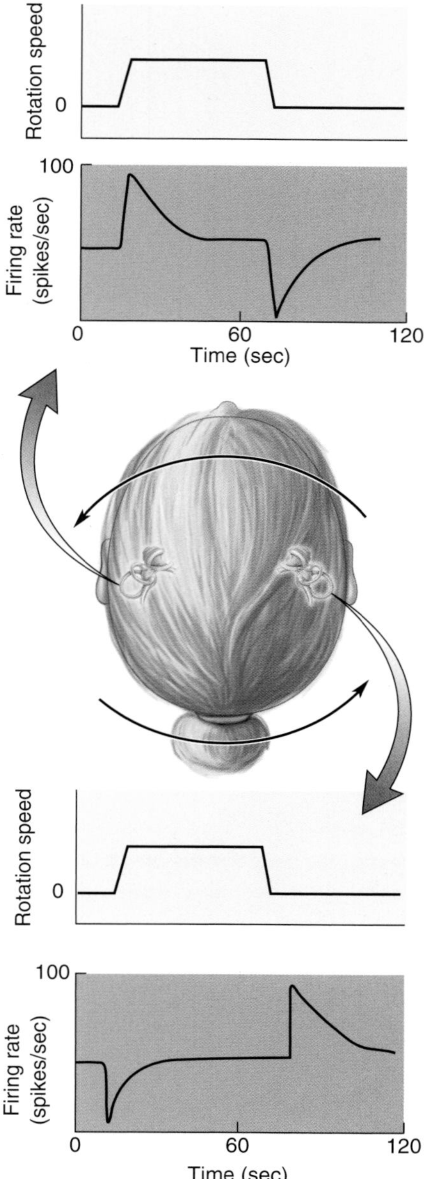

Figure 11.35
Push-pull activation of the semicircular canals. Head rotation excites hair cells in one horizontal semicircular canal and inhibits them in the other. The graphs show that long-lasting head rotation leads to adaptation of the firing in vestibular axons. When rotation is stopped, the vestibular axons from each side begin firing again, but with opposite patterns of excitation and inhibition.

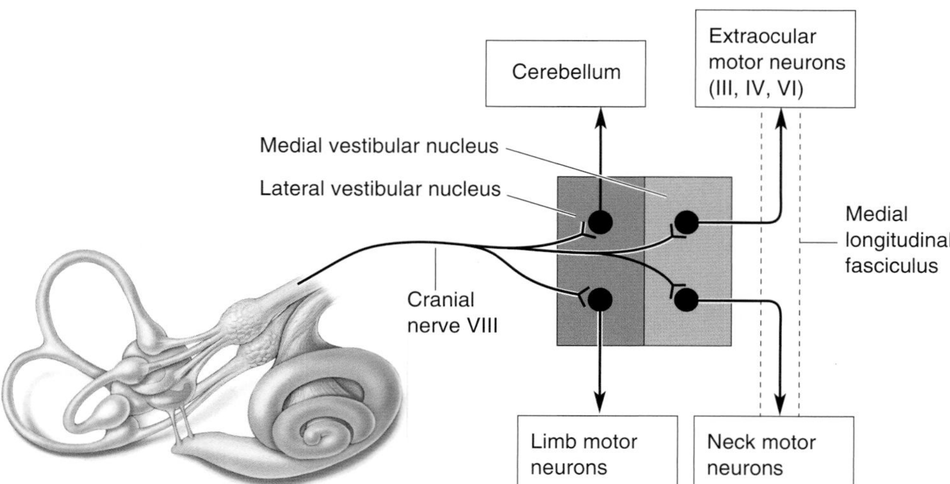

Figure 11.36
A summary of the central vestibular connections from one side.

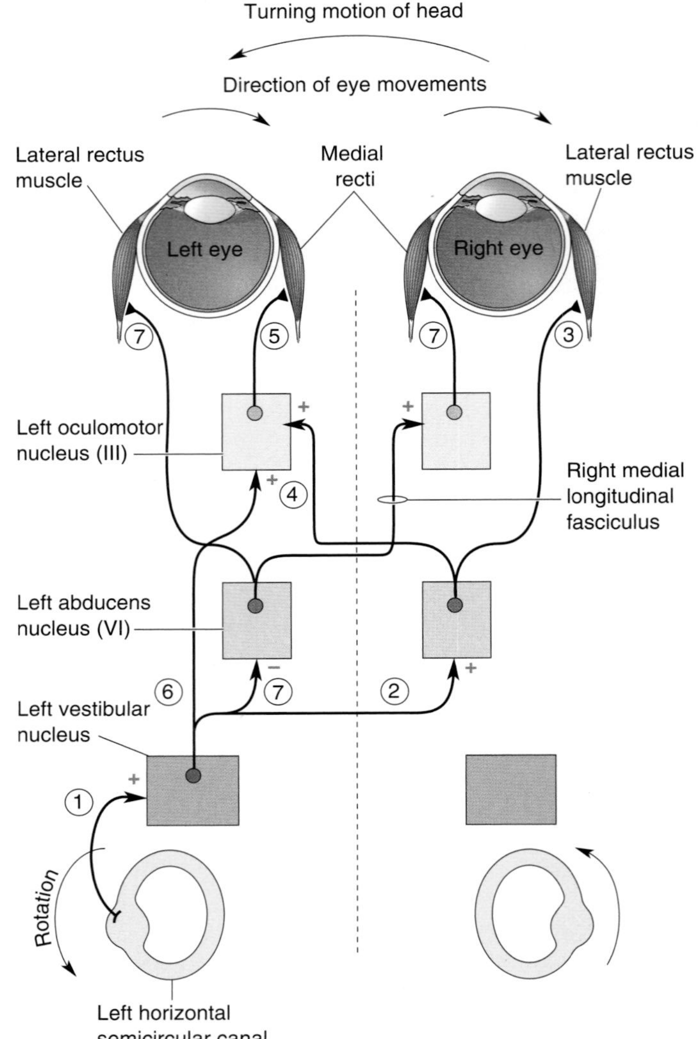

Figure 11.37
Vestibular connections mediating horizontal eye movements during the VOR. These pathways are active when the head suddenly turns to the left, causing the eyes to turn to the right. Excitatory connections are in green; inhibitory connections are in red.

extraocular muscles. Figure 11.37 shows only half of the horizontal component of this circuit and illustrates what happens when the head turns to the left and the VOR induces both eyes to turn right. Axons from the left horizontal canal innervate the left vestibular nucleus, which sends excitatory axons to the contralateral (right) cranial nerve VI nucleus (abducens nucleus). Motor axons from the abducens nucleus in turn excite the lateral rectus muscle of the right eye. Another excitatory projection from the abducens crosses the midline back to the left side and ascends (via the medial longitudinal fasciculus) to excite the left cranial nerve III nucleus (oculomotor nucleus), which excites the right medial rectus muscle of the left eye. Mission accomplished, so it would seem: Both eyes are turning right. To ensure speedy operation, however, the left medial rectus muscle also gets excited via a projection from the vestibular nucleus directly to the left oculomotor nucleus. Speed is also maximized by activating inhibitory connections to the muscles that oppose this movement (the lateral rectus and medial rectus in this case). To deal with head rotations in any direction, the complete VOR circuit includes similar connections between the right horizontal canal, the other semicircular canals, and the other extraocular muscles that control eye movements.

Vestibular Pathology

The vestibular system can be damaged in a variety of ways, e.g., by the toxicity of high doses of antibiotics such as streptomycin. People with bilateral lesions of the vestibular labyrinths have enormous trouble fixating on visual targets as they move about. Even the minute head pulsations due to blood pressure surges of heartbeats can be disturbing in some cases. When people with vestibular disturbances cannot stabilize an image on their moving retinas, they may also have the disconcerting feeling that the world is constantly moving around them. This can make walking and standing difficult. Compensatory adjustments come with time, as the brain learns to substitute more visual and proprioceptive cues to help guide smooth and accurate movements.

CONCLUDING REMARKS

Hearing and balance begin with nearly identical sensory receptors, the hair cells, which are exquisitely sensitive to deflections of their stereocilia. These movement detectors are surrounded by three types of inner ear apparatus that give them selectivity for three kinds of mechanical energy: periodic waves of air pressure (sound), rotational forces (head turns), and linear forces (head tilt). Except for the similarity in their transduction step and the fact that the hair cells of both are located in the inner ear, the auditory and vestibular systems are quite different. Auditory stimuli come mainly from the external environment, while the vestibular system senses only its own movements. Auditory and vestibular pathways are entirely separate, except perhaps at the highest levels of the cortex. Auditory information is often at the forefront of our consciousness, while vestibular sensation usually operates unnoticed to coordinate and calibrate our every movement.

We have followed the auditory pathways from the ear to cerebral cortex and seen the ways in which information about sound is transformed. Variations in the density of air molecules are converted to movement of the mechanical components of the middle and inner ear, which are transduced into neural responses. The structure of the ear and cochlea are highly specialized for the transduction of sound. This fact should not blind us, however, to the considerable similarities between the organization of the auditory system and that of other sensory systems. Many analogies can be made between the auditory and visual systems. In the sensory receptors of both systems, a spatial code is established. In the visual system, the code in the photoreceptors is retinotopic; the activity of a given photoreceptor indicates light at a particular location. The receptors in the auditory system establish a spatial code that is tonotopic because of the unique properties of the cochlea. In each system, the retinotopy or tonotopy is preserved as signals are processed in secondary neurons, in the thalamus, and finally in sensory cortex.

The convergence of inputs from lower levels produces neurons at higher levels that have more complex response properties. Combinations of LGN inputs give rise to simple and complex receptive fields in visual cortex; similarly in the auditory system, the integration of inputs tuned to various sound frequencies yields higher-level neurons that respond to complex combinations of frequencies. Another example of increasing visual complexity is the convergence of inputs from the two eyes, which yields binocular neurons that are important for depth perception. Analogously, in the auditory system, input from the two ears is combined to create binaural neurons, which are used for horizontal sound localization. These are just a few of the many similarities in the two systems. Principles governing one system can often help us understand other systems. Keep this in mind while reading about the somatic sensory system in Chapter 12, and you'll be able to predict some features of cortical organization based on the types of sensory receptors.

KEY TERMS

Introduction
audition (p. 351)
vestibular system (p. 351)

The Nature of Sound
frequency (p. 351)
hertz (Hz) (p. 352)
intensity (p. 352)

The Structure of the Auditory System
pinna (p. 353)
auditory canal (p. 353)
tympanic membrane (p. 353)
ossicle (p. 353)
oval window (p. 354)
cochlea (p. 354)
outer ear (p. 354)
middle ear (p. 354)
inner ear (p. 354)
medial geniculate nucleus (MGN) (p. 354)
primary auditory cortex (A1) (p. 354)

The Middle Ear
malleus (p. 355)
incus (p. 355)
stapes (p. 355)
Eustachian tube (p. 355)
attenuation reflex (p. 356)

The Inner Ear
round window (p. 357)
scala vestibuli (p. 358)
scala media (p. 358)
scala tympani (p. 358)
Reissner's membrane (p. 358)
basilar membrane (p. 358)
organ of Corti (p. 358)
tectorial membrane (p. 358)
helicotrema (p. 358)
perilymph (p. 358)
endolymph (p. 358)
hair cell (p. 360)
stereocilium (p. 360)
reticular lamina (p. 360)
inner hair cell (p. 362)
outer hair cell (p. 362)
spiral ganglion (p. 362)
auditory-vestibular nerve (p. 362)
cochlear amplifier (p. 367)

Central Auditory Processes
dorsal cochlear nucleus (p. 368)
ventral cochlear nucleus (p. 368)
superior olive (p. 368)
inferior colliculus (p. 368)
characteristic frequency (p. 371)

Encoding Sound Intensity and Frequency
tonotopy (p. 372)
phase locking (p. 373)
volley principle (p. 373)

Mechanisms of Sound Localization
duplex theory of sound localization (p. 375)

The Vestibular System
vestibular labyrinth (p. 385)
otolith organs (p. 385)
semicircular canals (p. 385)
macula (p. 387)
ampulla (p. 388)
vestibular nucleus (p. 390)
vestibulo-ocular reflex (VOR) (p. 391)

REVIEW QUESTIONS

1. How is the conduction of sound to the cochlea facilitated by the ossicles of the middle ear?
2. Why is the round window crucial for the function of the cochlea? What would happen to hearing if it suddenly didn't exist?
3. Why is it impossible to predict the frequency of a sound simply by looking at which portion of the basilar membrane is the most deformed?
4. Why would the transduction process in hair cells fail if the stereocilia as well as the hair cell bodies were surrounded by perilymph?
5. If inner hair cells are primarily responsible for hearing, what is the function of outer hair cells?
6. Why doesn't unilateral damage to the inferior colliculus or MGN lead to deafness in one ear?
7. What mechanisms function to localize sounds in the horizontal and vertical planes?
8. What symptoms would you expect to see in a person who had recently had a stroke affecting A1 unilaterally? How does the severity of these symptoms compare with the effects of a unilateral stroke involving V1?
9. What is the difference between nerve deafness and conduction deafness?
10. Each macula contains hair cells with kinocilia arranged in all directions. What is the advantage of this arrangement compared with arranging all of the cells in the same direction?
11. Imagine a semicircular canal rotating in two ways: around its axis (like a rolling coin) and end over end (like a flipped coin). How well would its hair cells respond in each case, and why?
12. How would you expect the functions of the otolith organs and the semicircular canals to change in the weightless environment of space?

CHAPTER 12

The Somatic Sensory System

INTRODUCTION

The somatic sensory system brings us some of life's most enjoyable experiences and some of its most aggravating. **Somatic sensation** enables our body to feel, to ache, to chill, and to know what its parts are doing. It is sensitive to many kinds of stimuli: the pressure of objects against the skin, the position of joints and muscles, distension of the bladder, and the temperature of the limbs and of the brain itself. When stimuli become so strong that they may be damaging, somatic sensation is also responsible for the feeling that is most offensive but vitally important—pain.

The somatic sensory system is different from other sensory systems in two interesting ways. First, its receptors are distributed throughout the body rather than being concentrated at small, specialized locations. Second, because it responds to many different kinds of stimuli, we can think of it as a group of at least four senses rather than a single one: the senses of touch, temperature, pain, and body position. In fact, those four can, in turn, be subdivided into many more. The somatic sensory system is really a catch-all name, a collective category for all of the sensations that are *not* seeing, hearing, tasting, smelling, and the vestibular sense of equilibrium. The familiar idea that we have only five senses is obviously too simple.

If something touches your finger, you can accurately gauge the place, pressure, sharpness, texture, and duration of the touch. If it is a pinprick, there is no mistaking it for a hammer. If the touch moves from your hand to your wrist and up your arm to your shoulder, you can track its speed and position. Assuming you are not looking, this information must be entirely described by the activity of the sensory nerves in your limb. A single sensory receptor can encode stimulus features such as intensity, duration, position, and sometimes direction. But a single stimulus usually activates many receptors. The task of the central nervous system (CNS) is to interpret the activity of the vast receptor array and use it to generate coherent perceptions.

In this chapter, we divide our discussion of somatic sensation into two main parts: the sense of touch and the sense of pain. As we will see, these different categories of somatic sensation depend on different receptors, different axonal pathways, and different regions of the brain. We also briefly discuss how we sense nonpainful changes in temperature. Discussion of the sense of body position, also called *proprioception*, is reserved for Chapter 13, in which we will explore how this type of somatic sensory information is used to control muscle reflexes.

TOUCH

The sensation of touch begins at the skin (Figure 12.1). The two major types of skin are called *hairy* and *glabrous* (hairless), as exemplified by the backs and palms of your hands. Skin has an outer layer, the *epidermis,* and an inner layer, the *dermis.* Skin performs an essential protective function and prevents the evaporation of body fluids into the dry environment we live in. But skin also provides our most direct contact with the world; indeed, skin is the largest sensory organ we have. Imagine the beach without the squish of sand between your toes, or consider *watching* a kiss instead of experiencing it yourself. Skin is sensitive enough that a raised dot measuring only 0.006 mm high and 0.04 mm wide can be felt when stroked by a fingertip. For comparison, a Braille dot is 167 times higher.

In this section, we see how a touch of the skin is transduced into neural signals, how these signals make their way to the brain, and how the brain makes sense of them.

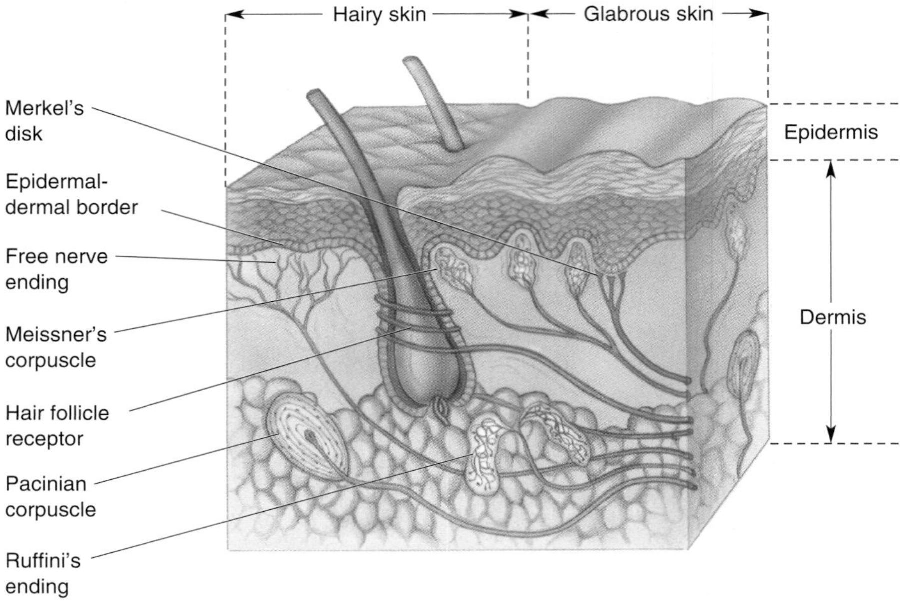

Figure 12.1
Somatic sensory receptors in the skin. Hairy and glabrous skin have a variety of sensory receptors in the dermal and epidermal layers. Each receptor has an axon, and except for free nerve endings, all of them have associated non-neural tissues.

Mechanoreceptors of the Skin

Most of the sensory receptors in the somatic sensory system are **mechanoreceptors**, which are sensitive to physical distortion such as bending or stretching. Present throughout the body, they monitor contact with the skin, as well as pressure in the heart and blood vessels, stretching of the digestive organs and urinary bladder, and force against the teeth. At the heart of each mechanoreceptor are unmyelinated axon branches. These axons have *mechanosensitive ion channels;* their gating depends on stretching or changes in tension of the surrounding membrane. Of the major types of known ion channels, mechanosensitive channels are the least understood.

The mechanoreceptors of the skin are shown in Figure 12.1. Most of them are named after the nineteenth century German and Italian histologists who discovered them. The largest and best studied receptor is the **Pacinian corpuscle**, which lies deep in the dermis and can be as long as 2 mm and almost 1 mm in diameter. *Ruffini's endings*, found in both hairy and glabrous skin, are slightly smaller than Pacinian corpuscles. *Meissner's corpuscles*, about one-tenth the size of Pacinian corpuscles, are in the ridges of glabrous skin (the raised parts of your fingerprints, for example). Located within the epidermis, *Merkel's disks* each consist of a nerve terminal and a flattened non-neural epithelial cell. In this case it may be that the epithelial cell is the mechanically sensitive part, because it makes a synapse-like junction with the nerve terminal. In *Krause end bulbs*, which lie in the border regions of dry skin and mucous membrane (around the lips and genitals, for example), the nerve terminals look like knotted balls of string.

Skin can be vibrated, pressed, pricked, and stroked, and its hairs can be bent or pulled. These are quite different kinds of mechanical energy, yet we can feel them all and easily tell them apart. Accordingly, we have mechanoreceptors that vary in their preferred stimulus frequencies, pressures, and receptive field sizes. Swedish neuroscientist Åke Vallbo and his colleagues developed methods to record from single sensory axons in the human arm so that they could simultaneously measure the sensitivity of mechanoreceptors in the hand *and* evaluate the perceptions produced by various mechanical stimuli (Figure 12.2a). When the stimulus probe was touched to the surface

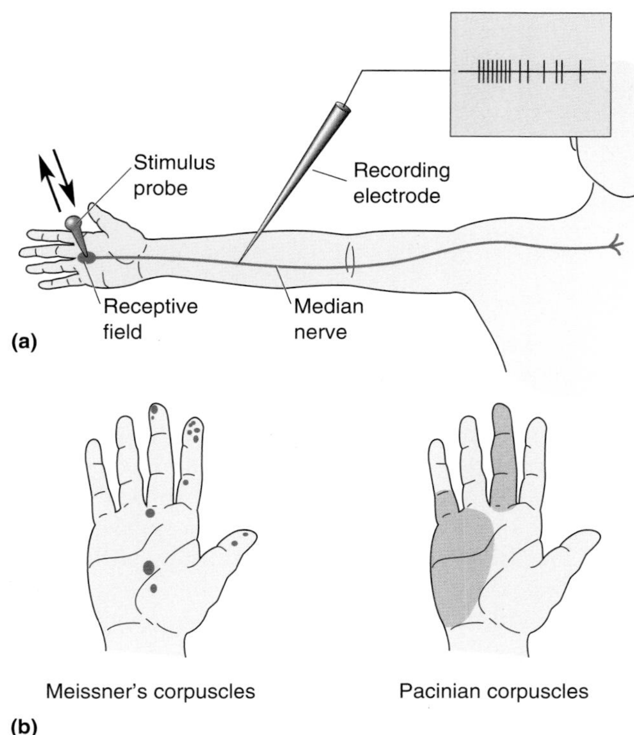

Figure 12.2
Testing the receptive fields of human sensory receptors. (a) By introducing a microelectrode into the median nerve of the arm, it is possible to record the action potentials from a single sensory axon and map its receptive field on the hand with a fine stimulus probe. **(b)** Results show that receptive fields are either relatively small, as in Meissner's corpuscles, or large, as in Pacinian corpuscles. (Source: Adapted from Vallbo and Johansson, 1984.)

of the skin and moved around, the receptive field of a single mechanoreceptor could be mapped. Meissner's corpuscles and Merkel's disks had small receptive fields only a few millimeters wide, while Pacinian corpuscles and Ruffini's endings had large receptive fields that could cover an entire finger or half the palm (Figure 12.2b).

Mechanoreceptors also vary in the persistence of their responses to long-lasting stimuli. If a stimulus probe is suddenly pressed against the skin within the receptive field, some mechanoreceptors, such as Meissner's and Pacinian corpuscles, tend to respond quickly at first but then stop firing, even though the stimulus continues; these receptors are said to be *rapidly adapting*. Other receptors, such as Merkel's disks and Ruffini's endings, are *slowly adapting*; they generate a more sustained response during a long stimulus. Figure 12.3 summarizes the receptive field size and adaptation rate for four mechanoreceptors of the skin.

Hairs do more than adorn our head and keep a dog warm in winter. Many hairs are part of a sensitive receptor system. To demonstrate this, brush just a single hair on the back of your arm with the tip of a pencil; it feels like an annoying mosquito. For some animals, hair is a major sensory system. Imagine a rat slinking confidently through dark passageways and alleys. The rat navigates in part by waving its facial *vibrissae* (whiskers) to sense the local environment and derive information about the texture, distance, and shape of nearby objects. Hairs grow from *follicles* embedded in the skin; each follicle is richly innervated by free nerve endings that either wrap around it or run parallel to it (Figure 12.1). There are several types of hair follicles, including some with erectile muscles (essential for mediating the strange sensations we know as goose pimples), and the details of their innervation differ. In all cases, the bending of the hair causes a deformation of the follicle and surrounding skin tissues. This, in turn, stretches, bends, or flattens the nearby nerve endings, which then increase or decrease their action potential

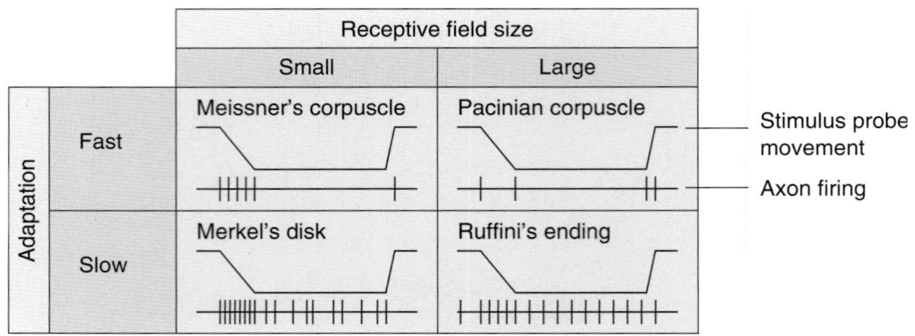

Figure 12.3
Variations among somatic sensory receptors of the skin in receptive field size and adaptation rate. (Source: Adapted from Vallbo and Johansson, 1984.)

firing frequency. The mechanoreceptors of hair follicles may be either slowly adapting or rapidly adapting.

The different mechanical sensitivities of mechanoreceptors mediate different sensations. Pacinian corpuscles are most sensitive to vibrations of about 200–300 Hz, while Meissner's corpuscles respond best around 50 Hz (Figure 12.4). Place your hand against a speaker while playing your favorite music loudly; you "feel" the music largely with your Pacinian corpuscles. If you stroke your fingertips across the coarse screen covering the speaker, each point of skin will hit the bumps at frequencies about optimal to activate Meissner's corpuscles. You feel this as a sensation of rough texture. Stimulation of even lower frequency can activate Ruffini's endings and Meissner's corpuscles, yielding a "fluttering" feeling.

Vibration and the Pacinian Corpuscle. The selectivity of a mechanoreceptive axon depends primarily on the structure of its special ending. For example, the Pacinian corpuscle has a football-shaped capsule with 20–70 concentric layers of connective tissue, arranged like the layers of an onion, with a nerve terminal in the middle (Figure 12.1). When the capsule is compressed, energy is transferred to the nerve terminal, its membrane is deformed, and mechanosensitive channels open. Current flowing through the channels generates a receptor potential, which is depolarizing (Figure 12.5a). If the depolarization is large enough, the axon will fire an action potential. But the capsule layers are slick, with viscous fluid between them. If the stimulus pressure is maintained, the layers slip past one another and transfer the stimulus energy in such a way that the axon terminal is no longer deformed and the receptor potential dissipates. When pressure is released, the events reverse themselves; the terminal depolarizes again and may fire another action potential.

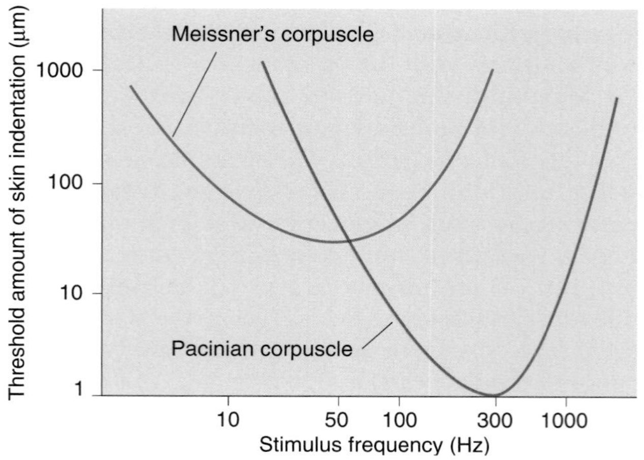

Figure 12.4
Frequency sensitivity of two rapidly adapting mechanoreceptors of the skin. Pacinian corpuscles are most sensitive to high-frequency stimuli, and Meissner's corpuscles are most sensitive to low-frequency stimuli. The skin was indented with a pressure probe at various frequencies during a recording from the nerve. The amplitude of the stimulus was increased until it generated action potentials; threshold was measured as the amount of skin indentation in micrometers (μm). (Source: Adapted from Schmidt, 1978.)

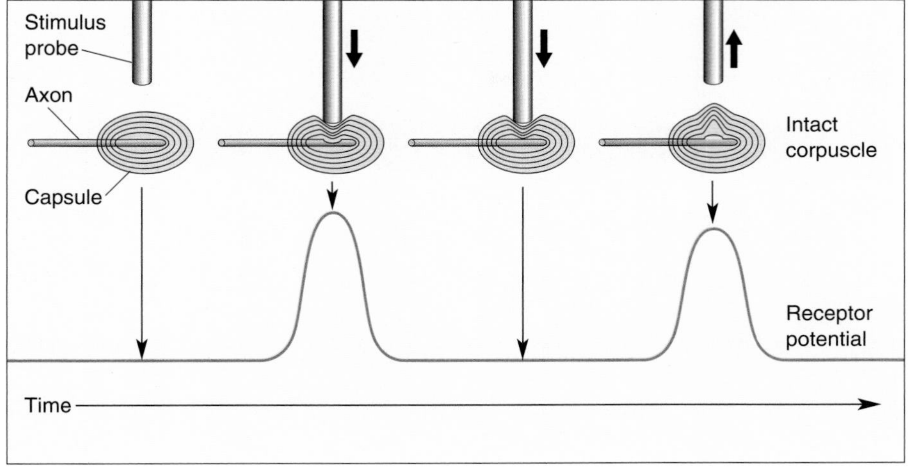

(a)

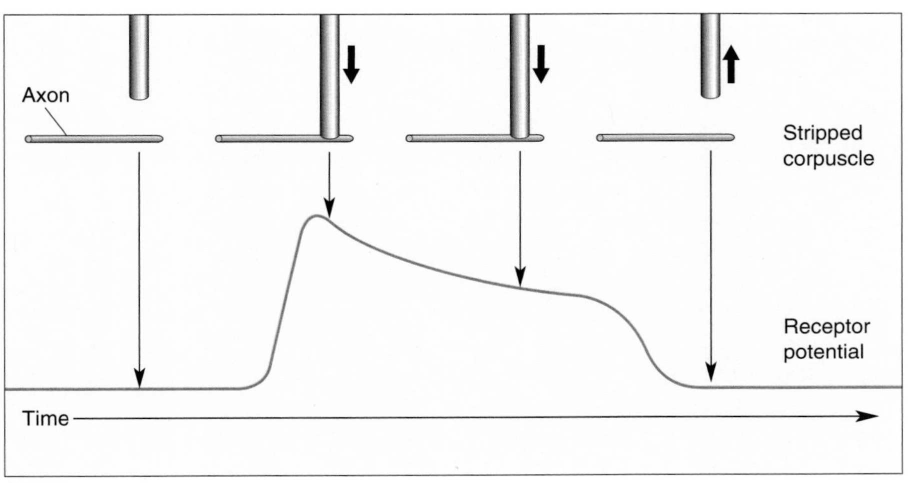

(b)

Figure 12.5

Adaptation in the Pacinian corpuscle. A single Pacinian corpuscle was isolated and stimulated by a probe that indented it briefly. The receptor potential was recorded from a nearby portion of the axon. **(a)** In the intact corpuscle, a large receptor potential was generated at the onset and offset of the stimulus; during maintained indentation, the receptor potential disappeared. **(b)** The onionlike encapsulation was dissected away, leaving a bare axon ending. When indented by the probe, a receptor potential was again generated, so the capsule is not necessary for mechanoreception. But while the normal corpuscle responded only to the onset or offset of a long indentation, the stripped version gave a much longer response; its adaptation rate was slowed. Apparently, it is the capsule that makes the corpuscle insensitive to low-frequency stimuli.

In the 1960s, Werner Loewenstein and his colleagues, working at Columbia University, stripped away the capsule from single corpuscles and found that the naked nerve terminal became much less sensitive to vibrating stimuli and much more sensitive to steady pressure (Figure 12.5b). Clearly, it is the layered capsule, not some property of the nerve ending itself, that makes the Pacinian corpuscle exquisitely sensitive to vibrating, high-frequency stimuli and almost unresponsive to steady pressure.

Two-Point Discrimination. Our ability to discriminate the detailed features of a stimulus varies tremendously across the body. A simple measure of spatial resolution is the two-point discrimination test. You can do this yourself with a paper clip bent into a U. Starting with the ends about an inch apart, touch them to the tip of a finger; you should have no problem telling that there are two separate points touching your finger. Then bend the wire to bring the points closer together, and touch them to your fingertip again. Repeat and see how close the points have to be before they feel like a single point. (This test is best done with two people, one testing and the other being tested without looking.) Now try it on the back of your hand, on your lips, on your leg, and on any other place that interests you. Compare your results with those shown in Figure 12.6.

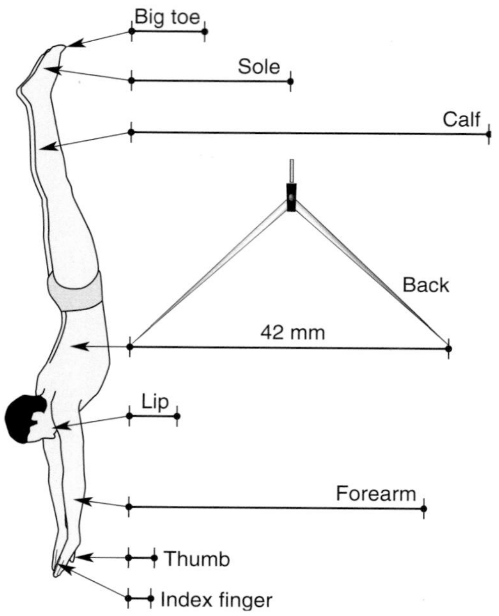

Figure 12.6
Two-point discrimination on the body surface. The pairs of dots show the minimum distance necessary to differentiate between two points touching you simultaneously. Notice the sensitivity of the fingertips compared with that of the rest of the body.

Two-point discrimination varies at least twentyfold across the body. Fingertips have the highest resolution. The dots of Braille are 1 mm high and 2.5 mm apart; up to six dots make a letter. An experienced Braille reader can scan an index finger across a page of raised dots and read about 600 letters per minute, which is roughly as fast as someone reading aloud. There are several reasons the fingertip is so much better than, for example, the elbow for Braille reading: (1) There is a much higher density of mechanoreceptors in the skin of the fingertip than on other parts of the body; (2) the fingertips are enriched in receptor types that have small receptive fields; (3) there is more brain tissue (and thus more raw computing power) devoted to the sensory information of each square millimeter of fingertip than elsewhere; and (4) there may be special neural mechanisms devoted to high-resolution discriminations.

Primary Afferent Axons

The skin is richly innervated by axons that course within the vast network of peripheral nerves on their way to the CNS (Figure 12.7). Axons bringing information from the somatic sensory receptors to the spinal cord or brain stem are the *primary afferent axons* of the somatic sensory system. The primary afferent axons enter the spinal cord through the dorsal roots; their cell bodies lie in the dorsal root ganglia (Figure 12.8).

Primary afferent axons have widely varying diameters, and their size correlates with the type of sensory receptor to which they are attached. Unfortunately, the terminology approaches absurdity here because the different sizes of axons are designated by two sets of names, using Arabic *and* Greek letters *and* Roman numerals. As shown in Figure 12.9, in order of decreasing size, axons from skin sensory receptors are usually designated Aα, Aβ, Aδ, and C; axons of similar size that innervate the muscles and tendons are called groups I, II, III, and IV. Sensory nerves from the skin lack the largest (Aα) group of axons. Group C (or IV) axons are, by definition, unmyelinated axons, while all the rest are myelinated.

An interesting and simple point is hidden in the many axon names. Recall that the diameter of an axon, together with its myelin, determines its speed of action potential conduction. The smallest axons, the so-called C fibers,

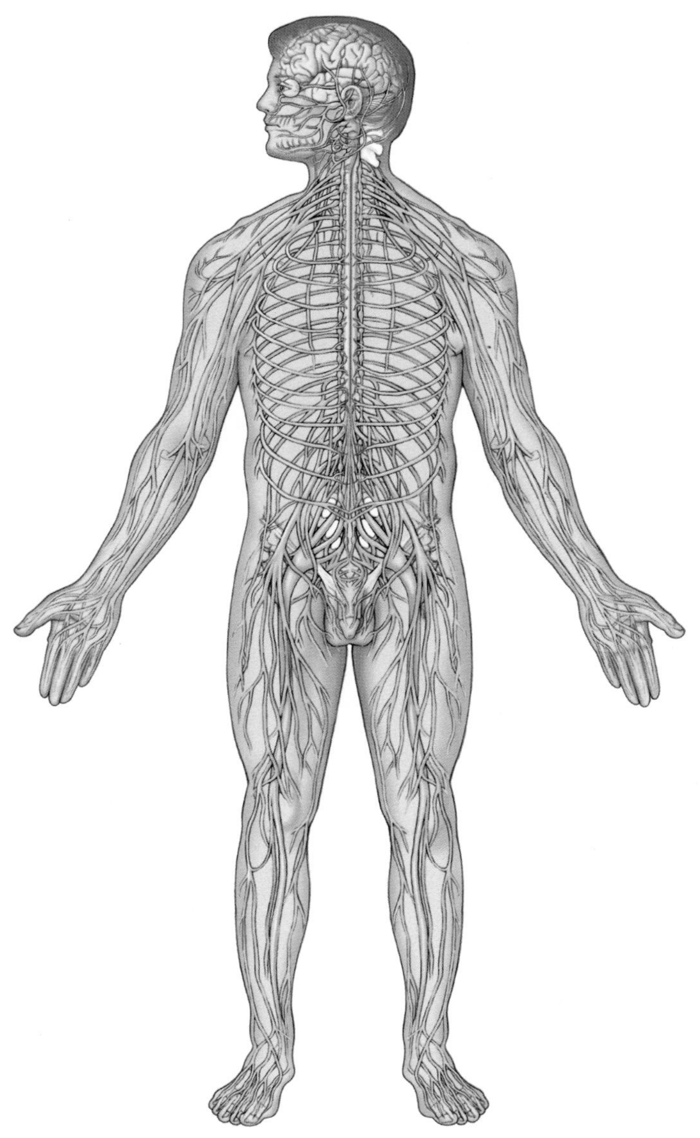

Figure 12.7
The peripheral nerves.

have no myelin and are less than about 1 μm in diameter. C fibers mediate pain and temperature sensation, and they are the slowest of axons, conducting at about 0.5–1 m/sec. To see how slow this is, take a big step, count to two, and then take another step. That's about how fast the action potentials travel along C fibers. On the other hand, touch sensations, mediated by the cutaneous mechanoreceptors, are conveyed by the relatively large Aβ axons, which can conduct at up to 75 m/sec. For comparison, consider that a good major league baseball pitcher can throw a fastball about 90 miles per hour, which is only 40 m/sec.

The Spinal Cord

Most peripheral nerves communicate with the CNS via the spinal cord, which is encased in the bony vertebral column.

Segmental Organization of the Spinal Cord. The arrangement of paired dorsal and ventral roots shown in Figure 12.8 is repeated 30 times down the

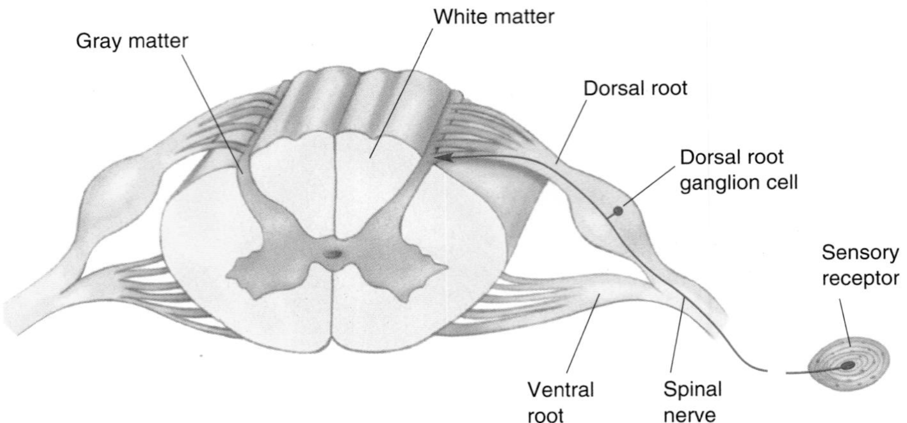

Figure 12.8
Structure of a segment of the spinal cord and its roots.

length of the human spinal cord. Each spinal nerve, consisting of dorsal root and ventral root axons, passes through a notch between the vertebrae ("back bones") of the spinal column. There are as many spinal nerves as there are notches between vertebrae. As shown in Figure 12.10, the 30 **spinal segments** are divided into four groups, and each segment is named after the vertebra adjacent to where the nerves originate: cervical (C) 1–8, thoracic (T) 1–12, lumbar (L) 1–5, and sacral (S) 1–5.

The segmental organization of spinal nerves and the sensory innervation of the skin are related. The area of skin innervated by the right and left dorsal roots of a single spinal segment is called a **dermatome**, and thus there is a one-to-one correspondence between dermatomes and spinal segments.

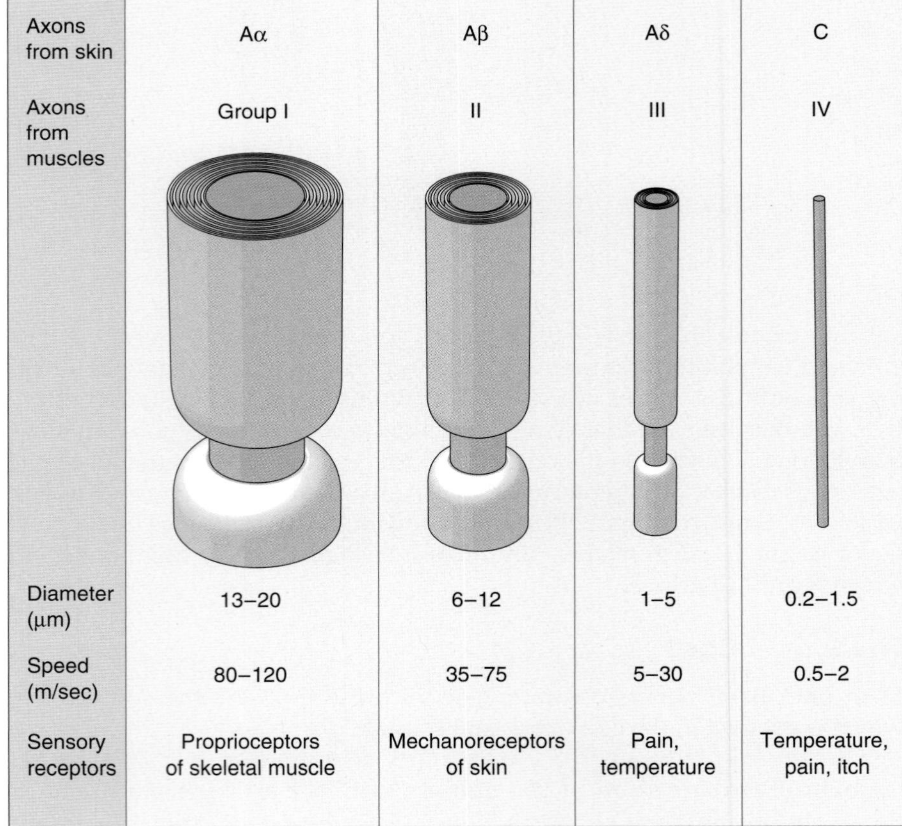

Axons from skin
Aα
Aβ
Aδ
C
Axons from muscles
Group I
II
III
IV
Diameter (μm)
13–20
6–12
1–5
0.2–1.5
Speed (m/sec)
80–120
35–75
5–30
0.5–2
Sensory receptors
Proprioceptors of skeletal muscle
Mechanoreceptors of skin
Pain, temperature
Temperature, pain, itch

Figure 12.9
Various sizes of primary afferent axons. The axons are drawn to scale but are shown 2000 times life size. The diameter of an axon is correlated with its conduction velocity and with the type of sensory receptor to which it is connected.

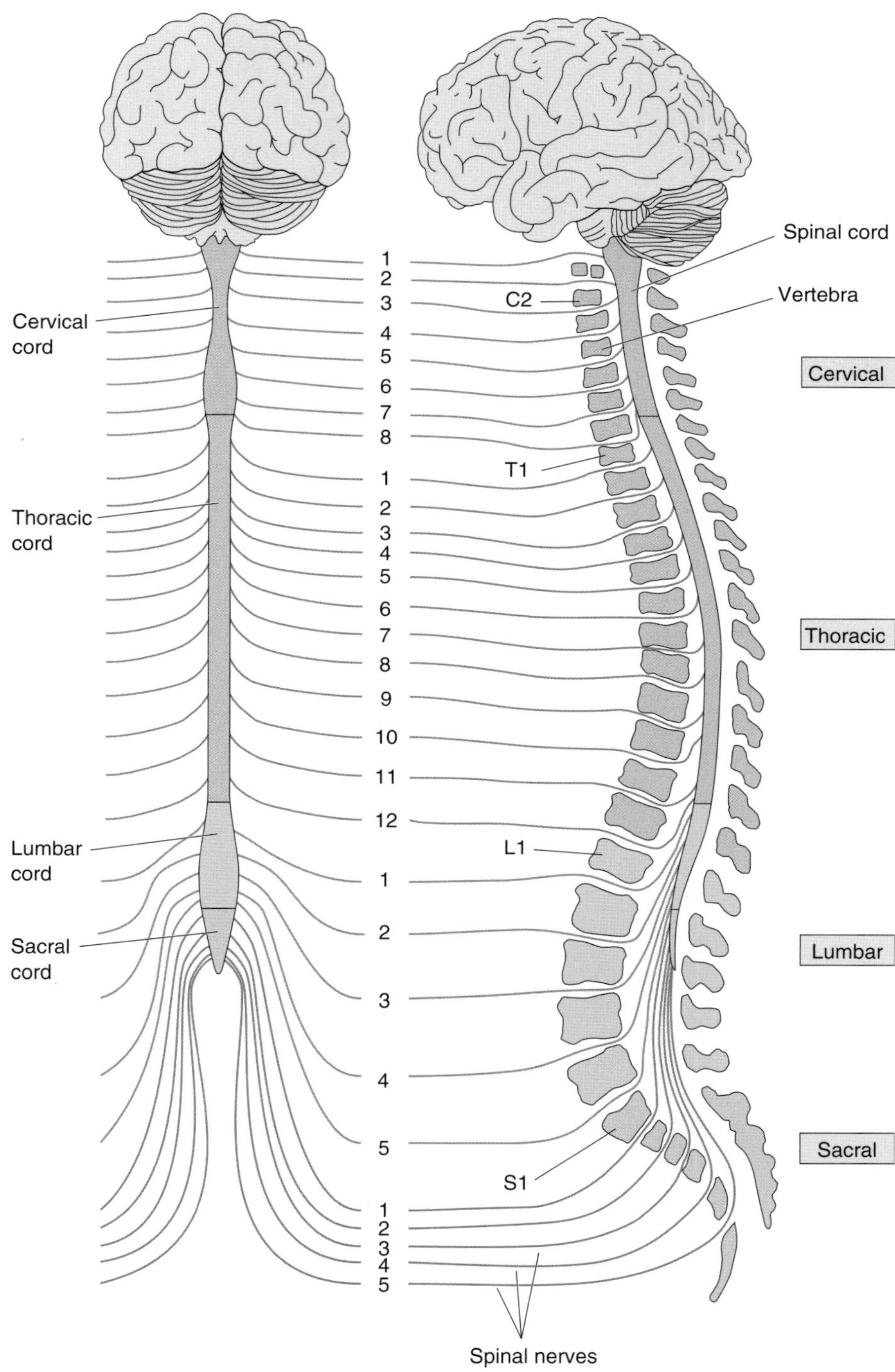

Figure 12.10
Segmental organization of the spinal cord. The spinal cord is divided into cervical, thoracic, lumbar, and sacral divisions (left). The right side shows the spinal cord within the vertebral column. Spinal nerves are named for the level of the spinal cord from which they exit and are numbered in order from rostral to caudal.

When mapped, the dermatomes delineate a set of bands on the body surface, as shown in Figure 12.11. The organization of the dermatomes is best revealed when one bends over to stand on both hands and feet (Figure 12.12). This organization presumably reflects our distant quadripedal ancestry.

When a dorsal root is cut, the corresponding dermatome on that side of the body does not lose all sensation. The residual somatic sensation is explained by the fact that the adjacent dorsal roots innervate overlapping areas. To lose all sensation in one dermatome, therefore, three adjacent dorsal roots must be cut. Skin innervated by the axons of one dorsal root, however, is plainly

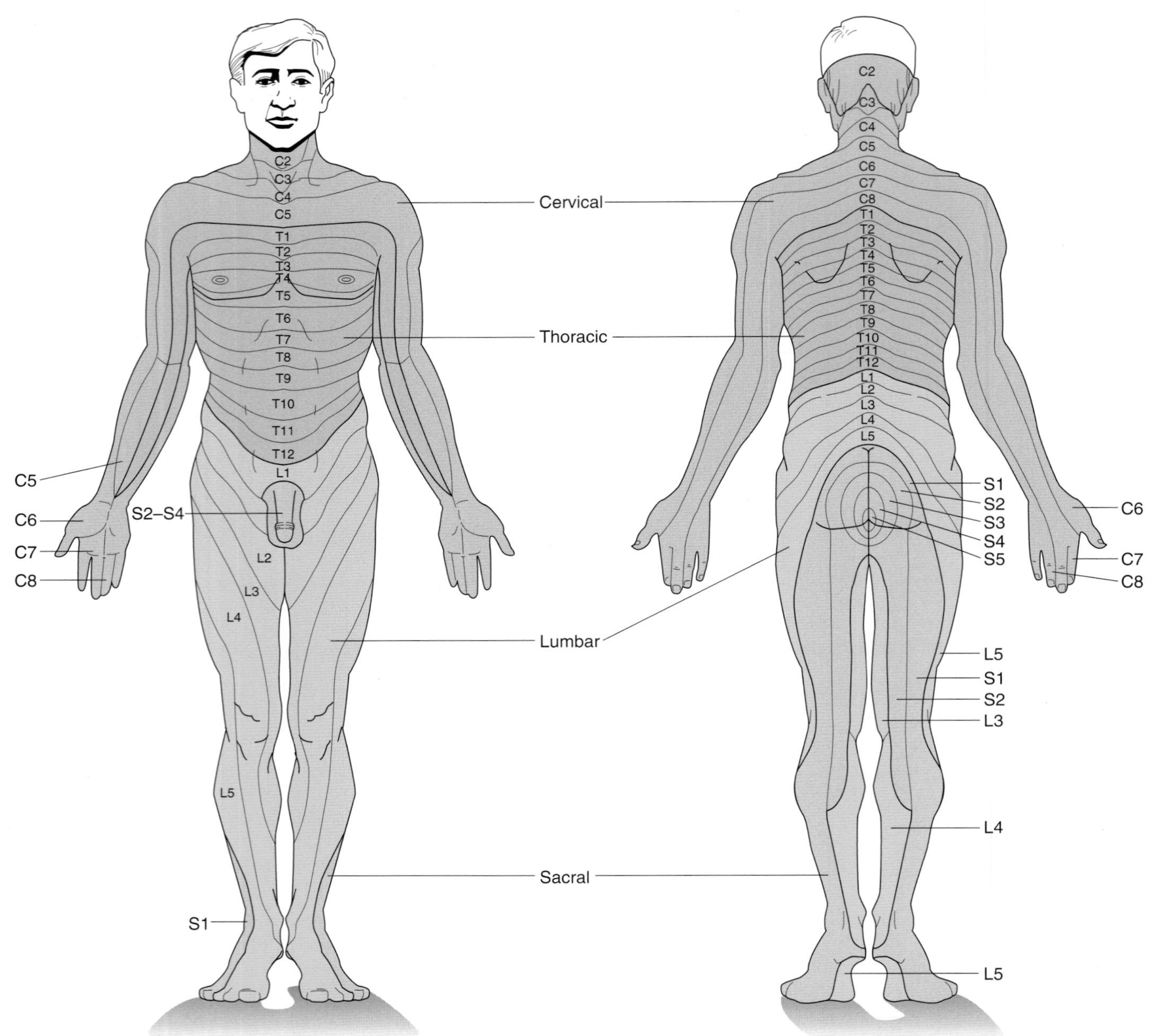

Figure 12.11
Dermatomes. This map shows the approximate boundaries of the dermatomes.

revealed by a condition called shingles, in which all of the neurons of a single dorsal root ganglion become infected with a virus (Box 12.1).

Figure 12.10 shows that the spinal cord in the adult ends at about the level of the third lumbar vertebra. The bundles of spinal nerves streaming down within the lumbar and sacral vertebral column are called the *cauda equina* (Latin for "horse's tail"). The cauda equina courses down the spinal column within a sac of dura, filled with CSF. In *lumbar puncture*, used to collect CSF for medical diagnostic tests, a needle is inserted into this CSF-filled cistern at the midline (this procedure is also called a spinal tap). If the needle is inserted a little off center, however, a nerve can be touched. Not surprisingly, this causes a sensation of sharp pain in the dermatome supplied by that nerve.

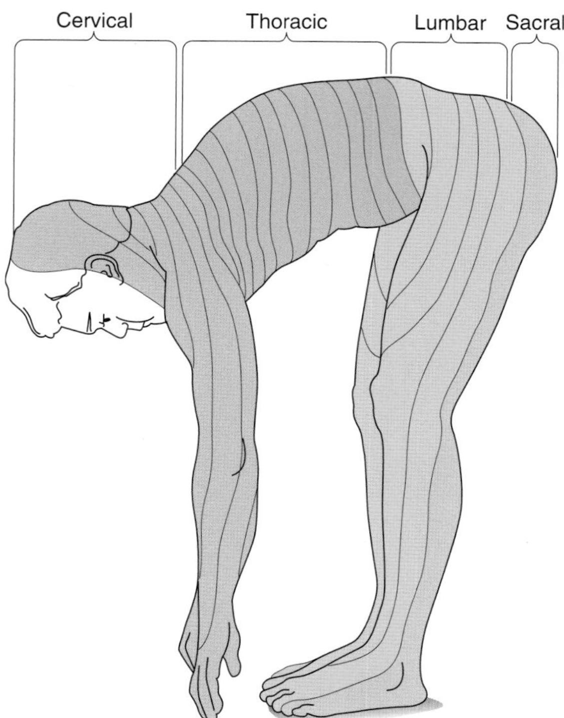

Figure 12.12
Dermatomes on all fours.

Sensory Organization of the Spinal Cord. The basic anatomy of the spinal cord was introduced in Chapter 7. The spinal cord is composed of an inner core of gray matter surrounded by a thick covering of white matter tracts that are often called columns. Each half of the spinal gray matter is divided into a *dorsal horn*, an *intermediate zone*, and a *ventral horn* (Figure 12.13). The neurons that receive sensory input from primary afferents are called *second-order sensory neurons*. Most of the second-order sensory neurons of the spinal cord lie within the dorsal horns.

The large, myelinated Aβ axons conveying information about a touch to the skin enter the dorsal horn and branch. One branch synapses in the deep part of the dorsal horn on second-order sensory neurons. These connections can initiate or modify a variety of rapid and unconscious reflexes. The other branch of the Aβ primary afferent axon ascends straight to the brain. This ascending input is responsible for perception, enabling us to form complex judgments about the stimuli touching the skin.

The Dorsal Column–Medial Lemniscal Pathway

Information about touch or vibration of the skin takes a path to the brain that is entirely distinct from that taken by information about pain and temperature. The pathway serving touch is called the **dorsal column–medial lemniscal pathway**, for reasons we will see in a moment. The organization of this pathway is summarized in Figure 12.14.

The ascending branch of the large sensory axons (Aβ) enters the ipsilateral **dorsal column** of the spinal cord, the white matter tract medial to the dorsal horn (Figure 12.13). The dorsal columns carry information about tactile sensation and limb position toward the brain. They are composed of primary sensory axons and second-order axons from neurons in the spinal gray matter. The axons of the dorsal column terminate in the **dorsal column nuclei**, which lie at the junction of the spinal cord and medulla. Consider that some of the axons terminating in the dorsal column nuclei at the base of your head

Box 12.1

OF SPECIAL INTEREST

Herpes, Shingles, and Dermatomes

As children, most of us were infected by the herpes zoster virus, commonly known as chickenpox. After a week or so covered with red, itchy spots on our skin, we recovered. Out of sight is not out of body, however. The virus remains in our primary sensory neurons, dormant but viable. Most people never hear from it again, but in some cases the virus revives decades later, wreaking havoc with the somatic sensory system. The result is *shingles*, a condition that can be agonizingly painful for months or even years. The reactivated virus increases the excitability of the sensory neurons, leading to very low thresholds of firing as well as spontaneous activity. The pain is a constant burning, sometimes a stabbing sensation, and the skin is exquisitely sensitive to any stimulus. Many people with shingles shun clothes because of their hypersensitivity. The skin itself becomes inflamed and blistered, then scaly—hence the name (Figure A).

Fortunately, the herpesvirus usually reactivates only in the neurons of one dorsal root ganglion. This means that the symptoms are restricted to the skin innervated by the axons of the affected dorsal root. In effect, the virus performs an anatomical labeling experiment for us by clearly marking the skin territory of one dermatome. Almost any dermatome may be involved, although the thoracic and facial areas are most common. Observations of many shingles patients and their infected areas were actually useful in mapping the dermatomes (Figure 12.11).

Neuroscientists are now learning to use herpesvirus and other viruses to their advantage. Viruses are useful research tools because they can be used to introduce new genes into neurons.

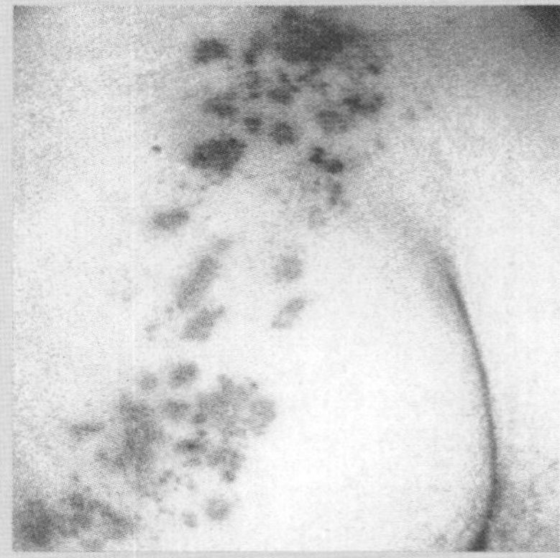

Figure A
Skin lesions caused by shingles, confined to the L4 dermatome on the left side. (Source: Fitzpatrick T, et al., 1997, p. 815.)

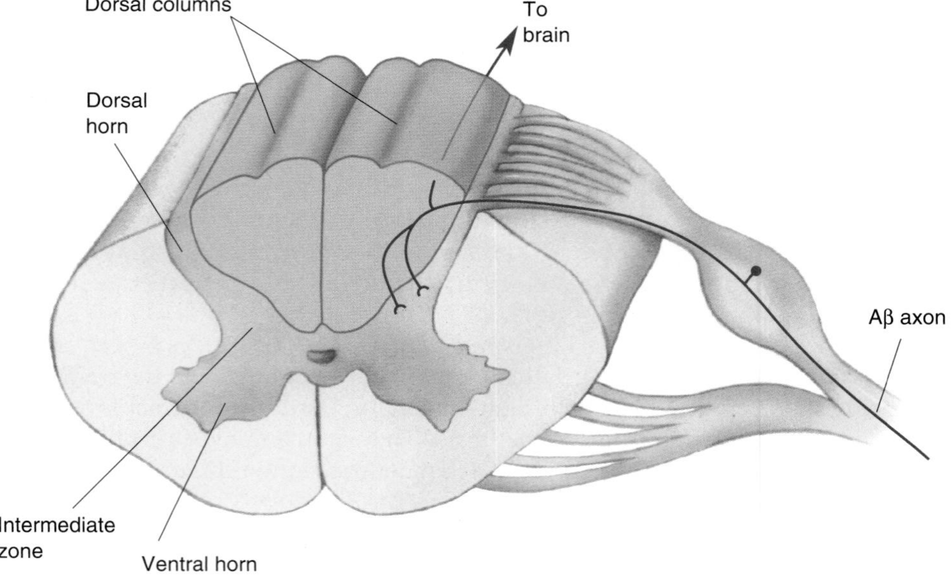

Figure 12.13
The route of the touch-sensitive Aβ axons in the spinal cord.

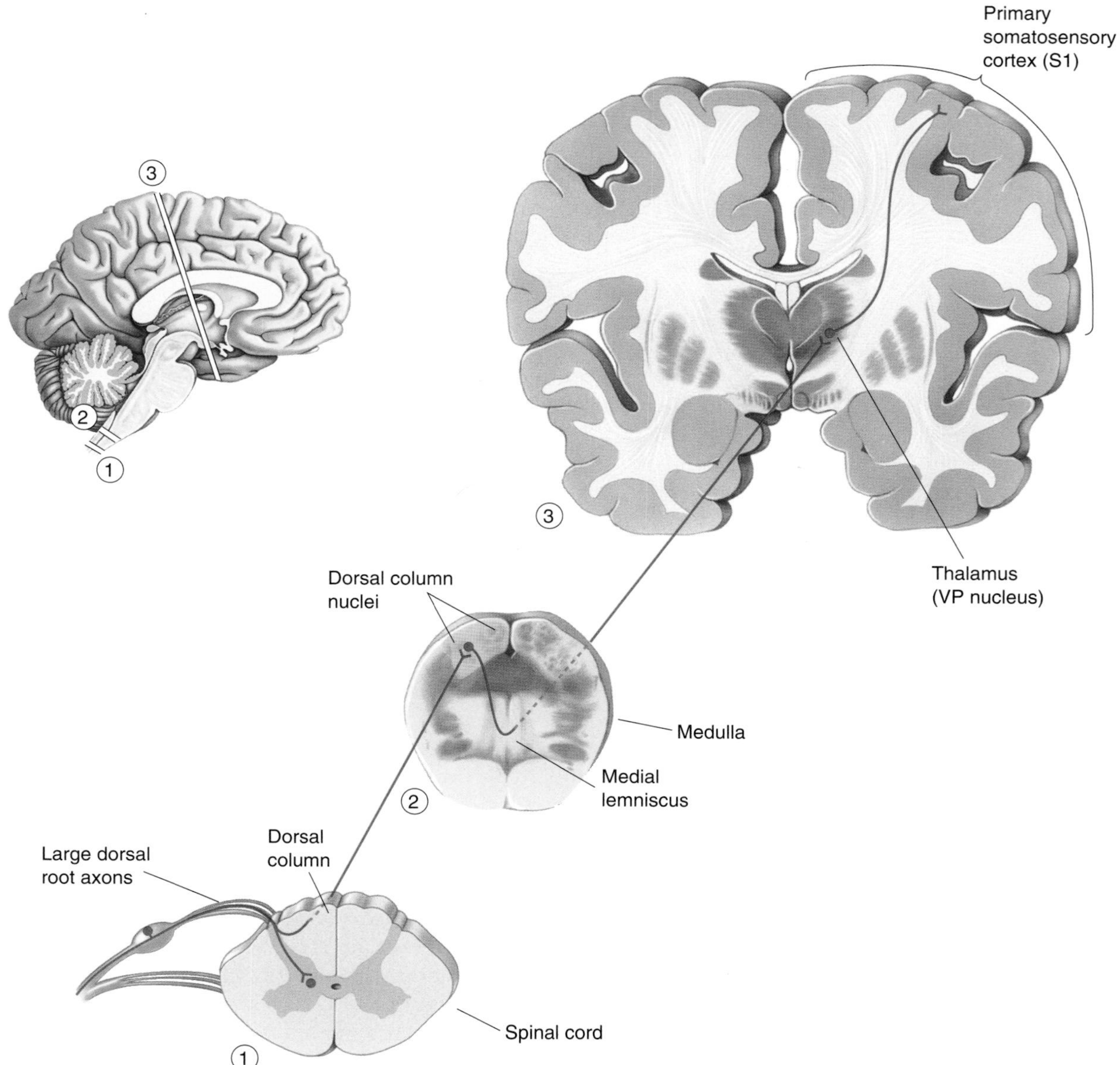

Figure 12.14

The dorsal column–medial lemniscal pathway. This is the major route by which touch and proprioceptive information ascends to the cerebral cortex.

originate as far away as the skin of your big toe! This is a fast, direct path that brings information from the skin to the brain without an intervening synapse.

At this point in the pathway, information is still represented ipsilaterally; i.e., touch information from the *right* side of the body is represented in the activity of cells in the *right* dorsal column nuclei. Axons from cells of the dorsal column nuclei, however, arch toward the ventral and medial medulla and decussate. *From this point onward, the somatic sensory system of one side of the brain is concerned with sensations deriving from the other side of the body.*

The axons of the dorsal column nuclei ascend within a conspicuous white matter tract called the **medial lemniscus**. The medial lemniscus rises through the medulla, pons, and midbrain, and its axons synapse upon neurons of the **ventral posterior (VP) nucleus** of the thalamus. Remember that no sensory information goes directly into the neocortex without first synapsing in the thalamus. Thalamic neurons of the VP nucleus then project to specific regions of **primary somatosensory cortex**, or **S1**.

It is tempting to assume that sensory information is simply transferred unchanged through nuclei in the brain stem and thalamus on its way to the cortex, with the actual processing taking place only in the cortex. In fact, this assumption is demonstrated by the term *relay nuclei*, which is often used to describe specific sensory nuclei of the thalamus such as the VP nucleus. Physiological studies prove otherwise, however. In both dorsal column and thalamic nuclei, considerable transformation of information takes place. As a general rule, information is altered every time it passes through a set of synapses in the brain. In particular, inhibitory interactions between adjacent sets of inputs in the dorsal column–medial lemniscal pathway enhance the responses to tactile stimuli (Box 12.2). As we see later in the chapter, some synapses in these nuclei can also change their strength, depending on their recent activity. Neurons of both the thalamus and the dorsal column nuclei are also controlled by input from the cerebral cortex. Accordingly, the output of the cortex can influence the input of the cortex!

The Trigeminal Touch Pathway

Thus far, we have described only the part of the somatic sensory system that enters the spinal cord. If this were the whole story, your face would be numb. Somatic sensation of the face is supplied mostly by the large **trigeminal nerves** (cranial nerve V), which enter the brain at the pons. (*Trigeminal* is from the Latin *tria*, "three," and *geminus*, "twin.") There are twin trigeminal nerves, one on each side, and each breaks up into three peripheral nerves that innervate the face, mouth areas, the outer two-thirds of the tongue, and the dura mater covering the brain. Additional sensation from the skin around the ears, nasal areas, and pharynx is provided by other cranial nerves: the facial (VII), glossopharyngeal (IX), and vagus (X).

The sensory connections of the trigeminal nerve are analogous to those of the dorsal roots. The large-diameter sensory axons of the trigeminal nerve carry tactile information from skin mechanoreceptors. They synapse onto second-order neurons in the ipsilateral trigeminal nucleus, which is analogous to a dorsal column nucleus (Figure 12.15). The axons from the trigeminal nucleus decussate and project into the medial part of the VP nucleus of the thalamus. From here, information is relayed to the somatosensory cortex.

Somatosensory Cortex

As with all other sensory systems, the most complex levels of somatosensory processing occur in the cerebral cortex. Most of the cortex concerned with the somatic sensory system is in the parietal lobe (Figure 12.16). Primary somatosensory cortex (S1) is easy to find in humans because it occupies an exposed cortical strip called the postcentral gyrus (right behind the central sulcus). Structurally, S1 consists of four distinct cortical areas, Brodmann's areas 3a, 3b, 1, and 2, counting from the central sulcus back. At the lateral end of S1, there is also a secondary somatosensory cortex (S2), which is revealed by pulling back the temporal lobe and peeking over the auditory cortex at the lower part of the parietal lobe. Finally, the **posterior parietal cortex,** consisting of areas 5 and 7, sits just posterior to S1.

Primary Somatosensory Cortex. S1 is the *primary* somatosensory cortex because (1) it receives dense inputs from the VP nucleus of the thalamus, (2) its neurons are very responsive to somatosensory stimuli (but not to other sen-

Box 12.2

BRAIN FOOD

Lateral Inhibition

Information is usually transformed as it is passed from one neuron to the next in a sensory pathway. One common transformation is the amplification of *differences* in the activity of neighboring neurons, also known as *contrast enhancement*. We have already seen this in retinal ganglion cell receptive fields (see Chapter 9). If all of the photoreceptors providing input to a ganglion cell are evenly illuminated, the cell hardly notices. If there is a contrast border—a *difference* in illumination—within the cell's receptive field, however, the cell's response is strongly modulated. Contrast enhancement is a general feature of information processing in sensory pathways, including the somatic sensory system. One general mechanism underlying contrast enhancement is *lateral inhibition*, whereby neighboring cells inhibit one another. Let's see how this works, using a simple model.

Consider the situation in Figure A. Dorsal root ganglion neurons a through i relay information to dorsal column nucleus neurons a′ through i′. Assume the output of the dorsal column nucleus cells is simply the presynaptic input multiplied by a synaptic gain factor of 1. If the input activity of cell a is 5, the output activity of cell a′ is also 5. This simple relay does nothing to enhance the difference between the more active neurons, d and f, and the others.

Now consider the situation in Figure B, with added inhibitory interneurons that project laterally to inhibit each cell's neighbor. Assume the synaptic gain of the excitatory synapses (white triangles) is 4, while the synaptic gain of the inhibitory synapses (black triangles) is −1. Calculate the activity of each cell by multiplying the input to each synapse by its synaptic gain and then summing the effect of all of the synapses on the cell. If you perform this calculation, you will see that there is significant contrast enhancement: The difference between the activity in cells d and f and their neighbors has been greatly amplified in the output of cells d′ and f′.

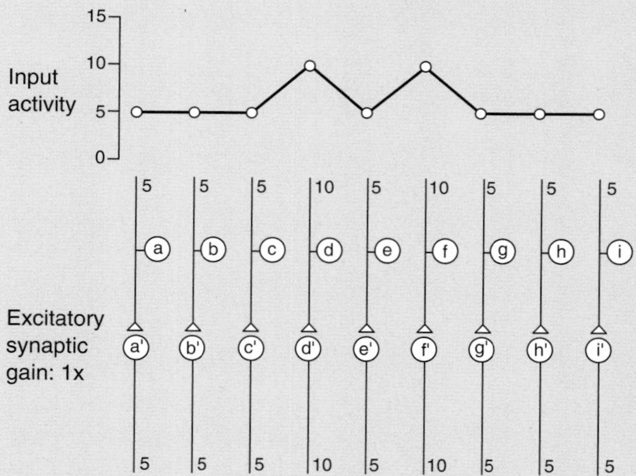

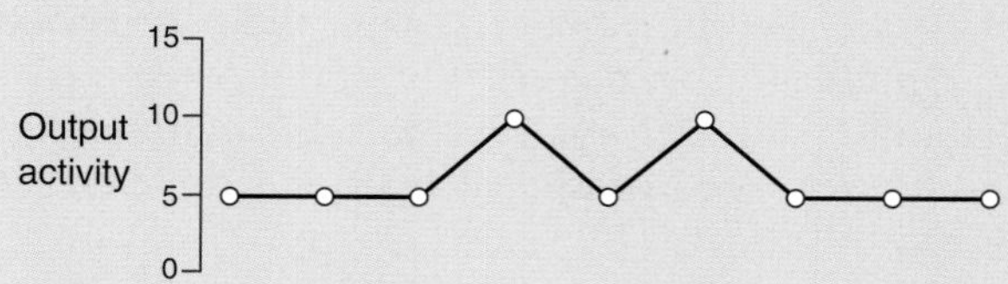

Figure A

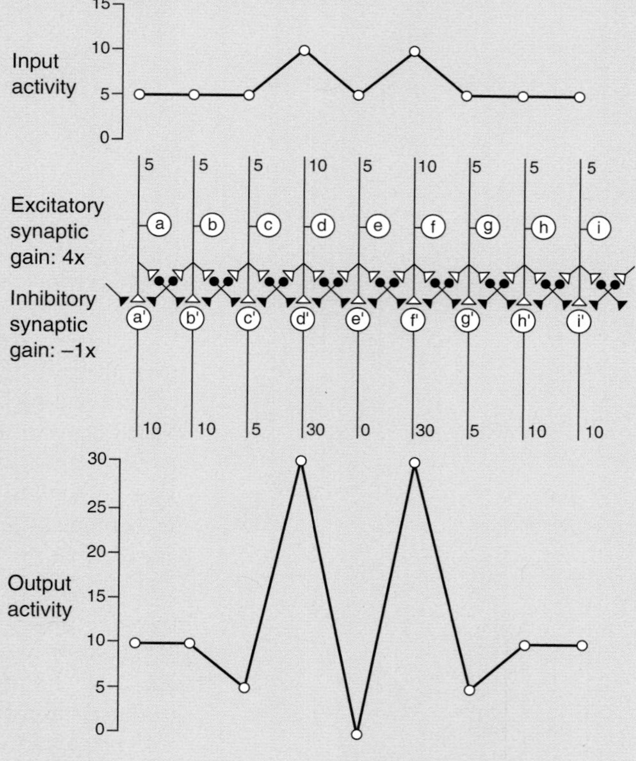

Figure B

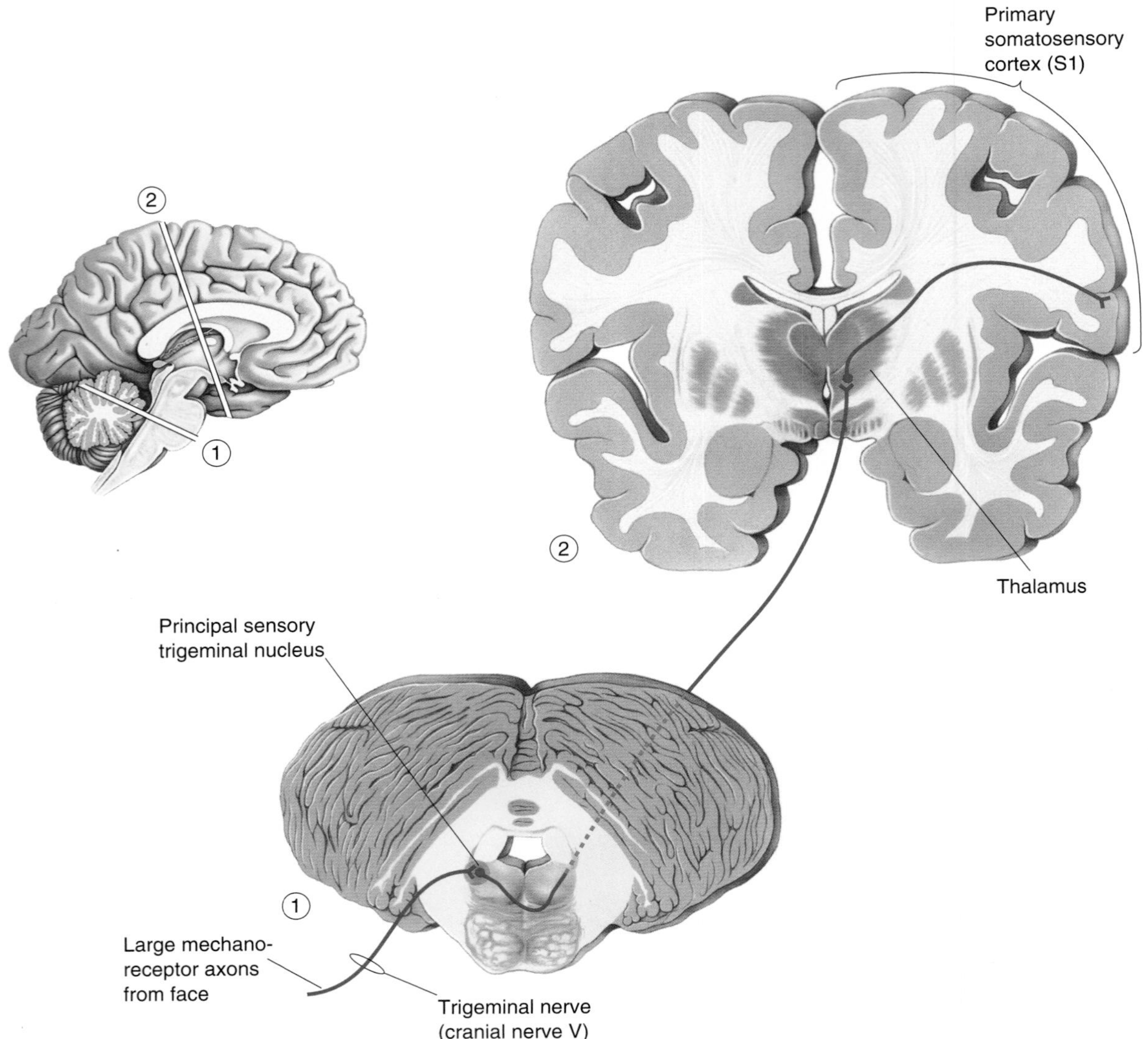

Figure 12.15
The trigeminal nerve pathway.

sory stimuli), (3) lesions in S1 impair somatic sensation, and (4) when electrically stimulated, it evokes somatic sensory experiences. Most inputs from the thalamus end in areas 3a and 3b, and these areas project to areas 1 and 2 and to S2; connections within the cortex are nearly always bidirectional, so axons also go back to areas 3a and 3b. These are more examples of cortical *association pathways* linking cortical areas, similar to the way area 17 is connected to area MT in the visual system.

The different areas of S1 have different functions. Area 3b is concerned mainly with the texture, size, and shape of objects. Its projection to area 1 sends mainly texture information, while its projection to area 2 emphasizes size and shape. Small lesions in area 1 or 2 produce predictable deficiencies in discrimination of texture, size, and shape.

Somatosensory cortex, like other areas of neocortex, is a layered structure. As is the case for visual and auditory cortex, the thalamic inputs to S1 terminate mainly in layer IV. The neurons of layer IV project, in turn, to the cells in the other layers. Another important similarity with other regions of cortex

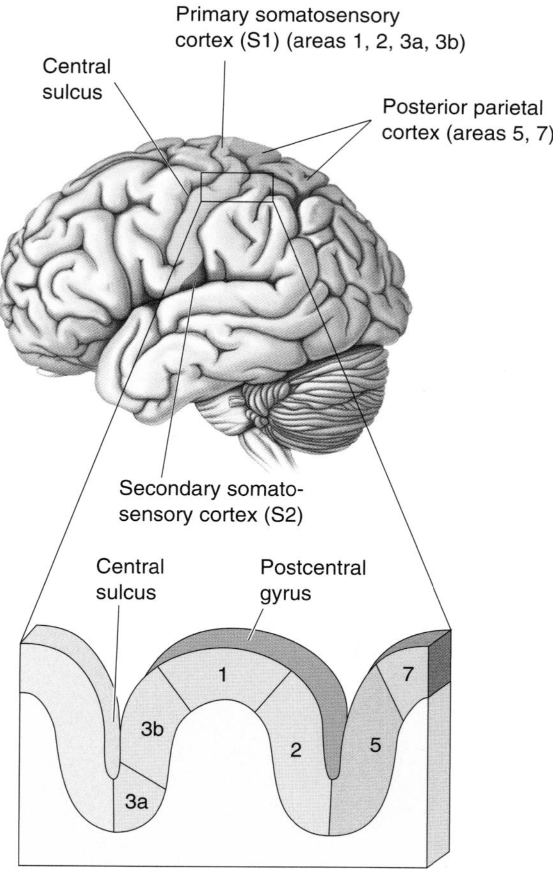

Figure 12.16
Somatic sensory areas of the cortex. All of the illustrated areas lie in the parietal lobe. The lower drawing shows that the postcentral gyrus contains S1, which consists of four cortical areas.

is that S1 neurons with similar inputs and responses are stacked vertically into columns that extend across the cortical layers (Figure 12.17). In fact, the concept of the cortical column, so beautifully elaborated by Hubel and Wiesel in visual cortex, was first described in somatosensory cortex by Johns Hopkins University scientist Vernon Mountcastle.

Cortical Somatotopy. Electrical stimulation of the S1 surface can cause somatic sensations in a specific part of the body. Systematically moving the stimulator around S1 will cause the sensation to move across the body. Canadian neurosurgeon Wilder Penfield, working at McGill University from the 1930s through the 1950s, actually used this method to map the cortex of neurosurgical patients. (These brain operations can be performed in awake patients with only local anesthesia of the scalp, because the brain tissue itself lacks the receptors of somatic sensation.) Another way to map the somatosensory cortex is to record the activity of a single neuron and determine the site of its somatosensory receptive field on the body. The receptive fields of many S1 neurons produce an orderly map of the body on the cortex. The mapping of the body's surface sensations onto a structure in the brain is called **somatotopy**. We have seen previously that the brain has maps of other sensory surfaces, such as the light-sensitive retina in the eye (*retinotopy*) and the frequency-sensitive cochlea in the inner ear (*tonotopy*).

Somatotopic maps generated by stimulating and recording methods are similar, roughly resembling a trapeze artist hanging upside down, his legs hooked over the top of the postcentral gyrus and dangling into the medial cortex between the hemispheres, and his head at the opposite, lower end of

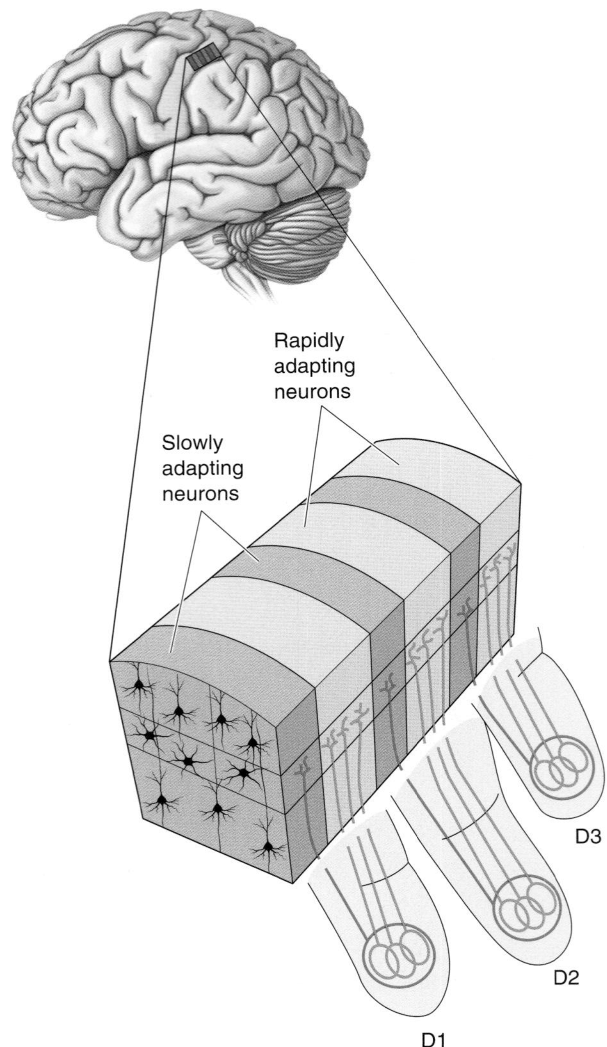

Figure 12.17
Columnar organization of S1's area 3b. Each finger (D1–D3) is represented by an adjacent area of cortex. Within the area of each finger representation are alternating columns of cells with rapidly adapting (green) and slowly adapting (red) sensory responses. (Source: Adapted from Kaas et al., 1981, Fig. 8.)

the gyrus (Figure 12.18). A somatotopic map is sometimes called a *homunculus* (from the Latin diminutive of "man"; the little man in the brain).

Two things are obvious about a somatotopic map. First, the map is not always continuous but can be broken up. Notice in Figure 12.18 that the representation of the hand separates that of the face and the head, while the genitals are mapped onto the most hidden part of S1, somewhere below the toes. Second, the map is not scaled like the human body. Instead, it looks like a caricature (Figure 12.19): The mouth, tongue, and fingers are absurdly large, while the trunk, arms, and legs are tiny. The relative size of cortex devoted to each body part is correlated with the *density* of sensory input received from that part. Size on the map is also related to the *importance* of the sensory input from that part of the body; information from your index finger is more useful than that from your elbow. The importance of touch information from our hands and fingers is obvious, but why throw so much cortical computing power at the mouth? Two likely reasons are that tactile sensations are important in the production of speech; and your lips and tongue (feeling, as well as tasting) are the last line of defense when deciding whether a morsel is delicious, nutritious food or something that could choke you, break your tooth, or bite back. As we will see in a moment, the importance of an input and the size of its representation in cortex also reflect how often it is used.

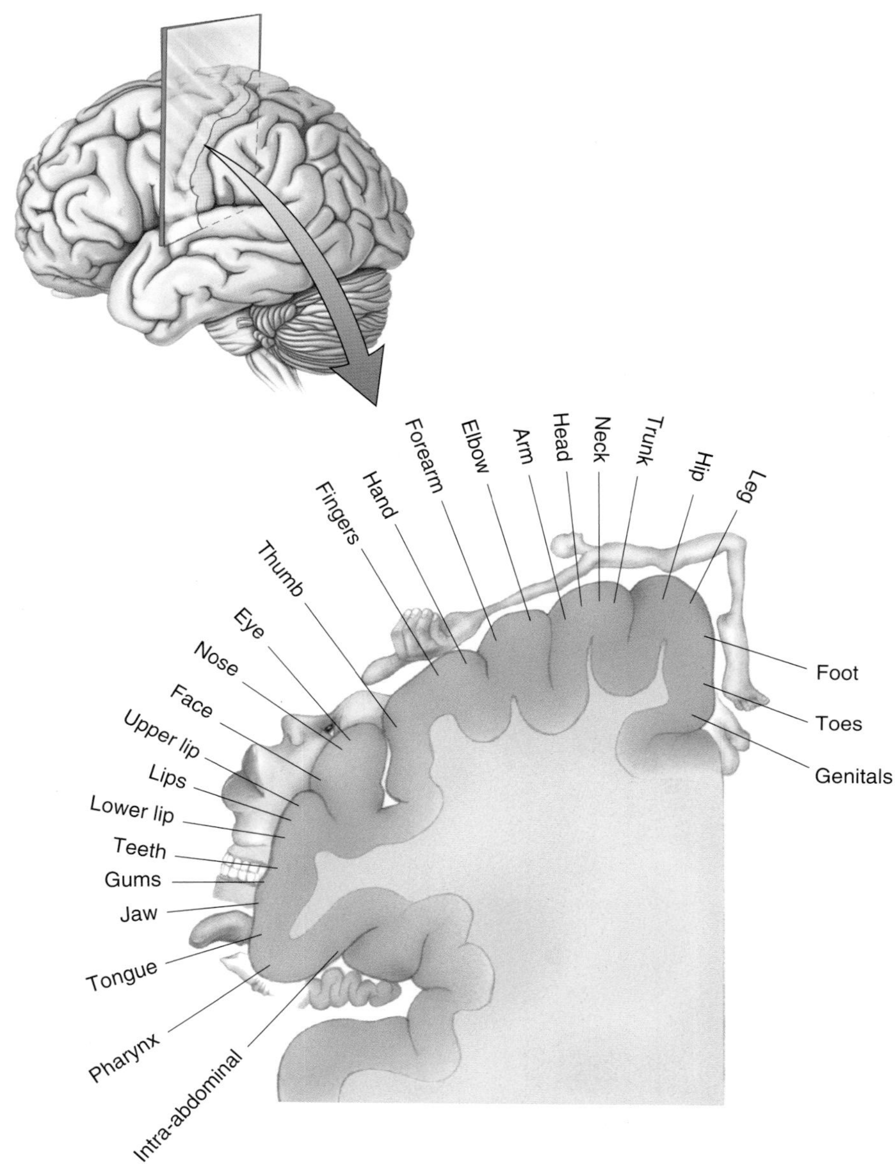

Figure 12.18
A somatotopic map of the body surface onto primary somatosensory cortex. This map is a cross section through the postcentral gyrus (shown at top). Neurons in each area are most responsive to the parts of the body illustrated above them. (Source: Adapted from Penfield and Rasmussen, 1952.)

The importance of a body part can vary greatly among species. For example, the large facial vibrissae of rodents receive a huge share of the territory in S1, while the digits of the paws receive relatively little (Figure 12.20). Remarkably, the sensory signals from each vibrissa follicle go to one clearly defined cluster of S1 neurons; such clusters are called *barrels*. The somatotopic map of rodent vibrissae is easily seen in thin sections of S1; the five rows of cortical barrels precisely match the five rows of facial vibrissae.

Somatotopy in the cerebral cortex is not limited to a single map. Just as the visual system builds multiple retinotopic maps, the somatic sensory system has several maps of the body. Figure 12.21 shows the detailed somatotopy of S1 in an owl monkey. Carefully compare the maps in areas 3b and 1; they map the same parts of the body, literally in parallel along adjacent strips of cortex. The two somatotopic maps are not identical but are mirror images, as an enlargement of the hand regions makes clear (Figure 12.21b). As described earlier, areas 3b and 1 are also concerned with different types of somatotopic information.

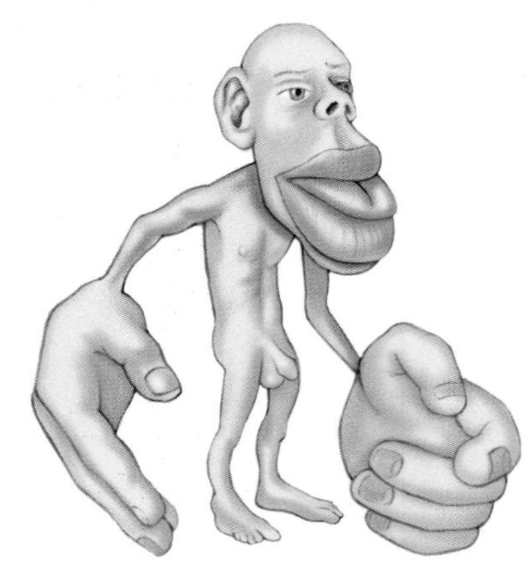

Figure 12.19
The homunculus.

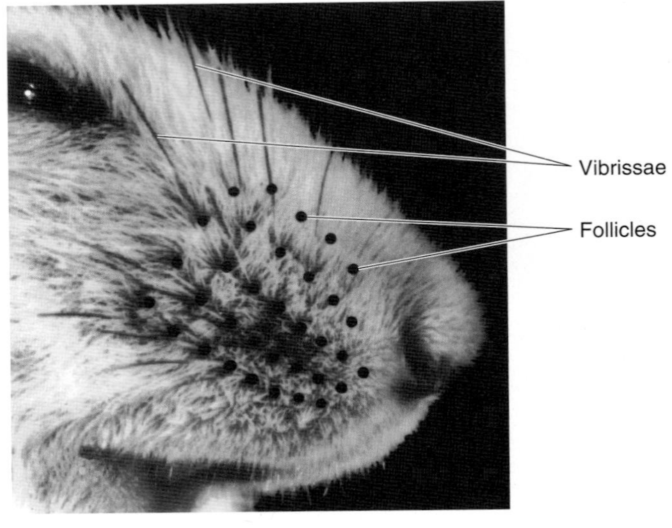

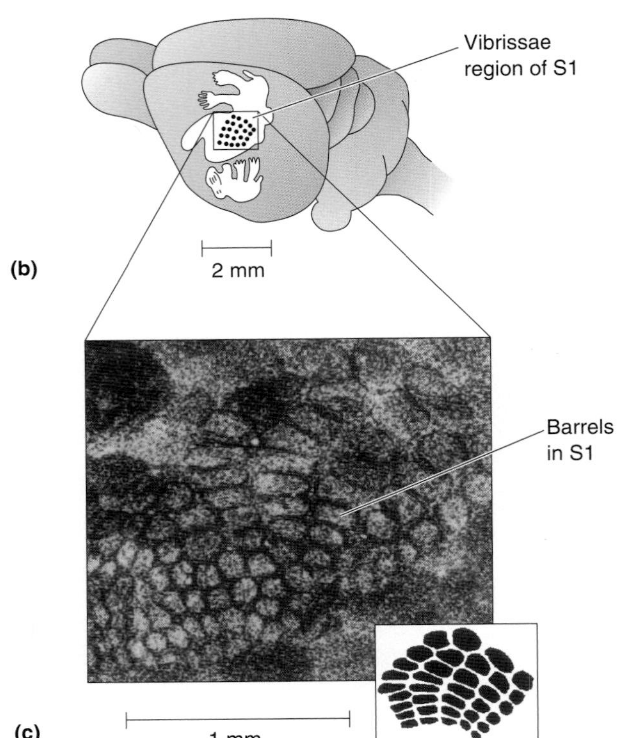

Figure 12.20
A somatotopic map of the facial vibrissae on mouse cerebral cortex. **(a)** The position of the major vibrissae on the face. **(b)** A somatotopic map within S1 of the mouse brain. **(c)** Barrels within S1. The cortex has been thinly sectioned parallel to the surface and Nissl-stained. The *inset* shows the pattern of barrels, laid out in five rows; compare with the five rows of vibrissae shown in part a. (Source: Adapted from Woolsey and Van der Loos, 1970.)

Cortical Map Plasticity. What happens to the somatotopic map in cortex when an input, such as the finger, is removed? Does the "finger area" of cortex simply go unused? Does it atrophy? Or is this tissue taken over by inputs from other sources? The answers to these questions may have important implications for the recovery of function after peripheral nerve injury. In the 1980s, University of California at San Francisco neuroscientist Michael Merzenich and his associates began a series of experiments to test the possibilities (Box 12.3).

The key experiments are summarized in Figure 12.22. First, the regions of S1 sensitive to stimulation of the hand in an adult owl monkey were carefully mapped with microelectrodes. Then, one finger (digit 3) on the hand was sur-

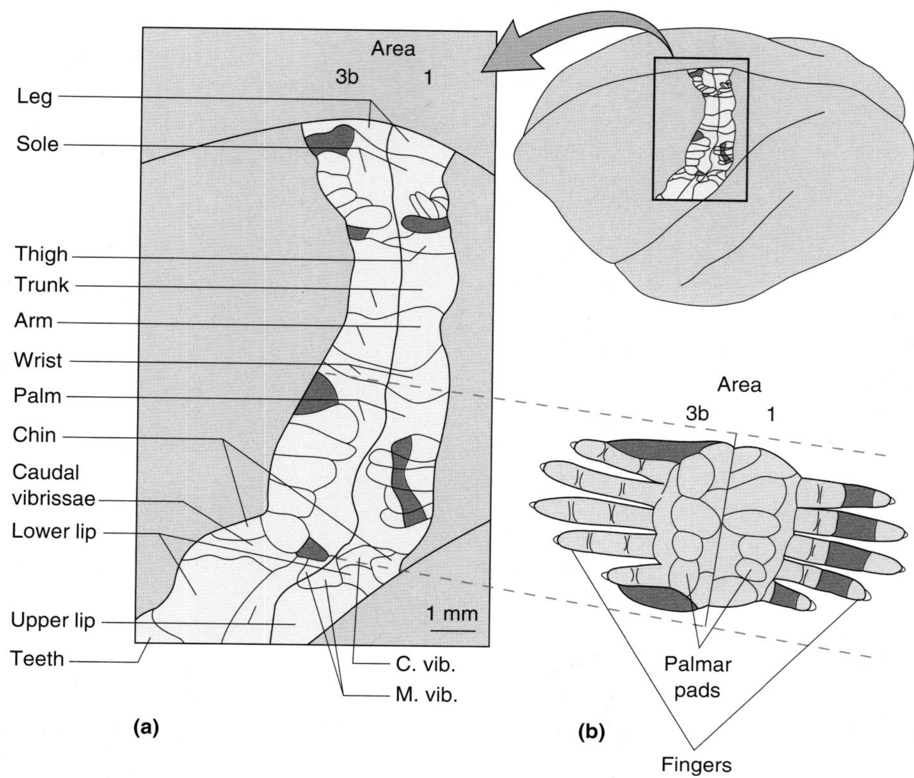

Figure 12.21
Multiple somatotopic maps in S1. Recordings were made from areas 3b and 1 of an owl monkey. **(a)** Results show that each area has its own somatotopic map. **(b)** Detailed examination of the hand area shows that the two maps are mirror images. Shaded regions indicate the dorsal surfaces of the hands and feet; unshaded regions, the ventral surfaces. (Source: Adapted from Kaas et al., 1981.)

gically removed. Several months later, the cortex was again mapped. The answer? The cortex originally devoted to the amputated digit now responded to stimulation of the adjacent digits (Figure 12.22c). There clearly had been a major rearrangement of the circuitry underlying cortical somatotopy.

In the amputation experiment, the cause of this map rearrangement was the absence of input from the missing digit. What happens when the input activity from a digit is *increased*? To answer this question, monkeys were trained to use selected digits to perform a task for which they received a food reward. After several weeks of this training, microelectrode mapping experiments showed that the representation of the stimulated digits had expanded in comparison with the adjacent, unstimulated ones (Figure 12.22d). These experiments reveal that cortical maps are dynamic, adjusting to the amount of sensory experience. Subsequent experiments in other areas of cortex (visual, auditory, motor) have shown that this type of map plasticity is widespread in the brain.

The findings of map plasticity in animals have led to a search for similar changes in the human brain. One interesting example comes from studies of amputees. A common experience among amputees is the perception of sensations coming from the missing limb when other body parts are touched. These "phantom limb" sensations are usually evoked by the stimulation of skin regions whose somatotopic representations border those of the missing limb. For example, feeling can be evoked in a phantom arm by stimulating the face. Functional brain imaging reveals that the cortical regions originally devoted to the missing limb are now activated by stimulating the face. While this plasticity may be adaptive in the sense that the cortex does not go unused, the mismatch between sensory stimulation and perception in amputees shows that it can lead to confusion on how signals from S1 should be interpreted.

Box 12.3

PATH OF DISCOVERY

The Power of Mutable Maps

BY MICHAEL MERZENICH

In the modern scientific era, most great insights come from the accumulation of information by many scientists. Virtually every great contemporary discovery is appropriately shared. Still, the great fun of neuroscience for me has come from those times, usually extending over several weeks or months, when a broad new scientific or medical vista opens up in front of my eyes because of experimental results achieved in my own laboratory.

One of those epiphanies came with the realization that the somatotopic map in the cortex of adult monkeys was completely reorganized in response to a peripheral nerve lesion. The dogma at the time was that sensory representations in the adult cortex were highly resistant to change in the adult brain. By studying the time course of reorganization following lesions, however, we discovered that extremely rapid changes were followed by dramatic, progressive, longer-term representational refinements. It was exciting for us to see very different details in the cortical representation of the skin surface, recorded over separate periods, in both reorganized and normal, always-innervated representational zones.

It took us a while to realize that with these latter observations, we had repeated in another form an experiment initially conducted by Karl Lashley and reported in 1923! We thought the changes recorded after peripheral input modification almost certainly derived from plasticity processes that normally shaped cortical representations and processing capabilities in the adult brain during learning. This idea quickly led to the direct demonstration that somatosensory cortical representations could be altered in predictable ways in adult primates by behaviorally engaging restricted skin surfaces. We found that behavioral training using sounds elicited an analogous reorganization in the tonotopic representation in the auditory cortex. Cortical representations were thereby demonstrated to be malleable reflections of sensory experience.

Michael Merzenich

This discovery eventually led to another wonderful epiphany, this time produced by studying a 6-year-old boy. The child had the language ability of a 30-month-old child, had very limited conversational skills, and was easily distracted. The problem seemed to stem from deficient complex signal-processing capabilities. He was given intensive listening training, using tools Bill Jenkins and I created based on what we had learned in our cortical plasticity experiments. Four weeks later, when tested in Dr. Paula Tallal's lab at Rutgers University, his language abilities had advanced to an age-appropriate level, his phonological processing abilities appeared to be more than adequate for developing normal reading, he was under attentional control, and he literally jabbered to anyone who would listen! With this single child, my colleagues and I realized that neuroscience-based training could change the educational outcomes and positively contribute to the self-esteem of millions of language- and reading-impaired children.

These periods in the life of one experimental scientist and his collaborators manifest what is the most remarkable quality of our brain: its capacity to develop and to specialize its own processing machinery, to shape its own abilities, and to enable, through hard brainwork, its own achievements.

While having more cortex devoted to a body part may not necessarily be beneficial to amputees, it apparently is to musicians. Violinists and other stringed instrument players must continually finger the strings with their left hand; the other hand, holding the bow, receives considerably less stimulation of individual fingers. Functional imaging of S1 shows that the amount of cortex devoted to the fingers of the left hand is greatly enlarged in string musicians. It is likely that this is an exaggerated version of a continuous remapping process that goes on in everyone's brain as their life experiences vary.

The mechanisms of these types of map plasticity are not understood. As

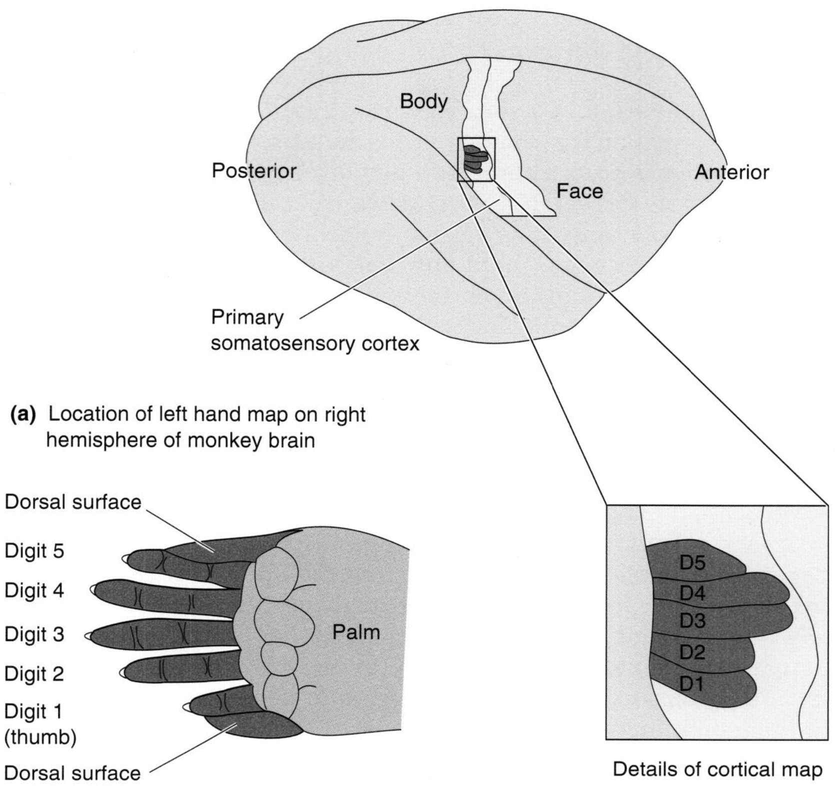

(a) Location of left hand map on right hemisphere of monkey brain

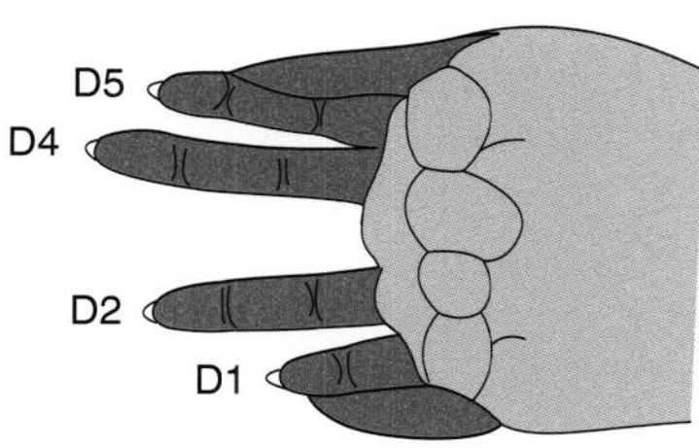

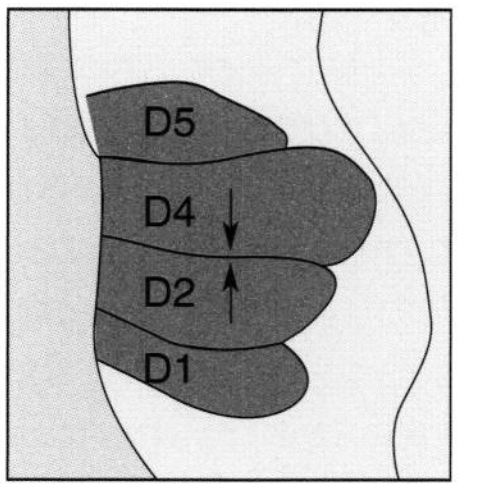

Details of cortical map

(b) Normal hand, palmar surface

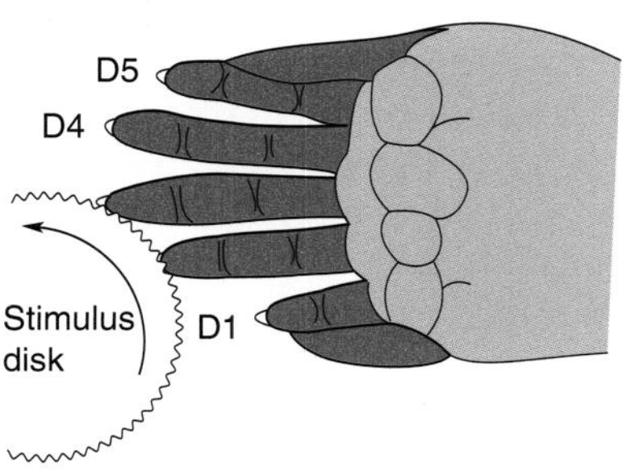

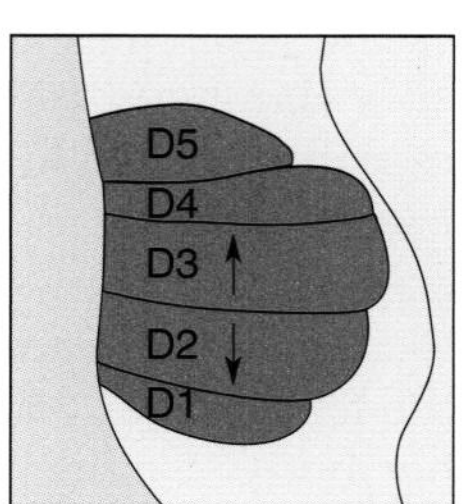

(c) Reorganization of cortical map after surgical removal of third finger (D3)

After reorganization of somatosensory cortex

(d) Reorganization of cortical map after discrimination training of two fingertips

Figure 12.22
Somatotopic map plasticity. (a, b) The fingers of the hand of an owl monkey are mapped onto the surface of S1 cortex. **(c)** If digit 3 is removed, over time the cortex reorganizes so that the representations of digits 2 and 4 expand. **(d)** If digits 2 and 3 are selectively stimulated, their cortical representations also expand.

we shall see in Chapter 24, however, they may be related to processes involved in learning and memory.

The Posterior Parietal Cortex. As we have seen, the segregation of different types of information is a general rule for the sensory systems, and the somatic sensory system is no exception. Information of different sensory types cannot remain separate forever, however. When we feel for a key in our pocket, we do not ordinarily sense it as a list of traits: a particular size and shape, textured and smooth edges, hard and smooth flat surfaces, a certain weight. Instead, without thinking much about it, we simply confirm with our fingers "key," as opposed to "coin" or "wad of old chewing gum." Separate aspects of a stimulus come effortlessly together as a meaningful object. We have a very poor understanding of how this occurs biologically within any sensory system, much less between sensory systems. After all, many objects have a distinct look, sound, feel, *and* smell, and the melding of these sensations is necessary for the complete mental image of something like your pet cat.

What we do know is that the character of neuronal receptive fields tends to change as information passes through the cortex and receptive fields enlarge. For example, neurons below the cortex and in cortical areas 3a and 3b are not sensitive to the direction of stimulus movement across the skin, but cells in areas 1 and 2 are. The stimuli that neurons prefer become increasingly complex. Certain cortical areas seem to be sites where simple, segregated streams of sensory information converge to generate particularly complex neural representations. When we discussed the visual system, we saw this in the complex receptive fields of area IT. The posterior parietal cortex is also such an area. Its neurons have large receptive fields, with stimulus preferences that are a challenge to characterize because they are so elaborate. Moreover, the area is concerned not only with somatic sensation but also with visual stimuli, movement planning, and even a person's state of attentiveness.

Damage to posterior parietal areas can yield some bizarre neurological disorders. Among these is **agnosia**, the inability to recognize objects even though simple sensory skills seem to be normal. People with *astereognosia* cannot recognize common objects by feeling them (a key, for example), though their sense of touch is otherwise normal and they may have no trouble recognizing the object by sight or sound. Deficits are often limited to the side contralateral to the damage.

Parietal cortical lesions may also cause a **neglect syndrome**, in which a part of the body or a part of the world (the entire visual field left of the center of gaze, for example) is ignored or suppressed and its very existence is denied (Figure 12.23). Neurologist Oliver Sacks described such a patient in his essay, "The Man Who Fell Out of Bed." After suffering a stroke that presumably damaged his cortex, the man insisted that someone was playing a macabre joke on him by hiding an amputated leg under his blanket. When he tried to remove the leg from his bed, he and the leg ended up on the floor. Of course, the leg in question was his own, still attached, but he was unable to recognize it as part of his body. A patient with neglect syndrome may ignore the food on half of his or her plate or attempt to dress only one side of his or her body. Neglect syndromes are most common following damage to the right hemisphere, and happily, they usually improve or disappear with time.

In general, the posterior parietal cortex seems to be essential for the perception and interpretation of spatial relationships, accurate body image, and the learning of tasks involving coordination of the body in space. This involves a complex integration of somatosensory information with that from other sensory systems, particularly the visual system.

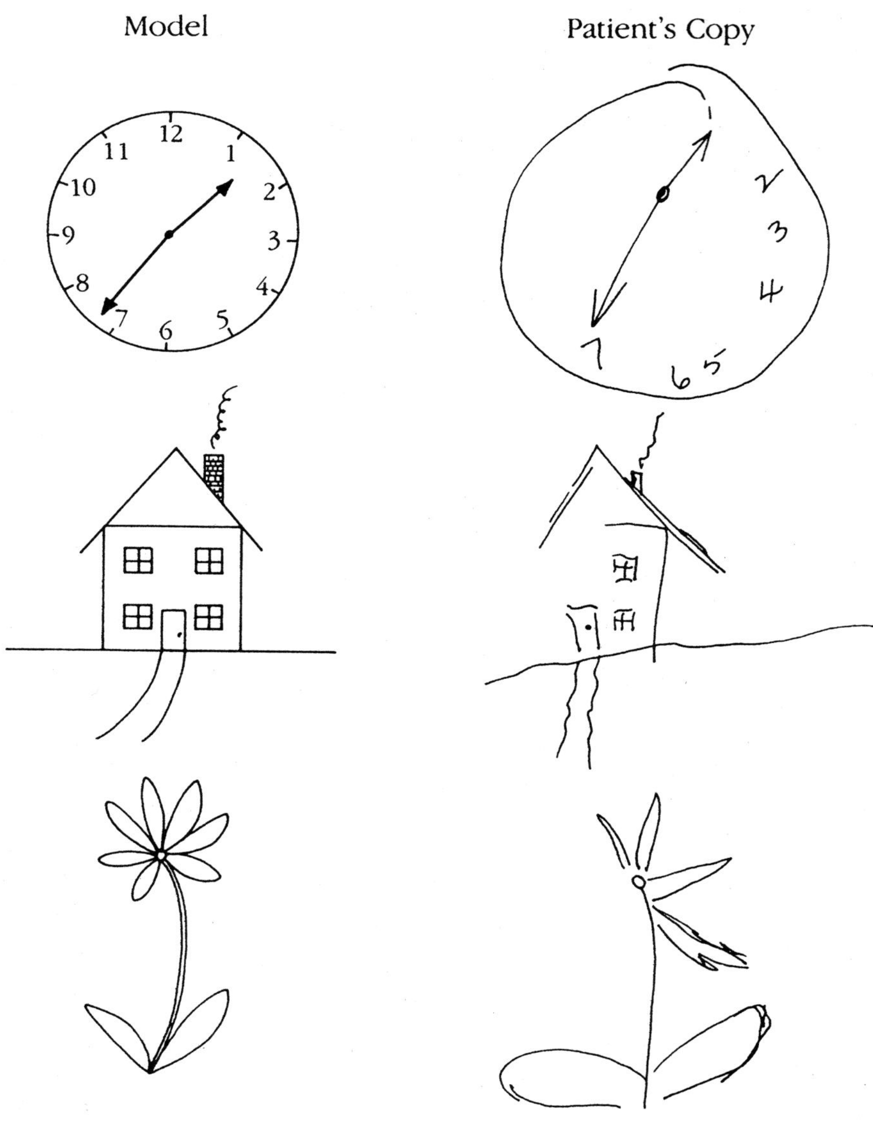

Figure 12.23
An example of a neglect syndrome. A patient who had had a stroke in the posterior parietal cortex was asked to copy the model drawing but was unable to reproduce many of the features on the left side of the model. (Source: Springer and Deutsch, 1989, p. 193.)

PAIN

Besides mechanoreceptors, somatic sensation depends strongly on **nociceptors**, the free, branching, unmyelinated nerve endings that signal that body tissue is being damaged or is at risk of being damaged. (The word is from the Latin *nocere*, "to hurt.") The information from nociceptors takes a path to the brain that is largely distinct from the path taken by mechanoreceptors; consequently, the subjective experience elicited by activation of these two pathways is different. Selective activation of nociceptors can lead to the conscious experience of pain. Nociception and pain are vital to life (Box 12.4).

Nociception and pain are not always the same thing, however. *Pain* is the feeling or perception of irritating, sore, stinging, aching, throbbing, miserable, or unbearable sensations arising from a part of the body. *Nociception* is the sensory process that provides the signals that trigger pain. While nociceptors may fire away wildly and continually, pain may come and go. The opposite may also happen. Pain may be agonizing even without activity in nociceptors. More than any other sensory system, the cognitive qualities of nociception can be controlled from within, by the brain itself.

Box 12.4

OF SPECIAL INTEREST

The Misery of Life Without Pain

Pain teaches us to avoid harmful situations. It elicits withdrawal reflexes from noxious stimuli. It exhorts us to rest an injured part of our body. Pain is vital. The most convincing arguments for this are the very rare people who are born without the sensation of pain. They go through life in constant danger of destroying themselves, because they do not realize the harm they are doing. Many of them die young.

A Canadian woman born with an indifference to painful stimuli had no other sensory deficits and was quite intelligent. Despite early training to avoid damaging situations, she developed progressive degeneration of her joints and spinal vertebrae, leading to skeletal deformation, degeneration, infection, and, finally, death at age 28.

Apparently, low levels of nociceptive activity are important during everyday tasks to tell us when a particular movement or prolonged posture is putting too much strain on our body. Even during sleep, nociception may be the prod that makes us toss and turn enough to prevent bedsores or skeletal strain.

People with a congenital absence of pain reveal that pain is a separate sensation, not simply an excess of the other sensations. Such people usually have a normal ability to perceive other somatic sensory stimuli. The causes can include the failure of peripheral nociceptors to develop and altered synaptic transmission in the pain-mediating pathways of the CNS. In any case, life without pain is *not* a blessing.

Nociceptors and the Transduction of Painful Stimuli

Nociceptors are activated by stimuli that have the potential to cause tissue damage. Tissue damage can result from strong mechanical stimulation, extremes in temperature, oxygen deprivation, and exposure to certain chemicals, among other causes. The membranes of nociceptors contain ion channels that are activated by these types of stimuli.

Consider as an example the events that accompany your stepping on a thumbtack (recall Chapter 3). The simple stretching or bending of the nociceptor membrane activates mechanically gated ion channels that cause the cell to depolarize and generate action potentials. In addition, damaged cells at the site of injury can release a number of substances that cause ion channels on nociceptor membranes to open. Examples of released substances are proteases (enzymes that digest proteins), adenosine triphosphate (ATP), and K^+ ions. Proteases can break down an abundant extracellular peptide called kininogen to form another peptide called *bradykinin*. Bradykinin binds to specific receptor molecules that activate ionic conductances in some nociceptors. Similarly, ATP causes nociceptors to depolarize by binding directly to ATP-gated ion channels. And as we learned back in Chapter 3, the elevation of extracellular $[K^+]$ directly depolarizes neuronal membranes.

Now consider leaning against a hot stove. Heat above 45°C burns tissue, and heat-sensitive ion channels in nociceptor membranes open at this temperature. Of course, we also have nonpainful sensations of warmth when the skin is heated at 37 to 45°C. These sensations depend on non-nociceptive thermoreceptors and their CNS connections, which we will discuss in a later section. But for now, note that the sensations of warmth and scalding are mediated by separate neural mechanisms.

Imagine you are a middle-aged runner on the last mile of a marathon. When your tissue oxygen levels do not meet the oxygen demand, your cells use anaerobic metabolism to generate ATP. A consequence of anaerobic metabolism is the release of lactic acid. The buildup of lactic acid leads to an excess of H^+ ions in the extracellular fluid, and these ions activate H^+-gated ion channels on nociceptors. This mechanism is responsible for the excruciating dull ache associated with very hard exercise.

A bee stings you. Your skin and connective tissue contain *mast cells*, a component of your immune system. Mast cells can be activated by exposure to foreign substances (e.g., bee venom), causing them to release histamine. Histamine binds to specific cell surface receptors on nociceptors and causes membrane depolarization. Histamine also causes blood capillaries to become leaky, which leads to swelling and redness at the site of an injury. Creams containing drugs that block histamine receptors (antihistamines) can be helpful for both the pain and the swelling.

Types of Nociceptors. The transduction of painful stimuli occurs in the free nerve endings of unmyelinated C fibers and lightly myelinated Aδ fibers. Most nociceptors respond to mechanical, thermal, and chemical stimuli and are therefore called *polymodal nociceptors*. Like mechanoreceptors, however, many nociceptors show selectivity in their responses to different stimuli. Thus, there are also *mechanical nociceptors* that show selective responses to strong pressure, *thermal nociceptors* that show selective responses to burning heat or extreme cold, and *chemical nociceptors* that show selective responses to histamine and other chemicals. Indeed, researchers recently discovered that the smallest of C fibers (conduction velocity of $\leq$0.5 m/sec) are selectively responsive to histamine and cause the perception of itch.

Nociceptors are present in most body tissues, including skin, bone, muscle, most internal organs, blood vessels, and the heart. They are notably absent in the brain itself, except for the meninges.

Hyperalgesia. Nociceptors normally respond only when stimuli are strong enough to damage tissue. But we all know that skin, joints, or muscles that have *already* been damaged or inflamed are unusually sensitive. A light, sympathetic mother's touch to a burned area of her child's skin may elicit howls of excruciating pain. This phenomenon is called **hyperalgesia**, and it is the most familiar example of our body's ability to control its own pain. Hyperalgesia can be a reduced threshold for pain, an increased intensity of painful stimuli, or even spontaneous pain. *Primary hyperalgesia* occurs within the area of damaged tissue, but tissues surrounding a damaged area may also become supersensitive by the process of *secondary hyperalgesia.*

Many different mechanisms appear to be involved in hyperalgesia, some in and around the peripheral receptors and others within the CNS. As noted earlier, a variety of substances are released when skin is damaged. A number of these chemicals modulate the excitability of nociceptors, making them more sensitive to thermal or mechanical stimuli. Examples of sensitizing chemicals are bradykinin, prostaglandins, and substance P (Figure 12.24).

Bradykinin was discussed earlier as one of the chemicals that directly depolarize nociceptors. In addition to this effect, bradykinin stimulates long-lasting intracellular changes that make heat-activated ion channels more sensitive. *Prostaglandins* are chemicals generated by the enzymatic breakdown of lipid membrane. While prostaglandins do not elicit overt pain, they do greatly increase the sensitivity of nociceptors to other stimuli. Aspirin and other nonsteroidal anti-inflammatory drugs are a useful treatment for hyperalgesia because they inhibit the enzymes required for prostaglandin synthesis.

Substance P is a peptide synthesized by the nociceptors themselves. Activation of one branch of a nociceptor axon can lead to the secretion of substance P by the other branches of that axon in the neighboring skin. Substance P causes vasodilation (swelling of the blood capillaries) and the release of his-

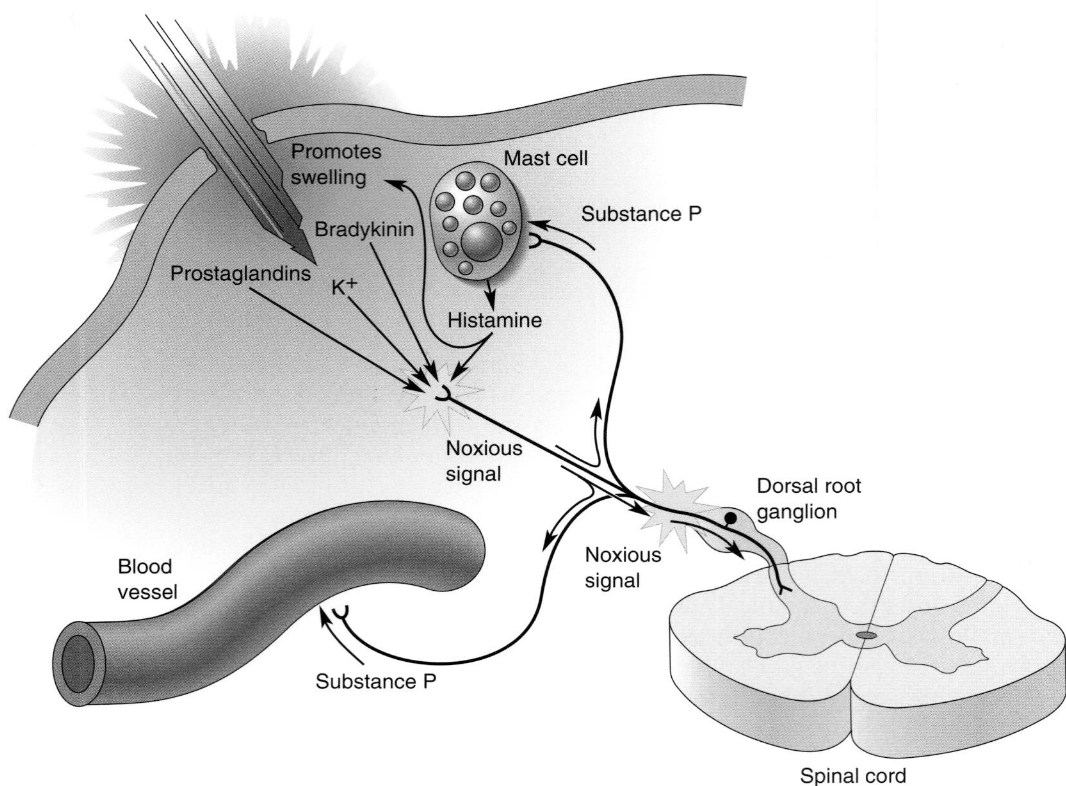

Figure 12.24
Peripheral chemical mediators of pain and hyperalgesia.

tamine from mast cells. Sensitization of other nociceptors around the site of injury by substance P is one cause of secondary hyperalgesia.

CNS mechanisms also contribute to secondary hyperalgesia. Following injury, the activation of mechanoreceptive Aβ axons by light touch can evoke pain. Thus, another mechanism of hyperalgesia involves cross-talk between the touch and pain pathways in the spinal cord.

Primary Afferents and Spinal Mechanisms

Aδ and C fibers bring information to the CNS at different rates because of differences in their action potential conduction velocities. Accordingly, activation of skin nociceptors produces two distinct perceptions of pain: a fast, sharp *first pain* followed by a duller, longer-lasting *second pain*. First pain is caused by the activation of Aδ fibers; second pain is caused by the activation of C fibers (Figure 12.25).

Like the Aβ mechanosensory fibers, the small-diameter fibers have their cell bodies in the segmental dorsal root ganglia, and they enter the dorsal horn of the spinal cord. The fibers branch immediately, travel a short distance up and down the spinal cord in a region called the *zone of Lissauer*, and then synapse on cells in the outer part of the dorsal horn in a region known as the **substantia gelatinosa** (Figure 12.26).

The neurotransmitter of the pain afferents is believed to be glutamate; as mentioned previously, however, these neurons also contain the peptide substance P (Figure 12.27). Substance P is contained within storage granules in the axon terminals (recall Chapter 5) and can be released by high-frequency trains of action potentials. Very recent experiments have shown that synap-

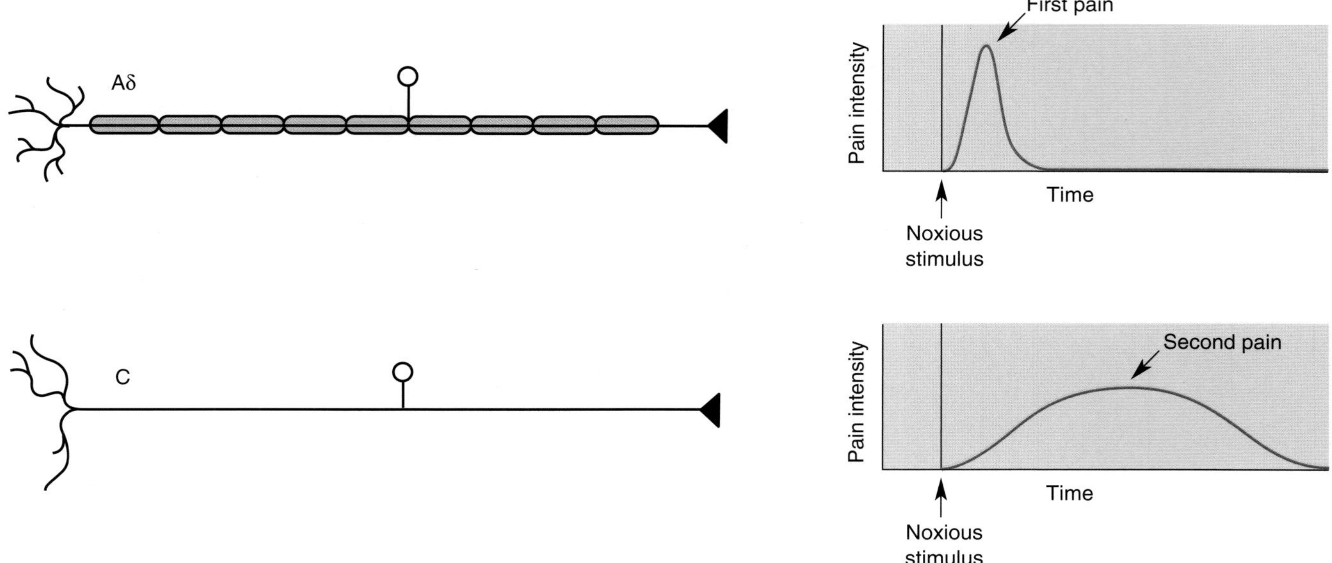

Figure 12.25
First and second pain. The first pain sensation registered by noxious stimulation is mediated by fast Aδ axons. The second, longer-lasting pain sensation is mediated by slow C fibers.

tic transmission mediated by substance P is required to produce moderate to intense pain. If you like spicy food, you should know that the active ingredient is *capsaicin,* which generates its piquant effect by causing the release of substance P from certain nociceptors in the mouth. If you cook spicy food, you should be forewarned about the searing pain that comes from rubbing your eyes with chili-stained (i.e., capsaicin-coated) fingers. Ironically, when applied in large quantities, capsaicin can cause **analgesia**, an absence of pain, because it depletes substance P from nerve terminals. This is one of the treatments for the pain associated with shingles (Box 12.1).

Nociceptor axons from the viscera enter the spinal cord by the same route as the cutaneous nociceptors. Within the spinal cord there is substantial mixing of information from these two sources of input. This cross-talk gives rise

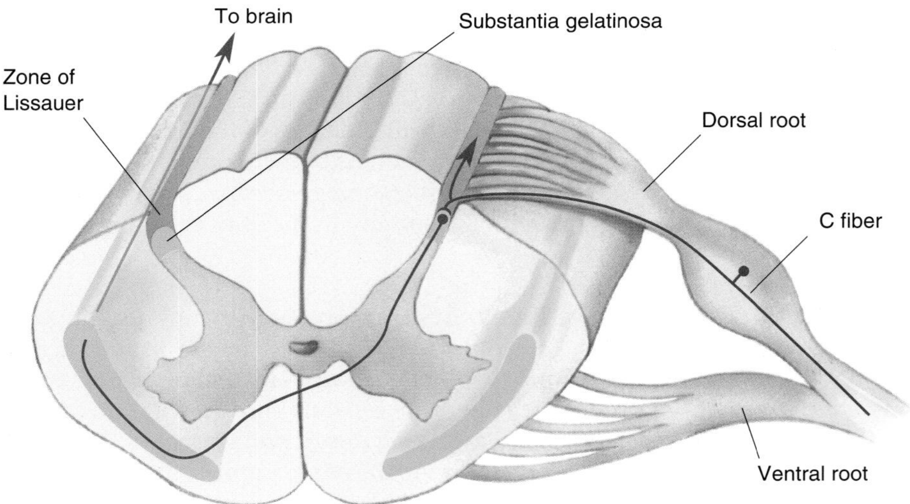

Figure 12.26
Spinal connections of nociceptive axons.

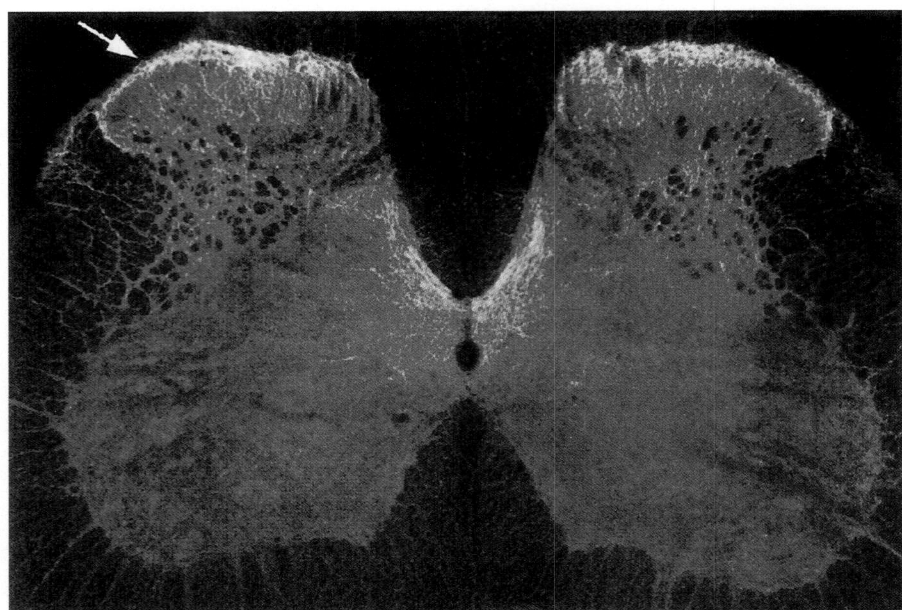

Figure 12.27
Immunocytochemical localization of substance P in the spinal cord. (Source: Mantyh et al., 1997.)

to the phenomenon of **referred pain**, wherein visceral nociceptor activation is perceived as a cutaneous sensation. The classic example of referred pain is angina, occurring when the heart fails to receive sufficient oxygen. The pain is often localized by the patient to the upper chest wall and the left arm. Another common example is the pain associated with appendicitis, which in its early stages is referred to the abdominal wall around the navel (Figure 12.28).

Ascending Pain Pathways

Let's briefly highlight the differences we've encountered between the touch and pain pathways. First, they differ with respect to their nerve endings in the skin. The touch pathway is characterized by specialized structures in the skin; the pain pathway has only free nerve endings. Second, they differ with respect to the diameter of their axons. The touch pathway is swift and uses thick myelinated Aβ fibers; the pain pathway is slow and uses thin, lightly myelinated Aδ fibers and unmyelinated C fibers. Third, they differ with respect to their connections in the spinal cord. Branches of the Aβ axons terminate in the deep dorsal horn; the Aδ and C fibers branch, run within the zone of Lissauer, and terminate within the substantia gelatinosa. As we will now see, the two pathways also differ substantially in the way they transmit information to the brain.

The Spinothalamic Pain Pathway. Information about pain and temperature in the body is conveyed from the spinal cord to the brain via the **spinothalamic pathway**. Unlike the dorsal column–medial lemniscal pathway, the axons of the second-order neurons *immediately decussate* and ascend through the *spinothalamic tract* running along the ventral surface of the spinal cord (compare Figures 12.13 and 12.26). As the name implies, the spinothalamic fibers project up the spinal cord and through the medulla, pons, and midbrain without synapsing until they reach the thalamus (Figure 12.29). As the spinothalamic axons journey through the brain stem, they eventually come to lie alongside the medial lemniscus, but the two groups of axons remain distinct from each other.

Figure 12.30 summarizes the different ascending pathways for touch and pain information. Information about touch ascends *ipsilaterally*, while information about pain and temperature ascends *contralaterally*. This organization can lead to a curious but predictable group of deficits when the nervous system is impaired. For example, if half of the spinal cord is damaged, certain deficits of mechanosensitivity occur on the same side as the spinal cord damage: insensitivity to light touch, the vibrations of a tuning fork on the skin, the position of a limb. Conversely, deficits in pain and temperature sensitivity show up on the side of the body *opposite* the cord damage. Other signs, such as motor deficiency and the exact map of sensory deficits, give additional clues about the site of spinal cord damage. For example, movements are impaired on the ipsilateral side. The constellation of sensory and motor signs following damage to one side of the spinal cord is called the *Brown-Séquard syndrome*.

The Trigeminal Pain Pathway. Pain and temperature information from the face and head takes a path to the thalamus that is analogous to the spinal path. The small-diameter fibers in the trigeminal nerve synapse first on second-order sensory neurons in the *spinal trigeminal nucleus* of the brain stem. The axons of these cells cross and ascend to the thalamus in the *trigeminal lemniscus*.

In addition to the spinothalamic and trigeminothalamic pathways, other closely related pain and temperature pathways send axons into a variety of structures at all levels of the brain stem, before they reach the thalamus. Some of these pathways are particularly important in generating sensations of slow, burning, agonizing pain, while others are involved in arousal.

The Thalamus and Cortex. The spinothalamic tract and trigeminal lemniscus axons synapse over a wider region of the thalamus than those of the me-

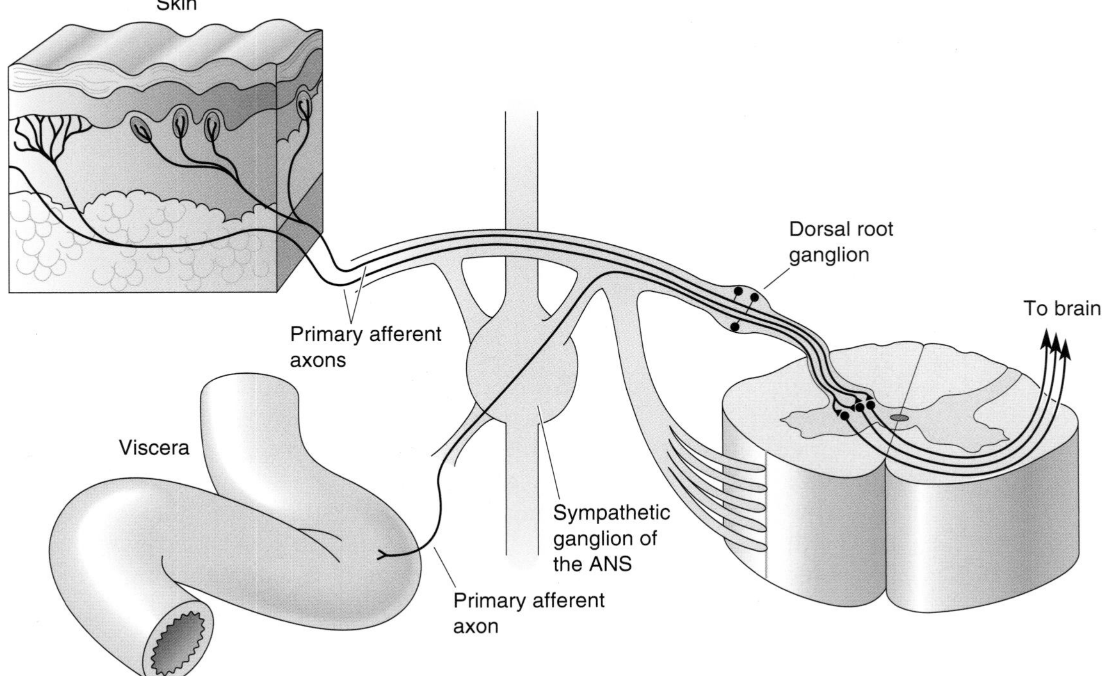

Figure 12.28
The convergence of nociceptor input from the viscera and the skin.

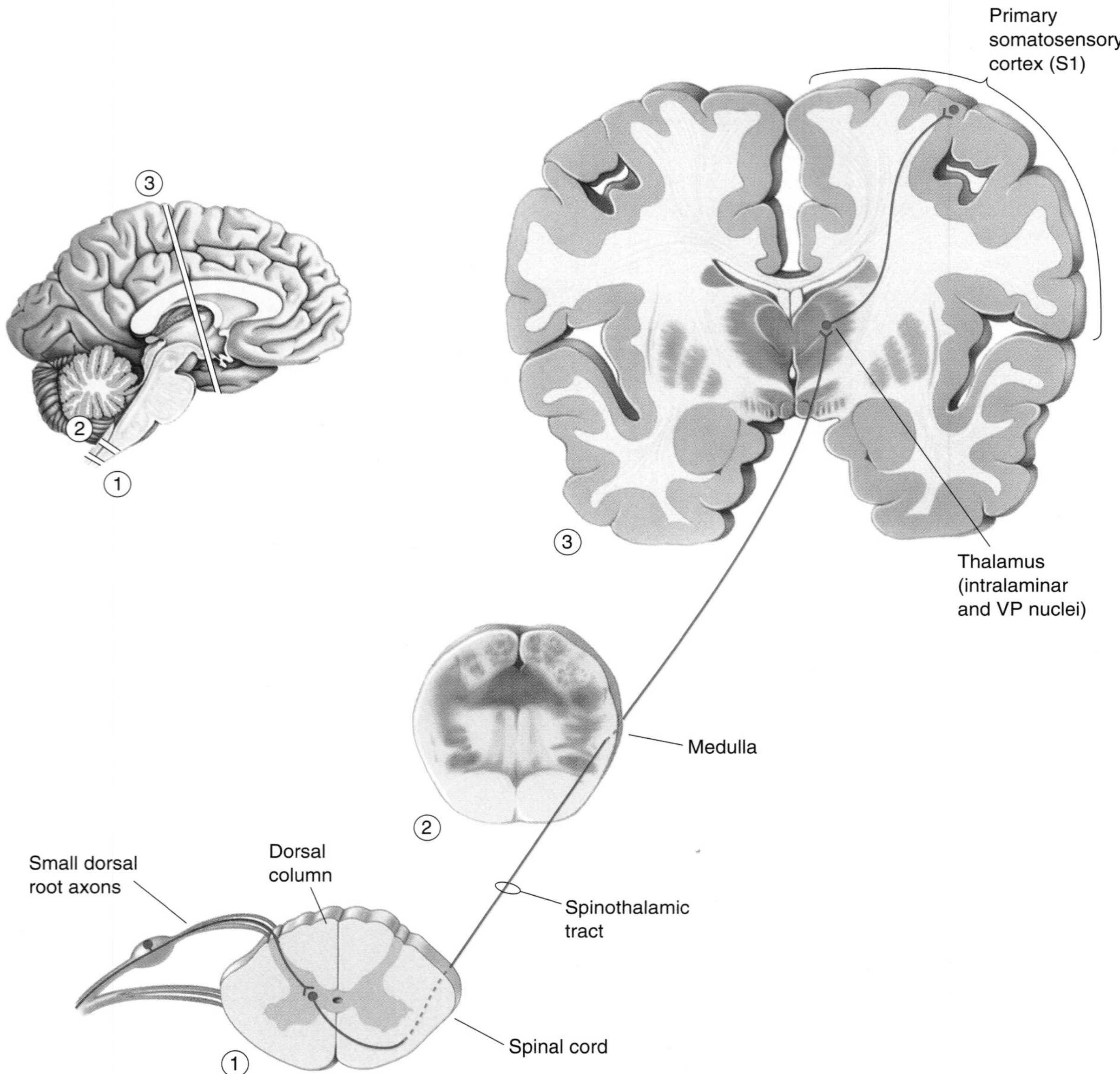

Figure 12.29

The spinothalamic pathway. This is the major route by which pain and temperature information ascend to the cerebral cortex.

dial lemniscus. Some of the axons terminate in the VP nucleus, just as the medial lemniscal axons do, but the touch and pain systems *still* remain segregated there by occupying separate regions of the nucleus. Other spinothalamic axons end in the small *intralaminar nuclei* of the thalamus (Figure 12.31). From the thalamus, pain and temperature information is projected to various areas of the cerebral cortex. As in the thalamus, this pathway covers a much wider territory than the cortical connections of the dorsal column–medial lemniscal pathway.

The Regulation of Pain

For many years it has been recognized that the perception of pain is variable. Depending on the concurrent level of nonpainful sensory input and the be-

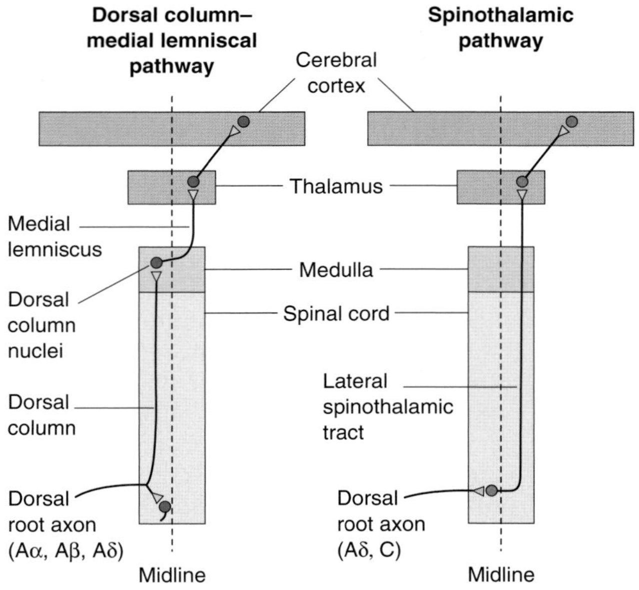

Figure 12.30
An overview of the two major ascending pathways of somatic sensation.

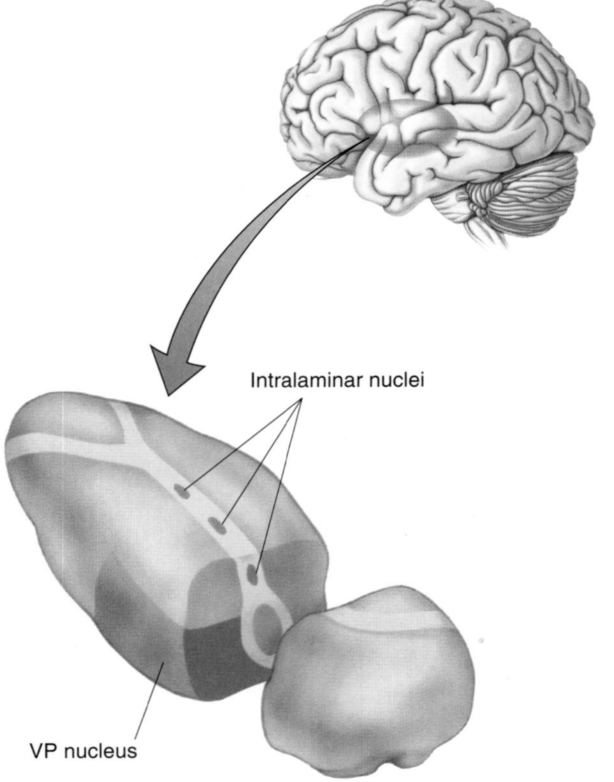

Figure 12.31
Somatic sensory nuclei of the thalamus. In addition to the VP nucleus, the intralaminar nuclei relay nociceptive information to a large expanse of the cerebral cortex.

havioral context, the same level of nociceptor activity can produce more or less pain. Understanding this modulation of pain is important because it may offer new strategies for the treatment of chronic pain, a condition that afflicts up to 20% of the adult population.

Afferent Regulation. We've already seen that light touch can evoke pain via the mechanisms of hyperalgesia. Pain evoked by activity in nociceptors, however, can also be *reduced* by simultaneous activity in low-threshold mechanoreceptors (Aβ fibers). Presumably, this is why it feels good to rub the skin around your shin when you bruise it. This may also explain an electrical treatment for some kinds of chronic, intractable pain. Wires are taped to the skin surface, and pain is suppressed when the patient simply turns on an electrical stimulator designed to activate large-diameter sensory axons.

In the 1960s, Ronald Melzack and Patrick Wall, then working at MIT, proposed a hypothesis to explain these phenomena. Their *gate theory of pain* proposes that certain neurons of the dorsal horn, which project an axon up the spinothalamic tract, are excited by both large-diameter sensory axons and unmyelinated pain axons. The projection neuron is also inhibited by an interneuron, and the interneuron is both *excited* by the large sensory axon and *inhibited* by the pain axon (Figure 12.32). By this arrangement, activity in the pain axon alone maximally excites the projection neuron, allowing nociceptive signals to rise to the brain. If the large mechanoreceptive axon fires concurrently, however, it activates the interneuron and suppresses nociceptive signals.

Descending Regulation. Stories abound of soldiers, athletes, and torture victims who sustained horrible injuries but apparently felt no pain. Strong emotion, stress, or stoic determination can powerfully suppress feelings of pain. Several brain regions have been implicated in pain suppression (Figure 12.33). One is a zone of neurons in the midbrain called the periventricular

Figure 12.32
Melzack and Wall's gate theory of pain.

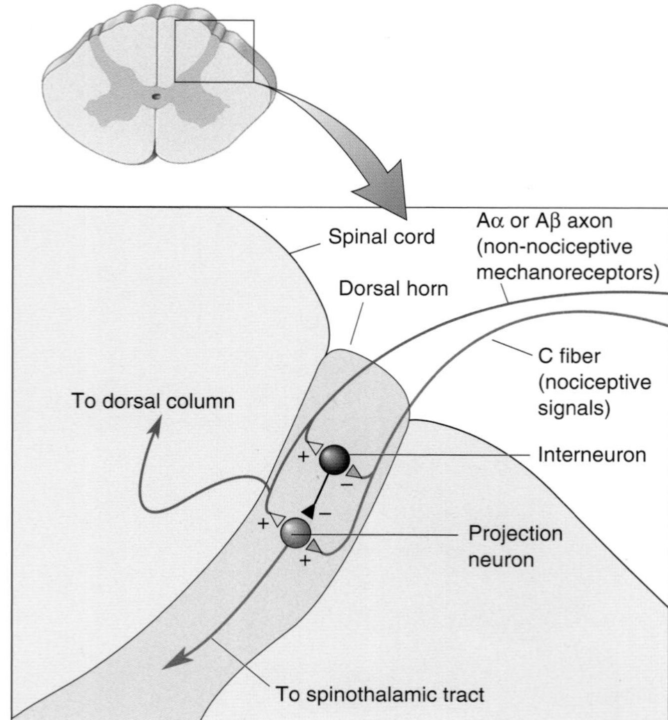

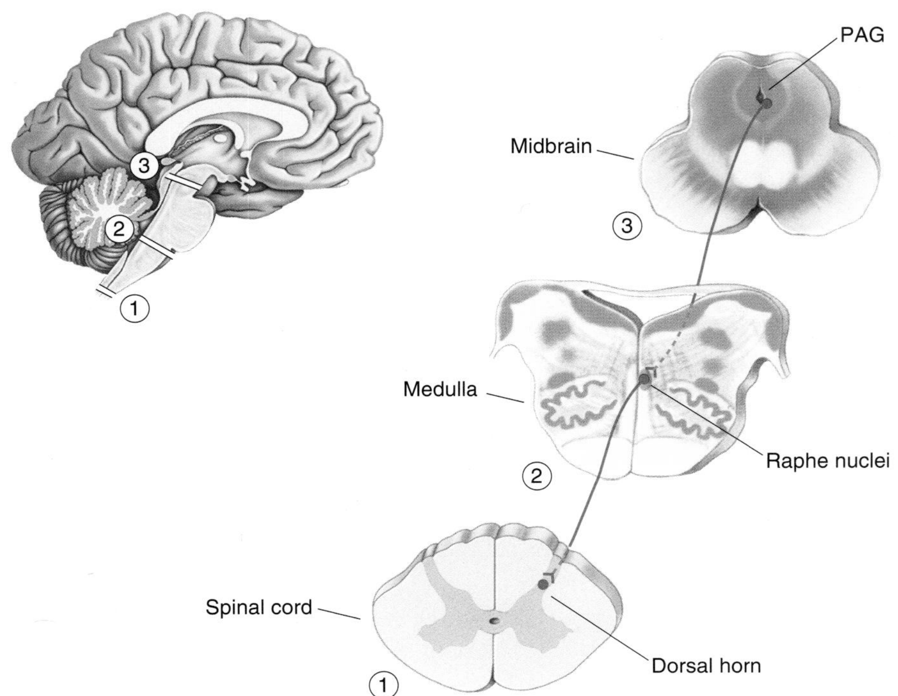

Figure 12.33
Descending pain-control pathways. A variety of brain structures, many of which are affected by behavioral state, can influence activity within the periaqueductal gray matter (PAG) of the midbrain. The PAG can influence the raphe nuclei of the medulla, which in turn can modulate the flow of nociceptive information through the dorsal horns of the spinal cord.

and **periaqueductal gray matter (PAG)**. Electrical stimulation of the PAG can cause profound analgesia that has sometimes been exploited clinically. The PAG normally receives input from several brain structures, many of them appropriate for transmitting signals related to emotional status. PAG neurons send descending axons into various midline regions of the medulla, particularly to the *raphe nuclei*, which use the neurotransmitter serotonin. These medullary neurons, in turn, project axons down to the dorsal horns of the spinal cord, where they can effectively depress the activity of nociceptive neurons.

The Endogenous Opioids. Opium was probably known to the ancient Sumerians around 4000 B.C. Their pictograph for the poppy roughly translates to "joy plant." By the seventeenth century, the therapeutic value of opium was undisputed. Opium, its active ingredients, and their analogues—including morphine, codeine, and heroin—are used and abused widely today in most cultures of the world. Opioids produce profound analgesia when taken systemically. They can also produce mood changes, drowsiness, mental clouding, nausea, vomiting, and constipation. The 1970s brought the stunning discoveries that opioids act by binding tightly and specifically to several types of **opioid receptors** in the brain and that the brain itself manufactures endogenous morphine-like substances, collectively called **endorphins**. Endorphins are relatively small proteins, or peptides.

Endorphins and their receptors are distributed widely in the CNS, but they are particularly concentrated in areas that process or modulate nociceptive information. Small injections of morphine or endorphins into the PAG, the raphe nuclei, or the dorsal horn can produce analgesia. Because this effect is prevented by administering the specific blocker of opioid receptors, *naloxone,* the injected drugs must have acted by binding to opioid receptors in those areas. Naloxone can also block the analgesic effects induced by electrically stimulating these areas. At the cellular level, endorphins exert multiple effects that include suppressing the release of glutamate from presynaptic

Box 12.5

OF SPECIAL INTEREST

Pain and the Placebo Effect

Trials to test the efficacy of a new drug often entail giving one group of subjects the drug and giving the other group an inert substance. Both groups of subjects believe they have been given the drug. Surprisingly, the inert substance is often reported to have the effect that patients are told to expect from the drug. The term *placebo* is used to describe such substances (the word is from the Latin meaning "I shall please"), and the phenomenon is called the *placebo effect.*

Placebos can be highly effective analgesics. It has been reported that a majority of patients suffering from postoperative pain get relief from an injection of sterile saline! Does this mean that these patients have only imagined their pain? Not at all. The opioid receptor antagonist naloxone can block the analgesic effect of the placebo, just as it antagonizes the effects of morphine, a true analgesic. Apparently, the belief that the treatment will work can be enough to cause activation of the endogenous pain relief systems of the brain. The placebo effect is a likely explanation for the success of other treatments for pain, such as acupuncture, hypnosis and, for children, a mother's loving kiss.

terminals and inhibiting neurons by hyperpolarizing their postsynaptic membranes. In general, extensive systems of endorphin-containing neurons in the spinal cord and brain stem prevent the passage of nociceptive signals through the dorsal horn and into higher levels of the brain, where the perception of pain is generated (Box 12.5).

TEMPERATURE

As is the case for the sense of touch and pain, nonpainful temperature sensations originate from receptors in the skin, and they depend on the neocortex for their conscious appreciation. Here we briefly describe how this system is organized.

Thermoreceptors

Because the rate of a chemical reaction depends on temperature, the functioning of all cells is sensitive to temperature. **Thermoreceptors,** however, are neurons that because of specific membrane mechanisms are exquisitely sensitive to temperature. For example, we can perceive changes in our average skin temperature of as little as 0.01°C. Temperature-sensitive neurons clustered in the hypothalamus and the spinal cord are important in the physiological responses that maintain stable body temperature, but it is the thermoreceptors in the skin that apparently contribute to our perception of temperature.

Although very little is known about the structure of thermoreceptive nerve endings, we do know that temperature sensitivity is not spread uniformly across the skin. You could take a small cold or warm probe and map your skin's sensitivity to temperature changes. Some spots about 1 mm wide are especially sensitive to *either* hot or cold but are not sensitive to both. The fact that the locations of hot and cold sensitivity are different demonstrates that separate receptors encode them. Also, small areas of skin between the hot and cold spots are relatively insensitive to temperature.

Warm receptors begin firing above about 30°C and increase their firing rate until about 45°C (Figure 12.34a). As temperature increases further, the firing rate actually falls off steeply. Recall that 45°C is just about the point where thermal nociceptors begin firing; it is also where perception changes from hot to scalding and tissues begin to burn. *Cold receptors* are relatively unresponsive at skin temperatures above about 35°C, but they fire faster over a broad range down to about 10°C. Below that temperature, firing ceases, and cold becomes a very effective anesthetic, as you know if you've ever had the pleasure of wading through an icy mountain stream. Strangely, some cold receptors also begin to fire when their temperature is raised *above* 45°C (Figure 12.34a). If such temperatures are applied to wide areas of skin, they are usually painful; but if they are restricted to small regions of skin innervated by a cold receptor, they produce a paradoxical feeling of cold. This emphasizes an important point: The CNS does not know *what* kind of stimulus (in this case, heat) caused the receptor to fire, but it continues to interpret all activity from its cold receptor as a response to cold.

As with mechanoreceptors, the responses of thermoreceptors adapt during long-duration stimuli. Figure 12.34b shows that a sudden drop in skin temperature causes a cold receptor to fire strongly, while it silences a warm receptor. After a few seconds at 32°C, however, the cold receptor slows its firing (but still fires faster than it did at 38°C), while the warm receptor speeds up slightly. Notice that a return to the original warm skin temperature causes opposite responses—transient silence of the cold receptor and a burst of ac-

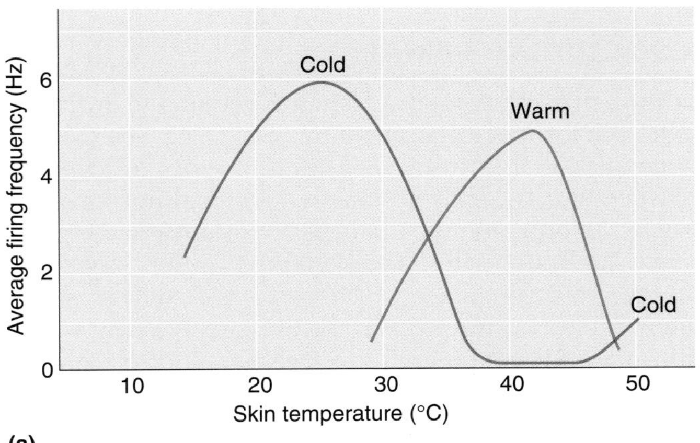

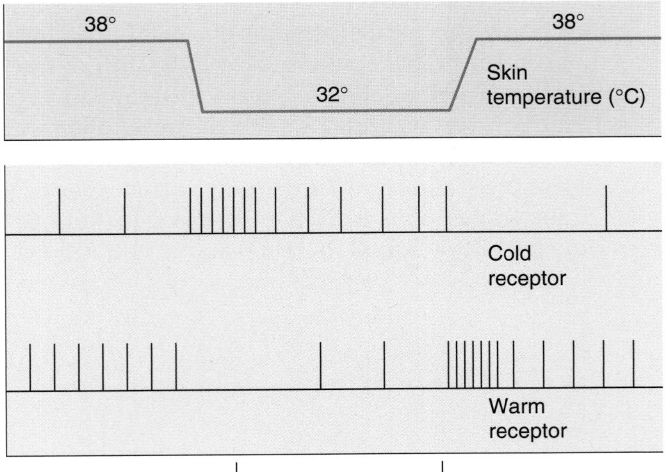

Figure 12.34
Various responses of thermoreceptors. **(a)** The action potential firing rate as a function of skin temperature for cold and warm receptors. The cold receptor fires not only at relatively low temperatures but also above about 45°C. **(b)** The responses of cold and warm receptors to a step reduction in skin temperature. Both receptors are most responsive to sudden changes in temperature, but they adapt over several seconds.

tivity in the warm receptor—before both return to their steady, adapted rates. Thus, the differences between the response rates of warm and cold receptors are greatest during and shortly after temperature changes. This often mirrors our perceptions of temperature.

Try a simple experiment. Fill two buckets with tap water, one cold and one hot (but not painfully hot). Then plunge your right hand for one minute into each of them in turn. Notice the striking sensations of hot and cold that occur with each change, but also notice how transient the sensations are. With thermoreception, as with most other sensory systems, it is the sudden *change* in the quality of a stimulus that generates the most intense neural and perceptual responses.

The Temperature Pathway

At this point you may be relieved to learn that the organization of the temperature pathway is virtually identical to that of the pain pathway already described. Cold receptors are coupled to Aδ and C fibers, while warm receptors are coupled only to C fibers. As we learned previously, small-diameter axons synapse within the substantia gelatinosa of the dorsal horn. The axons of second-order neurons immediately decussate and ascend in the contralateral spinothalamic tract. Thus, if the spinal cord is transected on one side, there is loss of temperature sensitivity (as well as pain) on the opposite side of the body, specifically in those regions of skin innervated by spinal segments below the cut.

CONCLUDING REMARKS

Although each sensory system has evolved to be the brain's interface with a different form of environmental energy, the systems are strikingly similar in organization and function. Different types of somatic sensory information are necessarily kept separate in the spinal nerves because each axon is connected to only one type of sensory receptor ending. Segregation of sensory types continues within the spinal cord and is largely maintained all the way to the cerebral cortex. In this way, the somatic sensory system repeats a theme common throughout the nervous system: Several flows of related but distinct information are passed in parallel through a series of neural structures. Mixing of these streams occurs along the way, but only judiciously, until higher levels of processing are reached in the cerebral cortex. We saw other examples of parallel processing of sensory information in the chemical senses, vision, and audition.

Exactly how the parallel streams of sensory data are melded into perception, images, and ideas remains the Holy Grail of neuroscience. Thus, perception of any handled object involves the seamless coordination of all facets of somatic sensory information. For example, the bird in hand is rounded, warm, soft, and light in weight, its heartbeat flutters against your fingertips, its claws scratch, and its textured wings brush against your palm. Somehow your brain knows it's a bird even without looking or listening and would never mistake it for a toad. In the chapters that follow, we describe how the brain begins to use sensory information to plan and coordinate movement.

KEY TERMS

Introduction
somatic sensation (p. 397)

Touch
mechanoreceptor (p. 398)
Pacinian corpuscle (p. 398)
spinal segment (p. 404)
dermatome (p. 404)
dorsal column–medial lemniscal pathway (p. 407)
dorsal column (p. 407)
dorsal column nucleus (p. 407)
medial lemniscus (p. 409)
ventral posterior (VP) nucleus (p. 409)
primary somatosensory cortex (S1) (p. 410)
trigeminal nerve (p. 410)
posterior parietal cortex (p. 410)
somatotopy (p. 413)
agnosia (p. 420)
neglect syndrome (p. 420)

Pain
nociceptor (p. 421)
hyperalgesia (p. 423)
substantia gelatinosa (p. 424)
analgesia (p. 425)
referred pain (p. 426)
spinothalamic pathway (p. 426)
periaqueductal gray matter (PAG) (p. 431)
opioid receptor (p. 431)
endorphin (p. 431)

Temperature
thermoreceptor (p. 432)

REVIEW QUESTIONS

1. Imagine rubbing your fingertips across a pane of smooth glass and then across a brick. What kinds of skin receptors help you distinguish the two surfaces? As far as your somatic sensory system is concerned, what is different about the two surfaces?
2. What purpose is served by the encapsulations around some sensory nerve endings in the skin?
3. If someone tossed you a hot potato and you caught it, which information would reach your CNS first: that the potato was hot or that it was relatively smooth? Why?
4. At what levels of the nervous system are *all* types of somatic sensory information represented on the contralateral side: the spinal cord, the medulla, the pons, the midbrain, the thalamus, the cortex?
5. What lobe of the cortex contains the main somatic sensory areas? Where are these areas relative to the main visual and auditory areas?
6. Where within the body can pain be modulated, and what causes its modulation?
7. Where in the CNS does information about touch, shape, temperature, and pain converge?
8. Imagine this experiment: Fill two buckets with water, one cold and one relatively hot. Fill a third bucket with water of an intermediate, lukewarm temperature. Put your left hand into the hot water and your right hand into the cold water, and wait one minute. Now quickly plunge both hands into the lukewarm water. Try to predict what sensations of temperature you will feel in each hand. Will they feel the same? Why?

CHAPTER

13

Spinal Control of Movement

INTRODUCTION

We are now ready to turn our attention to the system that actually gives rise to behavior. The motor system consists of all our muscles and the neurons that command them. The importance of the motor system was summarized by the pioneering English neurophysiologist Charles Sherrington in the Linacre lecture of 1924: "To move things is all that mankind can do . . . for such the sole executant is muscle, whether in whispering a syllable or in felling a forest." A moment's thought will convince you that the motor system is also incredibly complex. Behavior requires the coordinated action of various combinations of more than 750 muscles in a changing and often unpredictable environment.

Have you ever heard the expression "running around like a chicken with its head cut off"? It is based on the observation that complex patterns of behavior (running around the barnyard) can be generated without the participation of the brain. There exists a significant amount of circuitry within the spinal cord for the coordinated control of movements, particularly stereotyped (repetitive) ones such as those associated with locomotion. This point was established early in this century by Sherrington and his English contemporary Graham Brown, who showed that rhythmic movements could be elicited in the hind legs of cats and dogs long after their spinal cords had been severed from the rest of the central nervous system (CNS). Today's view is that the spinal cord contains certain *motor programs* for the generation of coordinated movements and that these programs are accessed, executed, and modified by descending commands from the brain. Thus, motor control can be divided into two parts: (1) the spinal cord's command and control of coordinated muscle contraction and (2) the brain's command and control of the motor programs in the spinal cord.

In this chapter, we explore the joints, skeletal muscles, and spinal motor neurons of the peripheral somatic motor system and how they communicate with each other. In Chapter 14, we will take a look at how the brain influences the activity of the spinal cord.

THE SOMATIC MOTOR SYSTEM

The muscles in the body may be divided into two broad categories, striated and smooth, based on their appearance under the microscope. But they are also distinct in other ways. **Smooth muscle** lines the digestive tract and arteries, among other things, and is innervated by nerve fibers from the autonomic nervous system (ANS) (discussed in Chapter 15). Smooth muscle plays a role in peristalsis (movement of material through the intestines) and the control of blood pressure. **Striated muscle** is further divisible into two types: cardiac and skeletal. **Cardiac muscle** is heart muscle, and it contracts rhythmically in the absence of any innervation. Innervation of the heart from the ANS functions to accelerate or slow down the heart rate. (Recall Otto Loewi's experiment in Chapter 5.)

Skeletal muscle, which constitutes the bulk of the muscle mass of the body, moves bones around joints and the eyes within the head, controls respiration and facial expression, and produces speech. Each skeletal muscle is enclosed in a connective tissue sheath that forms the tendons at the ends of the muscle. Within each muscle are hundreds of **muscle fibers**—the cells of skeletal muscle—and each fiber is innervated by a single axon branch from the CNS (Figure 13.1). Because skeletal muscle is derived embryologically from 33 paired somites (see Chapter 7), these muscles and the parts of the nervous system that control them are collectively called the **somatic motor system.** Here we focus our attention on this system because it is under vol-

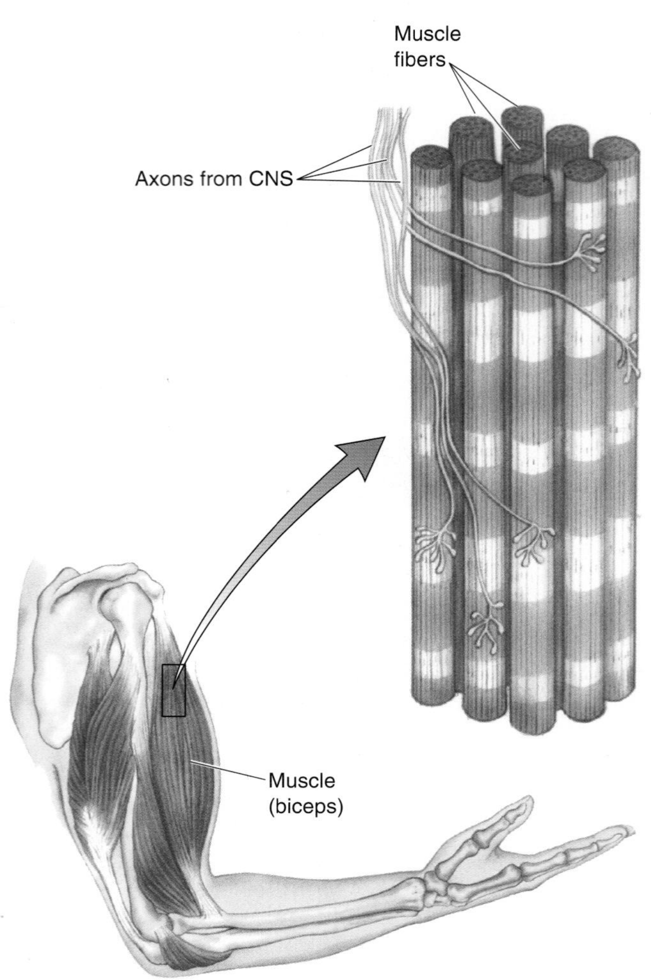

Figure 13.1
The structure of skeletal muscle. Each muscle fiber is innervated by a single axon.

untary control and is responsible for generating behavior. (The visceral motor system of the ANS will be discussed in Chapter 15.)

Consider the elbow joint (Figure 13.2). It is formed where the humerus, the upper arm bone, is bound by fibrous ligaments to the radius and ulna, the bones of the lower arm. The elbow joint functions like a hinge on a pocket knife. Movement in the direction that closes the knife is called **flexion**, and movement in the direction that opens the knife is called **extension**. The major muscle that causes flexion is the brachialis, whose tendons insert into the humerus at one end and into the ulna at the other. Two other muscles, the biceps brachii and the coracobrachialis (which lies under the biceps), cause flexion at this joint. Together, these muscles are called **flexors** of the elbow joint, and because the three muscles all work together, they are called **synergists** of one another. The two synergistic muscles that cause extension of the elbow joint are the triceps brachii and the anconeus; these two muscles are called **extensors**. Because the flexors and extensors pull on the joint in opposite directions, they are called **antagonists** to one another. Note that muscles only pull on a joint; they cannot push. Even the simple flexion of the elbow joint requires the coordinated contraction of the synergistic flexor muscles *and* the relaxation of the antagonistic extensor muscles.

Other terms to note about somatic musculature refer to the location of the joints they act on. The muscles that are responsible for movements of the trunk are called **axial muscles**; those that move the shoulder, elbow, pelvis,

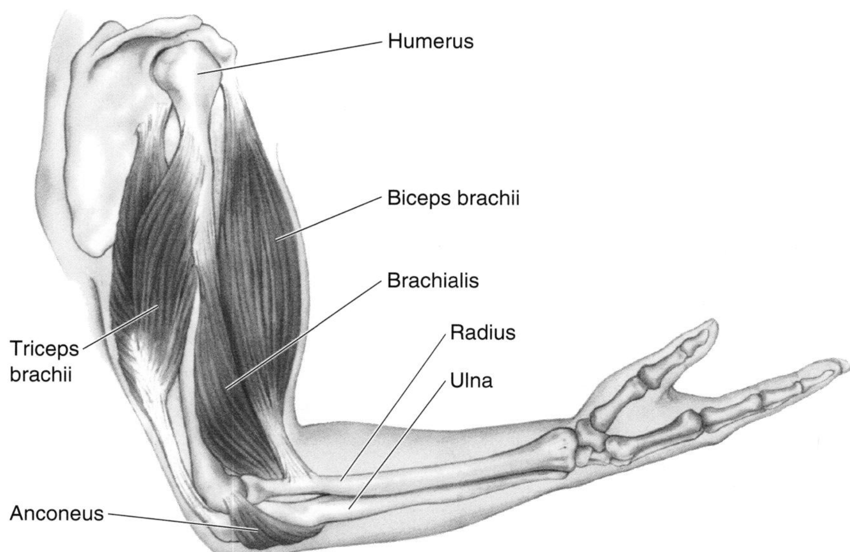

Figure 13.2
Major muscles of the elbow joint. The biceps and triceps brachii are antagonistic muscles. Contraction of the biceps brachii causes flexion, and contraction of the triceps brachii causes extension of the elbow.

and knee are called **proximal** (or **girdle**) **muscles**; and those that move the hands, feet, and digits (fingers and toes) are called **distal muscles.** The axial musculature is very important for maintaining posture; the proximal musculature is critical for locomotion; and the distal musculature, particularly of the hands, is specialized for the manipulation of objects.

THE LOWER MOTOR NEURON

The somatic musculature is innervated by the somatic motor neurons in the ventral horn of the spinal cord (Figure 13.3). These cells are sometimes called "lower motor neurons" to distinguish them from the higher-order "upper motor neurons" of the brain that supply input to the spinal cord. Remember that only the lower motor neurons directly command muscle contraction. Sherrington called these cells the *final common pathway* for the control of behavior.

Segmental Organization of Lower Motor Neurons

The axons of lower motor neurons bundle together to form ventral roots; each ventral root joins with a dorsal root to form a spinal nerve that exits the cord through the notches between vertebrae. Recall from Chapter 12 that there are as many spinal nerves as there are notches between vertebrae; in humans, this adds up to 30 on each side. Because they contain sensory and motor fibers, they are called *mixed spinal nerves.* The motor neurons that provide fibers to one spinal nerve are said to belong to a spinal segment, named for the vertebra where the nerve originates. The segments are cervical (C) 1–8, thoracic (T) 1–12, lumbar (1) 1–5, and sacral (S) 1–5 (see Figure 12.10).

Skeletal muscles are not distributed evenly throughout the body, nor are lower motor neurons distributed evenly within the spinal cord. For example, innervation of the more than 50 muscles of the arm originates entirely from spinal segments C3–Tl. Thus, in this region of the spinal cord, the ventral horns appear swollen to accommodate the large number of motor neurons that control the arm musculature (Figure 13.4). Similarly, spinal segments Ll–S3 have a swollen ventral horn because this is where the motor neurons controlling the leg musculature reside. Thus, we can see that the motor neurons that innervate distal and proximal musculature are found mainly in the

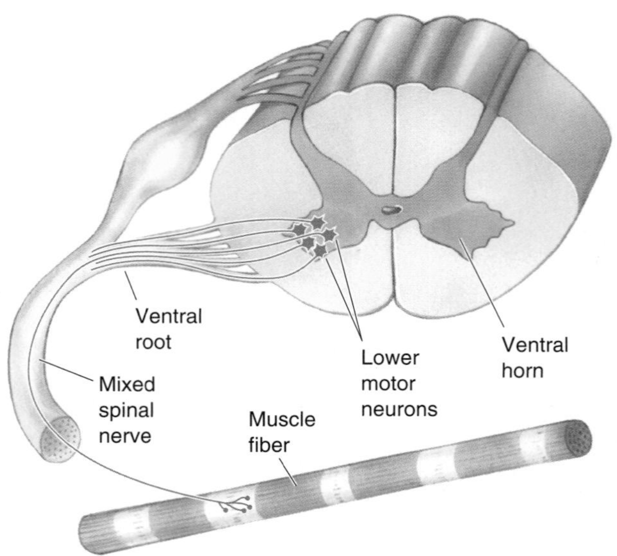

Figure 13.3
Muscle innervation by lower motor neurons. The ventral horn of the spinal cord contains motor neurons that innervate skeletal muscle fibers.

cervical and lumbar-sacral segments of the spinal cord, whereas those innervating axial musculature are found at all levels. The lower motor neurons are also distributed within the ventral horn at each spinal segment in a predictable way, depending on their function. The cells innervating the axial muscles are medial to those innervating the distal muscles, and the cells innervating flexors are dorsal to those innervating extensors (Figure 13.5).

Alpha Motor Neurons

Lower motor neurons of the spinal cord can be divided into two categories: alpha motor neurons and gamma motor neurons (the latter are discussed later in the chapter). **Alpha motor neurons** are directly responsible for the generation of force by muscle. One alpha motor neuron and all of the muscle fibers it innervates collectively make up the elementary component of motor control; Sherrington called it the **motor unit**. Muscle contraction results from the individual and combined actions of these motor units. The collection of alpha motor neurons that innervates a single muscle (e.g., the biceps brachii) is called a **motor neuron pool** (Figure 13.6).

Graded Control of Muscle Contraction by Alpha Motor Neurons. An alpha motor neuron communicates with a muscle fiber by releasing the neurotransmitter acetylcholine (ACh) at the neuromuscular junction, the specialized synapse between nerve and skeletal muscle (see Chapter 5). Because of the high reliability of neuromuscular transmission, the ACh released in response to one presynaptic action potential causes an excitatory postsynaptic potential (EPSP) in the muscle fiber (sometimes also called an *end-plate potential*) that is large enough to trigger one postsynaptic action potential. By mechanisms we will discuss in a moment, a postsynaptic action potential causes a twitch—a rapid sequence of contraction and relaxation—in the muscle fiber. A sustained contraction requires a continual barrage of action potentials. As for other types of synaptic transmission, high-frequency presynaptic activity causes temporal summation of the postsynaptic responses. Twitch summation increases the tension in the muscle fibers and smooths the contraction (Figure 13.7). The rate of firing of motor units is therefore one important way the CNS grades muscle contraction.

A second way the CNS grades muscle contraction is by recruiting addi-

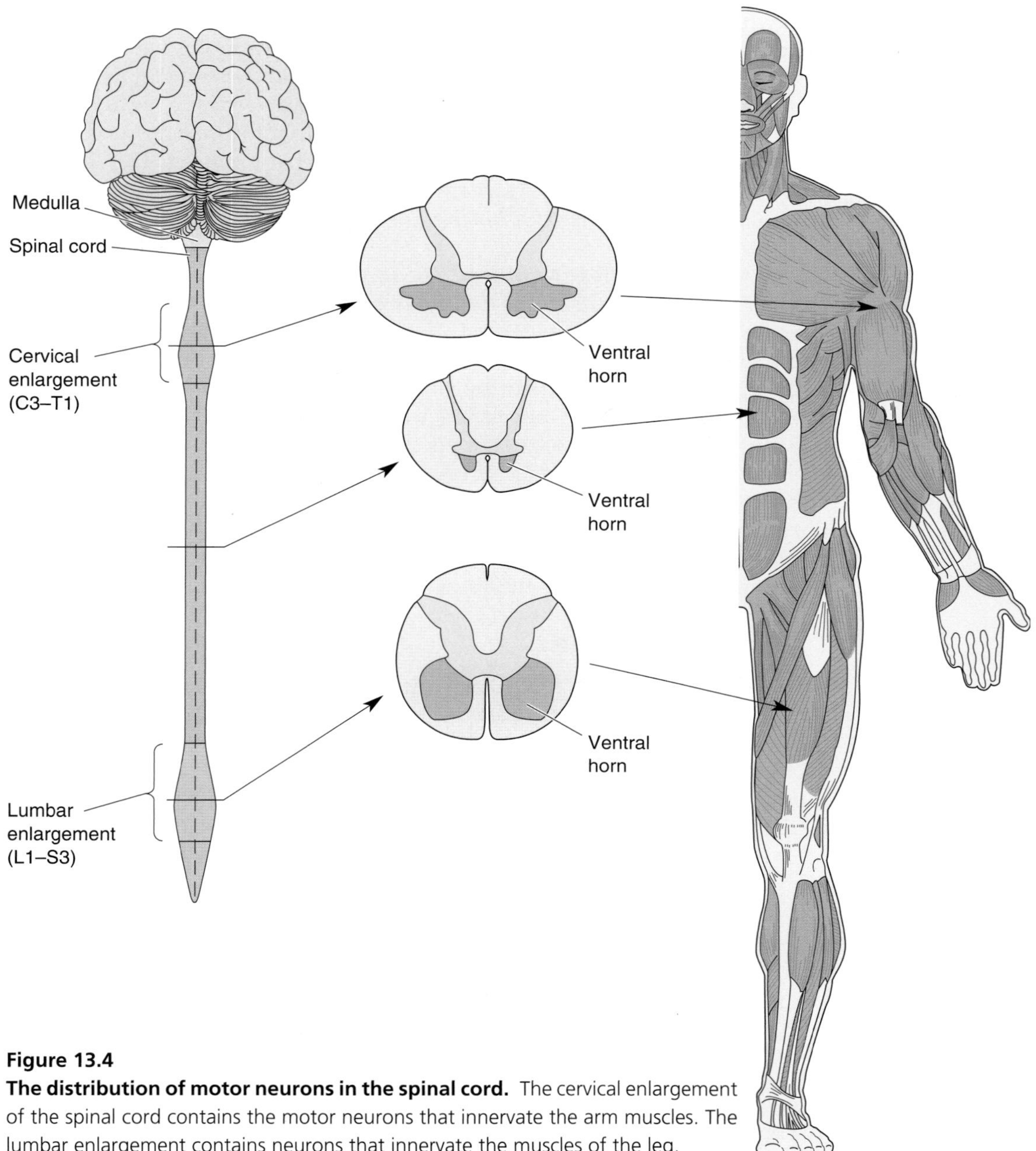

Figure 13.4
The distribution of motor neurons in the spinal cord. The cervical enlargement of the spinal cord contains the motor neurons that innervate the arm muscles. The lumbar enlargement contains neurons that innervate the muscles of the leg.

tional synergistic motor units. The extra tension provided by the recruitment of an active motor unit depends on how many muscle fibers are in that unit. In the antigravity muscles of the leg (muscles that oppose the force of gravity when standing upright), each motor unit tends to be quite large, with an innervation ratio of over 1000 muscle fibers per single alpha motor neuron. In contrast, the smaller muscles that control the movement of the fingers and the rotation of the eyes are characterized by much smaller innervation ratios, as few as 3 muscle fibers per alpha motor neuron. In general, the muscles with a large number of small motor units can be more finely controlled by the CNS.

Most muscles have a range of motor unit sizes, and these motor units are recruited in the order of smallest first, largest last. This orderly recruitment

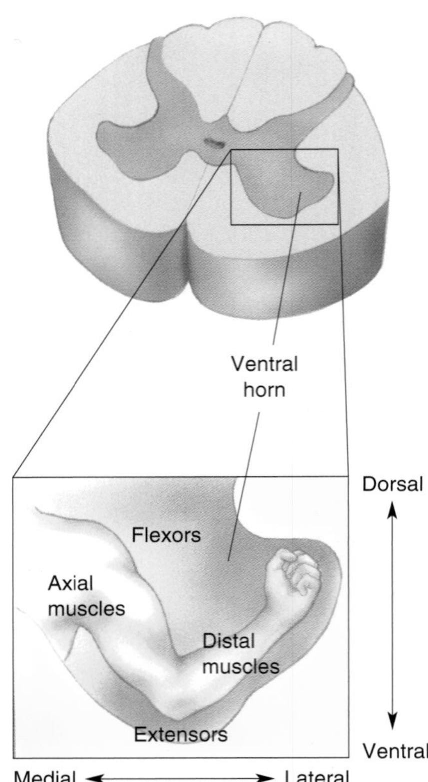

Figure 13.5
The distribution of lower motor neurons in the ventral horn. Motor neurons controlling flexors lie dorsal to those controlling extensors. Motor neurons controlling axial muscles lie medial to those controlling distal muscles.

explains why finer control is possible when muscles are under light loads than when they are under greater loads. Small motor units have small alpha motor neurons, and large motor units have large alpha motor neurons. Thus, one way that orderly recruitment may occur is if small neurons, as a consequence of the geometry of their soma and dendrites, are more easily excited by signals descending from the brain. The idea that the orderly recruitment of motor neurons is due to variations in alpha motor neuron size, first proposed in the late 1950s by Harvard University neurophysiologist Elwood Henneman, is called the *size principle.*

Inputs to Alpha Motor Neurons. The lower motor neurons are controlled by their synaptic inputs in the ventral horn. *There are only three sources of input to an alpha motor neuron* (Figure 13.8). The first source is dorsal root ganglion cells with axons that innervate a specialized sensory apparatus embedded within the muscle, called the *muscle spindle.* As we shall see, this input provides feedback about muscle length. The second source of input to an alpha motor neuron derives from upper motor neurons in the motor cortex and brain stem. This input, which is important for the initiation and control of voluntary movement, will be discussed in more detail in Chapter 14. The third and largest input to an alpha motor neuron derives from interneurons in the spinal cord. This input may be excitatory or inhibitory and is part of the circuitry that generates the spinal motor programs.

Types of Motor Units

If you have ever dined on chicken, you know immediately that not all muscle is the same; there is the dark meat of the leg and the white meat of the breast and wing. The appearance and taste of the various muscles differ because of the biochemistry of the constituent muscle fibers. The red (dark) muscle fibers are characterized by a large number of mitochondria and enzymes specialized for oxidative energy metabolism. These fibers are rela-

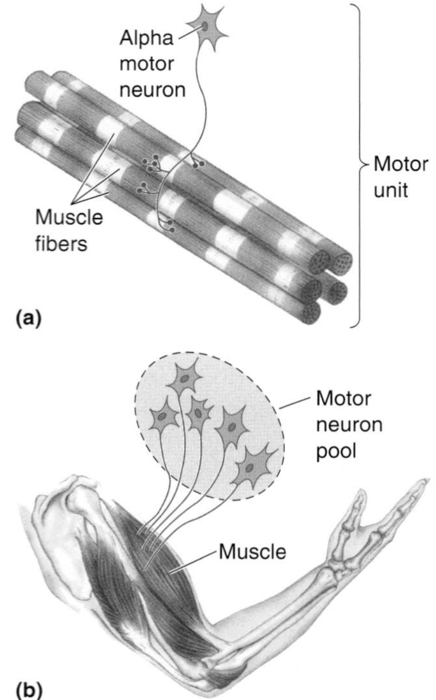

Figure 13.6
A motor unit and motor neuron pool. **(a)** The motor unit is an alpha motor neuron and all of the muscle fibers it innervates. **(b)** The motor neuron pool is all of the alpha motor neurons that innervate one muscle.

Figure 13.7
From muscle twitch to sustained contraction. A single action potential in an alpha motor neuron causes the muscle fiber to twitch. Summation of twitches causes a sustained contraction as the number and frequency of incoming action potentials increase.

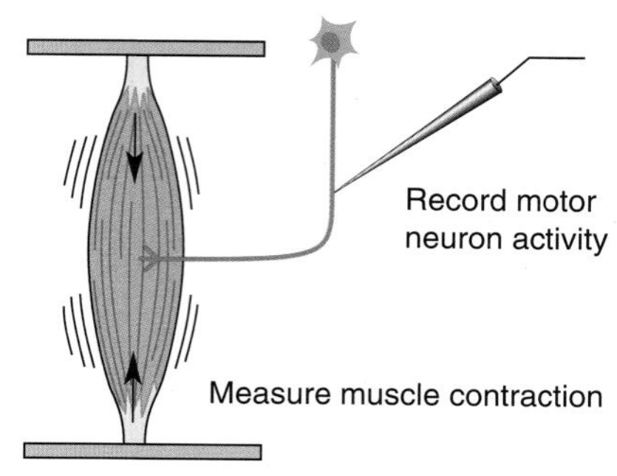

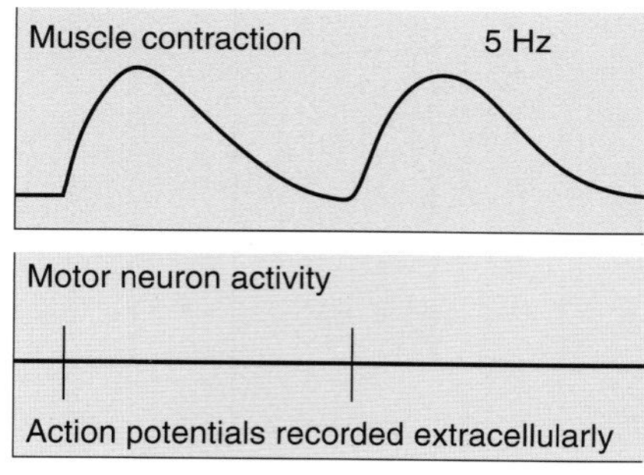

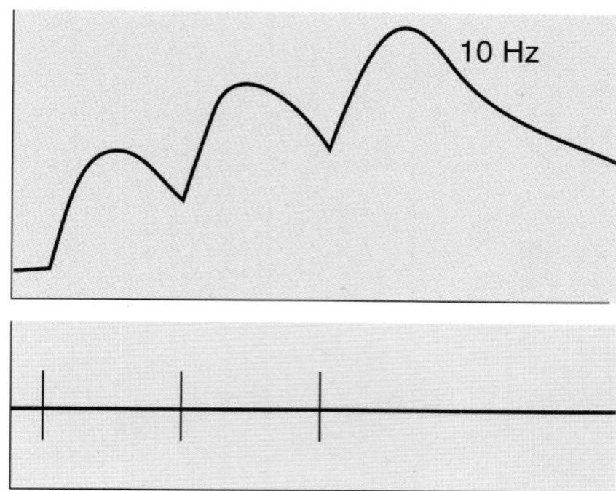

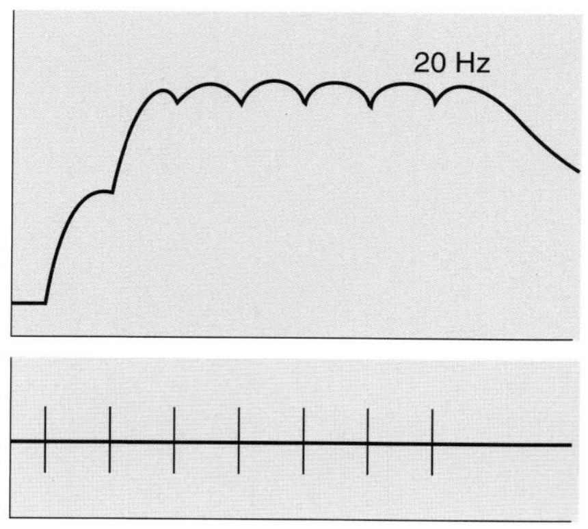

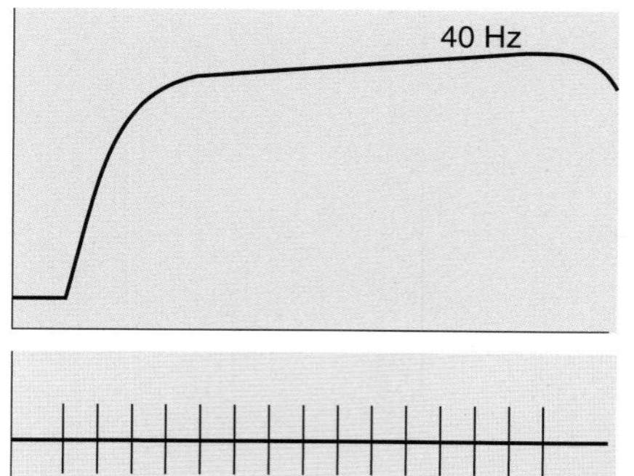

tively slow to contract but can sustain contraction for a long time without fatigue. They are typically found in the antigravity muscles of the leg and in the flight muscles of birds that fly (as opposed to chickens). In contrast, the pale (white) muscle fibers contain fewer mitochondria and rely mainly on anaerobic (without oxygen) metabolism. These fibers contract rapidly and powerfully, but they also fatigue rapidly. They are typical of muscles involved in escape reflexes, e.g., the jumping muscles of frogs and rabbits. In humans, the arm muscles contain a large number of white fibers.

Even though both types of muscle fiber can (and usually do) coexist in a given muscle, each *motor unit* contains muscle fibers of only a single type. Thus, **fast motor units** contain rapidly fatiguing white fibers, and **slow motor units** contain slowly fatiguing red fibers. Just as the muscle fibers of the two types of units differ, so do many of the properties of the alpha motor neurons. For example, the motor neurons of fast units are generally bigger and have larger-diameter, faster-conducting axons than the motor neurons of slow units. The firing properties of the two types of motor neuron also differ. Fast motor neurons tend to generate occasional high-frequency bursts of action potentials (30–60 impulses per second), whereas slow motor neurons are characterized by relatively steady, low-frequency activity (10–20 impulses per second).

Neuromuscular Matchmaking. The precise matching of particular motor neurons to particular muscle fibers raises an interesting question. Since we've been talking about chickens, let's pose the question this way: Which came first, the muscle fiber or the motor neuron? Perhaps during early embryonic development a matching of the appropriate axons with the appropriate muscle fibers occurs. Alternatively, we could imagine that the properties of the muscle are determined solely by the type of innervation it receives. If it receives a synaptic contact from a fast motor neuron, it becomes a fast

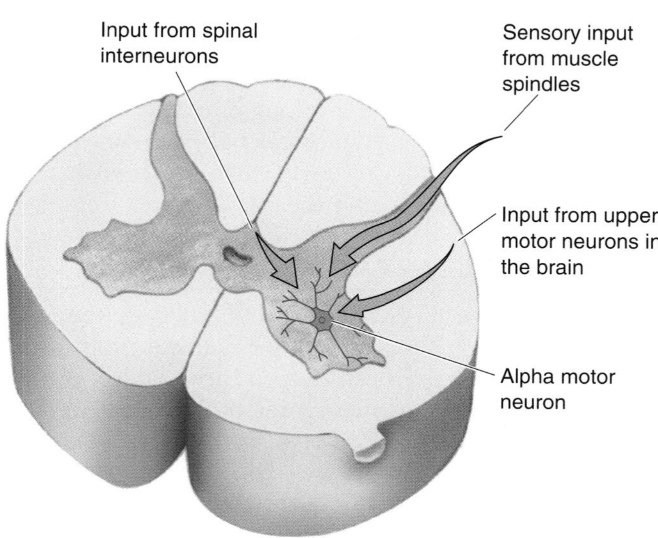

Figure 13.8
An alpha motor neuron and its three sources of input.

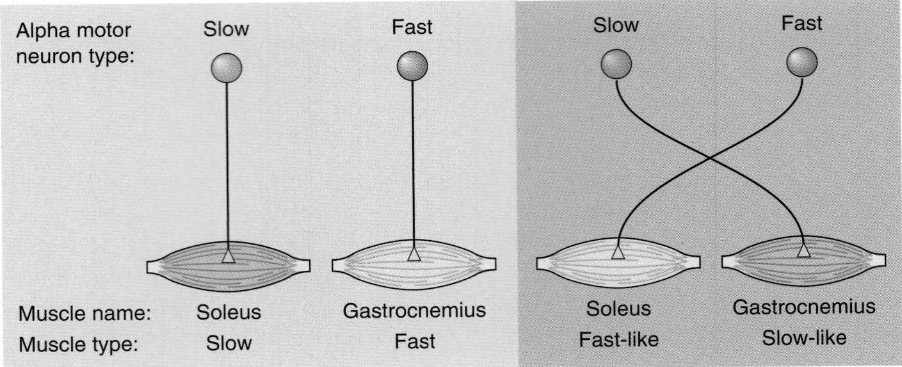

Figure 13.9
A crossed-innervation experiment. Forcing slow motor neurons to innervate a fast muscle causes the muscle to switch to assume slow properties.

fiber, and if it receives a synaptic contact from a slow motor neuron, it becomes a slow fiber.

John Eccles and his colleagues addressed this question in an experiment in which the normal innervation of a fast muscle was removed and replaced with a nerve that normally innervated a slow muscle (Figure 13.9). This procedure resulted in the muscle's taking on slow properties. These new properties included not only the type of contraction (slow, fatigue-resistant) but also a switch in much of the underlying biochemistry. This is a switch of muscle *phenotype*—its physical characteristics—because the types of proteins expressed by the muscle were altered by the new innervation. Work by Terje Lømo and his colleagues in Norway suggests that this switch in muscle phenotype can be induced simply by changing the activity in the motor neuron from a fast pattern (occasional bursts at 30–60 impulses per second) to a slow pattern (steady activity at 10–20 impulses per second). These findings are particularly interesting because they raise the possibility that *neurons* switch phenotype as a consequence of synaptic activity (experience) and that this may be a basis for learning and memory (discussed in detail in Chapters 23 and 24).

Besides the alterations imposed by *patterns* of motor neuron activity, muscle fibers are changed simply by varying the *absolute amount* of activity. A long-term consequence of increased activity (especially due to isometric exercise) is hypertrophy, or exaggerated growth, of the muscle fibers, as seen in bodybuilders. Conversely, prolonged inactivity leads to atrophy, or degeneration, of muscle fibers, which can happen when joints are immobilized in a cast following an injury. Clearly, there is an intimate relationship between the lower motor neuron and the muscle fibers it innervates (Box 13.1).

EXCITATION-CONTRACTION COUPLING

As we said, muscle contraction is initiated by the release of ACh from the axon terminals of the alpha motor neuron. As a result of the activation of nicotinic ACh receptors, ACh produces a large EPSP in the postsynaptic membrane. Because the membrane of the muscle cell contains voltage-gated sodium channels, this EPSP is sufficient to evoke an action potential in the muscle fiber (but see Box 13.2). This action potential triggers the release of Ca^{2+} ions from an organelle inside the muscle fiber, which leads to contraction of the fiber. Relaxation occurs when the Ca^{2+} levels are lowered by reuptake into the organelle. To understand this process, we must take a closer look at the muscle fiber.

Box 13.1

OF SPECIAL INTEREST

Amyotrophic Lateral Sclerosis

Amyotrophic lateral sclerosis (*ALS*) is an especially cruel disease. The first signs are muscle weakness and atrophy. Then, usually over the course of 3–5 years, all voluntary movement is lost—the abilities to walk, speak, swallow, and breathe are all taken away. Because the disease does not affect sensations, intellect, or cognitive function, patients are left to watch their body slowly waste away, keenly aware of what is happening. ALS is relatively rare, afflicting 1 in every 20,000 individuals. Still, an estimated 30,000 Americans are diagnosed with ALS. Perhaps its most famous victim was Lou Gehrig, a star baseball player with the New York Yankees, who died of ALS in 1936. ALS is still often called Lou Gehrig's disease.

Muscle weakness and paralysis are characteristics of motor unit damage. Indeed, the pathology associated with ALS is the degeneration of the large alpha motor neurons. The large neurons of the motor cortex that innervate alpha motor neurons are also affected, but curiously, other neurons in the CNS are spared. The selective damage to motor neurons explains the selective loss of motor functions in ALS patients.

The exact cause of ALS is unknown, but a few clues have emerged. A small percentage of ALS cases are genetic, and screens for the defective gene have pointed to a mutation affecting the enzyme *superoxide dismutase*. A toxic by-product of cellular metabolism is the negatively charged molecule O_2^-, called the superoxide radical. Superoxide radicals are extremely reactive and can inflict irreversible cellular damage. Superoxide dismutase is a key enzyme that causes superoxide radicals to lose their extra electrons, converting them back to oxygen. Thus, the loss of superoxide dismutase leads to a buildup of superoxide radicals and cellular damage, particularly in cells that are metabolically very active.

Another suspected cause of ALS is excitotoxicity. As we learned in Chapter 6, overstimulation by the excitatory neurotransmitter glutamate and closely related amino acids can cause the death of otherwise-normal neurons (see Box 6.2). Excitotoxicity has been implicated in the unusually high incidence of ALS on the island of Guam that occurred before World War II. It has been suggested that one environmental cause may have been the ingestion of cycad nuts, which contain an excitotoxic amino acid. In addition, research indicates that in ALS a glutamate transporter may be defective, thereby prolonging the exposure of active neurons to extracellular glutamate. Thus, the first drug approved by the U. S. Food and Drug Administration for the treatment of ALS was riluzole, a blocker of glutamate release. The drug treatment can slow the disease by a few months, but unfortunately the long-term outcome is the same.

There is still much to be learned about selective motor neuron loss in ALS. The slow accumulation of damage from the toxic by-products of cellular metabolism, however, probably occurs in all neurons and contributes to neuronal death over the lifespan of all people. This loss of neurons may not be inevitable. Vitamins C and E readily accept electrons from superoxide radicals and can be neuroprotective. Dietary supplementation with these vitamins may help stem the tide of neuronal loss that occurs in the brain as we grow older.

Muscle Fiber Structure

The structure of a muscle fiber is shown in Figure 13.10. Muscle fibers are formed early in fetal development by the fusion of muscle precursor cells, or myoblasts, which are derived from the mesoderm (see Chapter 7). This fusion leaves each cell with more than one cell nucleus, so individual muscle cells are said to be *multinucleated*. The fusion elongates the cells (hence the name "fiber"). Muscle fibers are delimited by an excitable cell membrane called the **sarcolemma**.

Within the muscle fiber are a number of cylindrical structures called **myofibrils**, which contract in response to an action potential sweeping down the sarcolemma. Myofibrils are surrounded by the **sarcoplasmic reticulum**, an extensive intracellular sac that stores Ca^{2+} ions (similar in appearance to the smooth endoplasmic reticulum (ER) of neurons; see Chapter 2). Action potentials sweeping along the sarcolemma gain access to sarcoplasmic reticu-

Box 13.2

OF SPECIAL INTEREST

Myasthenia Gravis

The neuromuscular junction is a synapse seemingly designed to work every time. A presynaptic action potential causes the contents of hundreds of synaptic vesicles to be released into the synaptic cleft. The liberated ACh molecules act at densely packed nicotinic receptors in the postsynaptic membrane, and the resulting EPSP is many times larger than what is necessary to trigger an action potential and twitch in the muscle fiber—normally, that is.

In the clinical condition *myasthenia gravis*, released ACh is far less effective, and neuromuscular transmission often fails. The name is derived from the Greek for "severe muscle weakness." The disorder is characterized by weakness and fatigability of voluntary muscles, typically including the muscles of facial expression, and it can be fatal if respiration is compromised. An unusual feature of myasthenia gravis is that the severity of the muscle weakness fluctuates, even over the course of a single day.

Myasthenia gravis is an *autoimmune disease*. For reasons we are only beginning to understand, the immune systems of afflicted individuals generate antibodies against their own nicotinic ACh receptors. The antibodies bind to the receptors, interfering with the normal actions of ACh at the neuromuscular junctions. In addition, the binding of antibodies to the receptors leads to secondary degenerative changes in the structure of the neuromuscular junctions that also make transmission less efficient.

An effective treatment for myasthenia gravis is the administration of drugs that inhibit the enzyme acetylcholinesterase (AChE). Recall from Chapters 5 and 6 that AChE breaks down ACh in the synaptic cleft. In low doses, AChE inhibitors can strengthen neuromuscular transmission by prolonging the life of released ACh. But the therapeutic window is narrow. As we saw in Box 5.5, too much ACh in the cleft leads to desensitization of the receptors and a block of neuromuscular transmission.

Another common treatment for myasthenia gravis involves suppression of the immune system, either with drugs or by surgical removal of the thymus gland. If managed carefully, the long-term prognosis is good for patients with this disease of the neuromuscular junction.

lum deep inside the fiber by way of a network of tunnels called **T tubules** (T for transverse). These are like inside-out axons; the lumen of each T tubule is continuous with the extracellular fluid.

Where the T tubule comes in close apposition to the sarcoplasmic reticulum, there is a specialized coupling of the proteins in the two membranes. A voltage-sensitive protein in the T tubule membrane is linked to a calcium channel protein in the sarcoplasmic reticulum. As illustrated in Figure 13.11, the arrival of an action potential in the T tubule membrane causes a conformational change in the voltage-sensitive membrane protein, which in essence "pulls the plug" on the calcium channel in the sarcoplasmic reticulum membrane. The resulting increase in intracellular free Ca^{2+} causes the myofibril to contract.

The Molecular Basis of Muscle Contraction

A closer look at the myofibril reveals how Ca^{2+} triggers contraction (Figure 13.12). The myofibril is divided into segments by disks called **Z lines** (so named because of their appearance when viewed from the side). A segment consists of two Z lines, and the myofibril in between is called a **sarcomere**. Anchored to each side of the Z lines is a series of bristles called **thin filaments**. The thin filaments from adjacent Z lines face one another but do not come in contact. Between and among the two sets of thin filaments are a series of fibers called **thick filaments**. Muscle contraction occurs when the thin

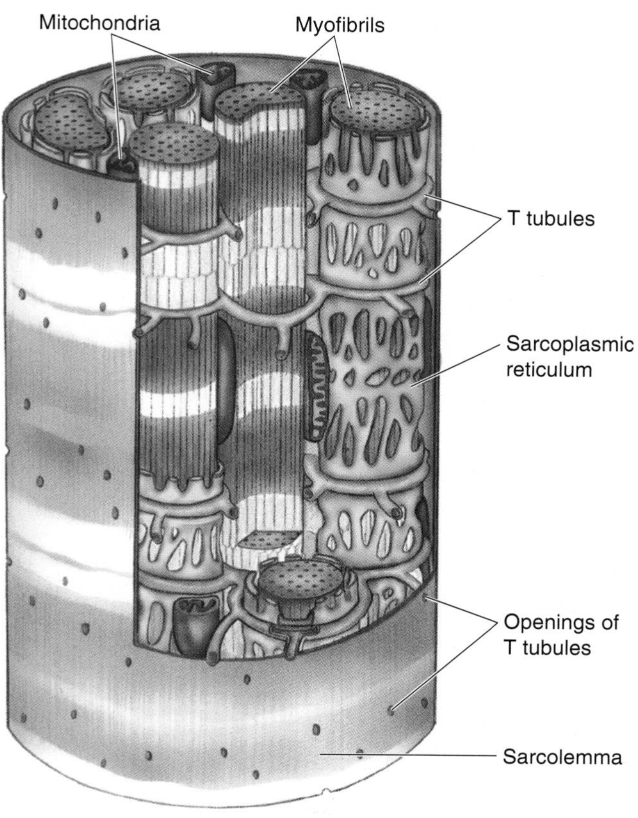

Figure 13.10
The structure of a muscle fiber. T tubules conduct electrical activity from the surface membrane into the depths of the muscle fiber.

filaments slide along the thick filaments, bringing adjacent Z lines toward one another. In other words, the sarcomere becomes shorter in length. This sliding-filament model of sarcomere shortening is shown in Figure 13.13.

The sliding of the filaments with respect to one another occurs because of the interaction between the major thick filament protein, **myosin**, and the major thin filament protein, **actin**. The exposed "heads" of the myosin molecules bind actin molecules and then undergo a conformational change that causes them to rotate (Figure 13.14). This rotation causes the thick filament to move with respect to the thin filament. At the expense of adenosine triphosphate (ATP), the myosin heads then disengage and "uncock" so that the process can repeat itself. Repeating this cycle enables the myosin heads to "walk" along the actin filament.

When the muscle is at rest, myosin cannot interact with actin because the myosin attachment sites on the actin molecule are covered by the protein **troponin**. Ca^{2+} initiates muscle contraction by binding to troponin, thereby exposing the sites where myosin binds to actin. Contraction continues as long as Ca^{2+} and ATP are available; relaxation occurs when the Ca^{2+} is sequestered by the sarcoplasmic reticulum. The reuptake of Ca^{2+} by the sarcoplasmic reticulum depends on the action of a calcium pump and hence also requires ATP.

We can summarize the steps of excitation-contraction coupling as follows:

Excitation

1. An action potential occurs in an alpha motor neuron axon.
2. ACh is released by the axon terminal of the alpha motor neuron at the neuromuscular junction.
3. Nicotinic receptor channels open and the postsynaptic sarcolemma depolarizes (EPSP).

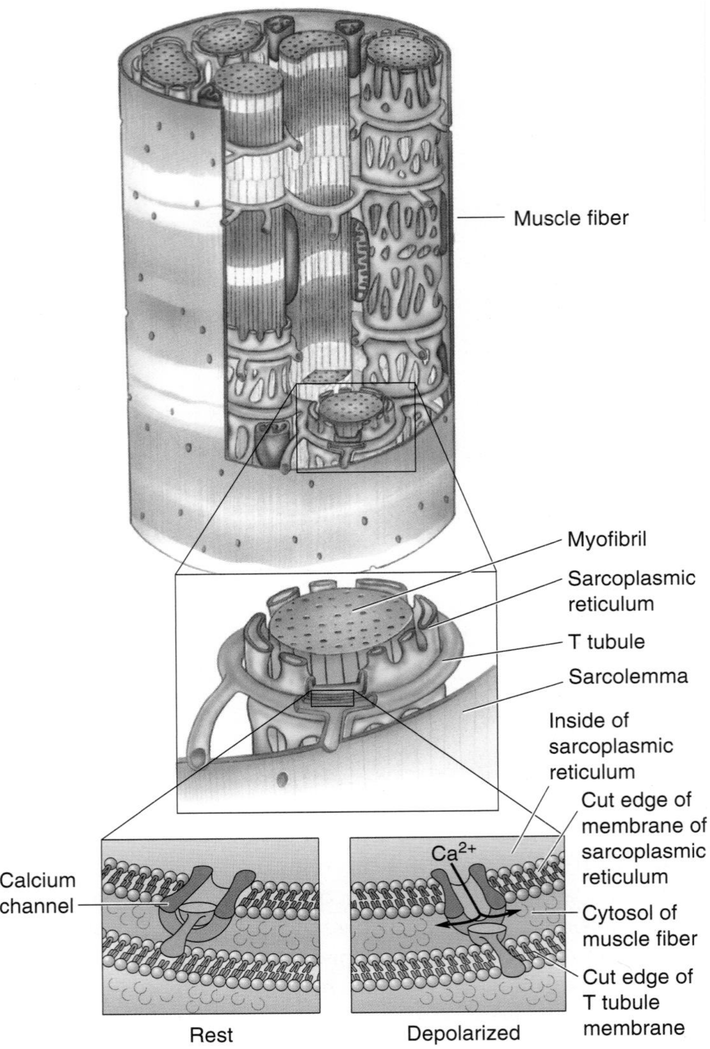

Figure 13.11
The release of Ca^{2+} from the sarcoplasmic reticulum. Depolarization of the T tubule membrane causes conformational changes in proteins that are linked to calcium channels in the sarcoplasmic reticulum, releasing stored Ca^{2+} into the cytosol of the muscle fiber.

4. Voltage-gated sodium channels open; an action potential is generated in the muscle fiber and sweeps down the sarcolemma.
5. Depolarization of the T tubules causes Ca^{2+} release from the sarcoplasmic reticulum.

Contraction
1. Ca^{2+} binds to troponin.
2. Myosin binding sites on actin are exposed.
3. Myosin heads bind actin.
4. Myosin heads rotate.
5. Myosin heads disengage at the expense of ATP.
6. The cycle continues as long as Ca^{2+} and ATP are present.

Relaxation
1. Ca^{2+} is sequestered by the sarcoplasmic reticulum by an ATP-driven pump.
2. Myosin binding sites on actin are covered by troponin.

You can now understand why death causes stiffening of the muscles, a condition known as *rigor mortis*. Starving the muscle cells of ATP both pre-

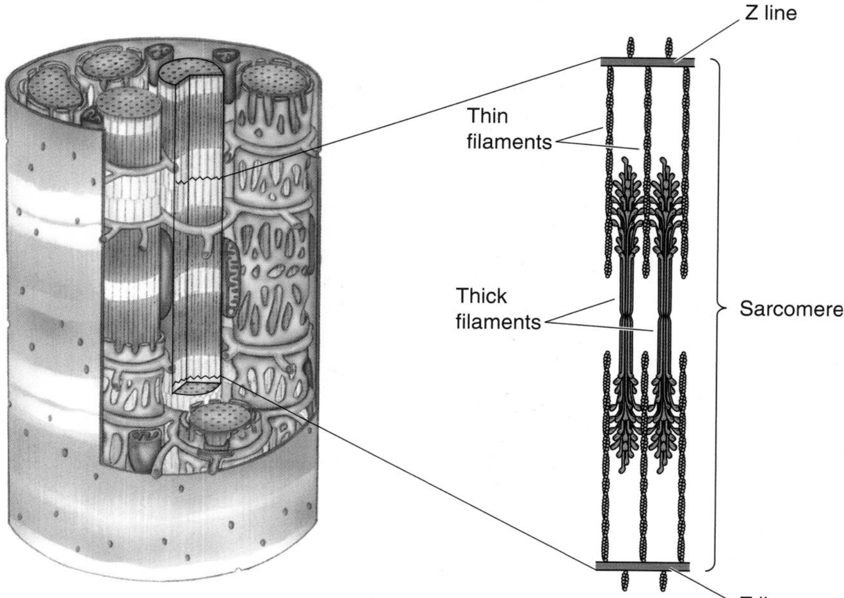

Figure 13.12
The myofibril: a closer look.

vents detachment of the myosin heads and leaves the myosin attachment sites on the actin filaments exposed for binding. The end result is the formation of permanent attachments between the thick and thin filaments.

Since the proposal of the sliding-filament model in 1954 by English physiologists Hugh and Andrew Huxley, there has been a tremendous amount of progress in identifying the detailed molecular mechanisms of excitation-contraction coupling in muscle. This progress has resulted from a multidisciplinary approach to the problem, with critical contributions made by the use of electron microscopy and biochemical and biophysical methods. The very recent application of molecular genetic techniques also has added important new information to our understanding of muscle function in both health and disease (Box 13.3).

SPINAL CONTROL OF MOTOR UNITS

We've traced the action potentials sweeping down the axon of the alpha motor neuron and seen how this causes contraction of the muscle fibers in the motor unit. Now we explore how the activity of the motor neuron is itself controlled. We begin with a discussion of the first source of synaptic input to the alpha motor neuron introduced earlier—sensory feedback from the muscles themselves.

Proprioception From Muscle Spindles

As already mentioned, deep within most skeletal muscles are specialized structures called **muscle spindles** (Figure 13.15). A muscle spindle consists of several types of specialized skeletal muscle fibers contained in a fibrous capsule. The middle third of the capsule is swollen, giving the structure the shape for which it is named. In this middle (equatorial) region, group Ia sensory axons wrap around the muscle fibers of the spindle. The spindles and their associated Ia axons, specialized for the detection of changes in muscle length (stretch), are examples of **proprioceptors**. These receptors are a component of the somatic sensory system that is specialized for "body sense," or **proprioception** (from the Latin *proprio*, "one's own," and *capio*, "take, understand"), which informs us about how our bodies are positioned and moving in space.

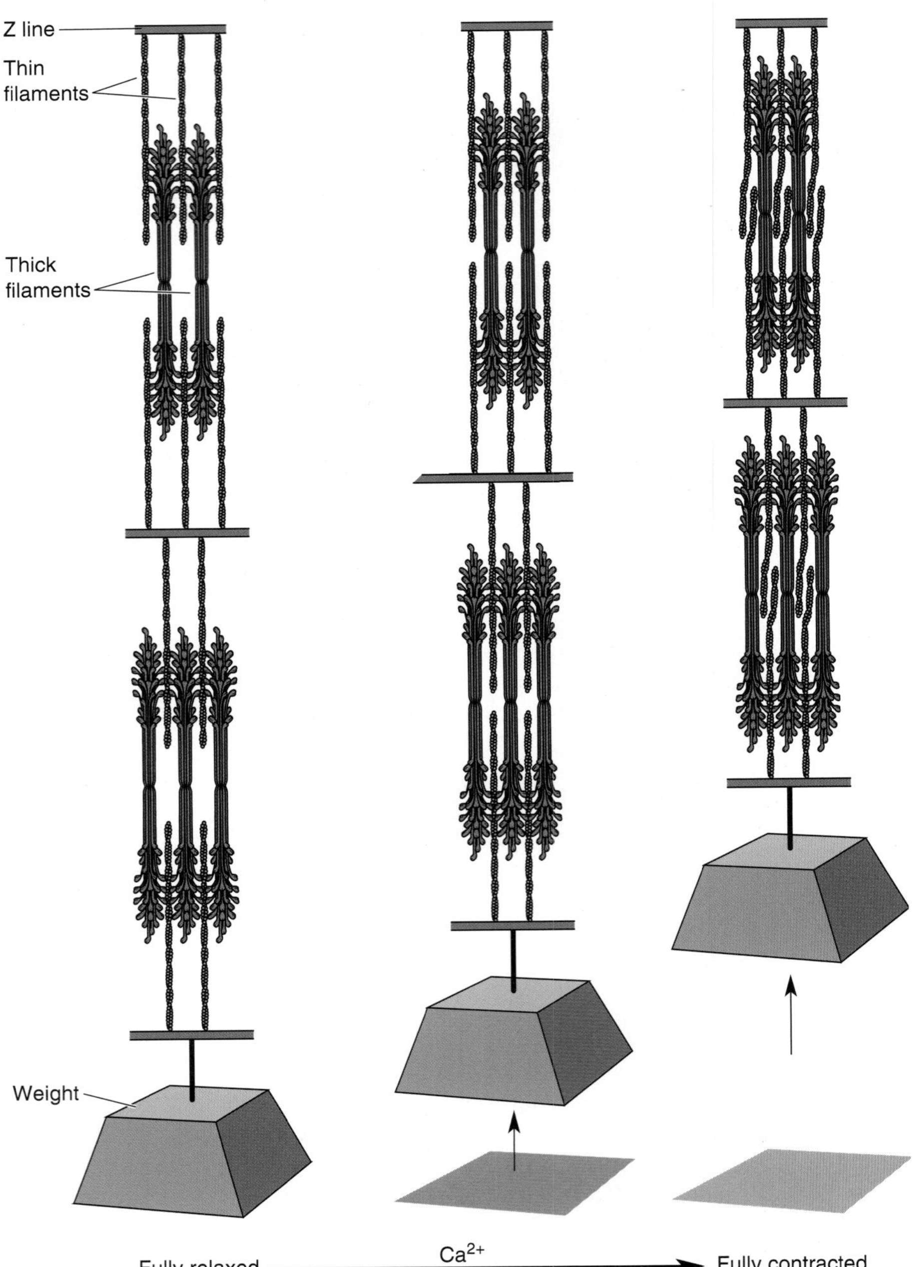

Figure 13.13
The sliding-filament model of muscle contraction. Myofibrils shorten when the thin filaments slide toward one another on the thick filaments.

Recall from Chapter 12 that group I axons are the thickest myelinated axons in the body, meaning that they conduct action potentials very rapidly. Within this group, the Ia axons are the largest and the fastest. Ia axons enter the spinal cord via the dorsal roots, branch repeatedly, and form excitatory synapses upon both interneurons and alpha motor neurons of the ventral horns. The Ia inputs are also very powerful. Neurophysiologist Lorne Mendell, working at Harvard with Henneman, was able to show that a single Ia axon synapses on virtually every alpha motor neuron in the pool innervating the same muscle that contains the spindle.

The Myotatic Reflex. The function of this sensory input to the spinal cord was first shown by Sherrington, who noted that when a muscle is pulled on, it tends to pull back (contract). The fact that this **myotatic reflex** (*myo* from

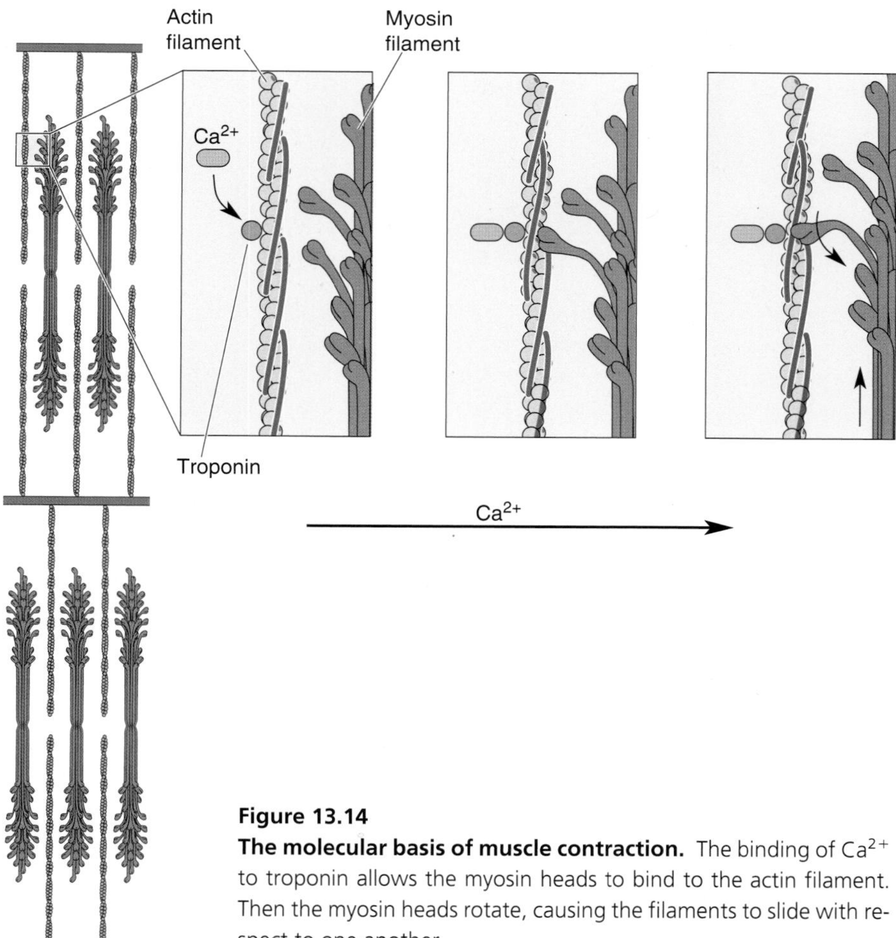

Figure 13.14
The molecular basis of muscle contraction. The binding of Ca^{2+} to troponin allows the myosin heads to bind to the actin filament. Then the myosin heads rotate, causing the filaments to slide with respect to one another.

the Greek for "muscle," *tatic* from the Greek for "stretch") involves sensory feedback from the muscle was shown by cutting the dorsal roots. Even though the alpha motor neurons were left intact, this procedure eliminated the stretch reflex and caused a loss of muscle tone. Sherrington deduced that the motor neurons must receive a continual synaptic input from the muscles. Later work showed that the discharge of Ia sensory axons is closely related to the length of the muscle. As the muscle is stretched, the discharge rate goes up; as the muscle is shortened and goes slack, the discharge rate goes down.

The Ia fiber and the alpha motor neurons on which it synapses constitute the *monosynaptic myotatic reflex arc*—"monosynaptic" because only one synapse separates the primary sensory input from the motor neuron output. Figure 13.16 shows how this reflex arc serves as an antigravity feedback loop. When a weight is placed on a muscle and the muscle starts to lengthen, the muscle spindles are stretched. The stretching of the equatorial region of the spindle leads to depolarization of the Ia axon endings because of the opening of mechanosensitive ion channels (see Chapter 12). The resulting increased action potential discharge of the Ia axons synaptically depolarizes the alpha motor neurons, which respond by increasing their action potential frequency. This causes the muscle to contract, thereby shortening it. The knee-jerk reflex that occurs when your doctor taps the tendon beneath your kneecap tests for the intactness of this reflex arc in the quadriceps muscle of your thigh (Figure 13.17).

Box 13.3

OF SPECIAL INTEREST

Duchenne Muscular Dystrophy

Muscular dystrophy describes a group of inherited disorders, all of which are characterized by muscle weakness. One type, *Duchenne muscular dystrophy*, afflicts boys before adolescence. First detected as a weakness of the legs, it usually puts its victim in a wheelchair by age 12. The disease continues to progress, and afflicted males typically do not survive past age 30.

The characteristic hereditary pattern of this disease, which afflicts only males but is passed on from their mothers, led to a search for a defective gene on the X chromosome. Major breakthroughs came in the late 1980s, when the defective region of the X chromosome was identified. Researchers discovered that this region contains the gene for the cytoskeletal protein *dystrophin*. Boys with Duchenne muscular dystrophy lack the mRNA encoding this protein. A milder form of the disease, called Becker muscular dystrophy, was found to be associated with an altered mRNA encoding only a portion of the dystrophin protein.

Dystrophin is a large protein that contributes to the muscle cytoskeleton lying just under the sarcolemma. It must not be an absolute requirement for muscle contraction, however, because movements in afflicted boys appear to be normal during the first few years of life. It is possible that the absence of dystrophin leads to secondary changes in the contractile apparatus, eventually resulting in muscle degeneration.

Whatever the normal function of dystrophin ultimately proves to be, it is clear that our models of excitation-contraction coupling are based on knowledge of only a fraction of the proteins that are normally expressed by neurons and muscle fibers. As more is learned about the various proteins in the membrane and cytosol, we can anticipate significant revisions of these models. Dystrophin is also concentrated in axon terminals in the brain, where it may contribute to excitation-secretion coupling.

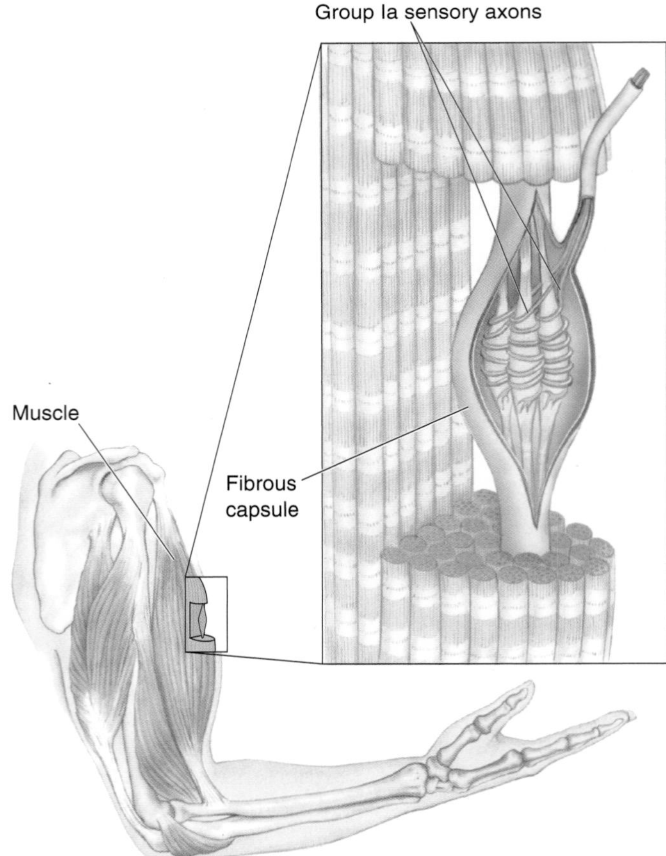

Figure 13.15
A muscle spindle and its sensory innervation.

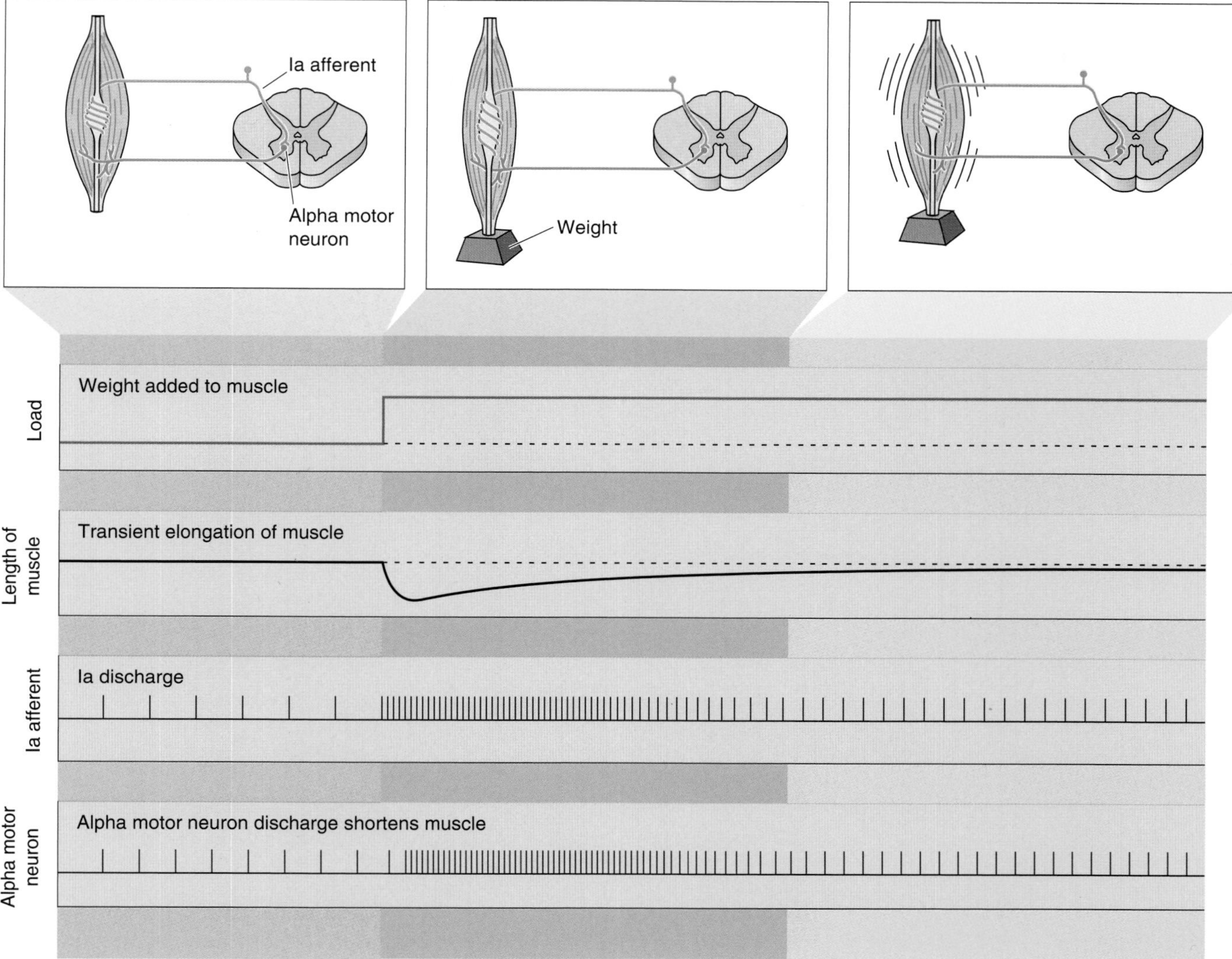

Figure 13.16
The myotatic reflex. This illustration shows the response of a Ia fiber and a motor neuron to the sudden addition of weight that stretches the muscle.

Gamma Motor Neurons

The muscle spindle contains modified skeletal muscle fibers within its fibrous capsule. These muscle fibers are called **intrafusal fibers** to distinguish them from the more numerous **extrafusal fibers**, which lie outside the spindle and form the bulk of the muscle. An important difference between the two types of muscle fibers is that only extrafusal fibers are innervated by alpha motor neurons. Intrafusal fibers receive their motor innervation by another type of lower motor neuron called a **gamma motor neuron** (Figure 13.18).

Imagine a situation in which a muscle contraction is commanded by an upper motor neuron. The alpha motor neurons respond, the extrafusal fibers contract, and the muscle shortens. The response of the muscle spindles is shown in Figure 13.19. If they went slack, the Ia fibers would become silent and the spindle would go "off the air," no longer providing information about muscle length. This does not happen, however, because the gamma motor neurons are also activated. Gamma motor neurons innervate the intrafusal muscle fiber at the two ends of the muscle spindle. Activation of

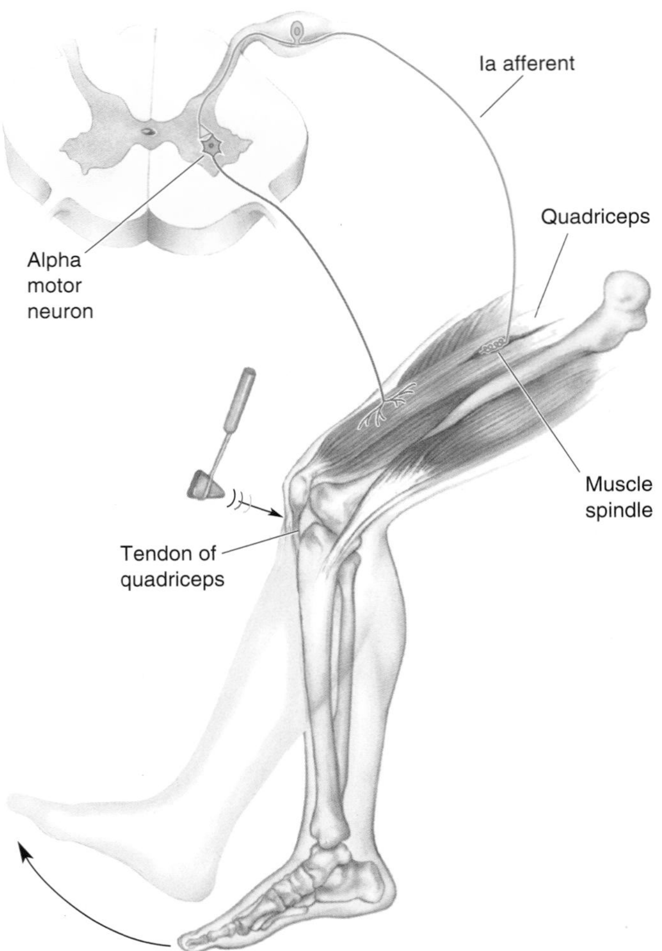

Figure 13.17
The knee-jerk reflex.

these fibers causes a contraction of the two poles of the muscle spindle, thereby pulling on the noncontractile equatorial region and keeping the Ia afferents active. Notice that the activation of alpha and gamma motor neurons has opposite effects on Ia output; alpha activation alone decreases Ia activity, while gamma activation alone increases Ia activity.

Recall from the earlier discussion that the monosynaptic myotatic reflex arc can be viewed as a feedback loop. The principles of feedback control systems are that a set point is determined (in this case, the desired muscle length), deviations from the set point are detected by a sensor (the Ia axon endings), and deviations are compensated for by an effector system (alpha motor neurons and extrafusal muscle fibers), returning the system to the set point. Changing the activity of the gamma motor neurons changes the set point of the myotatic feedback loop. This circuit, gamma motor neuron → intrafusal muscle fiber → Ia afferent → alpha motor neuron → extrafusal muscle fibers, is sometimes called the *gamma loop*.

Alpha and gamma motor neurons are simultaneously activated by descending commands from the brain. By regulating the set point of the myotatic feedback loop, the gamma loop provides additional control of alpha motor neurons and muscle contraction.

Proprioception From Golgi Tendon Organs

Muscle spindles are not the only source of proprioceptive inputs from the muscles. Another sensor in skeletal muscle is the **Golgi tendon organ**, which

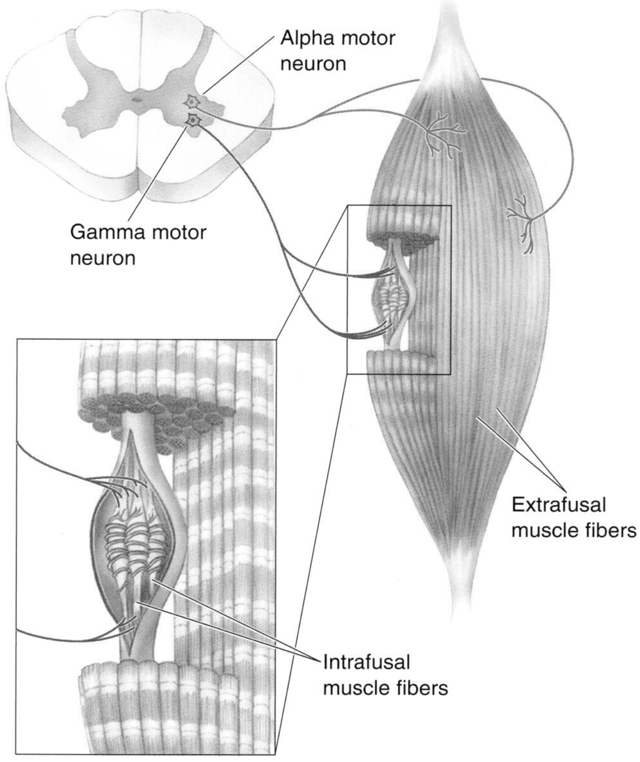

Figure 13.18
Alpha motor neurons, gamma motor neurons, and the muscle fibers they innervate.

acts like a strain gauge; that is, it monitors muscle tension, or the force of contraction. Golgi tendon organs are located at the junction of the muscle and the tendon and are innervated by group Ib sensory axons that are slightly smaller than the Ia axons innvervating the muscle spindles (Figure 13.20).

While spindles are situated *in parallel* with the muscle fibers, Golgi tendon organs are situated *in series* (Figure 13.21). This different anatomical arrangement is what distinguishes the types of information these two sensors provide the spinal cord: Ia activity from the spindle encodes *muscle length* information, while Ib activity from the Golgi tendon organ encodes *muscle tension* information.

The Ib afferents enter the spinal cord, branch repeatedly, and synapse on interneurons in the ventral horn. Some of these interneurons form inhibitory connections with the alpha motor neurons innervating the same muscle. This

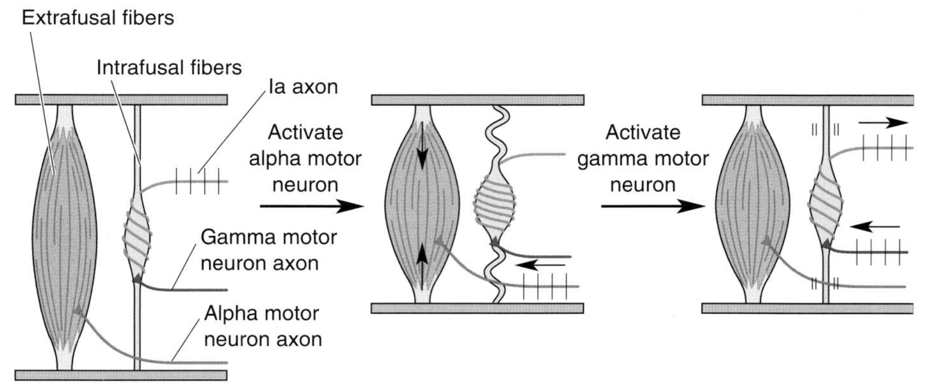

Figure 13.19
The function of gamma motor neurons. **(a)** Activation of alpha motor neurons shortens the extrafusal muscle fibers. If the muscle spindle becomes slack, it goes "off the air" and no longer reports the length of the muscle. **(b)** Activation of gamma motor neurons contracts the poles of the spindle, keeping it "on the air."

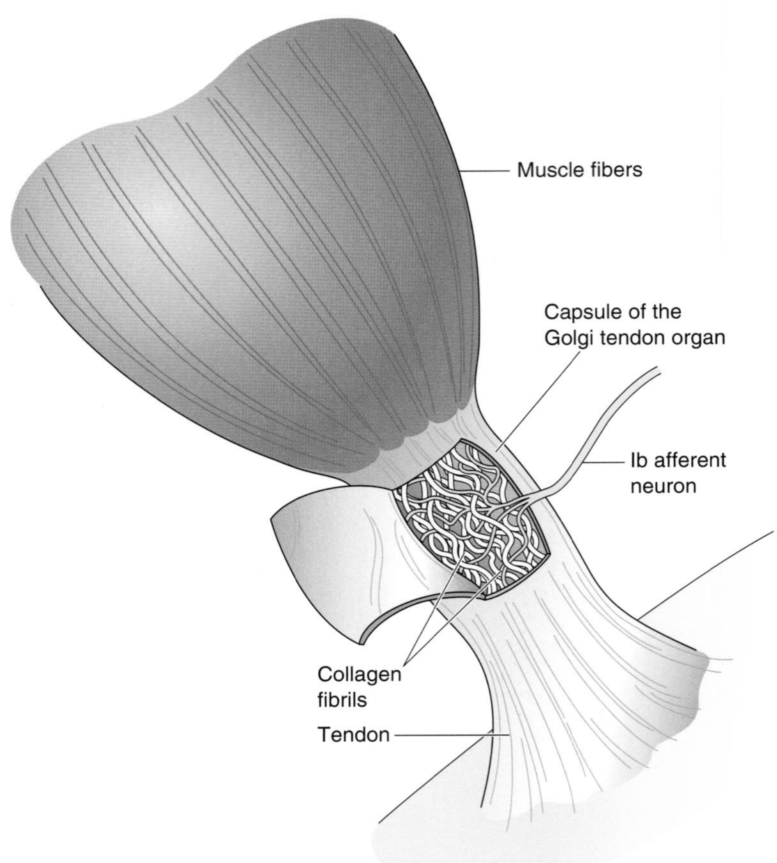

Figure 13.20
A Golgi tendon organ.

is the basis for another spinal reflex, the *reverse myotatic reflex* (Figure 13.22). In extreme circumstances, this reflex arc protects the muscle from overload. The normal function, however, is to regulate muscle tension within an optimal range. As muscle tension increases, the inhibition of the alpha motor neuron slows muscle contraction; as muscle tension falls, the inhibition of the alpha motor neuron is reduced, and muscle contraction increases. This type of proprioceptive feedback is thought to be particularly important for the

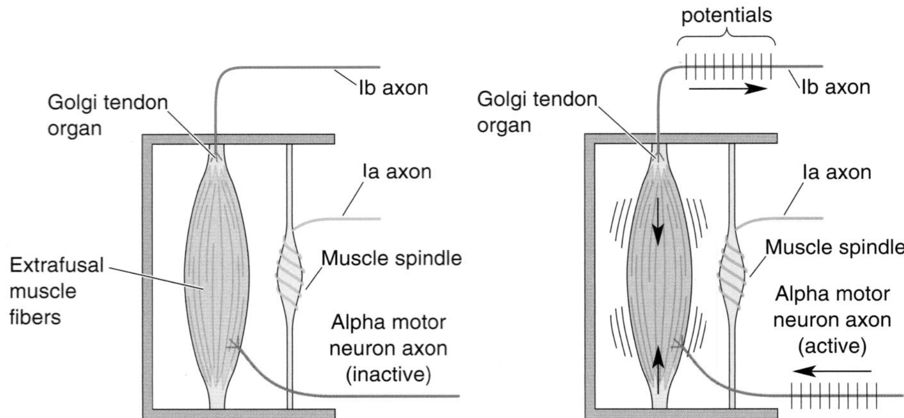

Figure 13.21
The organization of muscle proprioceptors. (a) Muscle spindles are arranged parallel to the extrafusal fibers; Golgi tendon organs lie in series between the muscle fibers and their points of attachment. **(b)** Golgi tendon organs respond to increased tension on the muscle and transmit this information to the spinal cord via type Ib sensory afferents. Because the activated muscle does not change length, the Ia afferents remain silent in this example.

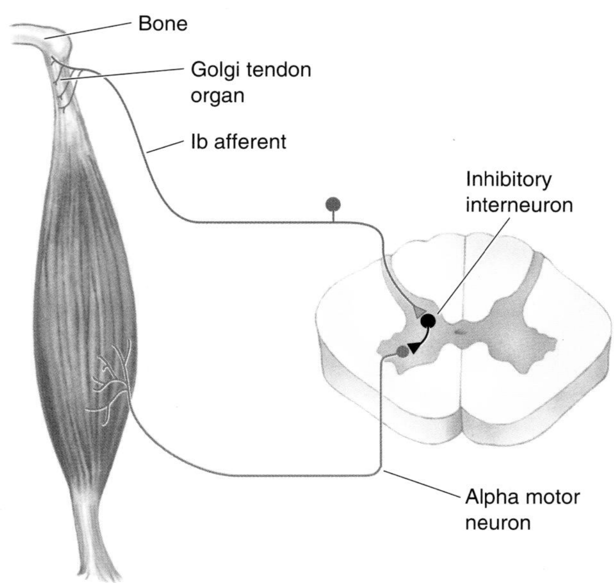

Figure 13.22
Circuitry of the reverse myotatic reflex.

proper execution of fine motor acts, such as the manipulation of fragile objects with the hands, which require a steady but not too powerful grip.

Proprioception From the Joints. We have focused on the proprioceptors that are involved in reflex control of the spinal motor neurons. Besides muscle spindles and Golgi tendon organs, however, various proprioceptive axons are present in the connective tissues of joints, especially within the fibrous tissue surrounding the joints (joint capsules) and ligaments. These mechanosensitive axons respond to changes in the angle, direction, and velocity of movement in a joint. Most are rapidly adapting, meaning that sensory information about a *moving* joint is plentiful, but nerves encoding the *resting* position of a joint are few. We are, nevertheless, quite good at judging the position of a joint even with our eyes closed. It seems that information from joint receptors is combined with that from muscle spindles and Golgi tendon organs and probably from receptors in the skin to estimate joint angle. Removal of one source of information can be compensated for by the use of the other sources. When an arthritic hip is replaced with a steel and plastic one, patients are still able to tell the angle between their thigh and their pelvis, despite the fact that all of their hip joint mechanoreceptors are sitting in a jar of formaldehyde in another room.

Spinal Interneurons

The actions of Ib inputs from Golgi tendon organs on alpha motor neurons are entirely *polysynaptic*—they are all mediated by intervening spinal interneurons. Indeed, most of the input to the alpha motor neurons comes from interneurons of the spinal cord. Spinal interneurons receive synaptic input from primary sensory axons, descending axons from the brain, and collaterals of lower motor neuron axons. The interneurons are themselves networked together in a way that allows coordinated motor programs to be generated in response to their many inputs.

Inhibitory Input. Interneurons play a critical role in the proper execution of even the simplest reflexes. Consider the myotatic reflex, for example. Compensation for the lengthening of one set of muscles, such as the flexors of the elbow, involves contraction of the flexors via the myotatic reflex, but it also requires the relaxation of the antagonist muscles, the extensors. This contraction of one set of muscles accompanied by the relaxation of the antagonist muscles is called **reciprocal inhibition**. Its importance is obvious; imagine how hard it would be to lift something if your own antagonist muscles were constantly opposing you. In the case of the myotatic reflex, reciprocal inhibition occurs because collaterals of the Ia afferents synapse on inhibitory spinal interneurons that contact the alpha motor neurons supplying the antagonist muscles (Figure 13.23).

Reciprocal inhibition is also used by descending pathways to overcome the powerful myotatic reflex. Suppose the flexors of the elbow are voluntarily commanded to contract. The resulting stretch of the antagonist extensor muscles activates their myotatic reflex arc, which strongly resists flexion of the joint. The descending pathways that activate the alpha motor neurons controlling the flexors, however, also activate interneurons, which inhibit the alpha motor neurons that supply the antagonist muscles.

Excitatory Input. Not all interneurons are inhibitory. An example of a reflex mediated in part by excitatory interneurons is the *flexor reflex* (Figure 13.24). This is a complex reflex arc used to withdraw a limb from an aversive stimulus, such as the withdrawal of your foot from the thumbtack as discussed in Chapter 3. This reflex is far slower than the myotatic reflex, indicating that a number of interneurons intervene between the sensory stimulus and the coordinated motor act. The pain fibers entering the spinal cord branch profusely and activate interneurons in several spinal segments. These cells eventually excite the alpha motor neurons that control all of the flexor muscles of the affected limb (and inhibitory interneurons are also recruited to inhibit the alphas that control the extensors).

You're walking along, and you step on a tack. Thanks to the flexor reflex, you (reflexively) yank your foot up. But where would that leave the rest of your body if nothing else happened? Falling down, most likely. Luckily, an additional component of the reflex is recruited: the activation of extensor muscles and the inhibition of flexors *on the opposite side.* This *crossed-extensor reflex* is used to compensate for the extra load imposed by limb withdrawal on the antigravity extensor muscles of the opposite leg (Figure 13.25). This is another example of reciprocal inhibition, but in this case, activation of the flexors on one side of the spinal cord is accompanied by inhibition of the flexors on the opposite side.

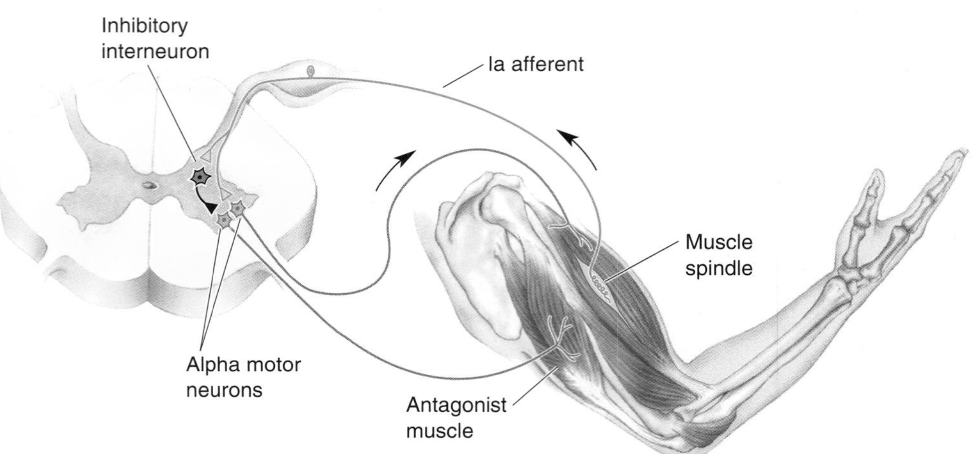

Figure 13.23
Reciprocal inhibition of flexors and extensors of the same joint.

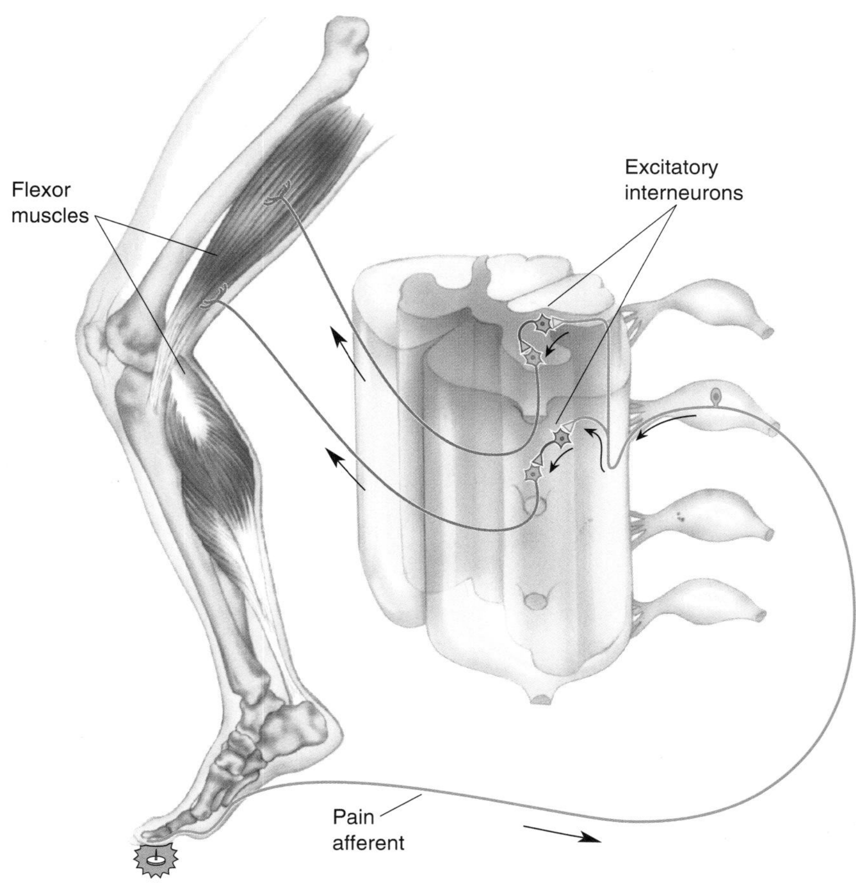

Figure 13.24
Circuitry of the polysynaptic flexor reflex.

The Generation of Spinal Motor Programs for Walking

The crossed-extensor reflex, in which one side extends as the other side flexes, seems to provide a building block for locomotion. When you walk, you alternately withdraw and extend your two legs. All that's lacking is a mechanism to coordinate the timing. In principle, this could be a series of descending commands from upper motor neurons. As we already suspect from our consideration of headless chickens' behavior, it seems likely, however, that this control is exerted from within the spinal cord. Indeed, a complete transection of a cat's spinal cord at the mid-thoracic level leaves the hind limbs capable of generating coordinated walking movements. The circuit for the coordinated control of walking must reside, therefore, within the spinal cord. In general, circuits that give rise to rhythmic motor activity are called **central pattern generators**.

How do neural circuits generate rhythmic patterns of activity? There is no single answer, and different circuits use different mechanisms. The simplest pattern generators, however, are single neurons whose membrane properties endow them with pacemaker properties. An interesting example comes from the work of Sten Grillner and his colleagues at the Nobel Institute for Neurophysiology in Stockholm, Sweden (Box 13.4). Based on the assumption that the spinal central pattern generators for locomotion in different species are variations on a plan that was established in a common ancestor, Grillner has focused on the mechanism for swimming in the lamprey, a jawless fish that has evolved slowly over the past 450 million years. The lamprey's spinal cord can be dissected and kept alive *in vitro* for several days. Electrical stimulation of the stumps of axons descending from the brain can generate alter-

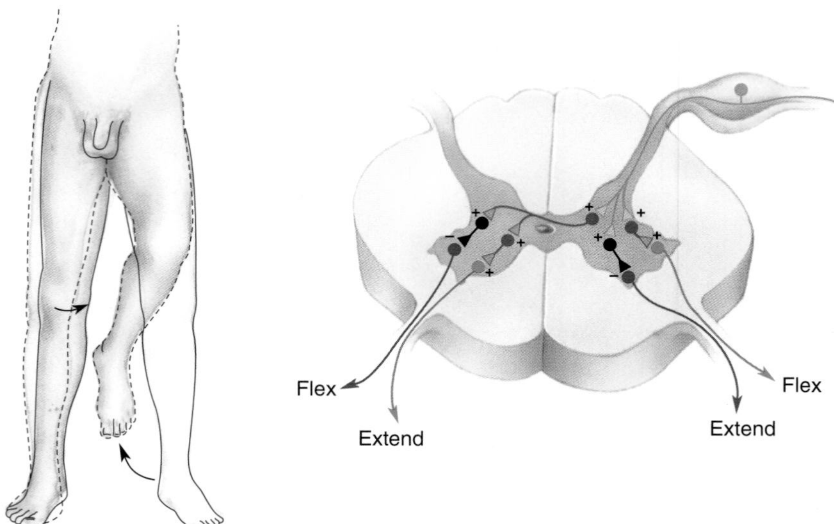

Figure 13.25
Circuitry of the crossed-extensor reflex.

nating rhythmic activity in the spinal cord, mimicking that which occurs during swimming. In an important series of experiments, Grillner showed that activation of *N*-methyl-D-aspartate (NMDA) receptors on spinal interneurons was sufficient to generate this locomotor activity.

Recall from Chapter 6 that NMDA receptors are glutamate-gated ion channels with two peculiar properties: (1) they allow more current to flow into the cell when the postsynaptic membrane is depolarized, and (2) they admit Ca^{2+} as well as Na^{+} into the cell. In addition to NMDA receptors, spinal interneurons possess Ca^{2+}-activated potassium channels. Now imagine the cycle that is initiated when NMDA receptors are activated by glutamate (Figure 13.26):

1. The membrane depolarizes.
2. Na^{+} and Ca^{2+} flow into the cell through the NMDA receptors.
3. Ca^{2+} activates potassium channels.
4. K^{+} flows out of the cell.
5. The membrane hyperpolarizes.
6. Ca^{2+} stops flowing into the cell.
7. The potassium channels close.
8. The membrane depolarizes and the cycle repeats.

It is easy to imagine how intrinsic pacemaker activity in spinal interneurons might act as the primary rhythmic driving force for sets of motor neurons that, in turn, command cyclic behaviors like walking. Pacemaker neurons are, however, not solely responsible for generating rhythms in vertebrates. They are embedded within interconnected circuits, and it is the combination of intrinsic pacemaker properties and synaptic interconnections that produces rhythm.

An example of a possible pattern-generating circuit for walking is shown in Figure 13.27. According to this scheme, walking is initiated when a steady input excites two interneurons that connect to the motor neurons controlling the flexors and extensors, respectively. The interneurons respond to a continuous input by generating bursts of outputs. The activities of the two interneurons alternate because they inhibit each other via other (inhibitory) interneurons. Thus, a burst of activity in one interneuron strongly inhibits the other. Then, using the spinal cord circuitry of the crossed-extensor reflex, the movements of the opposite limb can be coordinated so that flexion on one

Box 13.4

PATH OF DISCOVERY

Locomotion: From Molecules to Networks and Behavior

BY STEN GRILLNER

The nervous system contains a number of functional networks that can produce a broad repertoire of stereotypical motor patterns. I have been interested in this type of neural organization ever since I was a graduate student, when I realized it was possible to make these networks active in reduced preparations. In 1969, I performed some simple experiments demonstrating that animals with a transected spinal cord could make well-coordinated locomotor movements if injected with precursors of certain neurotransmitters. This meant that the spinal cord, devoid of all control from higher centers, could coordinate the limb movements used in walking and thus contained the necessary neuronal networks. This experiment had a major influence on the direction of my future research.

A few years earlier, G. N. Orlovsky and M. L. Shik, working in Moscow, showed that stimulation of certain areas of the brain stem could make an animal start to walk. The stronger the stimulation, the faster the animal would move. A simple control signal from the brain stem could thus make the networks in the spinal cord coordinate hundreds of muscles with the accurate timing required within each limb and between the different limbs. The same brain areas can be stimulated in birds, mammals, and fish to produce the locomotor activity characteristic of a given species. An overriding goal of my research has been to understand how this well-conserved system operates. How do molecules and cells interact to produce network activity? How does this activity translate into behavior?

I had to find an experimental model simple enough to provide answers to these technically demanding questions. I chose the lamprey, a jawless fish (cyclostome) with a simple nervous system containing fewer neurons than that of other vertebrates but retaining the same overall design (Figure A). Another advantage of the lamprey is that the brain stem or spinal cord can be isolated from the rest of the nervous system and kept alive *in vitro* for several days. When stimulated appropriately, this isolated part of the nervous system can generate the motor pattern underlying swimming. It thus attempts to perform an active "behavior." This preparation has enabled me, together with Peter Wallén and other colleagues, to analyze how a locomotor network operates.

Sten Grillner

Even this "simple" network is remarkably complex. It consists of glutamatergic excitatory and glycinergic inhibitory interneurons connected in an intricate pattern. The interneurons are turned on by glutamate released by axons descending from the brain stem. We discovered that NMDA receptors play a particularly important role. Working in concert with other types of ion channels, the NMDA receptors can convert a simple signal, such as an elevation in glutamate, into a periodic change in the voltage of the neuron. The blend of ion channels in each cell gives its particular electrical signature and accounts for its role in the network. It has been very interesting to see how the different factors that contribute to the burst pattern actually combine to ensure a safe and accurate control system. The network activity can be tuned to different environmental demands by different neurotransmitter systems (e.g., aminergic and peptidergic) that target specific ion channels in particular network neurons.

In a network, many processes occur in parallel: Interacting ion channels open and close in different parts of a cell, and different cells interact. It is practically impossible to fully comprehend what actually happens within a network like this. Modeling of cell and network function has therefore become an indispensable tool for testing whether the experimental results can account for the behavioral findings. Modern computers provide remarkable new possibilities: Nerve cells, networks, and muscles can be simulated and used to produce simulated behavior. In the case of the locomotor network, we can thus bridge the gap from molecule, cell, and synapse to network and actual behavior.

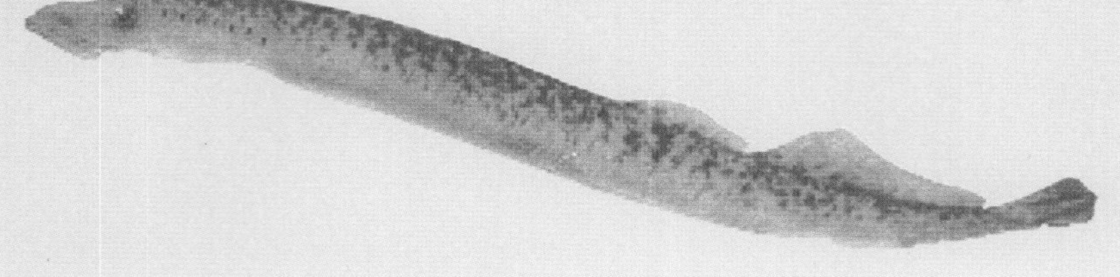

Figure A
A lamprey.

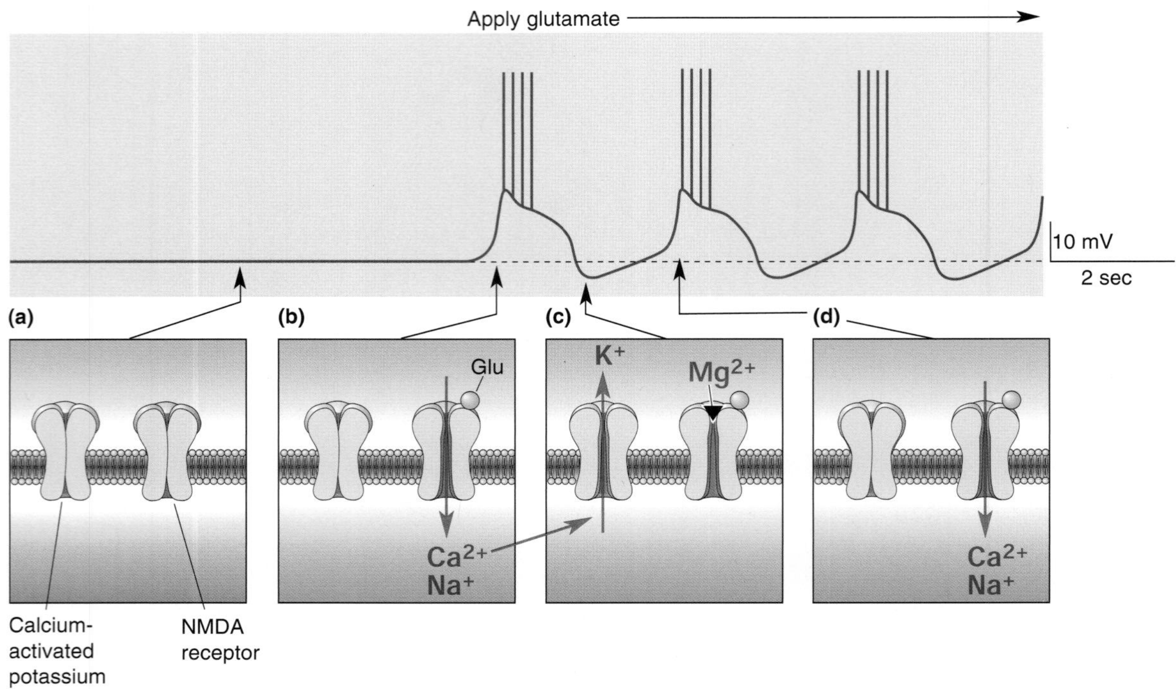

Figure 13.26

Rhythmic activity in a spinal interneuron. Some neurons respond to the activation of NMDA receptors with rhythmic depolarization. **(a)** In the resting state, the NMDA receptor channels and the calcium-activated potassium channels are closed. **(b)** Glutamate causes the NMDA receptors to open, the cell membrane to depolarize, and Ca^{2+} ions to enter the cell. **(c)** The rise in intracellular $[Ca^{2+}]$ causes the Ca^{2+}-activated potassium channels to open. K^+ ions leave the neuron, hyperpolarizing the membrane. The hyperpolarization allows Mg^{2+} ions to enter and clog the NMDA channel, arresting the flow of Ca^{2+}. **(d)** As $[Ca^{2+}]$ falls, the potassium channels close, resetting the membrane for another oscillation. (Source: Adapted from Wallén and Grillner, 1987.)

side is accompanied by extension on the other. By adding further interneuronal connections between the lumbar and cervical spinal segments, we can account for the swinging of the arms that accompanies walking.

Work on many vertebrate species, from lampreys to humans, has shown that locomotor activity in the spinal cord and its coordination depend on multiple mechanisms. Such complexity is not surprising when one considers the demands on the system, e.g., the adjustments necessary when one foot

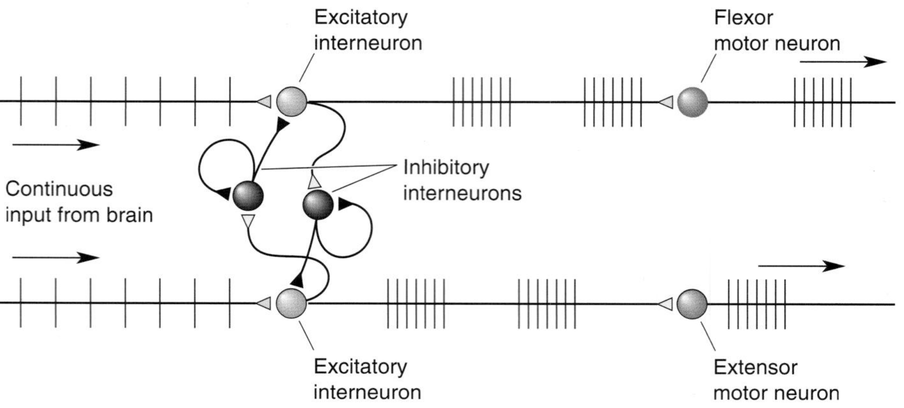

Figure 13.27

A possible circuit for rhythmic alternating activity.

strikes an obstacle while walking or the changes in output that are necessary to go from walking to trotting to running to jumping.

CONCLUDING REMARKS

We can draw several conclusions from this discussion of the spinal control of movement. First, a great deal has been learned about movement and its spinal control by working at various levels of analysis, ranging from biochemistry to biophysics to behavior. Indeed, a complete understanding, whether of excitation-contraction coupling or central pattern generation, requires knowledge derived from every approach. Second, sensation and movement are inextricably linked even at the lowest levels of the neural motor system. The normal function of the alpha motor neuron depends on direct feedback from the muscles themselves and indirect information from the tendons, joints, and skin. Third, the spinal cord contains an intricate network of circuits for the control of movement; it is far more than a conduit for somatic sensory and motor information.

Evidently coordinated and complex patterns of activity in these spinal circuits can be driven by relatively crude descending signals. This leaves the question of precisely what the upper motor neurons contribute to motor control—the subject of Chapter 14.

KEY TERMS

The Somatic Motor System
smooth muscle (p. 437)
striated muscle (p. 437)
cardiac muscle (p. 437)
skeletal muscle (p. 437)
muscle fiber (p. 437)
somatic motor system (p. 437)
flexion (p. 438)
extension (p. 438)
flexor (p. 438)
synergist muscle (p. 438)
extensor (p. 438)
antagonist muscle (p. 438)
axial muscle (p. 438)
proximal (girdle) muscle (p. 439)
distal muscle (p. 439)

The Lower Motor Neuron
alpha motor neuron (p. 440)
motor unit (p. 440)
motor neuron pool (p. 440)
fast motor unit (p. 443)
slow motor unit (p. 443)

Excitation-Contraction Coupling
sarcolemma (p. 445)
myofibril (p. 445)
sarcoplasmic reticulum (p. 445)
T tubule (p. 446)
Z line (p. 446)
sarcomere (p. 446)
thin filament (p. 446)
thick filament (p. 446)
myosin (p. 447)
actin (p. 447)
troponin (p. 447)

Spinal Control of Motor Units
muscle spindle (p. 449)
proprioceptor (p. 449)
proprioception (p. 449)
myotatic reflex (p. 450)
intrafusal fiber (p. 453)
extrafusal fiber (p. 453)
gamma motor neuron (p. 453)
Golgi tendon organ (p. 454)
reciprocal inhibition (p. 458)
central pattern generator (p. 459)

REVIEW QUESTIONS

1. What did Sherrington call the "final common pathway?" Why?
2. Define *motor unit* in one sentence. How does it differ from motor neuron pool?
3. Which is recruited first, a fast motor unit or a slow motor unit? Why?
4. When and why does rigor mortis occur?
5. Your doctor taps the tendon beneath your kneecap and your leg extends. What is the neural basis of this reflex? What is it called?
6. What is the function of gamma motor neurons?
7. Lenny, a character in Steinbeck's classic book *Of Mice and Men*, loved rabbits, but when he hugged them, they were crushed to death. Which type of proprioceptive input may Lenny have been lacking?

CHAPTER

14

Brain Control of Movement

INTRODUCTION

In Chapter 13, we discussed the organization of the peripheral somatic motor system: the joints, skeletal muscles, and their sensory and motor innervation. We saw that the final common pathway for behavior is the alpha motor neuron, that the activity of this cell is under the control of sensory feedback and spinal interneurons, and that reflex movements reveal the complexity of this spinal control system. In this chapter, we explore how the brain influences the activity of the spinal cord to command voluntary movements.

The central motor system is arranged as a hierarchy of control levels, with the forebrain at the top and the spinal cord at the bottom. It is useful to think of this motor control hierarchy as having three levels (Table 14.1). The highest level, represented by the association areas of neocortex and basal ganglia of the forebrain, is concerned with *strategy*: the goal of the movement and the movement strategy that best achieves the goal. The middle level, represented by the motor cortex and cerebellum, is concerned with *tactics*: the sequences of muscle contractions, arranged in space and time, required to smoothly and accurately achieve the strategic goal. The lowest level, represented by the brain stem and spinal cord, is concerned with *execution*: activation of the motor neuron and interneuron pools that generate the goal-directed movement and make any necessary adjustments of posture.

To appreciate the different contributions of the three hierarchical levels to movement, consider the actions of a baseball pitcher standing on the mound, preparing to pitch to a batter (Figure 14.1). The cerebral neocortex has information about precisely where the body is in space, based on vision, audition, somatic sensation, and proprioception. Strategies must be devised to move the body from the current state to one in which a pitch is delivered and the desired outcome (a swing and a miss) is attained. Several options—a curve ball, a fast ball, a knuckle ball, and so on—are available, and these alternatives are filtered through the basal ganglia and back to the cortex until a decision is made, based in large part on experience (e.g., "This batter hit a home run the last time I threw a fast ball"). The motor areas of cortex and the cerebellum then make the tactical decision (to throw a curve ball) and issue instructions to the brain stem and spinal cord. Activation of neurons in the brain stem and spinal cord then causes the movement to be executed. Properly timed activation of motor neurons in the cervical spinal cord generates a coordinated movement of the shoulder, elbow, wrist, and fingers. Simultaneously, brain stem input to the thoracic and lumbar spinal cord command the appropriate postural adjustments that keep the pitcher from falling over during the throw. In addition, brain stem motor neurons are activated to keep the pitcher's eyes fixed on the catcher, his target, as his head and body move about.

According to the laws of physics, the movement of a thrown baseball through space is *ballistic*, that is, a trajectory that cannot be altered. The movement of the pitcher's arm that throws the ball is also described as ballistic because it cannot be altered once initiated. This type of rapid voluntary movement is not under the same type of sensory feedback control that regulates

Table 14.1 The Motor Control Hierarchy

LEVEL	FUNCTION	STRUCTURES
High	Strategy	Association areas of neocortex, basal ganglia
Middle	Tactics	Motor cortex, cerebellum
Low	Execution	Brain stem, spinal cord

antigravity postural reflexes (see Chapter 13). The reason is simple: The movement is too fast to be altered by sensory feedback. But the movement does not occur in the absence of sensory information. Sensory information *before* the movement is initiated is crucial for determinating the starting positions of the limbs and body and for anticipating any changes in resistance during the throw. And sensory information *during* the movement is also important—not necessarily for the movement at hand but for improving subsequent similar movements.

The proper function of each level of the motor control hierarchy relies on sensory information so heavily that the motor system of the brain might properly be considered a *sensorimotor system*. At the highest level, sensory information generates a mental image of the body and its relationship to the environment. At the middle level, tactical decisions are based on the memory of sensory information from past movements. At the lowest level, sensory feedback is used to maintain posture, muscle length, and tension before and after each voluntary movement.

In this chapter, we investigate this hierarchy of motor control and how each level contributes to the control of the peripheral somatic motor system. We start by exploring the pathways that bring information to the spinal motor neurons. From there we will ascend to the highest levels of the motor hierarchy, and then we'll fill in the pieces of the puzzle that bring the different levels together. Along the way, we'll describe how pathology in specific parts of the motor system leads to particular movement disorders.

Figure 14.1
A baseball pitcher planning a pitch.

DESCENDING SPINAL TRACTS

How does the brain communicate with the motor neurons of the spinal cord? Axons from the brain descend through the spinal cord along two major groups of pathways (Figure 14.2). One is in the lateral column of the spinal cord, and the other is in the ventromedial column. Remember this rule of thumb: The **lateral pathways** are involved in voluntary movement of the dis-

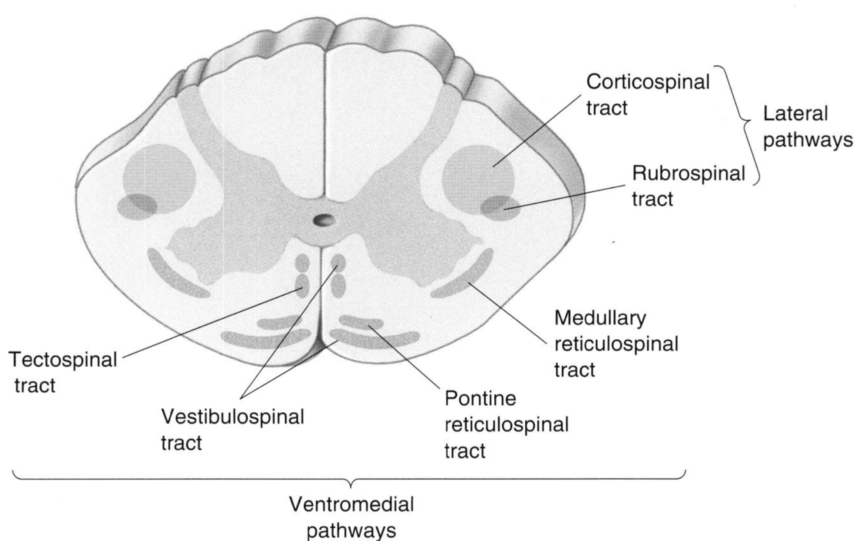

Figure 14.2
The descending tracts of the spinal cord. The lateral pathways, consisting of the corticospinal and rubrospinal tracts, control voluntary movements of the distal musculature. The ventromedial pathways, consisting of the reticulospinal, vestibulospinal, and tectospinal tracts, control postural muscles.

tal musculature and are under direct cortical control, and the **ventromedial pathways** are involved in the control of posture and locomotion and are under brain stem control.

The Lateral Pathways

The most important component of the lateral pathways is the **corticospinal tract** (Figure 14.3a). Originating in the neocortex, it is the longest and one of the largest central nervous system (CNS) tracts (10^6 axons). Two-thirds of the

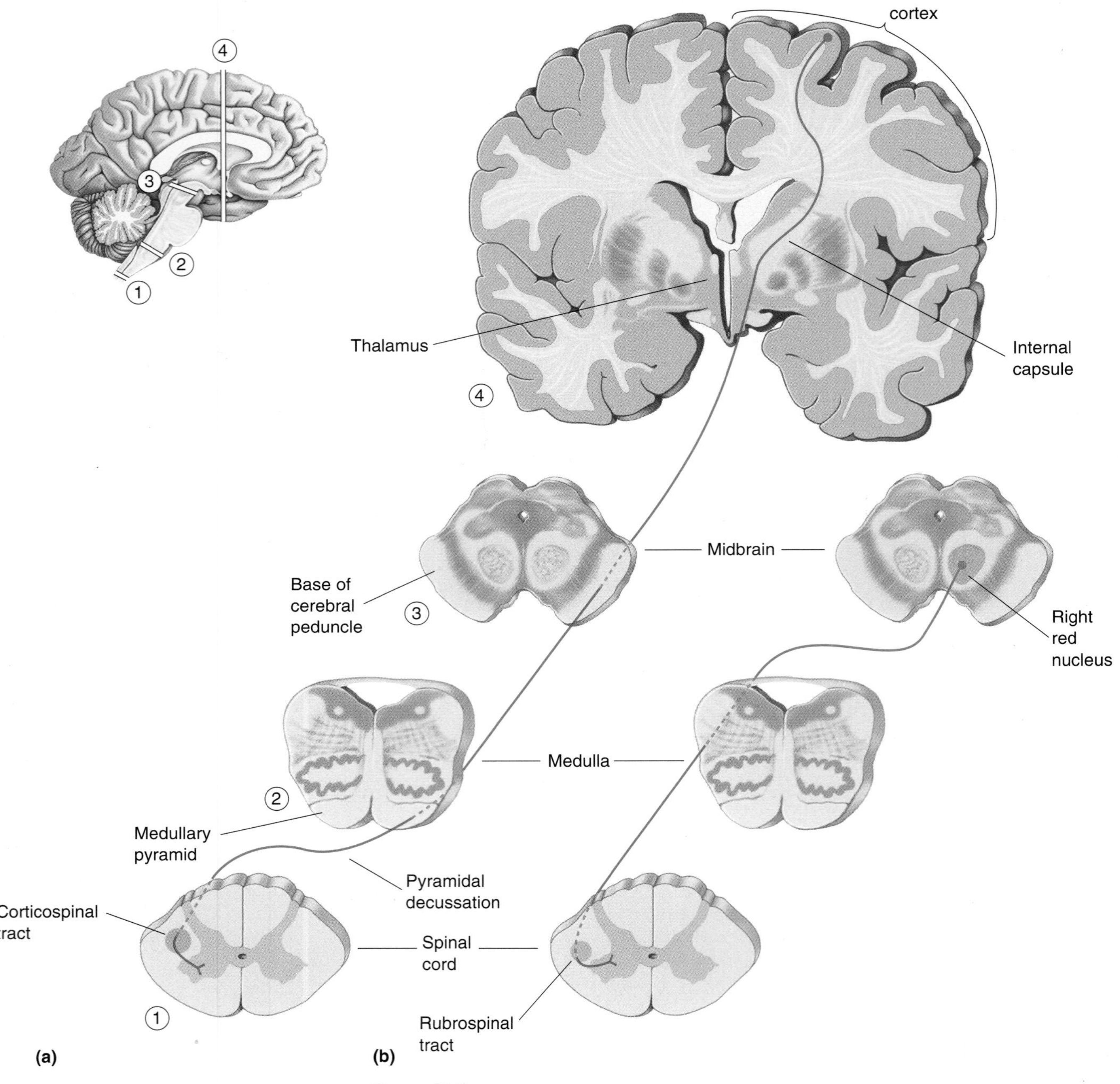

Figure 14.3
Origins and terminations of (a) the corticospinal tract and (b) the rubrospinal tract. These components of the lateral pathways control fine movements of the arms and fingers.

axons in the tract originate in areas 4 and 6 of the frontal lobe, collectively called **motor cortex**. Most of the remaining axons in the corticospinal tract derive from the somatosensory areas of the parietal lobe and serve to regulate the flow of somatosensory information to the brain (see Chapter 12). Axons from the cortex pass through the internal capsule, bridging the telencephalon and thalamus; course through the base of the *cerebral peduncle*, a large collection of axons in the midbrain; then pass through the pons and collect to form a tract at the base of the medulla. The tract forms a bulge running down the ventral surface of the medulla. When cut, the tract's cross section is roughly triangular, explaining why it is called the **pyramidal tract**.

At the junction of the medulla and spinal cord, the pyramidal tract decussates. This means that the *right* motor cortex directly commands the movement of the *left* side of the body, and the *left* motor cortex controls the muscles on the *right* side. As the axons cross, they collect in the lateral column of the spinal cord and form the lateral corticospinal tract. The corticospinal tract axons terminate in the dorsolateral region of the ventral horns and intermediate gray matter, where the motor neurons and interneurons that control the distal muscles, particularly the flexors, lie (see Chapter 13).

A much smaller component of the lateral pathways is the **rubrospinal tract**, which originates in the **red nucleus** of the midbrain, so named because of its distinctive pinkish hue in a freshly dissected brain (*rubro* is from the Latin for "red"). Axons from the red nucleus almost immediately decussate in the pons and join those in the corticospinal tract in the lateral column of the spinal cord (Figure 14.3b). A major source of input to the red nucleus is the very region of frontal cortex that also contributes to the corticospinal tract. Indeed, it appears that in the course of primate evolution, this indirect corticorubrospinal pathway has largely been replaced by the direct corticospinal path. Thus, while the rubrospinal tract contributes importantly to motor control in many mammalian species, in humans it appears to be reduced, most of its functions subsumed by the corticospinal tract.

The Effects of Lateral Pathway Lesions. In the late 1960s, Donald Lawrence and Hans Kuypers laid the foundation for the modern view of the function of the lateral pathways. Experimental lesions in both corticospinal and rubrospinal tracts in monkeys rendered them unable to make fractionated movements of the arms and hands; that is, they could not move their shoulders, elbows, wrists, and fingers independently. For example, they could grasp objects with their hands, but only by using all of their fingers at once. Voluntary movements were also slower and less accurate. Despite this, the animals could sit upright and stand with normal posture. By analogy, a human with a lateral pathway lesion can stand on the pitcher's mound but be unable to throw the ball accurately.

Lesions in the corticospinal tracts alone caused a movement deficit as severe as that observed after lesions in the lateral columns. Interestingly, however, many functions gradually reappeared over the months following surgery. In fact, the only permanent deficit was some weakness of the distal flexors and an inability to move the fingers independently. A subsequent lesion in the rubrospinal tract completely reversed this recovery. These results suggest that over time the corticorubrospinal pathway partially compensated for the loss of the corticospinal tract input.

Strokes that damage the motor cortex or the corticospinal tract are common in humans. Their immediate consequence can be paralysis on the contralateral side, but considerable recovery of voluntary movements may occur over time. As in Lawrence and Kuypers' lesioned monkeys, it is the fine, fractionated movements of the fingers that are least likely to recover.

The Ventromedial Pathways

The ventromedial pathways contain four descending tracts that originate in the brain stem and terminate among the spinal interneurons controlling proximal and axial muscles. These tracts are the vestibulospinal tract, the tectospinal tract, the pontine reticulospinal tract, and the medullary reticulospinal tract. The ventromedial pathways use sensory information about balance, body position, and the visual environment to reflexively maintain balance and body posture. We'll briefly discuss each pathway in turn.

The Vestibulospinal Tracts. The vestibulospinal and tectospinal tracts keep the head balanced on the shoulders as the body moves through space, and they turn the head in response to new sensory stimuli. The **vestibulospinal tracts** originate in the **vestibular nuclei** of the medulla, which relay sensory information from the vestibular labyrinth in the inner ear (Figure 14.4a). The *vestibular labyrinth* consists of fluid-filled canals and cavities in the temporal bone that are closely associated with the cochlea (see Chapter 11). The motion of the fluid in this labyrinth, which accompanies movement of the head, activates hair cells that signal the vestibular nuclei via cranial nerve VIII. One component of the vestibulospinal tracts projects bilaterally down the spinal cord and activates the cervical spinal circuits that control neck and back muscles and thus guide head movement. Stability of the head is important because the head contains our eyes, and keeping the eyes stable as our body

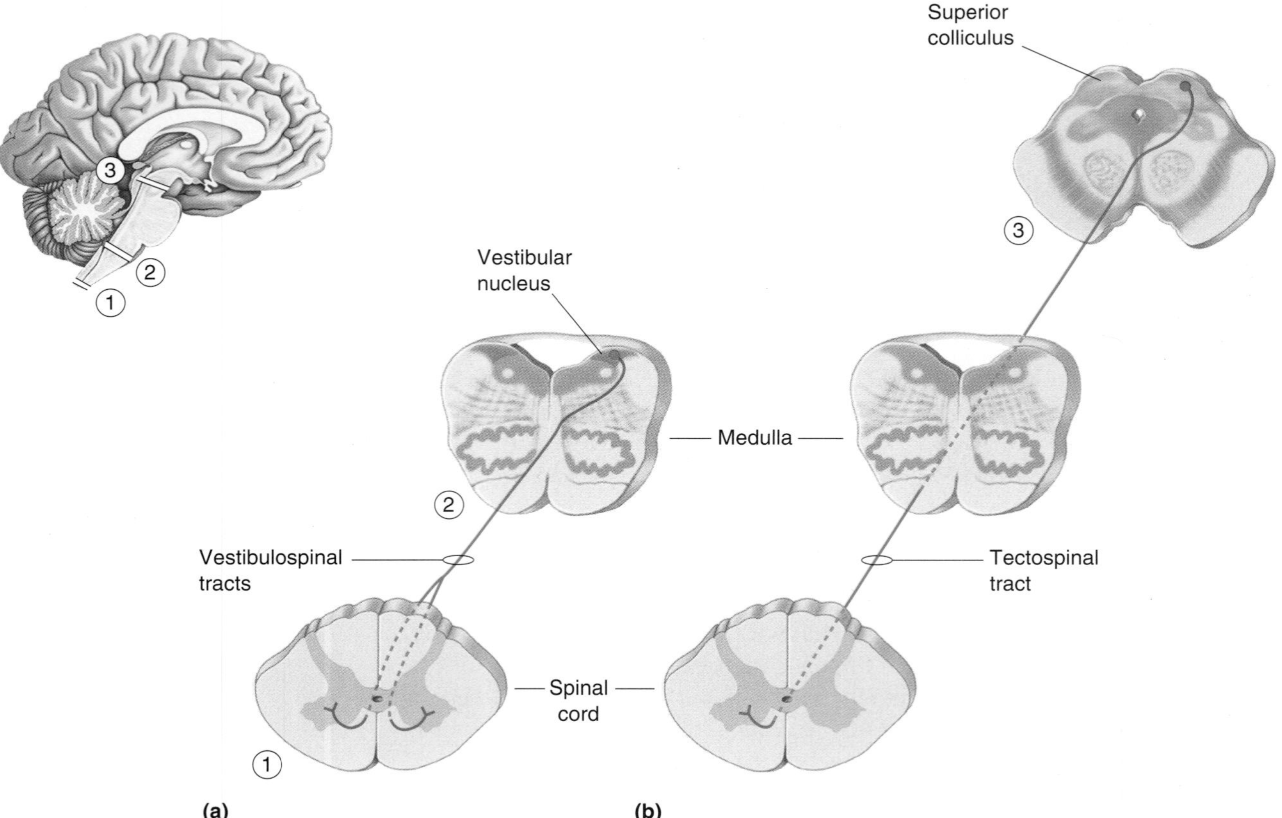

Figure 14.4
Origins and terminations of (a) the vestibulospinal tract and (b) the tectospinal tract. These components of the ventromedial pathway control the posture of the head and neck.

moves ensures that our image of the world remains stable. Another component of the vestibulospinal tracts projects ipsilaterally as far down as the lumbar spinal cord. It helps us maintain an upright and balanced posture by facilitating extensor motor neurons of the legs.

The Tectospinal Tract. The **tectospinal tract** originates in the superior colliculus of the midbrain, which receives direct input from the retina (Figure 14.4b). (Recall from Chapter 10 that optic tectum is another name for the superior colliculus.) Besides its retinal input, the superior colliculus receives projections from visual cortex, as well as afferents carrying somatosensory and auditory information. From this input, the superior colliculus constructs a map of the world around us; stimulation at one site in this map leads to an orienting response that directs the head and eyes to move so that the appropriate point of space is imaged on the fovea. Activation of the colliculus by the image of a runner sprinting toward second base, for example, would cause the pitcher to orient his head and eyes toward this important new stimulus.

The Pontine and Medullary Reticulospinal Tracts. The reticulospinal tracts arise mainly from the **reticular formation** of the brain stem, which runs the length of the brain stem at its core, just under the cerebral aqueduct and fourth ventricle. A complex meshwork of neurons and fibers, the reticular formation receives input from many sources and participates in many functions. For the purposes of our discussion of motor control, the reticular formation may be divided into two parts that give rise to two descending tracts: the pontine (medial) reticulospinal tract and the medullary (lateral) reticulospinal tract (Figure 14.5).

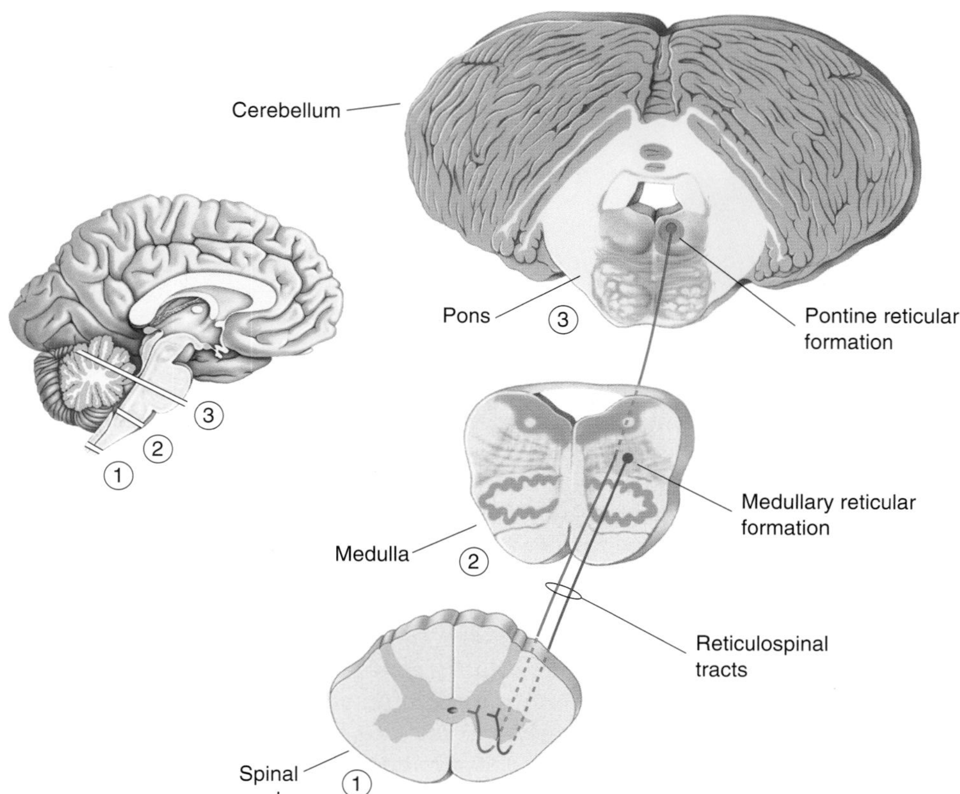

Figure 14.5
The pontine (medial) and medullary (lateral) reticulospinal tracts. These components of the ventromedial pathway control posture of the trunk and the antigravity muscles of the limbs.

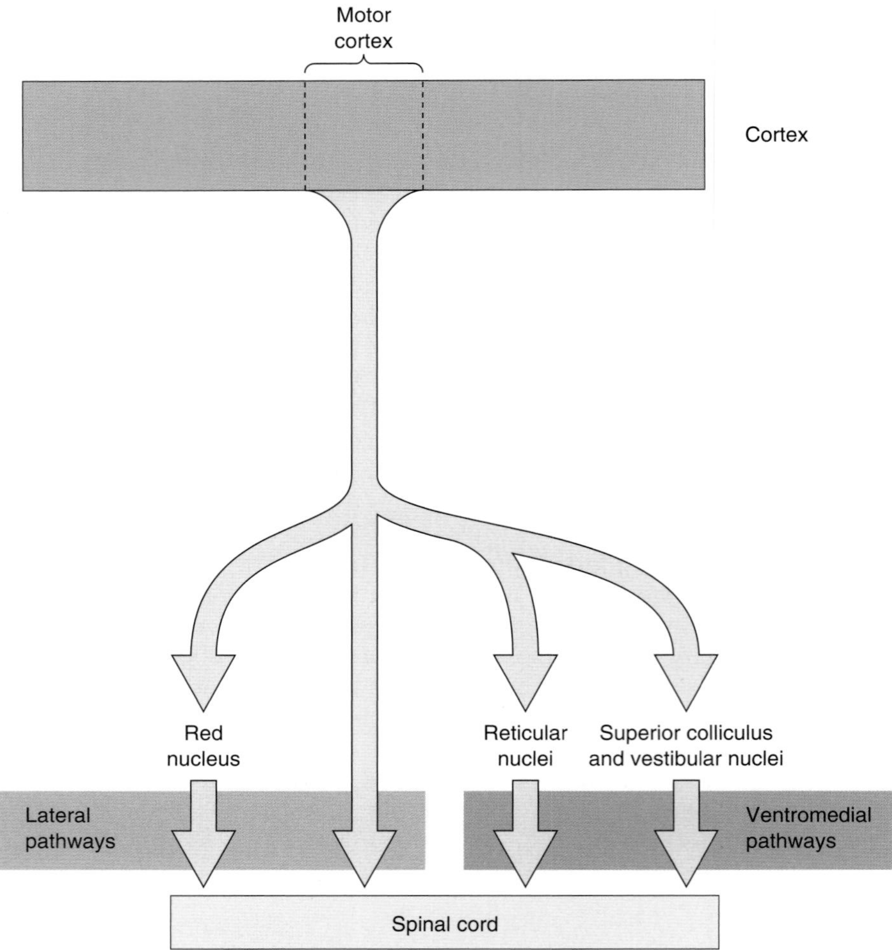

Figure 14.6
A summary of the major descending spinal tracts and their points of origin.

The **pontine reticulospinal tract** enhances the antigravity reflexes of the spinal cord. Activity in this pathway, by facilitating the extensors of the lower limbs, helps maintain a standing posture by resisting the effects of gravity. This type of regulation is an important component of motor control: Keep in mind that most of the time, the activity of ventral horn neurons maintains rather than changes muscle length and tension. The **medullary reticulospinal tract** has the opposite effect, however: It liberates the antigravity muscles from reflex control. Activity in both reticulospinal tracts is controlled by descending signals from the cortex. A fine balance between them is required as the pitcher goes from standing on the mound to winding up and throwing the ball.

Figure 14.6 provides a simple summary of the major descending spinal tracts. The ventromedial pathways originate from several regions of the brain stem and participate mainly in the maintenance of posture and certain reflex movements. Initiation of a voluntary ballistic movement, such as throwing a baseball, requires instructions that descend from the motor cortex along the lateral pathways. Motor cortex directly activates spinal motor neurons and liberates them from reflex control by communicating with the nuclei of the ventromedial pathways. Clearly, the cortex is key for voluntary movement and behavior, so we will now focus our attention there.

THE PLANNING OF MOVEMENT BY THE CEREBRAL CORTEX

Although cortical areas 4 and 6 are called motor cortex, the control of voluntary movement engages almost all of the neocortex. Goal-directed movement depends on knowledge of where the body is in space, where it intends to go, and the selection of a plan to get it there. Once a plan has been selected, it must be held in memory until the appropriate time. Finally, instructions to implement the plan must be issued. To some extent, these different aspects of motor control are localized to different regions of the cerebral cortex. In this section, we explore some of the cortical areas implicated in motor planning. Later we look at how a plan is converted into action.

Motor Cortex

The motor cortex is a circumscribed region of the frontal lobe. Area 4 lies just anterior to the central sulcus on the precentral gyrus, and area 6 lies just anterior to area 4 (Figure 14.7). The definitive demonstration that these areas constitute motor cortex in humans came from the work of Canadian neurosurgeon Wilder Penfield. Recall from Chapter 12 that Penfield electrically stimulated the cortex of patients who were undergoing surgery to remove bits of brain thought to be inducing epileptic seizures. The stimulation was used in an attempt to identify which regions of cortex were so important that they should be spared from the knife. In the course of these operations, Penfield discovered that weak electrical stimulation of area 4 in the precentral gyrus would elicit a twitch of the muscles in a particular region of the body on the contralateral side. Systematic probing of this region established that there is a somatotopic organization in the human precentral gyrus much like that seen in the somatosensory areas of the postcentral gyrus (Figure 14.8). Area 4 is now often called **primary motor cortex**, or **M1.**

The foundation for Penfield's discovery was laid nearly a century earlier by Gustav Fritsch and Eduard Hitzig, who in 1870 showed that stimulation of the frontal cortex of anesthetized dogs would elicit movement of the contralateral side of the body (see Chapter 1). Then, around the turn of the century, David Ferrier and Charles Sherrington discovered that the motor area

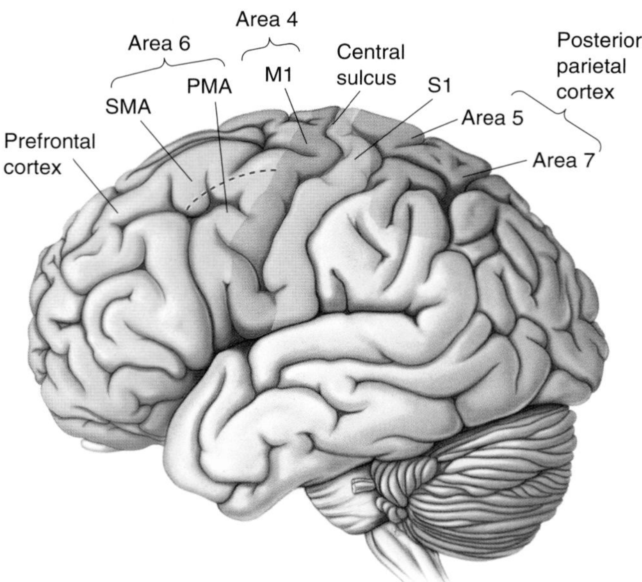

Figure 14.7
Areas of neocortex intimately involved in the planning and instruction of voluntary movement. Areas 4 and 6 constitute motor cortex.

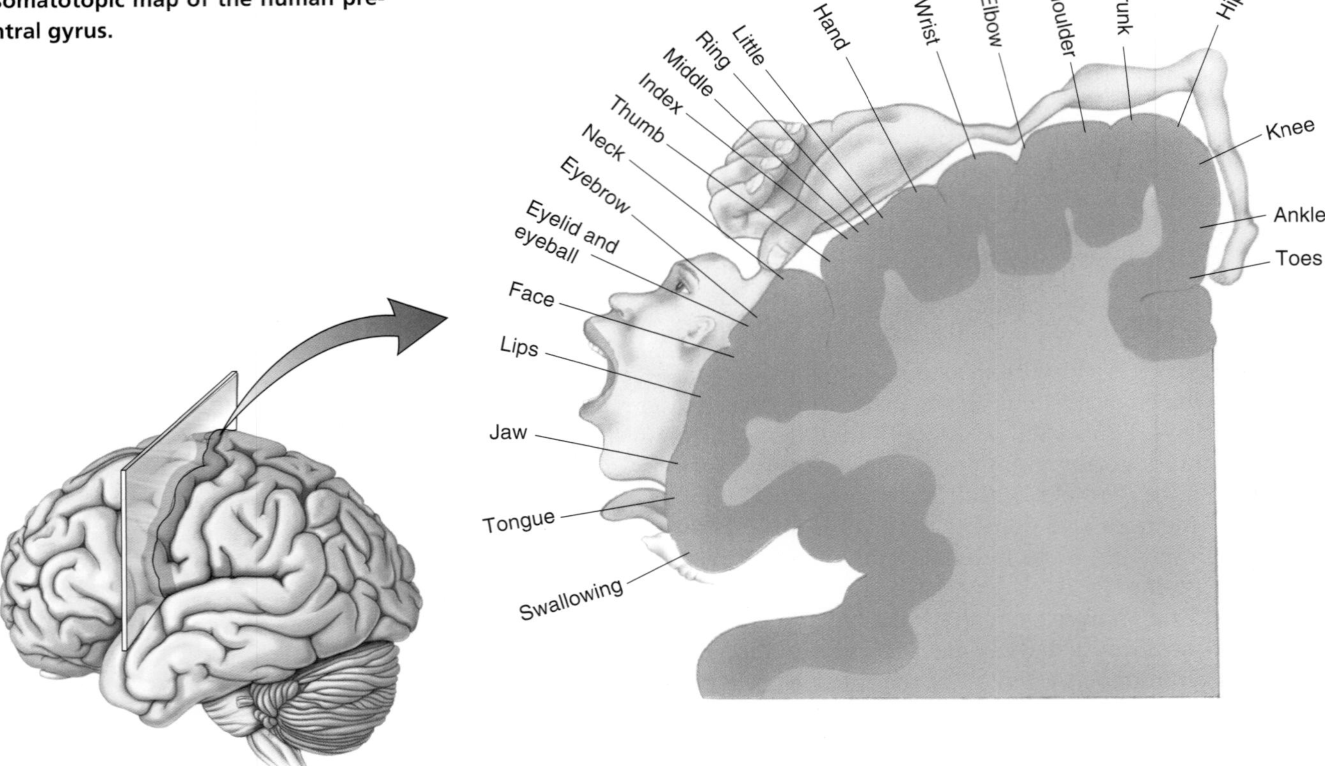

Figure 14.8
A somatotopic map of the human precentral gyrus.

in primates was in the precentral gyrus. By comparing the histology of this region in Sherrington's apes with that of the human brain, Australian neuroanatomist Alfred Walter Campbell concluded that cortical area 4 is motor cortex.

Campbell speculated that cortical area 6, just rostral to area 4, may be an area specialized for skilled voluntary movement. Penfield's studies 50 years later supported the conjecture that this was a "higher" motor area in humans by showing that electrical stimulation of area 6 could evoke complex movements of either side of the body. Penfield found two somatotopically organized motor maps in area 6: one in a lateral region he called the **premotor area**, or **PMA**, and one in a medial region called the **supplementary motor area**, or **SMA** (Figure 14.7). These two areas appear to perform similar functions but on different groups of muscles. While SMA sends axons that innervate distal motor units directly, PMA connects primarily with reticulospinal neurons that innervate proximal motor units.

The Contributions of Posterior Parietal and Prefrontal Cortex

Recall the baseball player standing on the mound, preparing to pitch. It should be apparent that before the detailed sequence of muscle contractions for the desired pitch can be calculated, the pitcher must have information about the current position of his body in space and how it relates to the positions of the batter and the catcher. This mental body image seems to be generated by somatosensory, proprioceptive, and visual inputs to the posterior parietal cortex.

Two areas are of particular interest in the posterior parietal cortex: area 5, a target of inputs from the primary somatosensory cortical areas 3, 1, and 2 (see Chapter 12), and area 7, a target of higher-order visual cortical areas such as MT (see Chapter 10). Recall that human patients with lesions in these areas of the parietal lobes, as can occur after a stroke, show bizarre abnormalities of body image and the perception of spatial relations. In its most extreme manifestation, the patient will simply ignore the side of the body and even the rest of the world opposite the parietal lesion.

The parietal lobes are extensively interconnected with regions in the anterior frontal lobes that in humans are thought to be important for abstract thought, decision making, and anticipating the consequences of action. These "prefrontal" areas, along with the posterior parietal cortex, represent the highest levels of the motor control hierarchy, where decisions are made about what actions to take and their likely outcome (a curve ball followed by a strike). Both the prefrontal and the parietal cortex send axons that converge on cortical area 6. Recall that areas 6 and 4 together contribute most of the axons to the descending corticospinal tract. Thus, area 6 lies at the junction where signals encoding *what* actions are desired are converted into signals that specify *how* the actions will be carried out.

This general view of higher-order motor planning received dramatic support in a series of studies on humans carried out by Danish neurologist Per Roland and his colleagues. They used positron emission tomography (PET) to monitor changes in the patterns of cortical activation that accompany voluntary movements (see Box 7.3). When the subjects were asked to perform a series of finger movements from memory, the following regions of cortex showed increased blood flow: the somatosensory and posterior parietal areas, parts of the prefrontal cortex (area 8), area 6, and area 4. These are the very regions of the cerebral cortex that are thought to play a role in generating the intention to move and converting that intention into a plan of action, as discussed earlier. Interestingly, when the subjects were asked only to rehearse the movement mentally without actually moving the finger, area 6 remained active but area 4 did not.

Neuronal Correlates of Motor Planning

Recent experimental work on monkeys further supports the idea that area 6 (SMA and PMA) plays an important role in the planning of movement, particularly complex movement sequences of the distal musculature. Using a method developed in the late 1960s by Edward Evarts at the National Institutes of Health, it has been possible to record the activity of neurons in the motor areas of awake, behaving animals (Box 14.1). Cells in SMA typically increase their discharge rates about a second before the execution of a hand or wrist movement, consistent with their proposed role in planning movement (recall Roland's findings in humans). An important feature of this activity is that it occurs in advance of the movements of *either* hand, suggesting that the supplementary areas of the two hemispheres are closely linked via the corpus callosum. Indeed, in both monkeys and humans, movement deficits observed following an SMA lesion on one side are particularly pronounced for tasks requiring the coordinated actions of the two hands, such as buttoning a shirt. In humans, a selective inability to perform complex (but not simple) motor acts is called *apraxia*.

You've heard the expression "Ready, set, go." The preceding discussion suggests that readiness ("Ready") depends on activity in the parietal and frontal lobes, along with important contributions from the brain centers that control levels of attention and alertness. "Set" may reside in the supplementary and premotor areas, where movement strategies are devised and held until they are executed. A good example is shown in Figure 14.9, based on the work of Michael Weinrich and Steven Wise at the National Institutes of

Box 14.1

OF SPECIAL INTEREST

Behavioral Neurophysiology

Showing that a brain lesion impairs movement and that brain stimulation elicits movement does not tell us how the brain *controls* movement. To address this problem, we need to know how the activity of neurons relates to different types of voluntary movement in the intact organism. PET scans and fMRI are extremely valuable in plotting out the distribution of activity in the brain as behaviors are performed, but they lack the resolution to track the millisecond-by-millisecond changes in the activity of individual neurons. The best method for this purpose is extracellular recording with metal microelectrodes (see Box 4.1). But how is this done in awake, behaving animals?

This problem was solved by Edward Evarts and his colleagues at the National Institutes of Health. They trained monkeys to perform simple tasks. When the monkeys performed the tasks successfully, they were rewarded with a sip of fruit juice. For example, to study the brain's guidance of hand and arm movements, the monkey might be trained to move its hand toward the brightest of several spots on a computer screen. Pointing to the correct spot earned it a juice reward. After training, the animals were anesthetized. In a simple surgical procedure, each monkey was fitted with a small headpiece so that a microelectrode could be introduced into the brain through a small opening in the skull. When the animals recovered from surgery, they showed no signs of discomfort from either the headpiece or the insertion of a microelectrode into the brain (recall from Chapter 12 that there are no nociceptors in the brain). Evarts and his colleagues then recorded the discharges of individual cells in the motor cortex as the animals made voluntary movements. They could then see how the neuron's response changed when the animal pointed to different spots on the screen.

This is an example of *behavioral neurophysiology*, the recording of cellular activity in the brain of awake, behaving animals. By altering the task that the animal performs, the same method can be applied to the investigation of a wide range of neuroscientific problems, including attention, perception, learning, and movement.

Some types of human neurosurgery are also done with the patient awake, at least during part of the procedure. By applying the techniques of behavioral neurophysiology to informed, consenting adults, we have also learned some fascinating information about uniquely human skills.

Health. They monitored the discharge of a neuron in PMA as a monkey performed a task requiring a specific arm movement to a target. The monkey was first given an *instruction stimulus* informing him what the target would be ("Get set, monkey!"), followed after a variable delay by a *trigger stimulus* informing the monkey that it was OK to move ("Go, monkey!"). Successful performance of the task (i.e., waiting for the "Go" signal and then making the movement to the appropriate target) was rewarded with a sip of juice. The neuron in PMA began firing if the instruction was to move the arm to the left, and it continued to discharge until the trigger stimulus came on and the movement was initiated. If the instruction was to move to the right, this neuron did not fire (presumably another population of PMA cells became active under this condition). Thus, the activity of this PMA neuron reported the direction of the upcoming movement and continued to do so until the movement was made. Although we do not yet understand the details of the coding taking place in SMA and PMA, the fact that neurons in these areas are selectively active well before movements are initiated is consistent with a role in planning the movement.

Now, let's consider our baseball player standing on the mound. The decision has been made to throw a curve ball, but the batter walks away from the plate to clean his cleats. The pitcher stands motionless on the mound, muscles tensed, waiting for the batter to return. The pitcher is "set"; a select population of neurons in the premotor and supplementary motor cortex (the cells

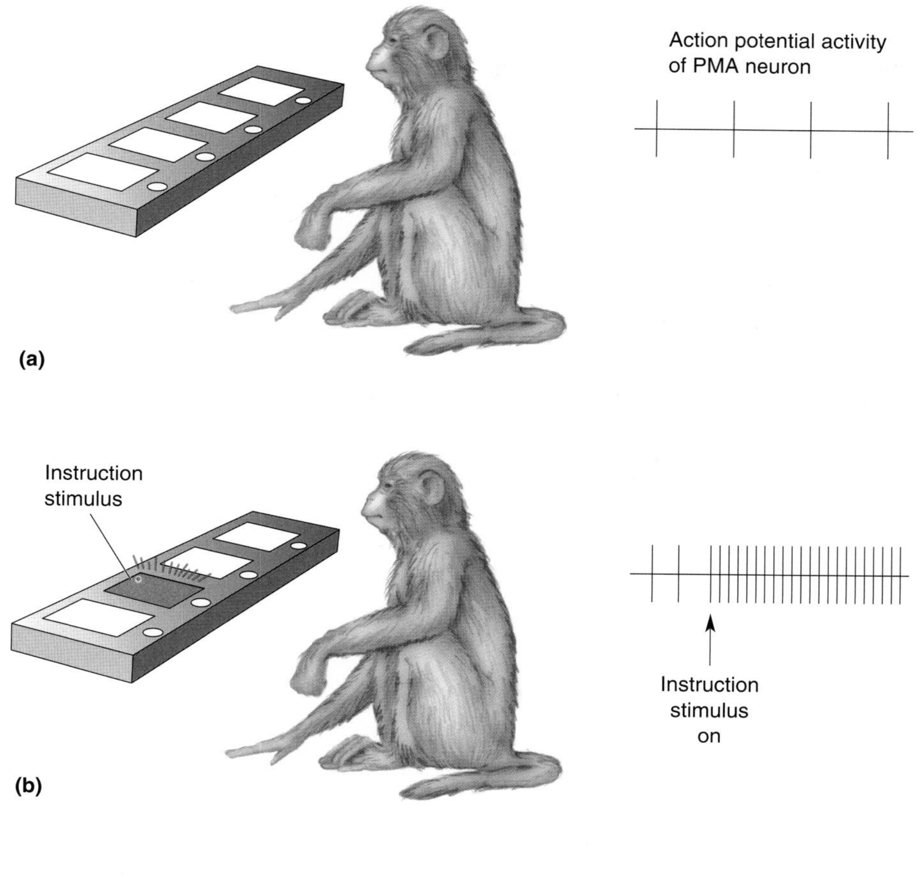

Figure 14.9

The discharge of a cell in the premotor area before a movement. (a) *Ready*: A monkey sits before a panel of lights. The task is to wait for an instruction stimulus that will inform him of the movement required to receive a juice reward, then perform the movement when a trigger stimulus goes on. The activity of a neuron in PMA is recorded during the task. **(b)** *Set*: The instruction stimulus occurs at the time indicated by the upward arrow, resulting in the discharge of the neuron in PMA. **(c)** *Go*: Shortly after the movement is initiated, the PMA cell ceases firing. (Source: Adapted from Weinrich and Wise, 1982.)

that are planning the curve ball movement sequence) are firing away in anticipation of the throw. Then the batter steps up to the plate, and an internally generated "Go" command is given. This command appears to be implemented with the participation of a major *subcortical* input to area 6, which is the subject of the next section. After that, we examine the origin of the "Go" command in the primary motor cortex.

THE BASAL GANGLIA

The major subcortical input to area 6 arises in a nucleus of the dorsal thalamus, the **ventral lateral nucleus** (**VL**). The input to this part of VL, called VLo, arises from the **basal ganglia** buried deep within the telencephalon. The

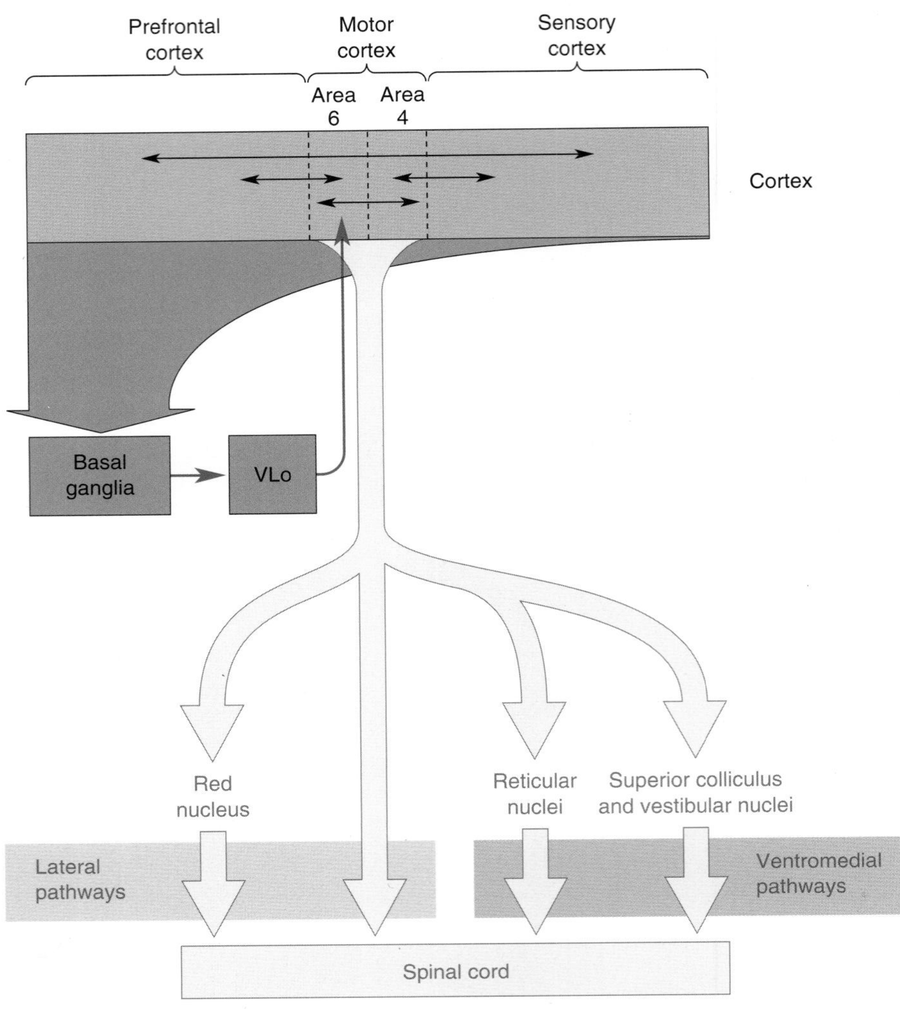

Figure 14.10
A summary of the loop from the cortex to the basal ganglia to the thalamus and back to area 6.

basal ganglia, in turn, are targets of the cerebral cortex, particularly the frontal, prefrontal, and parietal cortex. Thus, we have a loop in which information cycles from the cortex through the basal ganglia and thalamus and back to the cortex, particularly the SMA (Figure 14.10). One of the functions of this loop appears to be the selection and initiation of willed movements.

Anatomy of the Basal Ganglia

The basal ganglia consist of the **caudate nucleus**, the **putamen**, the **globus pallidus**, and the **subthalamic nucleus**. We can add the **substantia nigra**, a midbrain structure that is reciprocally connected with the basal ganglia of the forebrain (Figure 14.11). The caudate and putamen together are called the **striatum**, which is the target of the cortical input to the basal ganglia. The globus pallidus is the source of the output to the thalamus. The other structures participate in various side loops that modulate the direct path:

Cortex → Striatum → Globus pallidus → VLo → Cortex (SMA)

Through the microscope, the neurons of the striatum appear randomly scattered, with no apparent order such as that seen in the layers of the cortex. But this bland appearance hides a degree of complexity in the organization of the basal ganglia that we are only now beginning to appreciate. It appears that the basal ganglia participate in a large number of parallel circuits, only a few of which are strictly motor. Other circuits are involved in certain aspects

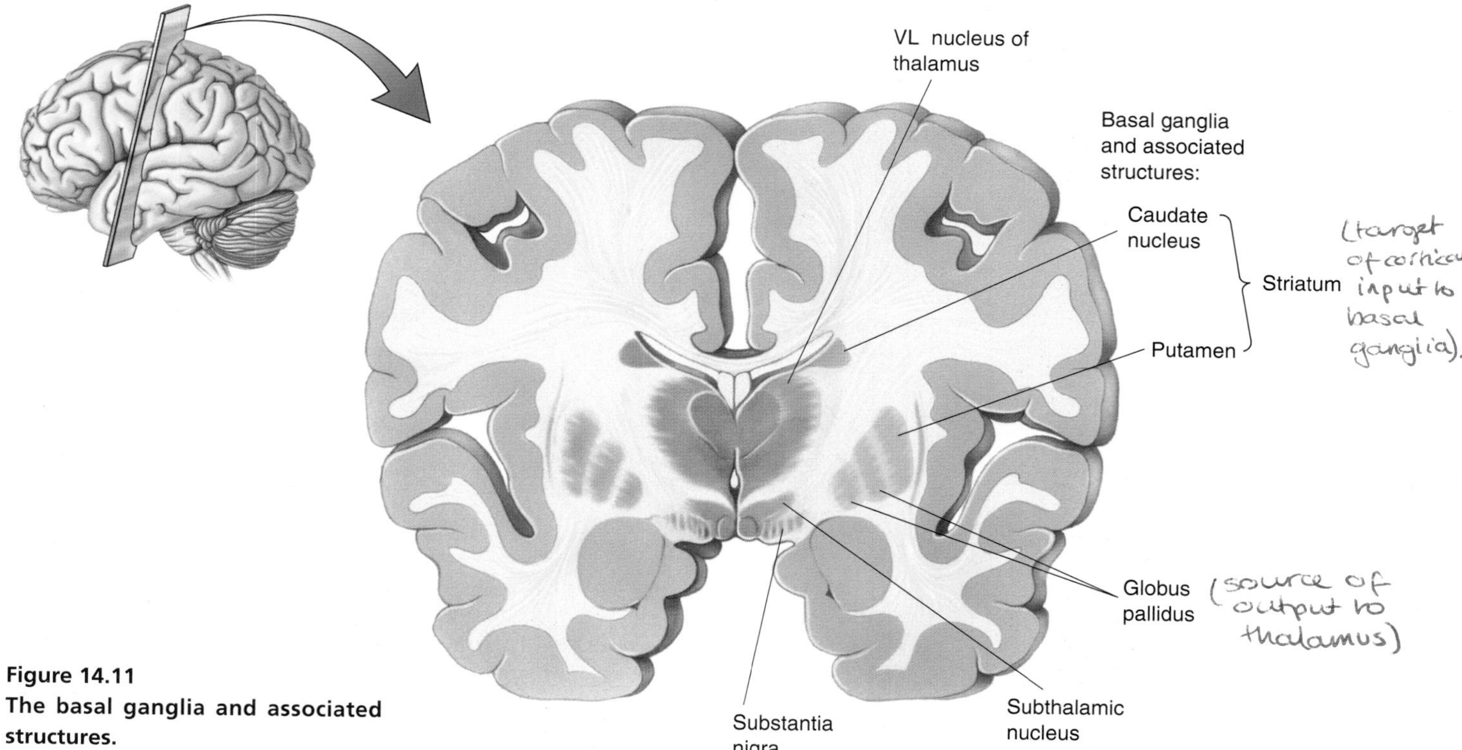

Figure 14.11
The basal ganglia and associated structures.

of memory and cognitive function. We will try to give a concise account of the motor function of the basal ganglia, simplifying a very complex and poorly understood part of the brain.

The Motor Loop

The most direct path in the motor loop through the basal ganglia originates with an excitatory connection from the cortex to cells in the putamen. The putamen cells make inhibitory synapses on neurons in the globus pallidus, which in turn make inhibitory connections with the cells in VLo. The thalamocortical connection from VLo to SMA is excitatory; it facilitates the discharge of movement-related cells in SMA. This direct motor loop is summarized in Figure 14.12.

The functional consequence of cortical activation of the putamen is excitation of SMA by VL. Let's work out how this happens. At rest, neurons in the globus pallidus are spontaneously active and therefore inhibit VL. Cortical activation (1) excites putamen neurons, which (2) inhibit globus pallidus neurons, which (3) release the cells in VLo from inhibition, allowing them to become active. The activity in VLo boosts the activity of SMA. Thus, this part of the circuit acts as a positive feedback loop that may serve to focus, or funnel, the activation of widespread cortical areas onto the supplementary motor area of cortex. We can speculate that the "Go" signal for an internally generated movement occurs when activation of SMA is boosted beyond some threshold amount by the activity reaching it through this basal ganglia "funnel."

Basal Ganglia Disorders. Support for the view that the direct motor loop through the basal ganglia functions to facilitate the initiation of willed movements has come from the study of several human diseases. According to one model, increased inhibition of the thalamus by the basal ganglia underlies *hypokinesia*, a paucity of movement, whereas decreased basal ganglia output leads to *hyperkinesia*, an excess of movement.

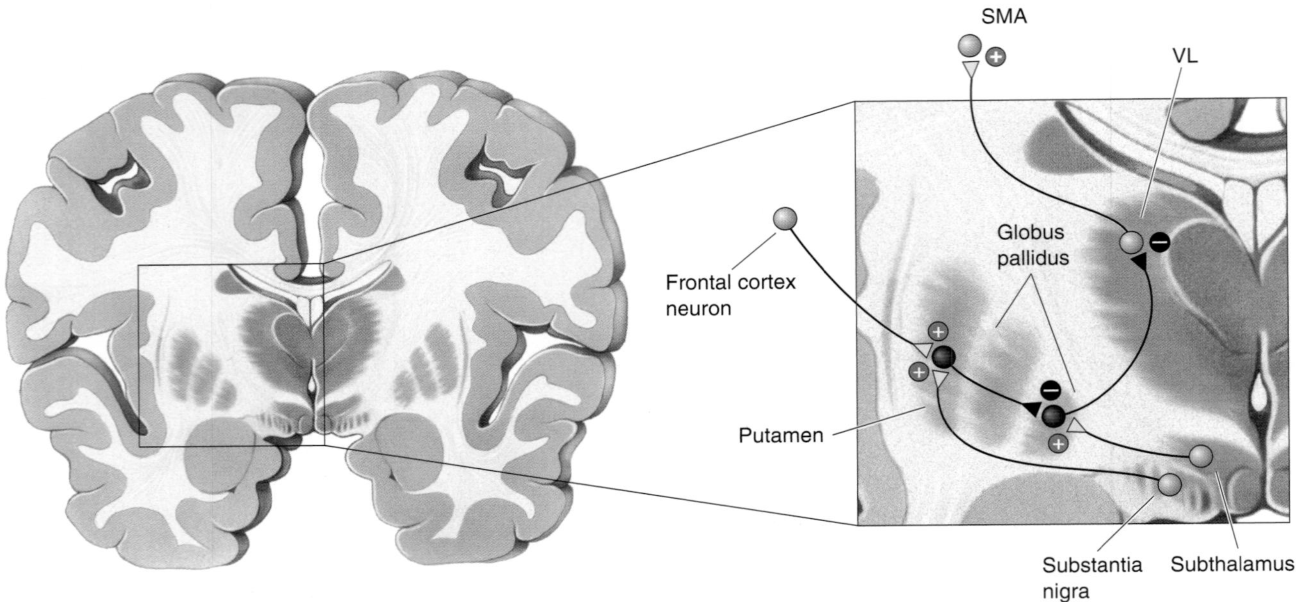

Figure 14.12
A wiring diagram of the basal ganglia motor loop. Synapses marked with a plus (+) are excitatory; those with a minus (−) are inhibitory.

Box 14.2

OF SPECIAL INTEREST

Do Neurons in Diseased Basal Ganglia Commit Suicide?

Several devastating neurological diseases involve the slow, progressive death of neurons. Most patients with Parkinson's disease have lost more than 80% of the dopamine-utilizing neurons in their substantia nigra (Figure A). Neurons of the striatum and other regions slowly degenerate in patients with Huntington's disease (Figure B). Why do these neurons die? Ironically, it may be that natural forms of cell death are involved. A process called *programmed cell death* is essential for normal brain development; certain neurons commit suicide as part of the "program" by which the nervous system forms (see Chapter 22). All cells have several "death genes" that trigger a set of enzymes that destroy cellular proteins and DNA. Some forms of cancer occur when normal programmed cell death is prevented and cells proliferate wildly. Some neurological diseases may result when programmed cell death is unnaturally activated.

Huntington's disease is caused by a dominant mutation in a gene that codes for a large brain protein called *huntingtin*. The normal molecule has a chain of 37 or fewer glutamines at one end, but in the mutated form this chain expands to as many as 150 glutamines. The abnormally long huntingtins aggregate; globs of them accumulate and trigger neuronal degeneration. The function of normal huntingtin is unknown, but it may counterbalance the triggers for programmed cell death. Thus, Huntington's disease may arise from normal processes of neuronal degeneration gone awry. Abnormally long chains of glutamines on other proteins are associated with several other inherited diseases that cause neural degeneration, and these expanded proteins may be the key to understanding why some neurons die prematurely.

Parkinson's disease is primarily a disease of aging, and the majority of cases occur after age 60. In 1976 and again in 1982, however, several relatively young drug abusers in Maryland and California developed the symptoms of severe Parkinson's disease over only a few days. This was extraordinary, because usually symptoms accumulate over many years.

Parkinson's disease epitomizes the first condition. Characterized by hypokinesia, it affects about 1% of people over age 50. Its symptoms include slowness of movement (bradykinesia), difficulty in initiating willed movements (akinesia), increased muscle tone (rigidity), and tremors of the hands and jaw that are most prominent at rest. The organic basis of Parkinson's disease is degeneration of substantia nigra inputs to the striatum (Box 14.2).

(Box 14.2, *continued*)

Astute medical detective work unraveled the cause of the addicts' affliction. Each had taken street versions of a synthetic narcotic that contained the chemical MPTP (1-methyl-4-phenyl-1,2,3,6-tetrahydropyridine). The incompetent basement chemists who synthesized the illegal drug tried to short-cut the procedure, thereby creating a chemical by-product that kills dopaminergic neurons. MPTP has since helped us understand Parkinson's disease better. We now know that MPTP is converted to MPP^+ (1-methyl-4-phenylpyridinium) in the brain; dopaminergic cells are selectively vulnerable to it because their membrane dopamine transporters mistake MPP^+ for dopamine, and they selectively accumulate this chemical Trojan horse. Once inside the cell, MPP^+ disrupts energy production in mitochondria, and the neurons apparently die because their ATP is depleted.

The effect of MPTP supports the idea that common forms of Parkinson's disease may be caused by chronic exposure to a slowly acting toxic chemical in the environment. Unfortunately, no one has identified such a toxin. Research has recently shown that MPTP can induce a form of programmed neuronal death in the substantia nigra. Dopaminergic neurons of Parkinsonian patients may degenerate for a similar reason. By understanding how and why neurons kill themselves, we may be able to devise strategies of suicide prevention that halt or avert a variety of terrible neurological diseases.

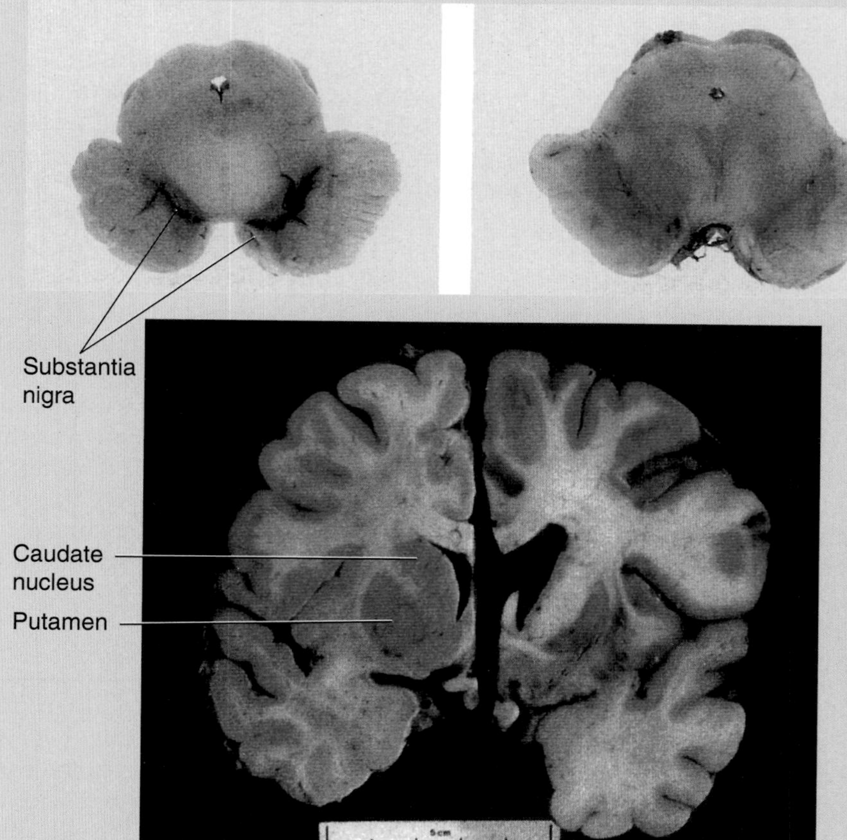

Figure A
Normal (left); Parkinson's disease (right)
(Source: Strange, 1992, Fig. 10.3.)

Figure B
Normal (left); Huntington's disease (right)
(Source: Strange, 1992, Fig. 11.2.)

These inputs use the neurotransmitter dopamine (DA), which normally facilitates the direct motor loop by activating cells in the putamen (which releases VLo from globus pallidus-induced inhibition). In essence, the depletion of dopamine closes the funnel that feeds activity to SMA via the basal ganglia and VLo.

Parkinson's disease can be treated by administration of the compound dopa (L-dihydroxyphenylalanine, introduced in Chapter 6), a precursor to dopamine. Dopa crosses the blood-brain barrier and boosts DA synthesis in the cells that remain alive in the substantia nigra, thus alleviating some of the symptoms. Dopa treatment, however, does not alter the course of the disease or the rate at which substantia nigra neurons degenerate. (We will return to the topic of dopamine neurons in Chapter 15.)

If Parkinson's disease lies at one end of the spectrum of basal ganglia disorders, Huntington's disease lies at the other. **Huntington's disease** is a hereditary, progressive, inevitably lethal syndrome characterized by hyperkinesia and *dyskinesias*, or abnormal movements; *dementia*, or impaired cognitive abilities; and a disorder of personality. The disease is particularly insidious because its symptoms usually do not appear until well into adulthood, often after the patient has passed the gene on to his or her children. People with Huntington's disease exhibit *chorea*, uncontrollable and purposeless movements with rapid, irregular flow and flicking motions of various parts of the body. The most obvious brain pathology is a profound loss of neurons in the caudate nucleus, putamen, and globus pallidus, with additional cell loss in the cerebral cortex and elsewhere (Box 14.2). The damage to structures in the basal ganglia and consequent loss of its inhibitory output to the thalamus seem to account for the disorders of movement in Huntington's patients. Cortical degeneration is primarily responsible for their dementia and personality changes.

Hyperkinesia can also result from other types of lesions that affect the basal ganglia. One example is a condition called **ballism**, which is characterized by violent, flinging movements of the extremities (somewhat like our baseball pitcher unintentionally winding up and throwing the ball while sitting in the dugout). As with Parkinson's disease, the cellular mechanisms associated with ballism are known: It is caused by damage to the subthalamic nucleus (usually resulting from an interruption of its blood supply caused by stroke). The subthalamic nucleus, part of another side loop within the basal ganglia, excites neurons in the globus pallidus that project to VLo. Remember that excitation of the globus pallidus inhibits VLo (Figure 14.12). Thus, a loss of excitatory drive to the globus pallidus facilitates VLo, in effect opening the funnel of activity to SMA.

In summary, the basal ganglia may facilitate movement by focusing activity from widespread regions of cortex onto SMA. Importantly, however, they also serve as a *filter* that keeps inappropriate movements from being expressed. We saw in Roland's PET studies that activity in SMA does not automatically trigger movement. The initiation of voluntary movement also requires activation of area 4, the subject of the next section.

THE INITIATION OF MOVEMENT BY PRIMARY MOTOR CORTEX

SMA is heavily interconnected with M1, cortical area 4 on the precentral gyrus. The designation of area 4 as primary motor cortex is somewhat arbitrary because this is not the only cortical area that contributes to the corticospinal tract or to movement. Nonetheless, it has been recognized since the time of Sherrington that this area has the lowest threshold for the elicitation of movement by electrical stimulation. In other words, stimulation intensities that are unable to evoke movement in other cortical areas are still effective in evoking movement when applied to area 4, meaning that area 4 has strong synaptic connections with the motor neurons. Focal electrical stimulation of area 4 evokes the contraction of small groups of muscles, and as we discussed earlier, the somatic musculature is mapped systematically in this area. This ribbon of cortex that stretches the full length of the precentral gyrus is sometimes also called the **motor strip**.

The Input-Output Organization of M1

The pathway by which motor cortex activates lower motor neurons originates in cortical layer V. Layer V has a population of pyramidal neurons, some of which can be quite large (soma diameters approaching 0.1 mm).The largest cells were first described as a separate class by Russian anatomist Vladimir Betz in 1874 and are therefore called *Betz cells*. The layer V pyramidal cells in M1 receive their inputs primarily from two sources: other cortical areas and the thalamus. The major cortical inputs originate in the areas adjacent to area 4: area 6, immediately anterior; and areas 3, 1, and 2, immediately posterior (Figure 14.7). The thalamic input to M1 arises mainly from another part of the ventral lateral nucleus called VLc, which relays information from the cerebellum. Besides projecting directly to the spinal cord, layer V pyramidal cells send axon collaterals to many subcortical sites involved in sensorimotor processing, especially the brain stem.

The Coding of Movement in M1

For a time it was thought that the motor cortex consisted of a detailed mapping of the individual muscles, such that the activity of a single pyramidal cell would lead to activity in a single motor neuron pool. The view that has emerged from more recent work, however, is that individual pyramidal cells can drive numerous motor neuron pools from a group of muscles involved in moving a limb toward a desired goal. Recordings from M1 neurons in behaving animals have revealed that a burst of activity occurs immediately before and during a voluntary movement and that this activity appears to encode two aspects of the movement, force and direction.

Considering that cortical microstimulation studies suggested the existence of a fine-grained movement map in M1, the discovery that the movement direction tuning of individual M1 neurons is rather broad came as a surprise. This breadth of tuning is shown clearly in a type of experiment devised by Apostolos Georgopoulos and his colleagues, then working at Johns Hopkins University. Monkeys were trained to move a joystick toward a small light whose position varied randomly around a circle. The M1 cells fired most vigorously during movement in one direction (180 degrees in the example in Figure 14.13a) but also discharged during movements that varied $\pm$ 45 degrees from the preferred direction. The coarseness in the directional tuning of the corticospinal neurons was certainly at odds with the high accuracy of the monkey's movements, suggesting that the direction of movement could not be encoded by the activity of individual cells that command movement in a single direction. Georgopoulos hypothesized that movement direction was encoded instead by the collective activity of a population of neurons.

To test the feasibility of this idea of population coding for movement direction, Georgopoulos and his colleagues recorded from more than 200 different neurons in M1; for each cell, they constructed a directional tuning curve such as is shown in Figure 14.13b. From these data, the researchers knew how vigorously each of the cells in the population responded during movement in each direction. The activity of each cell was represented as a *direction vector* pointing in the direction that was best for that cell; the length of

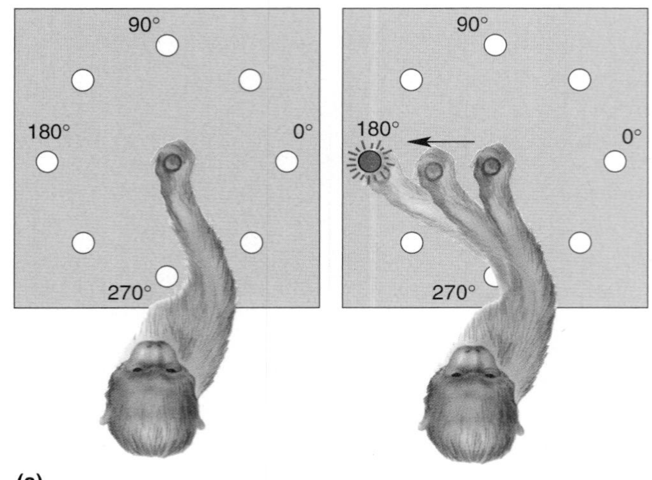

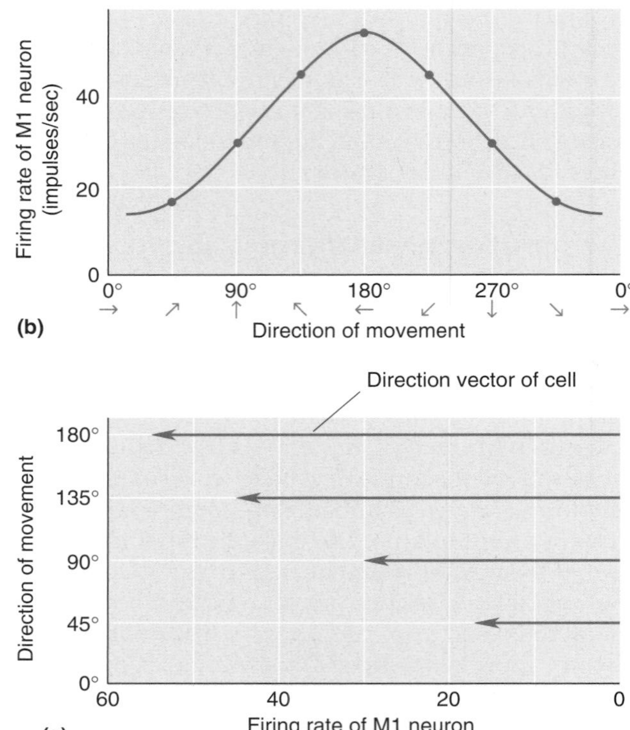

Figure 14.13

Responses of an M1 neuron during arm movements in different directions. (a) A monkey is instructed to move a handle toward a small light as responses of an M1 neuron are monitored. Having the monkey move in directions around the clock allows determination of the relationship between the cell's discharge rate and movement direction. **(b)** A tuning curve for an M1 neuron. This cell fires most during movements to the left. **(c)** Because the cell in part b responds best to leftward movement, it is represented by a vector pointing in that direction (the direction vector). The length of the vector is proportional to the firing rate of the cell. Notice that as the movement direction changes, the length of the direction vector changes. (Source: Adapted from Georgopoulos et al., 1982.)

the vector represented how active that cell had been during a particular movement (Figure 14.14). The vectors representing each cell's activity could be plotted together for each direction of movement, then averaged to yield what the researchers called a *population vector*. They found excellent agreement between this average vector, representing the activity of the entire population of M1 cells, and the actual direction of movement (Figure 14.15).

This work suggests three important conclusions about how M1 commands voluntary movement: (1) that much of the motor cortex is active for every movement, (2) that the activity of each cell represents a single "vote" for a particular direction of movement, and (3) that the direction of movement is determined by a tally and averaging of the votes registered by each cell in the population. Although this population-coding scheme remains hypothetical in M1, experiments on the superior colliculus by James McIlwain at Brown University and by David Sparks at the University of Alabama have shown conclusively that a population code is used by this structure to command precisely directed eye movements (Box 14.3).

The Malleable Motor Map. An interesting prediction that derives from this scheme for motor control is that the larger the population representing a type of movement, the finer the possible control. From the motor map shown in Figure 14.8, we would predict that finer control should be possible for the hands and the muscles of facial expression, and indeed this is normally the case. Of course, fine movements of other muscles can be learned with experience; consider the finger, wrist, elbow, and shoulder movements of an accomplished cellist. Does this mean that cortical cells in M1 can switch allegiance from participation in one type of movement to another as skills are learned? The answer appears to be yes. John Donoghue, Jerome Sanes, and their students at Brown University have presented evidence indicating that such plasticity of the adult motor cortex is possible. For example, in one series of experiments using cortical microstimulation in rats, they found that the region of M1

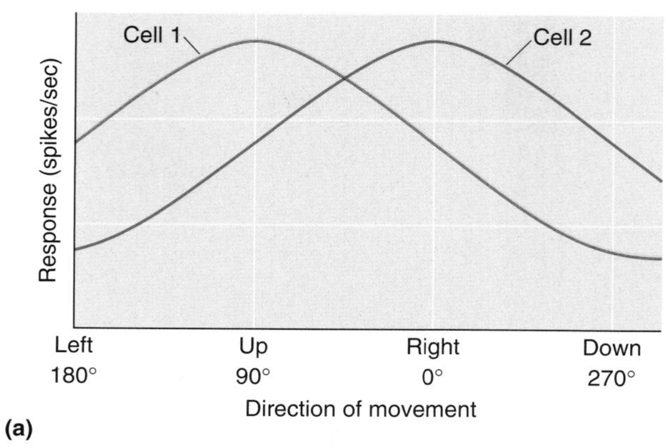

(a)

Movement direction | Direction vector for cell 1 | Direction vector for cell 2 | Population vector for cells 1 and 2

Right + =

Up + =

(b)

Figure 14.14
Direction vectors and population vectors. (a) The graph shows the tuning curves for two imaginary cells in the motor cortex (Figure 14.13). Both cells fire during movement in a wide range of directions, but cell 1 fires best when movement is upward, while cell 2 responds best when movement is from left to right. **(b)** The response of each cell can be represented as a direction vector. The direction vector always points in the preferred direction for the neuron, but the length varies with the number of action potentials fired by the cell during movement over a range of directions. For any given direction of movement, the direction vectors of the individual cells can be combined to yield a population vector, reflecting the strength of the response of both cells during this movement.

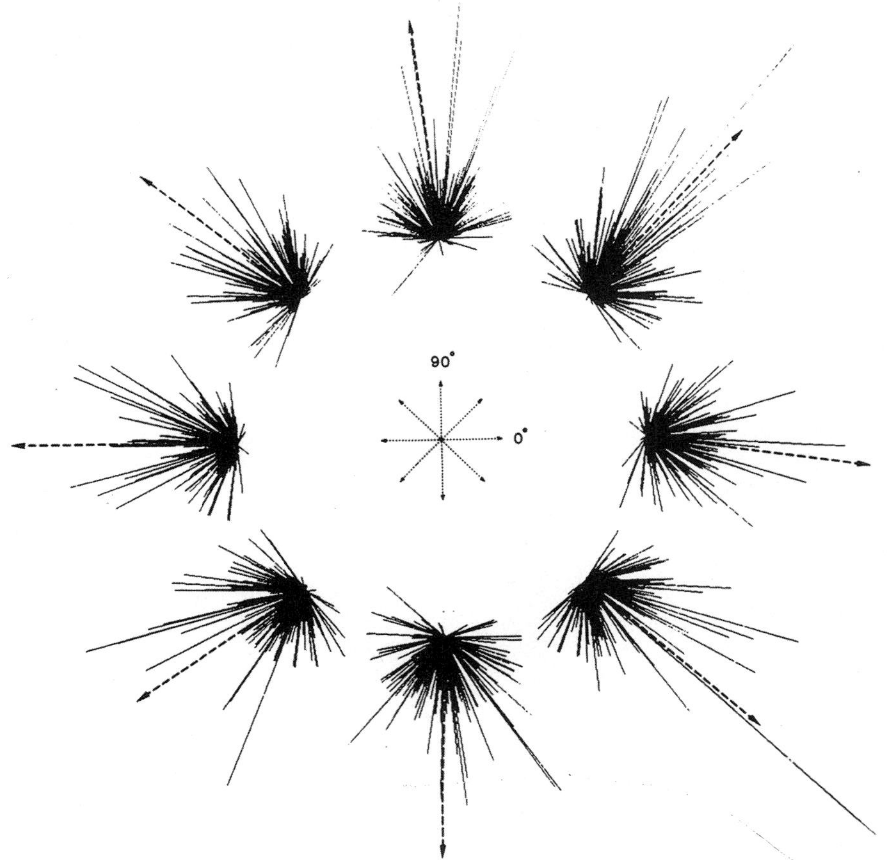

Figure 14.15
Predicting the direction of movement by population vectors. Each cluster of lines reflects the direction vectors of many cells in M1. Line length reflects the discharge rate of each cell during a movement in eight directions. For each direction, the individual direction vectors can be summed to yield the population vector (dashed arrows), which predicts the direction of movement. (Source: Georgopoulos et al., 1983.)

Box 14.3

PATH OF DISCOVERY

Distributed Coding in the Superior Colliculus

BY JAMES T. McILWAIN

During the 1960s and 1970s, it became relatively easy to record the electrical activity of single neurons in the brain. The power of this method led to the idea of the "feature-detector neuron," whose discharge was thought to announce the presence of the stimulus feature to which it was most sensitive. This view rarely tempted students of olfaction and gustation, because they had found that the discharge of single chemosensory neurons is highly ambiguous with respect to the identity of the stimulus. Not so in vision research, where the cells with the smallest spatial receptive fields and the most refined preferences for specific stimuli seemed most interesting. Any neuron responsive to a range of stimuli was regarded as crudely selective and unsuited to processes requiring high resolution.

Neurophysiologists who examined visual areas of the brain stem soon encountered a paradox. The receptive fields of cells in the superior colliculus turned out to be very large, yet this structure was clearly important for the execution of accurate saccadic eye movements, which change the direction of gaze to a stimulus of interest. The superior colliculus receives orderly input from both the retina and the visual cortex, and damage to it impairs an animal's ability to direct its gaze at novel stimuli. Focal electrical stimulation of the colliculus evokes saccades whose directions and amplitudes are correlated with the visual receptive fields of cells at the stimulus site. A small change in the position of the stimulating electrode results in small changes in the saccade's direction and amplitude. Certain collicular neurons discharge in association with saccadic eye movements, as if they are part of the control mechanism that specifies the dimensions of the move-

James T. McIlwain

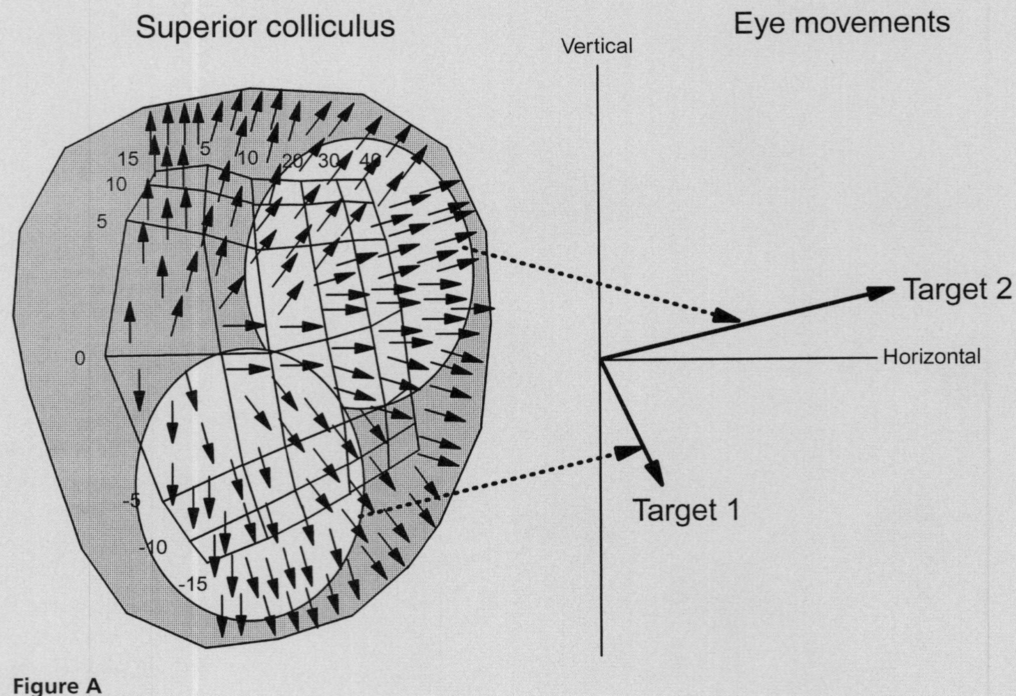

Figure A

(Box 14.3, *continued*)

ment. This activity occurs in association with saccades that terminate across a restricted zone of visual space called the *movement field* of the cell, by analogy to the sensory receptive field. How could such cells specify the target of a saccade with any accuracy if their movement fields and visual receptive fields are very large?

The answer began to emerge from experiments on the cat's superior colliculus in my laboratory at Brown University and from studies of the primate's superior colliculus by David Sparks, then at the University of Alabama. We asked the inverse of the traditional receptive field question or movement field question. From the sensory side, instead of asking where a point of light must be to activate a collicular neuron, we asked where in the superior colliculus are the cells that have the point in their receptive fields—that "see" the point. Similarly, the key consideration on the motor side is the location of the cells that discharge before a saccade to a given target, rather than the size of their individual movement fields. The analyses from both laboratories revealed that these regions of activity are widespread, occupying considerable fractions of the collicular tissue. As the stimulus or target location moves around in visual space, the corresponding patch of neural activity moves around in the superior colliculus.

A general idea of how a system of such neurons may encode a saccade is shown in Figure A. On the left, each arrow on the retinotopic map of the superior colliculus symbolizes the contribution of its location to the code for the direction of a saccade. The more closely packed the arrows, the stronger the signal from that region to brain stem circuits that shape the motor commands for the saccade. The distribution of the arrows is consistent with the effects of focal electrical stimulation. Thus, for example, stimulation at successively lower points in the lower half of the map (representing the lower visual field) leads to downwardly directed saccades of increasing amplitude. If the appearance of target 1 in the right side of the figure excited cells in the lower unshaded oval of the map, the collective activity of the cells would specify the downward and horizontal components needed for a saccade to the target. The appearance of target 2 would activate the upper unshaded area, assembling the components of the correct, upwardly directed saccade. In this model, changes in the position of the target modify the output signal to yield a saccade matching any position of the target.

The large size of the receptive fields and movement fields of collicular neurons means that information about the location of a visual point or a saccade target is distributed among many neurons. The model of Figure A shows how the active population of collicular neurons may indicate the target's position as a motor code. The simple and incomplete model shown here represents only one of several ideas about how the superior colliculus accomplishes its task. It seems certain, though, that target location and saccade measurements are encoded in the distribution of activity across a population of neurons.

that normally elicits whisker movements would evoke forelimb movements instead when the motor nerve that supplies the muscles of the snout was cut (Figure 14.16). These neuroscientists speculate that similar types of cortical reorganization may provide a basis for learning fine motor skills.

From the preceding discussion, we can imagine that when the time has come for our pitcher to wind up, the motor cortex generates a torrent of activity in the pyramidal tract. What could appear to be a discordant voice to the neurophysiologist recording from a single M1 neuron is part of a clear chorus of activity to the spinal motor neurons that generate the movement.

THE CEREBELLUM

It is not enough simply to command the muscles to contract. Throwing a curve ball requires a detailed *sequence* of muscle contractions, each one timed with great precision. Looking after this critical motor control function is the **cerebellum**, introduced in Chapter 7. Involvement of the cerebellum in this aspect of motor control is plainly revealed by cerebellar lesions; movements become uncoordinated and inaccurate, a condition known as **ataxia**.

Take this simple test. Lay your arms in your lap for a moment, then touch your nose with one finger. Try it again with your eyes closed. No problem,

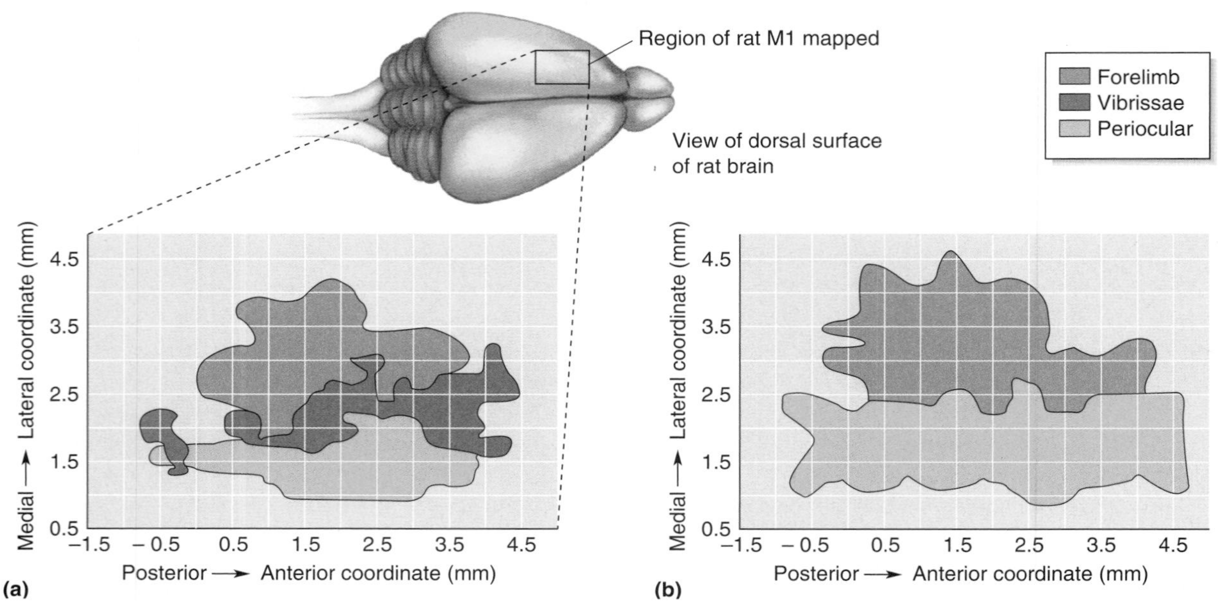

Figure 14.16

Malleable motor maps. (a) This map indicates motor cortex from a normal rat. **(b)** This map indicates a rat whose motor nerve that supplies the vibrissae (whiskers) was cut. The regions of cortex that used to evoke movement of the vibrissae now cause muscle movement in the forelimb or around the eyes (periocular). (Source: Adapted from Sanes and Donoghue, 1990.)

Box 14.4

OF SPECIAL INTEREST

Involuntary Movements, Normal and Abnormal

Raise your hand in front of your face and try to hold it as still as possible. You will see a very slight trembling of your fingers. This is called *physiological tremor*, a small, rhythmic oscillation of about 8–12 Hz (Figure A). It is perfectly normal, and there is nothing you can do to stop it short of resting your hand on the table. A variety of everyday circumstances—stress, anxiety, hunger, fatigue, fever, too much coffee—can increase the tremor.

As we discuss in this chapter, some neurological diseases lead to more dramatic involuntary movements with distinctive characteristics. Parkinson's disease is often associated with a large *resting tremor* of about 3–5 Hz. Movement is at its worst when the patient is not attempting to move. Strangely enough, the tremor immediately disappears during a voluntary movement. On the other hand, people with cerebellar damage have no abnormal tremor at rest but often show dramatic *intention tremor* when they try to move. Cerebellar tremor is an expression of ataxia, the uncoordinated contractions of the muscles used in the movement. For example, as the patient tries to move her finger from one point in space to another (Figure B) or track a pathway with her finger (Figure C), she makes large errors; in trying to correct an error, she makes more errors, and so on, as the finger wobbles toward its destination.

Huntington's disease causes *chorea* (from the Greek for "dance")—quick, irregular, involuntary, but relatively coordinated movements of the limbs, trunk, head, and face. Other types of basal ganglia disease can lead to *athetosis,* much slower, almost writhing movements of the neck and trunk. The unique properties of each abnormal movement can be very helpful in the diagnosis of neurological diseases, and they have also taught us volumes about the normal functions of the damaged parts of the brain.

right? Patients with cerebellar damage are often incapable of performing this simple task. Instead of smoothly and simultaneously moving the shoulder, elbow, and wrist to bring the finger to rest on the nose, they move each joint sequentially—first the shoulder, then the elbow, and finally the wrist. This is called *dyssynergia*, decomposition of synergistic multijoint movement. Another characteristic deficit shown by these patients is that their finger movement is *dysmetric*; they either come up short of the nose or shoot past it, poking themselves in the face. You may recognize these symptoms as similar to those that accompany ethanol intoxication. Indeed, the clumsiness that accompanies alcohol abuse is a direct consequence of the depression of cerebellar circuits (Box 14.4).

Anatomy of the Cerebellum

Cerebellar anatomy is shown in Figure 14.17. The cerebellum sits on stout stalks called peduncles that rise from the pons; the whole structure resembles

(Box 14.4, *continued*)

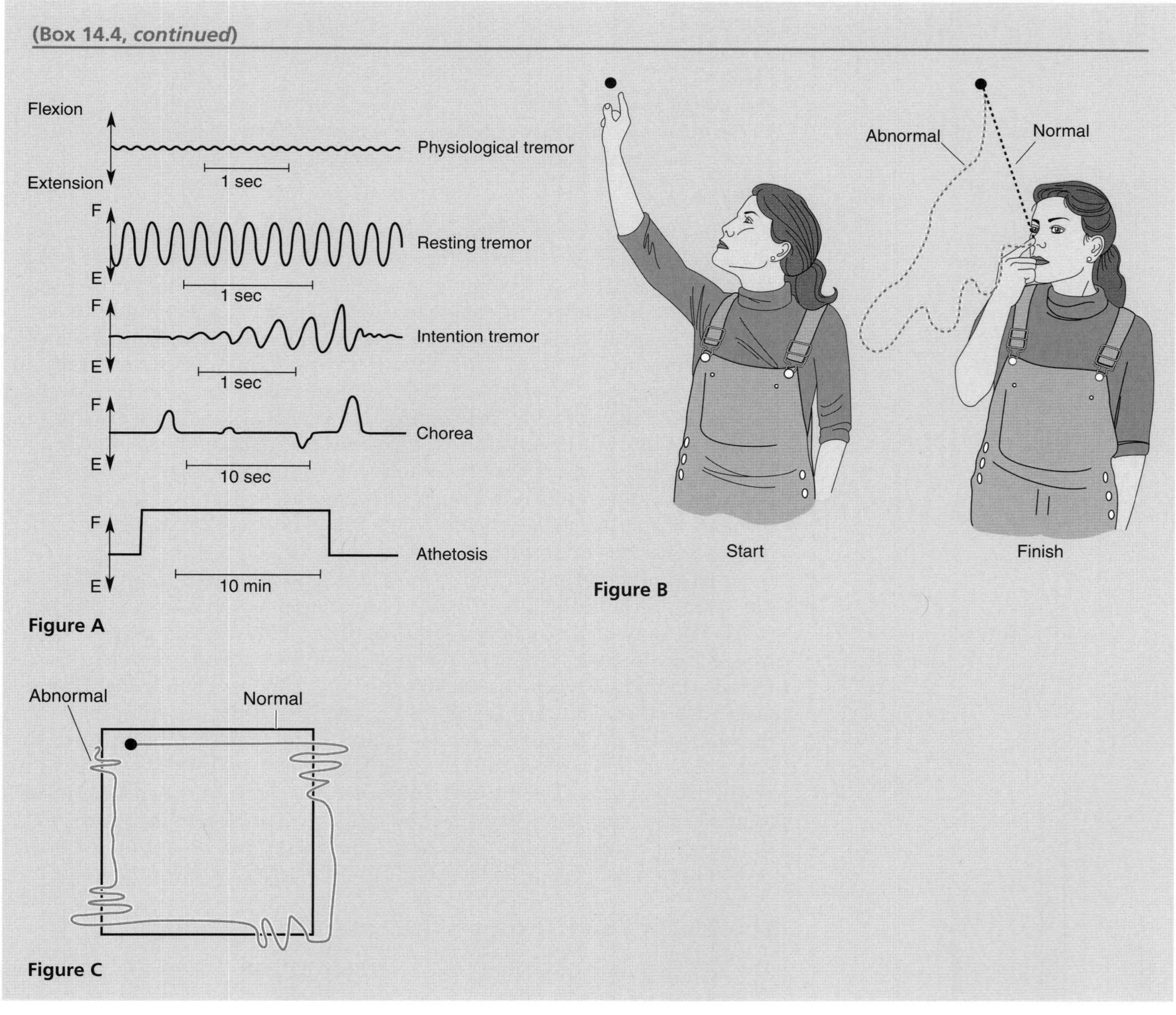

Figure A

Figure B

Figure C

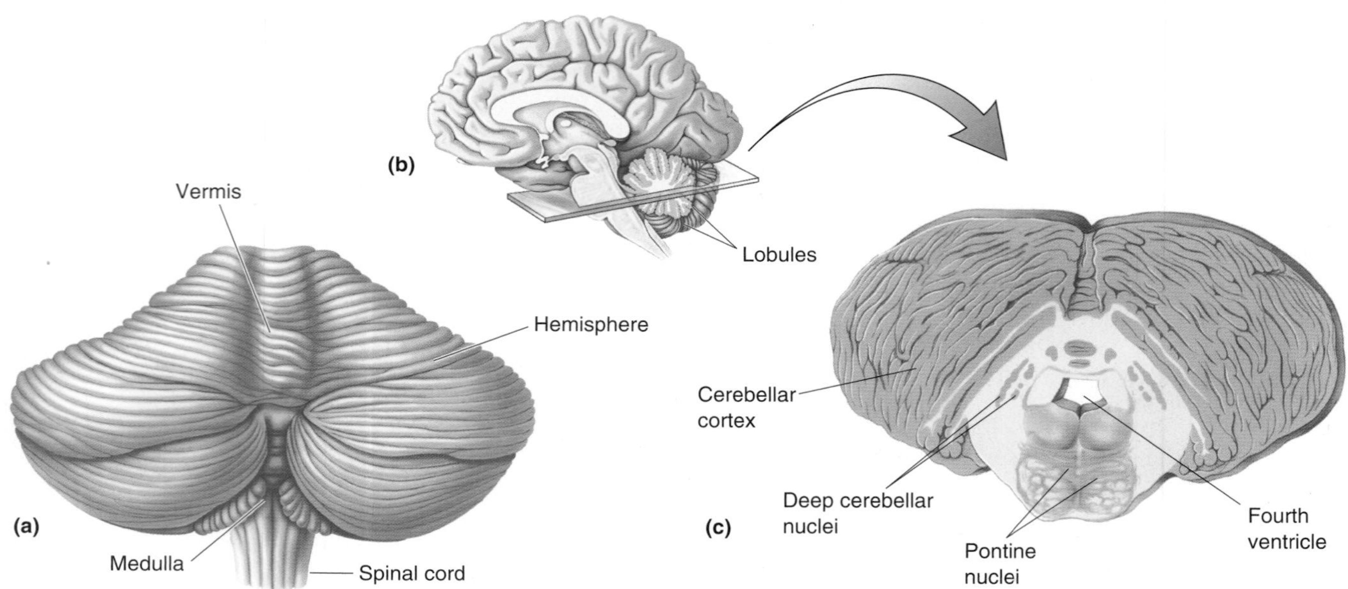

Figure 14.17

The cerebellum. (a) A dorsal view of the human cerebellum showing the vermis and hemispheres. **(b)** A midsagittal view of the brain, showing the lobules of the cerebellum. **(c)** A cross section of the cerebellum showing the cortex and deep nuclei.

a piece of broccoli or cauliflower. The visible part of the cerebellum is actually a thin sheet of cortex that is repeatedly folded. The dorsal surface is characterized by a series of shallow ridges called folia (singular: folium), which run transversely, that is, from side to side. In addition, deeper transverse fissures, revealed by making a sagittal slice through the cerebellum, divide the cerebellum into 10 lobules. Together, folia and lobules serve to greatly increase the surface area of the cerebellar cortex, as the gyri of the cerebrum do for the cerebral cortex. Because of the high density of neurons in its cortex, the cerebellum, which constitutes only about one-tenth of the total volume of the brain, contains more than *50%* of the total number of neurons in the CNS. Neurons are also embedded deep in the white matter of the cerebellum, forming the *deep cerebellar nuclei*, which relay most of the cerebellar cortical output to various brain stem structures.

Unlike the cerebrum, the cerebellum is not obviously split down the middle. At the midline, the folia appear to run uninterrupted from one side to the other. The only distinguishing feature of the midline is a bump that runs like a backbone down the length of the cerebellum. This midline region called the **vermis** (Latin for "worm") separates the two lateral **cerebellar hemispheres** from each other. The vermis and the hemispheres represent important functional divisions. The vermis sends output to the brain stem structures that contribute to the ventromedial descending spinal pathways, which, as already discussed, control the axial musculature. The hemispheres are related to other brain structures that contribute to the lateral pathways, particularly the cerebral cortex. For the purpose of illustration, we'll focus on the lateral cerebellum, which is particularly important for limb movements.

The Motor Loop Through the Lateral Cerebellum

The simplest circuit involving the lateral cerebellum constitutes yet another loop, shown schematically in Figure 14.18. Axons arising from layer V pyra-

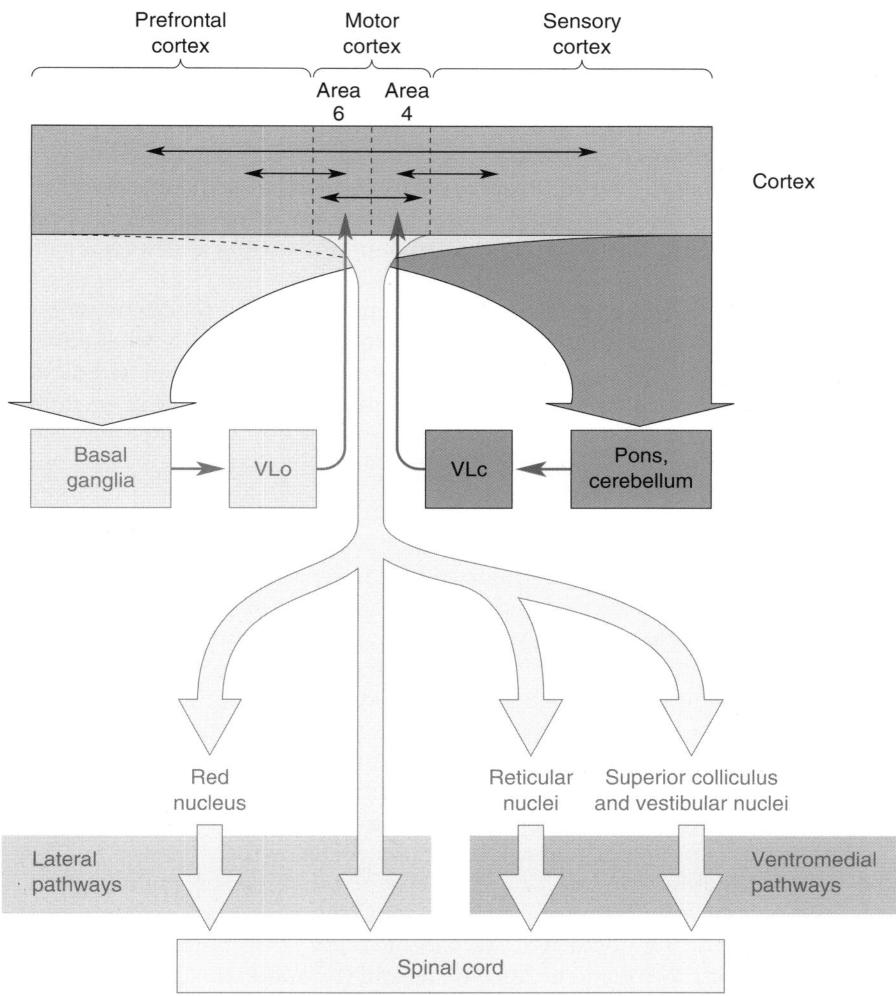

Figure 14.18
A summary of the motor loop through the cerebellum.

midal cells in the sensorimotor cortex—frontal areas 4 and 6, somatosensory areas on the postcentral gyrus, and the posterior parietal areas—form a massive projection to clusters of cells in the pons, the **pontine nuclei**, which in turn feed the cerebellum. To appreciate the size of this pathway, consider that the corticopontocerebellar projection contains about 20 million axons; that's *20 times* more than in the pyramidal tract. The lateral cerebellum projects back to the motor cortex via a relay in the ventral lateral nucleus of the thalamus (VLc).

From the effects of lesions in this pathway, we can deduce that it is critical for the proper execution of planned, voluntary multijoint movements. Indeed, once the signal for movement intent has been received by the cerebellum, the activity of this structure appears to instruct the primary motor cortex with respect to movement direction, timing, and force. For ballistic movements, these instructions are based entirely on predictions about their outcome (because such movements are too fast for feedback to be of much immediate use). Such predictions are based on experience; that is, they are learned. Therefore, the cerebellum is another important site for motor learning—it is a place where *what is intended* is compared with *what has happened.* When this comparison fails to meet expectations, compensatory modifications are made in certain cerebellar circuits.

Programming the Cerebellum. We will return to the details of cerebellar circuitry and how it is modified by experience in Chapter 24. But for now, think about the process of learning a new skill (e.g., downhill skiing, tennis, juggling, throwing a curve ball). Early on, you must concentrate on new movements, which you will perform in a disjointed and uncoordinated fashion. Practice makes perfect, however. As you master the skill, the movements become smooth, and eventually you perform them almost unconsciously. This process represents the creation of a new motor program that generates the appropriate movement sequences on demand without the need for conscious control.

We are reminded of a certain computer microchip manufacturer that advertises its product as the "computer inside." The cerebellum acts as the "brain inside" that unconsciously sees to it that the programs for skilled movement are executed properly and are adjusted whenever their execution fails to meet expectations.

CONCLUDING REMARKS

Let's return to the example of the baseball player one last time to put the pieces of the motor control puzzle together. Imagine the pitcher walking to the mound. The spinal circuits of the crossed extensor reflex are engaged and coordinated by descending commands on the ventromedial pathways. Extensors contract, flexors relax; flexors contract, extensors relax.

Once on the mound, the pitcher is joined by the umpire. Into his outstretched hand the umpire drops a new baseball. The added weight stretches the flexors of the arm. Group Ia afferents become more active and cause monosynaptic excitation of the motor neurons innervating the flexors. The muscles contract to hold the ball up against gravity.

The pitcher is now ready to pitch. His neocortex is fully engaged and active as he looks at the catcher for the hand signal that instructs which pitch is to be thrown. At the same time, the ventromedial pathways are working to maintain his standing posture. Although his body is still, the neurons of the ventral horns of the spinal cord are firing madly under the influence of the ventromedial pathways, keeping the extensors of the lower leg activated.

The catcher flashes the sign for the curve ball. The sensory information is communicated to the parietal and prefrontal cortex. These regions of cortex and area 6 begin planning the movement strategy.

The batter steps up to the plate and is ready. Activity cycling through the basal ganglia increases, triggering the initiation of the pitch. In response to this input, SMA activity increases and is followed immediately by the activation of M1. Now instructions are sweeping down the axons of the lateral pathways. The cerebellum, activated by the corticopontocerebellar inputs, uses these instructions to coordinate the timing of the descending activity so the proper sequence of muscle contractions can occur. Cortical input to the reticular formation leads to the release of the antigravity muscles from reflex control. Finally, lateral pathway signals engage the motor neurons and interneurons of the spinal cord, which cause the muscles to contract.

The pitcher winds up and throws. The batter swings. The ball sails over the left-field fence. The crowd jeers; the pitcher's manager curses; his team's owner frowns. Even as the cerebellum goes to work making adjustments for the next pitch, the player's body reacts. His face flushes; he sweats; he's angry and anxious. But these latter reactions are not the stuff of the somatic motor system. These are topics of Part III, The Brain and Behavior.

KEY TERMS

Descending Spinal Tracts
lateral pathway (p. 467)
ventromedial pathway (p. 468)
corticospinal tract (p. 468)
motor cortex (p. 469)
pyramidal tract (p. 469)
rubrospinal tract (p. 469)
red nucleus (p. 469)
vestibulospinal tract (p. 470)
vestibular nuclei (p. 470)
tectospinal tract (p. 471)
reticular formation (p. 471)
pontine reticulospinal tract (p. 472)
medullary reticulospinal tract (p. 472)

The Planning of Movement by the Cerebral Cortex
primary motor cortex (M1) (p. 473)
premotor area (PMA) (p. 474)
supplementary motor area (SMA) (p. 474)

The Basal Ganglia
ventral lateral nucleus (VL) (p. 477)
basal ganglia (p. 477)
caudate nucleus (p. 478)
putamen (p. 478)
globus pallidus (p. 478)
subthalamic nucleus (p. 478)
substantia nigra (p. 478)
striatum (p. 478)
Parkinson's disease (p. 481)
Huntington's disease (p. 482)
ballism (p. 482)

The Initiation of Movement by Primary Motor Cortex
motor strip (p. 483)

The Cerebellum
cerebellum (p. 487)
ataxia (p. 487)
vermis (p. 490)
cerebellar hemispheres (p. 490)
pontine nuclei (p. 491)

REVIEW QUESTIONS

1. List the components of the lateral and ventromedial descending spinal pathways. Which type of movement does each path control?
2. You are a neurologist presented with a patient who has the following symptom: an inability to independently wiggle the toes on the left foot, but with all other movements (walking, independent finger movement) apparently intact. You suspect a lesion in the spinal cord. Where?
3. PET scans can be used to measure blood flow in the cerebral cortex. What parts of the cortex show increased blood flow when a subject is asked to think about moving his or her right finger?
4. Why is dopa used to treat Parkinson's disease? How does it alleviate the symptoms?
5. Individual Betz cells fire during a fairly broad range of movement directions. How may they work together to command a precise movement?
6. Sketch the motor loop through the cerebellum. What movement disorders result from damage to the cerebellum?

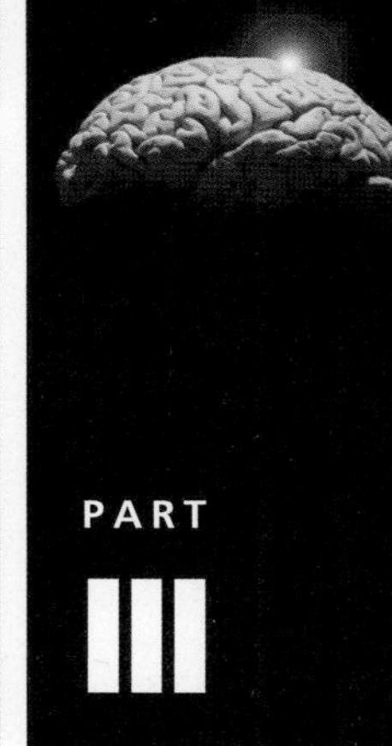

PART III

THE BRAIN AND BEHAVIOR

CHAPTER 15

Chemical Control of the Brain and Behavior

INTRODUCTION

It should be obvious by now that knowing the organization of synaptic connections is essential to understanding how the brain works. It's not from a love of Greek and Latin that we belabor the neuroanatomy! Most of the connections we have described are precise and specific. For example, for you to read these words, there must be a very fine-grained neural mapping of the light falling on your retina—how else could you see the dot on this question mark? The information must be carried centrally and precisely dispersed to many parts of the brain for processing, coordinated with control of the motor neurons that closely regulate the six muscles of each eye as it scans the page.

In addition to anatomical precision, point-to-point communication in the sensory and motor systems requires mechanisms that restrict synaptic communication to the cleft between the axon terminal and its target. It just wouldn't do for glutamate released in somatosensory cortex to activate neurons in motor cortex! Furthermore, transmission must be brief enough to allow rapid responses to new sensory inputs. Thus, at these synapses, only minute quantities of neurotransmitter are released with each impulse, and these molecules are quickly destroyed enzymatically or taken up by neighboring cells. The postsynaptic actions at transmitter-gated ion channels last only as long as the transmitter is in the cleft, a few milliseconds at most. Many axon terminals also possess presynaptic "autoreceptors" that detect the transmitter concentrations in the cleft and inhibit release if they get too high. These mechanisms ensure that this type of synaptic transmission is tightly constrained in both space and time.

The elaborate mechanisms that constrain point-to-point synaptic transmission bring to mind a telecommunications analogy. Telephone systems make possible very specific connections between one place and another; your mother in Tacoma can talk just to you in Providence, reminding you that her birthday was last week. The telephone lines can act like precise synaptic connections. The influence of one neuron (your mother) is targeted to a small number of other neurons (in this case, only you). The embarrassing message is limited to your ears only. For a real neuron in one of the sensory or motor systems discussed so far, its influence usually extends to the few dozen or few hundred cells it synapses on—a conference call, to be sure, but still relatively specific.

Now imagine your mother being interviewed on a television talk show, which is broadcast via a cable network. In this case, the widespread cable connections may allow her to tell millions of people that you forgot her birthday, and the loudspeaker in each television set will announce the message to anyone within earshot. Likewise, certain neurons communicate with hundreds of thousands of other cells. These widespread systems also tend to act relatively slowly, over seconds to minutes. Because of their broad, protracted actions, such systems in the brain can orchestrate entire behaviors, ranging from falling asleep to falling in love. Indeed, many of the behavioral dysfunctions collectively known as mental disorders are believed to result specifically from imbalances of certain of these chemicals.

In this chapter, we take a look at three components of the nervous system that operate in expanded space and time (Figure 15.1). One component is the *secretory hypothalamus*. By secreting chemicals directly into the bloodstream, the secretory hypothalamus can influence functions throughout both the brain and the body. A second component, controlled neurally by the hypothalamus, is the *autonomic nervous system* (*ANS*), introduced in Chapter 7. Through extensive interconnections within the body, the ANS simultaneously controls the responses of many internal organs, blood vessels, and glands. The third component exists entirely within the central nervous sys-

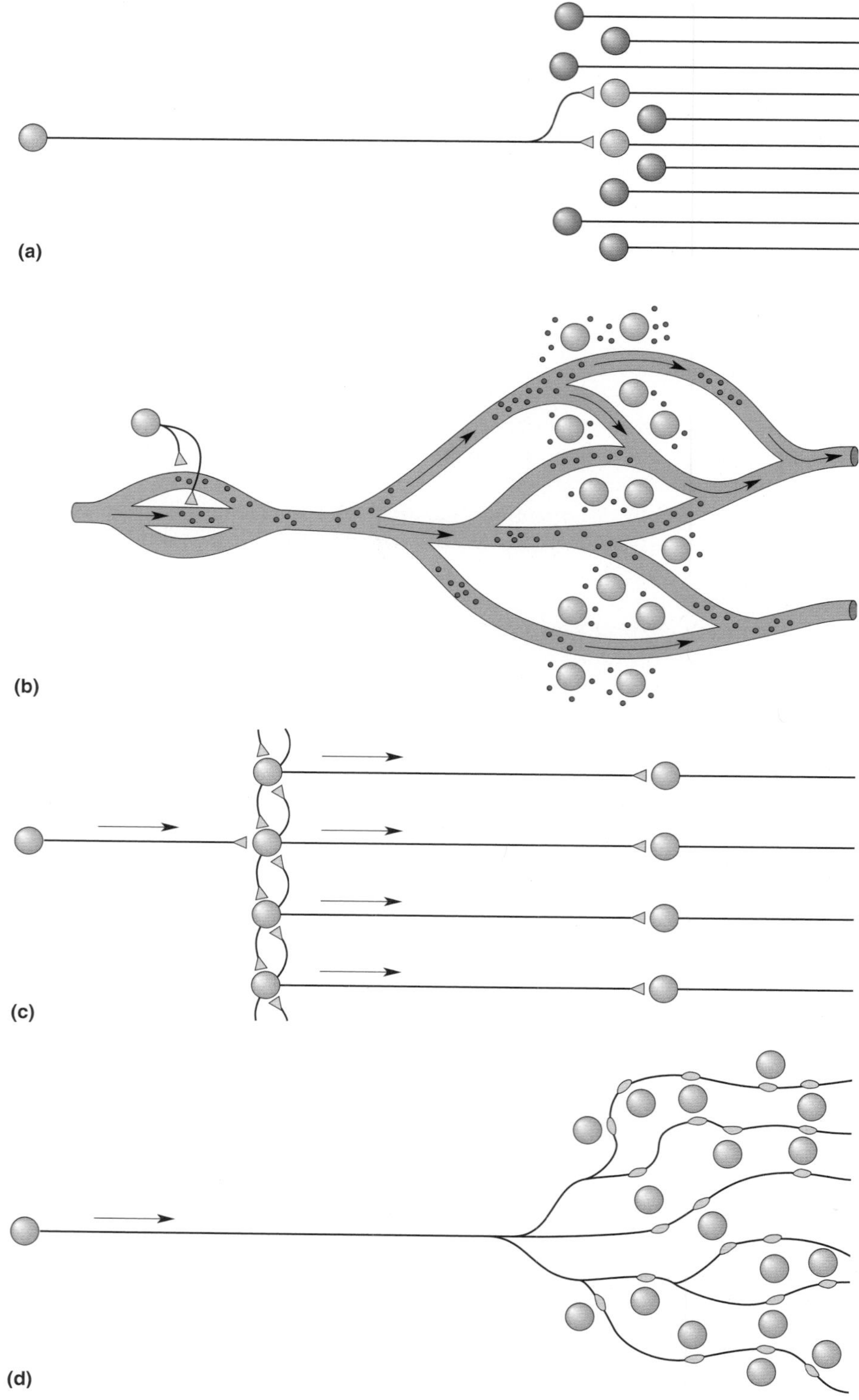

Figure 15.1

Patterns of communication in the nervous system. (a) Most of the systems we have discussed in this book may be described as point to point. The proper functioning of these systems requires restricted synaptic activation of target cells and signals of brief duration. In contrast, three other components of the nervous system act over great distances and for long periods. **(b)** Neurons of the secretory hypothalamus affect their many targets by releasing hormones directly into the bloodstream. **(c)** Networks of interconnected neurons of the ANS can work together to activate tissues all over the body. **(d)** Diffuse modulatory systems extend their reach with widely divergent axonal projections.

tem (CNS) and consists of several related cell groups that differ with respect to the neurotransmitter they use. All of these cell groups extend their spatial reach with highly divergent axonal projections and prolong their actions by using metabotropic postsynaptic receptors. Members of this component of the nervous system are called the *diffuse modulatory systems of the brain*. The diffuse systems are believed to regulate, among other things, the level of arousal and mood.

This chapter serves as a general introduction to these systems. In later chapters, we will see how they contribute to specific behaviors and brain states: motivation (Chapter 16), sexual behavior (Chapter 17), emotion (Chapter 18), sleep (Chapter 19), and psychiatric disorders (Chapter 21).

THE SECRETORY HYPOTHALAMUS

Recall from Chapter 7 that the hypothalamus sits below the thalamus, along the walls of the third ventricle. It is connected by a stalk to the pituitary gland, which dangles below the base of the brain, just above the roof of your mouth (Figure 15.2). Although this tiny cluster of nuclei makes up less than 1% of the brain's mass, the influence of the hypothalamus on body physiology is enormous. Let's take a brief tour of the hypothalamus and then focus on some of the ways in which it exerts its powerful influence.

An Overview of the Hypothalamus

The hypothalamus and dorsal thalamus are adjacent, but their functions are very different. As we saw in the previous seven chapters, the dorsal thalamus lies in the path of all of the point-to-point pathways whose destination is the neocortex. Accordingly, destruction of a small part of the dorsal thalamus can produce a discrete sensory or motor deficit: a little blind spot or a lack of

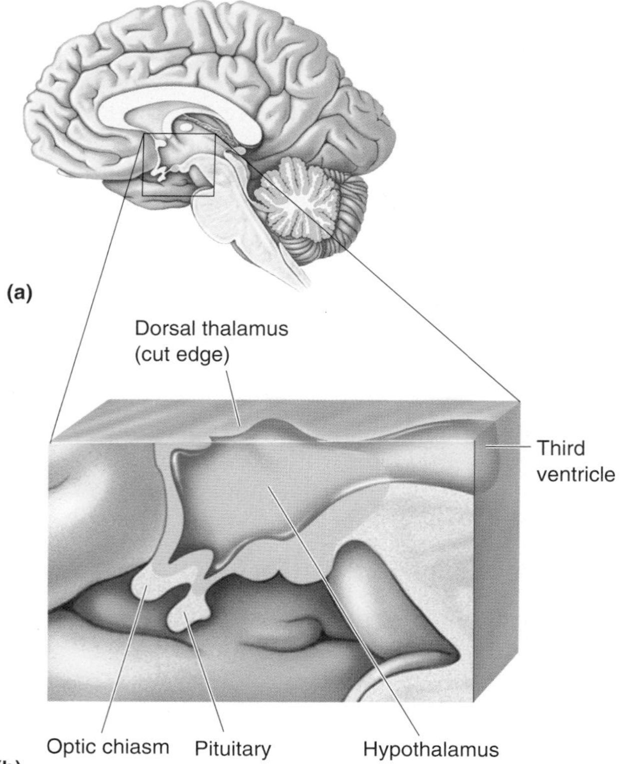

Figure 15.2
The location of the hypothalamus and pituitary. (a) A midsagittal section. **(b)** The hypothalamus forms the wall of the third ventricle and sits below the dorsal thalamus.

feeling on a portion of skin. In contrast, the *hypothalamus integrates somatic and visceral responses in accordance with the needs of the brain*. A tiny lesion in the hypothalamus can produce dramatic and often fatal disruptions of widely dispersed bodily functions.

Homeostasis. In mammals, the requirements for life include a narrow range of body temperatures and blood compositions. The hypothalamus regulates these levels in response to a changing external environment. This regulatory process is called **homeostasis**, the maintenance of the body's internal environment within a narrow physiological range.

Consider temperature regulation. Biochemical reactions in many cells of the body are fine-tuned to occur at about 37°C. A variation of more than a few degrees in either direction can be catastrophic. Temperature-sensitive cells in the hypothalamus detect variations in brain temperature and orchestrate the appropriate responses. For example, when you stroll naked through the snow, the hypothalamus issues commands that cause you to shiver (generating heat in the muscles), develop goose pimples (a futile attempt to fluff up your nonexistent fur—a reflexive remnant from our hairier ancestors), and turn blue (shunting blood *away from* the cold surface tissues to keep the sensitive core of the body warmer). In contrast, when you go for a jog in the tropics, the hypothalamus activates heat-loss mechanisms that make you turn red (shunting blood *to* the surface tissues, where heat can be radiated away) and sweat (cooling the skin by evaporation).

Other examples of homeostasis are the tight regulation of blood volume, pressure, salinity, acidity, and blood oxygen and glucose concentrations. The means by which the hypothalamus achieves these different types of regulation are remarkably diverse.

Structure and Connections of the Hypothalamus. Each side of the hypothalamus can be divided into three zones: lateral, medial, and periventricular (Figure 15.3). The lateral and medial zones have extensive connections with the brain stem and the telencephalon and regulate certain types of behavior, as we will see in Chapter 16. Here we are concerned only with the third zone, which actually receives much of its input from the other two.

The **periventricular zone** is so named because, with the exception of a thin finger of neurons that are displaced laterally by the optic tract (called the supraoptic nucleus), the cells of this region lie right next to the wall of the third ventricle. Within this zone exists a complex mix of neurons with various functions. One group of cells constitutes the *suprachiasmatic nucleus* (*SCN*), which lies just above the optic chiasm. These cells receive direct retinal innervation and function to synchronize circadian rhythms with the daily light-dark cycle (see Chapter 19). Other cells in the periventricular zone control the ANS and regulate the outflow of the sympathetic and parasympathetic innervation of the visceral organs. The cells in a third group, the *neurosecretory neurons*, extend axons down toward the stalk of the pituitary gland. These are the cells that now command our attention.

Pathways to the Pituitary

We have said that the pituitary dangles down below the base of the brain, which is true if the brain is lifted out of the head. In a living brain, the pituitary is gently held in a cradle of bone at the base of the skull. It deserves this special protection because it is the "mouthpiece" from which much of the hypothalamus "speaks" to the body. The pituitary has two lobes, posterior and anterior. The hypothalamus controls the two lobes in different ways.

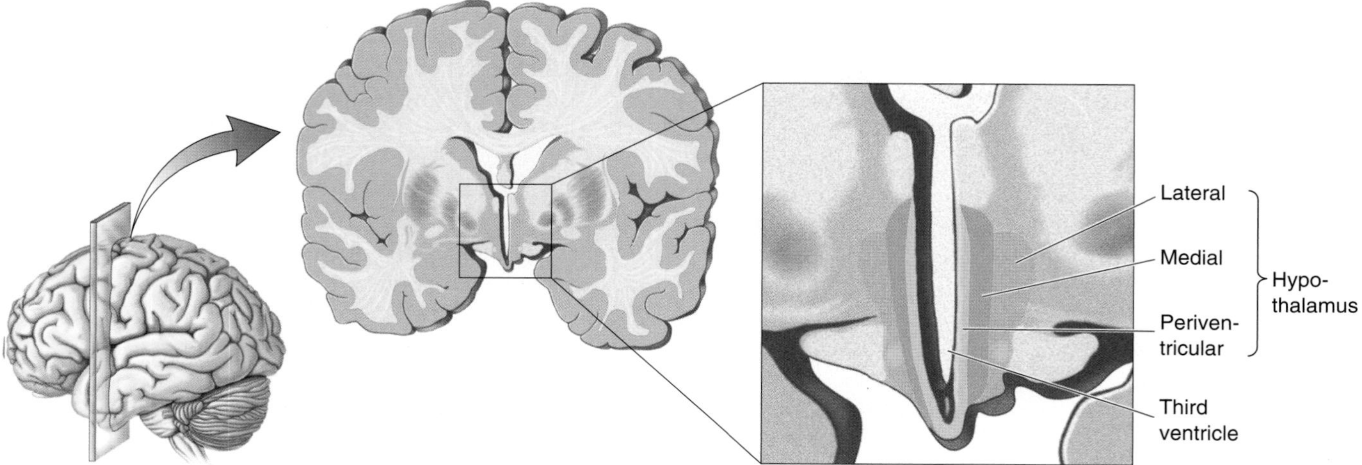

Figure 15.3
Zones of the hypothalamus. The hypothalamus is usually divided into three zones: lateral, medial, and periventricular. The periventricular zone receives input from the other zones, the brain stem, and the telencephalon. Neurosecretory cells in the periventricular zone secrete hormones into the bloodstream. Other periventricular cells control the ANS.

Hypothalamic Control of the Posterior Pituitary. The largest of the hypothalamic neurosecretory cells, **magnocellular neurosecretory cells**, extend axons around the optic chiasm, down the stalk of the pituitary, and into the posterior lobe (Figure 15.4). In the late 1930s, Ernst and Berta Scharrer, working at the University of Frankfurt in Germany, proposed that these neurons release chemical substances directly into the capillaries of the posterior lobe.

Figure 15.4
Magnocellular neurosecretory cells of the hypothalamus. Shown here is a midsagittal view of the hypothalamus and pituitary. Magnocellular neurosecretory cells secrete oxytocin and vasopressin directly into capillaries in the posterior lobe of the pituitary.

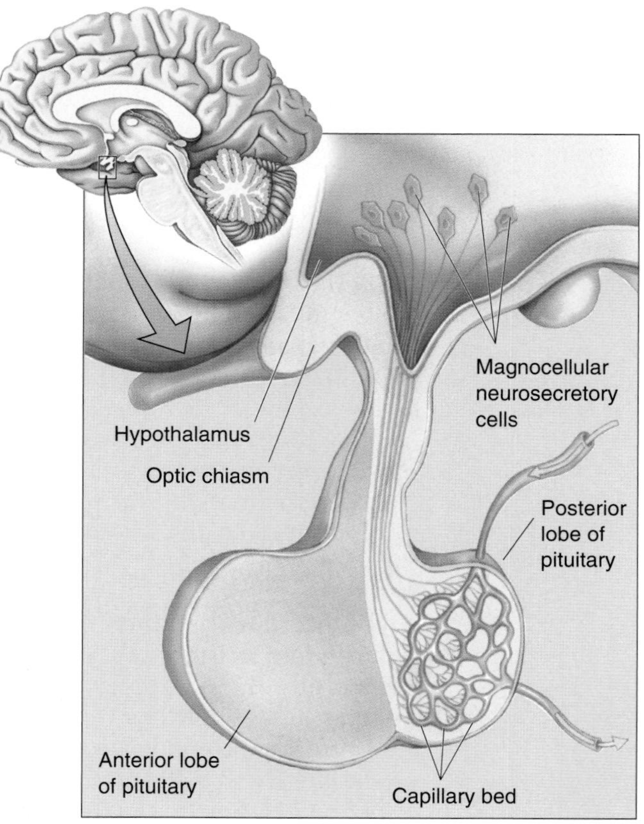

At the time, this was quite a radical idea. It was well known that chemical messengers called hormones were released by glands into the bloodstream, but no one anticipated that a neuron could act like a gland or that a neurotransmitter could act like a hormone. The Scharrers were correct, however. The substances released into the blood by neurons are now called **neurohormones**.

The magnocellular neurosecretory cells release two neurohormones, oxytocin and vasopressin, into the bloodstream. Both of these chemicals are peptides; each consists of a chain of nine amino acids. **Oxytocin**, released during the final stages of childbirth, causes the uterus to contract and facilitates the delivery of the newborn. It also stimulates the ejection of milk from the mammary glands. All lactating mothers know about the complex "letdown reflex," which involves the oxytocin neurons of the hypothalamus. Oxytocin release may be stimulated by the somatic sensations generated by a suckling baby. But the sight or cry of a baby (even someone else's) can also trigger the release of milk beyond the mother's conscious control. In each case, information about a sensory stimulus—somatic, visual, or auditory—reaches the cerebral cortex via the usual route, the thalamus, and the cortex ultimately stimulates the hypothalamus to trigger oxytocin release. The cortex can also suppress hypothalamic functions, as when anxiety inhibits the letdown of milk.

Vasopressin, also called *antidiuretic hormone* (*ADH*), regulates proper blood volume and salt concentration. When one is deprived of water, blood volume decreases and blood salt concentration increases. These changes are detected by pressure receptors in the cardiovascular system and salt concentration-sensitive cells in the hypothalamus, respectively. Vasopressin-containing neurons receive information about these changes and respond by releasing vasopressin, which acts directly on the kidneys and leads to water retention and reduced urine production.

When blood volume and pressure are low, communication between the brain and the kidneys occurs in both directions (Figure 15.5). The kidneys secrete the enzyme, *renin,* into the blood. Elevated renin sets off a sequence of biochemical reactions in the blood: *Angiotensinogen*, a large protein released from the liver, is converted by renin to *angiotensin I*, which breaks down further to form another small peptide hormone, *angiotensin II*. Angiotensin II has direct effects on the kidney and blood vessels, which help increase blood pressure. But angiotensin II in the blood is also detected by the *subfornical organ*, a part of the telencephalon that lacks a blood-brain barrier. Cells in the subfornical organ project axons into the hypothalamus, where they activate, among other things, the vasopressin-containing neurosecretory cells. In addition, however, the subfornical organ activates cells in the lateral area of the hypothalamus, producing an overwhelming thirst that motivates drinking behavior. It may be difficult to accept, but it's true: Our brain is, to a limited extent, controlled by our kidneys! This example also illustrates that the means by which the hypothalamus maintains homeostasis go beyond control of the visceral organs and can include the activation of entire behaviors. In Chapter 16, we will explore in more detail how the hypothalamus incites behavior.

Hypothalamic Control of the Anterior Pituitary. Unlike the posterior lobe, which is a part of the brain, the anterior lobe of the pituitary is a gland. The cells of the anterior lobe synthesize and secrete a wide range of hormones that regulate secretions from other glands throughout the body (together constituting the endocrine system). The pituitary hormones act on the gonads, the thyroid glands, the adrenal glands, and the mammary glands (Table 15.1). For this reason, the anterior pituitary was traditionally de-

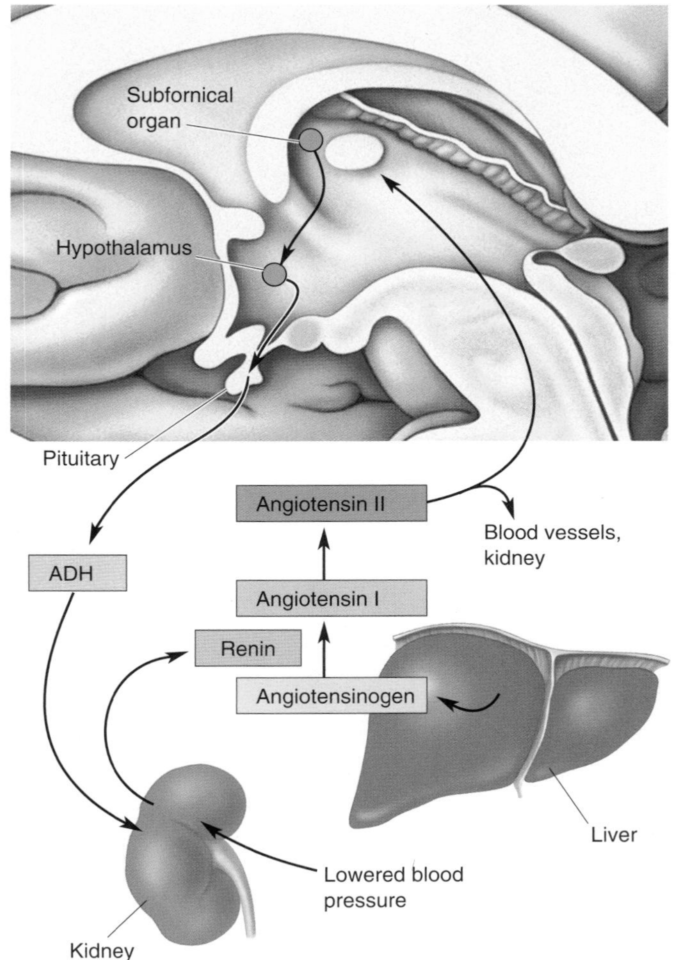

Figure 15.5
Communication between the kidneys and the brain. When blood volume or pressure is low, the kidney secretes renin into the bloodstream. Renin in the blood promotes the synthesis of the peptide angiotensin II, which excites the neurons in the subfornical organ. The subfornical neurons stimulate the hypothalamus, causing an increase in vasopressin (ADH) production and thirst.

scribed as the body's "master gland." But what controls the master gland? The secretory hypothalamus. *The hypothalamus itself is the true master gland of the endocrine system.*

The anterior lobe is under the control of neurons in the periventricular area, called **parvocellular neurosecretory cells**. These hypothalamic neurons do not extend axons all the way into the anterior lobe; instead, they communicate with their targets via the bloodstream (Figure 15.6). These neurons secrete **hypophysiotropic hormones** into a uniquely specialized capillary bed

Table 15.1 Hormones of the Anterior Pituitary

HORMONE	TARGET	ACTION
Follicle-stimulating hormone (FSH)	Gonads	Ovulation, spermatogenesis
Luteinizing hormone (LH)	Gonads	Ovarian, sperm maturation
Thyroid-stimulating hormone (TSH), also called thyrotropin	Thyroid	Thyroxin secretion (increases metabolic rate)
Adrenocorticotropic hormone (ACTH); also called corticotropin	Adrenal cortex	Cortisol secretion (mobilizes energy stores; inhibits immune system; other actions)
Growth hormone (GH)	All cells	Stimulation of protein synthesis
Prolactin	Mammary glands	Growth and milk secretion

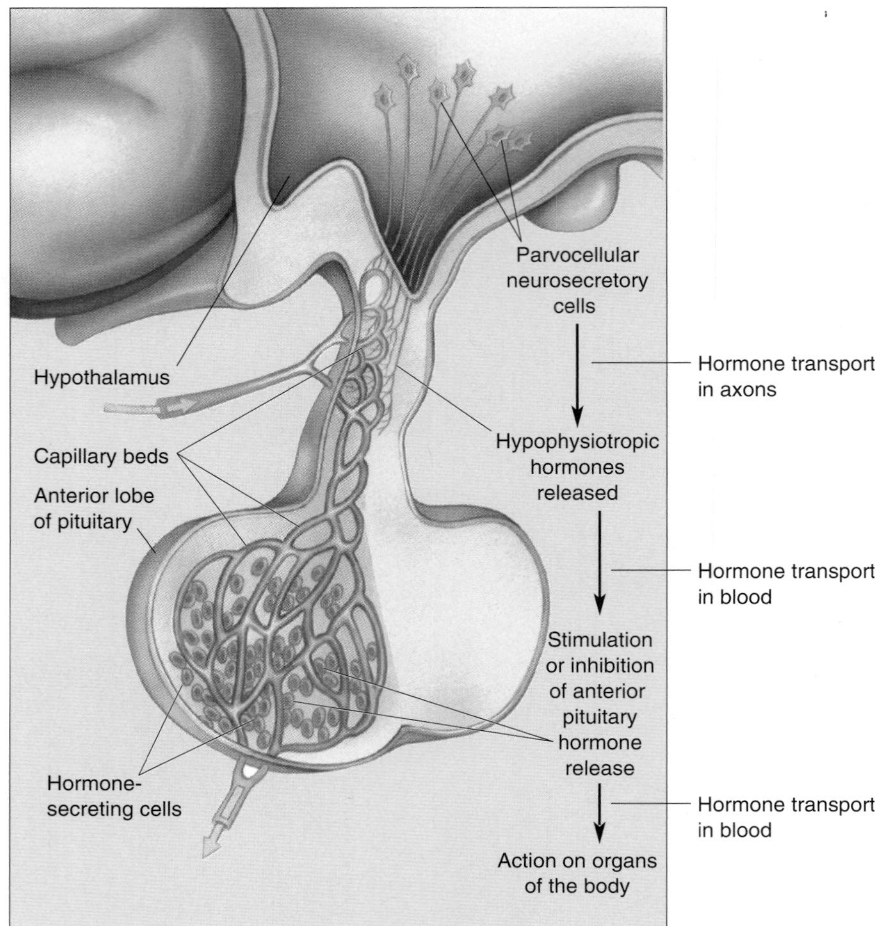

Figure 15.6
Parvocellular neurosecretory cells of the hypothalamus. Parvocellular neurosecretory cells secrete hypophysiotropic hormones into specialized capillary beds of the hypothalamo-pituitary portal circulation. These hormones travel to the anterior lobe of the pituitary, where they trigger or inhibit the release of pituitary hormones from secretory cells.

at the floor of the third ventricle. These tiny blood vessels run down the stalk of the pituitary and branch in the anterior lobe. This network of blood vessels is called the **hypothalamo-pituitary portal circulation.** Hypophysiotropic hormones secreted by hypothalamic neurons into the portal circulation travel downstream until they bind to specific receptors on the surface of pituitary cells. Activation of these receptors causes the pituitary cells either to secrete or to stop secreting hormones into the general circulation.

Regulation of the adrenal glands illustrates how this system works. Located just above the kidneys, the adrenal glands consist of two parts, a shell called the **adrenal cortex** and a center called the **adrenal medulla**. The adrenal cortex produces the steroid hormone **cortisol**, which when released into the bloodstream acts throughout the body to mobilize energy reserves and suppress the immune system, preparing us to carry on in the face of life's various stresses. In fact, a good stimulus for cortisol release is stress, ranging from physiological stress, such as a loss of blood, to positive emotional stimulation, such as falling in love, to psychological stress, such as anxiety over an upcoming exam. Parvocellular neurosecretory cells that control the adrenal cortex determine whether a stimulus is stressful or not (as defined by the release of cortisol). These neurons lie in the periventricular hypothalamus and release a peptide called *corticotropin-releasing hormone* (*CRH*) into the blood of the portal circulation. CRH travels the short distance to the anterior pituitary, where within about 15 seconds it stimulates the release of *corticotropin*, or *adrenocorticotropic hormone* (*ACTH*). ACTH enters the general cir-

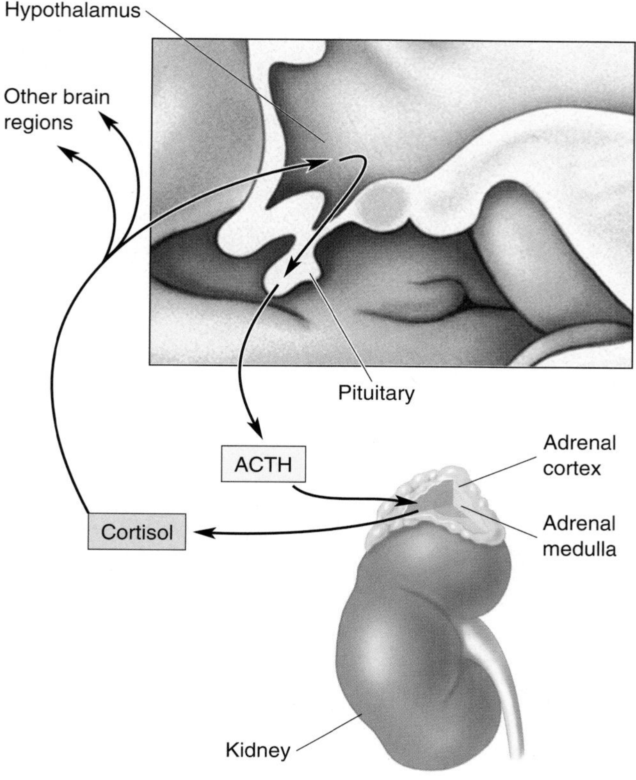

Figure 15.7
The stress response. During physiological, emotional, or psychological stimulation or stress, the periventricular hypothalamus secretes corticotropin-releasing hormone (CRH) into the hypothalamo-pituitary portal circulation. This triggers the release of adrenocorticotropic hormone (ACTH) into the general circulation. ACTH stimulates the release of cortisol from the adrenal cortex. Cortisol can act directly on hypothalamic neurons, as well as on other neurons elsewhere in the brain.

culation and travels to the adrenal cortex, where within a few minutes it stimulates cortisol release (Figure 15.7).

Blood levels of cortisol are, to some extent, self-regulated. Cortisol is a *steroid*; steroids are a class of biochemicals related to cholesterol. Thus, cortisol is a lipophilic ("fat-loving") molecule that dissolves easily in lipid membranes and readily crosses the blood-brain barrier. In the brain, cortisol interacts with specific receptors that lead to the inhibition of CRH release, thus ensuring that circulating cortisol levels do not get too high. Surprisingly, however, neurons with cortisol receptors are found widely distributed in the brain, not just in the hypothalamus. In these other CNS locations, cortisol has been shown to have significant effects on neuronal activity. Thus, the release of hypophysiotropic hormones by cells in the secretory hypothalamus can produce widespread alterations in the physiology of both the body and the brain (Box 15.1).

THE AUTONOMIC NERVOUS SYSTEM

Besides controlling the ingredients of the hormonal soup that flows in our veins, the periventricular zone of the hypothalamus also controls the **autonomic nervous system (ANS)**. The ANS is an extensive network of interconnected neurons that are widely distributed inside the body cavity. "Autonomic" is from the Greek *autonomia*, roughly meaning "independence"; autonomic functions are usually carried out automatically, without voluntary control. They are also highly coordinated functions. Imagine a sudden crisis. In a morning class, as you are engrossed in a crossword puzzle, the instructor unexpectedly calls you to the blackboard to solve an impossible-looking equation. You are faced with a classic fight-or-flight situation, and your body reacts accordingly, even as your conscious mind frantically considers whether to blunder through it or beg off in humiliation.

Box 15.1

OF SPECIAL INTEREST

Stress and the Brain

The brain creates biological stress in response to real or imagined stimuli. The many physiological responses associated with stress help protect the body and the brain from the dangers that triggered the stress in the first place. But stress in chronic doses can also have insidious harmful effects. Neuroscientists have only begun to sort out the relationships between stress, the brain, and brain damage.

Stress leads to the release of the steroid hormone cortisol from the adrenal cortex. Cortisol travels to the brain through the bloodstream and binds to receptors in the cytoplasm of many neurons. The activated receptors travel to the cell nucleus, where they stimulate gene transcription and ultimately protein synthesis. Steve Kerr, Philip Landfield, and their colleagues at the Bowman Gray School of Medicine in North Carolina have found that one consequence of cortisol's action is that neurons admit more Ca^{2+} through voltage-gated ion channels. This may be due to a direct change in the channels, or it may be indirectly caused by changes in the cell's energy metabolism. Whatever the mechanism, presumably in the short term, cortisol makes the brain better able to cope with the stress—perhaps by helping it figure out a way to avoid it!

But what about the effects of chronic, unavoidable stress? In Chapter 6, we learned that too much calcium can be a bad thing. If neurons become overloaded with calcium, they die (excitotoxicity). The question naturally arises: Can cortisol kill? Bruce McEwen and his colleagues at Rockefeller University and Robert Sapolsky and his colleagues at Stanford University have studied this question in the rat brain. They found that daily injections of corticosterone (rat cortisol) for several weeks caused dendrites to wither on many neurons with corticosterone receptors. A few weeks later, these cells started to die. A similar result was found when, instead of being given daily hormone injections, the rats were stressed every day.

Sapolsky's studies of baboons in Kenya further reveal the scourges of chronic stress. Baboons in the wild maintain a complex social hierarchy, and subordinate males steer clear of dominant males when they can. During one year when the baboon population boomed, local villagers caged many of the animals to prevent them from destroying their crops. Unable to escape the "top baboons" in the cages, many of the subordinate males died—not from wounds or malnutrition, but apparently from severe and sustained stress-induced effects. They had gastric ulcers, colitis, enlarged adrenal glands, and extensive degeneration of neurons in their hippocampus. Subsequent studies suggest that it is the direct effect of cortisol that damages the hippocampus. These effects of cortisol and stress resemble the effects of aging on the brain. Indeed, research has clearly shown that chronic stress causes premature aging of the brain.

In humans, exposure to the horrors of combat, sexual abuse, and other types of extreme violence can lead to post-traumatic stress disorder, with symptoms of heightened anxiety, memory disturbances, and intrusive thoughts. Imaging studies have consistently found degenerative changes in the brains of victims, particularly in the hippocampus. In Chapter 21 we will see that stress and the brain's response to it play a central role in several psychiatric disorders.

Your ANS triggers a host of physiological responses, including increasing heart rate and blood pressure, depressing digestive functions, and mobilizing glucose reserves. These responses are all produced by the **sympathetic division** of the ANS. Now imagine your relief as the class-ending bell suddenly rings, saving you from acute embarrassment and the instructor's anger. You settle back into your chair, breathe deeply, and read the clue for 24 down. Within a few minutes, your sympathetic responses decrease to low levels, and the functions of your **parasympathetic division** crank up again: Your heart rate slows and blood pressure drops, digestive functions work harder on breakfast, and you stop sweating.

You may not have moved out of your chair throughout this unpleasant event. Maybe you didn't even move your pencil. But your body's internal workings reacted dramatically. Unlike the *somatic motor system*, whose alpha motor neurons can rapidly excite skeletal muscles with pinpoint accuracy,

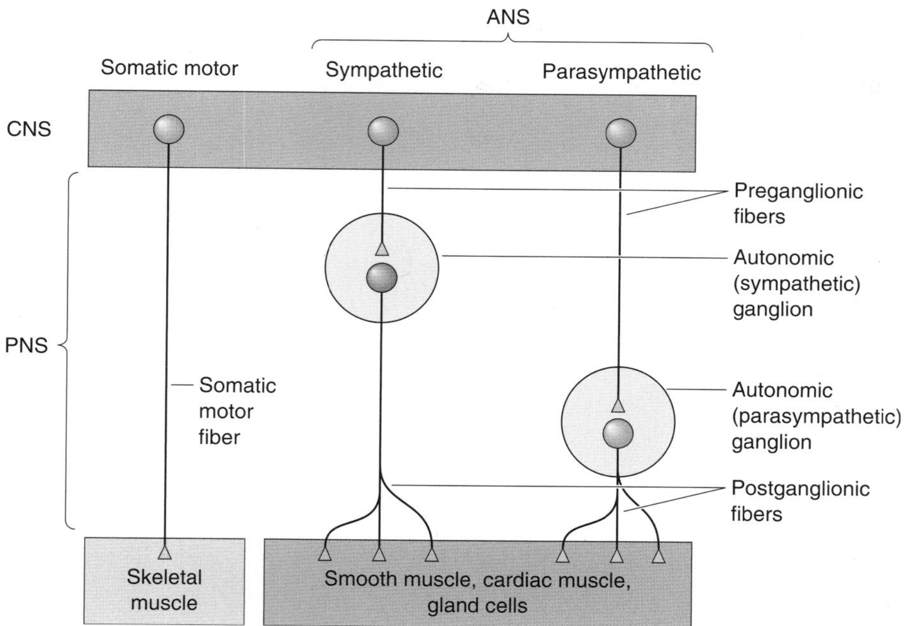

Figure 15.8
The organization of the three neural outputs of the CNS. The sole output of the somatic motor system is the lower motor neurons in the ventral horn of the spinal cord and the brain stem, which Sherrington called the final common pathway for the generation of behavior. But some behaviors, such as salivating, sweating, and genital responses, depend instead on the ANS. These visceral motor responses depend on the sympathetic and parasympathetic divisions of the ANS, whose lower motor neurons (i.e., postganglionic neurons) lie outside the CNS in autonomic ganglia. PNS indicates peripheral nervous system.

the actions of the ANS are typically multiple, widespread, and relatively slow. Therefore, the ANS operates in expanded space and time. In addition, unlike the somatic motor system, which can only excite its peripheral targets, the ANS balances synaptic excitation and inhibition to achieve widely coordinated and graded control.

ANS Circuits

Together, the somatic motor system and the ANS constitute the total neural output of the CNS. The somatic motor system has a single job: It innervates and commands skeletal muscle fibers. The ANS has the complex task of commanding *every other* tissue and organ in the body that is innervated. Both systems have upper motor neurons in the brain that send commands to lower motor neurons, which actually innervate the peripheral target structures. They have some interesting differences, however (Figure 15.8). The cell bodies of all somatic lower motor neurons lie within the CNS, in either the ventral horn of the spinal cord or the brain stem. The cell bodies of all autonomic lower motor neurons lie outside the CNS, within cell clusters called **autonomic ganglia**. The neurons in these ganglia are called **postganglionic neurons**. Postganglionic neurons are driven by **preganglionic neurons**, whose cell bodies are in the spinal cord and brain stem. Thus, the somatic motor system controls its peripheral targets via a *monosynaptic pathway*, while the ANS uses a *disynaptic pathway*.

Sympathetic and Parasympathetic Divisions. The sympathetic and parasympathetic divisions operate in parallel, but they use pathways that are quite distinct in structure and in their neurotransmitter systems. Preganglionic axons of the sympathetic division emerge only from the middle third of the spinal cord (thoracic and lumbar segments). In contrast, preganglionic axons of the parasympathetic division emerge only from the brain stem and the lowest (sacral) segments of the spinal cord, so the two systems complement each other anatomically (Figure 15.9). The preganglionic neurons of the sympathetic division lie within the *intermediolateral gray matter* of the spinal cord. They send their axons through the ventral roots to synapse

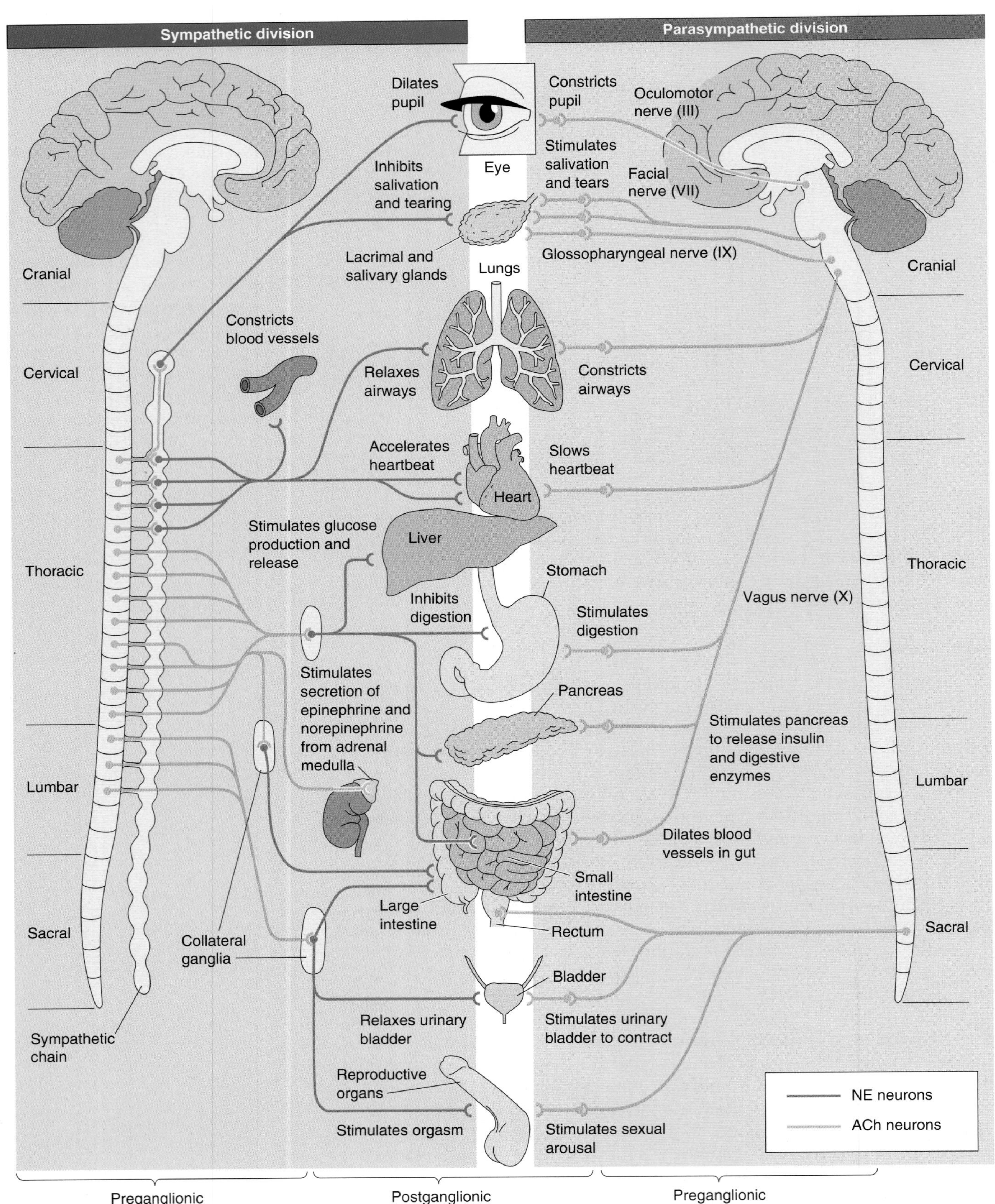
Sympathetic division
Parasympathetic division
Dilates pupil
Eye
Constricts pupil
Oculomotor nerve (III)
Inhibits salivation and tearing
Stimulates salivation and tears
Facial nerve (VII)
Lacrimal and salivary glands
Glossopharyngeal nerve (IX)
Cranial
Lungs
Constricts blood vessels
Relaxes airways
Constricts airways
Cervical
Accelerates heartbeat
Slows heartbeat
Heart
Stimulates glucose production and release
Liver
Thoracic
Stomach
Inhibits digestion
Stimulates digestion
Vagus nerve (X)
Stimulates secretion of epinephrine and norepinephrine from adrenal medulla
Pancreas
Stimulates pancreas to release insulin and digestive enzymes
Lumbar
Dilates blood vessels in gut
Small intestine
Large intestine
Rectum
Sacral
Collateral ganglia
Bladder
Relaxes urinary bladder
Stimulates urinary bladder to contract
Sympathetic chain
Reproductive organs
Stimulates orgasm
Stimulates sexual arousal
NE neurons
ACh neurons
Preganglionic neurons
Postganglionic neurons
Preganglionic neurons

on neurons in the ganglia of the **sympathetic chain**, which lies next to the spinal column, or in collateral ganglia within the abdominal cavity. Conversely, the preganglionic parasympathetic neurons sit within a variety of brain stem nuclei and the lower (sacral) spinal cord, and their axons travel within several cranial nerves as well as the nerves of the sacral spinal cord. The parasympathetic preganglionic axons travel much farther than the sympathetic axons, because the parasympathetic ganglia are typically located next to, on, or in their target organs (Figures 15.8 and 15.9).

The ANS innervates three types of tissue: glands, smooth muscle, and cardiac muscle. Thus, almost every part of the body is a target of the ANS, as shown in Figure 15.9. The ANS

- Innervates the secretory glands (salivary, sweat, tear, and various mucus-producing glands)
- Innervates the heart and blood vessels to control blood pressure and flow
- Innervates the bronchi of the lungs to meet the oxygen demands of the body
- Regulates the digestive and metabolic functions of the liver, gastrointestinal tract, and pancreas
- Regulates the functions of the kidney, urinary bladder, large intestine, and rectum
- Is essential to the sexual responses of the genitals and reproductive organs
- Interacts with the body's immune system

The physiological influences of the sympathetic and parasympathetic divisions generally oppose each other. The sympathetic division tends to be most active during a crisis, real or perceived. The behaviors related to it are summarized in the puerile (but effective) mnemonic used by medical students, called the four Fs: fight, flight, fright, and sex. The parasympathetic division facilitates various non–four-F processes, such as digestion, growth, immune responses, and energy storage. In most cases, the activity levels of the two ANS divisions are reciprocal: When one is high, the other tends to be low, and when one is low, the other tends to be high. The sympathetic division frenetically mobilizes the body for a short-term emergency at the expense of processes that keep it healthy over the long term. The parasympathetic division works calmly for the long-term good. Both cannot be stimulated strongly at the same time; their general goals are incompatible. Fortunately, neural circuits in the CNS inhibit activity in one division when the other is active.

Some examples will help to illustrate how the balance of activity in the sympathetic and parasympathetic divisions controls organ functions. The pacemaker region of the heart triggers each heartbeat without the help of neurons, but both divisions of the ANS innervate it and modulate it; sympa-

Figure 15.9

The chemical and anatomical organization of the sympathetic and parasympathetic divisions of the ANS. The preganglionic inputs of both divisions use ACh as a neurotransmitter. The postganglionic parasympathetic innervation of the visceral organs also uses ACh, but the postganglionic sympathetic innervation uses NE (except innervation of the sweat glands and vascular smooth muscle within skeletal muscle, which use ACh). The adrenal medulla receives preganglionic sympathetic innervation and, when activated, secretes epinephrine into the bloodstream. Note the pattern of innervation by the sympathetic division: Target organs in the chest cavity are contacted by postganglionic neurons originating in the sympathetic chain, and target organs in the abdominal cavity are contacted by postganglionic neurons originating in the collateral ganglia.

thetic activity results in an increase in the rate of beating, while parasympathetic activity slows it down. The smooth muscles of the gastrointestinal tract are also dually innervated, but the effect of each division is the opposite of its effect on the heart. Intestinal motility, hence digestion, is stimulated by parasympathetic axons and inhibited by sympathetic axons. Not all tissues receive innervation from both divisions of the ANS. For example, the blood vessels of the skin and the sweat glands are innervated only by excitatory sympathetic axons. Lacrimal (tear-producing) glands are excited only by parasympathetic input.

Another example of the balance of parasympathetic-sympathetic activity is the curious neural control of the male sexual response. Erection of the human penis is a hydraulic process. It occurs when the penis becomes engorged with blood, which is triggered and sustained by parasympathetic activity. The curious part is that orgasm and ejaculation are triggered by *sympathetic* activity. You can imagine how complicated it must be for the nervous system to orchestrate the entire sexual act; parasympathetic activity gets it going and keeps it going, but a shift to sympathetic activity is necessary to terminate it. Anxiety and worry, with their attendant sympathetic activity, tend to inhibit erection and promote ejaculation. Not surprisingly, impotence and premature ejaculation are common complaints of the overstressed male. (We will discuss sexual behavior further in Chapter 17.)

The Enteric Division. The "little brain," as the **enteric division** of the ANS is sometimes called, is a unique neural system embedded in an unlikely place: the lining of the esophagus, stomach, intestines, pancreas, and gallbladder. It consists of two complicated networks, each with sensory nerves, interneurons, and autonomic motor neurons, called the *myenteric* (or Auerbach's) *plexus* and *submucous* (or Meissner's) *plexus* (Figure 15.10). These networks control many of the physiological processes involved in the transport and digestion of food from oral to anal openings. The enteric system is not small; it contains about the same number of neurons as the entire spinal cord!

If the enteric division of the ANS qualifies as "brain" (which may be overstating the case), it is because it can operate with a great deal of independence. Enteric sensory neurons monitor tension and stretch of the gastrointestinal walls, the chemical status of stomach and intestinal contents, and hormone levels in the blood. The enteric interneuronal circuits use this information to control the activity levels of enteric output motor neurons, which govern smooth muscle motility, the production of mucous and digestive secretions, and the diameter of the local blood vessels. For example, consider a partially digested pizza making its way through the small intestine. The myenteric plexus ensures that lubricating mucus and digestive enzymes are delivered, that rhythmic (peristaltic) muscle action works to mix the pizza and enzymes thoroughly, and that intestinal blood flow increases to provide a sufficient fluid source and transport newly acquired nutrients to the rest of the body.

The enteric division is not entirely autonomous. It receives input indirectly from the "real" brain via axons of the sympathetic and parasympathetic divisions. These provide supplementary control and can supersede the functions of the enteric division in some circumstances. For example, the strong activation of the sympathetic nervous system that occurs during acute stress inhibits the enteric nervous system and digestive functions.

Central Control of the ANS. As we have said, the hypothalamus is the main regulator of the autonomic preganglionic neurons. Somehow this diminutive structure integrates the diverse information it receives about the body's status, anticipates some of its needs, and provides a coordinated set of both neu-

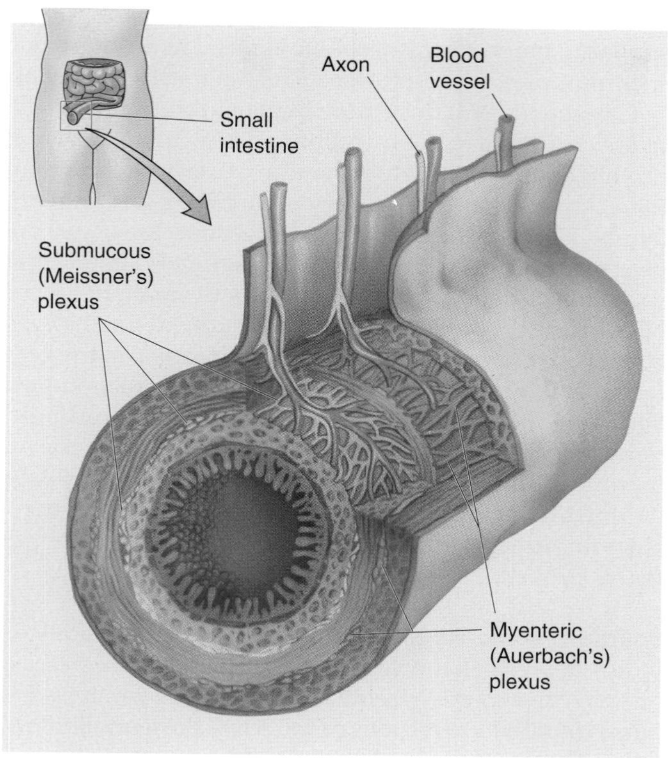

Figure 15.10
The enteric division of the ANS. This cross-sectional view of the small intestine shows the two networks of the enteric division, the myenteric plexus and the submucous plexus. Both contain visceral sensory and motor neurons that control the functions of the digestive organs.

ral and hormonal outputs. Essential to autonomic control are the connections of the periventricular zone to the brain stem and spinal cord nuclei that contain the preganglionic neurons of the sympathetic and parasympathetic divisions. The **nucleus of the solitary tract**, located in the medulla and connected with the hypothalamus, is another important center for autonomic control. In fact, some autonomic functions operate well even when the brain stem is disconnected from all structures above it, including the hypothalamus. The solitary nucleus integrates sensory information from the internal organs and coordinates output to the autonomic brain stem nuclei.

Neurotransmitters and the Pharmacology of Autonomic Function

Even people who have never heard the word *neurotransmitter* know what it means to "get your adrenaline flowing." (In the United Kingdom the compound is called adrenaline, while in the United States we call it epinephrine.) Historically, the ANS has probably taught us more than any other part of the body about how neurotransmitters work. Because the ANS is relatively simple compared with the CNS, we understand the ANS much better. In addition, neurons of the peripheral parts of the ANS are outside the blood-brain barrier, so all drugs that enter the bloodstream have direct access to them. The relative simplicity and accessibility of the ANS have led to a deeper understanding of the mechanisms of drugs that influence synaptic transmission.

Preganglionic Neurotransmitters. The primary transmitter of the peripheral autonomic neurons is acetylcholine (ACh), the same transmitter used at skeletal neuromuscular junctions. The preganglionic neurons of both sympathetic and parasympathetic divisions release ACh. The immediate effect is that the ACh binds to nicotinic ACh receptors (nAChRs), which are ACh-gated channels, and evokes a fast excitatory postsynaptic potential (EPSP) that usually triggers an action potential in the postganglionic cell. This is very similar to the mechanisms of the skeletal neuromuscular junction, and drugs such as curare, which block nAChRs in muscle, also block autonomic output.

Ganglionic ACh does more than neuromuscular ACh, however. It also activates muscarinic ACh receptors (mAChRs), which are metabotropic (G-protein-coupled) receptors that can cause both the opening and the closing of ion channels that lead to very slow EPSPs and inhibitory postsynaptic potentials (IPSPs). These slow mAChR events are usually not evident unless the preganglionic nerve is activated repetitively. In addition to ACh, some preganglionic terminals release a variety of small, neuroactive peptides such as *NPY* (*neuropeptide Y*) and *VIP* (*vasoactive intestinal polypeptide*). These also interact with G-protein-coupled receptors and can trigger small EPSPs that last for several minutes. The effects of peptides are modulatory; they do not usually bring the postsynaptic neurons to firing threshold, but they make them more responsive to the fast nicotinic effects when they do come along. Since more than one action potential is required to stimulate the release of these modulatory neurotransmitters, the pattern of firing in preganglionic neurons is an important variable in determining the type of postganglionic activity that is evoked.

Postganglionic Neurotransmitters. Postganglionic cells—the autonomic motor neurons that actually trigger glands to secrete, sphincters to contract or relax, and so on—use different neurotransmitters in the sympathetic and parasympathetic divisions of the ANS. Postganglionic parasympathetic neurons release ACh, but those of most parts of the sympathetic division use *norepinephrine* (*NE*). Parasympathetic ACh has a very local effect on its targets and acts entirely through mAChRs. In contrast, sympathetic NE often spreads far, even into the blood, where it can circulate widely.

The autonomic effects of a variety of drugs that interact with cholinergic and noradrenergic systems can be confidently predicted if you understand some of the autonomic circuitry and chemistry (Figure 15.9). In general, drugs that promote the actions of norepinephrine *or* inhibit the muscarinic actions of acetylcholine are *sympathomimetic*; they cause effects that mimic activation of the sympathetic division of the ANS. For example, *atropine*, an antagonist of mAChRs, produces signs of sympathetic activation, such as dilation of the pupils. This response occurs because the balance of ANS activity shifts toward the sympathetic division when parasympathetic actions are blocked. By contrast, drugs that promote the muscarinic actions of ACh *or* inhibit the actions of NE are *parasympathomimetic*; they cause effects that mimic activation of the parasympathetic division of the ANS. For example, *propranolol*, an antagonist of the β receptor for NE, slows the heart rate and lowers blood pressure. For this reason, propranolol is sometimes used to prevent the physiological consequences of stage fright.

But what about the familiar flow of adrenaline (i.e., epinephrine)? It is the compound released into the blood from the *adrenal medulla* when activated by preganglionic sympathetic innervation. Epinephrine is actually made from norepinephrine (noradrenaline in the United Kingdom), and its effects on target tissues are almost identical to those caused by sympathetic activation. Thus, the adrenal medulla is really nothing more than a modified sympathetic ganglion. You can imagine that as the adrenaline (epinephrine) flows, a coordinated, bodywide set of sympathetic effects kicks in.

THE DIFFUSE MODULATORY SYSTEMS OF THE BRAIN

Consider what happens when you fall asleep. The internal commands "You are becoming drowsy" and "You are falling asleep" are messages that must be received by broad regions of the brain. Dispensing this information requires neurons with a particularly widespread pattern of axons. The brain has several such collections of neurons, each using a particular neurotransmitter and making widely dispersed, diffuse, almost meandering connections. Rather than carrying detailed sensory information, these cells often perform regulatory functions, modulating vast assemblies of postsynaptic neurons, such as the cerebral cortex, the thalamus, and the spinal cord, so that they become more or less excitable, more or less synchronously active, and so on. Collectively, they are a bit like the volume, treble, and bass controls on a radio, which do not change the lyrics or melody of a song but dramatically regulate the impact of both. In addition, different systems appear to be essential for aspects of motor control, memory, mood, motivation, and metabolic state. Many psychoactive drugs affect these modulatory systems, and the systems figure prominently in current theories about the biological basis of certain psychiatric disorders.

Anatomy and Functions of the Diffuse Modulatory Systems

The **diffuse modulatory systems** differ in structure and function, yet they have certain principles in common:

- Typically, the core of each system has a small set of neurons (several thousand).
- Neurons of the diffuse systems arise from the central core of the brain, most of them from the brain stem.
- Each neuron can influence many others, because each one has an axon that may contact more than 100,000 postsynaptic neurons spread widely across the brain.
- The synapses made by many of these systems seem designed to release transmitter molecules into the extracellular fluid so that they can diffuse to many neurons rather than be confined to the vicinity of the synaptic cleft.

We'll focus on the modulatory systems of the brain that use either NE, serotonin (5-HT), dopamine (DA), or ACh as a neurotransmitter. Recall from Chapter 6 that all of these transmitters activate specific metabotropic (G-protein-coupled) receptors, and these receptors mediate most of their effects. For example, the brain has 10–100 times more mAChRs than ionotropic nAChRs. Because neuroscientists are still working hard to determine the exact functions of these systems in behavior, our explanations here will necessarily be vague.

The Noradrenergic Locus Coeruleus. Besides being a neurotransmitter in the peripheral ANS, NE is also used by neurons of the tiny **locus coeruleus** in the pons (from the Latin for "blue spot" because of the pigment in its cells). Each human locus coeruleus has about 12,000 neurons. We have two of them, one on each side.

A major breakthrough occurred in the mid-1960s, when Kjell Fuxe and his colleagues at the Karolinska Institute in Sweden developed a technique that enabled the catecholaminergic (noradrenergic and dopaminergic) neurons to be visualized selectively in histological sections prepared from the brain (Box 15.2). This analysis revealed that axons leave the locus coeruleus in several tracts but then fan out to innervate just about every part of the brain: all of the cerebral cortex, the thalamus and the hypothalamus, the olfactory bulb, the cerebellum, the midbrain, and the spinal cord (Figure 15.11). The locus coeruleus must make some of the most diffuse connections in the brain, con-

Box 15.2

PATH OF DISCOVERY

Illuminating the Catecholamines of the Brain

BY KJELL FUXE

It all began with my teacher Professor Nils-Åke Hillarp. I first met him in 1962, when he became chairman of my department at the Karolinska Institute in Stockholm, Sweden, where I was a medical student. He, together with Bengt Falck, had recently shown that formaldehyde gas could convert neuronal catecholamines (CA) and 5-HT into green and yellow fluorescent products, respectively. Now, for the first time, it was possible to visualize neurons that contain CA and 5-HT. Because of his creative, dynamic, generous, and warm personality, I became very enthusiastic about working with Hillarp to map the CA and 5-HT neurons in the CNS, which at that time remained a "black box" with regard to chemical signaling. My interest was further aroused by my neuropsychopharmacology teacher, Professor Arvid Carlsson at the University of Göteborg.

In the autumn of 1962, I started to set up the formaldehyde method with the invaluable help of Georg Thieme. Shortly afterward, another medical student, Annica Dahlström, joined in the project. The method was not reproducible at first, because of a lack of control of the water content in the paraformaldehyde. Nonetheless, Dahlström and I obtained a sufficient number of good formaldehyde reactions to cause unbelievable happiness. After 3 years of hard work, we were able to describe the various clusters of CA and 5-HT nerve cell groups in the brain stem (Figure A), as well as some of the various nerve terminal systems in the rat CNS. These results enabled me to complete my M.D. thesis in the spring of 1965. The mapping projects continued for a number of years, especially together with my colleague Tomas Hökfelt.

The next step was to test this hypothesis that the widespread monosynaptic CA and 5-HT pathways in the forebrain originate in the lower brain stem. To do this, I began a long-term and fruitful collaboration with Nils-Erik Anden and Knut Larsson at the University of Göteborg and with my colleagues Lars Olson and Urban Ungerstedt working in our department. We made lesions in the junction between the midbrain and the diencephalon and used combined biochemical and histochemical methods to detect changes. Suddenly, we were able to visualize the CA and 5-HT axon bundles in the brain stem, which were caused by a buildup of amine-containing secretory granules that were transported from the cell bodies to the stumps of the cut axons. This was a great moment. We were thrilled to see clearly the ascending CA and 5-HT axons in the medial forebrain bundle and tegmentum. Fluorescent nerve terminals disappeared in the forebrain rostral to the lesion, revealing that the brain stem was the source of CA and 5-HT axons in the cortex. In addition, by making spinal cord lesions, Dahlström and I obtained evidence for the existence of highly collateralized descending CA and 5-HT axons innervating the gray matter of the entire spinal cord.

Kjell Fuxe

In this way, through great teamwork, we obtained evidence that ascending and descending DA, NE, and 5-HT diffuse modulatory systems monosynaptically innervate large parts of the brain and spinal cord. The existence of such widespread projections, however, was questioned by many neuroanatomists. The standard method for tracing axons at that time had never given any indication that such a large projection exists from the lower brain stem to the cortex. This example illustrates the importance of new techniques for progress in the neurosciences. In a way, these studies represented the very beginning of chemical neuroanatomy. They have had a large influence on the way we think about the functions of neurotransmitters in the brain.

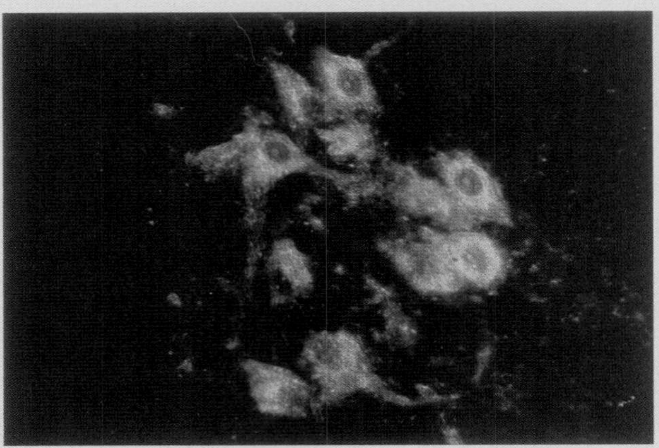

Figure A

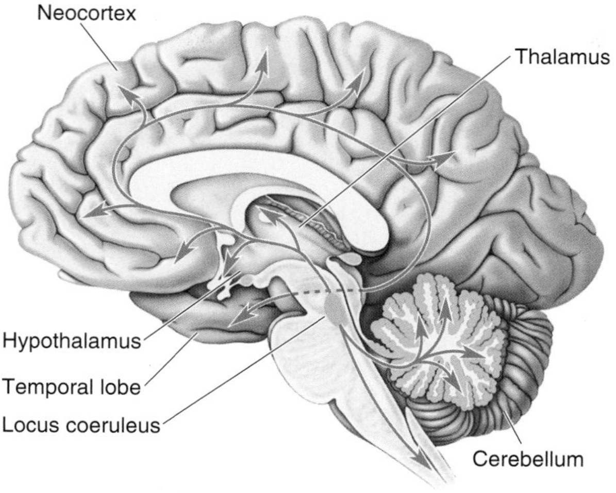

Figure 15.11
The noradrenergic diffuse modulatory system arising from the locus coeruleus. The small cluster of locus coeruleus neurons project axons that innervate vast areas of the CNS, including the spinal cord, cerebellum, thalamus, and cerebral cortex.

sidering that just one of its neurons can make more than 250,000 synapses and that it can have one axon branch in the *cerebral* cortex and another in the *cerebellar* cortex!

Locus coeruleus cells seem to be involved in the regulation of attention, arousal, and sleep-wake cycles, as well as learning and memory, anxiety and pain, mood, and brain metabolism. This makes it sound as if the locus coeruleus may run the whole show. But the key word is "involved," which can mean almost anything. For example, our heart, liver, lungs, and kidneys are also involved in every brain function, for without them, all behavior would fail utterly. Because of its widespread connections, the locus coeruleus can influence virtually all parts of the brain. But to understand its actual functions, we start by determining what activates its neurons. Recordings from awake, behaving rats and monkeys show that locus coeruleus neurons are best activated by new, unexpected, nonpainful sensory stimuli in the animal's environment. They are least active when the animals are not vigilant, just sitting around quietly, digesting a meal. The locus coeruleus may participate in a general arousal of the brain during interesting events in the outside world. Because NE can make neurons of the cerebral cortex more responsive to salient sensory stimuli, the locus coeruleus may function generally to increase brain responsiveness, speeding information processing by the point-to-point sensory and motor systems and making them more efficient.

The Serotonergic Raphe Nuclei. Serotonin-containing neurons are mostly clustered within the nine **raphe nuclei**. *Raphe* means "ridge" or "seam" in Greek, and indeed the raphe nuclei lie to either side of the midline of the brain stem. Each nucleus projects to different regions of the brain (Figure 15.12). Those more caudal, in the medulla, innervate the spinal cord, where they modulate pain-related sensory signals, as discussed in Chapter 12. Those more rostral, in the pons and midbrain, innervate most of the brain in much the same diffuse way as do the locus coeruleus neurons.

Similar to neurons of the locus coeruleus, raphe nuclei cells fire most rapidly during wakefulness, when an animal is aroused and active. Raphe neurons are quietest during sleep. The locus coeruleus and the raphe nuclei are part of a venerable concept called the *ascending reticular activating system*, which implicates the reticular "core" of the brain stem in processes that

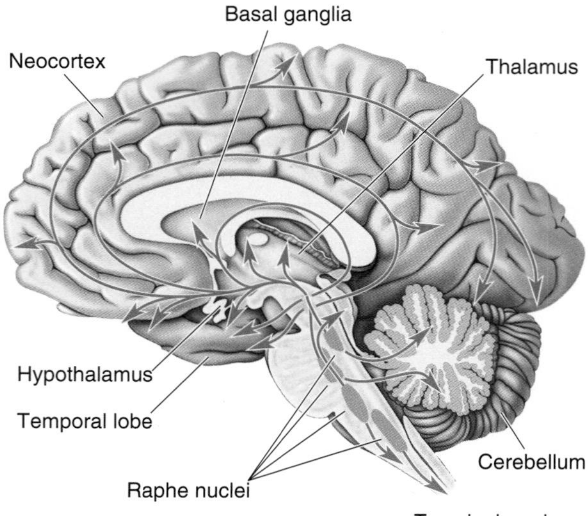

Figure 15.12
The serotonergic diffuse modulatory systems arising from the raphe nuclei. The raphe nuclei, which are clustered along the midline of the brain stem, project extensively to all levels of the CNS.

arouse and awaken the forebrain. This simple idea has been refined and redefined in countless ways since it was introduced in the 1950s, but its basic sense remains. Raphe neurons seem to be intimately involved in the control of sleep-wake cycles as well as the different stages of sleep. Several other transmitter systems are also involved in a coordinated way. We will discuss the involvement of the diffuse modulatory systems in sleep and wakefulness in Chapter 19.

Serotonergic raphe neurons have also been implicated in the control of mood and certain types of emotional behavior. We will return to serotonin and mood when we discuss clinical depression in Chapter 21.

The Dopaminergic Substantia Nigra and Ventral Tegmental Area. Although there are dopamine-containing neurons scattered throughout the CNS, including some in the retina, the olfactory bulb, and the periventricular hypothalamus, two closely related groups of dopaminergic cells have the characteristics of the diffuse modulatory systems (Figure 15.13). One of these arises in the *substantia nigra* in the midbrain. Recall from Chapter 14 that these cells project axons to the striatum (i.e., the caudate nucleus and the putamen), where they somehow facilitate the initiation of voluntary movements. Degeneration of the DA-containing cells in the substantia nigra is all that is necessary to produce the progressive, dreadful motor disorders of Parkinson's disease. This is not to say that we entirely understand the function of DA in motor control, but in general, it facilitates the initiation of motor responses by environmental stimuli.

The midbrain is also the origin of the other dopaminergic modulatory system, a group of cells that lie very close to the substantia nigra, in the *ventral tegmental area.* Axons from these neurons innervate a circumscribed region of the telencephalon that includes the frontal cortex and parts of the limbic system. (The limbic system will be described in Chapter 18.) This dopaminergic projection from the midbrain is sometimes called the *mesocorticolimbic dopamine system.* A number of functions have been ascribed to this complicated projection. For example, there is evidence that it is involved in a "reward" system that somehow assigns value to, or *reinforces,* certain adaptive behaviors (see Chapter 16). We will see in Chapter 18 that if rats (or humans) are given a chance to do so, they work to electrically stimulate this pathway.

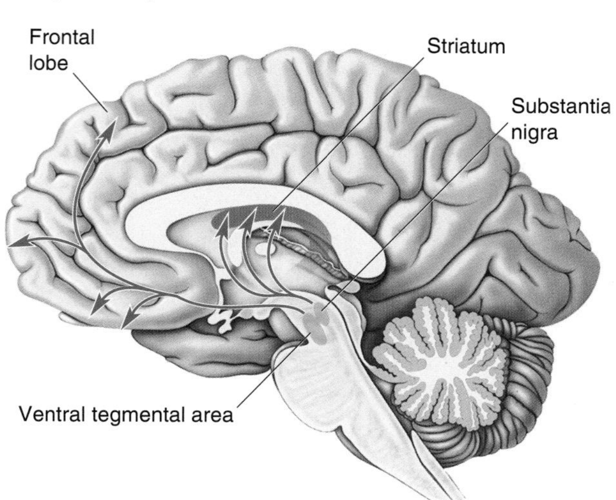

Figure 15.13
The dopaminergic diffuse modulatory systems arising from the substantia nigra and the ventral tegmental area. The substantia nigra and ventral tegmental area lie close together in the midbrain. They project to the striatum (caudate nucleus and putamen) and limbic and frontal cortical regions, respectively.

In addition, this projection has been implicated in psychiatric disorders, as we will discuss in Chapter 21.

The Cholinergic Basal Forebrain and Brain Stem Complexes. Acetylcholine is the familiar transmitter at the neuromuscular junction, at synapses in autonomic ganglia, and at postganglionic parasympathetic synapses. Cholinergic interneurons also exist within the brain—in the striatum and the cortex, for example. In addition, there are two major diffuse modulatory cholinergic systems in the brain, one of which is called the **basal forebrain complex**. It is a "complex" because the cholinergic neurons lie scattered among several related nuclei at the core of the telencephalon, medial and ventral to the basal ganglia. The best known of these are the *medial septal nuclei*, which provide the cholinergic innervation of the hippocampus, and the *basal nucleus of Meynert*, which provides most of the cholinergic innervation of the neocortex.

The function of the cells in the basal forebrain complex remains mostly unknown. But interest in this region has been fueled by the discovery that these are among the first cells to die during the course of Alzheimer's disease, which is characterized by a progressive and profound loss of cognitive functions. (There is, however, widespread neuronal death in Alzheimer's disease, and no specific link between the disease and cholinergic neurons has been established.) Like the noradrenergic and serotonergic systems, the cholinergic system has been implicated in regulating general brain excitability during arousal and sleep-wake cycles. The basal forebrain complex may also play a special role in learning and memory formation.

The second diffuse cholinergic system is called the *pontomesencephalotegmental complex*. These are ACh-utilizing cells in the pons and midbrain tegmentum. This system acts mainly on the dorsal thalamus, where, together with the noradrenergic and serotonergic systems, it regulates the excitability of the sensory relay nuclei. These cells also project up to the telencephalon, providing a cholinergic link between the brain stem and basal forebrain complexes. Figure 15.14 shows the cholinergic systems.

Drugs and the Diffuse Modulatory Systems

Psychoactive drugs, compounds with "mind-altering" effects, all act on the CNS, and most do so by interfering with chemical synaptic transmission.

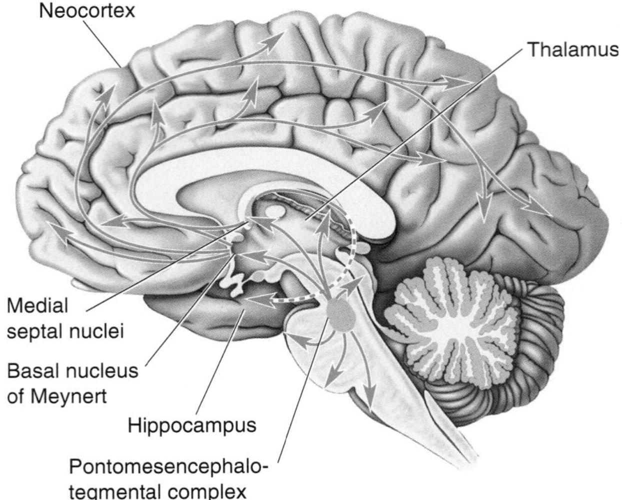

Figure 15.14
The cholinergic diffuse modulatory systems arising from the basal forebrain and brain stem. The medial septal nuclei and basal nucleus of Meynert project widely upon the cerebral cortex, including the hippocampus. The pontomesencephalotegmental complex projects to the thalamus and parts of the forebrain.

Many abused drugs act directly on the modulatory systems, particularly the noradrenergic, dopaminergic, and serotonergic systems.

Hallucinogens. The use of hallucinogens, drugs that produce hallucinations, goes back thousands of years. Hallucinogenic compounds are contained in a number of plants consumed as part of religious ritual, e.g., the *Psilocybe* mushroom by the Maya and the peyote cactus by the Aztec. The modern era of hallucinogenic drug use was unwittingly ushered in at the laboratory of Swiss chemist Albert Hofmann. In 1938, Hofmann chemically synthesized a new compound, *lysergic acid diethylamide*, or *LSD*. For 5 years, the LSD sat on the shelf. Then one day in 1943, Hofmann accidentally ingested some of the powder. His report of the effects attracted the immediate interest of the medical community. Psychiatrists began to use LSD in attempts to unlock the subconscious of mentally disturbed patients. Later the drug was discovered by intellectuals, artists, students, and the U. S. Defense Department, all of whom investigated its "mind-expanding" effects. (A chief advocate of LSD use was former Harvard psychologist Timothy Leary.) In the 1960s, LSD made its way to the street, where it remains today.

LSD is extremely potent. A dose sufficient to produce a full-blown hallucinogenic effect is only 25 μg (compared with a normal dose of aspirin at 650 mg, 25,000 times larger). Among the reported behavioral effects of LSD are a dreamlike state with heightened awareness of sensory stimuli, often with a mixing of perceptions such that sounds can evoke images, images can evoke smells, and so on.

The chemical structure of LSD (and the active ingredients of *Psilocybe* mushrooms and peyote) is very close to that of serotonin, suggesting that it acts on the serotonergic system. Indeed, LSD is a potent agonist at the serotonin receptors on the presynaptic terminals of neurons in the raphe nuclei. Activation of these receptors markedly inhibits the firing of raphe neurons. Thus, one known CNS effect of LSD is a reduction in the outflow of the brain's serotonergic diffuse modulatory system. It is worth noting in this regard that decreased activity of the raphe nuclei is also characteristic of dream-sleep (see Chapter 19).

Can we conclude that LSD produces hallucinations by silencing the brain's serotonin systems? If only drug effects on the brain were that simple!

Unfortunately, there are problems with this hypothesis. For one, silencing neurons in the raphe nuclei by other means—by destroying them, for example—does not mimic the effects of LSD in experimental animals. Furthermore, animals still respond to LSD as expected after their raphe nuclei have been destroyed.

In recent years, researchers have focused on direct LSD actions at serotonin receptors in the cerebral cortex. Current research suggests that LSD causes hallucinations by superseding the naturally modulated release of serotonin in cortical areas where perceptions normally are formed and interpreted.

Stimulants. In contrast to the uncertainties about hallucinogens and serotonin, it is clear that the powerful CNS stimulants *cocaine* and *amphetamine* both exert their effects at synapses made by dopaminergic and noradrenergic systems. Both drugs give users a feeling of increased alertness and self-confidence, a sense of exhilaration and euphoria, and a decreased appetite. Both are also sympathomimetic—they cause peripheral effects that mimic activation of the sympathetic division of the ANS: increased heart rate and blood pressure, dilation of the pupils, and so on.

Cocaine, which is extracted from the leaves of the coca plant, has been used by Andean Indians for hundreds of years. In the mid-nineteenth century, cocaine turned up in Europe and North America as the magic ingredient in a wide range of concoctions touted by their salesmen as having medicinal value (an example is Coca-Cola, which, as originally marketed in 1886 as a therapeutic agent, contained both cocaine and caffeine). Cocaine use fell out of favor early in the twentieth century, only to reemerge with a vengeance in the late 1960s as a recreational drug. Ironically, one of the main reasons for the rise in cocaine use during this period was the tightening of regulations against amphetamines. First chemically synthesized in 1887, amphetamines did not come into wide use until World War II, when they were taken by soldiers of both sides (particularly aviators) to sustain them in combat. Following the war, amphetamines became available as nonprescription diet aids, nasal decongestants, and pep pills. Regulations were finally tightened after recognition that amphetamines, like cocaine, are highly addictive and dangerous in large doses.

The neurotransmitters dopamine and norepinephrine are *catecholamines,* so named because of their chemical structure (see Chapter 6). The actions of catecholamines released into the synaptic cleft are normally terminated by specific uptake mechanisms. Cocaine and amphetamine both block this catecholamine uptake (Figure 15.15), although recent work suggests that cocaine targets DA reuptake more selectively; amphetamine blocks NE and DA reuptake *and* stimulates the release of DA. Thus, these drugs can prolong and intensify the effects of released DA or NE. Is this the means by which cocaine and amphetamine cause their stimulant effects? There is good reason for thinking so. For example, experimental depletion of brain catecholamines by use of synthesis inhibitors (such as α-methyltyrosine) will abolish the stimulant effects of both cocaine and amphetamine.

Besides having a similar stimulant effect, cocaine and amphetamine share another, more insidious behavioral action: psychological dependence, or addiction. Users develop powerful cravings for prolonging and continuing drug-induced pleasurable feelings. These effects are believed to result specifically from the enhanced transmission in the mesocorticolimbic dopamine system during drug use. Remember that this system may normally function to reinforce adaptive behaviors. By short-circuiting the system, these drugs instead reinforce drug-seeking behavior. Indeed, just as rats will work to

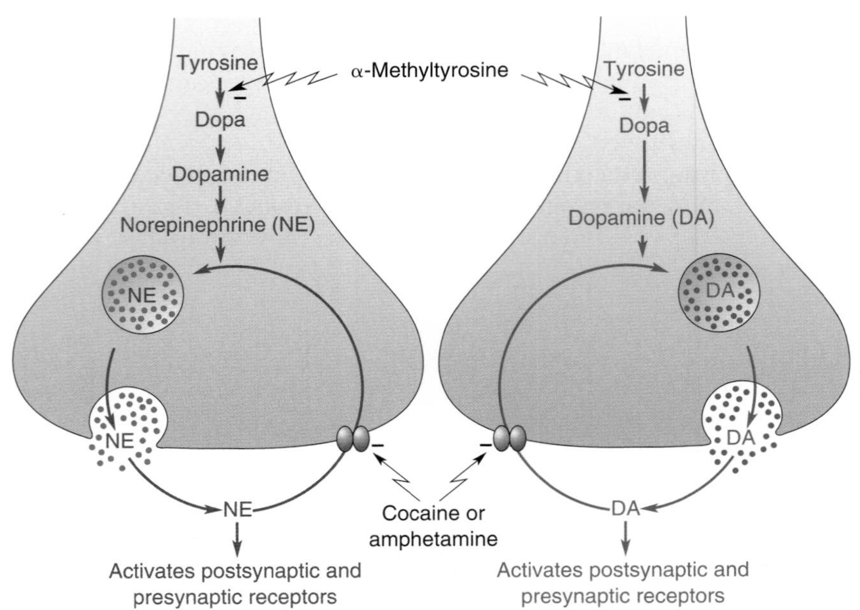

Figure 15.15
Stimulant drug action on the catecholamine axon terminal. On the left is a noradrenergic terminal, and on the right is a dopaminergic terminal. Both neurotransmitters are catecholamines synthesized from the dietary amino acid tyrosine. Dopa (3,4-dihydroxyphenylalanine) is an intermediate in the synthesis of both. The actions of NE and DA are usually terminated by uptake back into the axon terminal. Amphetamine and cocaine block this uptake, thereby allowing NE and DA to remain in the synaptic cleft longer.

electrically stimulate the mesocorticolimbic projection, they will also work to receive an injection of cocaine. The involvement of dopamine pathways in motivation and addiction will be discussed further in Chapter 16.

CONCLUDING REMARKS

In this chapter, we have examined three components of the nervous system that are characterized by the great reach of their influences. The secretory hypothalamus and ANS communicate with cells all over the body, and the diffuse modulatory systems communicate with neurons in many different parts of the brain. They are also characterized by the duration of their direct effects, which can range from minutes to hours. Finally, they are characterized by their chemical neurotransmitters. In many instances, the transmitter *defines* the system. For example, in the periphery, we can use the words "noradrenergic" and "sympathetic" interchangeably. The same thing goes for "raphe" and "serotonin" in the forebrain and "substantia nigra" and "DA" in the basal ganglia. These chemical idiosyncrasies have permitted interpretations of drug effects on behavior that are not possible with most other neural systems. Thus, we have a good idea where in the brain amphetamine and cocaine exert their stimulant effects and where in the periphery they act to raise blood pressure and heart rate.

At a detailed level, each of the systems discussed in this chapter performs different functions. But at a general level, they all *maintain brain homeostasis*: They regulate different processes within a certain physiological range. For example, the ANS regulates blood pressure within a range that is appropriate. Blood pressure variations optimize an animal's performance under different conditions. In a similar way, the noradrenergic locus coeruleus and serotonergic raphe nuclei regulate levels of consciousness and mood. These levels also vary within a range that is adaptive to the organism. In the next several chapters, we will encounter these systems again in the context of specific functions.

K E Y T E R M S

The Secretory Hypothalamus
homeostasis (p. 500)
periventricular zone (p. 500)
magnocellular neurosecretory cell (p. 501)
neurohormone (p. 502)
oxytocin (p. 502)
vasopressin (p. 502)
parvocellular neurosecretory cell (p. 503)
hypophysiotropic hormone (p. 503)
hypothalamo-pituitary portal circulation (p. 504)
adrenal cortex (p. 504)
adrenal medulla (p. 504)
cortisol (p. 504)

The Autonomic Nervous System
autonomic nervous system (ANS) (p. 505)
sympathetic division (p. 506)
parasympathetic division (p. 506)
autonomic ganglia (p. 507)
postganglionic neuron (p. 507)
preganglionic neuron (p. 507)
sympathetic chain (p. 509)
enteric division (p. 510)
nucleus of the solitary tract (p. 511)

The Diffuse Modulatory Systems of the Brain
diffuse modulatory system (p. 513)
locus coeruleus (p. 513)
raphe nuclei (p. 515)
basal forebrain complex (p. 517)

R E V I E W Q U E S T I O N S

1. Battlefield trauma victims who have lost large volumes of blood often express a craving to drink water. Why?
2. You've stayed up all night trying to meet a term paper deadline. You now are typing frantically, keeping one eye on the paper and the other on the clock. Describe in detail how the periventricular zone of the hypothalamus has orchestrated your body's physiological response to this stressful situation.
3. Why is the adrenal medulla often called a modified sympathetic ganglion? Why is the adrenal cortex not included in this description?
4. A number of famous athletes and entertainers have accidentally killed themselves by taking large quantities of cocaine. Usually, the cause of death is heart failure. How would you explain the peripheral actions of cocaine?
5. How do the diffuse modulatory and point-to-point synaptic communication systems in the brain differ? List four ways.
6. Under what behavioral conditions are the noradrenergic neurons of the locus coeruleus active? The noradrenergic neurons of the ANS?

CHAPTER

16

Motivation

INTRODUCTION

Behavior happens. But why? In Part II of this book, we discussed various types of motor responses. At the lowest level are unconscious reflexes initiated by sensory stimulation—dilation of the pupils when the lights go out, yanking back the foot from a thumbtack, and so on. At the highest level are willful movements initiated by the neurons of the frontal lobe, e.g., the finger movements that tap this text into the computer. Willful movements are incited to occur—or *motivated*—to satisfy a need. The motivation can be abstract (the "need" to go sailing on a warm and breezy summer afternoon), but it can also be quite concrete (the need to go to the bathroom when your bladder is full).

Motivation can be thought of as a driving force on behavior. By analogy, consider the driving force on sodium ions to cross the neuronal membrane (an odd analogy, perhaps, but not for a neuroscience text). As we learned back in Chapters 3 and 4, ionic driving force depends on a number of factors, including the concentration of the ion on either side of the membrane and the electrical membrane potential. Variations in driving force make transmembrane ionic current in a particular direction more or less likely. But the driving force alone does not determine whether the current flows; the transmembrane movement of ions also requires the appropriate gated ion channels to be opened and capable of conducting the current.

Of course, human behavior will never be described by anything as simple as Ohm's law. Still, it is useful to consider that the probability and direction of a behavior vary with the level of driving force to perform that behavior. And while motivation may be required for a certain behavior, it does not guarantee that behavior. The membrane analogy also allows us to highlight the fact that a crucial part of the control of behavior is appropriate gating of the expression of different motivated actions that have conflicting goals—tapping on the computer keyboard versus sailing, for example.

Neuroscience cannot yet concretely explain why the sailing expedition was abandoned in favor of writing this chapter. Considerable progress has been made, however, in understanding the motivation of behaviors that are basic to survival.

THE HYPOTHALAMUS, HOMEOSTASIS, AND MOTIVATED BEHAVIOR

Chapter 15 introduced the hypothalamus and homeostasis. Recall that homeostasis refers to the processes that maintain the internal environment of the body within a narrow physiological range. Although homeostatic reflexes occur at many levels of the nervous system, the hypothalamus plays a key role in the regulation of body temperature, fluid balance, and energy balance.

The hypothalamic regulation of homeostasis starts with sensory transduction. A regulated parameter (e.g., temperature) is measured by specialized sensory neurons, and deviations from the optimal range are detected by neurons concentrated in the periventricular zone of the hypothalamus. These neurons orchestrate an integrated response to bring the parameter back to its optimal value. The response generally has three components:

1. *Humoral response*: Hypothalamic neurons respond to sensory signals by stimulating or inhibiting the release of pituitary hormones into the bloodstream.
2. *Visceromotor response*: Neurons in the hypothalamus respond to sensory signals by adjusting the balance of sympathetic and parasympathetic outputs of the autonomic nervous system (ANS).

3. *Somatic motor response*: Hypothalamic neurons, particularly within the lateral hypothalamus, respond to sensory signals by inciting an appropriate somatic motor behavioral response.

You are cold, dehydrated, and depleted of energy. The appropriate humoral and visceromotor responses kick in automatically. You shiver, blood is shunted away from the body surface, urine production is inhibited, body fat reserves are mobilized, and so on. But the fastest and most effective way to correct these disturbances of brain homeostasis is to seek warmth or generate it by moving, to drink water, and to eat. These are examples of **motivated behaviors** generated by the somatic motor system, and they are incited to occur by the activity of the lateral hypothalamus. Our goal in this chapter is to explore the neural basis for this type of motivation. To illustrate, we will concentrate on a subject dear to our hearts: eating.

THE LONG-TERM REGULATION OF FEEDING BEHAVIOR

As you know, even a brief interruption in a person's oxygen supply can lead to serious brain damage and death. You may be surprised to learn that the brain's requirement for food in the form of glucose is no less urgent. Only a few minutes of glucose deprivation will lead to loss of consciousness, eventually followed by death if glucose is not restored. While the external environment normally provides a constant source of oxygen, the availability of food is less certain. Thus, complex internal regulatory mechanisms have evolved to store energy in the body so that it is available when needed. One primary reason we are motivated to eat is to keep these reserves at a level sufficient to ensure that there will not be an energy shortfall.

Energy Balance

The body's energy stores are replenished during and immediately after consuming a meal. This condition, in which the blood is filled with nutrients, is called the *prandial state* (from the Latin for "breakfast"). During this time, energy is stored in two forms: glycogen and triglycerides (Figure 16.1). Glycogen stores have a finite capacity, and they are found mainly in the liver and skeletal muscle. Triglyceride stores are found in adipose (fat) tissue, and they have a virtually unlimited capacity. The assembly of macromolecules such as glycogen and triglycerides from simple precursors is called **anabolism**, or anabolic metabolism.

During the fasting condition between meals, called the *postabsorptive state*, the stored glycogen and triglycerides are broken down to provide the body with a continuous supply of the molecules used as fuel for cellular metabolism (glucose for all cells; fatty acids and ketones for all cells other than neurons). The process of breaking down complex macromolecules is called **catabolism**, or catabolic metabolism; it is the opposite of anabolism. The system is in proper balance when energy reserves are replenished at the same average rate that they are expended. If the intake and storage of energy consistently exceed the usage, the amount of body fat, or *adiposity*, increases, eventually resulting in **obesity**. (The word *obese* is derived from the Latin word for "fat.") If the intake of energy consistently fails to meet the body's demands, loss of fat tissue occurs, eventually resulting in **starvation**. Figure 16.2 summarizes the concept of energy balance and body fat.

For the system to stay in balance, there must be some means of regulating feeding behavior, based on the size of the energy reserves and their rate of replenishment. In the past decade, we have seen breathtaking progress on the various means by which this regulation occurs—and none too soon, as eating disorders and obesity are widespread health problems. It is now apparent

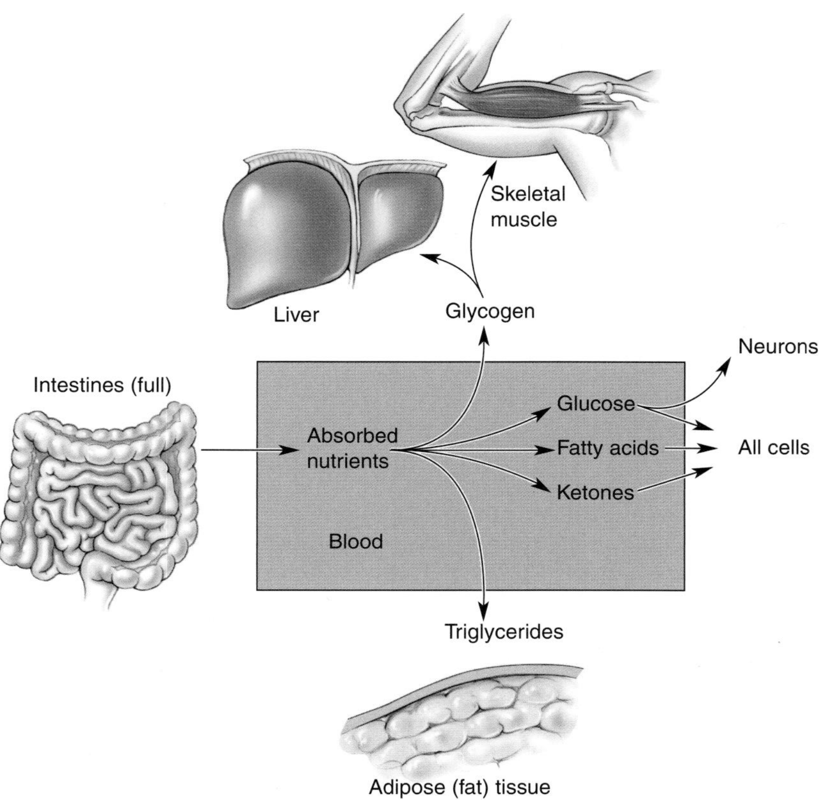

(a) Anabolism during the prandial state

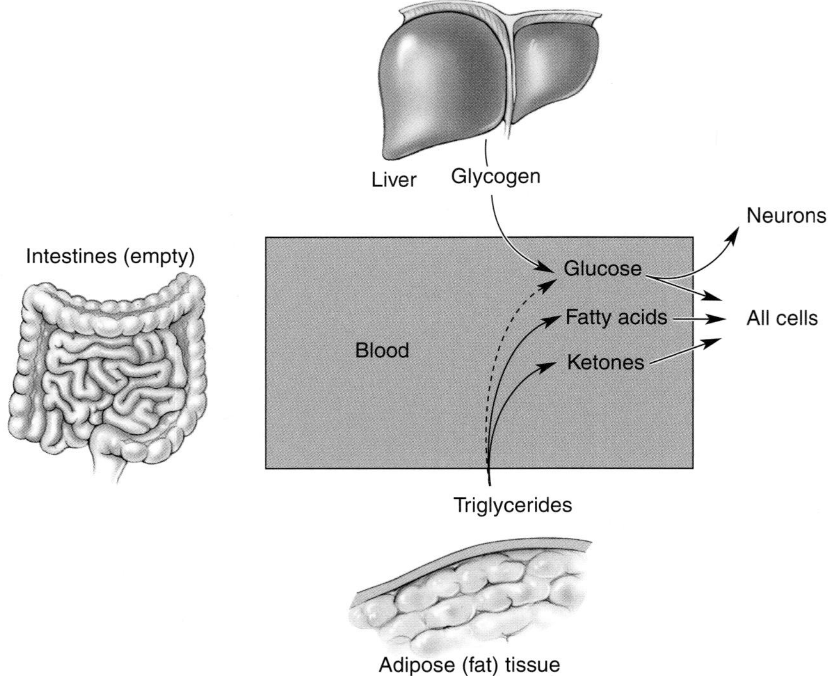

(b) Catabolism during the postabsorptive state

Figure 16.1

Loading and emptying the body's energy reserves. (a) After a meal, when we are in the prandial state, excess energy is stored as glycogen or as triglycerides. **(b)** Between meals, when we are in the postabsorptive state, the glycogen and triglycerides are broken down (catabolized) into smaller molecules that can be used as fuel by the cells of the body.

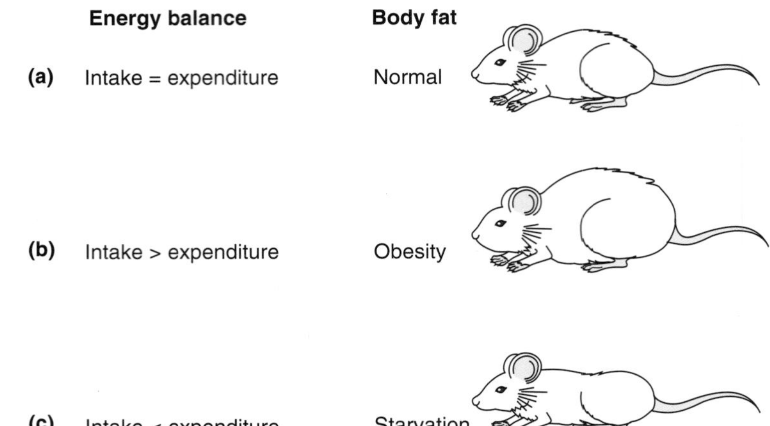

Figure 16.2
Energy balance and body fat. (a) Normal energy balance leads to normal adiposity. **(b)** Prolonged positive energy balance leads to obesity. **(c)** Prolonged negative energy balance leads to starvation.

that there are multiple regulatory mechanisms, some acting over a long time to maintain the body's fat reserves and others acting over a shorter time to regulate meal size and frequency. We begin our investigation by looking at long-term regulation.

Hormonal and Hypothalamic Regulation of Body Fat and Feeding

The study of the homeostatic regulation of feeding behavior has a long history, but the pieces of the puzzle are only now falling into place. As we will see, feeding is stimulated when neurons in the hypothalamus detect a drop in the level of a hormone released by fat cells. These hypothalamic cells are concentrated in the periventricular zone; the neurons that incite feeding behavior are in the lateral hypothalamus.

Body Fat and Food Consumption. If you've ever dieted, you do not have to be told that the body works hard to frustrate any efforts to alter adiposity. You are not alone. As illustrated in Figure 16.3, a rat can be induced to lose body fat by severely restricting its caloric intake. However, once free access to food is restored, the animal will overeat until the original level of body fat

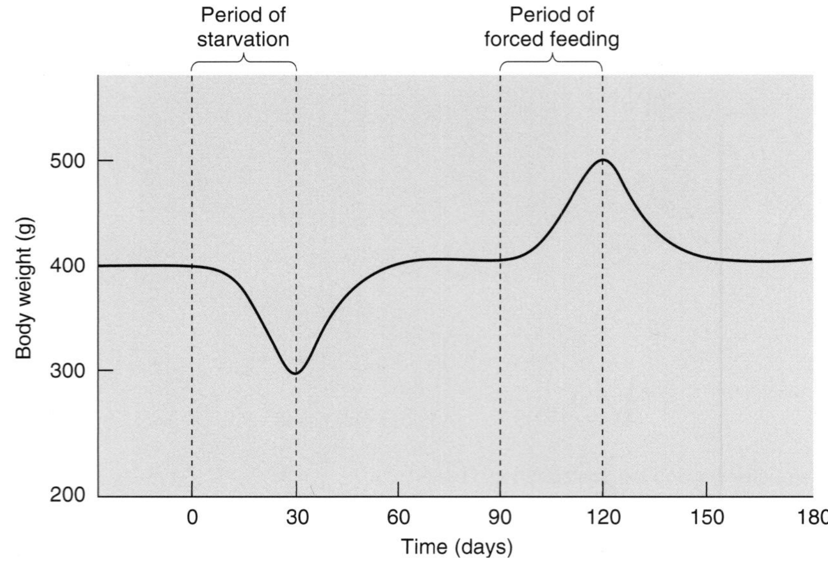

Figure 16.3
Maintenance of body weight around a set value. Body weight is normally stable. If an animal is force-fed, it will gain weight. The weight is lost, however, as soon as the animal can regulate its own food intake. Similarly, weight lost during a period of starvation is rapidly gained when food is freely available.

has fully returned. (Perhaps this sounds familiar.) But it also works the other way around. Animals force-fed to gain fat mass, once given the chance to regulate their own diet, will eat less until their fat levels return to normal. The rat's behavioral response obviously is not a reflection of vanity; it is a mechanism to maintain energy homeostasis. The idea that the brain monitors the amount of body fat and acts to "defend" this energy store against perturbations, first proposed in 1953 by British scientist Gordon Kennedy, is called the **lipostatic hypothesis**.

The coupling of fat to feeding behavior suggests that there must be communication from adipose tissue to the brain. Immediately, a bloodborne hormonal signal was suspected, and this suspicion was confirmed in the 1960s by Douglas Coleman and his colleagues at the Jackson Laboratories in Bar Harbor, Maine, working with genetically obese mice. The DNA of one strain of obese mouse lacks both copies of a gene called *ob* (these mice are therefore called *ob/ob* mice). Coleman hypothesized that the protein encoded by the *ob* gene is the hormone telling the brain that fat reserves are normal. Thus, in the *ob/ob* mice that lack this hormone, the brain is fooled into thinking that the fat reserves are low, and the animals are abnormally motivated to eat. To test this idea, researchers performed a parabiosis experiment. *Parabiosis* is the long-term anatomical and physiological union of two animals, as in Siamese twins. Surgical fusion can produce parabiotic animals sharing a common blood supply. Coleman and his colleagues found that when *ob/ob* animals were parabiotically joined with normal mice, their feeding behavior and obesity were greatly reduced, as if the missing hormone had been replaced (Figure 16.4).

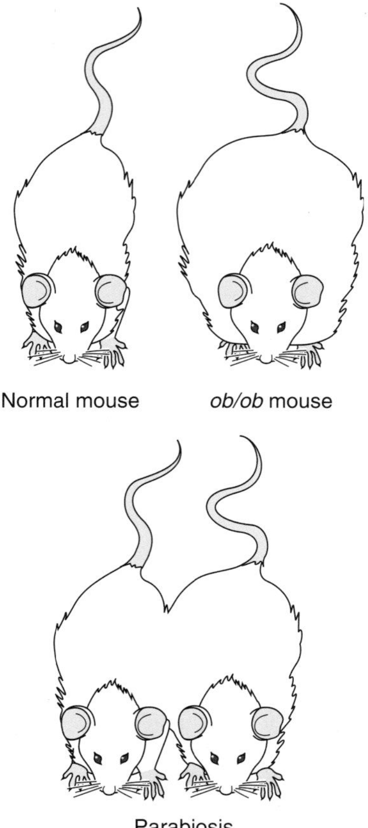

Figure 16.4
Regulation of body fat by a circulating hormone. If a genetically obese *ob/ob* mouse is surgically fused with a normal mouse so that bloodborne signals are now shared between the animals, the obesity of the *ob/ob* mouse is greatly moderated.

Then the search was on for the protein encoded by the *ob* gene. In 1994, a group of scientists led by Jeffrey Friedman at The Rockefeller University finally isolated the protein, which they called **leptin** (from the Greek for "slender") (Box 16.1). Treating *ob/ob* mice with leptin completely reverses the obesity and the eating disorder. The hormone leptin, released by adipocytes (fat cells), regulates body mass by acting directly on neurons of the hypothalamus.

The Hypothalamus and Feeding. A. W. Hetherington and S. W. Ranson of Northwestern University made the seminal discovery, published in 1940, that small lesions made on both sides of a rat's hypothalamus can have large effects on subsequent feeding behavior and adiposity. Bilateral lesions of the *lateral* hypothalamus caused **anorexia**, a severely diminished appetite for food. In contrast, bilateral lesions of the *ventromedial* hypothalamus caused the animals to overeat and become obese (Figure 16.5). This basic scenario applies to humans as well. Anorexia caused by damage to the lateral hypothalamus is commonly called **lateral hypothalamic syndrome**; the overeating and obesity caused by lesions to the ventromedial hypothalamus is called **ventromedial hypothalamic syndrome**.

For a time, the idea that the lateral hypothalamus was a "hunger center" acting in opposition to the ventromedial hypothalamus "satiety center" was popular. Thus, lesions of the medial or lateral hypothalamus bring the system out of balance. Destruction of the lateral hypothalamus leaves the animals inappropriately satiated, so they do not eat; destruction of the ventromedial hypothalamus leaves the animals insatiable, so they overeat. This "dual center" model, however, has proven to be simplistic. We now have a better idea why hypothalamic lesions affect body fat and feeding behavior—it has much to do with leptin signaling.

The Effects of Elevated Leptin Levels on the Hypothalamus. Although still sketchy in places, a picture of how the hypothalamus participates in body fat homeostasis is beginning to emerge. First, let's consider the response when leptin levels are high, as they are right after several days of "forced" holiday feasting.

Circulating leptin molecules, released into the bloodstream by adipocytes, activate leptin receptors on neurons of the **arcuate nucleus** of the hypothalamus, which lies near the base of the third ventricle (Figure 16.6). The arcuate neurons that are activated by a rise in blood leptin levels are characterized by a distinctive mix of peptide neurotransmitter molecules. Most appear to possess both *αMSH* and *CART*, and the levels of these peptides in the brain vary in proportion to the level of leptin in the blood. (To explain the alphabet soup: Peptides are often named by their first discovered function, and these names can lead to confusion when other roles are recognized. Therefore, it is usual to refer to neuropeptides simply by their abbreviations. For the record, αMSH stands for *alpha-melanocyte–stimulating hormone*, and CART stands for *cocaine- and amphetamine-regulated transcript*. Like other neurotransmitters,

Box 16.1

PATH OF DISCOVERY

Fat Chance

BY JEFFREY FRIEDMAN

My laboratory identified leptin, a hormone produced by fat tissue. Leptin acts on the brain to modulate food intake and functions as an afferent signal in a feedback loop that regulates weight. My route to this hormone was filled with a number of chance events that were in no way predictable at the time I started my career.

I was originally trained as a physician. In medical school and as a medical resident, I participated in some modest research studies. The first piece of work I completed related to the effects of dietary salt on the regulation of blood pressure. After completing this project, I excitedly submitted a paper for publication. I remember one of the reviews verbatim: "This paper should not be published in the *Journal of Clinical Investigation* or anywhere else." Fortunately, one of my mentors in medical school still thought I might have some aptitude for research. He suggested that I go to The Rockefeller University to work in a basic science research laboratory. I joined the lab of Dr. Mary-Jeanne Kreek to study the effects of endorphins in the development of narcotic addiction.

I was fascinated by the idea that endogenous molecules could alter behavior and emotional state. At The Rockefeller University, I met another scientist, Bruce Schneider. Bruce was studying cholecystokinin (CCK), a peptide hormone that is secreted by intestinal cells. CCK aids digestion by stimulating the secretion of enzymes from the pancreas and bile from the gallbladder. CCK had also been found in neurons of the brain, although its function there was less clear. In the late 1970s, it was shown that injections of CCK reduce food intake. This finding appealed to me as another example of how a single molecule can change behavior. One other fact also piqued my interest: There were indications that the levels of CCK were decreased in the genetically obese *ob/ob* mouse (Figure A). These mutant mice are massively obese as a consequence of a defect in a single gene. The mice eat excessively and weigh 3 to 5 times as much as normal mice. It was thus hypothesized that CCK functions as an endogenous appetite suppressant and that a deficiency of CCK caused the obesity evident in *ob/ob* mice. Fascinated by this possibility, I set out to establish the possible role of CCK in the pathogenesis of obesity in these animals. To do this I was going to need additional training in basic research, so I abandoned my plans to continue medical training in gastroenterology and instead entered the Ph.D. program at The Rockefeller University.

Jeffrey Friedman

As a Ph.D. student I worked in the laboratory of Jim Darnell, studying the regulation of gene expression in the liver and learning the basic tools of molecular biology. But I carried the functional role of these molecules depends on the circuits in which they participate.)

Now let's take a moment to consider the body's integrated response to excessive adiposity and high leptin levels. The *humoral response* consists of increased secretion of TSH (thyroid-stimulating hormone) and ACTH (adrenocorticotropic hormone) (see Table 15.1). These pituitary hormones act on the thyroid and adrenal glands and have the effect of raising the metabolic rate of cells throughout the body. The *visceromotor response* increases the tone of the sympathetic division of the ANS, which also raises the metabolic rate, in part by raising body temperature. The *somatic motor response* decreases feeding behavior. The αMSH and CART neurons of the arcuate nucleus project their axons directly to the regions of the nervous system that orchestrate this coordinated response (Figure 16.7).

The humoral response is triggered by the activation of neurons in the **paraventricular nucleus** of the hypothalamus, which in turn causes the release

(Box 16.1, *continued*)

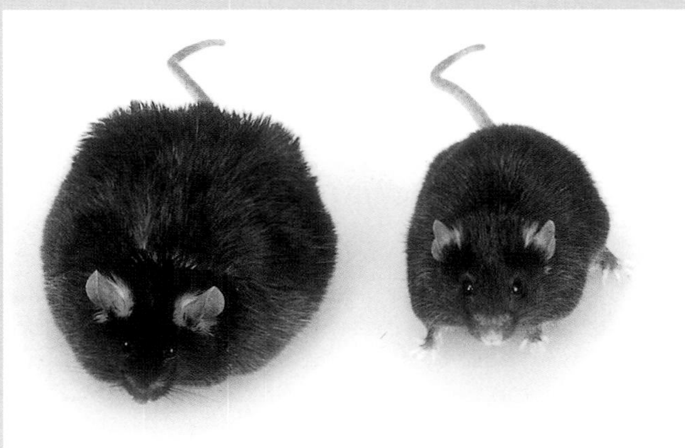

Figure A
Both of these mice have a defect in a gene called obese (*ob*). This mutation usually results in a marked increase in the amount of fat. Administration of the protein encoded by the *ob* gene, called leptin, reduced the body weight of the *ob* mice. After $4\frac{1}{2}$ weeks, the *ob* mouse on the left, which did not receive leptin, weighed approximately 67 g, while the mouse on the right, which received daily injections of leptin, weighed 35 g. Normal mice weigh approximately 24 g. Daily injections of leptin to *ob* mice reduced body weight via effects on food intake and energy expenditure, meaning the treated animals ate less and burned more calories. (Source: John Sholtis, The Rockefeller University, New York, NY.)

my interest in the *ob/ob* gene with me. At the end of my graduate studies, two colleagues and I successfully isolated the CCK gene from the mouse. One of the first studies we performed after sequencing the gene was to determine its chromosomal position. We found that the CCK gene was not on chromosome 6, where the *ob* mutation had been localized, which thus excluded defective CCK as the cause of the obesity. The question thus remained: What is the nature of the defective gene in *ob/ob* mice?

After receiving my Ph.D. in 1986, I became an assistant professor at The Rockefeller University and set out to answer this question. The culmination of what proved to be an 8-year odyssey was the identification of the *ob* gene in 1994. We now know that the *ob* gene encodes the hormone leptin. The discovery of this hormone, a singular event in my life, was absolutely exhilarating. The realization that nature had happened upon such a simple and elegant solution for regulating weight was the closest thing I have ever had to a religious experience. Subsequent studies revealed that injections of leptin dramatically decrease the food intake of mice and other mammals. My current studies now focus on the question that originally aroused my interest in this mutation: How is it that a single molecule—leptin—profoundly influences feeding behavior? An esteemed colleague of mine remarked recently that I had searched for the *ob* gene primarily so that I could approach the question I started with. It is as yet unclear whether I will succeed in understanding how a single molecule can influence a complex behavior.

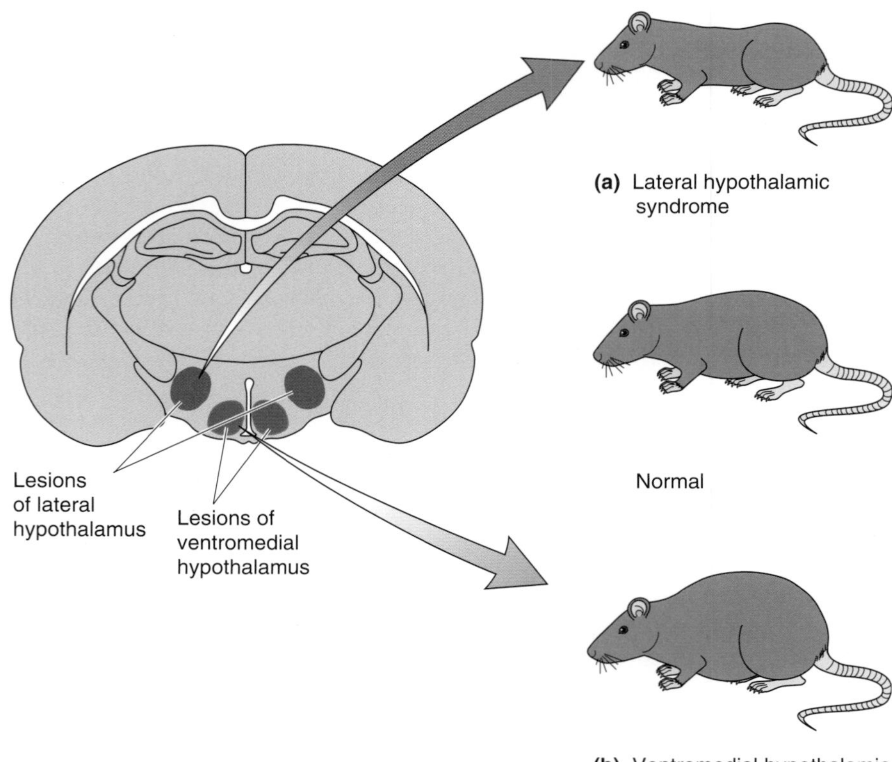

Figure 16.5
Altered feeding behavior and body weight resulting from bilateral lesions of the rat hypothalamus. (a) The lateral hypothalamic syndrome, characterized by anorexia, is caused by lesions of the lateral hypothalamus. **(b)** The ventromedial hypothalamic syndrome, characterized by obesity, is caused by lesions of the ventromedial hypothalamus.

of the hypophysiotropic hormones that regulate the secretion of TSH and ACTH from the anterior pituitary (see Chapter 15). The paraventricular nucleus also controls the activity of the sympathetic division of the ANS with direct axonal projections to neurons in the lower brain stem and to preganglionic neurons in the spinal cord. But there is also a direct path for arcuate control of the sympathetic response: The αMSH and CART neurons themselves project axons directly down to the intermediolateral gray matter of the spinal cord. Finally, feeding behavior is inhibited via connections of the arcuate nucleus neurons with cells in the lateral hypothalamus. We will take a closer look at the lateral hypothalamus in a moment.

The injection of αMSH or CART into the brain mimics the response to elevated leptin levels. Thus, these are said to be **anorectic peptides**—they diminish appetite. The injection of drugs that block the actions of these peptides increases feeding behavior. These findings suggest that αMSH and CART normally participate in the regulation of energy balance, in part by acting as the brain's own appetite suppressants.

The Effects of Decreased Leptin Levels on the Hypothalamus. In addition to turning off the responses mediated by αMSH and CART neurons, a *fall* in leptin levels actually *stimulates* another type of arcuate nucleus neuron. These neurons contain their own mix of peptides: *NPY* (*neuropeptide Y*) and *AgRP* (*agouti-related peptide*). The NPY and AgRP neurons of the arcuate nucleus also have connections with the paraventricular nucleus and the lateral hypothalamus (Figure 16.8). But the effects of these neuropeptides on energy balance are the opposite of the effects of αMSH and CART. NPY and AgRP *inhibit* the secretion of TSH and ACTH, activate the *parasympathetic* division of the ANS, and *stimulate* feeding behavior. They are therefore called **orexigenic peptides** (from the Greek for "appetite").

AgRP and αMSH literally are antagonistic. Both peptides bind to the same

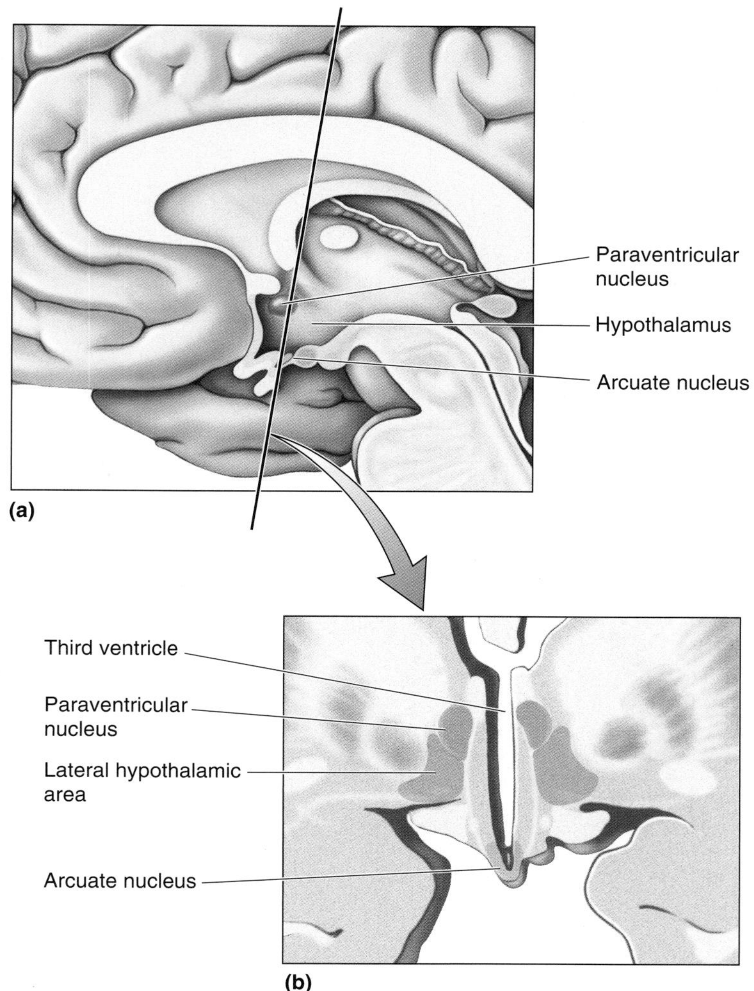

Figure 16.6
Hypothalamic nuclei important for the control of feeding. (a) A midsagittal view of the human brain, showing the location of the hypothalamus. **(b)** A coronal section of the human brain, showing, in part, three important nuclei for the control of feeding: the arcuate nucleus, the paraventricular nucleus, and the lateral hypothalamic area.

receptor, called the *MC4 receptor*. Whereas αMSH is the receptor's agonist, AgRP is a natural antagonist that blocks the stimulation of αMSH. Activating MC4 receptors on lateral hypothalamic neurons inhibits feeding; blocking the receptors stimulates feeding (Figure 16.9).

The Control of Feeding by Lateral Hypothalamic Peptides. We now come to the mysterious lateral hypothalamus, which appears to have a special role in motivating us to eat. Because this region of the brain is not organized into well-defined nuclei, it has the undistinctive name **lateral hypothalamic area** (Figure 16.6). As mentioned earlier, the first indication that the lateral hypothalamus is involved in motivating feeding behavior was that a lesion here caused animals to stop eating. Moreover, electrical stimulation of this area triggers feeding behavior, even in satiated animals. These basic findings apply to all mammals that have been examined (including humans). Unfortunately, interpreting these findings has proven to be difficult. One problem is that lesions and electrical stimulation not only affect the neurons with cell bodies in this region, they also affect many axonal pathways passing through the lateral hypothalamus. It is now apparent that both the neurons intrinsic to the lateral hypothalamus *and* the axons passing through the lateral hypothalamus contribute to the motivation of feeding behavior. For now, however, we will concentrate on the role of the neurons within the lateral hypothalamic area.

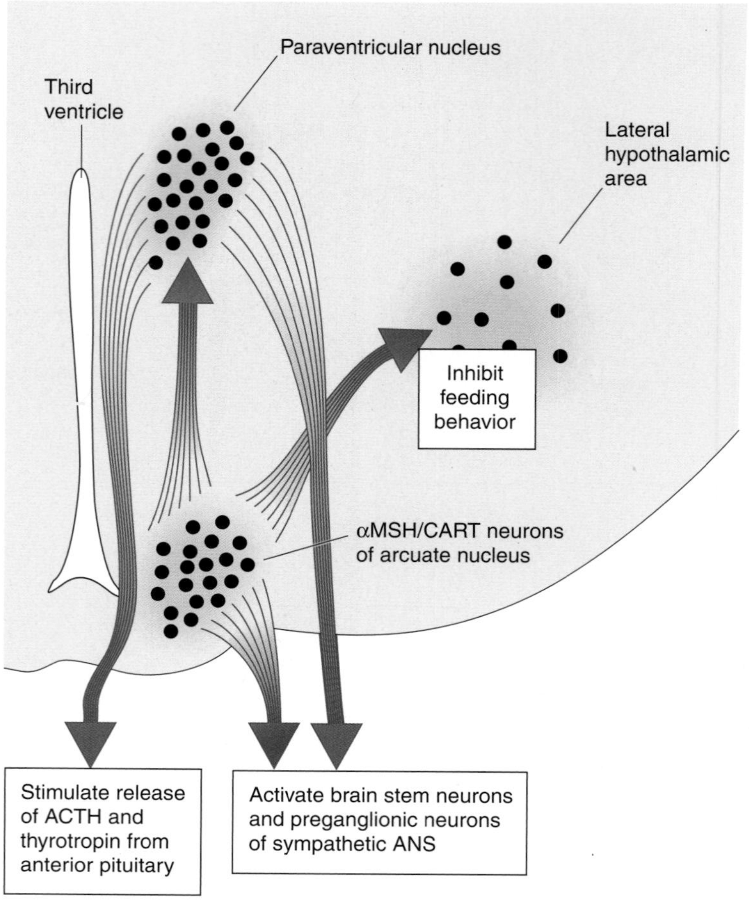

Paraventricular nucleus
Third ventricle
Lateral hypothalamic area
Inhibit feeding behavior
αMSH/CART neurons of arcuate nucleus
Stimulate release of ACTH and thyrotropin from anterior pituitary
Activate brain stem neurons and preganglionic neurons of sympathetic ANS

Figure 16.7
Response to elevated leptin levels. A rise in leptin levels in the blood is detected by neurons in the arcuate nucleus that contain the peptides αMSH and CART. These neurons project axons to the lower brain stem and spinal cord, the paraventricular nuclei of the hypothalamus, and the lateral hypothalamic area. Each of these connections contributes to the coordinated humoral, visceromotor, and somatic motor responses to increased leptin levels. (Source: Adapted from Sawchenko, 1998, p. 437.)

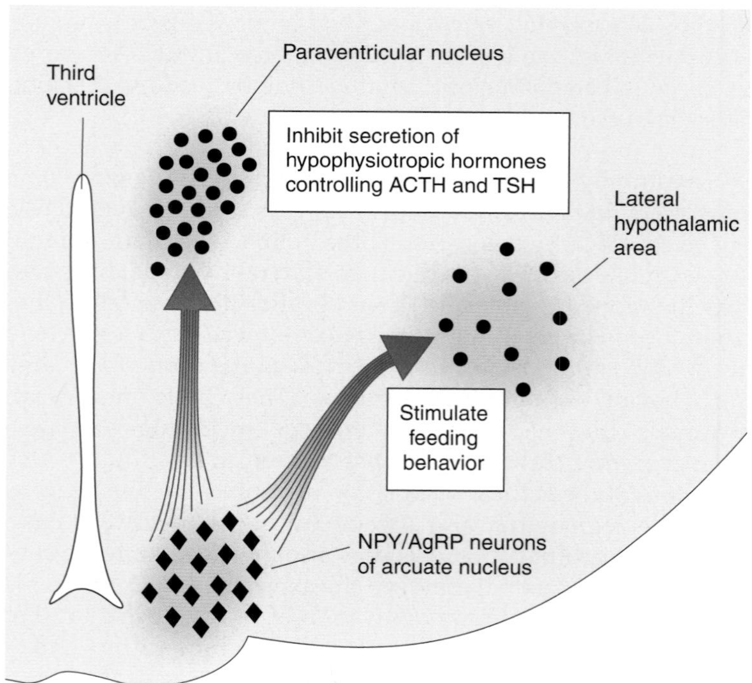

Figure 16.8
Response to decreased leptin levels. A reduction in blood levels of leptin is detected by neurons in the arcuate nucleus that contain the peptides NPY and AgRP. These arcuate nucleus neurons inhibit the neurons in the paraventricular nuclei that control the release of TSH and ACTH from the pituitary. In addition, they activate the neurons in the lateral hypothalamus that stimulate feeding behavior. Some of the activated lateral hypothalamic neurons contain the peptides melanin-concentrating hormone (MCH) and orexin.

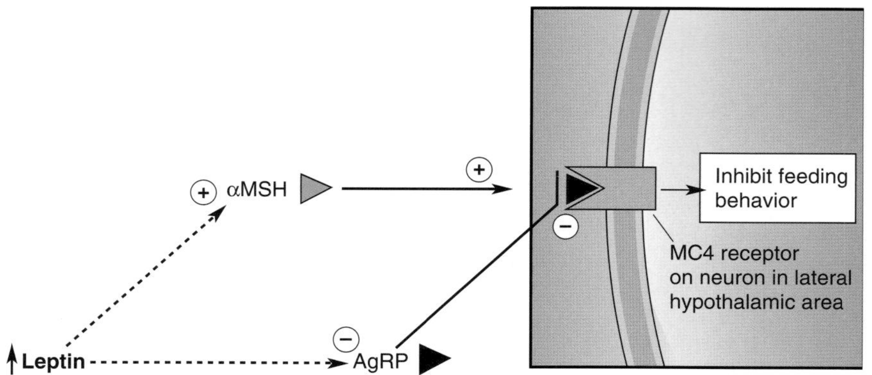

Figure 16.9
Competition for activation of the MC4 receptor. One way that αMSH, an anorectic peptide, and AgRP, an orexigenic peptide, exert opposite effects on metabolism and feeding behavior is via an interaction with the MC4 receptor on some hypothalamic neurons. While αMSH stimulates the MC4 receptor, AgRP blocks the action of αMSH at the receptor.

One group of neurons in the lateral hypothalamus that receives direct input from the leptin-sensitive cells of the arcuate nucleus has yet another peptide neurotransmitter, *MCH* (*melanin-concentrating hormone*). These cells have extremely widespread connections in the brain, including direct monosynaptic innervation of most of the cerebral cortex. The cortex participates in orga-

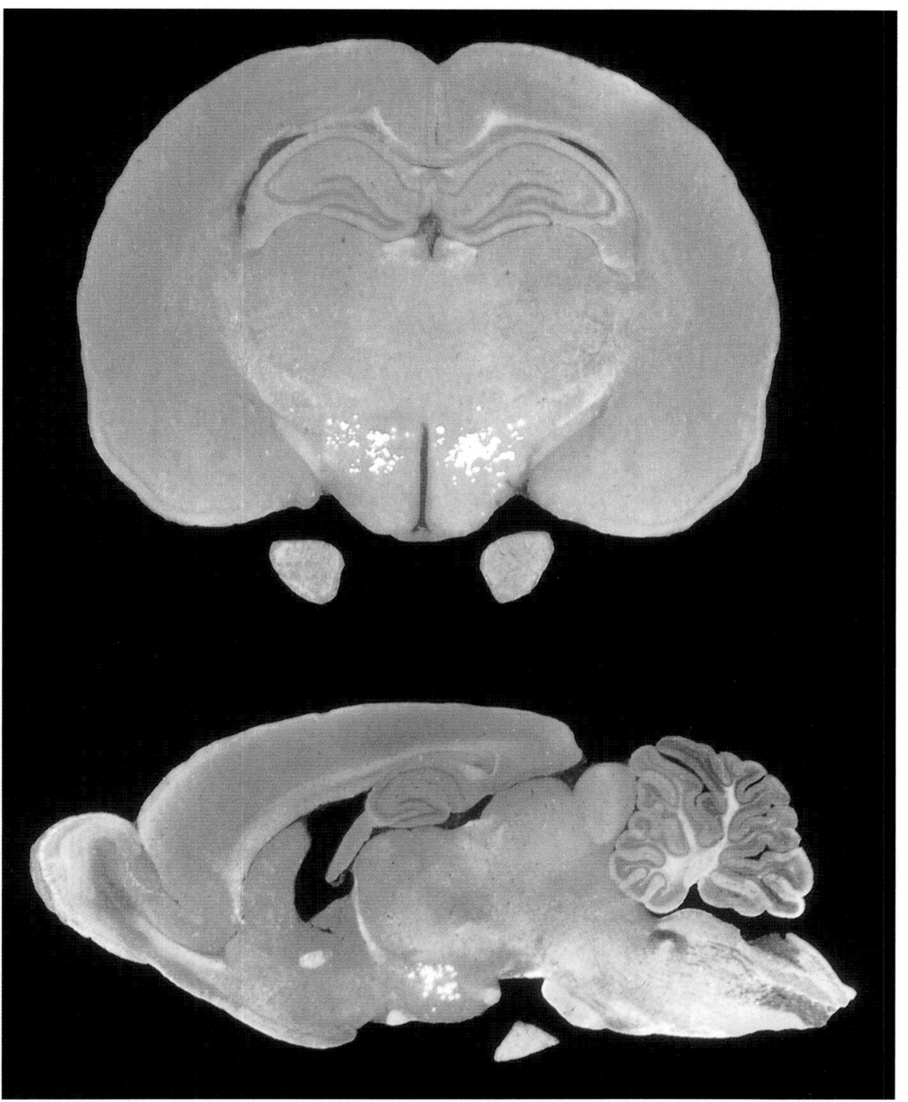

Figure 16.10
Neurons that stimulate feeding in the rat lateral hypothalamus. In this photomicrograph, neurons containing messenger RNA for orexin appear bright on a dark background. Orexin is one of the lateral hypothalamic peptides that stimulates feeding behavior. As we will see in Chapter 19, the same cells play an important role in the control of sleep and wakefulness. (Source: *Cell* 92 (4), February 22, 1998, cover image.)

Table 16.1 Anorectic and Orexigenic Peptides of the Hypothalamus

ANORECTIC PEPTIDES (INHIBIT FEEDING BEHAVIOR)			OREXIGENIC PEPTIDES (STIMULATE FEEDING BEHAVIOR)		
ABBREVIATION	FULL NAME	LOCATION	ABBREVIATION	FULL NAME	LOCATION
αMSH	Alpha-melanocyte-stimulating hormone	Arcuate nucleus	NPY	Neuropeptide Y	Arcuate nucleus
CART	Cocaine- and amphetamine-regulated transcript	Arcuate nucleus	AgRP	Agouti-related peptide	Arcuate nucleus
			MCH	Melanin-concentrating hormone	Lateral hypothalamic area
			—	Orexin	Lateral hypothalamic area

nizing and initiating goal-directed behaviors, such as raiding the refrigerator. The MCH system is in a strategic position to inform the cortex of leptin levels in the blood and, therefore, may contribute significantly to motivating the search for food. Supporting this idea, the injection of MCH into the brain stimulates feeding behavior. Moreover, mutant mice that lack this peptide exhibit reduced feeding behavior, have an elevated metabolic rate, and are lean.

A second population of lateral hypothalamic neurons with widespread cortical connections that contains another peptide called *orexin* (Figure 16.10) has recently been identified. These cells also receive direct input from the arcuate nucleus. As is the case for MCH and as the name suggests, orexin is an orexigenic peptide; i.e., it stimulates feeding behavior. The levels of both MCH and orexin rise in the brain when leptin levels in the blood fall.

To conclude this section, we briefly summarize hypothalamic responses to blood leptin levels (Table 16.1). Remember that leptin levels rise when body fat is increased and they fall when body fat is decreased:

- A *rise* in leptin levels increases αMSH and CART in arcuate nucleus neurons. These anorectic peptides act on the brain to inhibit feeding behavior and increase metabolism.
- A *fall* in leptin levels increases NPY and AgRP in the arcuate nucleus and MCH and orexin in the lateral hypothalamic area. These orexigenic peptides act on the brain to stimulate feeding behavior and decrease metabolism.

THE SHORT-TERM REGULATION OF FEEDING BEHAVIOR

Regulation of the tendency to seek and consume food by the body's levels of leptin is very important, but it is not the whole story. Even setting aside social and cultural factors (such as a mother's command: "Eat!"), we all know that the motivation to eat depends on how long it has been since the last meal and how much we ate then. Moreover, the motivation to continue eating once a meal starts depends on how much and what type of food has already been eaten. These are examples of what we are calling the short-term regulation of feeding behavior.

A useful way to think about this regulatory process is to imagine that the

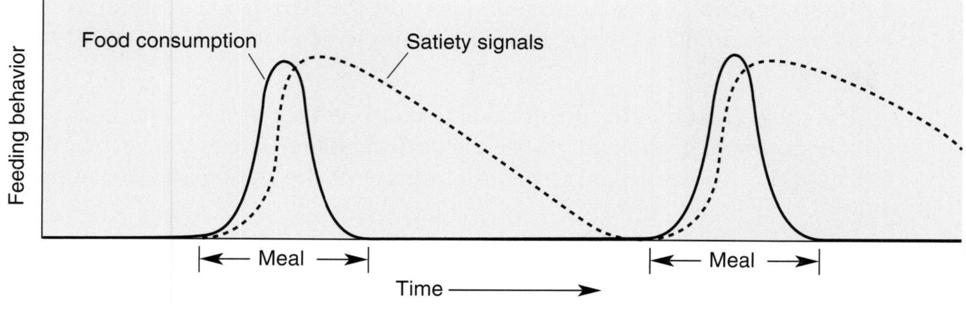

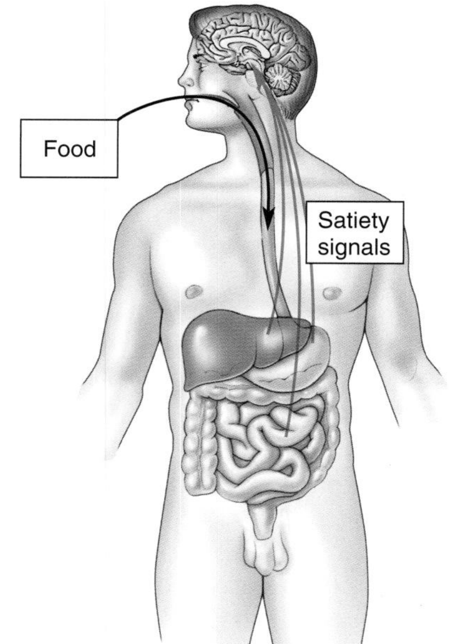

Figure 16.11
Hypothetical model for the short-term regulation of feeding behavior. This graph shows a possible means of regulating food consumption by satiety signals, which rise in response to feeding. When satiety signals are high, food consumption is inhibited. When the satiety signals fall to zero, the inhibition is eliminated, and food consumption ensues.

drive to eat, which may vary rather slowly with the rise and fall of leptin, is inhibited by **satiety signals** that occur when we eat and begin the digestive process (i.e., the prandial period). These satiety signals both terminate the meal and inhibit feeding for some time afterward. During this postabsorptive (fasting) period, the satiety signals slowly dissipate until the drive to eat again takes over (Figure 16.11). Let's use this model to explore the biological basis for the short-term regulation of feeding behavior.

Eating, Digestion, and Satiety

You have awakened in the morning after a long night's slumber. You come to the kitchen to find pancakes cooking on the stove; when they are ready, you enthusiastically eat them until you're satiated. Your body's reactions during this process can be divided into three phases: cephalic, gastric, and substrate (also called the intestinal phase).

1. *Cephalic phase*. The sight and the smell of the pancakes trigger a number of physiological processes that anticipate the arrival of breakfast. The parasympathetic and enteric divisions of the ANS are activated, causing the secretion of saliva into your mouth and digestive juices into your stomach.
2. *Gastric phase*. These responses grow much more intense when you start chewing, swallowing, and filling your stomach with food.

3. *Substrate phase*. As your stomach fills and the partially digested pancakes move into your intestines, nutrients begin to be absorbed into your bloodstream.

The meal ends by the concerted actions of several satiety signals, three of which are considered next: gastric distension, the release of the gastrointestinal peptide cholecystokinin, and the release of the pancreatic hormone insulin.

Gastric Distension. We all know what it is like to feel "full" after a big meal. As you might expect, stretching of the stomach wall is a powerful satiety signal. The stomach wall is richly innervated by mechanosensory axons, and most of these ascend to the brain via the **vagus nerve.** Recall from the Chapter 7 Appendix that the vagus (cranial nerve X) contains a mixture of sensory and motor axons, originates in the medulla, and meanders through much of the body cavity (*vagus* is from the Latin for "wandering"). The vagal sensory axons activate neurons in the **nucleus of the solitary tract** in the medulla. These signals inhibit feeding behavior.

You may have noticed that the nucleus of the solitary tract has been mentioned several times in various contexts. The gustatory nucleus, which receives direct sensory input from the taste buds (Chapter 8), is actually a subdivision of the nucleus of the solitary tract. The nucleus of the solitary tract is also an important center in the control of the ANS (Chapter 15). Now we find that the same nucleus receives visceral sensory input from the vagus nerve. It is easy to see how a nucleus with such widespread connections could serve as an important integration center in the control of feeding and metabolism. As you know, satiety induced by a full stomach can be delayed quite a while if what you are eating is tasty enough.

Cholecystokinin. In the 1970s, researchers discovered that administration of the peptide **cholecystokinin (CCK)** reduces frequency of eating and size of meals without sickening the experimental animals. CCK is present in some of the cells that line the intestines and in some of the neurons of the enteric nervous system. It is released in response to stimulation of the intestines by certain types of food, especially fatty ones. The major action of CCK as a satiety peptide is exerted on the vagal sensory axons. CCK acts synergistically with gastric distension to inhibit feeding behavior (Figure 16.12). Curiously, CCK, like many other gastrointestinal peptides, is also contained within selected populations of neurons within the CNS.

Insulin. Insulin is a vitally important hormone, released into the bloodstream by the β cells of the pancreas (Box 16.2). Although glucose is always readily transported into neurons, *glucose transport into the other cells of the body requires insulin*. This means that insulin is important for anabolic metabolism, during which glucose is transported into liver, skeletal muscle, and adipose cells for storage. Insulin is also necessary for catabolic metabolism, during which the glucose liberated from storage sites is taken up as fuel by the other cells of the body. Thus, the level of glucose in the blood is tightly regulated by the level of insulin: Blood glucose levels are elevated when insulin levels are reduced; blood glucose levels fall when insulin levels rise.

Insulin release by the pancreas is controlled in a number of ways (Figure 16.13). Consider your pancake breakfast. During the cephalic phase, when you are anticipating food, the parasympathetic innervation of the pancreas (delivered by the vagus nerve) stimulates the β cells to release insulin. In response, blood glucose levels fall slightly, and this change, detected by the neurons of the brain, increases your drive to eat, in part by activation of the

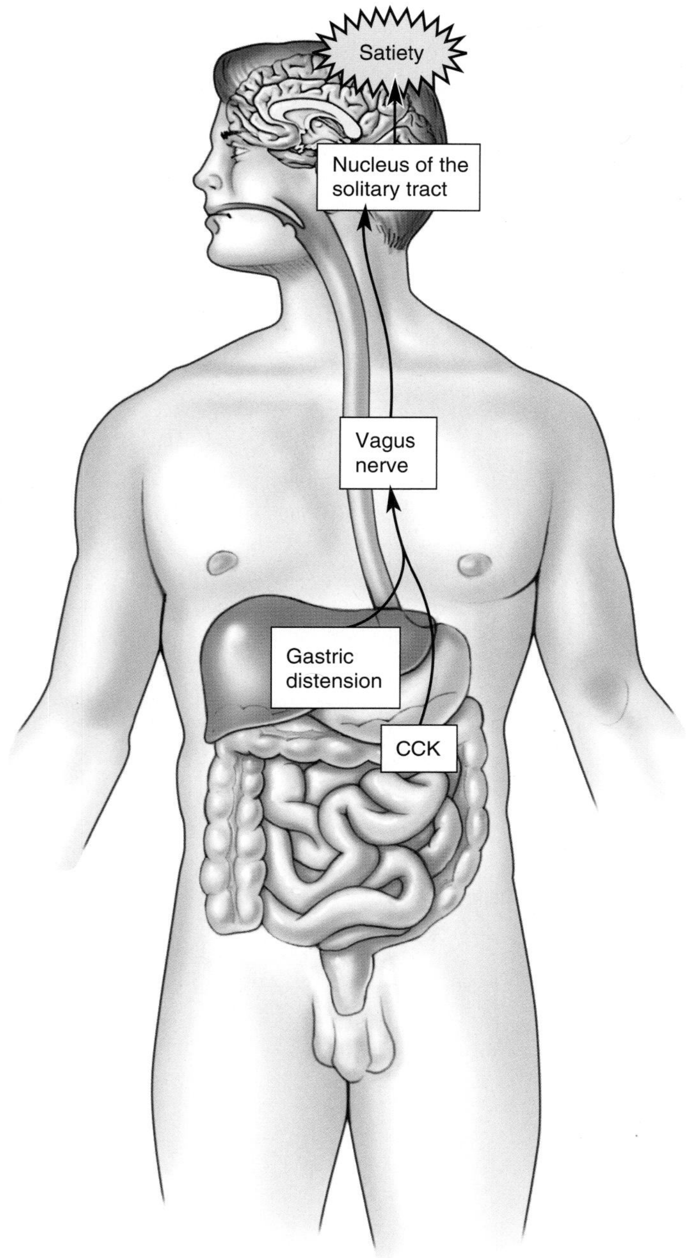

Figure 16.12
Synergistic action of gastric distension and CCK on feeding behavior. Both signals converge on axons in the vagus nerve that trigger satiety.

NPY neurons of the arcuate nucleus. During the gastric phase, when food enters your stomach, insulin secretion is stimulated further by gastrointestinal hormones, such as CCK. Insulin release is maximal when the food is finally absorbed in the intestines, and blood glucose levels rise, during the substrate phase. Indeed, the primary stimulus for insulin release is increased blood glucose levels. This rise in insulin, coupled with the elevated blood glucose levels, is a satiety signal; it causes you to stop eating.

In contrast to the other satiety signals we've discussed, which communicate with the brain mainly via the vagus nerve, bloodborne insulin acts to inhibit feeding behavior by acting directly on the arcuate and ventromedial nuclei of the hypothalamus. It appears that insulin acts in much the same way as leptin to regulate feeding behavior.

Box 16.2

OF SPECIAL INTEREST

Diabetes Mellitus and Insulin Shock

Insulin, released by the β cells of the pancreas, plays a pivotal role in maintaining energy balance. After a meal, glucose levels in the blood rise. To be used by the cells of the body, glucose must be shuttled across the plasma membrane by specialized proteins called glucose transporters. In all cells other than neurons, the insertion of glucose transporters into the membrane occurs when insulin binds to cell surface insulin receptors. Thus, for these cells to use or store the glucose, a rise in blood insulin levels must accompany the rise in blood glucose levels. In the clinical condition known as *diabetes mellitus*, defects in insulin production and release or in the cellular response to insulin prevent the normal reaction to elevated glucose. The consequence is elevated blood sugar levels (hyperglycemia), because the glucose absorbed from the intestines cannot be taken up by the cells of the body (other than neurons). The excess glucose passes to the urine, making it sweet. Indeed, the name *diabetes mellitus* is from the Latin for "siphoning honey."

An effective treatment for many types of diabetes mellitus is hypodermic injections of insulin. This treatment is not without risks, however. An overdose of insulin causes blood glucose levels to plummet (hypoglycemia), starving the neurons of the brain. The resulting condition, *insulin shock*, is characterized by sweating, tremor, anxiety, dizziness, and double vision. If it is not corrected promptly, these early signs are followed by delirium, convulsions, and loss of consciousness. The sudden neurological response to hypoglycemia illustrates how vital energy balance is for the normal functioning of the brain.

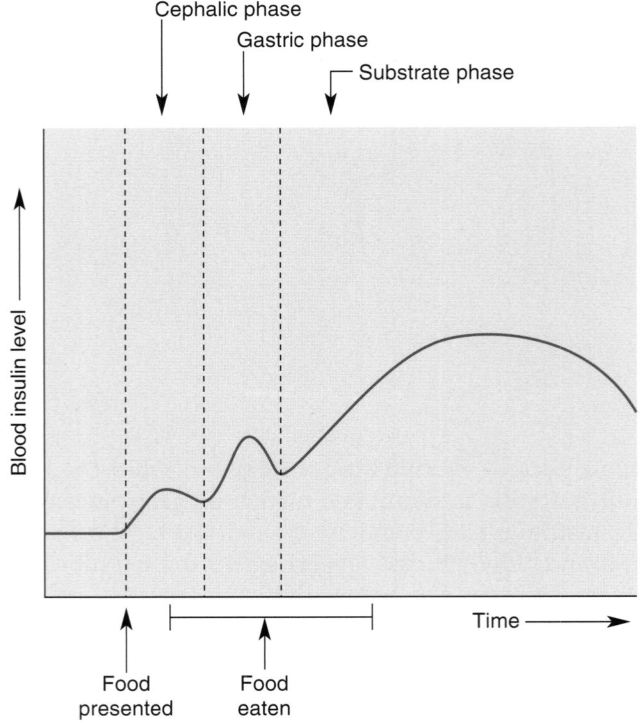

Figure 16.13
Changes in blood insulin levels before, during, and after a meal. (Source: Adapted from Woods and Stricker, 1999, p. 1094.)

WHY DO WE EAT?

We have talked about the signals that motivate feeding behavior, but we still have not discussed what that means in psychological terms. Obviously, we eat because we *like* food. This aspect of motivation is *hedonic*: It feels good, so we do it. We derive pleasure from the taste, smell, sight, and feel of the food and from the act of eating. We also eat because we are hungry and *want* food. This aspect of motivation can be considered as a *drive reduction*—satisfying a craving. A reasonable assumption is that "liking" and "wanting" are two aspects of a unified process; after all, we typically crave food that we like. Research on humans and animals suggests, however, that liking and wanting are mediated by separate circuits in the brain.

The Role of Dopamine in Motivation

As mentioned previously, electrical stimulation of the lateral hypothalamus triggers feeding behavior. To some extent, the feeding results from activation of the orexigenic neurons of the lateral hypothalamic area. Electrical stimulation also activates the dopaminergic axons of the *mesocorticolimbic dopamine system*, introduced in Chapter 15 (Figure 16.14). Strong evidence indicates that this dopaminergic system plays an important role in motivating behaviors, including feeding. For example, if the dopamine neurons are destroyed or dopamine receptors are blocked with drugs, electrical stimulation is much less effective in initiating feeding behavior.

For many years, this dopamine projection, from the ventral tegmental area to the basal forebrain area, was believed to serve hedonic reward—in other words, pleasure. The evidence behind this concept will be presented in more detail in Chapter 18. In the case of feeding, it was believed that dopamine was released in response to palatable foods, making the sensation pleasurable. Animals were motivated to seek palatable food for the hedonic reward—a squirt of dopamine in the forebrain.

This simple idea has been challenged in the past few years. Destruction of the dopamine axons passing through the lateral hypothalamus fails to reduce the hedonic responses to food, even though animals stop eating. If a tasty morsel is placed on the tongue of a rat that has sustained such a lesion, the animal will still behave as if the food evokes a pleasurable sensation (the rat equivalent of lip smacking), and the morsel will be consumed. The dopamine-depleted animal behaves as though it *likes* food but does not *want*

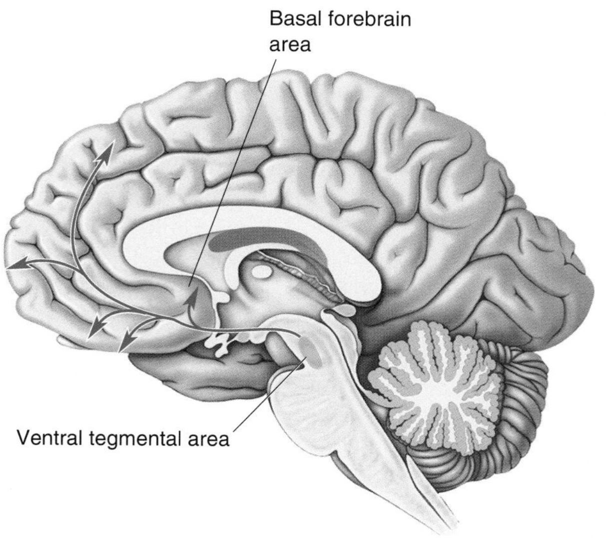

Figure 16.14
Mesocorticolimbic dopamine system. Animals are motivated to behave in ways that stimulate the release of dopamine in the basal forebrain area.

food. The animal apparently lacks the motivation to seek food, even though it seems to enjoy it when it is available. Conversely, stimulation of the dopamine axons in the lateral hypothalamus of normal rats appears to produce a craving for food without increasing the food's hedonic affect. Not surprisingly, recent research on the cravings associated with addiction (to drugs and alcohol, even to chocolate) has focused on the role of this dopaminergic pathway (Box 16.3).

Serotonin, Food, and Mood

Mood and food are connected. Consider how grouchy you are when you're on a restricted diet or the sense of well-being that accompanies a whiff and a bite of a freshly baked chocolate chip cookie. As mentioned in Chapter 15, one system in the brain involved in the control of mood uses serotonin as a neurotransmitter. Serotonin provides one of the links between food and mood.

Measurements of serotonin in the hypothalamus reveal that levels are low during the postabsorptive period, rise in anticipation of food, and spike during a meal, especially in response to carbohydrates (Figure 16.15). Serotonin is derived from the dietary amino acid tryptophan, and tryptophan levels in the blood vary with the amount of carbohydrate in the diet. The rise in blood tryptophan and brain serotonin is one likely explanation for the mood-elevating effects of a chocolate chip cookie. This effect of "carbos" on mood is particularly evident during periods of stress—possibly explaining the food-seeking behavior and weight gain of the typical first-year college student.

Drugs that elevate serotonin levels in the brain are powerful appetite suppressants. One of these drugs is dexfenfluramine (trade name Redux), which was used successfully as a treatment for human obesity. Unfortunately, the drug had toxic side effects that led to its withdrawal from the market in 1997.

Abnormalities in brain serotonin regulation are believed to be one factor that contributes to eating disorders. The defining characteristic of **anorexia nervosa** is the voluntary maintenance of body weight at an abnormally low level, while **bulimia nervosa** is characterized by frequent eating binges, often compensated for by forced vomiting. These disorders are also commonly accompanied by *depression*, a severe disturbance of mood that has been linked to lowered brain serotonin levels (we will discuss mood disorders in Chapter 21). The serotonin connection is clearest in the case of bulimia. In addition to depressing mood, lowered serotonin levels reduce satiety. Indeed, antidepressant drugs that act to elevate brain serotonin levels, such as fluoxetine (Prozac), are also an effective treatment for the majority of bulimia nervosa patients.

OTHER MOTIVATED BEHAVIORS

We have used eating and the regulation of energy balance to give you a fairly detailed picture of the brain mechanisms that incite behavior. The systems involved in motivating several other behaviors that are basic for survival have also been intensively studied. Although we will not cover these other systems in depth, a quick overview will show that the basic principles are the same as those for eating. We will see that the transduction of physiological stimuli in the blood occurs in specialized regions of the hypothalamus, that humoral and visceromotor responses are initiated by activation of the periventricular and medial hypothalamus, and that behavioral action depends on the lateral hypothalamus.

Box 16.3

OF SPECIAL INTEREST

Dopamine and Addiction

What do the drugs heroin, nicotine, and cocaine have in common? They act on different neurotransmitter systems in the brain—heroin on the opiate system, nicotine on the cholinergic system, and cocaine on the dopaminergic and noradrenergic systems—and they produce different psychoactive effects. However, all three drugs are highly addictive. This common quality is explained by the fact that they all act on the brain circuitry that motivates behavior—in this case, drug-seeking behavior. We can learn much about the brain mechanisms of motivation by studying drug addiction, and we can learn much about drug addiction by studying the brain mechanisms of motivation.

Rats, like humans, will self-administer drugs and will develop clear signs of drug dependence. Studies using microinfusions of drugs directly into the brain have mapped out the sites where the drugs cause addiction. In the case of heroin and nicotine, the key site of action is the ventral tegmental area, home of the dopamine neurons that project axons through the lateral hypothalamus to the forebrain. These dopaminergic neurons have both opiate and nicotinic acetylcholine receptors. In the case of cocaine, a key site of action is the nucleus accumbens, one of the major targets of the ascending dopaminergic axons in the forebrain (Figure A). Recall from Chapter 15 that cocaine prolongs the actions of dopamine at its receptors. Thus, these three drugs either stimulate dopamine release (heroin, nicotine) or enhance dopamine actions (cocaine) in the nucleus accumbens.

The exact role of dopamine in motivating behavior continues to be debated. Much evidence suggests however, that animals are motivated to perform behaviors that stimulate dopamine release in the nucleus accumbens and related structures. Behaviors associated with the delivery of drugs that stimulate dopamine release are therefore strongly reinforced. Chronic overstimulation of this pathway, however, causes a homeostatic response: The dopamine "reward" system is downregulated. This adaptation leads to the phenomenon of *drug tolerance*—it takes more and more of the drug to get the desired (or required) effect. Indeed, drug discontinuation in addicted animals is accompanied by a marked decrease in dopamine release and function in the nucleus accumbens. And, of course, one withdrawal symptom is the powerful craving for the discontinued drug.

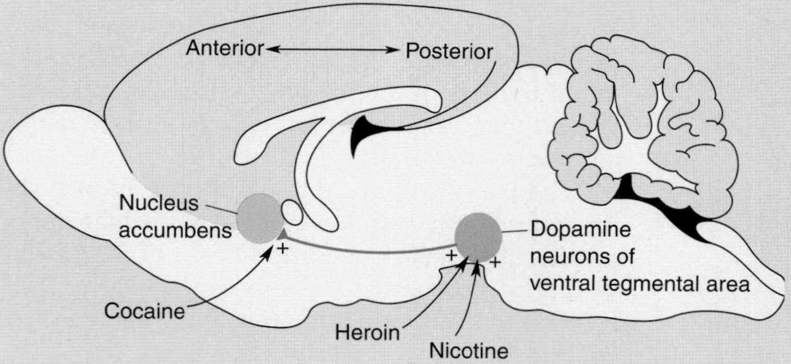

Figure A
Addictive drugs act on the dopaminergic pathway from the ventral tegmental area to the nucleus accumbens. (Source: Adapted from Wise, 1996, p. 248, Fig. 1.)

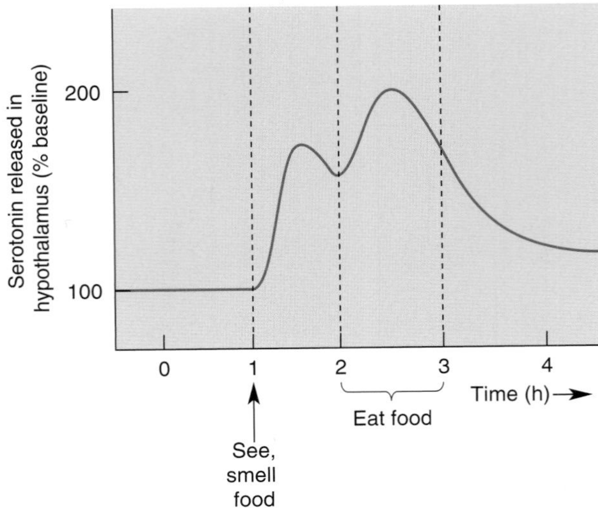

Figure 16.15
Changes in hypothalamic serotonin levels before and during a meal. The mood-elevating effects of eating are believed to be related to the release of serotonin in the brain. (Source: Adapted from Schwartz et al., 1990.)

Drinking

Two different physiological signals stimulate drinking behavior. As mentioned in Chapter 15, one of these is a decrease in blood volume, or *hypovolemia*. The other is an increase in the concentration of dissolved substances (solutes) in the blood, or *hypertonicity*. These two stimuli trigger thirst by different mechanisms.

Thirst triggered by hypovolemia is called **volumetric thirst**. In Chapter 15, we used the example of decreased blood volume to illustrate when and how vasopressin is released in the posterior pituitary by the magnocellular neurosecretory cells. Vasopressin (also called antidiuretic hormone, or ADH) acts directly on the kidneys to increase water retention and inhibit urine production. The release of vasopressin associated with volumetric thirst is triggered by two types of stimuli (Figure 16.16). First, a rise in blood levels of angiotensin II occurs in response to reduced blood flow to the kidneys (see Figure 15.5). The circulating angiotensin II acts on the neurons of the subfornical organ in the telencephalon, which in turn directly stimulate the magnocellular neurosecretory cells of the hypothalamus to release vasopressin. Second, mechanoreceptors in the walls of the major blood vessels and heart signal the loss of blood pressure that accompanies a loss of blood volume. These signals make their way to the hypothalamus via the vagus nerve and the nucleus of the solitary tract.

In addition to this humoral response, reduced blood volume (1) stimulates the sympathetic division of the ANS, which helps correct the drop in blood pressure by constricting arterioles, and (2) powerfully motivates animals to seek and consume water. Not surprisingly, the lateral hypothalamus has been implicated in inciting the behavioral response, although the details of this process are still poorly understood.

The other stimulus for thirst, hypertonicity of the blood, is sensed by neurons in yet another specialized region of the telencephalon lacking a blood-brain barrier, the **vascular organ of the lamina terminalis**, or **OVLT** (Figure 16.17). When the blood becomes hypertonic, water leaves cells by the process of osmosis. This loss of water is transduced by the OVLT neurons into a change in action potential firing frequency. The OVLT neurons (1) directly excite the magnocellular neurosecretory cells that secrete vasopressin and (2) stimulate **osmometric thirst**, the motivation to drink water when dehydrated. Lesions of the OVLT completely prevent the behavioral and humoral responses to dehydration (but not the responses to loss of blood volume).

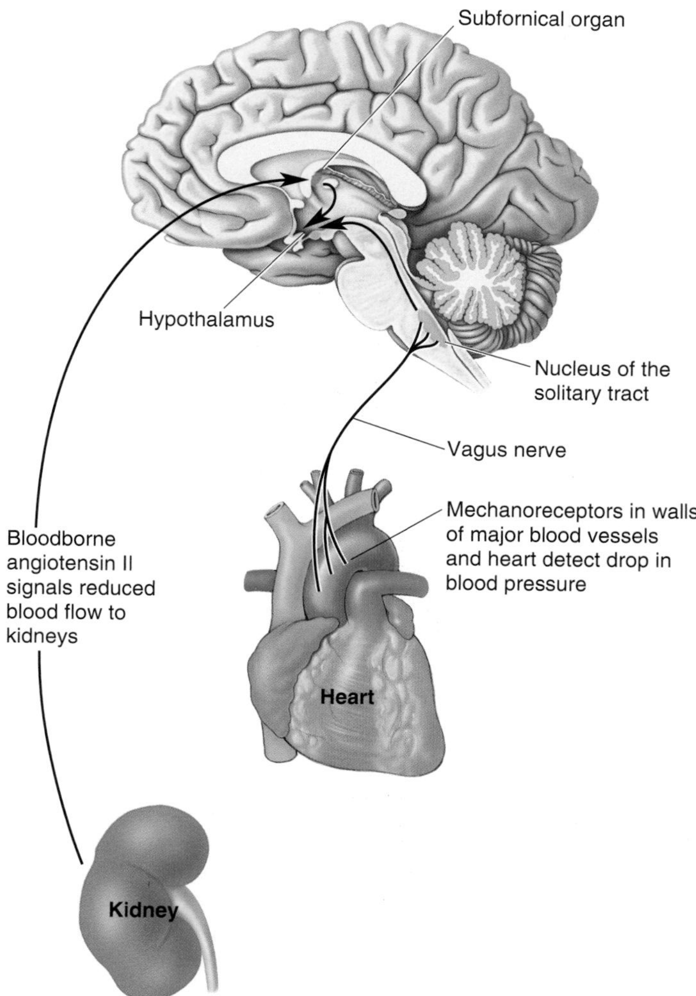

Figure 16.16
Pathways triggering volumetric thirst. Hypovolemia is detected in two ways. First, angiotensin II, released into the bloodstream in response to decreased blood flow to the kidneys, activates neurons in the subfornical organ. Second, mechanosensory axons in the vagus nerve, detecting a drop in blood pressure, activate neurons in the nucleus of the solitary tract. The subfornical organ and nucleus of the solitary tract relay this information to the hypothalamus, which orchestrates the coordinated response to reduced blood volume.

The motivation to drink, the secretion of vasopressin from the hypothalamus, and the retention of water by the kidneys normally go hand in hand. Selective loss of the vasopressin-secreting neurons of the hypothalamus, however, produces a curious condition called *diabetes insipidus,* in which the body works against the brain. As a consequence of the loss of vasopressin, the kidneys pass too much water from the blood to the urine. The resulting dehydration stimulates the strong motivation to drink water, but the water absorbed from the intestines passes quickly through the kidneys into the urine. Thus, diabetes insipidus is characterized by extreme thirst coupled with frequent excretion of a large amount of pale, watery urine. This condition can be treated by replacing the missing vasopressin.

Temperature Regulation

You are hot; you seek a cool place. You are cold; you seek warmth. We are all motivated to interact with our environment in such a way as to keep our bodies within a narrow range of temperatures. The need for such regulation is clear: The cells of the body are fine-tuned for a constant temperature, 37°C, and deviations from this temperature interfere with cellular functions.

Neurons that change their firing rate in response to small changes in temperature are found throughout the brain and spinal cord. The most important neurons for temperature homeostasis, however, are found clustered in the

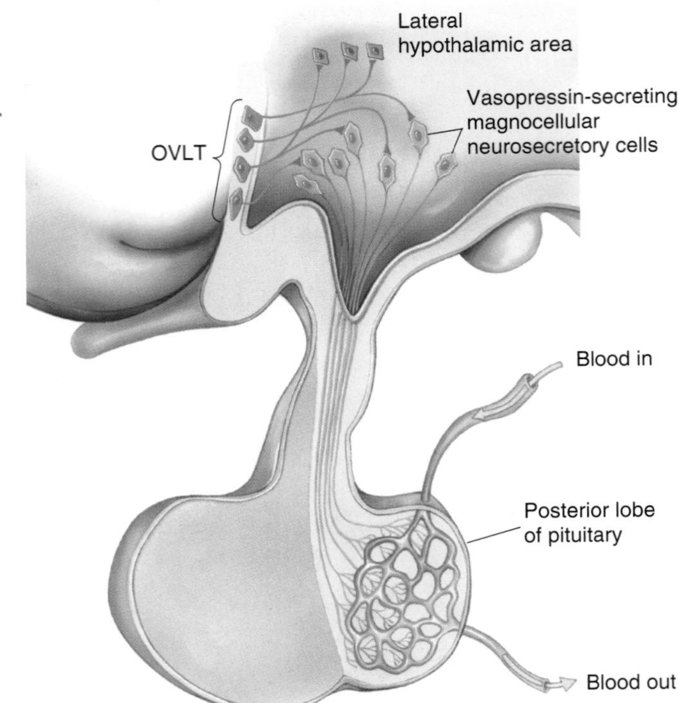

Figure 16.17
Hypothalamic response to dehydration. Blood becomes hypertonic when it loses water. Blood hypertonicity is sensed by neurons of the vascular organ of the lamina terminalis (OVLT). The OVLT activates magnocellular neurosecretory cells and cells in the lateral hypothalamus. The neurosecretory cells secrete vasopressin into the blood, and the neurons of the lateral hypothalamus trigger osmometric thirst.

anterior hypothalamus. These cells transduce small changes in blood temperature into changes in their firing rate. Humoral and visceromotor responses are subsequently initiated by neurons in the *medial preoptic area* of the hypothalamus; somatic motor (behavioral) responses are initiated by the neurons of the lateral hypothalamic area. Lesions in these different regions can selectively abolish different components of the integrated response.

A fall in temperature is detected by cold-sensitive neurons of the anterior hypothalamus. In response, TSH is released by the anterior pituitary. TSH stimulates the release of the hormone thyroxin from the thyroid gland, which causes a widespread increase in cellular metabolism. The visceromotor response is constricted blood vessels in the skin and piloerection (goose pimples). An involuntary somatic motor response is shivering, and of course, the other somatic response is to seek warmth.

Table 16.2 Hypothalamic Responses to Stimuli That Motivate Behavior

BLOODBORNE STIMULUS	SITE OF TRANSDUCTION	HUMORAL RESPONSE	VISCEROMOTOR RESPONSE	SOMATIC MOTOR RESPONSE
Eating signals				
↓ Leptin	Arcuate nucleus	↓ ACTH ↓ TSH	↑ Parasympathetic activity	Feeding
↓ Insulin	Arcuate nucleus	↓ ACTH ↓ TSH	↑ Parasympathetic activity	Feeding
Drinking signals				
↑ Angiotensin II	Subfornical organ	↑ Vasopressin	↑ Sympathetic activity	Drinking
↑ Blood tonicity	OVLT	↑ Vasopressin		Drinking
Thermal signals				
↑ Temperature	Medial preoptic area	↓ TSH	↑ Parasympathetic activity	Panting; seeking cold
↓ Temperature	Medial preoptic area	↑ TSH	↑ Sympathetic activity	Shivering; seeking warmth

A rise in temperature is detected by warm-sensitive neurons of the anterior hypothalamus. In response, metabolism is slowed by reducing TSH release, blood is shunted toward the body periphery to dissipate heat, and behavior is initiated to seek shade. In some species, an involuntary motor response is panting; in humans, it is sweating.

The strong parallels between the hypothalamic control of energy balance, water balance, and temperature should now be clear. In each case, specialized neurons detect variations in the regulated parameter. The hypothalamus orchestrates responses to these challenges, which always include adjustments in physiology and the stimulation of different types of behavior. Table 16.2 summarizes some of the hypothalamic responses we have discussed in this chapter.

CONCLUDING REMARKS

In the motor system chapters of Part II, we addressed the "how" questions of behavior: How do muscles contract? How is movement initiated? How are the actions of our different muscles coordinated? The discussion of motivation, however, asks a different question: Why? Why do we eat when our energy reserves become depleted? Why do we drink when we are dehydrated? Why do we seek warmth when our blood temperature falls?

Concrete answers to both the "how" and the "why" of behavior have been obtained in the body's periphery. Muscles contract because of the release of acetylcholine at the neuromuscular junction. We drink water because we are thirsty, and we are thirsty when angiotensin II levels rise in response to decreased blood flow to the kidneys. Answers are still lacking, however, where "how" and "why" come together—in the brain. In this chapter, we chose to focus on feeding behavior, in part because the trail leads farthest into the brain. The discovery of orexigenic peptide neurons in the lateral hypothalamus that respond to changes in leptin levels was a major breakthrough. We can at least begin to frame the question about how these neurons might act elsewhere in the brain to initiate feeding behavior. The next few years should be an exciting time of discovery. Advances in research will have a significant influence on how we interpret our own behavior and the behavior of those around us.

After reading about the bloodborne signals that motivate eating and drinking, you might begin to feel that indeed we are ruled by our hormones. While bloodborne signals do have a strong effect on the probability of specific types of behavior, however, we are not their slaves. Clearly, one of the great triumphs of human evolution is the ability to exert cognitive, cortical control over our more primitive instincts.

KEY TERMS

The Hypothalamus, Homeostasis, and Motivated Behavior
motivated behavior (p. 524)

The Long-Term Regulation of Feeding Behavior
anabolism (p. 524)
catabolism (p. 524)
obesity (p. 524)
starvation (p. 524)
lipostatic hypothesis (p. 527)
leptin (p. 527)
anorexia (p. 527)
lateral hypothalamic syndrome (p. 527)
ventromedial hypothalamic syndrome (p. 527)
arcuate nucleus (p. 528)
paraventricular nucleus (p. 529)
anorectic peptide (p. 530)
orexigenic peptide (p. 530)
lateral hypothalamic area (p. 531)

The Short-Term Regulation of Feeding Behavior
satiety signal (p. 535)
vagus nerve (p. 536)
nucleus of the solitary tract (p. 536)
cholecystokinin (CCK) (p. 536)
insulin (p. 536)

Why Do We Eat?
anorexia nervosa (p. 540)
bulimia nervosa (p. 540)

Other Motivated Behaviors
volumetric thirst (p. 542)
vascular organ of the lamina terminalis (OVLT) (p. 542)
osmometric thirst (p. 542)

REVIEW QUESTIONS

1. A surgical approach to reducing excessive body fat is liposuction, the removal of adipose tissue. Over time, however, body adiposity usually returns to precisely the same value as before surgery. Why does liposuction not work?
2. Bilateral lesions of the lateral hypothalamus lead to reduced feeding behavior. Name three types of neuron, distinguished by their neurotransmitter molecules, that contribute to this syndrome.
3. What neurotransmitter agonists and antagonists would you design to treat obesity? Consider drugs that could act on the neurons of the brain, as well as drugs that could act on the ANS.
4. Name one way the axons of the vagus nerve may stimulate feeding behavior and one way they may inhibit it.
5. What does it mean, in neural terms, to be addicted to chocolate? How could chocolate elevate mood?
6. Compare and contrast the functions of these three regions of the hypothalamus: the arcuate nucleus, the subfornical organ, and the vascular organ of the lamina terminalis.

CHAPTER

17

Sex and the Brain

INTRODUCTION

Without sex there is no human reproduction. And, without offspring, no species can survive. Those are the simple facts of life, and millions of years of evolution have given us a nervous system well-designed for survival. The drive to reproduce can be compared to the powerful motivation to eat or drink, which we discussed in Chapter 16. For survival's sake, life-maintaining functions, such as reproduction and eating, are not left entirely to the whims of conscious thought. Instead, they are regulated by subcortical structures, and thoughtful conscious control is provided by the cerebral cortex.

In this chapter, we will explore what is known about sex and the brain. Our goal is not a discussion of the birds and the bees—we will assume you have picked up the basics about human sexual behavior from your parents, teachers, friends, or cable TV. Instead, we will look at the neural machinery that makes reproduction possible. Sexual and reproductive behaviors are clearly different in men and women, but just how different are the brains of the two genders? Ultimately, the origin of *most* distinctions between males and females is the chromosomes provided by their parents. Under the guidance of certain genes, the human body produces a small number of sex hormones that have powerful effects on both the sexual differentiation of the body and adult sexual physiology and behavior. The reproductive organs (the ovaries and testes), which secrete sex hormones, are outside the nervous system, but they are activated by the brain. Recall from Chapter 15 that the hypothalamus controls the release of diverse hormones from the anterior pituitary. In the case of reproductive function, the hormones released by the anterior pituitary regulate secretions from the ovaries and testes.

Sex hormones have obvious effects on the human body, but they also influence the brain. At the scale of individual neurons, sex hormones have been found to influence the structure of dendrites. At a larger scale, structural differences between male and female brains have been reported over the past 100 years. We will look at these differences and try to distinguish those that are reliable from those that are not. Also, numerous studies indicate that men or women perform better on certain cognitive tasks. Are these gender differences significant? Are they the result of small brain dissimilarities, cultural upbringing, or some other factor?

Another point to consider is what it means to be male or female. Is gender determined by genetics, anatomy, or behavior? The answer is not simple; there are examples of gender identities that do not correlate with biological and behavioral factors. And what about sexual orientation? Is an attraction to members of the opposite or same sex determined by experiences in childhood or the structure of the brain? These are challenging questions that address how we perceive ourselves and others. We will examine the extent to which we can answer such questions by looking at the anatomy and physiology of the nervous system.

SEX AND GENDER

The words *sex* and *gender* both refer to the state of being male or female and are often used synonymously. Gender is medically defined by the many biological characteristics and qualities that distinguish between male and female, including the nature of the sex chromosomes, and the anatomy of the gonads and other genital organs. There are also behavioral and cultural implications of gender, and these start at birth. When a child is born, we ask the parents, "Is the baby a boy or a girl?" Regardless of one's ideas about equality of the sexes, the answer to this question leads to innumerable assumptions about the life experiences the child will have. In the case of adults, we don't inquire about someone's gender, as it is usually obvious from their appearance. However, identifying someone as female or male still involves many assumptions, because our ideas about gender involve a package of biological and behavioral traits. Gender-specific behaviors result from complex interactions among self-assessment, societal expectations, genetics, and hormones. These behaviors are related to **gender identity**—our perception of our own gender. In this section, we will discuss some of the genetic and developmental origins of gender.

The Genetics of Sex

Within the nucleus of every human cell, DNA provides a person's genetic blueprint—all the information needed to build an individual. The DNA is organized into 46 chromosomes: 23 from the father and 23 from the mother. Each of us has two versions of chromosome 1, 2, 3, and so on, conventionally numbered in order of decreasing size (Figure 17.1). The only exception to this pair system is the sex chromosomes, X and Y. Thus, it is usually stated that there are 44 autosomes (22 pairs of matching chromosomes) and two sex chromosomes. Females have two X chromosomes, one from each parent. Males have an X chromosome from the mother and a Y chromosome from the father. Therefore, the *female genotype* is denoted XX and the *male genotype* XY. These genotypes specify a person's *genetic sex*. Because the mother contributes an X chromosome to a child of either sex, gender is determined by the X or Y contribution from the father. In some nonhuman animals, such as birds, the mother's contribution determines the gender of the offspring.

The DNA molecules that make up chromosomes are some of the largest molecules known, and they contain genes, the basic units of hereditary information. The piece of DNA comprising a gene provides the unique information needed to construct a particular protein. Current estimates are that humans have roughly 100,000 genes.

As you can see in Figure 17.1, the X chromosome is significantly larger than the Y chromosome. Corresponding to this size difference, it is estimated that the X chromosome contains several thousand genes, whereas the Y chromosome may only contain 15 or so. You might joke that men appear to get genetically shortchanged, and in a sense, you are correct: There are serious medical consequences of the XY genotype. If there is a defective X gene in a

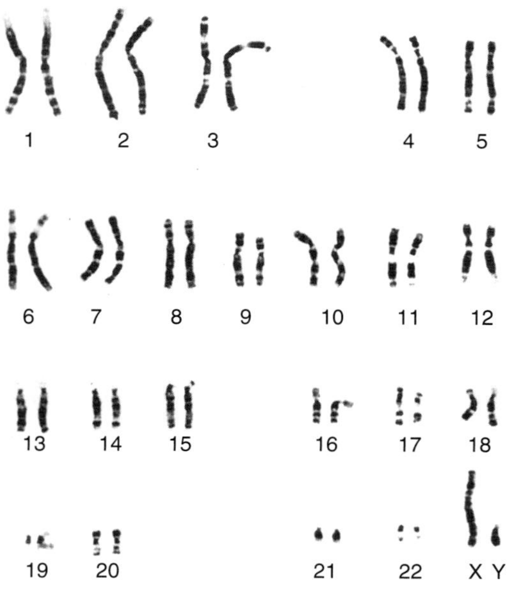

Figure 17.1
Human chromosomes. These 23 pairs of chromosomes are from a man. Notice how much smaller the Y chromosome is than the X chromosome. (Source: Yunis and Chandler, 1977.)

female, there may be no negative consequence if her other X gene is normal. However, any defect in the single X chromosome of a male can lead to a developmental defect. Such a defect is called an *X-linked disease*, and there are many. One example, discussed in Box 9.5, is red-green color blindness, which is relatively common in males. Other X-linked diseases that occur more often in men than women are hemophilia and Duchenne muscular dystrophy.

Compared to the X chromosome, the smaller Y chromosome has few genes and less diverse functions. Most importantly for sexual determination, it contains a gene called the **sex-determining region of the Y chromosome (SRY)**, which is responsible for the production of a protein called *testis-determining-factor* (*TDF*). A human with a Y chromosome and the SRY gene develops as a male, and without it, the person develops as a female. In rare individuals, there are too few or too many sex chromosomes, but gender is always determined by the presence or absence of the Y chromosome. Thus, an X or XXX individual is female and an XXY or XXYY individual is male.

The SRY gene was found to be located on the short arm of the Y chromosome in 1990 by Peter Goodfellow, Robin Lovell-Badge, and their colleagues at the Medical Research Council in London (Figure 17.2). If this bit of the Y chromosome is artificially incorporated into the DNA of an XX mouse, the mouse will develop as a male instead of a female. However, this doesn't mean that SRY is the only gene involved in sex determination, as SRY is known to regulate genes on other chromosomes. Also, male-specific physiology, such as sperm production, relies on other genes on the Y chromosome. Nonetheless, we will see shortly that expression of the SRY gene causes the development of the testes, and the hormones from the testes are largely responsible for making the male fetus develop differently from the female fetus.

Sexual Reproduction

Because it is so obvious that reproduction works by bringing together genetic information from two individuals of opposite gender, most people never question why this system came to be dominant. While sexual reproduction is the most common system and the one we are used to, it is not the only way

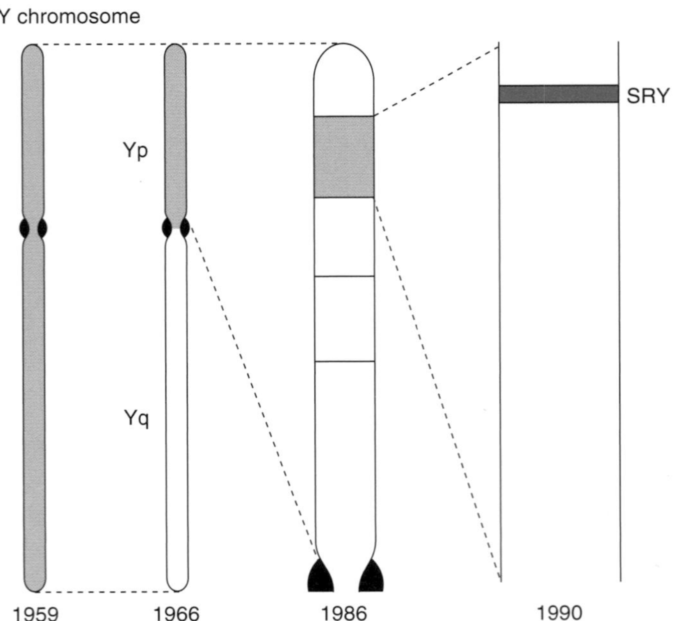

Figure 17.2
The location of the SRY gene on the Y chromosome. In 1959, researchers found that TDF depended on the Y chromosome, and in 1966, the important location was further restricted to the short (p) arm. Research in the 1980s established that TDF is coded by the SRY gene, a small segment near the tip of the short arm of the Y chromosome. (Source: Adapted from McLaren, 1990, p. 216.)

reproduction occurs. It may not surprise you that some bacteria and plants reproduce asexually. However, even more complex animals, including some insects and reptiles, can reproduce asexually by a process called parthenogenesis, in which an unfertilized sex cell grows into a new animal. Although it is not natural, recent reports of cloning mammals such as sheep demonstrate that, with human intervention, reproduction can occur asexually even in more complex animals. However, almost all complex animals use sexual reproduction, and asexual reproduction is limited primarily to simple organisms.

Why has sexual reproduction evolved in so many species? Before answering that question, it is worth noting that asexual reproduction provides for much faster expansion of a species. If every member of a species can reproduce, rather than having a bunch of unproductive males, population numbers increase much faster. Nonetheless, it is thought that sexual reproduction confers significant survival advantages. The asexual approach may be fine when the environment is stable, as each offspring is an exact copy of an adult that must at least have been healthy enough to reproduce. However, suppose there is a change in temperature, water availability, or food supply. If the change proves fatal to the parent, it will also wipe out one or many generations of children, precisely because they are the same as the adult. Similarly, if there is a disadvantageous genetic mutation, perhaps because of exposure to ultraviolet light, the degraded genes would be passed on to all future generations.

The great advantage of sexual reproduction is that it mixes genetic information, giving rise to a highly diverse population. By the chance mixture of genes, some children may be less robust than the parents, but there is the possibility that some will be stronger. A serious genetic mutation in a parent might not be a problem for the children if the other parent has a healthy gene that can take over. An important medical example of this is *cystic fibrosis*, a common incurable genetic disorder. Because of an abnormality in salt transport, cells produce thick mucus that clogs the lungs and can lead to fatal infections. The gene responsible for cystic fibrosis is on chromosome 7, and about 3% of all Americans are carriers because they have a defective gene on one of their copies of chromosome 7. If both parents are carriers, there is a 25% chance that the child will have cystic fibrosis. However, if the child inherits a defective gene from only one parent, he or she will not have the disease.

Sexual Development and Differentiation

Differences between males and females are everywhere, from body size and muscle development to endocrine function. We know it is ultimately the genetic sex of the child that determines its anatomical sex, or gender. But during development, when and how does the fetus differentiate into one sex or the other? How does the genotype of the child lead to the male or female development of the gonads? To answer this question, one must appreciate the unique situation of the gonads during development. Unlike organs such as the lung and liver, the rudimentary cells that develop into the gonads are not committed to a single developmental pathway. During the first 6 weeks of pregnancy, the gonads are in an indifferent stage that can develop into either ovaries or testes. The uncommitted gonads possess two key structures, the *Müllerian duct* and the *Wolffian duct* (Figure 17.3). If the fetus has a Y chromosome and an SRY gene, testosterone is produced, and the Wolffian duct develops into the male internal reproductive system. At the same time, the Müllerian duct is prevented from developing by a hormone called Müllerian-inhibiting factor. Conversely, if there is no Y chromosome and no upsurge of

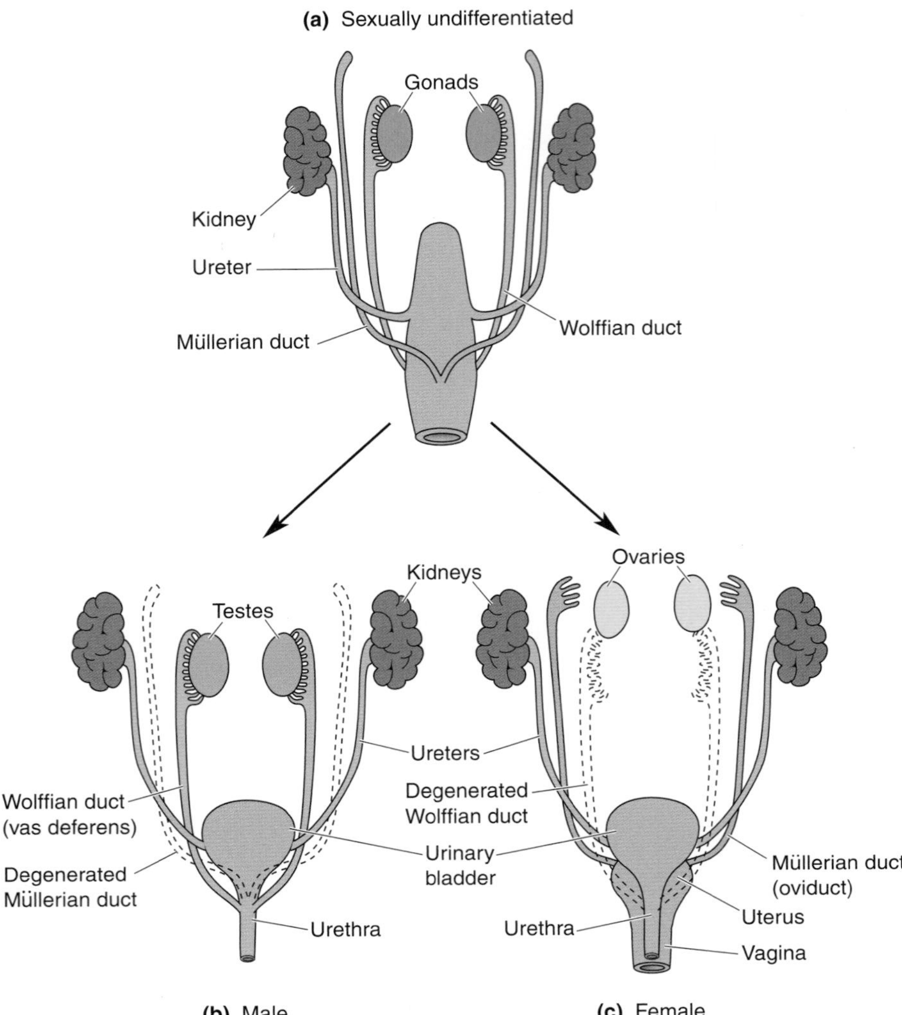

Figure 17.3
Development of the reproductive organs. (a) The undifferentiated urogenital system has both Müllerian and Wolffian ducts. **(b)** If there is an SRY gene, the Wolffian duct develops into the male reproductive organs. **(c)** If there is no SRY gene, the Müllerian duct develops into the female reproductive organs. (Source: Adapted from Gilbert, 1994, p. 759.)

testosterone, the Müllerian duct develops into the female internal reproductive system, and the Wolffian duct degenerates. The external genitals of both males and females develop from the same undifferentiated urogenital structures. This is why it is possible for a person to be born with genitals intermediate in form between those of typical males and females, a condition known as *hermaphroditism*.

THE HORMONAL CONTROL OF SEX

Hormones are chemicals, released into the bloodstream, that regulate physiological processes. The endocrine glands we are primarily interested in are the ovaries and testes, because they release sex hormones, and the pituitary because it regulates this release. The sex hormones are crucial to the development and function of the reproductive system and sexual behavior. The sex hormones are steroids (briefly mentioned in Chapter 15), and some of them are familiar, such as testosterone and estrogen. Steroids are molecules, synthesized from cholesterol, that have four carbon rings. Despite the warnings of the American Heart Association, cholesterol is good for something! Small alterations to the basic cholesterol structure have profound consequences for the effects of hormones. For example, testosterone is the most crucial hormone for male development, but it differs from the important female steroid estradiol in only a few places on the molecule.

The Principal Male and Female Hormones

Steroid sex hormones are often referred to as "male" or "female," but men also have "female" hormones and women also have "male" hormones. The designation reflects the fact that men have higher concentrations of **androgens,** or male hormones, and women have more **estrogens,** or female hormones. For example, *testosterone* is an androgen and *estradiol* is an estrogen. In the series of chemical reactions that lead from cholesterol to sex hormones, one of the principal female hormones, estradiol, is actually synthesized from the male hormone testosterone (Figure 17.4). This reaction takes place with the aid of an enzyme called *aromatase.*

Steroids act differently from other hormones because of their structure. Some hormones, such as vasopressin and oxytocin, are proteins and, as such, they cannot cross the lipid bilayer of a cell membrane. Hence, these hormones act at receptors with extracellular binding sites. In contrast, steroids are fatty and can easily pass through cell membranes and bind to receptors within the cytoplasm, giving them direct access to the nucleus and gene expression. Differences in the concentration of various receptors are responsible for steroid effects localized to different areas of the brain (Figure 17.5).

The testes are primarily responsible for the release of androgens, though small amounts are secreted in the adrenal glands and elsewhere. Testosterone is by far the most abundant androgen, and it is responsible for most masculinizing hormonal effects. Prenatally, elevated testosterone levels are essential for the development of the male reproductive system. Increases in testosterone much later, at puberty, are responsible for secondary sex characteristics, ranging from increased muscular development and facial hair in human males to the lion's mane. Oddly, for those with a genetic predisposition, testosterone is also responsible for baldness in men. Female concentrations of testosterone are roughly 10% of those found in males. Male testosterone levels vary during the course of the day due to innumerable factors, including stress, exertion, and aggression. It is not clear whether an increase in testosterone is a cause or an effect, but it is correlated with social challenges, anger, and conflict. Testosterone levels also rise in anticipation of sex or even fantasizing about it. One study found that testosterone levels increase if a male sports fan watches his team win, but levels decrease in fans of the losing team.

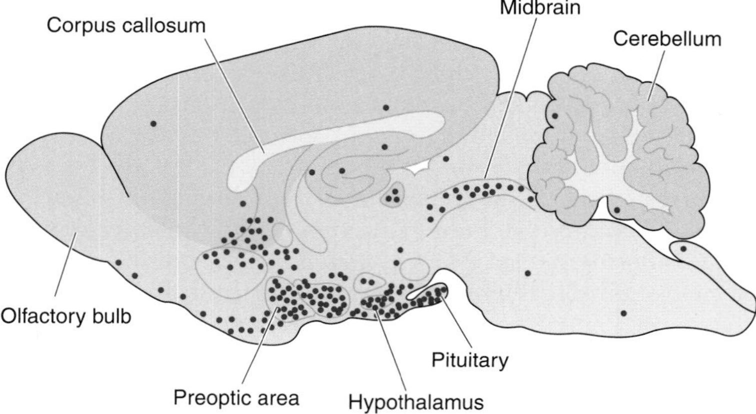

Figure 17.5
The distribution of estradiol receptors in a saggital section of the rat brain. High concentrations of these receptors are found in the pituitary and hypothalamus, including the preoptic area of the anterior hypothalamus. These brain areas are all involved in sexual and reproductive behaviors.

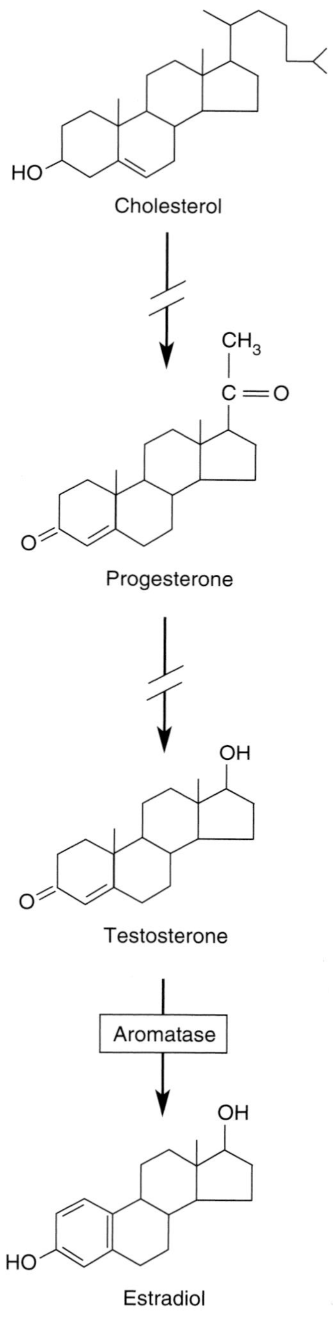

Figure 17.4
Cholesterol and synthesis of the principal sex steroid hormones. Broken arrows indicate where one or more intermediate reactions occurs. The enzyme aromatase directly converts testosterone into estradiol.

The principal female hormones are estradiol and *progesterone*, and they are secreted by the ovaries. As already mentioned, estradiol is an estrogen; progesterone is a member of a second class of female steroid hormones called *progestins*. Estrogen levels are quite low during childhood, but they increase dramatically at puberty and are responsible for the maturation of the female reproductive system and the development of breasts. As in the male, blood concentrations of sex hormones are quite variable in the female. However, whereas in men, fluctuations occur rapidly each day, in women, hormonal levels follow a regular 28-day cycle (discussed later in the chapter).

The Control of Steroid Sex Hormones by the Pituitary and Hypothalamus

The anterior pituitary gland secretes two hormones that are particularly important for normal sexual development and function in both women and men: **luteinizing hormone (LH)** and **follicle-stimulating hormone (FSH)**. These hormones are also called **gonadotropins**. LH and FSH are secreted by specialized cells scattered throughout the anterior pituitary that represent about 10% of the total cell population. Recall from Chapter 15 that the secretion of hormones from the anterior pituitary is under the control of hypophysiotropic hormones released by the hypothalamus. **Gonadotropin-releasing hormone (GnRH)** from the hypothalamus does what the name suggests, causing the release of LH and FSH from the pituitary. GnRH is also referred to as LHRH, for luteinizing hormone-releasing hormone, because it causes a much greater increase in LH than FSH. Neuronal activity in the hypothalamus is influenced by numerous psychological and environmental factors that indirectly affect the secretion of gonadotropins from the anterior pituitary.

In males, LH stimulates the testes to produce testosterone. FSH is involved in the maturation of sperm cells within the testes. Sperm maturation also requires testosterone, meaning that both LH and FSH play key roles in male fertility. Because there is cortical input to the hypothalamus, it is possible for psychological factors to decrease male fertility by inhibiting gonadotropin secretion and sperm production.

The chain of events from hypothalamic input to the secretion of gonadal hormones is illustrated in Figure 17.6. There is also neural input from the retina to the hypothalamus, which is responsible for changes in the release of GnRH based on daily variations in the levels of light. In some nonhuman species, strong seasonal variations in reproductive behavior and gonadotropin secretion occur. Light inhibits the production of the hormone *melatonin* in the pineal gland, increasing gonadotropin secretion because of the inhibitory effect of melatonin on gonadotropin release. By means of this circuit, reproductive activity can be influenced by the length of daylight during the course of the year, and offspring are born seasonally when they have the best chance of survival. In humans, there is also an inverse relationship between gonadotropin release and melatonin levels, but it is not yet known whether melatonin actually modulates reproductive behavior.

Hormonal Cycles and Feedback to the Brain. In females, LH and FSH (secreted by the anterior pituitary gland) cause the secretion of estrogens from the ovaries. In the absence of gonadotropins, the ovaries are inactive, which is the case throughout childhood. Cyclic variations in LH and FSH levels in adult females are responsible for periodic changes in the ovaries, and the timing and duration of LH and FSH secretion determine the nature of the reproductive cycle, or **menstrual cycle**.

The menstrual cycle illustrates the complex interplay between the brain

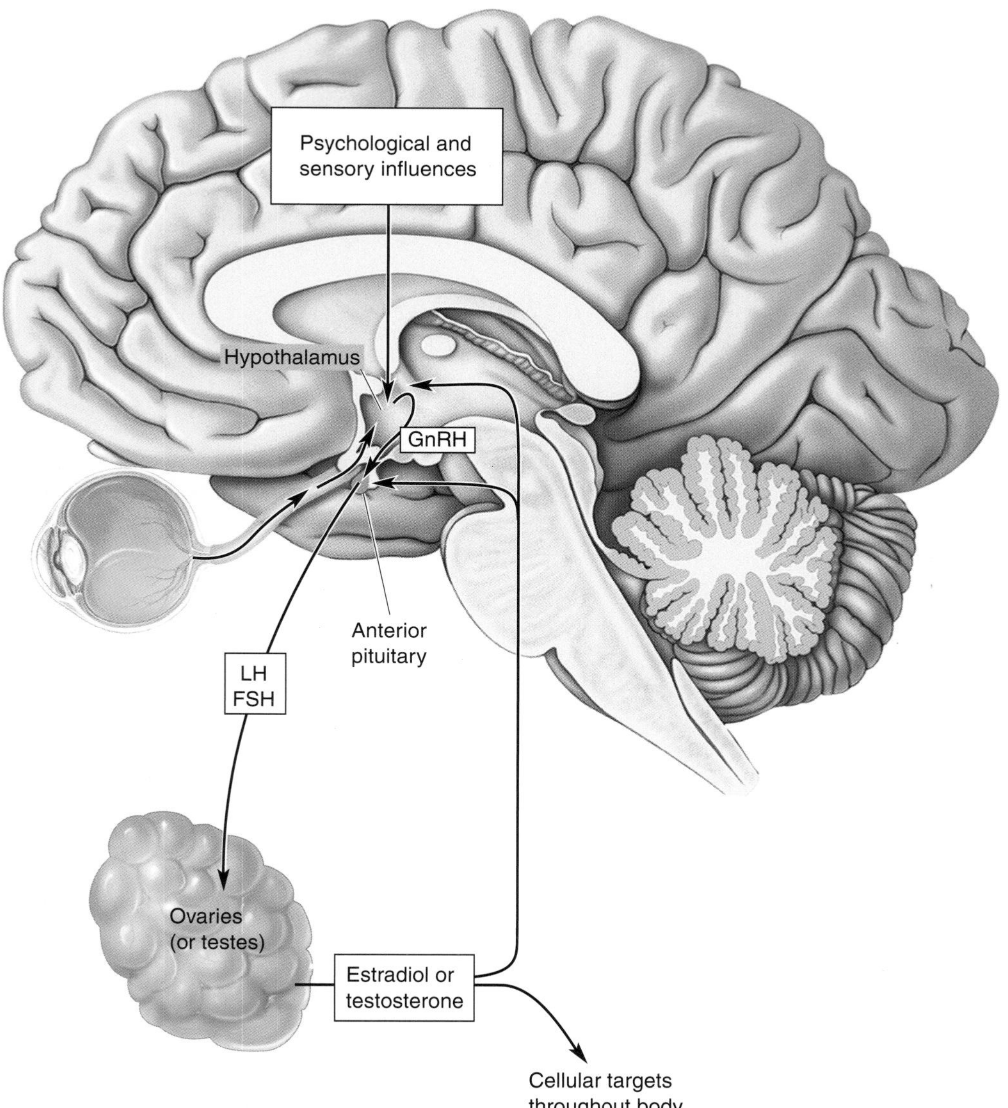

Figure 17.6
Bidirectional interactions between the brain and the gonads. The hypothalamus is influenced by both psychological factors and sensory information, such as light responses from the retina. GnRH from the hypothalamus regulates gonadotropin (LH and FSH) release from the anterior pituitary. The testes secrete testosterone and the ovaries secrete estradiol, as directed by the gonadotropins. The sex hormones have diverse effects on the body and also send feedback to the pituitary and hypothalamus.

and the sex hormones, as shown in Figure 17.7. At approximately the onset of menstruation, there is an increase in the secretion of both LH and FSH from the anterior pituitary (stage ①). In the ovaries, these hormones (particularly FSH) have the effect of increasing the growth of a small number of follicles, the cavities in the ovaries that enclose and maintain the ova (egg cells). The development of the follicles characterizes the *follicular phase* of the cycle, which lasts about 10 days. Cells in the growing follicles secrete a fluid that contains estrogens, and the rising estrogen levels in the bloodstream send feedback to the hypothalamus and pituitary (stage ②). One effect of this feedback is a decrease in LH release, and an increase in LH stores in the anterior

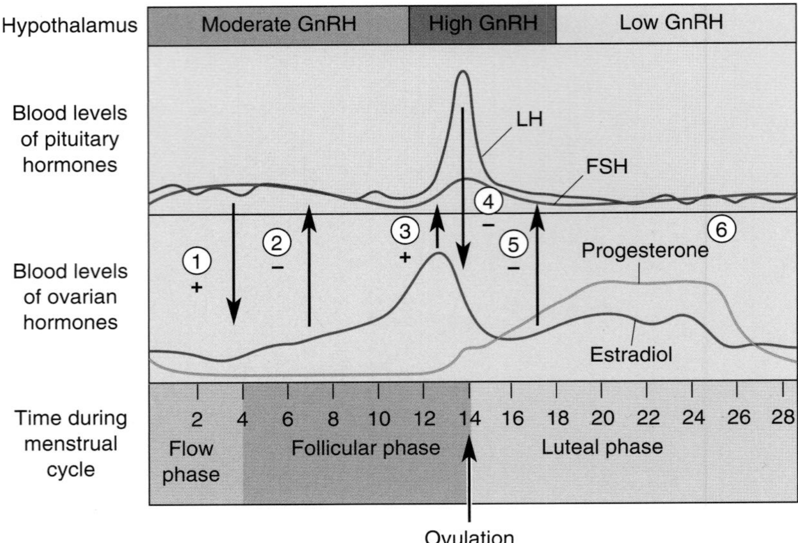

Figure 17.7
Negative and positive feedback between sex hormones and gonadotropins during the menstrual cycle. The cycle can be summarized in six stages, as described in the text.

pituitary in preparation for ovulation. Eventually, one follicle outgrows the others, and its higher production of estradiol inhibits the growth of the other follicles. Through the feedback to the hypothalamus and pituitary, a rapid increase in estradiol release from the dominant follicle causes a surge in GnRH and gonadotropins (stage ③). The marked increase in LH release from the pituitary both accelerates the growth of the follicle and modifies it, such that less estrogen is produced (stage ④). Now, about 14 days after menstruation begins, LH secretion from the anterior pituitary initiates ovulation. The follicle swells and ruptures, causing the release of the ovum.

After expulsion of the egg, the small cells that surround it undergo chemical changes in a process called luteinization, which depends on LH release from the pituitary. During this *luteal phase* of the cycle, estrogen and progesterone secreted by the luteinized cells send feedback to the hypothalamus, decreasing the release of LH and FSH from the anterior pituitary (stage ⑤). This feedback prevents the maturation of any other follicles in the ovaries. In the absence of fertilization of the egg, the luteinized cells degenerate after about 11 days, thereby terminating the negative feedback to the hypothalamus (stage ⑥). Gonadotropin secretion then increases, and a new cycle begins.

The duration of the follicular and luteal phases of the reproductive cycle vary significantly for different mammals. The phases are roughly equal in length in the primate menstrual cycle. In the **estrous cycle** of nonprimate mammals, such as rats and mice, the luteal phase is much shorter. In other estrous animals, such as dogs, cats, and farm animals, the phases are more equal in duration. Many estrous animals have only one cycle per year, usually in the spring. Presumably, this is timed to produce offspring when the weather and food supply are optimal. At the other extreme are animals such as rats, which are said to be *polyestrous* because they have short periods of estrus, or "heat," throughout the year.

THE NEURAL BASIS OF SEX-RELATED BEHAVIORS

Sexual behavior is a vast, complex, and provocative topic, ranging from the most mechanical and biological facts of copulation to the myriad cultural practices of human societies. Here we will touch on only fragments of the subject. We begin with genitals and the autonomic and spinal neurons that control them, then discuss various strategies of mating, and conclude with some research about brain mechanisms that are important for monogamy and parenting.

Reproductive Organs and Their Control

Despite the obvious structural differences between female and male reproductive organs, their neural regulation (to the extent it is understood) is surprisingly similar. Sexual arousal of adult men and women can result from both erotic psychological and sensory stimuli (including visual, olfactory, somatosensory—the things that arouse different people vary widely), and from direct tactile stimulation of the external sex organs. A full sexual response cycle consists of *arousal* followed by *plateau*, *orgasm*, and *resolution* phases. Although the duration of each phase can vary widely, the physiological changes associated with each one are relatively consistent. Neural control of the sexual response comes in part from the cerebral cortex—this is, after all, where erotic thoughts occur—but the spinal cord coordinates this brain activity with sensory information from the genitals and generates the critical outputs that mediate the sexual responses of the genital structures.

The major external and internal sex organs are shown in Figure 17.8. Research on the physiology of the human sexual response has tended to focus unduly on men, but we will try to summarize some of what is known about both sexes. Sexual arousal causes certain parts of the external genitals of both women and men to become engorged with blood, and thus to swell. In women, these structures include the *labia* and the *clitoris;* in men, it is primarily the *penis*. The external genitals are densely innervated by mechanoreceptors, particularly within the clitoris and the glans of the penis. Adequate stimulation of these sensory endings can, by itself, be enough to cause engorgement and erection. The best evidence that engorgement can be generated by a simple spinal reflex is that most men who have suffered a complete transection of the spinal cord at the thoracic or lumbar level can nevertheless generate an erection when their penis is mechanically stimulated. The mechanosensory pathways from the genitals are components of the somatosensory system (see Chapter 12), and their anatomy follows the usual pattern: Axons from mechanoreceptors in the penis and clitoris collect in the dorsal roots of the sacral spinal cord. They then send branches into the dorsal horns of the cord, and into the dorsal columns, through which they project toward the brain.

Engorgement and erection are controlled primarily by axons of the *parasympathetic* division of the autonomic nervous system (see Figure 15.9). Within the sacral spinal cord, the parasympathetic neurons can be excited by either mechanosensory activity from the genitals (which can directly trigger reflexive erection), or by axons descending from the brain (which account for responses mediated by more cerebral stimuli) (Figure 17.8). Engorgement of the clitoris and penis depend on dramatic changes in blood flow. Parasympathetic nerve endings are thought to release a potent combination of acetylcholine, vasoactive intestinal polypeptide (VIP), and nitric oxide (NO) directly into the erectile tissues. These neurotransmitters cause the relaxation of smooth muscle cells in the arteries and the spongy substance of the clitoris and penis. The usually flaccid arteries then become filled with blood, thereby distending the organs. (Sildenafil, a potent new drug with the trade name Viagra, is a treatment for erectile dysfunction that works by enhancing the effects of NO.) As the penis becomes longer and thicker, the spongy internal tissues swell against two thick, elastic outer coverings of connective tissue that give the erect penis its stiffness. In order to keep the organs sliding easily during copulation throughout the plateau phase, parasympathetic activity also stimulates the secretion of lubricating fluids from the woman's vaginal wall and from the man's bulbourethral gland.

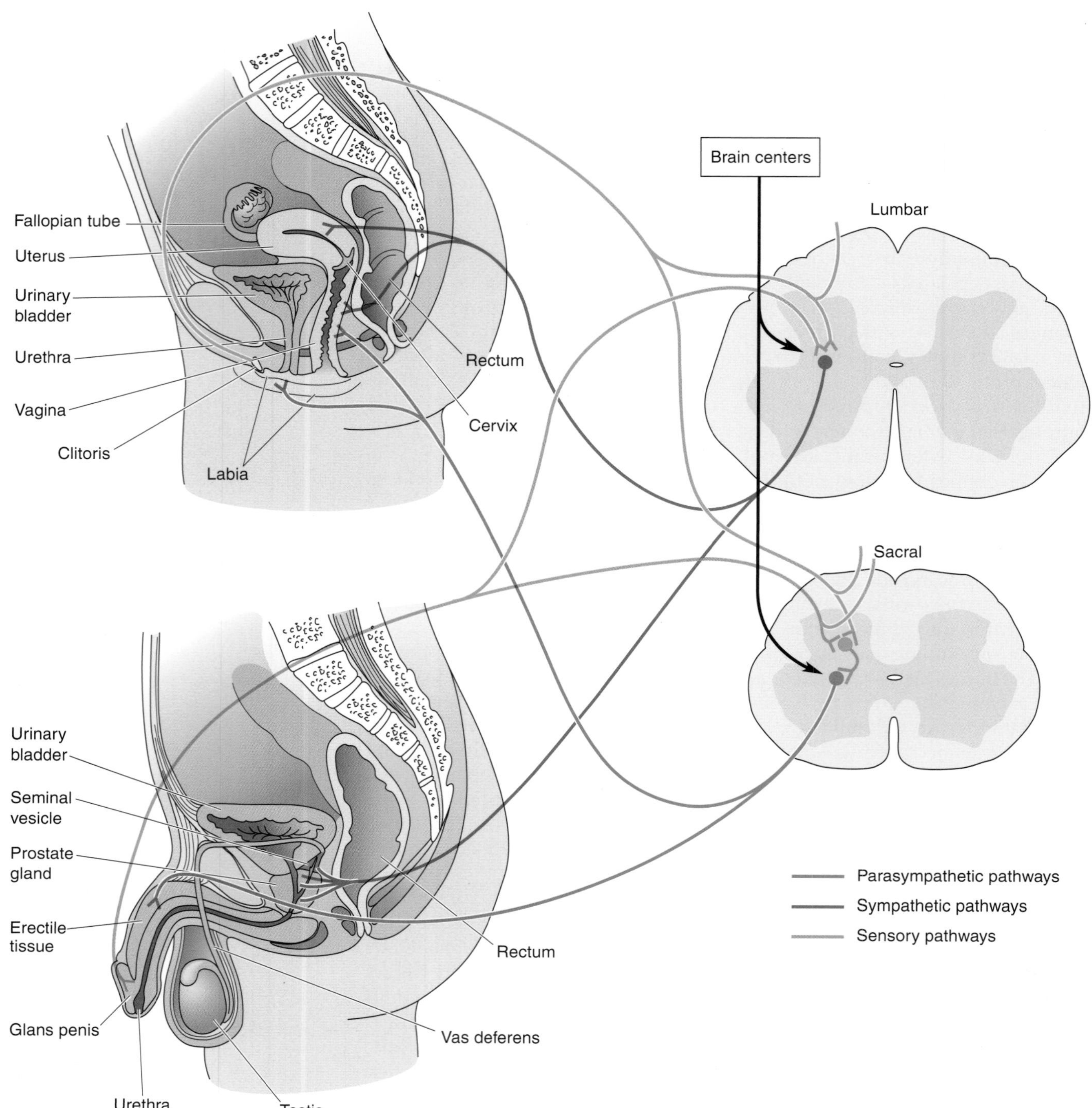

Figure 17.8
The neural control of human sex organs.

Completing the sexual response cycle requires activity from the *sympathetic* division of the ANS. As sensory axons, particularly from the penis or clitoris, become highly active, they, together with activity descending from the brain, excite sympathetic neurons in the thoracic and lumbar segments of the spinal cord (Figure 17.8). In men, the sympathetic efferent axons then trigger the process of *emission:* Muscular contractions move sperm from storage sites near the testes through two tubes called the *vas deferens*, combine the sperm with fluids produced by various glands, and propel the resulting mixture (called *semen*) into a larger tube called the *urethra*. Finally, during *ejaculation*, a series of coordinated muscular contractions expel the semen from the urethra, and this is usually coincident with the intense sensations of orgasm. In women, stimulation adequate to trigger orgasm probably also activates the sympathetic system. Sympathetic outflow causes the outer vaginal wall to thicken and, during orgasm itself, triggers a series of strong muscular contractions. Following an orgasm, some time must pass before another orgasm can be triggered in men. The orgasmic experience of women tends to be considerably more variable in frequency and intensity. The neural mechanisms underlying the sensations of sexual climax are entirely mysterious, in both sexes. The resolution phase, which ends the sexual response cycle, includes a draining of blood from the external genitals through veins, and a loss of erection and other signs and sensations of sexual excitement.

Mammalian Mating Strategies

Mammals practice a dazzling range of mating habits. Each one is a strategy that ultimately meets a single evolutionary objective: to maximize the survival of offspring and parental genes. But males and females are distinctly different in their reproductive biology, and this can lead to very different mating strategies within the same species.

Females of some species tend to be very selective when choosing a mate. An evolutionary explanation for this fact is that females must be discerning because they expend a huge amount of time and energy on their offspring. They have relatively few eggs, and they can incubate and nurture only a small number of young during their lifetimes. Once pregnant, they also have little choice about whether to stick with their embryos or abandon them. Female mammals of many species mate with a small number of males, often just one. Above all, they want the sperm they choose for their eggs to be of the best quality available, because this maximizes the chance that their offspring (and their genes) in turn will survive and reproduce. Courtship rituals often provide an opportunity for the female to assess the genetic fitness of a potential mate. (When we say that animals "want" the best sperm, or "assess" the fitness of a mate, we are using the words figuratively, and do not mean to suggest that they consciously reflect on the need to maximize the genetic fitness of their offspring. Nonhuman reproductive behaviors are largely driven by genetically determined characteristics of the brain, which has itself been shaped by natural selection.)

Males, on the other hand, often invest very little in reproduction; quick copulation and a small dose of sperm can be all it takes, since the male of many species provides no further service to his mate and their offspring. In that case, it is to the male's advantage to disseminate his sperm among as many fertile females as possible. This does not necessarily mean that all males get to mate promiscuously! On the contrary, in most species there is intense sexual competition among males, and most males *never* mate. Males, much more so than females, can also choose whether to stay or leave their mate and offspring. The evolutionary logic is that if they leave, they are free to mate with more females and thus enhance the frequency of their genes in the next generation; if they stay with one female, on the other hand, they may greatly increase the chances that their young will survive to reproductive age, but then they can't mate with as many females. It's a trade-off—so which is more beneficial for the male, to stay or to go? That depends on the species and the particular circumstances. For example, if an environment is particularly harsh or offspring are particularly fragile, the male's most effective strategy for passing on his genes to the next generation might be to forgo his wanderlust, and cooperate with his mate in raising the young.

Species variations in preferred mating systems seem to depend on the investment that males and females make in raising their offspring, although there are exceptions. Very common among mammals is **polygyny** (from the Greek for "many women"), in which the male mates with many females but the female usually mates with only one male. Polygynous mating (practiced by giraffes, montane voles, orangutans, and most other mammals) has a "one-night stand" quality to it, and the male never looks back to check on the outcome of his many liaisons. Sometimes polygyny takes the form of a harem—one male forms a lasting and exclusive association with a group of females—as practiced by gorillas, elephant seals, and a very small number of traditional human cultures. **Polyandry** ("many men"), in which one female mates with many males but the males mate with only that female, is a very rare system among mammals and vertebrates in general. One exception is the phalarope, a shorebird that breeds in the cold tundra. Polygyny and polyandry are both examples of *polygamy*, having more than one spouse. In **monogamy** ("one spouse"), a male and a female form a tightly bound relationship that includes exclusive (or nearly exclusive) mating with each other. Only about 3% of mammalian species are monogamous, although monogamy is practiced by roughly 12% of primate species (and 90% of bird species!).

If we survey various cultures and eras, we can find examples of nearly every type of mating system among humans. On balance, humans have a strong tendency toward (at least temporary) monogamy, although some cultures condone polygyny. Interestingly, even where polygyny is socially acceptable, most marriages are monogamous. Polyandry for reproductive purposes is rare, and most cultures have penalized women accused of it. Although there has been much speculation about the evolutionary explanations of human patterns, these ideas are difficult to evaluate. We must remember that our distant prehistory provided the environment and circumstances that shaped our current genetic makeup. Even the last few millennia are too brief to have had much influence on human genetics. However, human reproductive habits are strongly shaped by culture and society. Sorting out the precise influences of genetics and culture on complex behaviors is virtually impossible.

The Neurochemistry of Reproductive Behavior

Regardless of an animal's choice of reproductive strategy—staying faithful to a mate and devoted to one's children, or wandering promiscuously and abandoning offspring—complex social behaviors are required. It would be remarkable if the tendency to be monogamous or polygamous were controlled by a few simple brain chemicals. Yet recent work on mouselike rodents called voles suggests that certain well-known pituitary hormones do precisely that (at least in voles).

Perhaps you are wondering why neuroscientists study the sex lives of voles in the first place. The answer should be familiar by now: When you have a problem in biology, choosing just the right species to study can provide a shortcut to the solution. Voles are a wonderful natural experimental model, because closely related species of voles use very different reproductive behaviors. The prairie vole (*Microtus ochrogaster*) lives in the American grasslands, practicing solid "family values" (Figure 17.9). It is highly social and as reliably monogamous as any mammal known. After an intense period of initial matings, the male and female form a tight pair-bond and live together in one nest. The male will fiercely defend his mate, and both parents

Figure 17.9
A prairie vole.

cooperate in the long-term care of their young. By contrast, the montane vole (*Microtus montanus*) of the high meadows is asocial and promiscuous. Each one lives in an isolated nest, the males take no part in parenting, and the females care for their offspring only briefly before letting them fend for themselves in the world.

Because these two vole species are physically and genetically quite similar, relatively few biological factors might account for their disparate reproductive habits. Based on clues from previous studies of maternal and territorial behavior, research has focused on the roles of *vasopressin* and *oxytocin* in voles. Recall that these peptide hormones are synthesized in the hypothalamus and can be released into the bloodstream by neurosecretory terminals in the posterior pituitary gland (see Figure 15.4). Circulating vasopressin (also known as ADH) helps regulate water and salt levels in the body, mainly by affecting the kidneys; oxytocin stimulates smooth muscle, causing the uterus to contract during childbirth and milk to be let down during lactation. However, vasopressin and oxytocin are also released onto CNS neurons, and, like most signaling molecules, they bind to specific receptors scattered about the brain. As shown in Figure 17.10, maps of these receptors are strikingly different in the brains of prairie voles and montane voles, whereas maps of other types of neurotransmitter and hormone receptors are very sim-

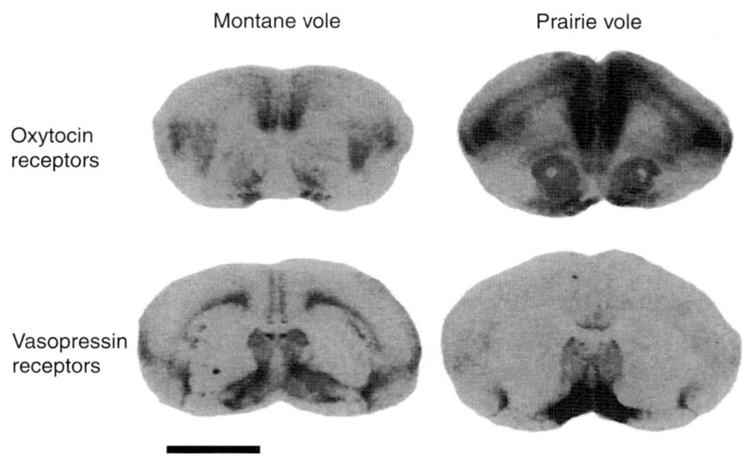

Figure 17.10
The distribution of oxytocin and vasopressin receptors in the brains of montane voles and prairie voles. The images are autoradiograms of coronal brain sections, and the regions with the highest receptor densities are the darkest. (Source: Young et al., 1998.)

ilar in the two species. The receptor differences correlate well with reproductive behavior even in other species of voles. Furthermore, the maps are plastic. When the female montane vole gives birth and assumes a maternal role (however briefly), her receptor maps temporarily change to resemble those of the prairie vole.

The distinctive maps of vasopressin and oxytocin receptors tell us that each hormone activates a different network of neurons in the monogamous and polygynous vole brains. This alone does not prove that the hormones have anything to do with sex-related behaviors. But together with the effects of the hormones and drugs that antagonize them, a strong case for cause and effect is made. When a pair of prairie voles copulates, levels of vasopressin (in males) and oxytocin (in females) rise sharply. Vasopressin antagonists given to a male prairie vole before mating prevent him from forming a pair-bond relationship, whereas oxytocin antagonists have no such effect. When a male is given vasopressin while he is exposed to a new female, he quickly forms a strong preference for her even without the intense mating that usually precedes pair-bonding. Oxytocin seems to be necessary for the female to establish a preference for her mate, while vasopressin has little effect.

The hormones are also involved in parenting habits: Vasopressin increases the male prairie vole's paternal proclivities, causing him to spend more time with his pups, whereas oxytocin stimulates maternal behaviors in the female. The research on voles suggests a very interesting hypothesis about the evolution of complex social behaviors. If genetic mutations cause a change in the anatomical distribution of a particular hormone's receptors, then that hormone may evoke an entirely new repertoire of behaviors. Consistent with this, giving vasopressin or oxytocin to the naturally promiscuous montane voles does not evoke the effects on pair-bonding and parenting seen in prairie voles, perhaps because they don't have receptors in the necessary places.

The vole story is a fascinating example of how brain chemicals can regulate critical behaviors. However, by now you are undoubtedly wondering: Does this have anything to do with human relationships, faithfulness, and love? We have only incomplete facts. There is some evidence from primates that vasopressin and oxytocin levels vary with sexual arousal, and that oxytocin facilitates nurturing behavior in females and sexual assertiveness in some males. This is a rapidly developing field of research, and worth watching closely.

Are oxytocin and vasopressin important for romantic and parental love in people? Maybe. It is too soon to tell.

WHY AND HOW MALE AND FEMALE BRAINS DIFFER

Sexual reproduction depends on a variety of individual and social behaviors—finding, attracting, and keeping a mate; copulating; giving birth; and nursing and nurturing the offspring—and in each case, the behavior of males and females is usually quite different. Since behavior depends on the structure and function of the nervous system, we can make the strong prediction that male and female brains are also somehow different—that is, they have **sexual dimorphisms** (from the Greek *dimorphos,* "having two forms"). Another good reason to expect that male and female brains differ is simply that male and female bodies differ. The body parts that are unique to each sex require neural systems that have evolved specifically to control them. For example, male rats have a particular muscle at the base of the penis, and their spinal cord has a small cluster of motor neurons that control that muscle; females lack both the muscle and the related motor neurons. Body size and general shape also vary with gender, and thus somatosensory and motor maps must adjust to fit them.

Sexual dimorphisms vary widely across species. Finding them in the brain is not always easy. In human brains, dimorphisms have so far proven to be small, subtle, few, and of unknown function. Differences between human male and female brains tend to vary along a continuum, with lots of overlap. A particular hypothalamic nucleus might be larger in women than men *on average*, but the size of the nucleus may vary so widely that many men have a larger nucleus than many women. By contrast, some nonhuman species have much more dramatic and explicable dimorphisms. In rodents, the trained eye can tell male from female brains with no ambiguity because of differences in their hypothalamus. The wide range of brain dimorphisms across species can sometimes be attributed to the remarkable variations in sexual behaviors. For example, in some songbird species, only males sing, and, not surprisingly, only males have large singing-related brain nuclei.

Another interesting feature of sexual behaviors is that they vary over time. For example, in some species, reproduction occurs only during a particular season, and mating may occur only during a specific phase of that season. Obviously, females of all species nurse their offspring only after birth and only temporarily. In most animals, but not in humans, sexual attractiveness and copulation occur only during certain phases of the estrous cycle. Sexually dimorphic changes in the brain are sometimes transient or cyclical, coinciding with the sexual behavior to which they are related. For example, in female rats the somatosensory cortex contains a sensory representation of the ventral skin surrounding the nipples; this representation expands dramatically but temporarily across the cortex during the time the mother rat nurses her young (Figure 17.11). This is an interesting example of somatosensory map plasticity (see Chapter 12).

In the rest of this section, we will describe sexual dimorphisms in the nervous systems of humans and other species, paying particular attention to examples that illuminate the relationships between the brain and behavior. We will also discuss some of the neurobiological mechanisms that generate these dimorphisms.

Sexual Dimorphisms of the Central Nervous System

Few dimorphic neural structures are related to their sexual functions in an obvious way. One structure that is related is the collection of spinal motor neurons that innervates the *bulbocavernosus* (*BC*) muscles surrounding the base of the penis. These muscles have a role in penile erection and help to eject urine. Both women and men have a BC muscle. In women, it surrounds the opening of the vagina and serves to constrict it slightly. The motor neuron pool controlling the BC muscles in humans is called *Onuf's nucleus*, and it is located in the sacral spinal cord. Onuf's nucleus is moderately dimorphic (there are more motor neurons in men than women) because the male BC muscles are larger than those of females.

The most distinct sexual dimorphisms in the mammalian brain are clustered around the third ventricle, within the *preoptic area of the anterior hypothalamus*. This region seems to have a role in reproductive behaviors. In rats, lesions of the preoptic area disrupt the estrous cycle in females and reduce the frequency of copulation in males. Histological sections of male and female preoptic areas from rats show an obvious difference: The aptly named *sexually dimorphic nucleus* (*SDN*) is 5–8 times larger in males than in females (Figure 17.12). The preoptic area of humans also has dimorphisms, but seeing them is more challenging. There are four clusters of neurons called the *interstitial nuclei of the anterior hypothalamus* (*INAH*). INAH-1 seems to be the human analogue of the rat SDN, but researchers disagree about whether INAH-1 is dimorphic. INAH-2 and INAH-3, however, are about twice as

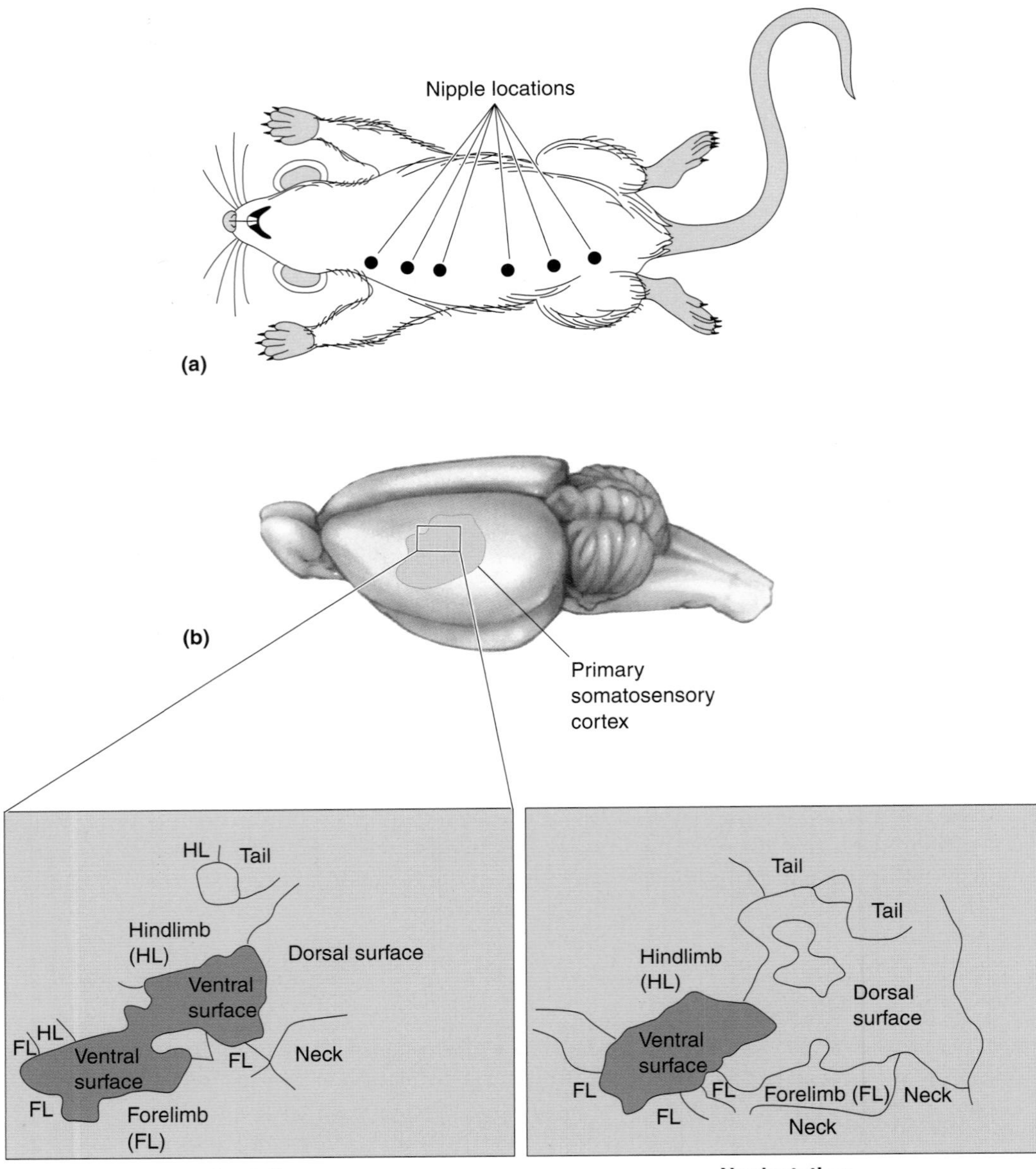

Figure 17.11

The effect of lactation on a sensory representation in the cortex. (a) The ventral skin of a nursing rat mother, showing the location of the nipples along one side. **(b)** The rat's brain (top) and primary somatosensory cortex. The boxed area, expanded below, illustrates how the cortical region that responds to the ventral skin around the nipples is enlarged in a postpartum lactating mother (left) compared with a postpartum nonlactating mother (right). Regions of somatosensory cortex subserving other regions of the body were not affected by the lactating state. (Source: Adapted from Xerri et al., 1994.)

large in men as in women. The evidence that these nuclei are involved in sexual behavior is thus far indirect. Various neurons of the medial preoptic area in male rhesus monkeys fire vigorously during specific phases of sexual behavior, including arousal and copulation. In addition, there may be subtle differences in the size of certain hypothalamic nuclei that correlate with sexual orientation in people.

Human brain dimorphisms outside the hypothalamus have been difficult

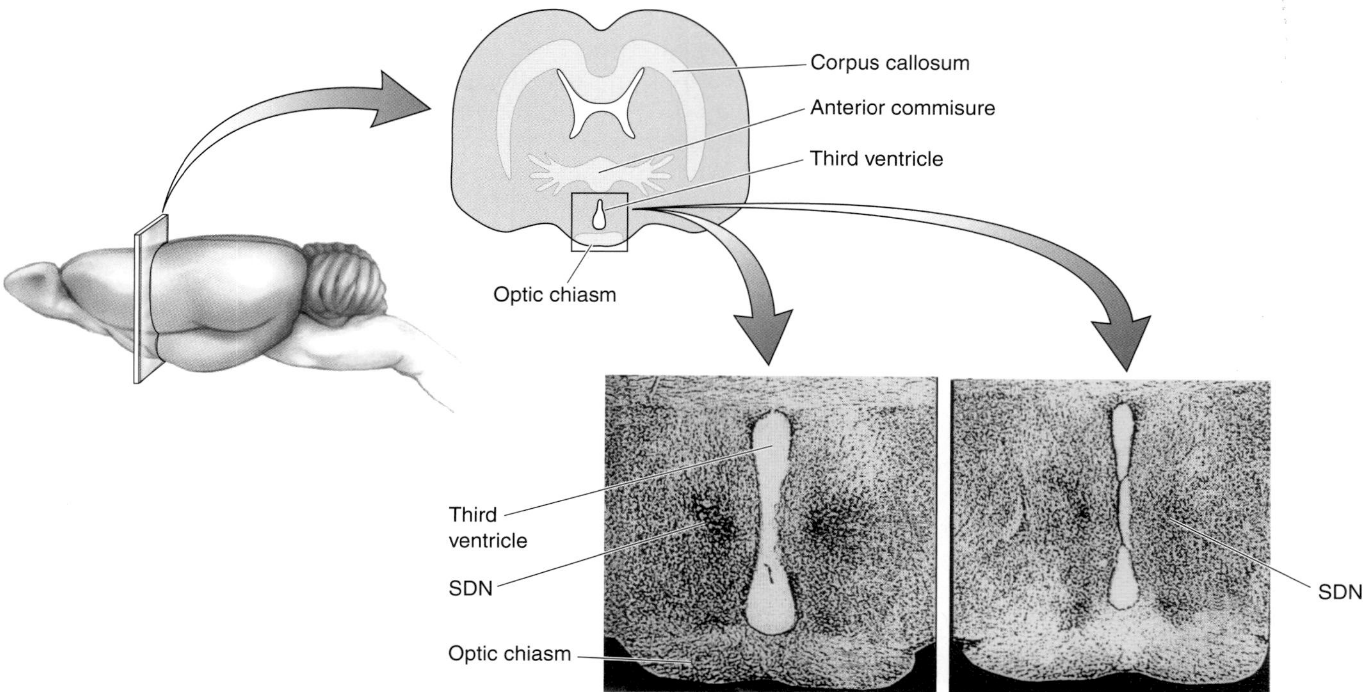

Figure 17.12
Sexual dimorphism in rats. The sexually dimorphic nucleus (SDN) in the hypothalamus of male rats (left) is much larger than the SDN in female rats (right). (Source: Adapted from Rosenzweig et al., 1999.)

to demonstrate. Although many sex differences have been claimed in the research literature, few have held up to the tests of replication by other neuroscientists. This is very tricky research, for numerous reasons: Dimorphisms tend to be quite small, while variability from one individual to the next is large; measurement techniques, particularly those that assess brain function, are often not very sensitive; the brain location or structure that should be measured is not always obvious; collecting enough samples, particularly of postmortem human brains, that are well-matched for age, history, and health can be difficult. We will mention only a few of the better studied and more interesting findings.

Many studies have suggested that the size and/or shape of the corpus callosum is (or is not) dimorphic. The corpus callosum, of course, is the large commissure that interconnects the two cerebral hemispheres. Measurements of the cross-sectional area of the callosum have been made on postmortem brains and, using MRI, on living brains. Some studies found that the callosum of males has, on average, a larger cross-sectional area than that of females; however, men's brains (and bodies) tend to be slightly larger than women's, and when overall brain size is taken into account, there may be no dimorphism in total callosal area. Several studies have claimed that the posterior end of the corpus callosum, called the *splenium,* is selectively larger in women than men. Not all researchers have seen this difference, however. One extensive recent report, which used measurements from MRIs of age-matched people, found no sex-related differences in the size of the corpus callosum but did describe a dramatic dimorphism in a bit of its shape. The splenium of adult women was more bulbous than that of men. No such sex differences were seen in the brains of children. Finally, one study of the other large bundle of axons that interconnects the hemispheres, the *anterior com-*

missure, found it to be about 12% larger in women (and homosexual men) than in heterosexual men.

Even if there is a dimorphism in the size or shape of the corpus callosum, what could it mean? We can only guess. The callosum has no obvious role in mediating specifically sex-related behaviors, but it is important for a variety of cognitive functions that involve coordinated activity between hemispheres (see Chapter 20). Observations of stroke patients in whom only one hemisphere has been damaged suggest that the functions of female brains may be less lateralized, that is, dependent on one cerebral hemisphere more than the other. But this conclusion, too, has been challenged.

Perhaps the most reliable conclusion we can draw about sexual dimorphisms in human brain structure is that there are so few of them. This probably should not come as a surprise, since the vast majority of women's and men's behaviors are very similar, if not indistinguishable. The gross anatomy of the brain provides only a crude view of the organization of the nervous system. If we are to see the reasons for sexually dimorphic behavior, we will need to look deeper into the patterns of neural connections, the neurochemistry of the brain, and the influence of sex-related hormones on neural development and function.

Sexual Dimorphisms of Cognition

Even if there are no major differences in the brain structures of men and women, we know that the brains work differently. Or do we? The idea that male and female brains endow people with different mental capabilities can be traced back thousands of years. The usual claim was that women were inferior in intelligence, and (surprise!) this opinion was held by men.

Ancient claims of large intellectual differences have not withstood the test of time. But modern studies report more subtle gender differences in cognitive abilities. An evolutionary explanation is sometimes offered: Men evolved as hunters and relied more on their abilities to navigate their environment. Women evolved the behavior of staying closer to home to care for children, making them more social and verbal. Regardless of whether this explanation is correct, do the brains really perform differently?

Numerous studies indicate that women are better at verbal tasks than men. Starting at around age 11, girls perform slightly better on tests of comprehension and writing, and this effect is sometimes said to extend through high school and beyond. Perhaps it reflects a difference in the rates of brain development in the two genders. Specific tasks at which women excel include naming objects of the same color, listing words beginning with the same letter, and verbal memory (Figure 17.13a).

(a) List words beginning with the letter B.

big, bag, bug, boy,
banana, bugle, bunny......

(b) Are these two shapes the same?

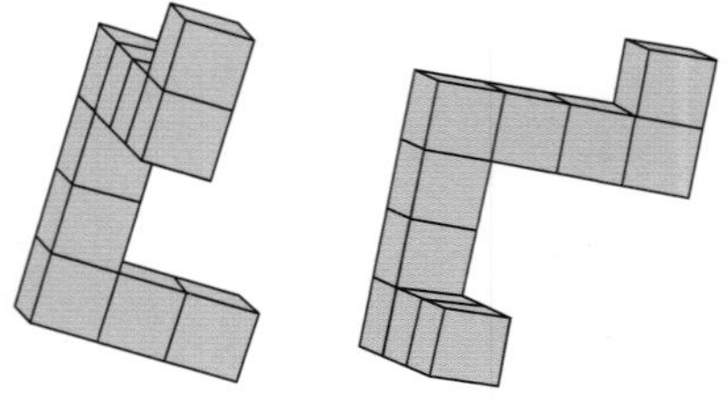

Figure 17.13
Cognitive tasks that slightly favor women or men. (a) Women may outperform men in listing words beginning with the same letter. **(b)** Men appear to be somewhat better at spatial rotation tasks, such as deciding whether two three-dimensional objects are the same. (Source: Adapted from Kimura, 1992, p. 120.)

Two cautionary comments. First, not all studies yield the same results. In some cases, there is no reported difference in performance between men and women, and occasionally men do better. Second, across large groups of people of both genders, there are huge differences in performance. But most of the variation is the result of differences *among individuals*, rather than being gender-specific. In other words, male and female performance on the verbal tasks could be plotted as two broad bell curves that would almost completely overlap. If you picked one man and one woman at random, and always guessed that the woman would be better at verbal tasks than the man, you would be wrong almost as often as right.

In other sorts of tasks, men are said to outperform women. Tasks that reportedly favor men include map reading, maze learning, and mathematical reasoning. Researchers speculate that these male advantages evolved from the days when men roamed large areas to hunt wild animals. One of the largest reported differences between the sexes is mental rotation of objects, a task that appears to favor men (Figure 17.13b). Gender differences in spatial tasks may favor men, but the cautions previously mentioned concerning verbal tasks also apply here. Not all studies even find a difference, and the variance among members of either sex is far greater than the average difference between men and women.

A common interpretation of gender-based differences in performance is that the distinctive hormonal environments of the male and female brains make them work somewhat differently. Perhaps there is a benefit or penalty associated with estrogens or androgens for each task. Consistent with this conjecture are reports that spatial reasoning in women correlates with the menstrual cycle, better performance being observed when estrogen levels are lowest. It has also been reported that administering testosterone enhances spatial performance in older men with low testosterone levels. However, cognition cannot be so simply related to hormones, as there is no reliable correlation between performance on verbal or spatial tasks and hormone levels. This doesn't mean that hormones do not affect cognitive function, but we must be cautious about overgeneralizing. In the past, the popular press has sometimes unfairly implied that each gender is poorly equipped to perform certain jobs because of hormones or genetic predisposition.

Many questions about gender differences remain. For example, how much of the reported differences are based on childhood or adult experiences? Typical males and females experience different things and they may, on average, develop slightly different skills. Even if it can be established conclusively that male and female brains are different, what is cause and what is effect? Naturally, scientists try to equate these sorts of things in their comparisons, but making a perfectly level playing field is difficult. The bottom line is that men and women perform about the same on almost all cognitive tasks. It is a challenge to future researchers to account for the small cognitive differences in verbal and spatial tasks in terms of hormones, neuroanatomy, and upbringing.

Sex Hormones, the Brain, and Behavior

Even if there are no gross anatomical dimorphisms, male and female brain circuitry must be somewhat different to account for gender-specific behaviors, whether they are singing by male birds or human sexual behavior. Sex hormones determine the sexual identity of the brain during early development. Recall that the kinds of sex hormones circulating in the blood are determined by the gonads, and the dimorphism of the gonads is ordinarily specified by our genes. As described earlier, people with a Y chromosome express a factor (testis-determining factor) that causes the undifferentiated gonads to become testes; people lacking a Y chromosome do not produce this factor, and their gonads differentiate into ovaries. The differentiation of testes or ovaries sets off a cascade of developmental events in the body. Most importantly for the sexual differentiation of the brain, the testes produce androgens, which trigger the masculinization of the nervous system by regulating the expression of a variety of sex-related genes. In the absence of androgens, the developing brain takes on different characteristics because some parts of it express a slightly different pattern of genes. There also seems to be a distinct feminization of the brain; in other words, the female brain is not simply one that missed out on androgenization.

There is nothing fundamentally unique about the brain's sensitivity to hormones. It is just one more body tissue waiting for a hormonal signal to decide its specific pattern of growth and development. Androgens provide a unitary signal for masculinization in the brain, just as in the various other tissues of the body that are sexually dimorphic. Steroids can influence neurons in two general ways (Figure 17.14). First, they can act quickly (within seconds or less) to alter membrane excitability, sensitivity to neurotransmitters, or neurotransmitter release. Steroids do this, in general, by directly binding to, and modulating the functions of, various enzymes, channels, and transmitter receptors. For example, certain metabolites (breakdown products) of progesterone bind to the inhibitory $GABA_A$ receptor and potentiate the amount of chloride current activated by GABA. The effects of these progesterone metabolites are quite similar to the sedative and anticonvulsant effects of the benzodiazepine class of drugs (see Figure 6.21). Second, steroids can diffuse across the outer membrane and bind to specific types of steroid receptors in the cytoplasm and nucleus. Receptors with bound steroid can either promote or inhibit the transcription of specific genes in the nucleus, a process that can take minutes to hours. Specific receptors exist for each type of sex hormone, and the distributions of each receptor type vary widely throughout the brain (Figure 17.5).

Ironically, testosterone does not cause the changes in gene expression responsible for masculinization of the male brain. Remember that testosterone is converted within neuronal cytoplasm into estradiol in a single chemical step, catalyzed by the enzyme aromatase (Figure 17.4). It is actually the estrogen, binding to estradiol receptors, that triggers masculinzation of the de-

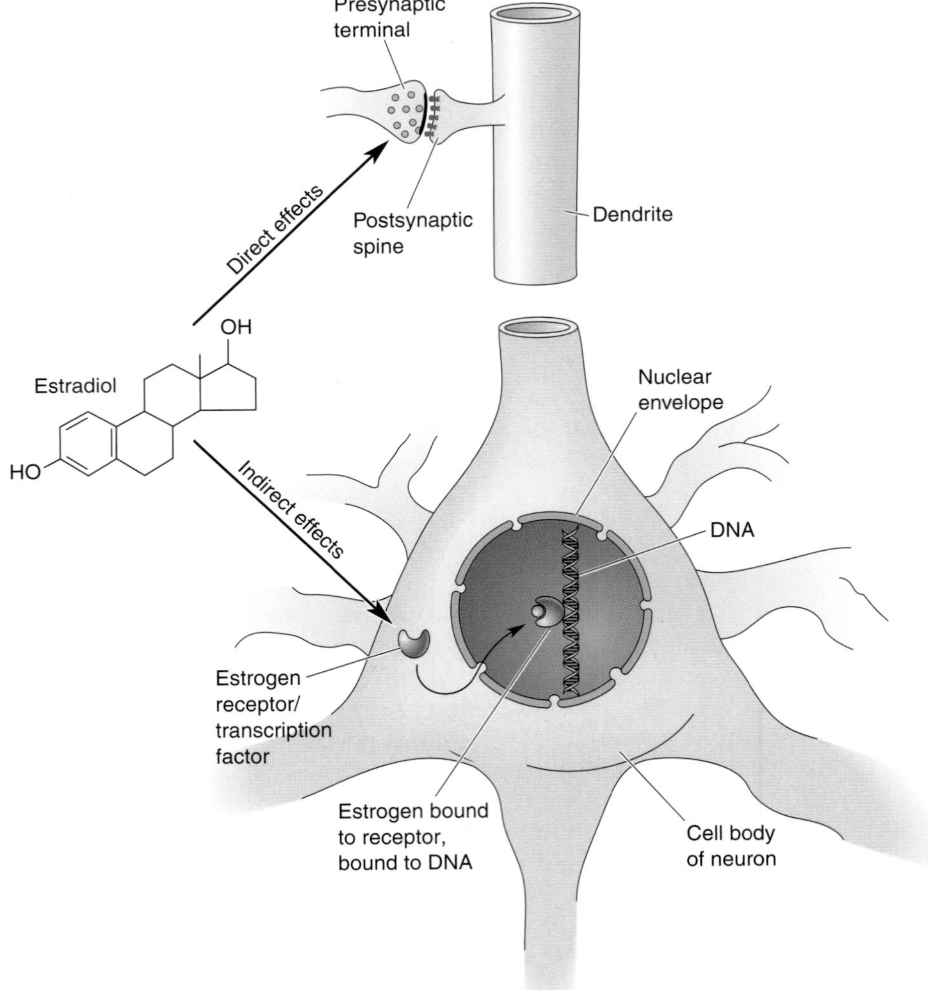

Figure 17.14
The direct and indirect effects of steroids on neurons. Steroids can directly affect transmitter synthesis, transmitter release, or postsynaptic transmitter receptors. They can indirectly influence gene transcription.

veloping nervous system. Since female gonads do not produce a surge of estrogens at early stages of development, female brains normally escape this steroid-triggered transformation.

Steroid hormones can exert effects on the body throughout life, but their influence early in development is often fundamentally different from their effect after an animal is mature. For example, the ability of testosterone to alter very young genitals and brain circuitry, leading to distinctly male genitals and masculine behaviors later in life, can be thought of as the **organizational effects** of the hormone. The hormone *organizes* the perinatal tissues in irreversible ways that allow it to generate male functions after sexual maturity has been reached. In order for a mature animal to express sexual behaviors fully, however, it is often necessary for steroid hormones to circulate *again* during periods of sexual activity, providing **activational effects** on the nervous system. Thus, for example, testosterone levels might surge in the body of a male songbird in the spring, *activating* changes in certain parts of his brain that are essential for normal reproductive behavior (Box 17.1). Activational effects are usually temporary.

Because hormones and not chromosomes directly determine the sexual characteristics of the nervous system, it is possible to have genetic males with female brains and genetic females with male brains. In all mammalian species studied, treatment with testosterone early in development leads to decreases in at least some features of adult female sexual behavior. Activating fully masculine behavior usually requires extended testosterone treatment before and after birth. If genetically female (XX) rats are exposed to testosterone during the few days around birth, then they will fail to elicit the typical female mating posture, called *lordosis*, when they reach maturity. Female guinea pigs treated *in utero* with enough testosterone to masculinize their external genitals will, as adults, energetically mount and attempt to mate with females in estrus. In a more natural experiment, when a cow carries twin calves that include both a male and a female, the female calf is exposed *in utero* to some testosterone produced by her male twin. As an adult the female, known as a freemartin, will invariably be infertile and behave more like a bull than a cow.

Some humans also experience mismatches between their chromosomes and sex hormones. For example, genetic males (XY) who carry a defective androgen receptor gene may have profound *androgen insensitivity*. The androgen receptor gene is on the X chromosome; males thus have only one copy of it, and males with the defective gene cannot produce functioning androgen receptors. These people develop normal testes and produce ample testosterone, but they appear outwardly quite female because their tissues cannot respond to androgen: They have a vagina, a clitoris, and labia, and at puberty they develop breasts and a female body shape. They do not menstruate, however, and remain infertile. Androgen-insensitive genetic males not only look like normal genetic females; they also behave like them. Even when they understand the circumstances of their biology, they prefer to call themselves women; they dress like women, and they choose men as their sex partners.

Occasionally, genetic females have a condition called *congenital adrenal hyperplasia* (*CAH*), which literally means overgrown adrenal glands present at birth. Although they are genetically female, because their adrenal glands secrete unusually large amounts of androgens, CAH females are exposed to abnormally high levels of circulating androgens early in their development. When they are born they have normal ovaries and no testes, but their external genitals are intermediate in size between a normal clitoris and a penis. Surgery and medications are the usual treatments after birth. Nevertheless, CAH girls (and their parents) are more likely to describe their behavior as ag-

Box 17.1

OF SPECIAL INTEREST

Bird Songs and Bird Brains

To our ears, the singing of birds may be simply a pleasant harbinger of spring, but for birds, it is part of the serious business of sex and reproduction. Singing is strictly a male talent for many species, performed for the purpose of attracting and keeping a mate and for warning off potential rivals. Studies of two bird species with different habits of reproduction and singing have revealed some fascinating clues about the control and diversity of sexual dimorphisms in the brain.

Zebra finches are popular pets, but their wild habitat is the harsh Australian desert. To breed successfully, birds require dependable sources of food, but in the desert, food comes only with sporadic and unpredictable rains. Zebra finches must therefore be ready and willing to breed whenever food and a mate are available, in any season. Wild canaries, on the other hand, live in the more predictable environment of the Azores and (where else?) the Canary Islands. They breed seasonally during spring and summer, and do not reproduce during fall and winter. The males of both species are passionate singers, but they differ greatly in the size of their repertoires. Zebra finches belt out one simple ditty all their lives, and cannot learn new ones. Canaries learn many elaborate songs, and they add new ones each spring. The different behaviors of zebra finches and canaries require different mechanisms of neural control.

The birds' sexually dimorphic behavior, singing, is generated by dramatically dimorphic neural structures. Birds sing by forcing air past a special muscularized organ called the syrinx, which encircles the air passage. The muscles of the syrinx are activated by motor neurons of the nucleus of cranial nerve XII, which are in turn controlled by a set of higher nuclei collectively called the *vocal control regions*, or *VCRs* (colored blue in Figure A). In zebra finches and canaries, VCR size is five or more times larger in males than in females.

As you might guess, the development of VCRs and singing behavior is under the strict control of steroid hormones. However, the very different seasonal requirements of zebra finches and canaries are paralleled by distinctly different modes of steroidal control. Zebra finches apparently require early doses of steroids to *organize* their VCRs, and later androgens to *activate* them. If a hatchling female zebra finch is exposed to testosterone or estradiol, its VCRs will be larger than those of normal females when it reaches adulthood. If the masculinized female is given more testosterone as an adult, its VCRs will grow larger still, and she will then sing like a male. Females that are not exposed to steroids when young are unresponsive to testosterone as adults.

By contrast, the song system in canaries seems to be independent of early steroid exposure, yet it bursts into full service each spring. If female canaries are given androgens for the first time as adults, they will begin singing within a few weeks. The androgens of males surge naturally each spring; their VCRs double in size as neurons grow larger dendrites and more synapses, and singing commences. Remarkably, *neurogenesis*, the birth of neurons, continues throughout adulthood in songbird brains, further contributing to the VCR circuitry during the mating season. By fall, male androgen levels drop, and the canary song system shrinks in size as his singing abates. In a sense, the male canary rebuilds much of his song control system anew each year as courtship begins. This may enable him to learn new songs more easily and, with his enlarged repertoire, gain some advantage in attracting a mate.

gressive and tomboyish. As adults, most CAH women are heterosexual, but, compared to other women, a higher percentage of CAH women are homosexual. Presumably, by analogy to the animal studies, prenatal exposure to high levels of androgens causes a somewhat male-like organization of certain brain circuits in CAH women. We have to be particularly cautious about drawing conclusions about the causes of human behavior, however (Box 17.2). It is very hard to determine whether masculine behavior of a CAH female is due entirely to early androgen exposure and male-like brain dimorphisms, whether her behavior is the product of subtle differences in the way she is treated by others (particularly parents faced with a child who has ambiguous genitals), or both.

(Box 17.1, *continued*)

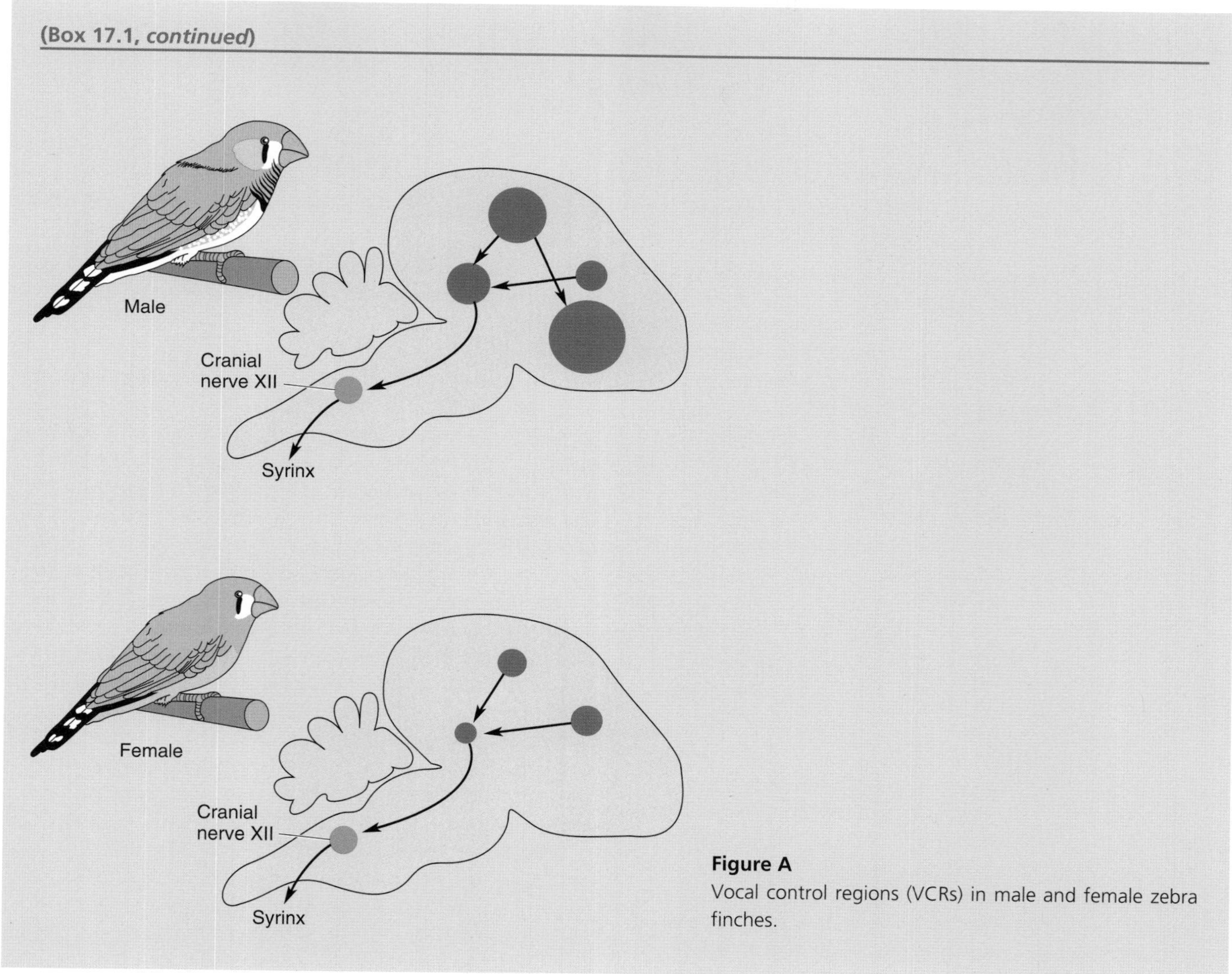

Figure A
Vocal control regions (VCRs) in male and female zebra finches.

The Activational Effects of Estrogens on Dendritic Spines

The estrous cycle provides a natural, repetitive surge of steroid hormones that have powerful effects on female reproductive tissues. These hormones also change brain circuitry. In 1990, Elizabeth Gould, Catherine Woolley, Bruce McEwen, and their colleagues, working at Rockefeller University, reported a fascinating example of a steroid's activational effects. They counted dendritic spines on neurons in the hippocampus of female rats, and found that the number of spines fluctuated dramatically during the 5-day estrous cycle. Spine density and estradiol levels peaked together, and treatment with injected estradiol also increased the numbers of spines in animals whose natural estradiol levels were kept low (Figure 17.15). Since spines are the major site of excitatory synapses on dendrites (see Chapter 2), this provides a possible explanation for the fact that hippocampal excitability also seems to track the estrous cycle. The hippocampus of experimental animals more easily generates seizures, for example, when estrogen levels increase (Figure 17.16).

Box 17.2

OF SPECIAL INTEREST

John/Joan and the Basis of Gender Identity

John was a normal healthy baby boy when he was born in 1965. But during a routine circumcision, an accident with an electrocautery device burned John's entire penis. The best efforts of John's parents to find help for the baby were fruitless, until they happened to watch a TV program about a physician who reported success with "gender transformation." The doctor's hypothesis was that, at birth, babies are essentially gender-neutral, and their identity as male or female is determined by their subsequent life experiences and identification with their anatomy. Because it was not possible to give John normal-appearing male genitals, it was recommended that the boy be castrated and undergo cosmetic surgery, followed by estrogen treatment at puberty, to turn him into a girl, Joan. Faced with a terrible decision, John's parents were eventually convinced that surgery combined with a female upbringing gave their child the best chance at a normal life.

In the physician's accounts of John's life after his transformation into Joan, it sounds as if the child adapted well and became a happy normal girl. The case even made it into the popular press, as evidenced by a 1973 article in *Time* magazine, stating that "This dramatic case . . . provides strong support . . . that conventional patterns of masculine and feminine behavior can be altered. It also casts doubt on the theory that major sex differences, psychological as well as anatomical, are immutably set by the genes at conception."[1] At that time, dramatic societal changes were taking place in the roles of men and women, and the success of John as a female appeared to confirm that society created gender identity as much as, or more than, biology.

Unfortunately, a follow-up report revealed that John's gender transformation was a disaster from the outset. According to John and his twin brother, John's behavior was always much more like other boys than girls. John rebelled at wearing girls' clothing and playing with traditional girls' toys. Despite cosmetic surgery and female indoctrination, as an adult, John said that he had suspected he was a boy as early as the second grade and imagined growing up to be a muscular man. Childhood can be hard on any kid who doesn't fit in. For John it was torture, because he was incessantly teased and ostracized. John knew nothing about the failed circumcision and subsequent surgery, nor the fact that he was genetically a male. However, as he got older, he was more attracted to girls than boys, and he expressed the opinion that he felt like a boy trapped in a girl's body. By the age of 14, after being on estrogen for 2 years, he looked increasingly like a girl, but he stopped living as one. John's father finally told him what had happened when he was young. John immediately requested sex-change hormonal therapy and surgery. For years, John has dealt with the overwhelming emotional problems resulting from his past. However, he is now married and has adopted his wife's children.

John's experience suggests that he had a "male brain" at birth, and no amount of hormonal therapy or female upbringing could change his gender identity. This provides an interesting contrast to cases of androgen insensitivity and congenital adrenal hyperplasia, in which gender identity appears at odds with the genetic sex. Clearly, gender identity involves a complex interplay between genetics, hormones, and life experiences.

[1] *Time,* January 8, 1973, p. 34.

Woolley and McEwen showed that it is indeed estradiol itself that triggers the increase in spine numbers, and that as hippocampal neurons grow more spines, they also grow more excitatory synapses. Furthermore, the new spines seem to have more of the NMDA type of glutamate receptor. This could explain why estradiol also enhances long-term synaptic plasticity in the hippocampus (see Chapter 24).

How does estradiol increase the numbers of hippocampal spines and excitatory synapses? Explanations in biology are often not simple and direct, especially where sex is concerned. Recent experiments by Diane Murphy, Menachem Segal, and their colleagues at the National Institutes of Health and the Weizmann Institute in Israel, suggest that the direct effect of estradiol in the hippocampus is to depress synaptic inhibition. Estradiol receptors in the hippocampus are primarily within inhibitory interneurons, which para-

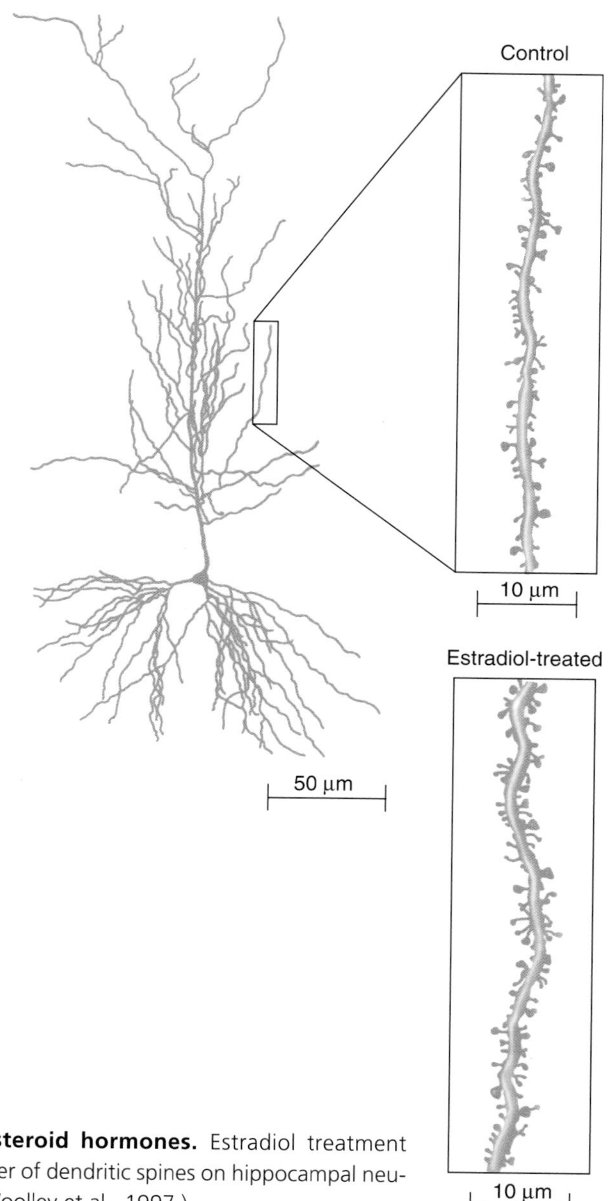

Figure 17.15
An activational effect of steroid hormones. Estradiol treatment causes an increase in the number of dendritic spines on hippocampal neurons. (Source: Adapted from Woolley et al., 1997.)

doxically are *not* the cells that grow more spines. However, estradiol causes the inhibitory cells to produce less GABA, their neurotransmitter, and therefore synaptic inhibition becomes less effective. Less inhibition, in turn, leads to an increase in general neural activity, which somehow triggers an increase in spines and excitatory synapses on the pyramidal cells.

Putting the pieces together, it seems that estradiol produces a hippocampus with less effective inhibitory synapses and additional excitatory circuitry. We can now only speculate about the significance of these results. Women with certain forms of epilepsy find that the frequency and severity of their seizures vary with their menstrual cycles. Seizures are most likely during the phase of the cycle with the highest ratio of estrogen to progesterone. Studies of experimental animals show that reducing inhibition and/or increasing excitation are very effective ways of provoking seizures—these are

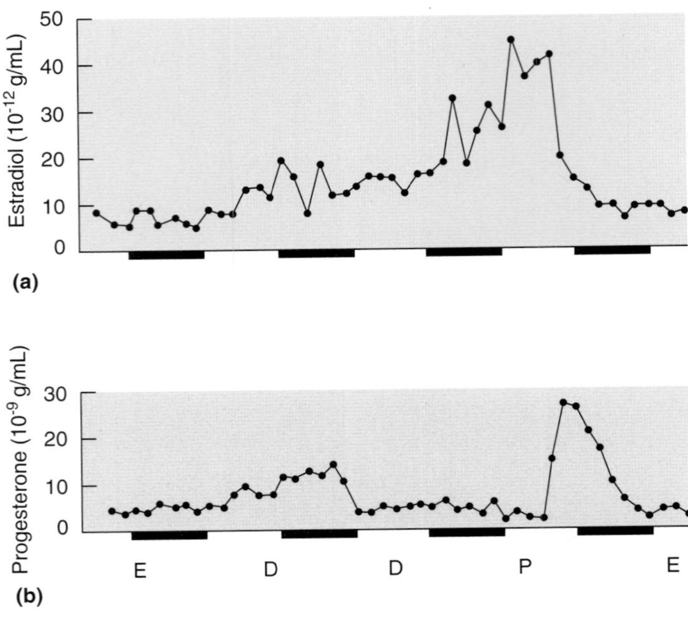

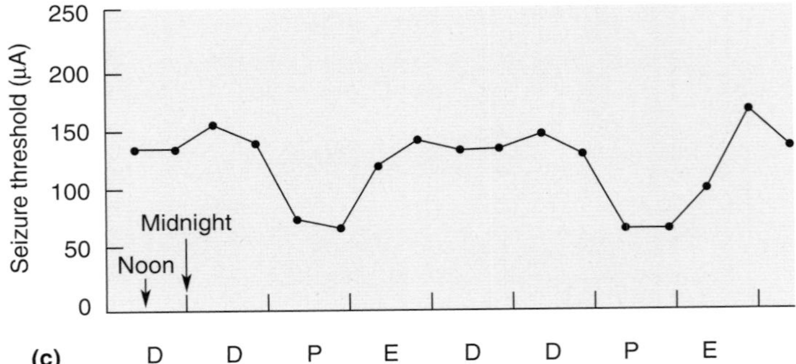

Figure 17.16
The correlation between fluctuations of steroid levels during the estrous cycle with hippocampal seizure threshold. The circulating levels of **(a)** estradiol, **(b)** progesterone, and **(c)** the threshold (in terms of stimulating current) for triggering a seizure in the hippocampus of a female rat during the estrous cycle. The phases of the estrous cycle are D, disestrous; P, proestrous; E, estrous. (Sources: a and b, Smith et al., 1995; c, Terasawa et al., 1968.)

precisely the effects of increased estradiol levels. A low level of progesterone may also contribute to seizure susceptibility; recall that progesterone facilitates the inhibitory effects of the $GABA_A$ receptor. But these effects of steroids certainly did not evolve in order to enhance epileptic seizures. Might they have a more adaptive function?

In rats, the hippocampus is particularly important for spatial memory and navigational skills. Woolley notes that the peak in hippocampal spine number coincides with the rat's peak in fertility. During that period the female actively seeks out mates, which may require the heightened spatial ability that might come with a more excitable, NMDA-receptor-filled hippocampus. Thus, the female rat's brain may fine-tune itself on a 5-day cycle in order to meet changing reproductive needs.

SEXUAL ORIENTATION

Although estimates vary widely, about 3% of the American population is homosexual. Homosexuality has been around a long time. There are references to it in the Bible and in ancient Greek literature and art. Why do some people seek sexual relationships with members of their own sex? Perhaps an equally valid question is why most people do not. There are many reasons why most people are drawn to the opposite sex: It is necessary for having children, most cultures strongly support heterosexual relationships, most cultures discourage homosexual relationships, it is the way our parents acted, and so on. Because heterosexuality is the norm, research into sexual orientation does *not* address the nebulous issue of why most people are attracted to the opposite sex; it seeks to find the cause of homosexuality. This line of research is fraught with religious and political overtones that sometimes make it difficult to be objective about the data.

Innumerable psychological studies comparing homosexual and heterosexual populations have been done. The results of this research often suggest that homosexual individuals are different from an early age. For example, statistically speaking, boys who prefer traditionally girls' toys as children are more likely to be gay as adults.

Although interesting, it is difficult to use the behavioral studies to answer the questions we want to address here: Are the brains of homosexual individuals different from those of heterosexuals? Is there a biological basis for sexual orientation? Of course, in some sense this has to be true, if we believe that all behavior is based on brain activity. But are the biological differences macroscopic, or are they embedded in the scattered thousands of unobservable synaptic connections that store a lifetime of experiences?

Hypothalamic Nuclei in Heterosexuals and Homosexuals

We saw already that there may be gender differences in the corpus callosum and some of the interstitial nuclei of the anterior hypothalamus (INAH). Some researchers have speculated that these differences are related to sexually dimorphic behavior. More recently, studies of the INAH suggest that there are differences between homosexual and heterosexual brains that might be related to sexual orientation. The best evidence for this comes from the research of Simon LeVay, then working at the Salk Institute (Box 17.3). LeVay first confirmed the earlier result of Laura Allen's work at UCLA, that INAH-3 is about twice as large in men as in women. By analyzing the brains of homosexual and heterosexual men, LeVay found that INAH-3 in gay men is only half the size of the nucleus in straight men (Figure 17.17). In other words, INAH-3 in homosexual men is similar in size to that of women. This result received a great deal of attention in the popular press. Some saw it as objective proof that homosexuality is biological rather than the outcome of a deficient childhood. However, as intriguing as this finding is, it is quite difficult to interpret in terms of complex human behavior. We know that destroying the homologue of INAH-3 in rats has little effect on sexual behavior. Also, even if the nucleus size is different in adults, how do we know whether this difference existed at birth, rather than being sculpted by childhood experiences? Clearly, LeVay's research is only a first step in the exploration of the biology of sexual orientation.

A Genetic Basis for Sexual Orientation?

One of the difficulties with LeVay's result is that homosexuality is only expressed in adults, and we don't know if any difference in INAH-3 existed before adulthood. A biological basis of sexual orientation might be more convincing if there were a homosexual/heterosexual brain difference that did not change with age. Is there a difference at birth that would allow us to reliably predict the sexual orientation a person would assume as an adult?

Dean Hamer and his colleagues at the National Cancer Institute have reported that there might be such a difference. Work leading up to Hamer's study found that the probability that a man would be homosexual was higher if other males in his family were homosexual. Moreover, this increased incidence was associated with cousins and uncles related through a man's

Box 17.3

PATH OF DISCOVERY

Queer Science

BY SIMON LEVAY

In 1991, I published a short report in *Science* describing a difference in the structure of the hypothalamus of gay and straight men. Although this report was the result of a "hobby project" (my lab's main focus was the visual system), it attracted more attention than all my other papers combined, and ended up drawing me away from active participation in science and into a more socially engaged role.

To understand the social significance of this kind of research, you have to recognize that people have wildly varying beliefs about the nature and causes of homosexuality, and that these views are strongly tied in with their attitudes toward gay people and how they should be treated. Some people believe that a person's sexual orientation—whether gay, bisexual, or straight—is a defining and durable aspect of his or her core identity. The majority of gay people, including myself, are in this camp. At the other extreme is the view that homosexuality is nothing more than a behavior pattern. According to a number of surveys, the latter view is correlated with anti-gay attitudes, and specifically with the belief that gay people can and should pull themselves together, get some therapy, find religion, or by some other means revert to a "normal" heterosexual state.

My finding was widely viewed as supporting the "born-that-way" theory of homosexuality. Thus, I found myself drawn into the culture wars that, in the early 1990s, revolved round gay people and gay rights. Should gays serve in the military? Should openly gay teachers be fired? Should gay people be covered by nondiscrimination statutes? Should gay couples be allowed to marry? To adopt children? It sometimes seemed as if the answers to all these questions depended on the size of a tiny speck of tissue in the anterior hypothalamus. So I spent (and still do spend) quite a bit of time speaking, writing, and testifying on the biology of sexual orientation and its relevance to public policy.

Of course, making the science relevant to social and legal issues in this way has its dangers. For one thing, the scientific findings are incomplete. Since my 1991 *Science* paper, there has been no independent verification (or refutation) of my findings. And the question of how the cell cluster I described comes to be different in homosexual and heterosexual men—as a result of different programs of fetal development, or perhaps different behaviors and experiences in adult life—has not been directly answered. Still, I think the totality of the evidence so far—human neuroanatomy, developmental studies in animals, genetic research, cognitive psychology, as well as research into attempts to convert gay people to heterosexuality—strongly supports the notion that a person's sexual orientation is a core personality trait, established at least in significant part through prenatal processes of brain differentiation.

Simon LeVay

Another possible danger is that the scientific findings might be used to "pathologize" homosexuality and to provide the possible tools for a "cure." The history of the field offers plenty of examples of such abuses, as I documented in my 1996 book, *Queer Science*. I have consistently presented the science in a neutral way. My *Science* paper, for example, spoke only of "differences," not of "normal" or "abnormal" structures. Yet right in the editorial pages of that same issue of *Science* was the headline "Missing Cells?"—as if the whole point of my study was to find out what is defective about homosexual brains.

Will people try to replace those "missing cells"? That's not an entirely implausible scenario. One research group has already shown that grafting neurons into the hypothalamus can modify another of its functions—the timing of circadian rhythms. In a 1997 biomedical thriller, *Albrick's Gold*, I explored the nightmarish consequences of an attempt to convert gay men to heterosexuality by means of hypothalamic cell implants.

Only one long-term strategy can prevent the abuse of neurobiological or genetic research in this field: creating a society in which homosexuality is seen as just another aspect of human diversity, where gay men and lesbians are respected as individuals and valued for the special gifts they bring to the world. The ultimate achievement of this goal will require, above all, the active engagement of gay people in all sectors of society. But the science itself, by moving the focus of attention from sex acts to sexual identity, contributes to that goal.

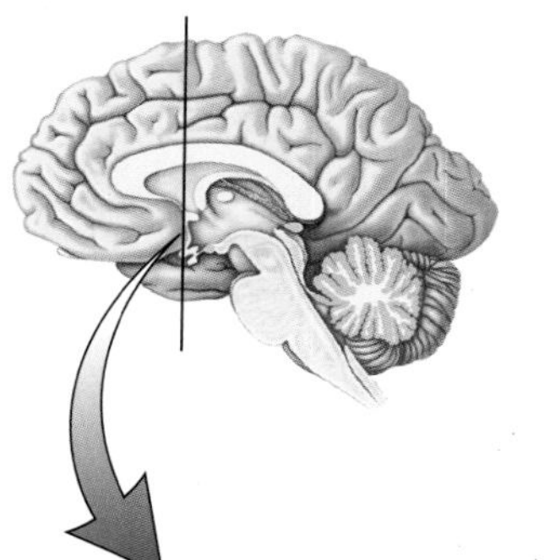

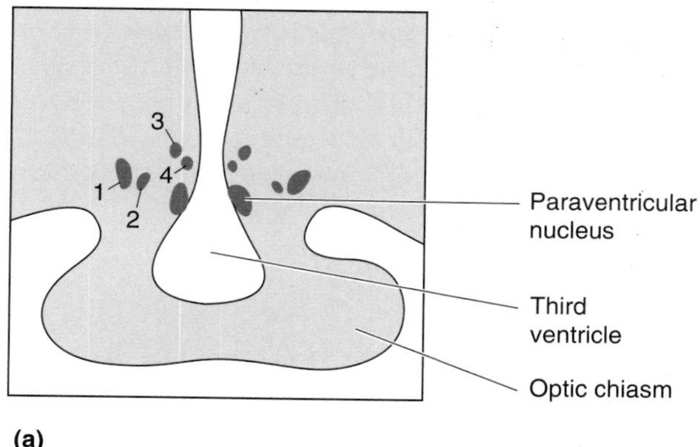

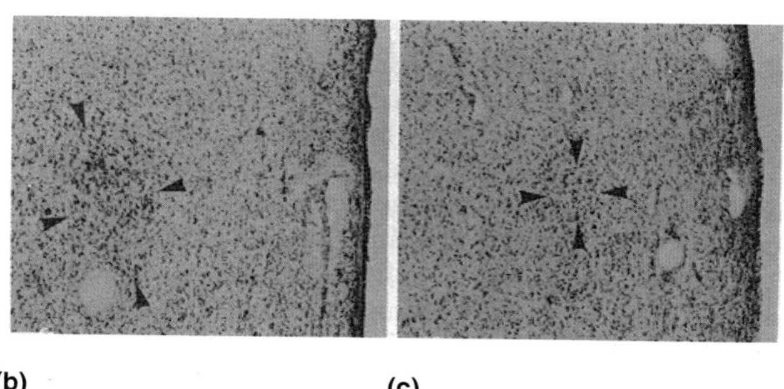

Figure 17.17

The location and size of INAH-3. (a) The location of the four INAH nuclei in the hypothalamus. **(b)** A micrograph of INAH-3 (arrows) in a heterosexual man. **(c)** A micrograph of INAH-3 in a homosexual man shows the nucleus to be smaller and the cells more scattered. (Source: Micrographs from LeVay, 1991, p. 1035.)

mother rather than father. This suggested there might be something on the X chromosome related to sexual orientation. In Hamer's study, DNA from the X chromosome was analyzed in 40 pairs of brothers (not twins) who were both homosexual. By the chance mixture of the mother's and father's DNA, any particular segment of DNA should be the same for two brothers only 50% of the time. Hamer found this to be true for most segments of DNA, except for one portion at the tip of the X chromosome that was much more likely to be the same in the gay brothers (Figure 17.18). The implication was that the DNA Hamer located might code for proteins that would somehow make the brain more likely to act in a homosexual manner.

Although Hamer's finding does not mean he located a "gay gene," it suggests there might be a genetic component to sexual orientation. However, it is clear that genetics are not the whole story. Most studies find that considerably fewer than 100% of identical twins share the same sexual orientation. Just as with LeVay's study, Hamer's report stirred up considerable controversy. Besides the political and religious debate about the study's implications, uncertainty has been introduced by other researchers who have been unable to replicate the original finding. At present, whether or not there is a neural or genetic basis for sexual orientation remains unclear.

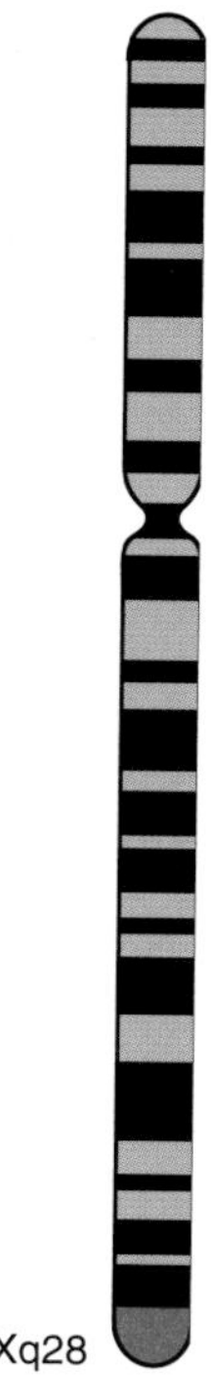

Figure 17.18

A homosexuality gene? At location 28, near the tip of the long (q) arm of the X chromosome, Hamer and his colleagues found a gene (Xq28) that appeared to correlate with homosexuality in brothers.

CONCLUDING REMARKS

The relationship between sex and the brain is one of the most slippery topics in neuroscience. Most adults think they know a lot about sex, but even issues that seem straightforward can be surprisingly complex and ambiguous, as in the definition of gender, for example.

The subject of sex and the brain is also complicated by the subtleties of the biological and cultural mechanisms that determine sexual behavior. Particularly in humans, the anatomical differences between the female and male nervous systems are not readily apparent, as indeed most human behavior is not distinctly masculine or feminine. Where small brain differences between the sexes occur, any adaptive purpose they may serve is not clear. And in no case is the neurobiological basis for sex differences in cognition known.

Nevertheless, the essential biological imperative—procreation—demands sex-specific behaviors, at least for mating and giving birth. For the most concretely sexual structures (such as the muscles and motor neurons controlling the penis or the sensory afferents innervating the clitoris), identifying some of the peripheral and spinal neural systems involved is fairly easy. The powerful role of sex hormones in sexual development and behavior is also clear. But the more complex aspects of sexual behavior and the brain systems that generate them are still quite mysterious.

We have touched on only a few of the issues that define the study of sex and the brain, and most of the basic questions remain unanswered. Scientific research about sex was long hampered by society's reluctance to talk openly about the subject, and these days, sexual politics tends to muddy the scientific waters. But sexual behavior is a defining feature of being human, and understanding its neural basis is a worthy challenge.

KEY TERMS

Sex and Gender
gender identity (p. 549)
sex-determining region of the Y chromosome (SRY) (p. 550)

The Hormonal Control of Sex
androgens (p. 553)
estrogens (p. 553)
luteinizing hormone (LH) (p. 554)
follicle-stimulating hormone (FSH) (p. 554)
gonadotropins (p. 554)
gonadotropin-releasing hormone (GnRH) (p. 554)
menstrual cycle (p. 554)
estrous cycle (p. 556)

The Neural Basis of Sex-Related Behaviors
polygyny (p. 560)
polyandry (p. 560)
monogamy (p. 560)

Why and How Male and Female Brains Differ
sexual dimorphism (p. 562)
organizational effect (p. 569)
activational effect (p. 569)

REVIEW QUESTIONS

1. Suppose you have just been captured by aliens who have landed on Earth to learn about humans. The aliens are all one gender, and they are curious about the two human genders. To earn your freedom, all you must do is tell them how to reliably distinguish males from females. What biological and/or behavioral tests do you tell them to conduct? Be sure to describe any exceptions that might violate your gender tests—you don't want the aliens to get angry!
2. Trace the chain of events that might lead psychological stress to reduce male sperm production and potency.
3. Figure 17.11 shows an interesting but unexplained observation: In the brain of a mother rat during periods of lactation, the size of the somatosensory cortex representing the skin around the nipples expands. Speculate about a likely mechanism for this phenomenon. Suggest a reason why such brain plasticity might be advantageous.
4. Estradiol is usually described as a female sex hormone, but it also plays a critical role in the early development of the male brain. Explain how this happens, and why the female brain is not similarly affected by estradiol at the same stage of development.
5. Where and how can steroid hormones influence neurons in the brain, at the cellular level?
6. Imagine that a research team has just claimed that a small and obscure nucleus in the brain stem, nucleus X, is sexually dimorphic and essential for certain "uniquely male" sexual behaviors. Discuss the kinds of evidence you would need to accept these claims about (a) the existence of a dimorphism, (b) the definitions of uniquely male behaviors, and (c) the involvement of nucleus X in these sexual behaviors.

CHAPTER

18 Brain Mechanisms of Emotion

INTRODUCTION

To appreciate the significance of emotions, just imagine life without them. Instead of the daily highs and lows we all experience, life would be a great empty plane of existence without significance. Without question, the expression of emotion is a large part of being human. Aliens in science fiction movies often look just like Earthlings, but they appear inhuman simply because they exhibit no emotions.

In this chapter, we explore the neural basis of emotion. Because it is difficult to study emotions by using the techniques for studying sensory and motor systems, it's a tricky subject. If you are studying a sensory system, you can present a stimulus and seek neurons that respond to it. You can manipulate the stimulus to determine the stimulus attributes (orientation, sound frequency, etc.) that are best for evoking a response. But how can this technique be used to study emotion? It is not a straightforward matter to study emotions in animals that cannot tell us their subjective feelings. What we observe are the behavioral manifestations of the internal emotions. Therefore, we must carefully distinguish between emotional *experience* and emotional *expression*. What we know about the brain mechanisms of emotion has been derived from a synthesis of animal studies on emotional expression and clinical cases that give us insight into human feelings.

In the simplest sense, the study of emotion can be reduced to an input-output problem. Most of the stimuli that evoke emotional responses come from our senses. The behavioral signs of emotion are controlled by the somatic motor system, the autonomic nervous system, and the secretory hypothalamus. Thus, one question we can ask is how incoming sensory stimuli lead to the behavioral and physiological responses indicative of emotional expression. The mechanisms of emotional experience are harder to grasp, but it is generally believed that the cerebral cortex plays a key role. The question is how sensory input or some internal signal ultimately leads to cortical activation characteristic of a specific emotion.

WHAT IS EMOTION?

Emotions—love, hate, disgust, joy, shame, envy, guilt, fear, anxiety, and so on—are feelings we all have at one time or another. But what precisely defines those feelings? Are they sensory signals from our body, diffuse patterns of activity in our cortex, or something else? Questions of this sort have proven to be surprisingly difficult to answer and have led to the development of various theories about exactly what emotions are.

Theories of Emotion

In the nineteenth century, several highly regarded scientists, including Darwin and Freud, considered the role of the brain in the expression of emotion (Figure 18.1). Careful observations of emotional expression in animals and humans and emotional experience in humans led to the development of theories relating expression and experience.

The James-Lange Theory. In 1884, the renowned American psychologist and philosopher William James proposed one of the first well-articulated theories of emotion. Danish psychologist Carl Lange proposed related ideas. Their theory, commonly known as the **James-Lange theory** of emotion, proposed that we experience emotion *in response to* physiological changes in our body. For example, we feel sad because we cry rather than cry because we are sad. Our sensory systems send information to our brain about our situation, and as a result our brain sends signals out to the body, changing muscle tone,

Figure 18.1
The emotional expression of a cat terrified by a dog. This drawing is from Darwin's book *The Expression of the Emotions in Man and Animals*. Darwin conducted one of the first extensive studies of emotional expression. (Source: Darwin, 1872/1955, p. 125.)

heart rate, and so on. The sensory systems react to the changes evoked by the brain, and it is this sensation that constitutes the emotion. According to James and Lange, the physiological changes *are* the emotion, and if they are removed, the emotion will go with them. This seems like a backward idea to many people today, as it did to many contemporaries of James and Lange. Until this theory was proposed, the commonly held conception was that an emotion is evoked by a situation and the body changes in response to the emotion. The James-Lange theory is the exact opposite.

Before you reject this theory as ridiculous, try one of the thought experiments suggested by James. Suppose you're boiling with anger about something that has just happened. Try to strip away all of the physiological changes associated with the emotion. Your pounding heart calms, your tense muscles relax, and your flushed face cools. It is hard to imagine maintaining rage in the absence of any physiological signs. In fact, this little experiment isn't much different from the technique used in meditation classes to relieve stress. Here is another example. You're on a first date with someone you are really attracted to, and you're swimming in a steaming soup of emotions that include happiness, love, lust, and anxiety. Then, poof! All at once the physiological signs of your infatuation are removed (like taking an imaginary cold shower). Are you still feeling the same emotional state? Probably not.

Even if it is true that emotion is intimately tied to physiological state, this doesn't mean that emotion cannot be felt in the absence of obvious physiological signs (a point even James and Lange would have conceded). But for strong emotions that are typically associated with physical change, there is a close relationship between emotion and its physiological manifestation, and it is not obvious which causes which.

The Cannon-Bard Theory. Although the James-Lange theory became popular in the early twentieth century, it soon came under attack. In 1927, American physiologist Walter Cannon published a paper containing several compelling criticisms of the James-Lange theory and went on to propose a new theory. Cannon's theory was modified by Philip Bard, and the **Cannon-**

Bard theory of emotion, as it came to be known, proposed that emotional experience can occur independently of emotional expression.

One of Cannon's arguments against the James-Lange theory was that emotions can be experienced even if physiological changes cannot be sensed. To support this claim, he offered the cases of animals he and others studied after transection of the spinal cord. Such surgery eliminated body sensations below the level of the cut, but it did not appear to abolish emotion. To the extent possible with muscular control of just the upper body or head, the animals still exhibited signs of emotions. Similarly, Cannon noted human cases in which a transected spinal cord did not diminish emotion. If emotional experience occurs when the brain senses physiological changes in the body, as the James-Lange theory proposed, then eliminating sensation should also eliminate emotions, and this did not appear to be the case.

A second observation of Cannon's that seems inconsistent with the James-Lange theory is that there is no reliable correlation between the experience of emotion and the physiological state of the body. For example, fear is accompanied by increased heart rate, inhibited digestion, and increased sweating. These same physiological changes, however, accompany other emotions, such as anger, and even nonemotional conditions of illness, such as fever. How can fear be a consequence of the physiological changes, when these same changes are associated with states other than fear?

Cannon's new theory focused on the idea that the thalamus plays a special role in emotional sensations. In this theory, sensory input is received by the cerebral cortex, which in turn activates certain changes in the body. But according to Cannon, this stimulus-response neural loop is devoid of emotion. Emotions are produced when signals reach the thalamus either directly from the sensory receptors or by descending cortical input. In other words, the character of the emotion is determined by the pattern of activation of the thalamus. An example may clarify the difference between this and the James-Lange theory. According to James and Lange, you feel sad because you cry; if you could prevent the crying, the sadness should go, too. In Cannon's theory, you don't have to cry to feel sad; there simply has to be the appropriate activation of your thalamus in response to the situation. The James-Lange and Cannon-Bard theories of emotion are compared in Figure 18.2.

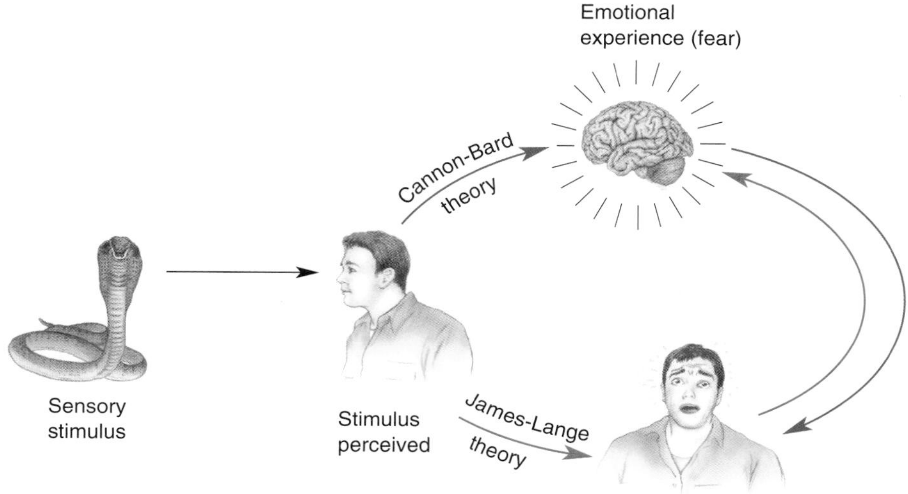

Figure 18.2
A comparison of the James-Lange and Cannon-Bard theories of emotion. In the James-Lange theory (red arrows), the man perceives the frightening animal and reacts. As a consequence of his body's response to the situation, he becomes afraid. In the Cannon-Bard theory (blue arrows), the frightening stimulus leads to the feeling of fear first, and then there is a reaction.

From Theory to Experimental Studies

It would take too much space to discuss all of the theories of emotion that have been proposed since the days of the James-Lange and Cannon-Bard theories. Subsequent work has demonstrated that each theory has merits as well as flaws. For instance, it has been shown, contrary to Cannon's statements, that fear and rage are associated with distinguishable physiological responses, even though they both activate the sympathetic division of the autonomic nervous system (ANS). Although this does not prove that these emotions are a result of distinct physiological responses, the responses are at least different.

Another interesting challenge to the Cannon-Bard theory that later studies demonstrated is that emotion is sometimes affected by damage to the spinal cord. In one study of adult men with spinal injuries, there was a correlation between the extent of sensory loss and reported decreases in emotional experiences. This fascinating result might be used to revive the James-Lange theory, that emotional experience depends on emotional expression, but other studies of people with spinal injuries have not always found a similar correlation. There is also some evidence that forcing oneself to *express* an emotion—such as smiling in order to *feel* happy—sometimes works. Perhaps experiencing some emotions does depend on behavioral manifestations, and experiencing others does not.

Ultimately, these questions can only be answered when we understand the neural basis for emotional experience. Although a complete understanding remains elusive, a valuable strategy has been to trace the pathways in the brain that link sensations (inputs) to the behavioral responses (outputs) that herald emotional experience. In the remainder of this chapter, we'll see that different emotions may depend on different neural circuits, but in many cases these circuits converge on the same parts of the brain.

THE LIMBIC SYSTEM CONCEPT

In previous chapters, we discussed how sensory information from peripheral receptors is processed along clearly defined, anatomically distinct pathways to the neocortex. The components of a pathway collectively constitute a *system*. For example, neurons located in the retina, lateral geniculate nucleus (LGN), and striate cortex work together to serve vision, so we say they are part of the visual system. Is there a system, in this sense, responsible for experiencing emotions? Beginning around 1930, some scientists argued that there is, and it came to be known as the limbic system. Shortly, we will discuss the difficulties of trying to define a single system for emotion. But first, let's examine the origin of the limbic system concept.

Broca's Limbic Lobe

In a paper published in 1878, French neurologist Paul Broca noted that on the medial surface of the cerebrum, all mammals possess a group of cortical areas that are distinctly different from the surrounding cortex. Using the Latin word for "border" (*limbus*), Broca named this collection of cortical areas the **limbic lobe** because they form a ring or border around the brain stem (Figure 18.3). According to this definition, the limbic lobe consists of the cortex around the corpus callosum, mainly in the cingulate gyrus, and the cortex on the medial surface of the temporal lobe, including the hippocampus. Broca did not write about the importance of these structures for emotion, and for some time they were thought to be primarily involved in olfaction. The word *limbic* and the structures in Broca's limbic lobe were, however, subsequently closely associated with emotion.

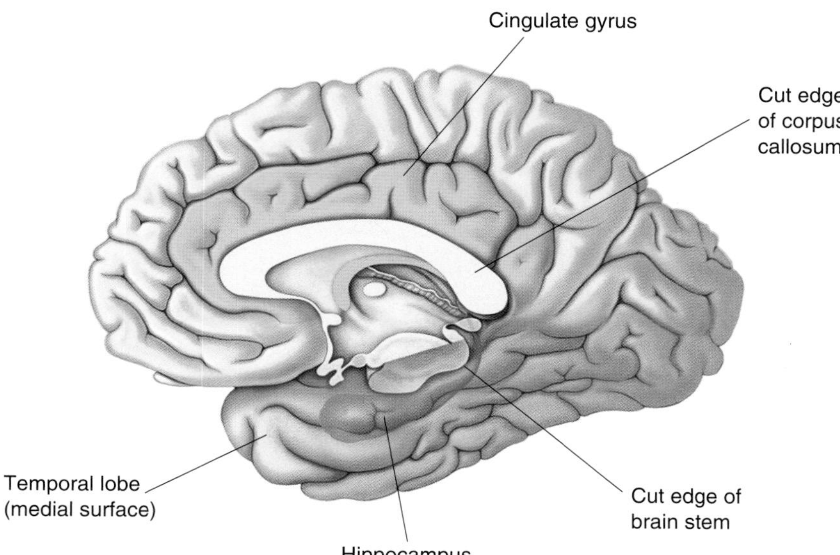

Figure 18.3
The limbic lobe. Broca defined the limbic lobe as the structures that form a ring around the brain stem and corpus callosum on the medial walls of the brain. The brain stem has been removed in the figure so that the medial surface of the temporal lobe is visible.

The Papez Circuit

By the 1930s, evidence suggested that a number of limbic structures are involved in emotion. Reflecting on the earlier work of Cannon, Bard, and others, American neurologist James Papez proposed that on the medial wall of the brain there is an "emotion system" that links the cortex with the hypothalamus. Figure 18.4 shows the group of structures that have come to be called the **Papez circuit**. Each is connected to another by a major fiber tract.

Papez believed, as do many scientists today, that the cortex is critically involved in the experience of emotion. Sometimes profound changes in emotional expression with little change in perception or intelligence follow damage to certain cortical areas (Box 18.1) Also, tumors near the cingulate cortex

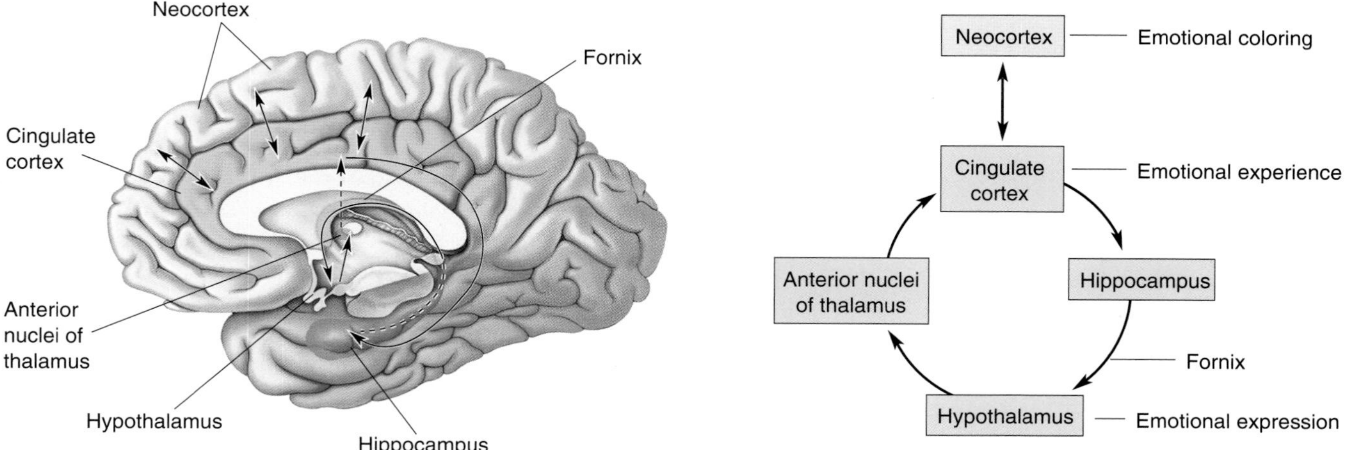

Figure 18.4
The Papez circuit. Papez believed that the experience of emotion was determined by activity in the cingulate cortex and, less directly, other cortical areas. Emotional expression was thought to be governed by the hypothalamus. The cingulate cortex projects to the hippocampus, and the hippocampus projects to the hypothalamus by way of the bundle of axons called the fornix. Hypothalamic effects reach the cortex via a relay in the anterior thalamic nuclei.

Box 18.1

OF SPECIAL INTEREST

The Amazing Case of Phineas Gage

One of the most amazing studies ever conducted on the brain's influence on emotional expression was actually an industrial accident. The unfortunate subject of this study was Phineas Gage, a 25-year-old foreman at a railroad construction site in Vermont. On September 13, 1848, while tamping explosive powder into a hole in preparation for blasting at the construction site, he made the mistake of looking away from what he was doing. His tamping iron hit a rock, and the powder exploded. The consequences of this act are described by Dr. John Harlow in an 1848 article entitled "Passage of an Iron Rod Through the Head." When the charge went off, it sent the meter-long, 6 kg iron rod into Gage's head just below his left eye. After passing through his left frontal lobe, the rod exited the top of Gage's head.

Incredibly, after being carried to an ox cart, Gage sat upright on the ride to a nearby hotel and walked up a long flight of stairs to go inside. When Harlow first saw Gage at the hotel, he commented that "the picture presented was, to one unaccustomed to military surgery, truly terrific" (p. 390). As you might imagine, the projectile destroyed a considerable portion of the skull and left frontal lobe, and Gage lost a great deal of blood. The hole through his head was more than 9 cm in diameter. Harlow was able to stick the full length of his index finger into the hole from the top of Gage's head, and also upward from the hole in his cheek. Harlow dressed the wound as best he could. Over the following weeks, considerable infection developed. No one would have been surprised if the man had died. But just about a month after the accident, he was out of bed and walking around town!

Harlow corresponded with Gage's family for many years and in 1868 published a second article, "Recovery From the Passage of an Iron Bar Through the Head," describing Gage's life after the accident. After Gage recovered from his wounds, he was apparently normal except for one thing: His personality was drastically and permanently changed. When he tried to return to his old job as construction foreman, the company found he had changed so much for the worse that they wouldn't rehire him. According to Harlow, before the accident Gage was considered "the most efficient and capable foreman. . . . He possessed a well-balanced mind, and was looked upon by those who knew him as a shrewd, smart business man, very persistent in executing all his plans of operation" (pp. 339–340). After the accident, Harlow described him as follows:

> He is fitful, irreverent, indulging at times in the grossest profanity (which was not previously his custom), manifesting but little deference for his fellows, impatient of restraint or advice when it conflicts with his desires, at times

are associated with certain emotional disturbances, including fear, irritability, and depression. Papez proposed that activity evoked in other neocortical areas by projections from the cingulate cortex adds "emotional coloring." We saw in Chapter 15 that the hypothalamus integrates the actions of the ANS. In the Papez circuit, the hypothalamus governs the behavioral expression of emotion. The hypothalamus and neocortex are arranged so that each can influence the other, thus linking the expression and experience of emotion. In the circuit, the cingulate cortex affects the hypothalamus via the hippocampus and fornix (the large bundle of axons leaving the hippocampus), whereas the hypothalamus affects the cingulate cortex via the anterior thalamus. The fact that communication between the cortex and hypothalamus is bidirectional means that the Papez circuit is compatible with both the James-Lange and Cannon-Bard theories of emotion.

While anatomical studies demonstrated that the components of the Papez circuit were interconnected as Papez had indicated, there was only suggestive evidence that each was involved in emotion. One reason Papez thought the hippocampus is involved in emotion is that it is affected by the virus re-

(Box 18.1, *continued*)

pertinaciously obstinate, yet capricious and vacillating, devising many plans of future operation, which are no sooner arranged than they are abandoned in turn for others appearing more feasible. . . . His mind was radically changed, so decidedly that his friends and acquaintances said he was "no longer Gage." (pp. 339–340)

There are no psychological test results to tell us what had happened to Phineas Gage's cognitive abilities. From Harlow's account of the man's life after the accident, however, it appears that Gage's personality was altered far more than his intelligence. Phineas went on to live for another 12 years, without direction.

No autopsy was performed on Gage. Gage's skull and the tamping iron have been preserved, however, in a museum at Harvard Medical School. Figure A is a sketch made by Harlow indicating the size of the tamping iron relative to Gage's skull. In 1994, Hanna and Antonio Damasio and their colleagues at the University of Iowa made new measurements of the skull and used modern imaging techniques to assess the damage to Gage's brain. Their reconstruction of the tamping iron's path is shown in Figure B. The iron rod severely damaged the cerebral cortex in both hemispheres, particularly the frontal lobes. It was this damage that caused Gage to behave like an ill-tempered child, giving expression to an endless succession of strong emotions. The significant increase in emotional behavior he exhibited suggested that the cerebral cortex normally plays an important role in regulating emotional expression.

Figure A
Relative size of Gage's skull and the tamping iron. (Source: Harlow, 1868, p. 347.)

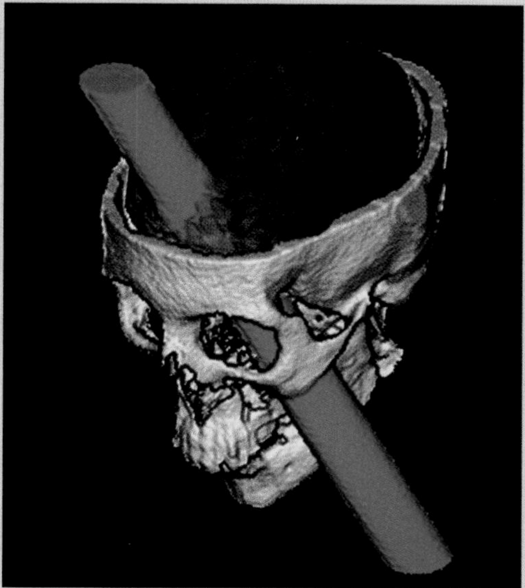

Figure B
Path of the tamping iron. (Source: Damasio et al., 1994, p. 1104.)

sponsible for rabies. An indication of rabies infection as well as an aid in its diagnosis is the presence of abnormal cytoplasmic bodies in neurons, especially in the hippocampus. Because rabies is characterized by hyperemotional responses such as exaggerated fear and aggressiveness, Papez reasoned that the hippocampus must be involved in normal emotional experience. Although little evidence related to the role of the anterior thalamus, clinical reports stated that lesions in this area led to apparent emotional disturbances, such as spontaneous laughing and crying.

You may have noticed the correlation between the elements composing both the Papez circuit and Broca's limbic lobe. Because of their similarity, the group of structures hypothetically responsible for the sensation and expression of emotion are often called the **limbic system**, even though Broca's anatomical notion of the limbic lobe originally had nothing to do with emotion. The term *limbic system* was popularized in 1952 by American physiologist Paul MacLean. According to MacLean, the evolution of a limbic system enabled animals to experience and express emotions and emancipated them from the stereotypical behavior dictated by their brain stem.

Difficulties With the Single Emotion System Concept

We have defined a group of interconnected anatomical structures roughly encircling the brain stem as a limbic system. Experimental work supports the hypothesis that some of the structures in Broca's limbic lobe and the Papez circuit play a role in emotion. On the other hand, some of the components of the Papez circuit, such as the hippocampus, are no longer thought to be important for the expression of emotion.

The critical point seems to be a conceptual one concerning the definition of an emotion system. What should we include in the system? For example, if a nucleus of cells in the brain plays a role in both emotion and some other function, such as olfaction or learning and memory, should it be included? Should we include every structure that contributes in any way to the experience or expression of any type of emotion? Part of the problem is the word *system*, which implies that the parts work together to perform a single common function. Given the diversity of our emotions, there is no compelling reason to think that only one system—rather than several—is involved. Conversely, solid evidence indicates that some structures involved in emotion are also involved in other functions; there is no one-to-one relationship between structure and function.

We are only beginning to learn how the experience and expression of emotion arise in the brain. Although the term *limbic system* is still commonly used, some question the utility of trying to define a single, discrete emotion system. Our approach is to focus on a few specific emotions for which there is strong evidence of the involvement of certain neural circuits.

FEAR AND ANXIETY

Suppose one night you're walking to a friend's house and you decide to take a shortcut. You become anxious when you find yourself on dark streets that are not particularly safe. You see a group of threatening-looking guys approaching you from the other end of the block, and now you are more than just anxious; you're frightened. Thinking about the anxiety and fear you would feel in this situation will remind you of the dramatic response the sympathetic division of your ANS can evoke. As discussed in Chapter 15, even before you have a behavioral reaction, your hypothalamus orchestrates a response in your ANS that affects virtually every part of your body, from increased heart rate and respiration to sweating. Generally, your level of anxiety and your body's response are proportional to the amount of perceived danger.

From a neurobiological standpoint, the question is how incoming sensory information leads to the behavioral and physiological responses associated with fear and anxiety. While we don't have the complete answer yet, compelling evidence suggests that a structure called the *amygdala*, which is buried in the temporal lobe, plays a key role. Before discussing recent studies of the amygdala, however, let's first look at a very influential experiment conducted in the 1930s that focused attention on the involvement of the temporal lobes in fear.

The Klüver-Bucy Syndrome

Shortly after Papez' proposal of an emotion circuit in the brain, neuroscientists Heinrich Klüver and Paul Bucy at the University of Chicago found that bilateral removal of the temporal lobes, or temporal *lobectomy*, in rhesus monkeys has a dramatic effect on the animals' responses to fearful situations. The surgery produces numerous bizarre behavioral abnormalities that Klüver and Bucy placed in five categories: psychic blindness, oral tendencies, hypermetamorphosis, altered sexual behavior, and emotional changes. This constellation of symptoms is called the **Klüver-Bucy syndrome**. Although our focus is the emotional disturbances, we briefly discuss the other symptoms so you can fully appreciate the syndrome.

Klüver and Bucy observed that the monkeys appeared to have *psychic blindness;* although they could see, they did not seem to recognize common objects or understand their meaning. They would pick up and examine every object within view and as the term *oral tendencies* implies, place each object in their mouth. They seemed to use their mouth instead of their eyes to identify each object. Klüver and Bucy put a variety of objects into the room with each monkey. Inedible items, such as a live mouse, a glass, and a nail, were examined by mouth and discarded. Pieces of food were examined the same way and then eaten. If a hungry monkey was shown a group of objects it had seen before, intermixed with food, the monkey would still go through the process of picking up each object for study. A normal hungry monkey in the same situation would make a beeline for the food. One thing this behavior seemed to show, in addition to psychic blindness and oral tendencies, was an irresistible compulsion to examine things. The animals appeared to be obsessed by the urge to run around and touch everything and place each found object in the mouth. This behavior is what Klüver and Bucy meant by the term *hypermetamorphosis*. There were also changes in sexual behavior. Some of the monkeys showed a strikingly increased interest in sex, including masturbation as well as heterosexual and homosexual acts. (The experimental notes of Klüver and Bucy read something like a script for an X-rated monkey movie.) These behaviors were not observed in normal monkeys.

The emotional changes in monkeys with Klüver-Bucy syndrome were most dramatically indicated by an apparent decrease in fear. For example, a normal wild monkey will avoid humans and other animals. In the presence of an experimenter, it will usually crouch in a corner and remain still; if approached, it will dash off to a safer corner. This sort of fear and excitement was not seen in the monkeys with bilateral temporal lobectomies. Experimental monkeys would not only approach and touch the human, they would even let the human stroke them and pick them up. Remember that these are otherwise wild monkeys. The same fearlessness was exhibited in the presence of other animals that monkeys normally fear. Even after approaching and being attacked by a natural enemy such as a snake, the monkey would go back and try to examine it again. You might think this behavior indicates stupidity or memory loss rather than fearlessness, but other evidence suggested that the monkeys actually experienced less fear. For example, there was a significant decrease in the vocalizations and facial expressions usually associated with fear. It appeared that both the normal experience and the normal expression of emotion were severely decreased by the temporal lobectomy.

Virtually all of the symptoms of the Klüver-Bucy syndrome reported in monkeys have also been seen in humans with temporal lobe lesions. In addition to visual recognition problems, oral tendencies, and hypersexuality, people appear to have "flattened" emotions.

The Amygdala

In interpreting the findings of Klüver and Bucy, we must keep in mind that a large amount of brain tissue was removed. Removal of the temporal lobes involves not only temporal cortex but also all of the subcortical structures in that area, including the amygdala and hippocampus. Some of the symptoms, especially psychic blindness, probably resulted from the removal of visual cortical areas in the temporal lobes (see Chapter 10). The emotional disturbances probably resulted, however, from destruction of the amygdala. Indeed, there is considerable evidence that the amygdala is involved in numerous aspects of emotion, not just fear.

Anatomy of the Amygdala. The **amygdala** lies in the pole of the temporal lobe just below the cortex on the medial side. Its name is derived from the Greek for "almond," because of its shape.

The amygdala is a complex of nuclei that are commonly divided into three groups: the **basolateral nuclei**, the **corticomedial nuclei**, and the central nucleus (Figure 18.5). Afferents to the amygdala come from a large variety of sources, including neocortex in all lobes of the brain and the hippocampal and cingulate gyri. Of particular interest here is the fact that information

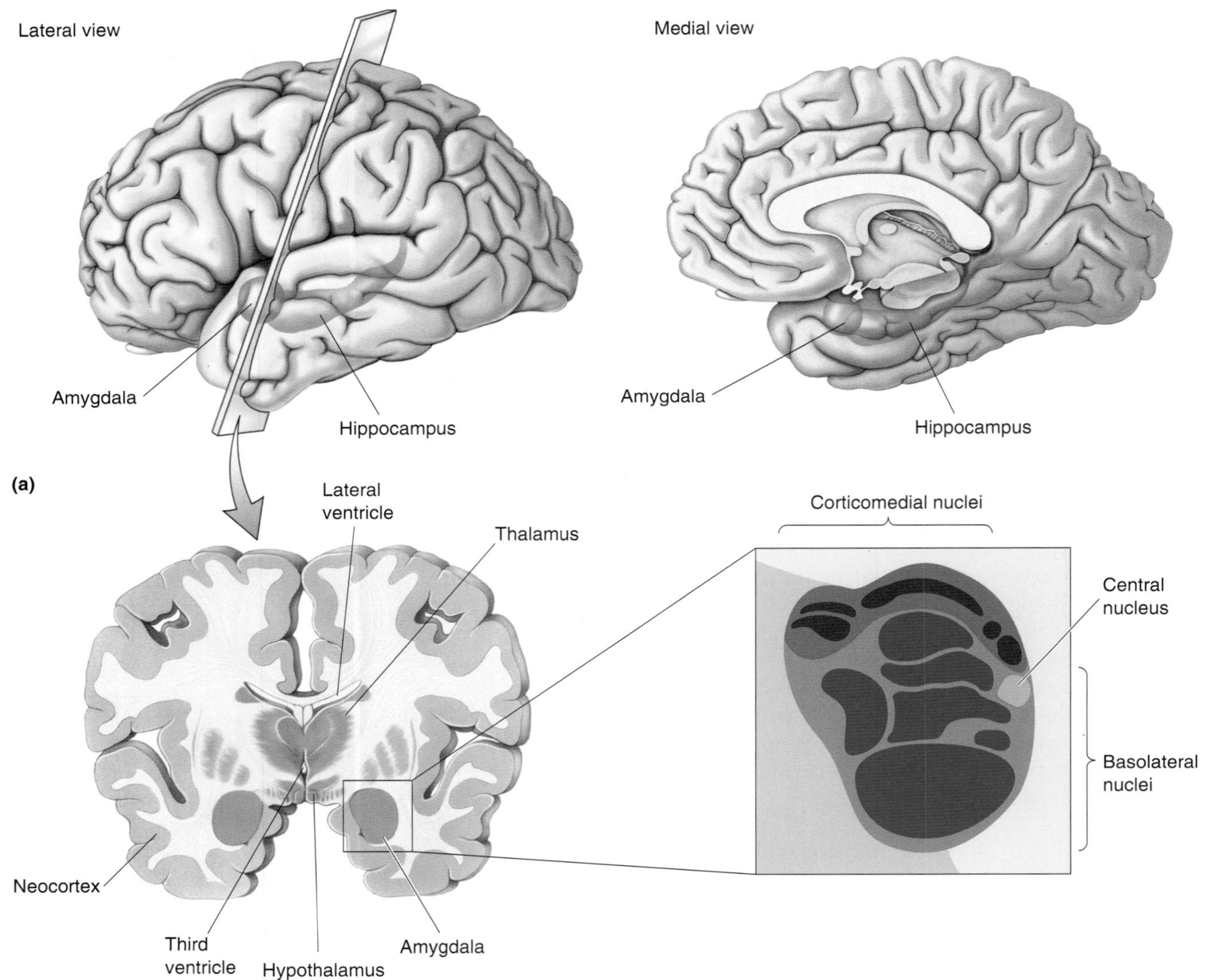

Figure 18.5
A cross section of the amygdala. (a) Lateral and medial views of the temporal lobe showing the location of the amygdala in relation to the hippocampus. **(b)** The brain is sectioned coronally to show the amygdala. The basolateral nuclei (surrounded by red) receive visual, auditory, gustatory, and tactile afferents. The corticomedial nuclei (surrounded by purple) receive olfactory afferents.

from all of the sensory systems feeds into the amygdala, particularly the basolateral nuclei. Each sensory system has a different projection pattern to the amygdala nuclei, and interconnections within the amygdala allow the integration of information from different sensory systems. Two major pathways, the *ventral amygdalofugal pathway* and the *stria terminalis*, connect the amygdala with the hypothalamus.

The Effects of Amygdala Destruction and Stimulation. It has been demonstrated in several species that lesions of the amygdala flatten emotion in a manner similar to the Klüver-Bucy syndrome. Amygdala damage in humans is associated with reduced emotionality. This appears to be a general effect extending to multiple emotions, but it is still possible that different emotions use different amygdala circuits. Bilateral amygdalectomy in animals can profoundly reduce fear and can affect aggression and memory. There are reports that rats so treated will approach a sedated cat and nibble its ear and that a wild lynx will become as docile as a house cat. The fearlessness is believed to result from the destruction of nuclei in the basolateral region of the amygdala. We will have more to say about amygdala involvement in aggression and memory a little later.

Very few cases of humans with selective damage to the amygdala have been documented. Ralph Adolphs and his colleagues at the University of Iowa studied a 30-year-old woman known as S. M., who had bilateral destruction of the amygdala resulting from Urbach-Wiethe disease. S. M. has normal intelligence and is perfectly able to identify people from photographs. But she has difficulty recognizing certain emotions expressed by people in the photographs. When asked to categorize the emotion expressed in a person's face, she normally describes happiness, sadness, and disgust. She is somewhat less likely to describe an angry expression as angry, and the most abnormal response is that she is unlikely to describe a fearful expression as afraid. It appears that the amygdala lesion selectively decreases her ability to recognize fear.

If removing the amygdala reduces the expression and recognition of fear, what happens if the intact amygdala is electrically stimulated? Depending on the site, amygdala stimulation can lead to a state of increased vigilance or attention. Stimulation of the lateral portion of the amygdala in cats can elicit a combination of fear and violent aggression. Electrical stimulation of the amygdala in humans has been reported to lead to anxiety and fear. Not surprisingly, the amygdala figures prominently in current theories about anxiety disorders, as we will see in Chapter 21.

A Neural Circuit for Learned Fear

Through socialization or painful experience, we all learn to avoid certain behaviors for fear of being hurt. If as a child you ever received a painful shock by pushing a paper clip into an electrical outlet, you probably never did it again. Memories associated with fear can be quickly formed and long-lasting. Although the amygdala is not thought to be a primary location for memory storage, it does seem to be involved in giving emotional content to memories.

A number of experiments suggest that neurons in the amygdala can "learn" to respond to stimuli associated with pain, and after such learning these stimuli evoke a fearful response. In an experiment performed by Bruce Kapp and his colleagues at the University of Vermont, rabbits were trained to associate the sound of a tone with mild pain. The researchers made use of the fact that a normal sign of fear in rabbits is a change in heart rate. An animal was placed in a cage, and at various times it would hear one of two tones. One tone was followed by a mild electrical shock to the feet through the metal floor of the cage; the other tone was benign. After training, Kapp's group found that the rabbit's heart rate developed a fearful response to the tone associated with pain but not to the benign tone. Prior to training, neurons in the central nucleus of the amygdala failed to respond to the tones used in the experiment. After training, however, neurons in the central nucleus of the amygdala responded to the shock-related tone (but not the benign one). Joseph LeDoux of New York University has shown that after this type of fear conditioning, amygdala lesions eliminate the learned visceral responses, such as the changes in heart rate and blood pressure. Michael Davis at Emory University has used the acoustic startle reflex in rats to work out the circuitry involved in fear responses (Box 18.2). This reflex is the behavioral response to a sudden, loud, scary sound, and a key component of the circuit underlying the reflex is the amygdala.

Recent research suggests that the role of the amygdala in learned fear, first studied in rabbits and rats, extends to humans. In one study, people were shown a number of visual stimuli and conditioned to expect a mild electrical shock when a particular visual stimulus was presented. This experiment was performed with the person in a functional magnetic resonance imaging (fMRI) machine so that brain activity could be monitored (see Box 7.2). The fMRI images show that the feared visual stimulus activated the amygdala significantly more than visual stimuli not associated with a shock (Figure 18.6a). Besides the amygdala, Figure 18.6a shows spots of increased activity in cerebral cortical areas associated with the conditioned stimulus. In a second fMRI study, activity in the cerebral cortex is even clearer. In this case, subjects viewed various visual stimuli and were conditioned to associate a loud, unpleasant tone with one of the pictures. In addition to amygdala activity, the visual stimulus associated with the loud aversive stimulus elicited prominent activity in the cingulate and insular cortex (Figure 18.6b). Presumably, the cortical activity is related to the emotional experience associated with the aversive stimulus.

Figure 18.7 shows a proposed circuit to account for learned fear. Auditory information is sent to the basolateral region of the amygdala, where, in turn, cells send axons to the central nucleus. Efferents from the central nucleus project to the hypothalamus, which can alter the state of the ANS, and to the periaqueductal gray matter in the brain stem, which can evoke behavioral reactions via the somatic motor system. The emotional experience is thought to be based on activity in the cerebral cortex.

ANGER AND AGGRESSION

In American culture, we are somewhat ambivalent about aggressiveness as a personality trait and about acts of violent aggression. We chastise some people who behave aggressively but praise others for aggressively pursuing their goals. Murder is considered a capital offense, but wartime killing is considered not only acceptable but honorable. Clearly, we distinguish different forms of aggression in humans. Likewise, we can categorize different types in animals. One animal may act aggressively toward another for many reasons: to kill for food, to defend offspring, to win a mate, to scare off a potential adversary. Though it has not been proven, there is some evidence that different types of aggression are regulated differently by the nervous system.

Aggression is a multifaceted behavior that is not a product of a single isolated system in the brain. One factor that influences aggression is the level of male sex hormones, or androgens (see Chapter 17). In animals there is a correlation between seasonal androgen levels and aggressive behavior. Consistent with one of the roles of androgens, injections of testosterone can make an immature animal more aggressive, and castration can reduce ag-

Box 18.2

PATH OF DISCOVERY

Startling Memories

BY MICHAEL DAVIS

When I was a freshman at Northwestern University, I took a course in philosophy and realized how exciting college was going to be. It had never dawned on me that I really didn't know for sure whether the sun would rise tomorrow! I became disenchanted with actually working in philosophy, however, because there didn't seem to be any rules. Psychology, on the other hand, also had great questions, but it had some rules. I began to take more courses in psychology and also one in art appreciation. When I was a sophomore, I had the grandiose idea that if I could study how the brain deals with vision, I might be able to create art like no one ever had before. I did a project in which I drew ambiguous forms and then had people rate them. Almost everyone said they liked the forms I liked and disliked the forms I disliked—even when I changed my ratings with a different group of subjects! I realized that psychology was more powerful than art, so I majored in psychology.

When I was a graduate student at Yale University, Allan Wagner and I studied startle habituation in rats. This happened quite by chance because it was useful for other studies going on in the lab at that time. We realized, however, this was an interesting measure of behavioral plasticity that might allow investigating mechanisms of learning and memory mediated by a known neural circuit in a complex vertebrate. Incredibly, it took my lab 17 years to figure out a startle pathway that we think has only three or four synapses! When we started, it was not possible to make a lesion that destroyed only cells, not fibers, passing through that area. So lots of leads turned out to be wrong. By eliciting startlelike responses electrically through implanted electrodes and then making small lesions through these electrodes, however, my students and I pieced together a probable "startle circuit."

I set up the fear-potentiated startle test because it measured learning by using the startle reflex as a probe to assess fear. I also hoped that knowing more about fear in rats would be relevant to studying the neural basis of fear and anxiety in people. After reading the work of Bruce Kapp on fear and the amygdala, my students found that lesions of this part of the brain produced a dramatic reduction in fear-potentiated startle and that activation of the amygdala greatly increased startle.

Michael Davis

Because we could elicit startlelike responses electrically from various points along the startle pathway, we used this technique to figure where fear ultimately alters neural transmission to increase startle. We reasoned that startle-like responses elicited electrically from points upstream from the site of modulation should be increased by fear, whereas responses elicited electrically from points downstream should not be. Based on these assumptions, we deduced that fear modulated transmission at the nucleus reticularis pontis caudalis, and when we put a retrograde tracer into this nucleus, we found lots of cells in the central nucleus of the amygdala—a joyous day in the lab!

Around that time, everyone was buzzing about the role of *N*-methyl-D-aspartate (NMDA) receptors in learning and memory. So our lab was very excited when we saw a total blockade of fear learning when we infused the NMDA receptor antagonist AP5 directly into the amygdala. We worked through the Christmas holidays to get the paper out. Now we are working on the intracellular processes that may be involved in long-term fear memories, as well as trying to figure out what parts of the brain are required for the suppression of fear—critical information for the treatment of anxiety disorders.

So I have devoted my entire scientific career to startling rats, even though my interest in startle began by chance. In fact, most important things in life happen quite by chance (where you are born, who your parents are, whom you marry, where you work). I never did get back to vision and art, where I thought I was headed. It has been a wonderful ride, however, and I can think of no better profession.

gressiveness. In humans the relationship is less clear, although some have claimed that there is a connection between testosterone levels and aggressive behavior in violent criminals. In any case, there is strong evidence for a neurobiological component to aggression, which is our focus here.

A useful distinction can be made between predatory aggression and af-

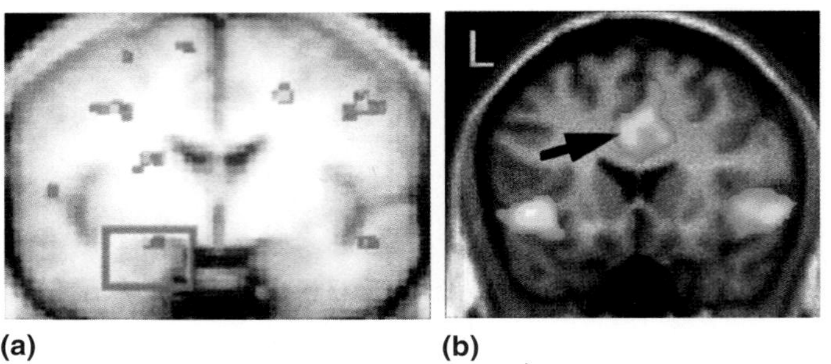

Figure 18.6
Human fMRI activity in response to aversive stimuli. (a) In this coronal brain section, the colored area within the green rectangle indicates that the amygdala was more active when a visual stimulus associated with an electrical shock was presented. Other sites more activated by the same stimulus were found in cerebral cortex. (Source: LaBar et al., 1998.) **(b)** The orange and yellow brain areas were more active in response to a visual stimulus associated with an aversive sound than to other visual stimuli. The arrow near the midline points to the cingulate cortex, and the symmetrical active areas on the sides are insular cortex. (Source: Büchel et al., 1998.)

fective aggression. **Predatory aggression** involves attacks against a member of a different species for the purpose of obtaining food, such as a lion hunting a zebra. Attacks of this type are typically accompanied by relatively few vocalizations, and they are aimed at the head and neck of the prey. Predatory aggression is not associated with high levels of activity in the sympathetic division of the ANS. **Affective aggression** is for show rather than to kill for food, and it involves high levels of activity in the sympathetic division of the ANS. An animal in this state will typically make vocalizations while adopting a threatening or defensive posture. A cat hissing and arching its back at the approach of a dog is a good example. The behavioral and physiological manifestations of both types of aggression must be mediated by the somatic motor system and the ANS, but the pathways must diverge at some point to account for the dramatic differences in the behavioral responses.

The Hypothalamus and Aggression

One of the first structures linked to aggressive behavior was the hypothalamus. Although some of the early experiments were crude by today's standards, they pointed the way for later studies.

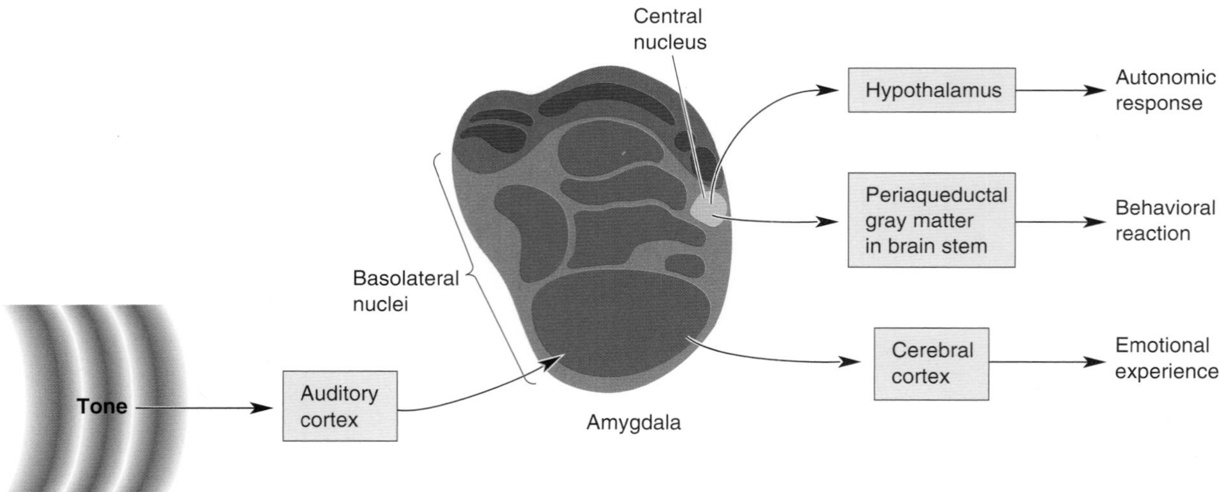

Figure 18.7
A neural circuit for learned fear. Through training, a sound becomes associated with pain. The presumed fear response is mediated by the amygdala. The emotional stimulus reaches the basolateral nuclei of the amygdala by way of the auditory cortex, and the signal is relayed to the central nucleus. Efferents from the amygdala project to the brain stem periaqueductal gray matter, causing the behavioral reaction to the stimulus and to the hypothalamus, resulting in the autonomic response. The experience of emotion presumably involves projections to the cerebral cortex.

Sham Rage. Experiments performed in the 1920s showed that a remarkable behavioral transformation took place in cats or dogs whose cerebral hemispheres have been removed. Animals that were not easy to provoke prior to the surgery would go into a state of violent rage with the least provocation after the surgery. For instance, a violent response might be produced by an act as mild as scratching a dog's back. This state was called **sham rage** because the animal demonstrated all of the behavioral manifestations of rage but in a situation that normally would not cause anger. It was also a sham in the sense that the animals would not actually attack as they normally might.

Perhaps these animals had every reason to be angry after such destructive surgery, but there is something to be learned from these experiments. While this extreme behavioral condition resulted from removing all of both cerebral hemispheres (telencephalon), the behavioral effect can be reversed by making the lesion just a little larger to include portions of the diencephalon, particularly the hypothalamus. Sham rage is observed if the anterior hypothalamus is destroyed along with the cortex, but it is not seen if the lesion is extended to include the posterior half of the hypothalamus (Figure 18.8). The implication is that the posterior hypothalamus may be particularly important for the expression of anger and aggression and that normally it is inhibited by the telencephalon. But we must bear in mind that the lesions were large, and something other than the posterior hypothalamus may have been destroyed with the larger lesion.

Electrical Stimulation of the Hypothalamus. In a series of pioneering studies begun in the 1920s, W. R. Hess at the University of Zurich investigated the behavioral effects of electrically stimulating the diencephalon. Hess made small holes in the skulls of anesthetized cats and implanted electrodes in the brain. After the animal awoke from the anesthesia, a small electrical current was passed through the electrodes, and behavioral effects were noted. Various structures were stimulated, but here we focus on the effects of stimulating different regions of the hypothalamus. The variety of responses from stimulating slightly different portions of the hypothalamus is amazing, considering that the hypothalamus is such a small part of the brain. Depending on where the electrode is placed, stimulation may cause the animal to sniff,

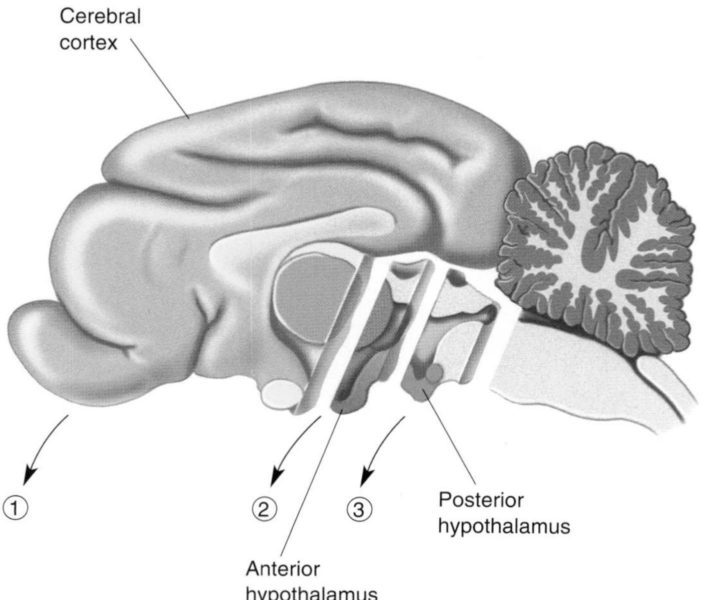

Figure 18.8
Brain transections and sham rage. ① If the cerebral hemispheres are removed and the hypothalamus is left intact, sham rage results. ① and ② A similar result is obtained if the anterior hypothalamus is removed in addition to the cerebral cortex. ①, ②, and ③ If the posterior hypothalamus is removed in addition to the anterior hypothalamus, sham rage does not result.

pant, eat, or express behaviors characteristic of fear or anger. These reactions illustrate the two primary functions of the hypothalamus discussed in Chapters 15 and 16: homeostasis and the organization of coordinated visceral and somatic motor responses. Responses related to emotional expression can include changes in heart rate, pupillary dilation, and gastrointestinal motility, to name a few. Because stimulation of some parts of the hypothalamus also elicits behavior characteristic of fear and rage, we hypothesize that the hypothalamus is an important component of the system normally involved in expressing these emotions.

The expression of rage Hess evoked by hypothalamic stimulation was similar to the sham rage seen in animals whose cerebral hemispheres had been removed. With a small application of electrical current, a cat will spit, growl, and fold its ears back, and its hair will stand on end. This complex and highly coordinated set of behaviors normally occurs when the cat feels threatened by an enemy. Sometimes, the cat suddenly runs as if fleeing an imaginary attacker. When the intensity of the stimulation is increased, the animal may make an actual attack, swatting with a paw or leaping onto an imaginary adversary. When the stimulation is stopped, the rage disappears as quickly as it started, and the cat may even curl up and go to sleep.

In a series of studies conducted at the Yale University Medical School in the 1960s, John Flynn found that affective aggression and predatory aggression could be elicited by stimulating different areas of the hypothalamus (Figure 18.9). Affective aggression (also known as a *threat attack*) was observed with stimulation of specific sites in the medial hypothalamus. Similar to the rage response that Hess reported, the animal would arch its back, hiss, and spit but would usually not attack the victim, such as a nearby rat. Predatory aggression (called by Flynn a *silent biting attack*) was evoked by stimulating parts of the lateral hypothalamus. While the back might be somewhat arched and the hair slightly on end, predatory aggression was not accompanied by the dramatic threatening gestures of affective aggression.

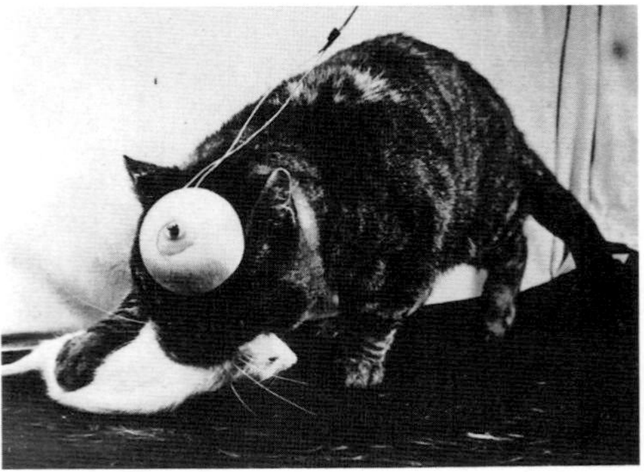

Figure 18.9
Rage reactions in cats with hypothalamic stimulation. (a) Stimulation of the medial hypothalamus produces affective aggression (threat attack). **(b)** Stimulation of the lateral hypothalamus evokes predatory aggression (silent biting attack). (Source: Flynn, 1967, p. 45.)

Nonetheless, in this "quiet attack," the cat would move swiftly toward a rat and viciously bite its neck.

The Midbrain and Aggression

There are two major pathways by which the hypothalamus sends signals involving autonomic function to the brain stem: the **medial forebrain bundle** and the **dorsal longitudinal fasciculus**. Axons from the lateral hypothalamus make up part of the medial forebrain bundle, and these project to the *ventral tegmental area* in the midbrain. Stimulation of sites within the ventral tegmental area can elicit behaviors characteristic of predatory aggression, just as stimulation of the lateral hypothalamus does. Conversely, lesions in the ventral tegmental area can disrupt offensive aggressive behaviors. One finding suggesting that the hypothalamus influences aggressive behavior via its effect on the ventral tegmental area is that hypothalamic stimulation will not evoke aggression if the medial forebrain bundle is cut. Interestingly, aggressive behavior is not entirely eliminated by this surgery, suggesting that this route is important when the hypothalamus is involved but that the hypothalamus need not always be involved.

The medial hypothalamus sends axons to the **periaqueductal gray matter** of the midbrain by way of the dorsal longitudinal fasciculus. Electrical stimulation of the periaqueductal gray matter can produce affective aggression, and lesions located there can disrupt this behavior.

The Amygdala and Aggression

In our discussion of the neurobiology of anxiety and fear, we saw that the amygdala plays a key role. The amygdala is also involved in aggressive behavior. In an experiment performed by American scientist Karl Pribram and his colleagues in 1954, amygdala lesions were shown to have a major effect on social interactions in a colony of eight male rhesus monkeys. Having lived together for some time, the animals had established a social hierarchy. The first intervention made by the investigators was to make bilateral amygdala lesions in the brain of the most dominant monkey. After this animal returned to the colony, it fell to the bottom of the hierarchy, and the previously next subordinate monkey became dominant. Presumably, the second monkey in the hierarchy discovered that the "top banana" had become more placid and less difficult to challenge. After an amygdalectomy was performed on the new dominant monkey, it likewise fell to the bottom of the hierarchy. This suggests that the amygdala is important for the aggression normally involved in maintaining a position in the social hierarchy.

Experiments in which amygdala subdivisions were electrically stimulated or destroyed suggest that it has multiple effects on aggressive behavior by way of its connections to the hypothalamus and other structures. Electrical stimulation of the basolateral nuclei produces affective aggression presumably through the effects of the efferents in the ventral amygdalofugal pathway on the hypothalamus and brain stem nuclei. Lesions in the basolateral nuclei reduce affective aggression.

The corticomedial nuclei have an inhibitory influence on aggression. Neurons in the corticomedial nuclei send axons through the stria terminalis to the hypothalamus. Lesions in either the corticomedial nuclei or the stria terminalis markedly increase predatory aggression. Therefore, we can infer that this portion of the amygdala may normally have an inhibitory effect on the hypothalamus, suppressing predatory aggression. Two hypothetical pathways for predatory aggression and affective aggression are summarized in Figure 18.10.

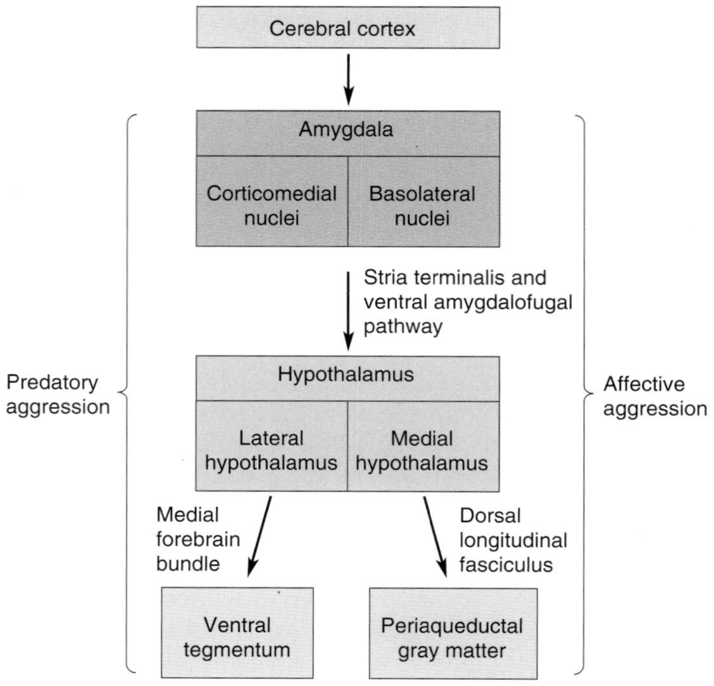

Figure 18.10
Two pathways for aggression. In this simplified scheme, one pathway extends from the corticomedial nuclei of the amygdala to the lateral hypothalamus and ventral tegmentum, controlling predatory aggression. Affective aggression is mediated by a pathway from the basolateral nuclei of the amygdala to the lateral hypothalamus and periaqueductal gray matter. Notice the overlap between these pathways and the circuit proposed for learned fear in Figure 18.7.

Surgery to Reduce Human Aggression. Recognizing that aggression is reduced by amygdalectomy in animals, some neurosurgeons thought it might be possible to modify aggressive behavior in violent humans similarly. It was thought by some that violent behavior frequently resulted from seizures in the temporal lobe. In a human amygdalectomy, electrodes are passed through the brain and down into the temporal lobe. By making neural recordings along the way and imaging the electrodes with X-rays, it is possible to get the tip of the electrode into the amygdala. Electrical current is passed through the electrode, or a solution is injected, to destroy all or part of the amygdala. Clinical reports claim considerable success in reducing aggressive asocial behavior, increasing the ability to concentrate, decreasing hyperactivity, and reducing seizures.

This type of brain surgery, a method of treating behavioral disorders, is called **psychosurgery**. Early in the twentieth century, treating severe disorders involving anxiety, aggression, or neuroses with psychosurgical techniques, including the frontal lobotomy, was a common practice (Box 18.3). Destroying a significant portion of the brain, however, is a drastic measure. Removing the amygdala and/or other structures is irreversible, and one never knows until after recovery whether there was inadvertent damage to cognitive or behavioral function. This clearly is a treatment of last resort.

Serotonin and Aggression

Studies indicate that the neurotransmitter serotonin may be involved in regulating aggression. Serotonin-containing neurons are located in the raphe nuclei of the brain stem, and they ascend in the medial forebrain bundle and project to the hypothalamus and various limbic structures involved in emotion (see Figure 15.12).

One link between serotonin and aggression comes from studies of induced aggression in rodents. Mice that are isolated in a small cage for 4 weeks frequently become hyperactive and extremely aggressive toward other mice. Although the isolation has no effect on the *level* of serotonin in the brain, there

Box 18.3

OF SPECIAL INTEREST

The Frontal Lobotomy

Ever since the discoveries by Klüver, Bucy, and others that brain lesions can alter emotional behavior, clinicians have attempted surgery as a means of treating severe behavioral disorders in humans. Probably no other brain operation has received as much media coverage as the frontal lobotomy. Today, it is difficult for many people to imagine that destroying a large portion of the brain was once thought to be therapeutic. Indeed, in 1949 the Nobel Prize in medicine was awarded to Dr. Egas Moniz for his development of the frontal lobotomy technique. Even stranger is the fact that Moniz was shot in the spine and partially paralyzed by a lobotomized patient—either a tragedy or poetic justice, depending on your point of view. One doesn't hear about lobotomies being performed anymore, but tens of thousands were performed following World War II.

Little theory supported the development of the lobotomy. In the 1930s, John Fulton and Carlyle Jacobsen of Yale University reported that frontal lobe lesions had a calming effect in chimpanzees. It has been suggested that frontal lesions have this effect because of the destruction of limbic structures and, in particular, connections with frontal and cingulate cortex. The guiding principle behind the surgery was something like this: The limbic system controls emotion; therefore, people with emotional problems may be helped by altering the system. In other words, a little emotion is a good thing, but too much is debilitating, and this may be surgically corrected.

A frightening variety of techniques were used to produce lesions in the frontal lobes. In the technique known as transorbital lobotomy, shown in Figure A, a knife was inserted through the thin bone at the top of the eye's orbit. The handle was then swung medially and laterally to destroy cells and interconnecting pathways. Thousands of people were lobotomized with this technique, sometimes called "ice pick psychosurgery," because it was so simple it could even be performed in the physician's office. Note that although it left no outward scars, the physician could not see what was being destroyed.

Frontal lobotomy reportedly had beneficial effects on people with a number of disorders, including psychosis, depression, and various neuroses. The effect of the surgery was described as a relief from anxiety and escape from thoughts that were unendurable. Only later did a pattern of less pleasant side effects emerge. While frontal lobotomy can be performed with little decrease in IQ or loss of memory, it does have other profound effects. The changes that appear to be related to the limbic system are a blunting of emotional responses and a loss of the emotional component of thoughts. In addition, it was commonly found that lobotomized patients developed "inappropriate behavior," or an apparent lowering of moral standards. Similar to Phineas Gage, patients had considerable difficulty planning and working toward goals. Lobotomized patients have trouble concentrating and are easily distracted.

Perhaps more to the point, with our modest understanding of the neural circuitry underlying emotion and other brain functions, it is hard to justify destroying a large portion of the brain. Fortunately, treatment with lobotomy decreased fairly rapidly, and today drug therapy is primarily used instead for serious emotional disorders.

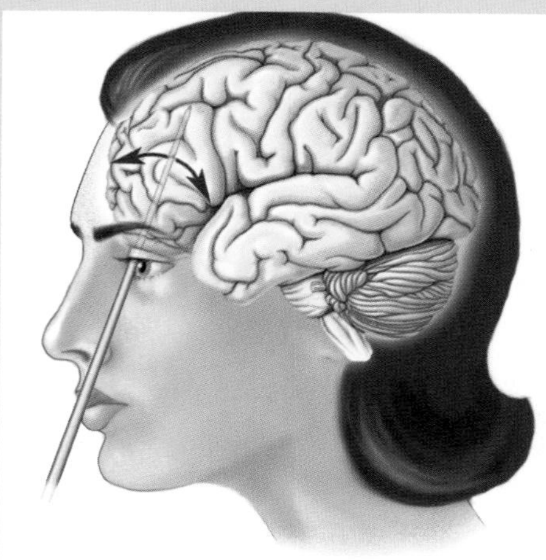

Figure A

is a decrease in the *turnover rate* (the rate of synthesis, release, and resynthesis) of this neurotransmitter. Moreover, this decrease is found only in the mice that later become unusually aggressive, not in those relatively unaffected by the isolation. Also, female mice typically do not become aggressive following isolation, and they show no decrease in serotonin turnover. Evidence indicates that drugs that block the synthesis or release of serotonin increase aggressive behavior. In one study, the drug PCPA (parachlorophenylalanine), which blocks serotonin synthesis, was administered, and the injected animals increased their attacks on others in their cage.

The relationship between serotonin and aggression is similar in primates that have been studied. For example, it was found that the dominance hierarchy in a colony of vervet monkeys could be manipulated by injecting animals with drugs that either increased or decreased serotonergic activity. The behavior of these animals was consistent: More aggression was associated with less serotonergic activity. There was, however, one interesting sociological twist—aggression did not correlate with dominance in the group. If the dominant male in the group was removed, the top position was taken by an animal with artificially *enhanced* serotonergic activity (i.e., one injected with a serotonin precursor or reuptake inhibitor who was *less* aggressive). Conversely, the injection of drugs that reduced serotonin function (serotonin antagonists) was correlated with animals becoming subordinate. The subordinate animals were actually significantly more likely to initiate aggression. It appeared that the less aggressive dominant male garnered his status by his skills in recruiting females to support his position.

Serotonin Receptor Knockout Mice. There are at least 14 serotonin receptor subtypes, and recombinant DNA techniques have been used to produce mice lacking certain types. They are called "knockout mice" because a normal gene—one that codes for a receptor, in this case—has been removed or mutated in a procedure called the *gene knockout technique*. Researchers have been particularly interested in the receptors 5-HT_{1A} and 5-HT_{1B}, which are found in high concentrations in the raphe nuclei.

We have mentioned that serotonin antagonists increase aggressiveness. Conversely, agonists of the 5-HT_{1A} and 5-HT_{1B} receptors are known to decrease anxiety and aggressiveness in mice. Based on these pharmacological results, one would predict that mice lacking the 5-HT_{1A} or 5-HT_{1B} receptor would be more anxious and aggressive than normal animals. This prediction has largely been found to be true. In one study, mice lacking the 5-HT_{1A} receptor were found to show little exploratory behavior. They would avoid open spaces, instead staying in covered protected areas. These results were interpreted as meaning that the mice lacking 5-HT_{1A} receptors were unusually anxious. In a second study, mice lacking the 5-HT_{1B} receptor were produced. The mutant mice lacking this receptor behaved normally and were not unusually aggressive living together as a group. But when they were put in a stressful situation, such as having a new mouse placed in their house, the mutants were far more aggressive. The anatomical location of the 5-HT_{1B} receptors is telling. In addition to the raphe nuclei, they are found in the amygdala, periaqueductal gray matter, and basal ganglia. The localization of the receptors in structures thought to be involved in aggressive behavior offers the potential for linking pharmacological studies of serotonin with the behavioral research we've examined.

REINFORCEMENT AND REWARD

In the early 1950s, James Olds and Peter Milner at the California Institute of Technology conducted an experiment in which a rat had an electrode implanted in its brain such that the brain could be locally stimulated at any time. The rat freely roamed about in a box about 3 feet square. Each time the rat went to a certain corner of the box, its brain was stimulated. After the first stimulation, the rat walked away but quickly returned and was again stimulated. Soon the rat was spending all of its time in the corner, apparently seeking the electrical stimulation. In a brilliant twist on this experiment, Olds and Milner set up a new box for the rat containing a lever that would deliver a brief stimulus to the brain when stepped on (Figure 18.11). At first the rat wandered about the box and stepped on the lever by accident, but before long it was pressing the lever repeatedly. This behavior is called **electrical self-stimulation**. Sometimes, the rats would become so involved in pressing the lever that they would shun food and water, stopping only after collapsing from exhaustion.

Thus, reinforcing brain stimulation was discovered, and it raises many questions. What structures must be stimulated to get this kind of reinforcement? Why does the rat repeatedly stimulate itself? Is the effect pleasurable? Does the rat feel something akin to the satisfaction it might receive from food or sex? Olds and Milner's unexpected finding led to many subsequent studies. It is hoped that deciphering the neural mechanisms underlying this artificial form of reinforcement will shed light on the physiological basis for normal reinforcing behaviors, such as eating, drinking, and sex (discussed in Chapters 16 and 17), and abnormal ones, such as addiction (Chapter 16).

Electrical Self-Stimulation and Reinforcement

It is not clear why the rats repeatedly pressed the lever in Olds and Milner's electrical self-stimulation experiments. One interpretation is that the rat got a positive feeling from the stimulation and therefore returned for more. Thus, the brain sites leading to reinforcing stimulation were called **pleasure centers**. That term, however, misrepresents the results of the experiments in two

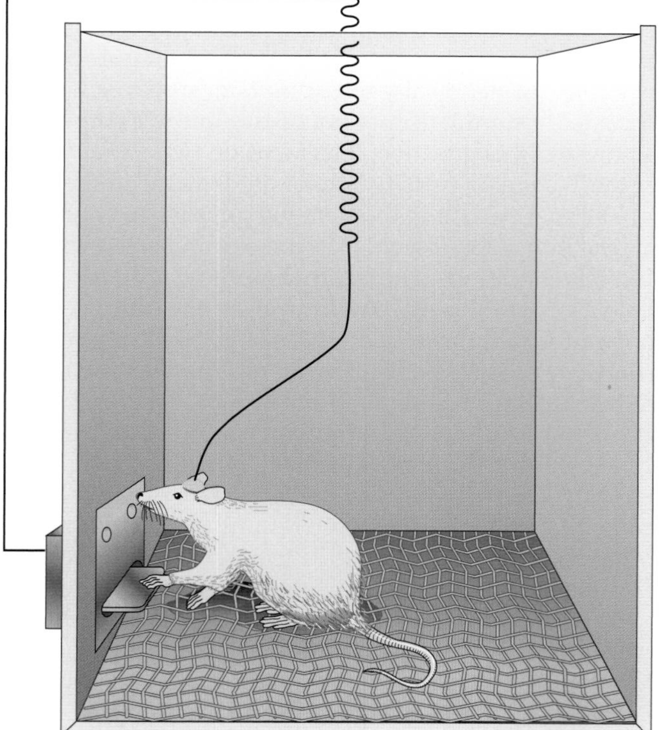

Figure 18.11
Electrical self-stimulation by a rat. When the rat presses on the lever, it receives a brief electrical current to an electrode in its brain.

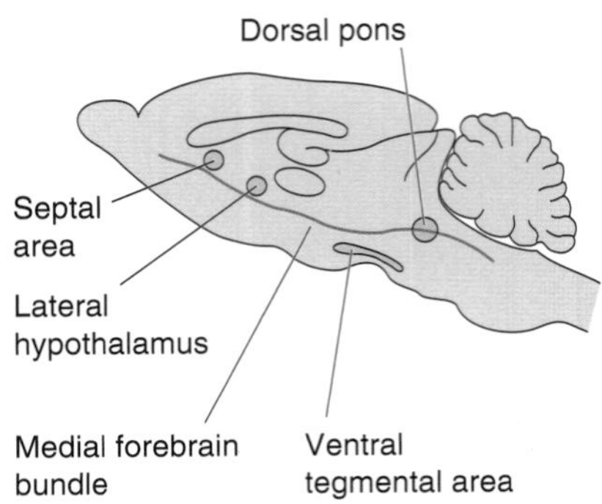

Figure 18.12
Self-stimulation sites in the rat brain. Rats will self-stimulate when electrodes are placed in the areas colored pink.

significant ways. First, we don't know that something like pleasure was experienced. Perhaps the self-stimulation makes the rat want more stimulation without its being pleasurable. We know that humans sometimes feel compelled to eat or drink alcohol even when these behaviors do not evoke a pleasurable feeling. Second, even if the stimulation is pleasurable, there may not be a particular "center" responsible for the reinforcement. Perhaps stimulation at a large number of diffusely scattered sites will produce self-stimulation, or maybe the stimulation affects bundles of axons rather than a nucleus of cells.

In the decades following Olds and Milner's discovery, many self-stimulation sites have been found in limbic structures and elsewhere in the rat's brain. These include the *septal area*, lateral hypothalamus, the medial forebrain bundle, the ventral tegmental area, and the dorsal pons (Figure 18.12). Stimulation of the medial forebrain bundle produces powerful reinforcement, and it has been studied more than any other structure. A smaller number of sites produce aversive behavior when electrically stimulated. For example, animals actively avoid behavior that results in stimulation of certain parts of their brain. They also learn to perform tasks that terminate stimulation at these sites. These "displeasure centers," or negative-reinforcement sites, are in medial portions of the hypothalamus and lateral parts of the midbrain tegmental area. Stimulation at these locations may evoke a negative feeling such as fear, or it may excite a pathway that is usually active in a negative-reinforcement situation, such as fleeing from a predator.

Brain Stimulation in Humans

To determine the sensations evoked by brain stimulation, it would be desirable to stimulate a person's brain and ask how it feels. Obviously, this is not normally feasible or ethical. But certain surgical procedures performed on humans require that they be alert during surgery and that their brain be electrically stimulated. As we have mentioned elsewhere in this book, the most common example is surgical treatment of severe epilepsy. When the brain is electrically stimulated, pleasurable feelings are sometimes evoked, so the term *pleasure center* may not be entirely inappropriate. In some cases when patients can electrically self-stimulate, however, the brain site they choose to repeatedly stimulate does not cause a pleasurable feeling. Why do they do it? To answer this question, let's focus on two patients studied by Robert Heath at the Tulane University School of Medicine in the 1960s.

The first patient had severe narcolepsy; he would abruptly go from being awake into a deep sleep. (Narcolepsy and sleep will be discussed in Chapter 19.) His condition significantly interfered with his life and made it difficult to hold a job. He was fitted with 14 electrodes in various areas of the brain in the hope of finding a self-stimulation site that might keep him alert. When he stimulated his hippocampus, he reported feeling mild pleasure. Stimulation of his midbrain tegmentum made him feel alert but unpleasant. The site he chose frequently to self-stimulate was the septal area of the forebrain (Figure 18.13). Stimulating this area made him more alert and gave him a good feeling, which he described as building up to orgasm. He reported that he would sometimes push the button over and over, trying unsuccessfully to achieve orgasm, ultimately ending in frustration. Despite the frustration, the most frequently self-stimulated site was associated with a positive feeling.

The second patient's case is a bit more complex. This person had electrodes implanted at 17 brain sites in the hope of learning something about the location of his severe epilepsy. He reported pleasurable feelings with stimulation of the septal area and the midbrain tegmentum. Consistent with the first case, septal stimulation was associated with sexual feelings. The mid-

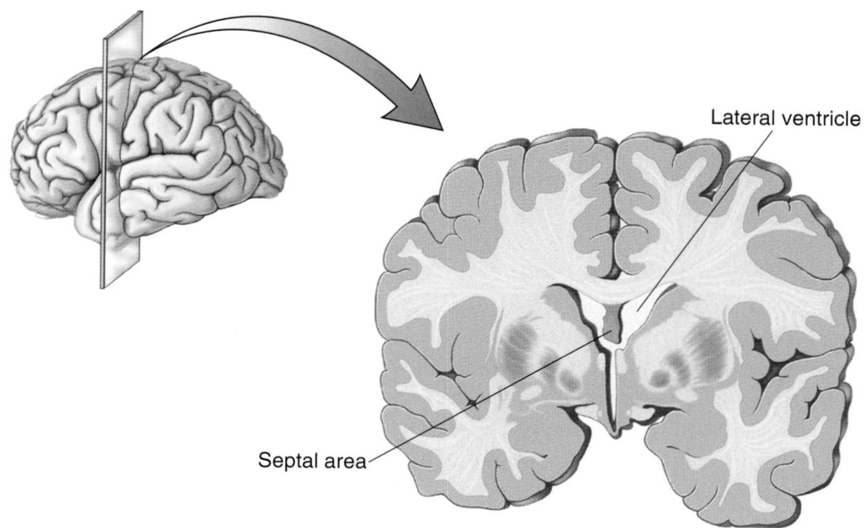

Figure 18.13
The septal area. The septal area is in the rostral forebrain below the lateral ventricle

brain stimulation gave him a "happy drunk" feeling. Other mildly positive feelings were produced by stimulation of the amygdala and caudate nucleus. Interestingly, the site that he most frequently stimulated was in the medial thalamus, even though stimulation here induced an irritable feeling, one that was less pleasurable than stimulation at other locations. The patient stated that the reason he stimulated this area the most was that it gave him the feeling he was about to recall a memory. He repeated the stimulation in a futile attempt to bring the memory into his mind, even though in the end this process proved to be frustrating.

These two cases and a host of others lead us to the conclusion that self-stimulation sites in the human brain are not always synonymous with pleasure. There is often some reward or anticipated reward associated with the stimulation, but the experience is not always pleasant.

Dopamine and Reinforcement

One proposed explanation for the large number of scattered self-stimulation sites is that they are all interconnected by a common pathway involved in normal reward behavior. It is interesting to note that high rates of self-stimulation are obtained with electrodes placed in the medial forebrain bundle and ventral tegmental area. As discussed in Chapter 15, cell bodies of dopaminergic neurons are located in the ventral tegmental area (and the substantia nigra) of the midbrain, and they send axons through the medial forebrain bundle to many areas of the brain (see Figure 15.13). The medial forebrain bundle also contains descending fibers that may conceivably carry reward signals to the ventral tegmental area.

Pharmacological evidence also links dopamine with reinforcement behavior. Dopamine agonists such as amphetamine increase the rate of self-stimulation. Injecting a rat with a drug that blocks dopamine receptors (e.g., haloperidol) decreases self-stimulation. But we must be cautious in interpreting this finding, because dopamine is also important for the brain's control of movement. We saw in Chapter 14 that degeneration of the substantia nigra causes depletion of dopamine, leading to the movement disorder known as Parkinson's disease. Therefore, it is conceivable that rats given dopamine antagonists simply cannot press a lever, although that does not appear to be the case.

The connection between dopamine and reinforcement should remind you of the role dopamine appears to play in motivated behaviors. We saw in

Chapter 16 that the destruction of ascending dopamine axons reduces food-seeking behavior, even though it does not reduce the hedonic response to food. We also saw in Box 16.3 that a variety of addictive drugs (heroin, nicotine, and cocaine) enhance the efficacy of dopamine and lead to its release in the nucleus accumbens. Thus, a common thread connecting reinforcement, motivation for feeding, and some types of drug addiction is that all of these behaviors are associated with activity in the mesocorticolimbic dopamine system.

While much evidence suggests that dopamine is involved in reinforcement and reward, the nature of this association is not clear. Indeed, certain experimental data are difficult to reconcile with the hypothesis that dopaminergic neurons in the medial forebrain bundle play a critical role. For instance, several studies have shown that cutting the medial forebrain bundle does not have the devastating effect on self-stimulation one might expect. Furthermore, self-stimulation sites are not identical to the brain sites that receive dopaminergic input from the midbrain. Thus, we have the somewhat unsettled situation in which pharmacological experiments suggest an important role for dopamine, but lesion studies, among others, raise concerns. Dopamine may be important in some situations, but it cannot simply be the reward transmitter for the brain.

CONCLUDING REMARKS

We have examined several neural pathways that appear to be involved in the experience and expression of emotion. Our approach has been somewhat the opposite of that taken by Papez in the 1930s. With good insight, Papez proposed a large emotion system composed of numerous structures along the midline of the brain. Although the role of each structure was far from clear, he knew that they were anatomically connected. Our discussion of the brain mechanisms of emotion focuses on a smaller number of structures. Rather than evaluating every component of the Papez circuit, we built up small circuits that experiments suggest are most clearly involved in emotion. Some structures, such as the amygdala and hypothalamus, appear to be important for several emotions.

This chapter contains much discussion of fear and aggression but no discussion of the physiology underlying many other emotions, such as hate and joy. The reason is simply the limitations of experimental technique. To study emotion, we must be able to reliably evoke that emotion in an animal and then find a way to describe the responses. This is why we are closer to understanding the neural circuitry involved in emotional expression than emotional experience. Even the studies of electrical self-stimulation in humans must be critically viewed because the effect of the stimulation is not simply a pleasant feeling, nor were these brains completely normal to begin with.

The experience and expression of emotions involve widespread activity in the nervous system, from the cerebral cortex to the ANS. Emotional reactions are the result of a complex interaction between sensory stimuli, brain circuitry, personal experiences, and the activity of neurotransmitter systems. In light of this complexity, we probably should not be surprised that humans exhibit a wide spectrum of emotional and mood disorders, as we will see when we discuss psychiatric disorders in Chapter 21.

Keep in mind for later chapters that the structures apparently involved in emotion also have other functions. For a considerable time after the limbic lobe was defined by Broca, it was thought to be primarily an olfactory system. And even though our perspective has changed a lot since Broca's time, parts of the brain involved in olfaction have been included in the definition of the limbic system. We will see in Chapter 23 that some of the limbic structures are also important for learning and memory. How is it that these structures evolved to perform so many functions? We don't know. There are interesting relationships between emotion, memory, and olfaction. Emotional events usually result in particularly enduring memories. Most people also feel that olfaction—a whiff of a certain perfume, for example—powerfully evokes memories.

KEY TERMS

What Is Emotion?
James-Lange theory (p. 581)
Cannon-Bard theory (p. 582)

The Limbic System Concept
limbic lobe (p. 584)
Papez circuit (p. 585)
limbic system (p. 587)

Fear and Anxiety
Klüver-Bucy syndrome (p. 589)
amygdala (p. 590)
basolateral nuclei (p. 590)
corticomedial nuclei (p. 590)

Anger and Aggression
predatory aggression (p. 594)
affective aggression (p. 594)
sham rage (p. 595)
medial forebrain bundle (p. 597)
dorsal longitudinal fasciculus (p. 597)
periaqueductal gray matter (p. 597)
psychosurgery (p. 598)

Reinforcement and Reward
electrical self-stimulation (p. 601)
pleasure center (p. 601)

REVIEW QUESTIONS

1. According to the James-Lange and Cannon-Bard theories of emotion, what is the relationship between the anxiety you would feel after oversleeping for an exam and your physical responses to the situation?
2. How have the definition of the limbic system and thoughts about its function changed since the time of Broca?
3. What procedures produce an abnormal rage reaction in an experimental animal? How do we know that the animals *feel* angry?
4. What changes in emotion were observed following temporal lobectomy by Klüver and Bucy? Of the numerous anatomical structures they removed, which is thought to be closely related to changes in temperament?
5. Why might performing bilateral amygdalectomy on a dominant monkey in a colony result in that monkey's becoming a subordinate?
6. Does the stimulation of "pleasure centers" always evoke pleasure? Why might a rat repeatedly self-stimulate a portion of its brain, even though the experience is not pleasurable?
7. What evidence is there that dopamine plays an important role in reinforcement? What evidence is there to the contrary? What behavioral abnormality was linked to dopamine in Chapter 15?
8. What assumptions about limbic structures underlie the surgical treatment of emotional disorders?
9. The drug fluoxetine (Prozac) is a serotonin-selective reuptake inhibitor. How might this drug affect a person's level of anxiety and aggression?

CHAPTER

19

Rhythms of the Brain

INTRODUCTION

The earth is a rhythmic environment. Temperature, precipitation, and daylight vary with the seasons; light and dark trade places each day; tides ebb and flow. To compete effectively and therefore survive, an animal's behavior must oscillate with the cadences of its environment. Brains have evolved a variety of systems for rhythmic control. Sleeping and waking are the most striking periodic behavior. But some rhythms controlled by the brain have much longer periods, as in hibernating animals, and many have shorter periods, such as the cycles of breathing, the steps of walking, the repetitive stages of one night's sleep, and the electrical rhythms of the cerebral cortex. The functions of some rhythms are obvious, while others are obscure, and some rhythms indicate pathology.

In this chapter, we explore selected brain rhythms, beginning with the fast ones and proceeding to the slow. The forebrain, especially the cerebral cortex, produces a range of rapid electrical rhythms that are easily measured and that closely correlate with interesting behaviors, including sleep. We discuss the electroencephalogram, or EEG, because it is the classical method of recording brain rhythms and is essential for studying sleep. Sleep is treated in detail because it is complex, ubiquitous, and so dear to our hearts. Finally, we summarize what is known about the timers that regulate the ups and downs of our hormones, temperature, alertness, and metabolism. Almost all physiological functions change according to daily cycles known as circadian rhythms. The clocks that time circadian rhythms are in the brain, calibrated by the sun via the visual system, and they profoundly influence our health and well-being.

THE ELECTROENCEPHALOGRAM

Very often the forest is more interesting than the trees. Similarly, we are often less concerned with the activities of single neurons than with understanding the activity of a large population of neurons. The **electroencephalogram** (**EEG**) is a measurement that enables us to glimpse the activity of the cerebral cortex. The roots of the EEG lie in work done by English physiologist Richard Caton in 1875. Using a primitive device sensitive to voltage, Caton made electrical recordings from the surface of dog and rabbit brains. The human EEG was first described by Austrian psychiatrist Hans Berger in 1929. Berger observed that waking and sleeping EEGs are distinctly different. Today, the EEG is used mostly to help diagnose certain neurological conditions, especially the seizures of epilepsy, and for research purposes, notably to study sleep.

Recording Brain Waves

Recording an EEG is relatively simple. The method is usually noninvasive, and it is painless. Countless people have slept through entire nights wearing EEG electrodes in the comfort of sleep research laboratories (Figure 19.1). The electrodes are wires taped to the scalp, along with conductive paste to ensure a low-resistance connection. As shown in Figure 19.2, some two-dozen electrodes are fixed to standard positions on the head and connected to banks of amplifiers and recording devices. Small-voltage fluctuations, usually a few tens of microvolts (μV) in amplitude, are measured between selected pairs of electrodes. Different regions of the brain—anterior and posterior, left and right—can be examined by selecting the appropriate electrode pairs. The typical EEG record is a set of many simultaneous squiggles, indicating voltage changes between pairs of electrodes.

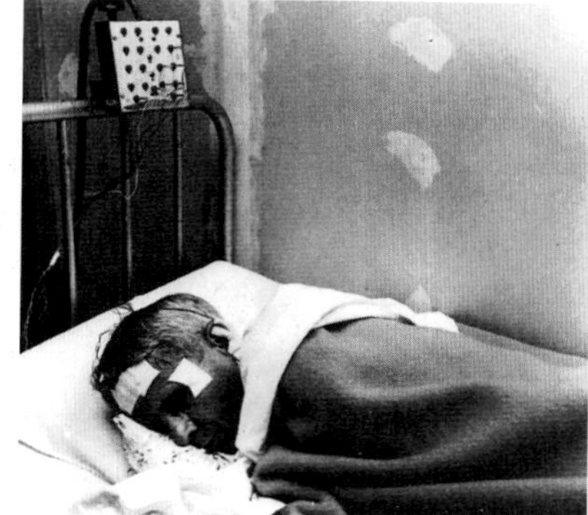

Figure 19.1
A subject in a sleep research study. This is American sleep researcher Nathaniel Kleitman, codiscoverer of REM sleep. The white patches on his head are pieces of tape holding EEG electrodes, and those next to his eyes hold electrodes that monitor his eye movements. (Source: Carskadon, 1993.)

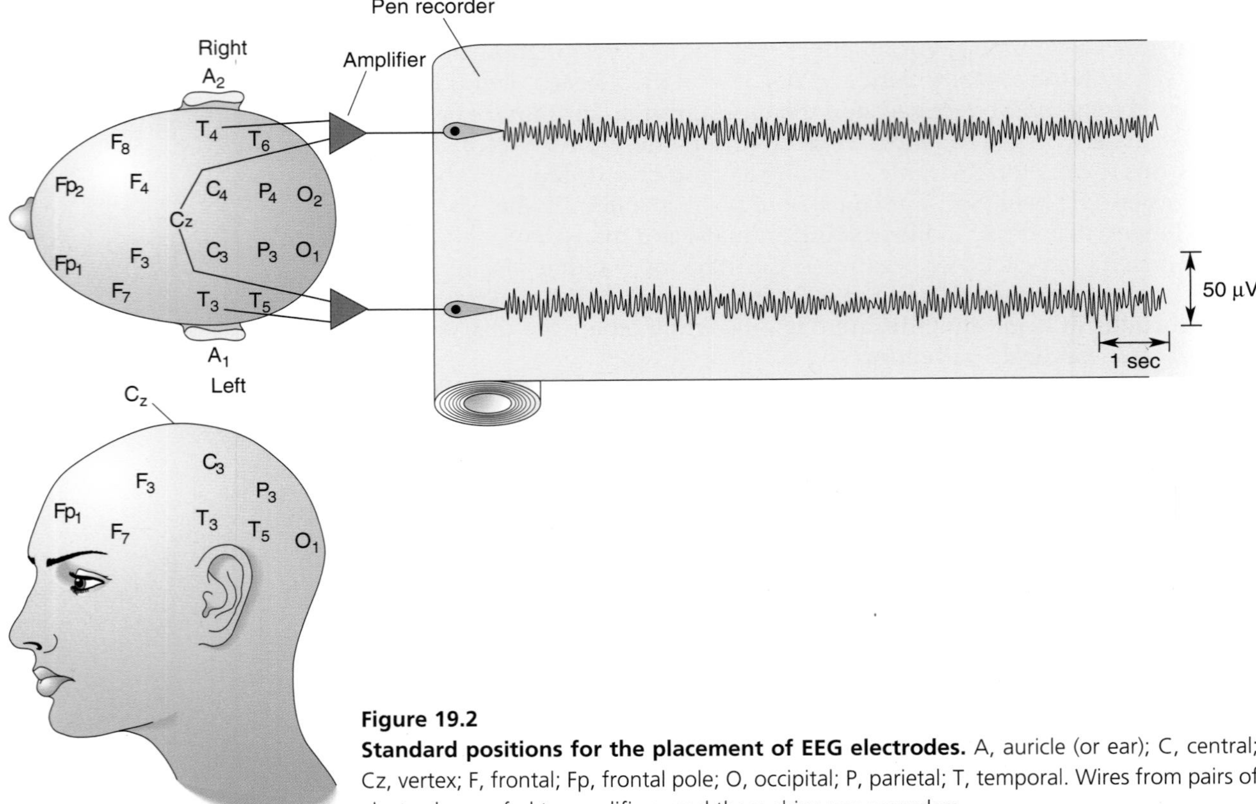

Figure 19.2
Standard positions for the placement of EEG electrodes. A, auricle (or ear); C, central; Cz, vertex; F, frontal; Fp, frontal pole; O, occipital; P, parietal; T, temporal. Wires from pairs of electrodes are fed to amplifiers, and these drive pen recorders.

What part of the nervous system generates the endless squiggles of an EEG? For the most part, an EEG measures the currents that flow during synaptic excitation of the dendrites of many pyramidal neurons in the cerebral cortex, which lies right under the skull and makes up 80% of the brain's mass. But the electrical contribution of any single cortical neuron is exceedingly small, and the signal must penetrate several layers of non-neural tissue, including the meninges, fluid, bones of the skull, and skin, to reach the electrodes (Figure 19.3). Therefore, it takes many thousands of underlying neurons, activated together, to generate an EEG signal big enough to see at all.

This has an interesting consequence: The amplitude of the EEG signal strongly depends on how *synchronous* is the activity of the underlying neurons. When a group of cells is excited simultaneously, the tiny signals sum to generate one large surface signal. But when each cell receives the same amount of excitation, but spread out in time, the summed signals are meager and irregular (Figure 19.4). Notice that in this case, the *number* of activated cells and the *total amount of excitation* may not have changed, only the timing of activity. If synchronous excitation of this group of cells is repeated again and again, the resulting EEG will consist of large, rhythmic waves. We often describe rhythmic EEG signals in terms of their relative amplitude, suggesting how synchronous the underlying activity is (although other factors may also contribute to amplitude).

EEG Rhythms

EEG rhythms vary dramatically and often correlate with particular states of behavior, such as level of attentiveness, sleeping, or waking, and pathology, such as seizures or coma. Figure 19.5 shows part of a normal EEG. The

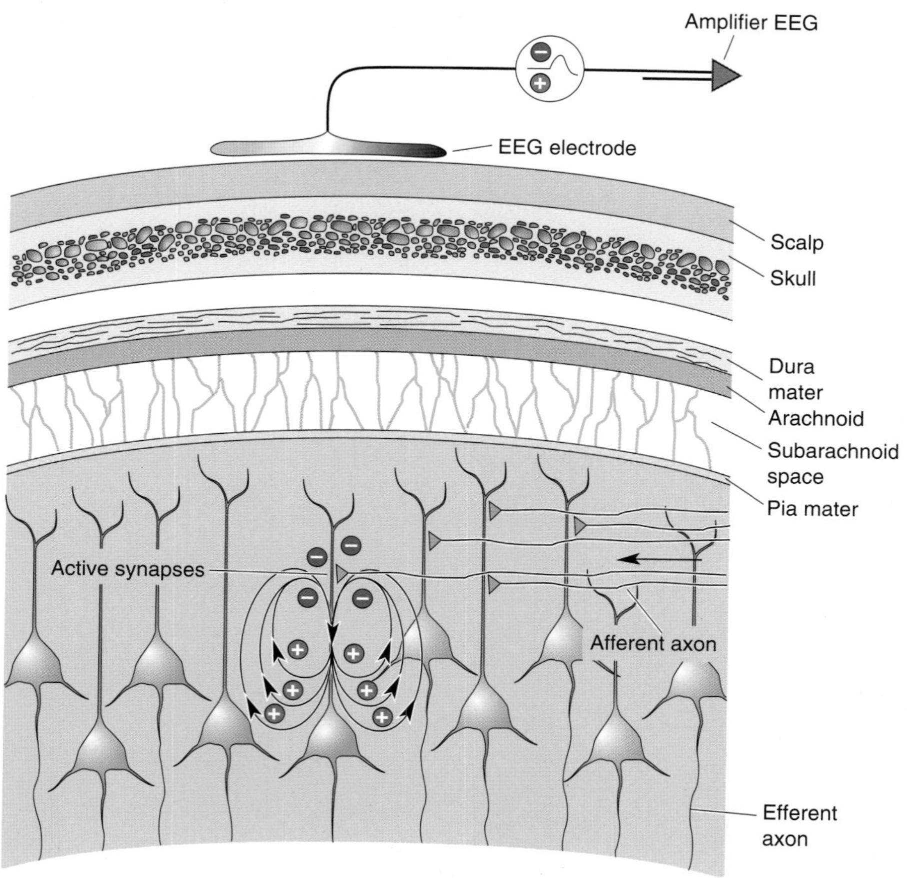

Figure 19.3
The generation of very small electrical fields by synaptic currents in pyramidal cells. In this case, the synapse is on the upper part of the dendrite. When the afferent axon fires, the presynaptic terminal releases glutamate, which opens numerous cation channels. Positive current flows into the dendrite, leaving a slight negativity in the extracellular fluid. Current spreads down the dendrite and escapes out of its deeper parts, leaving the extracellular fluid slightly positive at those sites. The EEG electrode, which is referred to a second electrode some distance away, measures this scene through thick layers of tissue. Only if thousands of cells contribute their small voltage does the signal become strong enough to see at the surface of the scalp. (The convention in EEG work is to plot the signals with negativity upward.)

rhythms are categorized by their frequency range, and each range is named after a Greek letter. *Beta* rhythms are fastest, greater than 14 Hz, and signal an activated cortex. *Alpha* rhythms are about 8–13 Hz and are associated with quiet, waking states. *Theta* rhythms are 4–7 Hz and occur during some sleep states. *Delta* rhythms are quite slow, less than 4 Hz, often large in amplitude, and are a hallmark of deep sleep.

While analysis of an EEG will never tell us *what* a person is thinking, it can help us know *whether* a person is thinking. In general, high-frequency, low-amplitude rhythms are associated with alertness and waking or the dreaming stages of sleep. Low-frequency, high-amplitude rhythms are associated with nondreaming sleep states and with the pathological state of coma. This is logical, because when the cortex is most actively engaged in processing information, whether generated by sensory input or by some internal process, the activity level of cortical neurons is relatively high but also relatively unsynchronized. In other words, each neuron or very small group of neurons is vigorously involved in a slightly different aspect of a complex cognitive task; it fires rapidly but not quite simultaneously with most of its neighbors. This leads to low synchrony, so EEG amplitude is low and beta waves dominate. By contrast, during deep sleep, cortical neurons are not engaged in information processing, and large numbers of them are phasically excited by a common, slow rhythmic input. In this case synchrony is high, so EEG amplitude is high.

Mechanisms and Meanings of Brain Rhythms

Electrical rhythms abound in the cerebral cortex. But how are they generated, and what functions do they perform? Let's take a look at each of these questions in turn.

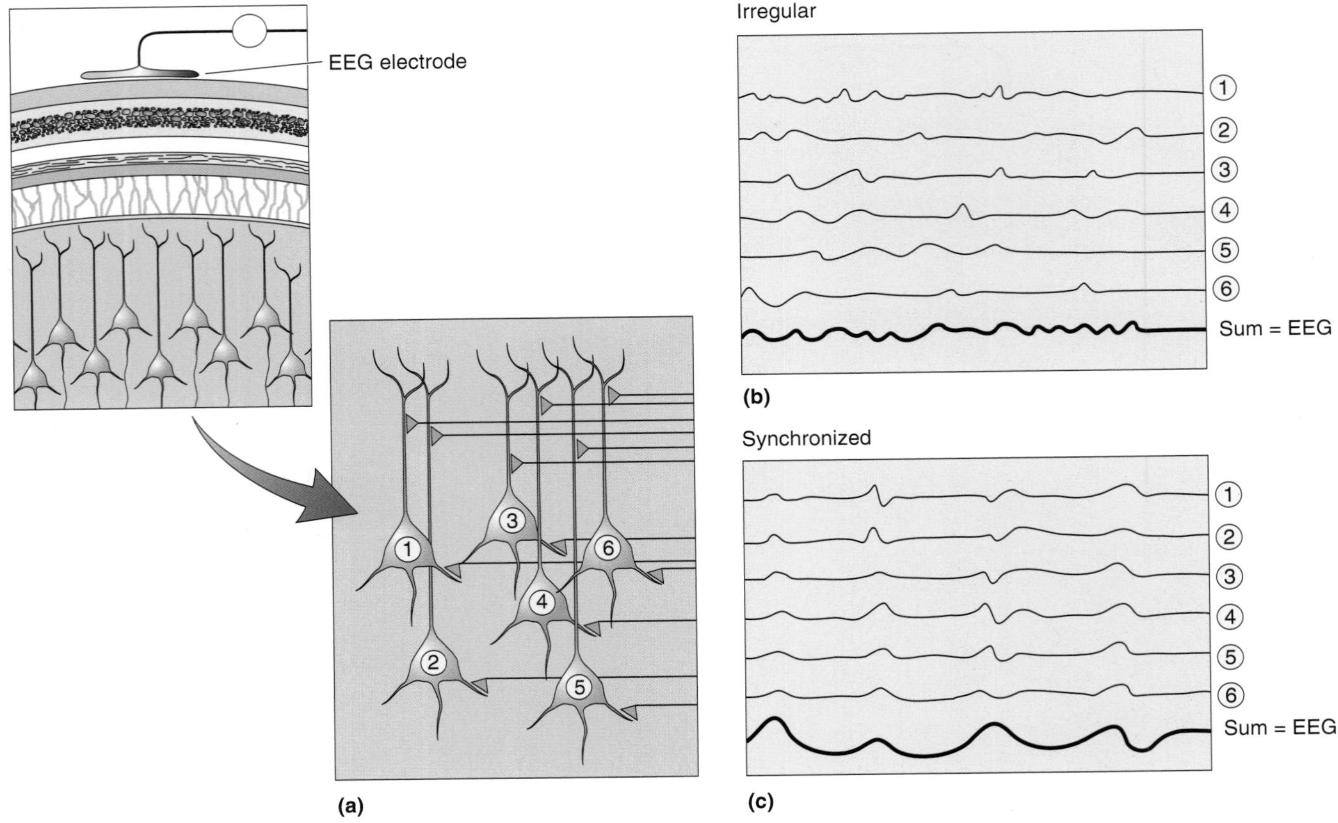

Figure 19.4

The generation of large EEG signals by synchronous activity. (a) In a population of pyramidal cells under an EEG electrode, each neuron receives many synaptic inputs. **(b)** If the inputs fire at irregular intervals, the pyramidal cell responses are not synchronized, and the summed activity detected by the electrode has small amplitude. **(c)** If the same number of inputs fire within a narrow time window so that the pyramidal cell responses are synchronized, the resulting EEG sum is much larger.

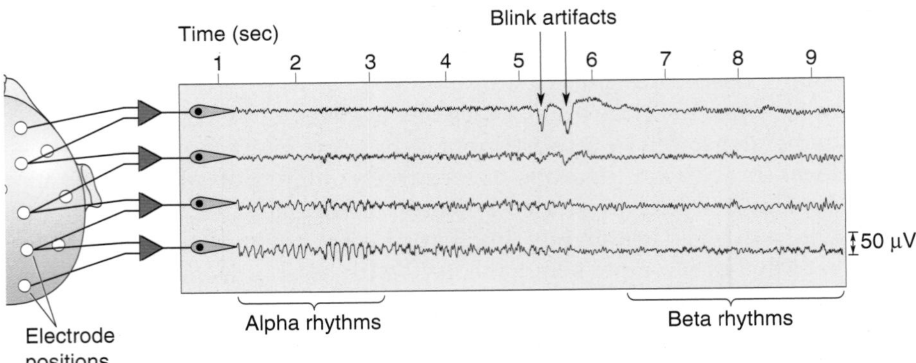

Figure 19.5

A normal EEG. The subject is awake and quiet, and recording sites are indicated at the left. The first few seconds show normal alpha activity, which has frequencies of 8–13 Hz and is largest in the occipital regions. About halfway through the recording, the subject opened his eyes, signaled by the large blink artifacts on the top traces (arrows), and alpha rhythms were suppressed.

The Generation of Synchronous Rhythms. The activity of a large set of neurons will produce synchronized oscillations in one of two fundamental ways: (1) They may all take their cues from a central clock, or *pacemaker*; or (2) they may share or distribute the timing function among themselves by exciting or inhibiting one another. The first mechanism is analogous to a leader and a band, with each musician playing in strict time to the baton's changing beat (Figure 19.6a). The second mechanism is more subtle, because the timing arises from the collective behavior of the cortical neurons themselves. Musically, it is more like a jam session (Figure 19.6b).

The concept can be easily demonstrated by a group of friends, even non-musical ones. Simply tell them to clap together, but give them no instructions about how fast to clap or whose beat to follow. Within one or two claps, they will all be clapping in synchrony. How? Listening and watching each other, they will adjust their clapping rates to match. A key factor is person-to-person interaction; in a network of neurons, these interactions occur via synaptic connections. People tend to clap within a narrow range of frequencies, so they don't have to adjust their timing far to clap in synchrony. Likewise, some neurons may fire at certain frequencies much more than others. This kind of collective, organized behavior can generate rhythms of impressive dimensions, which can move in space and time. Have you ever been part of a human wave in the stands of a sold-out football stadium?

Many circuits of neurons can generate rhythmic activity. A very simple one, consisting of only one excitatory and one inhibitory neuron, is shown in Figure 19.7.

Within the mammalian brain, rhythmic synchronous activity is usually coordinated by a combination of the pacemaker and the collective methods. The thalamus, with its massive input to all of the cortex, can act as a powerful pacemaker. Under certain conditions, thalamic neurons can generate very rhythmic action potential discharges. But how do thalamic neurons oscillate? Thalamic cells have a particular set of voltage-gated ion channels that allow each cell to generate rhythmic self-sustaining discharge patterns even when there is no external input to the cell (Figure 19.8). The rhythmic activity of each thalamic pacemaker neuron is synchronized with many other thalamic cells via a hand-clapping kind of collective interaction. Synaptic connections between excitatory and inhibitory thalamic neurons force each individual neuron to conform to the rhythm of the group. These coordinated rhythms are passed to the cortex by the thalamocortical axons, which excite cortical neurons. In this way, a relatively small group of centralized thalamic cells (acting as the band leader) can compel a much larger group of cortical cells (acting as the band) to march to the thalamic beat (Figure 19.9).

Some rhythms of the cerebral cortex do not depend on the thalamic pacemaker but rely instead on the collective, cooperative interactions of cortical neurons themselves. In this case, the excitatory and inhibitory interconnections of the neurons result in a coordinated synchronous pattern of activity that may remain localized or spread to encompass larger regions of cortex.

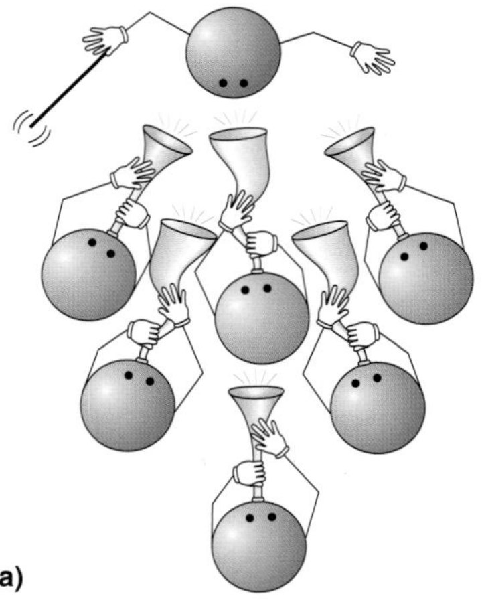
(a)

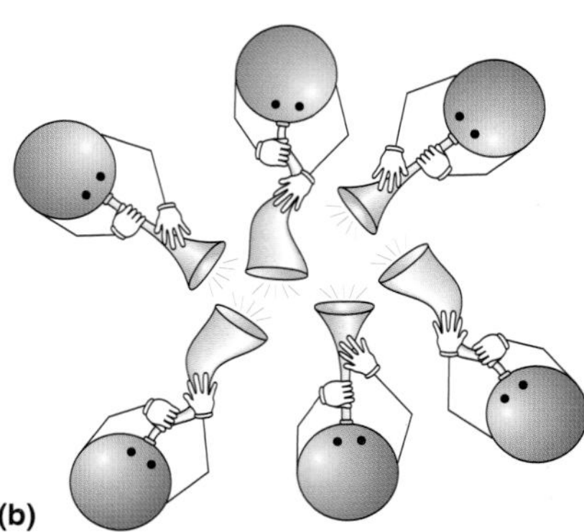
(b)

Figure 19.6
Two mechanisms of synchronous rhythms. Synchronous rhythms may **(a)** be led by a pacemaker or **(b)** arise from the collective behavior of all participants.

Functions of Brain Rhythms. Cortical rhythms are fascinating to watch in an EEG, and they parallel so many interesting human behaviors that we are compelled to ask: Why so many rhythms? More importantly, do they serve a purpose? There are no satisfactory answers yet. Ideas abound, but relevant data are scarce. One hypothesis for sleep-related rhythms is that they are the brain's way of disconnecting the cortex from sensory input. When you are awake, the thalamus allows sensory information to pass through it and be relayed up to the cortex. When you are asleep, thalamic neurons enter a self-generated rhythmic state that prevents organized sensory information from

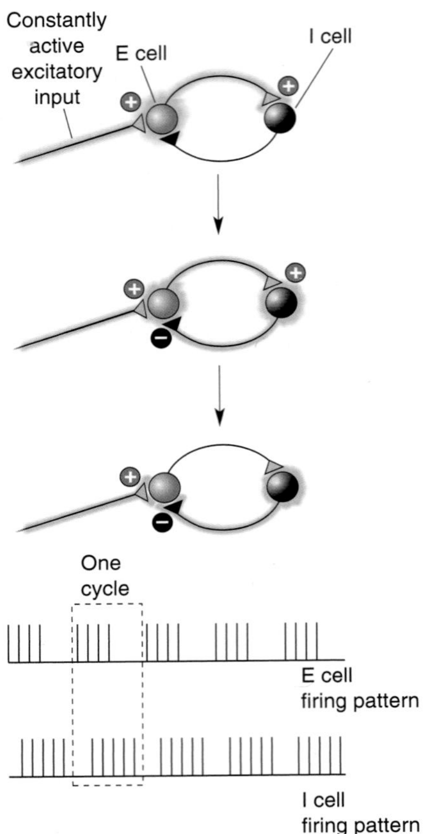

Figure 19.7
A two-neuron oscillator. One excitatory cell (E cell) and one inhibitory cell (I cell) synapse upon each other. As long as there is a constant excitatory drive, which does not have to be rhythmic, onto the E cell, activity tends to trade back and forth between the two neurons. One activity cycle through the network generates the pattern of firing shown in the box.

penetrating to the cortex. While this idea has intuitive appeal (most people do prefer to sleep in a dark, quiet environment), it does not explain why *rhythms* are necessary. Why not just steadily inhibit the thalamus and allow the cortex to rest quietly?

A function for fast rhythms in the awake cortex has recently been proposed. Recall from Chapter 10 that one scheme for understanding visual perception takes advantage of the fact that cortical objects responding to the same object are synchronously active. Walter Freeman, a neurobiologist at the University of California at Berkeley, has pioneered the idea that neural rhythms are used to coordinate activity between regions of the nervous system. Both sensory and motor systems of the awake brain often generate bursts of synchronous neural activity that give rise to EEG oscillations in the 30–80 Hz range. It may be that by momentarily synchronizing the fast oscillations generated by different regions of cortex, the brain binds together various neural components into a single perceptual construction. For example, when you are trying to catch a basketball, different groups of neurons that simultaneously respond to the specific shape, color, movement, distance, and even the significance of the basketball tend to oscillate synchronously. That the oscillations of these scattered groups of cells (those that together encode "basketballness") are highly *synchronous* somehow tags them as a meaningful group, distinct from other nearby neurons, thereby unifying the dis-

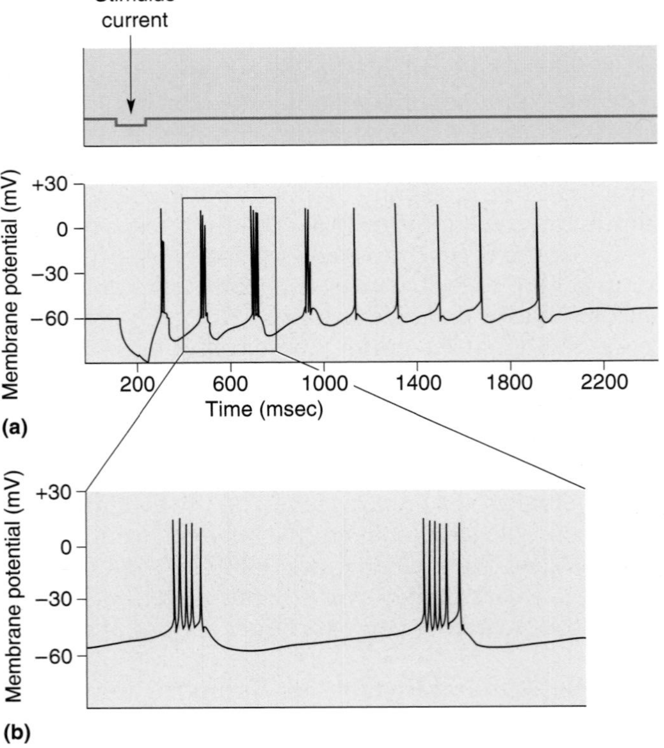

Figure 19.8
A one-neuron oscillator. At times during sleep states, thalamic neurons fire in rhythmic patterns that do not reflect their input. Shown here are intracellular recordings of membrane voltage in such a case. **(a)** A short pulse (less than 0.1 second) of stimulus current was applied, and the cell responded with almost 2 seconds of rhythmic firing, first with bursts at about 5 Hz and then with single spikes. **(b)** Two of the bursts expanded in time; each burst is a cluster of five or six action potentials. (Source: Adapted from Bal and McCormick, 1993, Fig. 2.)

jointed neural pieces of the "basketball puzzle." The evidence for this idea is indirect, far from proven, and understandably controversial.

For now, the functions of rhythms in the cerebral cortex are largely a mystery. One plausible hypothesis is that most rhythms have no direct function. Instead, they may be intriguing but unimportant by-products of the tendency for brain circuits to be strongly interconnected, with various forms of excitatory feedback. When something excites itself, whether it is an audio amplifier or the human stadium wave, it often leads to instability or oscillation. Feedback circuits are essential for the cortex to do all the marvelous things it does for us. Oscillations may be the unfortunate but unavoidable consequence, unwanted but tolerated by necessity. Even without a function, EEG rhythms provide us with a convenient window on the functional states of the brain.

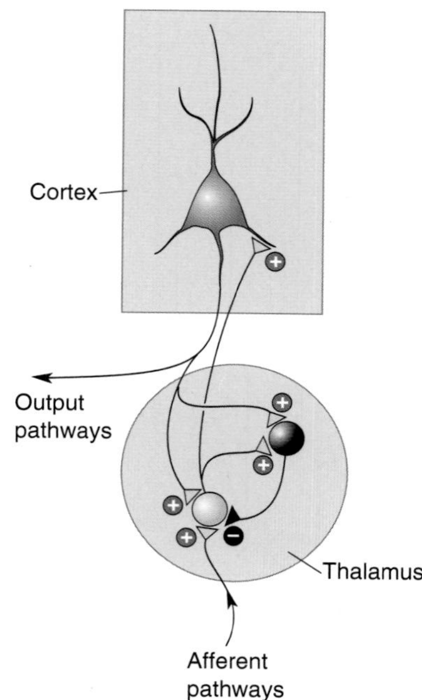

Figure 19.9
Rhythms in the thalamus drive rhythms in the cerebral cortex. The thalamus can generate rhythmic activity because of the intrinsic properties of its neurons and because of its synaptic interconnections. Yellow indicates a population of excitatory neurons, and black indicates a population of inhibitory neurons.

The Seizures of Epilepsy

Seizures, the most extreme form of synchronous brain activity, are always a sign of disorder. A **generalized seizure** involves the entire cerebral cortex of both hemispheres. A **partial seizure** involves only a circumscribed area of the cortex. In both cases, the neurons within the affected areas fire with a synchrony that never occurs during normal behavior. As a consequence, seizures are usually accompanied by very large EEG patterns. The cerebral cortex, probably because of its extensive feedback circuitry, is never far from the runaway excitation we know as a seizure. Isolated seizures are not uncommon during a lifetime, and 7–10% of the general population have at least one. When a person experiences repeated seizures, the condition is known as **epilepsy**. About 1% of the U.S. population (2.5 million people) has epilepsy.

Epilepsy is not so much a disease itself as a symptom of disease. Its causes can sometimes be identified, and they include tumors, trauma, metabolic dysfunction, infection, and vascular disease, but in many cases the cause of epilepsy is not known. It is highly unlikely that a single mechanism accounts for all seizures. Many forms of epilepsy show a genetic predisposition, and the genes responsible for numerous forms have been identified. These genes code for a diverse array of proteins, including ion channels, transporters, and receptors.

Research suggests that some seizures reflect an upset of the delicate balance of synaptic excitation and inhibition in the brain. Other seizures may be due to excessively strong or dense excitatory interconnections. Drugs that block gamma-aminobutyric acid (GABA) receptors are very potent *convulsants* (seizure-promoting agents). The withdrawal of chronic depressant drugs, such as alcohol or barbiturates, may also trigger seizures. A variety of drugs are useful in the suppression of seizures, and these *anticonvulsants* tend to counter excitability in various ways. For example, some act by prolonging the inhibitory actions of GABA (e.g., barbiturates and benzodiazepines), while others decrease the tendency for certain neurons to fire high-frequency action potentials (e.g., phenytoin and carbamazepine).

The behavioral features of a seizure depend on the neurons involved and the patterns of their activity. During most forms of generalized seizures, virtually all cortical neurons participate, so behavior is completely disrupted for many minutes. Consciousness is lost, while all muscle groups may be driven by tonic (i.e., ongoing) activity, or by clonic (i.e., rhythmic) patterns, or by both in sequence. *Absence seizures* occur during childhood; they consist of less than 30 seconds of generalized 3 Hz EEG waves accompanied by loss of consciousness. Strangely, the motor signs of an absence seizure are subtle—a fluttering of the eyelids or a twitching of the mouth.

Partial seizures can be fascinating to study and instructive anatomically. If they begin in a small area of motor cortex, they can cause clonic movement of part of a limb. In the late 1800s, British neurologist John Hughlings Jackson observed the progression of seizure-related movements across the body, looked for the scars in his patients' brains after they died, and correctly inferred the basic somatotopic map of the motor cortex (see Chapter 14). If seizures begin in a sensory area, they may trigger an abnormal sensation, or *aura*, such as an odd smell or sparkling lights. Most bizarre are the partial seizures that elicit better-formed auras such as *déjà vu* (the feeling that something has happened before) or hallucinations. These sometimes involve the cortex of the temporal lobes, including the hippocampus and amygdala, and impair memory, thought, and consciousness. Finally, partial seizures may spread uncontrollably and become generalized seizures.

SLEEP

Sleep and dreams are mysterious, even mystical to some people and a favorite subject of art and literature, philosophy, and science. Sleep is a powerful master. Each night we abandon our companions, our work, and our play and enter the cloister of sleep. We have only limited control over the decision; we can postpone sleep for a while, but eventually it overwhelms us. We spend about one-third of our lives sleeping and one-quarter of that time in a state of active dreaming.

Sleep is universal among higher vertebrates, perhaps among all animals. Prolonged sleep deprivation is devastating to proper functioning, at least temporarily, and in some animals (though not humans) it may even cause death. Sleep is essential to our lives—almost as important as eating and breathing. But why do we sleep? What purpose does it serve? Despite many years of research, the joke remains that the only thing we are sure of is that sleep overcomes sleepiness. But one of the wonderful things about science is that lack of consensus inspires a flourishing of hypotheses, and sleep research is no exception.

We can still describe what we cannot explain, and sleep has been richly studied. Let's begin with a definition: *Sleep is a readily reversible state of reduced responsiveness to and interaction with the environment.* (Coma and general anesthesia are not readily reversible and do not qualify as sleep.) In the sections that follow, we discuss the phenomenology and neural mechanisms of sleep and dreaming.

The Functional States of the Brain

During a normal day, you experience two very different and noticeable types of behavior: waking and sleeping. It is much less obvious that your sleep also has distinct phases. Several times during a night you enter a state called **rapid eye movement sleep**, or **REM sleep**, when your EEG looks more awake than asleep, your body (except for your eye muscles) is immobilized, and you conjure up the vivid, detailed illusions we call dreams. The rest of the time you spend in a state called **non-REM sleep**, in which the brain does not usually generate complex dreams. (Non-REM sleep is also sometimes called slow-wave sleep because of its domination by large, slow EEG rhythms.) These fundamental behavioral states—awake, non-REM sleep, and REM sleep—are produced by three distinct states of brain function (Table 19.1). Each state is also accompanied by large shifts in body function.

Non-REM sleep seems to be designed for rest. Muscle tension throughout the body is reduced, and movement is minimal. The body is *capable* of movement during non-REM sleep, but only rarely does the brain command it to move, usually to adjust the body's position. The temperature and energy

Table 19.1 Characteristics of the Three Functional States of the Brain

BEHAVIOR	AWAKE	NON-REM SLEEP	REM SLEEP
EEG	Low voltage, fast	High voltage, slow	Low voltage, fast
Sensation	Vivid, externally generated	Dull or absent	Vivid, internally generated
Thought	Logical, progressive	Logical, repetitive	Vivid, illogical, bizarre
Movement	Continuous, voluntary	Occasional, involuntary	Muscle paralysis; movement commanded by the brain but not carried out
Rapid eye movement	Often	Rare	Often

consumption of the body are lowered. Because of an increase in activity of the parasympathetic division of the autonomic nervous system, heart rate, respiration, and kidney function all slow down, and digestive processes increase. The brain also seems to rest. Its rate of energy use and the general firing rates of its neurons are at their lowest point of the day. The slow, large-amplitude EEG rhythms indicate that the neurons of the cortex are oscillating in relatively high synchrony, and experiments suggest that most sensory input cannot even reach the cortex. While there is no way to know for certain what people are thinking when they are asleep, studies indicate that mental processes also hit their daily low during the non-REM state. When awakened, people usually recall nothing or only very vague thoughts. Detailed, entertaining dreams are rare, although not absent, during non-REM sleep. William Dement, a prominent sleep researcher at Stanford University, characterizes non-REM sleep as *an idling brain in a movable body.*

In contrast, Dement calls REM sleep *an active, hallucinating brain in a paralyzed body.* REM sleep is dreaming sleep. Although the REM period accounts for only a small part of our sleep, it is the part most researchers get excited about (and this is the state that most excites the brain), probably because dreams are so intriguing and enigmatic. If you awaken someone during REM sleep, as Dement, Eugene Aserinsky, and Nathaniel Kleitman first did in the mid-1950s, the person will likely report visually detailed, lifelike episodes, often with bizarre story lines—the kinds of dreams we love to talk about and try to interpret.

The physiology of REM sleep is also bizarre. The EEG looks almost indistinguishable from that of an active, waking brain, with fast, low-voltage fluctuations. This is why REM sleep is sometimes called paradoxical sleep. In fact, the oxygen consumption of the brain (a measure of its energy use) is higher in REM sleep than when the brain is awake and concentrating on difficult mathematical problems. The paralysis that occurs during the REM state is almost a total loss of skeletal muscle tone, or **atonia**. Most of the body is actually *incapable* of moving! Respiratory muscles do continue to function, but just barely. The muscles controlling eye movement and the tiny muscles of the inner ear are the exceptions; these are strikingly active. With lids closed, the eyes occasionally dart rapidly back and forth. These bursts of rapid eye movement are the best predictors of vivid dreaming, and 90–95% of people awakened during or after them report dreams.

Physiological control systems are dominated by sympathetic activity during REM sleep. Inexplicably, the body's temperature control system simply quits, and core temperature begins to drift downward. Heart and respiration rates increase but become irregular. In healthy people, the clitoris and penis become engorged with blood and erect during REM sleep, although this usually has nothing to do with the sexual content of dreams. Overall, the brain seems to be doing everything except resting during REM sleep.

The Sleep Cycle

Even a good night's sleep is not a steady, unbroken journey. Sleep takes the brain through a repetitive roller-coaster ride of activity (Figure 19.10a). Sometimes the ride is wilder than at other times (Box 19.1). Roughly 75% of total sleep time is spent in non-REM sleep, and 25% is spent in REM sleep, with periodic cycles between these states throughout the night. Non-REM sleep is generally divided into four distinct stages. During a normal night, we slide through the stages of non-REM, into REM, then back through the non-REM stages, repeating the cycle about every 90 minutes. These cycles are examples of **ultradian rhythms**, which have faster periods than circadian rhythms.

EEG rhythms during the stages of sleep are shown in Figure 19.10b. As an average, healthy adult becomes drowsy and begins to sleep, he or she first enters stage 1 non-REM sleep. Stage 1 is transitional sleep, when the EEG alpha rhythms of relaxed waking become less regular and wane and the eyes make slow, rolling movements. Stage 1 is fleeting, usually lasting only a few minutes. It is also the lightest stage of sleep, meaning that we are most easily awakened during it. Stage 2 is slightly deeper and may last 5–15 minutes. Its characteristics include the occasional 8–14 Hz oscillation of the EEG called the *sleep spindle*, which is known to be generated by a thalamic pacemaker. In addition, a high-amplitude sharp wave called the K complex is observed. Eye movements almost cease. Next follows stage 3, and the EEG begins large-amplitude, slow delta rhythms. Eye and body movements are absent. Stage

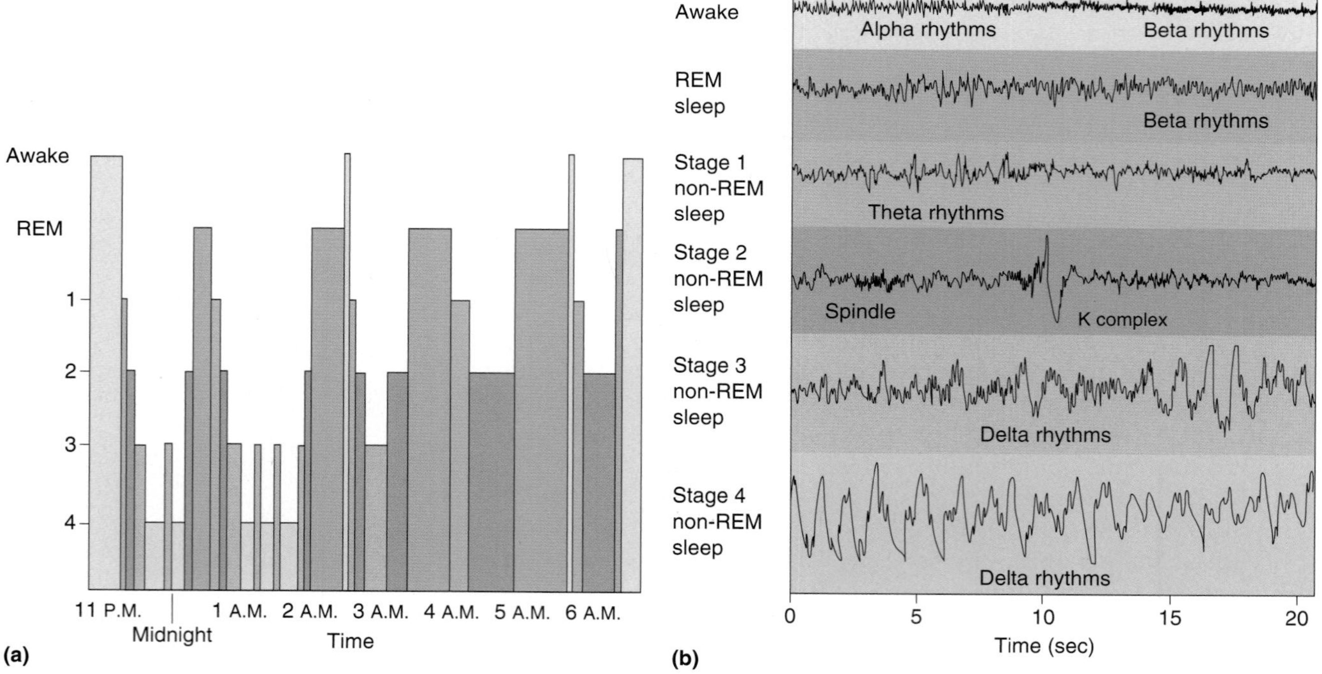

Figure 19.10

The stages of sleep through one night. (a) This graph represents falling asleep at 11 P.M. and entering non-REM sleep stage 1. The night's sleep progresses through the deeper stages of non-REM sleep, then into REM sleep. The cycle is repeated several times. Each cycle tends, however, to have shorter and shallower non-REM periods and longer REM periods. **(b)** EEG rhythms during the stages of sleep. (Source: Adapted from Horne, 1988, Fig. 1.1.)

Box 19.1

OF SPECIAL INTEREST

Walking, Talking, and Screaming in Your Sleep

Sleep is not always serene and stationary. Talking, walking, and screaming are common, and they usually occur during non-REM sleep. If this seems surprising, remember that REM sleep is accompanied by almost total body paralysis. You would therefore be incapable of walking or talking during REM sleep, even if your dream urged you to do it.

Sleepwalking, or *somnambulism*, peaks at about age 11. Although 40% of us were sleepwalkers as children, few of us sleepwalk as adults. Sleepwalking usually occurs during the first stage 4 non-REM period of the night. A full-blown sleepwalking incident involves open eyes and movement around the room, the house, or even outside, with enough awareness to avoid objects and climb stairs. Cognitive functioning and judgment are severely impaired. It is often difficult to awaken sleepwalkers because they are in deep, slow-wave sleep. The best treatment is a guiding hand back to bed. Sleepwalkers usually have no memory of the incident the next morning.

Almost everyone practices sleep talking, *somniloquy*, now and then. Unfortunately, sleeping speech is usually so garbled or nonsensical that a curious listener is disappointed by its emptiness.

More dramatic are *sleep terrors*, which are most common in children 5–7 years old. A girl screams in the middle of the night. Her parents rush to her bedside, frantic to know what has terrified her. The girl cries inconsolably, unable to explain her horrifying experience. After 10 agonizing minutes of shrieking and flailing, she finally sleeps quietly, leaving the parents shaken and baffled. The next morning she is bright and cheerful, with no recollection of the night's misadventure. Sleep terrors are distinctly different from nightmares, which are vivid, complex dreams, outwardly quiet, that occur during REM sleep. By contrast, sleep terrors begin in stage 3 or 4 of non-REM sleep, and the experience is not dreamlike but a feeling of uncontrollable panic, accompanied by greatly increased heart rate and blood pressure. They usually pass with age and are not a symptom of a psychiatric disorder.

4 is the deepest stage of sleep, with large EEG rhythms of 2 Hz or less. During the first cycle of sleep, stage 4 may persist for 20–40 minutes. Then sleep begins to lighten again, ascends to stage 2 for 10–15 minutes, and suddenly enters a brief period of REM sleep, with its fast EEG beta rhythms and sharp, frequent eye movements.

As the night progresses, there is a general reduction in the duration of non-REM sleep, particularly in stages 3 and 4, and an increase in the REM periods. Half of the night's REM sleep occurs during its last third, and the longest REM cycles may last 30–50 minutes. Still, there seems to be an obligatory refractory period of about 30 minutes between periods of REM; in other words, each REM cycle is followed by at least 30 minutes of non-REM sleep before the next REM period can begin.

What is a normal night's sleep? Your mother may have insisted that you need a "good 8 hours" of sleep each night. Research suggests that normal requirements vary widely among adults, from about 5 to 10 hours per night. The average length is about 7.5 hours, and the sleep duration of about 68% of young adults is between 6.5 and 8.5 hours. What is the proper length of sleep time for you? The best measure of successful sleep is the quality of your time awake. You need a certain amount of sleep to maintain a reasonable level of alertness. Too much daytime sleepiness can be more than annoying; it can be dangerous if it interferes with driving, for example. Because of the wide variations among individuals, you must decide for yourself how much sleep you need.

Why Do We Sleep?

All mammals, birds, and reptiles appear to sleep, although only mammals and some birds have a REM phase. Sleep time varies widely, from about 18 hours a day in bats and opossums to about 3 hours a day in horses and giraffes. Some people argue that a behavior as pervasive as sleep must have a critical function; otherwise, some species would have lost the need to sleep through evolution. Whatever the function, there is good reason to believe sleep is designed solely for the brain. A restful 8 hours in bed without sleep might allow your body to recover from physical exertion, but you would probably not be at your best mentally the next day.

Some animals apparently have more reason *not* to sleep than others do. Imagine living your entire life in deep or turbulent water, yet needing to breathe air every minute or so. Even a quick nap would be awkward at best. This is precisely the situation with dolphins, yet they sleep about as much as humans do. Remarkably, bottlenose dolphins sleep with only one cerebral hemisphere at a time: about 2 hours asleep on one side, then 1 hour awake on both sides, 2 hours asleep on the other side, and so on, for a total of 12 hours per night (Figure 19.11). (This gives new meaning to the phrase "half asleep.") Bottlenose dolphins seem to have no REM sleep. The blind Indus dolphin of Pakistan uses sonar to navigate through muddy, turbid, sweeping currents; and during monsoon season, it must never stop swimming or it will come to grief on the rocks and debris of the flooded estuary it calls home. Still, the Indus dolphin manages to sleep, snatching microsleeps 4–6 seconds long while continuing to swim slowly. Its many microsleeps add up to about 7 hours in a 24-hour day. Dolphins have evolved extraordinary sleep mechanisms that adapt them to a demanding environment. But the fact that dolphins are not sleepless reinforces our question: What is so important about sleeping?

No single theory of sleep function is widely accepted, but the most reasonable ideas fall into two categories: theories of *restoration* and theories of *adaptation*. The first category is a commonsense explanation: We sleep to rest and recover and to prepare to be awake again. The second category is less obvious: We sleep to keep ourselves out of trouble, to hide from predators when we are most vulnerable or from other harmful features of the environment, or to conserve energy.

If sleep is restorative, what is it restoring? Quiet rest is certainly not a sub-

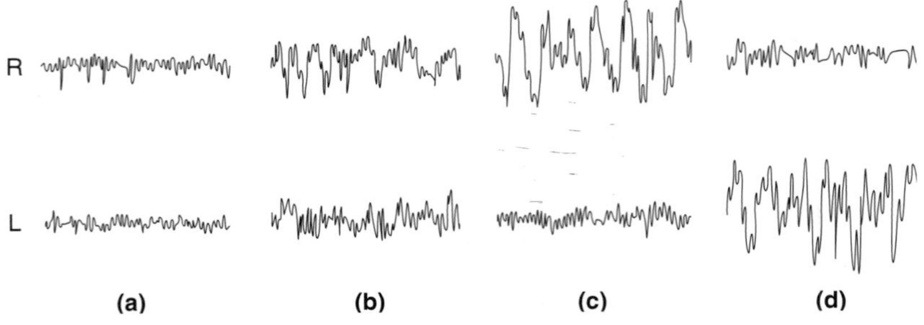

Figure 19.11
Sleep in the bottlenose dolphin. These EEG patterns were recorded from the right (R) and left (L) hemispheres. **(a)** High-frequency activity on both sides during alert wakefulness. **(b)** Low-amplitude delta rhythms of an intermediate sleep stage on the right side and, to some extent, on the left. **(c)** Large delta rhythms of deep sleep only on the right side, with fast activation on the left. **(d)** The patterns shift to opposite hemispheres some time later. (Source: Mukhametov, 1984, Fig. 1.)

stitute for sleep. Sleeping does something more than simple resting. Prolonged sleep deprivation may lead to serious physical and behavioral problems (Box 19.2). Unfortunately, no one has yet identified a particular physiological process that is clearly restored by sleep, an essential substance that is made or a toxin that is destroyed while sleeping. Sleep does prepare us to be effectively awake again. But does sleep renew us in the same way that eating and drinking do, by replacing essential substances, or the way the healing of a wound repairs damaged tissues? For the most part, evidence indicates that sleep is not a time of increased tissue repair for the body. It is possible, however, that brain regions such as the cerebral cortex can achieve some form of essential "rest" only during non-REM sleep.

Adaptation theories of sleep take many forms. Some large animals eat small animals; a stroll in the moonlight is far too risky for a squirrel living in owl and fox territory. The squirrel's best strategy may be to stay safely tucked away in its burrow during the night, and sleep is a good way to enforce such isolation. At the same time, sleep may be an adaptation for conserving energy. While sleeping, the body does just enough work to stay alive, core temperature drops, temperature regulation is depressed, and the rate of calories burned is kept low.

Functions of Dreaming and REM Sleep

In many ancient cultures, people believed that dreams were a window on some higher world and a source of information, guidance, power, or enlightenment. Perhaps they were right, but the collective wisdom of the past does

Box 19.2

OF SPECIAL INTEREST

The Longest All-Nighter

In 1963, Randy Gardner was a 17-year-old high school student with an ambitious idea for a San Diego Science Fair project. On December 28 he awoke at 6 A.M. to begin. When he finished 11 days (264 hours) later, he had broken the world's record for continuous wakefulness, under the continuous scrutiny of two friends and, during the last 5 days, fascinated sleep researchers. He had used no drugs, not even caffeine.

The experience was not pleasant. After 2 days without sleep, Randy became irritable, nauseated, had trouble remembering, and could not even watch television. By the fourth day he had mild delusions and overwhelming fatigue, and by the seventh day he had tremors, his speech was slurred, and his EEG no longer showed alpha rhythms. Fortunately, he did not become psychotic, despite the predictions of some "experts." On the contrary, on his last awake night, he beat one of his better-rested observers at an arcade baseball game, and he gave a coherent account of himself at a national press conference.

When he finally went to bed, he slept for almost 15 hours straight, then stayed awake 23 hours to wait for nightfall, and slept for another 10.5 hours. After the first sleep, his symptoms had mostly disappeared, and within a week, he was sleeping and behaving normally.

One of the most interesting things about Randy's ordeal is that there were no lasting harmful effects. The same is not true for some animals deprived of sleep. If rats are kept awake for long periods, they progressively lose weight while consuming much more food; they become weak, accumulate stomach ulcers and internal hemorrhages, and even die. They seem to suffer from an impairment of their ability to regulate body temperature and metabolic needs. Total sleep deprivation is not necessary. Prolonged REM sleep loss alone is detrimental. These results may imply that sleep does provide something physiologically essential.

not agree on exactly how to interpret the meaning of dreams. Today, we must take a step backward and first ask whether dreams even *have* meaning. Dreams are difficult to study. Obviously, we can't directly observe the dreams of someone else, and even the dreamer has access to them only after he or she has awakened and perhaps forgotten or distorted the experience. Because the phenomena of REM can be objectively measured, modern explanations of dreaming lean heavily on studies of REM sleep rather than of dreaming. But it is important to remember that the two are not synonymous. Some dreams can occur outside of REM sleep, and REM sleep has many peculiar features that have nothing to do with dreaming.

Do we need to dream? No one knows, but the body does seem to crave REM sleep. It is possible to deprive sleepers of REM sleep specifically, by waking them every time they enter the REM state; when they fall asleep a minute or two later, it is inevitably into a non-REM state, and they can accumulate an entire night of relatively pure non-REM sleep. As Dement first observed, after several days of this annoying treatment, sleepers attempt to enter the REM state much more frequently than normal. When they are finally allowed to sleep undisturbed, they undergo *REM rebound* and spend more time in REM proportional to the duration of their deprivation. Most studies have not found that REM deprivation causes any psychological harm during the daytime. Again, it is important not to interpret REM deprivation as dream deprivation, since during REM deprivation, dreams may continue to occur during sleep onset and during non-REM periods.

Sigmund Freud suggested many functions for dreams. For Freud, dreams were disguised wish-fulfillment, an unconscious way for us to express our sexual and aggressive fantasies, which are forbidden while we are awake. Bad dreams may help us conquer the anxiety-provoking events of life. Recent theories of dreaming are more biologically based. Allan Hobson and Robert McCarley of Harvard University propose an "activation-synthesis" hypothesis, which explicitly rejects the Freudian psychological interpretations. Instead, dreams, or at least some of their bizarre features, are seen as the associations and memories of the cerebral cortex that are elicited by the random discharges of the pons during REM sleep. Thus, the pontine neurons, via the thalamus, *activate* various areas of the cerebral cortex and elicit well-known images or emotions, and the cortex tries to *synthesize* the disparate images into a sensible whole. Not surprisingly, the "synthesized" dream product may be quite bizarre and even nonsensical because it is triggered by the semi-random activity of the pons. Evidence for the activation-synthesis hypothesis is mixed. It does predict the weirdness of dreams and their correlation with REM sleep, but it does not explain how random activity can trigger the complex and fluid stories that many dreams contain, nor how they can evoke dreams that recur night after night.

Many researchers have suggested that REM sleep and perhaps dreams themselves have an important role in memory. None of the evidence is definitive, but intriguing hints indicate that REM sleep somehow aids the integration or consolidation of memories. Depriving humans or rats of REM sleep can impair their ability to learn a variety of tasks. Some studies show an increase in the duration of REM sleep after an intense learning experience. In one study, Israeli neuroscientist Avi Karni and his colleagues trained people to identify the orientation of a small line in their peripheral visual field. The task was made difficult by presenting the visual stimulus for a very short time. With repeated practice over days, people got much better at this task; surprisingly, their performance also improved between evening and morning, after a night's sleep. Karni found that if people were deprived of REM sleep, their learning of the task did not improve overnight. Depriving them of non-REM sleep, however, actually enhanced their performance. Karni hypothesizes that this kind of memory requires time to strengthen and that REM sleep is particularly effective for this purpose.

You may have heard about sleep-learning, the notion that you can study for an exam by listening to a tape of the material while you blissfully snooze away. Sounds like a student's fantasy, right? Unfortunately, it is exactly that and no more. There is no scientific evidence for sleep-learning, and careful studies have shown that the very few things recalled the next morning were heard when the subjects briefly woke up. In fact, sleep is a profoundly amnesic state. Most of our dreams, for example, seem to be lost forever. Although we dream profusely during each of the four or five REM periods every night, we usually remember only the last dream before waking. Also, when we wake up to do something in the middle of the night, we have often forgotten the incident by morning.

At this point you are probably confused about the functions of dreaming and REM sleep. So are we. Unfortunately, there is not enough evidence to support or dismiss any of the theories we have discussed. There are also many other creative and plausible ideas that we do not have the space to present.

Neural Mechanisms of Sleep

Until the 1940s, it was generally believed that sleep was a passive process: Deprive the brain of sensory input, and it will fall asleep. When the sensory afferents to an animal's brain are blocked, however, the animal continues to have cycles of waking and sleeping. We now know that sleep is an active process that requires the participation of a variety of brain regions. As we saw in Chapter 15, wide expanses of the cortex are actually controlled by very small collections of neurons much deeper in the brain. These cells act like the switches or tuners of the forebrain, altering cortical excitability and gating the flow of sensory information into it. The full details of these control systems are complex and not fully understood. But we can summarize a few principles:

1. The neurons most critical to the control of sleeping and waking are part of the diffuse modulatory neurotransmitter systems.
2. The brain stem modulatory neurons using norepinephrine (NE) and serotonin fire during waking and enhance the awake state; some neurons using acetylcholine (ACh) enhance critical REM events, and other cholinergic neurons are active during waking.
3. The diffuse modulatory systems control the rhythmic behaviors of the thalamus, which in turn controls many EEG rhythms of the cerebral cortex; slow, sleep-related rhythms of the thalamus apparently block the flow of sensory information up into the cortex.
4. Sleep also involves activity in descending branches of the diffuse modulatory systems, e.g., active inhibition of motor neurons during dreaming.

There are three basic kinds of evidence for the localization of sleep mechanisms in the brain. Lesion data reveal changes in function after a part of the brain is removed, results of stimulation experiments identify changes following the activation of a brain region, and recordings of neural activity determine the relationship between that activity and different brain states.

Wakefulness and the Ascending Reticular Activating System. Lesions in the brain stem of humans can cause sleep and coma, suggesting that the brain stem has neurons whose activity is essential to keeping us awake. In the 1940s and 1950s, Italian neurophysiologist Giuseppe Moruzzi and his colleagues began to sort out the neurobiology of the brain stem's control of waking and arousal. They found that lesions in the midline structures of the brain stem caused a state similar to non-REM sleep, but lesions in the lateral tegmentum, which interrupted ascending sensory inputs, did not. Conversely, electrical stimulation of the midline tegmentum of the midbrain within the reticular formation transformed the cortex from the slow, rhythmic EEGs of non-REM sleep to a more alert and aroused state with an EEG similar to that of waking. Moruzzi called this ill-defined region of stimulation the *ascending reticular activating system* (mentioned in Chapter 15). This area is now much better defined anatomically and physiologically, and it is clear that Moruzzi's stimulation was affecting many different sets of ascending modulatory systems.

Several sets of neurons increase their firing rates in anticipation of awakening and during various forms of arousal. They include cells of the locus coeruleus, which contain norepinephrine, serotonin-containing cells of the raphe nuclei, acetylcholine-containing cells of the brain stem and basal forebrain, and neurons of the midbrain that use histamine as a neurotransmitter. Collectively, these neurons synapse directly on the entire thalamus, cerebral cortex, and other brain regions. The general effects of their transmitters are depolarization of neurons, an increase in their excitability, and suppression of rhythmic forms of firing. These effects are most clearly seen in the relay neurons of the thalamus (Figure 19.12).

Falling Asleep and the Non-REM State. Falling asleep involves a progression of changes over several minutes, culminating in the non-REM state. It is not clear what initiates non-REM sleep, but there is a general decrease in the firing rates of most brain stem modulatory neurons (those using NE, 5-HT [5-hydroxytryptamine], and ACh). Although most regions of the basal forebrain seem to promote alertness and arousal, a subset of its cholinergic neurons increases their firing rate with the onset of non-REM sleep and are silent during wakefulness.

Early stages of non-REM sleep include the EEG sleep spindles, described earlier, which are generated by the inherent rhythmicity of thalamic neurons. As non-REM sleep progresses, spindles disappear and are replaced by slow delta rhythms (less than 4 Hz). Delta rhythms also seem to be a product of thalamic cells, occurring when their membrane potentials become even more negative than during spindle rhythms (and much more negative than they are during waking). Synchronization of activity during spindle or delta rhythms is due to neural interconnections within the thalamus and possibly between the thalamus and cortex. Because of the strong two-way excitatory connections between the thalamus and cortex, rhythmic activity in one is often strongly and widely projected upon the other.

Mechanisms of REM Sleep. REM sleep is such a different state from non-REM sleep that we would expect some clear neural distinctions. Many cortical areas are at least as active during REM sleep as during waking. For example, neurons of the motor cortex fire rapidly and generate organized

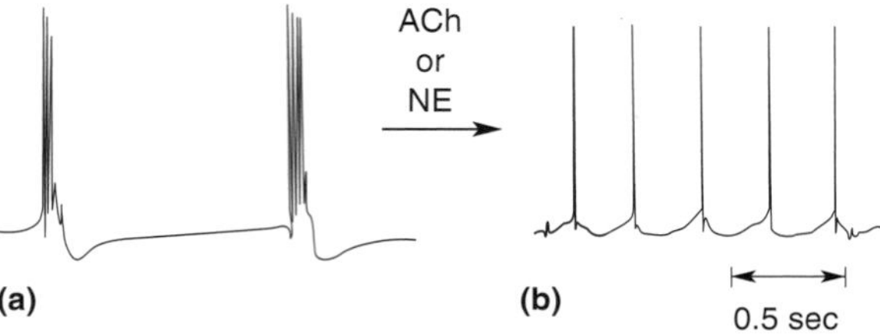

Figure 19.12
Modulating thalamic and cortical rhythmicity. In this case, either ACh *or* NE shifts cells from **(a)** an intrinsic burst-firing mode to **(b)** a single-spiking mode. This may be what happens during transitions from non-REM sleep to the waking state. (Source: Steriade et al., 1993, Fig. 5D.)

motor patterns that attempt to command the entire body but succeed only with a few muscles of the eye, the inner ear, and those essential for respiration. The elaborate dreams of REM sleep certainly require the cerebral cortex, but the cortex is not necessary for the *production* of REM sleep.

The use of positron emission tomography (PET) in the waking and sleeping human brain has provided fascinating glimpses into the activity patterns that distinguish waking from REM sleep and non-REM sleep. Figure 19.13a shows the difference in brain activity between REM sleep and waking. Some areas, including primary visual cortex, were equally active in the two states. But extrastriate cortical areas and portions of the limbic system were significantly more active during REM sleep. Conversely, regions of the frontal lobes were noticeably less active during REM. Figure 19.13b contrasts the brain activity in REM sleep and non-REM sleep. The primary visual cortex and a number of other areas are significantly less active during REM sleep, but extrastriate cortex is more active. These results paint an intriguing picture of what happens when we sleep. During REM sleep there is an explosion of extrastriate activity, presumably during the times when we dream. There is no corresponding activity increase in primary visual cortex, however, suggesting that the extrastriate excitation is internally generated. The emotional component to dreams may derive from the heightened limbic activation. The low activity in the frontal lobe suggests that high-level integration or interpretation of the extrastriate visual information may not take place, leaving us with a buzz of uninterpreted visual imagery.

The control of REM sleep, as with the other functional brain states, derives from diffuse modulatory systems in the core of the brain stem, particularly the pons. The firing rates of the two major systems of the upper brain stem, the locus coeruleus and the raphe nuclei, decrease to almost nothing with the onset of REM (Figure 19.14). There is, however, a concurrent sharp increase in the firing rates of ACh-containing neurons in the pons, and some evidence suggests that cholinergic neurons induce REM sleep. It is probably the action of ACh during REM sleep that causes the thalamus and cortex to behave so much like they do in the waking state.

Why don't we act out our dreams? The same core brain stem systems that control the sleep processes of the forebrain also actively inhibit our spinal motor neurons, preventing the descending motor activity from expressing itself as actual movement. This is clearly an adaptive mechanism, protecting us from ourselves. There are rare people, usually elderly men, who do seem to act out their dreams; they have a hazardous condition know as *REM sleep behavior disorder*. These people often sustain repeated injuries, and even their spouses have fallen victim to their nocturnal flailings. One man dreamed he was in a football game and tackled his bedroom bureau. Another imagined he was defending his wife from attack, when in fact he was beating her in her

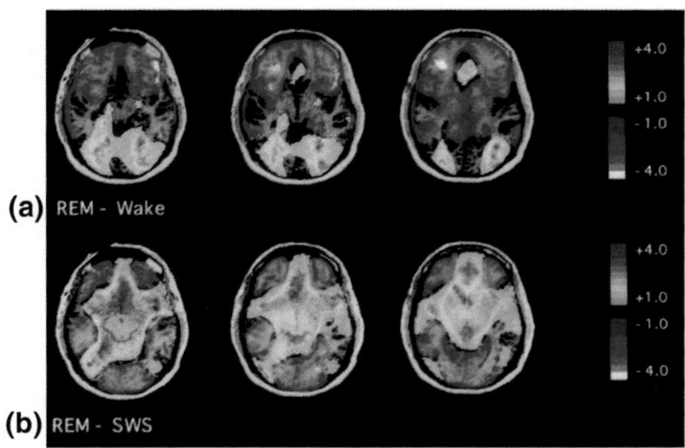

Figure 19.13

PET images of the waking and sleeping human brain. (a) These images show brain activity in three horizontal sections in which color represents changes in activity between REM sleep and waking (yellow and red areas are more active during REM sleep). The dark notch at the bottom (posterior) edge of the sections indicates that striate cortex is equally active in the two states. **(b)** In these three horizontal sections, REM sleep is compared with non-REM sleep (SWS). In REM, striate cortex is less active. (Source: Braun et al., 1998, Fig. 1.)

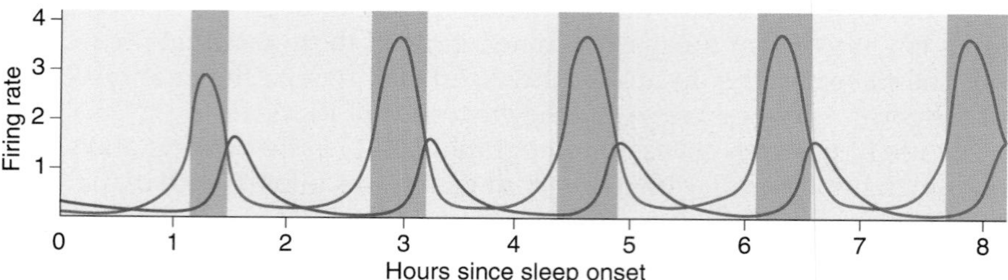

Figure 19.14
Control of the onset and offset of REM periods by brain stem neurons. This graph shows the relative firing rates of REM-associated neurons during a single night. Periods of REM sleep are green. REM-on cells are cholinergic neurons of the pons, and they increase their firing rates just before the onset of REM sleep (red line). REM-off cells are noradrenergic and serotonergic neurons of the locus coeruleus and raphe nuclei, respectively, and their firing rates increase just before the end of REM sleep (blue line). (Source: McCarley and Massaquoi, 1986, Fig. 4B.)

bed. The basis for this REM disorder seems to be disruption of the brain stem systems that normally mediate REM atonia. Experimental lesions in certain parts of the pons can cause a similar condition in cats. During REM periods, they may seem to chase imaginary mice or investigate invisible intruders. Disorders of REM control mechanisms may also underlie the problems of people with narcolepsy (Box 19.3).

Sleep-Promoting Factors. Sleepiness is one of the most familiar consequences of infectious diseases such as the common cold and the flu. There may be direct links between the immune response to infection and the regulation of sleep. Sleep researchers have searched intensively for a chemical in the blood or cerebrospinal fluid that promotes or even causes sleep. Many sleep-promoting substances have been identified in sleep-deprived animals. Most interact with the body's immune system. In the 1970s, physiologist John Pappenheimer of Harvard University identified a muramyl dipeptide in the spinal fluid of sleep-deprived goats that facilitated non-REM sleep. Muramyl peptides are usually produced only by the cell walls of bacteria, not brain cells, and they also cause fever and stimulate immune cells of the blood. It is not clear how they appear in spinal fluid, but they may be synthesized by bacteria in the intestines. Another sleep-promoting factor, interleukin-1, is synthesized by the brain in glia and in macrophages, cells throughout the body that scavenge foreign material. Interleukin-1 is also a peptide that stimulates the immune system.

A recent candidate as a sleep-promoting factor—adenosine—may appeal to the millions who drink coffee, tea, and cola. This small molecule acts as a neuromodulator at synapses throughout the brain. From ancient times, antagonists of adenosine receptors, such as caffeine and theophylline, have been used to keep people awake. Conversely, the administration of adenosine or its agonists increases sleep. In a number of brain areas that have been examined, extracellular adenosine levels are higher during waking than while sleeping, and the levels progressively increase during sustained waking periods. Adenosine levels gradually decrease during sleep.

How might adenosine promote sleep? Adenosine has an inhibitory effect on the diffuse modulatory systems for ACh, NE, and 5-HT that are more active in the awake brain. This suggests that sleep may be the result of a molecular chain reaction. Neural activity in the awake brain increases adenosine levels, and as these levels increase during waking hours, there is more inhi-

Box 19.3

OF SPECIAL INTEREST

Narcolepsy

Narcolepsy is a bizarre and disabling disturbance of sleeping and waking. Despite the sound of the name, it is not a form of epilepsy. It can include some or all of the following manifestations.

Excessive daytime sleepiness can be severe and often leads to unwanted "sleep attacks." *Cataplexy* is a sudden muscular paralysis while consciousness is maintained. In the middle of a normal day, sufferers suddenly collapse into a state similar to REM sleep. Cataplexy is often brought on by strong emotional expression, such as laughter or tears, or by surprise or sexual arousal, and it usually lasts less than a minute. *Sleep paralysis*, a similar loss of muscle control, occurs during the transition between sleeping and waking. It sometimes occurs in the absence of narcolepsy, and it can be very disconcerting; although conscious, a person may be unable to move or speak for several minutes. *Hypnagogic hallucinations* are graphic dreams, often frightening, that also accompany sleep onset and may follow sleep paralysis. Sometimes such dreams flow smoothly with real events that occurred just prior to falling asleep.

EEG monitoring reveals a distinct difference between narcoleptic and normal sleep. A narcoleptic person goes directly from waking into a REM phase, whereas normal adult sleepers always enter a long period of non-REM sleep first. Most of the symptoms of narcolepsy may be interpreted as an abnormal intrusion of the characteristics of REM sleep into waking.

The prevalence of narcolepsy varies greatly. For example, it affects about 1 in 1000–2000 people in the U.S. population, but only 1 in 500,000 in Israel. There is a genetic component to the disorder, and a high percentage of people with narcolepsy have a particular form of the human leukocyte antigen (HLA) gene. About 25% of the general population has the narcoleptic form of the HLA gene, however, and most do not develop narcolepsy. In the United States, first-degree relatives of people with narcolepsy have roughly a 1% chance of developing it themselves, compared with the 0.1% chance in people without an affected first-degree relative. Environmental factors presumably play an important role, along with genetics.

Narcolepsy occurs in goats, donkeys, ponies, and more than a dozen breeds of dogs. In dogs, genetics plays a more significant role than in humans. In 1999, Emmanuel Mignot, Seiji Nishino, and their research team at Stanford University found that canine narcolepsy is caused by a mutation of the orexin receptor 2 gene. Cell bodies containing orexin lie in the dorsolateral hypothalamus, and they project widely throughout the brain. Orexin is a brain peptide known to stimulate feeding behavior (see Chapter 16). It has been suggested that this effect of orexin actually results from a shift in the daily pattern of feeding rather than a simple augmentation of food intake. Although the relationship between orexin and narcolepsy is still unclear, increasing evidence indicates that there are complex interactions in the hypothalamus between mechanisms involved in periodic daily behaviors, including sleeping and eating.

Both cholinergic and catecholaminergic systems have been implicated in the abnormalities associated with narcolepsy. For example, even in normal animals, ACh agonists can induce REM sleep, but in narcoleptic animals, a significantly lower dose is needed to cause muscle atonia of the type seen in cataplexy. It has also been reported that pharmacologically increasing dopamine output from the ventral tegmental area induces cataplexy in narcoleptic animals but not in normal animals. As the output from this nucleus is associated with emotion and reward (see Chapter 18), it may explain the fact that emotions trigger cataplexy.

There is no known cure for narcolepsy, and treatment consists of trying to relieve the symptoms. Frequent naps, amphetamines, and a new drug called modafinil may help daytime sleepiness, while tricyclic antidepressant drugs (which have REM-suppressant effects) may reduce cataplexy.

bition of neurons in the modulatory systems associated with wakefulness. Increased suppression of the "wakeful" modulatory systems increases the likelihood that the brain will fall into the slow-wave synchronous activity characteristic of sleep. After sleep begins, adenosine levels slowly fall, and activity in the modulatory systems gradually increases until we wake up to start the cycle anew.

Why not just eliminate the adenosine loop and keep the brain awake all the time? We don't know the answer. One interesting hypothesis is that adenosine is a cue that the brain needs time for restoration. It has been proposed that glycogen, found in astrocytes, is an energy source for the brain that is called upon during the waking hours. If glycogen depletion were to increase adenosine release, this would close a simple negative feedback loop, initiating sleep, so that glycogen stores could be replenished. For now, this is simply a hypothesis, but it illustrates the molecular approach that research in this area is taking.

Gene Expression During Sleeping and Waking. Research into the neural function of sleep has benefited from studies at various levels of analysis, including sleep behavior, brain physiology, and the action of diffuse modulatory systems. The use of techniques from molecular neurobiology is relatively new. While the pieces do not all fit together quite yet, it is clear that the behavioral states of sleeping and waking are different even at the molecular level. For example, in the macaque monkey, most areas of cerebral cortex show higher rates of protein synthesis in deep sleep than in light sleep. In rats, levels of cyclic adenosine monophosphate (cAMP) in several brain areas have been found to be lower during sleeping than waking.

Recent research has demonstrated that sleeping, waking, and sleep deprivation are associated with differences in the expression of certain genes. In one study, Chiara Cirelli and Giulio Tononi of the Neurosciences Institute in San Diego examined the expression of thousands of genes in rats that were awake or asleep. Most genes were expressed at the same level in the two states. The genes that showed different levels of expression, however, may provide insight into what happens in the brain during sleep. Most of the genes that were more highly expressed in the awake brain fell into one of two groups. One group includes what are called *immediate early genes*, genes that code for transcription factors that affect the expression of other genes. Some of these genes appear to be related to changes in synaptic strength. The low expression of these genes during sleep may be associated with the fact that learning and memory formation are largely absent in this state. The second group of genes that are more highly expressed in the awake brain come from mitochondria. Increased expression of these genes may play a role in satisfying the higher metabolic demands of the awake brain.

CIRCADIAN RHYTHMS

Almost all land animals coordinate their behavior according to **circadian rhythms**, the daily cycles of lightness and darkness that result from the spin of the earth. (The term is from the Latin *circa*, "approximately," and *dies*, "day."). The precise schedules of circadian rhythms vary among species. Some animals are active during daylight hours, others only at night, and others mainly at the transitional periods of dawn and dusk. Most physiological and biochemical processes in the body also rise and fall with daily rhythms; body temperature, blood flow, urine production, hormone levels, hair growth, and metabolic rate all fluctuate (Figure 19.15). In humans, there is an approximate inverse relationship between the propensity to sleep and body temperature.

When the cycles of daylight and darkness are removed from an animal's environment, circadian rhythms continue on more or less the same schedule because the primary clocks for circadian rhythms are not astronomical (the sun and the earth) but biological in the brain. Brain clocks, like all clocks, are imperfect and require occasional resetting. Now and then you readjust your watch to keep it in sync with the rest of the world (or at least the time given on the radio). Similarly, external stimuli, such as light and dark, or daily temperature changes help adjust the brain's clocks to keep them synchronized

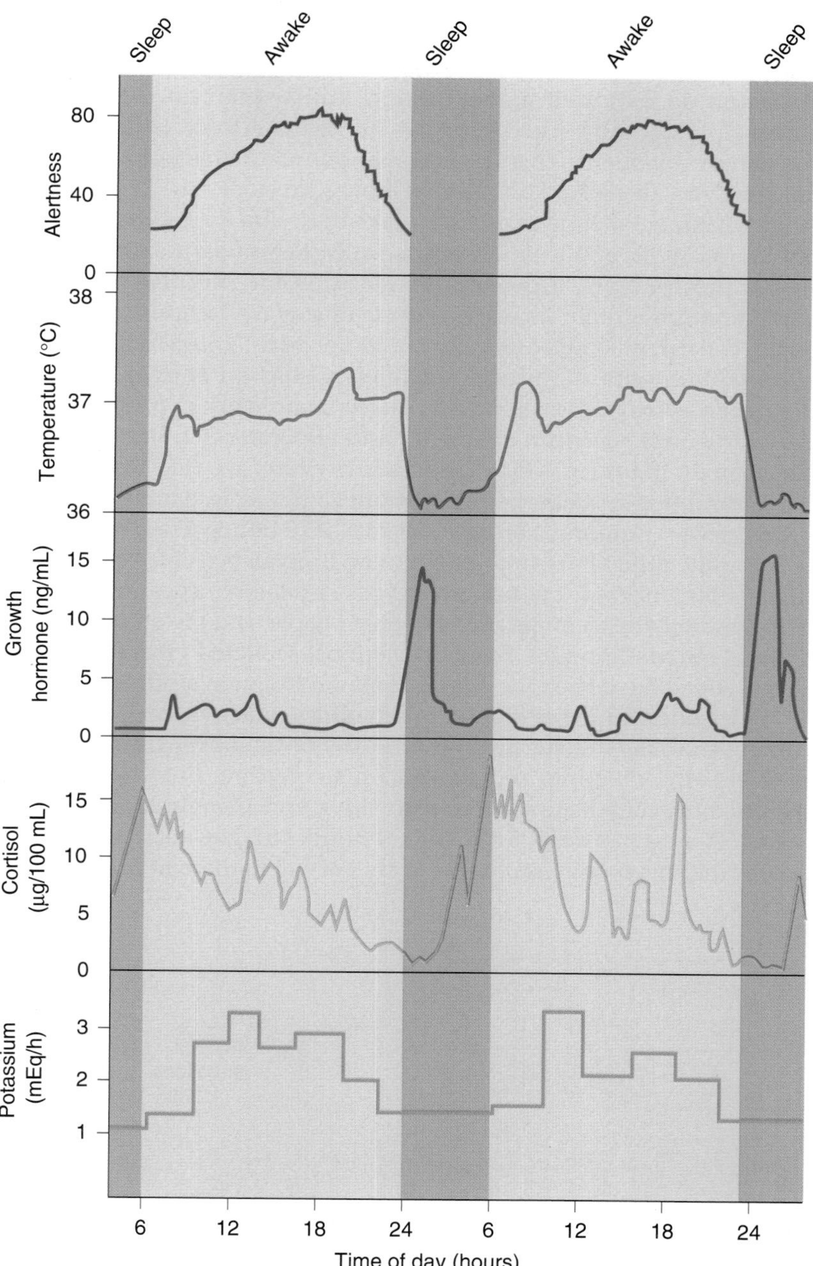

Figure 19.15
Circadian rhythms of physiological functions. Fluctuations over two consecutive days are shown here. Alertness and core body temperature vary similarly. Growth hormone and cortisol levels in the blood, however, are highest during sleep, although at different times. The bottom graph shows the excretion of potassium by the kidneys, which is highest during the day. (Source: Adapted from Coleman, 1986, Fig. 2.1.)

with the coming and going of the sunlight. Circadian rhythms have been well studied at both behavioral and cellular levels. Brain clocks are especially useful for studying the link between the activity of specific neurons and behavior.

Biological Clocks

The first evidence for a biological clock came from a brainless organism, the mimosa plant. The mimosa raises its leaves during the day and lowers them at night. It seemed obvious to many people that the plant simply reacts to sunlight, in some kind of reflex movement. In 1729, French physicist Jean Jacques d'Ortous de Mairan tested the obvious; he put some mimosa plants in a dark closet and found that they continued to raise and lower their leaves.

But a surprising new observation can still lead to a wrong conclusion. It was de Mairan's opinion that the plant was still somehow sensing the sun's movements, even in the darkness. More than a century later, Swiss botanist Augustin de Candolle showed that a similar plant in the dark moved its leaves up and down every 22 hours, rather than every 24 hours according to the sun's movement. This implied that the plant was not responding to the sun and very likely had an internal biological clock.

Environmental time cues (light and dark and variations in temperature and humidity) are collectively termed **zeitgebers** (German for "time givers"). In the presence of zeitgebers, animals become *entrained* to the day-night rhythm and maintain an activity cycle of exactly 24 hours. Obviously, even small, consistent errors of timing could not be tolerated for long. A 24.5-hour cycle would, within 3 weeks, completely shift an animal from daytime to nighttime activity. When mammals are completely deprived of zeitgebers, they settle into a rhythm of activity and rest that often has a period more or less than 24 hours, in which case their rhythms are said to *free-run*. In mice, the natural free-running period is about 23 hours; in hamsters it is close to 24 hours; and in humans it tends to be 24.5–25.5 hours (Figure 19.16).

It is quite difficult to separate a human from all possible zeitgebers. Even inside a laboratory, society provides many subtle time cues, such as the sounds of machinery, the comings and goings of people, and the on-off cycling of heating and air conditioning. Some of the most secluded environments are deep caves, which have been the sites for several isolation studies. When people in caves are allowed to set their own schedules of activity for months on end—waking and sleeping, turning lights on and off, and eating when they choose—they initially settle into roughly a 25-hour rhythm. But after days to weeks, their activity may begin to free-run with a surprisingly long period of 30–36 hours. They stay awake for about 20 hours straight, then sleep for about 12 hours, and this pattern seems perfectly normal to them at the time.

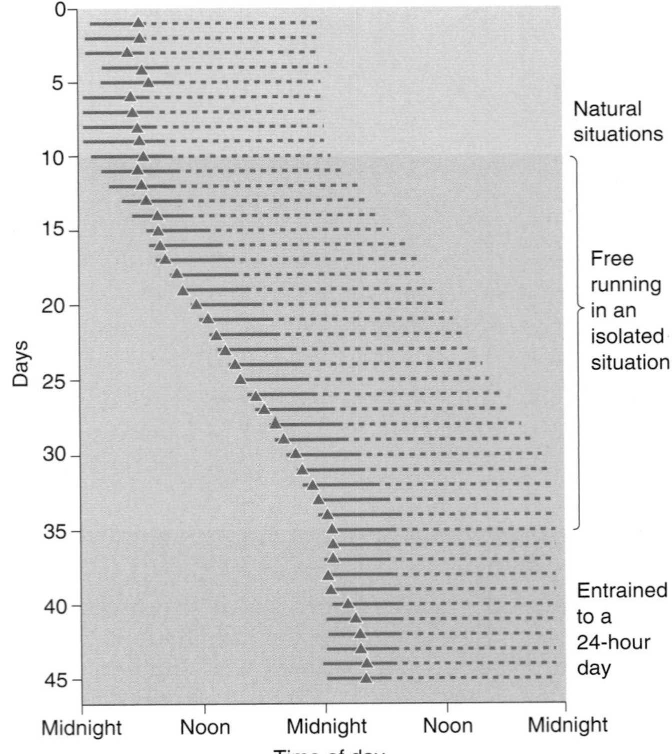

Figure 19.16
Circadian rhythms of sleep and wakefulness. Shown here is a daily plot of one person's sleep-wake cycles. Each horizontal line is a day; solid lines indicate sleep, and broken lines indicate waking. A triangle indicates the point of the day's lowest body temperature. The subject was first exposed to 9 days of natural 24-hour cycles of light and dark, noise and quiet, and air temperature. During the middle 25 days, all time cues were removed, but the subject was free to set his own schedule. Notice that the sleep-wake cycles remained stable, but each lengthened to about 25 hours. The subject was now *free-running*. Notice also that the low point of body temperature shifts from the end of the sleep period to the beginning. During the last 11 days, a 24-hour cycle of light and meals was reintroduced, the subject again entrained to a day-long rhythm, and body temperature gradually shifted back to its normal point in the sleep cycle. (Source: Adapted from Dement, 1976, Fig. 2.)

In isolation experiments, behavior and physiology do not always continue to cycle together. Recent studies have found that body temperature and other physiological measures change quite reliably over a 24-hour cycle, even if people are entrained on a 20-hour or 28-hour "day" with artificial lighting. This means that the rhythms of temperature and sleeping-waking, which are normally synchronized to a 24-hour period, become desynchronized. In the cave experiments, when people are allowed to set their own schedules, there can be even larger differences in the periods of behavioral and physiological cycles. Normally our lowest body temperature occurs shortly before we awaken in the morning, but when we are desynchronized, this temperature nadir can drift, first moving earlier into the sleep period and then into waking time. Sleep quality and waking comfort are impaired when cycles are desynchronized. One implication of this desynchronization is that the body has more than one biological clock, because sleeping-waking and temperature can cycle at their own pace, uncoupled from one another.

Desynchronization may occur temporarily when we travel and force our bodies suddenly into a new sleep-wake cycle. This is the familiar experience known as jet lag, and the best cure is bright light, which helps resynchronize our biological clocks.

The primary zeitgeber for mammals is the light-dark cycle. The mother's hormone levels may be the first zeitgeber for some mammals, however, already entraining their activity levels in the womb. In studies of various adult animals, effective zeitgebers have also included the periodic availability of food or water, social contact, environmental temperature cycles, and noise-quiet cycles. Although many of these are much less effective than light-dark cycles, they may be important for particular species in certain circumstances.

The Suprachiasmatic Nucleus: A Brain Clock

A biological clock that produces circadian rhythms consists of several components:

Light sensor → Clock → Output pathway

One or more input pathways are sensitive to light and dark, and they entrain the clock and keep its rhythm coordinated with the circadian rhythms of the environment. The clock itself continues to run and keep its basic rhythm even when the input pathway is removed. Output pathways from the clock allow it to control certain brain and body functions according to the timing of the clock.

Mammals have a tiny pair of neuron clusters in the hypothalamus that serves as a biological clock: the **suprachiasmatic nuclei** (**SCN**), introduced in Chapter 15. Each SCN has a volume of less than 0.3 mm^3, and its neurons are among the smallest in the brain. They are located on either side of the midline, bordering the third ventricle (Figure 19.17). When the SCN is stimulated electrically, circadian rhythms can be shifted in a predictable way. Removal of both nuclei abolishes the circadian rhythmicity of physical activity, sleeping and waking, and feeding and drinking (Figure 19.18). In hamsters, transplantation of a new SCN can restore rhythms within 2–4 weeks (Box 19.4). The brain's internal rhythms never return without an SCN. Lesions in the SCN do not abolish sleeping, however, and animals continue to coordinate their sleeping and waking with light-dark cycles if present. Sleep appears to be regulated by a mechanism other than the circadian clock that depends primarily on the amount and timing of prior sleep.

Because behavior is normally synchronized with light-dark cycles, there must also be a photosensitive mechanism for resetting the brain clock. The SCN accomplishes this via the retinohypothalamic tract: Axons from gan-

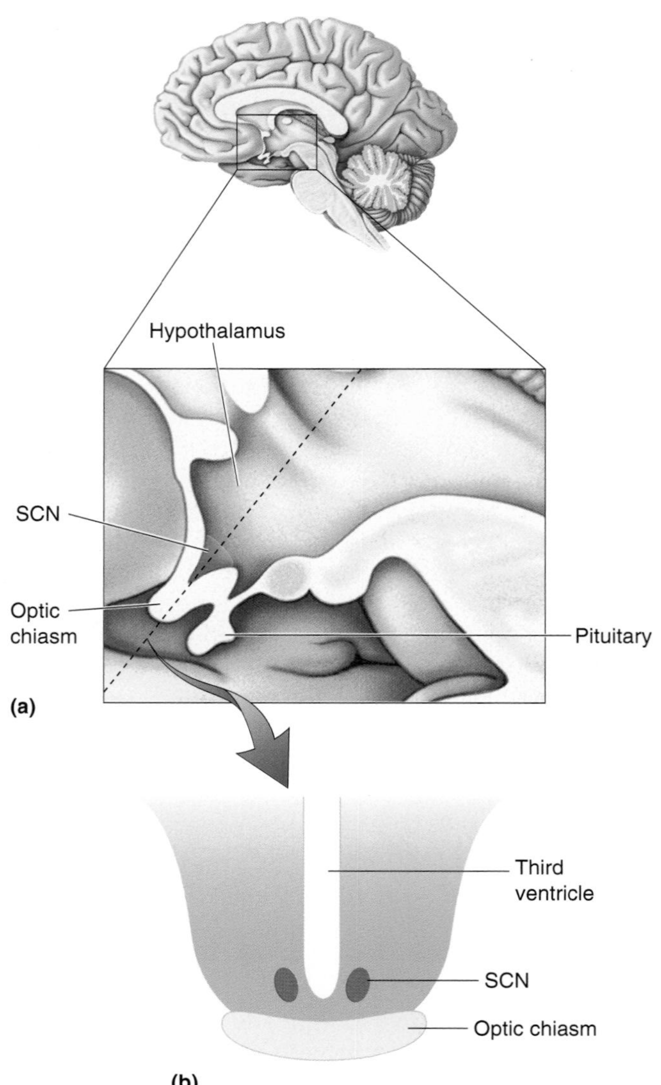

Figure 19.17
The human suprachiasmatic nuclei. There are two suprachiasmatic nuclei (SCN) within the hypothalamus, just above the optic chiasm and next to the third ventricle. **(a)** Sagittal view. **(b)** Frontal view, sectioned at the broken line in part a.

glion cells in the retina synapse directly on the dendrites of SCN neurons. This input from the retina is necessary and sufficient to entrain sleeping and waking cycles to night and day. When recordings are made from neurons of the SCN, many are indeed sensitive to light. Unlike the more familiar neurons of the visual pathways discussed in Chapter 10, SCN neurons have very large, nonselective receptive fields and respond to the luminance of light stimuli rather than their orientation or motion. Recent research suggests, surprisingly, that the retinal cells synchronizing the SCN may be neither rods nor cones. Eyeless mice cannot use light to reset their clocks, but mice with eyes lacking rods and cones can! No one knows what these mysterious photoreceptors involved in the SCN mechanism are, but a leading candidate is a protein in the retina called *cryptochrome*.

Output axons of the SCN mainly innervate nearby parts of the hypothalamus, but some also go to the midbrain and other parts of the diencephalon. Because almost all SCN neurons use GABA as their primary neurotransmitter, presumably they inhibit the neurons they innervate. It is not yet clear how the SCN sets the timing of so many important behaviors. Extensive lesions in the efferent SCN pathways disrupt circadian rhythms. In addition to the axonal output pathways, SCN neurons may rhythmically secrete the peptide neuromodulator vasopressin.

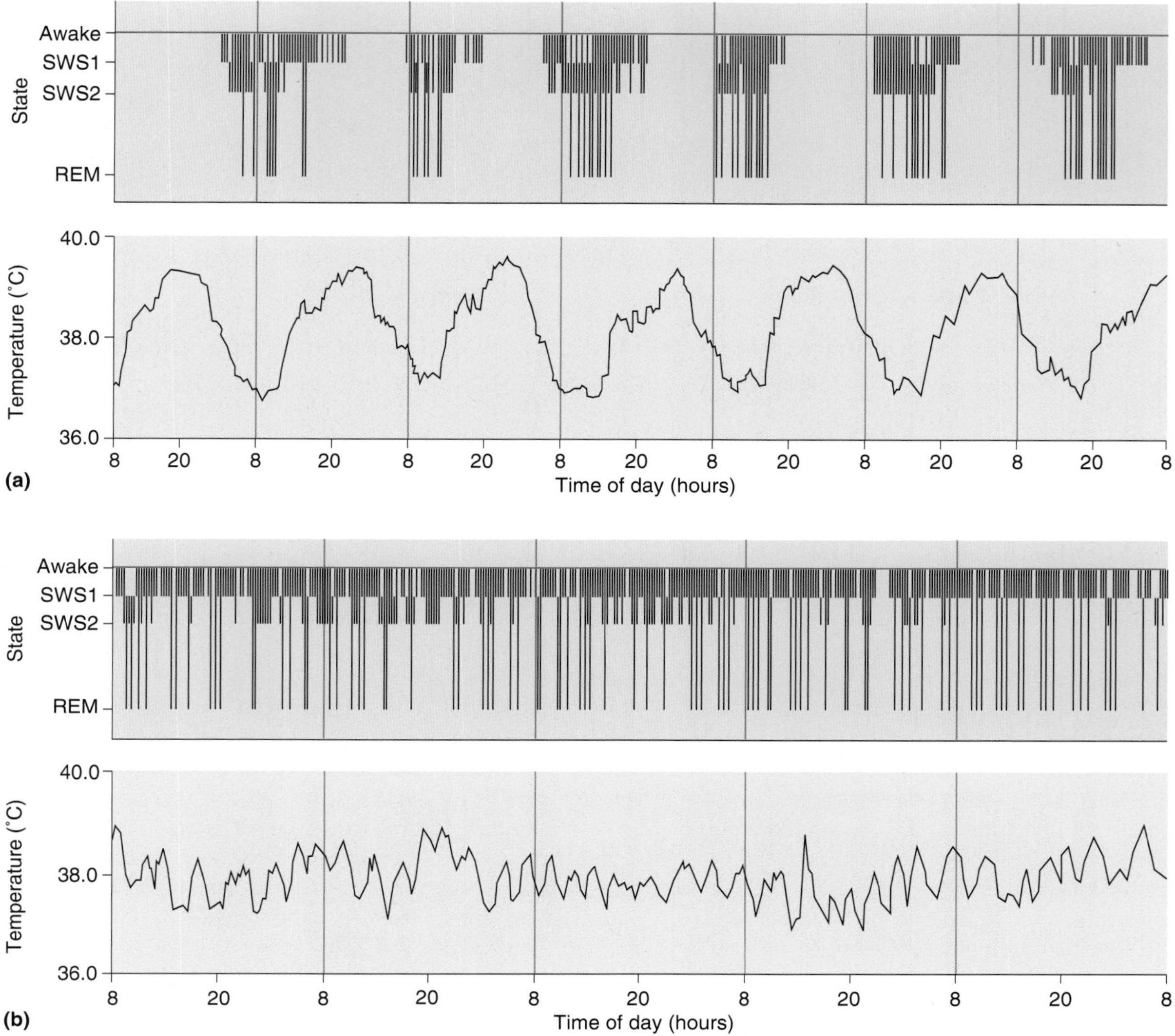

Figure 19.18

The SCN and circadian rhythms. (a) Normal squirrel monkeys kept in a constantly lit environment display circadian rhythms of about 25.5 hours. The graph shows the stages of waking and sleeping and concurrent variations in body temperature. The animals' activity states were defined as awake, two levels of non-REM sleep (SWS1 or SWS2), or REM sleep. **(b)** Circadian rhythms are abolished in monkeys that have lesions in both SCN and that are kept in a constant-light environment. Persistent high-frequency rhythms of both activity and temperature result from SCN lesions. (Source: Adapted from Edgar et al., 1993, Figs. 1, 3.)

SCN Mechanisms

How do neurons of the SCN keep time? We don't have the answer at the molecular level, but it's clear that each SCN cell is a minuscule clock. The ultimate isolation experiment has been simply to remove neurons from the SCN of a rat and grow them alone in a tissue culture dish, segregating them from the rest of the brain and from each other. Nevertheless, their rates of action potential firing, glucose utilization, vasopressin production, and protein synthesis continue to vary with rhythms of about 24 hours, just as they do in the intact brain (Figure 19.19). Other isolated mammalian neurons do not have circadian rhythmicity. SCN cells in culture cannot be entrained to light-dark cycles (input from the eyes is necessary for this), but their basic rhyth-

Box 19.4

OF SPECIAL INTEREST

Mutant Hamster Clocks

Golden hamsters are perfectionists of circadian timing. When placed in constant darkness, they continue sleeping and waking, running on their wheels, and eating and drinking over an average period of 24.1 hours, for weeks on end.

It was this dependability that made neuroscientists Martin Ralph and Michael Menaker, then working at the University of Oregon, take heed when one of the hamsters in their laboratory began punching in with 22.0-hour cycles during 3 weeks in the dark. This maverick male was bred with three females of unimpeachable circadian character (their free-running periods were 24.01, 24.03, and 24.04 hours—quite normal). When 20 pups from the three resulting litters were tested in the dark, their free-running periods were evenly split into two narrow groups. Half had periods of 24.0 hours, and half had periods of 22.3 hours. Further cross-breeding showed that the hamsters with the shorter circadian periods had one mutant copy of a gene (*tau*) that was dominant over their normal gene. After further breeding, Ralph and Menaker also found that animals with two copies of the mutant *tau* gene had free-running periods of only 20 hours!

Hamsters with mutant circadian rhythms provided a convincing way to answer a fundamental question: Is the SCN the brain's circadian clock? Ralph, Menaker, and their colleagues found that when both SCN of a hamster were ablated, rhythms were entirely lost. But rhythms could be restored to these ablated animals by simply transplanting a new SCN into their hypothalamus and waiting about 1 week.

The key finding was that hamsters receiving transplants adopted the circadian rhythm of the *transplanted* SCN, not the rhythm they were born with. In other words, if a genetically normal hamster with SCN lesions received an SCN from a donor with one copy of the mutant *tau* gene, it subsequently cycled at about 22 hours. If its transplanted SCN came from an animal with two mutant *taus*, it cycled at 20 hours. This is very compelling evidence that the SCN is the master circadian clock in the hamster brain and probably in our own brain as well.

Short circadian periods were often devastating to a mutant hamster's lifestyle when it was placed in normal, 24-hour light and dark cycles. A hamster's normal preference is to be active at night, but most *tau* animals could not completely entrain to the 24-hour rhythm. Instead, they found their activity periods continually shifting through various parts of the light-dark cycle.

A similar problem sometimes occurs in people, most often in the elderly. Because of an age-dependent shortening of the circadian rhythm, overwhelming sleepiness begins in early evening, and awakening comes at 3:00 or 4:00 in the morning. Some people are unable to entrain their sleep-wake cycles to a daily rhythm, and like the mutant hamsters, they find their activity cycles constantly shifting with respect to daylight.

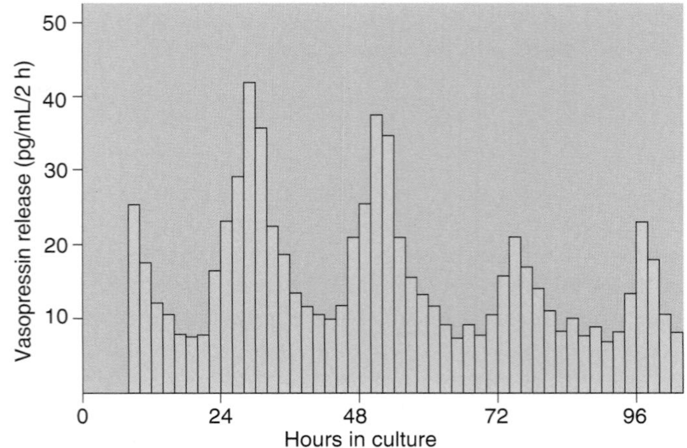

Figure 19.19
Circadian rhythms of the SCN isolated from the rest of the brain. Small portions of rat SCN were kept alive in tissue culture, and the rate of vasopressin release was monitored over several days. (Source: Adapted from Earnest and Sladek, 1986, Fig. 1.)

micity remains intact and expresses itself just as it does when an animal is deprived of zeitgebers.

SCN cells communicate their rhythmic message to the rest of the brain through efferent axons, using action potentials in the usual way, and rates of SCN cell firing vary with a circadian rhythm. Action potentials are not necessary, however, for SCN neurons to maintain their rhythm. When tetrodotoxin (TTX), a blocker of sodium channels, is applied to SCN cells, it blocks their action potentials but has no effect on the rhythmicity of their metabolism and biochemical functions. When the TTX is removed, action potentials resume firing with the same phase and frequency they had originally, before the TTX, implying that the SCN clock keeps running even without action potentials. SCN action potentials are like the hands of a clock; removing the clock's hands does not stop the clock from working, but it does make it difficult to read the time.

What is the nature of this clock that functions without action potentials? Recent research in a number of species indicates that it is a molecular cycle based on gene expression. Interestingly, the system used in humans is thought to be similar to those systems found in mice, fruit flies (*Drosophila*), and bread mold. In *Drosophila* and mice, the system involves clock genes known as *period* (*per*), *timeless* (*tim*), and *clock.* Although the details vary between species, the basic scheme is a negative feedback loop. Many of the details have been worked out in recent experiments performed by Joseph Takahashi and his colleagues at Northwestern University (Box 19.5). A clock gene is transcribed to produce mRNA that is translated into proteins. After a delay, the newly manufactured proteins send feedback and somehow interact with the transcription mechanism, decreasing gene expression. As a consequence of decreased transcription, less protein is produced, and gene expression again increases to start the cycle anew. This entire cycle takes about 24 hours; thus it is a circadian rhythm (Figure 19.20).

If each SCN neuron is a clock, there must be some mechanism to coordinate those thousands of cellular clocks so that the SCN as a whole gives a single, clear message about time to the rest of the brain. Light information from the retina certainly helps do this, but SCN neurons also communicate directly with each other. Even the coordination of rhythms *between* SCN cells, however, seems to be independent of action potentials and normal synaptic transmission, because TTX does not block them. Also, the SCN of the very young rat brain coordinates circadian rhythms perfectly well, even *before* it has developed any synapses. The nature of neuron-to-neuron communication within the SCN is not understood, but in addition to classical synapses, it may include other chemical signals, direct electrical interactions, or the participation of glia.

CONCLUDING REMARKS

Rhythms are ubiquitous in the mammalian central nervous system. They also span a broad range of frequencies, from nearly 100 Hz in the cortical EEG to once per year (0.00000003 Hz) for many seasonal behaviors, such as the autumn mating of deer, the winter hibernation of chipmunks, and the instinct that drives migrating swallows to return to Capistrano, California, every March 19. In 200 years, they have only missed the date twice. In some cases these rhythms are based on intrinsic brain mechanisms, in some they result from environmental factors, and in others, such as the SCN clock, they represent an interaction between a neural process and zeitgebers.

While the purpose of some rhythms is obvious, the functions of many neural rhythms are unknown. Indeed, some rhythms may have no function at all

Box 19.5

PATH OF DISCOVERY

Clock Genes

BY JOSEPH TAKAHASHI

I grew up in the tropics, surrounded by exotic plants, insects, and animals. My parents indulged my interest in nature by allowing me to have many unusual pets and by taking me on interesting forays into the jungles of Burma and trips to the Arabian Sea in Pakistan. So it was natural that upon entering college, I was drawn to the biology department. Two professors, Norman Meinkoth and Kenneth Rawson (Swarthmore College), sparked my first serious interest in biology. At the time, I was not even aware that one could practice biology as a profession; the default path for a biology major was medical school. I was not interested in genetics and molecular biology, but now, 30 years later, I can't imagine doing research without them.

I was introduced to circadian rhythms by Rawson, whose course had a remarkable field and laboratory section on the physiological basis of animal behavior. Later I did a senior thesis project with him on circadian rhythms in electric fish. Through Rawson I met Patricia DeCoursey and worked with her in marine biology. For graduate work, I studied with Michael Menaker at the University of Texas at Austin and the University of Oregon at Eugene. Menaker is a remarkable person who provided vast resources and an unusually stimulating environment for research. It was here that my appreciation for neuroscience, evolutionary biology, and large-scale experiments began.

The 1970s heralded the modern era of circadian biology on two fronts. Ron Konopka and Seymour Benzer (Caltech) had just discovered the locus of the *period* gene by finding the first *Drosophila* mutants affecting circadian rhythms. Robert Y. Moore (University of California, San Diego) and Irving Zucker (University of California, Berkeley) had just provided the first evidence that the suprachiasmatic nucleus (SCN) was the site of the circadian pacemaker in mammals. The SCN work directly influenced me as a graduate student. One of my thesis projects was to assess the effects of SCN lesions in birds. But it was Konopka's mutants that kept me awake at night. How could mutations in a single gene shorten, lengthen, or abolish circadian rhythms? In the late 1970s, the possibility of finding comparable "clock mutants" in mammals seemed unthinkable. I recall examining, with my friend Fred Davis, a menagerie of odd-looking hamsters collected from pet stores. Despite a multitude of coat color variants, their circadian rhythms were completely normal. Fred and I put this idea to rest for the time being.

Joseph Takahashi

Then in 1988, Martin Ralph and Menaker discovered the first single gene mutation to affect circadian rhythms in a mammal: the *tau* mutant hamster. This was exactly what Davis and I had hoped to find earlier. The *tau* hamster became an important tool (animal model) for physiological experiments, but because of the paucity of genetic resources in the hamster, finding the underlying gene was a daunting task. A search for alternative solutions eventually led to the idea, from conversations with my colleagues Fred Turek and Lawrence Pinto at Northwestern, that we should look for circadian mutants in the mouse. The Human Genome Project had just been launched, and it was clear that the mouse was the organism of choice for experimental mammalian genetics. Pinto made an important contact with William Dove and Alexandra Shedlovsky at the University of Wisconsin, Madison. Working with mice mutagenized in Madison, we began our first genetic screen for circadian rhythm mutants. Martha Vitaterna, a postdoctoral fellow, and I hit pay dirt in the first group of mice tested (mouse no. 25). The mutation lengthened circadian periodicity from 24 to 28 hours, with a complete loss of rhythmicity in constant conditions. We named the mutation *Clock*, an acronym for "*Circadian locomotor output cycles kaput.*"

In 1997, through the efforts of my laboratory working as a team, we cloned the *Clock* gene. *Clock* was significant because it provided an important entry point for molecular genetic analysis of mammalian circadian rhythms. Only 3 years later, at least nine genes are thought to be molecular components of the mammalian clock, and a striking level of conservation of the circadian mechanism has been found in animals.

Ultimately, genetics was the key to unlocking the secrets of the circadian clock. Many years later, I have come to appreciate those genetics and molecular biology classes I had to take at Swarthmore.

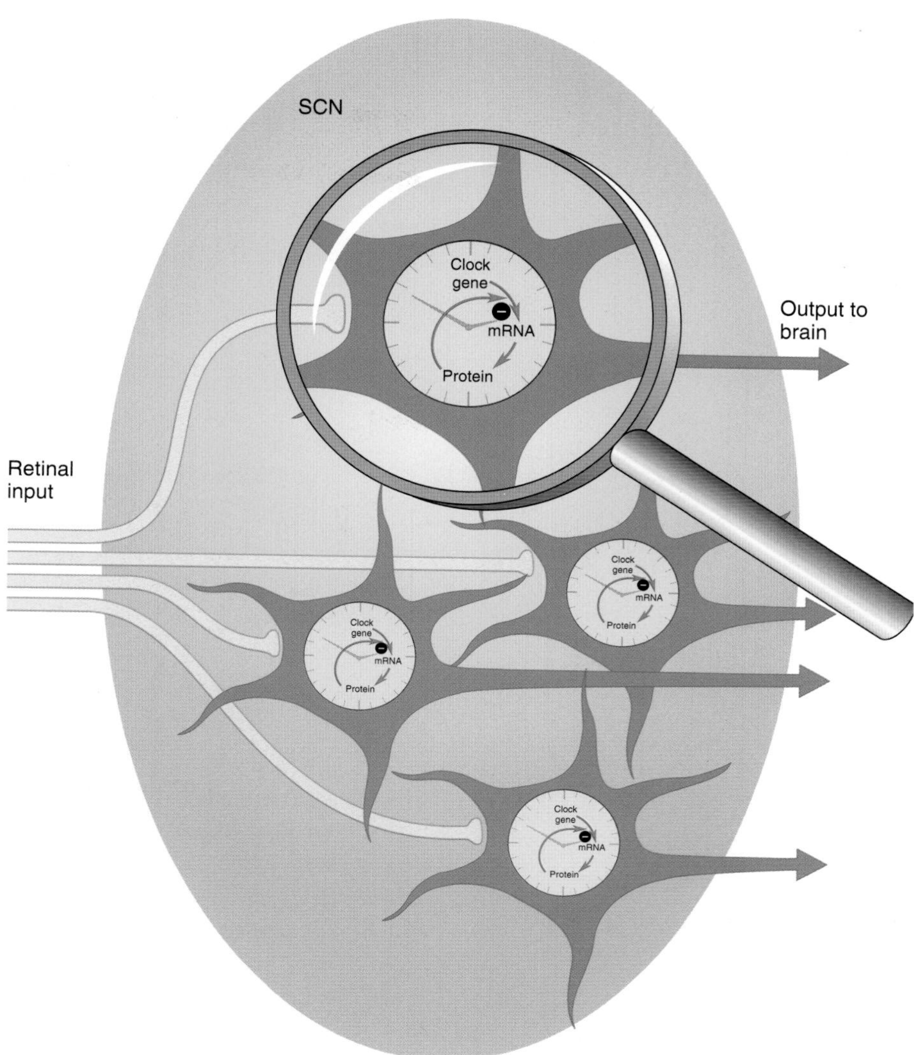

Figure 19.20
Clock genes. In the SCN, clock genes produce proteins that inhibit further transcription. Gene transcription and the firing rate of individual SCN neurons cycle up and down over 24 hours. The cycles of many cells are synchronized by light exposure (input from the retina).

but arise as a secondary consequence of neural interconnections that are essential for nonrhythmic purposes.

Among the most conspicuous yet inexplicable of brain rhythms is sleep. Sleep provides a fascinating set of problems for neuroscience. Unlike most studies of single ion channels, single neurons, or the systems mediating perception and movement, sleep research begins with profound ignorance about a most basic question: Why? We still admit ignorance about why we spend one-third of our lives sleeping, most of that time languid and vegetative and the rest of it paralyzed and hallucinating. Sleep and dreams may have no vital function, but they can be studied and enjoyed nevertheless. Ignoring the functional question will not be a satisfying approach for long, however. For most neuroscientists, asking "why" remains the deepest and most challenging problem of all.

KEY TERMS

The Electroencephalogram
electroencephalogram (EEG) (p. 607)
generalized seizure (p. 613)
partial seizure (p. 613)
epilepsy (p. 613)

Sleep
rapid eye movement sleep (REM sleep) (p. 614)
non-REM sleep (p. 614)
atonia (p. 615)
ultradian rhythm (p. 616)

Circadian Rhythms
circadian rhythm (p. 626)
zeitgeber (p. 628)
suprachiasmatic nucleus (SCN) (p. 629)

REVIEW QUESTIONS

1. Why do EEGs with relatively fast frequencies tend to have smaller amplitudes than EEGs with slower frequencies?
2. The human cerebral cortex is very large and must be folded extensively to fit within the skull. What do the foldings of the cortical surface do to the brain signals that are recorded by an EEG electrode at the scalp?
3. Sleep seems to be a behavior of every species of mammal, bird, and reptile. Does this mean that sleep performs a function essential for the life of these higher vertebrates? If you do not think so, what might be an explanation for the abundance of sleep?
4. An EEG during REM sleep is very similar to an EEG during the waking state. How do the brain and body in REM sleep *differ* from the brain and body during the waking state?
5. What is a likely explanation for the brain's relative insensitivity to sensory input during REM sleep as compared to the waking state?
6. The SCN receives direct input from the retina via the retinohypothalamic tract, and this is how light-dark cycles can entrain circadian rhythms. If the retinal axons were disrupted, what would be the likely effect on a person's circadian rhythms of sleeping and waking?
7. What differences would there be in the behavioral consequences of a free-running circadian clock versus no clock at all?

CHAPTER

20

Language and Attention

INTRODUCTION

One of the exciting aspects of current neuroscience research is the study of complex functions of the brain that would have seemed out of reach not long ago. In this chapter, we look at two complex facets of human behavior: attention and the use of language. Language and attention have been studied for many years by linguists and psychologists, and now the underlying brain processes are being examined by neuroscientists.

The use of language is a uniquely human behavior that has had a great effect on human society. Much of what we know about language processing is derived from studies of language deficits resulting from brain damage. Numerous aspects of language, including speech, comprehension, and naming, can be selectively disrupted, suggesting that language is processed in multiple, anatomically distinct stages. We will use studies of brain lesions, brain stimulation, and brain imaging to investigate components of language that appear to occur at different locations in the brain.

Humans are constantly shifting attention between stimuli in the environment. Think of the way you can voluntarily and selectively scrutinize faces at

Box 20.1

OF SPECIAL INTEREST

Is Language Unique to Humans?

To determine whether animals other than humans use language, one must be specific about what language is. If it is defined simply as communication, then most animals certainly do use language. But this definition misses the point. Human language is a remarkably complex, flexible, and powerful *system* for communication that involves the creative use of words according to the rules of a systematic grammar. Do other animals have anything similar? Actually, there are two questions: Do animals naturally use language? Can animals be taught human language?

Concerning the natural communication of animals, there is no evidence that they use a system even close to the complexity of that of humans. Chimpanzees and monkeys use a variety of sounds, gestures, and facial expressions to communicate alarm, warn adversaries, claim territory, and so on. Compared to humans, however, nonhuman primates have a very limited system used in stereotypical situations. Human language is a much more creative system. Limited only by the rules of grammar, it is effectively infinite. New word combinations and sentences are constantly being made, and the combinations have clear meaning according to the meaning of the individual words plus the rules used in arranging them. It is possible that further observations of nonhuman primates or dolphins might reveal that they combine their vocalizations in endlessly creative ways, but at present there is no evidence to support this claim.

The second question is whether animals can be taught human language. In the 1940s, several psychologists tried raising baby chimpanzees just like human children, including teaching them to speak. Despite extensive training, the chimps never learned to say more than a few words. This wasn't really a fair test though, because the position of the larynx in chimpanzees makes the animal unable to make the sounds of human speech. In more recent studies, animals have been trained to use either American Sign Language gestures or objects, such as plastic pieces in various shapes, to indicate words. Well-known examples are the chimpanzee named Washoe, trained by Allen and Beatrice Gardner, and the gorilla named Koko, trained by Francine Patterson. Without question the animals learn the meanings of gestures, but this only proves that the animals have good memories.

The big question is whether they combine the symbols in original ways according to some grammatical system in order to express new things. The answer is mired in controversy. Proponents of animal language report instances in which animals use combinations of symbols to describe a novel situation. For instance, after learning the sign for water and the sign for bird, one chimpanzee put the signs together as "water bird" when it was near a swimming swan. This may or may not indicate use of language. Opponents of animal language have argued that animals never really learn sign language. The chimpanzee may simply have made separate gestures for wa-a crowded airport to pick out grandma. Attention allows us to exert extra mental energy on the performance of selected tasks, and it has significant beneficial effects on performance. Physiological experiments on attention give us a dramatic view of brain function in which the receptive field properties of neurons change to suit the needs of ongoing behaviors. Recent studies using functional magnetic resonance imaging (fMRI) and positron emission tomography (PET) enable us to see the changes in human brain activity that result from increased attention.

LANGUAGE AND THE BRAIN

Our use of language—the fact that we have a brain sophisticated enough for language—is one of the key features that distinguishes humans from other animals (Box 20.1). More than just sounds, language is a system by which sounds, symbols, and gestures are used for communication.

It is estimated that there are about 10,000 languages and dialects throughout the world. Languages differ in many ways, such as the order in which nouns and verbs are arranged. But despite differences in syntax, from the

(Box 20.1, *continued*)

ter and bird. Part of the problem is that the animals frequently combine symbols in illogical or unintelligible ways and only occasionally come up with something like "water bird." Do the seemingly logical combinations of symbols reflect moments of language lucidity, or are they just coincidences that the researchers choose to emphasize? It is not easy to know, and more controlled experiments are ongoing. For now, it is probably prudent to take claims of animal language with a grain of salt.

In response to the question of whether language is unique to humans, the current answer appears to be that the creative system most of us think of as language *is* unique to the human species. To some people this conclusion may seem unlikely or even distasteful—how could *we* be the only species with language? But don't confuse language with intelligence. Language is not necessary for thought. Monkeys, as well as humans raised without any language, can do many things requiring abstract reasoning. Many creative people say they do some of their best thinking without words. Albert Einstein claimed that many of his ideas about relativity came from visually thinking of himself riding on a beam of light looking around at clocks and other objects. In any event, Fido probably does think, but he doesn't need language (Figure A).

Another thing people find hard to believe is that language could have evolved in humans without more primitive languages in other species. In his book *The Language Instinct*, Steven Pinker compares the evolution argument for language to a similar argument about the elephant's trunk. The trunk is obviously a prominent feature and is essential to the elephant's lifestyle. Nonetheless, there simply aren't living animals with lesser forms of the trunk. It seems to be unique to the elephant, just as language seems to be unique to humans.

Figure A

English spoken in the House of Lords to the dialects of aboriginal tribes around the globe, all languages convey the subtleties of human experience and emotion. Consider the fact that no mute tribe of people has ever been found, not even in the remotest corner of the world. Many believe this is a consequence of the fact that the human brain has evolved special language-processing systems. These systems are present in newborn babies; if a child grows up in a normal language environment, he or she inevitably learns to speak and understand language. Consistent with this idea, it has been found that children acquire language in a similar manner in all cultures.

The Discovery of Specialized Language Areas in the Brain

As in many other areas of neuroscience, only in the twentieth century has our understanding of a clear relationship between language and the brain emerged. Much of what we know about the importance of certain brain areas is derived from studies of aphasia. **Aphasia** is the partial or complete loss of language abilities following brain damage, often without the loss of cognitive faculties or the ability to move the muscles used in speech.

During the Greek and Roman empires, it was commonly thought that the tongue controlled speech and that speech disorders originated there rather than in the brain. If a head injury resulted in loss of speech, the treatment involved special gargles or massage of the tongue. By the sixteenth century, it had been noted that a person could suffer speech impairment without paralysis of the tongue. Despite this step forward, treatment still included such procedures as cutting the tongue, drawing blood, and applying leeches.

Around 1770, Johann Gesner published a relatively modern theory of aphasia, describing it as the inability to associate images or abstract ideas with their expressive verbal symbols. He attributed this loss to brain damage resulting from disease. Gesner's definition makes the important observation that in aphasia, cognitive ability may remain intact but some function specific to verbal expression is lost. Despite the incorrect association between skull shape and brain function (see Chapter 1) that Franz Joseph Gall and later phrenologists made, they, too, made an important observation about aphasia. They reasoned that cases of brain lesions, in which speech is lost but other mental faculties are retained, suggest that there is a specific region of the brain used for speech.

In 1825, based on many case studies, French physician Jean-Baptiste Bouillaud proposed that speech is specifically controlled by the frontal lobes. But it took another four decades before this idea was generally accepted. In 1861, Simon Alexandre Ernest Aubertin, Bouillaud's son-in-law, described the case of a man who had shot away his frontal skull bone in a failed suicide attempt. In treating this man, Aubertin discovered that if a spatula was pressed against the exposed frontal lobe while the man was speaking, his speech was immediately halted and did not resume until the pressure was released. He inferred that the pressure on the brain interfered with the normal function of a cortical area in the frontal lobe.

Broca's Area and Wernicke's Area. Also in 1861, French neurologist Paul Broca had a patient who was almost entirely unable to speak. Broca invited Aubertin to examine the brain, and they concluded that this man had a lesion in the frontal lobes. Perhaps because of a change in the scientific climate, Broca's case study appears to have swung popular opinion around to the idea that there is a language center in the brain. In 1863, Broca published a paper describing eight cases in which language was disturbed by damage to the frontal lobe in the left hemisphere. Additional similar cases, along with reports that speech was not disturbed by right hemispheric lesions, led Broca, in 1864, to propose that language expression is controlled by only one hemisphere, almost always the left. This view is supported by the results of a more modern procedure for assessing the role of the two hemispheres in language, called the **Wada procedure**, in which a single hemisphere of the brain is anesthetized (Box 20.2). In most cases, anesthesia of the left hemisphere, but not the right, disrupts speech.

Box 20.2

OF SPECIAL INTEREST

The Wada Procedure

A simple procedure used for studying the function of a single cerebral hemisphere is the *Wada procedure*, developed by John Wada at the Montreal Neurological Institute. A fast-acting barbiturate, such as Amytal Sodium, is injected into the carotid artery on one side of the neck (Figure A). The drug is preferentially carried in the bloodstream to the hemisphere ipsilateral to the injection, where it acts as an anesthetic for about 10 minutes. The effects are sudden and dramatic. Within a matter of seconds the limbs on the side of the body contralateral to the injection become paralyzed, and somatic sensation is lost.

By asking the patient to answer questions, one can assess his or her ability to speak. If the injected hemisphere is dominant for speech, the patient is completely unable to talk until the anesthesia wears off. If the injected hemisphere is not dominant, the person continues to speak throughout the procedure.

Table A shows that in 96% of right-handed people and 70% of left-handed people, the left hemisphere is dominant for speech. Because 90% of people are right-handed, this means that the left hemisphere is dominant for language in roughly 93% of people. While small but significant numbers of people with either handedness have a dominant right hemisphere, only in left-handers are bilateral representations of speech seen. In the Wada procedure, this is indicated when an injection into either hemisphere has some disruptive effect on speech, though specifics of the disruption may be different for the two hemispheres.

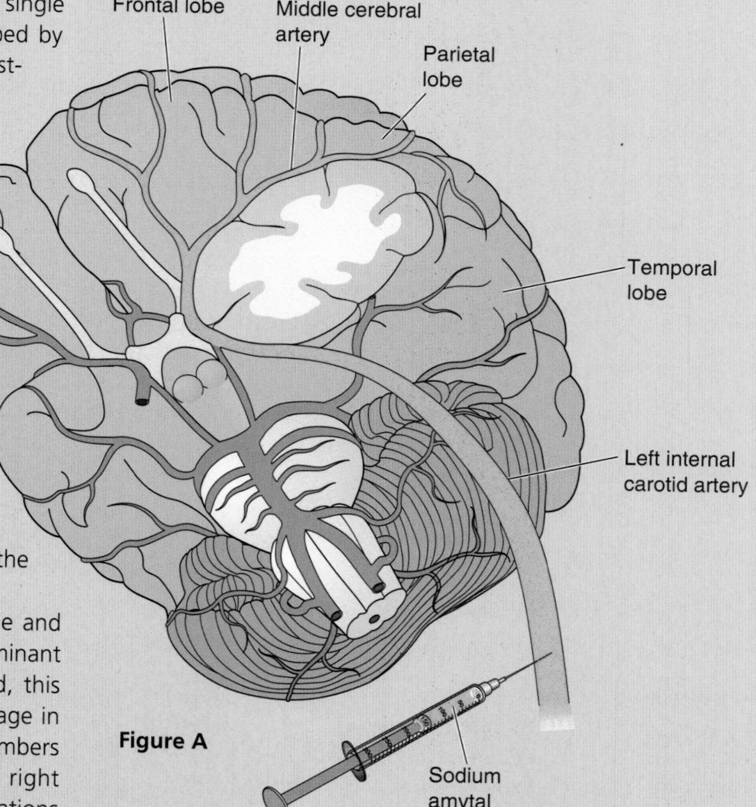

Figure A

Table A Hemispheric Control of Speech in Relation to Handedness

HANDEDNESS	NUMBER OF CASES	SPEECH REPRESENTATION (%) LEFT	BILATERAL	RIGHT
Right	140	96	0	4
Left	122	70	15	15

Source: Rasmussen and Milner, 1977, Table 1.

If one hemisphere is thought to be more heavily involved in a particular task, it is said to be *dominant*. The region of the dominant left frontal lobe that Broca identified as being critical for articulate speech has come to be called **Broca's area** (Figure 20.1). Broca's work has considerable significance because it was the first clear demonstration that brain functions can be anatomically localized.

In 1874, German neurologist Karl Wernicke reported that lesions in the left hemisphere in a region distinct from Broca's area also disrupted normal speech. Located on the superior surface of the temporal lobe between auditory cortex and the angular gyrus, this region is now commonly called **Wernicke's area** (Figure 20.1). The nature of the aphasia Wernicke observed is different from that associated with damage to Broca's area. Having established that there are two language areas in the left hemisphere, Wernicke and others proceeded to construct maps of language processing in the brain. Interconnections between auditory cortex, Wernicke's area, Broca's area, and the muscles required for speech were hypothesized, and different types of language disabilities were attributed to damage in different parts of this system.

Although the terms *Broca's area* and *Wernicke's area* are still commonly used, the boundaries of these areas are not clearly defined, and they appear to be quite variable from one person to the next. Furthermore, each area may be involved in more than one language function. This recent finding will make sense, however, only after we look at the aphasias produced by damage to Broca's and Wernicke's areas.

Types of Aphasia and Their Causes

As anyone who has ever tried to learn a foreign language knows, language is complicated. Trying to understand how such a complex process works in the brain is a challenge. The oldest technique for studying the relationship between language and the brain involves correlating functional deficits with lesions in particular brain areas. The occurrence of distinct types of aphasia, as shown in Table 20.1, suggests that language is processed in several stages at different locations in the brain. As demonstrated by the research of Sheila Blumstein at Brown University, brain lesions produce aphasias ranging from subtle alterations of speech to its complete elimination (Box 20.3).

Broca's Aphasia. The syndrome called **Broca's aphasia** is also known as motor or nonfluent aphasia, because the person has difficulty speaking even

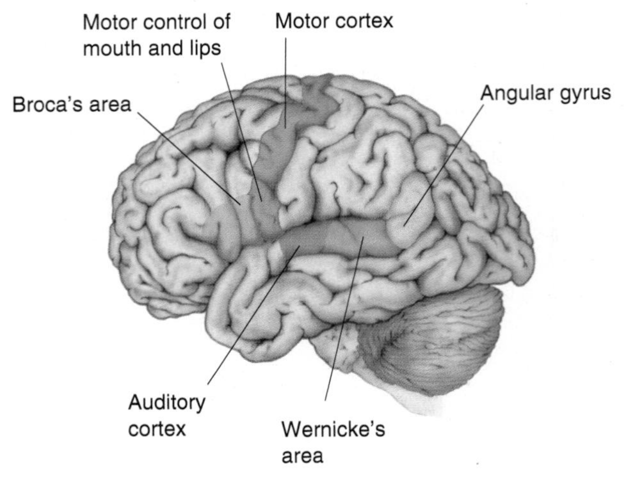

Figure 20.1
Key components of the language system in the left hemisphere. In the frontal lobe, Broca's area lies next to the area that controls the mouth and lips in motor cortex. Wernicke's area, on the superior surface of the temporal lobe, lies between auditory cortex and the angular gyrus.

Table 20.1 Characteristics of Types of Aphasia

TYPE OF APHASIA	SITE OF BRAIN DAMAGE	COMPREHENSION	SPEECH	IMPAIRED REPETITION	PARAPHASIC ERRORS
Broca's	Motor association cortex of frontal lobe	Good	Nonfluent, agrammatical	Yes	Yes
Wernicke's	Posterior temporal lobe	Poor	Fluent, grammatical, meaningless	Yes	Yes
Conduction	Arcuate fasciculus	Good	Fluent, grammatical	Yes	Yes
Global	Portions of temporal and frontal lobes	Poor	Very little	Yes	—
Transcortical motor area	Frontal lobe anterior to Broca's	Good	Nonfluent, agrammatical	No	Yes
Transcortical sensory	Cortex near the junction of temporal, parietal, and occipital lobes	Poor	Fluent, grammatical, meaningless	No	Yes
Anomic	Inferior temporal lobe	Good	Fluent, grammatical	No	No

though he or she can understand language heard or read. The case of David Ford is typical. Ford was a 39-year-old radio operator in the Coast Guard when he had a stroke. He remained an intelligent man, but he had little control over his right arm and leg (demonstrating that his lesion was in the left hemisphere). His speech was also abnormal, as the following discussion with psychologist Howard Gardner illustrates:

> *I asked Mr. Ford about his work before he entered the hospital.*
>
> *"I'm a sig . . . no . . . man . . . uh, well, . . . again." These words were emitted slowly, and with great effort. The sounds were not clearly articulated; each syllable was uttered harshly, explosively, in a throaty voice. With practice, it was possible to understand him, but at first I encountered considerable difficulty in this.*
>
> *"Let me help you," I interjected. "You were a signal . . . "*
>
> *"A signal man . . . right," Ford completed my phrase triumphantly.*
>
> *"Were you in the Coast Guard?"*
>
> *"No, er, yes, yes . . . ship . . . Massachu . . . chusetts . . . Coastguard . . . years." He raised his hands twice, indicating the number "nineteen."*
>
> *"Could you tell me Mr. Ford, what you've been doing in the hospital?"*
>
> *"Yes, sure. Me go, er, uh, P.T. nine o'cot, speech . . . two times . . . read . . . wr . . . ripe, er, rike, er, write . . . practice . . . get-ting better."*
>
> *"And have you been going home on weekends?"*
>
> *"Why, yes . . . Thursday, er, er, er, no, er, Friday . . . Bar-ba-ra . . . wife . . . and, oh, car . . . drive . . . purnpike . . . you know . . . rest and . . . tee-vee."*
>
> *"Are you able to understand everything on television?"*
>
> *"Oh, yes, yes . . . well . . . almost." Ford grinned a bit.*
>
> (Gardner, 1974, pp. 60–61.)

People with Broca's aphasia have difficulty saying anything, often pausing to search for the right word. The inability to find words is called **anomia** (literally meaning "no name"). Interestingly, there are certain "overlearned" things Broca aphasics can say without much hesitation, such as the days of the week. The hallmark of Broca's aphasia is a telegraphic style of speech in which mainly *content words* (nouns, verbs, and adjectives carrying content specific to the sentence) are used. For instance, when Mr. Ford was asked about being in the Coast Guard, his answer contained the words ship, Massachusetts, Coast Guard, and years but little else. Many *function words* (articles, pronouns, and conjunctions connecting the parts of the sentence

Box 20.3

PATH OF DISCOVERY

Foreign Accents and Native Tongues

BY SHEILA E. BLUMSTEIN

I have always been interested in the nature of human language. As a graduate student at Harvard, I studied formal linguistics, but after my first year of graduate school, I realized theoretical linguistics was not for me. Professor Roman Jacobson, one of the fathers of linguistics, taught a course in which he used cases of aphasia to illustrate points about language processing. Jacobson turned out to be a wonderful mentor who supported my research and encouraged me to learn more about aphasia. As a graduate student, I also had the good fortune to meet Harold Goodglass, the head of the Aphasia Research Center in Boston. He invited me to join the research team. I spent all of my time at the Veterans' Administration Hospital Medical Center, where the Aphasia Center was housed, and attended rounds and case conferences. Thus began my interest in and study of aphasia, which I have pursued for more than 30 years.

Our research today involves developing and testing theories about the nature of the underlying language deficits of Broca's and Wernicke's aphasia. Over the years, however, we have studied a diverse range of language abnormalities that result from brain damage. One of the most peculiar and interesting is the foreign accent syndrome. Imagine going to sleep one night and waking up the next day speaking with what sounds like a thick foreign accent. This is roughly what happens to these people; understandably, it is upsetting for them to discover that they no longer sound like themselves.

This is an extremely rare syndrome; fewer than 20 cases have been reported in the past 80 years, and very little is known about the neuropathology underlying the disorder. It was first reported by Arnold Pick in 1919. I was most influenced by a case described by Monrad-Krohn concerning a patient who suffered a brain lesion during World War II. This Norwegian speaker sounded as if she were German—not a safe thing in those times. I was skeptical when reading this case; perhaps the investigators didn't know how to listen as a linguist would. That attitude changed quickly when I encountered my first case of the foreign accent syndrome.

I met a woman with foreign accent syndrome at the Aphasia Center and was stunned. Born and bred in Boston, this patient had never traveled overseas or studied any foreign language in school. Yet after a stroke, she sounded quite foreign. The way the pitch of her voice raised at the end of sentences reminded me of French. When she was presented at rounds, the research team—consisting of neurologists, psychologists, speech pathologists, and linguists—all agreed that she had a foreign accent, but the opinions on the accent ranged from Swedish to French to German to Slavic. Rather than relying on my ear, I conducted an acoustical analysis of her speech. Different languages have particular sound patterns that can be quantified. I learned that she did not really have a foreign accent; she spoke in an unusual way because of a production speech deficit. An analogy is the way comedians imitate various accents: Careful analysis would probably reveal that they aren't correct foreign accents, but they sound that way to the listener. In other words, a foreign accent is in the ear of the beholder.

Sheila E Blumstein

I studied a second patient with the foreign accent syndrome who was quite different. Fortunately, we had a tape recording of this man's speaking voice before his stroke. After the stroke, he retained his normal English intonation, but his accent sounded rather Celtic to my ear.

Patients with the foreign accent syndrome often start with a mild Broca's aphasia that disappears. They also tend to have small lesions. Patients whose lesions are near those of patients with foreign accent syndrome, however, sometimes have more classic speech production deficits, such as slurred speech (dysarthria). These observations raise the question of whether the syndrome is one point on a continuum of speech deficits. One argument against this hypothesis is that different speech disorders have dramatically different symptoms. For example, patients with articulation problems make errors with consonants, while those with foreign accent syndrome have more problems with vowels.

One reason for studying aphasia is what the language consequences of discrete brain lesions teach us about normal language processing. Of course, nobody believes there is an accent or dialect box in the brain. Indeed, the diversity of brain lesions and language deficits warns us not to make quick superficial associations between structure and function. Yet the foreign accent syndrome and other aphasias provide clues about how speech production goes from an abstract representation to articulation. They teach us that there are probably dimensions that current theories of speech production do not capture.

grammatically) are left out (there are no ifs, ands, or buts). Also, verbs are frequently not conjugated. In the jargon of aphasia deficits, the inability to construct grammatically correct sentences is called *agrammatism*. There are some peculiar nuances to the agrammatical tendencies in Broca's aphasia. In the case of Ford, he could read and use the words "bee" and "oar" but had difficulty with the more common words "be" and "or." The problem isn't related to the sound but to whether or not the word is a noun. In a similar vein, Broca aphasics have difficulty repeating things spoken to them, though they tend to be better at familiar nouns such as "book" and "nose." Sometimes, they substitute incorrect sounds or words. Ford says "purnpike" for "turnpike." These are called *paraphasic errors*.

In contrast to the speech difficulties in Broca's aphasia, comprehension is generally quite good. In the earlier dialogue, Ford seems to understand the questions asked of him, and for the most part, he says he understands what he sees on television. In Gardner's study, Ford was able to answer simple questions, such as "Does a stone float on water?" More difficult questions demonstrated, however, that he did not have completely normal comprehension abilities. If he was told "The lion was killed by the tiger; which animal is dead?" or "Put the cup on top of the fork and place the knife inside the cup," he had difficulty understanding. This is probably related to the fact that he generally had trouble with the function words "by" in the first example and "on top of" in the second example.

Because the most obvious difficulty is in producing speech, Broca's aphasia is thought of as a language disturbance toward the motor end of the language system. Language is understood but not easily produced. While it is true that Broca aphasics are worse at speech than are other types of aphasics, several things suggest that there is more to the syndrome. As pointed out earlier, comprehension is generally good, but comprehension deficits can be demonstrated by tricky questions. Also, patients sometimes have considerable anomia, suggesting that they have problems "finding" words as well as making the appropriate sounds.

Wernicke suggested that the area damaged in Broca's aphasia contains memories for the fine series of motor commands required for articulating word sounds. Because Broca's area is near the part of motor cortex that controls the mouth and lips, there is an appealing logic to this idea. Wernicke's theory is still held by some, but there are other ways of looking at the problem. For instance, the difference in the aphasic's ability to use content words and function words suggests that Broca's area and nearby cortex may be specifically involved in making grammatical sentences out of words. This might explain why Gardner's Mr. Ford could produce sounds such as "bee" and "oar" when they represent content words, but not when the sounds represent the function words "be" and "or."

Wernicke's Aphasia. When Wernicke noted that superior temporal lesions could lead to aphasia, the syndrome he observed was quite distinct from Broca's aphasia. Indeed, Wernicke suggested that aphasia is of two general types. In Broca's aphasia, speech is disturbed but comprehension is relatively intact. In **Wernicke's aphasia**, speech is fluent but comprehension is poor. (Although these descriptions are oversimplified, they are useful for remembering the syndromes.)

Let's consider the case of Philip Gorgan, another patient studied by Gardner.

"What brings you to the hospital?" I asked the 72-year-old retired butcher 4 weeks after his admission to the hospital.

"Boy, I'm sweating, I'm awful nervous, you know, once in a while I get caught up, I can't mention the tarripoi, a month ago, quite a little, I've done a lot well, I impose a lot, while, on the other hand, you know what I mean, I have to run around, look it over, trebbin and all that sort of stuff."

I attempted several times to break in, but was unable to do so against this relentlessly steady and rapid outflow. Finally, I put up my hand, rested it on Gorgan's shoulder, and was able to gain a moment's reprieve.

"Thank you, Mr. Gorgan. I want to ask you a few—"

"Oh sure, go ahead, any old think you want. If I could I would. Oh, I'm taking the word the wrong way to say, all of the barbers here whenever they stop you it's going around and around, if you know what I mean, that is tying and tying for repucer, repuceration, well, we were trying the best that we could while another time it was with the beds over there the same thing . . . ."

(Gardner, 1974, pp. 67–68.)

Clearly, Mr. Gorgan's speech is altogether different from that of Mr. Ford. Gorgan's speech is fluent, and he has no trouble using function words as well as content words. If you didn't understand English, the speech would probably sound normal because of its fluency. The content, however, does not make much sense. It is a strange mixture of clarity and gibberish. Along with their far greater output of speech, compared with that of Broca aphasics, Wernicke aphasics also make far more paraphasic errors. Gorgan would sometimes use the correct sounds but in an incorrect sequence, such as "plick" instead of "clip." Occasionally he would stumble around the correct sound or word, as when, in another conversation, he called a piece of paper "piece of handkerchief, pauper, hand pepper, piece of hand paper." Interestingly, he would sometimes use an incorrect word but one categorically similar to the correct word, such as "knee" instead of "elbow."

Because of the stream of unintelligible speech, it is difficult to assess with speech alone whether the Wernicke aphasic comprehends what he hears or reads. Indeed, one of the intriguing things about Wernicke aphasics is that they frequently appear undisturbed by the sound of their own speech and the speech of others, even though they probably don't understand either. Comprehension is usually assessed by asking the patient to respond in a nonverbal manner. For instance, the patient could be asked to put object A on top of object B. Questions and commands of this sort quickly lead to the conclusion that Wernicke aphasics do not understand most instructions. They are completely unable to comprehend questions of the sort understood by Broca aphasics. When Gorgan was presented with commands written on cards ("Wave goodbye," "Pretend to brush your teeth"), he was often able to read the words but never acted as if he understood what the words meant.

Gorgan's strange speech was mirrored in his writing and his ability to play music. When Gardner gave him a pencil, he spontaneously took it and wrote "Philip Gorgan. This is a very good beautifyl day is a good day, when the wether has been for a very long time in this part of the campaning. Then we want on a ride and over to for it culd be first time . . . . " (p. 71). Likewise, when he sang or played the piano, pieces of the appropriate song were intermixed with musical gibberish, and he had a difficult time ending, just as in his speech.

Insight about the possible function of Wernicke's area is provided by its location on the superior temporal gyrus near primary auditory cortex. Wernicke's area may play a critical role in relating incoming sounds to their meaning. In other words, it is an area specialized for storing memories of the sounds that make up words. It has been suggested that Wernicke's area is a high-order area for sound recognition in the same sense that inferior temporal cortex is thought to be a high-order area for visual recognition. (Recall the psychic blindness observed in monkeys with temporal lesions in Chapter 18.) A sound recognition deficit would explain why Wernicke aphasics don't comprehend speech well. There must be more to Wernicke's area, however, to account for the odd speech patterns. Speech in Wernicke's aphasia suggests that Broca's area and the system responsible for speech production are running without control over content. The speech zooms along, swerving in every direction like a car with a sleepy driver at the wheel.

Aphasia and the Wernicke-Geschwind Model. Shortly after making his observations on what came to be called Wernicke's aphasia, Wernicke proposed a model for language processing in the brain. Later extended by Norman Geschwind at Boston University, it is known as the **Wernicke-Geschwind model.** The key elements in the system are Broca's area, Wernicke's area, the *arcuate fasciculus*, a bundle of axons connecting the two cortical areas, and the *angular gyrus*. The model also includes sensory and motor areas involved in receiving and producing language. To understand what the model entails, we'll consider the performance of two tasks.

The first task is the repetition of spoken words (Figure 20.2). When sounds of incoming speech reach the ear, the auditory system processes the sounds, and neural signals eventually reach auditory cortex. According to the model, the sounds are not understood as meaningful words until they are processed in Wernicke's area. To repeat the words, the brain passes word-based signals to Broca's area from Wernicke's area via the arcuate fasciculus. In Broca's area, the brain converts the words to a code for the muscular movements required for speech. Output from Broca's area is sent to the nearby motor cortical areas responsible for moving the lips, tongue, larynx, and so on.

The second task we'll consider is reading written text aloud (Figure 20.3). In this case, the incoming information is processed by the visual system through striate cortex and higher-order visual cortical areas. The visual signals are then passed to the angular gyrus at the junction of the occipital, parietal, and temporal lobes. In the cortex of the angular gyrus, it is assumed that a transformation occurs so that the output evokes the same pattern of activity in Wernicke's area as if the words were spoken rather than written. From this point, the processing follows the same progression as in the first example: Wernicke's area to Broca's area to motor cortex.

This model offers simple explanations for key elements of Broca's and Wernicke's aphasia. A lesion in Broca's area seriously interferes with speech production because the proper signals can no longer be sent to motor cortex. On the other hand, comprehension is relatively intact because Wernicke's

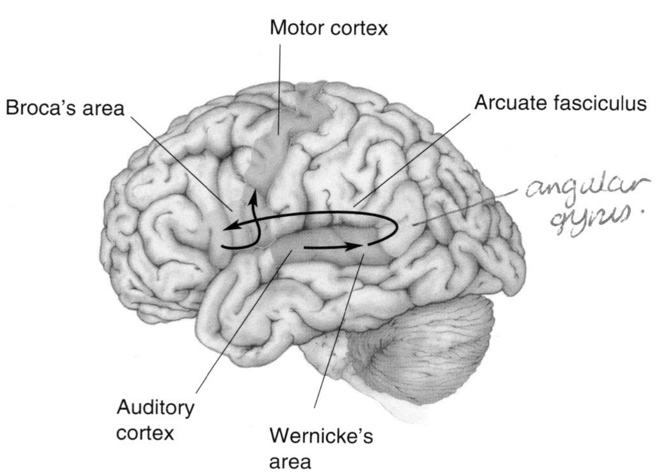

Figure 20.2
Repeating a spoken word, according to the Wernicke-Geschwind model.

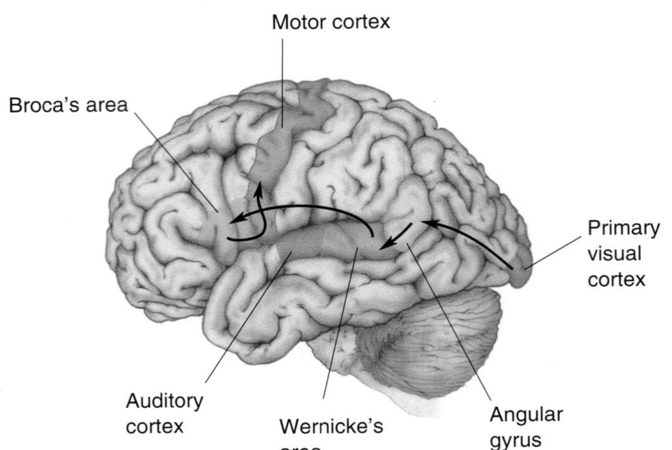

Figure 20.3
Repeating a written word, according to the Wernicke-Geschwind model.

area is undisturbed. A lesion in Wernicke's area produces great comprehension problems because this is the site of the transformation of sounds into words. The ability to speak is unaffected because Broca's area is still able to drive the muscles required for speech.

The Wernicke-Geschwind model has several errors and oversimplifications. For example, read words do not have to be transformed into a pseudoauditory response, as suggested in the reading task described earlier. In fact, visual information can reach Broca's area from visual cortex without making a stop at the angular gyrus. One of the dangers inherent in any model is overstating the significance of a given cortical area for a particular function. It has recently been found that the severity of Broca's and Wernicke's aphasias depends on how much cortex is damaged beyond the limits of Broca's and Wernicke's areas. Also, aphasia is influenced by damage to subcortical structures such as the thalamus and caudate nucleus, which are not in the model. If parts of cortex are surgically removed, the resulting language deficits are usually milder than the deficits resulting from stroke, which affects both cortical and subcortical structures.

Another important factor is that there is often significant recovery of language function after a stroke, and it appears that other cortical areas can sometimes compensate for what is lost. As in many neurological syndromes, young children recover extremely well, but even adults, especially left-handers, can show good recovery of function.

A final problem with the Wernicke-Geschwind model is that most aphasias involve both comprehension and speech deficits. Mr. Ford, with Broca's aphasia, had good comprehension but was confused by complex questions. Conversely, Mr. Gorgan, with Wernicke's aphasia, had several speech abnormalities in addition to a severe lack of comprehension. Therefore, in cortical processing, the sharp functional distinctions between regions as implied by the model do not exist. In response to these shortcomings, more complex models have been proposed, but they retain some of the features of the Wernicke-Geschwind model (Figure 20.4). Despite the problems with the Wernicke-Geschwind model, it continues to be clinically useful because of its simplicity and approximate validity.

Conduction Aphasia. The value of a model is not only its ability to account for previous observations but also its ability to predict. Wernicke showed that his language-processing model could predict that a form of aphasia would result from a lesion that disconnects Wernicke's area from Broca's area but leaves both areas themselves intact. In the Wernicke-Geschwind

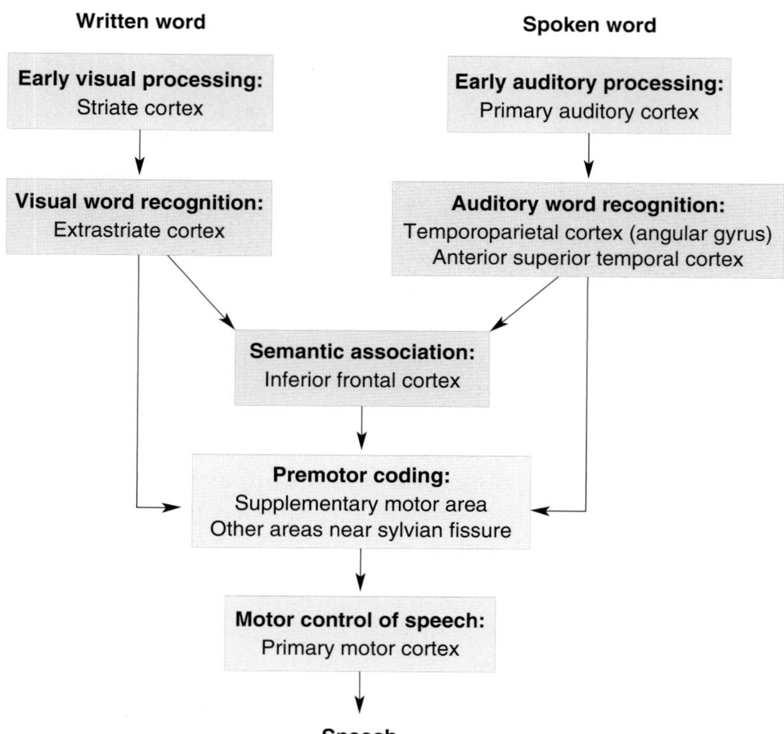

Figure 20.4
A model of language processing. Stages in the processes of repeating a written or spoken word are indicated. Below each stage is the cortical location where task-specific activity was observed with PET imaging. (Source: Adapted from Petersen et al., 1988, Fig, 3.)

model, a lesion in the fibers composing the arcuate fasciculus could accomplish this. In reality, such disconnection lesions usually involve damage to parietal cortex in addition to the arcuate fasciculus, but Broca's and Wernicke's areas are spared.

Sure enough, aphasia from such lesions was demonstrated and is now known as **conduction aphasia**. As the model predicts, based on the preservation of Broca's and Wernicke's areas, comprehension is good and speech is fluent. The patient is typically able to express himself through speech without difficulty. The deficit that chiefly characterizes conduction aphasia is difficulty in repeating words. In response to hearing a few words, the patient may attempt to repeat what was said, but the repetition will substitute words, omit words, and include paraphasic errors. Repetition is usually best with nouns and short common expressions, but it may fail entirely if the spoken words are function words, polysyllabic words, or nonsense sounds. Interestingly, a person with conduction aphasia comprehends sentences he or she reads aloud, even though what is said aloud contains many paraphasic errors. This is consistent with the idea that comprehension is good and the deficit occurs between regions involved in comprehension and speech.

One of the sad yet fascinating things about aphasia is the diversity of syndromes that occur following strokes. While the syndromes challenge any language model, each one offers a clue to our understanding of language processing. Characteristics of a few other aphasias are listed in Table 20.1.

Aphasia in Bilinguals and the Deaf. Cases of aphasia in bilingual people and deaf people provide fascinating insight about language processing in the brain. Suppose a person knows two languages before he or she has a stroke. Does the stroke produce aphasia in one language and not the other, or do both languages suffer equally? The answer depends on several factors, including the order in which the languages were learned, the fluency achieved in each language, and how recently the language was used. The consequences of a stroke are not always predictable, but language tends to be better preserved in the language learned more fluently and earlier in life. If the person learned two languages to equivalent levels of fluency at the same time, a lesion will probably produce similar deficits in both languages. If the languages were learned at different times in life, it is likely that one language will be affected more than the other. The implication is that the second language may make use of different, though overlapping populations of neurons than may the first.

The study of language deficits in those who are deaf and/or know sign language suggests that there is some universality to language processing in the brain. American Sign Language (ASL) uses hand gestures to express all of the ideas and emotions most of us convey with spoken language (Figure 20.5). Left hemispheric lesions in people who use sign language appear to cause language deficits similar to those seen in verbal aphasics. In some cases analogous to Broca's aphasia, comprehension is good, but the ability to "speak" through sign language is severely impaired. Importantly, the ability to move the hands is not impaired (i.e., the problem is not with motor control); rather, the deficit is specific to the use of hand movements for the expression of language.

There are also sign language versions of Wernicke's aphasia in which the patient signs fluently but with many mistakes, while having difficulty comprehending the signing of others. In one unusual case, a hearing man who was the child of deaf parents learned both sign language and verbal language. A left hemispheric stroke initially gave him global aphasia, but his condition significantly improved with time. The important observation is that his verbal and sign languages recovered together, as if overlapping brain areas were used. While there do appear to be sign language aphasias that are analogous to speech aphasias, there is also evidence that signing aphasia and speaking aphasia can be produced by left hemispheric lesions in somewhat different locations.

Lessons Learned From Split-Brain Studies

We have seen that damage to certain parts of the brain leads to a variety of aphasias. As the early work of Broca indicated, language is usually not handled equally by the two cerebral hemispheres. Some of the most valuable and fascinating findings on the language differences of the two hemispheres come from **split-brain studies** in which the hemispheres are surgically dis-

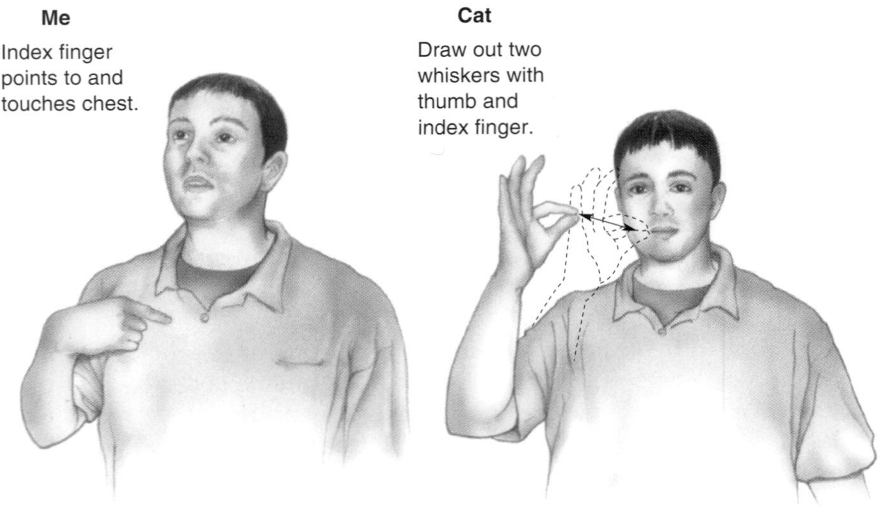

Me
Index finger points to and touches chest.
Cat
Draw out two whiskers with thumb and index finger.

Figure 20.5
"Speaking" in American Sign Language.

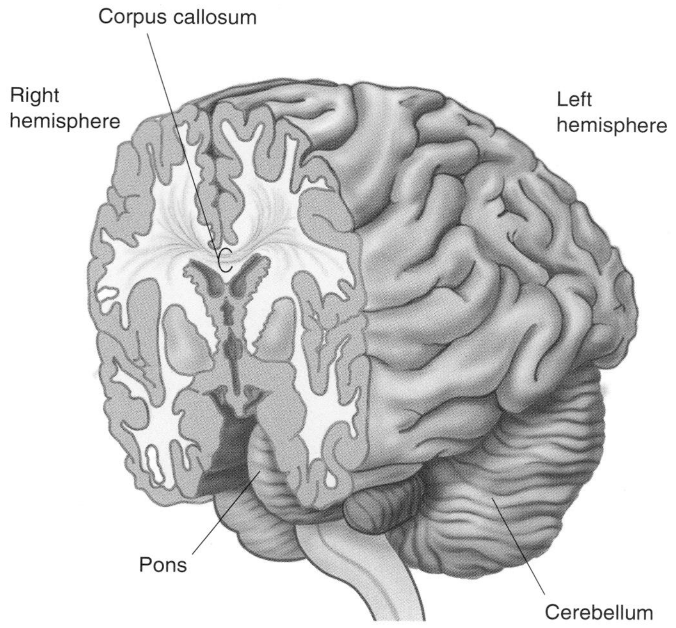

Figure 20.6
The corpus callosum. The corpus callosum is the largest bundle of axons providing communication between the cerebral hemispheres.

connected. Communication between the cerebral hemispheres is served by several bundles of axons known as commissures. Recall from Chapter 7 that the largest of these is the great cerebral commissure, also called the **corpus callosum** (Figure 20.6). The corpus callosum consists of about 200 million axons crossing between the hemispheres. Surely such a huge bundle of fibers must be of great importance. Surprisingly, until about 1950, researchers were unable to demonstrate any important role of the corpus callosum.

In split-brain procedures, the skull is opened and the axons making up the corpus callosum are severed (Figure 20.7). The hemispheres may retain some communication via the brain stem or smaller commissures if they aren't also severed, but most of the intercerebral communication is lost. In the 1950s, Roger Sperry and his colleagues at the University of Chicago and later at the California Institute of Technology performed a series of experiments on split-brain animals to explore the function of the corpus callosum and the separated cerebral hemispheres. Sperry's group confirmed earlier reports that cutting the corpus callosum in a cat or monkey has no noticeable effect on the animal's behavior. Temperament is unchanged, and the animal appears to be normal in coordination, reaction to stimuli, and ability to learn. In cleverly devised experiments, however, Sperry's group showed that the animals sometimes acted as if they had two separate brains.

Language Processing in Split-Brain Humans. Because split-brain monkeys did not appear to have any major deficits, surgeons felt they were justified in cutting the corpus callosum as a last resort in treating certain types of severe epilepsy in humans. They hoped to prevent the spread of epileptic activity from one hemisphere to the other. It may seem questionable to cut 200 million axons on the assumption that they are not very important, but the surgery is often beneficial in restoring a seizure-free life. Michael Gazzaniga, then at New York University, studied a number of these people. Gazzaniga initially worked with Sperry, and his techniques were modifications of those used with experimental animals.

One key methodological feature of studying split-brain humans involves careful control that visual stimuli are presented to only one cerebral hemi-

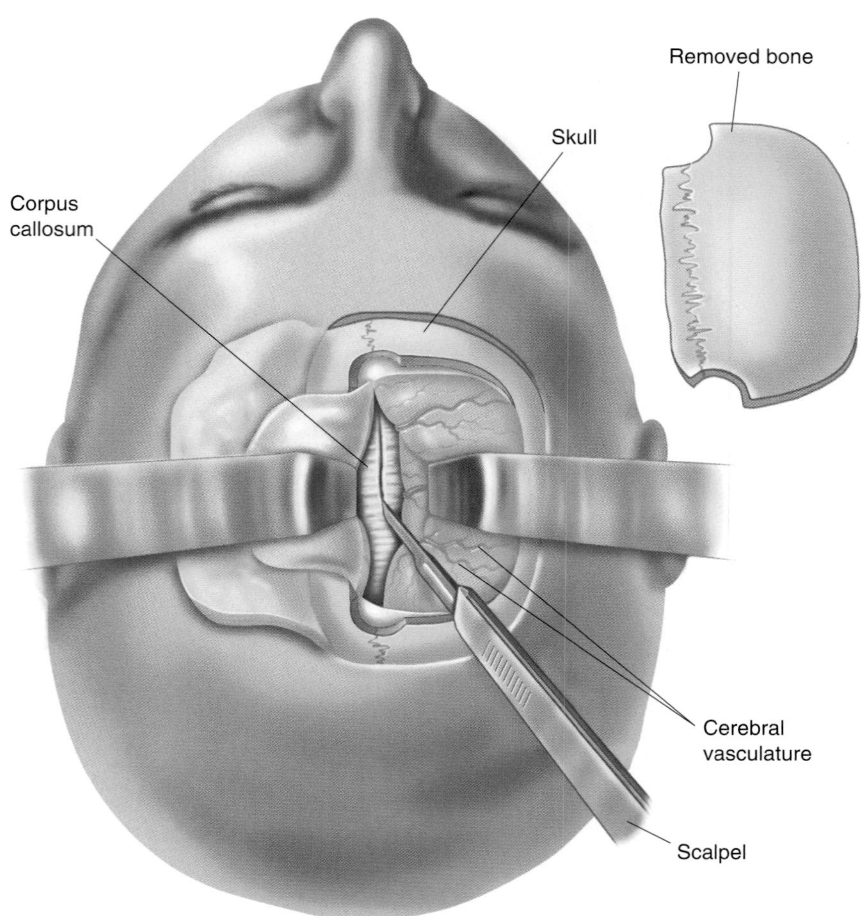

Figure 20.7
Split-brain surgery in a human. To reach the corpus callosum, a portion of the skull is removed and the cerebral hemispheres are retracted.

sphere. Gazzaniga did this by taking advantage of the fact that only the right hemisphere sees objects to the left of the point of fixation, and only the left hemisphere sees objects to the right, as long as the eyes can't move to bring the image onto the fovea (Figure 20.8). Pictures or words were flashed on for a fraction of a second using a device with a camera-like shutter. Because the images were presented for a shorter time than that required to move the eyes, the images were seen by only one hemisphere.

Although split-brain humans are normal in most every way, there is a striking asymmetry in their ability to verbalize answers to questions posed separately to the two hemispheres. For instance, numbers, words, and pictures presented only in the right visual field are repeated or described with no difficulty because the left hemisphere is usually dominant for language. Likewise, objects that can be manipulated only by the right hand (out of view of both eyes) can be described. These findings would be entirely unremarkable except for the fact that such simple verbal descriptions of sensory input are impossible for the right hemisphere.

If an image is shown only in the left visual field or an object is felt only by the left hand, the person will be unable to describe it and will usually say that nothing is there (Figure 20.9). An object could be covertly placed in a patient's left hand, and there would be no verbal indication of even noticing. This absence of response is a consequence (and demonstration) of the fact that the left hemisphere controls speech in most people. If you think about the implications for split-brain people, you'll realize that they have an unusual existence. Following their surgery, they are unable to describe anything to the left

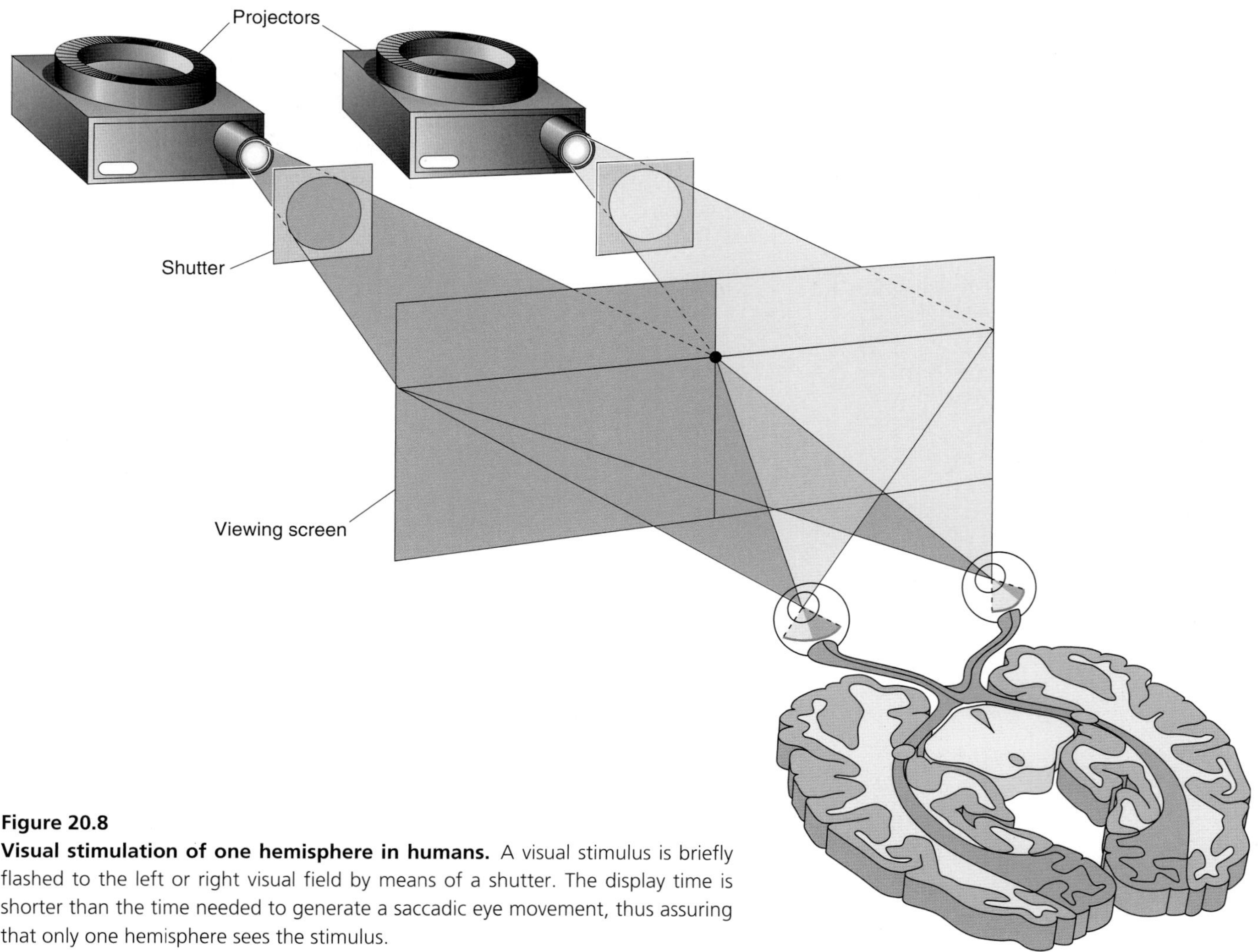

Figure 20.8
Visual stimulation of one hemisphere in humans. A visual stimulus is briefly flashed to the left or right visual field by means of a shutter. The display time is shorter than the time needed to generate a saccadic eye movement, thus assuring that only one hemisphere sees the stimulus.

of their visual fixation point: the left side of a person's face, the left side of the room, and so on. What is startling is that this doesn't seem to disturb the patients.

While there is a dramatic inability of the right hemisphere to speak, this does not mean it knows nothing of language. It can be demonstrated that the right hemisphere can read and understand numbers, letters, and short words as long as the required response is nonverbal. In one experiment, the right hemisphere is shown a word that is a noun. As already mentioned, the person will say he sees nothing. Of course, that's the talkative left hemisphere speaking, and it *didn't* see anything. But if he is urged to use his left hand to select a card containing a picture corresponding to the word he saw or to pick out an object by touch, he can do it (Figure 20.9). The right hemisphere cannot do this with more complex words or sentences, but the results clearly imply that the right hemisphere does have language comprehension.

A recent study conducted by Kathleen Baynes, Michael Gazzaniga, and their colleagues at the University of California at Davis suggests that the right hemisphere may sometimes also be able to write even if it can't speak. In most people, reading, speaking, and writing are all controlled by the left hemisphere. In a split-brain woman known as V. J., this was found not to be true. Words were flashed to either her left or her right hemisphere. Words

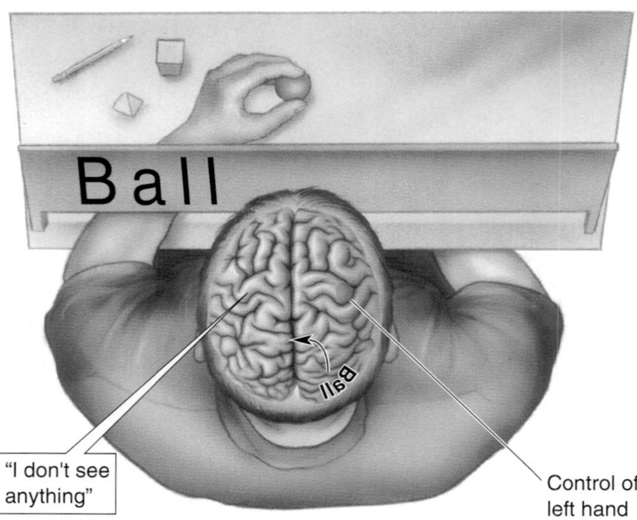

Figure 20.9
Demonstrating language comprehension in the right hemisphere. If a split-brain person sees a word in the left visual field, he will say he sees nothing. This is because the left hemisphere, which usually controls speech, did not see the word, and the right hemisphere, which saw the word, cannot speak. The left hand, which is controlled by the right hemisphere, can, however, pick out the object corresponding to the word by touch alone.

seen by her left hemisphere could be spoken but not written. Conversely, she could write but not speak words shown to her right hemisphere. While this separation of function may be an abnormal situation, the case of V. J. is particularly interesting because it indicates that there is not necessarily a single brain system for all aspects of language in one hemisphere.

Evidence also suggests that the right hemisphere understands complex pictures despite its inability to say so. In one experiment, a subject was shown a series of pictures in her left visual field, and at one point a nude photo appeared in the series. When the researcher asked what she saw, she said nothing, but then she began to laugh. She told the researcher that she didn't know what was funny, but that perhaps it was the machine used in the experiment.

The right hemisphere appears to be better than the left hemisphere at certain things. For instance, even though the split-brain patients were right-handed and thus their left hemispheres were much more practiced at drawing, the left hand controlled by the right hemisphere was better at drawing or copying figures containing three-dimensional perspective. The patients were also better at solving complex puzzles with their left hand. It has also been reported that the right hemisphere is somewhat better at perceiving nuances in sound.

In some of the split-brain studies, the two hemispheres initiated conflicting behaviors, apparently because they were thinking differently. In one task, a patient was asked to arrange a group of blocks to match a pattern on a small card. He was told to do this using only his right hand (left hemisphere), which is not generally good at this type of task. As the right hand struggled to arrange the blocks, the left hand (right hemisphere), which knew how to do it, reached in to take over. Only the restraint of the experimenter kept the left hand from pushing the right one out of the way to solve the puzzle. Another patient whom Gazzaniga studied would sometimes find himself pulling his pants down with one hand and pulling them up with the other. These bizarre behaviors make a strong case that there are two independent brains controlling the two sides of the body.

The results of these split-brain studies demonstrate that the two hemispheres can function as independent brains and that they have different language abilities. Although the left hemisphere is usually dominant for language, the right hemisphere has significant skills in comprehending language. Keep in mind that the split-brain studies test the ability of each hemisphere to perform on its own. Presumably in the intact brain, the callosum allows for synergistic interactions between the hemispheres for language and other functions.

Anatomical Asymmetry and Language

In the nineteenth century, there were reports of anatomical differences between the two hemispheres. For example, it was noted that the left Sylvian fissure is longer and less steep than the right (Figure 20.10). As recently as the 1960s, however, there was considerable doubt about the existence of significant cortical asymmetries. Because of the strikingly asymmetrical control of speech demonstrated by the Wada procedure, it would be interesting to know if the two cerebral hemispheres are anatomically different. Some of the first good quantitative data demonstrating hemispheric differences came from the work of Geschwind and his colleague Walter Levitsky. Initial observations were made on postmortem brains, but more recently the results were confirmed with magnetic resonance imaging (MRI) (see Box 7.2).

The most significant difference seen was in a region called the **planum temporale** on the superior surface of the temporal lobe (Figure 20.11). Based on measurements of 100 brains, Geschwind and Levitsky found that in about 65% of the brains, the left planum temporale was larger than the right, whereas in only about 10%, the right was larger. In some instances, the left area was more than five times larger than the right area. It has been suggested that the left planum temporale is larger because this hemisphere is dominant for speech. This hypothesis has not been proven, however. Interestingly, the asymmetry in this area is seen even in the human fetus, suggesting that it is not a developmental consequence of the use of the left hemisphere for speech. Indeed, apes also tend to have a larger left planum temporale. This suggests the possibility that if the planum temporale is related to language, speech became dominant in the left hemisphere because of a preexisting size difference.

It may have occurred to you that a functional human asymmetry more obvious than language is handedness. More than 90% of humans are right-handed and usually relatively uncoordinated with their left hand, implying that in some way the left hemisphere is specialized for fine motor control. Is this related to the left hemispheric dominance for language? The answer is not known, but it is interesting that humans are different from nonhuman primates in regard to handedness as well as language. While animals of many species show a consistent preference for using one hand over the other, there are typically equal numbers of left-handers and right-handers.

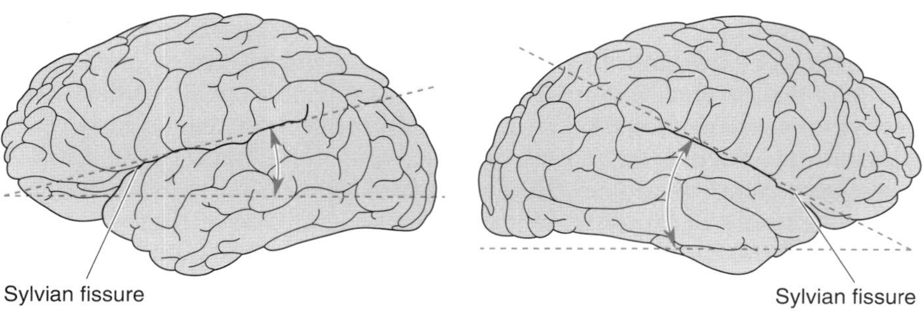

Figure 20.10
Asymmetry of the Sylvian fissure. In most right-handed people, the Sylvian fissure in the left hemisphere is longer and runs at a more shallow angle than the fissure in the right hemisphere. (Adapted from Geschwind, 1979, p. 192.)

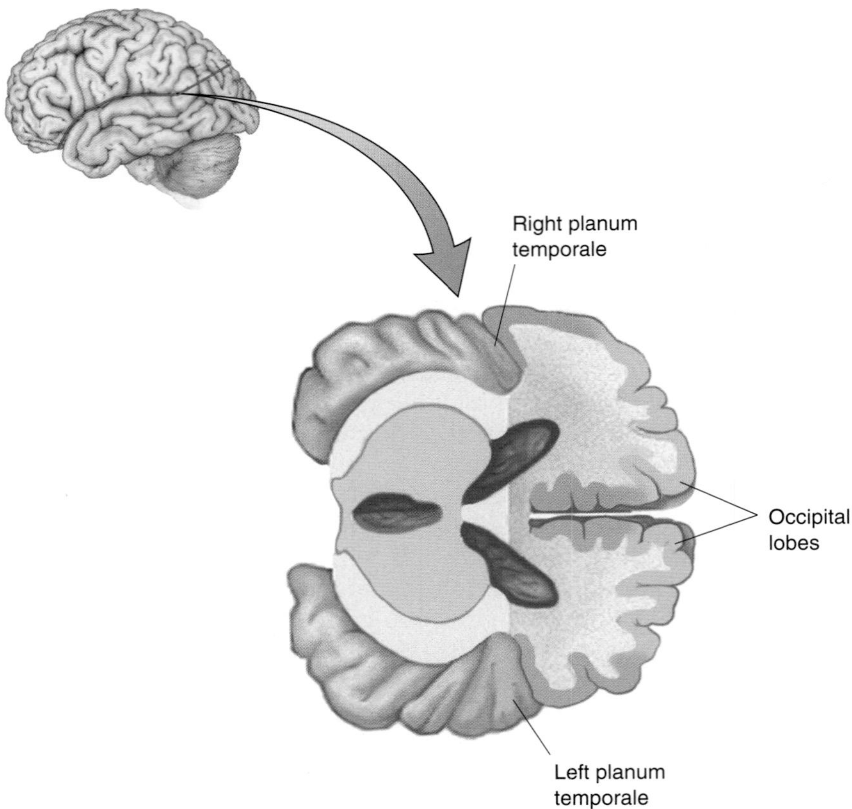

Figure 20.11
Asymmetry of the planum temporale. This region on the superior temporal lobe is usually significantly larger in the left hemisphere. (Source: Adapted from Geschwind and Levitsky, 1968, Fig. 1.)

Language Studies Using Brain Stimulation and PET Imaging

Until recently, the main way that language processing in the brain could be examined was by correlating language deficits with postmortem analysis of brain damage. Other techniques, however, enable researchers to study language function in the brains of living humans. Electrical brain stimulation and PET imaging are two of these techniques.

The Effects of Brain Stimulation on Language. At several points in this book, we have discussed the electrical brain stimulation studies of Wilder Penfield. Without general anesthesia, patients were able to report the effects of stimulation at different cortical sites. In the course of these experiments, Penfield noted that stimulation at certain locations affected speech. These effects fell into three main categories: vocalizations, speech arrest, and speech difficulties similar to aphasia.

Stimulation of motor cortex in the area that controls the mouth and lips caused immediate speech arrest (Figure 20.12). Such a response is logical because the activated muscles sometimes pulled the mouth to one side or clenched the jaw shut. Stimulation of motor cortex occasionally evoked cries or rhythmic vocalizations. Remember that these effects were found with electrical stimulation of motor cortex on either side of the brain. Penfield found three other areas where electrical stimulation interfered with speech, but these were only in the dominant left hemisphere. One of these areas appeared to correspond to Broca's area. If this area was stimulated while a person was speaking, either speech stopped entirely (with strong stimulation) or there was hesitation in the speech (with weaker stimulation). Some patients were unable to name objects they could name before and after the brain stim-

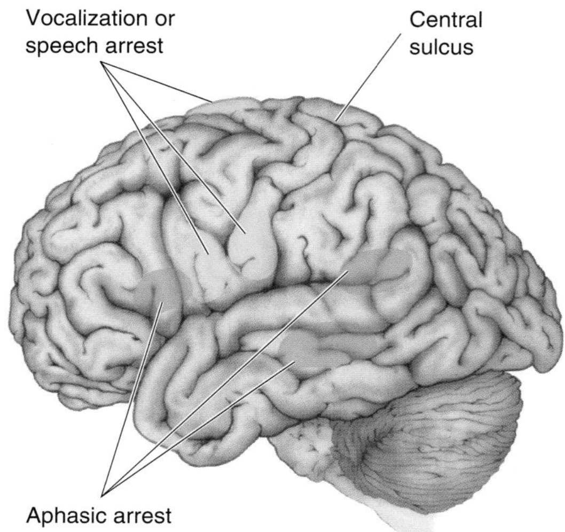

Figure 20.12
Sites where electrical brain stimulation affects language. Stimulation of motor cortex causes vocalizations or speech arrest by activating facial muscles. At other sites, stimulation causes an aphasic arrest in which language is agrammatical or anomia is observed. (Source: Adapted from Penfield and Rasmussen, 1950, Fig. 56.)

ulation. Occasionally, they would substitute an incorrect word. They apparently experienced a mild transient form of anomia. Confusion about words and speech arrest also occurred in some patients who had been stimulated at two other sites, one in the posterior parietal lobe near the Sylvian fissure and the other in the temporal lobe. These latter two areas were in the vicinity of the arcuate fasciculus and Wernicke's area, although they did not align perfectly with estimated locations of those areas.

It is somewhat reassuring to see that electrical stimulation selectively affects speech in the brain areas roughly corresponding to those responsible for aphasia. The consequences of stimulation are, however, surprisingly variable between nearby cortical sites and between subjects. In studies similar to those of Penfield, neurosurgeon George Ojemann at the University of Washington has found that the effects of stimulation can sometimes be quite specific. For example, stimulation of small parts of cortex at different locations can interfere with naming, reading, or repetition of facial movements (Figure 20.13). What makes these findings all the more intriguing is that different results are sometimes obtained at nearby stimulation sites and similar results can be obtained at distant sites. These results may indicate that the language areas in the brain are much more complex than is implied by the Wernicke-Geschwind model. Areas used in language are also more extensive than simply Broca's and Wernicke's areas, as they have been found to include other cortical areas, as well as parts of the thalamus and striatum. Within Broca's area and Wernicke's area there may be specialized regions, possibly on the scale of functional columns in somatosensory cortex or ocular dominance columns in visual cortex. It appears that the large language areas identified on the basis of aphasic syndromes may well encompass a good deal of finer structure.

PET Imaging of Language Processing. With the advent of modern imaging techniques, it has become possible to observe normal language processing. With PET, the level of neural activity in different parts of the brain is inferred from regional blood flow (see Box 7.3). In one study of language processing, researchers used PET imaging to observe the differences in brain activity between the sensory responses to words and the production of speech. They began by measuring cerebral blood flow with the subject at rest. They then had the person either listen to words being read or look at words presented on a monitor. By subtracting the levels of blood flow at rest from the levels during listening or seeing, they determined blood flow levels specifically corre-

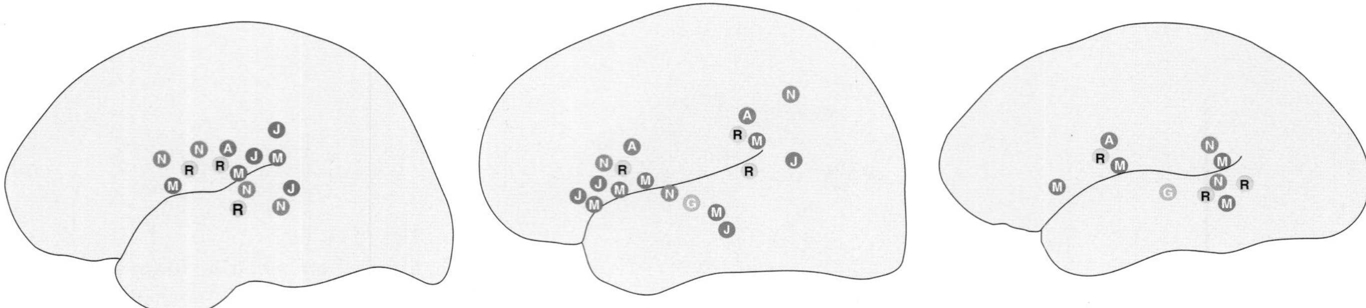

Figure 20.13
The effects of brain stimulation in three patients being treated for epilepsy. Patients were awake, and difficulties in speaking or reading were noted. N, naming difficulty with intact speech (anomia); A, arrest of speech; G, grammatical errors; J, jargon (fluent speech with frequent errors); R, failure to read; M, facial movement errors. (Source: Adapted from Ojemann and Mateer, 1979, Fig. 1.)

sponding to the activity evoked by the sensory input. The results are shown in the top half of Figure 20.14. Not surprisingly, the visual stimuli evoked increased brain activity in striate cortex and extrastriate cortex, and the auditory stimuli elicited activity in primary and secondary auditory cortex. The areas activated in extrastriate cortex and secondary auditory cortex did not respond, however, to visual and auditory stimuli that were not words. These areas may be specialized for the coding of words either seen or heard. The visual stimuli did not evoke noticeably increased activity in the area of the angular gyrus and Wernicke's area, as one would expect based on the Wernicke-Geschwind model.

Another task studied with PET imaging was repeating words. To know what words to repeat, subjects must perceive and process the words by either the visual or the auditory system. Thus, the brain activity seen in the repetition task should include a component associated with the basic perceptual process as well as a component associated with speech. To isolate the speech

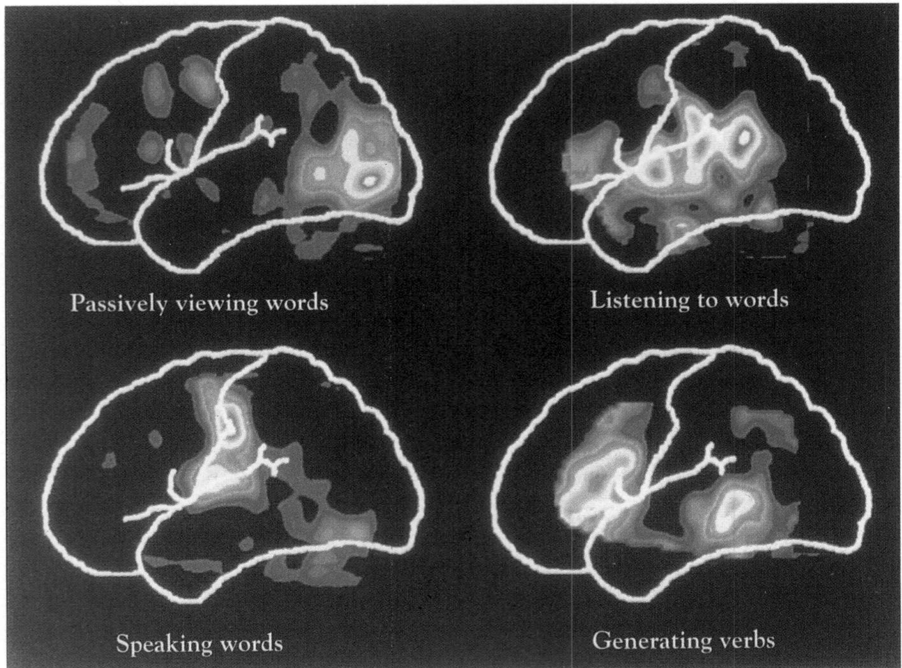

Figure 20.14
PET imaging of sensation and speech. Relative levels of cerebral blood flow are color coded. Red indicates the highest levels, and progressively lower levels are represented by orange, yellow, green, and blue. (Source: Posner and Raichle, 1994, p. 115.)

component, the response pattern previously obtained in the simple sensory task was subtracted. In other words, the image for "speaking words" equals an image corresponding to "repeating spoken words" minus an image corresponding to "listening to words." After the subtraction, the blood flow pattern indicated high levels of activity in primary motor cortex and the supplementary motor area (Figure 20.14, bottom left). There was also increased blood flow around the Sylvian fissure near Broca's area. The PET images showed such activity *bilaterally*, and it was even observed when the subjects were instructed to move their mouth and tongue without speaking. Since there is good evidence that Broca's area is unilateral, it may, for reasons unknown, not be showing up in these images.

The final task required the subjects to think a little. For each word presented, the subject had to state a use for the word (e.g., if "cake" were presented, the subject might say "eat"). To isolate activity specific to this verb-noun association task, the blood flow pattern obtained previously for speaking words was subtracted. Areas that were activated in the association task are located in the left inferior frontal area, the anterior cingulate gyrus, and the posterior temporal lobe (Figure 20.14, bottom right). The activity in frontal and temporal cortex is believed to be related to the performance of the word association task, whereas the activity in cingulate cortex is probably related to attention. Recent PET and fMRI studies indicate fascinating similarities and differences in language processing of spoken language, sign language, and Braille (Box 20.4).

The PET results shown in Figure 20.14 are consistent with the overall placement of language areas inferred from studies of aphasia. As in the brain stimulation studies we discussed, however, the PET results suggest that language processing involves more complex mechanisms than a simple interaction between two major language areas (Broca's and Wernicke's). Because language involves many skills, such as naming, articulation, use of grammar, and comprehension, this conclusion seems reasonable. An important question is whether there are separate systems involved in these various language functions. The model in Figure 20.4 may be a more realistic description of language processing than the Wernicke-Geschwind model, but there is clearly much to be learned. The hope is that further brain imaging studies will clarify language systems in the brain.

ATTENTION

Picture yourself at a crowded party where you are talking with a friend, numerous other conversations are going on around you, and loud music is playing. You are bombarded by sound from all directions. Yet somehow you're able to concentrate on the conversation you're having and ignore most of the other noise and talking. You are *paying attention* to the one conversation. Behind you, you hear someone mention your name, and you decide to eavesdrop. Without turning around, you start focusing your attention on this other conversation to find out what's being said about you.

The ability to select one conversation to listen to out of many going on at the same time is an example of attention. The act of differentially processing simultaneous sources of information is called *selective attention*. Attention can also be exerted in other sensory modalities. In the visual system, attention enables us to concentrate on one object over many others in our visual field. Interactions between modalities also occur. For example, if you are performing an attention-demanding visual task such as reading, you are relatively insensitive to incoming sounds. Clearly, attention has to do with preferential processing of sensory information. Amid all the sights, sounds, and tastes coming into our brain, we can selectively attend to some infor-

Box 20.4

OF SPECIAL INTEREST

Hearing Sight and Seeing Touch

The human brain is a remarkably adaptable organ, and some of the most dramatic examples of brain reorganization come from studies of human language processing. The top panel of Figure A is an fMRI showing brain areas activated when normally hearing English speakers read sentences in English. Areas in red are those most highly specifically activated by language, and areas in yellow are somewhat less activated. (Visual activity not associated with language has been subtracted.) There is significant activity in the standard language areas of the left hemisphere, including Broca's area and Wernicke's area, but little activity in the right hemisphere.

The center panel of Figure A is the activity seen when hearing English speakers saw sentences in American Sign Language (ASL). These people who did not understand ASL showed no specific activity after their response to meaningless hand signs was subtracted. On the other hand, the bottom panel of Figure A shows the response to ASL in deaf subjects who were raised with it as their sole language. The activation includes Broca's area and Wernicke's area in the left hemisphere, indicating that ASL uses the same language areas also used for spoken English in hearing subjects. Even more surprising is the great deal of activity in the right hemisphere. Remarkably, the superior temporal gyrus is active in response to ASL in deaf subjects, an area that would normally respond to spoken language in hearing subjects. As these areas are also active in hearing subjects who are bilingual in ASL, it appears that something about ASL recruits right hemisphere auditory areas in addition to the usual left hemisphere language areas.

A different form of brain reorganization is seen in blind subjects who read Braille. A system of writing that uses patterns of small bumps on paper to represent letters, Braille is read by scanning the fingertips across the bumps. As you would expect, Braille reading activates somatosensory cortex, but other brain areas activated are a real surprise. Figure B shows a PET image obtained from a person reading Braille. There is significant activity at the occipital pole of the brain (yellow)—unmistakably portions of visual cortex. Through a process of brain reorganization, the brains of these blind subjects use traditional visual areas to process Braille, similar in spirit to the auditory areas used for ASL in deaf subjects. (The chapters in Part IV explore the mechanisms by which sensory experience can affect brain organization and lead to learning and memory.)

mation and ignore the rest. Why, you may ask, do we do that? After all, if all this sensory information successfully makes its way into the brain, why not process all of it?

One possibility is that the brain simply cannot process all of the incoming sensory information simultaneously. For example, striate cortex in a macaque monkey occupies about 10% of cerebral cortex. By current estimates, there are more than 30 other visual areas, but many are much smaller than striate cortex. It is likely that these other areas cannot process as much information as Vl. If this is true, attention plays a key role in selecting what information should receive the limited processing resources of the brain.

Another possibility is that even if the brain could process all the sensory information at once, there might be a performance advantage to focusing on one thing at a time. Maybe the situation is analogous to juggling—it's a lot easier to keep one ball in the air than five. While it is not simply a problem of attention, attention-deficit/hyperactivity disorder demonstrates how critical attentional mechanisms are (Box 20.5).

Only recently have neuroscientists begun to explore the neural consequences of attention. We will look at some of these fascinating physiological studies. But first, let's examine a few key findings concerning the behavioral manifestations of attention. We'll focus on visual attention because that is where behavioral studies can best be related to physiological research.

(Box 20.4, *continued*)

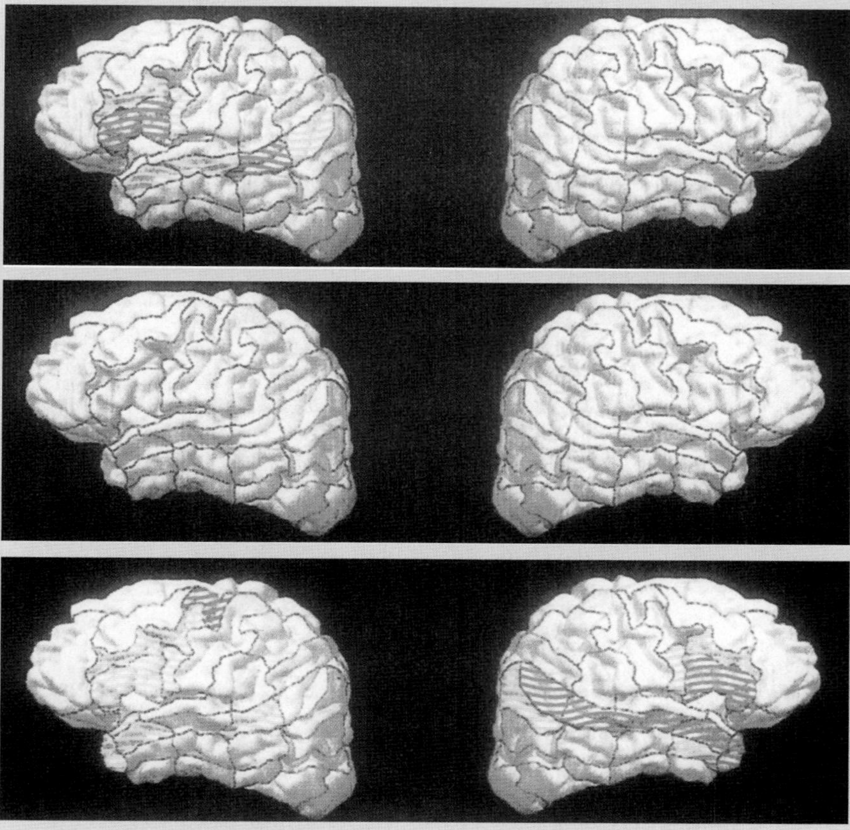

Figure A
Top: Written English, read by hearing subjects. Center: ASL seen by hearing subjects. Bottom: ASL seen by deaf subjects. (Source: Neville et al., 1998.)

Figure B
Reading Braille. (Source: Sadato et al., 1996.)

Behavioral Consequences of Attention

Under most conditions, if we want to scrutinize something visually, we move our eyes so that the object of interest is imaged on the fovea in each eye. Implicit in this behavior is the fact that most of the time we are paying attention to something imaged on our fovea. It is possible, however, to shift attention to objects imaged on parts of the retina outside the fovea. Shifting attention to some location on the retina enhances visual processing in several ways. Two ways we'll look at are enhanced detection and faster reaction times.

Attention Enhances Detection. Figure 20.15 shows an experiment for studying the effects of directing visual attention to different locations. During the entire course of an experiment, the observer fixated on a central

Box 20.5

OF SPECIAL INTEREST

Attention-Deficit/Hyperactivity Disorder

It is the last lecture of the school year, and you are failing miserably at focusing on the instructor as you stare longingly at the green grass and trees outside the window. At times we all have difficulty concentrating on our work, sitting still, and resisting the urge to play. But for millions of people, the syndrome that has come to be called *attention-deficit/hyperactivity disorder* (*ADHD*) routinely and seriously interferes with their ability to get things done.

The three traits commonly associated with ADHD are inattention, hyperactivity, and impulsiveness. Children normally exhibit these traits more than adults, but if the behaviors are significantly pronounced, ADHD may be diagnosed. By some estimates, 5–10% of all school-age children worldwide have ADHD, and the disorder interferes with schoolwork and interactions with classmates. Follow-up studies show that many people diagnosed with ADHD continue to exhibit some symptoms as adults.

We do not know what causes ADHD, but there are several clues. For example, it has been reported that in MRI scans, several brain structures, including prefrontal cortex and the basal ganglia, are smaller in boys with ADHD. It is not known whether these differences are behaviorally significant. The possible involvement of these structures is intriguing, however, because they have long been implicated in the regulation and planning of behavior. You might recall Phineas Gage from Box 18.1, who had great difficulty making and carrying out plans after a severe lesion to his prefrontal cortex.

Several pieces of evidence suggest that heredity plays a significant role in ADHD. Children of parents with ADHD are more likely to develop it than children of parents without ADHD, and a child is much more likely to have it if an identical twin does. Nongenetic factors, such as brain injury and premature birth, may also be involved. Several genes related to the function of dopaminergic neurons have been reported to be abnormal in people with ADHD. These include the D4 dopamine receptor gene, the D2 dopamine receptor gene, and the dopamine transporter gene. We have seen in several previous chapters how important dopaminergic transmission is for a variety of behaviors, so it will be a challenge to clarify dopamine's involvement in ADHD.

At present, aside from behavioral therapy, drugs such as Ritalin (methylphenidate), a mild central nervous system stimulant similar to amphetamine, are the most common treatment for ADHD. It also inhibits the dopamine transporter, increasing the postsynaptic effect of dopamine. In many children, Ritalin successfully decreases impulsiveness and inattention, although questions about the safety of long-term use remain.

point. The observer's task was to say whether a target stimulus was flashed on at a location to the left of the fixation point, to the right, or not at all. The task was difficult because the target presented to either side of the fixation point was a small circle and was flashed on for only about 15 msec. The experiment had several special procedures for identifying the effects of attention. Each trial began with the presentation of a cue stimulus at the fixation point. The cue was either a plus sign, an arrow pointing left, or an arrow pointing right. After the cue was extinguished, a variable delay period followed, during which the fixation point was seen. In half the trials there was no further stimulus, and in the other half a small circle was flashed on for 15 msec at either the left or the right position. A key element of the experiment is that the cue was used to direct attention. If the central cue was a plus sign, it was equally likely that a little circle would appear at either the left or the right. If the cue was a left arrow, it was four times as likely that a target would appear on the left than on the right. If the cue was a right arrow, it was four times as likely that a target would appear on the right than on the left. The observer had to keep his or her eyes pointing straight ahead, but to make the most correct responses on the difficult task of detecting the flashed circles, it would be advantageous to make use of the cue. For instance, if the cue was a

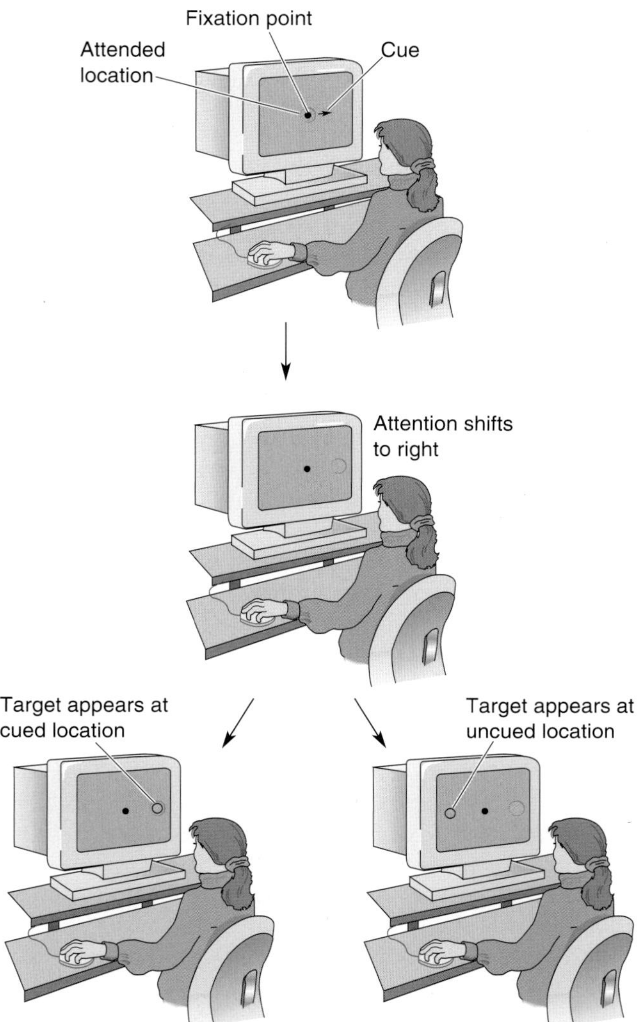

Figure 20.15
An experiment to measure the effect of attention on detection. While an observer maintains steady fixation, a cue directs her to shift her attention to one side of the computer screen. In each trial, the observer indicates whether a circular target is seen on either side of the screen.

left arrow, it would be beneficial to try and pay attention to the left target location more than the right.

For each of the subjects used in this experiment, the data collected consisted of the percentage of the time a circle was correctly detected. Because there was no target circle on half the trials, the observers could not get a high percentage correct by cheating (i.e., by saying there was always a target at the side where the arrow pointed). In trials in which the central cue was a plus sign, the observers detected a target stimulus on about 60% of the trials in which one was presented. When the cue was a right arrow, the observers detected a target stimulus at the right on about 80% of the trials in which one was presented there. When the cue pointed right, however, the observers detected a target stimulus at the left on only about 50% of the trials in which one was presented there. With the appropriate left-right reversal, the results were about the same with left arrows.

What do these data mean? To answer, we must imagine what one of the observers was doing. Evidently, the expectation of the observer based on the cues influenced his or her ability to detect the subsequent targets. It appears that the arrow cues caused the observer to shift his or her attention to the side where the arrow pointed, even though the observer's eyes did not move. Presumably, this shift of attention made it easier to detect the flashed targets than when the central cue was a plus sign. Conversely, the observer was less sensitive to the targets on the side of fixation opposite to where the arrow cue pointed. These results and those from many other similar experiments lead us to our first conclusion about the behavioral effects of attention: It makes things easier to detect. This is probably one of the reasons we can listen in on one conversation among many when we give it our attention.

Attention Speeds Reaction Times. Using a similar experimental technique has demonstrated that attention increases the speed of reactions in perceptual studies. In a typical experiment, an observer fixated on a central point on a computer screen, and target stimuli were presented to either the left or the right of the fixation point. But in this experiment, the observer was told to wait until he or she perceived a stimulus at either location and then to press a button. The researchers measured how long it took the observer to react to the presentation of a stimulus and press a button. Preceding the target was a cue stimulus, either a plus sign or an arrow pointing left or right. The arrows indicated the side to which a stimulus was more likely to appear, whereas the plus sign meant that either side was equally likely.

Results from this experiment demonstrated that an observer's reaction times were influenced by where the central cue directed the observer's attention. When the central cue was a plus sign, it took about 250–300 msec to press the button. When an arrow cue correctly indicated where a target would appear (e.g., right arrow and right target), reaction times were 20–30 msec faster. Conversely, when the arrow pointed in one direction and the target appeared at the opposite location, it took 20–30 msec longer to react to the target and press the button. The reaction time included time for transduction in the visual system, time for visual processing, time to make a decision, time to code for the finger movement, and time to press the button. Nonetheless, there was a small but reliable effect based on which direction the arrows directed the observer's attention (Figure 20.16). If we assume that attention to visual objects does not have a direct effect on visual transduction or motor coding, we are left with the hypothesis that attention can alter the speed of visual processing or the time to make a decision about pressing the button.

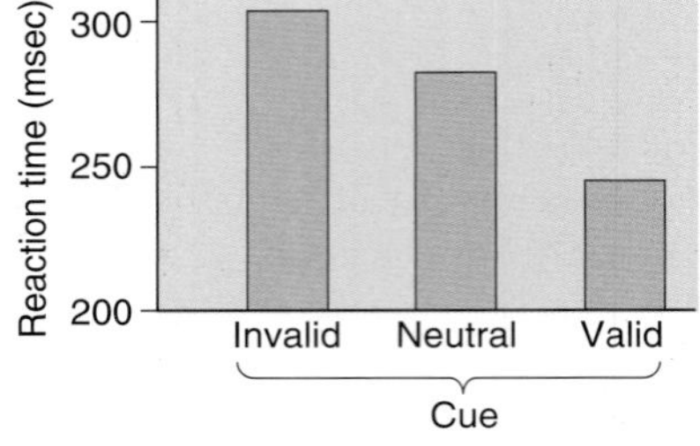

Figure 20.16
The effect of cueing on reaction time. In neutral cue trials, the cue was a plus sign, which gave no indication of the likely location of the following target. In valid cue trials, the arrow-shaped cue pointed to the location where the target appeared, speeding reactions to the targets. When the cue was invalid, pointing in a direction opposite to where the target later appeared, reaction times were slower. (Source: Adapted from Posner, Snyder, and Davidson, 1980, Fig. 1.)

Neglect Syndrome as an Attentional Disorder. In Chapter 12, we briefly discussed **neglect syndrome**, in which a person appears to ignore objects, people, and sometimes their own body to one side of their center of gaze. Some have argued that this syndrome is a unilateral deficit in attention. The manifestations of neglect syndrome can be so bizarre that they're hard to believe if not directly observed. In mild cases, the behavior may not be apparent from casual observation. But in severe cases, the patient acts as if half the universe no longer exists. He may shave only one side of his face, brush the teeth on only one side of his mouth, dress only one side of his body, and eat food from only one side of his plate.

Because neglect syndrome is less common following left hemispheric than right hemispheric damage, it has primarily been studied in regard to neglect of the left half of space as a result of damage to right cerebral cortex. In addition to neglecting objects to the left side, some patients exhibit denial. For instance, they may say that their left hand isn't really paralyzed or, in extreme cases, refuse to believe that a limb on their left side is part of their body. Refer to Figure 12.23 as a typical example of the distorted sense of space these patients have. If asked to make a drawing, they may crowd all of the features into the right half, leaving the left half blank. A particularly dramatic example is the paintings shown in Figure 20.17, made by an artist as he recovered from a stroke.

Figure 20.17
Self-portraits during recovery from a stroke that caused a neglect syndrome. Two months after suffering a stroke affecting parietal cortex on the right side, the artist made the upper left portrait. There is virtually no left side to the face in the painting. Three and a half months after the stroke (upper right), there is some detail on the left side but not nearly as much as on the right side. At 6 months (lower left) and 9 months (lower right) after the stroke, there is increasing treatment of the left side of the painting. (Source: Posner and Raichle, 1994, p. 152.)

If asked to close their eyes and point toward the midline of their body, patients with neglect syndrome typically point too far to the right, as if there has been a shrinkage of the left half. If blindfolded and asked to explore objects placed on a table before them, patients behave normally in exploring objects to the right but are haphazard about probing to the left. All of these examples point toward a problem in relating to the space around them.

Neglect syndrome is most commonly associated with lesions in posterior parietal cortex in the right hemisphere, but it has also been reported to occur following damage to right hemisphere prefrontal cortex, cingulate cortex, and other areas. It has been proposed that posterior parietal cortex is involved in attending to objects at different positions in extrapersonal space. If this is true, then neglect syndrome may be a disruption of the ability to shift attention. One piece of evidence supporting this hypothesis is that objects in the right visual field of patients with neglect syndrome are sometimes abnormally effective in capturing attention, and patients may have difficulty disengaging their attention from an object on this side.

It is not clear why neglect syndrome more often accompanies right hemispheric damage than left hemispheric damage. The right hemisphere appears to be dominant for understanding spatial relationships, and in split-brain studies, it has been shown to be superior at solving complex puzzles. This finding seems consistent with the greater loss of spatial sense after right hemispheric lesions. One hypothesis is that the left hemisphere attends to objects in the right visual field and the right hemisphere attends to objects in the left and right visual fields. While this would account for the asymmetrical effects of left and right hemispheric lesions, at present there is only suggestive evidence in support of such a hypothesis.

Physiological Effects of Attention

What is happening in the brain when we shift our attention to something? While it's conceivable that attention is strictly a high-level cognitive process, recent experiments demonstrate that the effects of attention can be observed in sensory areas such as visual cortex. But it is important to recognize a critical distinction. We primarily discuss experiments showing that brain activity changes when attention is devoted to a stimulus. While this activity change may *result* from attention, it may not have anything to do with the *control* of attention. How attention is allocated to particular objects is quite a different matter.

Functional MRI Studies of Attention to Location. A key observation made in behavioral studies of attention is that enhancements in detection and reaction time are selective for spatial location. When we know where an important stimulus is more likely to appear, we move our attention to it and process the sensory information with greater sensitivity and speed. A common analogy is that attention works like a spotlight, moving to illuminate objects of particular interest or significance. Recent experiments using fMRI of the human brain suggest that there may be selective changes in brain activity associated with spatial shifts in attention.

In one experiment, subjects in the fMRI machine viewed a stimulus consisting of colored line segments arranged in circular sectors at several distances from the central fixation point (Figure 20.18a). The subject was told which sector to pay attention to, and this location changed every 10 seconds. During the 10-second period, the color and orientation of the line segments changed every 2 seconds. Each time the line segments changed, the subject's task was to press one button if the lines were blue and horizontal or orange and vertical and a second button if the lines were blue and vertical or orange and horizontal. The reason for having the subjects perform this task was to force them to attend to a particular sector of the stimulus. Remember that subjects always kept their gaze fixed at the center of the stimulus.

The fascinating aspect of this experiment is what happens when the location of the sector being attended to changes. Figure 20.18b shows brain activity recorded with the attended sector at four locations at increasing distance from the fixation point. The areas of highest brain activity move away from the occipital pole as the attended sector moves out from the fovea. The pattern of brain activity shifts retinotopically, even though the visual stimuli are the same regardless of which sector is attended. The hypothesis is that these images show the neural effect of the spotlight of attention moving to different locations.

PET Imaging of Attention to Features. These fMRI findings appear consistent with the behavioral observation that visual attention can be moved independently of eye position. But attention involves more than just location. Imagine walking down a crowded city sidewalk in the winter looking for someone. Everyone is bundled in heavy coats, but you know your friend will be wearing a red hat. Mentally focusing on red makes it much easier to pick out your friend. Evidently, we are able to pay particular attention to visual features such as color to enhance our performance. Is there any reflection of this attention to features in brain activity? The answer has come from studies using PET in humans.

Steven Petersen and his colleagues at Washington University used PET while humans performed a same-different discrimination task (Figure 20.19). An image was flashed on a computer screen for about half a second; after a delay period, another image was flashed. Each image was composed of small

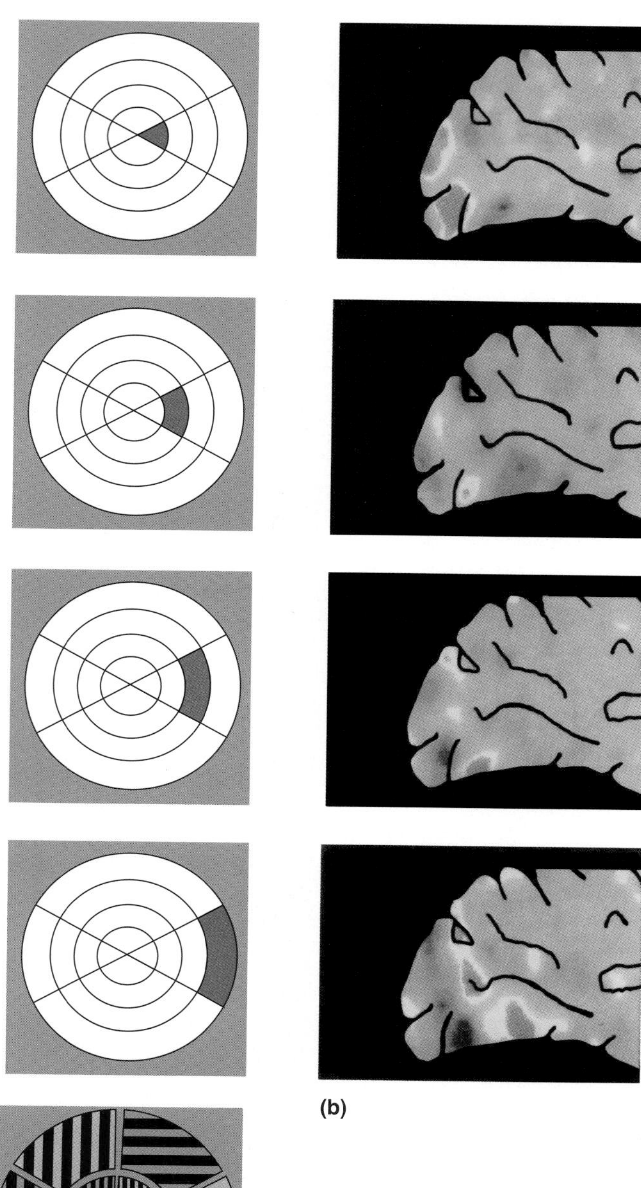

Figure 20.18
The spotlight of attention. (a) The visual stimulus. The stimulus (bottom frame) consisted of blue and orange line segments arranged into sectors radiating out from the central fixation point. The orientation and color of each sector changed every 2 seconds. The four bulls-eye patterns indicate in red the sector that a subject was instructed to attend to. **(b)** Enhanced activity in visual cortex. The visual stimuli elicited activity in multiple visual cortical areas, but patches of enhanced activity were associated with the attended sector. Enhanced activity is indicated in yellow and red. (Source: Courtesy of J. A. Brefczynski and E. A. DeYoe.)

elements that could vary in shape, color, and speed of motion. The task of the observer was to indicate whether the two successive images were the same or different. While the speed, shape, and color of the elements could all change between the two images, the observer was instructed to base same-different judgments on any one or all of the attributes. Accordingly, brain activity in response to identical stimuli could be measured when the person was attending to different features. Of course, the brain activity would simply indicate where neurons were activated by the visual patterns, unless something was added to the experiment.

To isolate the effect of attention, two versions of the experiment were con-

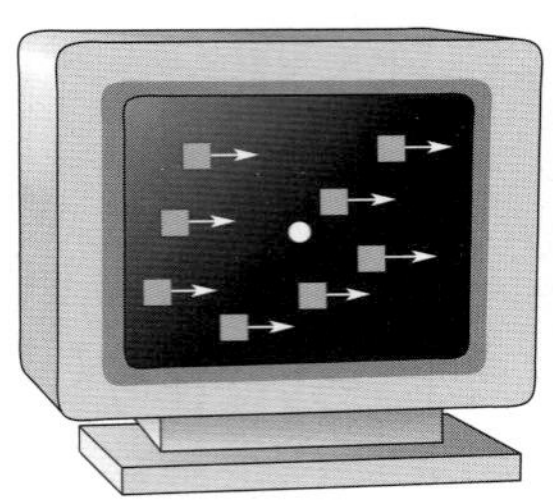

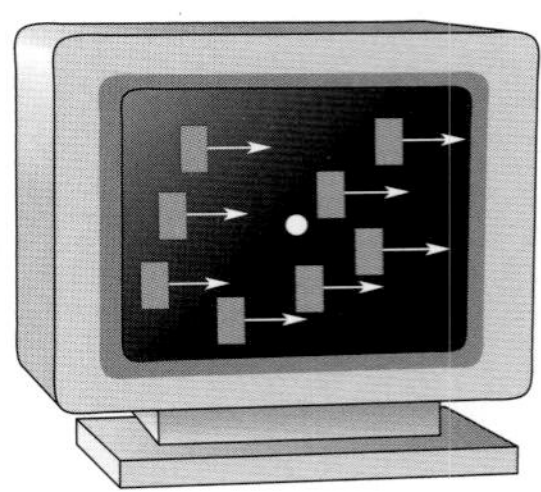

Figure 20.19
Same-different stimuli used for PET imaging. The observer sees two frames, each containing moving elements that can change in shape, color, and speed of motion. The observer responds by indicating whether the stimuli are the same or different.

ducted. In *selective-attention* experiments, subjects viewed the stimuli and performed the same-different task after being told to pay attention to just one of the features (shape, color, or speed). In *divided-attention* experiments, subjects simultaneously monitored all features and based their same-different judgments on changes in any feature. The researchers then subtracted the divided-attention responses from the selective-attention responses to obtain an image of changes in brain activity associated with attention to one feature.

Figure 20.20 illustrates the results. Different areas of cortex had higher activity when different attributes of the stimuli were being discriminated. For instance, whereas ventromedial occipital cortex was affected by attention in color and shape discrimination tasks, it was not affected in the speed discrimination task. Conversely, areas in parietal cortex were influenced by attention to the motion task but not the other tasks. While it is not possible to know with certainty which cortical areas were highlighted in these experiments, areas of heightened activity in the color and shape tasks may have corresponded to areas V4 and IT and other visual cortical areas in the temporal lobe. The area most affected by performing the motion task was near area MT. These effects of attention to different features are roughly consistent with the tuning properties of neurons in extrastriate visual areas discussed in Chapter 10.

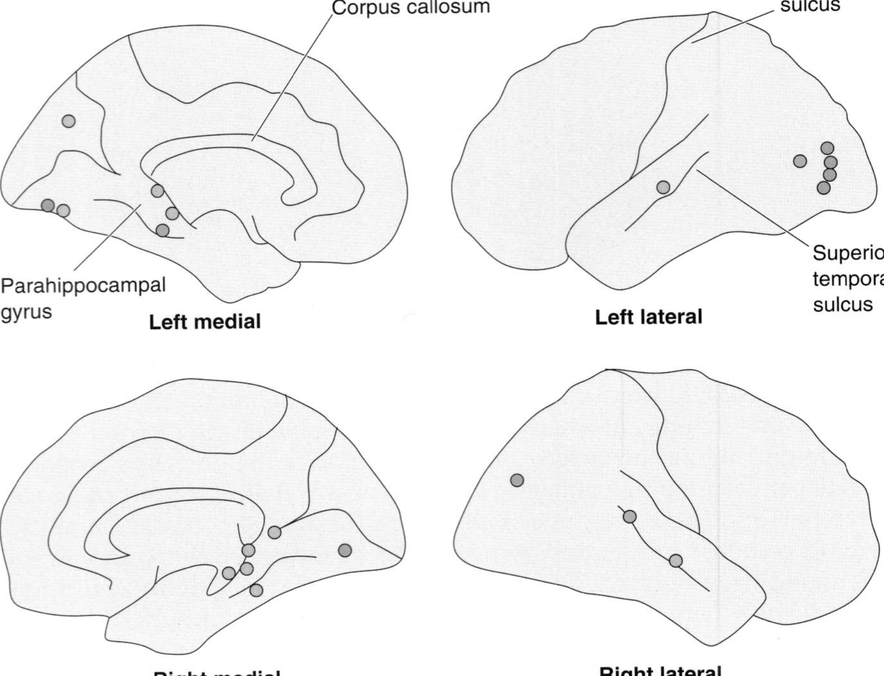

Figure 20.20
Feature-specific effects of visual attention. Symbols indicate where activity in PET images was higher in selective-attention experiments relative to divided-attention experiments. In selective-attention experiments, the same-different judgments were based on speed (green), color (blue), or shape (orange). (Source: Adapted from Corbetta et al., 1990, Fig. 2.)

The important point to learn from these and other PET experiments is that numerous cortical areas appear to be affected by attention and that the greatest attentional effects are seen in "late" rather than "early" areas in the visual system. Attention selectively increases brain activity, but the particular areas affected depend on the nature of the behavioral task performed. We'll now examine a couple of these areas in detail and see how studies in behaving monkeys have clarified the role of attention.

Enhanced Neuronal Responses in Parietal Cortex. The perceptual studies discussed earlier show that in carefully constructed experiments, attention can be moved independently of eye position. But what happens normally when you move your eyes? Let's assume that when you are scrutinizing an object focused on your fovea, your attention is also directed to the fovea. If a bright flash of light appeared in the periphery of your visual field, you normally would make a saccade toward the sudden flash so that you could look at it with your fovea. But what happened to your attention? Did it move before, during, or after the eye movement? Behavioral studies have shown that shifts in attention can occur in about 50 msec, whereas saccades take about 200 msec. Therefore, it is quite likely that attention is drawn to the flash before your eyes move.

The assumption that attention changes location prior to an eye movement underlies an experiment performed by neurophysiologists Robert Wurtz, Michael Goldberg, and David Robinson at the National Institutes of Health. They recorded neural activity at several locations in the brains of monkeys to determine whether, prior to eye movements, there was increased activity that might be related to attention. Assuming there is a relationship between attentional shifts and eye movements, it is logical to examine parts of the brain involved in generating saccades.

The researchers recorded from neurons in the posterior parietal cortex of monkeys while the animals performed a simple behavioral task (Figure 20.21). This cortical area is thought to be involved in directing eye movements, in part because electrical stimulation here evokes saccades. Parietal neurons have rather large visual receptive fields, perhaps encompassing 25% of the entire visual field. In the task, a monkey fixated on a spot on a computer screen; when a target appeared at a different location, it made a saccade to that location. In each trial the receptive field of a cortical neuron was located, and the target used in the behavioral task was positioned so that it appeared within the receptive field. As expected, the neuron was excited when the target flashed in its receptive field (Figure 20.22a). The key observation Wurtz and his colleagues made, however, was that the response of many parietal cortex neurons was significantly enhanced (a rapid burst of action potentials) when the animal subsequently made a saccade to the target (Figure 20.22b). This effect was spatially selective because the enhancement was not seen if a saccade was made to a location outside the receptive field (Figure 20.22c). This is an important point because it indicates that the brain was not simply generally more excitable.

The enhancement that occurred before the saccades is curious if the parietal neurons were simply responding to the visual stimuli. Why should the response to a target stimulus depend on an eye movement that happens long *after* the target is turned on? One explanation is that the increased activity prior to the saccades was a consequence of attention shifting to the location inside the receptive field of the neuron. Another possibility is that the enhanced response was a premotor signal related to coding for the subsequent eye movement, just as neurons in motor cortex fire before hand movements. To address this possibility, they performed a variation on the experiment in which the behavioral response was changed from a saccade to a hand move-

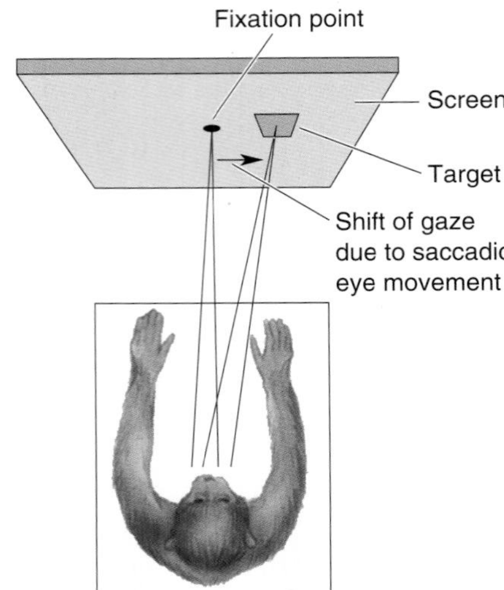

Figure 20.21
A behavioral task for directing a monkey's attention. While recordings are made from the posterior parietal cortex, the monkey fixates on a point on a computer screen. When a peripheral target appears (usually in a neuron's receptive field), the animal makes a saccade to the target. (Source: Adapted from Wurtz, Goldberg, and Robinson, 1982, p. 128.)

ment (Figure 20.22d). Again, there was an enhanced response to the target in the receptive field. This shows that the response enhancement was not a premotor signal for a saccade but rather was related to attention.

It isn't difficult to see how response enhancement of the kind observed in posterior parietal cortex could be involved in the behavioral effects of attention discussed earlier. That attention drawn to one location in the visual field by a cueing stimulus increases the response to other stimuli near that location could account for the spatially selective enhancement in the ability to detect a target. Likewise, it is conceivable that an increased response could lead to more rapid visual processing and ultimately faster reaction times, as seen in the perceptual experiments.

Receptive Field Changes in Area V4. In a fascinating series of experiments, Robert Desimone and his colleagues at the National Institutes of Mental Health have revealed surprisingly specific effects of attention on the response properties of neurons in visual cortical area V4. In one experiment, monkeys performed a same-different task with pairs of stimuli within the receptive fields of V4 neurons. As an example, suppose a particular V4 cell responded strongly to vertical and horizontal red bars of light in its receptive field but did not respond to vertical or horizontal green bars. The red bars were "effective" stimuli and the green bars were "ineffective" stimuli. While the monkey fixated, two stimuli (each either effective or ineffective) were briefly presented at different locations in the receptive field, and after a delay period, two more stimuli were presented at the same locations. In an experimental session, the animal was instructed to base its same-different judgments on the successive stimuli at one of the two locations within the receptive field. In other words, to perform the task, the animal had to pay attention to one location in the receptive field but not the other. The animal pushed a lever one way with its hand if the successive stimuli at the attended location were the same and the opposite way if the stimuli were different.

Consider what happened in a trial when effective stimuli appeared at the attended location and ineffective stimuli appeared at the other location (Figure 20.23a). Not surprisingly, the V4 neuron responded strongly in this situation because there were perfectly good effective stimuli in the receptive field. Suppose the monkey was then instructed to base its same-different judgments on the stimuli at the other location in the receptive field (Figure 20.23b). At this location, only green ineffective stimuli were presented. The

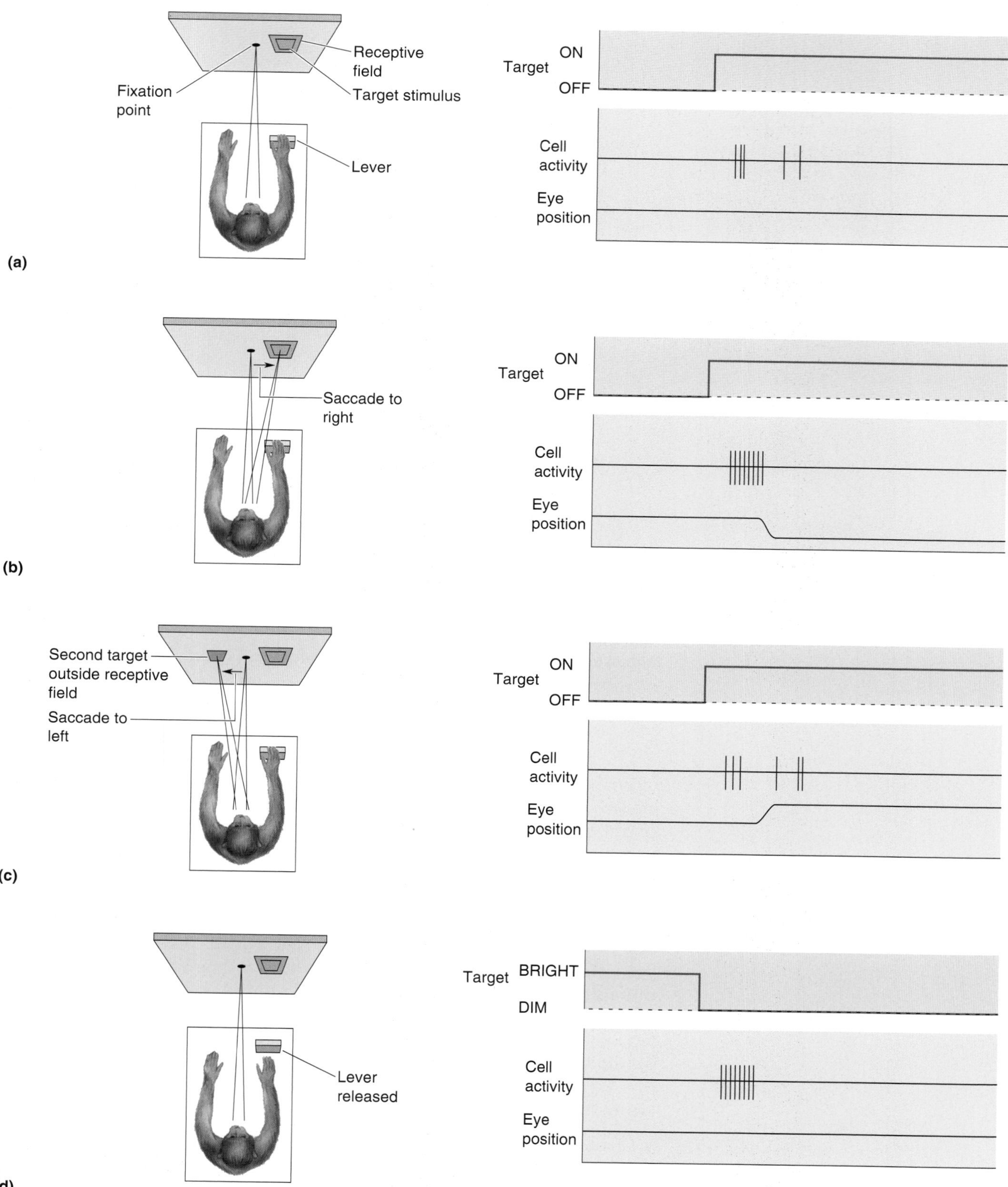

Figure 20.22

The effect of attention on the response of a neuron in posterior parietal cortex. (a) A neuron in posterior parietal cortex responds to a target stimulus in its receptive field. **(b)** The response is enhanced if the target presentation is followed by a saccade to the target. **(c)** The enhancement effect is spatially selective, as it is not seen if a saccade is made to a stimulus not in the receptive field. **(d)** Enhancement is also seen when the task requires the animal to release a hand lever when the peripheral spot dims. (Source: Adapted from Wurtz, Goldberg, and Robinson, 1982, p. 128.)

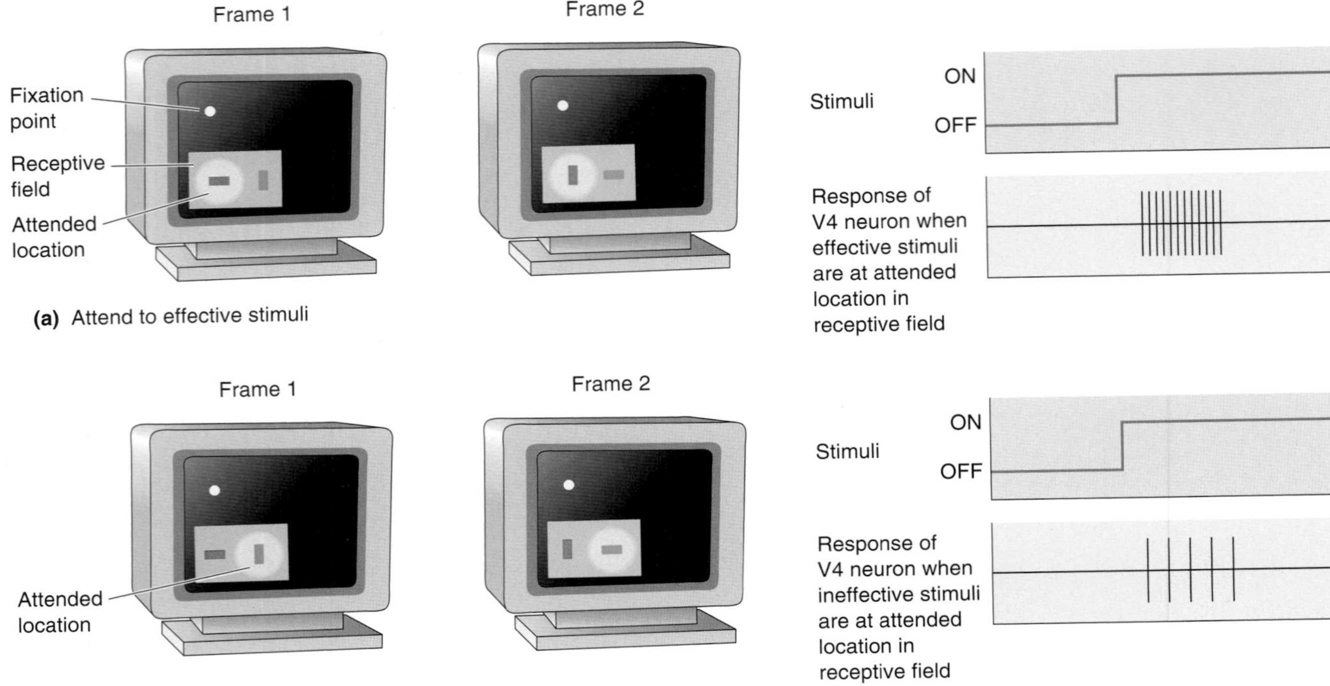

Figure 20.23
Stimuli used to study attentional effects on neurons in visual cortical area V4. The yellow circle indicates whether the monkey is attending to **(a)** the left or **(b)** the right location in the receptive field. For this neuron, red bars of light are effective in producing a response, and green bars are ineffective.

response of the neuron should have been the same as before because exactly the same stimuli were in the receptive field, right? Wrong! Even though the stimuli were identical, on average the responses of V4 neurons were less than half as great when the animal attended to the area in the neuron's receptive field containing the ineffective stimuli. It's as if the receptive field contracted around the attended area, decreasing the response to the effective stimuli at the unattended location. The location specificity of attention observed in this experiment may be directly related to that discussed earlier in the human perceptual experiments. Perceptually, detection is enhanced at attended locations relative to unattended locations. It's not too great a conceptual leap to imagine that the difference in ease of detection at the attended and unattended locations is based on the higher activity evoked by effective stimuli at the attended location.

How Is Attention Directed?

We have discussed the effects of attention on the responses of neurons in several cortical areas, and similar effects have been reported in other areas. But what is controlling attention? At present, there is no clear answer to this question. Certainly, the specificity observed in the physiological studies puts considerable demand on the neural mechanisms guiding attention.

One structure that has been studied for its possible role in guiding attention is the *pulvinar nucleus* of the thalamus. Several properties of the pulvinar make it interesting. For example, it has reciprocal connections with most visual cortical areas of the occipital, parietal, and temporal lobes, giving it the potential to modulate widespread cortical activity (Figure 20.24). Also, hu-

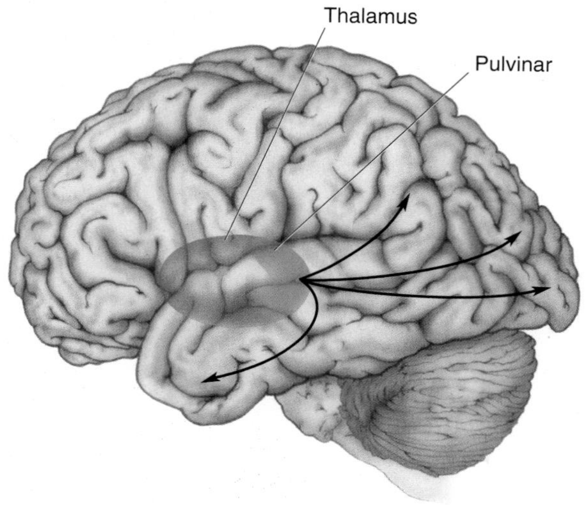

Figure 20.24
Pulvinar projections to the cortex. The pulvinar nucleus is in the posterior thalamus. It sends widespread efferents to areas of cerebral cortex, including areas V1, V2, MT, parietal cortex, and inferior temporal cortex.

mans with pulvinar lesions respond abnormally slowly to stimuli on the contralateral side, particularly when there are competing stimuli on the ipsilateral side. It has been proposed that such a deficit reflects a reduced ability to focus attention on objects in the contralateral visual field. A similar phenomenon has been observed in monkeys. When muscimol, an agonist of the inhibitory neurotransmitter gamma-aminobutyric acid (GABA), is unilaterally injected into the pulvinar, the activity of the neurons is suppressed. Behaviorally, the injection produces difficulty in shifting attention to contralateral stimuli, which seems similar to the effect of pulvinar lesions in humans. Interestingly, injection of the GABA antagonist bicuculline appears to facilitate shifting attention to the contralateral side.

There are several problems with the hypothesis that the pulvinar directs attention. Unilateral deactivation of the superior colliculus or posterior parietal cortex has behavioral effects similar to deactivation of the pulvinar. Also, bilateral pulvinar lesions do not have a large effect on attention, as would be expected if it were a critical part of an attention system.

Although from human and animal studies there is interesting evidence suggesting that the pulvinar is involved in attention, it is not any more clearly involved than the superior colliculus or the posterior parietal cortex. Perhaps these structures work together to guide attention. We must also consider, however, the possibility that there is no central control for attention but that attention is a property of the interactions between many areas.

CONCLUDING REMARKS

After study of the brain systems responsible for sensation and motor control, it is easy to forget that the human brain is much more than a passive information processor. In this chapter, we've explored several aspects of higher brain function that demonstrate the opposite. For example, through the process of selective attention, we dynamically allocate our mental energies to perform chosen tasks while ignoring other stimuli that might interfere. The effects of attention even include changes in receptive field properties that not long ago were considered static. A different sort of adaptability is seen in human language. Specialized language areas have evolved in the human brain that endow us with an incredibly flexible and creative system for communication.

Studies of language and attention represent a growing field of research known as *cognitive neuroscience*. Brain activity associated with such higher functions would be impossible to observe without recent technological advances. Attention has to be studied in awake animals, and this has become possible at the level of single neurons only with the development of painless techniques for recording from the brains of awake monkeys. Both attention and language can now be studied in the human brain by looking at the activity of large populations of neurons with PET and fMRI.

KEY TERMS

Language and the Brain
aphasia (p. 640)
Wada procedure (p. 641)
Broca's area (p. 642)
Wernicke's area (p. 642)
Broca's aphasia (p. 642)
anomia (p. 643)
Wernicke's aphasia (p. 645)
Wernicke-Geschwind model (p. 647)
conduction aphasia (p. 649)
split-brain study (p. 650)
corpus callosum (p. 651)
planum temporale (p. 655)

Attention
neglect syndrome (p. 664)

REVIEW QUESTIONS

1. How is it possible for a split-brain human to speak intelligibly if the left hemisphere controls speech? Isn't this inconsistent with the fact that the left hemisphere must direct motor cortex in both hemispheres to coordinate movements of the mouth?
2. What can you conclude about the normal function of Broca's area from the observation that there are usually some comprehension deficits in Broca's aphasia? Must Broca's area itself be directly involved in comprehension?
3. Pigeons can be trained to press one button when they want food and to press other buttons when they see particular visual stimuli. This means the bird can look around and "name" things it sees. How would you determine whether or not the pigeon is using a new language—"buttonese"?
4. What does the Wernicke-Geschwind language processing model explain? What data are inconsistent with this model?
5. What are the differences between the conscious states of a human with neglect syndrome and a split-brain human who can describe things only in the right visual field?
6. The following headline appeared in the newspaper: "Drunk Gets Nine Months in Violin Case." This phrase has two reasonable interpretations, and somewhere your brain activity must change when you switch interpretations. What kind of experiment would you use to look for changing activity correlated with the cognitive change? Where would you look?
7. How would you use fMRI or PET imaging to look for brain areas involved in directing selective attention in humans? What would you conclude if your experiment did not locate any such areas?
8. What neural mechanism or mechanisms could be responsible for the receptive field changes observed in area V4 in response to shifts in attention?

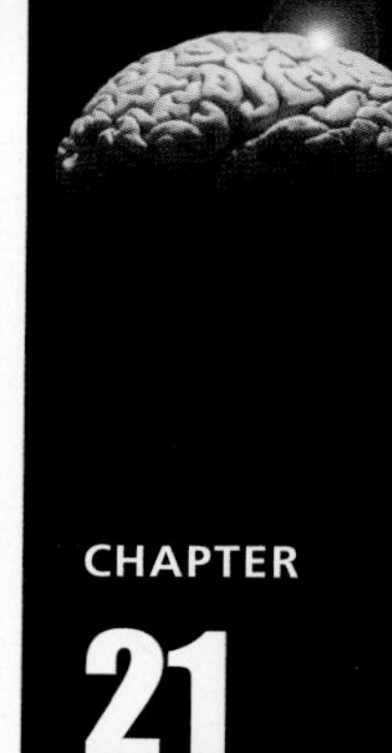

CHAPTER

Mental Illness

21

INTRODUCTION

MENTAL ILLNESS AND THE BRAIN

- Psychosocial Approaches to Mental Illness
- Biological Approaches to Mental Illness

ANXIETY DISORDERS

- A Description of Anxiety Disorders
  - Panic Disorder
  - Agoraphobia
    - Box 21.1 *Of Special Interest*: Agoraphobia With Panic Attacks
  - Obsessive-Compulsive Disorder
- Biological Bases of Anxiety Disorders
  - The Stress Response
  - Regulation of the HPA Axis by the Amygdala and Hippocampus
- Treatments for Anxiety Disorders
  - Psychotherapy
  - Anxiolytic Medications

MOOD DISORDERS

- A Description of Mood Disorders
  - Depression
  - Bipolar Disorder
    - Box 21.2 *Of Special Interest:* A Magical Orange Grove in a Nightmare
- Biological Bases of Mood Disorders
  - The Monoamine Hypothesis
  - The Diathesis-Stress Hypothesis
    - Box 21.3 *Path of Discovery*: Stress and Mood, by Charles Nemeroff
- Treatments for Mood Disorders
  - Electroconvulsive Therapy
  - Psychotherapy
  - Antidepressants
  - Lithium

SCHIZOPHRENIA

- A Description of Schizophrenia
- Biological Bases of Schizophrenia
  - Genes and the Environment
  - The Dopamine Hypothesis
  - The Glutamate Hypothesis
- Treatments for Schizophrenia

CONCLUDING REMARKS

INTRODUCTION

Neurology is a branch of medicine concerned with the diagnosis and treatment of nervous system disorders. We have discussed many neurological disorders in this book, ranging from multiple sclerosis to aphasia. In addition to the fact that they are fascinating in their own right, neurological disorders help illustrate the role of physiological processes in normal brain function—the importance of myelin for action potential conduction and the role of the frontal lobe in language, for example.

Psychiatry has a different focus. This branch of medicine is concerned with the diagnosis and treatment of disorders that affect the *mind*, or *psyche*. Aspects of brain function that are disturbed by mental illness—our fears, moods, and thoughts—were once thought to be beyond the reach of neuroscience. But as we have seen in the earlier chapters of Part III, many higher brain functions have begun to yield their secrets. There is now real hope that neuroscience will also solve the riddle of mental illness.

In this chapter, we discuss some of the most severe and prevalent psychiatric disorders: the anxiety disorders, the mood disorders, and schizophrenia. Once again we will see that a great deal can be learned about the nervous system by studying what happens when things go wrong.

MENTAL ILLNESS AND THE BRAIN

Human behavior is the product of brain activity, and the brain is the product of two interacting factors: heredity and environment. Obviously, one important determinant of your individualism is your complement of DNA, which, unless you have an identical twin, is unique. This means that physically your brain, like your fingerprint, is different from all others. A second factor that makes your brain unique is your history of personal experience. Experiences include trauma and disease, but as we saw in the case of somatosensory map plasticity (Chapter 12), the sensory environment itself can leave a permanent mark on the brain. (We'll return to this theme in Part IV, when we discuss development, learning, and memory.) Thus, despite the gross physical similarities you may share with a genetic twin, at a fine scale neither your brains nor your behaviors are identical. To complicate matters further, variations in genetic makeup and experience make the brain differentially susceptible to modification by subsequent experiences. It is these genetic and experiential variations, all ultimately expressed as physical changes in the brain, that give rise to the full range of behaviors exhibited by the human population.

Health and illness are relative, falling along a continuum of bodily function, and the same can be said for mental health and mental illness. While we all have our odd traits, the point at which an individual is said to be "mentally ill" is when the person has a diagnosable disorder of thought, mood, or behavior that causes distress or impaired functioning. An unfortunate legacy of our past ignorance about brain function is the common distinction drawn between "physical" and "mental" health. The philosophical roots of this distinction can be traced to Descartes' proposed separation of body and mind (see Chapter 1). Disorders of the body (which for Descartes included the brain) had an organic basis and were the concern of physicians and medicine. Disorders of the mind, on the other hand, were spiritual or moral and were the concern of clergy and religion. That most disorders of mood, thought, and behavior have, until very recently, remained resistant to biological explanations or treatments has reinforced this dichotomy.

Psychosocial Approaches to Mental Illness

An important advance in the secularization of mental illness was the emergence of the medical discipline of psychiatry, devoted to treating disorders of human behavior. Sigmund Freud (1856–1939) had an enormous influence on the new field, especially in the United States (Figure 21.1). Freud's theory of *psychoanalysis* was based on two major assumptions: (1) that much of mental life is unconscious (beyond awareness), and (2) that past experiences, particularly in childhood, shape how a person feels and responds throughout life. According to Freud, mental illness results when the unconscious and conscious elements of the psyche come into conflict. The way to resolve the conflict and to treat the illness is to help the patient unearth the secrets of the unconscious. Often these dark secrets are related to incidents (e.g., physical, mental, or sexual abuse) that occurred in childhood and were suppressed from consciousness.

A different theory of personality, championed by Harvard University psychologist B. F. Skinner (1904–1990), is based on the assumption that many behaviors are learned responses to the environment. *Behaviorism* rejects the notions of underlying conflicts and the unconscious and focuses instead on observable behaviors and their control by the environment. In Chapter 16, we

Figure 21.1
Sigmund Freud.

learned about some of the forces that motivate behavior. The probability of a type of behavior increases when it satisfies a craving or produces a pleasurable sensation (positive reinforcement), and it decreases when the consequences are deemed unpleasant or unsatisfactory (negative reinforcement). According to this theory, mental disorders may represent maladaptive behaviors that are learned. Treatment consists of active attempts to "unlearn" through behavior modification, either by introducing new types of behavioral reinforcement or by providing an opportunity to observe and recognize behavioral responses that are appropriate.

Such "psychosocial" approaches to treating mental illness have a sound neurobiological basis. The brain is structurally modified through learning and early experience, and these modifications alter behavioral responses. Treatment relies on *psychotherapy*, the use of verbal communication to help the patient. Of course, "talk therapy" is not appropriate for all mental disorders, any more than a particular type of antibiotic is appropriate for all infections. Until the recent revolution in biological psychiatry, however, variations in psychotherapy were the only tools available to psychiatrists. Moreover, despite the shift in "blame" away from one's moral character and toward early childhood experience, psychotherapy contributed to the stigma that mental illness (in contrast to physical illness) could be overcome by willpower alone. Freud himself recognized the shortcomings of psychotherapy, writing in 1920 that the "deficiencies in our [the psychoanalytic] description would probably vanish if we were already in a position to replace the psychological terms by physiological or chemical ones." Now, nearly a century later, neuroscience has advanced to a point where this goal seems attainable.

Biological Approaches to Mental Illness

A spectacular success in the early biological diagnosis and treatment of mental illness actually occurred in Freud's time. A major psychiatric disorder at the turn of the twentieth century, called *general paresis of the insane*, afflicted 10–15% of institutionalized psychiatric patients. The disorder had a progressive course, starting with symptoms of mania—excitement, euphoria, and grandiose delusions—and evolved to cognitive deterioration and, ultimately, paralysis and death. Initially blamed on psychological factors, the cause eventually was traced to infection of the brain with *Treponema pallidum*, the microorganism that causes syphilis. Once the cause was established, increasingly effective treatments quickly followed. By 1910, German microbiologist Paul Ehrlich had established that the drug arsphenamine could act as a "magic bullet," killing the *T. pallidum* in the blood without damaging its human host. Eventually, the antibiotic penicillin (discovered in 1928 by British microbiologist Alexander Fleming) was found to be so effective in killing *T. pallidum* that established brain infections could be eradicated. Thus, when penicillin became widely available by the end of World War II, a major psychiatric disorder was virtually eliminated.

A number of other mental illnesses can be traced directly to biological causes. For example, a dietary deficiency in niacin (a B vitamin) can cause agitation, impaired reasoning, and depression. The penetration of human immunodeficiency virus (HIV) into the brain causes progressive cognitive and behavioral impairments. Recently, a form of obsessive-compulsive disorder (discussed later in the chapter) has been linked to an autoimmune response triggered by streptococcal pharyngitis (strep throat) in children. Understanding the causes of these diseases will lead to treatments and, ultimately, cures of the associated mental disorders.

Of course, serious mental disorders also occur in well-nourished and infection-free individuals. Although the causes remain to be determined, it is safe to say that the roots of these disorders lie in altered brain anatomy, chemistry, and function. Next we explore the major psychiatric disorders and see how neuroscience has provided insight into their possible causes and contributed to their treatment.

ANXIETY DISORDERS

Fear is an adaptive response to threatening situations. As we learned in Chapter 18, fear is expressed by the autonomic fight-or-flight response, mediated by the sympathetic division of the autonomic nervous system (ANS) (see Chapter 15). Many fears are innate and species specific. A mouse does not have to be taught to fear a cat. But fear is also learned. One touch is usually all it takes to cause a horse to fear an electric fence. The adaptive value of fear is obvious. As the old aviation saying goes, "There are old pilots, and there are bold pilots, but there are no old bold pilots." But fear is not an appropriate or adaptive response in all circumstances. The inappropriate expression of fear characterizes the *anxiety disorders*, the most common of psychiatric disorders.

A Description of Anxiety Disorders

It has been estimated that in any given year, more than 15% of Americans will suffer from one of the recognized anxiety disorders listed in Table 21.1. Although they differ in terms of the real or imagined stimuli that evoke the anxiety and the behavioral responses the individual uses to attempt to reduce it, these disorders have in common the pathological expression of fear.

Panic Disorder. Panic attacks are sudden feelings of intense terror that occur without warning. The symptoms include palpitations, sweating, trembling, shortness of breath, chest pain, nausea, dizziness, tingling sensations, and chills or blushing. Most people report an overwhelming fear that they are dying or going crazy and flee from the place where the attack begins, often seeking emergency medical assistance. The attacks are short-lived, however, usually lasting less than 30 minutes. Panic attacks can occur in response to specific stimuli and can be a feature of a number of anxiety disorders, but

Table 21.1 Anxiety Disorders

NAME	DESCRIPTION
Panic disorder	Frequent panic attacks consisting of discrete periods with sudden onset of intense apprehension, fearfulness, or terror, often associated with feelings of impending doom
Agoraphobia	Anxiety about or the avoidance of places or situations from which escape might be difficult or embarrassing or in which help may not be available in the event of a panic attack
Obsessive-compulsive disorder	Characterized by obsessions, which cause marked anxiety or distress, and/or compulsions, which neutralize anxiety
Generalized anxiety disorder	At least 6 months of persistent and excessive anxiety and worry
Specific phobia	Clinically significant anxiety provoked by exposure to a specific feared object or situation, often leading to avoidance behavior
Social phobia	Clinically significant anxiety provoked by exposure to certain types of social or performance situations, often leading to avoidance behavior
Posttraumatic stress disorder	Reliving of an extremely traumatic event accompanied by symptoms of increased arousal and by avoidance of stimuli associated with the trauma

Source: Adapted from American Psychiatric Association, 1994.

they can also occur spontaneously. The condition that psychiatrists call *panic disorder* is characterized by recurring, seemingly unprovoked panic attacks and a persistent worry about having further attacks. About 2% of the population suffer from panic disorder, which is twice as common in women as in men. Its onset is most common after adolescence but before age 50. Half of the individuals who have panic disorder also have major depression (discussed later), and 25% of them become alcoholic or develop other substance abuse problems.

Agoraphobia. Severe anxiety about being in situations in which escape might be difficult or embarrassing is characteristic of *agoraphobia* (from the Greek for "fear of an open marketplace"). The anxiety leads to avoidance of situations irrationally perceived as threatening, such as being alone outside the home, in a crowd of people, in a car or airplane, or on a bridge or elevator. Agoraphobia is often one adverse outcome of panic disorder, as the situation in Box 21.1 describes. About 5% of the population are agoraphobic, with the incidence among women again being twice that of men.

Obsessive-Compulsive Disorder. Obsessions are recurrent, intrusive thoughts or impulses that are perceived as being inappropriate, grotesque, or forbidden. Common themes are thoughts of contamination with germs or body fluids, thoughts that the sufferer has unknowingly caused harm to someone, and violent or sexual impulses. These thoughts are recognized as being foreign, and they evoke considerable anxiety. Compulsions are repetitive behaviors or mental acts that are performed to reduce the anxiety associated with obsessions. Examples are repeated hand-washing, counting, and checking to make sure that something is not out of place. *Obsessive-compulsive disorder* (*OCD*) is not rare; it affects more than 2% of the population, with an equal incidence among men and women. OCD usually appears in young adult life, and the symptoms fluctuate in response to stress levels.

Biological Bases of Anxiety Disorders

A genetic predisposition for many anxiety disorders has been established, although the specific genes have not been identified. Other anxiety disorders appear to be rooted in stressful life events.

Fear is normally evoked by a threatening stimulus, called a *stressor*, and it is manifested by the stress response. As mentioned previously, the stimulus-response relationship can be strengthened by experience (recall the horse and the electric fence), but it can also be weakened. Consider, for example, the expert skier who no longer views a precipitous drop as fearful. A healthy person regulates the stress response through learning. The hallmark of the anxiety disorders is an inappropriate stress response either when a stressor is not present or when it is not immediately threatening. Thus, a key to understanding anxiety is to understand how the stress response is regulated by the brain.

The Stress Response. The stress response is the coordinated reaction to threatening stimuli. It is characterized by the following:

- Avoidance behavior
- Increased vigilance and arousal
- Activation of the sympathetic division of the ANS
- Release of cortisol from the adrenal glands

It should come as no surprise that the hypothalamus is centrally involved in orchestrating the appropriate humoral, visceromotor, and somatic motor

Box 21.1

OF SPECIAL INTEREST

Agoraphobia With Panic Attacks

To appreciate the distress and disruption caused by anxiety disorders, consider the following case history from Nancy C. Andreasen's book, *The Broken Brain*.

> Greg Miller is a 27-year-old unmarried computer programmer. When asked about his main problem, he replied, "I am afraid to leave my house or drive my car."
>
> The patient's problems began approximately one year ago. At that time he was driving across the bridge that he must traverse every day in order to go to work. While driving in the midst of the whizzing six-lane traffic, he began to think (as he often did) about how awful it would be to have an accident on that bridge. His small, vulnerable VW convertible could be crumpled like an aluminum beer can, and he could die a bloody, painful death or be crippled for life. His car could even hurtle over the side of the bridge and plunge into the river.
>
> As he thought about these possibilities, he began to feel increasingly tense and anxious. He glanced back and forth at the cars on either side of him and became frightened that he might run into one of them. Then he experienced an overwhelming rush of fear and panic. His heart started pounding and he felt as if he were going to suffocate. He began to take deeper and deeper breaths, but this only increased his sense of suffocation. His chest felt tight and he wondered if he might be about to die of a heart attack. He certainly felt that something dreadful was going to happen to him quite soon. He stopped his car in the far right lane in order to try to regain control of his body and his feelings. Traffic piled up behind with many honking horns, and drivers pulled around him yelling obscenities. On top of his terror, he experienced mortification. After about three minutes, the feeling of panic slowly subsided, and he was able to proceed across the bridge and go to work. During the remainder of the day, however, he worried constantly about whether or not he would be able to make the return trip home across the bridge without a recurrence of the same crippling fear.
>
> He managed to do so that day, but during the next several weeks he would begin to experience anxiety as he approached the bridge, and on three or four occasions he had a recurrence of the crippling attack of panic. The panic attacks began to occur more frequently so that he had them daily. By this time he was overwhelmed with fear and began to stay home from work, calling in sick each day. He knew that his main symptom was an irrational fear of driving across the bridge, but he suspected that he might also have some type of heart problem. He saw his family doctor, who found no evidence of any serious medical illness, and who told him that his main problem was excessive anxiety. The physician prescribed a tranquilizer for him and told him to try to return to work.
>
> For the next six months, Greg struggled with his fear of driving across the bridge. He was usually unsuccessful and continued to miss a great deal of work. Finally, he was put on disability for a few months and told by the company doctor to seek psychiatric treatment. Greg was reluctant and embarrassed to do this, and instead he stayed home most of the time, reading books, listening to records, playing chess on his Apple computer, and doing various "handy-man" chores around the house. As long as he stayed home, he had few problems with anxiety or the dreadful attacks of panic. But when he tried to drive his car, even to the nearby shopping center, he would sometimes have panic attacks. Consequently, he found himself staying home nearly all the time and soon became essentially housebound. (Source: Andreasen, 1984, pp. 65–66.)

responses (see Chapter 16). To get an idea of how this response is regulated, let's focus our attention on the humoral response, which is mediated by the **hypothalamic-pituitary-adrenal (HPA) axis** (Figure 21.2).

As we learned in Chapter 15, the hormone cortisol, a glucocorticoid, is released from the adrenal cortex in response to an elevation in the blood level of **adrenocorticotropic hormone (ACTH)**. ACTH is released by the anterior pituitary gland in response to **corticotropin-releasing hormone (CRH)**. CRH is released into the blood of the portal circulation by parvocellular neurosecretory neurons in the paraventricular nucleus of the hypothalamus. Thus, this arm of the stress response can be traced to activation of the CRH-

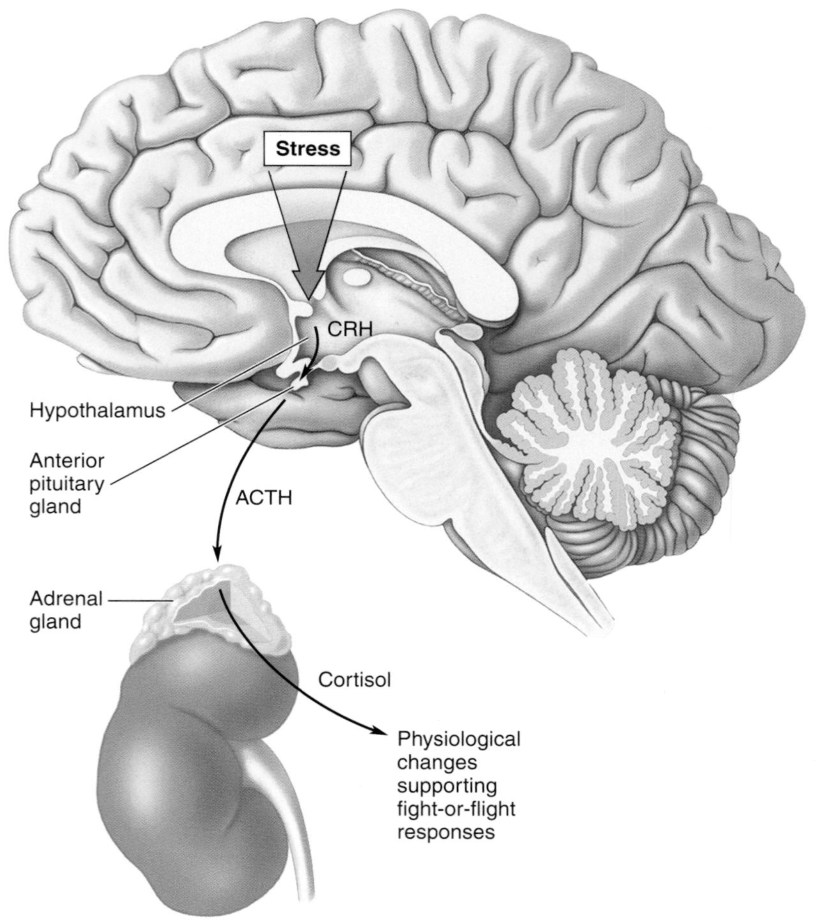

Figure 21.2
The hypothalamic-pituitary-adrenal axis. The HPA system regulates the secretion of cortisol from the adrenal gland in response to stress. CRH is the chemical messenger between the paraventricular nucleus of the hypothalamus and the anterior pituitary gland. ACTH released by the pituitary gland travels in the bloodstream to the adrenal gland lying atop the kidney, where it stimulates cortisol release. Cortisol contributes to the body's physiological response to stress.

containing neurons of the hypothalamus. Much can be learned about anxiety disorders by understanding how the activity of these neurons is regulated.

Regulation of the HPA Axis by the Amygdala and Hippocampus. The CRH neurons of the hypothalamus are regulated by two structures that were introduced in earlier chapters: the *amygdala* and the *hippocampus* (Figure 21.3). As we learned in Chapter 18, the amygdala is critical to fear responses. Sensory information enters the basolateral amygdala, where it is processed and relayed to neurons in the central nucleus. When the central nucleus of the amygdala becomes active, the stress response ensues (Figure 21.4). Inappropriate activation of the amygdala, as measured using functional magnetic resonance imaging (fMRI) (see Box 7.2), has been associated with some anxiety disorders. Downstream from the amygdala is a collection of neurons called the *bed nucleus of the stria terminalis*. The bed nucleus neurons activate the HPA axis and the stress response.

The HPA axis is also regulated by the hippocampus. Hippocampal activation suppresses rather than stimulates CRH release. The hippocampus contains numerous **glucocorticoid receptors** that respond to the cortisol released from the adrenal gland in response to HPA system activation. Thus, the hippocampus normally participates in the feedback regulation of the HPA axis by inhibiting CRH release and the subsequent release of ACTH and cortisol when circulating cortisol levels get too high. Continuous exposure to cortisol, however, as occurs during periods of chronic stress, can cause hippocampal neurons to wither and die in experimental animals (see Box 15.1). This de-

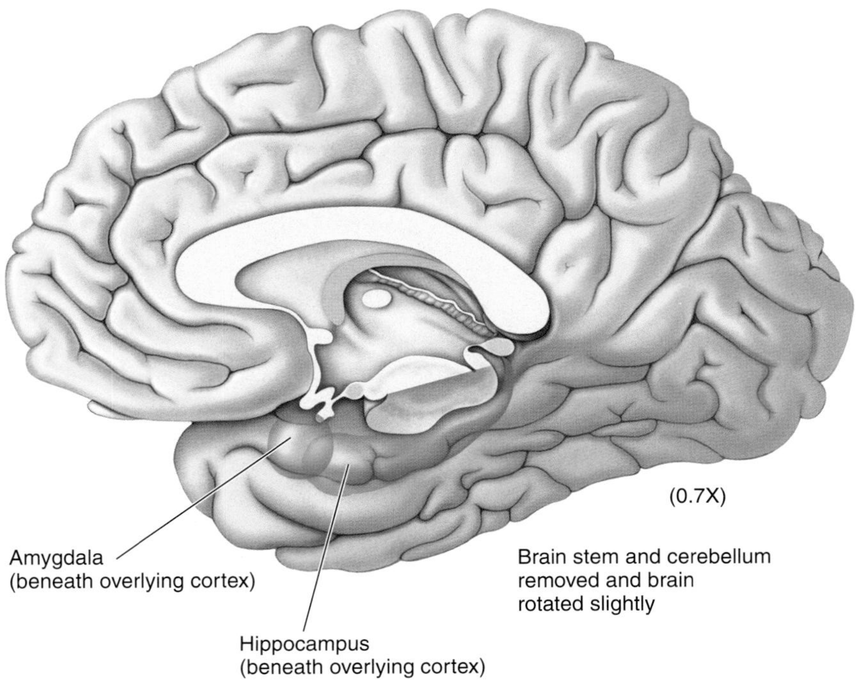

Figure 21.3
The location of the amygdala and hippocampus.

generation of the hippocampus sets off a vicious cycle in which the stress response becomes more pronounced, leading to even greater cortisol release and more hippocampal damage. Human brain imaging studies have shown a decrease in the volume of the hippocampus in some people who have post-traumatic stress disorder, an anxiety disorder that is triggered by exposure to inescapable stress (Table 21.1).

In summary, the amygdala and the hippocampus regulate the HPA system and the stress response in a push-pull fashion (Figure 21.5). Anxiety disorders have been related to both hyperactivity of the amygdala and dimin-

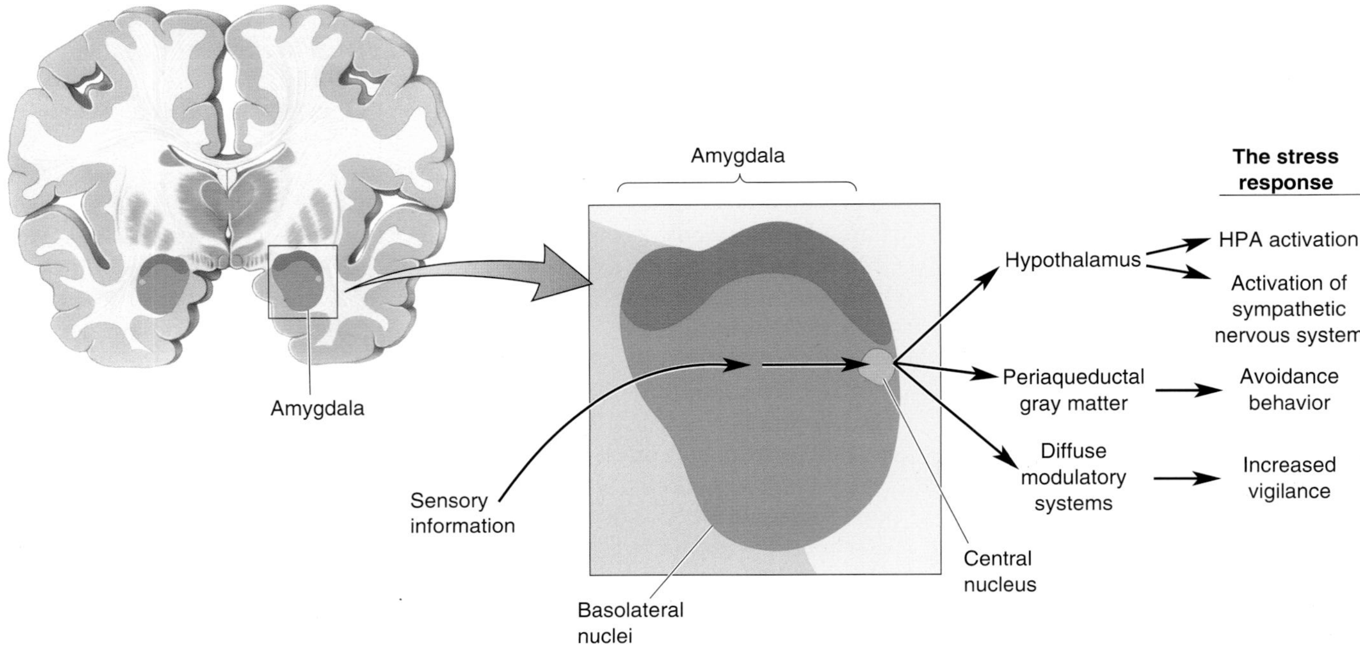

Figure 21.4
Control of the stress response by the amygdala. The amygdala receives ascending sensory information from the thalamus and descending inputs from the neocortex. This information is integrated by the basolateral nuclei and is relayed to the central nucleus. Activation of the central nucleus leads to the stress response.

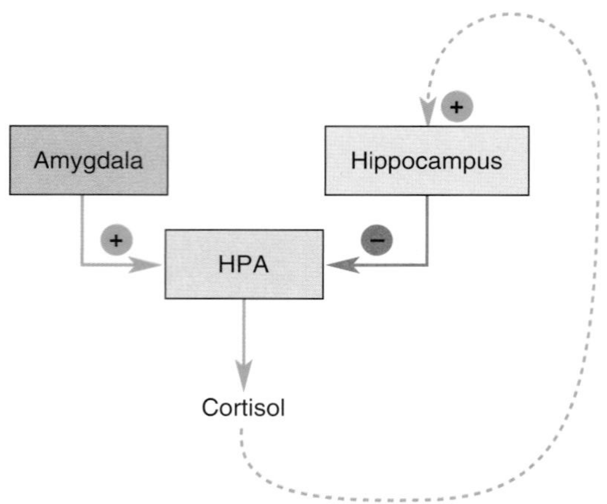

Figure 21.5
Push-pull regulation of the HPA axis by the amygdala and hippocampus. Amygdala activation stimulates the HPA system and the stress response (green lines). Hippocampal activation suppresses the HPA system (red line). The hippocampus has glucocorticoid receptors that are sensitive to circulating cortisol. Thus, the hippocampus is important in the feedback regulation of the HPA axis to prevent excessive cortisol release.

ished activity of the hippocampus. Keep in mind, however, that the amygdala and hippocampus both receive highly processed information from the neocortex. Indeed, another consistent finding in humans with anxiety disorders has been elevated activity of the prefrontal cortex.

Treatments for Anxiety Disorders

Several treatments for anxiety disorders are available. In many cases, patients respond well to psychotherapy and counseling; in other cases, specific medications are preferred.

Psychotherapy. Since there is a strong learning component to fear, psychotherapy can be an effective treatment for many of the anxiety disorders. The therapist gradually increases the exposure of the patient to the stimuli that produce anxiety, reinforcing the notion that the stimuli are not dangerous. At the neurobiological level, the aim of the psychotherapy is to alter connections in the brain so that the real or imagined stimuli no longer evoke the stress response.

Anxiolytic Medications. Drugs that reduce anxiety are described as anxiolytic (anxiety dissolving). All anxiolytic drugs act by altering chemical synaptic transmission in the brain. The major classes of drugs used in the treatment of anxiety disorders are the benzodiazepines and the serotonin-selective reuptake inhibitors.

Recall that gamma-aminobutyric acid (GABA) is an important inhibitory neurotransmitter in the brain. $GABA_A$ receptors are GABA-gated chloride channels that mediate fast inhibitory postsynaptic potentials (IPSPs) (see Chapter 6). The proper action of GABA is critical to the proper functioning of the brain: Too much inhibition results in coma, and too little results in seizures. In addition to its GABA binding site, the $GABA_A$ receptor contains sites where chemicals can act to powerfully modulate channel function. *Benzodiazepines* bind to one of these sites and make GABA much more effective in opening the channel and producing inhibition (Figure 21.6). The site on the receptor that binds benzodiazepines is believed to be used normally by a naturally occurring brain chemical, although the identity of the endogenous molecule has not been established. Benzodiazepines, of which Valium (diazepam) is perhaps the best known, are highly effective treatments for acute anxiety. Indeed, virtually all drugs that stimulate GABA actions are anxiolytic, including ethyl alcohol. A reduction in anxiety is likely to explain, at least in part, the widespread social use of alcohol. The anxiolytic effects of

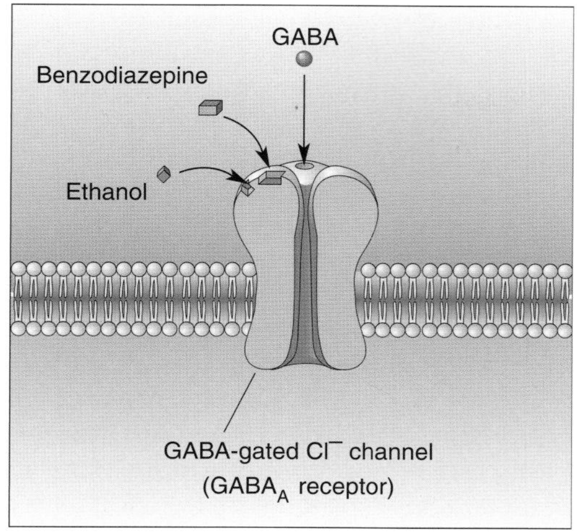

Figure 21.6
The action of benzodiazepine. Benzodiazepines bind to a site on the GABA$_A$ receptor that makes it much more responsive to GABA, the major inhibitory neurotransmitter in the forebrain.

alcohol are also an obvious reason that anxiety disorders and alcohol abuse often go hand in hand.

We may infer that the calming actions of the benzodiazepines are due to suppression of activity in the brain circuits used in the stress response. Benzodiazepine treatment may be required to restore normal function to these circuits. Indeed, a recent study of patients with panic disorder, using positron emission tomography (PET) (see Box 7.2), showed that the number of benzodiazepine binding sites was reduced in regions of frontal cortex that show hyperactive responsiveness during anxiety (Figure 21.7). These findings are exciting, not only because they may reveal the sites of benzodiazepine action in the brain but also because they suggest that an alteration in the endogenous regulation of GABA receptors is a cause of the anxiety disorder.

Serotonin-selective reuptake inhibitors (**SSRIs**), of which Prozac (fluoxetine) is the best known, are widely used in the treatment of mood disorders. The SSRIs are also highly effective for treating anxiety disorders, notably including OCD. Recall that serotonin is released throughout the brain by a diffuse modulatory system originating in the raphe nuclei of the brain stem. The actions of serotonin are mediated by G-protein-coupled receptors and are terminated by reuptake into the axon terminal. Thus, as the name implies, SSRIs prolong the actions of released serotonin at their receptors by inhibiting reuptake.

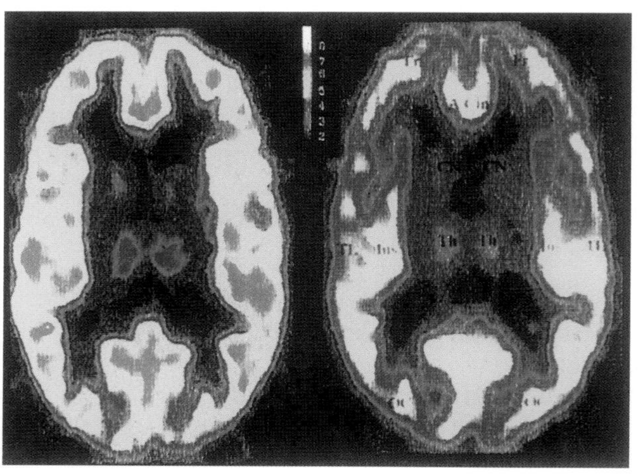

Figure 21.7
Diminished binding of radioactive benzodiazepine in a patient with panic disorder. PET scans in the horizontal plane of the brain of a healthy person (left) and the brain of a person suffering from panic disorder (right). The color code indicates the number of benzodiazepine binding sites in the brain (hot colors indicate more; cool colors indicate fewer). The frontal cortex at the top of the scan shows many fewer binding sites in the individual with panic disorder. (Source: Malizia et al., 1998, Fig. 1.)

Unlike the benzodiazepines, however, the anxiolytic actions of the SSRIs are not immediate. Therapeutic effects develop slowly over a period of weeks in response to regular daily dosing. This finding means that the immediate rise in extracellular serotonin caused by the SSRI is *not* responsible for the anxiolytic effect. Rather, the effect appears to be due to an adaptation of the nervous system to chronically elevated brain serotonin. We will return to a discussion of the actions of SSRIs when we discuss depression. In the context of anxiety disorders, however, it is very interesting that one adaptive response to SSRIs is an increase in the glucocorticoid receptors in the hippocampus. SSRIs may dampen anxiety by enhancing the feedback regulation of the CRH neurons in the hypothalamus (Figure 21.5).

Although the benzodiazepines and SSRIs have proven to be effective in treating a wide variety of anxiety disorders, novel drugs are being developed based on our new understanding of the stress response. One very promising drug target is the receptors for CRH. Not only is CRH used by hypothalamic neurons to control ACTH release from the pituitary, it is also used as a neurotransmitter in some of the central circuits involved in the stress response. For example, some neurons of the central nucleus of the amygdala contain CRH, and injections of CRH into the brain can produce the full-blown stress response and signs of anxiety. Thus, there is hope that antagonists of CRH receptors will be useful for the treatment of anxiety disorders.

MOOD DISORDERS

Affect is the medical term for emotional state or mood; **affective disorders** are disorders of mood. In a given year, more than 7% of the population will suffer from one of the mood disorders.

A Description of Mood Disorders

An occasional, brief feeling of depression—getting "the blues"—is a common response to life's events, such as suffering a loss or disappointment, and we can hardly call this a disorder. The affective disorder that psychiatrists and psychologists call *depression* is, however, something more prolonged and much more severe, characterized by a feeling that one's emotional state is no longer under one's control. Major depression can occur suddenly, often without obvious external cause, and if left untreated, usually lasts 4–12 months. Depression is a serious disease. It is a main precipitating cause of suicide, which claims more than 30,000 lives each year in the United States. Depression is also widespread. Perhaps as many as 20% of the population suffer a major, incapacitating episode of depression during their lifetime. In a subset of patients with *bipolar disorder*, bouts of depression are punctuated with emotional highs that can also be highly disruptive.

Depression. The mental illness known as **major depression** is the most common mood disorder, affecting 5% of the population in a year. The cardinal symptoms are lowered mood and decreased interest or pleasure in all activities. To be diagnosed with major depression, these symptoms must be present every day for at least 2 weeks and not be obviously related to bereavement. Other symptoms also occur:

- Loss of appetite or increased appetite
- Insomnia or hypersomnia
- Fatigue
- Feelings of worthlessness and guilt
- Diminished ability to concentrate
- Recurrent thoughts of death

Episodes of major depression rarely last longer than 2 years. Without treatment, however, depressions recur in 50% of cases, and after three or more episodes, the odds of recurrence increase to more than 70%. Another expression of depression afflicting 2% of the adult population is called *dysthymia*. Although milder than major depression, dysthymia has a chronic, "smoldering" course, and it seldom disappears spontaneously. Major depression and dysthymia are twice as common in women as in men.

Bipolar Disorder. Like major depression, **bipolar disorder** is a recurrent mood disorder. It consists of repeated episodes of mania or mixed episodes of mania and depression and therefore is also called *manic-depressive disorder*. **Mania** (derived from a French word for "crazed" or "frenzied") is a distinct period of abnormally and persistently elevated, expansive, or irritable mood. During the manic phase, other common symptoms include the following:

- Inflated self-esteem or grandiosity
- A decreased need for sleep
- Increased talkativeness or feelings of pressure to keep talking
- Flight of ideas or a subjective experience that thoughts are racing
- Distractibility
- Increased goal-directed activity

Another symptom is impaired judgment. Spending sprees, offensive or disinhibited behavior, promiscuity, and other reckless behaviors are common.

According to current diagnostic criteria, there are two types of bipolar disorder. Type I bipolar disorder, which occurs in about 1% of the population and equally among men and women, is characterized by the manic episodes just described, with or without incidents of major depression. Type II bipolar disorder, affecting about 0.6% of the population, is characterized by *hypomania*, a milder form of mania that is not associated with marked impairments in judgment or performance. Indeed, hypomania in some may take the form of a marked increase in efficiency, accomplishment, or creativity (Box 21.2). Type II bipolar disorder is, however, also always associated with episodes of major depression. When hypomania alternates with periods of depression that are not severe enough to warrant the description "major" (i.e., fewer symptoms and of shorter duration), the disorder is called *cyclothymia*.

Biological Bases of Mood Disorders

Like most other mental illnesses, mood disorders reflect the altered functioning of many parts of the brain at the same time. How else can we explain the coexistence of symptoms ranging from eating and sleeping disorders to loss of the ability to concentrate? For this reason, much attention has focused on the role of the diffuse modulatory systems, with their wide reach and diverse effects. In the past few years, however, disruption of the HPA system has also been implicated as playing an important role in depression. Let's take a closer look at the neurobiology of mood disorders.

The Monoamine Hypothesis. The first real indication that depression may result from a problem with the central diffuse modulatory systems came in the 1960s. The drug reserpine, introduced to control high blood pressure, caused psychotic depression in about 20% of patients. Reserpine depletes central catecholamines and serotonin by interfering with their loading into synaptic vesicles. Then it was found that another class of drugs introduced to treat tuberculosis caused a marked mood elevation. These drugs inhibit *monoamine oxidase* (*MAO*), the enzyme that destroys catecholamines and serotonin. Another piece of the puzzle fell into place when it was recognized

Box 21.2

OF SPECIAL INTEREST

A Magical Orange Grove in a Nightmare

Winston Churchill called it his "black dog."[1] The writer F. Scott Fitzgerald often found himself " . . . hating the night when I couldn't sleep and hating the day because it went toward night."[2] It was the most "terrible of all the evils of existence" for the composer Hector Berlioz.[3] They were speaking of their lifelong bouts with depression. From the Scottish poet Robert Burns to the American grunge rocker Kurt Cobain, extraordinarily creative people have suffered inordinately from affective disorders. Biographical studies of accomplished artists have been consistent and alarming; their estimated rates of major depression are about 10 times higher than in the general public, and rates of bipolar disorder may be up to 30 times higher.

Many artists have eloquently described their misfortunes. But can mood disorders actually reinforce great talent and creative productivity? Certainly most people with mood disorders are not artistic or unusually imaginative, and most artists are not manic-depressive. Artists with bipolar disorders can, however, sometimes draw vigor and inspiration from their condition. Edgar Allan Poe wrote of his cycles of depression and mania, "I am excessively slothful, and wonderfully industrious—by fits."[4] The poet Michael Drayton mused about "that fine madness . . . which rightly should possess a poet's brain."[5] Studies have suggested that hypomania can heighten certain cognitive processes, increase original and idiosyncratic thought, and even enhance linguistic skills. Manic states can also reduce the need for sleep, foster intense and obsessive concentration, create unmitigated self-confidence, and eliminate concern for social norms—just what you need, perhaps, to push the envelope of artistic creativity.

The poet's madness is much more often a scourge than an inspiration. For Robert Lowell, manic experiences were "a magical orange grove in a nightmare."[6] Virginia Woolf's husband described how "she talked almost without stopping for two or three days, paying no attention to anyone in the room or anything said to her."[7] It is hard to overstate the depths of melancholy that can accompany major depression. The suicide rate among accomplished poets is said to be 5 to 18 times higher than in the general population. Poet John Keats once wrote desperately, "I am in that temper that if I were under water I would scarcely kick to come to the top."[8] But when Keats' mood pitched the other way, he wrote most of his best poetry during a 9-month period in 1819 before dying of tuberculosis at age 25. Figure A shows how Robert Schumann's wildly fluctuating output of musical compositions coincided with the oscillations of his manic-depressive episodes.

The psychiatrist Kay Redfield Jamison has suggested that "depression is a view of the world through a dark glass, and mania is that seen through a kaleidoscope—often brilliant but fractured."[9] We are lucky now to have effective treatments for both conditions, for the dark glass and the kaleidoscope carry a heavy price.

[1] Quoted in Ludwig AM, *The Price of Greatness: Resolving the Creativity and Madness Controversy*. New York: Guilford Press, 1995, p. 174.
[2] F. Scott Fitzgerald, "The Crack-Up," in *The Crack-Up and Other Stories*. New York: New Directions, 1956, pp. 69-75.
[3] Hector Berlioz, *The Memoirs of Hector Berlioz*, trans. David Cairns. St. Albans, England: Granada, 1970, p. 142.
[4] Edgar Allan Poe, letter to James Russell Lowell, June 2, 1844, in *The Letters of Edgar Allan Poe*, Vol. 1, ed. John Wand Ostrom. Cambridge: Harvard University Press, 1948, p. 256.
[5] Michael Drayton, "To my dearly beloved Friend, Henry Reynolds, Esq.; of Poets and Poesy," lines 109-110, *The Works of Michael Drayton, Esq.*, vol. 4, London: W. Reeve, 1753.
[6] Ian Hamilton, *Robert Lowell: A Biography*. New York: Random House, 1982, p. 218.
[7] Leonard Woolf, *Beginning Again: An Autobiography of the Year 1911 to 1918*. New York: Harcourt Brace, 1964, pp. 172-173.
[8] Quoted by Kay Jamison in a presentation at the Depression and Related Affective Disorders Association/Johns Hopkins Symposium, Baltimore, Maryland, April 1997.
[9] Jamison KR, Manic-depressive illness and creativity. *Scientific American* 272: 62-67.

that the drug imipramine, introduced some years earlier as an antidepressant, inhibits the reuptake of released serotonin and norepinephrine, thus promoting their action in the synaptic cleft. From these observations came the hypothesis that mood is closely tied to the levels of released "monoamine" neurotransmitters—norepinephrine and/or serotonin—in the brain. According to this idea, called the **monoamine hypothesis of mood**

(Box 21.21, *continued*)

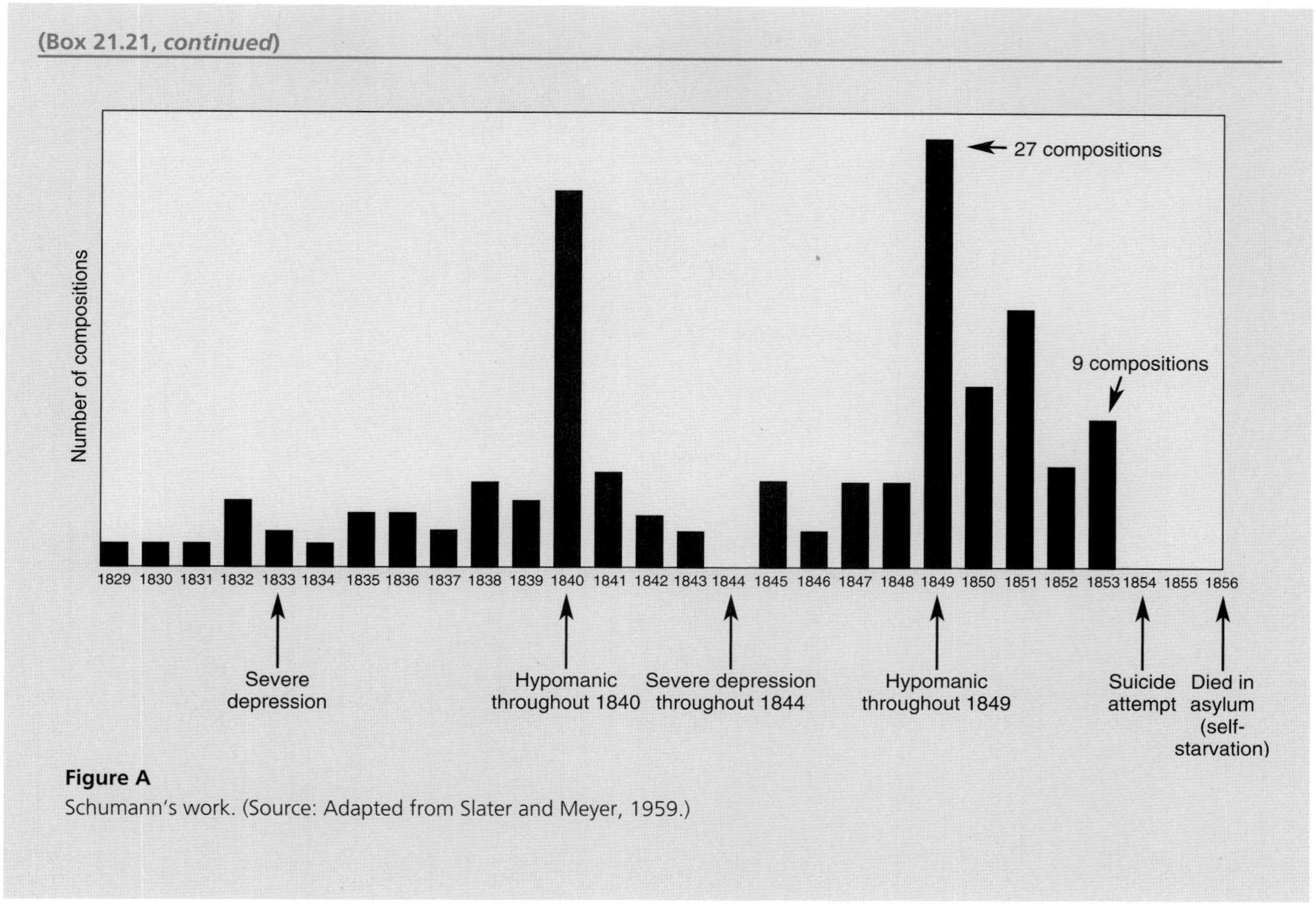

Figure A
Schumann's work. (Source: Adapted from Slater and Meyer, 1959.)

disorders, depression is a consequence of a deficit in one of these diffuse modulatory systems (Figure 21.8). Indeed, as we will see in a moment, many of the modern drug treatments for depression have in common enhanced neurotransmission at central serotonergic and/or noradrenergic synapses.

Unfortunately, there are problems with a straightforward equation between mood and modulator. Perhaps most striking is the clinical finding that the antidepressant action of all of these drugs takes several weeks to develop, even though they have almost immediate effects on transmission at the modulatory synapses. Another concern is that other drugs that raise norepinephrine levels in the synaptic cleft, such as cocaine, are not effective as antidepressants. A new hypothesis is that the effective drugs promote long-term adaptive changes in the brain that alleviate the depression. One adaptation occurs in the HPA axis, which has also been implicated in mood disorders.

The Diathesis-Stress Hypothesis. There is clear evidence that mood disorders run in families and that our genes predispose us to this type of mental illness. The medical term for a predisposition for a certain disease is *diathesis*. But it has also been established that early childhood abuse or neglect and life stresses are also important risk factors in the development of mood disorders in adults. In an attempt to bring these findings together, Charles Nemeroff

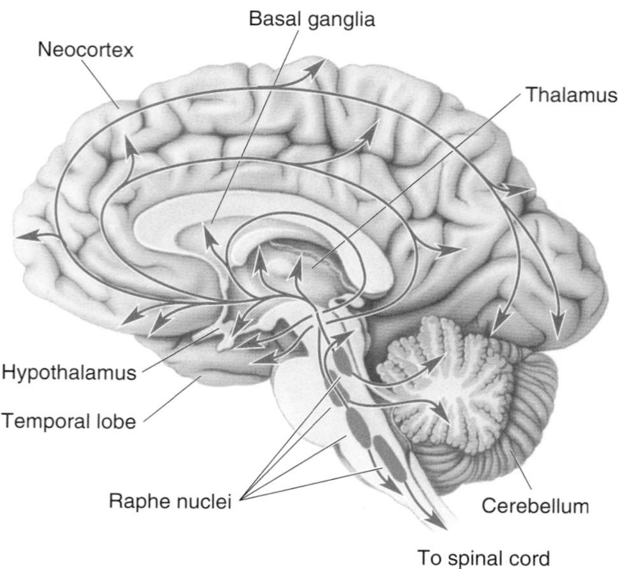

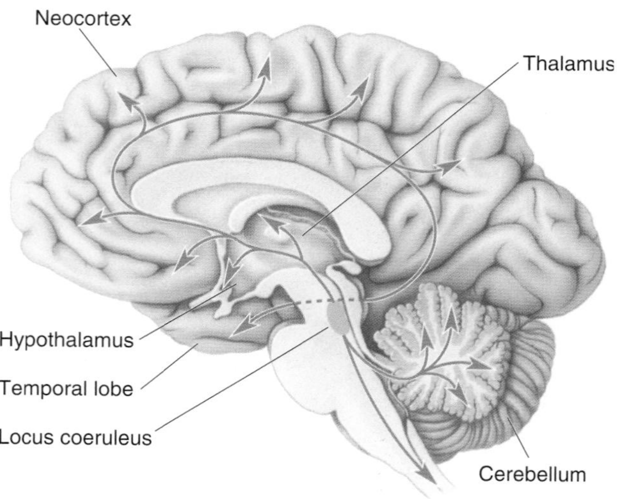

Figure 21.8
The diffuse modulatory systems implicated in mood disorders. (a) Serotonergic system. **(b)** Noradrenergic system.

and his colleagues at Emory University have proposed the **diathesis-stress hypothesis of mood disorders** (Box 21.3). According to this new idea, the HPA axis is the main site where genetic and environmental influences converge to cause mood disorders.

As we have seen, exaggerated activity of the HPA system is associated with anxiety disorders. But anxiety and depression often coexist; in fact, this comorbidity is the rule rather than the exception. Indeed, one of the most robust findings in all of biological psychiatry is hyperactivity of the HPA axis in severely depressed patients: Blood cortisol levels are elevated, as is the concentration of CRH in the cerebrospinal fluid. Could this hyperactive HPA system, with its resulting deleterious effects on brain function, be the *cause* of depression? Animal studies are highly suggestive. Injection of CRH into the brains of animals produces behavioral effects that are reminiscent of major depression: insomnia, decreased appetite, decreased interest in sex, and, of course, an increased behavioral expression of anxiety.

Box 21.3

PATH OF DISCOVERY

Stress and Mood

BY CHARLES NEMEROFF

When I received residency training in psychiatry and during my early years on faculty, the field was sharply divided between those who favored "psychological" causes of severe mental illness and those who viewed themselves as "biological" psychiatrists. In retrospect, considering the fact that the brain is the organ of the mind, such mind-brain dualism and nature-nurture controversies now seem simplistic and artificial.

At that time, in the 1970s and early 1980s, we knew relatively little about depression, its biology, or the risk factors that contributed to the likelihood of developing this devastating disorder. Although the effectiveness of antidepressant medications had clearly been demonstrated, the success of cognitive-behavior therapy in treating depression had not yet been realized. This was the environment in which I began my work on the HPA axis in general and CRH in particular.

During my training at the University of North Carolina at Chapel Hill, Dwight Evans and I confirmed and extended the observations of others, demonstrating hyperactivity of the HPA axis in depressed patients. Moreover, we found that the more severe the patient's depression, the greater the hyperactivity of the HPA axis. The activity of the HPA axis was assessed by measuring the adrenocortical hormone cortisol and the anterior pituitary hormone ACTH in blood samples from these patients. ACTH was known to be secreted in response to stress and to be a direct cause of secretion of cortisol from the adrenal cortex. The connection between the HPA axis, the major stress-responsive endocrine system in mammals, and depression was strengthened by the now well-replicated observation that stressful life events often precipitate depressive episodes in patients with major depression. We were struck by two seminal observations in 1983, one clinical and one experimental. That year, I relocated to Duke University Medical Center and with my colleague, Dr. K. Ranga R. Krishnan, cared for the majority of patients with mood disorders. We continued to document additional evidence of HPA axis hyperactivity (e.g., adrenal gland enlargement) in depressed patients and were struck by the remarkably high percentage of patients with major depression who reported horrendous personal histories of child abuse, physical and sexual, as well as severe emotional neglect and early loss. This was of considerable interest because psychoanalysts had long proposed a link between early traumatic experiences and the risk of developing depression in adulthood. The research finding that served as an impetus for much of our subsequent work was the demonstration that the newly discovered peptide CRH, which controlled the secretion of ACTH, was also higher in cerebrospinal fluid obtained by spinal tap from depressed patients than that obtained from patients with other psychiatric disorders and normal healthy volunteers.

Charles Nemeroff

This finding has been confirmed in several studies, and there is now postmortem evidence of marked CRH hypersecretion and increased CRH gene expression in depression. Indeed, when administered directly into the brain of laboratory animals, CRH produces many of the characteristic signs of depression, such as sleep disturbance, reduced appetite, and diminished sexual behavior. Thus, CRH, the brain's major chemical stress messenger, which coordinates the endocrine, autonomic, immune, and behavioral responses to stress, is now thought to be responsible for the endocrine pathology of depression and its major behavioral effects.

Two major research directions are under way. First, there is burgeoning evidence from both preclinical and clinical studies that early traumatic life experiences, such as maternal neglect or child abuse, is associated with persistent, long-lasting hyperactivity of CRH neurons. This is one likely mechanism by which early trauma produces vulnerability to depression in adulthood. The second major direction is the study of the effects of available and novel antidepressants to determine whether their mechanism of action is a reduction in CRH secretion, in addition to their effects on serotonergic and noradrenergic neurons. Most exciting, however, is the development of CRH receptor antagonists as novel antidepressants and anxiolytics.

To take observations from the bedside and clinic to the laboratory and conversely from the laboratory to the clinical arena is the most fun one could have—the best I believe in so-called translational research. Having a great team of students, fellows, junior and senior colleagues, and, of course, patients, all of whom continuously teach and challenge me, round out a rewarding professional life.

Recall that the activation of the hippocampal glucocorticoid receptors by cortisol normally leads to feedback inhibition of the HPA axis (Figure 21.5). In depressed patients, this feedback is disrupted, explaining why HPA function is hyperactive. A molecular basis for the diminished hippocampal response to cortisol is a decreased number of glucocorticoid receptors. What regulates glucocorticoid receptor number? In a fascinating parallel with the factors implicated in mood disorders, the answer is genes, monoamines, and early childhood experience.

Glucocorticoid receptors, like all proteins, are the product of gene expression. It has been shown in rats that early sensory experience regulates the amount of glucocorticoid receptor gene expression. Rats that received a lot of maternal care as pups express more glucocorticoid receptors in their hippocampus, less CRH in their hypothalamus, and reduced anxiety as adults. The maternal influence can be replaced by increasing the tactile stimulation of the pups. Tactile stimulation activates the ascending serotonergic inputs to the hippocampus, and the serotonin triggers a long-lasting increase in the expression of the glucocorticoid receptor gene. More glucocorticoid receptors make the animal better at "handling" stress as adults. The beneficial effect of experience, however, is restricted to a critical period of early postnatal life; stimulation of the rats as adults does not have the same effect.

Childhood abuse and neglect, in addition to genetic factors, are known to put people at risk for developing mood and anxiety disorders, and these animal findings suggest one cause. Elevations in brain CRH and decreased feedback inhibition of the HPA system may make the brain especially vulnerable to depression.

Treatments for Mood Disorders

Mood disorders are very common, and the burden they impose on human health and productivity is enormous. Fortunately, a number of highly effective treatments are available.

Electroconvulsive Therapy. It may surprise you to learn that one of the most effective treatments for depression and mania involves inducing seizure activity in the temporal lobes. In **electroconvulsive therapy** (**ECT**), electrical current is passed between two electrodes placed on the scalp. Localized electrical stimulation triggers seizure discharges in the brain, but the patient is given anesthesia and muscle relaxants to prevent violent movements during treatment. An advantage of ECT is that relief can occur quickly, sometimes after the first treatment session. This attribute of ECT is especially important when suicide risk is high. An adverse effect of ECT, however, is memory loss. As we will see in Chapter 23, temporal lobe structures, including the hippocampus, play a vital role in memory. ECT usually disrupts memories of events that occurred before treatment, extending back 6 months on average. In addition, ECT can temporarily impair the storage of new information.

The mechanism by which ECT relieves depression is unknown. As already mentioned, however, one temporal lobe structure affected by ECT is the hippocampus, which we have seen is involved in regulating CRH and the HPA axis.

Psychotherapy. Psychotherapy can be effective in treating mild to moderate cases of depression. The main goal of psychotherapy is to help depressed patients overcome negative views of themselves and their future. The neurobiological basis of the treatment has not been established, although we can infer that it relates to establishing cognitive neocortical control over the activity patterns in disturbed circuits.

Antidepressants. A number of highly effective pharmacological treatments are available for mood disorders. The most popular **antidepressant drugs** are (1) tricyclic compounds (so named because of their chemical structure), such as imipramine, which block the reuptake of both norepinephrine and serotonin; (2) SSRIs, such as fluoxetine, which act only on serotonin terminals; (3) norepinephrine-selective reuptake inhibitors, such as reboxetine; and (4) MAO inhibitors, such as phenelzine, which reduce the enzymatic degradation of serotonin and norepinephrine (Figure 21.9). All of these drugs elevate the levels of monoamine neurotransmitters in the brain, but as mentioned, their therapeutic actions take weeks to develop.

The adaptive response in the brain that is responsible for the clinical effectiveness of these drugs has not been established with certainty. Nonetheless, an intriguing finding is that clinically effective treatment with antidepressants dampens the hyperactivity of the HPA system in humans. Animal studies suggest that this effect is due to increased glucocorticoid receptor expression in the hippocampus, which occurs in response to a long-term elevation in serotonin. Recall that CRH plays a crucial role in the stress response of the HPA axis. New drugs that act as CRH receptor antagonists are under development for use as antidepressants.

Lithium. By now, you probably have formed the (correct) impression that until recently, most treatments for psychiatric disorders were discovered virtually by chance. For example, ECT was introduced initially in the 1930s as a treatment of last resort for psychotic behavior, based on the mistaken belief

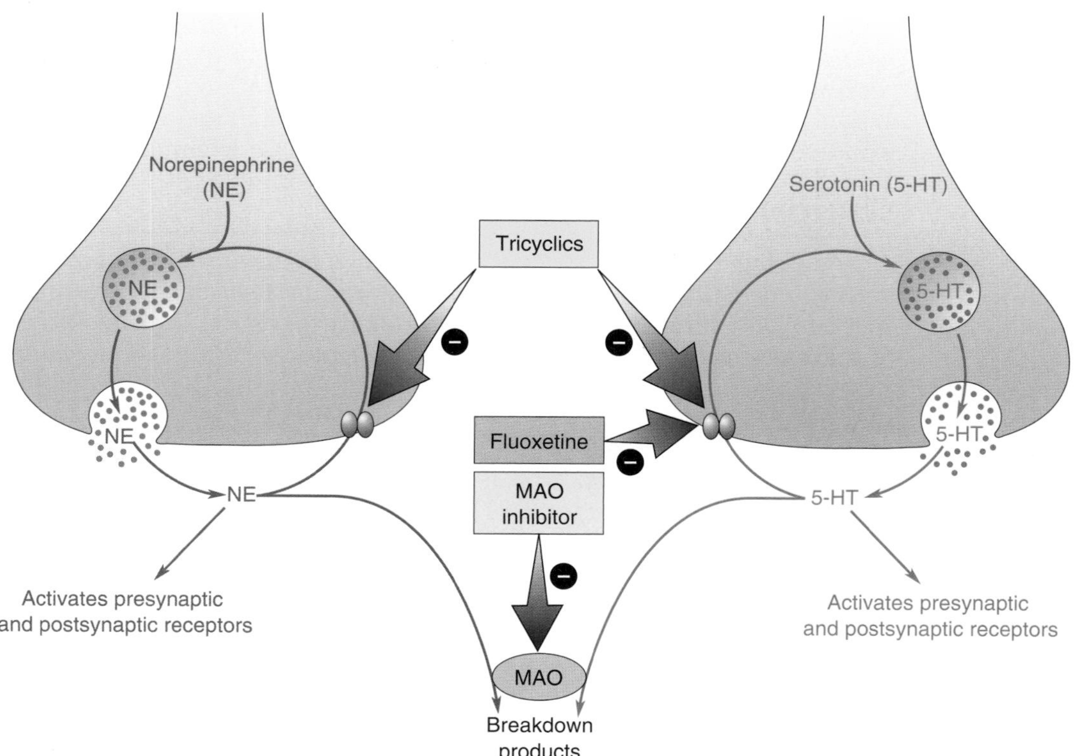

Figure 21.9
Antidepressant drugs and the biochemical life cycles of norepinephrine and serotonin. MAO inhibitors, tricyclics, and SSRIs are used as antidepressants. MAO inhibitors enhance the actions of NE and 5-HT by preventing their enzymatic destruction. Tricyclics enhance NE and 5-HT action by blocking uptake. SSRIs act the same way but are selective for serotonin.

that epilepsy and schizophrenia could not coexist in the same person. Only later was it shown to be an effective treatment for major depression, for reasons that still remain unknown.

"Enlightened serendipity" was again at work in the discovery of a highly effective treatment for bipolar disorder. Working in the 1940s, Australian psychiatrist John Cade was searching for psychoactive substances in the urine of manic patients. He injected guinea pigs with urine or urinary constituents and observed their behavioral effects. Cade wished to test the effect of uric acid, but he had difficulty getting it into solution. So instead, he used lithium urate because it dissolved easily and was readily available in the pharmacy. He observed, quite unexpectedly, that this treatment calmed the guinea pigs (he predicted the opposite effect). Because other lithium salts also produced the behavioral effect, he concluded that it was the lithium, not a constituent of urine, that was responsible. He went on to test lithium treatment on patients with mania, and amazingly, it worked. Subsequent studies showed that lithium is highly effective in stabilizing the mood of patients with bipolar disorder, not only by preventing the recurrence of mania but also by preventing episodes of depression (Figure 21.10).

Lithium affects neurons in many ways. In solution, it is a monovalent cation that passes freely through neuronal sodium channels. Inside the neuron, lithium prevents the normal turnover of phosphatidylinositol (PIP_2), a precursor for important second messenger molecules that are generated in response to activation of some G-protein-coupled neurotransmitter receptors (see Chapter 6). Lithium also interferes with the actions of adenylyl cyclase, important for generation of the second messenger cyclic adenosine monophosphate (cAMP), and glycogen synthase kinase, important for cellular energy metabolism. Why lithium is such an effective treatment for bipolar disorder, however, remains completely unknown. As with other antidepressants, the therapeutic effects of lithium require long-term use. The answer again appears to lie in an adaptive change in the central nervous system (CNS), but the nature of this change remains to be determined.

SCHIZOPHRENIA

Although their severity may be hard to comprehend, we all have some idea of what mood and anxiety disorders are like, because they are extremes in the spectrum of brain states that are part of normal experience. The same cannot be said for schizophrenia. This severe mental disorder distorts thoughts and

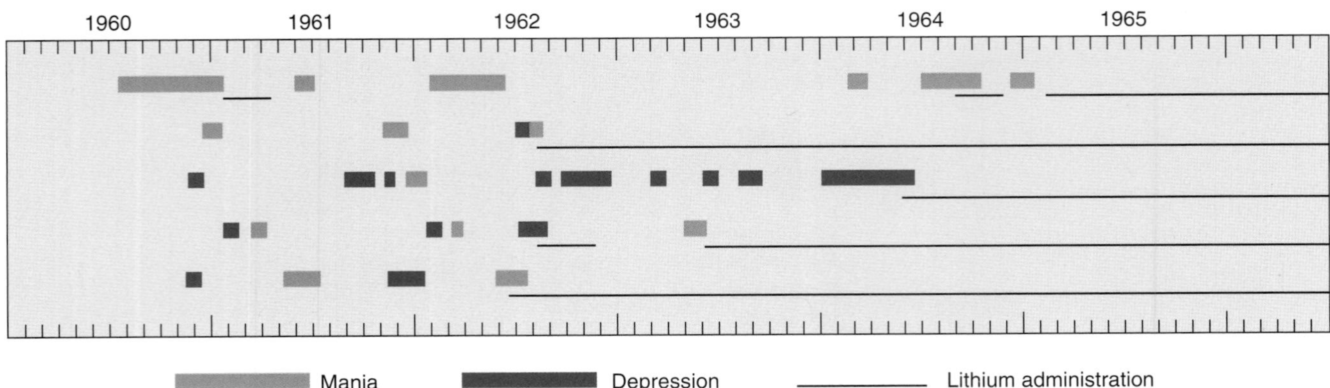

Figure 21.10
The mood-stabilizing effect of lithium treatment in five patients. (Source: Adapted from Barondes, 1993, p. 139.)

perceptions in ways that healthy people find difficult to understand. Schizophrenia is a major public health problem. It affects 1% of the population; more than 2 million people in the United States alone are affected.

A Description of Schizophrenia

Schizophrenia is characterized by a loss of contact with reality and a disruption of thought, perception, mood, and movement. The disorder typically becomes apparent during adolescence or early adulthood and usually persists for life. The name, introduced in 1911 by Swiss psychiatrist Eugen Bleuler, roughly means "divided mind," because of his observation that many patients seemed to oscillate between normal and abnormal states. There are, however, many variations in the manifestations of schizophrenia, including those that show a steadily deteriorating course. Indeed, it is still not clear whether what is called schizophrenia is a single disease or several.

The symptoms of schizophrenia fall into two categories: positive and negative. *Positive symptoms* reflect the presence of abnormal thoughts and behaviors, such as the following:

- Delusions
- Hallucinations
- Disorganized speech
- Grossly disorganized or catatonic behavior

Negative symptoms reflect the absence of responses that are normally present. These symptoms include the following:

- Reduced expression of emotion
- Poverty of speech
- Difficulty in initiating goal-directed behavior

According to current diagnostic criteria, schizophrenia may be classified into a number of types, based on the constellation of prominent symptoms that are present. *Paranoid schizophrenia* is characterized by a preoccupation with delusions organized around a theme, e.g., the belief that powerful adversaries are out to get them. These delusions are often accompanied by auditory hallucinations (hearing imaginary voices, for example) related to the same delusional theme. Of all patients with schizophrenia, those with the paranoid type have the best chance of rehabilitation. The outlook is less optimistic in cases of *disorganized schizophrenia*. The characteristics of this type of schizophrenia are lack of emotional expression (called a "flat affect"), coupled with disorganized behavior and incoherent speech. The speech may be accompanied by silliness and laughter that appear to have no relation to what is being said. This form of schizophrenia has a steadily worsening course without significant remissions. A third common type is *catatonic schizophrenia*, characterized by peculiarities of voluntary movement, such as immobility and stupor (catatonia), bizarre posturing and grimacing, and senseless, parrot-like repetition of words or phrases.

Biological Bases of Schizophrenia

Understanding the neurobiological basis for schizophrenia is one of the greatest challenges of neuroscience, because the disorder affects many of the traits that make us human: thought, perception, self-awareness. Although considerable progress has been made, we still have much more to learn.

Genes and the Environment. Schizophrenia runs in families. As shown in Figure 21.11, the likelihood of having the disorder varies in relation to the

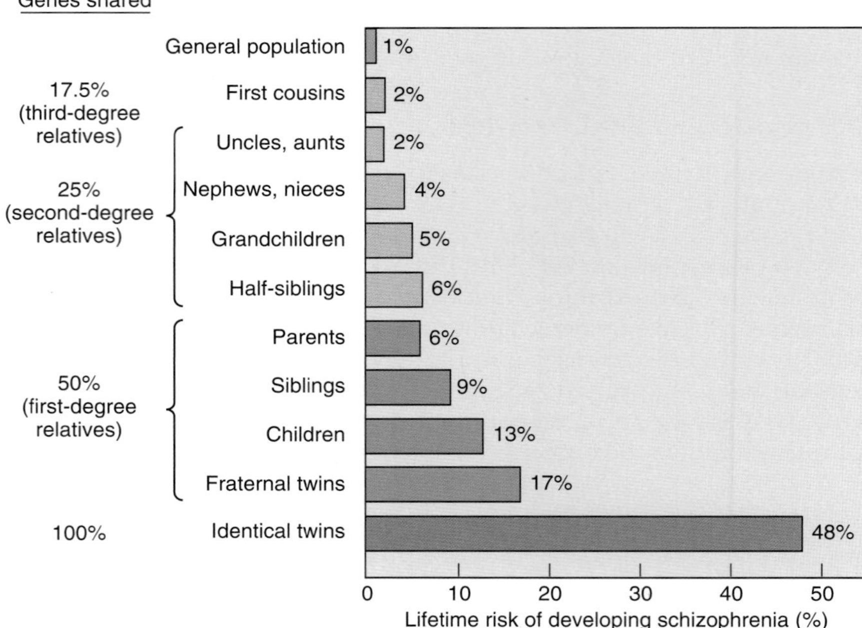

Figure 21.11
The familial nature of schizophrenia. The risk of schizophrenia increases with the number of shared genes, suggesting a genetic basis for the disease. (Source: Adapted from Gottesman, 1991, p. 96.)

number of genes that are shared with an affected family member. If your identical twin has schizophrenia, the probability is about 50% that you will also have it. The chances you will have schizophrenia decline as the number of genes you share with an affected family member decreases. These findings argue strongly for a genetic basis for the disorder. It seems that one or more faulty genes are to blame for schizophrenia.

Remember, however, that identical twins have exactly the same genes. So why, in 50% of cases, is one sibling spared when the other has schizophrenia? The answer must lie in the environment. In other words, faulty genes seem to make some people susceptible to environmental factors that cause schizophrenia. Viral infections during fetal and infant development have been implicated as possible causes, as has poor maternal nutrition. In addition, environmental stresses throughout life are known to exacerbate the course of the disorder. At present, however, neither the genes nor the exact environmental factors important for schizophrenia have been identified.

Schizophrenia is associated with physical changes in the brain. An interesting example appears in Figure 21.12. The figure shows brain scans of identical twins, one with schizophrenia and one without. Normally, identical twins have structurally identical brains. In this case, however, the brain of the schizophrenic sibling shows enlarged lateral ventricles, presumably reflecting the shrinkage of brain tissue around them. But such pronounced structural changes are not always apparent in the brains of schizophrenics. Current thinking is that the most important physical changes in their brains occur in the fine structure and function of cortical connections. As we will now see, particular attention has focused on alterations in chemical synaptic transmission mediated by dopamine and glutamate.

The Dopamine Hypothesis. Recall that dopamine is the neurotransmitter used by another of the diffuse modulatory systems (Figure 21.13). A link between schizophrenia and the *mesocorticolimbic dopamine system* has been made on the basis of two main observations. The first relates to the effects of amphetamine in otherwise healthy people. Remember from our discussion in Chapter 15 that amphetamine enhances neurotransmission at catecholamine-

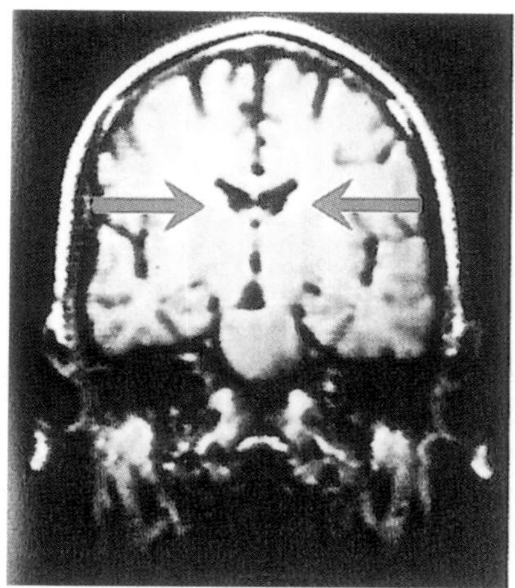

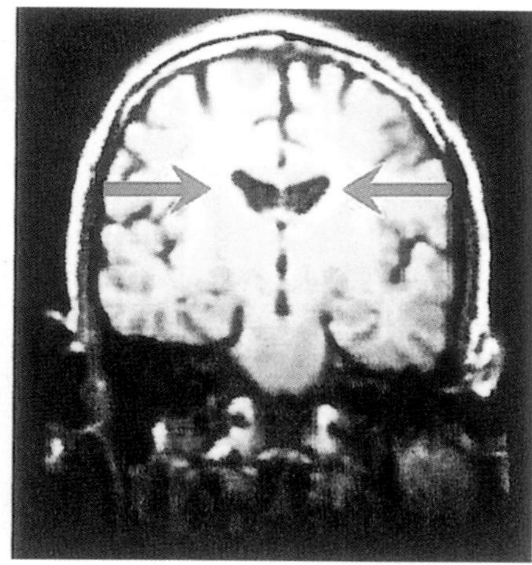

Figure 21.12

Enlargement of the lateral ventricles in schizophrenia. These MRIs are from the brains of identical twins. The sibling on the left was normal; the one on the right was diagnosed with schizophrenia. Notice the enlarged lateral ventricles in the schizophrenic sibling, indicating a loss of brain tissue. (Source: Barondes, 1993, p. 153.)

utilizing synapses and causes the release of dopamine. Amphetamine's normal stimulant action bears little resemblance to schizophrenia. Because of its addictive properties, however, users of amphetamines often risk taking more and more to satisfy their cravings. The resulting overdose can lead to a psychotic episode with positive symptoms that are virtually indistinguishable from those of schizophrenia. This suggests that psychosis is somehow related to too much catecholamine in the brain.

A second reason to associate dopamine with schizophrenia relates to the CNS effects of drugs that reduce the positive symptoms of the disorder. In the 1950s, it was discovered that the drug *chlorpromazine*, initially developed as an antihistamine, could prevent the positive symptoms in schizophrenia. It was later found that chlorpromazine and other related antipsychotic drugs, collectively called **neuroleptic drugs**, are potent blockers of dopamine receptors, specifically the D_2 receptor. When a large number of neuroleptics are examined, the correlation between the dosage effective for controlling

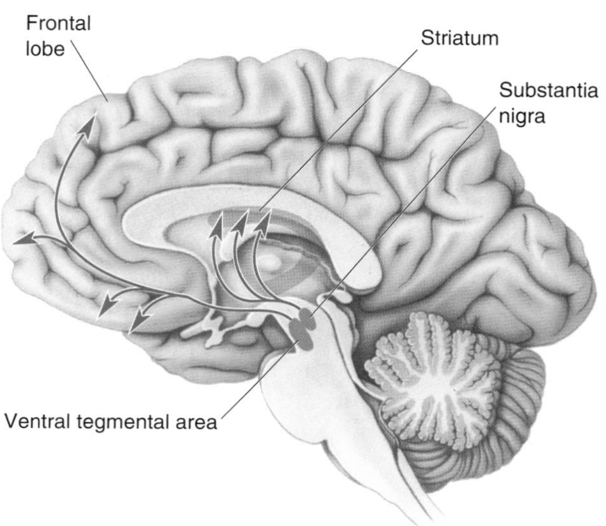

Figure 21.13

The dopaminergic diffuse modulatory systems of the brain. The mesocorticolimbic dopamine system arises in the ventral tegmental area. It has been implicated in the cause of schizophrenia. A second dopaminergic system arises from the substantia nigra and is involved in the control of voluntary movement by the striatum.

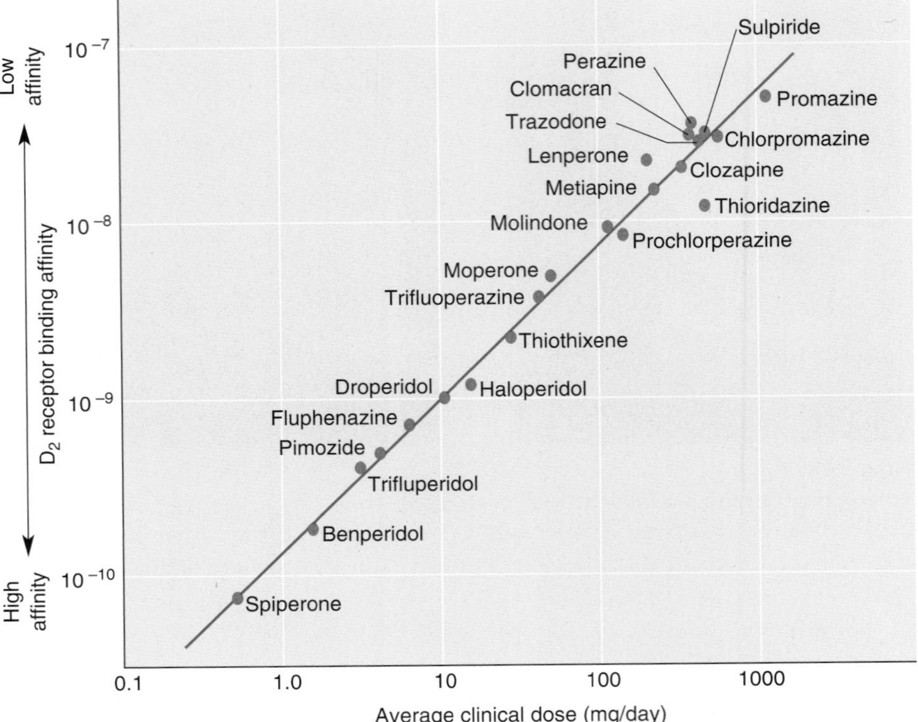

Figure 21.14
Neuroleptics and D_2 receptors. The neuroleptic dosages effective in controlling schizophrenia correlate well with the drugs' binding affinities for D_2 receptors. (Source: Adapted from Seeman, 1980.)

schizophrenia and their ability to bind to D_2 receptors is impressive (Figure 21.14). Indeed, these same drugs are effective in the treatment of amphetamine and cocaine psychoses. According to the **dopamine hypothesis of schizophrenia**, psychotic episodes in schizophrenia are triggered specifically by the activation of dopamine receptors.

Despite the tantalizing link between the positive symptoms of schizophrenia and dopamine, there seems to be more to the disorder than an overactive dopamine system. One indication is that newly developed antipsychotic drugs, such as *clozapine*, have little effect on D_2 receptors. These drugs are called *atypical neuroleptics*, indicating that they act in a novel way. The mechanism by which these compounds exert their neuroleptic effect has not been established with certainty, but an interaction with serotonin receptors is suspected.

The Glutamate Hypothesis. Another indication that there is more to schizophrenia than dopamine comes from the behavioral effects of *phencyclidine* (*PCP*). PCP was introduced in the 1950s as an anesthetic. Trials in humans were a failure, however, because many patients experienced adverse postoperative side effects, sometimes lasting for days, that included hallucinations and paranoia. Tragically, PCP is now a common drug of abuse, known on the street as "angel dust." PCP intoxication is accompanied by many of the symptoms of schizophrenia, both positive and negative. PCP has, however, no effect on dopaminergic transmission; rather, PCP affects synapses that use glutamate as a neurotransmitter.

Recall from Chapter 6 that glutamate is the main fast excitatory neurotransmitter in the brain and that *N*-methyl-D-aspartate (NMDA) receptors are one subtype of glutamate receptor. PCP acts by inhibiting NMDA receptors (Figure 21.15). Thus, according to the **glutamate hypothesis of schizophrenia**, the disorder reflects diminished activation of NMDA receptors in the brain.

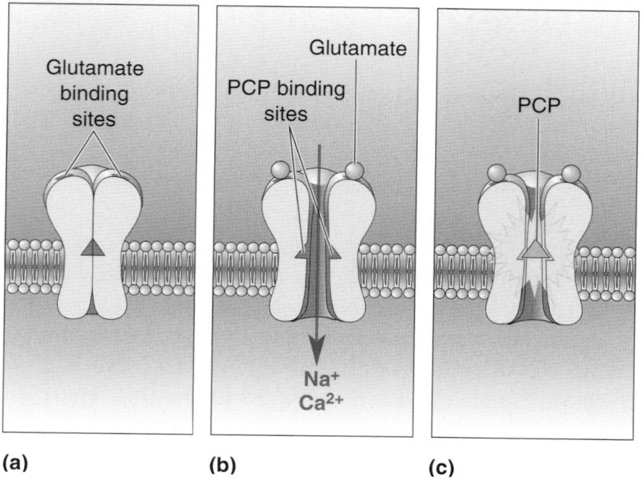

Figure 21.15
Blocking of the NMDA receptor by PCP. NMDA receptors are glutamate-gated ion channels. **(a)** In the absence of glutamate, the channels are closed. **(b)** In the presence of glutamate, the channel is open, exposing PCP binding sites. **(c)** NMDA channels are blocked when PCP enters and binds. The blockade of brain NMDA receptors by PCP produces effects on behavior that resemble the symptoms of schizophrenia.

To study the neurobiology of schizophrenia, neuroscientists have attempted to establish animal models of the disorder. Mice that have been genetically engineered to express fewer NMDA receptors display some behaviors that resemble human schizophrenia, including repetitive movements, agitation, and altered social interactions with other mice (Figure 21.16). Of course, we don't know whether the mutant mice feel paranoid or hear imaginary voices. But it is significant that the observed behavioral abnormalities can be ameliorated by treating the mice with either conventional or atypical neuroleptics.

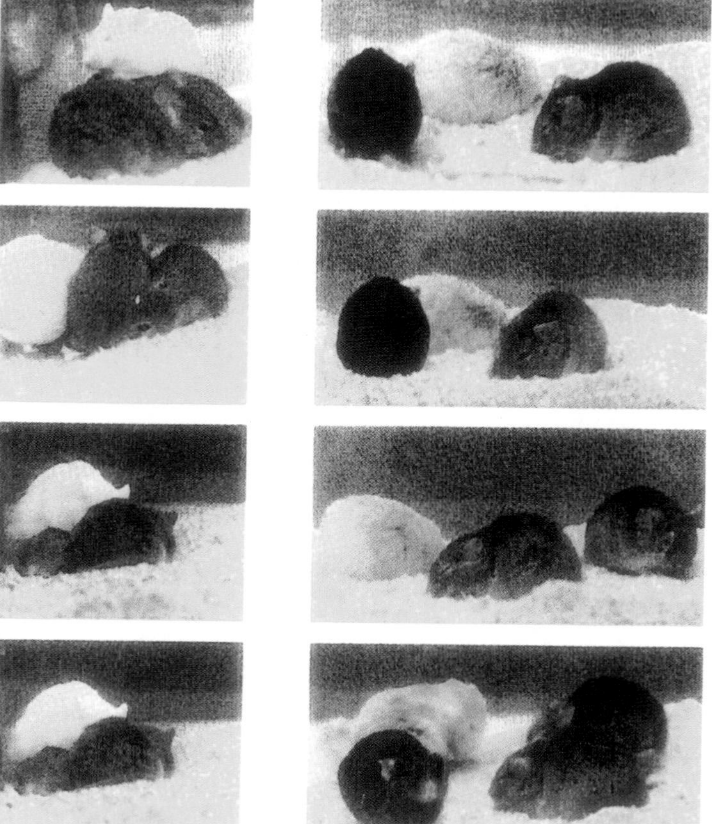

Figure 21.16
Social withdrawal in mutant mice with reduced numbers of NMDA receptors. The mice on the left have the normal number of NMDA receptors. Each photograph was taken 30 minutes apart for 2 hours to monitor social behavior. These mice tend to nest together. The mice on the right have been genetically altered to express fewer NMDA receptors. Notice that these mice tend to avoid social contact with one another. (Source: Mohn et al., 1999, p. 432.)

Treatments for Schizophrenia

The treatment of schizophrenia consists of drug therapy combined with psychosocial support. As we mentioned, the conventional neuroleptics, such as chlorpromazine and haloperidol, act at D_2 receptors. These drugs reduce the positive symptoms of schizophrenia in most patients. Unfortunately, the drugs also have numerous side effects relating to their actions on the dopaminergic input to the striatum that arises from the substantia nigra (see Chapter 14). Not surprisingly, the effects of blocking dopamine receptors in the striatum resemble Parkinson's disease; they include rigidity, tremor, and difficulty initiating movements. Chronic treatment with conventional neuroleptics also can result in the emergence of *tardive dyskinesia*, characterized by involuntary movements of the lips and jaw. Many of these side effects are avoided by using the atypical neuroleptics, such as clozapine and risperidone, because they do not act directly on the dopamine receptors in the striatum. These medications are also more effective against the negative symptoms of schizophrenia. The newest target for drug discovery is the NMDA receptor. The hope is that increasing NMDA receptor responsiveness in the brain, perhaps in combination with decreasing D_2 receptor activation, will further alleviate the symptoms of schizophrenia.

CONCLUDING REMARKS

Neuroscience has had a huge influence on psychiatry. Mental illness is now recognized to be the consequence of pathological modifications of the brain, and psychiatric treatments today are focused on correcting these changes. Just as important, neuroscience has changed the way people with mental illness are viewed by society. Suspicion of the mentally ill is slowly giving way to compassion. Mental illnesses today are recognized as diseases of the body, just like cancer or diabetes.

Despite remarkable progress in treating psychiatric disorders, our understanding of how these treatments work their magic on the brain is far from complete. In the case of drug treatments, we know with great precision about how chemical synaptic transmission is affected. But we do not know why, in many cases, the therapeutic effect of a drug takes weeks to emerge. Even less is known about how psychosocial treatments act on the brain. In general, the answer seems to lie in adaptive changes that occur in the brain in response to treatment.

We also do not know the causes of most mental disorders. It is clear that our genes either put us at risk or protect us. The environment, however, also plays an important role. Environmental stresses before birth may contribute to schizophrenia, and those after birth may precipitate depression. Not all environmental effects are bad, however. Appropriate sensory stimulation, especially in early childhood, apparently can produce adaptive changes that help us ward off mental illness later in life.

Psychiatric disorders and their treatment illustrate that our brains and behavior are influenced by experience, whether it is exposure to inescapable stress or to pharmacologically elevated levels of serotonin. Of course, much more subtle sensory experiences also leave their mark on the brain. In Part IV, we will explore how sensory experiences modify the brain during development, and during learning.

KEY TERMS

Anxiety Disorders
hypothalamic-pituitary-adrenal (HPA) axis (p. 681)
adrenocorticotropic hormone (ACTH) (p. 681)
corticotropin-releasing hormone (CRH) (p. 681)
glucocorticoid receptor (p. 682)
serotonin-selective reuptake inhibitor (SSRI) (p. 685)

Mood Disorders
affective disorders (p. 686)
major depression (p. 686)
bipolar disorder (p. 687)
mania (p. 687)
monoamine hypothesis of mood disorders (p. 688)
diathesis-stress hypothesis of mood disorders (p. 690)
electroconvulsive therapy (ECT) (p. 692)
antidepressant drug (p. 693)
lithium (p. 694)

Schizophrenia
schizophrenia (p. 695)
neuroleptic drug (p. 697)
dopamine hypothesis of schizophrenia (p. 698)
glutamate hypothesis of schizophrenia (p. 698)

REVIEW QUESTIONS

1. Why and where do benzodiazepines reduce anxiety?
2. Depression is often accompanied by bulimia nervosa, characterized by frequent eating binges. Where does the regulation of mood and appetite converge in the brain?
3. Snuggling with your mom as a baby might help you cope with stress better as an adult. Why?
4. What three types of drugs are used to treat depression? What do they have in common?
5. Psychiatrists often refer to the dopamine theory of schizophrenia. Why do they believe dopamine is linked to schizophrenia? Why must we be cautious in accepting a simple equation between schizophrenia and too much dopamine?

PART

IV

THE CHANGING BRAIN

CHAPTER 22

Wiring the Brain

INTRODUCTION

We have seen that most of the operations of the brain depend on remarkably precise interconnections among its 100 billion neurons. As an example, consider the precision in the wiring of the visual system, from retina to lateral geniculate nucleus (LGN) to cortex, shown in Figure 22.1. All retinal ganglion cells extend axons into the optic nerve, but only ganglion cell axons from the nasal retinas cross at the optic chiasm. Axons from the two eyes are mixed in the optic tract, but in the LGN they are sorted out again (1) by ganglion cell type, (2) by eye of origin (ipsilateral or contralateral), and (3) by retinotopic position. LGN neurons project axons into the optic radiations that travel via the internal capsule to the primary visual (striate) cortex. Here, they terminate (1) only in cortical area 17, (2) only in specific cortical layers (mainly layer IV), and (3) again according to cell type and retinotopic position.

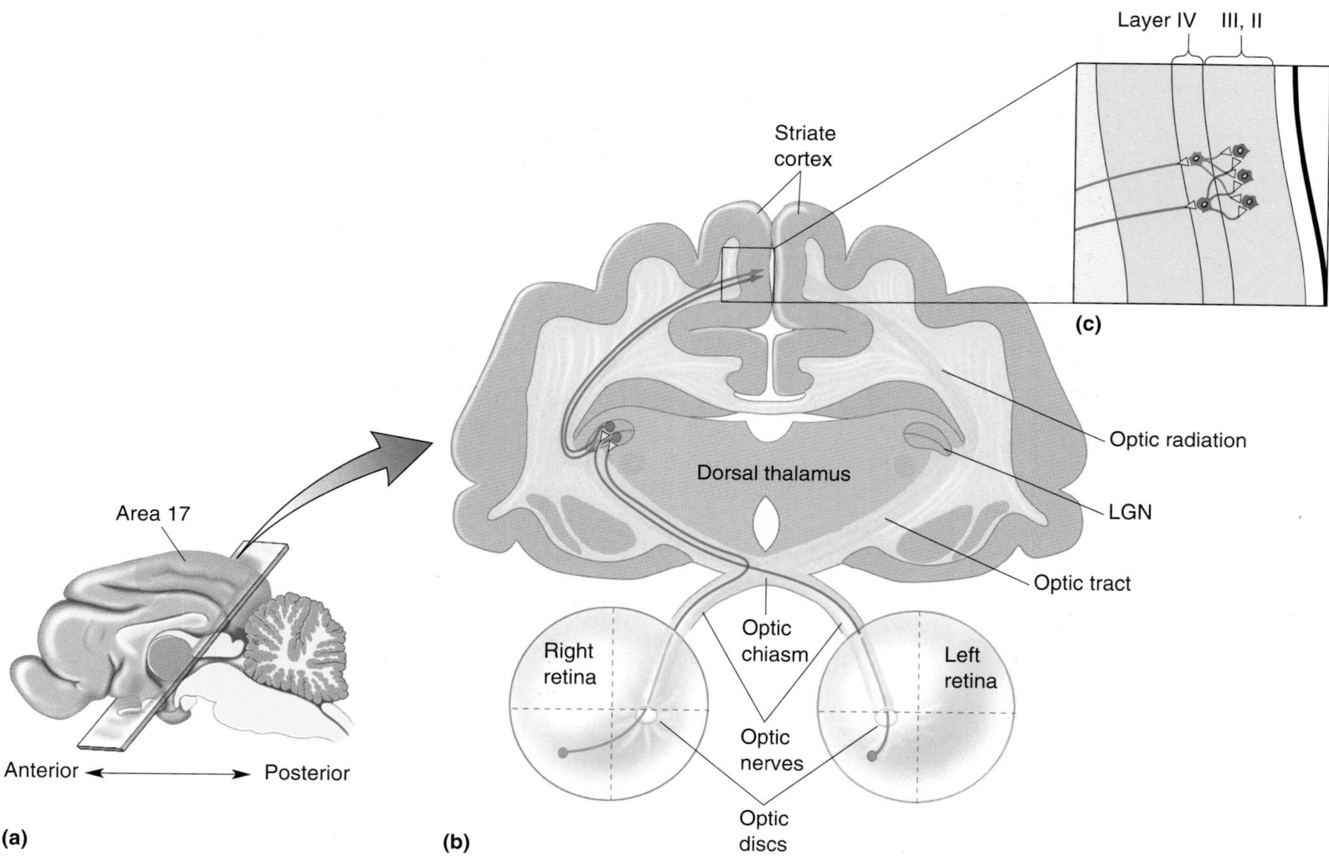

Figure 22.1
Components of the mature mammalian retinogeniculocortical pathway. (a) A midsagittal view of a cat brain showing the location of primary visual cortex (striate cortex, area 17). The line indicates a sectional plane, illustrated in part b, that reveals all of the components of the ascending visual pathway. **(b)** The temporal retina of the left eye and the nasal retina of the right eye project axons via the optic nerve and optic tract to the LGN of the left dorsal thalamus. Inputs from the two eyes remain segregated in separate layers at the level of this synaptic relay. LGN neurons project to striate cortex via the optic radiations. These axons terminate mainly in layer IV, where inputs serving the two eyes continue to be segregated. **(c)** The first site of major convergence of inputs from the two eyes is in the projection of layer IV cells onto cells in layer III.

Finally, the neurons in layer IV make very specific connections with cells in other cortical layers that are appropriate for binocular vision and are specialized to enable the detection of contrast borders. How did such precise wiring arise?

In Chapter 7, we looked at the embryological and fetal development of the nervous system to understand how it changed from a simple tube in the early embryo into the structures that we recognize in the adult as the brain and spinal cord. Here we take another look at brain development, this time to see how connections are formed and modified as the brain matures. Most of the wiring in the brain is specified by genetic programs that allow axons to detect the correct pathways and the correct targets. A small but important component of the final wiring, however, depends on sensory information about the world around us during early childhood. In this way, nurture and nature both contribute to the final structure and function of the nervous system. We use the central visual system as an example whenever possible, so you may want to review Chapter 10 before continuing.

THE GENESIS OF NEURONS

The first step in wiring the nervous system together is the generation of neurons. Consider as an example the striate cortex. In the adult, there are six cortical layers, and the neurons in each of these layers have characteristic appearances and connections that distinguish striate cortex from other areas. Neuronal structure develops in three major stages: cell proliferation, cell migration, and cell differentiation.

Cell Proliferation

Recall from Chapter 7 that the brain develops from the walls of the five fluid-filled vesicles. These fluid-filled spaces remain in the adult and constitute the ventricular system. Very early in development, the walls of the vesicles consist of only two layers: the ventricular zone and the marginal zone. The *ventricular zone* lines the inside of each vesicle, and the *marginal zone* faces the overlying pia. Within these layers of the telencephalic vesicle, a cellular ballet gives rise to all of the neurons and glia of the visual cortex. The choreography of cell proliferation is described next, and the five "positions" correspond to the circled numbers in Figure 22.2a:

1. A cell in the ventricular zone extends a process that reaches upward toward the pia.
2. The nucleus of the cell migrates upward from the ventricular surface toward the pial surface; the cell's DNA is copied.
3. The nucleus, containing two complete copies of the genetic instructions, settles back to the ventricular surface.
4. The cell retracts its arm from the pial surface.
5. The cell divides in two.

The fate of the newly formed *daughter cells* depends on a number of factors. Curiously, a ventricular zone precursor cell that is cleaved vertically during division has a different fate from that of one that is cleaved horizontally. After vertical cleavage, both daughter cells remain in the ventricular zone to divide again and again (Figure 22.2b). This mode of cell division predominates early in development to expand the population of neuronal precursors. Later in development, horizontal cleavage is the rule. In this case, the daughter cell lying farthest from the ventricular surface migrates away to take up its position in the cortex, where it will never divide again. The other daughter remains in the ventricular zone to undergo more divisions (Figure 22.2c).

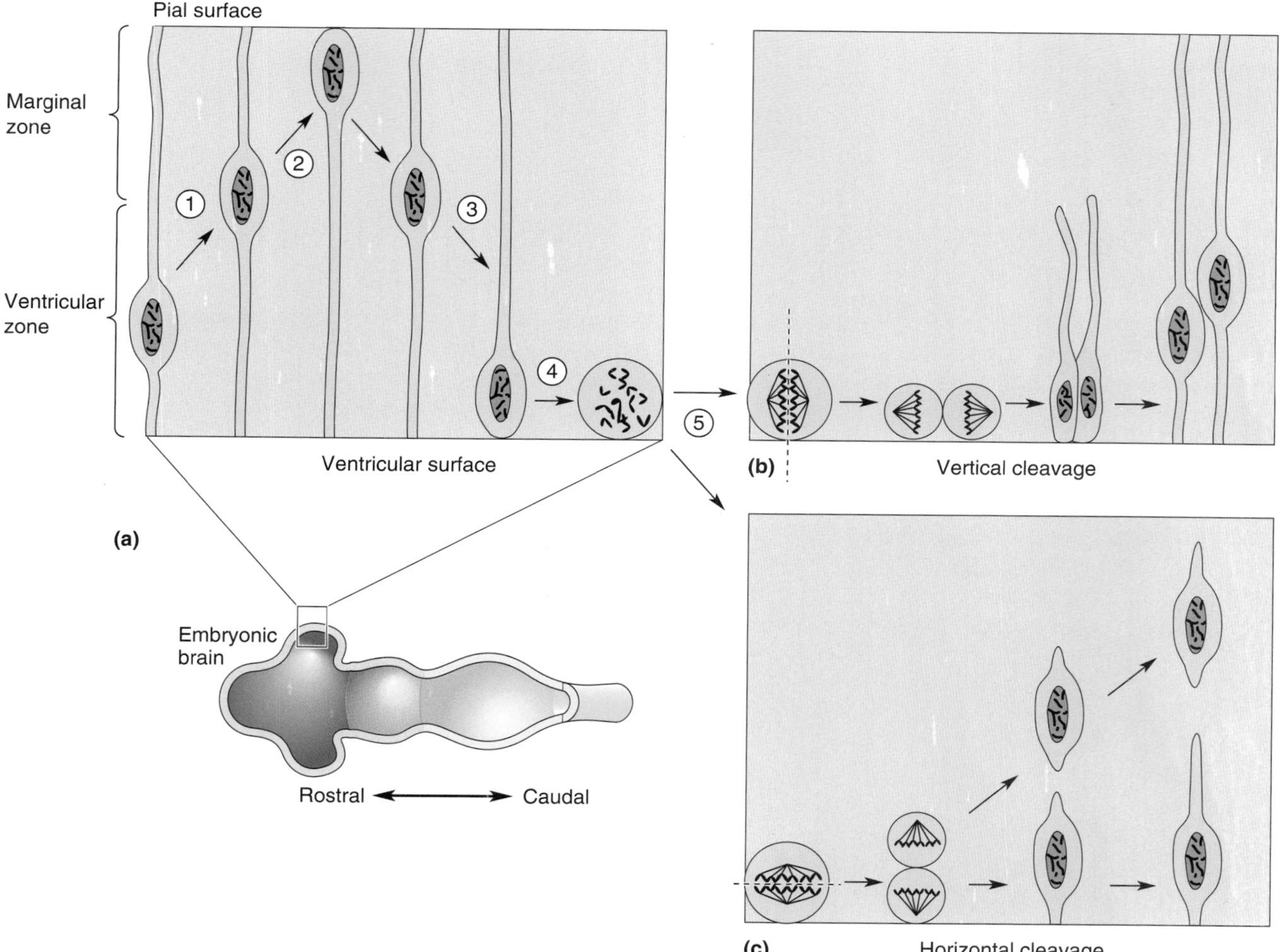

Figure 22.2

The choreography of cell proliferation. (a) The wall of the brain vesicles initially consists of only two layers, the marginal zone and the ventricular zone. Each cell performs a characteristic "dance" as it divides, shown here from left to right. The circled numbers correspond to the five positions described in the text. The fate of the daughter cells depends on the plane of cleavage during division. **(b)** After cleavage in the vertical plane, both daughters remain in the ventricular zone to divide again. **(c)** After cleavage in the horizontal plane, the daughter farthest away from the ventricle ceases any further division and migrates away.

Ventricular zone precursor cells repeat this pattern until all of the neurons of the cortex have been generated. In humans, most neocortical neurons are born between the fifth week and the fifth month of gestation (pregnancy), peaking at the astonishing rate of 250,000 new neurons per minute. Recent findings suggest that although most of the action is over well before birth, the adult ventricular zone retains some capacity to generate new neurons (Box 22.1). Once a daughter cell commits to a neuronal fate, however, it will never divide again.

How does the cleavage plane during cell division determine the cell's fate? Remember that all of our cells contain the same complement of DNA that we inherited from our parents, so every daughter cell has the same genes. The factor that makes one cell different from another is the specific genes used to generate messenger RNA (mRNA) and, ultimately, protein. Thus, cell fate is regulated by differences in *gene expression* during development. Gene ex-

Box 22.1

OF SPECIAL INTEREST

Neurogenesis in the Adult Neocortex

For many years, neuroscientists believed that neurogenesis—the generation of new neurons from precursor cells—was restricted to early brain development. New findings have challenged this view. It now appears that new neurons are continuously generated by neural precursor cells lining the ventricles in the adult brain.

Cell division requires the synthesis of DNA, which can be detected by feeding the cells chemically labeled DNA precursor molecules. Cells undergoing division at the time the precursor is available incorporate the chemical label into their DNA. In the mid-1980s, Fernando Nottebohm of Rockefeller University used this approach to prove that new neurons are generated in the brains of adult canaries, particularly in regions associated with song learning. This finding resurrected interest in adult neurogenesis in mammals, which John Altman and Gopal Das of the Massachusetts Institute of Technology first described in 1965. Research in the past few years by Fred Gage at the Salk Institute has established definitively that new neurons are generated in the adult rat hippocampus, a structure that is important for learning and memory (as we will see in Chapter 23). Interestingly, the number of new neurons in this region increases if the animal is exposed to an enriched environment, filled with toys and playmates. In addition, rats given the chance to have a daily run on an exercise wheel show enhanced neurogenesis. In both cases, the increased number of neurons correlates with enhanced performance on memory tasks that require the hippocampus. Hippocampal neurogenesis is not restricted to rats; Gage has found evidence that the same holds true for humans.

Can newly generated neurons be incorporated into the elaborate circuits of the primate neocortex? Elizabeth Gould and her colleagues at Princeton University addressed this question in 1999. They found that in adult macaque monkeys, cells are born in the ventricular zone, migrate through the white matter to the neocortex, and differentiate into neurons (Figure A). The new neurons appear to be added selectively to the association cortex of the temporal and frontal lobes. Primary sensory areas, such as striate cortex, had no new neurons.

Neurogenesis in the adult brain, unfortunately, is far too limited to repair CNS damage. The hope, however, is that understanding how adult neurogenesis is regulated—e.g., by the quality of the environment—will suggest ways it can be harnessed to promote regeneration after brain injury.

Migrating neuroblasts
Newly generated neurons
Neural precursor cells
Lateral ventricle

Figure A

pression is regulated by cellular proteins called *transcription factors*. If transcription factors or the "upstream" molecules that regulate them are unevenly distributed within a cell, then the cleavage plane can determine which factors are passed on to the daughter cells. For example, proteins called notch-1 and numb migrate to different poles of ventricular zone precursor cells (Figure 22.3). When the neuron divides vertically, the notch-1 and numb

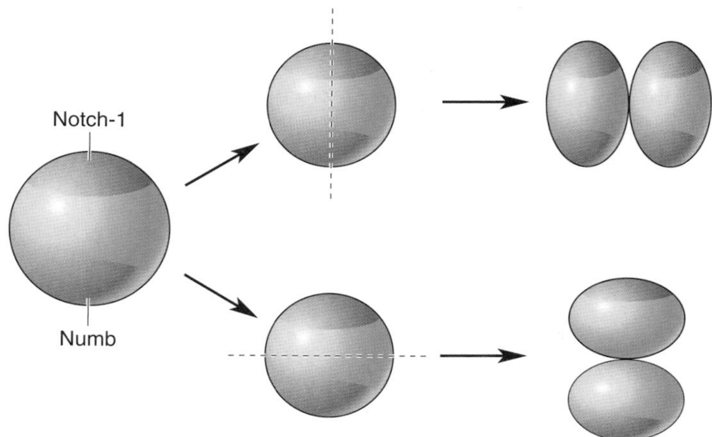

Figure 22.3
The distribution of cell constituents in precursor cells. The proteins notch-1 and numb are differentially distributed in the precursor cells of the developing neocortex. Vertical cleavage partitions these proteins equally in the daughters, but horizontal cleavage does not. Differences in the distribution of proteins in the daughters causes them to have different fates.

proteins are partitioned symmetrically. When the cell divides horizontally, however, notch-1 goes with the daughter that will migrate away, while numb remains with the cell that will divide again. Research suggests that notch-1, "unopposed" by numb, activates the gene expression that causes the cell to cease dividing and migrate away from the ventricular zone.

Mature cortical cells can be classified as glia or neurons, and the neurons can be further classified by the layer in which they reside, their dendritic morphology, and the neurotransmitter they use. Conceivably, this diversity could arise from different types of precursor cell in the ventricular zone. In other words, there could be one class of precursor cell that gives rise only to layer VI pyramidal cells, another that gives rise to layer V cells, and so on. But this is not the case. Multiple cell types, including neurons and glia, can arise from the same precursor cell. Because of this potential to give rise to many different types of tissue, these precursor cells are also called *neural stem cells.*

The ultimate fate of the migrating daughter cell is determined by a combination of factors, including the age of the precursor cell and its environment at the time of division. The first cells to migrate away from the ventricular zone are destined to reside in a layer called the **subplate**, which eventually disappears as development proceeds. The next cells to divide become layer VI neurons, followed by the neurons of layers V, IV, III, and II.

Cell Migration

Many daughter cells migrate by slithering along thin fibers that radiate from the ventricular zone toward the pia. These fibers are derived from specialized **radial glial cells**, providing the scaffold on which the cortex is built. The immature neurons, called **neuroblasts**, follow this radial path from the ventricular zone toward the surface of the brain (Figure 22.4). When cortical assembly is complete, the radial glial cells withdraw their radial processes. But not all migrating cells follow the path provided by the radial glial cells. About one-third of the neuroblasts wander horizontally on their way to the cortex.

The neuroblasts destined to become subplate cells are among the first to migrate away from the ventricular zone. Neuroblasts destined to become the adult cortex migrate next. They cross the subplate and form another cell layer called the **cortical plate**. The first cells to arrive in the cortical plate are those that will become layer VI neurons. Next come the layer V cells, followed by layer IV cells, and so on. Notice that each new wave of neuroblasts migrates right past those in the existing cortical plate. In this way, the cortex is said to be assembled *inside out* (Figure 22.5).

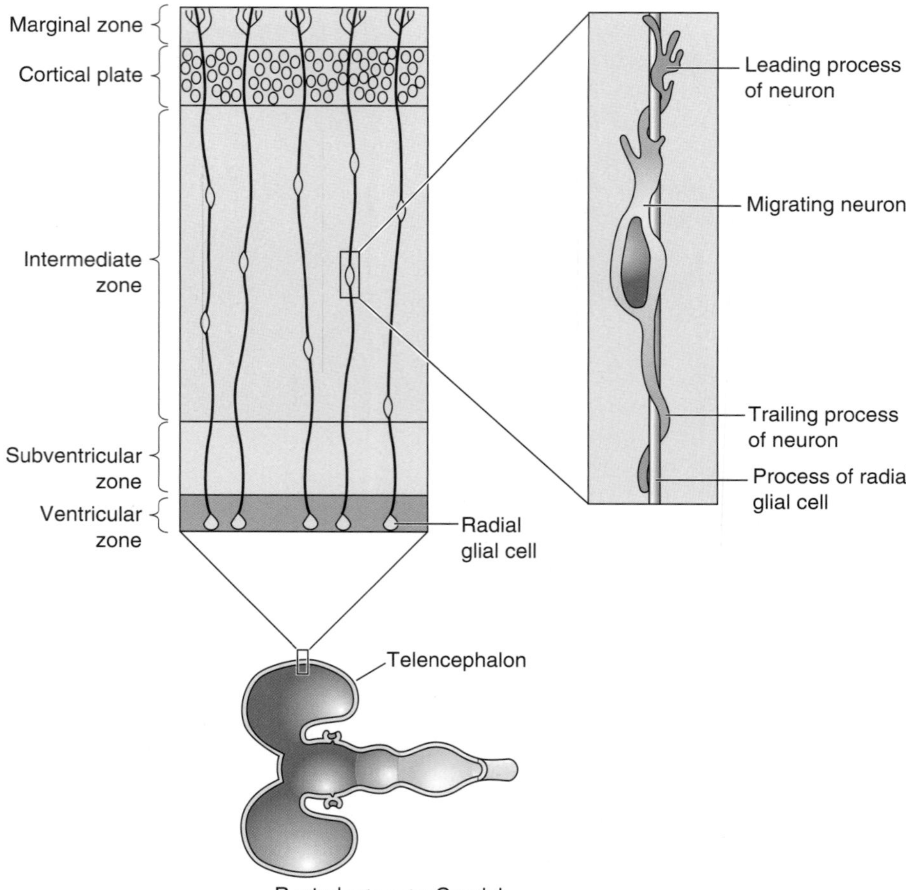

Figure 22.4
The migration of neuroblasts to the cortical plate. This is a schematic horizontal section through the brain early in development. Expanded views show neuroblasts crawling along the thin processes of the radial glia en route to the cortical plate, which forms just under the marginal zone.

Cell Differentiation

The process in which a cell takes on the appearance and characteristics of a neuron is known as *cell differentiation* (Figure 22.6). Differentiation is the consequence of a specific spatiotemporal pattern of gene expression. As we have seen, neuroblast differentiation begins as soon as the precursor cells divide with the uneven distribution of cell constituents. Further neuronal differentiation occurs when the neuroblast arrives in the cortical plate. Thus, layer V and VI neurons have differentiated into recognizable pyramidal cells even before layer II cells have migrated into the cortical plate.

Differentiation of the neuroblast into a neuron begins with the appearance of neurites sprouting off the cell body. At first these neurites all appear about the same, but soon one becomes recognizable as the axon, and the others become recognizable as dendrites. Differentiation occurs even if the neuroblast is removed from the brain and placed in tissue culture. For example, cells destined to become neocortical pyramidal cells often assume the same characteristic dendritic architecture in tissue culture. This means that differentiation is programmed well before the neuroblast arrives at its final resting place. The complexity of the dendritic trees is not entirely preprogrammed, however. The fine structure of axons and dendrites also depends on "environmental" factors in the cortex.

Differentiation of Cortical Areas

The neocortex is often described as a sheet of tissue. In reality, however, cortex is much more like a patchwork quilt with many structurally distinct areas

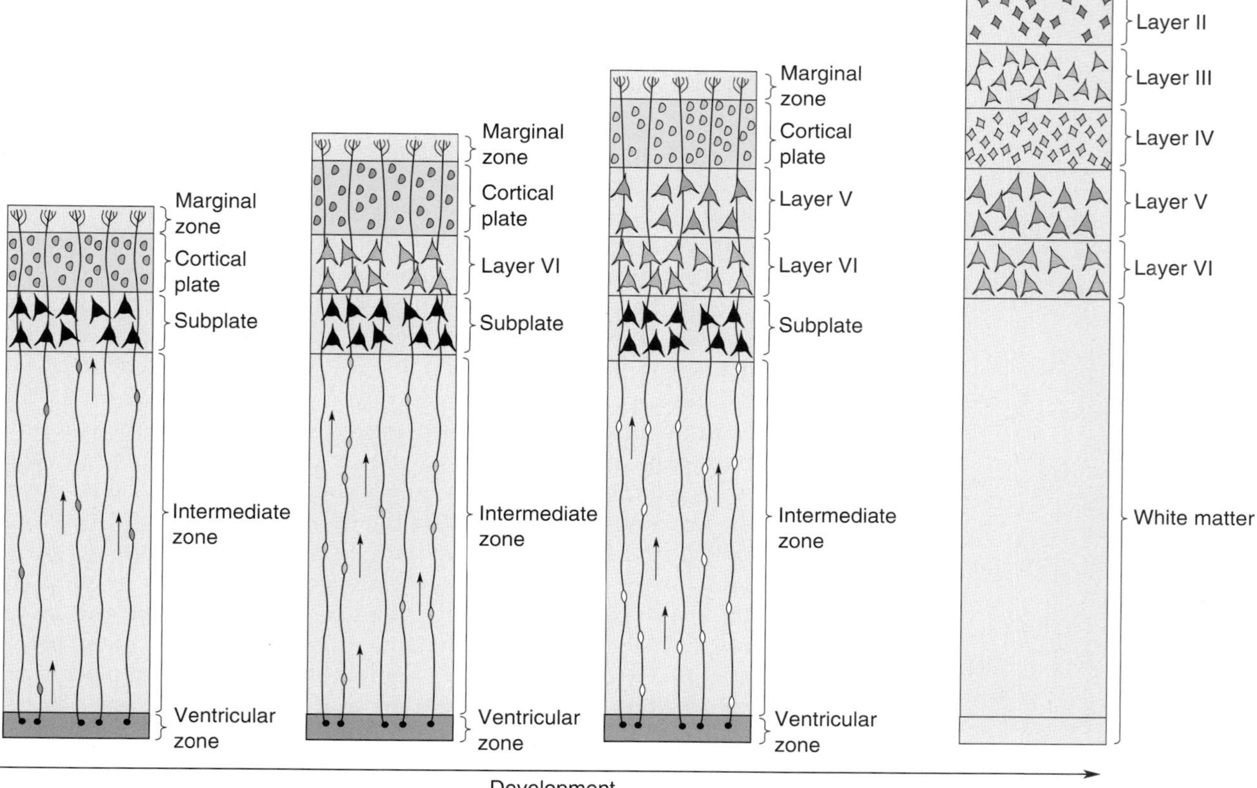

Figure 22.5
Inside-out development of the cortex. The first cells to migrate to the cortical plate are those that form the subplate. As these differentiate into neurons, the neuroblasts destined to become layer VI cells migrate past and collect in the cortical plate. This process repeats again and again until all layers of the cortex have differentiated. The subplate neurons then disappear.

stitched together. One of the consequences of human evolution was the creation of new neocortical areas that are specialized for increasingly sophisticated analysis. It is natural to wonder exactly how all of these areas arise during development.

As we have seen, cortical neurons are born in the ventricular zone and then migrate along radial glia to take up their final position in one of the cortical layers. Thus, it seems reasonable to conclude that cortical areas in the adult brain simply reflect an organization that is already present in the ventricular zone of the fetal telencephalon. According to this idea, the ventricular zone contains something like a microfilm record of the future cortex, which is projected onto the wall of the telencephalon as development proceeds.

The idea of such a cortical "protomap" was originally based on the assumption that migrating neuroblasts are precisely guided to the cortical plate by the network of radial glial fibers. If migration is strictly radial, then we may expect that all of the offspring of a precursor cell would migrate to exactly the same neighborhood of the cortex. As mentioned earlier, one-third of all neuroblasts stray considerable distances as they migrate toward the cortical plate. This finding initially seemed to be at odds with the protomap hypothesis.

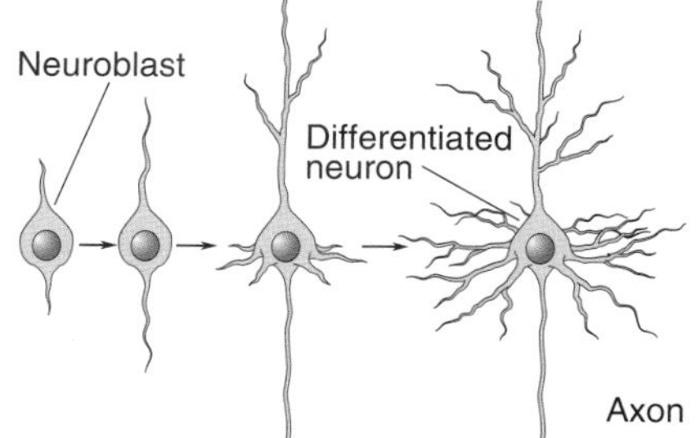

Figure 22.6
The differentiation of a neuroblast into a neuron.

Recall that cortical areas differ not only in terms of cytoarchitecture but also in terms of connections, particularly with the dorsal thalamus. Area 17 receives input from the LGN, area 3 receives input from the ventral posterior (VP) nucleus, and so on. What is the contribution of the thalamic input to the differentiation of cortical areas? Researchers Brad Schlaggar and Dennis O'Leary of the Salk Institute addressed this question in a novel way. In rats, the thalamic fibers wait in the cortical white matter and do not enter the cortex until a few days after birth. Schlaggar and O'Leary peeled off the parietal cortex in newborn rats and replaced it with a piece of occipital cortex so that the thalamic fibers from the VP nucleus were waiting under what would have been visual cortex. Remarkably, the fibers invaded the new piece of cortex, and it assumed the cytoarchitecture that is characteristic of the rodent somatosensory cortex (the "barrels"; see Figure 12.20). These results suggest that the thalamus is important for specifying the pattern of cortical areas.

But how did the appropriate thalamic axons come to lie in wait under the parietal cortex in the first place? This is beginning to sound like the chicken-and-egg problem! The answer apparently lies in the subplate. Subplate neurons, which have a more strictly radial migration pattern, attract the appropriate thalamic axons to different parts of the developing cortex: LGN axons to occipital cortex, VP nucleus axons to parietal cortex, and so on. The area-specific thalamic axons initially innervate distinct populations of subplate cells. When the overlying cortical plate grows to a sufficient size, the axons invade the cortex. The arrival of the thalamic axons causes the cytoarchitectural differentiation we recognize in the adult brain. Thus, the subplate layer of earliest-born neurons seems to contain the instructions for the assembly of the cortical quilt.

THE GENESIS OF CONNECTIONS

As neurons differentiate, they extend axons that must find their appropriate targets. Think of this development of long-range connections, or pathway formation, in the central nervous system (CNS) as occurring in three phases: pathway selection, target selection, and address selection. Let's understand the meaning of these terms in the context of the development of the visual pathway from the retina to the LGN, as shown in Figure 22.7.

Imagine for a moment that you must lead a growing retinal ganglion cell axon to the correct location in the LGN. First you travel down the optic stalk toward the brain. But soon you reach the optic chiasm at the base of the brain and must decide which fork in the road to take. You have three choices: You can enter the optic tract on the same side; you can enter the optic tract on the opposite side; or you can dive into the other optic nerve. The correct path depends on the location in the retina of your ganglion cell and on the cell type. If you came from the nasal retina, you would cross over at the chiasm into the contralateral optic tract; but if you came from the temporal retina, you would stay in the tract on the same side. And in no case would you enter the other optic nerve. These are examples of the decisions that must be made by the growing axon during *pathway selection*.

Having forged your way into the dorsal thalamus, you are confronted with a dozen possible targets, e.g., the medial geniculate nucleus. The correct choice, of course, is the *lateral* geniculate nucleus. This decision is called *target selection*.

But finding the correct target still isn't enough. You must now find the correct layer of the LGN. You must also make sure that you sort yourself out with respect to other invading retinal axons so that retinotopy in the LGN is established. These are examples of the decisions that must be made by the growing axon during *address selection*.

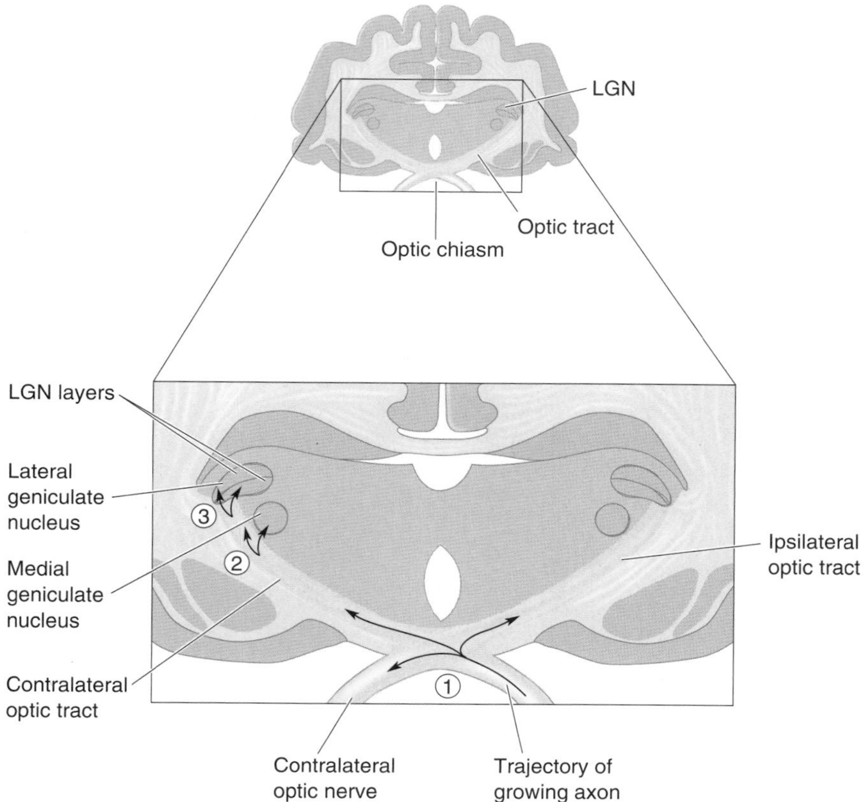

Figure 22.7
The three phases of pathway formation. The growing retinal axon must make several choices to find its correct target in the LGN. ① During pathway selection, the axon must choose the correct path. ② During target selection, the axon must choose the correct structure to innervate. ③ During address selection, the axon must choose the correct cells to synapse with in the target structure.

Each of the three phases of pathway formation depends critically on communication between cells. This communication occurs in several ways: direct cell-to-cell contact, contact between cells and the extracellular secretions of other cells, and communication between cells over a distance via diffusible chemicals. As the pathways develop, the neurons also begin to communicate via action potentials and synaptic transmission.

The Growing Axon

Once the neuroblast has migrated to take up its appropriate position in the nervous system, the neuron differentiates and extends the processes that will ultimately become the axon and dendrites. At this early stage, however, the axonal and dendritic processes appear quite similar and collectively are still called neurites. The growing tip of a neurite is called a **growth cone** (Figure 22.8).

The growth cone is specialized to identify an appropriate path for neurite elongation. The leading edge of the growth cone consists of flat sheets of membrane called *lamellipodia* that undulate in rhythmic waves like the wings of a stingray swimming along the ocean bottom. Extending from the lamellipodia are thin spikes called *filopodia,* which constantly probe the environment, moving in and out of the lamellipodia. Growth of the neurite occurs when a filopodium, instead of retracting, takes hold of the substrate (the surface on which it is growing) and pulls the advancing growth cone forward.

Obviously, axonal growth cannot occur unless the growth cone advances along the substrate. An important substrate consists of fibrous proteins that are deposited in the spaces between cells, the **extracellular matrix**. Growth occurs only if the extracellular matrix contains appropriate proteins. An example of a permissive substrate is the glycoprotein *laminin*. The growing ax-

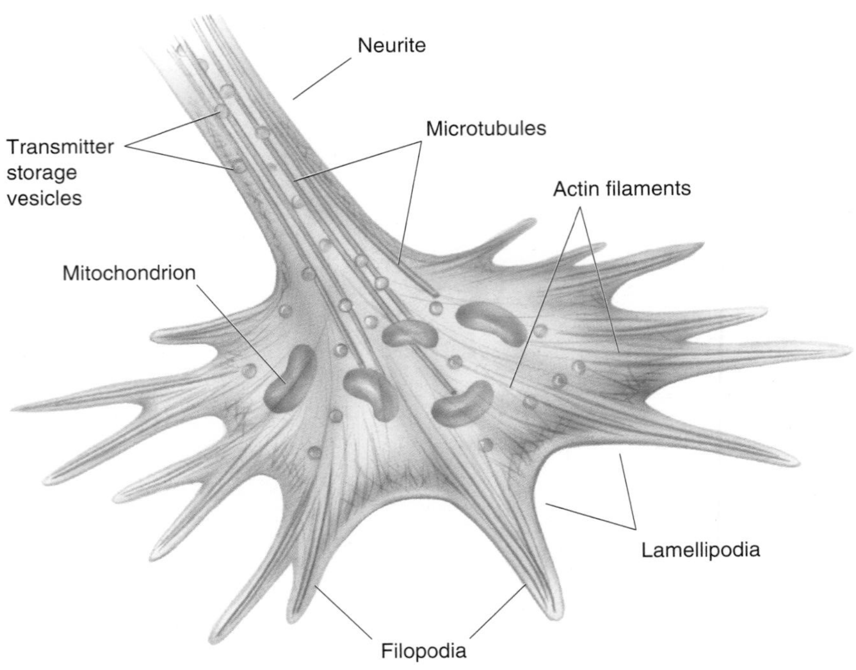

Figure 22.8
The growth cone.

ons express special surface molecules called *integrins* that bind laminin, and this interaction promotes axonal elongation. Permissive substrates, bordered by repulsive ones, can provide corridors that channel axon growth along specific pathways.

Travel down such molecular highways is also aided by **fasciculation**, a mechanism that causes axons that are growing together to stick together (Figure 22.9). Fasciculation is due to the expression of specific surface molecules called **cell adhesion molecules** (**CAMs**). The CAMs in the membrane of neighboring axons bind tightly to one another, causing the axons to grow in unison.

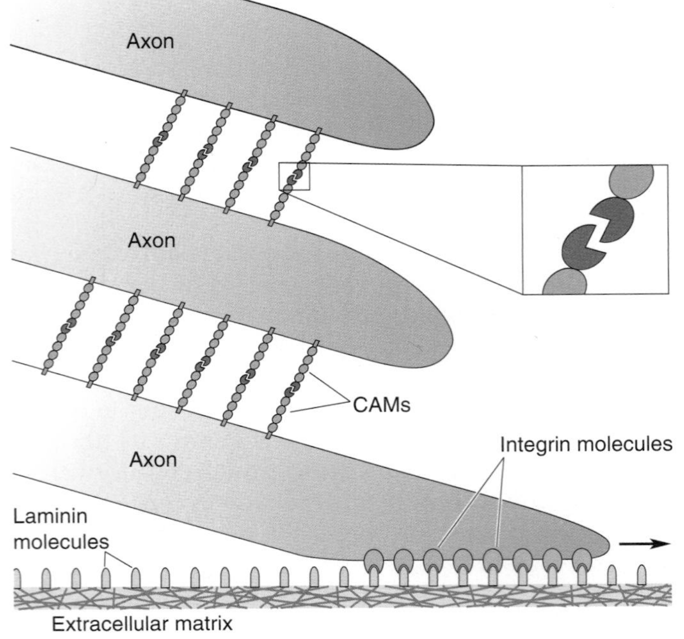

Figure 22.9
Fasciculation. The bottom axon grows along the molecular "highway" of the extracellular matrix. The other axons ride piggyback, sticking to one another by the interaction of cell adhesion molecules (CAMs) on their surfaces.

Axon Guidance

Wiring the brain is a formidable challenge when you consider the distances that many axons traverse in the mature nervous system. Remember, though, that distances are not nearly as great early in development, when the entire nervous system is no more than a few centimeters long. A common mode of pathway formation is the initial establishment of connections by *pioneer axons*. These axons "stretch" as the nervous system expands, and they are used to guide their later-developing neighbor axons to the same targets. Still, the question remains of how the pioneer axons grow in the correct direction along the correct paths to the correct targets. The answer appears to be that the trajectory of the axons is broken into short segments that may only be a few hundred microns long. Axons conclude a segment when they arrive at an intermediate target. The interaction of the axon and the intermediate target throws a molecular switch that sends the axons onward to another intermediate target. Thus, by "connecting the dots," the axons find their way to their final destination.

Guidance Cues. Growth cones differ in terms of the molecules they express on their membranes. Interactions of these cell surface molecules with *guidance cues* in the environment determine the direction and amount of growth. Guidance cues can be attractive or repulsive, depending on the receptors expressed by the axons.

A **chemoattractant** is a diffusible molecule that acts over a distance to attract growing axons toward their targets, as the aroma of freshly brewed java might attract the coffee lover. Although the existence of such chemoattractants was proposed more than a century ago by Cajal and has been inferred by many experimental studies since then, only very recently have attractant molecules been identified. An example is *netrin*, which is secreted by neurons in the ventral midline of the spinal cord (Figure 22.10). The gradient of netrin attracts the axons of dorsal horn neurons that will cross the midline to form the spinothalamic tract (see Figure 12.29). These axons possess netrin receptors, and the binding of netrin to the receptor spurs growth toward the source of netrin.

But this is only half the story. Once the decussating axons cross the midline, they must escape the powerful siren song of netrin. This escape is enabled by the action of *slit*, another protein secreted by midline cells. Slit is an example of a **chemorepellent**, a diffusible molecule that chases axons away. For slit to exert this action, however, the axon must express the slit receptor, a protein called *robo*, on its surface. The growth cones that are attracted to the midline by netrin express little robo and therefore are insensitive to repulsion by slit. Once they cross the midline, however, they encounter a signal that upregulates robo. Now slit repels the axons so they grow away from the midline.

This example shows how axons can be "pulled" and "pushed" by the coordinated actions of chemoattractants and chemorepellents. The trajectory of the axons to and from the midline is also constrained by the permissive substrates that are available for growth. In this example, the cells of the midline are an intermediate target—one of the "dots"—along the molecular highway that spans the midline. These cells serve to alternately attract and repel the growing axon as it crosses from one side of the CNS to the other.

Establishing Topographic Maps. Let's return to the example of the growing retinogeniculate axon (Figure 22.7). These axons grow along the substrate provided by the extracellular matrix of the ventral wall of the optic stalk. An important "choice point" occurs at the optic chiasm. Axons from the nasal

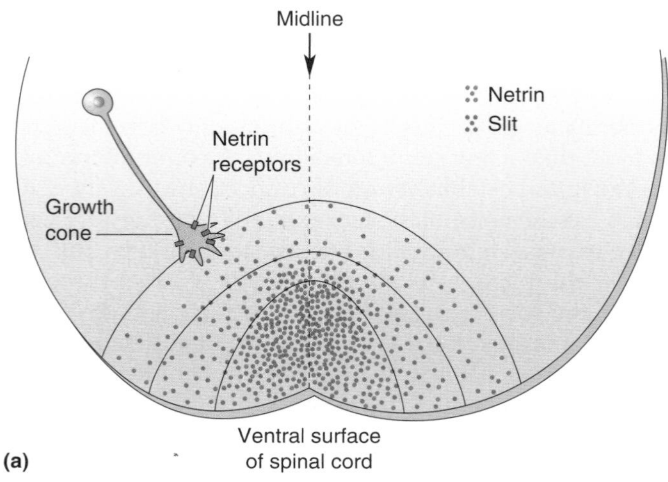

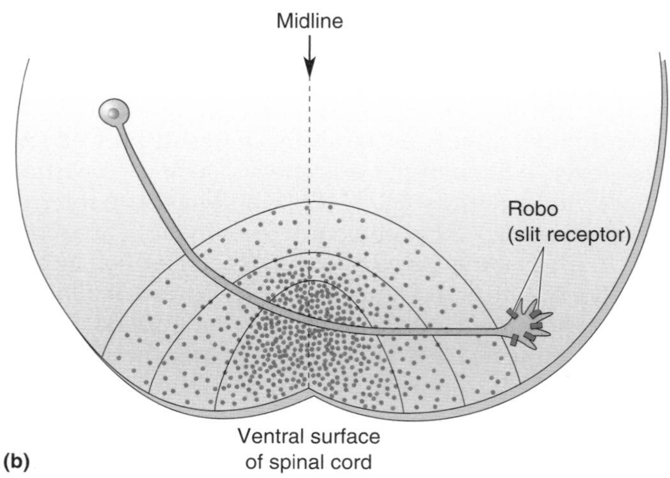

Figure 22.10
Chemoattraction and chemorepulsion. **(a)** Cells in the ventral midline of the spinal cord secrete the protein netrin. Axons with the appropriate netrin receptors are attracted to the region of highest netrin concentration. **(b)** Midline cells also secrete the protein slit. Axons that express the protein robo, the slit receptor, grow away from the region of highest slit concentration. Upregulation of robo by axons that cross the midline ensures that they keep growing away from the midline.

retina cross and ascend in the contralateral optic tract, while axons from the temporal retina remain in the ipsilateral optic tract. From our discussion so far, we can infer that nasal and temporal retinal axons must express different receptors to cues secreted at the midline.

Once the axons from the retinas are sorted out at the midline, they continue on to innervate targets such as the LGN and the superior colliculus. Sorting of the axons occurs again, this time to establish a retinotopic map in the target structure (see Figure 10.7). If we accept the notion that axons differ on the basis of their position in the retina (as they must to account for the partial decussation at the optic chiasm), we have a potential molecular basis for the establishment of retinotopy. This idea, that chemical markers on growing axons are matched with complementary chemical markers on their targets to establish precise connections, is called the **chemoaffinity hypothesis**.

In the 1940s, Roger Sperry, using the retinotectal projection in frogs, first tested this hypothesis in an important series of experiments at the California Institute of Technology. Recall from Chapter 10 that the tectum is the amphibian homologue of the mammalian superior colliculus. The tectum receives retinotopically ordered input from the contralateral eye and uses this information to organize movements in response to visual stimulation, such as lunging after a fly passing overhead. Thus, this system can be used to investigate the mechanisms that generate orderly maps in the CNS.

Another advantage of amphibians is that their CNS axons regenerate after being cut, which is not true for mammals (Box 22.2). Sperry took advantage of this property to investigate how the retinotopic map was established in the tectum. In one experiment, Sperry cut the optic nerve, rotated the eye 180 degrees in the orbit, and then allowed the upside-down nerve to regenerate. Despite the fact that the axons in the optic nerve were now scrambled from where they would occur naturally, the axons grew into the tectum to exactly the same sites that they occupied originally. Now, when a fly passed overhead, these frogs with rotated eyes lunged down instead of up!

What factors control the guidance of retinal axons to the correct part of the tectum? When the axons arrive at the tectum, they must grow along the membranes of tectal cells. The axons from the nasal retina cross the anterior part of the tectum and innervate the neurons in the posterior part. The axons of the temporal retina, in contrast, grow into the anterior tectum and stop there (Figure 22.11a). Why? Experiments have shown that the cell membranes of anterior and posterior tectal neurons differentially express factors that are permissive for the growth of nasal and temporal retinal axons. Nasal axons grow well on the substrate provided by both anterior and posterior tectal membranes. Temporal axons grow only on anterior tectal membranes, however; the posterior membranes are repulsive (Figure 22.11b). Recent research has led to the discovery that proteins called *ephrins* are one repulsive signal for temporal retinal axons. Specific ephrin molecules are expressed in a gradient across the surface of the tectum, with the highest levels found on posterior tectal cells. The ephrins interact with receptors that are expressed on the growing axons. The interaction of ephrin with its receptor inhibits further axonal growth, similar to the slit-robo interaction discussed earlier.

Although the story remains incomplete, it is clear that such gradients in the expression of guidance cues and their axonal receptors can impose considerable topographic order on the wiring of the retina to its targets in the brain.

Synapse Formation

When the growth cone comes in contact with its target, a synapse is formed. The details of the mechanisms of synapse formation in the CNS are still sketchy; most of what is known comes from studies of the neuromuscular junction. At the neuromuscular junction, the first step appears to be the induction of a cluster of postsynaptic receptors under the site of contact. This clustering is triggered by an interaction between proteins secreted by the growth cone and the target membrane. At the neuromuscular junction, one of these proteins, called *agrin*, is deposited in the extracellular space at the site of contact (Figure 22.12). The layer of proteins in this space is called the **basal lamina**. Agrin in the basal lamina binds to a receptor called *muscle-specific kinase*, or *MuSK*, in the muscle cell membrane. MuSK communicates with another molecule, *rapsyn*, which appears to act like a shepherd to gather the postsynaptic acetylcholine receptors (AChRs) at the synapse. The size of the flock of receptors is regulated by another molecule released by the axon, called *neuregulin*, which stimulates the receptor gene expression in the muscle cell.

The interaction between axon and target occurs in both directions, and the induction of a presynaptic terminal also appears to involve proteins in the basal lamina. Basal lamina factors provided by the target cell evidently can stimulate Ca^{2+} entry into the growth cone, which triggers neurotransmitter release. Thus, although the final maturation of synaptic structure may take a

Box 22.2

OF SPECIAL INTEREST

Why Don't Axons Regenerate in Our CNS?

Compared with other vertebrates, mammals are fortunate in many ways. We have computing power and behavioral flexibility that our distant aquatic cousins, the fish and amphibians, utterly lack. In one interesting respect, however, fish and frogs have a distinct advantage—the growth of axons in the adult CNS after injury. Cut the optic nerve in a frog, and it grows back. Do the same thing in a human, and the person is blind forever. Of course, our CNS axons do grow over long distances early in development. But something happens shortly after birth that makes the CNS, especially the white matter, a hostile environment for axon growth.

When an axon is cut, the distal segment degenerates because it is isolated from the soma. The severed tip of the proximal segment initially responds, however, by emitting growth cones. In the adult mammalian CNS, this growth is aborted. Not so in the mammalian PNS (peripheral nervous system). If you've ever had a deep cut sever a peripheral nerve, you know that eventually, over time, sensations can come back in the denervated skin. This happens because PNS axons are capable of regeneration over long distances.

Surprisingly, the critical difference between the mammalian PNS and the CNS is not the neurons. A PNS dorsal root ganglion cell axon regenerates fine in the peripheral nerve, but when it hits the environment of the CNS in the dorsal horn, growth ceases. Conversely, if a CNS alpha motor neuron axon is cut in the periphery, it grows back to its target. If it is cut in the CNS, no regeneration occurs. Thus, the critical difference seems to be the different environments of the CNS and PNS. Beginning in the early 1980s, Albert Aguayo and his colleagues at Montreal General Hospital tested this idea in a very important series of experiments. They showed that crushed optic nerve axons can grow long distances if they are given a peripheral nerve graft to grow along (Figure A).

What is different about peripheral nerves? One difference is the type of myelinating glial cell: oligodendroglia in the CNS and Schwann cells in the PNS (see Chapter 2). Experiments performed by Martin Schwab of the University of Zurich showed that CNS neurons grown in tissue culture extend axons along substrates prepared from Schwann cells but not from CNS oligodendroglia. This finding led to the search for glial factors that inhibit axon growth, and a molecule called nogo was finally identified early in 2000. Nogo apparently is released when oligodendroglia are damaged.

Antibodies raised against nogo neutralize the molecule's growth-suppressing activity. Schwab and his colleagues have injected the anti-nogo antibody (called IN-1) into adult rats after spinal cord injury. This treatment enabled about 5% of the severed axons to regenerate—a modest effect, perhaps, but sufficient for the animals to show a remarkable functional recovery. The same antibodies have also been used to localize nogo in the nervous system. The protein is made by oligodendroglia in mammals but not in fish, and it is not found in Schwann cells.

One of the last steps in wiring the mammalian brain is wrapping the young axons in myelin. This has the beneficial effect of speeding action potential conduction, but it comes with a heavy cost—the inhibition of axon growth after injury. In the twentieth century, neurologists accepted the lack of axon regeneration in the adult CNS as a dismal fact of life. The recent understanding of molecules with the power to stimulate or inhibit CNS axon growth, however, offers hope for the twenty-first century that treatments can be devised to promote axon regeneration in the damaged human brain and spinal cord.

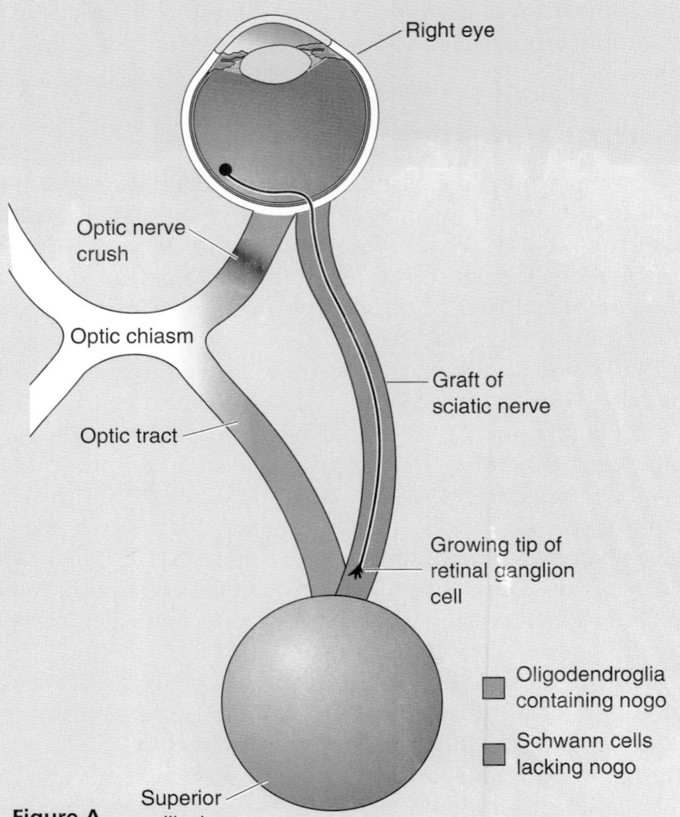

Figure A

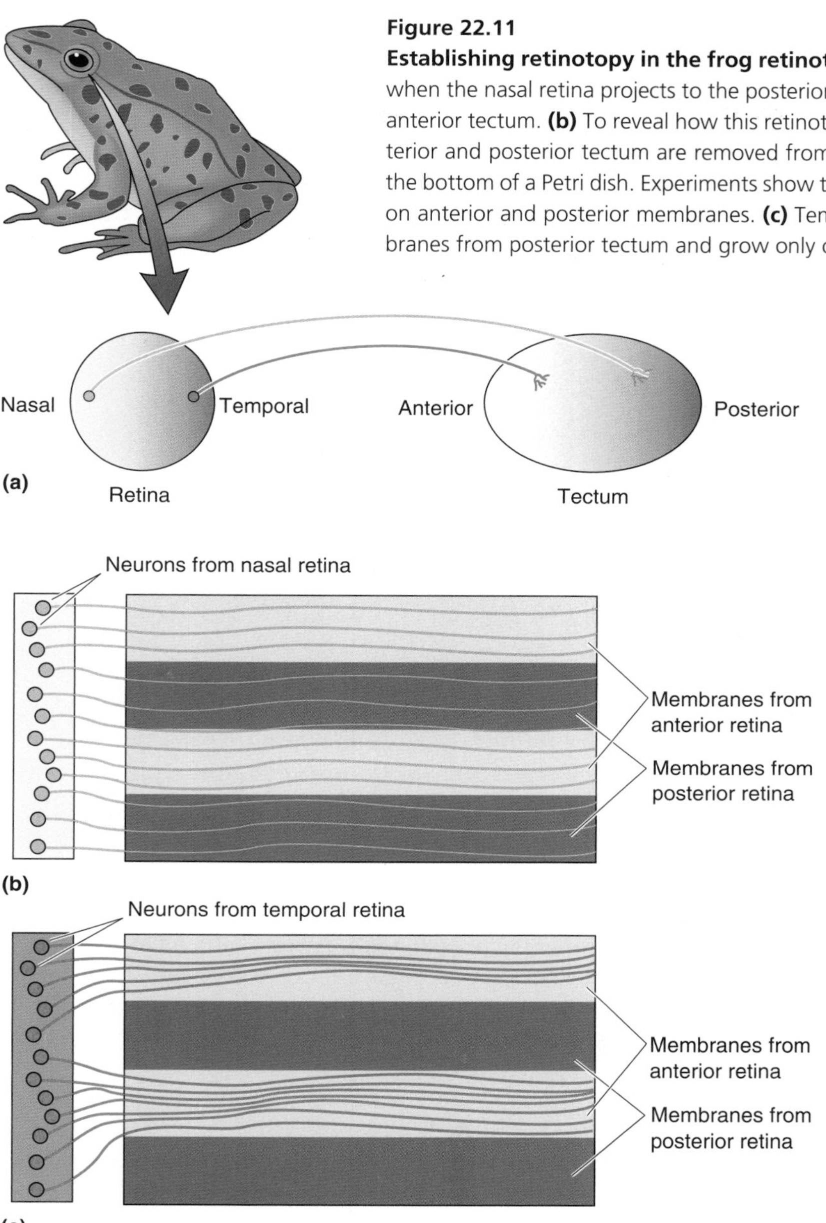

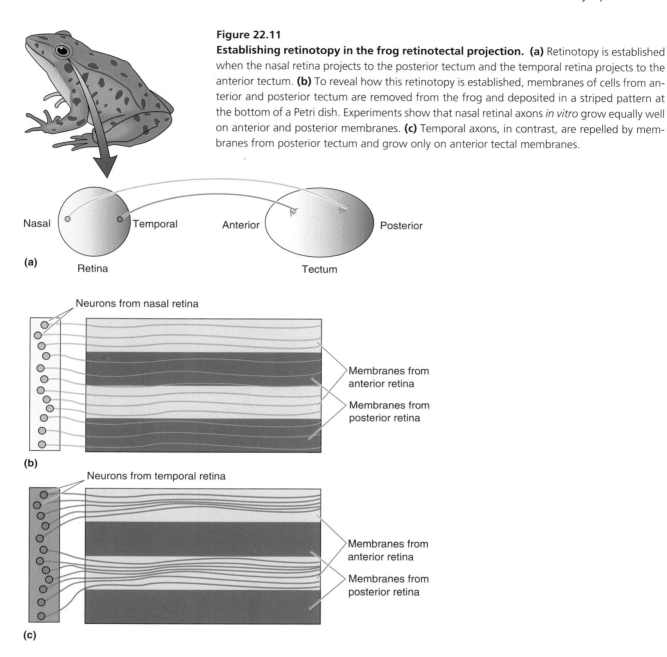

Figure 22.11
Establishing retinotopy in the frog retinotectal projection. (a) Retinotopy is established when the nasal retina projects to the posterior tectum and the temporal retina projects to the anterior tectum. **(b)** To reveal how this retinotopy is established, membranes of cells from anterior and posterior tectum are removed from the frog and deposited in a striped pattern at the bottom of a Petri dish. Experiments show that nasal retinal axons *in vitro* grow equally well on anterior and posterior membranes. **(c)** Temporal axons, in contrast, are repelled by membranes from posterior tectum and grow only on anterior tectal membranes.

matter of weeks, rudimentary synaptic transmission appears very rapidly after contact is made. Besides mobilizing transmitter, Ca^{2+} entry into the axon also triggers changes in the cytoskeleton that cause it to assume the appearance of a presynaptic terminal and to adhere tightly to its postsynaptic partner.

THE ELIMINATION OF CELLS AND SYNAPSES

These mechanisms of pathway formation are sufficient to establish considerable order in the connections of the fetal brain. For example, in the visual system these mechanisms ensure that (1) retinal axons reach the LGN, (2) geniculate axons reach layer IV of striate cortex, and (3) both of these sets of axons form synapses in their target structures in proper retinotopic order. But the job of wiring together the nervous system isn't finished yet. During a long period of development, from before birth all the way through adolescence,

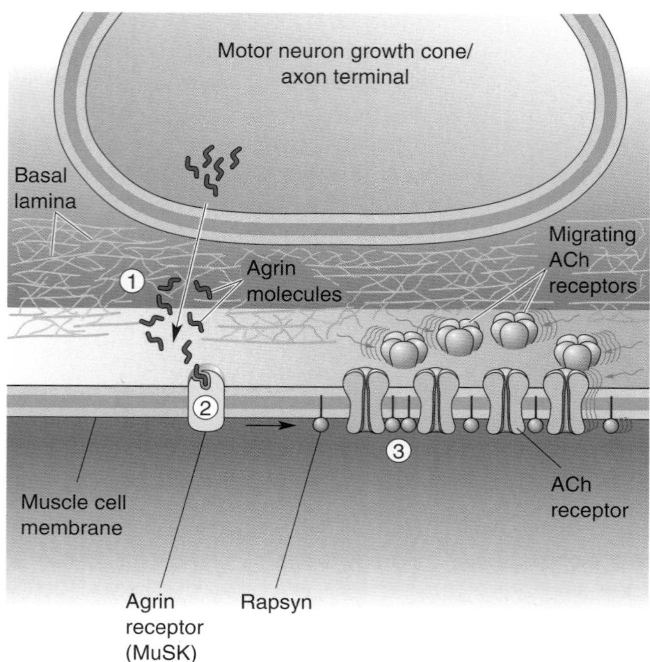

Figure 22.12
Steps in the formation of a neuromuscular synapse. ① The growing motor neuron secretes the protein agrin into the basal lamina. ② Agrin interacts with MuSK in the muscle cell membrane. This interaction leads to ③ the clustering of ACh receptors in the postsynaptic membrane via the actions of rapsyn.

these connections are refined. It may come as a surprise that one of the most significant refinements is a large-scale *reduction* in the numbers of all those newly formed neurons and synapses.

Cell Death

Entire populations of neurons are eliminated during pathway formation, a process known as *programmed cell death*. After axons have reached their targets and synapse formation has begun, there is a progressive decline in the number of presynaptic axons and neurons. Cell death reflects competition for **trophic factors**, life-sustaining substances that are provided in limited quantities by the target cells. This process is believed to produce the proper match in the number of presynaptic and postsynaptic neurons (Figure 22.13).

A peptide called **nerve growth factor** (**NGF**) was the first trophic factor to be identified in the 1940s by Italian biologist Rita Levi-Montalcini. NGF is produced by the targets of axons in the sympathetic division of the autonomic nervous system (ANS). Levi-Montalcini and Stanley Cohen found that the injection of antibodies to NGF into newborn mice resulted in total degeneration of the sympathetic ganglia. NGF, produced and released by the target tissue, is taken up by the sympathetic axons and transported retrogradely, where it acts to promote neuronal survival. Indeed, if axoplasmic transport is disrupted, the neurons die despite the release of NGF by the target tissue. Their pioneering work earned Levi-Montalcini and Cohen the 1986 Nobel Prize.

NGF is one of a family of related trophic proteins collectively called the **neurotrophins**. Family members include the proteins *NT-3*, *NT-4*, and *brain-derived neurotrophic factor* (*BDNF*), which is important for the survival of visual cortical neurons. Neurotrophins act at specific cell surface receptors. Most of the receptors are neurotrophin-activated protein kinases, called *trk receptors*, that phosphorylate tyrosine residues on their substrate proteins (recall phosphorylation from Chapter 6). This phosphorylation reaction stimulates a second messenger cascade that ultimately alters gene expression in the cell's nucleus.

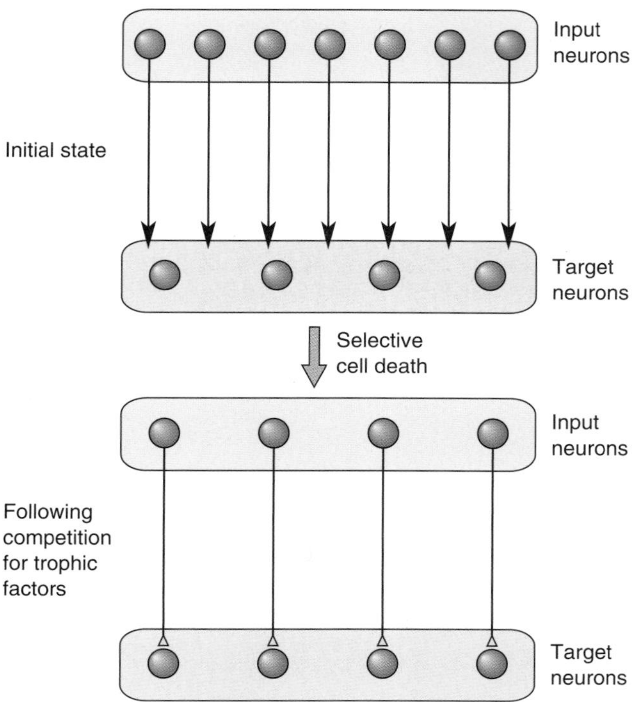

Figure 22.13
Matching inputs with targets by selective cell death. The input neurons are believed to compete with one another for limited quantities of trophic factors produced by the target neurons.

The description of cell death during development as "programmed" reflects the fact that it is actually a consequence of genetic instructions to self-destruct. Neurotrophins save neurons by switching off this genetic program. Expression of the cell death genes causes neurons to die by a process called **apoptosis**, the systematic disassembly of the neuron. Apoptosis differs from *necrosis*, which is accidental cell death resulting from injury to cells. Research on neuronal cell death is proceeding at a rapid pace, fueled by the promise of rescuing dying neurons in neurodegenerative disorders such as Alzheimer's disease (see Box 2.2) and amyotrophic lateral sclerosis (see Box 13.1).

Changes in Synaptic Capacity

Each neuron can receive on its dendrites and soma a finite number of synapses. This number is the *synaptic capacity* of the neuron. Throughout the nervous system, synaptic capacity peaks early in development and then declines as the neurons mature. For example, in the striate cortex of all species examined so far, the synaptic capacity of immature neurons exceeds that of adult cells by about 50%. In other words, visual cortical neurons in the infant brain receive one and a half times as many synapses as do the neurons in adults.

When do cortical neurons lose all those synapses? Yale University scientists Jean-Pierre Bourgeois and Pasko Rakic conducted a detailed study to address this question in the striate cortex of the macaque monkey. They discovered that synaptic capacity was remarkably constant in the striate cortex from infancy until puberty. During the subsequent adolescent period, however, synaptic capacity declined sharply—by almost 50% in just over 2 years. A quick calculation revealed the following startling fact: The loss of synapses in the primary visual cortex during adolescence occurs at an average rate of *5000 per second*. (No wonder adolescence is such a trying time!)

Once again, the neuromuscular junction has provided a useful model system for the study of synaptic elimination. Initially, a muscle fiber may receive

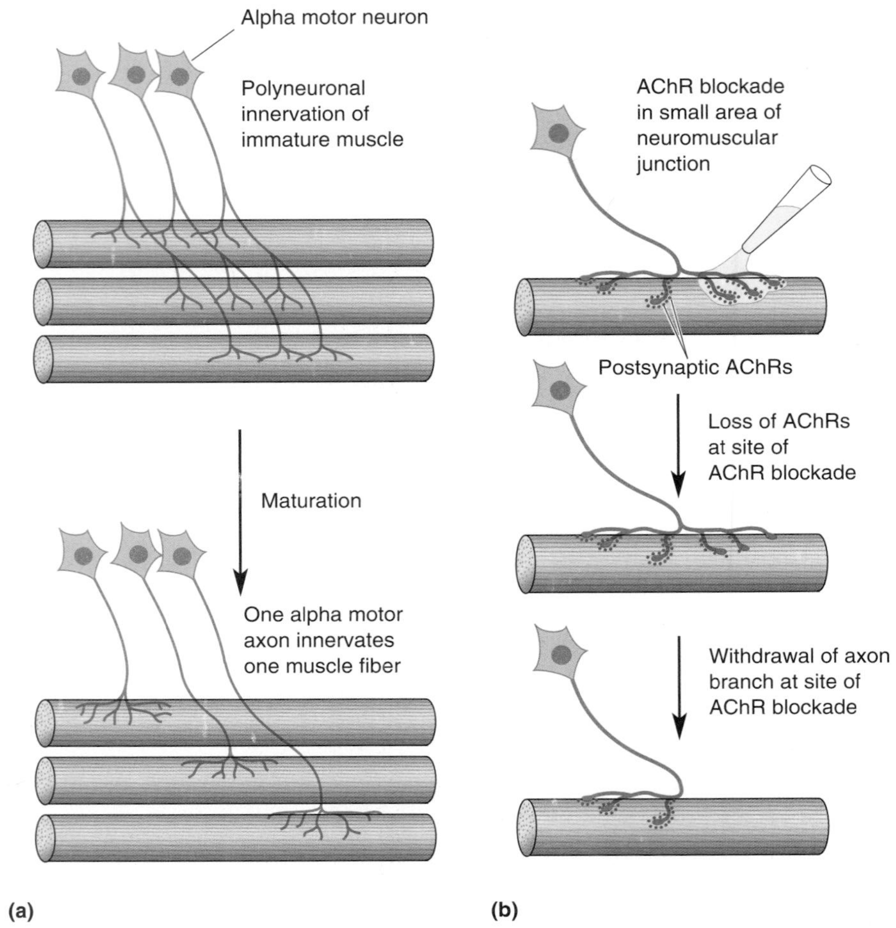

Figure 22.14
Synapse elimination. (a) Initially, each muscle fiber receives inputs from several alpha motor neurons. Over the course of development, all inputs but one are lost. **(b)** Normally, postsynaptic AChR loss precedes the withdrawal of the axon branch. Simply blocking a subset of receptors with α-bungarotoxin can also stimulate synapse elimination.

input from several motor neurons. Eventually, however, this polyneuronal innervation is lost, and each muscle fiber receives synaptic input from a single alpha motor neuron (Figure 22.14a). This process is regulated by electrical activity in the muscle. Silencing the activity of the muscle fiber leads to a retention of polyneuronal innervation, while stimulation of the muscle accelerates the elimination of all but one input.

Careful observations have revealed that the first change during synapse elimination is the loss of postsynaptic AChRs, followed by the disassembly of the presynaptic terminal and retraction of the axon branch. What causes the receptors to disappear? The answer appears to be insufficient receptor activation in an otherwise active muscle. If receptors are partially blocked with α-bungarotoxin (see Box 5.5), they are internalized and the overlying axon terminal withdraws (Figure 22.14b). If *all* of the AChRs are blocked, however, the synapses remain because the muscle is also silent. As we will see in a moment, a similar process appears to occur during the refinement of connections in the CNS.

ACTIVITY-DEPENDENT SYNAPTIC REARRANGEMENT

Imagine a neuron that has a synaptic capacity of 12 synapses and receives inputs from 3 presynaptic neurons, A, B, and C (Figure 22.15). One possible arrangement is that each of the 3 presynaptic neurons provides 4 synapses apiece. Another arrangement is that neurons A and C provide 6 synapses apiece and neuron B provides none. A change from one such pattern of

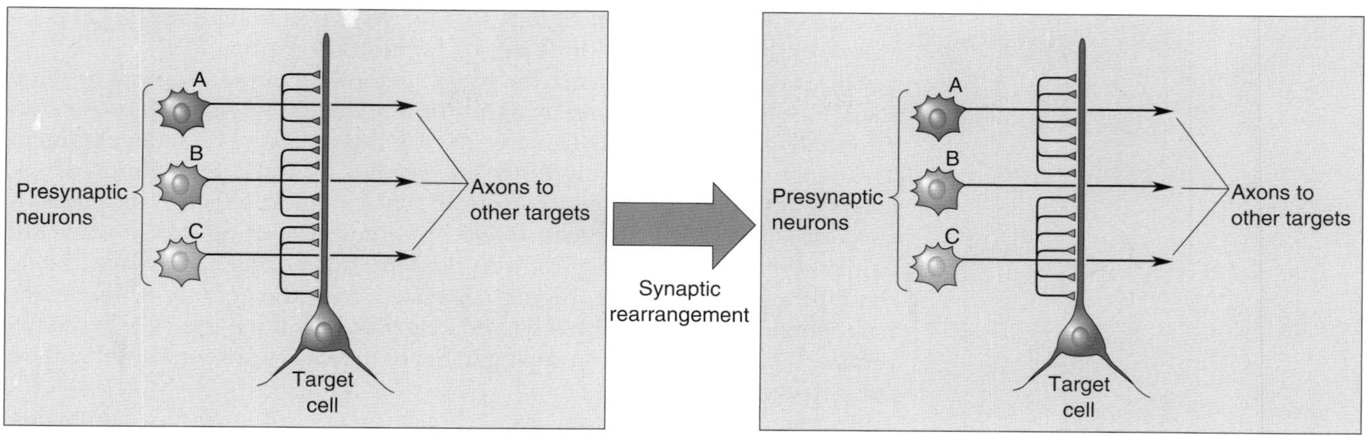

Figure 22.15

Synaptic rearrangement. The target cell receives the same number of synapses in both cases, but the innervation pattern has changed.

synapses to another is called *synaptic rearrangement.* There is abundant evidence for widespread synaptic rearrangement in the immature brain.

Synaptic rearrangement is the final step in the process of address selection. Unlike most of the earlier steps of pathway formation, *synaptic rearrangement occurs as a consequence of neural activity and synaptic transmission.* In the visual system, some of this activity-dependent shaping of connections occurs prior to birth in response to spontaneous neuronal discharges. Significant activity-dependent development occurs after birth, however, profoundly influenced by sensory experience during childhood. Thus, we will find that the ultimate performance of the adult visual system is determined to a significant extent by the quality of the visual environment during the early postnatal period. In a very real sense, *we learn to see during a critical period of postnatal development.*

Macaque monkeys and cats have typically been used as models for studies of activity-dependent visual system development because, like humans, both of these species have good binocular vision. The neuroscientists who pioneered this field were none other than David Hubel and Torsten Wiesel, who, as you will recall from Chapter 10, also laid the foundation for our current understanding of the central visual system in the adult brain. In 1981, they shared the Nobel Prize with Roger Sperry.

Synaptic Segregation

A feature of the development of both the LGN and the striate cortex, first discovered in the macaque monkey by Pasko Rakic, is that the inputs that carry information from the two eyes initially converge onto the same target cells. Then, over several weeks, these inputs segregate out into the eye-specific LGN layers and ocular dominance columns we learned about in Chapter 10. We will discuss segregation of synaptic inputs in the LGN and striate cortex in turn.

Segregation of Retinal Axons in the LGN. The first axons to reach the LGN are usually those from the contralateral retina, and they spread out to occupy the entire nucleus. Somewhat later, the ipsilateral projection arrives and intermingles with the axons of the contralateral eye. Then the axons from the two eyes segregate into the eye-specific domains that are characteristic of the adult nucleus. Silencing retinal activity with TTX (tetrodotoxin) prevents this process of segregation (recall that TTX blocks action potentials). What is the source of the activity, and how does it orchestrate segregation?

Since segregation occurs in the womb before the photoreceptors develop, the activity cannot be driven by light stimulation. Rather, it appears that ganglion cells are spontaneously active during this period of fetal development. This activity is not random, however. Studies by Carla Shatz and her colleagues at Stanford University indicate that ganglion cells fire in quasisynchronous "waves" that spread across the retina. The origin of the wave and its direction of propagation may be random, but during each wave, the activity in a ganglion cell is highly correlated with the activity of its nearest neighbors. And because these waves are generated independently in the two retinas, the activity patterns arising in the two eyes are not correlated with respect to each other.

Segregation is thought to depend on a process of synaptic stabilization whereby only retinal terminals are retained that are active at the same time as their postsynaptic LGN target neuron. This hypothetical mechanism of synaptic plasticity was first articulated by Canadian psychologist Donald Hebb in the 1940s. Consequently, synapses that can be modified in this way are called **Hebb synapses**, and synaptic rearrangements of this sort are called **Hebbian modifications**. According to this hypothesis, whenever a wave of retinal activity drives a postsynaptic LGN neuron to fire action potentials, the synapses between them are stabilized (Figure 22.16). Because the activity from the two eyes does not occur at the same time, the inputs compete on a "winner-takes-all" basis until one input is retained and the other is eliminated. Stray retinal inputs in the inappropriate LGN layer are the losers be-

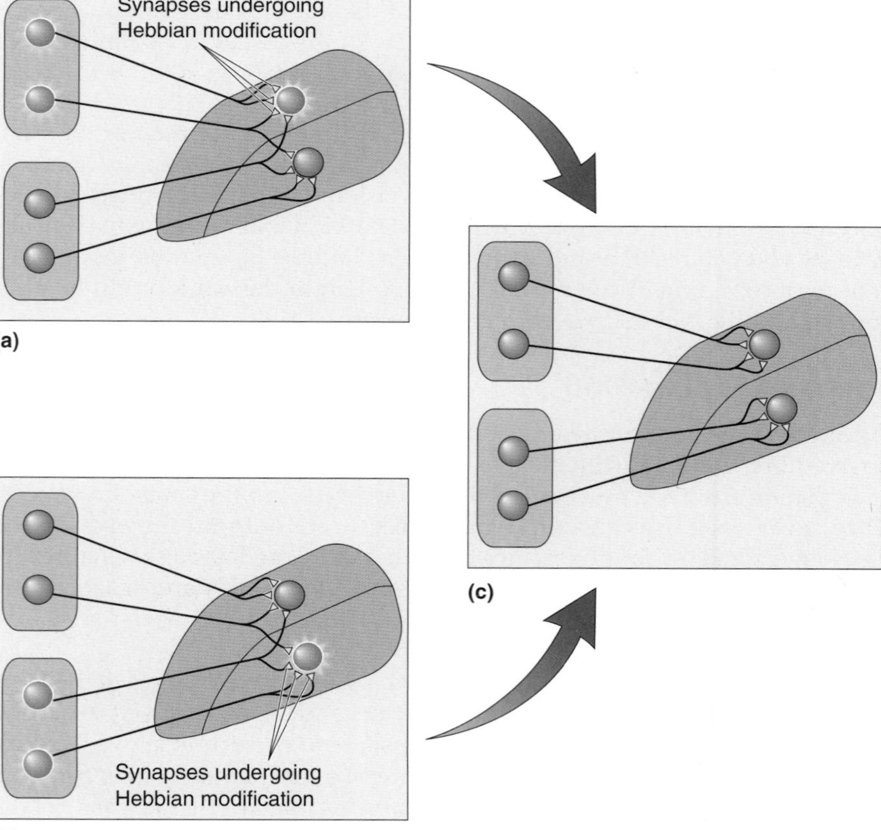

Figure 22.16
Plasticity at Hebb synapses. Two target neurons in the LGN have inputs from different eyes. Inputs from the two eyes initially overlap and then segregate under the influence of activity. **(a)** The two input neurons in one eye (top) fire at the same time. This is sufficient to cause the top LGN target neuron to fire but not the bottom one. The active inputs onto the active target undergo Hebbian modification and become more effective. **(b)** This is the same situation as in part a, except that now the two input neurons in the other eye (bottom) are active simultaneously, causing the bottom target neuron to fire. **(c)** Over time, neurons that fire together wire together. Notice also that input cells that fire out of sync with the target lose their link.

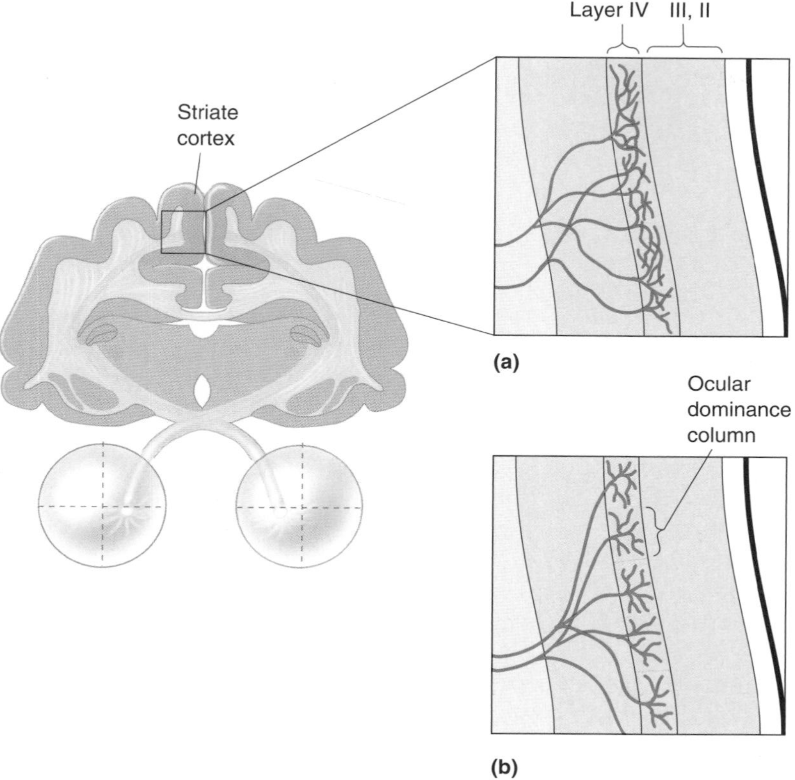

Figure 22.17
Segregation of ocular dominance columns in cat striate cortex. (a) Initially the inputs from the LGN serving the two eyes (different colors) are intermingled in layer IV. **(b)** Over the course of fetal and early postnatal development, the inputs from the two eyes segregate into ocular dominance columns in layer IV.

cause their activity does not consistently correlate with the strongest postsynaptic response, which is evoked by the activity of the other eye. In a moment, we will explore some potential mechanisms for such correlation-based synaptic modification.

Segregation of LGN Inputs in the Striate Cortex. Similar to the situation in the LGN, the afferents serving the two eyes are initially intermingled in cortical layer IV and then segregate under the influence of activity (Figure 22.17). In the macaque monkey, this segregation of ocular dominance columns is completed by birth, while in the cat it persists for several postnatal weeks.

The appearance of segregation does not mean that the axons lose their ability to grow and retract, however. "Plasticity" of ocular dominance columns after birth can be dramatically demonstrated by an experimental manipulation used by Wiesel and Hubel, called **monocular deprivation**, in which one eyelid is sealed closed (Box 22.3). If monocular deprivation is begun shortly after birth, the striking result is that the "open-eye" columns expand in width while the "closed-eye" columns shrink (Figure 22.18). Moreover, these effects of monocular deprivation can be reversed simply by closing the previously open eye and opening the previously closed eye. The result of this "reversed-occlusion" manipulation is that the shrunken ocular dominance columns of the formerly closed eye expand and the expanded columns of the formerly open eye shrink. Thus, LGN axons and their synapses in layer IV are highly dynamic even after birth. This type of synaptic rearrangement is not just activity dependent, it is *experience dependent* because it relies on the quality of the sensory environment.

The plasticity of ocular dominance columns does not occur throughout life. Hubel and Wiesel found that if the deprivation is begun later in life, these

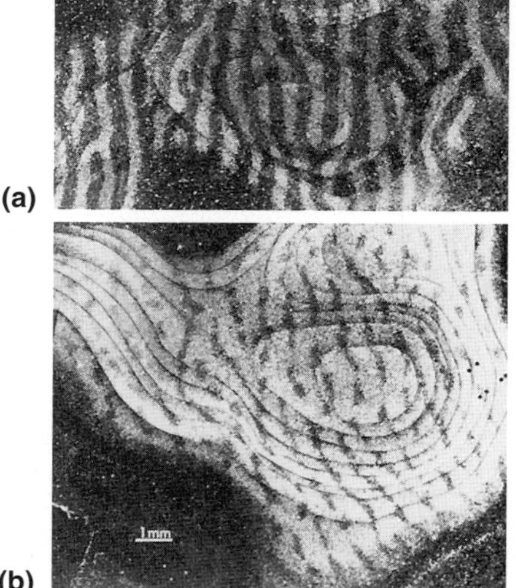

Figure 22.18
Modification of ocular dominance stripes after monocular deprivation. Tangential sections through layer IV of macaque monkey striate cortex illuminated to show the distribution of radioactive LGN terminals serving one eye. **(a)** A normal monkey. **(b)** A monkey that had been monocularly deprived for 22 months, starting at 2 weeks of age. The nondeprived eye had been injected, revealing expanded ocular dominance columns in layer IV. (Source: Wiesel, 1982, p. 585.)

Box 22.3

PATH OF DISCOVERY

Exploring Visual Cortical Plasticity

BY TORSTEN WIESEL

My background is a bit unusual. My father was the director of a mental hospital in Sweden, and since we lived on the hospital grounds, I learned as a young boy to interact with individuals with behavioral problems. It was probably this experience that led me to medical school and fostered my interest in neuroscience. After medical school and some clinical experience in child psychiatry, I became painfully aware of how little was known about higher brain functions; ever since, my focus has been on basic research in systems neuroscience and brain development.

I had the great luck in 1955 to be invited as a postdoctoral student to the laboratory of Stephen Kuffler, who was then at Johns Hopkins Medical School. He had just published his classical work on the receptive field of cat retinal ganglion cells, and he became my mentor and role model. My second stroke of luck came when David Hubel joined the laboratory in 1958. The two of us embarked on an exhilarating 20-year journey from the retina to the lateral geniculate body to the primary visual cortex, analyzing the structure and function of this central visual pathway at the level of single cells.

The cat and monkey visual cortex revealed some very intriguing information as we probed single neurons with David's tungsten microelectrode. To our amazement, we discovered that a given cortical cell responded to contours of a specific orientation; moreover, cells with the same orientation preference were grouped together into what we called orientation columns, using a term first formulated by Vernon Mountcastle to describe the organization of cells in the somatosensory cortex. Also, although most cells were binocular—i.e., they could be driven by input from either eye—most showed a clear preference for one eye or the other; such cells were grouped into what we termed ocular dominance columns.

Torsten Wiesel

We started to wonder how such intricate cellular properties developed and how cells with similar properties aggregated into columns. Was visual experience necessary? David and I both had clinical backgrounds and knew, for instance, that children born with cataracts never recovered full vision

anatomical effects are not observed in layer IV. Thus, a **critical period** exists for this type of structural modification. In the macaque monkey, the critical period for anatomical plasticity in layer IV lasts until about 6 weeks of age. At the end of the critical period, the LGN afferents apparently lose their capacity for growth and retraction and, in a sense, are cemented in place.

There are many critical periods during development—specific times when developmental fate is influenced by the environment (Box 22.4). In the visual cortex, the end of the critical period for anatomical plasticity in layer IV does not spell the end of the influence of visual experience on cortical development. The synapses outside layer IV in striate cortex remain modifiable by experience until adolescence.

Synaptic Convergence

Although the streams of information from the two eyes are segregated in the LGN and layer IV of striate cortex, eventually they must be combined to give rise to binocular vision. The anatomical basis for binocular vision is the convergence of inputs from layer IV cells serving the right and left eyes onto cells

(Box 22.3, *continued*)

after the cataracts were removed when the child was 3–5 years of age. Indeed, such cases were often mentioned in the debate on the relative contributions of nature (heredity) and nurture (experience) in the development of brain functions.

Our initial approach to these developmental questions was simply to record from single cortical cells in newborn animals that had no visual experience. We found that these cells were remarkably similar to cells in the adult animal; they responded to specific orientations, showed precise binocular responses, and were organized into columns. We concluded that, at least in the monkey and the cat, the central visual pathway was wired up and ready to function at birth.

But if central visual circuits developed in the absence of visual experience, why did children with congenital cataracts fail to see properly when their cataracts were removed? How could we investigate whether abnormal visual experience early in life led to changes in the visual brain? We at first contemplated raising animals in the dark but quickly concluded that we lacked the resources and patience for such an elaborate procedure. Instead, we decided to adapt a standard clinical practice and occlude one eye by lid suture, leaving the other eye open as a control. This approach of closing the lids at various ages and for different lengths of time became our standard procedure for studying the effects of visual deprivation in kittens and neonatal monkeys.

From a series of experiments we learned that early in life, there is a critical period during which the neural connections present at birth can be lost or modified by visual deprivation. In the monkey, the occlusion of one eye for a few days during the first few weeks after birth caused marked and permanent changes in the visual cortex: Few cells responded to stimulation of the previously occluded eye. Instead, there was nearly complete dominance of vision by the open eye, accompanied by dramatic changes in the ocular dominance columns: The open eye took control of cortical territory normally occupied by the closed eye.

The animal's susceptibility to these changes declined with age, so that the primate was resistant to the effects of lid closure at 6–12 months after birth. This would explain the inability to restore vision in a child who was born with a cataract that was not removed until 3–5 years of age; it also would explain the converse finding that vision is completely restored when a cataract acquired as an adult is removed.

These studies demonstrated that early in life, neuronal pathways are highly sensitive to environmental influences. We should not conclude, however, that circuits in the visual system and other parts of the brain cannot be modified after the critical period and throughout adulthood. In classic experiments with adult monkeys, Michael Merzernich showed that the representation of fingers in the somatosensory cortex can be modified dramatically through experience; thus, we imagine that learning to play the piano even as an adult can change brain circuits connected to our hands. Indeed, understanding plasticity in terms of the neural basis of learning and memory is among the central challenges for future generations of neuroscientists.

in layer III. These are among the last connections to be specified during the development of the retinogeniculocortical pathway. Again, experience-dependent synaptic rearrangement plays a major role in this process.

Binocular connections are formed and modified under the influence of the visual environment during infancy and early childhood. Unlike segregation of eye-specific domains, which evidently depends on *asynchronous* patterns of activity spontaneously generated in the two eyes, the *maintenance of binocular receptive fields depends on correlated patterns of activity that arise from the two eyes as a consequence of vision*. This has been demonstrated clearly by experiments that bring the patterns of activity from the two eyes out of register. For example, monocular deprivation, which replaces patterned activity in one eye with random activity, profoundly disrupts the binocular connections in striate cortex. Neurons outside of layer IV, which normally have binocular receptive fields, respond only to stimulation of the nondeprived eye after even a brief period of monocular deprivation. This change in the binocular organization of the cortex is called an **ocular dominance shift** (Figure 22.19).

These effects of monocular deprivation are not merely a passive reflection of the anatomical changes in layer IV discussed earlier. An ocular dominance

Box 22.4

BRAIN FOOD

The Critical Period Concept

A critical period of development may be defined as a time in which intercellular communication alters a cell's fate. The concept is usually credited to the experimental embryologist Hans Spemann. Working around the turn of the twentieth century, Spemann showed that transplantation of a piece of early embryo from one location to another often caused the "donor" tissue to take on the characteristics of the "host," but only if transplantation took place during a well-defined period. Once the transplanted tissue had been induced to change its developmental fate, the outcome could not be reversed. The intercellular communication that altered the physical characteristics (the phenotype) of the transplanted cells was shown to be mediated both by contact and by chemical signals.

The term took on new significance with respect to brain development as a consequence of the work of Konrad Lorenz in the mid-1930s. Lorenz was interested in the process by which young graylag geese come to be socially attached to their mother. He discovered that in the absence of the mother the social attachment could occur instead to a wide variety of moving objects, including Lorenz himself (Figure A). Once imprinted on an object, the goslings would follow it and behave toward it as if it were their mother. Lorenz used the word "imprinting" to suggest that this first visual image was somehow permanently etched in the young bird's nervous system. Imprinting was also found to be limited to a finite window of time (the first 2 days after hatching), which Lorenz called the "critical period" for social attachment. Lorenz himself drew the analogy between this process of imprinting the external environment on the nervous system and the induction of tissue to change its developmental fate during critical periods of embryonic development.

This work had a tremendous influence in the field of developmental psychology. The very terms *imprinting* and *critical period* suggested that changes in the behavioral phenotype caused by early sensory experience were permanent and irreversible later in life, much like the determination of tissue phenotype during embryonic development. Numerous studies extended the critical period concept to aspects of mammalian psychosocial development. The fascinating implication was that the fate of neurons and neural circuits in the brain depended on the experience of the animal during early postnatal life. It is not difficult to appreciate why research in this area took on social as well as scientific significance.

By necessity, the effects of experience on neuronal fate must be exercised by neural activity generated at the sensory epithelia and communicated by chemical synaptic transmission. The idea that synaptic activity can alter the fate of neuronal connectivity during CNS development first received solid neurobiological support from the study of mammalian visual system development, beginning with the experiments of Hubel and Wiesel. Using anatomical and neurophysiological methods, they found that visual experience or lack thereof was an important determinant of the state of connectivity in the central visual pathways and that this environmental influence was restricted to a finite period of early postnatal life. A great deal of work has been devoted to the analysis of experience-dependent plasticity of connections in the visual system. Thus, it is an excellent model for illustrating the principles of critical periods in nervous system development.

Figure A
Konrad Lorenz with graylag geese. (Source: Nina Leen/TimePix.)

shift, reflecting changes in intracortical connections, occurs in response to monocular deprivation initiated in the macaque monkey at 1 year of age, well beyond the period of susceptibility of LGN axonal arbors. The plasticity of these intracortical connections is, however, also limited to a critical period of postnatal life. For example, in the cat, in which it has been studied in greatest detail, ocular dominance plasticity peaks at about 1 month of age and then progressively declines to a very low level by 3–4 months (Figure 22.20). The critical period for modification of binocular connections in mice and rats lasts

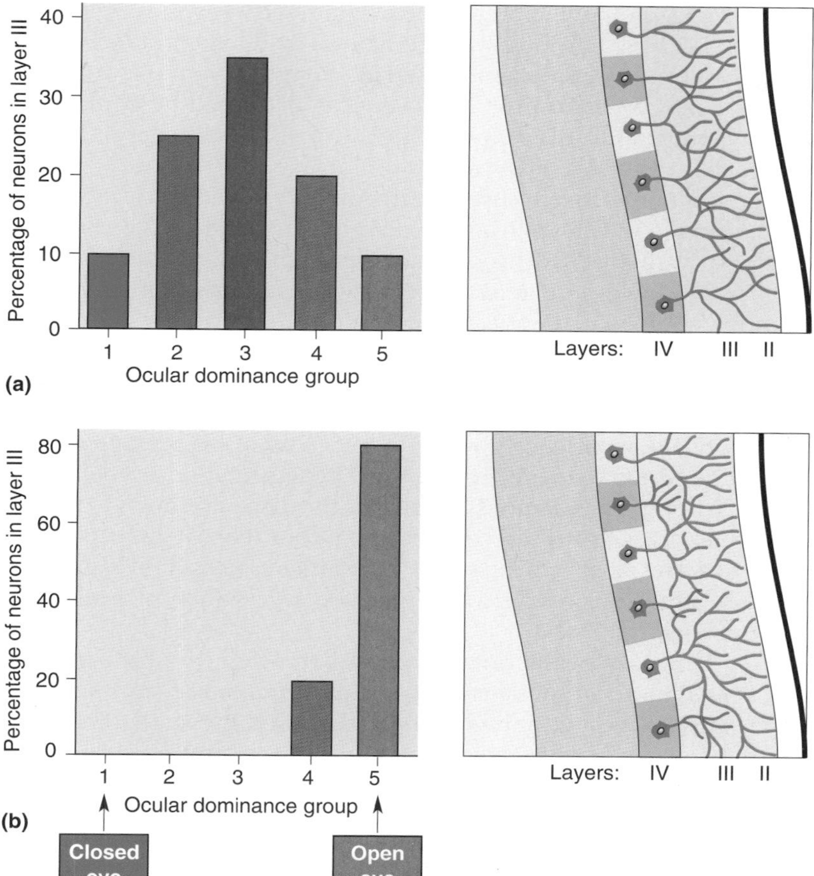

Figure 22.19
The ocular dominance shift. These ocular dominance histograms were constructed after electrophysiological recording from neurons in the striate cortex of **(a)** normal cats and **(b)** cats that had been monocularly deprived early in life. The bars show the percentage of neurons in layer III in each of five ocular dominance categories. Cells in groups 1 and 5 are activated by stimulation of either the left or the right eye, respectively, but not both. Cells in group 3 are activated equally well by either eye. Cells in groups 2 and 4 are binocularly activated but show a preference for either the left or the right eye, respectively. The histogram in part **a** reveals that the majority of neurons in the visual cortex of a normal animal are driven binocularly. The histogram in part **b** shows that a period of monocular deprivation leaves few neurons responsive to the deprived eye. The illustrations to the right of the histograms show the likely anatomical basis for this change in physiology.

until 5 weeks of age. In the macaque monkey, the critical period extends to about age 2 years, and estimates are that this plasticity ends in humans at about age 10. These critical periods coincide with the times of greatest growth of the head and eyes in these species. Thus, it is believed that the plasticity of binocular connections is normally required for maintaining good binocular vision throughout this period of rapid growth. The hazard associated with such activity-dependent fine-tuning is that these connections are also highly susceptible to deprivation.

Synaptic Competition

As you well know, a muscle that is not used regularly will atrophy and lose strength; hence the saying, "use it or lose it." Is the disconnection of activity-deprived synapses simply a consequence of disuse? This does not appear to be the case in striate cortex, because the disconnection of a deprived eye input requires that the open-eye inputs be active. Rather, a process of **binocu-**

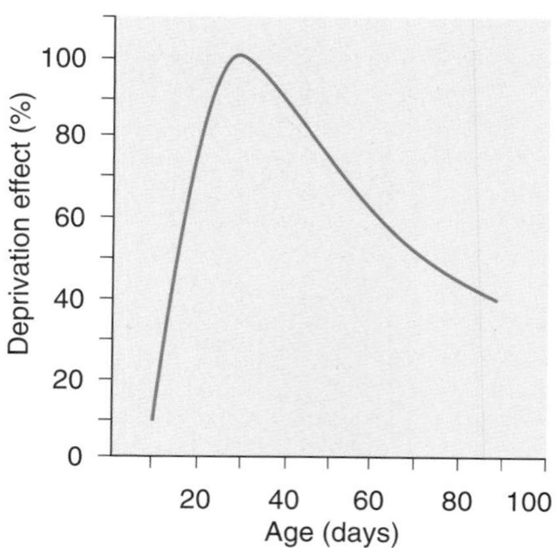

Figure 22.20
The critical period for plasticity of binocular connection in the kitten. Plotted here is the sensitivity of binocular connections in cat striate cortex to monocular deprivation at different postnatal ages. The deprivation effect is the percentage of neurons in area 17 whose responses are dominated by stimulation of the nondeprived eye. The critical period in cats begins at about 3 weeks and ends at about 3 months. (Source: Adapted from Olson and Freeman, 1980.)

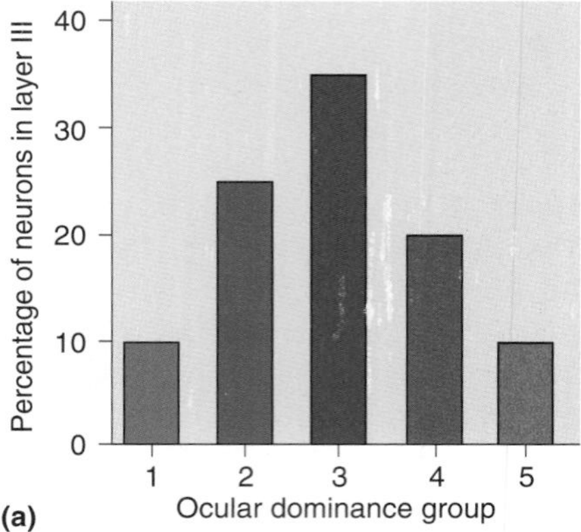

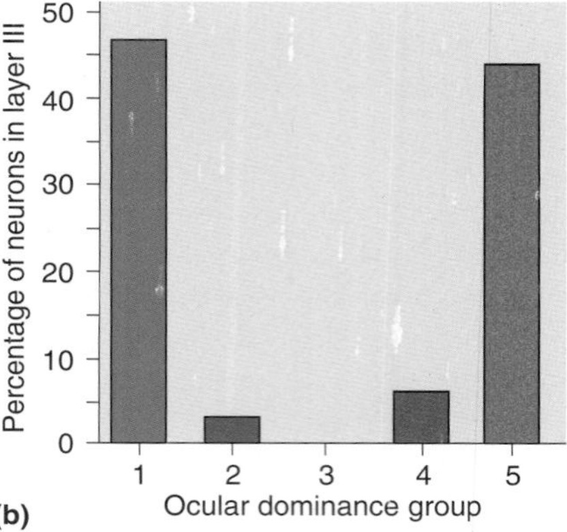

Figure 22.21

The effects of strabismus on cortical binocularity. (a) An ocular dominance histogram from a normal animal like that in Figure 22.19a. **(b)** In this case, the eyes have been brought out of alignment by cutting one of the eye muscles. After a brief period of strabismus, binocular cells are almost completely absent. The cells in visual cortex are driven by either the right or the left eye but not by both.

lar competition in which the inputs from the two eyes actively compete for synaptic control of the postsynaptic neuron evidently occurs. If the activity of the two eyes is correlated and equal in strength, the two inputs are retained on the same cortical cell. If this balance is disrupted by depriving one eye, however, the more active input somehow displaces the deprived synapses or causes them to be less effective.

Competition in visual cortex is demonstrated by the effects of **strabismus**, a condition in which the eyes are not perfectly aligned (i.e., they are "cross-eyed" or "wall-eyed"). This common visual disorder in humans can result in the permanent loss of stereoscopic vision. Experimental strabismus is produced by surgically or optically misaligning the two eyes, thereby causing visually evoked patterns of activity from the two eyes to arrive out of sync in the cortex. If you press gently with a finger alongside one eye, you can see the consequences of misalignment of the two eyes. As a result of this manipulation of visual experience, there is a total loss of binocular receptive fields, even though the two eyes retain equal representation in the cortex (Figure 22.21). This is a clear demonstration that the disconnection of inputs from one eye occurs as the result of competition rather than disuse (the two eyes are equally active, but for each cell, a "winner takes all"). If produced early enough, strabismus can also sharpen the segregation of ocular dominance columns in layer IV.

The changes in ocular dominance and binocularity after deprivation have clear behavioral consequences. An ocular dominance shift after monocular deprivation leaves the animal visually impaired in the deprived eye, and the loss of binocularity associated with strabismus eliminates stereoscopic depth perception. Neither of these effects is irreversible, however, if corrected early enough in the critical period. The clinical lesson is clear: Congenital cataracts or ocular misalignment must be corrected in early childhood, as soon as surgically feasible, to avoid permanent visual disability.

Modulatory Influences

With increasing age, there appear to be additional constraints on the forms of activity that cause modifications of cortical circuits. Before birth, spontaneous bursts of retinal activity are sufficient to orchestrate aspects of address selection in the LGN and cortex. After birth, an interaction with the visual environment is critically important. But even visually driven retinal activity may be insufficient for modifications of binocularity during this critical period. Increasing experimental evidence indicates that such modifications also require the animals to pay attention to visual stimuli and use vision to guide behavior. For example, modifications of binocularity following monocular stimulation do not occur when animals are kept anesthetized, even though it is known that cortical neurons respond briskly to visual stimulation under this condition. These and related observations have led to the proposal that synaptic plasticity in the cortex requires the release of extraretinal "enabling factors" that are linked to behavioral state (level of alertness, for example).

Some progress has been made in identifying the physical basis of these enabling factors. Recall that a number of diffuse modulatory systems innervate the cortex (see Chapter 15). These include the noradrenergic inputs from the locus coeruleus and the cholinergic inputs from the basal forebrain. The effects of monocular deprivation have been studied in animals in which these modulatory inputs to striate cortex were eliminated. This was found to cause a substantial impairment of ocular dominance plasticity outside of cortical layer IV, even though transmission in the retinogeniculocortical pathway was apparently normal (Figure 22.22).

The mechanism for modulating synaptic plasticity by acetylcholine (ACh)

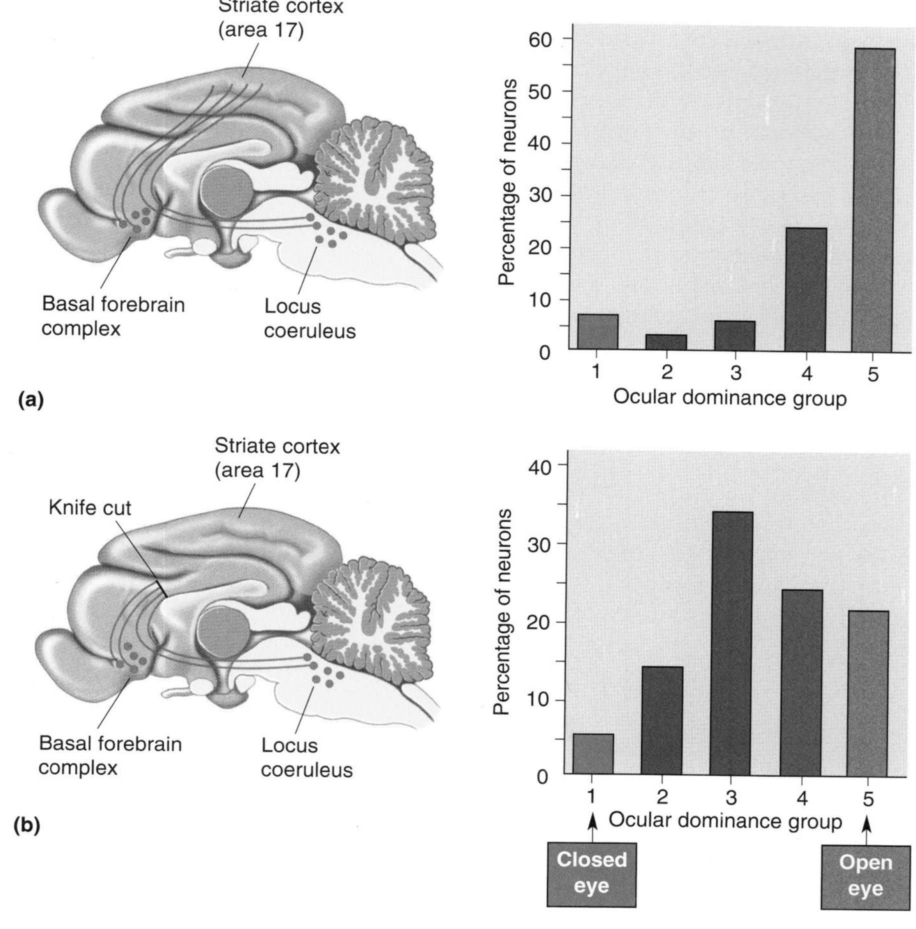

Figure 22.22
The dependence of plasticity of binocular connections on modulatory inputs. **(a)** A midsagittal view of a cat brain showing the trajectory of two modulatory inputs to striate cortex. One arises in the locus coeruleus and uses norepinephrine as a transmitter, and the other arises in the basal forebrain complex and uses ACh as a transmitter. The activity of both of these inputs is related to levels of attention and alertness. If these systems are intact, monocular deprivation produces the expected ocular dominance shift shown in the histogram at the right. **(b)** The result of depleting the cortex of these modulatory inputs: Monocular deprivation has little effect on the binocular connections in striate cortex. (Source: Adapted from Bear and Singer, 1986.)

and norepinephrine remains to be determined. It is known, however, that these neurotransmitters increase the excitability of neurons, making cortical neurons more likely to respond to visual stimulation with action potentials.

ELEMENTARY MECHANISMS OF CORTICAL SYNAPTIC PLASTICITY

Synapses form in the absence of any electrical activity. As we have seen, however, the awakening of synaptic transmission during development plays a vital role in the final refinement of connections. Based on the analysis of experience-dependent plasticity in the visual cortex and elsewhere, we can formulate two simple "rules" for synaptic modification:

1. When the presynaptic axon is active and at the same time the postsynaptic neuron is *strongly activated* under the influence of other inputs, then the synapse formed by the presynaptic axon is strengthened. This is another way of stating Hebb's hypothesis, mentioned previously. In other words, *neurons that fire together wire together.*
2. When the presynaptic axon is active and at the same time the postsynaptic neuron is *weakly activated* by other inputs, then the synapse formed by the presynaptic axon is weakened. In other words, *neurons that fire out of sync lose their link.*

The key appears to be *correlation*. Remember that in most locations in the CNS, including the visual cortex, a single synapse has little influence on the firing rate of the postsynaptic neuron. To be "heard," the activity of the synapse must be correlated with the activity of many other inputs converging on the same postsynaptic neuron. When the activity of the synapse consistently correlates with a strong postsynaptic response (and, therefore, the activity of many other inputs), the synapse is retained and strengthened. When synaptic activity consistently fails to correlate with a strong postsynaptic response, the synapse is weakened and eliminated. In this way, synapses are "validated" on the basis of their ability to participate in the firing of their postsynaptic partner.

What mechanisms are responsible for such correlation-based synaptic modifications? The answer requires knowledge of the mechanisms of excitatory synaptic transmission in the brain.

Excitatory Synaptic Transmission in the Immature Visual System

Glutamate is the transmitter at all of the modifiable synapses we have discussed (retinogeniculate, geniculocortical, and corticocortical), and it activates several subtypes of postsynaptic receptors. Recall from Chapter 6 that neurotransmitter receptors may be classified into two broad categories: G-protein-coupled, or metabotropic, receptors; and transmitter-gated ion channels (Figure 22.23). Postsynaptic glutamate-gated ion channels allow the pas-

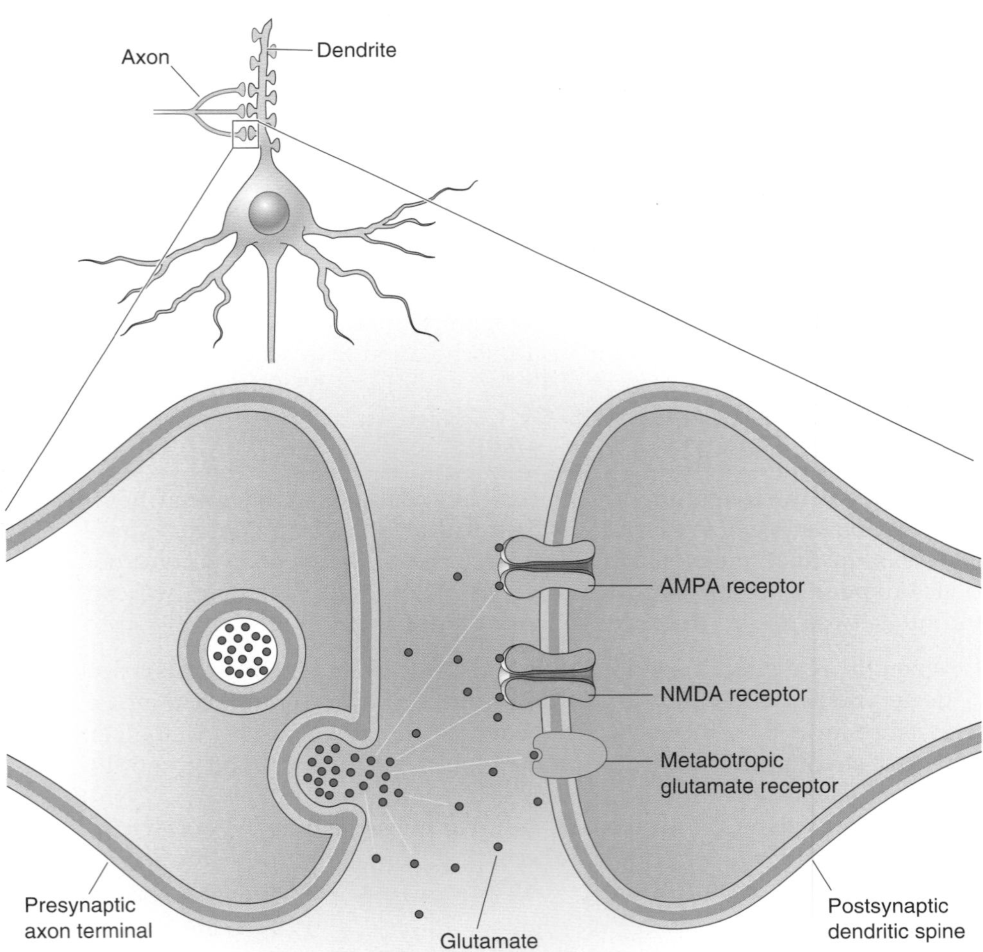

Figure 22.23
Glutamate receptors at excitatory synapses.

sage of positively charged ions into the postsynaptic cell and may be further classified as either AMPA (α-amino-3-hydroxy-5-methyl-4-isoxazole propionic acid) receptors or NMDA (*N*-methyl-D-aspartate) receptors. AMPA and NMDA receptors are colocalized at many synapses.

An NMDA receptor has two unusual features that distinguish it from an AMPA receptor (Figure 22.24). First, NMDA receptor conductance is voltage gated because of the action of Mg^{2+} at the channel. At the resting membrane potential, the inward current through the NMDA receptor is interrupted by the movement of Mg^{2+} ions into the channel, where they become lodged. As the membrane is depolarized, however, the Mg^{2+} block is displaced from the channel, and current is free to pass into the cell. Thus, substantial current through the NMDA receptor channel requires the concurrent release of glutamate by the presynaptic terminal *and* depolarization of the postsynaptic membrane. The other distinguishing feature of an NMDA receptor is that its channel conducts Ca^{2+} ions. Therefore, *the magnitude of the Ca^{2+} flux passing through the NMDA receptor channel specifically signals the level of presynaptic and postsynaptic coactivation.*

Curiously, when a glutamatergic synapse first forms, only NMDA receptors appear in the postsynaptic membrane. As a consequence, released glutamate at a single synapse evokes little response when the postsynaptic membrane is at the resting potential. Such "silent" synapses announce their presence only when enough of them are active at the same time to cause suf-

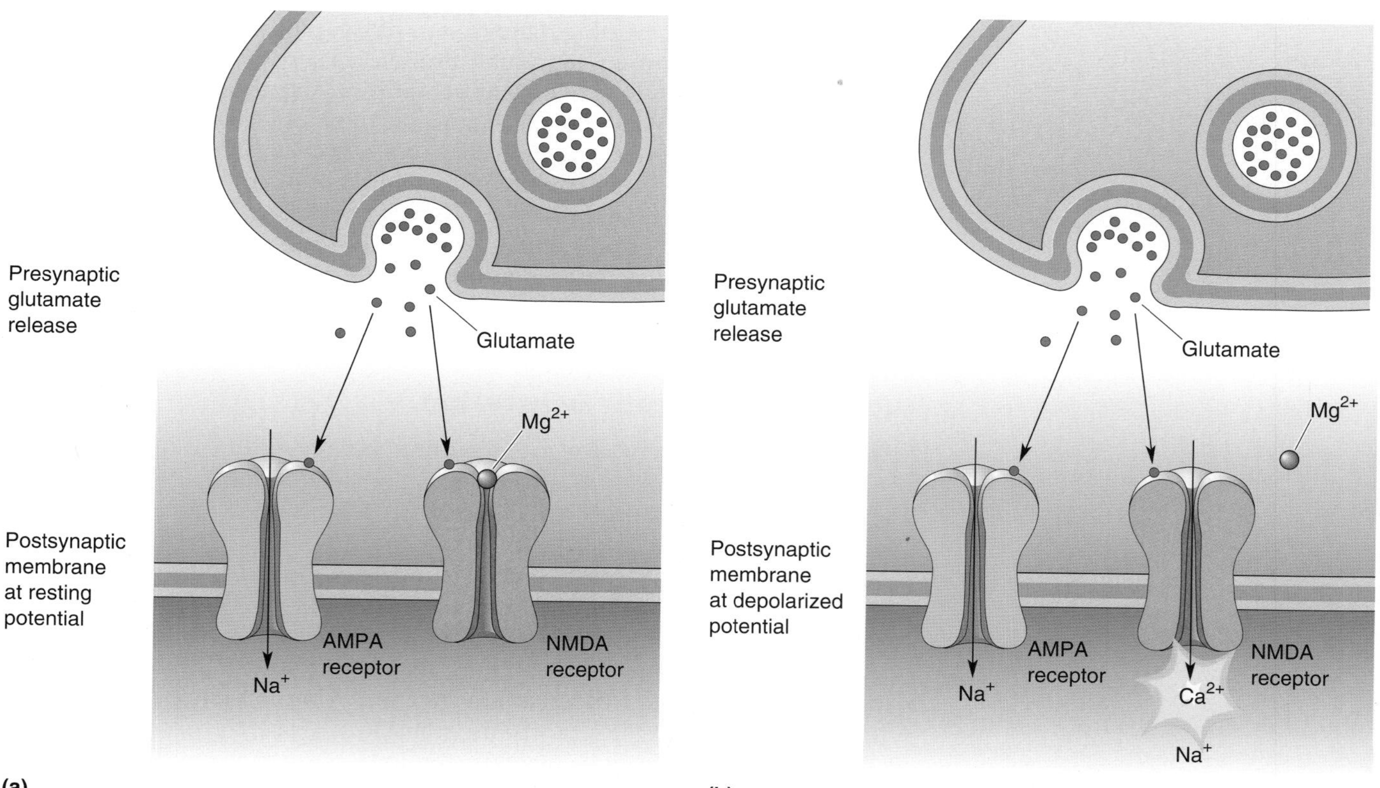

Figure 22.24

NMDA receptors activated by simultaneous presynaptic and postsynaptic activity. (a) Presynaptic activation causes the release of glutamate, which acts on postsynaptic AMPA receptors and NMDA receptors. At the negative resting membrane potential, the NMDA receptors pass little ionic current because they are blocked with Mg^{2+} ions. **(b)** If glutamate release coincides with depolarization sufficient to displace the Mg^{2+} ions, Ca^{2+} enters the postsynaptic neuron via the NMDA receptor. Hebbian modification could be explained if the Ca^{2+} admitted by the NMDA receptor triggered enhanced synaptic effectiveness.

ficient depolarization to relieve the Mg^{2+} block of the NMDA receptor channels. In other words, "silent" synapses "speak" only when there is highly correlated activity—the necessary condition for synaptic enhancement during development.

Long-Term Synaptic Potentiation

Perhaps NMDA receptors serve as Hebbian detectors of simultaneous presynaptic and postsynaptic activity, and Ca^{2+} entry through the NMDA receptor channel triggers the biochemical mechanisms that modify synaptic effectiveness. Tests of this hypothesis have been performed by electrically stimulating axons to monitor the strength of synaptic transmission before and after an episode of strong NMDA receptor activation (Figure 22.25a and b). Results consistently indicate that a consequence of strong NMDA receptor activation is a strengthening of synaptic transmission, called **long-term potentiation** (**LTP**).

What accounts for LTP of the synapse? One consequence of the strong NMDA receptor activation and the resulting flood of Ca^{2+} ions into the postsynaptic dendrite is the insertion of new AMPA receptors into the synaptic membrane (Figure 22.25c). Such "AMPAfication" of the synapse makes transmission stronger. In addition to this change in the complement of glutamate receptors, recent evidence suggests that the potentiated synapses can actually split in half, forming distinct sites of synaptic contact.

Cortical neurons grown in tissue culture form synapses with one another and become electrically active. The immature synapses contain clusters of NMDA receptors but few AMPA receptors. Consistent with the idea that LTP is a mechanism for synaptic maturation, electrically active synapses gain AMPA receptors over the course of development in culture (Figure 22.26). This change fails to occur, however, if NMDA receptors are blocked with an antagonist. Thus, the strong activation of NMDA receptors that occurs when presynaptic and postsynaptic neurons fire together appears to account, at least in part, for why they wire together during visual system development. (We will discuss LTP and its molecular basis further in Chapter 24.)

Long-Term Synaptic Depression

Neurons that fire out of sync lose their link. In the case of strabismus, for example, synapses whose activity fails to correlate with that of the postsynaptic cell are weakened and then eliminated. What mechanism is responsible for this form of synaptic plasticity?

In principle, weak coincidences could be signaled by lower levels of NMDA receptor activation and less Ca^{2+} influx. Indeed, experiments suggest that the lower level of Ca^{2+} admitted under these conditions triggers an opposite form of synaptic plasticity, **long-term depression** (**LTD**), whereby the active synapses are decreased in effectiveness. One consequence of LTD induction is a loss of AMPA receptors from the synapse, the exact opposite of LTP. We do not yet know whether the long-term consequence of LTD is synapse elimination, but it is an appealing hypothesis. Recall that at the neuromuscular junction, the loss of postsynaptic receptors stimulates the physical retraction of the presynaptic axon.

How are presynaptic and postsynaptic correlations used to refine synaptic connections in the visual system? The data accumulated to date suggest that the maintenance of some connections formed during development depends on their success in evoking an NMDA receptor-mediated response beyond some threshold level. Failure to achieve this threshold leads to disconnection. Both processes depend on activity originating in the retina, NMDA receptor activation, and postsynaptic Ca^{2+} entry.

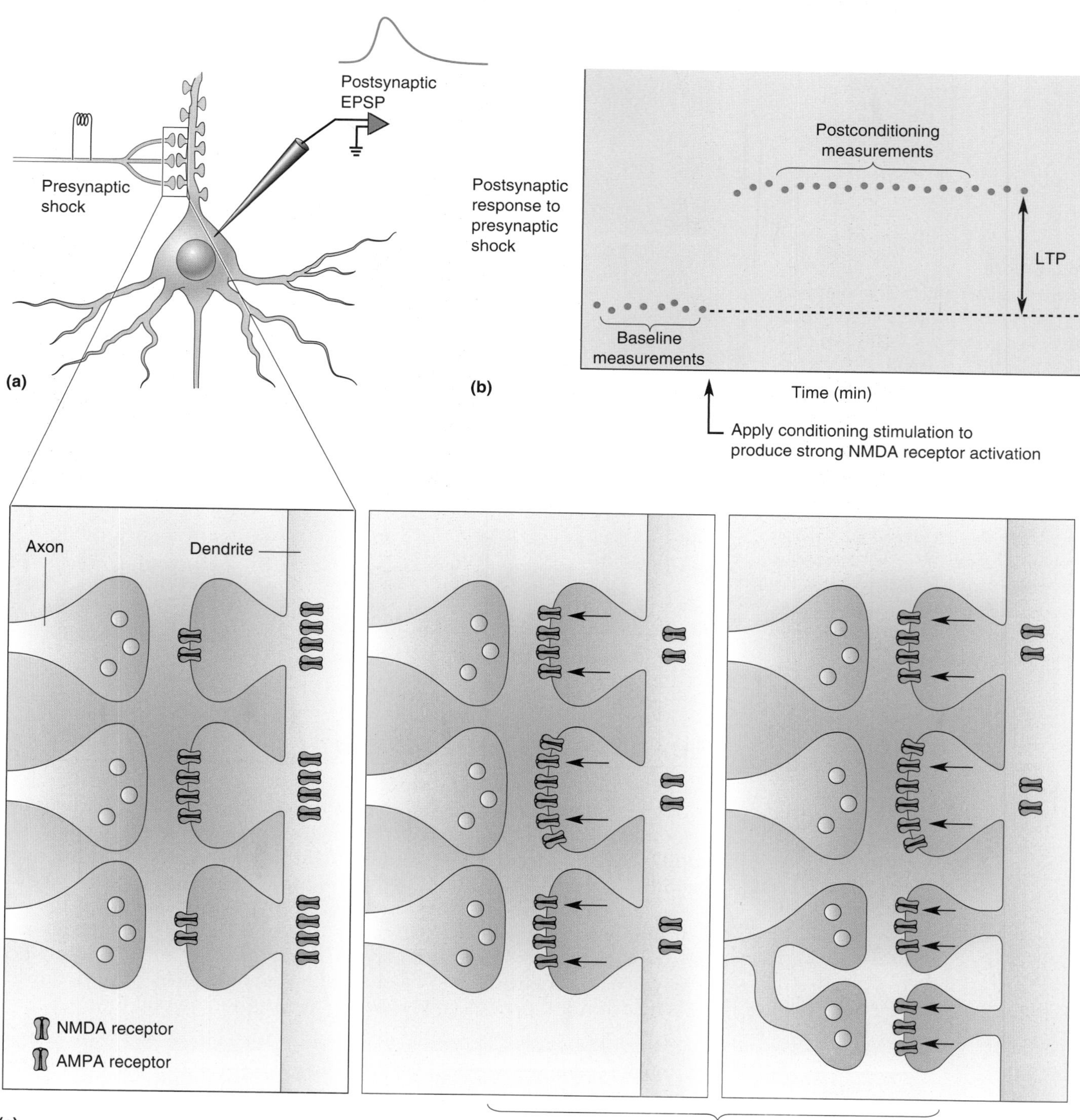

Figure 22.25

The lasting synaptic effects of strong NMDA receptor activation. (a) An experiment in which presynaptic axons are stimulated electrically to evoke an action potential and microelectrode recordings of the resulting EPSPs are made from the postsynaptic neuron. **(b)** A graph showing how the strength of synaptic transmission is changed by strong NMDA receptor activation. The conditioning stimulation consists of depolarizing the postsynaptic neuron via current injection through the microelectrode when the synapses are repeatedly stimulated. LTP is the resulting enhancement of synaptic transmission. **(c)** LTP at many synapses is associated with the insertion of AMPA receptors into synapses that previously had none. A longer-term consequence is the splitting of potentiated synapses to form new sites of contact.

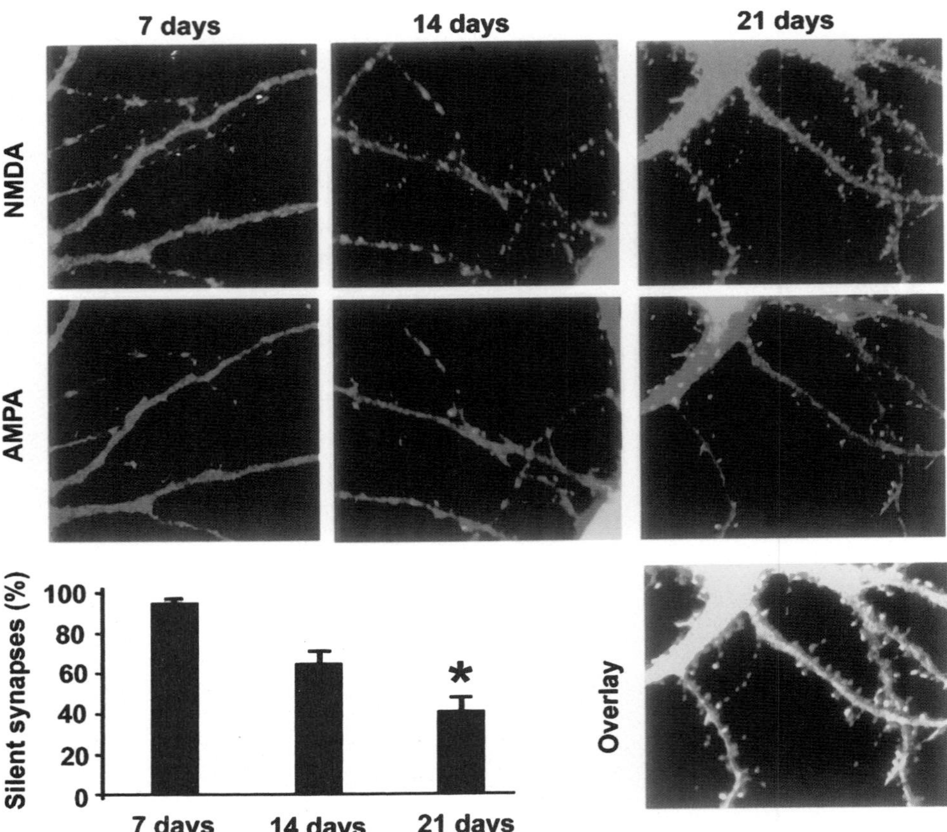

Figure 22.26
Silent synapses and their maturation. The dendrites of neurons in cell culture have been stained at different ages using antibodies that recognize NMDA receptors (red) and AMPA receptors (green). When the images are combined, the regions where receptor distributions overlap appear yellow. Notice, however, the large number of red dendritic spines. These postsynaptic regions, which contain only NMDA receptors, reflect "silent" synapses. The graph shows how the percentage of silent synapses decreases during development. (Source: Courtesy of D. Liao and R. L. Huganir, Johns Hopkins University.)

WHY DO CRITICAL PERIODS END?

Although plasticity of visual connections persists in the adult brain, the range over which this plasticity occurs constricts with increasing age. Early in development, gross rearrangements of axonal arbors are possible, while in the adult, plasticity appears to be restricted to local changes in synaptic efficacy. In addition, the adequate stimulus for evoking a change also appears to be increasingly constrained as the brain matures. An obvious example is the fact that simply patching one eye will cause a profound alteration in the binocular connections of the superficial layers during infancy, but by adolescence this type of experience fails to cause a lasting alteration in cortical circuitry.

Why do critical periods end? Here are three current hypotheses:

1. *Plasticity diminishes when axon growth ceases.* We've seen that there is a period of several weeks when the geniculate arbors can contract and expand within layer IV under the influence of visual experience. Thus, a factor that limits the critical period in layer IV could be a loss of the capability for changes in axon length, which in turn may be due to changes in the extracellular matrix or the myelination of the axons by oligodendroglia.
2. *Plasticity diminishes when synaptic transmission matures.* The end of a critical period may reflect changes in the elementary mechanisms of synaptic plasticity. Evidence indicates that glutamate receptors change during postnatal development. For example, the activation of metabotropic glutamate receptors stimulates very different postsynaptic responses in striate cortex during the critical period when binocular connections are most susceptible to monocular deprivation. In addition, the molecular compo-

sition and properties of NMDA receptors change during the course of the critical period. Accordingly, the properties of LTP and LTD vary with age, and at some synapses, they seem to disappear altogether.

3. *Plasticity diminishes when cortical activation is constrained.* As development proceeds, certain types of activity may be filtered by successive synaptic relays to the point where they no longer activate NMDA receptors or other elementary mechanisms sufficiently to trigger plasticity. As mentioned previously, ACh and norepinephrine facilitate synaptic plasticity in the superficial cortical layers, perhaps simply by enhancing polysynaptic intracortical transmission. A decline in the effectiveness of these neurotransmitters or a change in the conditions under which they are released may contribute to the decline in plasticity. Indeed, evidence suggests that supplementing the adult cortex with norepinephrine can restore some degree of modifiability.

Evidence also indicates that intrinsic inhibitory circuitry is late to mature in the striate cortex. Consequently, patterns of activity that may have gained access to modifiable synapses in superficial layers early in postnatal development may be tempered by inhibition in the adult. Consistent with the idea that inhibition regulates the duration of the critical period, recent research in mice has shown that genetic manipulations that accelerate the maturation of GABAergic inhibition in visual cortex also shorten the critical period for ocular dominance plasticity. Conversely, manipulations that slow the development of inhibition can prolong the critical period.

Why critical periods end is important. Synaptic modification and rewiring of circuitry provide the capacity for some recovery of function after CNS damage. Such recovery is, however, disappointingly limited in the adult brain. On the other hand, the recovery of function after brain injury can be nearly 100% in the immature nervous system when synaptic rearrangements are widespread. Thus, an understanding of how plasticity is regulated during normal development may suggest ways to promote recovery from damage later in life.

CONCLUDING REMARKS

We have seen that the generation of circuitry during brain development occurs mostly prior to birth and is guided by cell-to-cell communication through physical contact and by diffusible chemical signals. Nonetheless, while most of the "wires" find their proper place before birth, the final refinement of synaptic connections, particularly in the cortex, occurs during infancy and is influenced by the sensory environment. Although we have focused on the visual system, other sensory and motor systems are also readily modified by the environment during critical periods of early childhood. In this way, our brain is a product not only of our genes but also of the world in which we grow up.

The end of developmental critical periods does not signify an end to experience-dependent synaptic plasticity in the brain. It is possible that some higher cortical areas in the frontal and temporal lobes exhibit in adults plasticity as robust as that seen in area 17 during infancy. In any case, the environment must modify the brain throughout life at some level, or there would be no basis for memory formation. In the next two chapters, we will take a close look at the neurobiology of learning and memory. We will see that the mechanisms of synaptic plasticity proposed to account for learning bear a close resemblance to those believed to play a role in synaptic rearrangement during development.

KEY TERMS

The Genesis of Neurons
subplate (p. 709)
radial glial cell (p. 709)
neuroblast (p. 709)
cortical plate (p. 709)

The Genesis of Connections
growth cone (p. 713)
extracellular matrix (p. 713)
fasciculation (p. 714)
cell adhesion molecule (CAM) (p. 714)
chemoattractant (p. 715)
chemorepellent (p. 715)
chemoaffinity hypothesis (p. 716)
basal lamina (p. 717)

The Elimination of Cells and Synapses
trophic factor (p. 720)
nerve growth factor (NGF) (p. 720)
neurotrophin (p. 720)
apoptosis (p. 721)

Activity-Dependent Synaptic Rearrangement
Hebb synapse (p. 724)
Hebbian modification (p. 724)
monocular deprivation (p. 725)
critical period (p. 726)
ocular dominance shift (p. 727)
binocular competition (p. 729)
strabismus (p. 730)

Elementary Mechanisms of Cortical Synaptic Plasticity
long-term potentiation (LTP) (p. 734)
long-term depression (LTD) (p. 734)

REVIEW QUESTIONS

1. What do we mean by saying that the cortex develops inside out?
2. Describe the three phases of pathway formation. In which phase (or phases) does neural activity play a role?
3. What are three ways that Ca^{2+} ions are thought to contribute to the processes of synapse formation and rearrangement?
4. How are the elimination of polyneuronal innervation of a muscle fiber and the segregation of retinal terminals in the LGN similar? How do these processes differ?
5. Not long ago, when a child was born with strabismus, the defect was usually not corrected until after adolescence. Today, surgical correction is always attempted during early childhood. Why? How does strabismus affect the connections in the brain, and how does it affect vision?
6. Children are often able to learn several languages apparently without effort, while most adults must struggle to master a second language. From what you know about brain development, why would this be true?

CHAPTER

23 Memory Systems

INTRODUCTION

It is difficult to study the brain without developing a sense of awe about how well it works. The sensory, motor, limbic, and modulatory systems contain billions of individual neural elements with enormous numbers of interconnections. As we saw in Chapter 22, the development of these connections is an orderly process that follows certain rules. Wiring the brain isn't magic—at least not entirely. But as impressive and orderly as prenatal development is, we are by no means complete creatures when we are born. From the moment we take our first breath, and possibly before, sensory stimuli modify our brain and influence our behavior. Indeed, one of the main goals of the first 20 years of life is to learn the skills we need to survive in the world. We learn an enormous number of things, some straightforward (snow is cold), and others more abstract (an isosceles triangle has two sides of equal length).

There is a close relationship between what we called experience-dependent brain development in Chapter 22 and what we call learning in this chapter. Visual experience during infancy is essential for the normal development of the visual cortex, but it also allows us to recognize an image of our mother's face. Visual development and learning probably use similar mechanisms, but at different times and in different cortical areas. Viewed in this way, learning and memory are the lifelong adaptations of brain circuitry to the environment. They enable us to respond appropriately to situations we have experienced before. This example also illustrates that a particular form of information is likely to be stored in the parts of the brain that normally process that type of information. Thus, we can expect different parts of the brain to participate in different types of memory.

In this chapter, we discuss the anatomy of memory, i.e., the different parts of the brain involved in storing particular types of information. In Chapter 24, we will focus on the elementary synaptic mechanisms that can store information in the brain.

TYPES OF MEMORY AND AMNESIA

Learning is the acquisition of new information or knowledge. **Memory** is the retention of learned information. We learn and remember many things, and it is important to appreciate that these various things may not be processed and stored by the same neural hardware. No single brain structure or cellular mechanism accounts for all learning. Moreover, the way in which information of a particular type is stored may change over time.

Declarative and Nondeclarative Memory

Psychologists have studied learning and memory extensively, and they have distinguished what appear to be different types. A useful distinction for our purposes is between declarative memory and nondeclarative memory.

During the course of our lives, we learn many facts—the capital of Thailand is Bangkok; the coyote will never catch the roadrunner. We also store memories of life's events—"I had Cheerios for breakfast"; "I heard a boring chemistry lecture yesterday." Memory for facts and events is called **declarative memory** (Figure 23.1). Declarative memory is what we usually mean by the word "memory" in everyday usage, but we actually remember many other things. These **nondeclarative memories** fall into several categories. The type we are most concerned with here is **procedural memory**, or memory for skills, habits, and behaviors. We learn to play the piano, throw a Frisbee, or tie our shoes, and somewhere that information is stored in our brain. In Chapter 18, we discussed learned fear, another type of nondeclara-

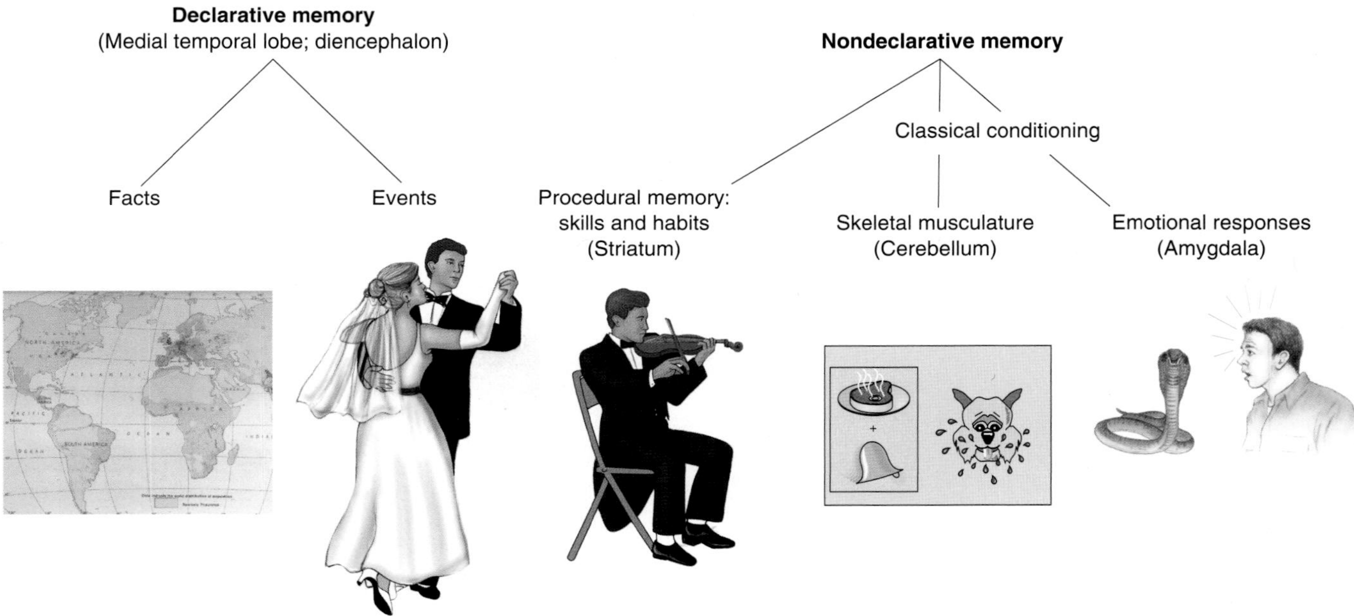

Figure 23.1
Types of declarative and nondeclarative memory. Brain structures thought to be involved in each type of memory are indicated. (This does not represent all types of memory.)

tive memory that involves the amygdala. Other forms of nondeclarative memory will be discussed in Chapter 24.

Generally, declarative memories are available for conscious recollection, and nondeclarative memories are not. The tasks we learn and the reflexes and emotional associations we have formed operate smoothly, however, without conscious recollection. As the old saying goes, you never forget how to ride a bicycle. You may not explicitly remember the day you first rode a two-wheeler on your own (the declarative part of the memory), but your brain remembers what to do when you're on one (the procedural part of the memory). Nondeclarative memory is also frequently called *implicit memory*, because it results from direct experience, and declarative memory is often called *explicit memory*, because it results from more conscious effort.

Another distinction is that declarative memories are often easy to form and are easily forgotten. By contrast, forming nondeclarative memories tends to require repetition and practice over a longer period, but these memories are less likely to be forgotten. Think of the difference between memorizing the capitals of foreign countries and learning to ski. While there is no clear limit to the number of declarative memories the brain can store, there can be great diversity in the ease and speed with which new information is acquired. Studies of humans with abnormally good memories suggest that the limit on the storage of declarative information is remarkably high (Box 23.1).

Long-Term and Short-Term Memory

Long-term memories are those that you can recall days, months, or years after they were stored. But not all memories make it into long-term storage. What did you have for dinner last night? You probably have no problem remembering. But how about dinner a week ago? Chances are that this memory has faded completely. Thus, it is useful to distinguish short-term memory (last night's dinner) from long-term memory. **Short-term memories** last on the order of seconds to hours and are vulnerable to disruption. For exam-

Box 23.1

OF SPECIAL INTEREST

An Extraordinary Memory

In the 1920s, a man named Sherashevsky came to see the Russian psychologist Aleksandr Luria. Thus began a 30-year study of the uncommon memory of this man Luria simply called S. Luria's fascinating description of this study is contained in the book *The Mind of a Mnemonist*. Luria initially studied S. by giving him conventional tests, such as memorizing lists of words, numbers, or nonsense syllables. He'd read the list once and then ask S. to repeat it. Much to Luria's surprise, he couldn't come up with a test S. could not pass. Even when 70 words were read in a row, S. could repeat them forward, backward, and in any other order. During the many years they worked together, Luria never found a limit to S.'s memory. In tests of his retention, S. demonstrated that he remembered lists he had previously seen even 15 years earlier!

How did he do it? S. described several factors that may have contributed to his great memory. One was his unusual sensory response to stimuli—he retained vivid images of things he saw. When shown a table of 50 numbers, he claimed it was easy to later read off numbers in one row or along the diagonal because he simply had to call up a visual image of the entire table. Interestingly, when he occasionally made errors in recalling tables of numbers written on a chalkboard, they appeared to be "reading" errors rather than memory errors. For instance, if the handwriting was sloppy, he would mistake a 3 for an 8 or a 4 for a 9. It was as if he was seeing the chalkboard and numbers all over again when he was recalling the information.

Another interesting aspect of S.'s sensory response to stimuli was a powerful form of *synesthesia*, a phenomenon in which sensory stimuli evoke sensations usually associated with different stimuli. For example, when S. heard a sound, in addition to hearing, he would see splashes of colored light and perhaps have a certain taste in his mouth.

After learning that his memory was unusual, S. left his job as a reporter and became a professional stage performer—a mnemonist. To remember huge lists of numbers or tables of words given by members of the audience trying to stump him, he complemented his lasting sensory responses to stimuli and his synesthesia with memory "tricks." To remember a long list of items, he made use of the fact that each item evoked some sort of visual image. As the list was read or written, S. imagined himself walking through his home town; as each item was given, he placed its evoked image along his walk—the image evoked by item 1 by the mailbox, the image for item 2 by a bush, and so on. To recall the items, he walked the same route and picked up the items he had put down. Though we may not have the complex synesthetic sensations of S., this ancient technique of making associations with familiar objects is one we all can use.

Not everything about S.'s memory was to his advantage. While the complex sensations evoked by stimuli helped him remember lists of numbers and words, they interfered with his ability to integrate and remember more complex things. He had trouble recognizing faces because each time a person's expression changed, he would also "see" changing patterns of light and shade, which would confuse him. He also wasn't very good at following a story read to him. Rather than ignoring the exact words and focusing on the important ideas, S. was overwhelmed by an explosion of sensory responses. Imagine how bewildering it would be to be bombarded by constant visual images evoked by each word, plus sounds and images evoked by the tone of voice of the person reading the story. No wonder S. had trouble!

S. also experienced the inability to forget. This became a particular problem when he was performing as a professional mnemonist and was asked to remember things written on a chalkboard. He would see things that had been written there on many occasions. Although he tried various tricks to forget old information, such as mentally erasing the board, nothing worked. Only by the strength of his attention and by actively telling himself to let information slip away was he able to forget. It was as if the effort most of us use for remembering and the ease with which we forget were reversed for S.

We don't know the neural basis for S.'s remarkable memory. Perhaps he lacked the same sort of segregation most of us have between sensations in different sensory systems. This may have contributed to an uncommonly strong multimodal coding of memories. Maybe his synapses were more malleable than normal. Unfortunately, we'll never know.

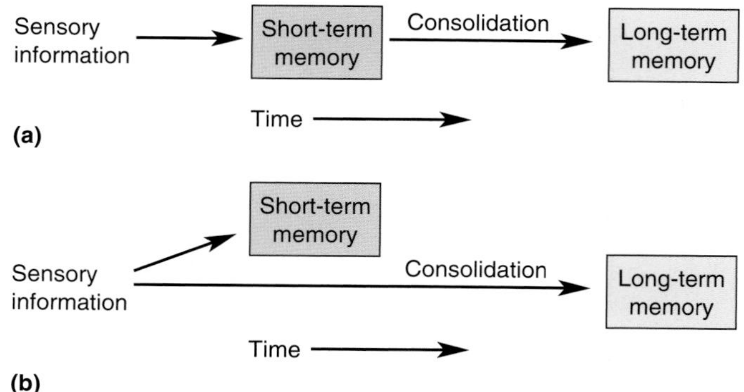

Figure 23.2
Short-term and long-term memory. Sensory information can be temporarily held in short-term memory, but permanent storage in long-term memory requires consolidation. **(a)** Information may be consolidated from short-term memory. **(b)** Alternatively, information processing necessary for consolidation may occur separately from short-term memory.

ple, short-term memory can be erased by head trauma or electroconvulsive shock. But the same treatments do not affect long-term memories (e.g., childhood memories), which were stored long ago. These observations have led to the idea that memories are stored in short-term memory and gradually converted into a permanent form via a process called **memory consolidation**. Memory consolidation, however, does not necessarily require short-term memory as an intermediary; the two types of memory may exist in parallel (Figure 23.2).

Short-term memory often requires holding information in mind. When someone tells you his or her phone number, you can retain it for a limited time by repeating the number to yourself. If the number is too long (e.g., a phone number with extra numbers for a foreign country), you may have trouble remembering it at all. Short-term memory is commonly studied by measuring a person's *digit span*, the maximum number of randomly chosen numbers a person can repeat after hearing a list read. The normal digit span is seven plus or minus two numbers.

Interestingly, there are reports of humans with cortical lesions who have normal short-term memory for information coming from one sensory system (e.g., they can remember as many written numbers as other people can) but a profound deficit when information comes from a different sense (e.g., they cannot remember more than one number spoken to them). These different digit spans in different modalities are consistent with the notion of multiple temporary storage areas in the brain.

Amnesia

As we all know, in daily life, forgetting is nearly as common as learning. But certain diseases and injuries to the brain cause **amnesia**, a serious loss of memory and/or the ability to learn. Concussion, chronic alcoholism, encephalitis, brain tumor, and stroke can all disrupt memory. You've probably seen a movie or television show in which a person undergoes a great trauma and wakes up the next day not knowing who he or she is and not remembering the past. That kind of absolute amnesia for past events and information is actually quite rare. It is more common for trauma to cause limited amnesia along with other nonmemory deficits. If amnesia is not accompanied by any other cognitive deficit, it is known as *dissociated amnesia* (i.e., the memory problems are dissociated from any other problems). We will focus on cases of dissociated amnesia because a clear relationship can be drawn between memory deficits and brain injury.

Following trauma to the brain, memory loss may manifest itself in two ways: retrograde amnesia and anterograde amnesia (Figure 23.3). **Retrograde amnesia** is characterized by memory loss for events prior to the

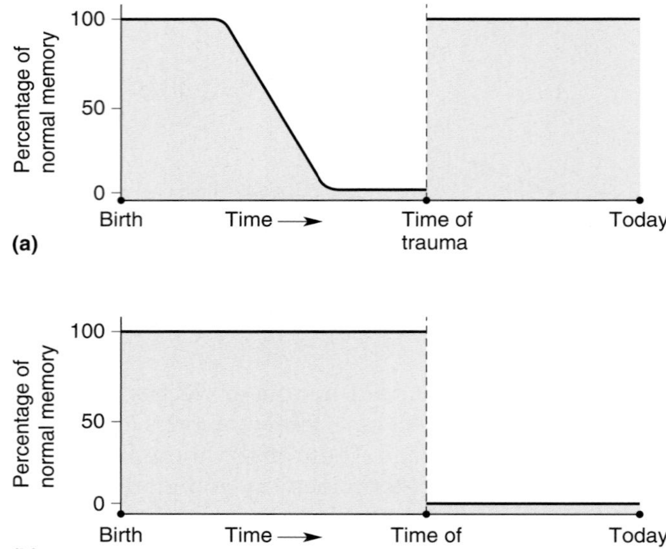

Figure 23.3
Amnesia produced by trauma to the brain. (a) In retrograde amnesia, events for a period prior to the trauma are forgotten, but memories from the distant past and the period following the trauma are intact. **(b)** In anterograde amnesia, events prior to the trauma can be remembered, but there are no memories for the period following the trauma.

trauma. In other words, you forget things you already knew. In severe cases, there may be complete amnesia for all declarative information learned before the trauma. More often, retrograde amnesia follows a pattern in which events in the months or years preceding the trauma are forgotten, but memory is increasingly strong for older memories. This is quite different from **anterograde amnesia**, the inability to form new memories following brain trauma. If the anterograde amnesia is severe, a person may be completely incapable of learning anything new. In milder cases, learning may be slower and require more repetition than normal. In clinical cases, there is often a mixture of retrograde and anterograde amnesias of different degrees of severity.

An example will help clarify. Consider the case of a 45-year-old man who had brain trauma at age 40. If he has severe retrograde amnesia, he cannot remember many things that occurred before the injury. If he has severe anterograde amnesia, he can remember nothing from age 40 until the present time.

A form of amnesia that involves a much shorter period is *transient global amnesia*. In this case, a sudden onset of anterograde amnesia lasts only for a period of minutes to days, accompanied by retrograde amnesia for recent events preceding the attack. This type of amnesia can result from brief cerebral ischemia, in which the blood supply to the brain is temporarily reduced, or concussion to the head from trauma, such as a car accident or a hard blow while playing football. There have been reports of transient global amnesia brought on by physical stress, drugs, cold showers, and even sex, presumably because all of these affect cerebral blood flow. Many cases have been linked to use of the antidiarrheal drug clioquinol (which has been taken off the market). In one case, a college student visiting Paris took clioquinol on the fifth day of a six-day holiday, during which she traveled with friends. When she got home, she discovered she couldn't remember her vacation at all! The memories for this short period were absent years later, but otherwise there was no lasting effect of the clioquinol. While we don't know exactly what causes transient global amnesia, it is probably a consequence of temporary blood deprivation to structures essential for learning and memory. Other forms of temporary amnesia can be caused by disease, brain trauma, and environmental toxins (Box 23.2).

Box 23.2

OF SPECIAL INTEREST

Forgettable Fish

Along the mid-Atlantic coast of the United States lurks a mysterious killer that has left millions of dead fish washed up on the shore. More sinister to humans, in the 1990s fishermen began to show up in doctors' offices with memory loss that was sometimes severe. The source of the trouble appears to be a unicellular microorganism called *Pfiesteria piscicida* and a number of its algal cousins. *Pfiesteria* is a dinoflagellate; at certain stages in its life cycle, it swims by means of a whiplike flagellum. *Pfiesteria* releases a neurotoxin that stuns fish, making them unable to escape. The dinoflagellates then swim to the fish and consume them.

Other dinoflagellates produce neurotoxins, but *Pfiesteria* is unusual in several ways. For example, the dinoflagellate *Gambierdiscus toxicus*, more commonly known as ciguatera, also produces a neurotoxin, ciguatoxin, that accumulates in fish that feed on coral reefs. Unlike *Pfiesteria,* however, ciguatera does not kill, and the fish continue to grow and are caught by fishermen. When the fish are consumed, the ciguatoxin binds to the Na^+ channels of people who eat the fish, increasing membrane excitability. Aside from nausea and diarrhea, the toxin produces a broad array of neurological symptoms, including numbness and tingling around the mouth, intense itching, and a reversal of the sensation of hot and cold.

Pfiesteria's modus operandi is different, as people do not become sick from ingesting the neurotoxin. Illness results from the seemingly innocuous activities of touching water containing the neurotoxin or inhaling it in the air. The fishermen, scientists, and others exposed to *Pfiesteria* experience a frightening array of neurological symptoms, including memory loss, difficulty concentrating, disorientation, and confusion. On visiting the doctor, patients tell of driving in the car and being unable to remember where they were going and what they were going to do. Some were even unable to remember their name or do simple arithmetic. The memory loss was sometimes so severe that it was confused with Alzheimer's disease. Fortunately, the symptoms generally subsided over the course of weeks or months after exposure to the toxin was stopped.

At present, the specific neurotoxins released by *Pfiesteria* that are responsible for the amnesia and other symptoms have not been isolated. Rats injected with solutions containing the neurotoxin have, however, been shown to have severe learning and memory impairment. Aside from identifying the toxin and the manner in which it attacks neurons is the question of why the sudden *Pfiesteria* outbreaks occur. The leading hypothesis is that water pollution from human sewage and fertilizers provided a rich source of food for the microbes, leading to tremendous growth in the *Pfiesteria* population.

THE SEARCH FOR THE ENGRAM

Now let's turn our attention to the parts of the brain that are involved in memory storage. The physical representation or location of a memory is called an **engram**, also known as a *memory trace*. When you learn the meaning of a word in French, where is this information stored—where is the engram? The technique most often used to answer this type of question is the time-honored experimental ablation method of Marie-Jean-Pierre Flourens (see Chapter 1).

Lashley's Studies of Maze Learning in Rats

In the 1920s, American psychologist Karl Lashley conducted experiments to study the effects of brain lesions on learning in rats. Well aware of the cytoarchitecture of the neocortex, Lashley set out to determine whether the engram resided in particular association areas of cortex (see Chapter 7), as was widely believed at the time.

In a typical experiment, he trained a rat to run through a maze to get a food reward (Figure 23.4a). On the first trial, the rat was slow getting to the food

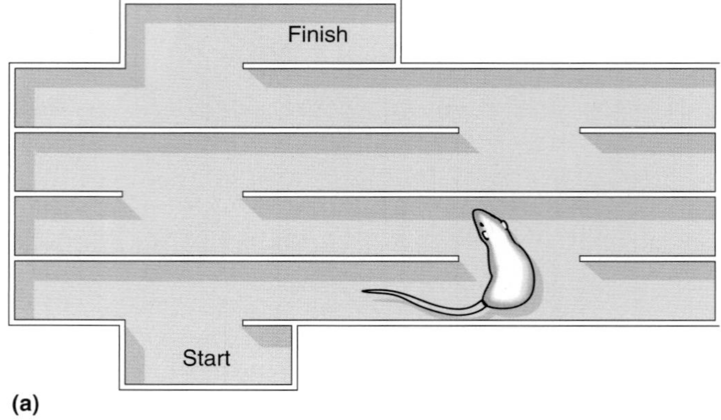

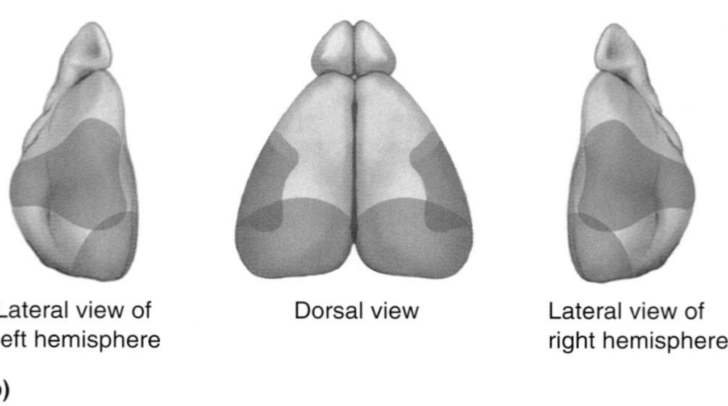

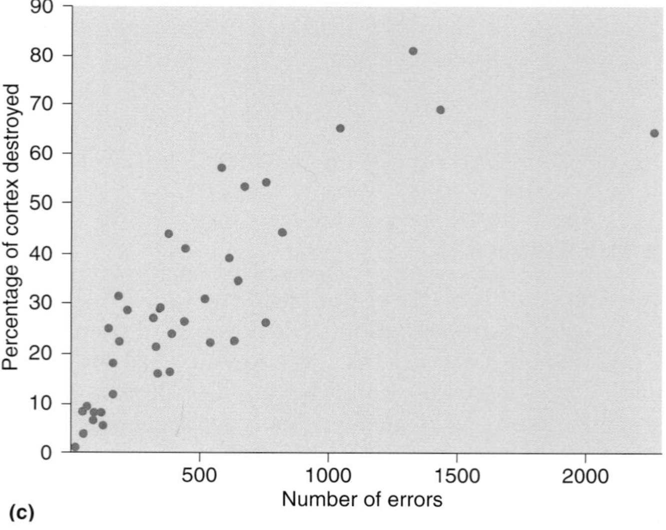

Figure 23.4
The effects of cortical lesions on maze performance. (a) The rat is trained to run through a maze from start to finish without entering blind alleys. **(b)** Three cortical lesions are shown in blue, yellow, and pink. **(c)** The greater the percentage of cortex destroyed, the more errors the rats make while they learn to run the maze. The number of errors shown is cumulative across trials, suggesting that rats with larger lesions had difficulty remembering which arms of the maze were blind alleys. (Source: Adapted from Lashley, 1929, Fig. 2, 16, and 23.)

because it would enter blind alleys and have to turn around. After running through the same maze repeatedly, the rat learned to avoid blind alleys and go straight to the food. Lashley was investigating how performance on this task was affected by lesions in the rat's cortex (Figure 23.4b). He found that rats given brain lesions before learning needed more trials to run the maze without going down blind alleys. The lesions seemed to interfere with their ability to learn.

In another group, a lesion was made after a rat had learned to run the maze without making mistakes. With the lesion, the rat made mistakes and went down blind alleys it had previously learned to avoid. It appeared that the lesion damaged or destroyed the memory for how to reach the food.

What was the effect of the size and location of the lesion? Interestingly, Lashley found that the severity of the deficits caused by the lesions (both learning and remembering) correlated with the *size* of the lesions (Figure 23.4c) but was apparently unrelated to the *location* of the lesion within the cortex. These findings led him to speculate that all cortical areas contribute equally to learning and memory; it is simply a matter of getting poorer performance on the maze task as the lesion gets bigger and the ability to remember the maze worsens. If true, this would be a very important finding because it implies that engrams are based on neural changes spread throughout the cortex rather than being localized to one area. The main problem with this interpretation is that it goes beyond what can safely be concluded from the data.

Notice how large the lesions are in Figure 23.4b. Perhaps one of the reasons Lashley did not find a difference in the effects of lesions at various locations is that his lesions were so large that they each damaged several cortical areas involved in learning the maze task. Another problem was that the rats might solve the maze in several different ways—by sight, feel, and smell—and the loss of one memory might be compensated for by another.

Subsequent research has proven Lashley's conclusions to be incorrect. All cortical areas do not contribute equally to memory. Nonetheless, he was correct that memories are distributed. Lashley had an important and lasting influence on the study of learning and memory because he led other scientists to consider ways in which memories might be distributed among the vast number of neurons of the cerebral cortex.

Hebb and the Cell Assembly

Lashley's most famous student was Donald Hebb, introduced in Chapter 22. Hebb reasoned that it was crucial to understand how external events are represented in the activity of the brain before one can hope to understand how and where these representations are stored. In a remarkable book published in 1949 entitled *The Organization of Behavior*, Hebb proposed that the internal representation of an object consists of all of the cortical cells that are activated by the external stimulus. Hebb called this group of simultaneously active neurons a **cell assembly** (Figure 23.5). Hebb imagined that all of these cells were reciprocally interconnected. The internal representation of the object was held in short-term memory as long as activity reverberated through the connections of the cell assembly. Hebb further hypothesized that if activation of the cell assembly persisted long enough, consolidation would occur by a "growth process" that made these reciprocal connections more effective (i.e., neurons that fired together would wire together; recall Hebb synapses from Chapter 22). Subsequently, if only a fraction of the cells of the assembly were activated by a later stimulus, as in Figure 23.5, the now-powerful reciprocal connections would cause the whole assembly to become active again, thus recalling the entire internal representation of the external stimulus—in this case, a circle.

Hebb's important message was that the engram (1) could be widely distributed among the connections that link the cells of the assembly and (2) could include the same neurons that are involved in sensation and perception. Destruction of only a fraction of the cells of the assembly would not be expected to eliminate the memory, which may explain Lashley's results. Hebb's ideas stimulated the development of neural network computer models. Although his original assumptions had to be modified slightly, these

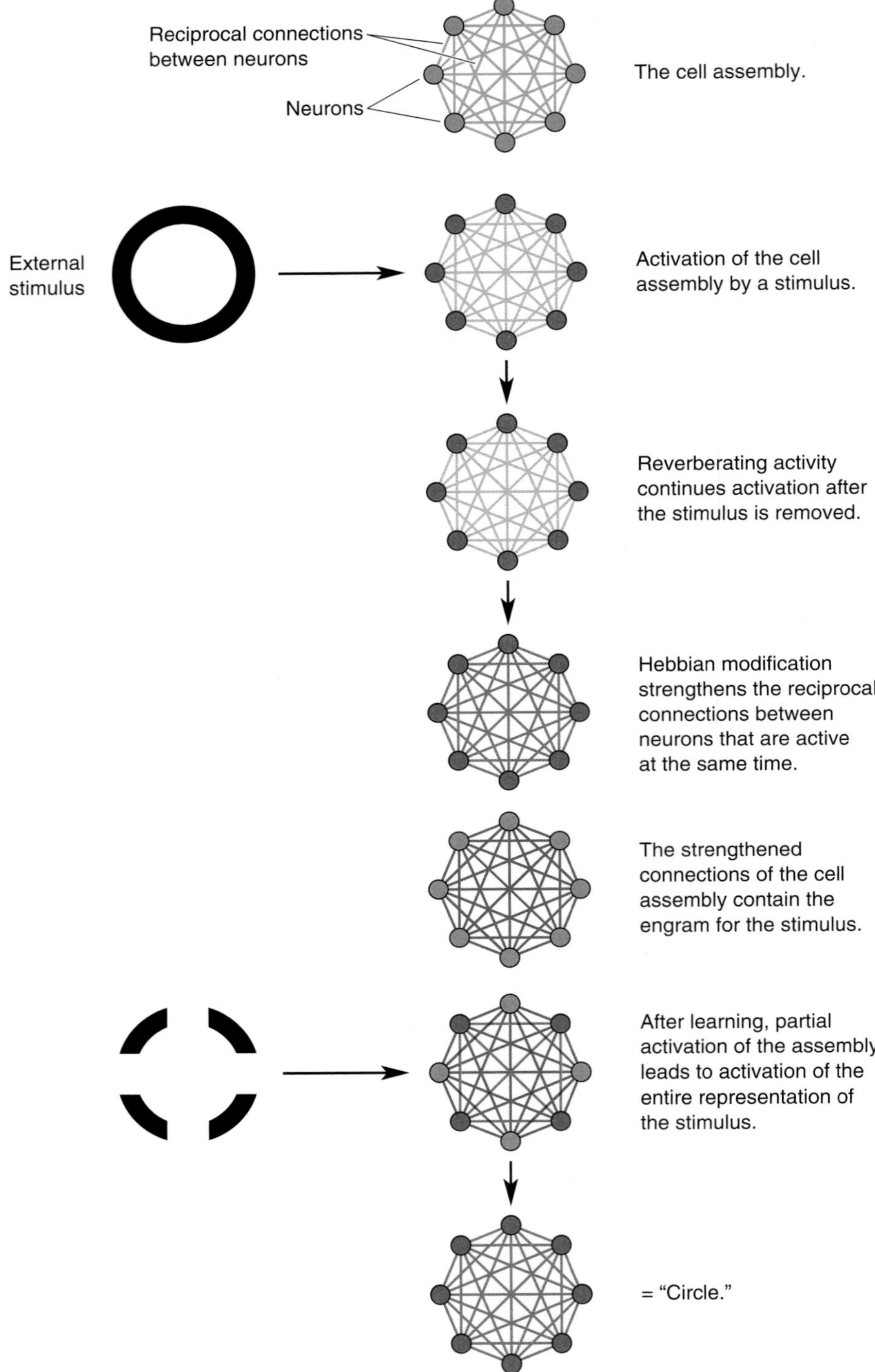

Reciprocal connections between neurons
Neurons
The cell assembly.
External stimulus
Activation of the cell assembly by a stimulus.
Reverberating activity continues activation after the stimulus is removed.
Hebbian modification strengthens the reciprocal connections between neurons that are active at the same time.
The strengthened connections of the cell assembly contain the engram for the stimulus.
After learning, partial activation of the assembly leads to activation of the entire representation of the stimulus.
= "Circle."

Figure 23.5
Hebb's cell assembly and memory storage.

models have successfully reproduced many features of human memory (Box 23.3).

Localization of Declarative Memories in the Neocortex

According to Hebb, if an engram is based on information from only one sensory modality, it should be possible to localize it within the regions of cortex that serve this modality. For example, if the engram relies only on visual in-

Box 23.3

BRAIN FOOD

A Model of a Distributed Memory

Historically, most of the progress in neuroscience research has come from experimental studies. But today, theoretical neuroscience is playing an increasing role, and the use of computational models of neural systems is widespread. In some cases, a model can provide insights into the workings of a system that are otherwise hard to gain. One area in which models have been helpful is the study of systems for learning and memory. Let's look at a model for distributed memory storage.

To keep things simple, we'll examine a nervous system consisting of three sensory neurons (the inputs) and three postsynaptic neurons (the outputs). The inputs will represent patterns of activity in visual afferents in response to the faces of three people, Eric, Kyle, and Kenny. Ignoring the complexities of visual processing, we'll assume each of these three inputs excites all three output neurons—A, B, and C (Figure A, part a). Before the system learns who Eric, Kyle, and Kenny are, each of the three output cells responds at roughly the same level to each of the inputs (Figure A, part b). There is no way to tell from the outputs of cells A, B, and C which face was the stimulus at a given time.

Now let's imagine that after being exposed to the three inputs repeatedly, the synaptic effectiveness, or strength, changes. One possibility is that after learning, Eric's face activates only cell A, Kyle's face activates only cell B, and Kenny's face activates only cell C (Figure A, part c). We can reliably tell which face is present at any time by looking to see which output neuron is active. All of the relevant information that allows the system to recognize the face—the memory—is stored in the synaptic weights. Part c represents a nondistributed memory because all of the information about Eric, for example, is stored in the single synapse with output cell A. The system can recognize that Eric is present without even bothering to look at the outputs of cells B and C.

In an alternative system, learning the three faces would again alter the synaptic weights, but none of them would be zero (Figure A, part d). The synaptic changes that store the memories can make the inputs more or less effective; memory formation does not involve only increases in synaptic strength. This is a *distributed memory* system because the memory of each face is stored in three synapses. In a real nervous system, many thousands of synapses may be involved. Recognition of one of the input faces requires comparing the strength of activity across all of the output neurons—the memory is "distributed." One reason the distributed system seems more realistic is that recordings from cortical neurons do not suggest individual neurons are specifically responsive to every image we recognize. Presumably, human recognition is based on the relative activation of thousands of neurons.

An attractive feature of the distributed memory system is its relative immunity to catastrophic memory loss if output neurons die. In the nondistributed scheme, recognition of Kenny totally depends on the response of cell C. If this cell dies, say goodbye to Kenny. Contrast this with the distributed system. If cell B is lost, we can still recognize Kyle by comparing the responses of cells A and C. The more neurons and synapses that are involved in the distributed memory, the lower the consequence of losing any single cell. This relative immunity to the effects of cell loss is a great advantage. Neurons in the human brain die every day, and it is probably because of the distributed nature of memory that we don't suddenly lose memories for people and events.

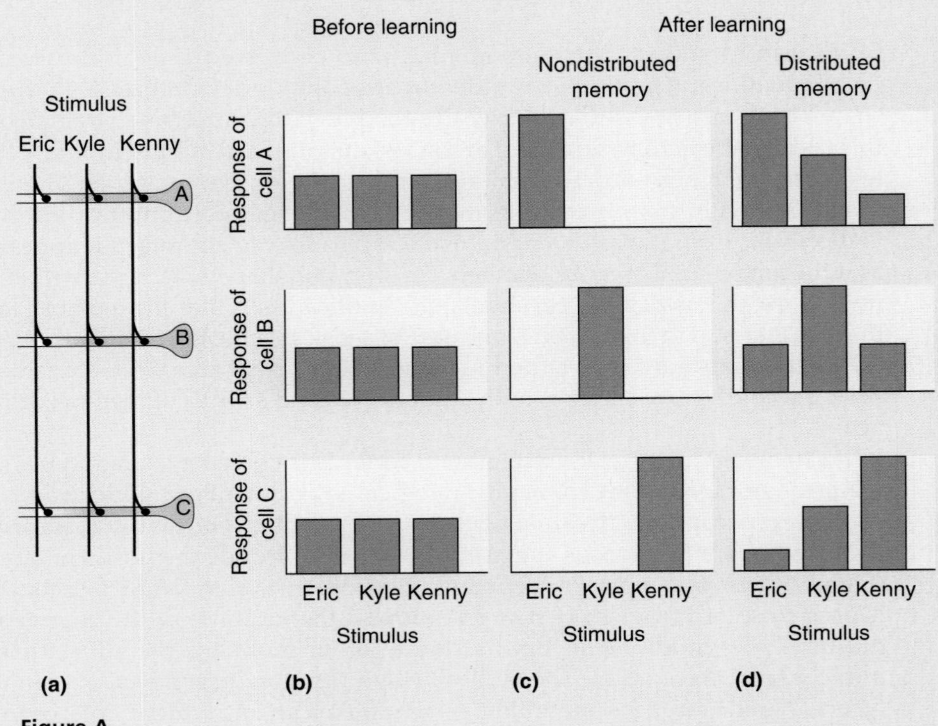

Figure A

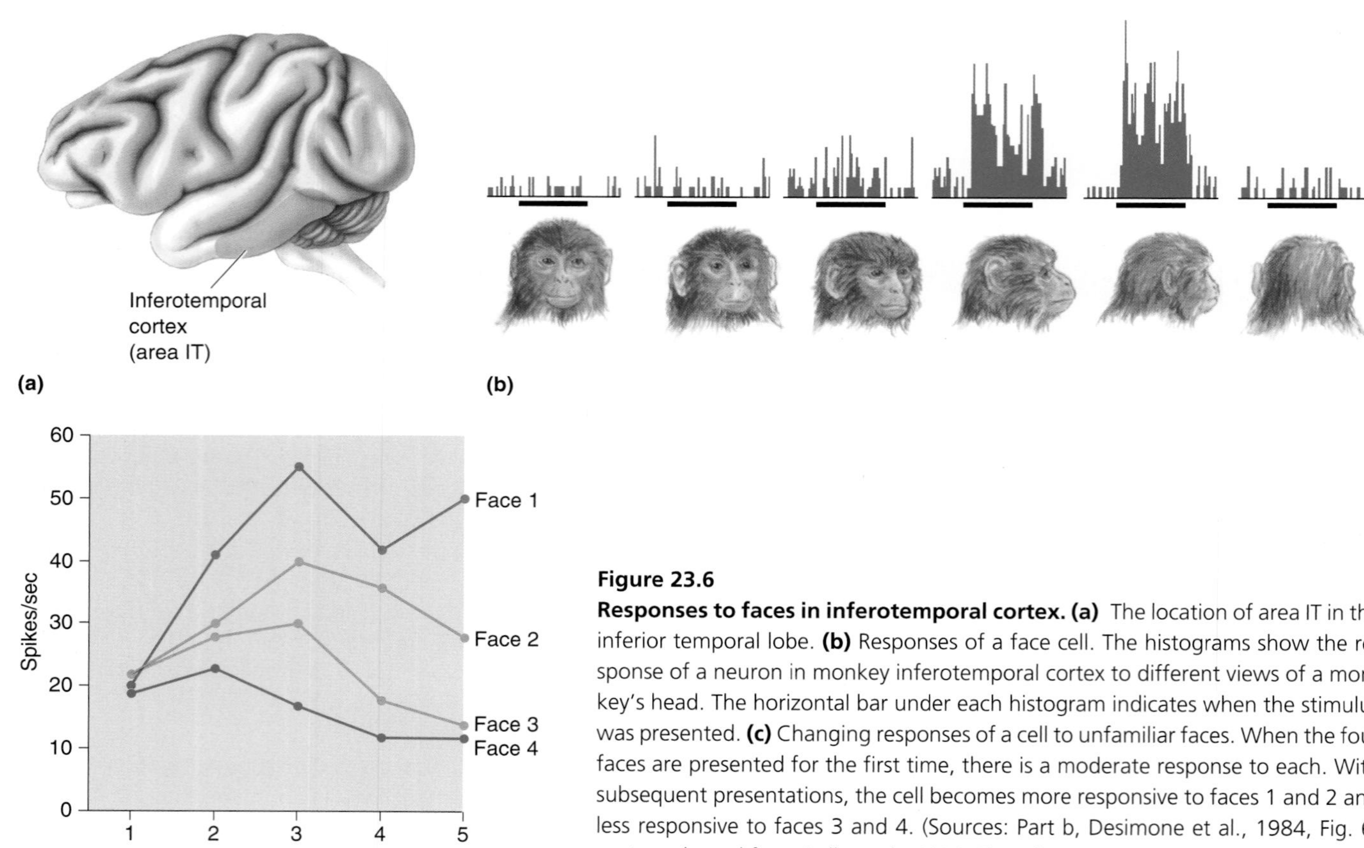

Figure 23.6
Responses to faces in inferotemporal cortex. (a) The location of area IT in the inferior temporal lobe. **(b)** Responses of a face cell. The histograms show the response of a neuron in monkey inferotemporal cortex to different views of a monkey's head. The horizontal bar under each histogram indicates when the stimulus was presented. **(c)** Changing responses of a cell to unfamiliar faces. When the four faces are presented for the first time, there is a moderate response to each. With subsequent presentations, the cell becomes more responsive to faces 1 and 2 and less responsive to faces 3 and 4. (Sources: Part b, Desimone et al., 1984, Fig. 6; part c, adapted from Rolls et al., 1989, Fig. 1.)

formation, then we expect it to reside within the visual cortex. Studies of visual discrimination in monkeys are consistent with this proposal.

Studies in Monkeys. Macaque monkeys can be trained to perform visual discriminations (e.g., they can differentiate pairs of objects based on their shapes and learn to associate one with a food reward). After being trained on this task and becoming proficient at it, a lesion is made in inferotemporal cortex (area IT), a high-order visual area in the inferior temporal lobe (Figure 23.6a). With this brain lesion, the animal can no longer perform the discrimination task, even though its basic visual capacities remain intact. It appears as if the animal no longer remembers the stimulus shape associated with the reward. As in Lashley's experiments, the implication is that the memory for the task is stored in the cortex. In the case of a task specific to vision, however, the memory appears to be stored in a high-order visual area. Stated another way, inferotemporal cortex is both a visual area and an area of memory storage.

Further evidence that inferotemporal cortex is involved in certain types of memory storage comes from physiological experiments in which the response properties of individual neurons are examined. For example, recordings made from IT neurons suggest that they may encode memories of faces (Figure 23.6b). In a typical experiment, the monkey is alert, and an electrode is used to record from an IT neuron. Initially, the response of the neuron to multiple presentations of familiar faces (other monkeys the subject frequently sees) is recorded. The face cell responds more to some faces than to others. When new faces unfamiliar to the monkey are introduced, interesting neuronal responses result, as shown in Figure 23.6c. The first time new faces are seen, the cell responds at about the same moderate level to all of them. With a couple of additional exposures, however, the response changes such that some faces evoke a significantly greater response than others do. The cell is becoming selective in its response to these new stimuli as we watch it. With continued presentation of the same group of faces, the response of the neuron to each pattern becomes more stable. We can speculate that the response selectivity of this neuron and others is part of a distributed code for the representation—the memory—of many faces. This dynamic aspect of responses in area IT supports Hebb's view that the brain can use cortical areas for both the processing of sensory information and the storage of memories.

Studies in Humans. When we discussed visual processing beyond striate cortex in Chapter 10, we saw, from functional magnetic resonance imaging (fMRI) studies in humans, results that indicate a small portion of the brain is particularly activated by faces. Other fMRI experiments have shown responses specific to a range of types of objects. For example, in one study, subjects were presented with various pictures of birds and cars. Everyone can recognize a few bird species or car models, but bird-watchers and car buffs are experts at discriminating the subtleties. When bird experts and car experts view pictures of both birds and cars, the brain responses are different. The bird experts have areas of extrastriate visual cortex that are significantly more activated by images of birds than by other objects, such as cars. Conversely, car experts show an especially strong response to pictures of cars (Figure 23.7).

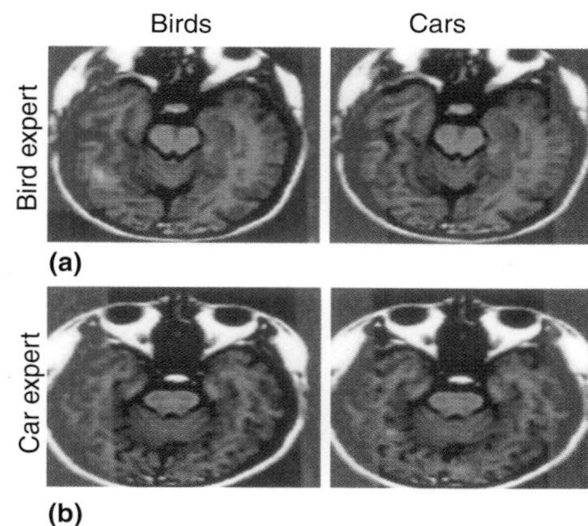

Figure 23.7
Functional MRI recordings of brain activity in bird and car experts. (a) Bird experts showed greater activity (red) in extrastriate visual areas in response to birds than other objects. **(b)** Extrastriate cortex in car experts was more activated by images of cars. (Source: Gauthier et al., 2000.)

The meaning of these activity patterns is hotly debated. One possibility is that the different activity patterns in the two types of experts reflect highly developed specialized processing of visual features needed to classify particular examples. Obviously, birds and cars differ in many ways—feathers, sheet metal, and so on. Another interpretation is that extrastriate visual areas encode memories for birds or cars (or even faces, as we saw in Chapter 10). Rather than focusing on the uncertainty about whether the responses in this type of experiment are sensory or memory, we note that the results are in line with Hebb's idea that the same area of cortex can subserve both functions.

Electrical Stimulation of the Human Temporal Lobes

One of the most intriguing and controversial studies implicating the neocortex of the temporal lobe in the storage of declarative memory traces involved electrical stimulation of the human brain. In Chapters 12 and 14, we discussed the work of Wilder Penfield in which patients, as part of surgical treatment for severe epilepsy, had their brain electrically stimulated at numerous locations prior to ablation of the seizure-prone region. Stimulation of somatic sensory cortex caused the patient to feel tingling in regions of skin, whereas stimulation of motor cortex caused a certain muscle to twitch.

Electrical stimulation of the temporal lobe occasionally produced more complex sensations than those obtained with stimulation in other brain areas. In a number of cases, Penfield's patients described sensations that sounded like hallucinations or recollections of past experiences. This is consistent with reports that epileptic seizures of the temporal lobes can evoke complex sensations, behaviors, and memories. Here is part of Penfield's account of one operation:

At the time of operation, stimulation of a point on the anterior part of the first temporal convolution on the right caused him [the patient] to say, "I feel as though I were in the bathroom at school." Five minutes later, after negative stimulations elsewhere, the electrode was reapplied near the same point. The patient then said something about "street corner." The surgeon asked him, "where" and he replied,

"South Bend, Indiana, corner of Jacob and Washington." When asked to explain, he said he seemed to be looking at himself—at a younger age.

(Penfield, 1958, p. 25)

Another patient reported similar flashbacks. When her temporal cortex was stimulated, she said, "I think I heard a mother calling her little boy somewhere. It seemed to be something that happened years ago." With stimulation at another location, she said, "Yes, I hear voices. It is late at night around the carnival somewhere—some sort of traveling circus. . . . I just saw lots of big wagons that they use to haul animals in."

Are these people reexperiencing events from earlier in their life because memories are evoked by the electrical stimulation? Does this mean that memories are stored in the neocortex of the temporal lobe? Those are tough questions. One interpretation is that the sensations are recollections of past events. That such elaborate sensations were obtained only when the temporal lobe was stimulated suggests that the temporal lobe may play a special role in memory storage. Other aspects of the findings, however, do not clearly support the hypothesis that engrams are being electrically activated. For instance, in some of the cases in which the stimulated part of the temporal lobe was removed, the memories that had been evoked by stimulation in that area could be evoked by stimulation somewhere else. In other words, the memory had not been "cut out." Also, it is important to appreciate that complex sensations were reported by only a minority of the patients, and all of these patients had abnormal cortex associated with their epilepsy.

There is no way to prove with certainty whether the complex sensations evoked by temporal lobe stimulation are recalled memories. Clearly, however, the consequences of temporal lobe stimulation and temporal lobe seizures can be qualitatively different from stimulation of other areas of the neocortex. We will next take a closer look at the structure of the temporal lobes and the elements within them that are strongly implicated in learning and memory.

THE TEMPORAL LOBES AND DECLARATIVE MEMORY

The temporal lobe lies under the temporal bone, so named because the hair of the temples is often the first to go gray with the passage of time (*tempus* is Latin for "time"). The association of the temporal lobe with time was fortuitous, in that considerable evidence points to this region of the brain as being particularly important for the recording of past events. The temporal lobes contain the temporal neocortex, which may be a site of long-term memory storage. Also within the temporal lobe are the hippocampus and other structures, which are critical for the formation of declarative memories.

The Effects of Temporal Lobectomy

If the temporal lobe is particularly important for learning and memory, one would expect removing both temporal lobes to have a profound effect on these functions. Recall from the discussion of the Klüver-Bucy syndrome in Chapter 18 that bilateral temporal lobectomy does seem to have some effect on memory. The monkeys with temporal lobectomy that Klüver and Bucy studied had a peculiar way of interacting with their environment in addition to a host of other abnormalities. The monkeys explored their room by placing objects in their mouth. If an object was edible, they would eat it; if it wasn't, they would drop it. Their behavior suggested, however, that they did not have basic perceptual deficits. In the words of Klüver and Bucy, they exhibited "psychic blindness"—even though they could see things, they did not appear to understand with their eyes what the objects were. They would repeatedly go back to the same inedible objects, put them in their mouth, then toss them aside. This problem with object recognition is probably related to memory function in the temporal lobe.

A Human Case Study: H. M. A renowned case of amnesia resulting from temporal lobe damage provides further evidence for the importance of this region in memory. A man known by his initials, H. M., has been extensively studied. H. M. had minor epileptic seizures beginning around age 10, and as he aged, they became more serious generalized seizures involving convulsions, tongue biting, and loss of consciousness. Although the cause of the seizures is not known, they may have resulted from damage sustained in an accident in which he was knocked off his bicycle at age 9 and lay unconscious for 5 minutes. After graduating from high school he got a job, but despite heavy medication with anticonvulsants, his seizures increased in frequency and severity to the point that he was unable to work. In 1953, at age 27, H. M. had an operation in which an 8 cm length of medial temporal lobe was bilaterally excised, including cortex, the underlying amygdala, and the anterior two-thirds of the hippocampus, in a last-ditch attempt to assuage the seizures (Figure 23.8). The surgery was successful in alleviating the seizures.

The removal of much of the temporal lobes had little effect on H. M.'s perception, intelligence, or personality. H. M. is still alive, and in nearly every way, he appears to be normal. This normal appearance, however, belies the reality that the surgery left him with amnesia so profound that he is incapable of performing basic human activities. H. M. has partial retrograde amnesia for the years preceding the operation. Much more serious, however, is his extreme anterograde amnesia. While he can remember a great deal about his childhood, he is unable to remember someone he met 5 minutes earlier. Dr. Brenda Milner of the Montreal Neurological Institute has worked with H. M. for more than 40 years, but she has to introduce herself to him every time they meet (Box 23.4). Milner has said it appears that H. M. forgets events as quickly as they occur. If he is told to remember a number and is then distracted, he not only forgets the number, he also forgets that he was asked to remember one. H. M. does not live in the same home he did in 1953, so he cannot find his way around his own neighborhood. He always underestimates his age, and he is unable to recognize a current picture of himself.

To be clear about the nature of H. M.'s amnesia, we must contrast what was lost with what is retained. He remembers his childhood, so long-term memories formed before the surgery and his ability to recollect past events were not destroyed. His short-term memory is also normal. For instance, with constant rehearsal he can remember a list of six numbers, although any interruption will cause him to forget. He simply has an extreme inability to form new declarative memories. Importantly, he is able to learn new tasks (i.e., form new *procedural* memories). For example, he was taught to draw by looking at his hand in a mirror, a task that takes a good deal of practice for anyone. The odd thing is that he has learned to perform new tasks, despite the fact that he has no recollection of the specific experiences in which he was taught to do them (the declarative component of the learning).

The characteristics of H. M.'s amnesia reinforce the idea that the neuroanatomy and neural mechanisms underlying procedural and declarative memory and short-term and long-term memory are not the same. In our search to understand the role of the medial temporal lobe in learning and memory, we are led to focus on the processing and consolidation of new declarative memories.

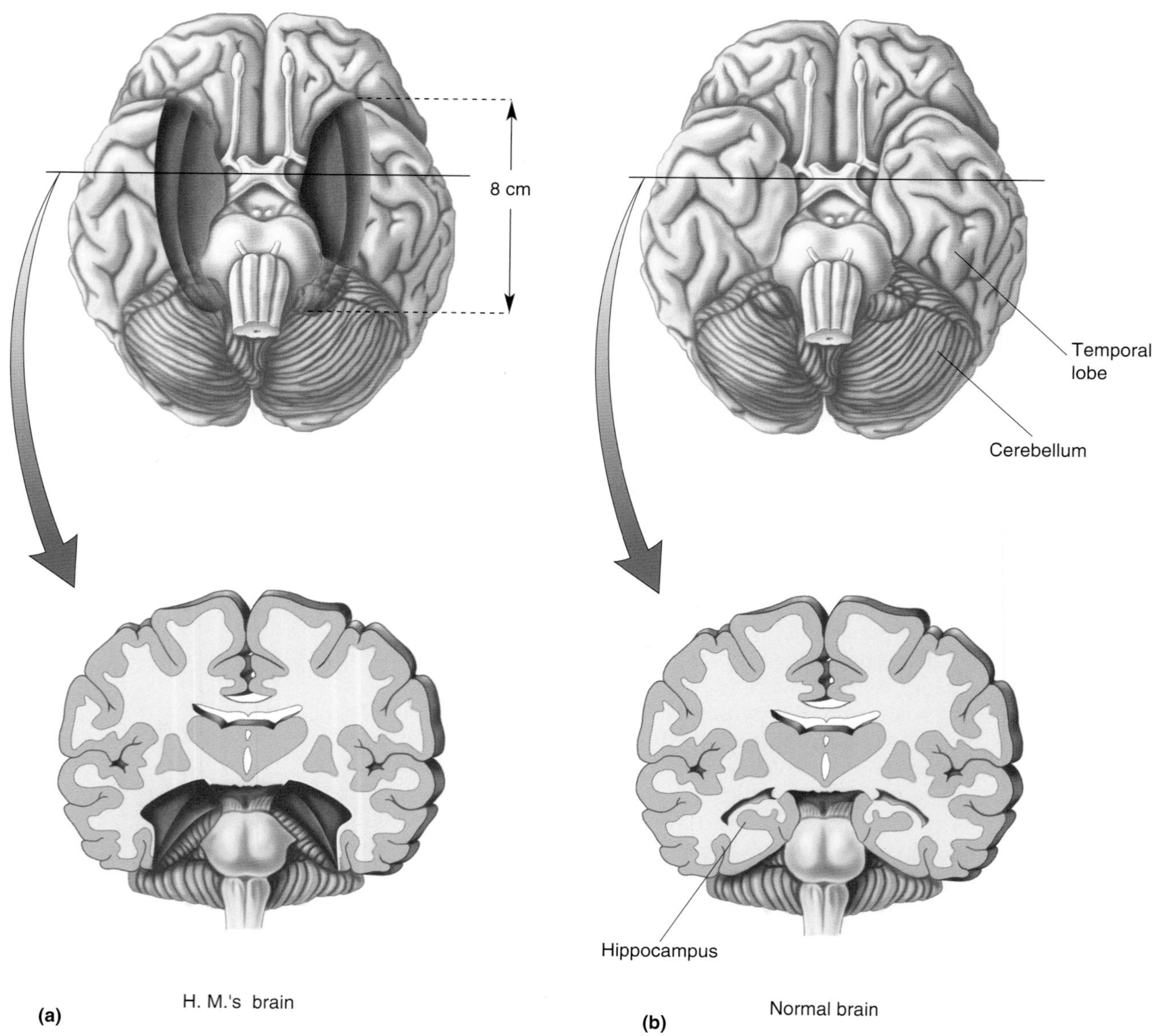

Figure 23.8

The brain lesion in patient H. M. that produced severe anterograde amnesia. (a) The medial temporal lobe was removed from both hemispheres in H. M.'s brain to alleviate severe epileptic seizures. **(b)** A normal brain showing the location of the hippocampus and cortex that were removed from H. M.'s brain. (Source: Adapted from Scoville and Milner, 1957, Fig. 2.)

The Medial Temporal Lobes and Memory Processing

In the medial temporal lobe, a group of interconnected structures appears to be of great importance for declarative memory consolidation. The key structures are the hippocampus, the nearby cortical areas, and the pathways that connect these structures with other parts of the brain (Figure 23.9). As we saw in Chapter 7, the **hippocampus** is a folded structure situated medial to the lateral ventricle. Ventral to the hippocampus are three important cortical regions that surround the rhinal sulcus: the **entorhinal cortex**, which occupies the medial bank of the rhinal sulcus; the **perirhinal cortex**, which occupies the lateral bank; and the **parahippocampal cortex**, which lies lateral to the

Box 23.4

PATH OF DISCOVERY

Discovering Memory in the Medial Temporal Lobe With H. M.

BY BRENDA MILNER

There was nothing in my background to predict a career in science. My parents' life was dominated by music, and it was a bitter disappointment that I had no musical aptitude. Instead, I studied psychology and eventually became a graduate student under Donald Hebb at McGill University. My research on the effects of temporal lobe lesions was conducted with patients who had undergone brain surgery for the relief of epilepsy. I was impressed, as no one could fail to be, by the experience of being present in the gallery of the operating room while Wilder Penfield stimulated the exposed cortex in awake patients and elicited reports of complex hallucinatory experiences. Penfield was convinced that he had excited part of the neural substrate of past experience conceived by him as a continuous record. To me, as an experimental psychologist, this tape recorder notion of memory seemed implausible.

Because most of Penfield's operations, which were in one hemisphere only, caused little behavioral change, we were shocked to discover that in two of the patients, unilateral resection of the medial structures of the temporal lobe led to severe anterograde amnesia. We hypothesized that in each case there had been a preexisting but preoperatively undetected atrophic lesion in the unoperated hemisphere. We were subsequently contacted by the neurosurgeon William Scoville, who said he had seen a similar memory disturbance in a patient on whom he had performed a bilateral medial temporal lobe resection. This was patient H. M., who is now very well known. Penfield asked me if I would like to go down to Hartford to study the patient, and that's how it started.

H. M.'s life had been totally disrupted by serious epileptic seizures. Prior to Scoville's surgery, he was on near-toxic doses of all kinds of medication. His graduation from high school came very late. He couldn't keep a job. He had no social life; he had nothing.

Removal of the medial temporal lobe didn't affect H. M.'s reasoning ability, his ability to repeat a short series of digits, or any such tasks. Incredibly, I found that he could retain a three-digit number for at least 15 minutes by continuous rehearsal, combining and recombining the digits according to an elaborate mnemonic scheme. Only when H. M. puts something out of his mind and turns to something else is the first thing lost. Say I had been working with him all morning, just the two of us, and then I'd go out for lunch. I would come back and walk by H. M. in the waiting room, and he would have no recognition of me, even though he is a very polite man. Such results appear to support the distinction between a primary memory process with a rapid decay and an overlapping secondary process (impaired in H. M.) by which the long-term storage of information is achieved.

Brenda Milner

To my surprise, H. M. had no trouble learning a mirror-drawing task I administered. Although his learning was normal, at the end of the last trial he had no idea that he had ever performed the task before. This was learning without any sense of familiarity. Nowadays we are well aware that such dissociations are possible following a discrete brain lesion, but for me at the time, it was quite astonishing. It was also early evidence of the existence of more than one memory system in the brain.

Our findings with H. M. had a mixed reception during that period, largely because monkeys with similar bilateral lesions performed normally on visual discrimination learning tasks. An important breakthrough came in 1978, when Mortimer Mishkin demonstrated a severe deficit in monkeys with bilateral medial temporal lobe lesions on a one-trial task of object recognition memory. This is, of course, what we should have predicted from H. M.'s failure on single-trial, nonverbal delayed matching tasks. It was a real discovery—the reason the monkeys with lesions exactly like H. M.'s had apparently succeeded where H. M. had failed was that they were learning the task in a different way, as a procedure over hundreds of trials.

As I look back over the past 50 years, it seems to me that I have had a lot of luck in being in the right place at the right time but also enough tenacity of purpose not to be discouraged when the going got rough. I am also grateful for my curiosity, which led me to wish to delve deeper into phenomena that caught my eye and which keeps me going to this day.

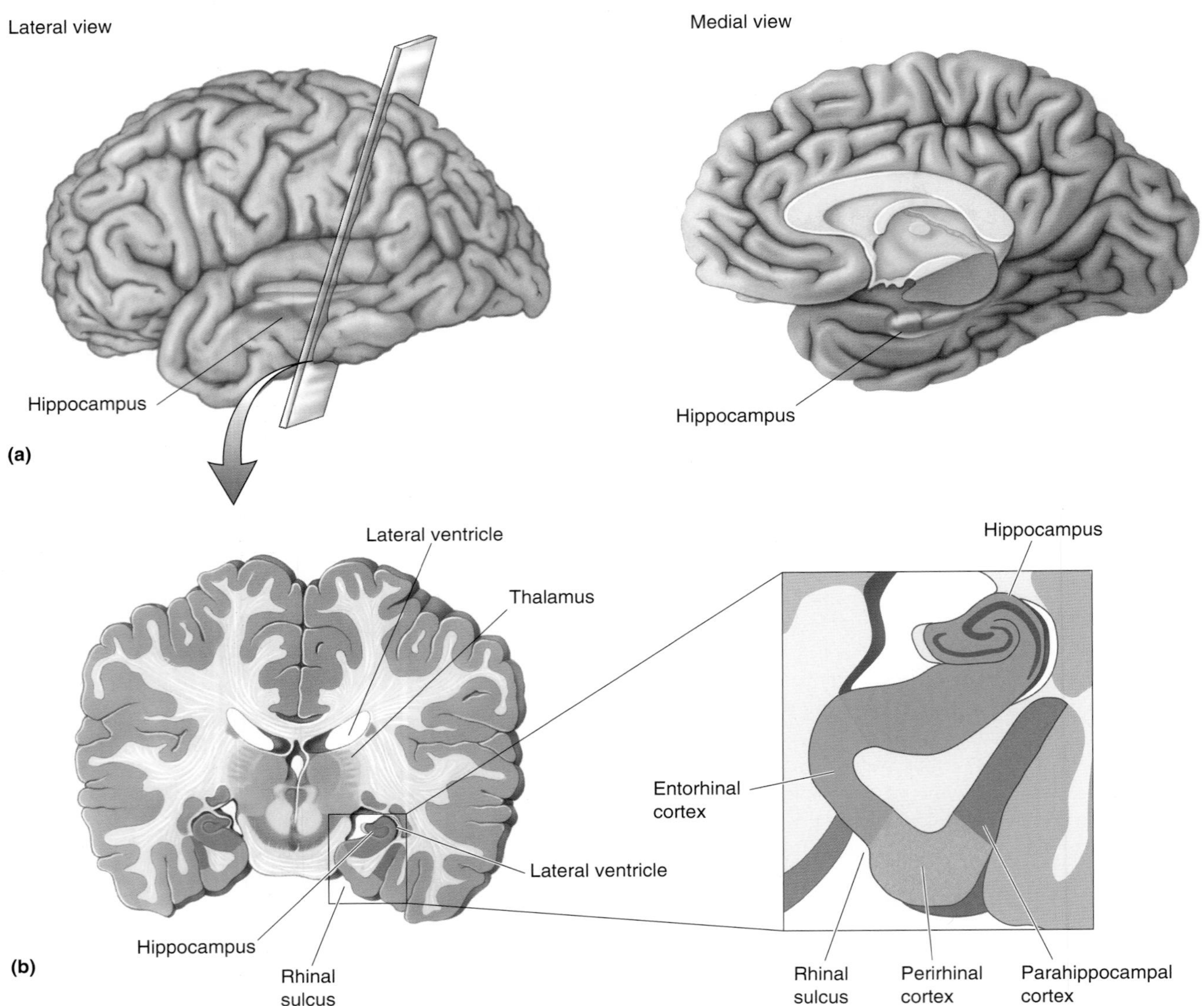

Figure 23.9

Structures in the medial temporal lobe involved in declarative memory formation. (a) Lateral and medial views show the location of the hippocampus in the temporal lobe. **(b)** The brain is sectioned coronally to show the hippocampus and cortex of the medial temporal lobe.

rhinal sulcus. (We'll refer to entorhinal cortex and perirhinal cortex collectively as rhinal cortex.)

Inputs to the medial temporal lobe come from the association areas of the cerebral cortex, containing highly processed information from all sensory modalities (Figure 23.10). For instance, inferotemporal visual cortex projects to the medial temporal lobe, but low-order visual areas such as striate cortex do not. This means that the input contains complex representations, perhaps of behaviorally important sensory information, rather than responses to simple features such as light-dark borders. Input first reaches the rhinal and parahippocampal cortex before being passed to the hippocampus. A major output pathway from the hippocampus is the **fornix**, which loops around the thalamus before terminating in the hypothalamus.

An Animal Model of Human Amnesia. The effects of temporal lobectomy, particularly the amnesia of H. M., make a strong case that one or more struc-

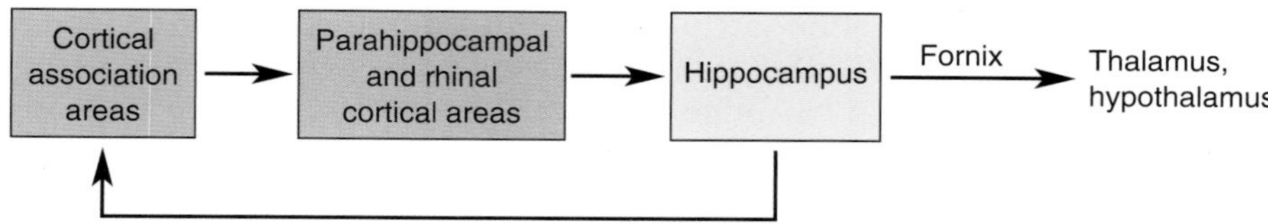

Figure 23.10
Information flow through the medial temporal lobe.

tures in the medial temporal lobe are essential for the formation of declarative memories. If these structures are damaged, severe anterograde amnesia results. There has been intensive investigation into the particular structures in the medial temporal lobe that are essential for memory formation. For the most part, these studies use the experimental ablation technique to assess whether the removal of some part of the temporal lobe affects memory.

Because the macaque monkey brain is similar in many ways to the human brain, macaques are frequently studied to further our understanding of human amnesia. The monkeys are most often trained to perform a task called **delayed non-match to sample (DNMS)** (Figure 23.11). In this type of experiment, a monkey faces a table that has several small wells in its surface. It first sees the table with one object on it covering a well. The object may be a wooden block or a chalkboard eraser (the sample stimulus). The monkey is trained to displace the object so that it can get a food reward in the well under the object. After the monkey gets the food, a screen is put down to prevent the monkey from seeing the table for some period (the delay interval). Finally, the animal gets to see the table again, but now there are two objects on it: One is the same as before, and another is new. The monkey's task is to displace the new object (the non-matching object) to get another food reward in a well below it. Normal monkeys are relatively easy to train on the non-matching task and get very good at it, probably because it exploits their natural curiosity for novel objects. With delays between the two stimulus presentations of anywhere from a few seconds to 10 minutes, the monkey

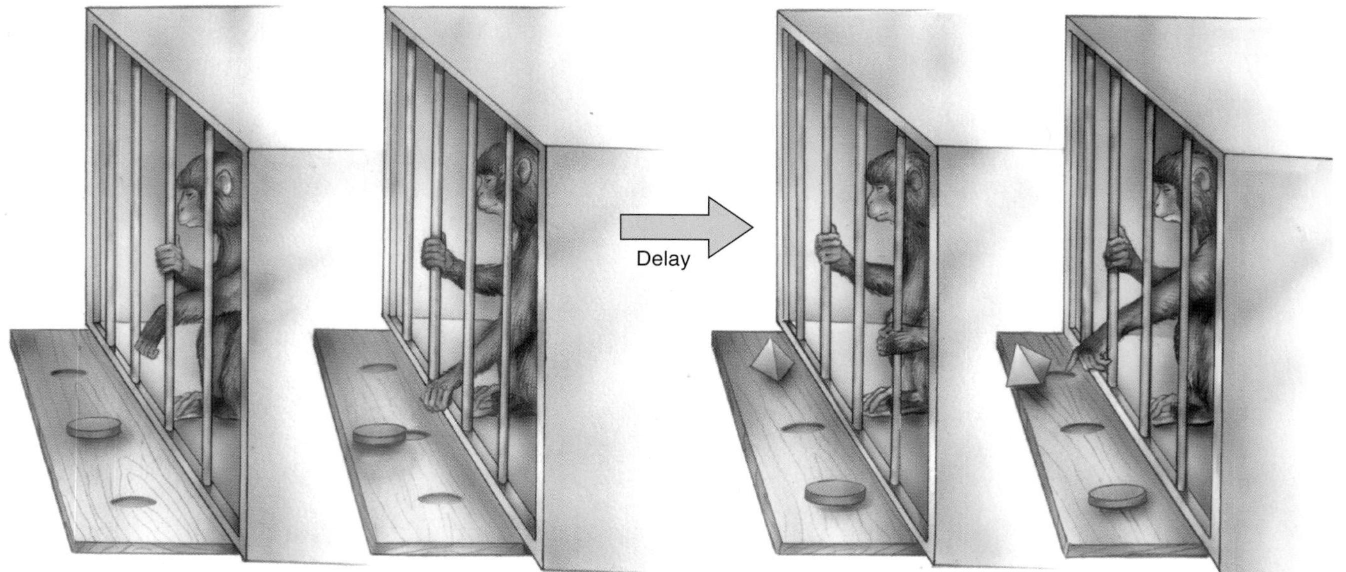

Figure 23.11
The DNMS task. A monkey first displaces a sample object to obtain a food reward. After a delay, two objects are shown, and recognition memory is tested by having the animal choose the object that does not match the sample. (Source: Adapted from Mishkin and Appenzeller, 1987, p. 6.)

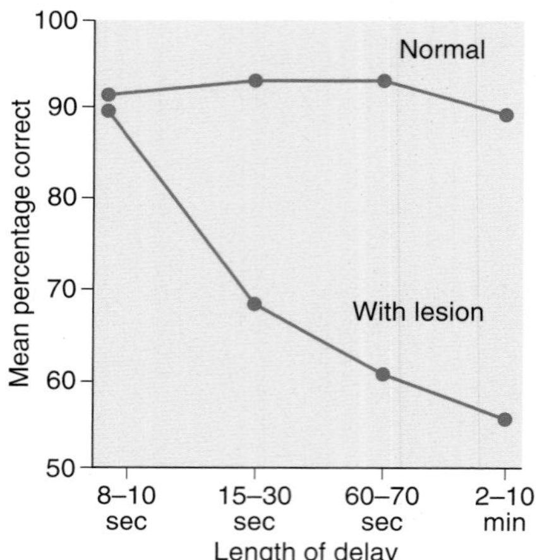

Figure 23.12

The effect of medial temporal lesions on DNMS performance. The Y axis shows the percentage of correct choices made by monkeys as a function of the length of the delay interval. The performance of normal monkeys is compared with that of monkeys with large bilateral medial temporal lesions. (Source: Adapted from Squire, 1987, Fig. 49.)

correctly displaces the non-matching stimulus on about 90% of the trials. Memory required in the DNMS task has been called **recognition memory** because it involves the ability to judge whether a stimulus has been seen before. It is probably reasonable to assume that if the animal can perform this task with a delay several minutes long, it must have formed a long-term memory for the sample object.

In the early 1980s, experiments performed by Mortimer Mishkin and his colleagues at the National Institute of Mental Health and Larry Squire and his coworkers at the University of California, San Diego, demonstrated that severe deficits on the DNMS task result from bilateral medial temporal lesions in macaque monkeys. Performance was close to normal if the delay between the sample stimulus and the two test stimuli was short (a few seconds). This is very important because it indicates that the monkey's perception was still intact after the ablation and it remembered the DNMS procedure. But when the delay was increased from a few seconds to a few minutes, the monkey made more and more errors choosing the non-matching stimulus (Figure 23.12). With the lesion, the animal was no longer as good at remembering what the sample stimulus was in order to choose the other object. Its behavior suggests that it forgot the sample stimulus if the delay was too long. The deficit in recognition memory produced by the lesion was not modality specific, since this deficit was also observed if the monkey was allowed to touch but not see the objects.

The monkeys with medial temporal lesions appeared to provide a good model of human amnesia. As with H. M., the amnesia involved declarative rather than procedural memory, it seemed more anterograde than retrograde, and long-term memory was more severely impaired than short-term memory. Note that the lesions that were originally found to produce recognition memory deficits in monkeys were quite large. They included the hippocampus, amygdala, and rhinal cortex. At one time it was thought that the key structures damaged in the lesions were the hippocampus and amygdala. It has now been shown, however, that selective amygdala lesions have no effect on this form of memory and lesions of the hippocampus alone produce only relatively mild amnesia. For example, Squire studied a man known by the initials R. B. who had bilateral hippocampal damage as a result of oxygen deprivation during surgery. Although R. B. clearly had difficulty forming new memories, this anterograde amnesia was not nearly as severe as that seen in H. M. The most severe memory deficits result from damage to the perirhinal cortex.

Together with the hippocampus, the cortex in and around the rhinal sulcus evidently performs a critical transformation of the information coming from association cortex. One hypothesis is that these medial temporal structures consolidate the memory into cortex. It is also possible, however, that they serve as an essential intermediate processing stage that involves something other than consolidation. These cortical regions may also do something else. In H. M. and possibly in R. B., there was some retrograde amnesia. Perhaps the cortex of the medial temporal lobes stores memories temporarily before they are ultimately transferred to the neocortex elsewhere for more permanent storage.

The Diencephalon and Memory Processing

Lesions in the medial temporal lobe can produce profound amnesia, but other lesions can also disrupt memory. Outside the temporal lobe, one of the brain regions most associated with memory and amnesia is the diencephalon.

The three regions of the diencephalon implicated in the processing of recognition memory are the anterior and dorsomedial nuclei in the thalamus and the mammillary bodies in the hypothalamus. Recall that a major output of the hippocampal formation is a bundle of axons making up the fornix. Most of these axons project to the mammillary bodies (Figure 23.13). Neurons in the mammillary bodies project to the anterior nucleus of the thalamus. This circuit from hippocampus to hypothalamus to anterior nucleus and then to cingulate cortex should sound familiar; it is half of the Papez circuit we discussed in Chapter 18. The dorsomedial nucleus of the thalamus also receives input from temporal lobe structures, including the amygdala and inferotemporal neocortex, and it projects to virtually all of frontal cortex.

Large midline thalamic lesions in monkeys produce relatively severe deficits on the DNMS task. These lesions damage the anterior and dorsomedial nuclei of the thalamus, producing retrograde degeneration in the mammillary bodies. Bilateral lesions limited to either the dorsomedial nuclei or the anterior nuclei produce significant but milder deficits. The limited data available suggest that rather mild deficits are seen after lesions in only the mammillary bodies.

A Human Case Study: N. A. Reports for many years have suggested a relationship between human amnesia and damage to the diencephalon. A particularly dramatic example is the case of a man known as N. A. In 1959, at age 21, N. A. was a radar technician in the U. S. Air Force. One day he was sitting down assembling a model in his barracks while behind him a roommate played with a miniature fencing foil. N. A. turned at the wrong moment and was stabbed. The foil went through his right nostril, taking a leftward course into his brain. Many years later, when computed tomography was performed, the only obvious damage was a lesion in his left dorsomedial thalamus, though there may have been other damage.

After his recovery, N. A.'s cognitive ability was normal but his memory was impaired. He had retrograde amnesia of about 2 years and relatively severe anterograde amnesia. While he could remember some faces and events

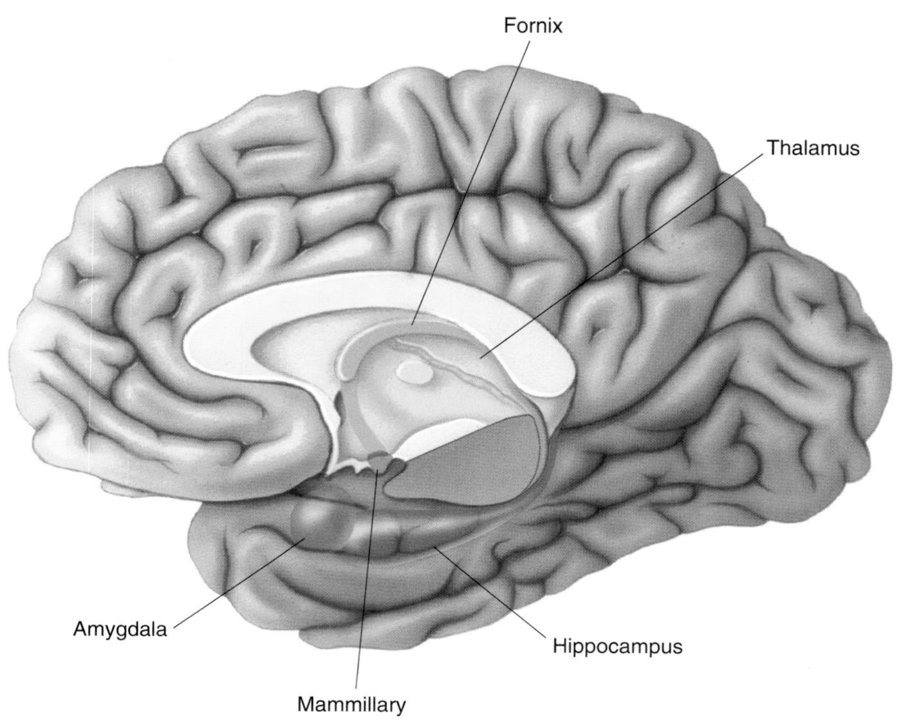

Figure 23.13
Components of the diencephalon involved in memory. The thalamus and mammillary bodies receive afferents from structures in the medial temporal lobe.

from the years following his accident, even these memories were sketchy. He had difficulty watching television because during commercials he'd forget what was happening on the show. In a sense, he lived in the past by preferring to wear old familiar clothes and keeping his hair in a crewcut.

Although N. A.'s amnesia was less severe than H. M.'s, the quality was strikingly similar. There was preservation of short-term memory, recollection of old memories, and general intelligence. Along with difficulty forming new declarative memories, he had retrograde amnesia for years preceding the accident that produced the amnesia. The similarities in the effects of medial temporal and diencephalic lesions suggest that these interconnected areas are part of a system serving the common function of memory consolidation.

Korsakoff's Syndrome. Further support for a role of the diencephalon in memory comes from Korsakoff's syndrome. Usually resulting from chronic alcoholism, **Korsakoff's syndrome** is characterized by confusion, confabulations, severe memory impairment, and apathy. As a result of poor nutrition, alcoholics may develop a thiamin deficiency, which can lead to such symptoms as abnormal eye movements, loss of coordination, and tremors. This condition can be treated with supplemental thiamin. If left untreated, thiamin deficiency can lead to structural brain damage, which does not respond to thiamin treatment. The structural damage produces Korsakoff's syndrome. Although not all cases of Korsakoff's syndrome are associated with damage to the same parts of the brain, there are usually lesions in the dorsomedial thalamus and the mammillary bodies.

In addition to anterograde amnesia, Korsakoff's syndrome can involve more severe retrograde amnesia than observed in N. A. and H. M. There is no strong correlation between the severity of anterograde amnesia and retrograde amnesia in Korsakoff's syndrome. This is consistent with the other studies of amnesia we've discussed, suggesting that the mechanisms involved in consolidation (disrupted in anterograde amnesia) are largely distinct from processes used to recall memories (disrupted in retrograde amnesia). Based on a small number of cases such as that of N. A., researchers suspect that anterograde amnesia associated with diencephalic lesions results from damage to the thalamus and mammillary bodies. Although it is not clear which damage is responsible for the retrograde amnesia, in addition to diencephalic lesions, Korsakoff's patients sometimes have damage to the cerebellum, brain stem, and neocortex.

Memory Functions of the Hippocampus

One important role of the medial temporal lobes is in declarative memory processing or consolidation. Research on the rat hippocampus suggests, however, that it is involved in memory function for a diverse range of tasks. Before we discuss these tasks, let's look at what is known about the physiology of the hippocampus and the effects of lesions.

The Effects of Hippocampal Lesions in Rats. A role for the hippocampus in memory has been demonstrated by experiments in which rats are trained to get food in a *radial arm maze*, devised by David Olton and his colleagues at Johns Hopkins University. This apparatus consists of arms, or passageways, radiating from a central platform (Figure 23.14a). If a normal rat is put in such a maze, it explores until it finds the food at the end of each arm. With practice, the rat becomes efficient at finding all of the food, going down each arm of the maze just once (Figure 23.14b). To run through the maze without going twice into one of the arms, the rat uses visual or other cues around the

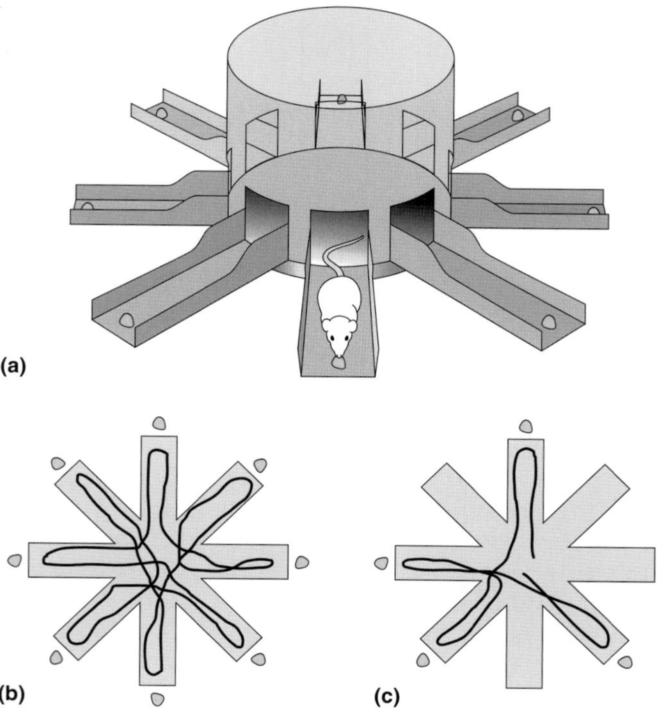

Figure 23.14
Following a rat through a radial arm maze. (a) An eight-arm radial maze. **(b)** The path of a rat through a maze in which all of the arms contain food. **(c)** If a rat learns that four of the eight arms never contain food, it will ignore these and follow a path to only the baited arms. (Source: Parts b and c adapted from Cohen and Eichenbaum, 1993, Fig. 7.4.)

maze to remember where it has already been. The type of memory used to retain information about which arms have been visited was called working memory by Olton. More generally, **working memory** refers to the retention of information needed to guide ongoing behaviors.

If the hippocampus is destroyed before the rat is put in the maze, performance differs from normal in an interesting way. In one sense, rats with lesions seem normal; they learn to go down the arms of the maze and eat the food placed at the end of each arm. But unlike normal rats, they never learn to do this efficiently. Rats with hippocampal lesions go down the same arms more than once, only to find no food after the first trip, and they leave other arms containing food unexplored for an abnormally long time. It appears that the rats can learn the task in the sense that they go down the arms in search of food. But they cannot seem to remember which arms they've already been down.

A variation on the radial arm experiment illustrates an important subtlety in the deficit produced by destroying the hippocampus. Instead of placing food at the end of all of the arms of the maze, food is placed only at the ends of certain arms and never in the other ones. After a bit of practice, a normal rat learns to avoid going down the arms that never contain food (Figure 23.14c). At the same time, the rat learns to get the food in the other arms efficiently, i.e., entering each arm just once. How do rats with hippocampal lesions do on this task? Interestingly, just like normal rats, they are able to learn to avoid the arms that never contain food. But they still are not able to get the food from the other arms without wasting time going down the same arms more than once. Isn't this a little odd? How can we argue that the lesion disrupts the ability to learn the locations of arms that have already been entered, even though the rat can learn to avoid arms that never contain food? Evidently, the key to making sense of these findings is that the information about the no-food arms is always the same each time the rat goes in the maze, whereas the information about which arms the rat has already entered requires working memory and varies from one trial to the next.

Place Cells. In a fascinating series of experiments begun in the early 1970s, John O'Keefe and his colleagues at University College London showed that many neurons in the hippocampus selectively respond when a rat is in a particular location in its environment. Suppose we have a microelectrode implanted in the hippocampus of a rat while it scurries about inside a large box. At first the cell is quiet, but when the rat moves into the northwest corner of the box, the cell starts firing. When it moves out of the corner, the firing stops; when it returns, the cell starts firing again. The cell responds only when the rat is in that one portion of the box (Figure 23.15a). This location, which evokes the greatest response, is called the neuron's *place field*. We try recording from another hippocampal cell, and it too has a place field, but this one fires only when the rat goes to the center of the box. For obvious reasons, these neurons are called **place cells**.

In some ways, place fields are similar to the receptive fields of neurons in the sensory systems. For instance, the location of the place field is related to sensory input such as visual stimuli in the environment. In our experiment with the rat in the box, we could paint images above the four corners, such as a star above the northwest corner, a happy face above the southeast corner, and so on. Consider a cell that responds only when the rat is in the northwest corner of the box near the painted star. Suppose we take the rat out of the box and blindfold it. We then secretly go back and rotate the box 180 degrees so that now the northwest corner has the happy face and the southeast corner has the star. Will the cell we were previously studying respond when the animal is in the northwest corner, or will it respond in the corner where the star

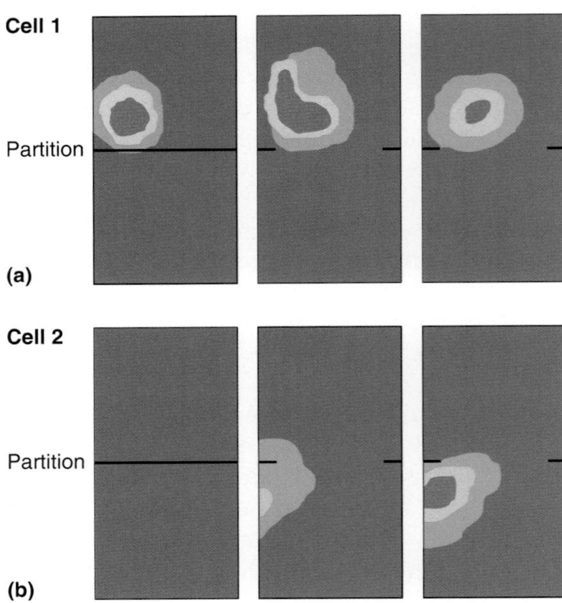

Figure 23.15
Place cells in the hippocampus. A rat explores a small box for 10 minutes (left panels). Then a partition is removed so the rat can explore a larger area (center and right panels). **(a)** Color coding indicates the area in the box where one place cell in the hippocampus responds: red, large response; yellow, moderate response; blue, no response. This cell has a place field in the smaller upper box; when the partition is removed, it stays in the same location. **(b)** In this case, an electrode is next to a cell in the hippocampus that does not respond when the animal is in the smaller upper box (left). In the first 10 minutes after the partition is removed, the cell also does not respond (center). But after another 10 minutes, a place field develops in the new larger box (right). (Source: Adapted from Wilson and McNaughton, 1993, Fig. 2.)

is now located (the southeast corner)? We put the rat back in the box and take off the blindfold. It starts exploring, and the neuron becomes active when the rat goes to the corner near the star. This demonstrates that at least under some conditions, the response is based on visual stimuli.

While place cells are similar to receptive fields in some ways, there are also major differences. For instance, once the animal has become familiar with the box with the images painted in each corner, a neuron continues to fire when the rat goes into the northwest corner even if we turn off the lights so that the animal cannot see the location markers. Evidently, the responses of place cells are related to where the animal *thinks* it is. If there are obvious visual cues, such as the star and happy face, the place fields are based on these cues. But if there are no cues, e.g., because the lights are out, the place cells are still location specific as long as the animal has had enough time to explore the environment and develop a sense of where it is.

Performance in the radial arm maze, discussed earlier, may utilize these place cells, which code for location. Of particular importance in this regard is the finding that the place fields are dynamic. For instance, let's say we first let a rat explore a small box and we determine the place fields of several cells. Then we cut a hole in a side of the box so the animal can explore a larger area. Initially, there are no place fields outside the smaller box. But after the rat has explored its new expanded environment, some cells develop place fields outside the smaller box (Figure 23.15b). These cells seem to *learn* in the sense that they alter their receptive fields to suit the environment. It's easy to imagine how these sorts of cells could be involved in remembering arms already visited in the radial arm maze. And if they are involved in running the maze, it certainly makes sense that performance is degraded by destroying the hippocampus.

Whether there are place cells in the human brain is not known. Positron emission tomography (PET) studies show, however, that the human hippocampus is activated in situations involving virtual or imagined navigation through the environment. In one experiment, subjects were positioned in a PET machine to view a video game on a computer monitor. They could navigate a virtual town in the game by using buttons for forward movement, backward movement, and turning (Figure 23.16a). After the subjects learned their way around the virtual town, their brain activity was recorded while they navigated from an arbitrary starting point to a chosen destination. In a control condition, the person moved through the virtual environment from the same start to finish locations, but there were arrows in the town that always pointed them in the correct direction. In this condition, they did not have to think about how to navigate.

Figure 23.16b shows the difference in brain activity between the navigation condition and the control condition with directional arrows. When the person had to navigate the environment, there was increased activation of the right hippocampus and the left tail of the caudate. The asymmetry in the activation of the left and right hemispheres is an interesting observation that has been repeatedly made, but our primary point is that the hippocampus is particularly active in this spatial navigation task with humans, just as it is in rats. The caudate activation is thought to reflect movement planning. Similar hippocampal activation was observed in an interesting study in which the brains of experienced taxi drivers were imaged as they imagined themselves driving to a destination through the complex streets of London.

Spatial Memory, Working Memory, and Relational Memory. Our discussion of the hippocampus to this point may make it seem that its role is easily defined. First, we saw that performance in a radial arm maze, which requires memory for the locations of arms already explored, is disrupted by hip-

(a)

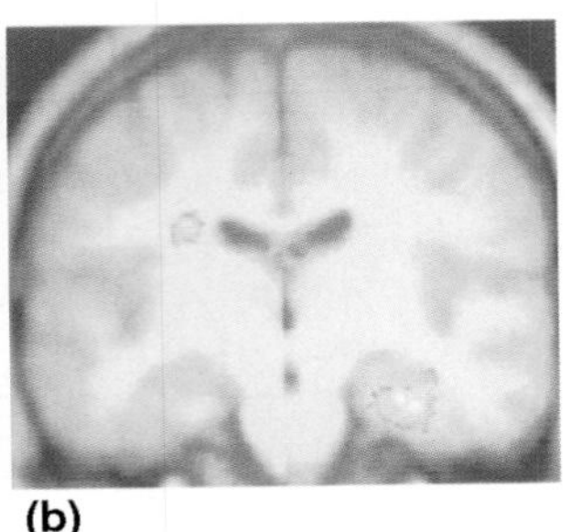
(b)

Figure 23.16
Activity in the human brain related to spatial navigation. (a) A virtual town was shown on a computer monitor, and subjects in a PET machine used buttons to navigate the virtual environment. **(b)** In this coronal slice, increased brain activity associated with spatial navigation was observed in the right hippocampus and left tail of the caudate (yellow). (Source: Maguire et al., 1998, Fig. 1.)

pocampal lesions. Second, the responses of neurons in the hippocampus, the place cells, suggests that these neurons are specialized for location memory. This is consistent with O'Keefe's hypothesis that the hippocampus is specialized for creating a spatial map of the environment. In one sense it is undeniable that the hippocampus, at least in rats, plays an important role in spatial memory. Others argue, however, that this is not the best description of what the hippocampus does. In Olton's original studies using the radial arm maze, he described the result of hippocampal lesions as a deficit in working memory. The rats were not able to retain recently acquired information concerning arms already explored. Thus, working memory may be one aspect of hippocampal function. This would explain why the rats with lesions could avoid going down arms that never contained food but still not remember which arms they had recently visited. Presumably after training, the information about no-food arms was saved in long-term memory, but working memory was still required to avoid the arms where food had already been retrieved.

A more recent hypothesis attempting to integrate a range of experimental findings has been put forward by behavioral neuroscientists Neal Cohen at the University of Illinois, Howard Eichenbaum at Boston University, and their colleagues. Cohen and Eichenbaum describe the function of the hippocampus, in conjunction with other structures of the medial temporal lobe, as involving relational memory. The basic idea of **relational memory** is that highly processed sensory information comes into the hippocampus and nearby cortex, and processing occurs, leading to the storage of memories in a manner that ties together or relates all of the things happening at the time the memory was stored. For example, as you read this book, you may form memories relating multiple things: specific facts, illustrations that catch your eye, interesting passages, the arrangement of material on the page, and information about the sounds or events going on around you as you read. You may have had the experience of searching for a particular passage in a book by searching for a page that looks a certain way that you remember. It is also a common event that remembering one thing (such as the theme song to an old television show) brings back a flood of related facts (the characters in the show, your living room at home, the friends you watched with, and so on). Interconnectedness is a key feature of declarative memory storage.

To navigate its environment, a rat could use either a mental map of space or relational memories associated with environmental cues. The distinction between the spatial map and relational memory hypotheses is illustrated in Figure 23.17. If there is a spatial map, one would expect place fields to be ordered in the hippocampus as the locations are in space, much like the retinotopic receptive fields in visual cortex. Experts disagree about the extent to which hippocampal place cells provide such an organized map of the entire area around the animal. In a relational memory scheme, neurons encode information about place as a series of simple associations between nearby ob-

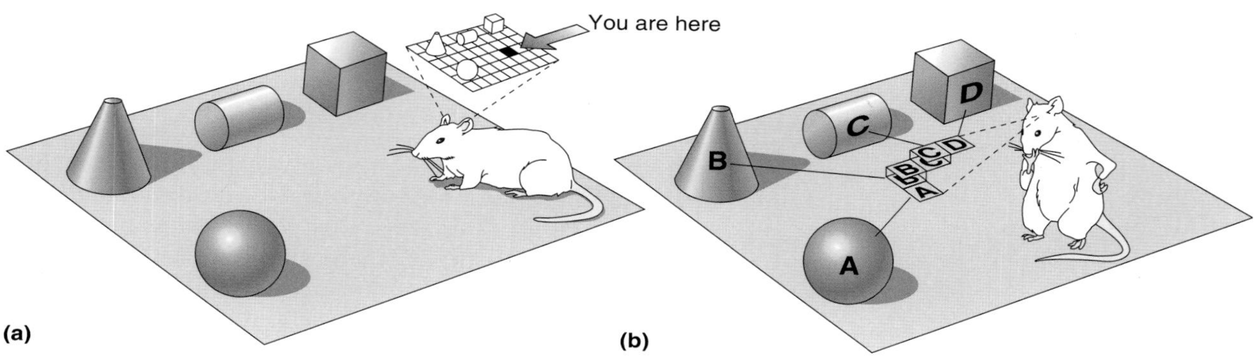

Figure 23.17

Spatial navigation based on (a) a spatial map and (b) a series of relational memories. (Source: Adapted from Eichenbaum et al., 1999.)

jects and concurrent sounds and smells. For example, in Figure 23.17b, "ball A is below cone B" would be one memory, and "cone B is to the left of cylinder C" would be another. Strung together, such relational memories could provide an understanding of the layout of the environment without having a complete, organized map in the hippocampus.

Whether or not relational memory is the principal function of the hippocampus has not been settled. It is a useful concept, however. In the maze experiments, a case can be made that the rat performs the task by storing memories of the arms it has been down in terms of the sensory cues in the room and the time a given arm was entered. Evidence also suggests that place cells may encode relational information of types other than spatial location. A simple example is that the responses of neurons with place fields are sometimes also affected by other factors, such as the speed or direction the rat is moving.

The responses of hippocampal neurons are also sometimes determined by nonspatial factors. This was demonstrated in an experiment in which Eichenbaum, Cohen, and their colleagues trained rats to discriminate odors. At one end of the rat's cage were two ports putting out two odors for the animal to sniff (Figure 23.18). For each pair of odors, the animal was trained to go toward the port releasing one odor and avoid the other port. The researchers found that some neurons in the hippocampus became selectively responsive for certain pairs of odors. Moreover, the neurons were particular about which odor was at which port—they would respond strongly with odor 1 at port A and odor 2 at port B, but not with the odors switched to the opposite ports. This indicates that the response of the hippocampal neurons relates the specific odors, their spatial locations, and the fact that they are pre-

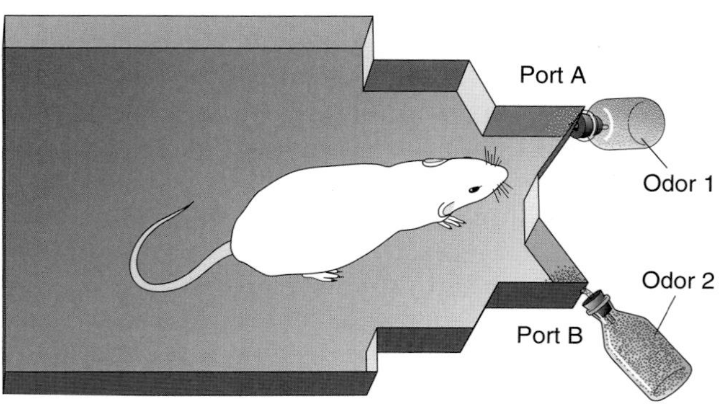

Figure 23.18

An odor discrimination experiment used to study relational memory. For various combinations of odors, rats were trained to move toward a port emitting one odor and avoid the other port. (Source: Adapted from Eichenbaum et al., 1988, Fig. 1.)

sented separately or together. It was also shown that hippocampal lesions produce deficits on this discrimination task.

THE STRIATUM AND PROCEDURAL MEMORY

Thus far, we have focused on the brain systems involved in the formation and retention of declarative memories, partly because declarative information is what we commonly mean when we say we remember something. In addition, the neural basis of nondeclarative memory is complex because different types seem to involve different brain structures. As indicated in Figure 23.1, various kinds of nondeclarative memory are thought to involve different parts of the brain. As an example of nondeclarative memory, we will take a look at evidence supporting the involvement of the striatum in habit learning and procedural memory.

Recall from Chapter 14 that the basal ganglia are important for the control of voluntary movements. Two elements of the basal ganglia are the caudate nucleus and the putamen, and together they form the **striatum**. The striatum sits at a key location in the motor loop, receiving input from frontal and parietal cortex and sending output to thalamic nuclei and cortical areas involved in movement. Several lines of evidence in studies of rodents and humans suggest that the striatum is critical for the procedural memory involved in forming behavioral habits.

Rodent Recordings and Lesions in the Striatum

The amnesia experienced by H. M. is surprising, in part, because he is able to learn new habits despite his complete inability to form new declarative memories. Indeed, this is one of the most compelling reasons for hypothesizing that procedural memory uses distinct circuitry. In the monkey model of amnesia, we saw that the formation of new declarative memories could be disrupted by making small lesions in the rhinal cortex of the medial temporal lobe. Such a lesion has relatively little effect on procedural memory, which raises an obvious question: Are there comparable lesions that disrupt procedural memory without affecting declarative memory? In rodents, lesions to the striatum have this effect.

In one study, rats had to learn two versions of the radial arm maze task. The first was the standard version in which the rat must move as efficiently as possible to retrieve the food from each of the baited arms of the maze. In the second version, small lights were illuminated above one or more arms containing food, and the unlit arms had no food. The lights could be turned on or off at any time. In this case, optimal performance meant that the animal kept returning to retrieve food from lit arms as long as they were lit and avoided arms that were never lit. The standard maze task was designed to require the use of declarative memory. The "light" version of the task was intended to draw on procedural memory because of the consistent association between the presence of food and illuminated lights. The rat does not have to remember which arms it has already explored; it must simply form a habit based on the association that light correlates with food. The rat's performance on the light task is analogous to the habits H. M. was able to form, such as mirror drawing.

Performance on the two versions of the radial arm maze task was affected in markedly different ways by two types of brain lesions. If the hippocampal system was damaged (in this case, by a lesion in the fornix that sends hippocampal output), performance was degraded on the standard maze task but was relatively unaffected on the light version. Conversely, a lesion in the striatum impaired performance of the light task but had little effect on the

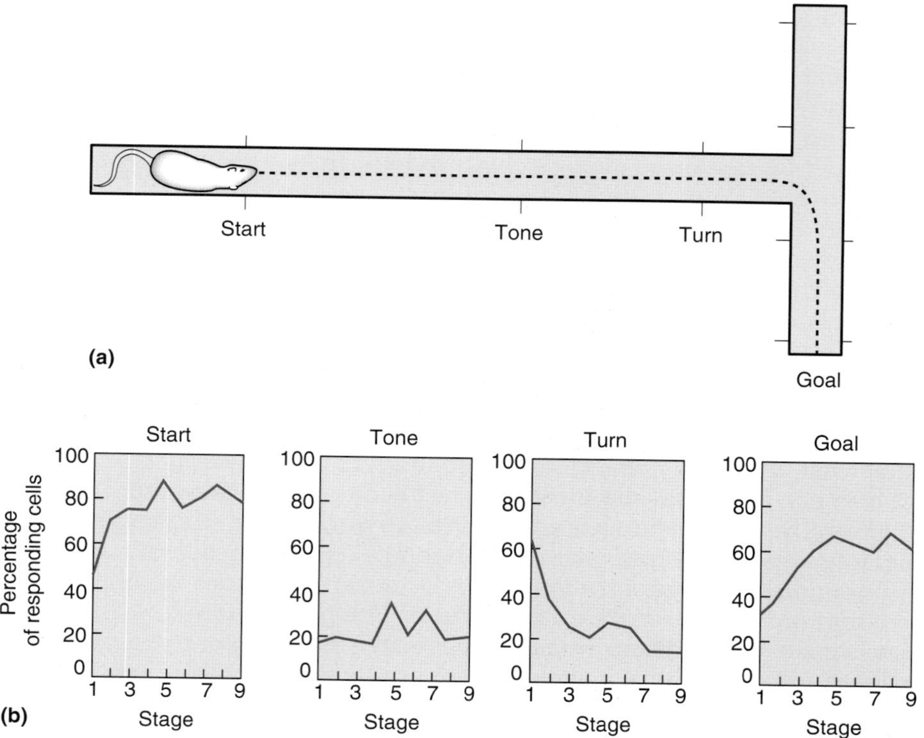

Figure 23.19
Changing responses in rat striatum during the learning of a habit. (a) The rat began at the end of the long arm of a T-maze and turned left or right depending on the pitch of a tone. **(b)** The percentage of neurons that responded at several stages of the task: start position, tone sounding, turning into reward arm, and reaching the reward. Over the stages of learning, more cells responded to the start and the goal and fewer to the turn. (Source: Adapted from Jog et al., 1999.)

standard task. This "double dissociation" of the lesion site and the behavioral deficit suggests that the striatum is part of a procedural memory system but is not crucial for the formation of declarative memories.

Recordings made from the rat striatum in other experiments show that neural responses change as the animals learn a procedure associated with a food reward. In a simple T-maze task, rats were placed at the end of one arm of the T and, as they moved away from the arm's end, a tone sounded (Figure 23.19a). A low tone instructed the animal to turn left into the next arm of the T in order to get a chocolate snack, and a high tone instructed the animal to turn right for the reward. Figure 23.19b shows the percentage of neurons that responded when the animal was in several stages of the task: start position, tone sounding, turning into reward arm, and reaching the reward. When the rats first performed the task, the highest percentage of neurons responded when the animals turned into the reward arm. As training progressed, however, this percentage decreased significantly. As the procedure was mastered, increasingly more neurons became responsive at the start and completion of the task. Also, increasing numbers of neurons responded during more than one stage of the task. One possible interpretation of these changes in response patterns is that they reflect the formation of a habit for which the striatum codes a sequence of behaviors initiated in the T-maze situation. At present, this is only a hypothesis, but it is intriguing because of the connectivity of the striatum—taking in highly processed sensory information and sending out signals involved in motor responses.

Habit Learning in Humans and Nonhuman Primates

Studies with monkeys indicate that the effects of selective brain lesions are comparable in rodents and primates. In primates, there is a similar dissociation between the effects of lesions to the hippocampal system and those to the striatum. As we have already seen, lesions to the medial temporal lobe significantly impair performance on the DNMS task that uses declarative memory. Consider, however, another task in which the animal repeatedly sees two visual stimuli, such as a square and a cross, and it must learn to associate a food reward with only the cross. This sort of habit learning is relatively unaffected by medial temporal lesions. This preservation of habit learning in monkeys is analogous to the rat's ability to retrieve food consistently associated with light above an arm of a maze, even after a fornix lesion.

In monkeys, lesions that involve the striatum or connections to it have quite different effects from medial temporal lesions. When the striatum is damaged, there is no effect on the performance of the DNMS task, demonstrating both that declarative memory formation is still possible and that the animal can discriminate visual stimuli. But when the striatum is damaged, the animal is unable to form the habit of always retrieving food associated with one visual stimulus over another. Repeated exposure to this fixed stimulus-reward situation just doesn't seem to sink in. Thus, there appear to be somewhat distinct anatomical systems for declarative memory and procedural memory, and behaviors such as learned habits utilize the striatum.

Several diseases in humans attack the basal ganglia, and certain effects on memory appear consistent with the striatum's role in procedural memory. For example, Huntington's disease kills neurons throughout the brain, but the striatum is a focus of the attack. Patients with Huntington's disease have been shown to have difficulty learning tasks in which a motor response is associated with a stimulus. Although these people generally have motor dysfunction, difficulty in learning the stimulus-response habit does not correlate with the severity of the motor deficits, suggesting that it is an independent consequence of the disease.

Further evidence for the involvement of the striatum in habit learning comes from comparisons of patients with Parkinson's disease with amnesic patients. As we saw in Chapter 14, Parkinson's disease is characterized by degeneration of the substantia nigra inputs to the striatum. In one study, people were tested on two tasks. In the first task, patients saw one, two, or three of four possible cues in one of fourteen possible combinations. They then had to guess whether this combination was arbitrarily associated with a prediction of sunny or rainy weather (Figure 23.20a). For each patient, the experimenter assigned different probabilities that various cues were associated with sun or rain. By being told when they guessed correctly or incorrectly about the predicted weather, the patients slowly built up an association between the cues and the weather. The idea behind this task was that it draws on the formation of a stimulus-response habit. In the second task, declarative memory was tested by having patients answer multiple-choice questions about the appearance of the cues and the computer screen.

The Parkinson's patients had great difficulty learning the weather forecasting task (Figure 23.20b) but performed at normal levels on the declarative memory questionnaire (Figure 23.20c). Conversely, amnesic patients had no trouble learning the weather classification, but they performed significantly worse than either Parkinson's patients or normal controls on the questionnaire. These results suggest that the striatum in humans may play a role in procedural memory as part of a system distinct from the medial temporal system used for declarative memory.

THE NEOCORTEX AND WORKING MEMORY

As we have seen, there may be a working memory function of the hippocampus, demonstrated by rats retrieving food in a radial arm maze. Here we'll take a closer look at two locations in the neocortex that appear to be involved in working memory: the prefrontal cortex and the lateral intraparietal area (area LIP).

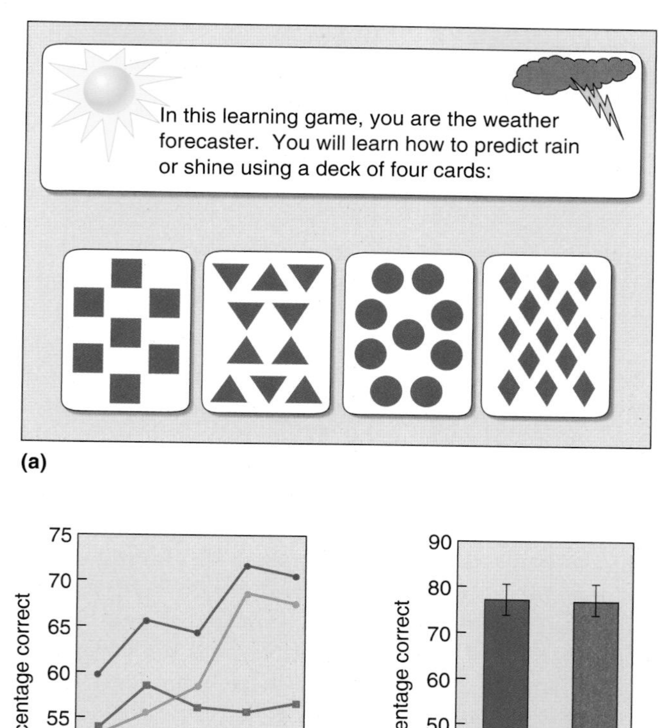

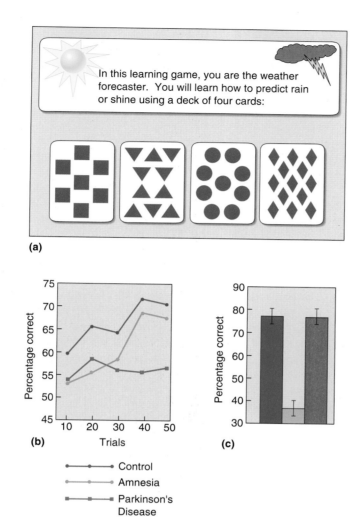

Figure 23.20
The performance of patients with amnesia and Parkinson's disease on two memory tasks. (a) Four cue cards were presented in various combinations associated with the icons indicating sun or rain. Based on repeated exposure to the combinations, patients had to learn to predict sun or rain by inferring the associations. **(b)** With successive trials, control subjects and amnesic patients improved on the association task. Parkinson's patients showed little improvement. **(c)** On a test of declarative memory formation (a questionnaire), Parkinson's patients performed similarly to control subjects, and amnesia patients were greatly impaired. (Source: Adapted from Knowlton et al., 1996.)

The Prefrontal Cortex and Working Memory

One of the most obvious anatomical differences between primates (especially humans) and other mammals is that primates have a large frontal lobe. The rostral end of the frontal lobe, the **prefrontal cortex**, is particularly highly developed (Figure 23.21). Compared with the functions of sensory and motor cortical areas, the function of prefrontal cortex is relatively poorly understood. But because it is so well developed in humans, it is often assumed that prefrontal cortex is responsible for those characteristics, such as self-awareness and the capacity for complex planning and problem solving, that distinguish us from other animals. One reason for thinking that prefrontal cortex may be involved in learning and memory is that it is interconnected with the medial temporal lobe and diencephalic structures previously discussed (Figure 23.22).

Some of the first evidence suggesting that the frontal lobe is important for learning and memory came from experiments performed in the 1930s using a *delayed-response task*. In this task, a monkey first sees food being placed in a well below one of two identical covers in a table. A delay period follows, during which the animal cannot see the table. Finally, the animal is allowed to see the table again and receives the food as a reward if it chooses the correct well. Large prefrontal lesions seriously degrade performance on this delayed-response task and other tasks including a delay period. Moreover, the

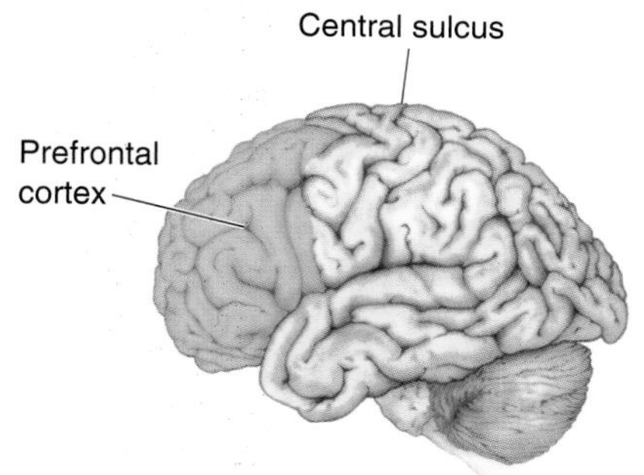

Figure 23.21
Prefrontal cortex. The prefrontal cortex at the rostral end of the frontal lobe receives afferents from the medial dorsal nucleus of the thalamus.

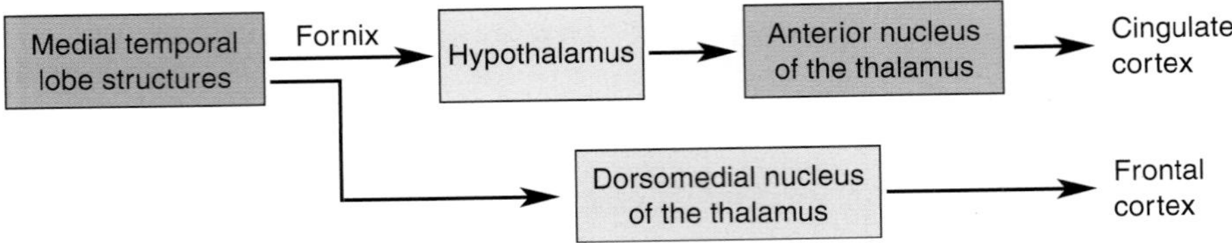

Figure 23.22
Connections between cortical association areas and the medial temporal lobe. Prefrontal cortex and cingulate cortex receive afferents from structures of the medial temporal lobe.

monkeys perform increasingly poorly as the delay period is lengthened. These results imply that prefrontal cortex plays some important role in memory.

Experiments conducted more recently suggest that prefrontal cortex is involved with working memory for problem solving and the planning of behavior. One piece of evidence comes from the behavior of humans with lesions in prefrontal cortex. These people usually perform better on simple memory tasks, such as recalling information after a delay period, than do those with medial temporal lesions. In more complex tasks, however, humans with prefrontal damage show marked deficits. Recall the case of Phineas Gage, discussed in Chapter 18. Having sustained severe frontal lobe damage (an iron bar passing through the head qualifies as severe), Gage had a difficult time maintaining a course of behavior after his injury. Although he could carry out behaviors appropriate for various situations, he had difficulty planning and organizing these behaviors, perhaps because of the damage to his frontal lobe.

A task that brings out the problems associated with prefrontal cortical damage is the Wisconsin card-sorting test. A person is asked to sort a deck of cards having a variable number of colored geometric shapes on them (Figure 23.23). The cards can be sorted by color, shape, or number of symbols, but at the beginning of the test, the subject isn't told which category to use. By putting cards into stacks and being informed when errors occur, however, the subject learns what the current sorting category is. After ten correct card placements are made, the sorting category is changed, and the routine starts over again. To perform well on this test, the person must use memory of previous cards and errors made to plan the next card placement. People with

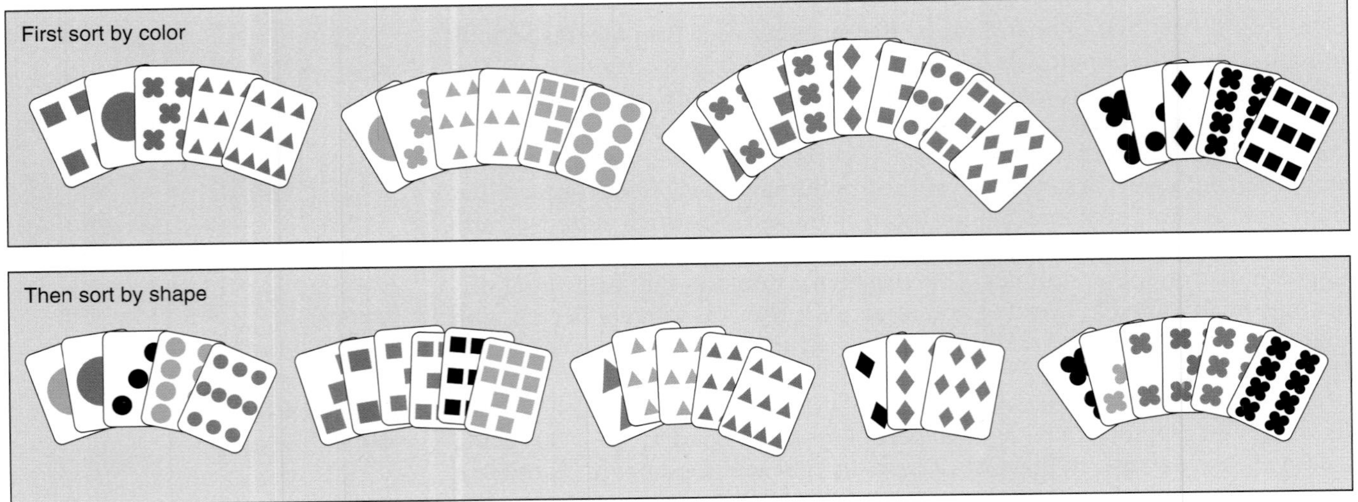

Figure 23.23
The Wisconsin card-sorting test. Cards containing various numbers of colored symbols must first be sorted by color. After a string of correct responses is made, the sorting category is changed to shape.

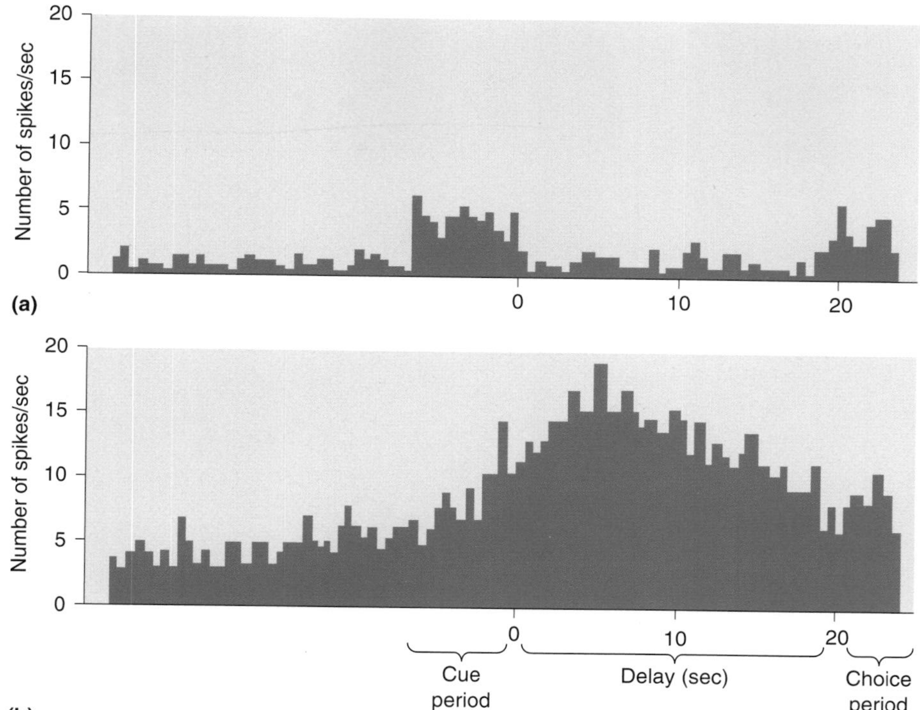

Figure 23.24
Neural responses in monkey prefrontal cortex. The two histograms show the activity of cells in prefrontal cortex recorded while the animal performed a delayed-response task. During a cue period of 7 seconds, within the view of the monkey food is placed in one of two wells. During the delay period, the animal cannot see the food wells; after the delay, it is allowed to choose a well to receive a food reward (the choice period). **(a)** This cell responds when the well is first baited with food and again when the animal exposes the well to get the food. **(b)** This cell responds most strongly during the delay period, when there is no visual stimulus. (Source: Adapted from Fuster, 1973, Fig. 2.)

prefrontal lesions have great difficulty on this task when the sorting category is changed; they continue to sort according to a rule that no longer applies. It appears that they have difficulty using recent information (i.e., data in working memory) to change their behavior.

The same sort of deficit is seen in other tasks. For example, a person with a prefrontal lesion may be asked to trace a path through a maze drawn on a piece of paper. While the patient understands the task, he or she repeatedly makes the same mistakes, returning to blind alleys. In other words, these patients do not learn from their recent experience in the same way that a normal person does. And when they make mistakes, they may have difficulty returning to an earlier point in the maze and instead start over from the beginning.

The neurons in prefrontal cortex have a variety of response types, some of which may reflect a role in working memory. Figure 23.24 shows two response patterns obtained while a monkey performed a delayed-response task. The neuron in the top trace responded while the animal saw the location of the food, was unresponsive during the delay interval, and responded again when the animal made a choice (Figure 23.24a). The response of the neuron simply correlates with the presentation of the stimuli. Perhaps more interesting is the response pattern of the other neuron, which increased its firing rate only during the delay interval (Figure 23.24b). This cell was not directly activated by the stimuli in the first or second interval in which it saw the food wells. The increased activity during the delay period may be related to the retention of information needed to make the correct choice after the delay (i.e., working memory).

Lateral Intraparietal Cortex (Area LIP) and Working Memory

In recent years, several other cortical areas have been found to contain neurons that appear to retain working memory information. In Chapter 14, we

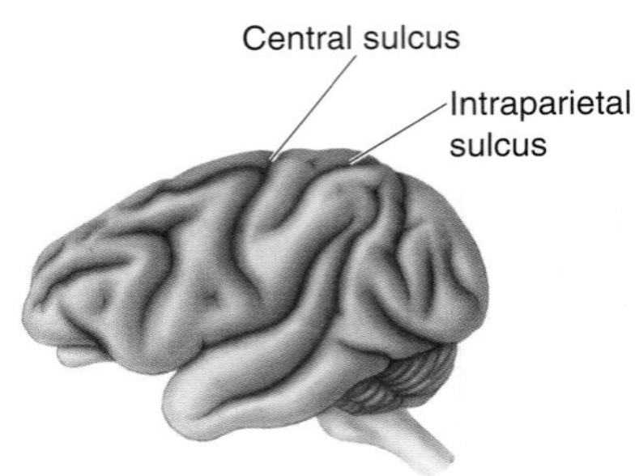

Figure 23.25
LIP buried in the intraparietal sulcus. Working memory responses are observed in some neurons in this area involved in visually guided behaviors.

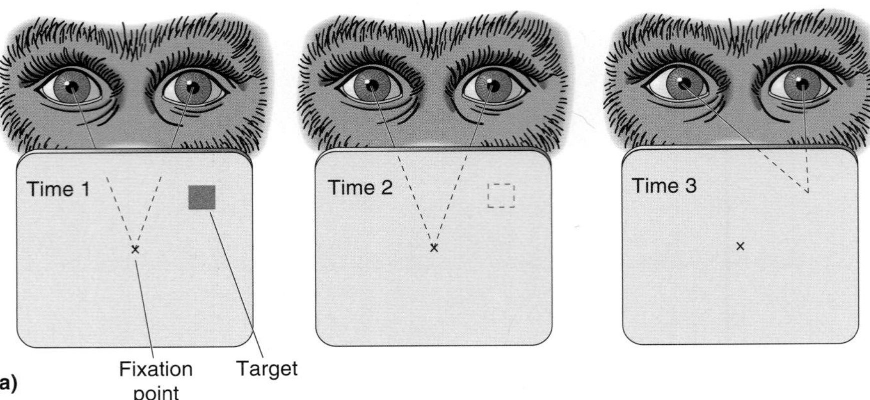

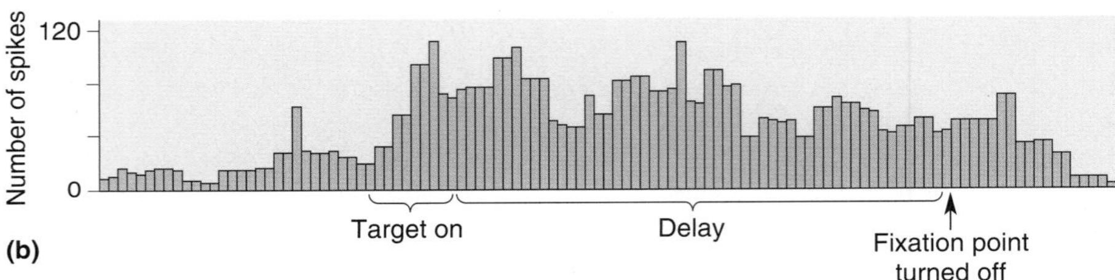

Figure 23.26
The delayed-saccade task. (a) After fixating on a central point, a target goes on and off at a peripheral location. There is a delay period after the target goes off in which the monkey must continue to fixate on the central point. At the end of the delay period, the fixation point goes off, and the animal knows to move its eyes to the remembered location of the target. **(b)** The histogram shows the response of an LIP neuron. The neuron begins firing when the target is presented and continues firing until after the fixation point is gone and the saccadic eye movement begins. (Source: Adapted from Goldman-Rakic, 1992, Fig. p. 113, and Gnadt and Andersen, 1988, Fig. 2.)

saw an example in area 6 (see Figure 14.9). Another example is provided by **lateral intraparietal cortex** (**area LIP**), buried in the intraparietal sulcus (Figure 23.25). Because electrical stimulation here elicits saccades, area LIP is thought to be involved in guiding eye movements. The responses of many neurons suggest that they are involved in a type of working memory. This pattern is evident in a *delayed-saccade task* in which the animal fixates on a spot on a computer screen and a target is briefly flashed at a peripheral location (Figure 23.26a). After the target disappears, there is a variable delay period. At the end of the delay period, the fixation spot goes off, and the animal's eyes make a saccadic movement to the remembered location of the target. The response of an LIP neuron while a monkey performs this task is shown in Figure 23.26b. The neuron begins firing shortly after the peripheral target is presented; this seems like a normal stimulus-evoked response. But the cell keeps firing throughout the delay period in which there is no stimulus, until saccadic movements finally occur. Further experiments using this delayed-saccade task suggest that the response of the LIP neuron is temporarily holding information that will be used to produce the saccades.

Other areas in parietal and temporal cortex have been shown to have analogous working memory responses. These areas seem to be modality specific, just as the responses in area LIP are specific to vision. This is consistent with the clinical observation that there are distinct auditory and visual working memory deficits in humans produced by cortical lesions.

CONCLUDING REMARKS

A clear message from the topics discussed in this chapter is that learning and memory are not confined to a single place in the brain. It is not the case that a small number of specialized "memory cells" store our life experiences and learned behaviors independent from the rest of brain function. This is logical, considering that behavioral adaptability is critical for survival and that the required brain adaptations involve sensory systems (you recognize what you see over there as an angry dog that might bite you), motor control systems (the combination of muscle contractions required to keep you atop a unicycle), and systems more specialized for storing facts. In our effort to understand the learning process itself, we are still at an early stage. Declarative memory depends heavily on the hippocampus and related structures, procedural memory involves the striatum, and working memory traces are found in many brain locations.

But how do different parts of the brain interact so that we learn? We've said that engrams may exist in temporal lobe neocortex, among other places. But what is the physiological basis for the memory storage? When we try to remember a phone number, an interruption can make us forget, suggesting that memories are initially held in a particularly fragile form. Long-term memory is much more robust, however; it can survive interruption, anesthesia, and life's normal bumps and traumas. Partly because of this robustness, it is thought that memories are ultimately stored in structural changes in neocortex. Our brain is constantly undergoing rewiring, to a certain degree, to adapt to life's experiences. The nature of these structural changes in the brain, which underlie learning and memory, are the subject of Chapter 24.

KEY TERMS

Types of Memory and Amnesia
learning (p. 740)
memory (p. 740)
declarative memory (p. 740)
nondeclarative memory (p. 740)
procedural memory (p. 740)
long-term memory (p. 741)
short-term memory (p. 741)
memory consolidation (p. 743)
amnesia (p. 743)
retrograde amnesia (p. 743)
anterograde amnesia (p. 744)

The Search for the Engram
engram (p. 745)
cell assembly (p. 747)

The Temporal Lobes and Declarative Memory
hippocampus (p. 754)
entorhinal cortex (p. 754)
perirhinal cortex (p. 754)
parahippocampal cortex (p. 754)
fornix (p. 756)
delayed non-match to sample (DNMS) (p. 757)
recognition memory (p. 758)
Korsakoff's syndrome (p. 760)
working memory (p. 761)
place cell (p. 762)
relational memory (p. 764)

The Striatum and Procedural Memory
striatum (p. 766)

The Neocortex and Working Memory
prefrontal cortex (p. 769)
lateral intraparietal cortex (area LIP) (p. 772)

REVIEW QUESTIONS

1. If you try to recall how many windows there are in your house by mentally walking from room to room, are you using declarative memory, procedural memory, or both?
2. What evidence indicates that declarative and nondeclarative memory use separate circuits?
3. What abilities and disabilities do you think a person completely lacking short-term memory would have?
4. Why did Lashley conclude that all cortical areas contribute equally to learning and memory? Why was this conclusion later called into question?
5. What evidence indicates that long-term memories are stored in neocortex?
6. If you were using a microelectrode to record from the brain and you suspected that a neuron you encountered was involved in storing long-term memories, how would you test that hypothesis?
7. If a neuron in visual cortex responds to faces, how can you determine whether it is a high-level visual neuron or a neuron involved in storing memories for faces?
8. What are place cells, and where are they found? In what ways are the response characteristics of place cells different from the receptive fields of sensory neurons?
9. What role does the hippocampus play in spatial memory, working memory, and relational memory?
10. What is working memory, and in what brain areas have neural correlates of working memory been observed?

CHAPTER 24

Molecular Mechanisms of Learning and Memory

INTRODUCTION

An important first step in understanding the neurobiology of memory is identifying *where* different types of information are stored. As we saw in Chapter 23, answers to this question are beginning to yield to basic neuroscientific research. An equally important question, however, concerns *how* information is stored. As Hebb pointed out and as research using computer models has amply confirmed, memories can result from subtle alterations in synapses, and these alterations can be widely distributed in the brain. This helps narrow the search for a physical basis of memory—synaptic modifications—but it also raises a dilemma. The synaptic modifications that underlie memory may be too small and too widely distributed to be observed and studied experimentally.

These considerations inspired some researchers to study the nervous systems of simple invertebrate animals for insights about the molecular mechanisms of memory. Although such animals obviously aren't going to remember the definition of an isosceles triangle, they do display certain types of simple procedural memory. Invertebrate research has clearly shown that Hebb was right: Memories *can* reside in synaptic alterations. Moreover, it has been possible to identify some of the molecular mechanisms that lead to this synaptic plasticity. Although nonsynaptic changes that may account for some types of memory have also been found, research on invertebrates leaves little doubt that the synapse is an important site of information storage.

It would be disappointing if the best we could do was understand how a sea slug learns. We really want to know how our own brains form memories. Hope for this understanding now comes from the study of synaptic plasticity in regions of the mammalian brain associated with different types of memory. Theoretical work and invertebrate studies point to lasting alterations in the effectiveness of synapses in the brain as a basis for memory. Therefore, researchers have used electrical stimulation of the brain to produce measurable synaptic alterations whose mechanisms can be studied. One of the interesting conclusions of this research is that the mechanisms of activity-dependent synaptic plasticity in the adult brain have much in common with those operating during development to ensure that the brain is wired properly. Another conclusion is that the basic molecular machinery operating in the mammalian brain to alter synaptic effectiveness is quite similar to that which is responsible for memory formation in invertebrates.

There is a growing sense of optimism among neuroscientists that we may soon know the physical basis of memory. This investigation has benefited from combined approaches of researchers in disciplines ranging from psychology to molecular biology. In this chapter, we take a look at some of their discoveries.

PROCEDURAL LEARNING

Recall that *declarative memories* have a certain ethereal quality. They are easily formed and easily forgotten, and they may result from small modifications of synapses that are widely distributed in the brain. These characteristics make this type of memory particularly challenging to study at the synaptic level. *Procedural memories,* however, have characteristics that make them more amenable to investigation. Besides the fact that these memories are particularly robust, they can be formed along simple reflex pathways that link sensations to movements. Procedural learning involves learning a motor response (procedure) in reaction to a sensory input. It is typically categorized as two types: nonassociative learning and associative learning.

Nonassociative Learning

Nonassociative learning describes the change in the behavioral response that occurs over time in response to a single type of stimulus. There are two types: habituation and sensitization.

Habituation. Suppose you live in a dorm where there is a single telephone in the hall. When the phone rings, you run to answer it. But every time you do, the call is for someone else. Over time, you stop reacting to the ringing of the phone and eventually no longer even hear it. This type of learning, **habituation**, is learning to ignore a stimulus that lacks meaning (Figure 24.1a). You are habituated to a lot of stimuli. Perhaps as you read this sentence, cars and trucks are passing by outside, a dog is barking, your roommate is playing the Limp Bizkit for the hundredth time—and all this goes on without your really noticing. You have habituated to these stimuli.

Sensitization. Suppose you're walking down the sidewalk on a well-lit city street at night, and suddenly there is a blackout. You hear footsteps behind you, and though normally this wouldn't disturb you, now you nearly jump out of your skin. Car headlights appear, and you react by stepping away from the street. The strong sensory stimulus (the blackout) caused **sensitization**, learning to intensify your response to all stimuli, even ones that previously evoked little or no reaction (Figure 24.1b).

Associative Learning

During **associative learning**, we form associations between events. Two types are usually distinguished: classical conditioning and instrumental conditioning.

Classical Conditioning. This type of learning was discovered and characterized in dogs by the famous Russian physiologist Ivan Pavlov around the turn

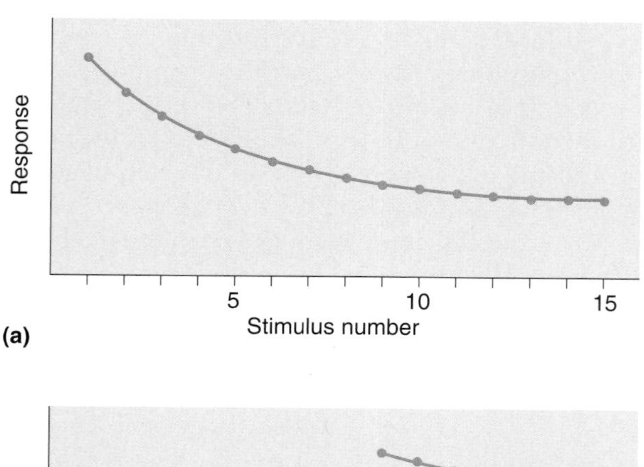

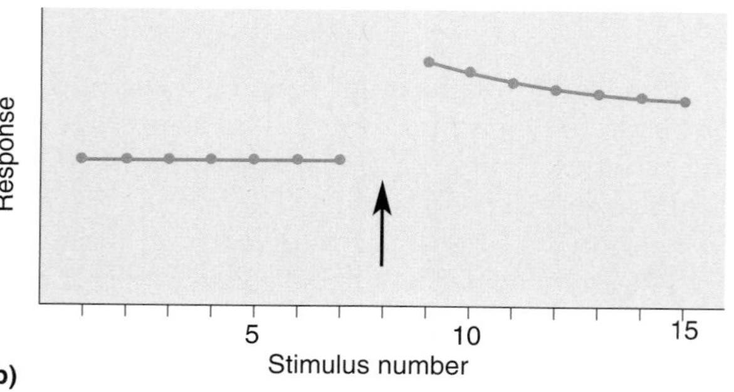

Figure 24.1
Types of nonassociative learning. (a) In habituation, repeated presentation of the same stimulus produces a progressively smaller response. **(b)** In sensitization, a strong stimulus (arrow) results in an exaggerated response to all subsequent stimuli.

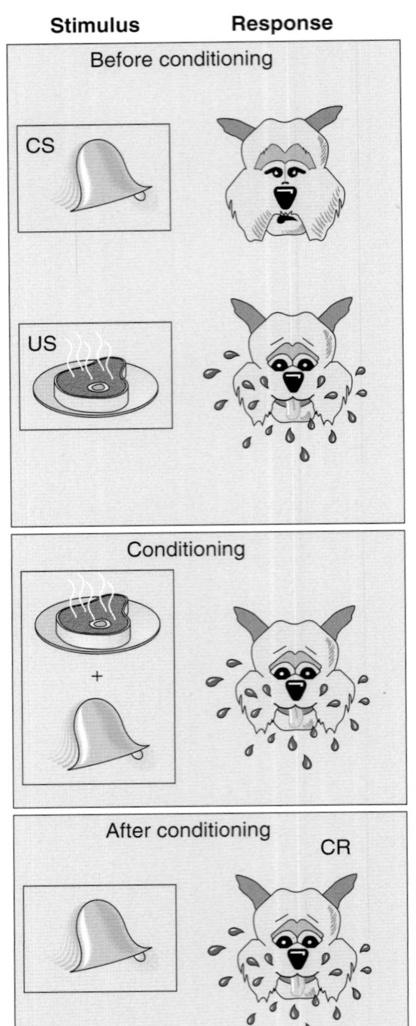

Figure 24.2
Classical conditioning. Prior to conditioning, the sound of a bell, the conditioned stimulus (CS), elicits no response, in sharp contrast to the response elicited by the sight of a piece of meat, the unconditioned stimulus (US). Conditioning entails pairing the sound of the bell with the sight of the meat. The dog learns that the bell predicts the meat.

of the twentieth century. **Classical conditioning** involves associating a stimulus that evokes a measurable response with a second stimulus that normally does not evoke this response. The first type of stimulus, the one that normally evokes the response, is called the *unconditioned stimulus* (*US*) because no training (conditioning) is required for it to yield a response. In Pavlov's experiments, the US was the sight of a piece of meat, and the response was salivation by the dog. The second type of stimulus, the one that normally does not evoke this same response, is called the *conditioned stimulus* (*CS*) because this one requires training (conditioning) before it will yield this response. In Pavlov's experiments, the CS was an auditory stimulus, such as the sound of a bell. Training consisted of repeatedly *pairing* the sound with the presentation of the meat. After many of these pairings the meat was withheld, and the animal salivated to the sound alone. The dog had learned an association between the sound (CS) and the presentation of meat (US); the dog learned that the sound *predicted* the meat. The learned response to the CS is called the *conditioned response* (*CR*) (Figure 24.2).

There are certain timing requirements in successful classical conditioning. Conditioning occurs if the US and CS are presented simultaneously or if the CS precedes the US by a short interval. If the CS precedes the US by too much, however, the conditioning is much weaker or absent. Conditioning typically does not occur if the CS follows the US. The synaptic mechanism of classical conditioning must account for these stringent timing requirements.

Instrumental Conditioning. This type of associative learning was discovered and studied by Columbia University psychologist Edward Thorndike early in the last century. In **instrumental conditioning**, an individual learns to associate a response, a motor act, with a meaningful stimulus, typically a reward such as food. As an example, consider what happens when a hungry rat is placed in a box with a lever that dispenses food. In the course of exploring the box, the rat bumps the lever and out pops a piece of food. After this happy accident occurs a few more times, the rat learns that pressing the lever leads to a food reward. The rat works the lever and eats the food until it is no longer hungry. This should sound familiar. Recall from previous chapters the experiments in which monkeys were trained to make responses as subtle as saccadic eye movements to receive a juice reward. Behavioral neurophysiology makes use of instrumental learning. The reward need not be food or drink, either. Rats will lever-press for a cocaine reward or for electrical stimulation of the medial forebrain bundle. Instrumental conditioning also occurs if a response, instead of evoking a rewarding stimulus, prevents an aversive stimulus, such as a foot shock.

As in classical conditioning, a predictive relationship is learned during instrumental conditioning. In classical conditioning, it is learned that one stimulus (CS) predicts another stimulus (US). In instrumental conditioning, it is learned that a particular behavior is associated with a particular consequence. Also like classical conditioning, timing is important. Successful instrumental conditioning requires the stimulus to occur shortly after the response.

Because motivation plays such a large part in instrumental conditioning (after all, only a hungry rat will lever-press for a food reward), the underlying neural circuits are considerably more complex than those involved in simple classical conditioning. Therefore, we leave instrumental conditioning to focus on the simpler forms of learning and memory for which mechanisms have been identified in the nervous system of an invertebrate animal.

SIMPLE SYSTEMS: INVERTEBRATE MODELS OF LEARNING

Over the course of neuroscience history, a large menagerie of invertebrate creatures has been used for experimentation. You are already familiar with the squid and the contribution of its giant axon and giant synapse to our understanding of cellular neurophysiology (see Chapters 4 and 5). Other useful invertebrate species include cockroaches, flies, bees, leeches, and nematodes. The reason is that invertebrate nervous systems offer some important experimental advantages:

- *Small nervous systems:* Neurons in invertebrate nervous systems number as few as a thousand; the human brain has roughly 10 million times as many.
- *Large neurons:* Many neurons in invertebrates are very large, making them easy to study electrophysiologically.
- *Identifiable neurons:* Neurons in invertebrates can be catalogued by size, location, and electrophysiological properties and can therefore be identified from individual to individual.
- *Identifiable circuits:* Identifiable neurons make the same connections with one another from individual to individual.
- *Simple genetics:* The small genomes and rapid life cycles of some invertebrates (such as flies and nematodes) make them ideal for studies of the genetic and molecular biological basis of learning.

Invertebrates can be particularly useful for analysis of the neural basis of behavior. Granted, the behavioral repertoire of the average invertebrate is limited. Nonetheless, many invertebrate species exhibit the simple forms of learning we discussed above: habituation, sensitization, and classical conditioning. One species in particular has been used to study the neurobiology of learning, the sea slug *Aplysia californica* (Figure 24.3).

Nonassociative Learning in *Aplysia*

If someone blows gently at your eye, you blink. Eventually, however, you habituate, assuming the air puff is not painful. Similarly, if a jet of water is squirted onto a fleshy region of *Aplysia* called the siphon, the siphon and the gill retract (Figure 24.4). This is called the *gill withdrawal reflex*. Like the eye blink, the gill withdrawal reflex displays habituation after repeated presentation of the water jet. In a pioneering series of experiments beginning in the 1960s, Eric Kandel and his colleagues at Columbia University set out to determine where this procedural memory resides and how it is formed.

Habituation of the Gill Withdrawal Reflex. Sensory information from the siphon travels along a nerve until it enters a region of the *Aplysia* nervous system called the abdominal ganglion (Figure 24.5). Here the information is distributed to motor neurons and interneurons. One of the motor neurons that

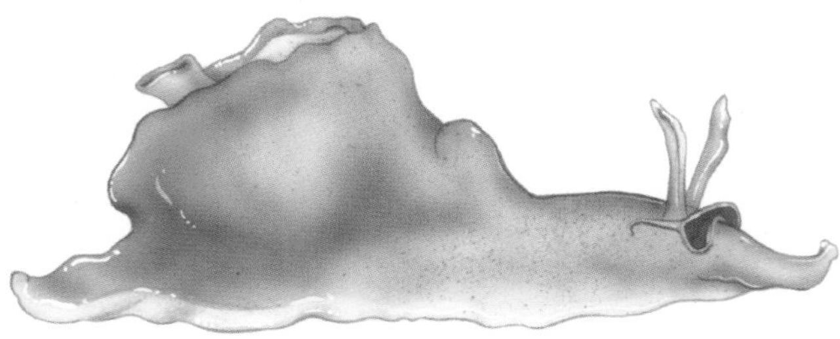

Figure 24.3
Aplysia californica. One cool sea slug used for neurobiological studies of learning and memory.

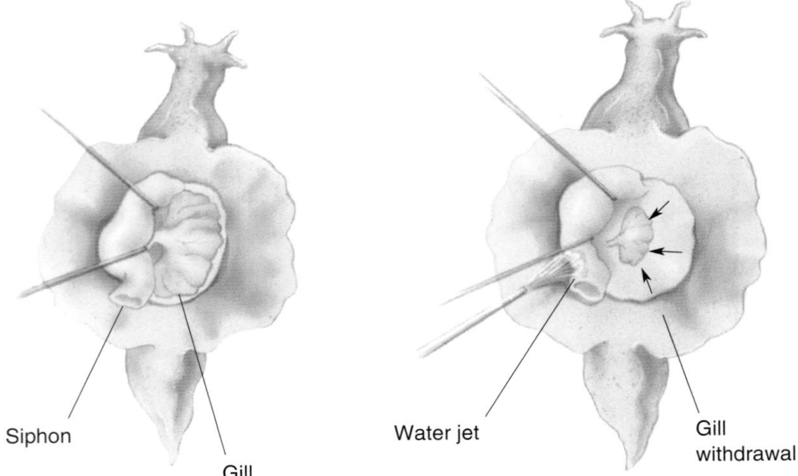

Figure 24.4
The gill withdrawal reflex in *Aplysia*.

receives direct monosynaptic sensory input from the siphon is identified as L7, and this cell innervates the muscles that produce gill withdrawal. Therefore, research has focused on understanding how this simple monosynaptic reflex arc changes during habituation.

The first important question concerns where the habituation occurs. Figure 24.6 shows a simplified wiring diagram of the gill withdrawal reflex. Repeated stimulation of the siphon skin leads to progressively less contraction of the gill withdrawal muscles. The change underlying this habituation could occur (1) at the sensory nerve endings in the skin, making them less sensitive to the squirt of water; (2) at the muscle, making it less responsive to synaptic stimulation by the motor neuron; or (3) at the synapse between the sensory neuron and the motor neuron. The first possibility was ruled out by making microelectrode recordings from the sensory neuron as habituation occurred. The sensory neuron continued to fire action potentials in response to stimulation of the skin, even as the motor response decreased. Similarly, the third possibility was ruled out by electrically stimulating the motor neuron (L7) and showing that it always evoked the same amount of muscle contraction. This left the second possibility: that habituation occurs at the synapse joining the sensory input to the motor neuron. Indeed, simply giving repetitive electrical stimulation to the sensory neuron was sufficient to

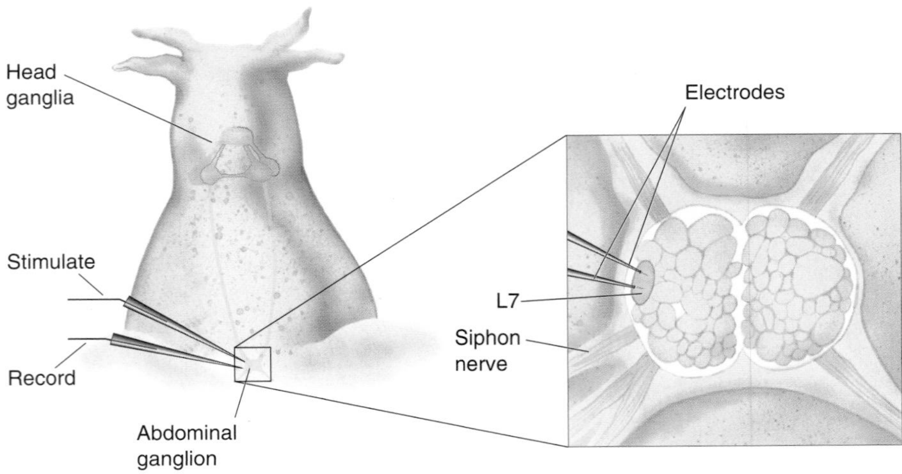

Figure 24.5
The abdominal ganglion of *Aplysia*. The gill withdrawal reflex involves neurons within the abdominal ganglion that can be dissected and studied electrophysiologically.

cause a progressive decrease in the size of the excitatory postsynaptic potential (EPSP) (Figure 24.7).

Identifying a synaptic modification as a possible basis for habituation still does not reveal where the synapse is modified. Let's consider the possibilities. After habituation, there could be (1) less neurotransmitter release by the presynaptic axon or (2) decreased postsynaptic responsiveness to the transmitter (perhaps caused by fewer receptors). Recall from Chapter 5 that transmitter molecules are released in discrete packets called *quanta*, believed to correspond to the contents of individual synaptic vesicles. Vincent Castelluci and Kandel performed a sophisticated quantal analysis of transmission at this synapse and concluded that *after habituation, fewer quanta per action potential are released*. The sensitivity of the postsynaptic cell to neurotransmitter did not change. In other words, habituation of the gill withdrawal reflex is associated with a *presynaptic* modification.

Why is neurotransmitter release reduced after repeated stimulation of the sensory nerve terminal? Recall the sequence of events that couples the arrival of the action potential in the terminal with the release of neurotransmitter. A critical step is the entry of Ca^{2+} into the terminal, which is mediated by voltage-gated calcium channels. It appears that in the nerve terminal of the sensory neuron, these channels become progressively and persistently less effective when they are repeatedly opened. It is still unknown exactly why this is so, but the consequence of reduced presynaptic Ca^{2+} entry per action potential is decreased transmitter release.

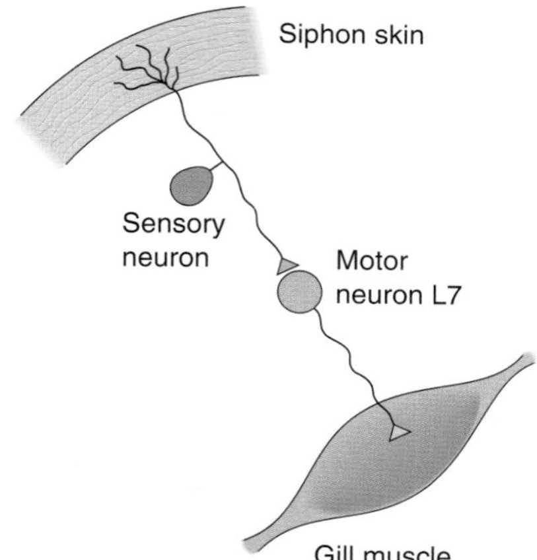

Figure 24.6
A simple wiring diagram for the gill withdrawal reflex. The sensory neuron that detects stimuli applied to the skin of the siphon synapses directly on the motor neuron that causes the gill to withdraw.

Sensitization of the Gill Withdrawal Reflex. To cause sensitization of the gill withdrawal reflex, Kandel and his colleagues applied a brief electrical shock to the head of *Aplysia*. This resulted in exaggerated gill withdrawal in response to stimulation of the siphon. Using the strategy outlined earlier, they identified a site of plasticity showing changes that correlate with the behavior. Once again, it turned out to be a modification of transmitter release in the sensory nerve terminal.

How might stimulation of the head lead to sensitization of the gill withdrawal reflex? Understanding the answer to this question requires that we add a third neuron to the wiring diagram (Figure 24.8). This third cell, L29, is activated by the head shock and makes a synapse *on the axon terminal of the sensory neuron.* The neurotransmitter released by L29 is believed to be serotonin (5-HT). Serotonin sets in motion a molecular cascade that sensitizes the sensory axon terminal so that it lets in more Ca^{2+} per action potential.

The serotonin receptor on the sensory axon terminal is a G-protein-coupled metabotropic receptor. Stimulation of this receptor leads to the pro-

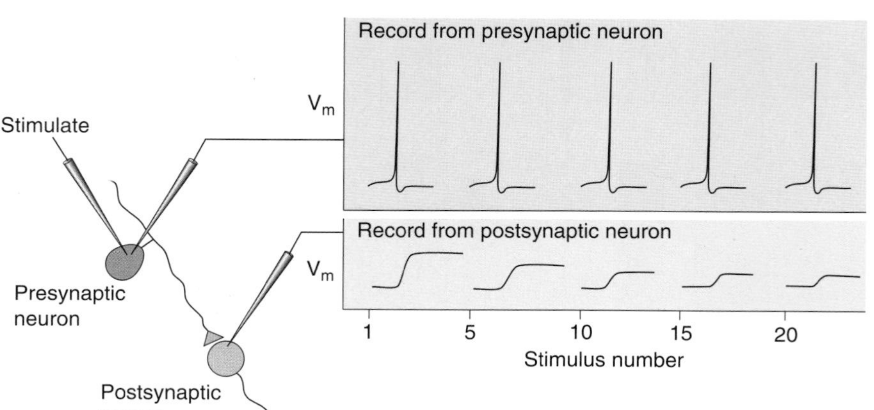

Figure 24.7
Habituation at the cellular level. Repeated electrical stimulation of a sensory neuron leads to a progressively smaller EPSP in the postsynaptic motor neuron.

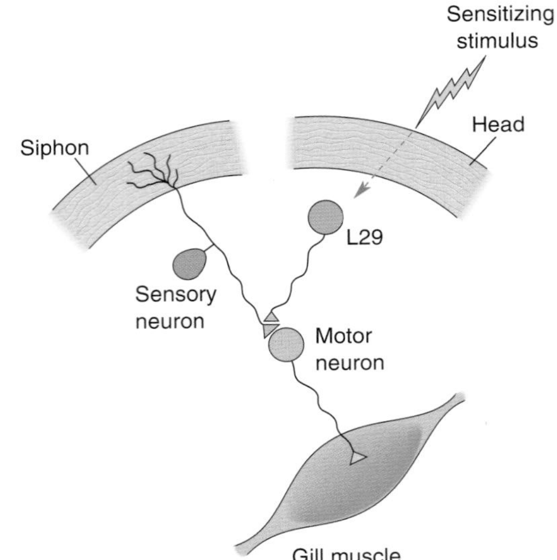

Figure 24.8
A wiring diagram for sensitization of the gill withdrawal reflex. A sensitizing stimulus to the head indirectly activates an interneuron, L29, which makes an axoaxonic synapse on the terminal of the sensory neuron.

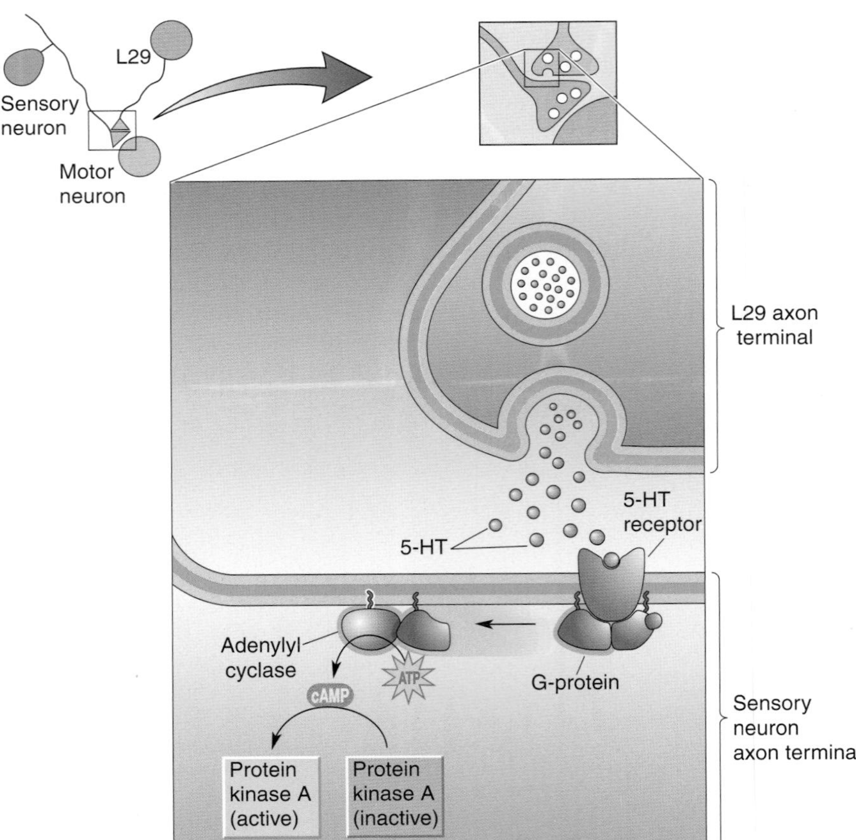

Figure 24.9
A mechanism for sensitization of the gill withdrawal reflex. Serotonin released by L29 in response to the head shock leads to G-protein-coupled activation of adenylyl cyclase in the sensory axon terminal. Activation of this enzyme leads to the production of cAMP, which in turn activates protein kinase A. Protein kinase A attaches phosphate groups to a potassium channel, causing it to close.

duction of intracellular second messengers. In the case of the *Aplysia* sensory nerve terminal, this second messenger is *cAMP* (cyclic adenosine monophosphate) produced from ATP (adenosine triphosphate) by the enzyme *adenylyl cyclase* (Figure 24.9). Recall from Chapters 5 and 6 that cAMP activates *protein kinase A*, and this enzyme phosphorylates (attaches phosphate groups to) various proteins. In the sensory nerve terminal, one of these proteins is a potassium channel, and phosphorylation of this channel causes it to close. The closure of potassium channels in the axon terminal leads to a prolongation of the presynaptic action potential. This results in more Ca^{2+} entry through voltage-gated Ca^{2+} channels during the action potential, and therefore, more quanta of neurotransmitter are released (Figure 24.10).

Associative Learning in *Aplysia*

In the 1980s, researchers discovered that *Aplysia* also could be classically conditioned. Again, the response measured was the withdrawal of the gill. The US was a strong shock to the tail. The CS was stimulation of the siphon that was so gentle that it did not cause much of a response. It was discovered that if stimulation of the tail (US) was *paired* with stimulation of the siphon (CS), the subsequent response to the siphon alone (CR) was much greater than what could be accounted for by sensitization (Figure 24.11). As in Pavlov's experiments, timing was critical. Conditioning occurred only if the CS

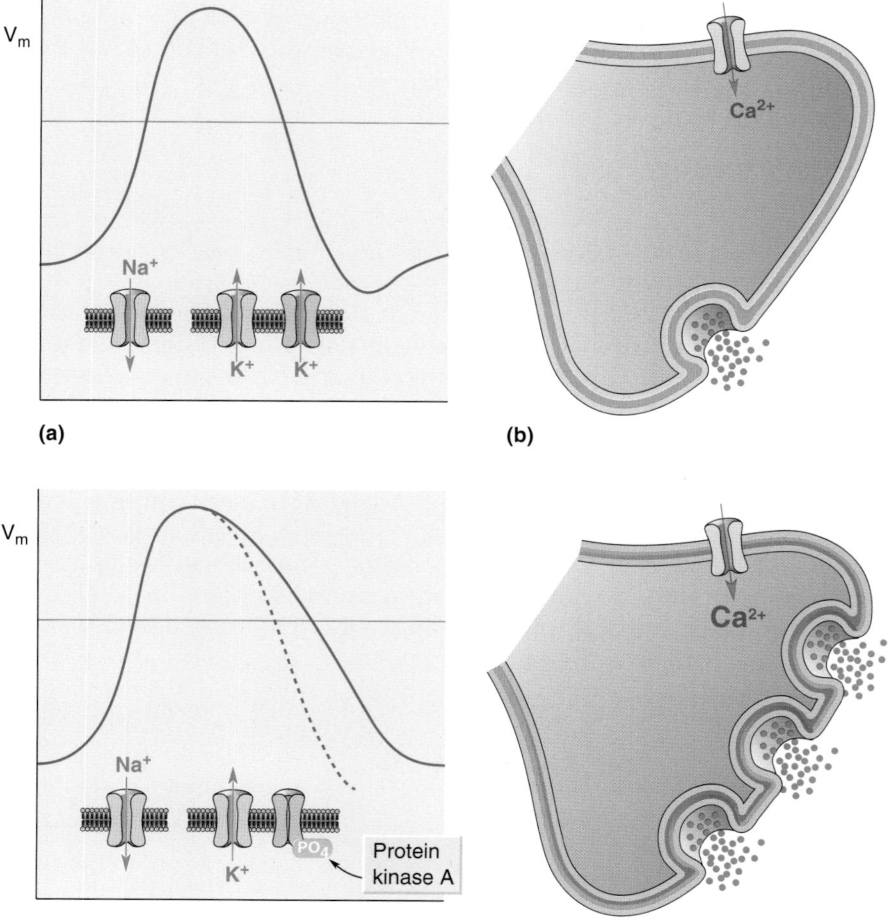

Figure 24.10
The effect of decreased potassium conductance in the sensory axon terminal. **(a)** The trace shows membrane voltage changes during an action potential. The rising phase is caused by the opening of voltage-gated sodium channels, and the falling phase is caused by the closing of the sodium channels and the opening of potassium channels. In the axon terminal, voltage-gated Ca^{2+} channels stay open as long as the membrane voltage exceeds a threshold value, indicated by the red line. **(b)** The resulting entry of Ca^{2+} stimulates the release of neurotransmitter. **(c)** A decrease in K^+ conductance after sensitization prolongs the action potential. **(d)** The voltage-gated Ca^{2+} channels stay open longer, thereby admitting more Ca^{2+} into the terminal. This causes more transmitter to be released per action potential.

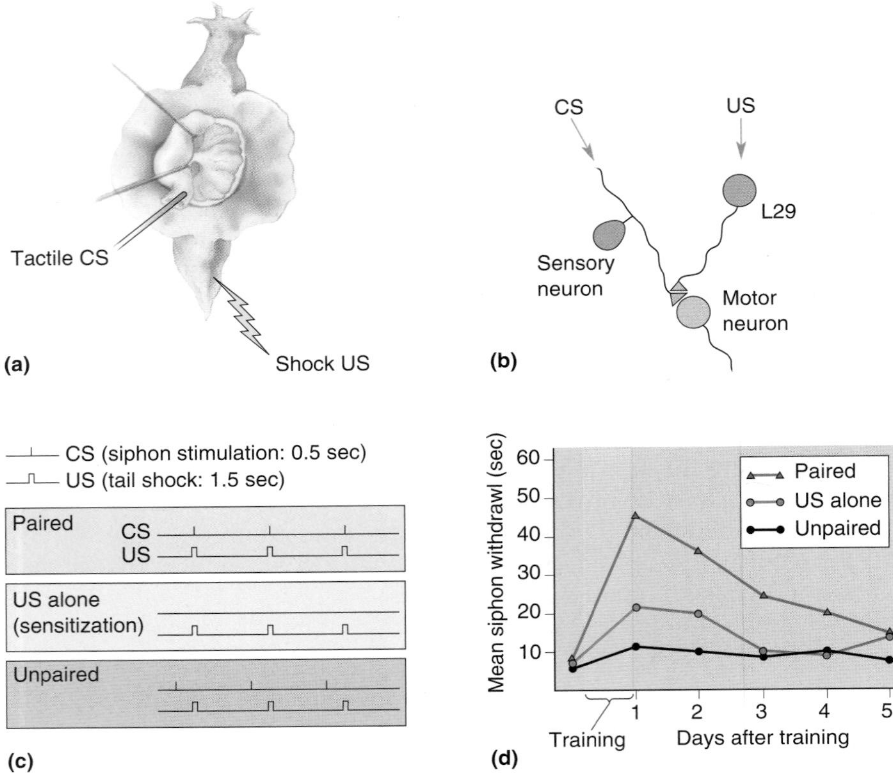

Figure 24.11
Classical conditioning in *Aplysia*. (a) A gentle water jet to the siphon is the CS. A shock to the tail is the US. The measured response is the withdrawal of the gill. **(b)** The wiring diagram for classical conditioning. The US activates the same serotonergic cell (L29) that is activated during sensitization. **(c)** Timing of the CS and US during three different types of training. **(d)** Plotted here is the magnitude of the gill withdrawal in response to the CS. After pairing (classical conditioning), the animal withdraws the gill in response to the CS, which before training was ineffective in eliciting a response.

(siphon stimulus) preceded the US (tail stimulus) by no more than 0.5 second.

Once again, a critical modification that involves increased neurotransmitter release occurs at the synapse between the sensory and motor neuron. To understand how this comes about, let's focus on what is happening at that synapse during the CS-US pairing. At the *cellular* level, the CS is represented by the arrival of an action potential in the sensory axon terminal, and the US is represented by the release of serotonin by L29 (activated by the tail shock). At the *molecular* level, the CS is represented by the influx of Ca^{2+}, and the US is represented by the G-protein-coupled activation of the enzyme adenylyl cyclase in the terminal (Table 24.1). Remember that adenylyl cyclase generates cAMP. In 1991, Thomas Abrams and his colleagues at Columbia University made a key discovery: *In the presence of elevated [Ca^{2+}], adenylyl cyclase churns out more cAMP* (Figure 24.12). More cAMP means more activation of protein kinase A, more phosphorylation of potassium channels, and therefore the release of more transmitter molecules. Thus, in the case of classical conditioning of the gill withdrawal reflex, adenylyl cyclase may serve as a detector of CS-US coincidence. According to this hypothesis, *learning* occurs

Table 24.1 Cellular and Molecular Correlates of Classical Conditioning in *Aplysia*

	LEVEL OF ANALYSIS		
STIMULUS	BEHAVIORAL	CELLULAR	MOLECULAR
CS	Touch to siphon	Presynaptic action potential	Presynaptic influx of Ca^{2+}
US	Shock to tail	Release of serotonin	G-protein activation of adenylyl cyclase in presynaptic terminal

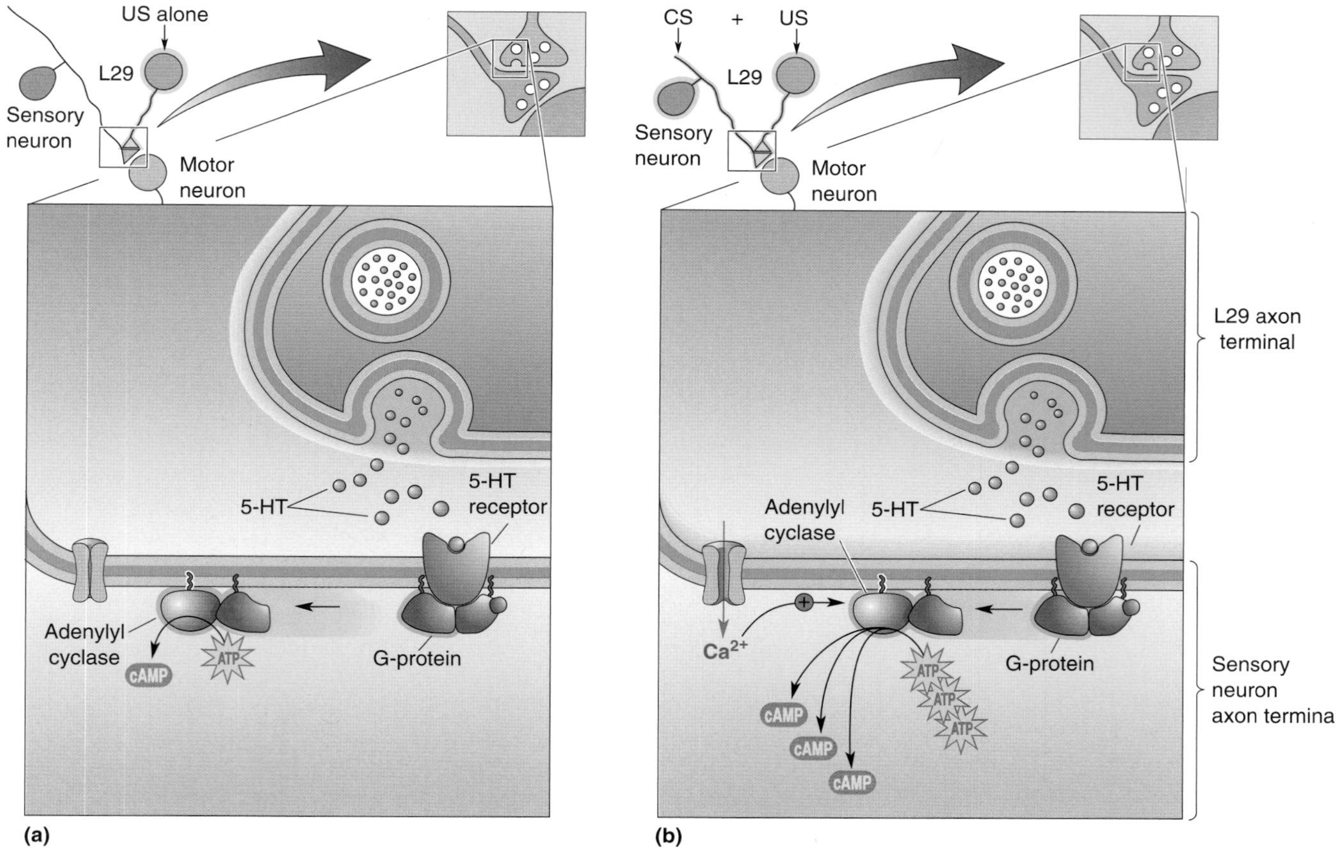

Figure 24.12

The molecular basis for classical conditioning in *Aplysia*. (a) The US alone leads to activation of the motor neuron (via an interneuron, not shown) and to sensitization of the sensory input by the same mechanism illustrated in Figures 24.9 and 24.10. **(b)** Pairing the CS and the US causes greater activation of adenylyl cyclase than either stimulus does by itself because the CS admits Ca^{2+} into the presynaptic terminal. The Ca^{2+} (by interacting with a protein called calmodulin, not shown) increases the response of adenylyl cyclase to G-proteins.

when a presynaptic Ca^{2+} pulse coincides with (or just precedes) the G-protein-coupled activation of adenylyl cyclase, which stimulates the production of a lot of cAMP. *Memory* occurs when potassium channels are phosphorylated and neurotransmitter release is enhanced.

Pavlov would have been surprised by these results. He assumed that this type of associative memory required the sophisticated neural hardware of the cerebral cortex. Even a sea slug can be classically conditioned, however, and many of the defining characteristics of the conditioning can be accounted for by the properties of a single type of enzyme.

A great deal can be learned about the neurobiology of behavior in invertebrates, but none of the *Aplysia* studies has uncovered the cellular basis of learning, only some *cellular correlates* of learning. Although the studied synaptic changes can accompany behavioral learning, they may not be essential for learning. In other words, there may be, and in some instances definitely are, additional mechanisms that contribute to habituation, sensitization, and classical conditioning. Thousands of neurons in the *Aplysia* nervous system are active during even the simplest reflexes, and synaptic changes related to learning are likely to be widely distributed among them. Nonetheless, the invertebrate studies have been invaluable for identifying candidate molecular mechanisms for learning and memory. We return to invertebrate studies again when we discuss the molecular basis of memory consolidation.

VERTEBRATE MODELS OF LEARNING

Let's summarize what we've learned from the invertebrate studies about the possible neural basis of memory:

- Learning and memory can result from modifications of synaptic transmission.
- Synaptic modifications can be triggered by the conversion of neural activity into intracellular second messengers.
- Memories can result from alterations in existing synaptic proteins.

Keep these points in mind as we explore different types of activity-dependent synaptic plasticity in the mammalian brain.

Synaptic Plasticity in the Cerebellar Cortex

In Chapter 14, we discussed the cerebellum in the context of motor control. We concluded that the cerebellum was an important site for motor learning, a place where corrections are made when the outcome of movements fails to meet expectations. Because these corrections are believed to be made by modifications of synaptic connections, the cerebellum has become a model system for the study of the synaptic basis of learning in the mammalian brain.

Anatomy of the Cerebellar Cortex. Particularly interesting is the **cerebellar cortex**, the sheet of tissue that lies just under the surface of the cerebellum (Figure 24.13a). The cortex consists of two layers of neuronal cell bodies, the *Purkinje cell layer* and the *granule cell layer*, and these are separated from the pial surface by a *molecular layer* that is largely devoid of somata. **Purkinje cells**, named for the Czech neuroanatomist who first described them in 1837, have a number of interesting features. First, their dendrites extend only into the molecular layer, where they branch like a fan, flattened in one plane. Second, Purkinje cell axons synapse on neurons in the deep cerebellar nuclei, which are the major output cells of the cerebellum. Thus, Purkinje cells are in a powerful position to modify the output of the cerebellum. Third, Purkinje cells use gamma-aminobutyric acid (GABA) as a neurotransmitter, so their influence on cerebellar output is inhibitory.

Purkinje cell dendrites are directly contacted by one of the two major inputs to the cerebellum. This input arises in a nucleus of the medulla called the **inferior olive**, which integrates information from muscle proprioceptors. Axons from the inferior olive are called **climbing fibers** because they twist around the Purkinje cell dendrites like a vine on the branches of a tree. Each Purkinje cell receives input from only one inferior olive cell, but this input is very powerful. A single climbing fiber axon makes hundreds of excitatory synapses on the dendritic tree of its target Purkinje neuron. An action potential in a climbing fiber generates an exceptionally large EPSP that always strongly activates the postsynaptic Purkinje cell.

The second major input to the cerebellum arises from a variety of brain stem cell groups, notably the pontine nuclei that relay information from the cerebral neocortex. These inputs, called **mossy fibers**, synapse on **cerebellar granule cells**, which form a layer just below the Purkinje cells (Figure 24.13b). Granule cells are very small, very tightly packed, and very numerous. (Estimates are that granule cells may make up half the total number of neurons in the brain.) Cerebellar granule cells give rise to axons that ascend

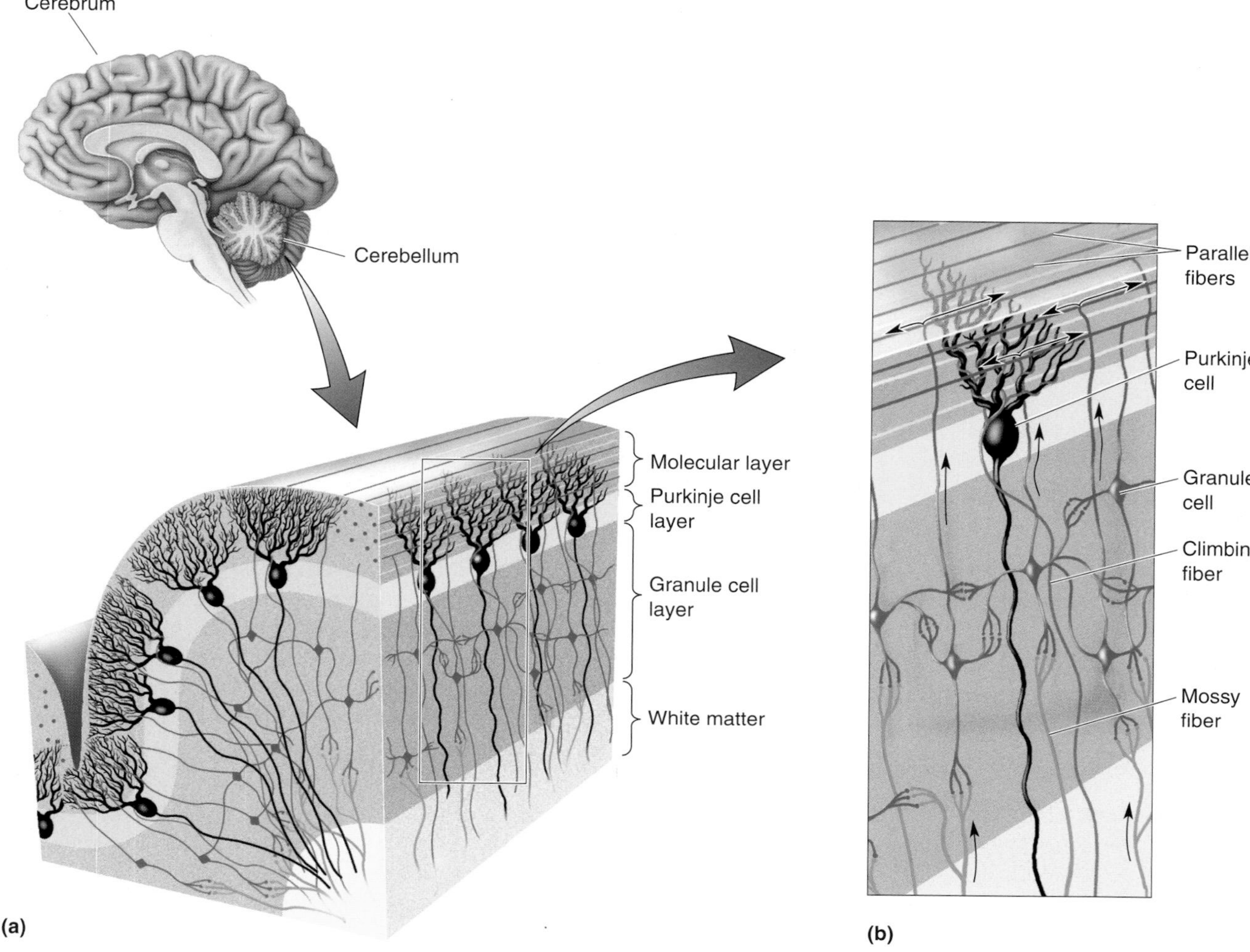

Figure 24.13
The structure of the cerebellar cortex. (a) A view of the cortex showing the organization of the granule cell, Purkinje cell, and molecular layers. **(b)** The major inputs to the Purkinje cells are parallel fibers arising from cerebellar granule cells and climbing fibers arising from the inferior olive. The major input to the granule cells is the mossy fibers, arising from neurons in the pontine nuclei.

into the molecular layer, where they branch like a T. Each branch, called a **parallel fiber**, runs straight for several millimeters in a direction that intersects the plane of the Purkinje cell dendrites at a right angle, like wires passing a telephone pole. Therefore, a single parallel fiber has only a brief encounter with any one Purkinje cell, but along its length it encounters many. Although a Purkinje cell receives only one synapse from each passing parallel fiber, however, it does so from as many as 100,000 fibers.

This unusual convergence of parallel and climbing fiber inputs onto Purkinje cell dendrites has fascinated neuroscientists ever since Cajal originally described it. In the early 1970s, James Albus at the Goddard Space Flight Center in Greenbelt, Maryland, suggested a way that this arrangement could serve motor learning. He proposed (1) that the climbing fiber input carries error signals indicating that a movement has failed to meet expectations and (2) that corrections are made by adjusting the effectiveness of the parallel fiber inputs to the Purkinje cell. The idea that motor learning is served by plasticity of the parallel fiber–Purkinje cell synapse actually was proposed a few years earlier by David Marr, then at Cambridge University in England. What is now called the **Marr-Albus theory of motor learning** specifically predicts plasticity of the parallel fiber synapse *if it is active at the same time as the climbing fiber input to the postsynaptic Purkinje cell.*

Long-Term Depression in the Cerebellar Cortex. Masao Ito and his colleagues at the University of Tokyo directly tested the prediction of the Marr-Albus theory. To monitor the effectiveness of parallel fiber synapses on Purkinje cells, they applied brief electrical stimulation to the parallel fibers and measured the EPSP in the Purkinje cell. Then, to induce synaptic plasticity, they paired stimulation of the climbing fibers with stimulation of the parallel fibers. Remarkably, they found that after this pairing procedure, activation of the parallel fibers alone resulted in a smaller postsynaptic response in the Purkinje cell. This type of modification could last at least an hour and therefore was termed **long-term depression**, or **LTD** (Figure 24.14).

An important property of LTD is that it occurs only in parallel fiber synapses that are active at the same time as the climbing fibers. Other parallel fiber synapses that were not stimulated in conjunction with the climbing fibers did not exhibit the plasticity. This property, that only the active inputs show the synaptic plasticity, is called **input specificity**.

Immediately we can see similarities between cerebellar LTD and classical conditioning in *Aplysia*. In *Aplysia* there is an input-specific synaptic modification when presynaptic activity in the sensory axon (event 1) occurs at the same time as stimulation of the sensory axon terminal with serotonin (event 2). In the cerebellar cortex, there is an input-specific modification when activation of the parallel fiber–Purkinje cell synapse (event 1) occurs at the same time as depolarization of the postsynaptic Purkinje cell by the climbing fiber

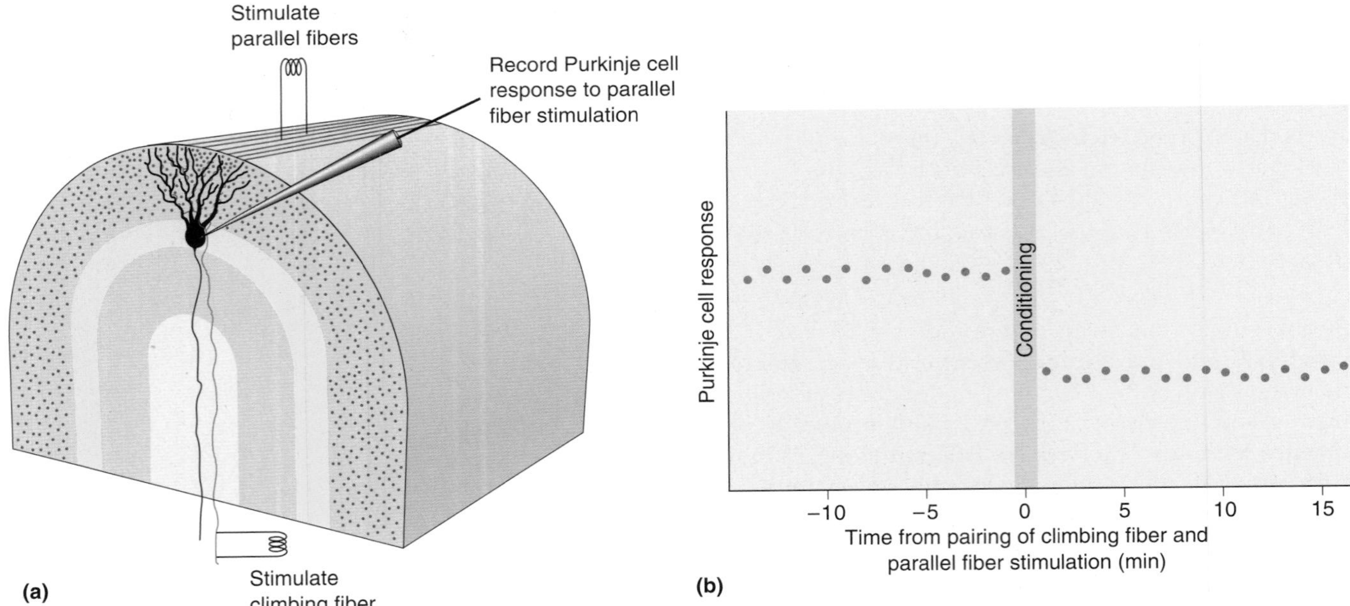

Figure 24.14
Cerebellar long-term depression. (a) The experimental arrangement for demonstrating LTD. The magnitude of the Purkinje cell response to stimulation of a "beam" of parallel fibers is monitored. Conditioning involves pairing parallel fiber stimulation with climbing fiber stimulation. **(b)** A graph of an experiment performed in this way. After the pairing, there is a long-term depression of the response to parallel fiber stimulation.

(event 2). The site of convergence of events 1 and 2 is different, however, in the two systems. In *Aplysia*, convergence occurs in the presynaptic axon terminal, while in the cerebellar cortex, convergence apparently occurs in the Purkinje cell dendrite. Another difference is the nature of the synaptic change. After classical conditioning, the sensory-motor synapse becomes *more* effective, while after pairing parallel and climbing fiber activity, the parallel fiber–Purkinje synapse becomes *less* effective.

As with *Aplysia*, now we must ask what side of the parallel fiber–Purkinje synapse is modified. Recall that in *Aplysia*, the modifications were always presynaptic alterations in transmitter release. But in the cerebellum, the synaptic modification was discovered to be postsynaptic. Specifically, LTD was found to result from a decrease in the postsynaptic response to glutamate released by the parallel fibers. The glutamate receptor that mediates excitatory transmission at this synapse is called the *AMPA* (α-amino-3-hydroxy-5-methyl-4-isoxazole propionic acid) *receptor*. Recent research suggests that the postsynaptic cell internalizes AMPA receptors following LTD induction, thus making the synapse less sensitive to glutamate.

Mechanisms of Cerebellar LTD. To show how paired stimulation of the climbing and parallel fibers leads to LTD, we focus on the Purkinje cell dendrite where these signals converge. First, what is so important about activating the climbing fiber synapses on the Purkinje cell? Recall that this is a very powerful input, causing a large EPSP that always stimulates the Purkinje cell to fire an action potential. In addition to activating voltage-gated sodium channels (causing the action potential), however, this depolarization is powerful enough to activate voltage-gated calcium channels in the membrane of the Purkinje cell dendrites. Thus, climbing fiber activation is associated with a surge of Ca^{2+} into the Purkinje cell dendrite. To assess the importance of this Ca^{2+} signal in the induction of LTD, a substance that binds Ca^{2+} and prevents rises in $[Ca^{2+}]$, called a *Ca^{2+} chelator*, was injected into the Purkinje cell. This treatment blocked LTD. This and related experiments have led to the conclusion that the critical signal provided by climbing fiber activation is a surge in $[Ca^{2+}]$ in the Purkinje cell dendrite.

What is special about activation of the parallel fibers? The transmitter released after parallel fiber activation is glutamate, and as we have said, one of the postsynaptic receptors is the AMPA receptor. The AMPA receptor is the channel that mediates the EPSP by allowing Na^+ ions into the Purkinje cell dendrite. There is a second type of glutamate receptor postsynaptic to the parallel fibers, however. This is a *metabotropic glutamate receptor* that is coupled via a G-protein to the enzyme *phospholipase C*. Activation of this enzyme leads to the production of a second messenger (diacylglycerol), which activates *protein kinase C* (Figure 24.15).

The evidence, to date, indicates that LTD is caused when *three* intracellular signals occur at the same time: (1) a rise in $[Ca^{2+}]_i$ due to climbing fiber activation, (2) a rise in $[Na^+]_i$ due to AMPA receptor activation, and (3) activation of protein kinase C due to metabotropic receptor activation. Precisely what happens next is still under investigation, but it certainly involves the phosphorylation of proteins by protein kinase C, and the end result is a decrease in the number of AMPA receptor channels in the postsynaptic membrane. In this model, *learning* occurs when rises in $[Ca^{2+}]_i$ and $[Na^+]_i$ coincide with the activation of protein kinase C. *Memory* occurs when AMPA channels are internalized and excitatory postsynaptic currents are depressed.

If we make a leap of faith and assume that LTD plays a role in motor learning (not yet proven), we see that (1) learning and memory can result from modifications of synaptic transmission, (2) the conversion of neural activity

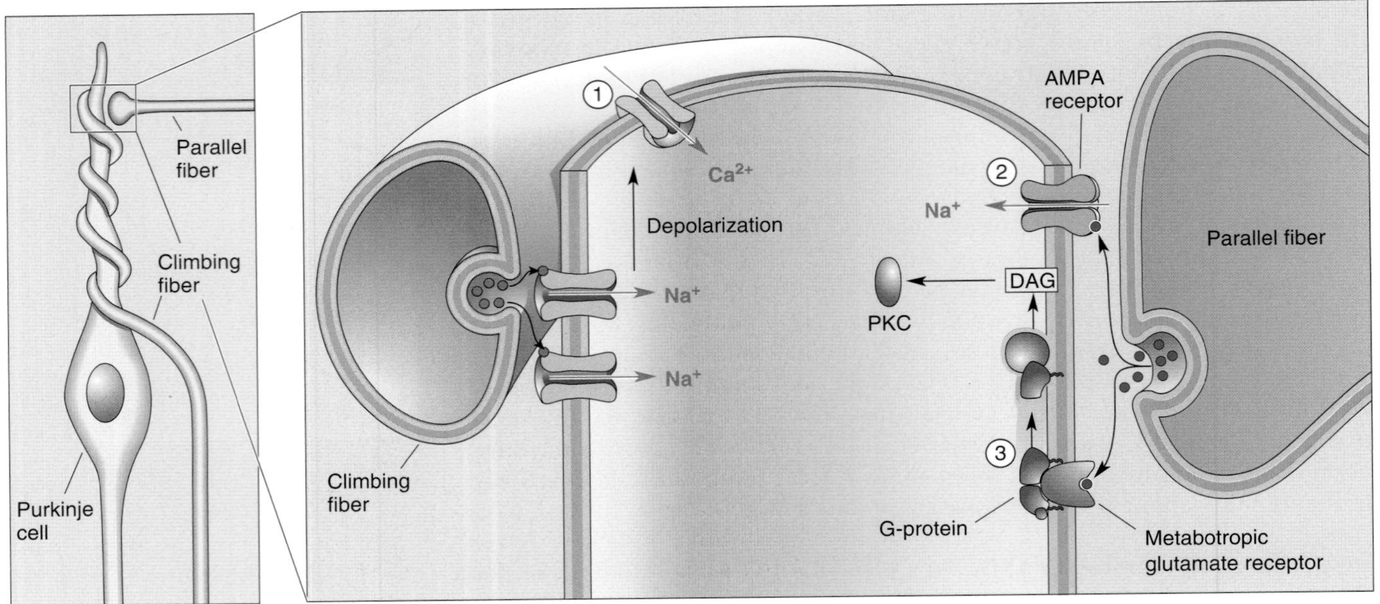

Figure 24.15
A mechanism of LTD induction in the cerebellum. ① Climbing fiber activation strongly depolarizes the Purkinje cell dendrite, which leads to the activation of voltage-gated Ca^{2+} channels. ② Parallel fiber activation leads to Na^{+} entry through AMPA receptors, and ③ the generation of diacylglycerol (DAG) via stimulation of the metabotropic receptor. DAG activates protein kinase C (PKC).

Figure 24.16
A comparison of the events leading to classical conditioning in *Aplysia* and to LTD in the cerebellum.

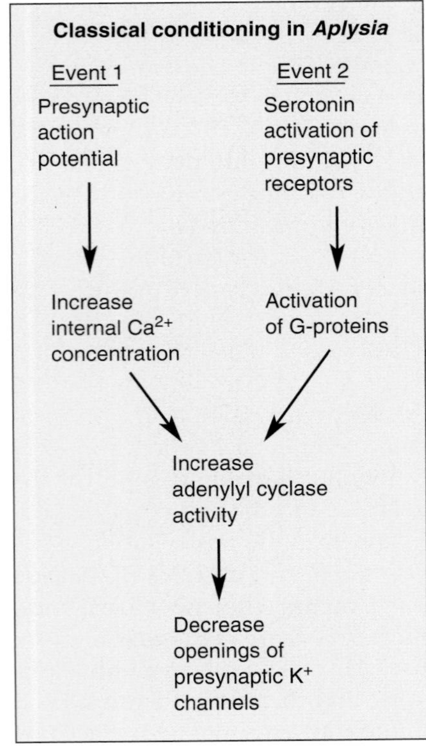

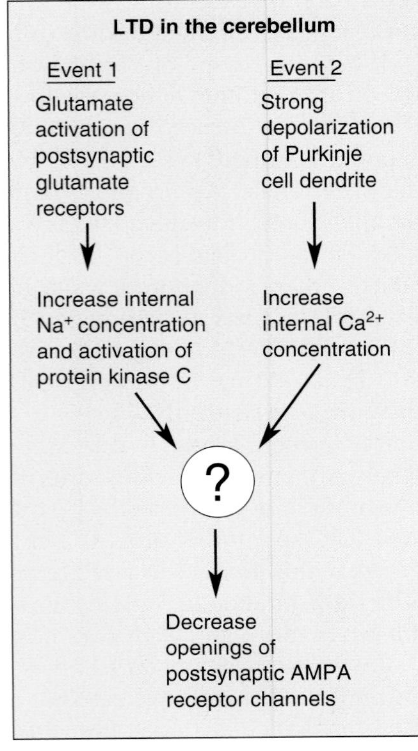

into intracellular second messengers can trigger synaptic modifications, and (3) memories can result from alterations in existing synaptic proteins. Does this sound familiar? Figure 24.16 compares the mechanisms of classical conditioning in *Aplysia* and LTD in the cerebellum.

Synaptic Plasticity in the Hippocampus and Neocortex

Recall from Chapter 23 that declarative memory, the type of memory you'll rely on to pass your next exam, involves the neocortex and structures in the medial temporal lobe, including the hippocampus. In 1973, Timothy Bliss and Terje Lømo, working together in Norway, made an important discovery about the hippocampus. They found that brief high-frequency electrical stimulation of an excitatory pathway to the hippocampus produced a long-lasting enhancement in the strength of the stimulated synapses. This effect is now known as **long-term potentiation**, or **LTP** (first introduced in Chapter 22). More recently, a form of LTD has also been described in the hippocampus. LTP and LTD are considered by many to hold the secret for how declarative memories are formed in the brain. Let's explore LTP and LTD where they are usually studied, in the hippocampus.

Anatomy of the Hippocampus. The hippocampus, illustrated in Figure 24.17, consists of two thin sheets of neurons folded onto each other. One sheet is called the **dentate gyrus,** and the other sheet is called **Ammon's horn**. Ammon's horn has four divisions, of which we need concern ourselves with only two: **CA3** and **CA1** (CA stands for *cornu Ammonis*, Latin for "Ammon's horn").

Recall from Chapter 23 that a major input to the hippocampus is the *entorhinal cortex*. Entorhinal cortex sends information to the hippocampus by way of a bundle of axons called the **perforant path**. Perforant path axons synapse on neurons of the dentate gyrus. Dentate gyrus neurons give rise to axons (also called mossy fibers) that synapse on cells in CA3. The CA3 cells give rise to axons that branch. One branch leaves the hippocampus via the fornix. The other branch, called the **Schaffer collateral**, forms synapses on the neurons of CA1. These connections are summarized in Figure 24.17.

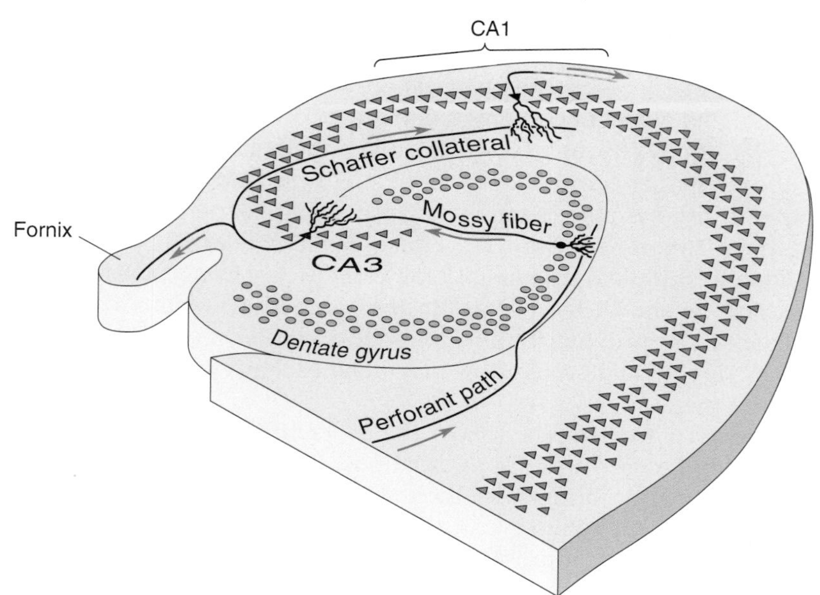

Figure 24.17
Some microcircuits of the hippocampus.

Because of its very simple architecture and organization, the hippocampus is an ideal place to study synaptic transmission in the mammalian brain. In the late 1960s, it was discovered that the hippocampus actually could be removed from the brain (usually in experimental animals) and cut up like a loaf of bread and that the resulting slices could be kept alive *in vitro* for many hours. In such a *brain slice preparation*, fiber tracts can be stimulated electrically and synaptic responses recorded. Because cells in the slice can be observed, stimulating and recording electrodes can be positioned with the precision previously reserved for invertebrate preparations. This preparation has greatly facilitated the study of LTP and LTD.

Properties of LTP in CA1. Although LTP was first demonstrated at the perforant path synapses on the neurons of the dentate gyrus, today most of the experiments on the mechanism of LTP are performed on the Schaffer collateral synapses on the CA1 pyramidal neurons in brain slice preparations.

In a typical experiment, the effectiveness of the Schaffer collateral synapse is monitored by giving a bundle of presynaptic axons a brief electrical stimulus and then measuring the size of the resulting EPSP in a postsynaptic CA1 neuron (Figure 24.18). This method is similar to the way parallel fiber responses are monitored in the cerebellar cortex. Usually, such a test stimulation is given every minute or so for 15–30 minutes to ensure that the baseline response is stable. Then, to induce LTP, the same axons are given a **tetanus**, which is a brief burst of high-frequency stimulation (typically 50–100 stimuli at a rate of 100/sec). Usually, this tetanus induces LTP, and subsequent test stimulation evokes an EPSP that is much greater than it was during the initial baseline period. In other words, the tetanus has modified the stimulated synapses so that they are more effective. Other synaptic inputs onto the same neuron that did not receive tetanic stimulation do not show LTP. Therefore, like cerebellar LTD, hippocampal LTP is *input specific*.

One remarkable feature of this plasticity is that it can be induced by a brief tetanus, lasting less than a second, consisting of stimulation at frequencies well within the range of normal axon firing. A second remarkable feature of LTP is its longevity. LTP induced in CA1 of awake animals can last many weeks, possibly even a lifetime. No wonder this form of synaptic plasticity has attracted interest as a candidate mechanism for declarative memory.

Subsequent research has shown that high-frequency stimulation is not an absolute requirement for LTP. Rather, what is required is that *synapses be active at the same time that the postsynaptic CA1 neuron is strongly depolarized*. To achieve the necessary depolarization with a tetanus, (1) synapses must be stimulated at frequencies high enough to cause temporal summation of the EPSPs, and (2) enough synapses must be active simultaneously to cause significant spatial summation of EPSPs. This second requirement is called **cooperativity** because coactive synapses must cooperate to produce enough depolarization to cause LTP. Thus, like cerebellar LTD, LTP results when synaptic stimulation (event 1) coincides with strong postsynaptic depolarization (event 2). In contrast to the cerebellum where a single powerful synapse can provide the critical depolarization, in the hippocampus, adequate depolarization requires that many excitatory synapses be active at the same time.

Consider for a moment how the cooperativity property of hippocampal LTP may be used to form associations. Imagine a hippocampal neuron receiving synaptic inputs from three sources, A, B, and C. Initially, no single input is strong enough to evoke an action potential in the postsynaptic neuron. Now imagine that inputs A and B repeatedly fire at the same time. Because of spatial summation, inputs A and B are now capable of firing the postsy-

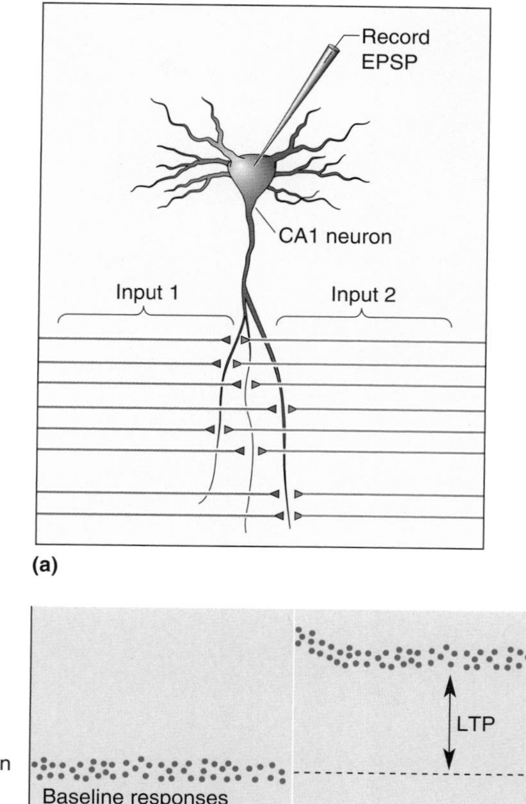

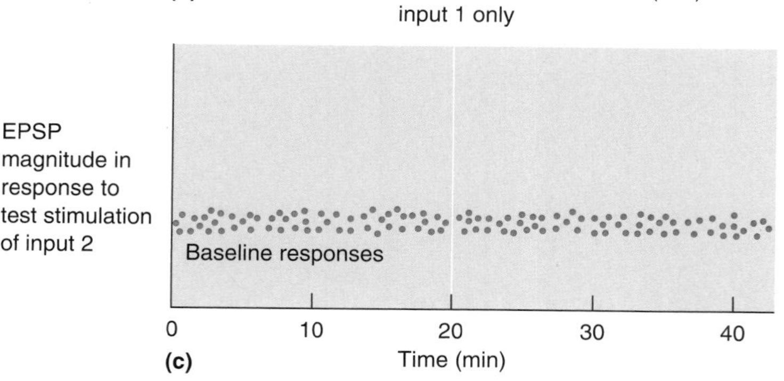

Figure 24.18
Long-term potentiation in CA1. (a) The response of a CA1 neuron is monitored as two inputs are alternately stimulated. LTP is induced in input 1 by giving this input a tetanus. **(b)** The graph shows a record of the experiment. The tetanus to input 1 (arrow) yields a potentiated response to stimulation of this input. **(c)** LTP is input specific, so there is no change in the response to input 2 after a tetanus to input 1.

naptic neuron and of causing LTP. Only the active synapses are potentiated, and these, of course, are the synapses belonging to inputs A and B. Now, because of potentiation of their synapses, *either* input A *or* input B can fire the postsynaptic neuron (but not input C). Thus, LTP has caused an association of inputs A and B. In this way, the sight of a rose may be associated with the smell of a rose (they often occur at the same time) but never with the smell of an onion (Figure 24.19).

Speaking of associations, remember the idea of a Hebb synapse introduced in Chapter 22 to account for aspects of visual development? LTP in CA1 is Hebbian: Inputs that fire together wire together.

Mechanisms of LTP in CA1. Glutamate receptors mediate excitatory synaptic transmission in the hippocampus. As is the case at the parallel

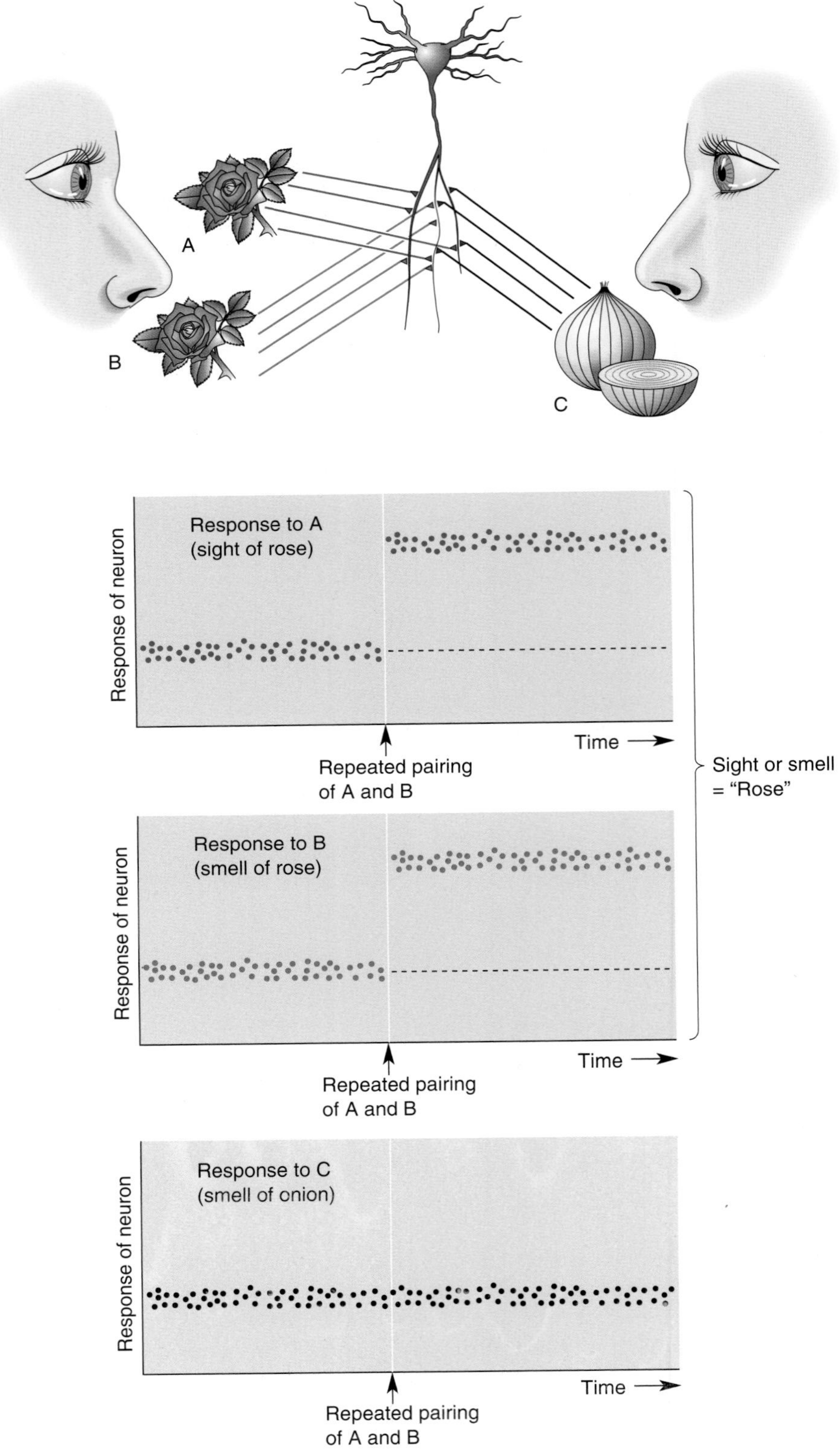

Figure 24.19

A rose is a rose, but it is not an onion. Because the sight and smell of the rose occur at the same time, the inputs carrying this information to a neuron may undergo LTP, thus forming an association between the two stimuli.

fiber–Purkinje cell synapse in the cerebellum, Na^+ ions passing through the AMPA subclass of glutamate receptor are responsible for the EPSP at the Schaffer collateral–CA1 pyramidal cell synapse. Unlike the cerebellum, however, CA1 neurons also have postsynaptic *NMDA* (*N*-methyl-D-aspartate) *receptors*. Recall that these glutamate receptors have the unusual property that they conduct Ca^{2+} ions, but only when glutamate binds *and* the postsynaptic membrane is depolarzed enough to displace Mg^{2+} ions that clog the channel (see Figure 22.24). Thus, Ca^{2+} entry through the NMDA receptor specifically signals when presynaptic and postsynaptic elements are active at the same time (Box 24.1).

Considerable evidence now links this rise in postsynaptic $[Ca^{2+}]_i$ to the induction of LTP. For example, LTP induction is prevented if NMDA receptors are pharmacologically inhibited or if rises in postsynaptic $[Ca^{2+}]_i$ are prevented by the injection of a Ca^{2+} chelator into the postsynaptic neuron. The rise in $[Ca^{2+}]_i$ activates two protein kinases: *protein kinase C* and *calcium-calmodulin-dependent protein kinase II*, also known as CaMKII (pronounced "cam K two"). Pharmacological inhibition of either kinase blocks the induction of long-term potentiation.

Following the rise in postsynaptic $[Ca^{2+}]_i$ and the activation of the kinases, however, the molecular trail that leads to a potentiated synapse gets harder

Box 24.1

BRAIN FOOD

Synaptic Plasticity: Timing Is Everything

When enough synapses are active at the same time, the postsynaptic neuron is depolarized sufficiently to fire an action potential. Donald Hebb proposed that each individual synapse grows a little stronger when it successfully participates in the firing of the postsynaptic neuron. The phenomenon of LTP comes close to satisfying Hebb's ideal. The synapse gets stronger when the glutamate released by the presynaptic terminal binds to postsynaptic NMDA receptors *and* the postsynaptic membrane is depolarized strongly enough to displace Mg^{2+} from the NMDA receptor channel.

Is there a role for postsynaptic action potentials in this "strong" depolarization? Action potentials are generated in the soma in response to depolarization of the membrane beyond threshold. Because this happens far from the synapses on the dendritic tree, it was assumed for a time that the actual occurrence of the spike was not important for the mechanism of synaptic potentiation. The important thing was the strong depolarization in the dendrite because of summed synaptic currents, which coincidentally was also usually sufficient to evoke a postsynaptic action potential.

While it remains true that the key is strong postsynaptic depolarization, researchers have recently taken another look at the role of the postsynaptic spike in LTP. This new attention resulted from the discovery that action potentials generated in the soma can actually back-propagate into the dendrites of some cells. Thus, Henry Markram, Bert Sakmann (of patch clamp fame; see Box 4.2), and their colleagues at the Max Planck Institute investigated what happens when a postsynaptic spike is generated via a microelectrode at various time intervals before or after an EPSP. Remarkably, they found that if an EPSP is followed by a postsynaptic action potential within about 50 msec, the synapse potentiates. Nothing happens in response to the spike or the EPSP alone—LTP results specifically from the precise timing of EPSP and spike, just as Hebb suggested.

What accounts for the LTP-promoting effect of a back-propagating action potential? The answer, of course, is strong depolarization. NMDA receptors have a high affinity for glutamate, so the transmitter remains bound to the receptor for many tens of milliseconds. This bound glutamate does nothing, however, if the postsynaptic membrane is not depolarized strongly, because the channel is clogged with Mg^{2+}. The timely occurrence of the action potential is sufficient to awaken these dormant channels by ejecting the Mg^{2+}. Then, as long as glutamate is still bound to the receptor, Ca^{2+} will enter the cell and trigger the mechanism of LTP.

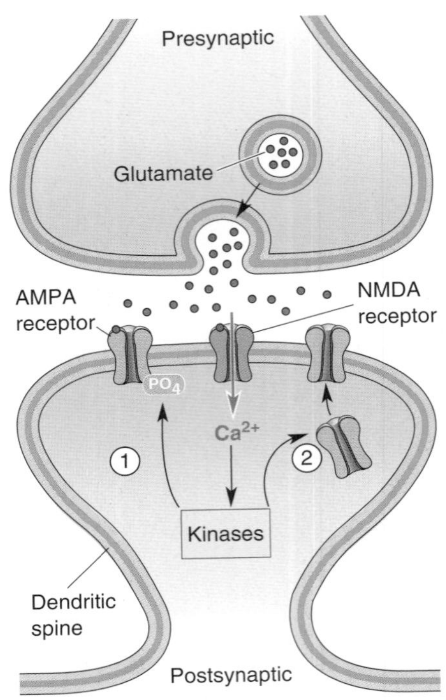

Figure 24.20

Routes for the expression of LTP in CA1. Ca^{2+} entering through the NMDA receptor activates protein kinases. This can cause LTP by ① changing the effectiveness of existing postsynaptic AMPA receptors or ② stimulating the insertion of new AMPA receptors.

to follow. Current research suggests that this trail may actually branch (Figure 24.20). One path appears to lead toward an increased effectiveness of existing postsynaptic AMPA receptors by way of phosphorylation. Phosphorylation of the AMPA receptor, by either protein kinase C or CaMKII, leads to a change in the protein that increases the ionic conductance of the channel. The other path leads to the insertion of entirely new AMPA receptors into the postsynaptic membrane. According to a current model, vesicular organelles studded with AMPA receptors lie in wait near the postsynaptic membrane. In response to CaMKII activation, the vesicle membrane fuses with the postsynaptic membrane, and the new AMPA receptors are thereby delivered to the synapse.

Evidence also indicates that synaptic structure changes following LTP. In particular, postsynaptic dendritic spines appear to bud and form new synaptic contacts with axons. Thus, following LTP, a single axon can make multiple synapses on the same postsynaptic neuron, which is not the normal pattern in CA1. This sprouting of synapses increases not only the responsive postsynaptic surface but also the probability that an action potential in the axon will trigger presynaptic glutamate release.

Long-Term Depression in CA1. We have seen that information can be stored either as a decrease in synaptic effectiveness (cerebellar LTD) or as an increase in synaptic effectiveness (hippocampal LTP). There is no reason, however, these two types of synaptic plasticity cannot reside in one location and contribute to information storage. In fact, hippocampal responses can be modified in both directions. Remember the place cells in Chapter 23? When the spatial environment changed, *two* changes occurred in the place cells: They became *more* responsive to a new location and *less* responsive to an old location. Similar changes in response selectivity are observed in many regions of the cerebral cortex when information is learned (see, for example, Figure 23.6). Neural network models suggest that the experience-dependent changes in selectivity of individual neurons reflect synaptic modifications that, when distributed over many neurons, store information (see Box 23.3). From this perspective, memories are encoded as specific *patterns* of synaptic change, with some synapses growing stronger and others growing weaker.

Recall Hebb's theory that a synapse grows stronger, or potentiates, when the activity of that synapse correlates with the strong activation of the postsynaptic neuron by other converging inputs. An extension of Hebb's theory designed to account for bidirectional (i.e., up and down) regulation of synaptic strength is the **BCM theory**, named for its authors, Elie Bienenstock, Leon Cooper, and Paul Munro, working at Brown University. According to the BCM theory, synapses that are active when the postsynaptic cell is only weakly depolarized by other inputs undergo LTD instead of LTP. This idea inspired a search for LTD in CA1 with stimuli that were designed to evoke a modest postsynaptic response being used. Soon researchers discovered that prolonged tetanic stimulation of the Schaffer collaterals at low frequencies (1–5 Hz) produces input-specific LTD (Figure 24.21).

Surprisingly, LTD, like LTP, is triggered by postsynaptic Ca^{2+} entry through the NMDA receptor. But how can the same signal, Ca^{2+} entry through the NMDA receptor, trigger *both* LTP *and* LTD? The key difference lies in the *level* of NMDA receptor activation (Figure 24.22). When the postsynaptic neuron is only weakly depolarized, the partial blocking of the NMDA receptor channels by Mg^{2+} prevents all but a trickle of Ca^{2+} into the postsynaptic neuron. On the other hand, when the postsynaptic neuron is strongly depolarized, the Mg^{2+} block is displaced entirely, and Ca^{2+} floods into the postsynaptic neuron. These different types of Ca^{2+} response selectively activate different types of enzymes. Instead of the kinases that are ac-

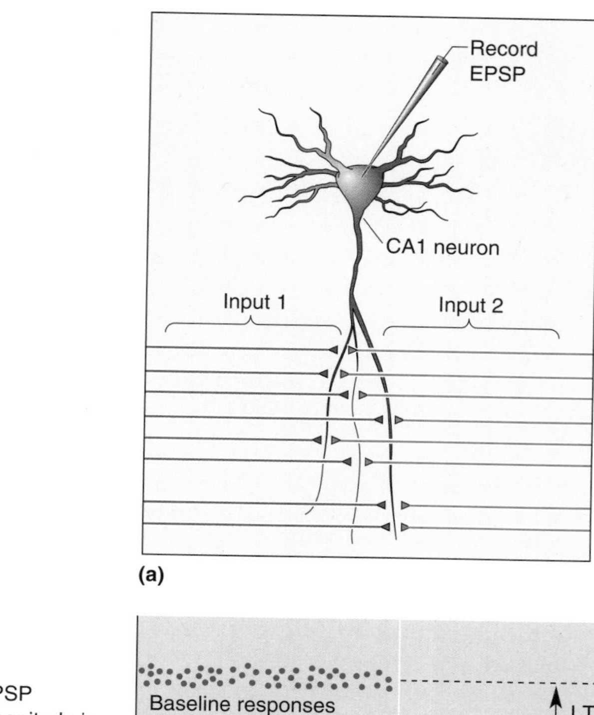

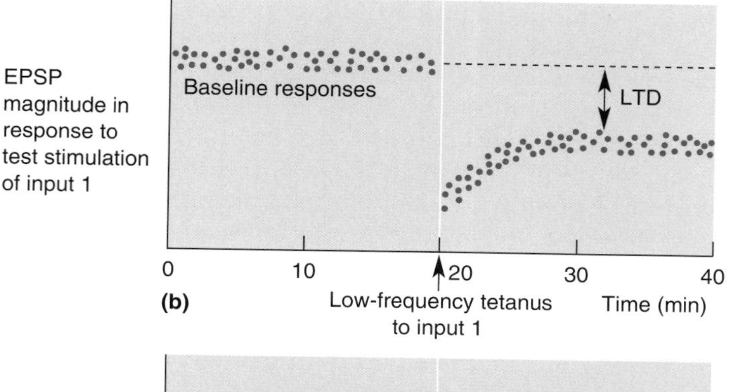

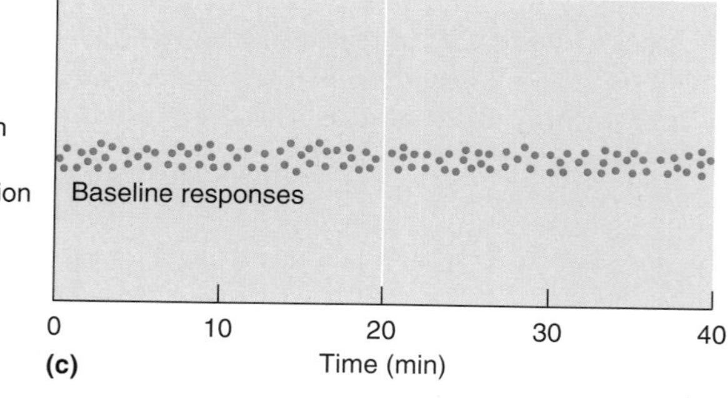

Figure 24.21
Hippocampal long-term depression. **(a)** The response of a CA1 neuron is monitored as two inputs are alternately stimulated. LTD is induced in input 1 by giving this input a 1 Hz tetanus. **(b)** The graph shows a record of the experiment. The low-frequency tetanus to input 1 (arrow) yields a depressed response to stimulation of this input. **(c)** LTD is input specific, so there is no change in the response to input 2 after a tetanus to input 1.

tivated by high $[Ca^{2+}]_i$, modest and prolonged elevations in $[Ca^{2+}]_i$ activate *protein phosphatases*, enzymes that pluck phosphate groups off proteins. Therefore, if LTP is putting phosphate groups on, LTD apparently is taking them off. Indeed, biochemical evidence now indicates that AMPA receptors are dephosphorylated in response to stimulation that induces LTD (Figure 24.23). Moreover, the induction of hippocampal LTD can also be associated with the internalization of AMPA receptors at the synapse (similar to cerebellar LTD). Thus, LTP and LTD appear to reflect the bidirectional and symmetrical regulation of postsynaptic AMPA receptors.

LTP, LTD, and Memory. LTP and LTD have attracted a lot of interest because theoretical work shows that these mechanisms of synaptic plasticity

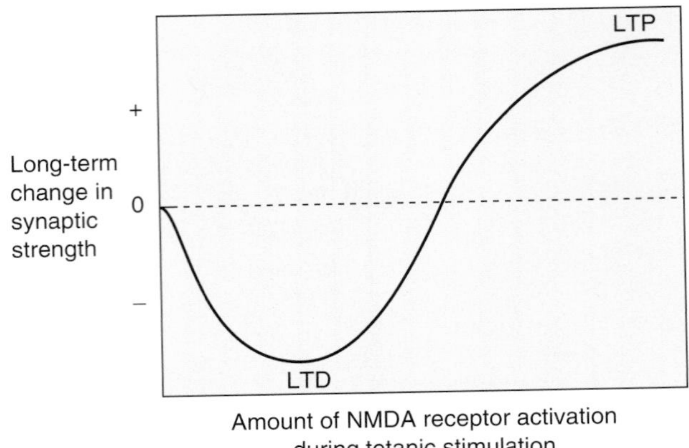

Figure 24.22
NMDA receptor activation and bidirectional synaptic plasticity. The long-term change in synaptic transmission is graphed as a function of the level of NMDA receptor activation during conditioning stimulation.

can contribute to the formation of declarative memories. But these forms of plasticity must be demonstrated in the neocortex, not just the hippocampus, because neocortex is the likely site of declarative long-term memory. Fortunately, recent research indicates that the types of NMDA-receptor-dependent synaptic plasticity that have been characterized in the hippocampus also occur in the neocortex (Figure 24.24). It appears that synaptic plasticity throughout the cerebral cortex may be governed according to the same rules and may use the same mechanisms. Interestingly, these mechanisms closely resemble those that have been identified to play a role in the development of cortical connections (see Chapter 22).

But what evidence links LTP and LTD to memory? So far, all we've described is a possible neural basis for a memory of having one's brain electrically stimulated. The most useful approach has been to see whether the molecules involved in LTP and LTD are also involved in learning and memory. A test of spatial memory in rats, called the **Morris water maze**, is normally used because proper performance on this task is known to depend on the hippocampus (Figure 24.25). In this test, a rat is placed in a pool filled with milky water. Submerged just below the surface in one location is a small platform that allows the rat to escape. A naive rat placed in the water will swim around until it bumps into the hidden platform, and then it will climb

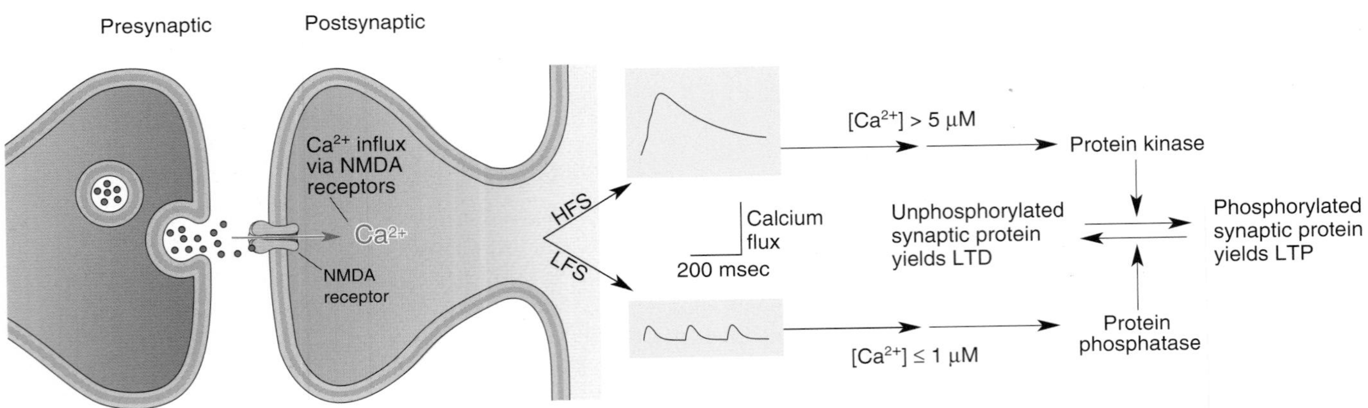

Figure 24.23
A model for how Ca^{2+} can trigger both LTP and LTD in the hippocampus. High-frequency stimulation (HFS) yields LTP by causing a large elevation of $[Ca^{2+}]$. Low-frequency stimulation (LFS) yields LTD by causing a smaller elevation of $[Ca^{2+}]$. (Source: Adapted from Bear and Malenka, 1994, Fig. 1.)

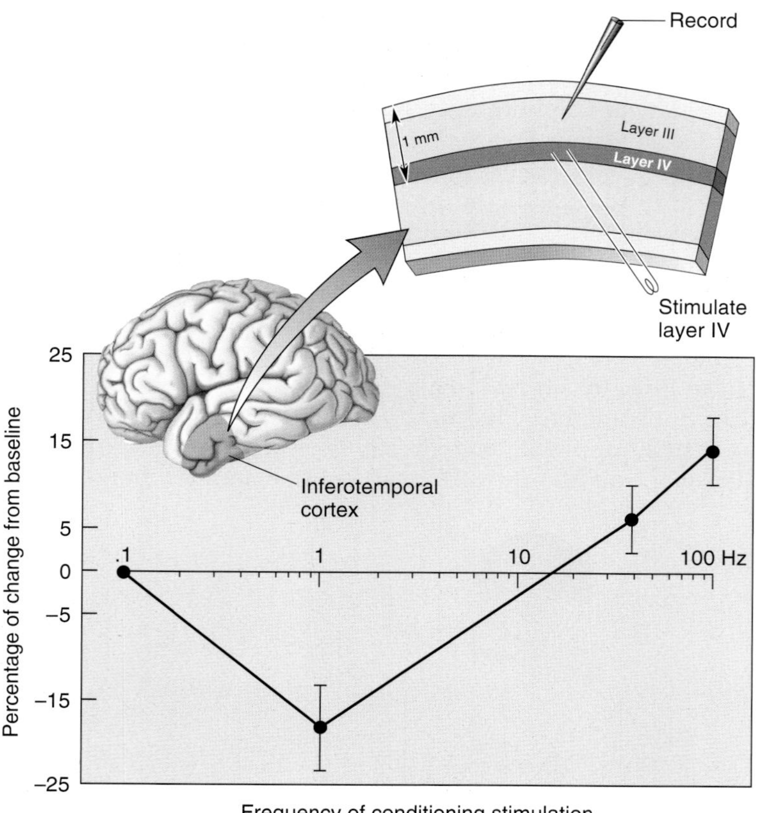

Figure 24.24
Bidirectional synaptic modifications in the human inferotemporal cortex. Slices of human temporal cortex, removed during the course of surgery to gain access to deeper structures, were maintained *in vitro*. Synaptic responses were monitored following various types of tetanic stimulation. As in rat CA1, stimulation of 1 Hz produced LTD, while 100 Hz stimulation produced LTP. (Source: Adapted from Chen et al., 1996.)

onto it. Normal rats quickly learn where the platform is and on subsequent trials swim straight to it. Moreover, once they have figured out what to search for, rats put in a new maze learn the task much faster. But rats with bilateral hippocampal damage never seem to figure out the game or remember the location of the platform.

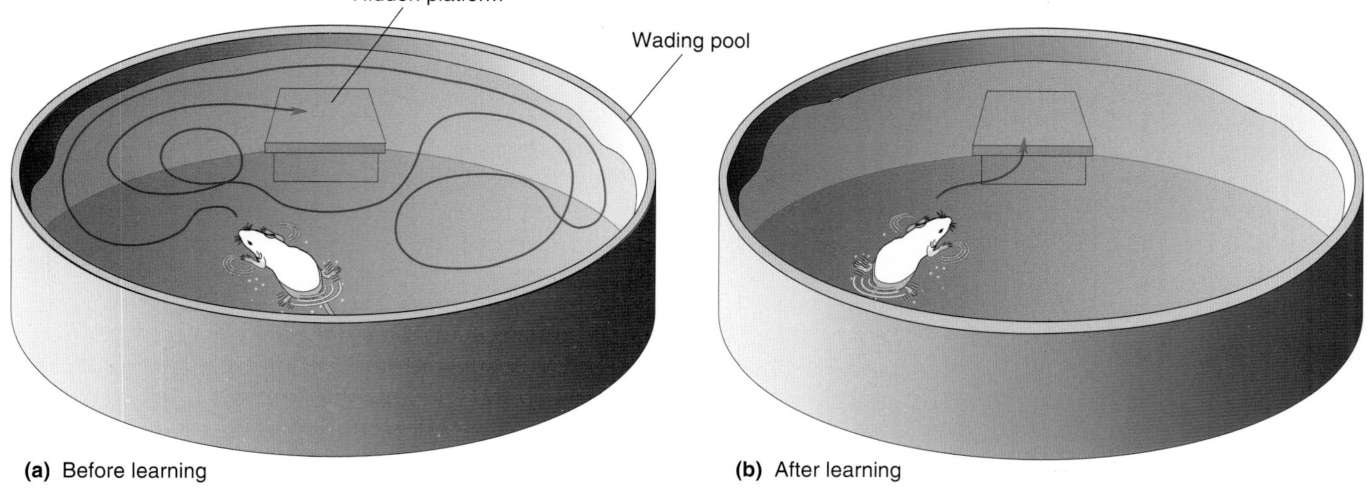

Figure 24.25
The Morris water maze. (a) The trajectory a rat might take to find a hidden platform the first time the rat is placed in the pool. **(b)** After repeated trials, the rat knows where the platform is, and swims straight to it. Learning this task requires the activation of NMDA receptors in the hippocampus.

Studies of LTP and LTD in the hippocampus point to the NMDA receptor as a key molecule for synaptic modification. To assess the possible role of hippocampal NMDA receptors in the learning of the maze, during the late 1980s, University of Edinburgh psychologist Richard Morris (inventor of the maze) and his colleagues injected an NMDA-receptor blocker into the hippocampus of rats that were being trained in the water maze. Unlike normal animals, these rats failed to learn the rules of the game or the location of the submerged platform. This finding provided the first evidence that NMDA receptor-dependent processes play a role in memory.

Susumu Tonegawa of the Massachusetts Institute of Technology introduced a revolutionary new approach to the molecular basis of learning and memory. Tonegawa, who switched to neuroscience after winning the 1987 Nobel Prize for his research in immunology, recognized that molecules and behavior could be connected by manipulating the genes of experimental animals. This approach had already been tried with success in simple organisms like fruit flies (Box 24.2) but not in mammals. In their first experiment,

Box 24.2

OF SPECIAL INTEREST

Memory Mutants

Of the several hundred thousand proteins manufactured by a neuron, some may be more important than others when it comes to learning. It is even possible that some proteins are *uniquely* involved in learning and memory. Needless to say, we could gain considerable insight into the molecular basis of learning and learning disorders if such hypothetical memory molecules could be identified.

Recall that each protein molecule is the readout of a segment of DNA called a gene. One way to identify a memory protein is to delete genes one at a time and see if specific learning deficits result. This is precisely the strategy that Seymour Benzer, Yadin Dudai, and their colleagues at the California Institute of Technology tried using the fruit fly *Drosophila melanogaster*. *Drosophila* has long been a favorite species of geneticists, but one might reasonably question to what extent a fruit fly learns. Fortunately, *Drosophila* can perform the same tricks that *Aplysia* has mastered: habituation, sensitization, and classical conditioning. For example, fruit flies can learn that a particular odor predicts a shock. They demonstrate this memory after training by flying away when the odor is presented. The strategy is to produce mutant flies by exposing them to chemicals or X-rays. They are then bred and screened for behavioral deficits. The first mutant displaying a fairly specific learning deficit, described in 1976, was called Dunce. Other memory-deficient mutants were later described and given vegetable names such as Rutabaga and Cabbage. The next challenge was to identify exactly which proteins had been deleted. It turned out that all three of these memory mutants lacked particular enzymes in cAMP-associated signaling pathways. Perhaps this isn't too surprising, considering the key role this second messenger system plays in procedural memory in *Aplysia*.

In these early *Drosophila* studies, the mutations were made at random, followed by extensive screening, first to find a learning deficit and then to determine exactly which gene was missing. More recently, however, genetic engineering techniques have made it possible to make very specific deletions of known genes, not only in *Drosophila* but also in mammals. Thus, for example, in 1992, Susumu Tonegawa, Alcino Silva, and their colleagues at MIT were able to isolate and delete one subunit (α) of the calcium-calmodulin-dependent protein kinase II in mice (Box 24.3). Experiments already suggested that this enzyme is critical for the induction of long-term potentiation. Sure enough, these mice have a clear deficit in LTP in the hippocampus and the neocortex. And when tested in the Morris water maze, they were found to have a severe memory deficit. Thus, these mice are memory mutants, just like their distant cousins Dunce, Rutabaga, and Cabbage.

Are we to conclude that the missing proteins in these mutants are the elusive memory molecules? No. All of these mutants show other behavioral deficits besides memory. At present, we can conclude only that animals growing up without these proteins are unusually poor learners. These studies do underscore, however, the critical importance of specific second messenger pathways in translating a fleeting experience into a lasting memory.

Tonegawa, Alcino Silva, and their colleagues "knocked out" (deleted) the gene for one subunit (α) of CaMKII and found parallel deficits in hippocampal LTP and memory. Since then, many mouse genes have been manipulated with the aim of assessing the role of LTP and LTD mechanisms in learning. Although the jury is still out, it does appear that LTP, LTD, and learning have many common requirements.

Despite the power of the genetic approach, it has some serious limitations. Loss of a function, such as LTP or learning, may be a secondary consequence of developmental abnormalities caused by growing up without a particular protein. Moreover, since the protein is missing in all cells that normally express it, pinpointing where and how a molecule contributes to learning can be difficult. For these reasons, researchers have attempted to devise ways to restrict their genetic manipulations to specific times and specific locations. In one interesting example of this approach, Tonegawa, Joe Tsien, and their colleagues found a way to restrict the genetic deletion of NMDA receptors to the CA1 region, starting when the animals were about 3 weeks of age (Box 24.3). These animals show a striking deficit in LTP, LTD, and water maze performance, thus revealing an essential role for CA1 NMDA receptors in this type of learning.

If too little hippocampal NMDA receptor activation is bad for learning and memory, what would happen if we boosted the number of NMDA receptors? Joe Tsien, now at Princeton University, recently generated mice that make extra NMDA receptors. These animals show *enhanced* learning ability in some tasks. Taken together, the pharmacological and genetic studies show that hippocampal NMDA receptors play a key role not only in synaptic modification, such as LTP and LTD, but also in learning and memory.

THE MOLECULAR BASIS OF LONG-TERM MEMORY

Several model systems have shown that memory can result from experience-dependent alterations in synaptic transmission. In almost every case, synaptic transmission is modified as a result of changing the number of phosphate groups that are attached to proteins in the synaptic membrane. In the case of sensitization and classical conditioning in *Aplysia*, this occurs at certain potassium channels in the presynaptic membrane. In the case of LTD and LTP in the mammalian central nervous system, it is very likely that this occurs at the postsynaptic AMPA receptor.

Adding phosphate groups to a protein can change synaptic effectiveness and form a memory, but only as long as the phosphate groups remain attached to that protein. Phosphorylation as a long-term memory mechanism is problematic for two reasons:

1. Phosphorylation of a protein is not permanent. Over time, the phosphate groups are removed, which erases the memory.
2. Protein molecules themselves are not permanent. Most proteins in the brain have a life of less than 2 weeks and undergo a continual process of replacement. Memories tied to changes in individual protein molecules cannot be expected to survive this rate of molecular turnover.

Thus, we must consider the mechanisms that might convert what initially is a change in synaptic protein phosphorylation to a form that can last a lifetime.

Persistently Active Protein Kinases

Phosphorylation of synaptic proteins and memory could be maintained if the kinases—the enzymes that attach phosphate groups to proteins—were made to stay "on" all the time. Normally the kinases are tightly regulated and are

Box 24.3

PATH OF DISCOVERY

Of Mice and Memories

BY SUSUMU TONEGAWA

I have always been fascinated by the so-called mind-brain problem. To what extent can we understand the mind by studying the brain? For much of my research career, my expertise was in molecular biology and immunology, and I knew almost nothing about neuroscience. My *virtual* interest in neuroscience began to be transformed to a *real* interest when Alcino Silva joined my lab as a postdoctoral fellow in the late 1980s. At that time our lab was working on problems in immunology, and we were making genetically altered mice. While discussing possible research projects with Alcino, I realized that he too had a hidden interest in neuroscience. As far as we knew, no one had attempted to apply powerful genetic engineering technology to neuroscience.

To decipher events occurring in the brain that underlie a cognitive phenomenon, one must select an experimental method that uses a whole, live animal. It seemed clear that knockout mice had the potential to be a very powerful tool. In knockout mice, a specific gene is deleted, so any cognitive or behavioral deficits observed in these mice relative to normal mice can directly or indirectly be attributed to the lack of a single gene product. A similar approach had been used by others to study the molecular basis of invertebrate behaviors, but work along these lines had never been applied to vertebrates, certainly not to mammals.

We decided to study learning and memory as the cognitive function and chose the α form of CaMKII as the target for knockout. In 1992, this work led to the report of the first knockout mice in neuroscience. It was shown that deletion of the α-CaMKII gene causes a deficit in LTP at Schaffer collateral–CA1 synapses and an impairment of spatial learning. Even before publication, however, we were aware of the limitations of this approach in conventional knockout mice. These limitations were primarily because the gene of interest is deleted in the entire animal throughout the animal's life. Although no obvious developmental defects were observed in the α-CaMKII knockout mice, more subtle defects could not be excluded. Furthermore, the universal absence of the protein in question (in this case, α-CaMKII) certainly did not permit the establishment of a causal relationship between CA1 LTP and spatial learning.

When Joe Tsien, another postdoctoral fellow, arrived in my lab in 1993, we started to develop a second-generation gene knockout technology that would restrict the deletion of a specific gene to a limited region of the brain. After 2 years of hard work, Joe generated several lines of mice whose genetic changes were restricted to specific neurons in the forebrain. We were thrilled to find that in one of these lines of mice, changes were restricted to the CA1 region of hippocampus. Joe went on to use this line to produce a new mouse strain in which deletion of the NMDA receptor gene, NR1, was restricted to the CA1 pyramidal cells. Joe and Pato Huerta, another postdoctoral fellow, demonstrated that in these mice, LTP and LTD are deficient at Schaffer collateral–CA1 synapses but are normal elsewhere, such as perforant path dentate granular cell synapses. They also showed that the mutant mice are deficient in hippocampus-dependent learning. Furthermore, a collaboration with Matthew Wilson's laboratory across the hall demonstrated that these mutant animals could not form normal place fields in CA1. All of these data provided very strong evidence for Hebb's hypothesis on the synaptic basis of memory formation.

Susumu Tonegawa

I informed Alcino that we seemed to have developed a technology to restrict NR1 gene knockout to the CA1 area, and the ever cheerful and enthusiastic Alcino exclaimed "Wow! God given! Congratulations!" CA1 is a major anatomical region in the hippocampus that had been implicated in memory formation, and the Schaffer collateral–CA1 synapse is the most intensively studied synapse in the brain. To set the experimental goal of targeting a gene knockout to a much broader forebrain region and obtain a mouse line in which the NR1 gene deletion is restricted to just one type of synapse may indeed be regarded as a God-given gift.

Can we develop technology that would target gene manipulation to other specific brain areas or cell types? Recent studies in our laboratory indicate that this is indeed possible. With the addition of techniques that would permit reversible temporal control of gene manipulation, rodent behavioral genetics is poised to become a powerful approach for the dissection of mechanisms underlying cognition and behavior.

"on" only in the presence of a second messenger. But what if learning changed these kinases so they no longer required the second messenger? The relevant synaptic proteins would remain phosphorylated all the time.

Recent evidence suggests that some kinases can become independent of their second messengers. Let's consider as an example the changes that occur in one protein kinase during LTP in the hippocampus.

CaMKII and LTP. Recall that Ca^{2+} entry into the postsynaptic cell and activation of CaMKII are required for the induction of LTP in CA1. Research has shown that CaMKII stays "on" long after $[Ca^{2+}]_i$ has fallen back to a low level.

CaMKII consists of ten subunits arranged in a rosette pattern. Each subunit catalyzes the phosphorylation of substrate proteins in response to a rise in Ca^{2+}-calmodulin. How might CaMKII be switched permanently on? The answer requires some knowledge of how this enzyme is normally regulated (Figure 24.26). Each subunit is built like a pocket knife, with two parts connected by a hinge. One part, the *catalytic* region, performs the phosphorylation reaction. The other part is the *regulatory* region. Normally, in the absence of the appropriate second messenger, the knife is closed and the catalytic region is covered by the regulatory region. This keeps the enzyme "off." The normal action of the second messenger (Ca^{2+}-calmodulin) is to pry the knife open, but only as long as the messenger is present. When the messenger is removed, the molecule usually snaps shut, and the kinase turns "off" again. After LTP, however, it appears that the knife fails to close completely in the α subunits of CaMKII. The exposed catalytic region continues to phosphorylate CaMKII substrates.

How is the hinge of the protein kinase molecule kept open? The answer lies in the fact that CaMKII is an *autophosphorylating protein kinase*; i.e., each subunit within the CaMKII molecule can be phosphorylated by a neighboring subunit. The consequence of subunit phosphorylation is that the hinge stays open. If the initial activation of CaMKII by Ca^{2+}-calmodulin is sufficiently strong, autophosphorylation will occur faster than dephosphorylation, and the molecule will be switched "on." Persistent activity of CaMKII may contribute to the maintenance of synaptic potentiation, for example, by keeping the postsynaptic AMPA receptors phosphorylated. The general idea that an autophosphorylating kinase could store information at the synapse, initially proposed by John Lisman at Brandeis University, is called the **molecular switch hypothesis**.

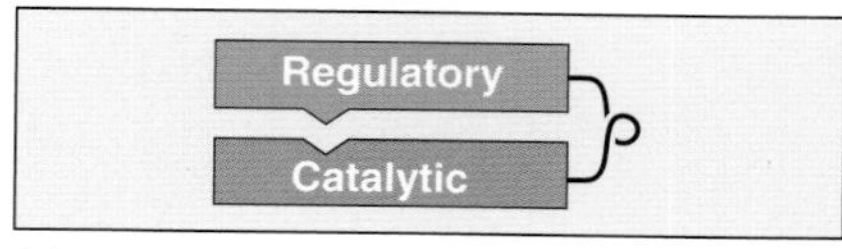

(a)

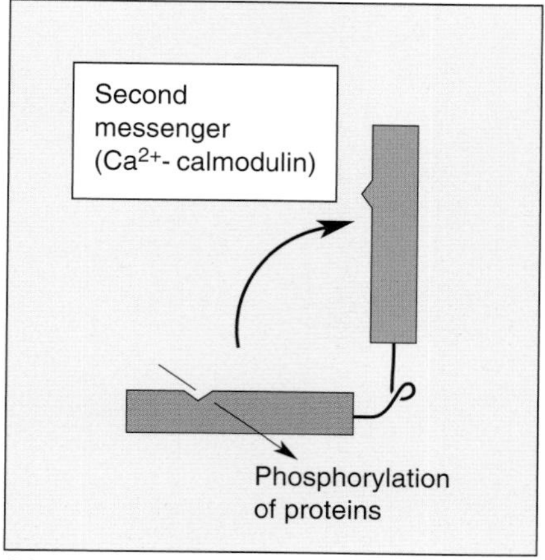

(b)

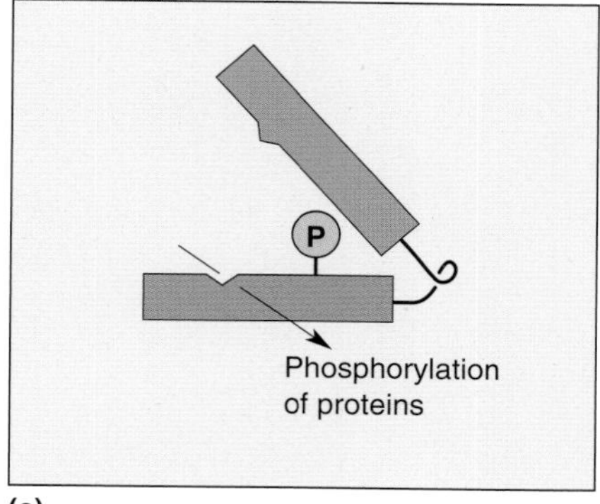

(c)

Figure 24.26
The regulation of CaMKII. (a) The hinge-like subunit of CaMKII is normally "off" when the catalytic region is covered by the regulatory region. **(b)** The hinge opens upon activation of the molecule by Ca^{2+}-bound calmodulin, freeing the catalytic region to add phosphate groups (P) to other proteins. **(c)** A large elevation of Ca^{2+} can cause phosphorylation of one subunit by another (autophosphorylation), which enables the catalytic region to stay "on" permanently.

Protein Synthesis

Persistent kinases are likely to contribute to the maintenance of a synaptic modification, but apparently only for a limited time (minutes to hours). After that, a requirement for long-term memory is the synthesis of new protein. This protein is used to assemble new synapses.

Protein Synthesis and Memory Consolidation. The possible role of new protein synthesis in memory has been investigated extensively since the introduction of drugs in the 1960s that selectively inhibit the assembly of protein from messenger RNA (mRNA). Protein synthesis inhibitors can be injected into the brains of experimental animals as they are trained to perform a task, and deficits in learning and memory can be assessed. These studies reveal that if brain protein synthesis is inhibited at the time of training, the animals learn normally but fail to remember when tested days later. A deficit in long-term memory is also often observed if the inhibitors are injected shortly after training. The memories become increasingly resistant to the inhibition of protein synthesis, however, as the interval between the training and the injection of inhibitor is increased. These findings indicate a requirement for new protein synthesis during the period of *memory consolidation,* when short-term memories are converted to long-term ones.

Very similar effects of protein synthesis inhibitors have been observed in the model systems we've been discussing. In *Aplysia,* the repeated application of sensitizing stimuli causes a form of long-term memory that can last many days. The application of protein synthesis inhibitors at the time of training has no effect on sensitization measured hours later, but it completely blocks the development of the long-term memory. Likewise, inhibiting protein synthesis at the time of a tetanus has no effect on the induction of LTP in the hippocampus. Instead of lasting days to weeks, however, the synaptic potentiation gradually disappears over a few hours.

How are we to interpret these results? From what we have learned so far, memory formation appears initially to involve the rapid modification of existing synaptic proteins. These modifications, perhaps with the help of persistently active kinases, work against the factors that would erase our memory (such as molecular turnover). It's a losing battle unless a new protein arrives at the modified synapse and converts the temporary change in the synapse to a more permanent one.

Cyclic AMP Response Element Binding Protein and Memory. What regulates the protein synthesis that is required for memory consolidation? The first step in protein synthesis is the generation of an mRNA transcript of a gene (see Figure 2.8). This process of gene expression is regulated by *transcription factors* in the nucleus. One transcription factor is the **cAMP response element binding protein (CREB).** CREB is a protein that binds to specific segments of DNA, called *cAMP response elements (CREs),* and regulates the expression of neighboring genes (Figure 24.27). There are two forms of CREB: CREB-2 re-

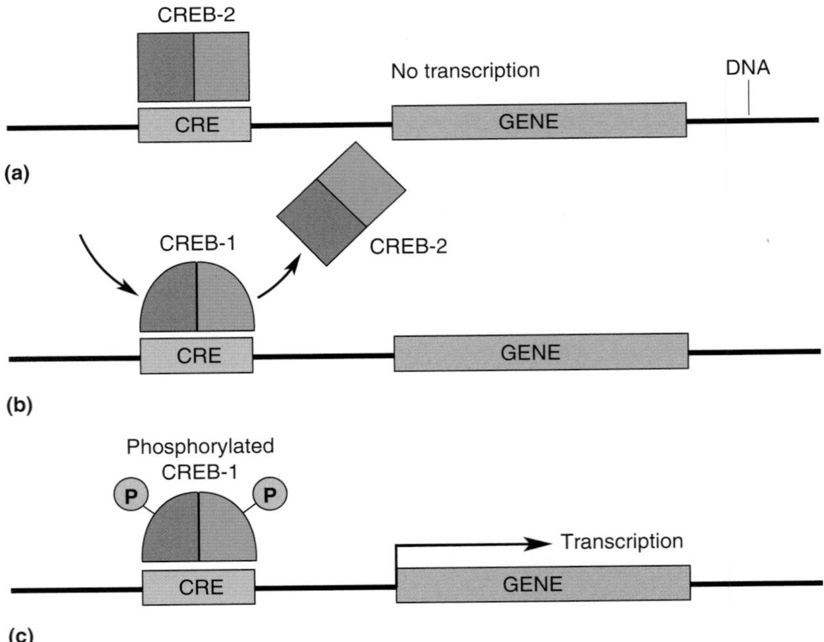

Figure 24.27
The regulation of gene expression by CREB. Shown here is a piece of DNA containing a gene whose expression is regulated by the interaction of a CREB protein with a CRE on the DNA. **(a)** CREB-2 functions as a repressor of gene expression. **(b)** CREB-1, an activator of gene expression, can displace CREB-2. **(c)** When CREB-1 is phosphorylated by protein kinase A and other kinases, transcription can ensue.

presses gene expression when it binds to the CRE; CREB-1 activates transcription, but only when it is phosphorylated by protein kinase A. In a seminal study published in 1994, Tim Tully and Jerry Yin at Cold Spring Harbor Laboratory showed that CREB regulates the gene expression required for memory consolidation in the fruit fly *Drosophila melanogaster* (Box 24.2).

In their first series of experiments, Tully and Yin bred *Drosophila* that would make extra copies of the fly's version of CREB-2 (called dCREBb) when the animal was warmed up (a miracle of fly genetic engineering that is not possible in mammals). This manipulation repressed all gene expression that is regulated by the CREs and blocked memory consolidation in a simple associative memory task. Thus, CREB-regulated gene expression is critical for memory consolidation in flies. More interesting, however, is what they found when they generated flies that could make extra copies of fly CREB-1 (called dCREBa). Now tasks that would take normal flies many trials to learn could be remembered after a single training trial. These mutant flies had a perfect memory. And these results are not peculiar to flies; CREB has been implicated in regulating the consolidation of sensitization in *Aplysia*, as well as long-term potentiation and spatial memory in mice.

As you know, not all experiences are remembered equally. Some, particularly those with strong emotional content, are seared permanently into our memories. (Think of your first romance.) Others remain with us only briefly and then fade away. The modulation of gene expression by CREB offers a molecular mechanism that can control the strength of a memory.

Structural Plasticity and Memory. How does the synapse make use of the timely occurrence of gene expression and the arrival of a new protein? In the case of synapses that are strengthened after learning, it appears that this protein is used to construct new synapses. This has been best demonstrated in *Aplysia*, in which it is possible to examine identified synapses made by the sensory neuron in the gill withdrawal reflex. Long-term (but not short-term) sensitization *doubles* the number of synapses made by this neuron. Furthermore, this increase in synapse number decays at the same rate as the long-term memory. Thus, in this system, long-term memory is associated with the formation of new synapses, and forgetting is associated with a loss of these synapses.

Do similar structural changes occur in the mammalian nervous system after learning? This problem is difficult to solve experimentally because of the complexity of the mammalian brain and the distributed nature of memory. One approach has been to compare brain structure in animals that have had ample opportunity to learn with that of animals that have had little chance to learn. Thus, putting a laboratory rat in a complex environment filled with toys and playmates (other rats) has been shown to increase by about 25% the number of synapses per neuron in the occipital cortex.

Structural changes after learning need not be confined to *increases* in synapse number, however. For example, long-term habituation of the gill withdrawal reflex in *Aplysia* is associated with a *decrease* (by one-third) in the number of synapses made by the sensory neuron. And recent research suggests that the conditioning stimulation that yields long-term depression in the cerebellar cortex also decreases the number of parallel fiber–Purkinje cell synapses.

There are limits to structural plasticity in the adult brain, however. As we discussed in Chapter 22, large changes in brain circuitry are generally confined to critical periods of early life. The growth and retraction of most axons in the adult central nervous system are restricted to no more than a few tens of micrometers. But it is now very clear that the end of a critical period does not necessarily signify an end to changes in the structure of axon terminals or the effectiveness of their synapses.

CONCLUDING REMARKS

Learning and memory can occur at synapses. Regardless of the species, brain location, and memory type, many of the underlying mechanisms appear to be universal. Events are represented first as changes in the electrical activity of the brain, then as second messenger molecules, and next as modifications of existing synaptic proteins. These temporary changes are converted to permanent ones—and long-term memory—by altering the structure of the synapse. In many forms of memory, this entails synthesis of new protein and assembly of new microcircuits. In other forms of memory, existing circuits may be disassembled. In either case, learning requires many of the same mechanisms that were used to refine brain circuitry during development.

One universal feature is the involvement of Ca^{2+} ions. Clearly, calcium does far more than build strong bones and teeth. Not only is it critical for neurotransmitter secretion and muscle contraction, it is involved in nearly every form of synaptic plasticity. Because it is a charge-carrying ion *and* a potent second messenger substance, Ca^{2+} has a unique ability to couple electrical activity directly with long-term changes in the brain.

Can basic neuroscience research take us from ions to intelligence? From calcium to cognition? If your performance on the next exam is a fair measure of intelligence and cognition and if synaptic plasticity truly is the basis of declarative memory, it would appear that the answer is yes.

KEY TERMS

Procedural Learning
nonassociative learning (p. 777)
habituation (p. 777)
sensitization (p. 777)
associative learning (p. 777)
classical conditioning (p. 778)
instrumental conditioning (p. 778)

Vertebrate Models of Learning
cerebellar cortex (p. 786)
Purkinje cell (p. 786)
inferior olive (p. 786)
climbing fiber (p. 786)
mossy fiber (p. 786)
cerebellar granule cell (p. 786)
parallel fiber (p. 787)
Marr-Albus theory of motor learning (p. 788)
long-term depression (LTD) (p. 788)
input specificity (p. 788)
long-term potentiation (LTP) (p. 791)
dentate gyrus (p. 791)
Ammon's horn (p. 791)
CA3 (p. 791)
CA1 (p. 791)
perforant path (p. 791)
Schaffer collateral (p. 791)
tetanus (p. 792)
cooperativity (p. 792)
BCM theory (p. 796)
Morris water maze (p. 798)

The Molecular Basis of Long-Term Memory
molecular switch hypothesis (p. 803)
cAMP response element binding protein (CREB) (p. 804)

REVIEW QUESTIONS

1. Outline the steps involved in the presynaptic release of neurotransmitter. Why would the closure of a potassium channel in the presynaptic axon terminal change the amount of Ca^{2+} entering and the amount of neurotransmitter released?

2. Rabbits can be classically conditioned to blink in response to a tone. This is accomplished by repeatedly pairing the tone with an air puff to the eye. Richard Thompson and his colleagues at Stanford University have made the following observations: Learning fails to occur and the memory is wiped out if the cerebellum is surgically removed; the air puff activates cells in the inferior olive; the tone activates cerebellar mossy fibers. Using your knowledge of synaptic plasticity in the cerebellum, propose a mechanism for classical conditioning in the rabbit.

3. In Figure 24.16, the mechanisms of classical conditioning in *Aplysia* and LTD in the cerebellar cortex are compared. Expand this comparison to include LTP in the hippocampus. What would events 1 and 2 be? How do these signals converge to affect a common intracellular process? How is the synaptic change expressed?

4. What property of the NMDA receptor makes it well suited to detect coincident presynaptic and postsynaptic activity? How may Ca^{2+} entering through the NMDA receptor possibly trigger both LTP and LTD in CA1 and neocortex?

5. People given electroconvulsive shocks to the head forget events that occurred shortly before the shock. One interpretation of this retrograde amnesia is that short-term memory and/or the consolidation into long-term memory of these events was disrupted by the shock. How may protein synthesis be involved in the conversion of short-term into long-term memory?

6. In H. M and R. B. (see Chapter 23), destruction of the hippocampus appears to have impaired the mechanism that fixes new memories in the neocortex. Propose a mechanism involving CREB explaining why this may be true.

GLOSSARY

A1 *See* primary auditory cortex.

absolute refractory period The period of time, measured from the onset of an action potential, during which another action potential cannot be triggered.

accommodation The focusing of light by changing the shape of the eye's lens.

acetylcholine (ACh) An amine that serves as a neurotransmitter at many synapses in the peripheral and central nervous systems, including the neuromuscular junction.

ACTH *See* adrenocorticotropic hormone.

actin A cytoskeletal protein in all cells and the major thin filament protein in a muscle fiber; causes muscle contraction by specific chemical interactions with myosin.

action potential A brief fluctuation in membrane potential caused by the rapid opening and closing of voltage-gated ion channels; also known as spike, nerve impulse, or discharge. Action potentials sweep like a wave along axons to transfer information from one place to another in the nervous system.

activational effect The ability of a hormone to activate reproductive processes or behaviors in the mature organism.

active zone A presynaptic membrane differentiation that is the site of neurotransmitter release.

adenosine triphosphate (ATP) The molecule that is the cell's energy source. The hydrolysis of ATP to produce adenosine diphosphate (ADP) releases energy that fuels most of the biochemical reactions of the neuron. ADP is converted back to ATP in the mitochondria.

adenylyl cyclase An enzyme that catalyzes the conversion of adenosine triphosphate to cyclic adenosine monophosphate, a second messenger.

adrenal cortex The outer segment of the adrenal gland; releases cortisol when stimulated by the pituitary adrenocorticotropic hormone.

adrenal medulla The inner segment of the adrenal gland, innervated by preganglionic sympathetic fibers; releases epinephrine.

adrenaline A catecholamine neurotransmitter synthesized from norepinephrine; also called epinephrine.

adrenocorticotropic hormone (ACTH) A hormone released by the anterior pituitary in response to corticotropin-releasing hormone; stimulates the release of cortisol from the adrenal gland.

affective aggression A threatening or defensive form of aggression accompanied by vocalizations and a high level of activity in the autonomic nervous system.

affective disorder A psychiatric mood disorder, such as major depression or bipolar disorder.

afferent An axon coursing toward and innervating a given structure. *See also* efferent.

after-hyperpolarization The hyperpolarization that follows strong depolarization of the membrane; the last part of an action potential, also called undershoot.

agnosia The inability to recognize objects, even though simple sensory skills appear to be normal; most commonly caused by damage to posterior parietal areas of the brain.

alpha motor neuron A neuron that innervates the extrafusal fibers of skeletal muscle.

amacrine cell A neuron in the retina of the eye that projects neurites laterally in the inner plexiform layer.

amino acid A chemical building block of protein molecules, containing a central carbon atom, an amino group, a carboxyl group, and a variable R group.

Ammon's horn A layer of neurons in the hippocampus that sends axons into the fornix.

amnesia A severe loss of memory or the ability to learn. *See also* anterograde amnesia, retrograde amnesia.

AMPA receptor A subtype of glutamate receptor; a glutamate-gated ion channel that is permeable to Na^+ and K^+.

ampulla The bulge along a semicircular canal, which contains the hair cells that transduce rotation.

amygdala An almond-shaped nucleus in the anterior temporal lobe, thought to be involved in emotion and certain types of learning and memory.

anabolism The biosynthesis of organic molecules from nutritive precursors; also called anabolic metabolism. *See also* catabolism.

analgesia The absence of normal sensations of pain.

androgens Male sex steroidal hormones, the most important of which is testosterone.

anion A negatively charged ion. *See also* cation.

anomia The inability to find words.

anorectic peptide A neuroactive peptide that acts to inhibit feeding behavior; examples are cholecystokinin, alpha-melanocyte-stimulating hormone (αMSH), and cocaine- and amphetamine-regulated transcript peptide (CART).

anorexia A state of negative energy balance in which energy expenditure exceeds energy intake.

anorexia nervosa A psychiatric disorder characterized by an obsession with food, an intense fear of gaining weight, and voluntary maintenance of weight at below normal levels.

ANS *See* autonomic nervous system.

antagonist muscle A muscle that acts against another at the same joint.

anterior A direction meaning toward the nose, or rostral.

anterograde amnesia The inability to form new memories.

anterograde transport Axoplasmic transport from the soma to the axon terminal.

antidepressant drug A drug that treats the symptoms of depression; examples are tricyclics, monoamine oxidase (MAO) inhibitors, and serotonin-selective reuptake inhibitors.

aphasia A partial or complete loss of language abilities following brain damage. *See also* Broca's aphasia, conduction aphasia, Wernicke's aphasia.

apoptosis A mechanism of orderly, genetically programmed cell death.

aqueous humor The fluid between the cornea and the lens of the eye.

arachnoid membrane The middle of the three meninges that cover the surface of the central nervous system.

arcuate nucleus A nucleus in the periventricular area of the hypothalamus containing a large number of neurons sensitive to changes in leptin levels, contributing to the regulation of energy balance.

area 17 Primary visual cortex.

area IT An area of neocortex, on the inferior surface of the temporal lobe, that is part of the ventral visual processing stream and contains neurons with responses to complex objects, including faces.

area MT An area of neocortex, at the junction of the parietal and temporal lobes, that receives input from primary visual cortex and appears to be specialized for the detection of stimulus movement; also called V5.

area V4 An area of neocortex, anterior to striate cortex, that is in the ventral visual processing stream and appears to be important for both shape perception and color perception.

aspinous neuron A neuron lacking dendritic spines.

associative learning The learning of associations between events; two types are usually distinguished: classical conditioning and instrumental conditioning.

astrocyte A glial cell in the brain that supports neurons and regulates the extracellular ionic and chemical environment.

ataxia Abnormally uncoordinated and inaccurate movements, often associated with cerebellar dysfunction.

atonia The absence of muscle tone.

ATP *See* adenosine triphosphate.

attenuation reflex The contraction of muscles in the middle ear, resulting in a reduction in auditory sensitivity.

audition The sense of hearing.

auditory canal A channel leading from the pinna to the tympanic membrane; the entrance to the internal ear.

auditory-vestibular nerve Cranial nerve VIII consisting of axons projecting from the spiral ganglion to the cochlear nuclei.

autonomic ganglia Peripheral ganglia of the sympathetic and parasympathetic divisions of the autonomic nervous system.

autonomic nervous system (**ANS**) A system of central and peripheral nerves that innervates the internal organs, cardiovascular system, and glands; also called the visceral peripheral nervous system. The ANS consists of sympathetic, parasympathetic, and enteric divisions.

autoradiography A method for visualizing sites of radioactive emissions in tissue sections.

autoreceptor A receptor in the membrane of a presynaptic axon terminal that is sensitive to the neurotransmitter released by that terminal.

axial muscle A muscle that controls movements of the trunk of the body.

axon A neurite specialized to conduct nerve impulses, or action potentials, normally away from the soma.

axon collateral A branch of an axon.

axon hillock A swelling of the axon where it joins the soma.

axon terminal The end region of an axon, usually a site of synaptic contact with another cell; also called terminal bouton or presynaptic terminal.

axoplasmic transport The process of transporting materials down an axon.

ballism A movement disorder caused by damage to the subthalamus, characterized by violent, flinging movements of the extremities.

barbiturate A class of drugs having sedative, general anesthetic, and anticonvulsant effects; act in part by binding to gamma-aminobutyric acid-A receptors and prolonging their inhibitory actions.

basal forebrain complex Several cholinergic nuclei of the telencephalon, including the medial septal nuclei and basal nucleus of Meynert.

basal ganglia A collection of associated cell groups in the basal forebrain, including the caudate nucleus, putamen, globus pallidus, and subthalamus.

basal lamina A layer of proteins in the space between a nerve terminal and the muscle cell it innervates.

basal telencephalon The region of the telencephalon lying deep in the cerebral hemispheres.

basilar membrane A membrane separating the scala tympani and scala media in the cochlea in the inner ear.

basolateral nuclei A group of nuclei in the basolateral amygdala whose axons constitute the ventral amygdalofugal pathway.

BCM theory A theory proposing that synapses are bidirectionally modifiable. Synaptic potentiation results when presynaptic activity correlates with a strong postsynaptic response, and synaptic depression results when presynaptic activity correlates with a weak postsynaptic response. An extension of the Hebb synapse concept, proposed by Bienenstock, Cooper, and Munro at Brown University. *See also* Hebb synapse, Hebbian modification.

benzodiazepine A class of drugs having antianxiety, sedative, muscle-relaxing, and anticonvulsant effects; act by binding to gamma-aminobutyric acid-A receptors and prolonging their inhibitory actions.

binocular competition A process believed to occur during the development of the visual system whereby the inputs from the two eyes actively compete to innervate the same cells.

binocular receptive field The receptive field of a neuron that responds to stimulation of either eye.

binocular visual field The portion of the visual field viewed by both eyes.

bipolar cell In the retina, a cell that connects photoreceptors to ganglion cells.

bipolar disorder A psychiatric mood disorder characterized by episodes of mania, sometimes interspersed with episodes of depression; also called manic-depressive disorder.

bipolar neuron A neuron with two neurites.

blob A collection of cells, mainly in primary visual cortical layers II and III, characterized by a high level of the enzyme cytochrome oxidase.

blob channel The visual information-processing channel that passes through the parvocellular and koniocellular layers of the lateral geniculate nucleus and converges on the blobs of striate cortical layer III; believed to process information about color.

blood-brain barrier A specialization of the walls of brain capillaries that limits the movement of blood-borne substances into the extracellular fluid of the brain.

brain The part of the central nervous system contained in the skull, consisting of the cerebrum, cerebellum, brain stem, and retinas.

brain stem The diencephalon, midbrain, pons, and medulla. (Some anatomists exclude the diencephalon.)

Broca's aphasia A language disturbance in which a person has difficulty speaking or repeating words but can understand language; also known as motor or nonfluent aphasia.

Broca's area A region of the frontal lobe associated with Broca's (motor) aphasia when damaged.

bulimia nervosa A psychiatric disorder characterized by large, uncontrolled eating binges followed by compensatory behavior, such as forced vomiting.

bundle A collection of axons that run together but do not necessarily have the same origin and destination.

CA1 A region of Ammon's horn in the hippocampus that receives input from the neurons of CA3.

CA3 A region of Ammon's horn in the hippocampus that receives input from the neurons of the dentate gyrus.

CAM *See* cell adhesion molecule.

calcium-calmodulin-dependent protein kinase (CaMK) A protein kinase activated by elevations of internal Ca^{2+} concentration.

calcium pump An ion pump that removes cytosolic Ca^{2+} ions.

cAMP *See* cyclic adenosine monophosphate.

Cannon-Bard theory A theory of emotion proposing that emotional experience is independent of emotional expression and is determined by the pattern of thalamic activation.

capsule A collection of axons that connect the cerebrum with the brain stem.

catabolism The breaking down of complex nutrient molecules into simpler molecules; also called catabolic metabolism. *See also* anabolism.

catecholamines The neurotransmitters dopamine, norepinephrine, and epinephrine.

cation A positively charged ion. *See also* anion.

caudal A direction toward the tail, or posterior.

caudate nucleus A part of the basal ganglia in the basal forebrain, involved in motor control.

CCK *See* cholecystokinin.

cell assembly The group of simultaneously active neurons that represents an object held in memory.

cell body The central region of the neuron containing the nucleus; also called soma or perikaryon.

cell adhesion molecule (CAM) A molecule on the cell surface that causes cells to adhere to one another.

center-surround receptive field A visual receptive field with a circular center region and a surround region forming a ring around the center; stimulation of the center produces a response opposite that generated by stimulation of the surround.

central nervous system (CNS) The brain (including the retinas) and spinal cord. *See also* peripheral nervous system.

central pattern generator A neural circuit that gives rise to rhythmic motor activity.

central sulcus The sulcus in the cerebrum that divides the frontal lobe from the parietal lobe.

cerebellar cortex A sheet of gray matter lying just under the pial surface of the cerebellum.

cerebellar granule cell A neuron in the cerebellar cortex that receives input from mossy fibers and gives rise to parallel fibers that innervate Purkinje cells.

cerebellar hemispheres The lateral regions of the cerebellum.

cerebellum A structure derived from the rhombencephalon, attached to the brain stem at the pons; an important movement control center.

cerebral aqueduct A canal filled with cerebrospinal fluid within the midbrain.

cerebral cortex The layer of gray matter that lies just under the surface of the cerebrum.

cerebral hemispheres The two sides of the cerebrum, derived from the paired telencephalic vesicles.

cerebrospinal fluid (CSF) In the central nervous system, the fluid produced by the choroid plexus that flows through the ventricular system to the subarachnoid space.

cerebrum The largest part of the forebrain; also called telencephalon.

cGMP *See* cyclic guanosine monophosphate.

characteristic frequency The sound frequency to which a neuron in the auditory system gives its greatest response.

chemical synapse A synapse in which presynaptic activity stimulates the release of neurotransmitter, which activates receptors in the postsynaptic membrane.

chemoaffinity hypothesis The hypothesis that chemical markers on growing axons are matched with complementary chemical markers on their targets.

chemoattractant A diffusible molecule that acts over a distance to attract growing axons.

chemoreceptor Any sensory receptor selective for chemicals.

chemorepellent A diffusible molecule that acts over a distance to repel growing axons.

cholecystokinin (CCK) A peptide found within some neurons of the central and peripheral nervous systems and in some endothelial cells lining the upper gastrointestinal tract; a satiety signal that inhibits feeding behavior, in part, by acting on axons of the vagus nerve that respond to gastric distension.

cholinergic Describing neurons or synapses that produce and release acetylcholine.

chromosome A structure in the cell nucleus containing a single linear thread of DNA.

ciliary muscle A muscle that controls the shape of the eye's lens.

circadian rhythm Any rhythm with a period of about one day.

classical conditioning A learning procedure used to associate a stimulus that evokes a measurable response with another stimulus that normally does not evoke this response.

climbing fiber The axon of an inferior olive neuron that innervates a Purkinje cell of the cerebellum.

CNS *See* central nervous system.

cochlea A spiral bony structure in the inner ear that contains the hair cells that transduce sound.

cochlear amplifier Outer hair cells, including the motor proteins in the outer hair cell membrane, responsible for amplifying the displacements of the basilar membrane in the cochlea.

cochlear nucleus *See* dorsal cochlear nucleus, ventral cochlear nucleus.

color-opponent cell A cell in the visual system with an excitatory response to wavelengths of light of one color and an inhibitory response to wavelengths of another color; the color pairs that cancel each other are red-green and blue-yellow.

commissure Any collection of axons that connect one side of the brain with the other side.

complex cell A type of visual cortical neuron that has an orientation-selective receptive field without distinct "on" and "off" subregions.

concentration gradient A difference in concentration from one region to another. Ionic concentration gradients across the neuronal membrane help determine the membrane potential.

conductance *See* electrical conductance.

conduction aphasia A type of aphasia associated with damage to the arcuate fasciculus, characterized by good comprehension and speech but difficulty repeating words.

cone photoreceptor A photoreceptor in the retina containing one of three photopigments that are maximally sensitive to different wavelengths of light. Cones are concentrated in the fovea, specialized for daytime vision, and responsible for all color vision. *See also* rod photoreceptor.

conjunctiva The membrane that folds back from the eyelids and attaches to the sclera of the eye.

contralateral A direction, the opposite side of midline.

cooperativity The property of long-term potentiation, reflecting the requirement that many inputs be active at the same time during a tetanus to induce long-term potentiation. *See also* long-term potentiation.

cornea The transparent external surface of the eye.

coronal plane An anatomical plane of section that divides the nervous system into anterior and posterior parts.

corpus callosum The great cerebral commissure consisting of axons connecting the cortex of the two cerebral hemispheres.

cortex Any collection of neurons that forms a thin sheet, usually at the brain's surface.

cortical module The chunk of cerebral cortex that is necessary and sufficient to analyze one discrete point in a sensory surface.

cortical plate A cell layer of the immature cerebral cortex containing undifferentiated neurons.

cortical white matter A collection of axons lying just below the cerebral cortex.

corticomedial nuclei A group of nuclei in the medial amygdala whose axons constitute the stria terminalis.

corticospinal tract The tract that originates in the neocortex and terminates in the spinal cord; involved in the control of voluntary movement.

corticotropin-releasing hormone (CRH) A hormone released by neurons in the paraventricular nucleus of the hypothalamus; stimulates the release of adrenocorticotropic hormone from the anterior pituitary.

cortisol A steroid hormone released by the adrenal cortex; mobilizes energy reserves, suppresses the immune system, and has direct actions on some central nervous system neurons.

cranial nerves Twelve pairs of nerves that arise from each side of the brain stem, numbered from anterior to posterior. Cranial nerve I is actually the olfactory tract, and cranial nerve II is the optic nerve; both are parts of the central nervous system. Cranial nerves III–XII, which are in the peripheral nervous system, perform many diverse functions.

CRH *See* corticotropin-releasing hormone.

critical period A limited period of time when a particular aspect of brain development is sensitive to a change in the external environment.

cyclic adenosine monophosphate (cAMP) A second messenger formed from adenosine triphosphate by the action of the enzyme adenylyl cyclase.

cyclic AMP response element binding protein (CREB) A protein that binds to specific regions of DNA (cyclic AMP response elements) and functions to regulate gene transcription; a key regulator of protein synthesis-dependent memory consolidation.

cyclic guanosine monophosphate (cGMP) A second messenger formed from guanosine triphosphate by the action of the enzyme guanylate cyclase.

cytoarchitectural map A map, usually of cerebral cortex, based on cytoarchitectural differences.

cytoarchitecture The arrangement of neuronal cell bodies in various parts of the brain.

cytochrome oxidase A mitochondrial enzyme concentrated in cells that form the blobs in primary visual cortex.

cytoplasm Cellular material contained by the cell membrane, including the organelles but excluding the nucleus.

cytoskeleton The internal scaffolding, consisting of microtubules, neurofilaments, and microfilaments, that gives a cell its characteristic shape.

cytosol The watery fluid inside the cell.

DA *See* dopamine.

DAG *See* diacylglycerol.

Dale's principle The idea that a neuron has a unique identity with respect to neurotransmitter.

dark adaptation The process by which the retina becomes more sensitive to light in dim light.

dark current The inward sodium current that occurs in photoreceptors in the dark.

declarative memory Memory for facts and events.

decussation A crossing of an axonal pathway from one side of the central nervous system to the other.

delayed non-match to sample (DNMS) A behavioral task in which animals are trained to displace one of two alternative objects that does not match a previously seen sample object.

dendrite A neurite specialized to receive synaptic inputs from other neurons.

dendritic spine A small sac of membrane that protrudes from the dendrites of some cells and receives synaptic input.

dendritic tree All of the dendrites of a single neuron.

dentate gyrus A layer of neurons in the hippocampus that receives input from the entorhinal cortex.

depolarization A change in membrane potential, taking it from the value at rest (e.g., −65 mV) to a less negative value (e.g., 0 mV).

dermatome A region of skin innervated by the pair of dorsal roots from one spinal segment.

deoxyribonucleic acid *See* DNA.

diacylglycerol (DAG) A second messenger molecule formed by the action of phospholipase C on the membrane phospholipid phosphatidylinositol-4,5-bisphosphate. DAG activates the enzyme protein kinase C.

diathesis-stress hypothesis of mood disorders A hypothesis suggesting that depression is caused by a combination of genetic predisposition and environmental stress.

diencephalon A region of the brain stem derived from the prosencephalon. Diencephalic structures include the thalamus and the hypothalamus.

differentiation During embryonic development, the process by which structures become more elaborate and specialized.

diffuse modulatory system One of several systems of central nervous system neurons that project widely and diffusely onto large areas of the brain and use modulatory neurotransmitters, including dopamine, norepinephrine, serotonin, and acetylcholine.

diffusion The temperature-dependent movement of molecules from regions of high concentration to regions of low concentration, resulting in a more even distribution.

diopter A unit of measurement for refractive power of the eye; the reciprocal of the focal distance.

direction selectivity The property of cells in the visual system that respond only when stimuli move within a limited range of directions.

distal muscle A muscle that controls the hands, feet, or digits.

DNA (deoxyribonucleic acid) A double-stranded molecule constructed from four nucleic acids that contains the genetic instructions for a cell.

DNMS *See* delayed non-match to sample.

dopa A chemical precursor of dopamine and the other catecholamines.

dopamine (DA) A catecholamine neurotransmitter synthesized from dopa.

dopamine hypothesis of schizophrenia A hypothesis suggesting that schizophrenia is caused by excessive activation of D_2 receptors in the mesocorticolimbic dopamine system in the brain.

dorsal A direction, toward the back.

dorsal cochlear nucleus A nucleus in the medulla that receives afferents from the spiral ganglion in the cochlea.

dorsal column A white matter tract on the dorsal side of the spinal cord, carrying touch and proprioceptive axons to the brain stem.

dorsal column nuclei A pair of nuclei located in the posterior medulla; target of dorsal column axons, mediating touch and proprioceptive input from the limbs and trunk.

dorsal column–medial lemniscal pathway An ascending somatic sensory pathway that mediates information about touch, pressure, vibration, and limb proprioception.

dorsal horn The dorsal region of the spinal cord containing neuronal cell bodies.

dorsal longitudinal fasciculus A bundle of axons reciprocally connecting the hypothalamus and midbrain periaqueductal gray matter.

dorsal root A bundle of sensory axons that emerges from a spinal nerve and attaches to the dorsal side of the spinal cord.

dorsal root ganglion A collection of cell bodies of the sensory neurons that are part of the somatic peripheral nervous system. There is one dorsal root ganglion for each spinal nerve.

duplex theory of sound localization The principle that two schemes function in sound localization: interaural time delay at low frequencies and interaural intensity difference at high frequencies.

dura mater The outermost of the three meninges that cover the surface of the central nervous system.

EEG *See* electroencephalogram.

efferent An axon originating in and coursing away from a given structure. *See also* afferent.

electrical conductance The relative ability of an electrical charge to migrate from one point to another, represented by the symbol g and measured in siemens (S). Conductance is the inverse of resistance and is related to electrical current and voltage by Ohm's law.

electrical current The rate of movement of electrical charge, represented by the symbol I and measured in amperes (amp).

electrical potential The force exerted on an electrically charged particle, represented by the symbol V and measured in volts; also called voltage or potential difference.

electrical resistance The relative inability of an electrical charge to migrate from one point to another, represented by the symbol R and measured in ohms (Ω). Resistance is the inverse of conductance and is related to electrical current and voltage by Ohm's law.

electrical self-stimulation Electrical stimulation that an animal can voluntarily deliver to a portion of its brain.

electrical synapse A synapse in which electrical current flows directly from one cell to another via a gap junction.

electroencephalogram (EEG) A measurement of electrical activity generated by the brain and recorded from the scalp.

electroconvulsive therapy (ECT) A treatment for major depression that consists of eliciting electrical seizure activity in the brain.

endocytosis The process by which a bit of the cell membrane is pinched off, internalized, and converted to an intracellular vesicle. *See also* exocytosis.

endolymph The fluid that fills the scala media in the cochlea of the inner ear, containing high K^+ and low Na^+ concentrations.

endorphin One of many endogenous opioid peptides with actions similar to those of morphine; present in many brain structures, particularly those related to pain.

engram The physical representation or location of a memory; also called memory trace.

enteric division A division of the autonomic nervous system that innervates the digestive organs; consists of the myenteric and submucous plexuses.

entorhinal cortex A cortical region in the medial temporal lobe that occupies the medial bank of the rhinal sulcus; provides input to the hippocampus.

ependymal cell A type of glial cell that provides the lining of the brain ventricular system.

epilepsy A chronic brain disorder characterized by recurrent seizures.

epinephrine A catecholamine neurotransmitter synthesized from norepinephrine; also called adrenaline.

EPSP *See* excitatory postsynaptic potential.

EPSP summation A simple form of synaptic integration whereby excitatory postsynaptic potentials combine to produce a larger postsynaptic depolarization.

equilibrium potential *See* ionic equilibrium potential.

estrogens Female steroidal hormones, the most important of which are estradiol and progesterone.

estrous cycle The female reproductive cycle in most

nonprimate mammals in which there are periodic episodes of estrus, or "heat."

Eustachian tube An air-filled tube connecting the middle ear to the nasal cavities.

excitable membrane Any membrane capable of generating action potentials. The membrane of axons and muscle cells is excitable.

excitatory postsynaptic potential (EPSP) Depolarization of the postsynaptic membrane potential by the action of a synaptically released neurotransmitter.

exocytosis The process whereby material is released from an intracellular vesicle into the extracellular space by fusion of the vesicle membrane with the cell membrane. *See also* endocytosis.

extension The direction of movement that opens a joint.

extensor A muscle that causes extension when it contracts.

extracellular matrix The network of fibrous proteins deposited in the space between cells.

extrafusal fiber The muscle fiber in skeletal muscle that lies outside muscle spindles and receives innervation from alpha motor neurons.

extraocular muscle A muscle that moves the eye in the orbit.

falling phase The part of an action potential characterized by a rapid fall of membrane potential from positive to negative.

fasciculation A process in which axons growing together stick to one another.

fast motor unit A motor unit with a large alpha motor neuron innervating rapidly contracting and rapidly fatiguing white muscle fibers.

5-HT *See* serotonin.

flexion The direction of movement that closes a joint.

flexor A muscle that causes flexion when it contracts.

follicle-stimulating hormone (FSH) A hormone secreted by the anterior pituitary; its diverse roles include the growth of follicles in the ovaries and the maturation of sperm in the testes.

forebrain The region of the brain derived from the rostral primary embryonic brain vesicle; also called prosencephalon. Forebrain structures include the telencephalon and the diencephalon.

fornix A bundle of axons that originates in the hippocampal formation, loops around the thalamus, and terminates in the diencephalon.

fourth ventricle The cerebrospinal fluid-filled space within the hindbrain.

fovea The pit or depression in the retina at the center of the macula; in humans, the fovea contains only cone photoreceptors and is specialized for high-acuity vision.

frequency The number of waves or other discrete events per second, expressed in hertz (Hz).

frontal lobe The region of the cerebrum lying anterior to the central sulcus under the frontal bone.

FSH *See* follicle-stimulating hormone.

GABA *See* gamma-aminobutyric acid.

GABAergic Describing neurons or synapses that produce and release gamma-aminobutyric acid.

gamma-aminobutyric acid (GABA) An amino acid synthesized from glutamate; the major inhibitory neurotransmitter in the central nervous system.

gamma motor neuron A motor neuron that innervates intrafusal muscle fibers.

ganglion A collection of neurons in the peripheral nervous system.

ganglion cell A cell in the retina that receives input from the bipolar cells and sends an axon into the optic nerve.

ganglion cell layer The layer of the retina closest to the center of the eye, containing ganglion cells.

gap junction A specialized junction where a narrow gap between two cells is spanned by protein channels (connexons) that allow ions to pass directly from one cell to another.

gating A property of many ion channels, making them open or closed in response to specific signals such as membrane voltage or the presence of neurotransmitters.

gender identity A person's perception of his or her maleness or femaleness.

gene expression The process of transcribing the information from a gene into messenger RNA; a gene is a segment of DNA carrying the instructions for a single protein.

generalized seizure Pathologically large and synchronous neural activity that spreads to encompass the entire cerebral hemispheres.

globus pallidus A part of the basal ganglia in the basal forebrain; involved in motor control.

glomerulus A cluster of neurons in the olfactory bulb that receives input from olfactory receptor neurons.

glucocorticoid receptor A receptor activated by cortisol released from the adrenal gland.

glutamate (Glu) An amino acid; the major excitatory neurotransmitter in the central nervous system.

glutamate hypothesis of schizophrenia A hypothesis suggesting that schizophrenia is caused by the reduced activation of NMDA (*N*-methyl-D-aspartate) receptors in the brain.

glutamatergic Describing neurons or synapses that produce and release glutamate.

glycine (Gly) An amino acid; an inhibitory neurotransmitter at some locations in the central nervous system.

GnRH See gonadotropin-releasing hormone.

Goldman equation A mathematical relationship used to predict membrane potential from the concentrations and membrane permeabilities of ions.

Golgi apparatus An organelle that sorts and chemically modifies proteins that are destined for delivery to different parts of the cell.

Golgi stain A method of staining brain tissue that shows neurons and all of their neurites; named for its discoverer, Italian histologist Camillo Golgi (1843–1926).

Golgi tendon organ A specialized structure within the tendons of skeletal muscle that senses muscle tension.

Golgi type I neuron A neuron in the brain with a long axon that transfers information from one region of the brain to another.

Golgi type II neuron A neuron in the brain with a short axon that does not extend beyond the vicinity of the cell body.

gonadotropins Hormones secreted by the anterior pituitary that regulate the release of androgens and estrogens from the testes and ovaries.

gonadotropin-releasing hormone (GnRH) A hypophysiotropic hormone secreted by the hypothalamus; regulates the release of luteinizing hormone and follicle-stimulating hormone form the anterior pituitary.

G-protein A membrane-enclosed protein that binds guanosine triphosphate (GTP) when activated by a membrane receptor. Active G-proteins can stimulate or inhibit other membrane-enclosed proteins.

G-protein-coupled receptor A membrane protein that activates G-proteins when it binds neurotransmitter.

gray matter A generic term for a collection of neuronal cell bodies in the central nervous system.

Gray's type I synapse A chemical synapse in the central nervous system with asymmetrical membrane differentiations.

Gray's type II synapse A chemical synapse in the central nervous system with symmetrical membrane differentiations.

growth cone The specialized tip of a growing neurite.

gustation The sense of taste.

gustatory nucleus A nucleus in the brain stem that receives primary taste input.

gyrus A bump or bulge lying between the sulci of the cerebrum (plural: gyri).

habituation A type of nonassociative learning leading to decreased behavioral responses to repeated stimulation.

hair cell An auditory cell that transduces sound into a change in membrane potential or a vestibular cell that transduces head movements into a change in membrane potential.

Hebb synapse A synapse that exhibits Hebbian modifications.

Hebbian modification An increase in the effectiveness of a synapse caused by the simultaneous activation of presynaptic and postsynaptic neurons.

helicotrema A hole at the apex of the cochlea in the inner ear that connects the scala tympani to the scala vestibuli.

hertz (Hz) The unit of frequency equivalent to cycles per second.

hindbrain The region of the brain derived from the caudal primary embryonic brain vesicle; also called rhombencephalon. Hindbrain structures include the cerebellum, pons, and medulla.

hippocampus A region of the cerebral cortex lying adjacent and medial to the olfactory cortex. In humans, the hippocampus is in the temporal lobe and may play a special role in learning and memory.

histology The microscopic study of the structure of tissues.

homeostasis The balanced functioning of physiological processes and maintenance of an organism's internal environment within a narrow range.

horizontal cell A cell in the retina of the eye that projects neurites laterally in the outer plexiform layer.

horizontal plane An anatomical plane of section that divides the nervous system into dorsal and ventral parts.

HPA *See* hypothalamic-pituitary-adrenal axis.

Huntington's disease A hereditary, progressive, inevitably fatal condition characterized by dyskinesias, dementia, and personality disorder and associated with profound degeneration of neurons in the basal ganglia and cerebral cortex.

hyperalgesia A reduced threshold for pain, an increased response to painful stimuli, or a spontaneous pain that follows localized injury.

hypophysiotropic hormone A peptide hormone, such as corticotropin-releasing hormone or gonadotropin-releasing hormone, released into the blood by the parvocellular neurosecretory cells of the hypothalamus, which stimulates or inhibits the secretion of hormones from the anterior pituitary.

hypothalamic-pituitary-adrenal (HPA) axis A system that regulates the release of cortisol from the adrenal gland. Dysfunction of the HPA system has been implicated in anxiety disorder and mood disorders.

hypothalamo-pituitary portal circulation A system of blood vessels that carries hypophysiotropic hormones from the hypothalamus to the anterior pituitary.

hypothalamus The ventral part of the diencephalon, involved in the control of the autonomic nervous system and the pituitary gland.

immunocytochemistry An anatomical method that uses antibodies to study the location of molecules within cells.

incus An ossicle in the middle ear whose shape somewhat resembles an anvil.

inferior colliculus A nucleus in the midbrain from which all ascending auditory signals project to the medial geniculate nucleus.

inferior olive A nucleus of the medulla that gives rise to climbing fiber input to the cerebellar cortex.

inhibitor A drug or toxin that blocks the normal action of a protein or a biochemical process.

inhibitory postsynaptic potential (**IPSP**) A change in the postsynaptic membrane potential by the action of a synaptically released neurotransmitter, making the postsynaptic neuron less likely to fire action potentials.

inner ear The cochlea, which is part of the auditory system, plus the labyrinth, which is part of the vestibular system.

inner hair cell An auditory cell located between the modiolus and the rods of Corti; the primary transducer of sound into an electrochemical signal.

inner nuclear layer The layer of the retina of the eye containing the cell bodies of bipolar, horizontal, and amacrine cells.

inner plexiform layer The layer of the retina of the eye, located between the ganglion cell layer and the inner nuclear layer, that contains the neurites and synapses between bipolar cells, amacrine cells, and ganglion cells.

innervation The provision of synaptic input to a cell or collection of cells.

inositol-1,4,5-triphosphate (**IP_3**) A second messenger molecule formed by the action of phospholipase C on the membrane phospholipid phosphatidylinositol-4,5-bisphosphate. IP_3 causes the release of Ca^{2+} from intracellular stores.

input specificity A property of some forms of synaptic plasticity such that only the active synapses onto a neuron are modified.

***in situ* hybridization** A method for localizing strands of messenger RNA within cells.

insulin A hormone released by the β cells of the pancreas; regulates blood glucose levels by controlling the expression of glucose transporters in the plasma membrane of non-neuronal cells.

instrumental conditioning A learning procedure used to associate a response, such as a motor act, with a stimulus reward, such as food.

intensity The amplitude of a wave. Sound intensity is the amplitude of the pressure differences in a sound wave that perceptually determines loudness.

internal capsule A large collection of axons that connects the telencephalon with the diencephalon.

internal resistance The resistance to electrical current flow longitudinally down a cable or neurite, represented by the symbol r_i.

interneuron Any neuron that is not a sensory or motor neuron; also used to describe a central nervous system neuron whose axon does not leave the structure in which it resides.

intrafusal fiber The specialized muscle fiber within a muscle spindle that receives motor innervation from gamma motor neurons.

ion An atom or molecule that has a net electrical charge because of a difference in the number of electrons and protons.

ion channel A membrane-spanning protein that forms a pore that allows the passage of ions from one side of the membrane to the other.

ion pump An enzyme that transports ions across a membrane at the expense of metabolic energy.

ion selectivity A property of ion channels that are selectively permeable to some ions and not to others.

ionic driving force The difference between the real membrane potential, V_m, and the ionic equilibrium potential, E_{ion}.

ionic equilibrium potential The electrical potential difference that exactly balances an ionic concentration gradient, represented by the symbol E_{ion}; also known as equilibrium potential.

IP_3 *See* inositol-1,4,5-triphosphate.

ipsilateral A direction, the same side of the midline.

IPSP *See* inhibitory postsynaptic potential.

iris The circular, pigmented muscle that controls the size of the pupil in the eye.

James-Lange theory A theory proposing that the subjective experience of emotion is a consequence of physiological changes in the body.

kainate receptor A subtype of glutamate receptor; a glutamate-gated ion channel that is permeable to Na^+ and K^+.

Klüver-Bucy syndrome A constellation of symptoms, including psychic blindness, oral tendencies, hypermetamorphosis, altered sexual behavior, and emotional changes that are observed in humans and monkeys after bilateral temporal lobectomy.

koniocellular LGN layer A layer of the lateral geniculate nucleus containing very small cells, lying just ventral to each magnocellular and parvocellular layer.

Korsakoff's syndrome A neurological syndrome resulting from chronic alcoholism, characterized by confusion, confabulations, apathy, and amnesia.

large, dense-core vesicle *See* secretory granule.

lateral A direction, away from the midline.

lateral geniculate nucleus (**LGN**) A thalamic nucleus that relays information from the retina to the primary visual cortex.

lateral hypothalamic area A poorly defined region of the hypothalamus that has been implicated in the motivation of behavior.

lateral hypothalamic syndrome Anorexia associated with lesions of the lateral hypothalamic area.

lateral intraparietal cortex (area LIP) A cortical area buried in the intraparietal sulcus.

lateral pathway Axons in the lateral column of the spinal cord that are involved in the control of voluntary movements of the distal musculature and are under direct cortical control.

lateral ventricle The cerebrospinal fluid-filled space within each cerebral hemisphere.

layer of photoreceptor outer segments The layer of the retina farthest from the center of the eye, containing the light-sensitive elements of the photoreceptors.

learning The acquisition of new information or knowledge.

lemniscus A tract that meanders through the brain like a ribbon.

length constant A parameter used to describe how far changes in membrane potential can passively spread down a cable such as an axon or a dendrite, represented by the symbol λ. The length constant λ is the distance at which the depolarization falls to 37% of its original value; λ depends on the ratio of membrane resistance (r_m) to internal resistance (r_i).

lens The transparent structure lying between the aqueous humor and the vitreous humor that enables the eye to adjust its focus to different viewing distances.

leptin A protein hormone released by adipocytes (fat cells) that communicates with neurons of the arcuate nucleus of the hypothalamus.

LGN *See* lateral geniculate nucleus.

ligand-binding method A method that uses radioactive receptor ligands (agonists or antagonists) to locate neurotransmitter receptors.

light adaptation The process by which the retina becomes less sensitive to light in bright light conditions.

limbic lobe The hippocampus and cortical areas bordering the brain stem in mammals, which Broca proposed as a distinct lobe of the brain.

limbic system A group of structures, including those in the limbic lobe and Papez circuit, that are anatomically interconnected and are probably involved in emotion, learning, and memory.

lipostatic hypothesis A hypothesis proposing that body fat is maintained homeostatically at a specific level.

lithium An element, existing in solution as a monovalent cation, that is effective in the treatment of bipolar disorder.

locus A small, well-defined group of cells.

locus coeruleus A small nucleus located bilaterally in the pons; its neurons use norepinephrine as their neurotransmitter and project widely upon all levels of the central nervous system.

long-term depression (**LTD**) A long-lasting decrease in the effectiveness of synaptic transmission that follows certain types of conditioning stimulation.

long-term memory Information storage that is relatively permanent and does not require continual rehearsal.

long-term potentiation (**LTP**) A long-lasting enhancement of the effectiveness of synaptic transmission that follows certain types of conditioning stimulation.

luteinizing hormone (LH) A hormone secreted by the anterior pituitary; its diverse roles include the stimulation of testosterone production in males and the facilitation of follicle development and ovulation in females.

M1 Primary motor cortex, area 4.

macula (1) In the eye, a yellowish spot in the middle of the retina with relatively few large blood vessels; contains the fovea. (2) In the ear, a sensory epithelium in the otolith organs whose hair cells transduce head tilt and acceleration.

magnocellular channel (**M channel**) The visual information-processing channel that begins with the M-type retinal ganglion cells and leads to layer IVB of striate cortex; believed to process information about visual movement.

magnocellular LGN layer A layer of the lateral geniculate nucleus receiving synaptic input from the M-type retinal ganglion cells.

magnocellular neurosecretory cell A large neuron of the periventricular and supraoptic nuclei of the hypothalamus that projects to the posterior pituitary and secretes oxytocin or vasopressin into the blood.

major depression An affective disorder characterized by prolonged, severe impairment of mood; may include anxiety, sleep disturbances, and other physiological disturbances.

malleus An ossicle in the middle ear attached to the tympanic membrane; shaped somewhat like a hammer.

mania An elevated, expansive, or irritable mood that is characteristic of bipolar disorder.

Marr-Albus theory of motor learning The theory that parallel fiber synapses on Purkinje cells are modified when their activity coincides with climbing fiber activity.

mechanoreceptor Any sensory receptor selective for mechanical stimuli, such as hair cells of the inner ear, various receptors of the skin, and stretch receptors of skeletal muscle.

medial A direction, toward the midline.

medial forebrain bundle A large bundle of axons coursing through the hypothalamus carrying efferents from the dopaminergic, noradrenergic, and serotonergic neurons in the brain stem and fibers interconnecting the hypothalamus, limbic structures, and midbrain tegmental area.

medial geniculate nucleus (MGN) A relay nucleus in

the thalamus through which all auditory information passes on its way from the inferior colliculus to auditory cortex.

medial lemniscus A white matter tract of the somatic sensory system, carrying axons from dorsal column nuclei to the thalamus.

medulla oblongata The part of the hindbrain caudal to the pons and cerebellum.

medullary reticulospinal tract A tract originating in the medullary reticular formation and terminating in the spinal cord; involved in the control of movement.

membrane differentiation A dense accumulation of protein adjacent to and within the membrane on either side of a synaptic cleft.

membrane potential The voltage across a cell membrane; represented by the symbol V_m.

membrane resistance The resistance to electrical current flow across a membrane; represented by the symbol r_m.

memory The retention of learned information.

memory consolidation The process of storing new information into long-term memory.

meninges Three membranes that cover the surface of the central nervous system: the dura mater, arachnoid membrane, and pia mater (singular: meninx).

menstrual cycle The female reproductive cycle in primates.

messenger RNA (mRNA) A molecule constructed from four nucleic acids that carries the genetic instructions for the assembly of a protein from the nucleus to the cytoplasm.

metabotropic receptor A G-protein-coupled receptor whose primary action is to stimulate an intracellular biochemical response.

MGN *See* medial geniculate nucleus.

microelectrode A probe used to measure the electrical activity of cells. Microelectrodes have a very fine tip and can be fashioned from etched metal or glass pipettes filled with electrically conductive solutions.

microfilament A polymer of the protein actin, forming a braided strand 5 nm in diameter; a component of the cytoskeleton.

microglial cell A type of cell that functions as a phagocyte in the nervous system to remove debris left by dead or dying neurons and glia.

microionophoresis A method of applying drugs and neurotransmitters in very small quantities to cells.

microtubule A polymer of the protein tubulin, forming a straight, hollow tube 20 nm in diameter. Microtubules, a component of the cytoskeleton, play an important role in axoplasmic transport.

midbrain The region of the brain derived from the middle primary embryonic brain vesicle; also called mesencephalon. Midbrain structures include the tectum and the tegmentum.

middle ear The tympanic membrane plus the ossicles.

midline The line that bisects the nervous system into right and left halves.

midsagittal plane An anatomical plane of section through the midline that is perpendicular to the ground. A section in the midsagittal plane divides the nervous system into right and left halves.

miniature postsynaptic potential A change in postsynaptic membrane potential caused by the action of neurotransmitter released from a single synaptic vesicle.

mitochondrion An organelle responsible for cellular respiration. Mitochondria generate adenosine triphosphate, using the energy produced by the oxidation of food.

modulation A term used to describe the actions of neurotransmitters that do not directly evoke postsynaptic potentials but modify the cellular response to excitatory postsynaptic potentials and inhibitory postsynaptic potentials generated by other synapses.

molecular switch hypothesis The idea that protein kinases can be switched "on" by autophosphorylation to a state in which they no longer require the presence of a specific second messenger to be active. Such persistently active kinases may hold the memory of an episode of strong synaptic activation. Initially proposed by John Lisman at Brandeis University.

monoamine hypothesis of mood disorders A hypothesis suggesting that depression is a consequence of a reduction in the levels of monoamine neurotransmitters, particularly serotonin and norepinephrine, in the brain.

monocular deprivation An experimental manipulation that deprives one eye of normal vision.

monogamy Mating behavior in which two individuals form a tightly bound relationship that includes exclusive or nearly exclusive mating with each other.

Morris water maze A task used to assess spatial memory in which a rodent must swim to a hidden platform below the surface of a pool of water.

mossy fiber The axon of a pontine neuron that innervates cerebellar granule cells.

motivated behavior Behavior that is incited to achieve a goal.

motor cortex Cortical areas 4 and 6, which are directly involved in the control of voluntary movement.

motor end-plate The postsynaptic membrane at the neuromuscular junction.

motor neuron A neuron that synapses on a muscle cell and causes muscle contraction.

motor neuron pool All of the alpha motor neurons innervating the fibers of a single skeletal muscle.

motor strip A name for area 4 on the precentral gyrus.

motor unit One alpha motor neuron and all of the muscle fibers it innervates.

mRNA *See* messenger RNA.

M-type ganglion cell A type of ganglion cell in the retina characterized by a large cell body and dendritic arbor, a transient response to light, and no sensitivity to different wavelengths of light; also called M cell.

Müllerian duct A structure in the embryonic gonads that develops into the female internal reproductive system.

multipolar neuron A neuron with three or more neurites.

muscarinic receptor A subtype of acetylcholine receptor that is G-protein-coupled.

muscle fiber A multinucleated skeletal muscle cell.

muscle spindle A specialized structure within skeletal muscles that senses muscle length; provides sensory information to neurons in the spinal cord via group Ia afferents; also called stretch receptor.

myelin A membranous wrapping around axons provided by oligodendroglia in the central nervous system and Schwann cells in the peripheral nervous system.

myofibril A cylindrical structure within a muscle fiber that contracts in response to an action potential.

myosin A cytoskeletal protein in all cells and the major thick filament protein in a muscle fiber; causes muscle contraction by chemical interaction with actin.

myotatic reflex A reflex that leads to muscle contraction in response to muscle stretch; mediated by the monosynaptic connection of a group Ia afferent from a muscle spindle with an alpha motor neuron innervating the same muscle.

NE *See* norepinephrine.

neglect syndrome A neurological disorder in which a part of the body or a part of the visual field is ignored or suppressed; most commonly associated with damage to posterior parietal areas of the brain.

neocortex The cerebral cortex, with six or more layers of neurons, found only in mammals.

Nernst equation A mathematical relationship used to calculate an ionic equilibrium potential.

nerve A bundle of axons in the peripheral nervous system.

nerve growth factor (**NGF**) A neurotrophin required for survival of the cells of the sympathetic division of the autonomic nervous system; also important for aspects of central nervous system development.

neural crest The primitive embryonic peripheral nervous system consisting of neural ectoderm that pinches off laterally as the neural tube forms.

neural tube The primitive embryonic central nervous system consisting of a tube of neural ectoderm.

neurite A thin tube extending from a neuronal cell body; the two types are axons and dendrites.

neuroblast An immature neuron prior to cell differentiation.

neurofilament A type of intermediate filament found in neurons. Neurofilaments, which measure 10 nm in diameter, are an important component of the neuronal cytoskeleton.

neurohormone A hormone released by neurons into the bloodstream.

neuroleptic drug An antipsychotic drug used to treat schizophrenia; examples are chlorpromazine and clozapine.

neuromuscular junction A chemical synapse between a spinal motor neuron axon and a skeletal muscle fiber.

neuron doctrine The concept that the neuron is the elementary functional unit of the brain and that neurons communicate with each other by contact, not continuity.

neuronal membrane The barrier, about 5 nm thick, that separates the inside of a nerve cell from the outside; consists of a phospholipid bilayer with proteins embedded in it; encloses the intracellular organelles and vesicles.

neuropharmacology The study of the effects of drugs on nervous system tissue.

neurotransmitter A chemical that is released by a presynaptic element upon stimulation and activates postsynaptic receptors.

neurotrophin A member of a family of related neuronal trophic factors, including nerve growth factor and brain-derived neurotrophic factor.

neurulation The formation of the neural tube from the neural ectoderm during embryonic development.

NGF *See* nerve growth factor.

nicotinic ACh receptor A class of acetylcholine-gated ion channel found in various locations, notably at the neuromuscular junction.

Nissl stain A class of basic dyes that stain the somata of neurons; named for its discoverer, German histologist Franz Nissl (1860–1919).

nitric oxide (NO) A gas produced from the amino acid arginine that serves as an intercellular messenger.

NMDA receptor A subtype of glutamate receptor; a glutamate-gated ion channel that is permeable to Na^+, K^+, and Ca^{2+}. Inward ionic current through the *N*-methyl-D-aspartate receptor is voltage-dependent because of a magnesium block at negative membrane potentials.

nociceptor Any receptor selective for potentially harmful stimuli; may induce sensations of pain.

node of Ranvier A space between two consecutive myelin sheaths where an axon comes in contact with the extracellular fluid.

nonassociative learning A change in the behavioral response that occurs over time in response to a single type of stimulus; the two types are habituation and sensitization.

nondeclarative memory Memory for skills, habits, emotional responses, and some reflexes.

nonM-nonP ganglion cell A ganglion cell in the retina that is not of the M type or P type, based on cell morphology and response properties. Of the variety of cell types in this category, some are known to be sensitive to the wavelength of light.

non-REM sleep A stage of sleep characterized by large, slow electroencephalogram waves, a paucity of dreams, and some muscle tone.

noradrenergic Describing neurons or synapses that produce and release norepinephrine.

norepinephrine (NE) A catecholamine neurotransmitter synthesized from dopamine; also called noradrenaline.

nucleus (1) The roughly spherical organelle in the cell body containing the chromosomes. (2) A general term used to describe a clearly distinguishable mass of neurons, usually deep in the brain.

nucleus of the solitary tract A brain stem nucleus that receives sensory input and uses it to coordinate autonomic function via its outputs to other brain stem and forebrain nuclei and to the hypothalamus.

obesity A state of positive energy balance in which energy intake exceeds energy expenditure.

occipital lobe The region of the cerebrum lying under the occipital bone.

ocular dominance column A region of striate cortex receiving information predominantly from one eye.

ocular dominance shift A change in visual cortex interconnections that makes more neurons responsive to one eye or the other.

OFF bipolar cell A bipolar cell of the retina that depolarizes in response to dark (light OFF) in the center of its receptive field.

Ohm's law The relationship between electrical current (I), voltage (V), and conductance (g): $I = gV$. Because electrical conductance is the inverse of resistance (R), Ohm's law may also be written: $V = IR$.

olfaction The sense of smell.

olfactory bulb A bulb-shaped brain structure derived from the telencephalon that receives input from olfactory receptor neurons.

olfactory cortex The region of the cerebral cortex connected to the olfactory bulb and separated from the neocortex by the rhinal fissure.

olfactory epithelium A sheet of cells lining part of the nasal passages that contains olfactory receptor neurons.

oligodendroglial cell A glial cell that provides myelin in the central nervous system.

ON bipolar cell A bipolar cell of the retina that depolarizes in response to light (light ON) in the center of its receptive field.

opioid receptor A membrane protein that selectively binds natural (e.g., endorphin) and synthetic (e.g., morphine) opioid substances.

optic chiasm The structure in which the right and left optic nerves converge and partially decussate (cross) to form the optic tracts.

optic disk The location on the retina where optic nerve axons leave the eye; also called optic nerve head.

optic nerve The bundle of ganglion cell axons that passes from the eye to the optic chiasm.

optic radiation A collection of axons coursing from the lateral geniculate nucleus to the visual cortex.

optic tectum A term used to describe the superior colliculus, particularly in nonmammalian vertebrates.

optic tract A collection of retinal ganglion cell axons stretching from the optic chiasm to the brain stem. Important targets of the optic tract are the lateral geniculate nucleus and superior colliculus.

orexigenic peptide A neuroactive peptide that stimulates feeding behavior; examples are neuropeptide Y (NPY), agouti-related peptide (AgRP), melanin-concentrating hormone (MCH), and orexin.

organ of Corti An auditory receptor organ that contains hair cells, rods of Corti, and supporting cells.

organelle A membrane-enclosed structure inside a cell; examples are the nucleus, mitochondrion, endoplasmic reticulum, and Golgi apparatus.

organizational effect The ability of a hormone to influence the prenatal development of sex organs and the brain.

orientation column A column of visual cortical neurons stretching from layer II to layer VI that responds best to the same stimulus orientation.

orientation selectivity The property of a cell in the visual system that responds to a limited range of stimulus orientations.

osmometric thirst The motivation to drink water as a result of an increase in blood tonicity.

ossicle One of three small bones in the middle ear.

otolith organ The utricle or the saccule, organs of the vestibular labyrinth in the inner ear that transduce head tilt and acceleration.

outer ear The pinna plus the auditory canal.

outer hair cell An auditory receptor cell located farther from the modiolus than the rods of Corti.

outer nuclear layer The layer of the retina of the eye containing the cell bodies of the photoreceptors.

outer plexiform layer The layer of the retina of the eye between the inner nuclear layer and the outer nuclear layer; contains the neurites and synapses between photoreceptors, horizontal cells, and bipolar cells.

oval window A hole in the bony cochlea at which movement of the ossicles is transferred to movement of the fluids in the cochlea.

overshoot The part of an action potential when the membrane potential is more positive than 0 mV.

oxytocin A small peptide hormone released from the posterior pituitary by magnocellular neurosecretory cells; stimulates uterine contractions and the let-down of milk from mammary glands.

Pacinian corpuscle A mechanoreceptor of the deep skin, selective for high-frequency vibrations.

PAG *See* periaqueductal gray matter.

Papez circuit A circuit of anatomical structures interconnecting the hypothalamus and cortex, which Papez proposed to be an emotion system.

papilla A small protuberance on the surface of the tongue that contains taste buds.

parahippocampal cortex A cortical region in the medial temporal lobe that lies lateral to the rhinal sulcus.

parallel fiber The axon of a cerebellar granule cell that innervates Purkinje cells.

parallel processing The idea that different stimulus attributes are processed by the brain in parallel, using distinct pathways.

parasympathetic division A division of the autonomic nervous system that maintains heart rate and respiratory, metabolic, and digestive functions under normal conditions; its peripheral axons emerge from the brain stem and sacral spinal cord. *See also* sympathetic division.

paraventricular nucleus A region of the hypothalamus involved in the regulation of the autonomic nervous system and in controlling the secretion of thyroid-stimulating hormone and adrenocorticotropic hormone from the anterior pituitary.

parietal lobe The region of the cerebrum lying under the parietal bone.

Parkinson's disease A movement disorder caused by damage to the substantia nigra, characterized by paucity of movement, difficulty in initiating willed movement, and resting tremor.

partial seizure Pathologically large and synchronous neural activity that remains localized to a relatively small region of the brain.

parvocellular LGN layer A layer of the lateral geniculate nucleus receiving synaptic input from the P-type retinal ganglion cells.

parvocellular neurosecretory cell A small neuron of the medial and periventricular hypothalamus that secretes hypophysiotropic peptide hormones into the hypothalamo-pituitary portal circulation to stimulate or inhibit the release of hormones from the anterior pituitary.

parvocellular-interblob channel (**P-IB channel**) The visual information-processing channel that begins with the P-type retinal ganglion cells and leads to the interblob regions of striate cortical layer III; believed to process information about object shape.

patch clamp A method that enables an investigator to hold constant the membrane potential of a patch of membrane while current through a small number of membrane channels is measured.

PDE *See* phosphodiesterase.

peptide bond The covalent bond between the amino group of one amino acid and the carboxyl group of another.

peptidergic Describing neurons or synapses that produce and release peptide neurotransmitters.

perforant path The axonal pathway from the entorhinal cortex to the dentate gyrus of the hippocampus.

periaqueductal gray matter (PAG) A region surrounding the cerebral aqueduct in the core of the midbrain, with descending pathways that can inhibit the transmission of pain-causing signals.

perikaryon The central region of the neuron containing the nucleus; also called soma or cell body.

perilymph The fluid that fills the scala vestibuli and scala tympani in the cochlea in the inner ear, containing low K^+ and high Na^+ concentrations.

peripheral nervous system (**PNS)** The parts of the nervous system other than the brain and spinal cord. The PNS includes all the spinal ganglia and nerves, cranial nerves III–XII, and the autonomic nervous system.

perirhinal cortex A cortical region in the medial temporal lobe that occupies the lateral bank of the rhinal sulcus. Lesions to this area in humans produce profound anterograde amnesia.

periventricular zone A hypothalamic region that lies most medially, bordering the third ventricle.

phase locking The consistent firing of an auditory neuron at the same phase of a sound wave.

pheromone An olfactory stimulus used for chemical communication between individuals.

phosphodiesterase (**PDE**) An enzyme that breaks down the cyclic nucleotide second messengers cyclic adenosine monophosphate and cyclic guanosine monophosphate.

phospholipase C (**PLC**) An enzyme that cleaves the membrane phospholipid phosphatidylinositol-4,5-bisphosphate to form the second messengers diacylglycerol (DAG) and inositol triphosphate (IP_3).

phospholipid bilayer The arrangement of phospholipid molecules that forms the basic structure of the cell membrane. The core of the bilayer is lipid, creating a barrier to water and to water-soluble ions and molecules.

phosphorylation A biochemical reaction in which a phosphate group (PO_4^{2-}) is transferred from adenosine triphosphate to another molecule. Phosphorylation of proteins by protein kinases changes their biological activity.

photoreceptor A specialized cell in the retina that transduces light energy into changes in membrane potential.

pia mater The innermost of the three meninges that cover the surface of the central nervous system.

P-IB channel *See* parvocellular-interblob channel.

pinna The funnel-shaped outer ear consisting of cartilage covered by skin.

place cell A neuron in the rat hippocampus that responds only when the animal is in a certain region of space.

planum temporale An area on the superior surface of the temporal lobe that is frequently larger in the left than the right hemisphere.

PKA *See* protein kinase A.

PKC *See* protein kinase C.

PLC *See* phospholipase C.

pleasure center A name originally given to any self-reinforcing site in the brain, regardless of whether the electrical stimulation evoked pleasure.

PNS *See* peripheral nervous system.

polyandry Mating behavior in which one female mates with more than one male.

polygyny Mating behavior in which one male mates with more than one female.

polypeptide A string of amino acids held together by peptide bonds.

polyribosome A collection of several ribosomes, floating freely in the cytoplasm.

pons The part of the rostral hindbrain that lies ventral to the cerebellum and the fourth ventricle.

pontine nuclei The clusters of neurons that relay information from the cerebral cortex to the cerebellar cortex.

pontine reticulospinal tract A tract originating in the pontine reticular formation and terminating in the spinal cord, involved in the control of movement.

population coding The representation of sensory, motor, or cognitive information by activity distributed over a large number of neurons. An example is color, which is encoded by the relative activity of the types of retinal cones.

posterior A direction, toward the tail, or caudal.

posterior parietal cortex The posterior region of the parietal lobe, mainly Brodmann's areas 5 and 7, involved in visual and somatosensory integration and attention.

postganglionic neuron A peripheral neuron of the sympathetic and parasympathetic divisions of the autonomic nervous system; its cell body lies in autonomic ganglia, and its axons terminate on peripheral organs and tissues.

postsynaptic density A postsynaptic membrane differentiation that is the site of neurotransmitter receptors.

predatory aggression Attack behavior, often with the goal of obtaining food, accompanied by few vocalizations and low activity of the autonomic nervous system.

prefrontal cortex The cortical area at the rostral end of the frontal lobe that receives input from the dorsomedial nucleus of the thalamus.

preganglionic neuron A neuron of the sympathetic and parasympathetic divisions of the autonomic nervous system; its cell body lies in the central nervous system (spinal cord or brain stem), and its axons extend peripherally to synapse on postganglionic neurons in the autonomic ganglia.

premotor area (PMA) The lateral part of cortical area 6, involved in the control of voluntary movement.

primary auditory cortex Brodmann's area 41, on the superior surface of the temporal lobe; also called A1.

primary gustatory cortex The area of neocortex that receives taste information from the ventroposterior medial nucleus.

primary motor cortex Brodmann's area 4, located on the precentral gyrus; the region of cortex that, when weakly stimulated, elicits localized muscle contractions; also called M1.

primary sensory neuron A neuron specialized to detect environmental signals at the body's sensory surfaces.

primary somatosensory cortex Brodmann's areas 3a, 3b, 1, and 2, located in the postcentral gyrus; also called S1.

procedural memory Memory for skills and behavior.

proprioception The sensation of body position and movement using sensory signals from muscles, joints, and skin.

proprioceptor A sensory receptor from the muscles, joints, and skin that contributes to proprioception.

protein kinase A class of enzyme that phosphorylates proteins, a reaction that changes the conformation of the protein and its biological activity.

protein kinase A (PKA) A protein kinase activated by the second messenger cyclic adenosine monophosphate.

protein kinase C (PKC) A protein kinase activated by the second messenger diacylglycerol.

protein phosphatase An enzyme that removes phosphate groups from proteins.

protein synthesis The assembly of protein molecules in the cell's cytoplasm according to genetic instructions.

proximal (girdle) muscle A muscle that controls the shoulder or pelvis.

P-type ganglion cell A type of ganglion cell in the retina characterized by a small cell body and dendritic arbor, a sustained response to light, and sensitivity to different wavelengths of light; also called P cell.

pupil The opening that allows light to enter the eye and strike the retina.

pupillary light reflex The reflex involving retinal inputs to brain stem neurons that control the iris causes the diameter of the pupil to become larger in dim light and smaller in bright light.

Purkinje cell A cell in the cerebellar cortex that projects an axon to the deep cerebellar nuclei.

putamen A part of the basal ganglia in the basal forebrain; involved in motor control.

pyramidal cell A neuron characterized by a pyramid-shaped cell body and elongated dendritic tree; found in the cerebral cortex.

pyramidal tract A tract running along the ventral medulla that carries corticospinal axons.

quantal analysis A method of determining how many vesicles release neurotransmitter during normal synaptic transmission.

radial glial cell A glial cell in the embryonic brain extending a process from the ventricular zone to the surface of the brain, along which process immature neurons and glia migrate.

raphe nuclei Clusters of serotonergic neurons that lie along the midline of the brain stem from the midbrain to the medulla and project diffusely upon all levels of the central nervous system.

rapid eye movement sleep (REM sleep) A stage of sleep characterized by low-amplitude, high-frequency electroencephalogram waves, vivid dreams, rapid eye movements, and atonia.

rate-limiting step In the series of biochemical reactions that leads to the production of a chemical, the one step that limits the rate of synthesis.

receptive field The region of a sensory surface (retina, skin) that, when stimulated, changes the membrane potential of a neuron.

receptor (1) A specialized protein that detects chemical signals, such as neurotransmitters, and initiates a cellular response. (2) A specialized cell that detects environmental stimuli and generates neural responses.

receptor agonist A drug that binds to a receptor and activates it.

receptor antagonist A drug that binds to a receptor and inhibits its function.

receptor potential A stimulus-induced change in the membrane potential of a sensory receptor.

receptor subtype One of several receptors to which a neurotransmitter binds.

reciprocal inhibition The process whereby the contraction of one set of muscles is accompanied by the relaxation of antagonist muscles.

recognition memory Memory required to perform a delayed non-match to sample task.

red nucleus A cell group in the midbrain involved in the control of movement.

referred pain Pain that is perceived as coming from a site other than its true origin. Nociceptor activation within visceral organs is typically perceived as pain originating in skin or skeletal muscle.

refraction The bending of light rays that can occur when they travel from one transparent medium to another.

Reissner's membrane The cochlear membrane in the inner ear that separates the scala vestibuli from the scala media.

relational memory A type of memory in which all of the events occurring at a given time are stored in a manner linking them.

relative refractory period The period of time following an action potential during which more depolarizing current than usual is required to achieve threshold.

REM sleep *See* rapid eye movement sleep.

resistance *See* electrical resistance.

resting membrane potential The membrane potential, or membrane voltage, maintained by a cell when it is not generating action potentials; also called resting potential. Neurons have a resting membrane potential of about −65 mV.

reticular formation A region of the brain stem ventral to the cerebral aqueduct and fourth ventricle; involved in many functions, including the control of posture and locomotion.

reticular lamina A thin sheet of tissue that holds the tops of hair cells in the organ of Corti.

retina A thin layer of cells at the back of the eye that transduces light energy into neural activity.

retinofugal projection A pathway that carries information from the retina to the visual cortex.

retinotectal projection A collection of axons coursing from the retina to the superior colliculus.

retinotopy The topographic organization of visual pathways in which neighboring cells on the retina feed information to neighboring cells in a target structure.

retrograde amnesia Memory loss for events prior to an illness or brain trauma.

retrograde messenger Any chemical messenger that communicates information from the postsynaptic side of a synapse to the presynaptic side.

retrograde transport Axoplasmic transport from an axon terminal to the soma.

rhodopsin The photopigment in rod photoreceptors.

ribosome A cellular organelle that assembles new proteins from amino acids according to the instructions carried by messenger RNA.

rising phase The first part of an action potential; characterized by a rapid depolarization of the membrane.

rod photoreceptor A photoreceptor in the retina containing rhodopsin and specialized for low light levels. *See also* cone photoreceptor.

rostral A direction, toward the nose, or anterior.

rough endoplasmic reticulum (rough ER) A membrane-enclosed cellular organelle with ribosomes attached to its outer surface; a site of synthesis for proteins destined to be inserted into membrane or to be enclosed by membrane.

round window A membrane-covered hole in the bony cochlea of the inner ear that is continuous with the scala tympani in the cochlea.

rubrospinal tract A tract originating in the red nucleus and terminating in the spinal cord; involved in the control of movement.

S1 *See* primary somatosensory cortex.

sagittal plane An anatomical plane of section that is parallel to the midsagittal plane.

saltatory conduction Propagation of an action potential down a myelinated axon.

sarcolemma The outer cell membrane of a muscle fiber.

sarcomere The contractile element between Z lines in a myofibril; contains the thick and thin filaments that slide along one another to cause muscle contraction.

sarcoplasmic reticulum An organelle within a muscle fiber that stores Ca^{2+} and releases it when stimulated by an action potential in T tubules.

satiety signal A factor that reduces the drive to eat without causing sickness; examples are gastric distension and cholecystokinin released by the intestinal cells in response to food.

scala media A chamber in the cochlea that lies between the scala vestibuli and the scala tympani.

scala tympani A chamber in the cochlea that runs from the helicotrema to the round window.

scala vestibuli A chamber in the cochlea that runs from the oval window to the helicotrema.

Schaffer collateral An axon of a CA3 neuron that innervates neurons in CA1 of the hippocampus.

schizophrenia A mental disorder characterized by a loss of contact with reality; fragmentation and disruption of thought, perception, mood, and movement; delusions, hallucinations, and disordered memory.

Schwann cell A glial cell that provides myelin in the peripheral nervous system.

sclera The tough outer wall of the eyeball; the white of the eye.

SCN *See* suprachiasmatic nucleus.

second messenger A short-lived chemical signal in the cytosol that can trigger a biochemical response. Second messenger formation is usually stimulated by a first messenger (a neurotransmitter or hormone) acting at a G-protein-coupled cell surface receptor. Examples of second messengers are cyclic adenosine monophosphate, cyclic guanosine monophosphate, and inositol-1,4,5-triphosphate.

second messenger cascade A multistep process that couples activation of a neurotransmitter receptor to activation of intracellular enzymes.

secretory granule A spherical membrane-enclosed vesicle about 100 nm in diameter containing peptides intended for secretion by exocytosis; also called large, dense-core vesicle.

semicircular canal A component of the vestibular labyrinth in the inner ear that transduces head rotation.

sensitization A type of nonassociative learning leading to an intensified response to all stimuli.

sensory map A representation of sensory information within a neural structure that preserves the spatial organization of that information established on the sensory organ. Examples are retinotopic maps in the superior colliculus, lateral geniculate nucleus, and visual cortex, where neurons in specific places respond selectively to stimulation of specific parts of the retina.

serotonergic Describing neurons or synapses that produce and release serotonin.

serotonin (5-HT) An amine neurotransmitter, 5-hydroxytryptamine.

serotonin-selective reuptake inhibitor (SSRI) A drug, such as fluoxetine (Prozac), that prolongs the actions of synaptically released serotonin by preventing reuptake; used to treat depression and obsessive-compulsive disorder.

sex-determining region of the Y chromosome (SRY) A gene on the Y chromosome responsible for the production of testis-determining factor; essential for normal for male development.

sexual dimorphism A sex-related difference in structure or behavior.

sham rage Behavior produced by brain lesions; a display of great anger in a situation that would not normally cause anger.

short-term memory Retention of information about recent events or facts that is not yet consolidated into long-term memory.

shunting inhibition A form of synaptic inhibition in which the main effect is to reduce membrane resistance, thereby shunting depolarizing current generated at excitatory synapses.

simple cell A cell found in primary visual cortex, having an elongated orientation-selective receptive field with distinct ON and OFF subregions.

skeletal muscle The type of muscle that is under voluntary control; derived from the mesodermal somites.

slow motor unit A motor unit with a small alpha motor neuron innervating slowly contracting and slowly fatiguing red muscle fibers.

SMA *See* supplementary motor area.

smooth endoplasmic reticulum (smooth ER) A membrane-enclosed cellular organelle that is heterogeneous and performs different functions in different locations.

smooth muscle A type of muscle in the digestive tract, arteries, and related structures; innervated by the autonomic nervous system and not under voluntary control.

sodium-potassium pump An ion pump that removes intracellular Na^+ and concentrates intracellular K^+, using adenosine triphosphate as its energy source.

soma The central region of the neuron containing the nucleus; also called the cell body or perikaryon.

somatic motor system The skeletal muscles and the parts of the nervous system that control them.

somatic PNS The part of the peripheral nervous system that innervates the skin, joints, and skeletal muscles.

somatic sensation The senses of touch, temperature, body position, and pain.

somatotopy The topographic organization of somatic sensory pathways in which neighboring receptors in the skin feed information to neighboring cells in a target structure.

spatial summation The combining of excitatory postsynaptic potentials generated at more than one synapse on the same cell. *See also* temporal summation.

spike-initiation zone A region of the neuronal membrane where action potentials are normally initiated; characterized by a high density of voltage-gated sodium channels.

spinal canal The cerebrospinal fluid-filled space within the spinal cord.

spinal cord The part of the central nervous system in the vertebral column.

spinal nerve A nerve attached to the spinal cord that innervates the body.

spinal segment One set of dorsal and ventral roots plus the portion of spinal cord related to them.

spinothalamic pathway An ascending somatic sensory pathway traveling from the spinal cord to the thalamus via the lateral spinothalamic columns that mediates information about pain, temperature, and some forms of touch.

spiny neuron A neuron with dendritic spines.

spiral ganglion A collection of cells in the modiolus of the cochlea that receives input from hair cells and sends output to the cochlear nuclei in the medulla via the auditory nerve.

split-brain study An examination of behavior in animals or humans that have had the cerebral hemispheres disconnected by cutting the corpus callosum.

SSRI *See* serotonin-selective reuptake inhibitor.

stapes An ossicle in the middle ear attached to the oval window that somewhat resembles a stirrup.

stellate cell A neuron characterized by a radial, starlike distribution of dendrites.

stereocilium A hairlike cilium attached to the top of a hair cell in the inner ear.

strabismus A condition in which the eyes are not perfectly aligned.

striate cortex Primary visual cortex, Brodmann's area 17; also called V1.

striated muscle A type of muscle with a striated, or striped, appearance; two categories are skeletal and cardiac.

striatum A collective term for the caudate nucleus and putamen; involved in the initiation of willed movements of the body; plays a role in procedural memory.

subplate A layer of cortical neurons lying below the cortical plate early in development; when the cortical plate has differentiated into the six layers of the neocortex, the subplate disappears.

substantia A group of related neurons deep within the brain, usually with less distinct borders than those of nuclei.

substantia gelatinosa A thin dorsal part of the dorsal horn of the spinal cord that receives input from unmyelinated C fibers; important in the transmission of nociceptive signals.

substantia nigra A cell group in the midbrain that uses dopamine as a neurotransmitter and innervates the striatum.

subthalamic nucleus A part of the basal ganglia in the basal forebrain; involved in motor control.

sulcus A groove in the surface of the cerebrum running between neighboring gyri (plural: sulci).

superior colliculus A structure in the tectum of the midbrain that receives direct retinal input and directs saccadic eye movements.

superior olive A nucleus in the caudate pons that receives afferents from the cochlear nuclei and sends efferents to the inferior colliculus; also called superior olivary nucleus.

supplementary motor area (SMA) The medial part of cortical area 6; involved in the control of voluntary movement.

suprachiasmatic nucleus (SCN) A small nucleus of the hypothalamus just above the optic chiasm that receives retinal innervation and synchronizes circadian rhythms with the daily light-dark cycle.

sympathetic chain A series of interconnected sympathetic ganglia of the autonomic nervous system, adjacent to the vertebral column, that receive input from preganglionic sympathetic fibers and project postganglionic fibers to target organs and tissues.

sympathetic division A division of the autonomic nervous system that in fight-or-flight situations activates a variety of physiological responses, including increased heart rate, respiration, blood pressure, and energy mobilization and decreased digestive and reproductive functions; its peripheral axons emerge from the thoracic and lumbar spinal cord. *See also* parasympathetic division.

synapse The region of contact where a neuron transfers information to another cell.

synaptic cleft The region separating the presynaptic and postsynaptic membranes.

synaptic transmission The process of transferring information from one cell to another at a synapse.

synaptic vesicle A membrane-enclosed structure, about 50 nm in diameter, containing neurotransmitter and found at a site of synaptic contact.

synergist muscle A muscle that contracts with other muscles to produce movement in one direction.

taste bud A cluster of cells, including taste receptor cells, in papillae of the tongue.

taste receptor cell A modified epithelial cell that transduces taste stimuli.

TDF *See* testis-determining factor.

tectorial membrane A sheet of tissue that hangs over the organ of Corti in the cochlea.

tectospinal tract A tract originating in the superior colliculus and terminating in the spinal cord; involved in the control of head and neck movement.

tectum The part of the midbrain lying dorsal to the cerebral aqueduct.

tegmentum The part of the midbrain lying ventral to the cerebral aqueduct.

telencephalon A region of the brain derived from the prosencephalon. Telencephalic structures include the paired cerebral hemispheres that contain cerebral cortex and the basal telencephalon.

temporal coding The representation of information by the timing of action potentials rather than by their average rate.

temporal lobe The region of the cerebrum lying under the temporal bone.

temporal summation The combining of excitatory postsynaptic potentials generated in rapid succession at the same synapse. *See also* spatial summation.

terminal arbor Branches at the end of an axon terminating in the same region of the nervous system.

terminal bouton The end region of an axon, usually a site of synaptic contact with another cell; also called axon terminal.

testis-determining factor (TDF) A protein of critical importance for the development of the testes in the fetus.

tetanus A type of repetitive stimulation.

tetrodotoxin (TTX) A toxin that blocks Na^+ permeation through voltage-gated sodium channels, thereby blocking action potentials.

thalamus The dorsal part of the diencephalon, highly interconnected with the cerebral neocortex.

thermoreceptor A sensory receptor selective for temperature changes.

thick filament A part of the cytoskeleton of a muscle cell containing myosin, lying between and among thin filaments and sliding along them to cause muscle contraction.

thin filament A part of the cytoskeleton of a muscle cell containing actin, anchored to Z lines and sliding along thick filaments to cause muscle contraction.

third ventricle The cerebrospinal fluid-filled space within the diencephalon.

threshold A level of depolarization sufficient to trigger an action potential.

tonotopy The systematic organization within an auditory structure on the basis of characteristic frequency.

tract A collection of central nervous system axons having a common site of origin and a common destination.

transcription The process of synthesizing a messenger RNA molecule according to genetic instructions encoded in DNA.

transducin The G-protein that couples rhodopsin to the enzyme phosphodiesterase in rod photoreceptors.

transduction The transformation of sensory stimulus energy into a cellular signal, such as a receptor potential.

translation The process of synthesizing a protein molecule according to genetic instructions carried by a messenger RNA molecule.

transmitter-gated ion channel A membrane protein forming a pore that is permeable to ions and gated by neurotransmitter.

transporter A membrane protein that transports neurotransmitters, or their precursors, across membranes to concentrate them in either presynaptic cytosol or synaptic vesicles.

trigeminal nerve Cranial nerve V; attaches to the pons and carries primarily sensory axons from the head, mouth, and dura mater and motor axons of mastication.

trophic factor Any molecule that promotes cell survival.

troponin A protein that binds Ca^{2+} in a muscle cell and thereby regulates the interaction of myosin and actin.

T tubule A membrane-enclosed tunnel running within a skeletal muscle fiber that links excitation of the sarcolemma with the release of Ca^{2+} from the sarcoplasmic reticulum.

TTX *See* tetrodotoxin.

tympanic membrane A membrane at the internal end of the auditory canal that moves in response to variations in air pressure; also called eardrum.

ultradian rhythm Any rhythm with a period significantly less than one day.

undershoot The part of an action potential when the membrane potential is more negative than at rest; also called after-hyperpolarization.

unipolar neuron A neuron with a single neurite.

V1 Primary visual cortex, or striate cortex.

vagus nerve Cranial nerve X, arising from the medulla and innervating the viscera of the thoracic and abdominal cavities; a major source of preganglionic parasympathetic visceromotor axons.

vascular organ of the lamina terminalis (OVLT) A specialized region of the hypothalamus containing neurons that are sensitive to the tonicity of the blood; these neurons activate the magnocellular neurosecretory cells that release vasopressin into the blood and trigger osmometric thirst.

vasopressin A small peptide hormone released from the posterior pituitary by magnocellular neurosecretory cells; promotes water retention and decreased urine production by the kidney; also called antidiuretic hormone.

ventral A direction, toward the belly.

ventral cochlear nucleus A nucleus in the medulla that receives afferents from the spiral ganglion in the cochlea.

ventral horn The ventral region of the spinal cord containing neuronal cell bodies.

ventral lateral (VL) nucleus A nucleus of the thalamus that relays information from the basal ganglia and cerebellum to the motor cortex.

ventral posterior (VP) nucleus The main thalamic relay nucleus of the somatic sensory system.

ventral posterior medial (VPM) nucleus The part of the ventral posterior nucleus of the thalamus that receives somatosensory input from the face, including afferents from the tongue.

ventral root A bundle of motor neuron axons that emerges from the ventral spinal cord and joins sensory fibers to form a spinal nerve.

ventricular system The cerebrospinal fluid-filled spaces inside the brain, consisting of the lateral ventricles, the third ventricle, the cerebral aqueduct, and the fourth ventricle.

ventromedial hypothalamic syndrome Obesity associated with lesions of the lateral hypothalamic area.

ventromedial pathway Axons in the ventromedial column of the spinal cord that are involved in the control of posture and locomotion and are under brain stem control.

vermis The midline region of the cerebellum.

vestibular labyrinth A part of the inner ear specialized for the detection of head motion; consists of the otolith organs and semicircular canals.

vestibular nucleus A nucleus in the medulla that receives input from the vestibular labyrinth of the inner ear.

vestibulo-ocular reflex (VOR) A reflexive movement of the eyes stimulated by rotational movements of the head; stabilizes the visual image on the retinas.

vestibulospinal tract A tract originating in the vestibular nuclei of the medulla and terminating in the spinal cord; involved in the control of movement and posture.

visceral PNS The part of the peripheral nervous system that innervates the internal organs, blood vessels, and glands; also called autonomic nervous system.

vision The sense of sight.

visual acuity The ability of the visual system to distinguish between two nearby points.

visual angle A way to describe distance across the retina; an object that subtends an angle of 3.5 degrees will form an image on the retina that is 1 mm across.

visual field The total region of space that is viewed by both eyes when the eyes are fixated on a point.

visual hemifield The half of the visual field to one side of the fixation point.

vitreous humor The jellylike substance filling the eye between the lens and the retina.

VL nucleus *See* ventral lateral nucleus.

volley principle The idea that high sound frequencies are represented in the pooled activity of a number of neurons, each of which fires in a phase-locked manner.

voltage The force exerted on an electrically charged particle, represented by the symbol V and measured in volts; also called electrical potential or potential difference.

voltage clamp A device that enables an investigator to hold the membrane potential constant while transmembrane currents are measured.

voltage-gated calcium channel A membrane protein forming a pore that is permeable to Ca^{2+} ions and gated by depolarization of the membrane.

voltage-gated potassium channel A membrane protein forming a pore that is permeable to K^+ ions and gated by depolarization of the membrane.

voltage-gated sodium channel A membrane protein forming a pore that is permeable to Na^+ ions and gated by depolarization of the membrane.

volumetric thirst The motivation to drink water as a result of a decrease in blood volume.

VOR *See* vestibulo-ocular reflex.

VP nucleus *See* ventral posterior nucleus.

VPM nucleus *See* ventral posterior medial nucleus.

Wada procedure A procedure in which one cerebral hemisphere is anesthetized to enable testing of the function of the other hemisphere.

Wernicke-Geschwind model A model for language processing involving interactions between Broca's area and Wernicke's area with sensory and motor areas.

Wernicke's aphasia A language disturbance in which speech is fluent but comprehension is poor.

Wernicke's area An area on the superior surface of the

temporal lobe between auditory cortex and the angular gyrus; associated with Wernicke's aphasia when damaged.

white matter A generic term for a collection of central nervous system axons. When a freshly dissected brain is cut open, axons appear white.

Wolffian duct A structure in the embryonic gonads that develops into the male internal reproductive system.

working memory Information storage that is temporary, limited in capacity, and requires continual rehearsal.

Young-Helmholtz trichromacy theory The theory that the brain assigns colors based on a comparison of the readout of the three types of cone photoreceptors.

Z line A band delineating sarcomeres in a myofibril of a muscle fiber.

zeitgeber Any environmental cue, such as the light-dark cycle, that signals the passage of time.

REFERENCES AND SUGGESTED READINGS

Chapter 1

Alt KW, et al. 1997. Evidence for stone age cranial surgery. *Nature* 387:360.

Clarke E, Dewhurst K. 1972. *An Illustrated History of Brain Function*. Los Angeles: University of California Press.

Clarke E, O'Malley C. 1968. *The Human Brain and Spinal Cord*, 2nd ed. Los Angeles: University of California Press.

Corsi P, ed. 1991. *The Enchanted Loom*. New York: Oxford University Press.

Crick F. 1994. *The Astonishing Hypothesis: The Scientific Search for the Soul*. New York: Macmillan.

Finger S. 1994. *Origins of Neuroscience*. New York: Oxford University Press.

National Academy of Sciences Institute of Medicine. 1991. Science, medicine, and animals. Washington, DC: National Academy Press.

US Office of Science and Technology Policy, Subcommittee on Brain and Behavioral Sciences. 1991. Report to Congress: Maximizing human potential.

Worden F, Swazey J, Adelman G, eds. 1975. *The Neurosciences: Paths of Discovery*. Cambridge, MA: MIT Press.

Chapter 2

Alberts B, Bray D, Lewis J, Raff M, Roberts K, Watson JD. 1994. *Molecular Biology of the Cell*, 3rd ed. New York: Garland.

Bick, K, Amaducci, L, Pepeu, G. 1987. *The Early Story of Alzheimer's Disease*. New York: Raven.

DeFelipe J, Jones EG. 1998. *Cajal on the Cerebral Cortex*. New York: Oxford University Press.

Finger S. 1994. *Origins of Neuroscience*. New York: Oxford University Press.

Goedert. 1996. Molecular dissection of the neurofibrillary lesions of Alzheimer's disease. *Cold Spring Harbor Symposia on Quantitative Biology*, Vol. LXI. Cold Spring Harbor, NY: Cold Spring Harbor Laboratory Press.

Grafstein B, Forman DS. 1980. Intracellular transport in neurons. *Physiological Reviews* 60:1167–1283.

Hall Z. 1992. *An Introduction to Molecular Neurobiology*. Sunderland, MA: Sinauer.

Hammersen F. 1980. *Histology*. Baltimore: Urban & Schwarzenberg.

Harris KM, Stevens JK. 1989. Dendritic spines of CA1 pyramidal cells in the rat hippocampus: serial electron microscopy with reference to their biophysical characteristics. *Journal of Neuroscience* 9:2982–2997.

Hubel DH. 1988. *Eye, Brain and Vision*. New York: Scientific American Library.

Levitan I, Kaczmarek L. 1997. *The Neuron: Cell and Molecular Biology*, 2nd ed. New York: Oxford University Press.

Peters A, Palay SL, Webster H deF. 1991. *The Fine Structure of the Nervous System*, 3rd ed. New York: Oxford University Press.

Purpura D. 1974. Dendritic spine "dysgenesis" and mental retardation. *Science* 20:1126–1128.

Robertson JD. 1957. New observations on the ultrastructure of the membranes of frog peripheral nerve fibers. *Journal of Biophysical and Biochemical Cytology* 3:1043–1048.

Widnell CC, Pfenniger KH, eds. 1990. *Essential Cell Biology*. Baltimore: Williams & Wilkins.

Chapter 3

Doyle DA, Cabral JM, Pfuetzner RA, Kuo A, Gulbis JM, Cohen SL, Chait BT, MacKinnon R. 1998. The structure of the potassium channel: molecular basis of K^+ conduction and selectivity. *Science* 280:69–77.

Hille B. 1992. *Ionic Channels of Excitable Membranes*, 2nd ed. Sunderland, MA: Sinauer.

Jan L, Jan YN. 1997. Cloned potassium channels from eukaryotes and prokaryotes. *Annual Review of Neuroscience* 20:91–123.

Levitan I, Kaczmarek L. 1997. *The Neuron: Cell and Molecular Biology*, 2nd ed. New York: Oxford University Press.

Li M, Unwin N, Staufer KA, Jan YN, Jan L. 1994. Images of purified *Shaker* potassium channels. *Current Biology* 4:110–115.

MacKinnon R. 1995. Pore loops: an emerging theme in ion channel structure. *Neuron* 14:889–892.

Miller C. 1988. *Shaker* shakes out potassium channels. *Trends in Neurosciences* 11:185–186.

Nicholls J, Martin A, Wallace B. 1993. *From Neuron to Brain*, 3rd ed. Sunderland, MA: Sinauer.

Ransom BR, Goldring S. 1973. Slow depolarization in cells presumed to be glia in cerebral cortex of cat. *Journal of Neurophysiology* 36:869–878.

Sanguinetti MC, Spector PS. 1997. Potassium channelopathies. *Neuropharmacology* 36: 755–762.

Shepard G. 1994. *Neurobiology*, 3rd ed. New York: Oxford University Press.

Stoffel M, Jan LY. 1998. Epilepsy genes: excitement traced to potassium channels. *Nature Genetics* 18:6–8.

Chapter 4

Agmon A, Connors BW. 1992. Correlation between intrinsic firing patterns and thalamocortical synaptic responses of neurons in mouse barrel cortex. *Journal of Neuroscience* 12:319–329.

Armstrong CM, Hille B. 1998. Voltage-gated ion channels and electrical excitability. *Neuron* 20:371–380.

Connors B, Gutnick M. 1990. Intrinsic firing patterns of diverse neocortical neurons. *Trends in Neurosciences* 13:99–104.

Hardman JG, Goodman Gilman A, Limbird LE, ed. 1996. *Goodman and Gilman's The Pharmacological Basis of Therapeutics*, 9th ed. New York: McGraw-Hill.

Hille B. 1992. *Ionic Channels of Excitable Membranes*, 2nd ed. Sunderland, MA: Sinauer.

Hodgkin A. 1976. Chance and design in electrophysiology: an informal account of certain experiments on nerves carried out between 1942 and 1952. *Journal of Physiology* (London) 263:1–21.

Huguenard J, McCormick D. 1994. *Electrophysiology of the Neuron*. New York: Oxford University Press.

Levitan I, Kaczmarek L. 1997. *The Neuron: Cell and Molecular Biology*, 2nd ed. New York: Oxford University Press.

Llinás R. 1988. The intrinsic electrophysiological properties of mammalian neurons: insights into central nervous system function. *Science* 242:1654–1664.

Neher E. 1992. Nobel lecture: Ion channels or communication between and within cells. *Neuron* 8:605–612.

Neher E, Sakmann B. 1992. The patch clamp technique. *Scientific American* 266:28–35.

Nicholls J, Martin A, Wallace B. 1993. *From Neuron to Brain*, 3rd ed. Sunderland, MA: Sinauer.

Noda M, Ikeda T, Suzuki H, et al. 1994. Primary structure of *Electrophorus electricus* sodium channel deduced from cDNA sequence. *Nature* 312:121–127.

Shepherd G. 1994. *Neurobiology*, 3rd ed. New York: Oxford University Press.

Sigworth FJ, Neher E. 1980. Single Na^+ channel currents observed in cultured rat muscle cells. *Nature* 287:447–449.

Unwin N. 1989. The structure of ion channels in membranes of excitable cells. *Neuron* 3:665–676.

Chapter 5

Bloedel JR, Gage PW, Llinás R, Quastel DM. 1966. Transmitter release at the squid giant synapse in the presence of tetrodotoxin. *Nature* 212:49–50.

Capogna M, Gahwiler BH, Thompson SM. 1966. Calcium-independent actions of α-latrotoxin on spontaneous and evoked synaptic transmission in the hippocampus. *Journal of Neurophysiology* 76:3149–3158.

Colquhoun D, Sakmann B. 1998. From muscle endplate to brain synapses: a short history of synapses and agonist-activated ion channels. *Neuron* 20:381–387.

Cooper JR, Bloom FE, Roth RH. 1996. *The Biochemical Basis of Neuropharmacology*, 7th ed. New York: Oxford University Press.

Fatt P, Katz B. 1951. An analysis of the endlate potential recorded with an intracellular electrode. *Journal of Physiology* (London) 115:320–370.

Ferro-Novick S, Jahn R. 1994. Vesicle fusion from yeast to man. *Nature* 370:191–193.

Furshpan E, Potter D. 1959. Transmission at the giant motor synapses of the crayfish. *Journal of Physiology* (London) 145:289–325.

Hanson PI, Heuser JE, Jahn R. 1997. Neurotransmitter release—four years of SNARE complexes. *Current Opinion in Neurobiology* 7:310–315.

Heuser J, Reese T. 1973. Evidence for recycling of synaptic vesicle membrane during transmitter release at the frog neuromuscular junction. *Journal of Cell Biology* 57:315–344.

Heuser J, Reese T. 1977. Structure of the synapse. In: Brookhart JM, Mountcastle VB, eds. *Handbook of Physiology—Section 1. The Nervous System, Vol. I. Cellular Biology of Neurons*. Bethesda, MD: American Physiological Society, pp. 261–294.

Kelly R. 1993. Storage and release of neurotransmitters. *Cell* 72:43–53.

Levitan I, Kaczmarek L. 1997. *The Neuron: Cell and Molecular Biology*, 2nd ed. New York: Oxford University Press.

Llinás R, Sugimori M, Silver RB. 1992. Microdomains of high calcium concentration in a presynaptic terminal. *Science* 256: 677–679.

Loewi O. 1953. *From the Workshop of Discoveries*. Lawrence: University of Kansas Press.

Matthews G. 1996. Neurotransmitter release. *Annual Review of Neuroscience* 19:219–233.

Matthews R. 1995. *Nightmares of Nature*. London: HarperCollins.

Neher E. 1998. Vesicle pools and Ca^{2+} microdomains: new tools for understanding their roles in neurotransmitter release. *Neuron* 20:389–399.

Neher E, Sakmann B. 1992. The patch clamp technique. *Scientific American* 266:44–51.

Rajendra S, Schofield PR. 1995. Molecular mechanisms of inherited startle syndromes. *Trends in Neurosciences* 18:80–82.

Shepherd G. 1994. *Neurobiology*, 3rd ed. New York: Oxford University Press.

Sherrington C. 1906. *Integrative action of the nervous system*. New Haven: Yale University Press.

Südhof TC. 1995. The synaptic vesicle cycle: a cascade of protein-protein interactions. *Nature* 375:645–653.

Südhof TC. 1998. RAB3 and synaptotagmin: the yin and yang of synaptic membrane fusion. *Annual Review of Neuroscience* 21:75–95.

Unwin N. 1993. Neurotransmitter action: opening of ligand-gated ion channels. *Cell* 72:31–41.

Weber T, Zemelman BV, McNew JA, Westermann B, Gmachl M, Parlati F, Söllner TH, Rothman JE. 1998. SNAREpins: minimal machinery for membrane fusion. *Cell* 92:759–772.

Chapter 6

Amara SG, Arriza JL. 1993. Neurotransmitter transporters: three distinct gene families. *Current Opinion in Neurobiology* 3:337–344.

Attwell D, Mobbs P. 1994. Neurotransmitter transporters. *Current Opinion in Neurobiology* 4:353–359.

Brezina V, Weiss KR. 1997. Analyzing the functional consequences of transmitter complexity. *Trends in Neurosciences* 20:538–543.

Changeux J-P. 1993. Chemical signaling in the brain. *Scientific American* 269:58–62.

Colquhoun D, Sakmann B. 1998. From muscle endplate to brain synapses: a short history of synapse and agonist-activated ion channels. *Neuron* 20:381–387.

Cooper JR, Bloom FE, Roth RH. 1996. *The Biochemical Basis of Neuropharmacology*, 7th ed. New York: Oxford University Press.

Feldman RS, Meyer JS, Quenzer LF. 1997. *Principles of Neuropsychopharmacology*, Sunderland, MA: Sinauer.

Gilman AG. 1995. Nobel lecture: G proteins and regulation of adenylyl cyclase. *Bioscience Report* 15:65–97.

Gudermann T, Schöneberg T, Schultz G. 1997. Functional and structural complexity of signal transduction via G-protein-couple receptors. *Annual Review of Neuroscience* 20:399–427.

Hall Z. 1992. *An Introduction to Molecular Neurobiology*. Sunderland, MA: Sinauer.

Hille B. 1992. *Ionic Channels of Excitable Membranes*, 2nd ed. Sunderland, MA: Sinauer.

Howlett AC. 1995. Pharmacology of cannabinoid receptors. *Annual Review of Pharmacology and Toxicology* 35:607−634.

Matsuda LA. 1997. Molecular aspects of cannabinoid receptors. *Critical Reviews in Neurobiology* 11:143–166.

McGeer P, Eccles J, McGeer E. 1987. *Molecular Neurobiology of the Mammalian Brain*, 2nd ed. New York: Plenum.

Nicoll R, Malenka R, Kauer J. 1990. Functional comparison of neurotransmitter receptor subtypes in the mammalian nervous system. *Physiological Reviews* 70:513–565.

Nicholls J, Martin A, Wallace B. 1993. *From Neuron to Brain*, 3rd ed. Sunderland, MA: Sinauer.

North RA, Barnard EA. 1997. Nucleotide receptors. *Current Opinion in Neurobiology* 7:346–357.

Rabow LE, Russek SJ, Farb DH. 1995. From ion channels to genomic analysis: recent advances in $GABA_A$ receptor research. *Synapse* 21:189–274.

Rosenmund C, Stern-Bach Y, Stevens CF. 1998. The tetrameric structure of a glutamate receptor channel. *Science* 280:1596–1599.

Siegel G, Agranoff B, Albers R, Molinoff P, eds. 1994. *Basic Neurochemistry: Molecular, Cellular and Medical Aspects*, 5th ed. New York: Raven Press.

Snyder S. 1986. *Drugs and the Brain*. New York: W.H. Freeman.

Stella N, Schweitzer P, Piomelli D. 1997. A second endogenous cannabinoid that modulates long-term potentiation. *Nature* 388:773–778.

Walker JM, Huang SM, Strangman NM, Tsou K, Sanudo-Pena MC. 1999. Pain modulation by release of the endogenous cannabinoid anandamide. *Proceedings of the National Academy of Sciences USA* 96:12198–12203.

Zigmond MJ, Bloom FE, Landis SC, Roberts JL, Squire LR. 1999. *Fundamental Neuroscience*. San Diego: Academic Press.

Chapter 7

Butterworth CE, Bendich A. 1996. Folic acid and the prevention of birth defects. *Annual Review of Nutrition* 16:73–97.

Creslin E. 1974. Development of the nervous system: a logical approach to neuroanatomy. *CIBA Clinical Symposium* 26:1–32.

Frackowick RSJ. 1998. The functional architecture of the brain. *Daedalus* 127:105–130.

Gilbert SF. 1997. *Developmental Biology*. Sunderland, MA: Sinauer.

Glubegoric N, Williams TH. 1980. *The Human Brain: A Photographic Guide*. Philadelphia: Lippincott.

Kaas JH. 1995. The evolution of neocortex. *Brain, Behavior and Evolution* 46:187–196.

Krubitzer L. 1995. The organization of neocortex in mammals: Are species really so different? *Trends in Neurosciences* 18:408–418.

Nauta W, Feirtag M. 1986. *Fundamental Neuroanatomy*. New York: W.H. Freeman.

Northcutt RG, Kaas JH. 1995. The emergence and evolution of mammalian neocortex. *Trends in Neurosciences* 18:373–375.

Posner MI, Raichle M. 1994. *Images of Mind*. New York: Scientific American Library.

Povinelli DJ, Preuss TM. 1995. Theory of mind: evolutionary history of a cognitive specialization. *Trends in Neurosciences* 18:414–424.

Smith JL, Schoenwolf GC. 1997. Neurulation: coming to closure. *Trends in Neurosciences* 20:510–517.

Watson C. 1995. *Basic Human Neuroanatomy: an Introductory Atlas*, 5th ed. New York: Little, Brown & Co, 1995.

Chapter 8

Belluscio L, Gold GH, Nemes A, Axel R. 1998. Mice deficient in G(olf) are anosmic. *Neuron* 20:69–81.

Bernstein IL. 1991. Flavor aversion. In *Smell and Taste in Health and Disease*, eds. Getchell TV, Doty RL, Bartoshuk LM, Snow JB, eds. New York: Raven Press, pp. 417–428.

Buck LB. 1996. Information coding in the vertebrate olfactory system. *Annual Review of Neurosciences* 19:517–554.

Buck LB, Axel R. 1991. A novel multigene family may encode odorant receptors: a molecular basis for odor recognition. *Cell* 65:175–187.

Corey DP, Roper SD. 1992. *Sensory Transduction*. New York: Rockefeller University Press.

Dorries KM. 1998. Olfactory coding: time in a model. *Neuron* 20:7–10.

Dulac C. 1997. How does the brain smell? *Neuron* 19:477–480.

Engen T. 1991. *Odor Sensation and Memory*. New York: Praeger.

Garcia J, Ervin FR, Koelling RA. 1966. Learning with prolonged delay of reinforcement. *Psychonomic Science* 5:121–122.

Getchell TV, Doty RL, Bartoshuk LM, Snow JB. 1991. *Smell and Taste in Health and Disease*. New York: Raven Press.

Hildebrand JG, Shepherd GM. 1997. Mechanisms of olfactory discrimination: converging evidence for common principles across phyla. *Annual Review of Neuroscience* 20:595–631.

Kauer JS. 1991. Contributions of topography and parallel processing to odor coding in the vertebrate olfactory pathway. *Trends in Neurosciences* 14:79–85.

Kinnamon SC, Margolskee RF. 1996. Mechanisms of taste transduction. *Current Opinion in Neurobiology* 6:506–513.

Laurent G. 1997. Olfactory processing: maps, time and codes. *Current Opinion in Neurobiology* 7:547–553.

Laurent G, Wehr M, Davidowitz H. 1996. Temporal representations of odors in an olfactory network. *Journal of Neuroscience* 16:3837–3847.

Liman ER. 1996. Pheromone transduction in the vomeronasal organ. *Current Opinion in Neurobiology* 6:487–493.

Lindemann B. 1997. Chemoreception: tasting the sweet and the bitter. *Current Biology* 10:1234–1237.

Mombaerts P, Wang F, Dulac C, Chao SK, Nemes A, Mendelsohn M, Edmondson J, Axel R. 1996. Visualizing an olfactory sensory map. *Cell* 87:675–686.

Nakamura T, Gold GH. 1987. A cyclic nucleotide-gated conductance in olfactory receptor cilia. *Nature* 325:442–444.

Ressler J, Sullivan SL, Buck LB. 1993. A zonal organization of odorant receptor gene expression in the olfactory epithelium. *Cell* 73:597–609.

Roper SD. 1992. The microphysiology of peripheral taste organs. *Journal of Neuroscience* 12:1127–1134.

Sato T. 1980. Recent advances in the physiology of taste cells. *Progress in Neurobiology* 14:25–67.

Stern K, McClintock MK. 1998. Regulation of ovulation by human pheromones. *Nature* 392:177–179.

Stewart RE, DeSimone JA, Hill DL. 1997. New perspectives in gustatory physiology: transduction, development, and plasticity. *American Journal of Physiology* 272:C1-C26.

Stopfer M, Bhagavan S, Smith BH, Laurent G. 1997. Impaired odour discrimination on desynchronization of odour-encoding neural assemblies. *Nature* 390:70–74.

Chapter 9

Barlow H. 1953. Summation and inhibition in the frog's retina. *Journal of Physiology* (London) 119:69–78.

Baylor DA. 1987. Photoreceptor signals and vision. *Investigative Ophthalmology and Visual Science* 28:34–49.

Curcio CA, Sloan KR, Kalina RE, Hendrickson AE. 1990. Human photoreceptor topography. *Journal of Comparative Neurology* 292:497–523.

Daw NW. 1972. Color-coded cells in goldfish, cat, and rhesus monkey. *Investigative Ophthalmology* 11:411–417.

Dowling JE, Werblin FS. 1971. Synaptic organization of the vertebrate retina. *Vision Research Suppl.* 3:1–15.

Dowling JE. 1987. *The Retina: An Approachable Part of the Brain*. Cambridge, MA: Harvard University Press.

Enroth-Cugell C, Robson JG. 1984. Functional characteristics and diversity of cat retinal ganglion cells. *Investigative Ophthalmology and Visual Science* 25:250–257.

Fesenko EE, Kolesnikov SS, Lyubarsky AL. 1985. Induction by cyclic GMP of cationic conductance in plasma membrane of retinal rod outer segment. *Nature* 313:310–313.

Kuffler S. 1953. Discharge patterns and functional organization of the mammalian retina. *Journal of Neurophysiology* 16:37–68.

McIlwain JT. 1996. *An Introduction to the Biology of Vision*. Cambridge, England: Cambridge University Press.

Moses RA, Hart WM, eds. 1987. *Adler's Physiology of the Eye*, 3rd ed. St. Louis: Mosby.

Nathans J. 1989. The genes for color vision. *Scientific American* 260:42–49.

Nathans J. 1994. In the eye of the beholder: visual pigments and inherited variation in human vision. *Cell* 78:357–360.

Neitz J, Jacobs GH. 1986. Polymorphism of the long-wavelength cone in normal human colour vision. *Nature* 323:623–625.

Newell FW. 1965. *Ophthalmology*. St. Louis: Mosby.

Rodieck RW. 1998. *The First Steps in Seeing*. Sunderland, MA: Sinauer.

Schwab L. 1987. *Primary Eye Care in Developing Nations*. New York: Oxford University Press.

Schnapf JL, Baylor DA. 1987. How photoreceptor cells respond to light. *Scientific American* 256:40–47.

Wässle H. Boycott B. 1991. Functional architecture of the mammalian retina. *Physiological Reviews* 71:447–480.

Watanabe M, Rodieck RW. 1989. Parasol and midget ganglion cells of the primate retina. *Journal of Comparative Neurology* 289:434–454.

Chapter 10

Barlow H. 1972. Single units and sensation: a neuron doctrine for perceptual psychology? *Perception* 1:371–394.

Gauthier I, Tarr MJ, Anderson AW, Skudlarski P, Gore JC. 1999. Activation of the middle fusiform "face area" increases with expertise in recognizing novel objects. *Nature Neuroscience* 2:568–573.

Grafstein B, Laureno R. 1973. Transport of radioactivity from eye to visual cortex in the mouse. *Experimental Neurology* 39:44–57.

Gray C, Singer W. 1989. Stimulus-specific neuronal oscillations in orientation columns of cat visual cortex. *Proceedings of the National Academy of Sciences USA* 86:1698–1702.

Hendry S, Yoshioka T. 1994. A neurochemically distinct third channel in the macaque dorsal lateral geniculate nucleus. *Science* 264:575–577.

Hubel D. 1982. Explorations of the primary visual cortex, 1955–78 (Nobel lecture). *Nature* 299:515–524.

Hubel D. 1988. *Eye, Brain, and Vision*. New York: W. H. Freeman.

Hubel D, Wiesel T. 1962. Receptive fields, binocular interaction and functional architecture in the cat's visual cortex. *Journal of Physiology (London)*160:106–154.

Hubel D, Wiesel T. 1968. Receptive fields and functional architecture of monkey striate cortex. *Journal of Physiology (London)* 195: 215–243.

Hubel D, Wiesel T. 1977. Functional architecture of the macaque monkey visual cortex (Ferrier lecture). *Proceedings of the Royal Society of London*. Series B 198:1–59.

Horibuchi S, ed. 1994. *Stereogram*. Tokyo: Shogakukan.

Julesz B. 1971. *Foundations of Cyclopean Perception*. Chicago: University of Chicago Press.

LeVay S, Wiesel TN, Hubel DH. 1980. The development of ocular dominance columns in normal and visually deprived monkeys. *Journal of Comparative Neurology* 191:1–51.

Livingstone M, Hubel D. 1984. Anatomy and physiology of a color system in the primate visual cortex. *Journal of Neuroscience* 4: 309–356.

Martin K. 1994. A brief history of the "feature detector." *Cerebral Cortex* 4:1–7.

Movshon JA, Adelson EH, Gizzi MS, Newsome WT. 1985. The analysis of moving visual patterns. In Chagas C, Gattass R, Gross CG, eds. *Study Group on Pattern Recognition Mechanisms*. Vatican City, Italy: Pontificia Academia Scientiarum.

Palmer SE. 1999. *Vision Science: Photons to Phenomenology*. Cambridge, MA: MIT Press.

Salzman C, Britten K, Newsome W. 1990. Cortical microstimulation influences perceptual judgments of motion detection. *Nature* 346:174–177.

Sereno MI, Dale AM, Reppas JB, Kwong KK, Belliveau JW, Brady TJ, Rosen BR, Tootell RBH. 1995. Borders of multiple visual areas in humans revealed by functional magnetic resonance imaging. *Science* 268:889–893.

Singer W, Gray CM. 1995. Visual feature integration and the temporal correlation hypothesis. *Annual Review of Neuroscience* 18:555–586.

Ts'o DY, Frostig RD, Lieke EE, Grinivald A. 1990. Functional organization of primate visual cortex revealed by high resolution optical imaging. *Science* 249:417–420.

Tyler C, Clarke MB. 1990. The autostereogram. *Proceedings of the International Society for Optical Engineering* 1256:182–197.

Zeki S. 1993. *A Vision of the Brain*. London: Blackwell Scientific.

Zihl J, Cramon D von, Mai N. 1983. Selective disturbance of movement vision after bilateral brain damage. *Brain* 106:313–340.

Chapter 11

Ashmore JF, Kolston PJ. 1994. Hair cell based amplification in the cochlea. *Current Opinion in Neurobiology* 4:503–508.

Baloh RW, Honrubia V. 1979. *Clinical Neurophysiology of the Vestibular System*. Philadelphia: FA Davis.

Brandt T. 1991. Man in motion: historical and clinical aspects of vestibular function. A review. *Brain* 114:2159–2174.

Goldberg JM. 1991. The vestibular end organs: morphological and physiological diversity of afferents. *Current Opinion in Neurobiology* 1:229–235.

Hudspeth AJ. 1983. The hair cells of the inner ear. *Scientific American* 248:54–64.

Hudspeth AJ. 1997. How hearing happens. *Neuron* 19:947–950.

Hudspeth AJ. 1997. Mechanical amplification of stimuli by hair cells. *Current Opinion in Neurobiology* 7:480–486.

Joris PX, Smith PH, Yin TCT. 1998. Coincidence detection in the auditory system: 50 years after Jeffress. *Neuron* 21: 1235–1238.

Köppl C. 1995. Otoacoustic emissions as an indicator for active cochlear mechanics: a primitive property of vertebrate auditory organs. In *Advances in Hearing Research*, ed. Manley GA, Klump GM, Koppl C, Fastl H, Oeckinghaus H. Singapore: World Scientific, pp. 207–216.

Middlebrooks JC, Green DM. 1991. Sound localization by human listeners. *Annual Review of Psychology* 42:135–159.

Moore BCJ. 1989. *An Introduction to the Psychology of Hearing*. San Diego: Academic Press.

Nobili R, Mammano F, Ashmore J. 1998. How well do we understand the cochlea? *Trends in Neurosciences* 21:159–167.

Oertel D. 1997. Encoding of timing in the brain stem auditory nuclei of vertebrates. *Neuron* 19:959–962.

Powers NL, Salvi RJ, Want J, Spongr V, Qiu CX. 1995. Elevation of auditory thresholds of spontaneous cochlear oscillations. *Nature* 375:585–587.

Rose JE, Hind JE, Anderson DJ, Brugge JF. 1971. Some effects of stimulus intensity on response of auditory nerve fibers in the squirrel monkey. *Journal of Neurophysiology* 24: 685–699.

Ruggero MA, Rich NC. 1996. Furosemide alters organ of Corti mechanics: evidence for feedback of outer hair cells upon the basilar membrane. *Journal of Neuroscience* 11: 1057–1067.

Sento S, Ryugo DK. 1989. Endbulbs of Held and spherical bushy cells in cats: morphological correlates with physiological properties. *Journal of Comparative Neurology* 280:553–562.

Simmons JA. 1989. A view of the world through the bat's ear: the formation of acoustic images in echolocation. *Cognition* 33: 155–199.

Suga N. 1995. Processing of auditory information carried by species-specific complex sounds. In *The Cognitive Neurosciences*, ed. Gazzaniga MS. Cambridge, MA: MIT Press, pp. 295–314.

Trussell LO. 1997. Cellular mechanisms for preservation of timing in central auditory pathways. *Current Opinion in Neurobiology* 7:487–492.

Ulfendahl M, Flock Å. 1998. Outer hair cells provide active tuning in the organ of Corti. *News in Physiological Sciences* 13:107–117.

von Békésy G. 1960. *Experiments in Hearing*, ed. and trans. Wever EG. New York: McGraw-Hill.

Zenner H-P, Gummer AW. 1996. The vestibular system. In *Comprehensive Mammalian Physiology. From Cellular Mechanisms to Integration, Vol. 1*, ed. Greger R, Windhorst U. Berlin: Springer-Verlag, pp. 697–710.

Chapter 12

Brown A. 1989. Somatic sensation: peripheral aspects. In *Textbook of Physiology*, eds. Patton HD, Fuchs AF, Hille B, Scher AM, Steiner R. Philadelphia: Saunders, pp. 298–313.

Cao Y, Mantyh P, Carlson E, Gillespie A, Epstein C. 1998. Primary afferent tachykinins are required to experience moderate to intense pain. *Nature* 392:90–394.

Cesare P, McNaughton P. 1997. Peripheral pain mechanisms. *Current Opinion in Neurobiology* 7:493–499.

Chen R, Corwell B, Yaseen Z, Hallett M, Cohen L. 1998. Mechanisms of cortical reorganization in lower-limb amputees. *Journal of Neuroscience* 18(9):3443–3450.

DeFelipe C, Huerrero J, O'Brien J, Palmer J, Doyle C. 1998. Altered nociception, analgesia and aggression in mice lacking the receptor for substance P. *Nature* 392:394–397.

Elbert T, Pantev C, Wienbruch C, Rockstroh B, Taub E. 1995. Increased cortical representation of the fingers of the left hand in string players. *Science* 270:305–306.

Fields HL. 1987. *Pain*. New York: McGraw-Hill.

Fields H, Basbaum A. 1994. Central nervous system mechanisms of pain modulation. In *Textbook of Pain*, eds, Wall P, Melzack R. Edinburgh: Churchill Livingstone, pp. 243–257.

Fitzpatrick T, et al. 1997. *Color Atlas and Synopsis of Clinical Dermatology*, 3rd ed. New York: McGraw-Hill.

Jenkins WM, Merzenich MM, Ochs MT, Allard T, Guic-Robles E. 1990. Functional reorganization of primary somatosensory cortex in adult owl monkeys after behaviorally controlled tactile stimulation. *Journal of Neurophysiology* 63:82–104.

Johnson KO, Hsiao SS. 1992. Neural mechanisms of tactile form and texture perception. *Annual Review of Neuroscience* 15:227–250.

Kaas J. 1990. Somatosensory system. In *The Human Nervous System*, ed. Paxinos G. San Diego: Academic Press, pp. 813–844.

Kass J. 1998. Phantoms of the brain. *Nature* 391:331–333.

Kaas SH, Nelson RH, Sur M, Merzenich MM. 1981. Organization of somatosensory cortex in primates. In *The Organization of the Cerebral Cortex*, eds. Schmitt FO, Worden FG, Adelman G, Dennis SG. Cambridge, MA: MIT Press, pp. 237–262.

Mantyh PW, Rogers SD, Honore P, Allen BJ, et al. 1997. Inhibition of hyperalgesia by ablation of lamina I spinal neurons expressing the substance P receptor. *Science* 278:275–279.

McMahon S, Koltzenburg M. 1992. Itching for an explanation. *Trends in Neurosciences* 15:497–501.

Melzack R, Wall P. 1983. *The Challenge of Pain*. New York: Basic Books.

Merzenich MM, Nelson RJ, Stryker MP, Cynader MS, Schoppman A. 1984. Somatosensory cortical map changes following digit amputation in adult monkeys. *Journal of Comparative Neurology* 224:591–605.

Meyer R, Campbell J, Raja S. 1994. Peripheral neural mechanisms of nociception. In *Textbook of Pain*, eds. Wall P, Melzack R. Edinburgh: Churchill Livingstone, pp. 13–44.

Penfield W, Rasmussen T. 1952. *The Cerebral Cortex of Man*. New York: Macmillan.

Sacks O. 1985. *The Man Who Mistook His Wife for a Hat and Other Clinical Tales*. New York: Summit.

Sadato N, Pascual-Leone A, Grafman J, Ibanez V, Delber M-P. 1996. Activation of the primary visual cortex by Braille reading in blind subjects. *Nature* 380:526–527.

Schmelz M, Schmidt R, Bickel A, Hardwerker H, Torebjork H. 1997. Specific C-receptors for itch in human skin. *Journal of Neuroscience* 17(20):8003–8008.

Schmidt RF. 1978. *Fundamentals of Sensory Physiology*. New York: Springer–Verlag.

Springer SP, Deutsch G. 1989. *Left Brain, Right Brain*. New York: W.H. Freeman.

Taddese A, Nah S-Y, McCleskey E. 1995. Selective opioid inhibition of small nociceptive neurons. *Science* 270:1366–1369.

Treede R-K, Magerl W. 1995. Modern concepts of pain and hyperalgesia: beyond the polymodal c-nociceptor. *News in Physiological Sciences* 10:216–228.

Vallbo Å, Johansson R. 1984. Properties of cutaneous mechanoreceptors in the human hand related to touch sensation. In *Human Neurobiology* 3: 3–14.

Vallbo Å. 1995. Single-afferent neurons and somatic sensation in humans. In *The Cognitive Neurosciences*, ed. Gazzaniga M. Cambridge, MA: MIT Press, pp. 237–251.

Wall P. 1994. The placebo and the placebo response. In *Textbook of Pain*, eds. Wall P, Melzack R, Edinburgh: Churchill Livingstone, pp. 1297–1308.

Woolsey TA, Van Der Loos H. 1970. The structural organization of layer IV in the somatosensory region (S1) of mouse cerebral cortex: the description of a cortical field composed of discrete cytoarchitectonic units. *Brain Research* 17:205–242.

Chapter 13

Brown RH. 1995. Amyotrophic lateral sclerosis: recent insights from genetics and transgenic mice. *Cell* 80:687–692.

Brown T. 1911. The intrinsic factors in the act of progression in the mammal. *Proceedings of the Royal Society of London*. Series B, 84:308–319.

Buller A, Eccles J, Eccles R. 1960. Interactions between motoneurons and muscles in respect to the characteristic speeds of their responses. *Journal of Physiology* (London) 150:417–439.

Drachman DB. 1994. Myasthenia gravis. *New England Journal of Medicine* 330:1797–1810.

Evarts E, Wise S, Bousfield D, eds. 1985. *The Motor System in Neurobiology*. New York: Elsevier.

Ghez C. 1991. Muscles: effectors of the motor system. In *Principles of Neural Science*, eds. Kandel E, Schwartz J, Jessel T. New York: Elsevier.

Grillner S, Ekeberg Ö, El Manira A, Lansner A, Parker D, et al. 1998. Intrinsic function of a neuronal network: a vertebrate central pattern generator. *Brain Research Reviews* 26:184–197.

Henneman E, Somjen G, Carpenter D. 1965. Functional significance of cell size in spinal motoneurons. *Journal of Neurophysiology* 28:560–580.

Huxley A, Niedergerke R. 1954. Structural changes in muscle during contraction. Interference microscopy of living muscle fibres. *Nature* 173:971–973.

Huxley H, Hanson J. 1954. Changes in cross-striations of muscle during contraction and stretch and their structural interpretation. *Nature* 173:973–976.

Lømo T, Westgaard R, Dahl H. 1974. Contractile properties of muscle: control by pattern of muscle activity in the rat. *Proceedings of the Royal Society of London*. Series B, 187:99–103.

Mendell L, Henneman E. 1968. Terminals of single Ia fibers: distribution within a pool of 300 homonymous motor neurons. *Science* 160:96–98.

Rowland L. 1988. Clinical concepts of Duchenne muscular dystrophy: the impact of molecular genetics. *Brain* 111:479–495.

Sherrington C. 1947. *The Integrative Action of the Nervous System*, 2nd ed. New Haven: Yale University Press.

Stein PSG, et al., eds. 1997. *Neurons, Networks, and Motor Behavior*. Cambridge, MA: MIT Press.

Wallen P, Grillner S. 1994. *N*-Methyl-D-aspartate receptor-induced, inherent oscillatory activity in neurons active during fictive locomotion in the lamprey. *Journal of Neuroscience* 7:2745–2755.

Chapter 14

Brooks V. 1986. *The Neural Basis of Motor Control*. New York: Oxford University Press.

Campbell A. 1905. *Histological Studies on the Localization of Cerebral Function*. Cambridge, England: Cambridge University Press.

The cerebellum: development, physiology and plasticity. 1998. *Trends in Neurosciences* 21:367–419 (special issue).

Donoghue J, Sanes J. 1994. Motor areas of the cerebral cortex. *Journal of Clinical Neurophysiology* 11:382–396.

Feigin A. 1998. Advances in Huntington's disease: implications for experimental therapeutics. *Current Opinions in Neurology* 11:357–362.

Feldman RS, Meyer JS, Quenzer LF. 1997. Parkinson's disease. In *Principles of Neuropharmacology*. Sunderland, MA: Sinauer, pp. 861–887.

Georgopoulos A, Caminiti R, Kalaska J, Massey J. 1983. Spatial coding of movement:

a hypothesis concerning the coding of movement direction by motor control populations. *Experimental Brain Research Suppl.* 7:327–336.

Georgopoulos A, Kalaska J, Caminiti R, Massey J. 1982. On the relations between the direction of two-dimensional arm movements and cell discharge in primate motor cortex. *Journal of Neuroscience* 2:1527–1537.

Langston JW. 1998. Epidemiology versus genetics in Parkinson's disease: progress in resolving an age-old debate. *Annals of Neurology* 44: S45–52.

Langston JW, Palfreman J. 1995. *The Case of the Frozen Addicts*. New York: Pantheon.

Lawrence D, Kuypers H. 1968. The functional organization of the motor system in the monkey: I. The effects of bilateral pyramidal lesions. *Brain* 91:1–14.

Lawrence D, Kuypers H. 1968. The functional organization of the motor system in the monkey: II. The effects of lesions of the descending brain-stem pathways. *Brain* 91:15–36.

Lee C, Rohrer W, Sparks D. 1988. Population coding of saccadic eye movements by neurons in the superior colliculus. *Nature* 332:357–360.

McIlwain JT. 1991. Distributed spatial coding in the superior colliculus: a review. *Visual Neuroscience* 6:3–13.

Porter R, Lemon R. 1993. *Corticospinal Function and Voluntary Movement.* Oxford, England: Clarendon Press.

Roland P, Larsen B, Lassen N, Skinhøf E. 1980. Supplementary motor area and other cortical areas in organization of voluntary movements in man. *Journal of Neurophysiology* 43:118–136.

Roland PE, Zilles K. 1996. Functions and structures of the motor cortices in humans. *Current Opinion in Neurobiology* 6:773–781.

Sanes JN, Donoghue JP. 1997. Static and dynamic organization of motor cortex. *Advances in Neurology* 73:277–296.

Strange, PG. 1992. *Brain Biochemistry and Brain Disorders*. New York: Oxford University Press.

Weinrich M, Wise S. 1982. The premotor cortex of the monkey. *Journal of Neuroscience* 2:1329–1345.

Wichmann T, DeLong MR. 1996. Functional and pathophysiological models of the basal ganglia. *Current Opinion in Neurobiology* 6:751–758.

Zigmond MJ, Bloom FE, Landis SC, Roberts JL, Squire LR. 1999. *Fundamental Neuroscience*. San Diego: Academic Press.

Chapter 15

Aghananian GK, Marek GJ. 1999. Serotonin and hallucinogens. *Neuropsychopharmacology* 2 (Suppl):16S–23S.

Appenzeller O. 1990. *The Autonomic Nervous System: An Introduction to Basic and Clinical Concepts*. New York: Elsevier.

Barajas-López C, Huizinga JD. 1993. New transmitters and new targets in the autonomic nervous system. *Current Opinion in Neurobiology* 3:1020–1027.

Bloom FE, Lazerson A, Hofstader L. 1985. *Brain, Mind, and Behavior*. New York: W.H. Freeman.

Caroll CR. 1985. *Drugs in Modern Society*. Dubuque, IA: WC Brown.

Cooper JR, Bloom FE, Roth RH. 1991. *The Biochemical Basis of Neuropharmacology*, 6th ed. New York: Oxford University Press.

Jänig W, McLachlan EM. 1992. Characteristics of function-specific pathways in the sympathetic nervous system. *Trends in Neurosciences* 15:475–481.

Kerr DS, Campbell LW, Hao S-Y, Landsfield PW. 1989. Corticosteroid modulation of hippocampal potentials: increased effect with aging. *Science* 245:1505–1509.

Kerr DS, Campbell LW, Applegate MD, Brodish A, Landsfield PW. 1991. Chronic stress-induced acceleration of electrophysiologic and morphometric biomarkers of hippocampal aging. *Journal of Neuroscience* 11:1316–1324.

Koob GF. 1992. Drugs of abuse: anatomy, pharmacology and function of reward pathways. *Trends in Pharmacological Sciences* 13:177–184.

McEwen BS, Schmeck HM. 1994. *The Hostage Brain*. New York: Rockefeller University Press.

Sapolsky RM. 1994. *Why Zebras Don't Get Ulcers: A Guide to Stress, Stress-Related Diseases, and Coping*. New York: W.H. Freeman.

Sapolsky RM, Krey LC, McEwen BS. 1986. The neuroendocrinology of stress and aging: the glucocorticoid cascade hypothesis. *Endocrine Reviews* 7:284–301.

Snyder SH. 1986. *Drugs and the Brain*. New York: Scientific American Books.

Swanson LW. 1987. The hypothalamus. In *Handbook of Chemical Neuroanatomy*, eds. Björkland A, Hökfelt T, Swanson LW. New York: Elsevier, pp. 1–124.

Watanabe Y, Gould E, McEwen BS. 1992. Stress induces atrophy of apical dendrites of hippocampal CA3 pyramidal neurons. *Brain Research* 588:341–345.

Chapter 16

Berridge KC. 1996. Food reward: brain substrates of wanting and liking. *Neuroscience and Biobehavioral Reviews* 20:1–25.

Berridge KC, Robinson TE. 1998. What is the role of dopamine in reward: hedonic impact, reward learning, or incentive salience? *Brain Research Review* 28:308–367.

Flier JS, Maratos-Flier E. 1998. Obesity and the hypothalamus: novel peptides for new pathways. *Cell* 92:437–440.

Friedman JM. 1997. The alphabet of weight control. *Nature* 385:119–120.

Hoebel BG. 1997. Neuroscience and appetitive behavior research: 25 years. *Appetite* 29:119–133.

Koob GF, Sanna PP, Bloom FE. 1998. Neuroscience of addiction. *Neuron* 21: 467–476.

Sawchenko PE. 1998. Toward a new neurobiology of energy balance, appetite, and obesity: The anatomists weigh in. *Journal of Comparative Neurology* 402:435–441.

Schwartz DH, Hernandez L, Hoebel BG. 1990. Serotonin release in lateral and medial hypothalamus during feeding and its anticipation. *Brain Research Bulletin* 25:797–802.

Stricker EM, Verbalis JG. 1999. Water intake and body fluids. In *Fundamental Neuroscience*, eds. Zigmond MJ, et al. New York: Academic Press, pp. 1111–1125.

Wise RA. 1996. Neurobiology of addiction. *Current Opinion in Neurobiology* 6:243–251.

Woods SC, Seeley RJ, Porte D, Schwartz MW. 1998. Signals that regulate food intake and energy homeostasis. *Science* 280:1378–1382.

Woods SC, Stricker EM. 1999. Food intake and metabolism. In *Fundamental Neuroscience*, eds. Zigmond MJ, et al. New York: Academic Press, pp. 1091–1109.

Chapter 17

Allen LS, Richey MF, Chai YM, Gorski RA. 1991. Sex differences in the corpus callosum of the living human being. *Journal of Neuroscience* 11:933–942.

Alvarez-Buylla A, Kirn JR. 1997. Birth, migration, incorporation, and death of vocal control neurons in adult songbirds. *Journal of Neurobiology* 33:585–601.

Berne RM, Levy MN. 1998. *Physiology*. St. Louis: Mosby.

Blum D. 1997. *Sex on the Brain: The Biological Differences Between Men and Women*. New York: Viking.

Breedlove SM. 1994. Sexual differentiation in the human nervous system. *Annual Review of Psychology* 45:389–418.

Diamond J. 1997. *Why Is Sex Fun? The Evolution of Human Sexuality*. New York: Basic Books.

Fausto-Sterling A. 1992. *Myths of Gender: Biological Theories About Women and Men*. New York: Basic Books.

Gilbert SF. 1994. *Developmental Biology*, 4th ed. Sunderland, MA: Sinauer.

Gould E, Wooley CS, Frankfurt M, McEwen BS. 1990. Gonadal steroids regulate spine density on hippocampal pyramidal cells in adulthood. *Journal of Neuroscience* 10: 1286–1291.

Hamer DH, Hu S, Magnuson VL, Hu N, Pattatucci AM. 1993. A linkage between DNA markers on the X chromosome and male sexual orientation. *Science* 261:321–327.

Kimura D. 1992. Sex differences in the brain. *Scientific American* 267:119–125.

Kimura D. 1996. Sex, sexual orientation and sex hormones influence human cognitive function. *Current Opinion in Neurobiology* 6:259–263.

Koopman P, Gubbay J, Vivian N, Goodfellow P, Lovell-Badge R. 1991. Male development of chromosomally female mice transgenic for Sry. *Nature* 351:117–121.

LeVay S. 1991. A difference in hypothalamic structure between heterosexual and homosexual men. *Science* 253:1034–1037.

LeVay S. 1993. *The Sexual Brain*. Cambridge, MA: MIT Press.

McEwen BS. 1976. Interactions between hormones and nerve tissue. *Scientific American* 235:48–58.

McEwen BS. 1999. Permanence of brain sex differences and structural plasticity of the adult brain. *Proceedings of the National Academy of Sciences, USA* 96:7128–7130.

McEwen BS, Davis PG, Parsons BS, Pfaff DW. 1979. The brain as a target for steroid hormone action. *Annual Review of Neuroscience* 2: 65–112.

McLaren A. 1990. What makes a man a man? *Nature* 346:216–217.

Murphy DD, Cole NB, Greenberger V, Segal M. 1998. Estradiol increases dendritic spine density by reducing GABA neurotransmission in hippocampal neurons. *Journal of Neuroscience* 18:2550–2559.

Nottebohm F, Arnold AP. 1976. Sexual dimorphism in vocal control areas of the songbird brain. *Science* 194:211–213.

Rosenzweig MR, Leiman AL, Breedlove SM. 1999. *Biological Psychology*, 2nd ed. Sunderland, MA: Sinauer.

Sinclair AH, Berta P, Palmer MS, Hawkins JR, Griffths BL, Smith MJ, Foster JW, Frischauf AM, Lovell-Badge R, Goodfellow PN. 1990. A gene from the human sex-determining region encodes a protein with homology to a conserved DNA-binding motif. *Nature* 346:240–242.

Wooley CS. 1999. Effects of estrogen in the CNS. *Current Opinion in Neurobiology* 9:349–354.

Wooley CS, Weiland N, McEwen BS, Schwartzkroin PA. 1997. Estradiol increases the sensitivity of hippocampal CA1 pyramidal cells to NMDA receptor-mediated synaptic input: correlation with dendritic spine density. *Journal of Neuroscience* 17:1848–1859.

Xerri C, Stern JM, Merzenich MM. 1994. Alterations of the cortical representation of the rat ventrum induced by nursing behavior. *Journal of Neuroscience* 14:1710–1721.

Young LJ, Wang Z, Insel TR. 1998. Neuroendocrine bases of monogamy. *Trends in Neurosciences* 21:71–75.

Chapter 18

Adolphs R, Tranel D, Damasio H, Damasio A. 1994. Impaired recognition of emotion in facial expressions following bilateral damage to the human amygdala. *Nature* 372:669–672.

Aggleton JP. 1993. The contribution of the amygdala to normal and abnormal emotional states. *Trends in Neurosciences* 16: 328–333.

Bard P. 1934. On emotional expression after decortication with some remarks on certain theoretical views. *Psychological Reviews* 41:309–329.

Broca P. 1878. Anatomie comparé de circonvolutions cérébrales. Le grand lobe limbique et la scissure limbique dans la série des mamniféres. *Revue d'Anthropologie* 1:385–498.

Büchel C, Morris J, Dolan RJ, Friston KJ. 1998. Brain systems mediating aversive conditioning: an event-related fMRI study. *Neuron* 20:947–957.

Cannon WB. 1927. The James-Lange theory of emotion. *American Journal of Psychology* 39:106–124.

Damasio H, Grabowski T, Frank R, Galaburda AM, Damasio AR. 1994. The return of Phineas Gage: clues about the brain from the skull of a famous patient. *Science* 264:1102–1105.

Darwin C. 1872/1955. *The Expression of the Emotions in Man and Animals*. New York: The Philosophical Library. Original edition published in 1872.

Davis M. 1992. The role of the amygdala in fear and anxiety. *Annual Review of Neuroscience* 15:353–375.

Edwards DH, Kravitz EA. 1997. Serotonin, social status and aggression. *Current Opinion in Neurobiology* 7:812–819.

Flynn JP. 1967. The neural basis of aggression in cats. *In Neurophysiology and Emotion*, ed. Glass DC. New York: Rockefeller University Press.

Fulton JF. 1951. *Frontal Lobotomy and Affective Behavior. A Neurophysiological Analysis*. New York: Norton.

Gallagher M, Chiba AA. 1996. The amygdala and emotion. *Current Opinion in Neurobiology* 6:221–227.

Harlow JM. 1848. Passage of an iron rod through the head. *Boston Medical and Surgical Journal* 39:389–393.

Harlow JM. 1868. Recovery from the passage of an iron bar through the head. *Publication of the Massachusetts Medical Society* 2:329–347.

Heath RG. 1963. Electrical self-stimulation of the brain in man. *American Journal of Psychiatry* 120:571–577.

Heisler LK, Chu HM, Brennan TJ, Danao JA, Bajwa P, Parsons LH, Tecott LH. 1998. Elevated anxiety and antidepressant-like responses in serotonin 5-HT1A receptor mutant mice. *Proceedings of the National Academy of Science USA* 95:15049–15054.

Hess WR. 1954. *Diencephalon: Autonomic and Extrapyramidal Functions*. New York: Grune & Stratton.

Jacobsen CF, Wolf JB, Jackson TA. 1935. An experimental analysis of the functions of the frontal association areas in primates. *Journal of Nervous and Mental Disease* 82:1–14.

James W. 1884. What is an emotion? *Mind* 9:188–205.

Julius D. 1998. Serotonin receptor knockouts: a moody subject. *Proceedings of the National Academy of Science USA* 95:15153–15154.

Kalin NH. 1993. The neurobiology of fear. *Scientific American* 268:94–101.

Kapp BS, Pascoe JP, Bixler MA. 1984. The amygdala: a neuroanatomical systems approach to its contributions to aversive conditioning. In *Neuropsychology of Memory*, eds. Butler N, Squire LR. New York: Guilford.

Klüver H, Bucy PC. 1939. Preliminary analysis of functions of the temporal lobes in monkeys. *Archives of Neurology and Psychiatry* 42:979–1000.

LaBar KS, Gatenby JC, Gore JC, LeDoux JE, Phelps EA. 1998. Human amygdala activation during conditioned fear acquisition and extinction: a mixed-trial fMRI study. *Neuron* 20:937–945.

Lange CG. 1887. *Uber Gemuthsbewegungen*. Liepzig: T. Thomas.

LeDoux JE. 1994. Emotion, memory and the brain. *Scientific American* 270:50–57.

MacLean PD. 1955. The limbic system ("visceral brain") and emotional behavior. *Archives of Neurology and Psychiatry* 73: 130–134.

Olds J, Milner P. 1954. Positive reinforcement produced by electrical stimulation of the septal area and other regions of the rat brain. *Journal of Comparative Physiological Psychology* 47:419–427.

Papez JW. 1937. A proposed mechanism of emotion. *Archives of Neurology and Psychiatry* 38:725–743.

Pribram KH. 1954. Towards a science of neuropsychology (method and data). *In Current Trends in Psychology and the Behavioral Sciences*, ed. Patton RA. Pittsburgh: University of Pittsburgh Press.

Raleigh MJ, McGuire MT, Brammer GL, Pollack DB, Yuwiler A. 1991. Serotonergic mechanisms promote dominance acquisition in adult vervet monkeys. *Brain Research* 559:181–190.

Ramboz S, Oosting R, Amara DA, Kung HF, Blier P, Mendelsohn M, Mann JJ, Brunner D, Hen R. 1998. *Proceedings of the National Academy of Science USA* 95:14476–14481.

Saudou F, Amara DA, Dierich A, LeMeur M, Ramboz S, Segu L, Buhot M, Hen R. 1994. Enhanced aggressive behavior in mice lacking 5-HT_{1B} receptor. *Science* 265:1875–1878.

Schultz W. 1997. Dopamine neurons and their role in reward mechanisms. *Current Opinion in Neurobiology* 7:191–197.

Thompson JG. 1988. *The Psychobiology of Emotions*. New York: Plenum Press.

Tinbergen N, et al. 1965. *Animal Behavior*. New York: Time Inc. (Nina Leen/TimePix.)

Wise RA, Rompre P. 1989. Brain dopamine and reward. *Annual Review of Psychology* 40: 191–225.

Chapter 19

Aldrich MS. 1986. The neurobiology of narcolepsy-cataplexy. *Progress in Neurobiology* 41:533–541.

Aréchiga H. 1993. Circadian rhythms. *Current Opinion in Neurobiology* 3:1005–1010.

Bal T, McCormick DA. 1993. Mechanisms of oscillatory activity in guinea-pig nucleus reticularis thalami *in vitro*: a mammalian pacemaker. *Journal of Physiology* (London) 468:669–691.

Borbely AA, Tonini G. 1998. The quest for the essence of sleep. *Daedalus* 127:167–196.

Braun AR, Balkin TJ, Wesensten NJ, Gwadry F, Carson RE, Varga M, Baldwin P, Belenky G, Herscovitch P. 1998. Dissociated pattern of activity in visual cortices and their projections during human rapid eye movement sleep. *Science* 279:91–95.

Carskadon MA. ed. 1993. *Encyclopedia of Sleep and Dreaming*. New York: Macmillan.

Cirelli C, Tonini G. 1998. Differences in gene expression between sleep and waking as revealed by mRNA differential display. *Molecular Brain Research* 56:293–305.

Coleman RM. 1986. *Wide Awake at 3:00 A.M. by Choice or by Chance*? New York: W.H. Freeman.

Czeisler CA, Duffy JF, Shanahan TL, Brown EN, Mitchell JF, Rimmer DW, Ronda JM, Silva EJ, Allan JS, Emens JS, Dijk D, Kronauer RE. 1999. Stability, precision, and near-24-hour period of the human circadian pacemaker. *Science* 284:2177–2181.

Dement WC. 1976. *Some Must Watch While Some Must Sleep*. San Francisco: San Francisco Book Company.

Earnest DJ, Sladek CD. 1987. Circadian vasopressin release from perfused rat suprachiasmatic explants in vitro: effects of acute stimulation. *Brain Research* 422:398–402.

Edgar DM, Dement WC, Fuller CA. 1993. Effect of SCN lesions on sleep in squirrel monkeys: evidence for opponent processes in sleep-wake regulation. *Journal of Neuroscience* 13:1065–1079.

Freeman W. 1991. The physiology of perception. *Scientific American* 264:78–85.

Gekakis N, Staknis D, Nguyen HB, Davis FC, Wilsbacher LD, King DP, Takahashi JS, Weitz CJ. 1998. Role of the CLOCK protein in the mammalian circadian mechanism. *Science* 280:1564–1568.

Gray CM. 1994. Synchronous oscillations in neuronal systems: mechanisms and functions. *Journal of Computational Neuroscience* 1:11–38.

Hobson JA. 1993. Sleep and dreaming. *Current Opinion in Neurobiology* 10:371–382.

Karni A, Tanne D, Rubenstein BS, Akenasy JJM, Sagi D. 1994. Dependence on REM sleep of overnight performance of a perceptual skill. *Science* 265:679–682.

Klein DC, Moore RY, Reppert SM, eds. 1991. *Suprachiasmatic Nucleus: The Mind's Clock*. New York: Oxford University Press.

Lamberg L. 1994. *Bodyrhythms: Chronobiology and Peak Performance*. New York: Morrow.

Lin L, Faraco J, Li R, Kadotani H, Rogers W, Lin X, Qiu X, de Jong PJ, Nishino S, Mignot E. 1999. The sleep disorder canine narcolepsy is caused by a mutation in the hypocretin (orexin) receptor 2 gene. *Cell* 98:365–376.

McCarley RW, Massaquoi SG. 1986. A limit cycle reciprocal interaction model of the REM sleep oscillator system. *American Journal of Physiology* 251: R1011.

McCormick DA, Thierry B. 1997. Sleep and arousal: thalamocortical mechanisms. *Annual Review of Neuroscience* 20:185–216.

Moruzzi G. 1964. Reticular influences on the EEG. *Electroencephalography and Clinical Neurophysiology* 16:1.

Mukhametov LM. 1984. Sleep in marine mammals. In: *Sleep Mechanisms*, eds. Borbély AA, Valatx JL. Munich: Springer-Verlag, pp. 227–238.

Pappenheimer JR, Koski G, Fencl V, Karnovsky ML, Krueger J. 1975. Extraction of sleep-promoting factor S from cerebrospinal fluid and from brains of sleep-deprived animals. *Journal of Neurophysiology* 38:1299–1311.

Porkka-Heiskanen T, Strecker RE, Thakkar M, Bjorkum AA, Greene RW, McCarley RW. 1997. Adenosine: a mediator of the sleep-inducing effects of prolonged wakefulness. *Science* 276:1265–1268.

Ralph MR, Foster RG, Davis FC, Menaker M. 1990. Transplanted suprachiasmatic nucleus determines circadian period. *Science* 247:975–978.

Ralph MR, Menaker M. 1988. A mutation of the circadian system in golden hamsters. *Science* 241:1225–1227.

Steriade M, McCormick DA, Sejnowski TJ. 1993. Thalamocortical oscillations in the sleeping and aroused brain. *Science* 262:679–685.

Winson J. 1993. The biology and function of rapid eye movement sleep. *Current Opinion in Neurobiology* 3:243–248.

Chapter 20

Baynes K, Eliassen JC, Lutsep HL, Gazzaniga MS. 1988. Modular organization of cognitive systems masked by interhemispheric integration. *Science* 280:902–905.

Binder JR, Frost JA, Hammeke TA, Cox RW, Rao SM, Prieto T. 1997. Human brain language areas identified by functional magnetic resonance imaging. *Journal of Neuroscience* 17:353–362.

Brefczynski JA, DeYoe EA. 1999. A physiological correlate of the "spotlight" of visual attention. *Nature Neuroscience* 2:370–374.

Corbetta M, Miezin FM, Dobmeyer S, Shulman GL, Petersen SE. 1990. Attentional modulation of neural processing of shape, color, and velocity in humans. *Science* 248:1556–1559.

Crick R. 1994. *The Astonishing Hypothesis: The Scientific Search for the Soul*. New York: Scribner's.

Damasio AR, Damasio H. 1992. Brain and language. *Scientific American* 267:88–95.

Fromkin V, Rodman R. 1992. *An Introduction to Language*. New York: Harcourt Brace Jovanovich.

Gardner H. 1974. *The Shattered Mind*. New York: Vintage Books.

Gardner RA, Gardner B. 1969. Teaching sign language to a chimpanzee. *Science* 165: 664–672.

Gazzaniga MS. 1970. *The Bisected Brain*. New York: Appleton-Century-Crofts.

Geschwind N, Levitsky W. 1968. Human-brain: left-right asymmetries in temporal speech region. *Science* 161:186–187.

Moran J, Desimone R. 1985. Selective attention gates visual processing in the extrastriate cortex. *Science* 229:782–784.

Neville HJ, Bavelier D, Corina D, Rauschecker J, Karni A, Lalwani A, Braun A, Clark V, Jezzard P, Turner R. 1998. Cerebral organization for language in deaf and hearing subjects: biological constraints and effects of experience. *Proceedings of the National Academy of Science USA* 95:922–929.

Ojemann G, Mateer C. 1979. Human language cortex: localization of memory, syntax, and sequential motor-phoneme identification systems. *Science* 205:1401–1403.

Patterson FG. 1978. The gestures of a gorilla: language acquisition in another pongid. *Brain and Language* 5:56–71.

Penfield W, Rasmussen T. 1950. *The Cerebral Cortex of Man*. New York: Macmillan.

Petersen SE, Fox PT, Posner MI, Mintum M, Raichle ME. 1988. Positron emission tomographic studies of the cortical anatomy of single-word processing. *Nature* 331:585–589.

Pinker S. 1994. *The Language Instinct*. New York: Morrow.

Posner MI, Petersen SE. 1990. The attention system of the human brain. *Annual Review of Neuroscience* 13:25–42.

Posner MI, Raichle M. 1994. *Images of Mind*. New York: Scientific American Library.

Posner MI, Snyder CRR, Davidson BJ. 1980. Attention and the detection of signals. *Journal of Experimental Psychology General* 109:160–174.

Rasmussen T, Milner B. 1977. The role of early left-brain injury in determining lateralization of cerebral speech functions. *Annals of New York Academy of Sciences* 299:355–369.

Sadato N, Pascual-Leone A, Grafman J, Ibanez V, Deiber M, Dold G, Hallett M. 1996. Activation of the primary visual cortex by Braille reading in blind subjects. *Nature* 380:526–528.

Sperry RW. 1964. The great cerebral commissure. *Scientific American* 210:42–52.

Wurtz RH, Goldberg ME, Robinson DL. 1982. Brain mechanisms of visual attention. *Scientific American* 246:124–135.

Chapter 21

Andreasen NC. 1984. *The Broken Brain*. New York: Harper Collins.

American Psychiatric Association. 1994. *Diagnostic and Statistical Manual of Mental Disorders*, 4th ed. Washington, DC: American Psychiatric Association.

Barondes SH. 1993. *Molecules and Mental Illness*. New York: W.H. Freeman.

Davidson RJ, Abercrombie H, Nitschke JB, Putnam K. 1999. Regional brain function, emotion and disorders of emotion. *Current Opinion in Neurobiology* 9:228–234.

Gottesman II. 1991. *Schizophrenia Genesis*. New York: W.H. Freeman.

Heuser I. 1998. The hypothalamic-pituitary-adrenal system in depression. Anna-Monika-Prize paper. *Pharmacopsychiatry* 31:10–13.

Liu D, Diorio J, Tannenbaum B, Caldji C, Francis D, et al. 1997. Maternal care, hippocampal glucocorticoid receptors, and hypothalamic-pituitary-adrenal responses to stress. *Science* 277:1659–1662.

Malizia AL, Cunningham VJ, Bell CJ, Liddle PF, Jones T, et al. 1998. Decreased brain GABA(A)-benzodiazepine receptor binding in panic disorder: preliminary results from a quantitative PET study. *Archives General Psychiatry* 55:715–720.

Mohn AR, Gainetdinov RR, Caron MG, Koller BH. 1999. Mice with reduced NMDA receptor expression display behaviors related to schizophrenia. *Cell* 98:427–436.

Nemeroff CB. 1998. The neurobiology of depression. *Scientific American* 278(6):42–49.

Satcher D. 1999. *Mental Health: A Report of the Surgeon General*. Washington, DC: US Government Printing Office.

Seeman P. 1980. Brain dopamine receptors. *Pharmacological Reviews* 32:229–313.

Slater E, Meyer A. 1959. Contributions to a pathology of the musicians. *Confinia Psychiatrica* 2:65–94.

Chapter 22

Balice-Gordon RJ, Lichtman JW. 1994. Long-term synapse loss induced by focal blockade of postsynaptic receptors. *Nature* 372: 519–524.

Bear MF, Kleinschmidt A, Gu Q, Singer W. 1990. Disruption of experience-dependent synaptic modifications in striate cortex by infusion of an NMDA receptor antagonist. *Journal of Neuroscience* 10:909–925.

Bear MF, Rittenhouse CD. 1999. Molecular basis for ocular dominance plasticity. *Journal of Neurobiology* 41:83–91.

Bear MF, Singer W. 1986. Modulation of visual cortical plasticity by acetylcholine and noradrenaline. *Nature* 320:172–176.

Bourgeois J, Rakic P. 1993. Changes of synaptic density in the primary visual cortex of the macaque monkey from fetal to adult stage. *Journal of Neuroscience* 13:2801–2820.

Chenn A, Braisted JE, McConnel SK, O'Leary DDM. 1999. Development of the cerebral cortex: mechanisms controlling cell fate, laminar and areal patterning, and axonal connectivity. In *Molecular and Cellular Approaches to Neural Development*, eds. Cowan MW, Jessell TM, Zipursky SL. New York: Oxford University Press.

Dudek SM, Bear MF. 1989. A biochemical correlate of the critical period for synaptic modification in the visual cortex. *Science* 246:673–675.

Ghosh A, Carnahan J, Greenberg M. 1994. Requirement for BDNF in activity-dependent survival of cortical neurons. *Science* 263:1618–1623.

Goldberg JL, Barres BA. 2000. Nogo in nerve regeneration. *Nature* 403:369–370.

Goodman C, Shatz C. 1993. Developmental mechanisms that generate precise patterns of neuronal connectivity. *Cell* 72:77–98.

Gould E, Reeves AJ, Graziano MSA, Gross CG. 1999. Neurogenesis in the neocortex of adult primates. *Science* 286:548–552.

Harris WC, Holt CE. 1999. Slit, the midline repellent. *Nature* 398:462–463.

Huang ZJ, Kirkwood A, Pizzorusso T, Porciatti V, Morales B, Bear MF, Maffei L, Tonegawa S. 1999. BDNF regulates the maturation of inhibition and the critical period of plasticity in mouse visual cortex. *Cell* 98: 39–55.

Ip N, Yancopoulos G. 1994. Neurotrophic factor receptors: just like other growth factor receptors. *Current Opinion in Neurobiology* 4:400–405.

Katz LC, Shatz CJ. 1996. Synaptic activity and the construction of cortical circuits. *Science* 274:1133–1138.

Kempermann G, Gage FH. 1999. New nerve cells for the adult brain. *Scientific American* 280:48–53.

Kennedy T, Serafini T, Torre JDL, Tessier-Lavigne M. 1994. Netrins are diffusible chemotropic factors for commissural axons in the embryonic spinal cord. *Cell* 78:425–435.

LeVay S, Stryker MP, Shatz CJ. 1978. Ocular dominance columns and their development in layer IV of the cat's visual cortex: a quantitative study. *Journal of Comparative Neurology* 179:223–244.

Liao D, Zhang X, O'Brien R, Ehlers MD, Huganir RL. 1999. Regulation of morphological postsynaptic silent synapses in developing hippocampal neurons. *Nature Neuroscience* 2:37–43.

McConnel SK. 1995. Constructing the cerebral cortex: Neurogenesis and fate determination. *Neuron* 15:761–768.

Meister M, Wong R, Baylor D, Shatz C. 1991. Synchronous bursts of action potentials in ganglion cells of the developing mammalian retina. *Science* 252:939–943.

Olson CR, Freeman RD. 1980. Profile of the sensitive period for monocular deprivation in kittens. *Experimental Brain Research* 39:17–21.

Rakic P. 1981. Development of visual centers in the primate brain depends on binocular competition before birth. *Science* 214:928–931.

Schlagger BL, O'Leary DD. 1991. Potential of visual cortex to develop an array of functional units unique to somatosensory cortex. *Science* 252:1556–1560.

Sperry R. 1963. Chemoaffinity in the orderly growth of nerve fiber patterns and connections. *Proceedings National Academy Science USA* 4:703–710.

Tessier-Lavigne M, Goodman CS. 1996. The molecular biology of axon guidance. *Science* 274:1123–1133.

Tinbergen N, et al. 1965. *Animal Behavior*. New York: Time Inc. (Nina Leen/TimePix.)

Walsh C, Cepko C. 1992. Widespread dispersion of neuronal clones across functional regions of the cerebral cortex. *Science* 255:434.

Wiesel T. 1982. Postnatal development of the visual cortex and the influence of the environment. *Nature* 299:583–592.

Chapter 23

Bear MF. 1996. A synaptic basis for memory storage in the cerebral cortex. *Proceedings of the National Academy of Sciences USA* 93:13453–13459.

Burkholder JM. 1999. The lurking perils of pfiesteria. *Scientific American* 281:42–49.

Cohen NJ, Eichenbaum H. 1993. *Memory, Amnesia, and the Hippocampal System*. Cambridge, MA: MIT Press.

Desimone R, Albright TD, Gross CG. Bruce C. 1984. Stimulus-selective properties of inferior temporal neurons in the macaque. *Journal of Neuroscience* 4:2051–2062.

Dudai Y. 1989. *The Neurobiology of Memory*. New York: Oxford University Press.

Eichenbaum H, Dudchenko P, Wood E, Shapiro M, Tanila H. 1999. The hippocampus, memory, and place cells: is it spatial memory or a memory space? *Neuron* 23:209–226.

Eichenbaum H, Fagan H, Mathews P, Cohen NJ. 1988. Hippocampal system dysfunction and odor discrimination learning in rats: impairment or facilitation depending on representational demands. *Behavioral Neuroscience* 102:331–339.

Fuster JM. 1973. Unit activity in prefrontal cortex during delayed-response performance: neuronal correlates of transient memory. *Journal of Neurophysiology* 36:61–78.

Fuster JM. 1995. *Memory in the Cerebral Cortex*. Cambridge, MA: MIT Press.

Gauthier I, Skularski P, Gore JC, Anderson AW. 2000. Expertise for cars and birds recruits brain areas involved in face recognition. *Nature Neuroscience* 3:191–197.

Gnadt JW and Andersen RA. 1988. Memory related motor planning activity in posterior parietal cortex of macaque. *Experimental Brain Research* 70:216–220.

Goldman-Rakic P. 1992. Working memory and the mind. *Scientific American* 267:111–117.

Grattan LM, Oldach D, Perl TM, Lowitt MH, Matuszak DL, Dickson C, Parrott C, Shoemaker RC, Kauffman CL, Wasserman MP, Hebel JR, Charache P, Morris JG. 1998. Learning and memory difficulties after environmental exposure to waterways containing toxin-producing *Pfiesteria* or *Pfiesteria*-like dinoflagellates. *Lancet* 352:532–539.

Hebb DO. 1949. *The Organization of Behavior: A Neuropsychological Theory*. New York: Wiley.

Jog MS, Kubota Y, Connolly CI, Hillegaart, Graybiel AM. 1999. Building neural representations of habits. *Science* 286:1745–1749.

Knowlton BJ, Mangels JA, Squire LR. 1996. A neostriatal habit learning system in humans. *Science* 273:1399–1402.

Lashley KS. 1929. *Brain Mechanisms and Intelligence*. Chicago: University of Chicago Press.

Luria A. 1968. *The Mind of a Mnemonist*. Cambridge, MA: Harvard University Press.

Maguire EA, Burgess N, Donnett JG, Frackowiak RS, Frith CD, O'Keefe J. 1998. Knowing where and getting there: a human navigation network. *Science* 280:921–924.

Mishkin M, Appenzeller T. 1987. *The Anatomy of Memory*. Scientific American 256:80–89.

O'Keefe JA. 1979. Place units in the hippocampus of the freely moving rat. *Experimental Neurology* 51:78–109.

O'Keefe JA, Nadel L. 1978. *The Hippocampus as a Cognitive Map*. London: Oxford University Press.

Olton DS, Samuelson RJ. 1976. Remembrance of places passed: spatial memory in rats. *Journal of Experimental Psychology* 2:97–116.

Penfield W. 1958. *The Excitable Cortex in Conscious Man*. Liverpool: Liverpool University Press.

Rolls ET, Baylis GC, Hasselmo ME, Nalwa V. 1989. The effect of learning on the face selective responses of neurons in the cortex in the superior temporal sulcus of the monkey. *Experimental Brain Research* 76: 153–164.

Scoville WB, Milner B. 1957. Loss of recent memory after bilateral hippocampal lesions. *Journal of Neurology, Neurosurgery, and Psychiatry* 20:11–21.

Squire LR. 1987. *Memory and Brain*. New York: Oxford University Press.

Wilson MA, McNaughton BL. 1993. Dynamics of the hippocampal ensemble code for space. *Science* 261:1055–1058.

Zola-Morgan S, Squire LR, Amaral DG, Suzuki WA. 1989. Lesions of perirhinal and parahippocampal cortex that spare the amygdala and hippocampal formation produce severe memory impairment. *Journal of Neuroscience* 9:4355–4370.

Chapter 24

Bailey CH, Kandel ER. 1993. Structural changes accompanying memory storage. *Annual Review of Neuroscience* 55:397–426.

Bear MF. 1996. A synaptic basis for memory storage in the cerebral cortex. *Proceedings of the National Academy of Sciences USA* 93:13453–13459.

Bliss TVP, Collingridge GL. 1993. A synaptic model of memory: long-term potentiation in the hippocampus. *Nature* 361:31–39.

Bourne HR, Nicoll R. 1993. Molecular machines integrate coincident synaptic signals. *Neuron* 10:65–75.

Carew TJ, Sahley CL. 1986. Invertebrate learning and memory: from behavior to molecules. *Annual Review of Neuroscience* 9:435–487.

Castellucci VF, Kandel ER. 1974. A quantal analysis of the synaptic depression underlying habituation of the gill-withdrawal reflex in *Aplysia*. *Proceedings of the National Academy of Sciences USA* 77:7492–7496.

Chen, WR, Lee S, Kato K, Spencer DD, Shepherd GM, Williamson A. 1996. Long-term modifications of synaptic efficacy in the human inferior and middle temporal cortex. *Proceedings of the National Academy of Sciences USA* 93:8011–8015.

Davis HP, Squire LR. Protein synthesis and memory. 1984. *Psychological Bulletin* 96: 518–559.

Grant SGN, Silva AJ. 1994. Targeting learning. *Trends in Neurosciences* 17:71–75.

Greenough WT, Bailey CH. 1988. The anatomy of memory: convergence of results across a diversity of tests. *Trends in Neurosciences* 11:142–147.

Kirkwood A, Dudek SD, Gold JT, Aizenman CD, Bear MF. 1993. Common forms of synaptic plasticity in hippocampus and neocortex in vitro. *Science* 260:1518–1521.

Linden DJ, Connor JA. 1993. Cellular mechanisms of long-term depression in the cerebellum. *Current Opinion in Neurobiology* 3:401–406.

Lisman JE, Fallon JR. 1999 What maintains memories? *Science* 283:339–340.

Malenka RC, Nicoll RA. 1999. Long-term potentiation–a decade of progress? *Science* 285:1870–1874.

Markram H, Lubke J, Frotscher M, Sakmann B. 1997. Regulation of synaptic efficacy by coincidence of postsynaptic APs and EPSPs. *Science* 275:213–215.

Morris RGM, Anderson E, Lynch GS, Baudry M. 1986. Selective impairment of learning and blockade of long-term potentiation by an *N*-methyl-D-aspartate receptor antagonist, AP5. *Nature* 319:774–776.

Morris RGM, Kandel ER, Squire LR. 1988. The neuroscience of learning and memory: cells, neural circuits and behavior. *Trends in Neurosciences* 11:125–127.

Otani S, Abraham WC. 1989. Inhibition of protein synthesis in the dentate gyrus, but not the entorhinal cortex, blocks maintenance of long-term potentiation in rats. *Neuroscience Letters* 106:175–180.

Pavlov IP. 1927. *Conditioned Reflexes: An Investigation of the Physiological Activity of the Cerebral Cortex.* London: Oxford University Press.

Schwartz JH. 1993. Cognitive kinases. *Proceedings of the National Academy of Sciences USA* 90:8310–8313.

Tang YP, Shimizu E, Dube GR, Rampon C, Kerchner GA, Zhuo M, Liu G, Tsien JZ. 1999. Genetic enhancement of learning and memory in mice. *Nature* 401:63–69.

Thompson RF, Krupa DJ. 1994. Organization of memory traces in the mammalian brain. *Annual Review of Neuroscience* 17:519–549.

Thorndike EL. 1911. *Animal Intelligence: Experimental Studies.* New York: Macmillan.

Tsien JZ, Huerta PT, Tonegawa S. 1996. The essential role of hippocampal CA1 NMDA receptor-dependent synaptic plasticity in spatial memory. *Cell* 87:1327–1338.

Yin JC, Tully T. 1996. CREB and the formation of long-term memory. *Current Opinion in Neurobiology* 6:264–268.

INDEX

Page numbers in *italics* followed by *f* denote figures; those followed by *t* denote tables.

B

C

D

E

N

O

P

Q

R

T

U

V

W

Y

Z